Temperature/Temperature Difference

$T(°F) = 1.8T(°C) + 32$
$T(°C) = [T(°F) - 32]/1.8$
$T(K) = T(°C) + 273.15$
$T(R) = T(°F) + 459.67$
$T(R) = 1.8\,T(K)$
$\Delta T(K) = \Delta T(°C)$
$\Delta T(R) = \Delta T(°F)$

Thermal Conductivity

$1\ \text{Btu/h} \cdot \text{ft} \cdot °F = 1.731\ \text{W/m} \cdot \text{K}$

Velocity

1 mph $= 1\ \text{mile/hour}$
$= 1.466\ \text{ft/s}$
$= 1.609\ \text{km/h}$
1 m/s $= 2.237\ \text{mph}$
$= 3.6\ \text{km/h}$
1 knot $= 1.151\ \text{mph}$

Viscosity (Dynamic)

1 centipoise $= 0.001\ \text{N} \cdot \text{s/m}^2$
1 poise $= 0.1\ \text{N} \cdot \text{s/m}^2$
$1\ \text{N} \cdot \text{s/m}^2 = 2.089 \times 10^{-2}\ \text{lbf s/ft}^2$
$1\ \text{N} \cdot \text{s/m}^2 = 2419.1\ \text{lbm/ft} \cdot \text{h}$

Viscosity (Kinematic)

$1\ \text{m}^2/\text{s} = 10.76\ \text{ft}^2/\text{s}$
1 stoke $= 10^{-4}\ \text{m}^2/\text{s}$

Volume

$1\ \text{m}^3 = 10^6\ \text{cm}^3$
$= 10^6\ \text{cc}$
$= 264.17\ \text{gal (US)}$
1 US gal $= 231\ \text{in.}^3$
$= 3.7854 \times 10^{-3}\ \text{m}^3$
$= 0.1337\ \text{ft}^3$
$= 128\ \text{fl ounces}$

1 UK gal $= 4.54609\ \text{L}$
1 L $= 1\ \text{liter}$
$= 10^{-3}\ \text{m}^3$
$= 0.2642\ \text{US}$
$= 0.0353\ \text{ft}^3$
$1\ \text{ft}^3 = 1728\ \text{in.}^2$
1 fl ounce $= 29.5735\ \text{cm}^3$

Volumetric Flow Rate

1 US gal/min $= 2.228 \times 10^{-3}\ \text{ft}^3/\text{s}$
$= 6.309 \times 10^{-5}\ \text{m}^3/\text{s}$
$1\ \text{ft}^3/\text{s} = 2.832 \times 10^{-2}\ \text{m}^3/\text{s}$
$1\ \text{ft}^3/\text{min} = 0.000472\ \text{m}^3/\text{s}$

Physical Constants

Standard acceleration of gravity (sea level)

$g = 32.174\ \text{ft/s}^2$
$= 9.80665\ \text{m/s}^2$

Standard Atmospheric pressure

1 atm $= 1.01325 \times 10^5\ \text{Pa}$
$= 101.325\ \text{kPa}$
$= 14.696\ \text{lbf/in.}^2 = 14.696\ \text{psia}$
$= 1.01325\ \text{bar}$

Universal gas constant

$\overline{R} = 8.31434\ \text{kJ/kmol} \cdot \text{K}$
$= 1.9858\ \text{Btu/lbmol} \cdot \text{R}$
$= 1545.35\ \text{ft} \cdot \text{lbf/lbmol} \cdot \text{R}$
$= 10.73\ \text{psia} \cdot \text{ft}^3/\text{lbmol} \cdot \text{R}$

Stefan-Boltzman constant

$\sigma = 5.670 \times 10^{-8}\ \text{W/m}^2 \cdot \text{K}^4$
$= 0.1714 \times 10^{-8}\ \text{Btu/h} \cdot \text{ft}^2 \cdot \text{R}^4$

INTRODUCTION TO THERMAL AND FLUIDS ENGINEERING

INTRODUCTION TO THERMAL AND FLUIDS ENGINEERING

DEBORAH A. KAMINSKI
MICHAEL K. JENSEN

Rensselaer Polytechnic Institute

WILEY

JOHN WILEY & SONS, INC.

Acquisitions Editor   *Joseph Hayton*
Senior Production Editor   *Norine M. Pigliucci / Sandra Dumas*
Senior Marketing Manager   *Jenny Powers*
Senior Design Manager   *Kevin Murphy*
New Media Editor   *Thomas Kulesa*
Production Management Services   *Hermitage Publishing Services*

This book was set in Times Roman by Hermitage Publishing Services and printed and bound by Courier Kendallville. The cover was printed by Courier Kendallville.

This book is printed on acid free paper. ∞

ISBN 0-471-26873-0
WIE ISBN 0-471-45236-X

Printed in the United States of America

10 9 8 7 6 5

PREFACE

Historically, thermal engineering has been somewhat arbitrarily divided into thermodynamics, fluid mechanics, and heat transfer due to specialization that has occurred in the profession. In recent years there has been renewed interest in teaching the field in a more integrated way. Traditional introductory textbooks in these three disciplines each approach 1000 pages in length, include many topics that are seldom covered in one- or two-semester disciplinary courses, and address many advanced topics that are not appropriate for introductory courses. Our experience teaching these subjects at Rensselaer indicated that many students failed to see the connections between the three topics; subsequently, we introduced a two-course sequence that presented the three topics in an integrated manner. Students responded well to the new approach and their understanding improved. To further aid our students, we saw a need to write a textbook reflecting the integrated approach.

This textbook is a fresh approach to the teaching of thermal and fluids engineering as an integrated subject. Our objectives are to:

- present appropriate material at an introductory level on thermodynamics, heat transfer, and fluid mechanics
- develop governing equations and approaches in sufficient detail so that the students can understand how the equations are based on fundamental conservation laws and other basic concepts
- explain the physics of processes and phenomena with language and examples that students have seen and used in everyday life
- integrate the presentation of the three subjects with common notation, examples, and homework problems
- demonstrate how to solve any problem in a systematic, logical manner

Features

An integrated approach: As can be illustrated in countless engineering systems, specific applications may need only thermodynamics, heat transfer, or fluid mechanics. However, many other applications require the integration of principles and tools from two or more of these disciplines. We use unifying themes to tie the text together so that boundaries between disciplines become transparent. For example, the first law is presented with a common notation and format as it applies in thermodynamics, fluid mechanics, and heat transfer. By necessity, topics are introduced in the context of their disciplines. However, examples and problems are given that illustrate how the three disciplines are integrated in practice.

An emphasis on problem solving: Students learn by problem solving, and the text features a rich collection of example problems (over 150) and end-of-chapter exercises (over 850). The problems range from the simple (to illustrate one concept or point) to the complex (to show the need for integration, synthesis of topics and tools, and the use of a logical problem solution approach). Some of the example problems are industrially relevant; these example problems and other practical engineering applications are used throughout the text to provide motivation to the students, to illustrate where and when certain equations and topics are needed, and to demonstrate the power and utility of basic concepts. Other problems, which relate to the student's personal experience and to established technologies, are used to develop physical understanding. Finally, many types of tried-and-true problems, which have been staples in thermal-fluids curricula for many years, are incorporated. The example problems include, at the beginning of the solution, a discussion of the thought process used to arrive at a solution procedure. This teaches the student to first focus on a global understanding of the solution (that is, an identification of all the tasks and parameters needed and a path to follow) instead of immediately looking up properties or applying an equation. In addition, assumptions are given in the context of the solution rather than at the beginning of the example.

A flexible organization: It is important for students to have good motivation for studying a subject and to be able to place in context the concepts and tools presented. Thus, Chapter 1 is an introduction that describes numerous engineering applications that require thermodynamics, heat transfer, and/or fluid mechanics, as well as basic concepts and definitions used throughout the book. The next three chapters (Chapter 2, The First Law; Chapter 3, Thermal Resistances; Chapter 4, Fundamentals of Fluid Mechanics) are intended to give the student an introduction to the three disciplines so that reasonable problems can be presented and solved early in the book. The remaining chapters delve into the topics in more detail and rigor, and integrated examples and problems are given.

The text is suitable for a single semester introduction to the subject or a two-semester sequence of courses. The approach is appropriate for both majors and non-majors. The text is designed to support a wide variety of syllabi and course structures. After Chapters 2–4, there are multiple paths through the book depending on the curricular needs (see the figure below). Chapter 2, which focuses on the first law, is absolutely essential to all students.

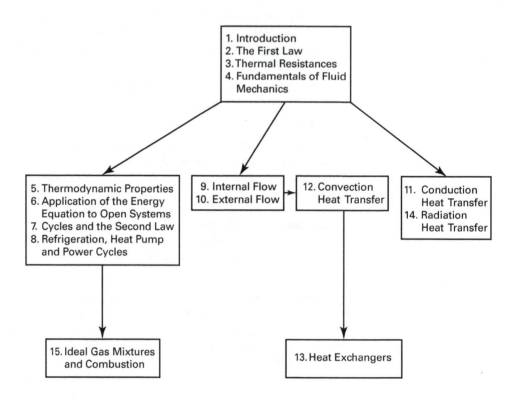

In Chapter 3, heat transfer is introduced with a strong emphasis on thermal resistances. This chapter includes some of the most useful concepts in heat transfer. Chapter 4 presents the conservation equations of mass, momentum, and energy in open systems and is the gateway to further study in both thermodynamics and fluid mechanics. After Chapter 4, four chapters on thermodynamics are presented; Chapters 5–8 focus on thermodynamic properties, steady flow devices, and thermodynamic cycles. While these are essential topics in some engineering disciplines, (e.g., mechanical and aeronautical engineering) they are much less important in others (e.g., electrical and civil engineering). If the course is designed for electrical engineering students, the instructor could skip Chapters 5–8 and proceed directly to Chapter 9 on internal flow. By design, Chapters 2 through 4 present enough of the rudiments of thermodynamics to allow students access to fluid mechanics (Chapters 9 and 10) and heat transfer (Chapters 11–14) without further thermodynamic study. On the textbook website, www.wiley.com/college/kaminski, supplementary material is given that expands or extends information given in the first 14 chapters. Chapter 15 (Ideal Gas Mixtures and Combustion) presents material on ideal gas mixtures, psychrometrics, and basic combustion calculations. Each chapter ends with a concise and useful summary of the important concepts and equations developed in that chapter.

Supporting Material: The textbook has a solution manual to all end-of-chapter exercises; the solutions are complete and detailed. Note that we have included problems (and appendices) in both SI and English units. While we would have preferred to use only SI units, in the United States there are still many companies and industries that resist change; we believe that students should be exposed to both unit systems because neither we nor the students know where they will be working once they graduate. Many reference tables of fluid and solid properties are given in the appendices.

The text is augmented with web-based material to extend the coverage of topics. The sections that are available on the text website are included in the Table of Contents. Titles for these sections also appear in the text at the appropriate locations with a reference to the web address: www.wiley.com/college/kaminski. The material on the website is optional and is not necessary to preserve the continuity of the material in the printed text. End-of-chapter problems based on the web material are included in the Problems section at the end of each chapter and are designated by WEB after the number.

Acknowledgments

Valuable suggestions, criticisms, and comments have been made by numerous individuals. We greatly appreciate the time and effort the following people gave when they reviewed early versions of this text, and we thank them for their help in improving the quality of this text:

J. Iwan D. Alexander, Case Western Reserve University

Brian M. Argrow, University of Colorado, Boulder

Louay M. Chamra, Mississippi State University

Fan-Bill Cheung, Pennsylvania State University

Chan Ching, McMaster University

Kirk Christensen, University of Missouri-Rolla

Benjamin T. F. Chung, University of Akron

S.A. Sherif, University of Florida

S. C. Yao, Carnegie Mellon University

We also would like to thank our editor, Joseph Hayton, at John Wiley & Sons and all the other contributors of that organization who have ensured the successful completion of this project.

Finally, we thank our spouses, Chris and Lois, and our children for their encouragement, support, and patience throughout the process of creating this book.

One last note: Debbie won the coin toss and got first billing.

Deborah A. Kaminski
Michael K. Jensen

CONTENTS

NOMENCLATURE

A	Area, m^2, ft^2
a	Acceleration, m/s^2, ft/s^2
$A, B, C, \ldots$	Stoichiometric coefficients
$AFR, \overline{AFR}$	Air-fuel ratio, molar air-fuel ratio
B	Momentum, $kg \cdot m/s$, $lbm \cdot ft/s$
Bi	Biot number $= hL_{char}/k$
BWR	Back work ratio
C	Heat capacity rate $= \dot{m}c_p$, W/K, Btu/h $\cdot$ °F
C_D	Drag coefficient $= F_D/(\rho \mathcal{V}^2 A/2)$
C_f	Skin friction coefficient $= \tau_w/(\rho \mathcal{V}^2/2)$
C_L	Lift coefficient $= F_L/(\rho \mathcal{V}^2 A/2)$
COP	Coefficient of performance
c	Specific heat, J/kg $\cdot$ K, Btu/lbm $\cdot$ R
$c_p, \overline{c}_p$	Constant pressure specific heat, J/kg $\cdot$ K, Btu/lbm $\cdot$ R; molar specific heat, J/kmol $\cdot$ K, Btu/lbmol $\cdot$ R
$c_v, \overline{c}_v$	Constant volume specific heat, J/kg $\cdot$ K, Btu/lbm $\cdot$ R, molar specific heat, J/kmol $\cdot$ K, Btu/lbmol $\cdot$ R
D, d	Diameter, m, ft
E	Total energy $= me$, kJ, Btu
E	Radiative emissive power, W/m^2, Btu/h $\cdot$ ft^2
e	Specific energy, kJ/kg, Btu/lbm
F	Force, N, lbf
F	Temperature difference correction factor for use with heat exchanger analysis
f	Friction factor $= \Delta P / \left[\left(\rho \mathcal{V}^2/2 \right) (L/D) \right]$
Fo	Fourier number $= \alpha t / L_{char}^2$
$F_{i \rightarrow j}$	Radiation view factor
G	Irradiation, W/m^2, Btu/h $\cdot$ ft^2
g	Acceleration of gravity, m/s^2, ft/s^2
Gz	Graetz number $= RePrD/L$
Gr	Grashof number $= g\beta\rho^2 \Delta T L_{char}^3/\mu^2$
H	Total enthalpy $= mh = U + PV$, kJ, Btu

h	Heat transfer coefficient, $W/m^2 \cdot K$, Btu/h $\cdot$ ft^2 $\cdot$ °F
$h, \overline{h}$	Specific enthalpy $= u + Pv$, kJ/kg, Btu/lbm; molar specific enthalpy, kJ/kmol, Btu/lbmol
h_L	Head loss, m, ft
h_P	Pump head, m, ft
h_T	Turbine head, m, ft
$\Delta \overline{h}_f^o$	Enthalpy of formation, kJ/kmol, Btu/lbmol
$\Delta \overline{h}_C^o$	Enthalpy of combustion, kJ/kmol, Btu/lbmol
HV	Heating value, kJ/kg, Btu/lbm
HHV	Higher heating value, kJ/kg, Btu/lbm
I	Current, A
I	Moment of inertia, m^4, ft^4
J	Radiosity, W/m^2, Btu/h $\cdot$ ft^2
K	Loss coefficient
k	Thermal conductivity, W/m $\cdot$ K, Btu/ft $\cdot$ h $\cdot$ °F
k	Specific heat ratio $= c_p/c_v$
KE	Total kinetic energy $= m\mathcal{V}^2/2$, kJ, Btu
ke	Specific kinetic energy, $\mathcal{V}^2/2$, kJ/kg, Btu/lbm
L	Length, m, ft
L^*	Corrected length, m, ft
LHV	Lower heating value, kJ/kg, Btu/lbm
m	Mass, kg, lbm
$\dot{m}$	Mass flowrate, kg/s, lbm/s
M	Molecular weight, kg/kmol, lbm/lbmol
M	Moment, N·m, ft $\cdot$ lbf
n	Number of moles, kmol, lbmol
$\dot{n}$	Molar flow rate, kmol/s, lbmol/s
NTU	Number of Transfer Units $= UA/C_{min}$

Nu Nusselt number $= hL_{char}/k$

P Pressure, kPa, psia

p Perimeter

P_r Relative pressure

PE Total potential energy $= mgz$, kJ, Btu

pe Specific potential energy $= gz$, kJ/kg, Btu/lbm

Pr Prandtl number $= \mu c_p/k$

ΔP Pressure difference, kPa, psi

Q Heat, kJ, Btu

$\dot{Q}$ Heat transfer rate, W, Btu/h

q Heat transfer per unit mass, kJ/kg, Btu/lbm

q'' Heat flux, W/m^2, Btu/h $\cdot$ ft^2

q''' Volumetric heat generation rate, W/m^3, Btu/h $\cdot$ ft^3

$\overline{R}$ Universal gas constant $= 8.314$ kJ/kmol $\cdot$ K $= 1545$ ft $\cdot$ lbf/lbmol $\cdot$ R $= 1.986$ Btu/lbmol $\cdot$ R

R Resistance, K/W, °F $\cdot$ h/Btu

r_c Cutoff ratio

r_p Pressure ratio

r_v Volume (compression) ratio

Ra Rayleigh number $= GrPr$

Re Reynolds number $= \rho \mathcal{V} L_{char}/\mu$

Ri Richardson number $= Gr/Re^2$

R'' Fouling factor, m$^2 \cdot$ K/W, h $\cdot$ ft$^2 \cdot$ R/Btu

S Conduction shape factor, m, ft

S Total entropy $= ms$, kJ/K, Btu/R

$s, \overline{s}$ Specific entropy, kJ/kg $\cdot$ K, Btu/lbm $\cdot$ R, molar specific entropy, J/kmol $\cdot$ K, Btu/lbmol $\cdot$ R

$s^o, \overline{s}^o$ Specific entropy tabulated for ideal gas at temperature T and pressure of one atmosphere, kJ/kg $\cdot$ K, Btu/lbm $\cdot$ R, molar specific entropy, J/kmol $\cdot$ K, Btu/lbmol $\cdot$ R

$\dot{S}_{gen}$ Entropy generation rate, W/K, Btu/h $\cdot$ R

SG Specific gravity

T Temperature, K, °C, R, °F

T_{film} Film temperature $= (T_s + T_\infty)/2$, K, °C, R, °F

t Time, s

t Thickness, m, ft

$\Im$ Torque, N $\cdot$ m, ft $\cdot$ lbf

U Total internal energy $= mu$, kJ, Btu

U Overall heat transfer coefficient, W/m$^2 \cdot$ K, Btu/h $\cdot$ ft$^2 \cdot$ R

$u, \overline{u}$ Specific internal energy, kJ/kg, Btu/lbm, molar specific internal energy, kJ/kmol, Btu/lbmol

V Total volume, m^3, ft^3

v Specific volume, m^3/kg, ft^3/lbm

$\mathcal{V}$ Velocity, m/s, ft/s

$\dot{V}$ Volume flowrate, m^3/s, ft^3/s

v_r Relative volume

W Work, kJ, Btu

w Work per unit mass, kJ/kg, Btu/lbm

$\dot{W}$ Power, W, Btu/h

X Mass fraction

x Quality (mass fraction of vapor in two-phase mixture)

Y Mole fraction

Z Compressibility factor

z Elevation, m, ft

GREEK LETTERS

α Thermal diffusivity, $k/\rho c_p$, m^2/s, ft^2/s

α Absorptivity

β Coefficient of volume expansion

δ Boundary layer thickness

ε Heat exchanger effectiveness

ε Emissivity

ε Roughness height, m, ft

ε_f Fin efficiency

η Efficiency

η Non-dimensional distance from wall in boundary layer

η_o Overall surface efficiency

μ	Dynamic viscosity, kg/m · s, lbm/ft · h
ν	Kinematic viscosity $= \mu/\rho$, m^2/s, ft^2/s
ξ	Voltage, V
ρ	Reflectivity
$\rho, \overline{\rho}$	Density, kg/m^3, molar density kmol/m^3
σ	Stefan-Boltzmann constant $= 5.67 \times 10^{-8}$ W/m^2 ·K$^4 = 0.171 \times 10^{-8}$ Btu/h · ft^2 · R^4
τ	Transmissivity
τ	Non-dimensional time, Fourier number $= \alpha t / L_{char}^2$
τ	Shear stress, N/m^2, lbf/in^2.
ϕ	Relative humidity
ω	Angular velocity, rad/s, deg/s
ω	Specific or absolute humidity, kg water vapor/kg dry air, lbm water vapor/lbm dry air

SUBSCRIPTS

1, 2, 3,...	Locations or times
∞	Free stream or far from a surface
λ	Wavelength
a	Air
$A, B, C,...$	Locations
act	Actual
atm	Atmosphere
avg	Average
b	Bulk
b	Base
b	Black body
buoy	Buoyancy
C	Centroid
C	Compressor
C	Cold
Carnot	Carnot cycle
cf	Counterflow
char	Characteristic
cond	Conduction
conv	Convection
cv	Control volume

crit	Critical
D	Diameter
D	Drag
DB	Dry bulb
e	Exit
ent	Entrance
f	Saturated liquid
f	Fin
f	Fluid
fg	Difference between saturated vapor and saturated liquid
F	Fuel
g	Saturated vapor
gen	Generation
H	High
H	Hot
h	Hydraulic
HP	Heat pump
I	Irreversible
i	Initial
i	Inside
i	Inlet
$i, j, k,...$	Index
in	Input
L	Lift
L	Loss
L	Low
lam	Laminar
LM	Log mean
m, mean	Mean
m, mix	Mixture
max	Maximum
min	Minimum
o	Outlet
o	Outside
opt	Optimum
out	Output
P	Pump
P	Products of reaction
pf	Parallel flow
r	Relative

R	Reactants
R	Resultant
R	Reversible
rad	Radiation
Ref	Refrigerator
ref	Reference
s	Isentropic process
s	Surface
sat	Saturated
surr	Surroundings
T	Turbine
t	Temperature or thermal
tot	Total
turb	Turbulent

v	Vapor
w	Wall
WB	Wet bulb
wetted	Portion of wall touched by fluid
x	Cross-section
x, y, z	Coordinate directions

SUPERSCRIPTS

° (circle)	Standard reference state
$'$	Quantity per unit length
$''$	Quantity per unit area
$'''$	Quantity per unit volume
¯ (over bar)	Quantity per unit mole
· (over dot)	Quantity per unit time

INTRODUCTION TO THERMAL AND FLUIDS ENGINEERING

1.1 OVERVIEW OF THERMAL AND FLUIDS SYSTEMS

In thermal–fluids systems, the focus is on energy: its use, conversion, or transmission in one form or another. For example, consider a few of the energy flows in a car. Gasoline is stored in a tank until its energy is needed to move the vehicle from one place to another, and then the gasoline is pumped from the tank to the engine. In the engine the fuel is burned, and some of the released chemical energy is converted to useful mechanical power to propel the car. Mechanical power is also extracted to drive: the water pump used in the engine-cooling system; the alternator to provide electrical power for the CD player, lights, cooling fan motor, and fuel pump; and the air-conditioning system.

Cars of the 21st century are dramatically improved over those of the early 1900s. The advances in engineering are the result of improved technical knowledge and the systematic application of this knowledge. The intelligent use of basic thermal and fluids engineering principles has improved the design of cars and other thermal–fluids systems as diverse as buildings, window air conditioners, oil refineries, electrical power plants, computers, airplanes, wind turbines, water distribution systems, plastic injection molding machines, and metal processing plants (Figure 1-1).

To analyze these systems, one, two, or three energy disciplines are needed, separately or in combination. These disciplines are:

Thermodynamics   The study of energy use and transformations from one form to another and the physical properties of substances (solids, liquids, gases) involved in energy use or transformation

Heat transfer   The study of energy flow that is caused by a temperature difference

Fluid mechanics   The study of fluids (liquids, gases) at rest or in motion and the interactions between a solid and a fluid either flowing past or acting on the solid in some manner

We can use the automobile to illustrate how these three subjects must be used together and separately. To begin an analysis, we must decide what aspect of the car we want to study. Is it the engine, the radiator where heat is removed from the engine coolant and released into the atmosphere, the water pump, the fuel supply system (pump, fuel lines, fuel injector), the air-conditioning system, or the passenger compartment? Do we want to examine the water-cooling system to determine what is needed to pump water through the engine-cooling system, the heat transfer from the water to the air flowing through the radiator, the conversion of the chemical energy in the gasoline to mechanical power in the engine, the energy contained within the exhaust gases, the refrigerant flow in the air-conditioning system, or the air flow through the air-conditioning system into the passenger compartment? Clearly, we need to identify carefully what we want to study.

FIGURE 1-1   Typical thermal–fluids systems: (a) office building, (b) wind turbines, (c) F-22 fighter aircraft, (d) room air conditioner, (e) desktop computer. (*Sources*: (a) Jon Riley/Getty Images, Inc., (b) John Turp/Getty Images, Inc., (c) Matt Ottosen/Getty Images, Inc., (d) Jessie Jean/GettyImages, Inc., (e) Don Farrall/Getty Images, Inc.)

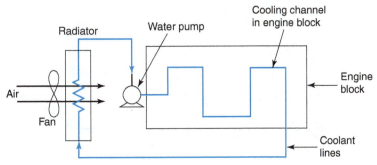

FIGURE 1-2 Schematic of engine water-cooling system.

Let us consider several of these car systems or subsystems. The water-cooling system (Figure 1-2) includes four main components: a water pump, the engine block, the radiator, and the radiator fan. Pipes connect the first three components, and there are water passages inside the engine block. A thermodynamic analysis of the engine would tell us how much heat must be removed from the engine block by the water and rejected by the water in the radiator to the air flowing through the radiator. Heat transfer analysis would tell us the number and size of passages needed in the engine block to remove the heat and would permit us to determine the necessary size of the radiator. Fluid mechanics would help us determine the pressure that must be produced by the water pump to overcome resistance to flow in the water passages, pipes, and water side of the radiator and by the fan to overcome the flow resistance on the air side of the radiator. Fluid mechanics also would tell us the power required to drive both the water pump and the fan.

Perhaps our focus is on a single piston–cylinder assembly in the engine, an idealized drawing of which is given in Figure 1-3. Thermodynamics would tell us how much energy might be extracted from a given amount of fuel. Fluid mechanics would be used to determine how effectively fresh fuel–air mixtures are inducted into the piston–cylinder through the intake valves and expelled through the exhaust valves. Heat transfer would be used to determine the energy loss from the hot combustion gases to the cooler cylinder walls.

We can examine another system—a house—to illustrate a different way in which the three governing disciplines must be used together (Figure 1-4). Consider the systems needed

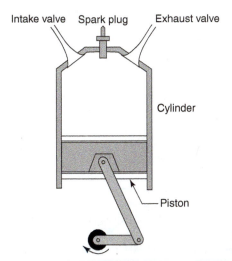

FIGURE 1-3 Schematic drawing of piston–cylinder assembly.

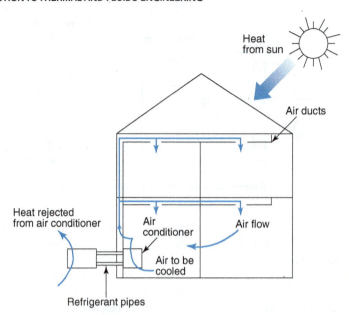

FIGURE 1-4 Schematic drawing of a house and energy subsystems.

to maintain a comfortable environment inside the house. By analyzing the construction of the walls and roof, we learn what heat transfer can tell us about how much heat will enter the house on a hot summer day. A thermodynamic analysis will tell us the size of the air conditioner needed to maintain the temperature and relative humidity inside the house. Fluid mechanics will tell us the size of the fan required to push the air through the air conditioner and the ducts needed to distribute the cooled air throughout the house.

Perhaps we want to focus on the air conditioner itself (Figure 1-5). This device is composed of two heat exchangers, a compressor, and a valve across which the refrigerant expands. A thermodynamic analysis would tell us how much electric power is needed to obtain a desired amount of cooling. A heat transfer analysis would tell us how big to build the two heat exchangers. The fluid dynamics analysis would tell us how big the pipes connecting the components must be, as well as the needed compressor characteristics.

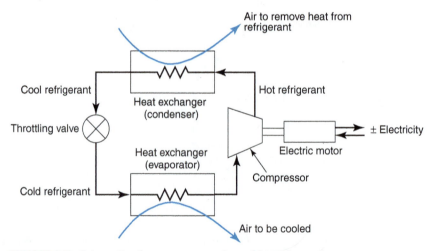

FIGURE 1-5 Schematic of vapor-compression refrigeration cycle.

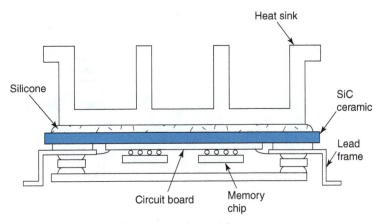

FIGURE 1-6   Sketch of a computer heat sink.

On a smaller scale, consider a cooling system used in computers (Figure 1-6). The computer chips and power supply must be cooled so that the reliability of the system is not compromised. One common cooling approach is to force air through the computer case to remove unwanted heat. A thermodynamic analysis of each chip, power supply, and component can tell us the amount of energy that must be removed from inside the computer case so that the temperature of the air surrounding electronic components will not exceed a given level. Heat transfer analysis will tell us what heat sink designs, fins, or other cooling techniques are needed to maintain the components at a safe temperature. Fluid dynamics will tell us the fan size needed to draw air through an air filter and blow the air over all the components and out of the computer case.

1.2 THERMAL AND FLUIDS SYSTEMS ANALYSIS AND ENGINEERING

Additional descriptions, similar to those given in the previous section, can be given of large industrial systems (e.g., power plants, oil refineries, chemical processing plants) and industrial processes (e.g., heat treatment of metals, food preparation) that would illustrate how thermodynamics, heat transfer, and fluid dynamics are all needed in their design. Indeed, it is the integrated use of these three disciplines that is required for rational and complete analysis of many systems.

Whenever an engineer is given an assignment to design a new device or troubleshoot an existing process or predict the performance of a system, the objective of the investigation must be very clear. Likewise, which aspect of the device, process, or system on which to focus must be well defined. A systematic approach to the whole analysis is needed. Figure 1-7 shows a flow diagram of the steps engineers typically take when analyzing and engineering a thermal fluids system. We always begin with the physical system (e.g., the engine of a car). That is reality. The engineer's job is, first, to translate the physical system into a physical model and, second, to describe the physical model with a mathematical model. The actual physical system may be so complex that it is impossible to fully describe each part and/or process. However, an engineer must obtain an answer, a solution, so he/she must use assumptions and experience to simplify the system sufficiently so that it can be modeled. Once a physical model is developed, then physical laws that govern the process (e.g., a force, momentum, mass, or energy balance) are used to create the

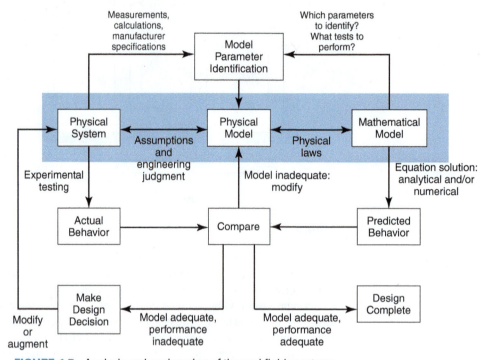

FIGURE 1-7 Analysis and engineering of thermal fluids systems.

(Adapted from K. Craig, "Is anything really new in mechatronics education?" *IEEE Robotics & Automation Magazine*, Vol. 8, No. 2, pp. 12-19, 2001. ©2001 IEEE. Used with permission.)

mathematical model, which is solved for the quantity being sought. The steps involved in defining the object being studied and identifying the processes involved in the investigation aid in quantifying/identifying the terms in the physical laws governing the analysis.

Once the mathematical model is developed and solved, the design and analysis loop would be closed by (ideally) comparing the model predictions with experimental data obtained from the actual operating device. If the model and experimental data agree sufficiently well, then the design would be complete. However, if there were disagreement, then the model would need to be modified or, perhaps, the measurements made in the experiment checked to ensure that valid data had been obtained. As shown in the figure, design and analysis are not a sequential process. Feedback and revisions are very common.

This textbook focuses on the tasks included in the shaded box in Figure 1-7. In the sections and chapters to follow, we show how to reduce a physical system to a physical model, and we show how the three primary disciplines—thermodynamics, heat transfer, and fluid dynamics—are used to organize thinking and to develop mathematical models. By necessity, topics are introduced in the context of their disciplines. However, examples and problems illustrating how the three disciplines are integrated in practice are given.

1.3 THERMODYNAMICS

Thermodynamics can be considered the unifying idea for the solution of thermal–fluids system problems. The governing concepts are: conservation of mass, conservation of energy (also called the first law of thermodynamics), and the second law of thermodynamics. Before

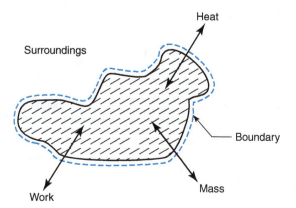

Surroundings

Heat

Boundary

Work

Mass

FIGURE 1-8 A system interacting with the surroundings.

we can discuss these concepts, we must set up a system and terminology for approaching the subject logically. We begin with identifying what we want to study.

The object we analyze is called a *system* (Figure 1-8). The region in space that contains the system is called the *control volume*. For example, the system may be an entire car or just its engine or only one piston–cylinder assembly in the engine. We may identify a complete oil refinery to examine. Likewise, we may wish to examine a particular pump or a heat exchanger in the refinery. A more detailed need may require us to determine what is occurring in one tube inside a heat exchanger. Whatever we want to examine, it is important to specify carefully what that system is. We do this by drawing a *boundary* (sometimes called a *control surface*) around the object. This line can follow the actual surface of the object or it might follow an imaginary path around the device or assembly of devices. Everything inside this line is the system; everything outside this line is the *surroundings*. Our analysis is dictated by the choice of the boundary, and several different boundaries might be chosen. One system boundary may have some advantages over another. Nevertheless, correct application of governing principles will result in identical results being obtained from the analyses. The choice of the boundary helps to establish what processes are involved and to quantify terms in the physical laws governing the process.

In thermodynamics, we can identify three types of systems. A *closed system* (Figure 1-9a) is one in which no mass crosses the boundary. Energy in any form can pass through the boundary. For example, suppose we want to determine how long it would take to boil water in a pan on a stove. We add a fixed mass of water to the pan and cover it with a perfectly sealing lid. (Ignore the air in the pan.) We identify the boundary as the inside surface of the pan and lid, and the system is only the water. We now turn on the stove. Heat transfer from the gas flame raises the temperature of the water until it begins to boil. Because of the lid, the amount of water (mass) in the system does not change; it is the same mass as at the beginning of the heating. A slightly more involved example could be a piston–cylinder assembly, similar to what is used in an engine. We assume there is perfect sealing between the piston and the cylinder and between the inlet and exhaust valves and cylinder head, so that no gas can escape from the assembly. We define the boundary to follow the walls of the cylinder and the top of the piston, so that the system is only the gas contained in the piston–cylinder assembly. Heat is added, and the piston moves because of the temperature increase in the gas. In both of these examples, the mass of the system is fixed. The volume of the first system (pot of water) is constant; the volume of the second system (piston–cylinder assembly) changes. Heat crosses the boundary in both systems. In the second system, mechanical work also crosses the boundary. (From physics, mechanical work, W, is defined as a force operating through a distance, and a force operates on the

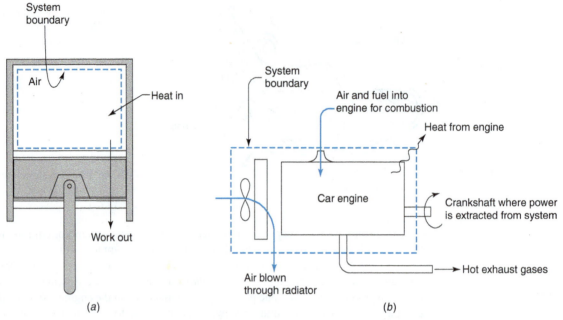

FIGURE 1-9   Examples of (a) a closed system and (b) an open system.

piston-face force due to the pressure in the cylinder.) All this information may be needed to analyze these two systems.

An *open system* is one in which both mass and energy can pass through the boundary. For example, consider an energy balance on a car engine. We define the boundary as shown on Figure 1-9b. Mechanical power produced by the engine crosses the boundary at the crankshaft. Energy leaves the system at the radiator, and there is heat transfer to the surroundings at other locations on the engine. Air enters the system through the intake manifold, and hot gases leave through the exhaust pipe. Hence, mass, heat, and work all cross the boundary. It should be noted that work and heat are defined *only* at boundaries.

Another example is an energy balance on a computer. The cooling fan draws air into the case. The air is heated by the electrical energy dissipation in the electronic components and then is blown out of the case. Electricity (a form of work) crosses the boundary to run the computer. In addition, the case itself may be hotter than the surroundings, and heat transfer occurs from the case.

The third type of system is an *isolated system*. Neither mass nor energy crosses its boundary. Consider a mixing process as shown in Figure 1-10. Two tanks are connected by a pipe in which a closed valve is placed. Each tank contains a gas at a given temperature and pressure. When the valve is opened, the gases mix and attain a common pressure and temperature. With a boundary drawn around both tanks and the connecting pipe, no mass crosses the boundary. Likewise, we could insulate the system so there is no heat transfer,

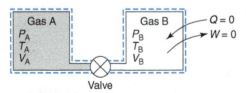

FIGURE 1-10   Example of an isolated system.

and we do not do any work on the system either. Hence, the system does not interact with its surroundings in any way. This system could be a simplified physical model of a chemical processing step in a chemical plant, in which fixed amounts of the gases are mixed together.

The various devices described above undergo some sort of *process*. The water in the pan is heated. Power is extracted from the expanding air in the piston–cylinder assembly. Heat is transferred from the electronic components in a computer to the air flowing over them and is blown into the room surrounding the computer. *A process occurs whenever some property of a system changes or if there is an energy or mass flow across the boundary of the system.* In the boiling water example, the properties that change are the temperature of the water and the total energy in the water. Because the properties of interest are different between the start and finish of the process (at different times), this is called an *unsteady* (or *transient*) process (Figure 1-11). In the computer-cooling example, both mass and energy (heat and electrical work) flow across the boundary. The property of interest may be the temperature of the air. The air temperature changes with location (from inlet to exit) but does not vary with time at either inlet or exit. This is called a *steady* process.

An electric power plant has an impressive assembly of pumps, turbines, heat exchangers, pipes, valves, controls, and so on. How would we start an analysis of such a complex installation? A simpler device to analyze may be one of the turbines used in the power plant (Figure 1-12). Again, the question is: How would we start an analysis of such a device? While the photograph of the turbine is interesting, an engineer must translate this picture into something that can be used in an analysis. In addition, the engineer must organize any and all information about the system being analyzed. One of the simplest ways to accomplish both tasks is to draw a *schematic diagram* of the system. The purpose of a schematic diagram is to show the relationship and/or interactions among the various pieces of equipment, flow streams, and energy transfers. A schematic of a power plant is shown in Figure 1-13. This drawing does not show the actual physical layout or size of the equipment. It shows only the relationships between parts. In addition, information about the equipment,

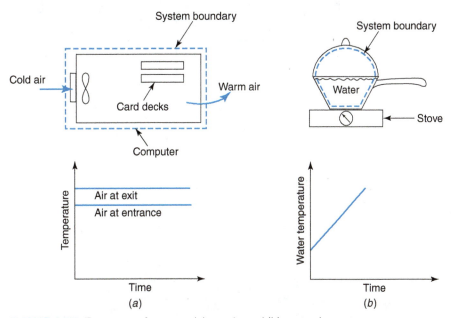

FIGURE 1-11 Two types of process: (a) steady, and (b) unsteady.

FIGURE 1-12 Gas turbine. (*Source*: Mason Morfit/ Getty Images, Inc.)

flows, or operating conditions is shown. Note that each piece of information is uniquely identified with a variable name and subscript. For example, pressures are specified at five locations, which are indicated with a subscripted number. A schematic of a gas turbine is shown in Figure 1-14. In this schematic, the boundary is indicated, mass and energy flows across this boundary are noted, and data are uniquely identified.

Many problems are too difficult to solve with all their real complexity. However, an engineer is expected to analyze the problem and obtain a reasonable or an approximate solution. Again, consider the internal combustion engine described above and the analysis of a single piston–cylinder assembly. We used the sentence: "We assume there is perfect

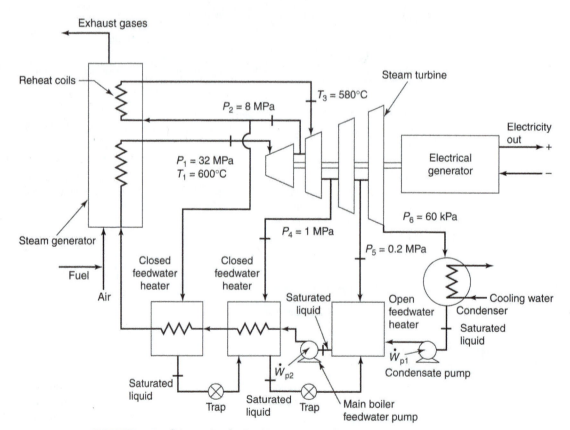

FIGURE 1-13 Schematic of a Rankine power cycle.

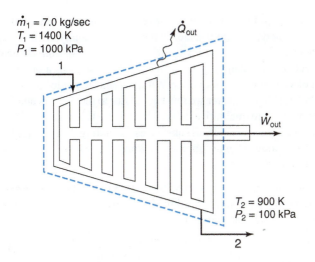

$\dot{m}_1 = 7.0$ kg/sec
$T_1 = 1400$ K
$P_1 = 1000$ kPa

1

$\dot{Q}_{out}$

$\dot{W}_{out}$

$T_2 = 900$ K
$P_2 = 100$ kPa

2

FIGURE 1-14   Schematic of a gas turbine.

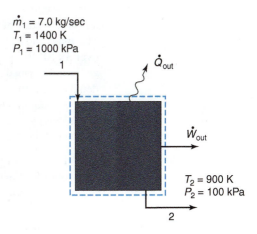

$\dot{m}_1 = 7.0$ kg/sec
$T_1 = 1400$ K
$P_1 = 1000$ kPa

1

$\dot{Q}_{out}$

$\dot{W}_{out}$

$T_2 = 900$ K
$P_2 = 100$ kPa

2

FIGURE 1-15   Blackbox representation of the gas turbine given in Figure 1-14.

sealing between the piston and the cylinder and between the inlet and exhaust valves and cylinder head, so that no gas can escape from the assembly." If we did not make this assumption, then we could not analyze this problem in a simple manner. We would need to (somehow) estimate the amount of gas that leaks past the piston. An *assumption* is used to simplify a problem. It can be considered a limitation or a restriction on the general applicability of the result obtained. Assumptions must be reasonable and justifiable. It would be all too easy to assume away the whole problem. Hence, the task of the engineer is to make enough *appropriate* assumptions to render the problem solvable, but not so many as to invalidate the result because the simplified system is too far from the actual situation.

In thermodynamics, we use what is called a *blackbox* analysis. Once we define the boundary around the object of our analysis, we infer characteristics of the system or what is happening inside the system by accounting for all the processes that take place across the boundary. That is, we account for mass flows into or out of the system, the energy flowing along with this mass flow, any mechanical work or power that crosses the boundary in either direction, and heat transfer into or out of the system across the boundary. Figures 1-14 and 1-15 are schematics of a gas turbine. The analysis of these two systems would be identical, even though they hardly resemble the gas turbine shown in Figure 1-12.

The identification of the object, the definition of a boundary, the making of assumptions, the drawing of a schematic, and the recognition of the processes involved all are intended to aid you in analyzing a system. It helps immensely if we visualize the system and physically interpret or describe what is occurring. The task of analysis is much simpler if we take time at the beginning to think about what is going on, rather than jumping in and writing with little forethought.

1.4 HEAT TRANSFER

Heat is transferred wherever there is a temperature difference between two points in a substance, whether that substance is a solid, liquid, gas, or plasma. Three types of heat transfer can occur—*conduction, convection,* and *radiation*—but regardless of the mode of heat transfer, a temperature difference drives the process. The amount or rate of heat transfer depends on the magnitude of the *thermal resistance* between the two points. For many systems, only one mode of heat transfer is needed in an analysis. In others, two or all three modes of heat transfer may be involved; this is called *multimode* heat transfer. The magnitude of heat transfer can vary from the 1–2 W typical in a computer chip to over 3×10^9 W in an electric power plant boiler. While thermodynamics uses the blackbox analysis described above, in heat transfer we must get closer to the process and look at more details of the process. Below are qualitative descriptions of the three modes of heat transfer.

Conduction heat transfer occurs in all substances, including solids, liquids, and gases and is energy transfer due to molecular vibrations within the material. A few examples of conduction heat transfer are:

- In a northern environment in the winter, the inside of a house is warmer than the outside. Energy is lost by conduction through the walls, but the loss is minimized with the use of insulating materials (Figure 1-16).

- In cold weather, people wear coats to stay warm. Body heat is conducted through the coat material out to the air. The coat is designed to minimize conduction.

- The temperature of a computer chip must be maintained below a specified temperature to ensure chip reliability. A heat sink (see Figure 1-6) is mounted on a chip to conduct away unwanted thermal energy (due to electrical power dissipation in the chip) that could impair its operation. Heat then is removed from the heat sink by air blowing over it.

- Large hydroelectric dams are constructed of concrete. The curing (or drying) of concrete is an exothermic reaction; that is, when concrete dries, it produces heat. Thermal expansion could crack the dam if too large a temperature nonuniformity occurred. Hence, a conduction heat transfer analysis is used to estimate the temperature distribution in the dam, and this information is used with a stress analysis in the dam design.

- Some machine tools are built from exotic metals that must have specific material properties, including a very hard surface and a softer core. When hot steel is removed from a furnace, the metal is quenched (cooled rapidly) at a specified cooling rate. The gradients in the properties depend, among other things, on the size of the grain structure in the solid. Grain growth and, hence, the material properties depend on the rate of cooling. A conduction heat transfer analysis can predict the temperature variations in the solid as a function of time.

Convective heat transfer occurs whenever a moving fluid (liquid or gas) flows past a solid surface that is at a temperature different from the fluid. A few examples of convective heat transfer are:

- When you run hot water over your hands, your skin temperature rises. Convective heat transfer from the hot water to your cooler hands causes the temperature rise.

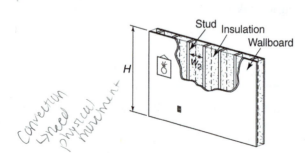

FIGURE 1-16 A typical wall construction.

- In the manufacturing of optical fibers, a long thin filament of glass is drawn continuously from a high-temperature furnace. The molten glass must be cooled before the fiber can be coated with a protective seal. This is accomplished by blowing cold gas over the fiber.

- In the winter, houses often have drafts of cold air along the floor. Heat transfer from a warm house to the cold outside air causes a decrease in the air temperature near the inside wall. Due to this cooling, the density of the air near the wall increases, and buoyancy causes this air to flow downward. Hotter air from near the ceiling replaces the cooled air, and a circulation cell is formed. This moving air past the solid surface results in *natural convection heat transfer* (also called *free convection heat transfer*). "Natural" means that buoyancy forces induce flow.

Whereas convection and conduction require some sort of material for heat transfer to occur, *radiation heat transfer* can occur in the presence of a vacuum or in the presence of a transparent or semitransparent solid, liquid, gas, or plasma. A few examples of radiation heat transfer are:

- On a clear summer day, the interior of a car with all its windows closed will have a much higher temperature than the outside air. Solar energy passes through the car windows (Figure 1-17), is absorbed by the interior seats, and then is reemitted. However, the reemitted energy cannot pass through the glass as easily as the solar energy. Hence, the trapped energy raises the air temperature. This is called the greenhouse effect.

- Concerns about global warming revolve around an energy balance and the greenhouse effect on the earth. The sun supplies radiant heat (solar energy) to the earth. How much radiant heat passes through the atmosphere to the earth from the sun or from the earth to outer space depends on the radiation characteristics of the atmosphere, which is changed by its chemical composition.

- Infrared radiant heaters are often used in industrial drying or curing processes to maintain product quality and to save energy. The radiant heat given off is similar to the radiant heat given off by a campfire, a white-hot sheet of metal as it is removed from a furnace, or the sun.

- Laser machining of materials is a technique in which precise contouring of surfaces can be obtained through the selective application of radiant energy. The laser beam heats and vaporizes the material being machined.

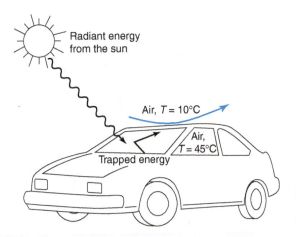

FIGURE 1-17 The greenhouse effect causes high temperatures inside the passenger cabin.

Heat transfer is intimately coupled to thermodynamics through the first law of thermodynamics. Thermodynamics concepts might be used for an overall analysis—a sort of standing back and looking at the big picture from a distance. Heat transfer analysis requires moving closer in to look at a process in more detail, and information developed with these concepts is then used in the thermodynamic analysis.

1.5 FLUID MECHANICS

Fluid mechanics is often divided into two general areas, one associated with fluids at rest—*hydrostatics*—and the other addressing relative motion between a fluid and a solid surface—*fluid dynamics*. In each case, we deal with a substance—a *fluid*—that will deform or change shape if a *shear* (or *tangential*) force is applied to it, no matter how small this force is. A fluid will not necessarily deform if we apply a *normal* force to it. One way to visualize this is to consider a stack of 500 sheets of paper. If we push down normally (perpendicularly) on the stack with our finger, nothing moves. However, if we lay our hand on the stack of paper and push sideways (parallel to the sheets), the sheets will slide over each other and the stack changes shape.

Hydrostatics (or *fluid statics*) deals with forces exerted by a stationary fluid on a solid surface. A few examples follow:

- The Monterey Bay Aquarium is a 326,000-gallon tank in which hundreds of fish from all over the world are displayed. To design the frames and support structure around the viewing windows, and to help determine the required window thickness, hydrostatics is used to calculate the forces on the window. In another example, the forces exerted on a dam (Figure 1-18) must be calculated so that the strength required to hold back the reservoir is engineered into the dam.

- Many systems have internal pressures different from that outside. Examples include aircraft flying at high altitudes, spacecraft, submarines, pipelines, helium tanks, and so on. Forces acting on the surfaces separating the two pressures can be calculated using hydrostatic principles.

- Hydraulic systems used in car and aircraft brakes, car hoists, and other hydraulic machinery employ hydrostatics principles to calculate forces and the amplification of these forces.

Fluid dynamics deals with the forces needed to push a fluid inside a conduit or past a solid surface. A few examples follow:

- Car manufacturers advertise how aerodynamically efficient their vehicles are. Fluid mechanics principles are used to estimate the drag forces on a car and to suggest ways

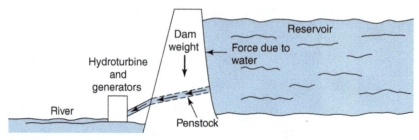

FIGURE 1-18   Schematic of a hydroelectric power plant.

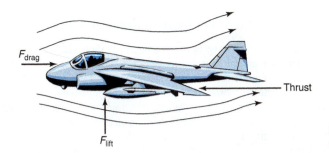

FIGURE 1-19 Aerodynamic forces acting on an airplane.

to modify the car body shape. Likewise, plane builders need to know both the drag and the lift forces acting on their designs (Figure 1-19) so that wings and fuselages are appropriate for their needs and engines can be specified accordingly. Likewise, golf ball manufacturers use fluid dynamics principles to design the dimples for the best performance of the ball.

- If the dam in Figure 1-18 is used for hydroelectric power generation, a pipe (the penstock) conveys the water from the reservoir to the water turbine, which extracts energy from the flowing water. Fluid mechanics principles and techniques are used to calculate the size of the penstock, the water turbine, and the power that can be extracted from the flowing water.

- Home and car air conditioners have fans that blow air through their cooling coils. The cooled air is circulated into the conditioned space. How "big" a fan is needed depends on the flow rate of air desired, the flow path, and the resistance to flow present in the flow path.

- In cities, utilities supply water to countless buildings of every size over a wide geographic area (Figure 1-20). Many kilometers of piping and countless valves are used. Efficient distribution of water, the pressures required, the pipe thickness, and the power required to drive the pumps are determined with fluid dynamics.

Fluid mechanics is coupled to thermodynamics through the conservation of mass, the first law of thermodynamics, and the second law of thermodynamics. In addition, conservation of momentum is used for some fluids problems. As with heat transfer, fluid mechanics concepts often require an up-close examination of details, and this information then is used in a larger view of a process.

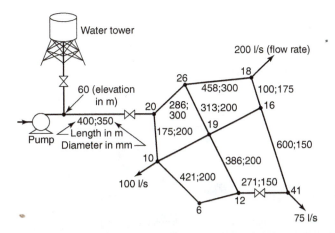

FIGURE 1-20 Water distribution pipeline network in a neighborhood.

EXAMPLE 1-1 Illustration of system identification and analysis

Consider the electric hot water heater in a house, as shown in the figure below. Because the water is at a temperature greater than its surroundings, there is heat transfer at all times from the tank, and the heat transfer rate depends on the temperature of the water inside. If the temperature in the tank falls below the thermostat set point, the heater is turned on, and it is turned off when the water temperature reaches the set point. At different times during the day, water may or may not flow through the water tank, and the electric heating element may or may not be activated. For this hot water heater, identify three different control volumes, and describe the processes, mass flows, and energy flows that occur during different operating modes throughout the day.

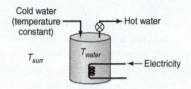

Approach:

Examining the hot water heater and the functions and processes involved in its operation, we can identify three distinct systems to analyze: the electric heater, the water in the tank, and the water plus the heater. For these chosen systems, we then can describe the processes involved. When hot water is withdrawn and sufficient cold water enters, the heater is turned on. When the hot water is shut off, the water temperature in the tank is below the thermostat set point, so the heater remains on. Finally, when the water reaches the set point, the heater is turned off. Heat conducts through the insulation around the tank and convects into the surrounding air until the water temperature falls below the set point again, and the heater is reactivated.

Assumptions:

Solution:

a) Define a control volume around *only* the electric heater, as shown in the figure below.

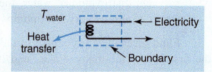

When the heater is on, electricity (an energy flow we call work) crosses the boundary into the system. The electrical resistance in the heater converts the electricity to thermal energy (Joulean heating), and the heater temperature increases to a value greater than that of the water. Because of the temperature difference, there is heat transfer from the heater to the water across the system boundary. No mass crosses the boundary, so this is a closed system. If we consider the heater immediately after the heater is turned on, the system would be transient (changing with time) because the temperature of the electric heater increases with time. If we consider the heater after it has been on for a long time, then the system would be steady because no system property (temperature, mass, energy content) changes with time.

b) Define a control volume around *only* the water in the hot water tank, as shown below.

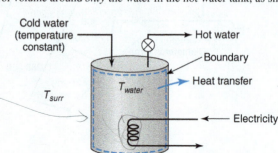

If someone in the house is using hot water, then we have an open system, because mass (water) crosses the boundary in two places. Energy flows along with the water flows. There is heat transfer from the hot water through the insulated tank wall to the surroundings, because the water temperature would be much higher than that of the surroundings. However, if the tank is heavily insulated, we might simplify the problem by ignoring this heat transfer [A1].

A1. No heat transfer from the tank to the air.

When hot water is removed from the tank and cold water enters, we would have a transient system, because the average temperature level of the water in the tank would decrease with time. If the hot water temperature drops below the temperature set on the thermostat, then the electric heater would turn on, and there would be heat transfer from the heater to the water; again, this is a transient process.

Generally, electric heaters are not large enough to raise the water temperature to the thermostat setting while water flows through the tank continuously. Therefore, after a long time period with hot water being drawn out of the tank, cold water being added, and the electric heater operating, the hot water outlet temperature would reach a constant temperature, and we would have a steady-state, open system.

If no hot water is withdrawn from the tank, then no mass crosses the boundary, and we have a closed system. Heat transfer from the tank to the surroundings would occur, and the temperature of the water in the tank would drop. Hence, this would be a transient system. If the water temperature dropped sufficiently, then the electric heater would turn on to raise the water temperature. The operating heater results in a second heat transfer process, and the system would still be transient.

c) Define a control volume around the electric heater and the water in the tank.

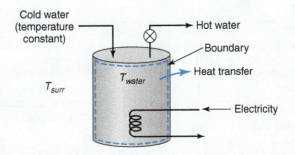

If hot water is being used, then this is an open system, because mass crosses the boundary; energy flows along with the two water flows. With the electric heater turned on, there is *only one* heat transfer process, which is from the hot water to the surroundings because of the temperature difference across the boundary. There is *no* heat transfer from the electric heater to the water, because that energy flow is *not across a boundary*. The electricity crosses the boundary and must be taken into account, as we did in part a. We identify electricity crossing a boundary as work.

Depending on how the hot water heater is operated, such as described in part b, the system defined as the water and electric heater could also operate as a closed system (no water withdrawn). Likewise, the system could be transient or steady state.

Comments:

For this simple device, the choice of the boundary will affect what we analyze, what processes occur, and how we will need to account for the energy and/or mass flows. As shown, we could have a transient or a steady-state system, heat transfer or no heat transfer, and an open or a closed system. The choice of a boundary is usually dictated by what is sought from the analysis. As long as you are careful with your analysis, the chosen boundary will have no effect on the final answer.

SUMMARY

In countless engineered systems, some aspect of thermodynamics, heat transfer, and fluid mechanics is used. Only one discipline might be needed for a specific application, or the principles and tools from all three disciplines might be required in the development of a reasonable solution to a design and/or analysis of a system. To design any of the above examples or to model or investigate their performance, three steps are always required: (1) The problem must be given thought, information organized, and a solution approach considered. (2) Fundamental concepts, equations, and definitions must be used. (3) The properties of the substances used in the problem must be evaluated. In the following chapters of this book, both the specific disciplines and integration of the concepts are presented, such that thermal and fluids engineering problems can be solved. A problem-solving approach is discussed, and methods to evaluate properties are given.

SELECTED REFERENCES

ÇENGEL, Y. A., and M. A. BOLES, *Thermodynamics: An Engineering Approach*, 3rd ed., McGraw-Hill, New York, 1998.

FOX, R. W., and A. T. McDONALD, *Introduction to Fluid Mechanics*, 5th ed., Wiley, New York, 1999.

INCROPERA, F. P., and D. P. DEWITT, *Fundamentals of Heat and Mass Transfer*, 5th ed., Wiley, New York, 2001.

MORAN, M. J., and H. N. SHAPIRO, *Fundamentals of Engineering Thermodynamics*, 5th ed., Wiley, New York, 2003.

MUNSON, B. R., D. F. YOUNG, and T. H. OKIISHI, *Fundamentals of Fluid Mechanics*, 4th ed., Wiley, New York, 2002.

SONNTAG, R. E., C. BORGNAKKE, and G. J. VAN WYLEN, *Fundamentals of Thermodynamics*, 5th ed., Wiley, New York, 1998.

THOMAS, L. C., *Heat Transfer*, 2nd ed., Capstone, Tulsa, 2000.

PROBLEMS

P1-1   For the following systems, define a control volume and state whether the system is open or closed and steady or unsteady. Identify any and all heat transfer, energy flows, mass flows, and energy transformations.

a. Rocket

b. Pot of boiling water with no lid

c. Portable space heater with fan

d. The jet airplane in Figure 1-1c

e. The house in Figure 1-4

P1-2   Describe some of the thermal–fluids systems in a typical residence, define a boundary, and describe the energy and/or mass flows associated with them.

P1-3   For the following four systems, define a control volume, state whether the system is steady or unsteady, is open or closed, has constant volume or changing volume, and has constant fluid density or changing fluid density. Also, identify all heat transfer, energy flows, and mass flows.

a. Swimming pool being filled (Choose one control volume as the whole pool; then choose a second control volume, one surface of which follows the surface of the rising water.)

b. Helium tank being filled

c. Helium balloon being filled

P1-4   A thermal solar energy system consists of a solar collector on the roof of a house, a hot water storage tank to store hot water, a heat exchanger through which the hot water passes, a fan that blows air through the heat exchanger to heat the house, and a pump to circulate water through the complete system. Define several different control volumes around different individual pieces of equipment or collections of equipment, and identify whether the control volume is steady or unsteady, open or closed; what heat transfer, energy flows, mass flows, and energy transformations occur; and whether the volume is constant or varying.

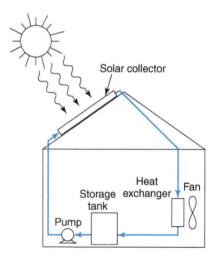

P1-5   In hydroelectric plants, electric power is generated from the flow of water from a reservoir, such as shown in Figure 1-18. The water flows continuously with a seemingly endless supply. How is the water replenished? Where does the energy in the water come from that is converted to electrical power?

P1-6 The radiator of a car is a heat exchanger. Energy from the hot water that flows through the heat exchanger is transferred to the cooler air that also flows through the radiator. For the three control volumes defined below, state whether the system is steady or unsteady, open or closed, and what heat transfer, energy flows, mass flows, and energy transformations occur.

a. Water

b. Air

c. Complete heat exchanger

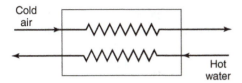

P1-7 An acorn is planted in the ground. After many years, the acorn grows into a mighty oak tree. Define a system, and describe the processes involved. Where did the mass in the tree come from?

P1-8 A Rankine cycle power plant is shown schematically in Figure 1-13. For the control volumes defined below, state whether the system is steady or unsteady, open or closed, and what heat transfer, energy flows, mass flows, and energy transformations occur.

a. Electric generator

b. Steam generator

c. Complete turbine

d. All the equipment shown

P1-9 A vapor-compression refrigeration cycle, similar to what is used in air-conditioning systems, is shown schematically in Figure 1-5. For the control volumes defined below, state whether the system is steady or unsteady, open or closed, and what heat transfer, energy flows, mass flows, and energy transformations occur.

a. Electric motor

b. Refrigerant flowing through condenser

c. Complete condenser

d. Throttling valve

e. All the equipment shown

P1-10 A hot cup of coffee is placed on a tabletop to cool. Define a control volume, and state whether the system is steady or unsteady, open or closed, and what heat transfer, energy flows, mass flows, and energy transformations occur.

P1-11 The water in a canal lock is at the downstream river level and the gates are opened. A boat enters the lock, and the downstream gates are closed. A valve is opened, and water from upstream flows into the lock, raising the boat. After the water reaches the upstream river level, the upstream gates are opened, and the boat travels upstream. Finally, the first valve is closed and a second valve is opened, allowing the water in the lock to

drain to the downstream river level. Another boat arrives from downstream, and the process is repeated. Neglect the energy required to open and close the gates and valves. Where does the energy come from to raise the boat?

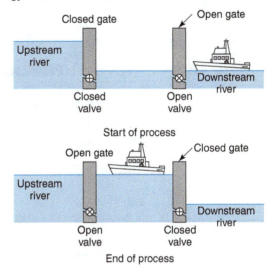

P1-12 A closed pan of cold water is placed on a burner of an electric stove, which is already turned on. For the control volumes defined below, state whether the system is steady or unsteady, open or closed, and what heat transfer, energy flows, mass flows, and energy transformations occur.

a. Pan of water

b. Burner

c. Pan of water plus burner

P1-13 Water from a home swimming pool is pumped through a filter and returned to the pool. If the system is all the water in the pool and filter, is this an open or closed system? If the system is just the water in the filter, is this an open or closed system?

P1-14 Wind turbine systems, such as shown in Figure 1-1b, consist of a wind turbine, an electric generator connected to the wind turbine, and a power line connecting the generator either to the electrical grid or to battery storage. In a steady wind, for the control volumes defined below, state whether the system is steady or unsteady, open or closed, and what heat transfer, energy flows, mass flows, and energy transformations occur.

a. Wind turbine

b. Battery

c. Electric generator

d. Wind turbine, electric generator, and electrical grid

e. Wind turbine, electric generator, and battery

P1-15 Global warming has been in the news much in recent years. Define an appropriate control volume to study this system and state whether it is steady or unsteady, open or closed, and what heat transfer, energy flows, mass flows, and energy transformations occur.

CHAPTER 2

THE FIRST LAW

2.1 THE FIRST LAW OF THERMODYNAMICS

2.40

The central organizing idea of thermodynamics is the principle of conservation of energy. This one idea is vital to understanding an enormous range of processes. In the absence of nuclear reactions, in which mass is converted to energy, total energy is always conserved under all circumstances, regardless of the form of energy. Conservation of energy is so important in thermodynamics that it is called the *first law of thermodynamics.*

In this chapter, the first law for a **closed system** will be introduced. As defined earlier, a closed system consists of a fixed amount of mass. No mass enters or leaves the system. In a closed system, the first law may be expressed as

$$\Delta E = Q - W \tag{2-1}$$

where ΔE is the change in all forms of energy stored in the system, Q is the net energy that is added to the system in the form of heat, and W is the net energy that leaves the system in the form of work. Eq. 2-1 applies to a process that takes place over a finite time interval. The quantity Q is the net heat that is added during this time interval, and the quantity W is the net work done during the time interval; Q and W could be positive or negative depending on the direction of the net energy flow of each quantity. The change in stored energy, ΔE, is the difference between the energy of the system at the end of the process and the energy of the system at the start of the process.

A simple schematic that illustrates the first law for a closed system is shown in Figure 2-1. The system is the mass contained within the dotted line. Heat and work are forms of energy that cross the system boundary, while E is a form of energy that is stored within the system boundary. The first law is a balance among these various forms of energy. It states that:

$$\begin{pmatrix} change\ in \\ energy \end{pmatrix} = \begin{pmatrix} energy \\ entering \end{pmatrix} - \begin{pmatrix} energy \\ leaving \end{pmatrix}$$

The energy, E, stored in the system consists of three components: kinetic energy, potential energy, and internal energy. Kinetic energy, KE, is due to the velocity of the system and has a magnitude given by

$$KE = \frac{1}{2}m\mathcal{V}^2 \tag{2-2}$$

where m is the total mass and $\mathcal{V}$ is the magnitude of the velocity of the system relative to an inertial reference frame. In this text the magnitude of the velocity vector will always be designated as $\mathcal{V}$ to distinguish it from volume, V. The change in kinetic energy during a process may be expressed as

$$\Delta KE = \frac{1}{2}m\mathcal{V}_2^2 - \frac{1}{2}m\mathcal{V}_1^2 \tag{2-3}$$

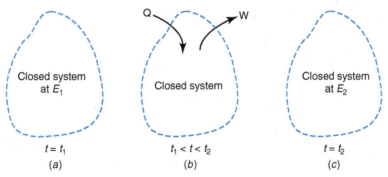

FIGURE 2-1 The stored energy in this closed system changes from E_1 to E_2 as time goes from t_1 to t_2 ($\Delta E = E_2 - E_1$). Q is the net energy entering as heat between t_1 and t_2, while W is the net energy leaving as work between t_1 and t_2.

Potential energy, PE, is due to the elevation of the system in a gravitational field and is given by

$$PE = mgz \tag{2-4}$$

where g is the acceleration of gravity and z is the elevation above a reference plane. The change in potential energy during a process is

$$\Delta PE = mgz_2 - mgz_1 \tag{2-5}$$

Internal energy, U, is energy stored at a molecular or atomic level. There is no simple expression for internal energy that applies to all cases. In single-phase materials such as solids, liquids, or gases, the internal energy depends primarily on the temperature. Internal energy is also stored in chemical bonds and in the attractive forces between the molecules of solids and liquids.

If kinetic, potential, and internal energy are substituted into Eq. 2-1, the result is:

$$\Delta KE + \Delta PE + \Delta U = Q - W \tag{2-6}$$

where ΔKE is the change in kinetic energy, ΔPE is the change in potential energy, and $\Delta U = U_2 - U_1$ is the change in internal energy of the system. Much of the discussion in this chapter and the next is devoted to explaining each of the five forms of energy in Eq. 2-6 and showing how they interact in a wide variety of applications. But, before formal definitions and detailed explanations are given, a few examples of the use of Eq. 2-6 are presented. This approach develops an intuitive understanding of the first law and will be a useful introduction to the study of thermal and fluid systems.

Consider a gas contained in a piston–cylinder assembly as shown in Figure 2-2. We define the closed system as the gas. Its boundary is indicated by a dotted line. The piston, on which a weight rests, is free to rise or fall. At the start of the process, the gas is at temperature T_1. When heat is added, the gas in the cylinder expands, and the temperature of the gas increases to T_2.

In this process, there is no change in kinetic or potential energy. Therefore, the first law becomes

$$\Delta U = Q - W$$

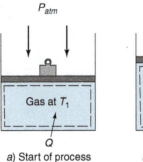
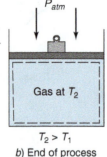

a) Start of process

b) End of process

$T_2 > T_1$

FIGURE 2-2 Conversion of heat into work and internal energy in a piston–cylinder assembly.

As the gas expands, it does work in lifting the weight and in pushing against atmospheric pressure. The work occurs at the boundary, specifically at the boundary between the gas and the face of the piston. Heat is also transferred at a boundary, that is, at the bottom surface of the piston–cylinder assembly as shown in Figure 2-2. During the process, the temperature of the gas rises. Higher-temperature gas has more internal energy than lower-temperature gas. Unlike heat and work, the internal energy is stored throughout the volume of the gas. In effect, the added heat has been converted into work, which leaves the system, and internal energy, which is stored in the system.

In this example, the heat added to the system has been considered to be a positive quantity. Conversely, if heat were removed, then the heat would be a negative quantity. This sign convention will be used throughout the text:

- heat transfer to a system is positive
- heat transfer from a system is negative

Note that if no heat transfer occurs, the system is called **adiabatic**. Work is also subject to a sign convention, which is

- work done by a system is positive
- work done on a system is negative

These conventions are arbitrary. If we had defined work done by a system as negative, then the minus sign on the right hand side of Eq. 2-1 would become a plus sign. A good way to remember the sign convention is to think of an automobile engine. Heat is **transferred to** the engine during combustion of the gasoline. Work is **done by** the engine to drive the wheels. In our convention, both are positive quantities. When you worked a statics problem, you often had to assume the direction of a force on a free-body diagram. For heat and work, we always assume a consistent direction and let the sign of the quantity tell us the actual direction.

In Figure 2-3, the example system from Figure 2-2 is modified to include a paddlewheel and heat loss to the surroundings. The boundary is defined as the surface that covers the inside of the cylinder and also encloses the paddlewheel blades. Thus the system is all the gas in the cylinder. The paddlewheel does work on the gas by stirring it. If we assume the gas is at a temperature higher than the surrounding air and the sides of the cylinder are not thermally insulated, then heat is lost from the gas through the wall. There are two heat interactions shown in Figure 2-3. The quantity Q_1 is the heat added through the bottom of the cylinder Q_{in}, and the quantity Q_2 is the heat lost through the side walls Q_{out}. Both of these heats cross the boundary, but at different locations. There are also two work interactions. The work W_1 is the work done by the paddlewheel on the gas W_{in}. This

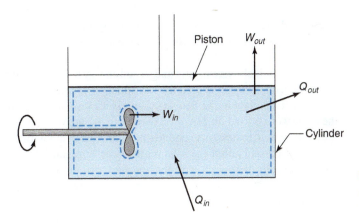

FIGURE 2-3 Gas in a piston–cylinder assembly with a paddlewheel.

work occurs at the boundary of the gas, where it contacts the paddlewheel surfaces. The work W_2 is the work done by the gas pushing against the piston and elevating it W_{out}. This work also crosses the boundary, but at a different location.

In Eq. 2-1, Q is the **net heat added to** the system and W is the **net work done by** the system. The net heat is the sum of all the individual heat interactions across the boundary, that is

$$Q = \sum_n Q_i$$

where n is the number of interactions. The net work is the sum of all the individual work interactions, or

$$W = \sum_n W_i$$

For example, suppose 8 kJ of heat are added through the bottom of the cylinder in Figure 2-3 and 2 kJ of heat are lost through the sides. Then the net heat transfer is

$$Q = \sum_n Q_i = Q_1 + Q_2 = (8 \text{ kJ}) + (-2 \text{ kJ}) = 6 \text{ kJ}$$

Note that Q_2 is negative because heat is leaving the system. Further suppose that the paddlewheel does 5 kJ of work on the gas and the gas does 9 kJ of work in raising the piston. Then, the net work is

$$W = \sum_n W_i = W_1 + W_2 = (-5 \text{ kJ}) + (9 \text{ kJ}) = 4 \text{ kJ}$$

In this case W_1 is negative because the paddlewheel is doing work on the system, while W_2 is positive because it is work done by the system on the piston and surroundings. The change in internal energy for this process is

$$\Delta U = Q - W = 6 \text{ kJ} - 4 \text{ kJ} = 2 \text{ kJ}$$

In an actual physical system, there would be additional heat loss from the gas into the blades of the paddlewheel. To account for this heat loss, an additional term would have to be included in the equation.

We describe heat, work, and stored energy with units of energy. Two systems of units will be employed in this text, the Standard International System (SI), used throughout the world, and the British system, used primarily in the United States. Practicing engineers typically must be capable of using either system. In the SI unit system, energy has units of joules (J).

In the British system, heat and stored energy are typically measured in British thermal units, or Btu. Work is ordinarily measured in ft-lbf, where lbf means "pounds-force." Unfortunately, mass in the British system is also usually measured in "pounds." Although the same word is used, in fact these two "pounds" are very different. In this text, the symbol lbm will be used to designate pounds-mass and the symbol lbf will designate pounds-force. More will be said about the relation between lbm and lbf later in the chapter. Although work is usually measured in ft-lbf and heat and stored energy are usually measured in Btu, in fact, both ft-lbf and Btu are units of energy. Units of Btu can be converted into ft-lbf using

$$1 \text{ Btu} = 778.169 \text{ ft-lbf}$$

EXAMPLE 2-1 Compression with heat transfer and shaft work

A gas is contained in a piston–cylinder assembly. The gas is compressed when 670 J of work are done on it. Over the same time period, a paddlewheel does 182 J of work on the gas and the internal energy decreases by 201 J. How much heat has been transferred during this process? Was the gas heated or cooled?

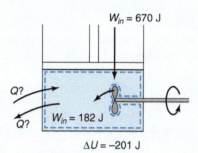

Approach:

Define the gas as the control volume. Use the first law, assuming kinetic and potential energies are negligible. Pay careful attention to the signs of all terms. Calculate the net heat transfer. If this is positive, the gas has been heated. If it is negative, then the gas has been cooled.

Assumptions:

A1. Kinetic energy is negligible.
A2. Potential energy is negligible.

Solution:

The system is the gas in the cylinder. Assuming no kinetic or potential energy changes [A1][A2], the first law is

$$\Delta U = Q - W$$

The change in internal energy may be written as

$$\Delta U = U_2 - U_1$$

where U_2 is the internal energy at the end of the process and U_1 is the internal energy at the start of the process. Because the internal energy decreases, $U_2 < U_1$ and ΔU is negative. Thus

$$\Delta U = -201 \text{ J}$$

Both the piston and the paddlewheel do work *on* the gas. By the sign convention, work done *on* a system is negative. The net work in this process is

$$W = -(670 + 182)\,\text{J}$$
$$= -852\,\text{J}$$

The first law may be rearranged to

$$Q = \Delta U + W$$

Then,

$$Q = -201 + (-852)$$
$$Q = -1053\,\text{J}$$

Because the heat transfer is negative, the system has been cooled.

The next application of the first law involves both kinetic and potential energy. Consider the motion of a bicycle and rider, as shown in Figure 2-4. Initially, the bicycle is at rest at a point near the top of a hill. The rider releases the brakes and rolls down the hill without pedaling. After reaching the bottom, the bike climbs the next hill, gradually slowing down. We analyse this motion for several different cases. In case 1, let us imagine a perfect world in which there is no friction and no aerodynamic drag. From the first law,

$$\Delta KE + \Delta PE + \Delta U = Q - W$$

The system is the bicycle and the rider. No work is done, either by the rider in pedaling the bicycle or by the bicycle in overcoming air resistance. We assume the bike and rider are at atmospheric temperature at the start of the process. If there is no frictional heating as the bike moves, then there is no change in temperature of any of the moving parts of the bicycle. Because the temperature is unchanged, the internal energy of the system remains constant and no heat is transferred between the bike or rider and the surroundings. Under these circumstances, the first law reduces to

$$\Delta KE + \Delta PE = 0$$

Suppose the bike starts at rest at an elevation, z_1, above the bottom of the hill. Then, in the perfect world of case 1, the bike rolls down the hill and travels up the next hill, coming to rest at elevation z_2. In this process the kinetic energy is zero at both the start and the finish,

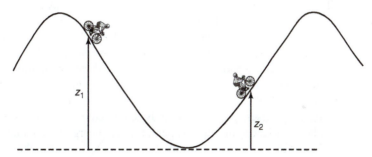

FIGURE 2-4   Motion of a bicycle in hilly terrain.

so ΔKE is zero. The first law then becomes

$$\Delta PE = 0 = mgz_2 - mgz_1$$

Clearly, $z_1 = z_2$. In case 1, the bike comes to rest at exactly the same elevation at which it started.

In case 2, the bike again starts at rest and rolls down one hill and up the next while the rider does not pedal. However, in this case, the real effects of friction and aerodynamic drag will be considered. The rolling friction of the tires against the pavement and the friction within the wheel bearings both contribute to a localized rise in temperature and, thus, in internal energy of the system. During the motion, some of this internal energy is transferred to the surroundings in the form of heat. Because friction always results in a temperature rise, ΔU will be positive in magnitude. Heat, Q, leaves the system, so it is negative. The bike and rider must overcome the drag force exerted by the air. In acting against this drag, the system does work on the surrounding air. Work done by a system is positive. Note that the work again occurs at the boundary, this time at the outer surface of rider and machine. The first law for case 2 is

$$\Delta KE + \Delta PE + \Delta U = Q - W$$

The bike begins and ends at rest, so ΔKE is zero and

$$\Delta PE = -\Delta U + Q - W$$

As described above, ΔU is positive, Q is negative, and W is positive. It follows that

$$\Delta PE < 0$$

Furthermore,

$$\Delta PE = mgz_2 - mgz_1 < 0$$

or

$$z_2 < z_1$$

As shown in Figure 2-4, z_2 is the elevation at the end of the bike's trajectory and z_1 is the elevation at the beginning. The first law predicts that z_2 will be less than z_1, in accordance with physical experience.

In case 3, the bike and rider start at rest at the top of the hill, roll down the first slope, and start to climb the next hill. This time, the rider sees that the bike will not rise to the top of the hill and pedals for a short time. The bike comes to rest at exactly the same elevation as it started. Now there are several new terms in the first law. Within the rider's body, stored chemical energy is converted into muscular energy, and work is done by the rider's feet in pressing against the pedals. (If you have ever ridden a bicycle, you know that this *is* work). The work done by the rider in operating the pedals is transmitted through the bicycle drivetrain and is manifested as work done by the tires against the pavement. The conversion of stored chemical energy into muscular energy is not 100% efficient; some of the chemical energy is converted into internal energy of the rider's body, and the body temperature rises. Because the body temperature has risen, some heat is given off from the body surface to the environment. Internal energy is contained in the chemical bonds of the glucose that is burned within the body. The chemical change results in a reduction in internal energy of the rider's body.

The first law is

$$\Delta KE + \Delta PE + \Delta U = Q - W$$

In case 3, the bicycle starts and stops at rest and the net change in elevation is zero. Therefore, the first law becomes

$$\Delta U = Q - W$$

When each of these terms is expanded to show all the interactions, we get

$$\Delta U_1 + \Delta U_2 + \Delta U_3 = Q_1 + Q_2 - (W_1 + W_2)$$

where

$\Delta U_1 = $ internal energy increase due to friction in the tires, wheel bearings, and drivetrain components

$\Delta U_2 = $ internal energy increase from the chemical reaction, which causes a rise in the rider's body temperature

$\Delta U_3 = $ internal energy reduction due to chemical reaction (the products of reaction have less internal energy than the reactants)

$Q_1 = $ heat leaving the surface of the bicycle from parts heated by friction

$Q_2 = $ heat leaving the surface of the rider's body

$W_1 = $ work done in overcoming air resistance

$W_2 = $ work done by the tires against the pavement

Note that the work done by the feet on the pedals is not included in this equation. Work in the first law is energy that *crosses* the boundary of the system. The system here is the rider *and* the bicycle. The feet contacting the pedals are internal to the system. Likewise, the transmission of power from the chain to the gears is work internal to the system. The boundary contacts the air and the pavement. Thus only the terms for overcoming air resistance and friction between the tires and pavement appear as work in the first law.

The actual calculation of the heat, work, and internal energy terms in this example requires a rather broad knowledge of thermal and fluids engineering. The calculation of drag, for example, depends on a knowledge of external flow, which is described in Chapter 10. The calculation of heat leaving the surface of the bike and rider requires the information presented in Chapter 12. The chemistry of the reaction within the rider's body can be understood using concepts in Chapter 15. Because so much knowledge is required for most real-world situations, the examples and problems presented in this text are generally simplified. Nevertheless, they capture the essence of the phenomena under study and include the most important effects and interactions.

An additional application of the first law is shown in Figure 2-5. When electric power is first supplied to a resistive element on an electric stove, the temperature of the element rises. We define the system as the resistive element. Electrical work is done on the element. This work is converted into heat, which leaves the system, and internal energy, which acts to raise the resistor temperature. Eventually, the temperature of the element reaches steady state, and heat from the element is used to cook food.

The examples in this introduction are just a small sample of all the possible applications of the first law. As you can see, the applications are quite diverse and can involve

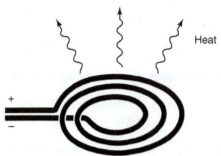

FIGURE 2-5   Conversion of work into energy and heat in a stove's resistive element.

many different forms of work and internal energy. There are also several different modes of heat transfer, for example, conduction, convection, and radiation. These are described qualitatively in the next section and will be briefly treated quantitatively in Chapter 3. The first law is a very powerful tool in understanding engineering systems and is one of the truly great ideas of all time. The first law will be used extensively throughout the text.

EXAMPLE 2-2   Kinetic and potential energy in the first law

A diver runs down a diving board, jumps into the air, lands on the board, depresses the end by 0.4 m, and then is launched into the air. The end of the board, when undeflected, is 4 m above the surface of the water. The board does 900 J of work on the diver, who has a mass of 59 kg. With what velocity does the diver strike the surface of the water? (Neglect aerodynamic drag and velocity parallel to the surface of the water.)

Approach:

Define the system as the diver, and divide the problem into two segments. Consider first the motion of the diver between the end of the board and the high point of the trajectory. Use the first law to find the maximum height attained. Next, consider the motion from the high point to the surface of the water. Again apply the first law, this time to calculate the velocity.

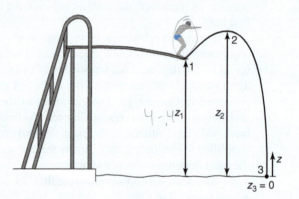

Assumptions: ### Solution:

We define the diver as the system under study. From point 1 to point 2, the first law for the diver is

$$\Delta KE + \Delta PE + \Delta U = Q - W$$

A1. The diver is at the air temperature.

There is no change in internal energy and no heat transferred [A1]. The first law therefore becomes

$$(KE_2 - KE_1) + (PE_2 - PE_1) = -W$$

At point 1, the diver has just decelerated to zero velocity and is now changing direction and beginning to accelerate upward. At point 1, the direction of motion changes and the velocity is instantaneously zero. At point 2, the diver is again changing direction and the velocity (in the z direction) is also instantaneously zero. We neglect components of velocity in the horizontal direction [A2]. Therefore,

A2. Horizontal velocity is small compared to vertical velocity.

$$KE_1 = KE_2 = 0$$

and the first law reduces to

$$PE_2 - PE_1 = -W$$

A3. Aerodynamic drag is negligible.

Work is done **on** the system, that is, on the diver by the board. We neglect any work done by aerodynamic drag [A3], so

$$W = -900 \text{ J}$$

At point 1, the board is deflected 0.4 meters below its undeflected position of 4 m, so $z_1 = 3.6$ m. The first law then becomes

$$mgz_2 - mgz_1 = -W$$

$$(59 \text{ kg}) \left(9.8 \, \frac{\text{m}}{\text{s}^2}\right) (z_2 - 3.6) \text{ m} \left(\frac{1 \text{ N s}^2}{1 \text{ kg·m}}\right) \left(\frac{1 \text{ N·m}}{1 \text{ J}}\right) = -(-900 \text{ J})$$

or

$$z_2 = 5.15 \text{ m}$$

In the second part of the analysis, the diver accelerates from point 2 to point 3, which is at the surface of the water. No heat is transferred and no work is done; therefore, the first law becomes

$$\Delta KE + \Delta PE = 0$$

$$(KE_3 - KE_2) + (PE_3 - PE_2) = 0$$

or

$$\frac{1}{2}m(V_3^2 - V_2^2) + mg(z_3 - z_2) = 0$$

Solving for V_3

$$V_3 = \sqrt{2g(z_2 - z_3) + V_2^2}$$

Substituting values,

$$V_3 = \sqrt{2\left(9.8 \, \frac{\text{m}}{\text{s}^2}\right)(5.15 - 0) \text{ m} + 0}$$

$$V_3 = 10 \text{ m/s}$$

2.2   HEAT TRANSFER

Heat is defined as the transfer of energy due to a temperature difference. There are two fundamental modes of heat transfer, conduction and radiation. **Conduction heat transfer** occurs in solids, liquids, and gases. In a solid, molecules vibrate about their equilibrium positions. The higher the temperature of the solid, the more energetic the vibrations. Now imagine two solids at different temperatures coming into contact. An example might be an ice cube and the back of your neck. At the surface of contact, molecules in the hot solid—your neck—vibrate vigorously, whereas the molecules in the ice cube vibrate less energetically. Because the surfaces are in contact, the vigorous molecules in the hot solid excite the sluggish molecules in the cold solid. As a result, energy transfers from the hot solid to the cold solid, and this process is called **heat transfer**. As your neck loses heat to the ice cube, the molecules in your skin vibrate less vigorously, the skin cools precipitously, and your entire organism is likely to react.

As stated earlier, heat is energy that enters or leaves a system at a boundary. In the last example, the boundary is the plane of contact between the ice cube and the skin. In conduction, heat flows across the boundary because of a temperature difference. For conduction to occur, molecules in the hot and cold substances must be in close proximity. Heat is not stored in a system. Rather, it is energy in motion across the boundary of a system.

Radiation heat transfer, the second fundamental mode of heat transfer, is energy transfer via electromagnetic waves. Radiation can occur in gases, liquids, and solids. As an example, consider the radiation from a campfire to a tired hiker. In the molecules of the hot gas that makes up the flame, electrons fall to lower energy levels and emit photons as a result. You can see some of these photons, that is, those in the visible range of wavelengths. The combusting gas in the flame also emits photons in the infrared range. The photons propagate through the air, and the infrared photons are absorbed on the skin of the hiker's outstretched hands. The absorbed photons raise electrons in the skin to higher energy levels and the hands become warm. Radiation differs from conduction in one important way. Photons can travel through a vacuum, so radiation does not require a transmission medium. This is why we can feel radiation from the sun. Conduction, on the other hand, always occurs within a medium or between two media in contact.

Conduction and radiation are the two fundamental modes of heat transfer. However, when we consider conduction in the presence of a moving fluid, we generally call this **convective heat transfer**. When you fan yourself on a hot day, you benefit from convection heat transfer. The heat from your face is conducted into nearby air and the air temperature increases. With the fan, you create a flow that displaces the warm air and replaces it with cool air. This is the essence of the convection process.

Heat transfer is a major topic in this text. Chapter 3 will introduce a quantitative description of heat transfer.

2.3   INTERNAL ENERGY

When energy is added to or removed from a system, changes in system properties occur. For example, if energy (either heat or work) is added to a copper block, its temperature will rise. On a microscopic level, the energy that flows into the block causes more energetic vibrations of the copper molecules, and temperature depends on these vibrations. On the other hand, energy addition does not always result in a temperature increase. If heat is transferred to a block of ice at 0°C, it melts into liquid water, but its temperature does not change. In this case, the heat transfer to the ice breaks the bonds in the solid structure and causes the solid to liquify.

In order to understand changes such as these, the concept of internal energy was invented. Internal energy is energy stored in the material. It can take many forms. In a solid, energy is stored in the vibrations of the atoms about their equilibrium positions. It is also stored in the interatomic and/or intermolecular bonds that hold a solid crystal in place. When energy flows into a solid due to heat transfer, that energy is stored as internal energy. The increased internal energy is manifested either in increased vibrations or in a change of phase. If vibrations increase, temperature rises; if bonds are broken, temperature remains constant.

In addition to these examples, there are many other forms of internal energy. The chemical bonds between the atoms in a molecule contain internal energy. In a gas, the translation of the molecules, the rotation of the molecules about their centers of mass, and the internal vibrations of the molecules all contribute to the internal energy. Internal energy is also stored in the nuclei of atoms.

2.4 SPECIFIC HEAT OF IDEAL LIQUIDS AND SOLIDS

Consider a tank partially filled with liquid, as shown in Figure 2-6. When heat is added to the tank, as in Figure 2-6a, the temperature will rise. The temperature will also rise if work is done on the liquid, even if there is no heat transfer. This situation is depicted in Figure 2-6b, where shaft work is done on a liquid in a heavily insulated tank, so that it can be considered adiabatic (no heat transfer). The temperature rise can be determined by experiment. Suppose, for example, we add 5 J of heat to the liquid shown in Figure 2-6a, and the temperature rises 12°C. If we add 5 J of shaft work as shown in Figure 2-6b, the temperature again would rise 12°C.

These two cases can be analyzed with the first law. Define the system as the liquid in the tank. In both cases, there is no change in the liquid kinetic or potential energy. The first law then becomes

$$\Delta U = Q - W$$

In case (a), no work is done, and the first law reduces to

$$\Delta U = Q = 5\,\text{J}$$

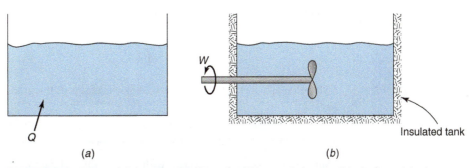

(a)   (b)

FIGURE 2-6 In case (a), heat is added to a liquid in a tank. In case (b), shaft work is done on a liquid in an insulated (adiabatic) tank. In this case, we also assume insignificant heat transfer between the liquid and the paddlewheel blades.

In case (b), no heat is transferred. Also note that work is done *on* the system in this case. Therefore, work is negative and the first law becomes

$$\Delta U = -W = -(-5 \text{ J}) = 5 \text{ J}$$

In each case, the internal energy increases by 5 J while the temperature increases by 12°C. If the mass, m, of liquid in the tank is doubled, then the internal energy must increase by 10 J to produce a temperature rise of $\Delta T = 12$°C. These ideas may be expressed as

$$\Delta U \propto m\Delta T$$

Now suppose a differential amount of heat is added to the liquid and the internal energy changes by a differential amount dU. Let dT be the differential temperature rise that results. Then

$$dU \propto m\,dT \qquad (2\text{-}7)$$

We can turn Eq. 2-7 into an equality if a proportionality constant is introduced. Experiments show that, in the general case, the proportionality constant is a function of temperature. It is common to designate this proportionality constant as $c(T)$, so that Eq. 2-7 becomes

$$dU = mc(T)\,dT \qquad (2\text{-}8)$$

where $c(T)$ is called the **specific heat** of the liquid. Eq. 2-8 also holds true for solids. In fact, it applies to so-called **ideal liquids and solids**. An ideal liquid or solid is one that is **incompressible**. By definition, an incompressible substance has a constant volume per unit mass. Most solids and liquids can be considered incompressible for ordinary ranges of temperature and pressure. For example, the pressure on an open tank of water at atmospheric pressure can be doubled and the change in volume of the water will be virtually imperceptible. If a solid block is heated, its volume will expand slightly. The expansion is usually so small as to be negligible, so that we can idealize the solid as "incompressible."

In some cases, the specific heat does not vary significantly with temperature and may be regarded as constant. Then, integrating Eq. 2-8 between states 1 and 2:

$$\int_1^2 dU = \int_1^2 mc\,(T)\,dT$$

If $c(T)$ is not a function of temperature, it may be removed from the integral to give

$$\int_1^2 dU = mc \int_1^2 dT$$

where m, also a constant, has likewise been removed. Integration results in

$$\boxed{\Delta U = mc\Delta T} \qquad \text{constant specific heat} \qquad (2\text{-}9)$$

Suppose the specific heat varies with temperature. Then, integrating Eq. 2-8:

$$\int_1^2 dU = \int_1^2 mc(T)\,dT$$

Performing the integral on the left and removing the constant, m, from the integral on the right gives

$$U_2 - U_1 = m \int_1^2 c(T)\,dT$$

If an expression for $c(T)$ is known, that expression can be substituted into the above equation and the integral evaluated. In many cases, however, an equation for $c(T)$ is not available. As an approximation, an average value of specific heat, c_{avg}, representative of the specific heat during the complete process, will often produce good results. With this approach

$$U_2 - U_1 = m \int_1^2 c_{avg}\,dT$$

or, evaluating the integral,

$$U_2 - U_1 = mc_{avg}(T_2 - T_1) \tag{2-10}$$

where

$$c_{avg} = c(T_{avg})$$

and

$$T_{avg} = \frac{T_1 + T_2}{2}$$

Eq. 2-10 will be exactly correct if specific heat varies linearly with temperature. Specific heat has units of J/kg · K or Btu/lbm · °F.

Specific heat is one of many useful quantities in thermal–fluids engineering that is determined by experimental measurement. The appendices contain numerous tables of data for a variety of thermophysical and thermodynamic properties, including specific heat. Tables A-1 through A-17 present results in SI units and Tables B-1 through B-16 give the corresponding quantities in British units. Specific heats for many solids are given in Tables A-2 through A-5 and in Tables B-2 through B-5. Values of specific heat for some liquids are included in Tables A-6 and B-6.

EXAMPLE 2-3 **Specific heat of a solid**

A 0.5-kg steel ball is dropped from a height of 60 m. It becomes embedded in the ground. Estimate the temperature rise of the ball just after impact.

Approach:

The first law will be used to find the potential energy change and to relate this to the change in internal energy of the ball. Then the relation between internal energy and temperature, that is,

$$\Delta U = mc\Delta T$$

will be used to find the temperature rise.

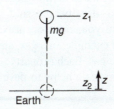

Assumptions:

A1. The ball begins at the air temperature.

A2. The ball rises in temperature rapidly after impact and does not exchange heat with the soil.

A3. Aerodynamic drag is negligible.

A4. The specific heat of the ball is constant.

Solution:

Define the ball as the system under study. The first law is

$$\Delta KE + \Delta PE + \Delta U = Q - W$$

In this process, the ball starts at rest and ends at rest. While it does have kinetic energy during flight, the *change* in kinetic energy between start and finish is zero. We assume the ball begins at the same temperature as the environment [A1], so no heat is transferred during flight. In addition, we would like to calculate the temperature just after impact, before the ball has time to exchange heat with the surrounding soil [A2]. These last two points imply that $Q = 0$. The aerodynamic drag on the ball is small and can be neglected [A3]; therefore, no work is done on the ball by the atmosphere and $W = 0$. This leaves

$$\Delta PE + \Delta U = 0$$

or

$$(PE_2 - PE_1) + (U_2 - U_1) = 0$$

where points 1 and 2 are shown on the figure. If we assume that the specific heat is constant [A4],

$$mg(z_2 - z_1) + mc(T_2 - T_1) = 0$$

Rearranging,

$$T_2 - T_1 = \frac{g(z_1 - z_2)}{c}$$

Using the specific heat from Table A-2 (in the appendix),

$$T_2 - T_1 = \frac{\left[9.81\,\frac{m}{s^2}\right]\left[60 - 0\right]m}{\left[0.235\frac{kJ}{kg\cdot °C}\right]\left[\frac{1000\,J}{1\,kJ}\right]\left[\frac{1\frac{kg\cdot m}{s^2}}{1\,N}\right]\left[\frac{1\,N\cdot m}{1\,J}\right]} = 2.50\,°C$$

Note that T_2, the temperature at the end of the process when the ball is embedded in the ground, is greater than T_1, the initial ball temperature, as expected.

2.5 FUNDAMENTAL PROPERTIES

Before introducing more applications of the first law, certain fundamental properties of a system must be discussed. To calculate expansion work, for example, we need to understand pressure. To calculate internal energy, temperature is required. Therefore, this section focuses on three of the most basic quantities used in understanding thermal and fluid processes—density, pressure, and temperature. Although each is undoubtedly already familiar to you, there are aspects of these properties that you may never have encountered. For example, in a gas, any one of these properties may vary as a function of location within the gas. Each property also has an atomic scale interpretation. Imagining events on the atomic scale helps to develop an intuitive understanding of many thermal and fluids processes.

2.5.1 Density

Density is defined as the mass per unit volume, or

$$\rho = \frac{m}{V}$$

To measure density, a volume is chosen and the mass of material within that volume is determined. This approach is adequate to provide an average density over the total volume, V. For many applications, assuming an average or constant density is sufficient. There are, however, many important processes that involve variable density, that is, density that varies with location in a volume. An example occurs in natural convection heat transfer. Suppose a roast turkey has just been taken out of an oven and placed on a countertop. Heat flows from the hot surface of the bird into the air, raising the air temperature in the immediate vicinity of the turkey. The density of the air decreases locally. Colder air above the turkey is now heavier (more dense). This cold air is dragged downward by gravity, displacing the hot air next to the bird, and the hot air rises. Throughout this process, density varies continuously with location in the air.

To allow for spatial variation, density may be written as:

$$\rho = \lim_{\Delta V \to \varepsilon} \frac{m}{\Delta V}$$

where m is the mass within the differential volume, ΔV, and ε is very small. This definition allows us to specify the density at a point within the material. Note, however, that as the volume becomes very small, it may contain only three or four molecules and the size of the volume could theoretically make a difference in the measurement of density. Figure 2-7 shows a plot of the density of a gas as a function of the volume chosen. At very low volumes, the density measurement is uncertain because it depends on how many molecules happen to be included in the measurement volume. At very high volumes, the density measurement may vary because the density is not homogeneous (constant) in this large volume. But, for most substances, there is a stable asymptotic value for density somewhere between the ultrasmall volumes and the rather large volumes. When we use the asymptotic value, we are

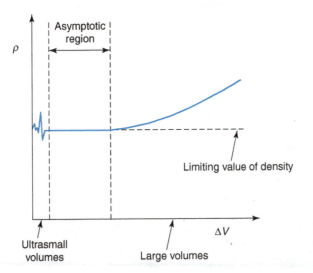

FIGURE 2-7 Density of a gas as a function of the volume chosen for measurement.

said to be making a **continuum** assumption. In a continuum, we characterize the material as if it were infinitely divisible and not composed of discrete molecules. The continuum assumption is applicable in almost all ordinary circumstances. An exception occurs in very high altitude plane flights, where the gas molecules are so far apart that the distance between molecules is significant compared to the size of the solid structures near the gas. So-called rarefied gases will not be treated in this text. Rather, all substances will be modeled using the continuum approximation. Note that, for many substances, density is a function of two other properties—pressure and temperature—which are discussed in the next two sections.

We describe density with units of mass per unit volume. In the SI unit system, density has the units of kg/m^3. In the British system, density is typically measured in lbm/ft^3.

2.5.2 Pressure

On a macroscopic level, pressure is something we feel as a force acting on our bodies. A diver who swims deep underwater feels pressure that can hurt the ears and constrict the chest. On a molecular level, pressure results from the combined motion of many molecules. For example, in a gas at rest, molecules travel incessantly in random directions with a range of velocities. This motion was first definitively detected by R. Brown in 1827. Such molecular motion accounts for the behavior of a dust particle suspended in air. If the particle is small and light enough, it will not fall due to gravity but will dance about randomly as it is jostled by collisions with the moving air molecules.

If a flat plate is inserted into a gas, then the gas molecules will strike the plate and bounce off. The collisions of the molecules with the plate impart a force to the plate. The integrated effect of all the collisions of the molecules against the plate is observed macroscopically as the pressure. Thus, pressure, P, is defined as a force per unit area, that is,

$$P = \frac{F}{A} \qquad (2\text{-}11)$$

As with density, we use the continuum approximation so that we can define pressure as a function that varies continuously throughout a gas or liquid. Because the force in this definition is due to the motion of the molecules, and this motion has no preferred direction, pressure in a fluid is independent of direction. It is a scalar rather than a vector quantity.

If a solid surface is placed in a gas or liquid, pressure exerts a force that is normal (perpendicuar) to the surface because the resultant force from all the collisions of molecules is in the normal direction. This can be seen in Figure 2-8, which shows the paths of two molecules colliding with a surface. Imagine that molecule A strikes the surface and imparts a force F_1 to it. Because the molecules in the fluid are moving in random directions, for every molecule A moving toward the right, there will be a molecule B moving toward the left. Molecule B imparts a force F_2 to the surface. The components of forces F_1 and F_2 parallel to the surface will cancel out, and the resultant net force will be in the normal direction.

The units of pressure in the British system are pounds-force per square inch ($lbf/in.^2$), also known as "psi." In the SI system, pressure is measured in pascals (Pa). By definition, a pascal is a newton per square meter, or

$$1 \text{ Pa} = 1 \frac{N}{m^2}$$

Recall that the newton is a unit of force.

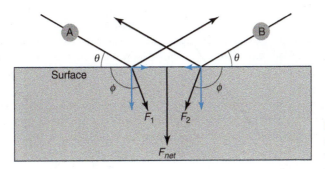

FIGURE 2-8 A pair of molecules in a fluid colliding with a surface.

There are many instruments available to measure pressure, including manometers, piezoelectric crystals, McLeod gauges, barometers, Bourdon tubes, and many others. Some of these devices measure not the actual pressure but rather the pressure relative to atmospheric pressure. As a result, it is common to make a distinction between the so-called absolute pressure, P_{abs}, and the gage pressure, P_g, where

$$P_g = P_{abs} - P_{atm}$$

and P_{atm} is the atmospheric pressure or ambient pressure surrounding the pressure gauge.

In the British system of units, gage pressure is indicated by "psig" to distinguish it from absolute pressure, which is called "psia" or simply "psi." For example, if an engineer reports that the pressure is 3 psig, then the pressure is 3 pounds-force per square inch above atmospheric pressure. In this text, "psi" will always mean "psia." The SI system has no special designation for gage pressure; generally, pressure is absolute pressure, unless otherwise stated.

2.5.3 Temperature

We all have experience with hot and cold objects. Although it is usually easy to sense that one object is hotter than another, it is difficult to specify the precise temperatures involved. Fortunately, all substances have characteristics that vary with temperature. Any one of these could be used to specify a temperature scale, though practical considerations preclude many. One substance that can be applied to temperature measurement is liquid mercury. A mercury thermometer is shown in Figure 2-9. Liquid mercury expands when it is heated and contracts when it is cooled. The height, L, of the liquid in the small-diameter bore of the glass tube is related to the temperature of the mercury in the bulb. One can define a temperature scale by scribing equally spaced marks on the glass and considering these as degrees of temperature.

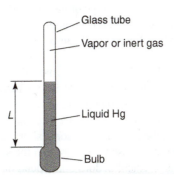

FIGURE 2-9 A mercury thermometer.

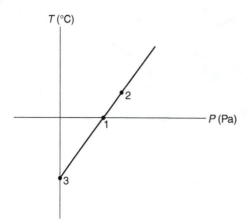

FIGURE 2-10 Temperature–pressure relationship for a constant volume of gas.

The earliest practical thermometer was a mercury-in-glass device of this type developed in 1715 by Gabriel Fahrenheit. His work built upon that of Isaac Newton, who had proposed an oil-filled thermometer in 1701. Newton selected the freezing point of water as the zero in his temperature scale. He selected body temperature as the second fixed point and divided the scale into 12 parts, in typical British fashion. When Fahrenheit developed his more accurate thermometer, he depressed the freezing point of water by adding salt and used this as the zero point in his scale. He introduced eight times as many divisions as Newton, so that body temperature became 96°. Later recalibration of the Fahrenheit scale led to the familiar body temperature of 98.6°. On the Fahrenheit scale, pure water freezes at 32°F and boils at 212°F.

Another temperature scale was introduced in the 18th century by Anders Celsius. On the Celsius scale, water freezes at 0°C and boils at 100°C. Both the Fahrenheit and Celsius temperature scales are widely used today, and both will be used in this text.

One of the problems with using the mercury thermometer to define temperature is that the definition depends on the properties of a single substance, mercury. A more universal definition would be desirable. The situation improves if gases are used to define temperature. Imagine that a fixed amount of gas is contained in a rigid tank. Figure 2-10 shows the measured relationship between temperature and pressure for a gas in this tank. Point 1 on this figure is the freezing point of water and point 2 is the boiling point. The values of pressure depend on the amount of gas in the tank. Smaller amounts of gas will lead to lower pressures. Experiments show that if the pressures are low enough, then the ratio P_2/P_1 will approach 1.3661 for all gases. At these low pressures, the relationship between temperature and pressure is linear. Therefore, it is possible to extrapolate the line to zero pressure, shown as point 3 in the figure. The temperature at point 3 is −273.15°C. This temperature is independent of the type of gas in the thermometer and the actual pressure of the gas, as long as the pressure is low enough.

The fixed point 3 in Figure 2-10 can be used to define a new temperature scale, the gas temperature scale. On this scale, point 3 is assigned the value of zero, water freezes at 273.15, and it boils at 373.15. In modern times, this scale has been slightly modified and designated the Kelvin temperature scale. Using the behavior of gases to define a temperature scale has some advantages over using the expansion of mercury. The scale is dependent on the properties of gases in general rather than on the properties of the single substance, mercury. Nevertheless, the gas scale has some shortcomings. At low enough temperatures, all gases, even helium, condense to liquids, and the scale is no longer usable except as an extrapolation. There is a scale that does not depend on the properties of any substance, the so-called thermodynamic scale. A description of the modern thermodynamic temperature scale is given later in the text.

The Kelvin scale is derived from the Celsius scale. The corresponding scale derived from the Fahrenheit scale is the Rankine scale. All four of the scales are frequently used in modern engineering practice. The relationships among them are

$$T(°F) = 1.8T(°C) + 32$$
$$T(°C) = [T(°F) - 32]/1.8$$
$$T(K) = T(°C) + 273.15$$
$$T(R) = T(°F) + 459.67$$
$$T(R) = 1.8T(K)$$

The Kelvin and the Rankine scales, are both **absolute temperature** scales, while the Celsius and the Fahrenheit scales are **relative** scales. The observed behavior of gases shown in Figure 2-10 must be described using an absolute scale rather than a relative one.

Although temperature is a very familiar concept, it is somewhat subtle. Certain common experiences involving temperature can lead us to incorrect conclusions and confuse our physical intuition. For example, if a barefoot person steps out of bed in the morning, a plush carpet will feel much "warmer" than a ceramic tile floor. However, both the carpet and the tile are at the same temperature. The tile feels cooler because heat is conducted from the sole of the foot into the tile at a high rate, while heat is conducted into the insulating carpet at a low rate. The foot contacting the tile cools quickly, and thus the tile seems cooler.

Even Newton was confused about the distinction between temperature and heat (as was everyone else in the scientific community of his time). Newton used the same Latin word, *calor*, to signify both temperature and heat. Temperature and heat are also distinctly different from internal energy, although the untrained observer may confuse these three ideas.

Before proceeding with the study of thermal and fluids systems, it is important to have a very clear idea of the distinction among temperature, heat, and internal energy. Temperature is a property of a system, heat is energy in motion across the boundary of a system, and internal energy is energy stored in a system or substance. The following three examples will help to clarify these distinctions. In each example, one of the three quantities remains constant while the other two change.

Example 1 is the compression of a gas in a well-insulated piston–cylinder assembly. Such a process is called adiabatic, which means that no heat is added or removed. Although no heat is added, the gas temperature rises due to the compression as work is converted into internal energy. In this case, both temperature and internal energy increase, but heat transferred is zero.

Example 2 is the heating of a glass of water containing ice cubes. The system is the liquid water and the ice. As heat is added, the ice cubes begin to melt. The system, however, remains at the melting temperature of the ice, that is, at 0°C. This process is **isothermal**, meaning that the temperature does not change during the process. Because bonds between the molecules in the ice crystals are broken during the melting, the internal energy of the system increases. The internal energy per unit mass of liquid water is greater than the internal energy per unit mass of ice. As a result of the melting, there is more liquid and less ice in the system, so the internal energy of the system increases. In this process, heat is added, internal energy increases, but temperature remains constant.

In **Example 3**, an electric current passes through a very-well-insulated rod. The current does work on the rod. However, the rod is assumed to be perfectly insulated, so no heat is transferred and the system is adiabatic. The temperature and the internal energy of the rod rise in this process, even though no heat is transferred.

2.6   IDEAL GASES

In the discussion of thermometry, we noted the linear relationship between pressure and temperature for a low-pressure gas in a rigid tank. When the temperature rises, the pressure does as well. You may be familiar with this phenomenon in automobile tires. If you measure the pressure when the tire is cold and later measure it after the car has been driven for some distance, you will find that the pressure has risen. Friction between the tire and the road caused the temperature rise. Motorists are warned not to remove air from a hot tire. The manufacturer has accounted for the rise in pressure due to frictional heating and has specified the correct tire pressure when the tire is cold.

We can describe the behavior of gases from a molecular viewpoint. In a gas, individual molecules are in motion in random directions. Each molecule of mass, m, and velocity, $\mathcal{V}$, has a kinetic energy given by $m\mathcal{V}^2/2$. Temperature is related to the average kinetic energy of the molecules in the gas. At higher temperatures, the molecules move faster on average, and at lower temperatures, the molecules move slower. Imagine a rigid container with a fixed number of molecules of gas. Recall that the pressure on the side of the container is due to the impact of molecules against the side. If the temperature rises, the molecules travel faster and more force is imparted on impact. Pressure is by definition a force per unit area, so the pressure increases. Thus higher temperatures lead to higher pressures.

What happens if the number of molecules in the container is decreased? If the temperature stays constant, the molecules travel just as fast. However, with fewer molecules striking each unit area of the container, the pressure will decrease.

The last important parameter characterizing the behavior of gases is volume, V. Suppose that the container is not rigid, but flexible. Imagine the volume is decreased to half its original size without changing the temperature or the number of molecules. The molecules move at the same average speed, but they are likely to strike the walls of the container more often. This increased collision rate leads to higher pressure.

All of these phenomena are embodied in a relationship called the **ideal gas law**:

$$PV = n\overline{R}T \tag{2-12}$$

where n is the number of moles of gas and $\overline{R}$ is the universal gas constant. The temperature in the ideal gas law must be expressed on an absolute scale, that is, as either degrees Kelvin or Rankine. The pressure must be expressed as absolute pressure, not gage pressure.

The ideal gas law applies to all gases for some range of temperature and pressure. However, it is an approximation to real gas behavior and will sometimes give inaccurate results. More accurate relationships among P, T, and V have been developed, but none are as simple and useful as the ideal gas law, and they will not be discussed in this text.

The ideal gas law of Eq. 2-12 involves n, the number of moles. A mole is a fixed number of molecules of material. Thus two moles of oxygen will have the same number of molecules as two moles of helium. While dealing with moles is very useful in chemistry, mass units are more convenient when no chemical reaction is taking place. Moles are related to mass through the molecular weight, M, that is,

$$m = nM \qquad (2\text{-}13)$$

The molecular weights in Tables A-1 and B-1 are given without units. They are ratios of the masses of molecules of different substances. Oxygen has a molecular weight of 31.999, while hydrogen has a molecular weight of 2.016. Thus an oxygen molecule is about 16 times more massive than a hydrogen molecule. A reference to one mole of oxygen usually means 31.999 grams of oxygen. Hence, the molecular weight, M, may be thought of as the number of grams per mole and may be assigned the units g/mol.

Difficulties can arise, however, if mass units other than grams are in use. An engineer, for example, might refer to one mole of oxygen and mean 31.999 kg of oxygen, rather than 31.999 g of oxygen. Since we will use different unit systems in this text, we will always specify the type of mole meant. In this spirit, the molecular weight for oxygen can be written as:

$$M = 31.999 \ \frac{\text{g}}{\text{mol}} = 31.999 \ \frac{\text{kg}}{\text{kmol}} = 31.999 \ \frac{\text{lbm}}{\text{lbmol}}$$

There are additional forms of the ideal gas law useful in engineering. Substituting Eq. 2-13 into the ideal gas law (Eq. 2-12) gives

$$PV = \frac{m\bar{R}T}{M}$$

Recall that the definition of density is mass per unit volume, or

$$\rho = \frac{m}{V}$$

With this substitution, the ideal gas law can be written

$$P = \frac{\rho\bar{R}T}{M}$$

Values of the universal gas constant, $\bar{R}$, are:

$$\bar{R} = 8.31434 \ \text{kJ}/(\text{kmol}\cdot\text{K})$$
$$= 1.9858 \ \text{Btu}/(\text{lbmol}\cdot\text{R})$$
$$= 1545.35 \ \text{ft}\cdot\text{lbf}/(\text{lbmol}\cdot\text{R})$$
$$= 10.73 \ \text{psia}\cdot\text{ft}^3/(\text{lbmol}\cdot\text{R})$$

It is important to reemphasize that both temperature and pressure in the ideal gas law must be expressed on absolute scales. Temperature must be given in degrees Rankine or Kelvin.

One of the most important gases that we encounter in engineering practice is air. Strictly speaking, the ideal gas law applies to pure substances, that is, substances that are composed of a single chemical species. Air is a mixture of gases, including nitrogen, oxygen, carbon dioxide, and other constituents. Experiments show that air can often be treated as an ideal gas as long as an appropriate value for the molecular weight is used. This molecular weight is a weighted average of the molecular weights of all the gases that make up air (mixtures of ideal gases are discussed in Chapter 15). The value of the molecular weight of air is included in Tables A-1 and B-1.

There is also one other form of the ideal gas law that is commonly used in thermodynamics. It involves the specific volume. By definition, the **specific volume**, v, is the volume per unit mass, or

$$v = \frac{V}{m} = \frac{1}{\rho}$$

The word *specific* will be used in this text to mean "per unit mass." When the specific volume is used in the ideal gas law, it becomes

$$\boxed{Pv = \frac{\overline{R}T}{M}}$$

EXAMPLE 2-4 Ideal gas law

Calculate the mass of the air in a typical residential living room of size 8 ft by 12 ft. The ceiling is 8 ft high. Assume the air is at a uniform temperature of 70°F and a pressure of 1 atm.

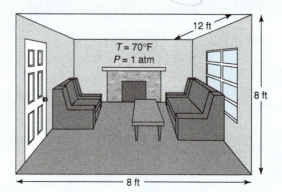

Approach:

We define the closed system as the air in the room and assume that air can be treated as an ideal gas. The size of the room is known, so volume can be calculated from the ideal gas law. Be certain to convert temperature to Rankine.

Assumptions:

A1. Air at 70°F and 1 atm may be considered an ideal gas.

A2. The volume occupied by the furniture is small.

Solution:

The system is the air in the room. The ideal gas law [A1] may be rearranged to the form

$$m = \frac{PVM}{\overline{R}T}$$

The molecular weight of air may be found in Table B-1. The mass can be calculated from [A2]:

$$m = \frac{(1\,\text{atm})\left[(8)(12)(8)\ \text{ft}^3\right]\left[28.97\ \dfrac{\text{lbm}}{\text{lbmol}}\right]}{\left[10.73\ \dfrac{\text{psia}\cdot\text{ft}^3}{\text{lbmol}\cdot\text{R}}\right]\left[\dfrac{1\ \text{atm}}{14.7\ \text{psia}}\right](70 + 460)\ \text{R}}$$

$$m = 57.5\ \text{lbm}$$

Note that the temperature has been converted to degrees Rankine. One must never use degrees Fahrenheit in the ideal gas law.

EXAMPLE 2-5   Force balance on a piston

Oxygen is contained in a cylinder fitted with a piston. The cylinder has a height of 8 cm and a diameter of 3 cm. The piston is held in place by a weight of mass 13.4 kg, as shown in the figure below. The mass of the oxygen is 0.1 g. Calculate the temperature of the oxygen, in °C. Assume that the piston itself has negligible mass and atmospheric pressure is 101 kPa.

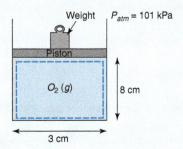

Approach:

We define a closed system that encompasses only the gas contained in the piston–cylinder assembly and assume that oxygen can be treated as an ideal gas. The temperature can be found from the ideal gas law. We simply need to rearrange the equation

$$PV = \frac{m\overline{R}T}{M}$$

and solve for temperature. The mass is given in the problem statement, and the volume is easily calculated. The constants $\overline{R}$ and M can be found in Table A-1. Pressure is a little more problematic. To find the pressure, a fundamental force balance on the piston will be needed. Then the pressure can be found from its definition:

$$P = \frac{F}{A}$$

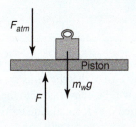

Assumptions:

A1. The mass of the piston is insignificant.

Solution:

To find the pressure of the oxygen, draw a free-body diagram of the piston and weight, as shown in the figure. Assume that the mass of the piston is very small [A1]. Three forces act on the piston. The pressure of the atmosphere exerts a downward force, gravity exerts a downward force on the weight, and the pressure of the oxygen exerts an upward force. Therefore, the upward force, F, due to the oxygen is given by

$$F = F_{atm} + m_w g$$

where m_w is the mass of the weight. Using the definition of pressure as a force per unit area, this equation can be rewritten in terms of pressure as

$$PA = P_{atm} A + m_w g$$

where A is the area of the piston. Solving for the oxygen pressure gives:

$$P = P_{atm} + \frac{m_w g}{A} = P_{atm} + \frac{m_w g}{\pi \left[\dfrac{D}{2}\right]^2}$$

Substituting values

$$P = 101 \text{ kPa} \left[\frac{1000 \text{ Pa}}{1 \text{ kPa}}\right] + \frac{(13.4 \text{ kg}) \left[9.81 \dfrac{\text{m}}{\text{s}^2}\right]}{\pi \left[\dfrac{3 \text{ cm}}{2} \dfrac{1 \text{ m}}{100 \text{ cm}}\right]^2} = 2.87 \times 10^5 \text{ Pa}$$

A2. Oxygen may be considered to be an ideal gas at these conditions.

Now that pressure is known, the temperature can be found from the ideal gas law [A2] rearranged as

$$T = \frac{PVM}{m\overline{R}}$$

Substituting values gives

$$T = \frac{\left(2.87 \times 10^5 \text{ Pa}\right) \left[\pi \left[\dfrac{0.03}{2}\right]^2 (0.08)\right] \text{m}^3 \left(32.0 \dfrac{\text{kg}}{\text{kmol}}\right)}{\left[0.1 \text{ g}\right] \left[\dfrac{1 \text{ kg}}{1000 \text{ g}}\right] \left[8.314 \dfrac{\text{kJ}}{\text{kmol·K}}\right] \left[\dfrac{1000 \text{ J}}{1 \text{ kJ}}\right]}$$

$$T = 625 \text{ K} = 356°\text{C}$$

Note that consistent units must be used throughout the calculation. For example, grams were changed to kg, and kJ were changed to J. This is necessary to obtain a dimensionally correct result. Units can be a major cause of error in working problems. Because there are many important points that must be understood, the next section deals explicitly with unit systems.

2.7   UNIT SYSTEMS

Two unit systems are used in this text, the SI system and the British system. The approach taken to ensure that the units are correct will be different in each case. In the SI system, all values will be converted into the appropriate SI unit. For example, the quantity kJ will be converted into J, and the quantity atm will be converted into Pa. Thus the units will form a consistent set, and it will be easy to determine the units of each variable in an equation. This approach works very well in the SI system, because most quantities are routinely expressed in SI units or in units that are simple multiples of SI units. For example, in Example 2-5, this equation arose:

$$P = P_{atm} + \frac{m_w g}{A}$$

Every term in the equation must have the same units. How do we know that the second term on the right-hand side really has units of pressure? If we use the correct SI unit for each quantity in the equation, then the units would be:

$$\text{Pa} = \text{Pa} + \frac{\text{kg} \frac{\text{m}}{\text{s}^2}}{\text{m}^2}$$

Since $1\,\text{N} = 1\,\text{kg·m/s}^2$, this may also be written as:

$$\text{Pa} = \text{Pa} + \left[\frac{\text{kg} \frac{\text{m}}{\text{s}^2}}{\text{m}^2} \right] \left[\frac{\text{N}}{\text{kg} \frac{\text{m}}{\text{s}^2}} \right] = \text{Pa} + \frac{\text{N}}{\text{m}^2}$$

Furthermore, $1\,\text{Pa} = 1\,\text{N/m}^2$, so the equation is dimensionally correct. Note that the equation would require conversion factors if we entered mass in g or pressure in kPa. A list of the correct SI units for quantities that will be commonly used in this text is given in Table 2-1. Only the unit for mass, the kilogram, has a prefix. All others have no prefix.

When performing calculations in the British system, a different approach will be followed in this text. In the British system, the commonly used units are often not part of a consistent set. For example, the consistent set of units includes feet and seconds. But we will often express flow rate in gallons per hour rather than ft^3/s. It is laborious to always convert the British units into a consistent set, and this approach is rarely followed in engineering practice.

Most conversion factors in the British system are straightforward, but there is one that is a little odd. In the British system, mass is often expressed in "pounds" and so is force. As noted above, we will always distinguish between "pounds-mass" as lbm and "pounds-force" as lbf. The consistent set of units includes ft, s, and lbf, but does not include lbm. In the British system, the consistent unit for mass is the slug. To see how slugs are related to the other units, consider Newton's second law,

$$F = ma$$

In consistent British units, acceleration is expressed in ft/s^2, and force is expressed in lbf. If slug is the consistent unit for mass, then

$$1\,\text{lbf} = 1\,\frac{\text{slug} \cdot \text{ft}}{\text{s}^2}$$

How are lbm related to slugs? By definition, earth's gravity exerts one pound force on a mass of one pound mass at the surface of the earth. From Newton's second law applied

TABLE 2-1 **Consistent Units of the SI System**

Quantity	SI Unit	Description
Mass	kg	Kilogram
Length	m	Meter
Time	s	Second
Energy, heat, work	J	Joule
Power, rate of heat transfer	W	Watts
Pressure	Pa	Pascals
Temperature	K	Kelvin
Current	A	Amperes
Voltage	V	Volts

to a gravitational force,

$$F = mg$$

The acceleration of gravity, g, at the earth's surface is 32.174 ft/s^2. How many slugs are pulled by gravity with a force of one pound force at the earth's surface? Rearranging the second law

$$m = \frac{F}{g}$$

Setting $F = 1$ lbf and using the value of g gives

$$m = \frac{1 \text{ lbf}}{32.174 \frac{\text{ft}}{\text{s}^2}} = \frac{1}{32.174} \text{ slug}$$

A mass of 1/32.174 slug is pulled on by gravity with a force of 1 lbf at the surface of the earth; therefore,

$$1 \text{ lbm} = \frac{1}{32.174} \text{ slug}$$

The slug is not a very popular unit and is not recommended in this text. It would be awkward to convert mass into slugs and then use the relation

$$1 \text{ lbf} = 1 \frac{\text{slug} \cdot \text{ft}}{\text{s}^2}$$

every time a lbm unit is encountered. Instead, we will use

$$1 \text{ lbm} = \left(\frac{1}{32.174} \text{ slug} \right) \left(\frac{1 \text{ lbf}}{1 \frac{\text{slug} \cdot \text{ft}}{\text{s}^2}} \right)$$

which can be shortened to

$$1 \text{ lbm} = \left(\frac{1 \text{ lbf}}{32.174 \frac{\text{ft}}{\text{s}^2}} \right) \tag{2-14}$$

In this text, we will never mention slugs again. Instead, we will make frequent use of Eq. 2-14 as a unit conversion factor. Note that in SI units, the equivalent statement is

$$1 \text{ kg} = \left(\frac{1 \text{ N}}{1 \frac{\text{m}}{\text{s}^2}} \right)$$

Thus the difference between the two systems is that the British system uses earth's gravity as the acceleration and the SI system uses 1 m/s^2, which is not earth's gravity. (In SI units, earth's gravity is 9.81 m/s^2.)

It is plain from Eq. 2-14 that lbm and lbf are *not* interchangable. One must always check units carefully in equations that contain both lbm and lbf. Other conversion factors commonly used in the British system will be introduced as needed throughout the text. Conversion factors for both the British system and the SI system are listed on the front inside cover of the text.

2.8 WORK

The first law is a relationship among work, heat, and changes in stored energy. There are many different kinds of work, including shaft work, expansion work, and electrical work. In this section, quantitative expressions for several of the most common types of work are developed.

Consider a body moving in a straight line because of an applied force, as shown in Figure 2-11. In this figure, the velocity vector and the force vector are both in the same direction. The general case in which the velocity vector and the force vector are in different directions is not usually important in thermal and fluids engineering and will not be discussed here. An example of a practical situation in which the net force vector is in the direction of motion is shown in Figure 2-12, where a crate is being lifted against the force of gravity.

To get an intuitive feel for work, first consider motion with a constant force. In Figure 2-13, a body has moved from s_1 to s_2 under the action of a constant force with magnitude F. The work done is the product of the applied force and the distance traveled, or

$$W = F(s_2 - s_1)$$

Now apply this equation to the situation shown in Figure 2-14, where a person is pushing a piano across a floor. The farther the distance traveled (i.e., larger values of $s_2 - s_1$), the greater the work done. Likewise, the greater the force needed (e.g., for a heavy piano), the more the work done. Notice that if a force is applied but the piano does not move, no work is done.

In the general case, the applied force may not be constant. Consider a variable force that acts in the same direction as s (Figure 2-15). The work done in moving the body from s_1 to s_2 is the sum of all the differential contributions to the work as the body moves along

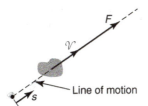

FIGURE 2-11 A body moving in the direction s.

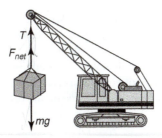

FIGURE 2-12 Force and motion in the same direction.

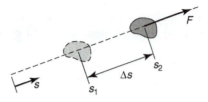

FIGURE 2-13   A body moving under the action of a constant force.

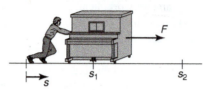

FIGURE 2-14   Work is done in moving a piano across a floor.

the path. Each differential contribution of work is given by $F(s)\,ds$, so the sum of all those contributions is

$$W = \int_{s_1}^{s_2} F(s)\,ds \qquad (2\text{-}15)$$

2.8.1   Compression and Expansion Work

When a gas expands or is compressed, work is done. For example, consider the gas-filled piston–cylinder assembly shown in Figure 2-16. We define the gas as a closed system. The gas exerts a force on the bottom of the piston. Imagine that the piston rises a differential distance dx due to this force. The work done is

$$\boxed{W = \int F\,dx} \qquad (2\text{-}16)$$

The force on the piston is related to the pressure of the gas. We assume the gas is in **equilibrium** at the start of the process. Equilibrium implies a state in which there are no imbalances in forces, temperatures, pressures, phases, or chemical composition; that is, all properties are uniform throughout the volume of the gas. In equilibrium, the pressure is the same at every location in the gas, and there is no driving force causing the gas to flow.

Now imagine heat is transferred to the system and the piston rises. The pressure in the cylinder is due to the collision of molecules with the surface of the piston. As the piston moves, these molecules must adjust to a new equilibrium state with the new position of the piston. If the piston moves slowly compared to the speed of the molecules, the molecular adjustment takes place very rapidly and the expansion process can be imagined as a succession of equilibrium states. This is the so-called **quasi-equilibrium process**, one that passes through a set of equilibrium states. On the other hand, if the piston moves very rapidly, there will be a delay before the molecules can catch up to the piston and reestablish equilibrium; this process would not be quasi-equilibrium. For example, a sudden gas expansion into a vacuum would not be a quasi-equilibrium process.

During a quasi-equilibrium process, the force on the piston is given by

$$F = PA$$

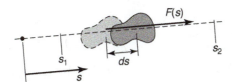

FIGURE 2-15 A body moving under the action of a variable force.

where A is the area of the piston. With this substitution, the expansion work during a quasi-equilibrium process becomes

$$W = \int F \, dx = \int PA \, dx$$

Referring again to Figure 2-16, the process begins at time zero at position x; after a change in time Δt, the new position is $x + \Delta x$. Note that $A dx$ is the differential change in volume, or

$$A \, dx = dV$$

Thus the expansion work becomes

$$\boxed{W = \int P \, dV} \tag{2-17}$$

This expression is valid for any closed system undergoing a quasi-equilibrium process that has occured over a time interval Δt. The units of work are force times distance. In the SI unit system, work has the dimensions of joules, where

$$1 \text{ joule} = (1 \text{ newton})(1 \text{ meter})$$

In the British system, the units of work are feet times pounds-force, or ft-lbf.

From calculus, we know that the integral of a function is the area under the plot of that function. This idea can be used to visualize the amount of work done in a quasi-equilibrium process. Because

$$W = \int P \, dV$$

when pressure is plotted against volume, the work is the area under the curve, as shown shaded in Figure 2-17. Work depends on how pressure varies with volume. For example, in Figure 2-18, the end points of the process, points 1 and 2, are the same as in Figure 2-17, but the pressure–volume curve is different. The amount of work done in Figure 2-18

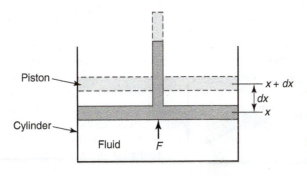

FIGURE 2-16 Gas expanding against a piston.

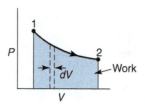

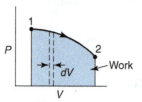

FIGURE 2-17 Work is the area under the curve of P versus V.

FIGURE 2-18 Work for an alternate path between points 1 and 2, which are the same points as in Figure 2-17.

is greater than that done in Figure 2-17. The work depends on the path taken between the end points, not solely on values at the end points.

It is easy to see intuitively how the work depends on the path by considering the work done in pushing a crate across the floor. In Figure 2-19, path A is longer than path B. More work is needed to push the crate against friction along path A than along path B.

When work is expressed as a differential quantity, special care is needed. It is tempting to write

$$dW = P\,dV \qquad \text{caution—meaningless equation}$$

However, this expression could be misleading. Suppose we decided to integrate both sides of the equation to get

$$\int_1^2 dW = \int_1^2 P\,dV \qquad \text{caution—meaningless equation}$$

If the left-hand side is integrated and evaluated at the limits, then

$$W_2 - W_1 = \int_1^2 P\,dV \qquad \text{caution—meaningless equation}$$

This is a meaningless equation and should *not* be used, because it is incorrect to assign values to W_1 and W_2 at points 1 and 2. We can show that if we do assign a value to W_1, then an absurd conclusion is obtained. (Remember that points 1 and 2 are the same in Figure 2-17 and Figure 2-18.) Let W_1 have a value of 7 J at point 1 in Figure 2-17. Then W_1 would also be 7 J at point 1 in Figure 2-18. Let the area under the curve in Figure 2-17 be 3 J and the area under the curve in Figure 2-18 be 5 J. Then, according to the last equation,

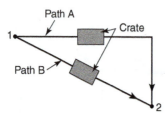

FIGURE 2-19 Top view of the motion of a crate being pushed along a floor against friction.

W_2 would be $7+3 = 10$ J for the path shown in Figure 2-17, *but* W_2 would be $7+5 = 12$ J for the path shown in Figure 2-18. For two different processes, two different values of W_2 are obtained. There is no unique value of W_2, thus demonstrating that the idea of work at a point is absurd. Writing the differential dW implies that work does have a meaning at a point. To avoid this problem, it is common to write

$$\boxed{\delta W = P\,dV}$$

(2-18)

where δW is called an **inexact differential**. This notation serves to remind us that the left-hand side cannot be integrated and evaluated at the limits.

EXAMPLE 2-6 Work in a two-step quasi-equilibrium process

Carbon dioxide is slowly heated from an initial temperature of 50°C to a final temperature of 500°C. The process occurs in two steps. In the first step, pressure varies linearly with volume; in the second step, pressure is constant, as shown in the figure below. The initial pressure, P_1, is 100 kPa and the final pressure, P_3, is 150 kPa. The temperature, T_2, at the end of the first step is 350°C. If the mass of CO_2 is 0.044 kg, calculate the total work done.

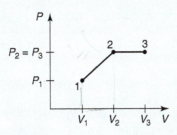

Approach:

Because the process is slow, we may assume it is a quasi-equilibrium process. The work done is then

$$W = \int_{V_1}^{V_3} P\,dV$$

The integral is the area under the curve; therefore, by inspection of the graph, the work is

$$W = \left(\frac{P_1 + P_2}{2}\right)(V_2 - V_1) + P_2\,(V_3 - V_2)$$

The values of pressure are given in the problem statement, and the volume may be calculated from the ideal gas law:

$$PV = \frac{m\overline{R}T}{M}$$

Assumptions:

A1. The process is slow and is considered to be quasi-equilibrium.

Solution:

For this quasi-equilibrium process [A1], work is given by

$$W = \int_{V_1}^{V_3} P\,dV$$

The integral is the area under the curve; therefore, by inspection of the figure,

$$W = \left(\frac{P_1 + P_2}{2}\right)(V_2 - V_1) + P_2\,(V_3 - V_2)$$

A2. Carbon dioxide behaves as an ideal gas for these temperatures and pressures.

To find the volume at state 1, use the ideal gas law [A2] in the form

$$V_1 = \frac{m\bar{R}T_1}{P_1 M}$$

Substituting values:

$$V_1 = \frac{(0.044\ \text{kg}) \left[8.314\ \dfrac{\text{kJ}}{\text{kmol·K}}\right] (50 + 273)\ \text{K} \left(\dfrac{1000\ \text{J}}{1\ \text{kJ}}\right)}{(100\ \text{kPa}) \left[44 \dfrac{\text{kg}}{\text{kmol}}\right] \left[\dfrac{1000\ \text{Pa}}{1\ \text{kPa}}\right]}$$

$$V_1 = 0.0268\ \text{m}^3$$

where the value of molecular weight, M, for carbon dioxide has been taken from Table A-1. To find V_2, again apply the ideal gas law:

$$V_2 = \frac{m\bar{R}T_2}{P_2 M}$$

$$V_2 = \frac{(0.044\ \text{kg}) \left[8.314\ \dfrac{\text{kJ}}{\text{kmol·K}}\right] (350 + 273)\ \text{K} \left[\dfrac{1000\ \text{J}}{1\ \text{kJ}}\right]}{(150\ \text{kPa}) \left[44 \dfrac{\text{kg}}{\text{kmol}}\right] \left[\dfrac{1000\ \text{Pa}}{1\ \text{kPa}}\right]}$$

$$V_2 = 0.0345\ \text{m}^3$$

By a similar calculation, $V_3 = 0.0428\ \text{m}^3$. The work may now be calculated as

$$W = \left(\frac{100 + 150}{2}\right)\ \text{kPa}\ (0.0345 - 0.0268)\ \text{m}^3 \left(\frac{1000\ \text{Pa}}{1\ \text{kPa}}\right)$$

$$+ (150\ \text{kPa})\ (0.0428 - 0.0345)\ \text{m}^3 \left(\frac{1000\ \text{Pa}}{1\ \text{kPa}}\right)$$

$$W = 2200\ \text{J} = 2.2\ \text{kJ}$$

2.8.2   Electrical Work

Work is also done in electrical systems. Consider the electrical circuit shown in Figure 2-20. The battery contains a solution of positive and negative ions. The two terminals of the battery are constructed of two different materials, for example, zinc and copper. The negative ions are attracted to one of the terminals and combine chemically with the atoms in that terminal. Conversely, the positive ions move toward the other terminal and combine chemically there. An electrostatic force acts on the ions to move them this distance through the battery, and there is work associated with this force. The differential work required to move a differential charge dq through the battery is

$$\boxed{\delta W = \xi\, dq} \tag{2-19}$$

where ξ is the so-called electromotive force. Note that ξ is not actually a force and does not have the units of force. Its units are work/charge, or joule/coulomb, which is the volt. Eq. 2-19 is not an expression of force through a distance; instead, it is the definition of

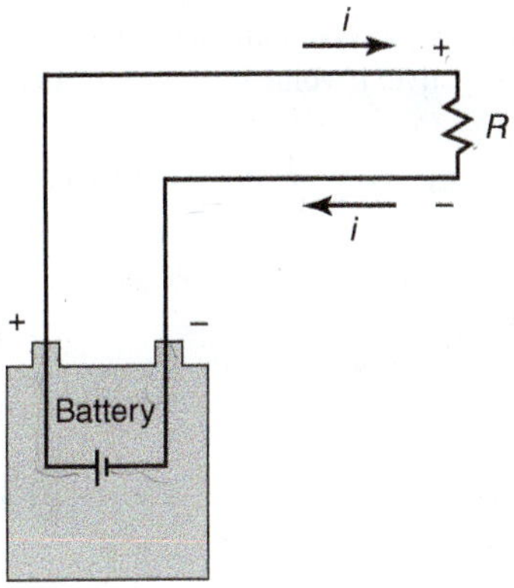

FIGURE 2-20   A simple resistive circuit.

electromotive force. The work is calculated by other means, and Eq. 2-19 is used to find ξ, the voltage of the battery.

The rate of flow of charge with time is current. Eq. 2-19 may be expressed in terms of current as

$$\delta W = \xi \frac{dq}{dt}\, dt = \xi I\, dt$$

Integrating over time gives the work done as

$$W = \int \xi I\, dt$$

EXAMPLE 2-7   Electrical work

In the simple circuit shown in Figure 2-20, the battery has a voltage of 10 volts and the resistor has a resistance of 25 Ω. In the span of five minutes, how much work is done by the battery on the resistor?

Approach:

The work will be calculated from

$$W = \int \xi I\, dt$$

Because neither current nor voltage vary with time, this integral is very easy to evaluate. The voltage is given and the current can be calculated from Ohm's law, which is:

$$\xi = IR$$

where R is the resistance of the resistor.

Assumptions: **Solution:**

The current in the circuit is, from Ohm's law,

$$I = \frac{\xi}{R} = \frac{10\ \text{V}}{25\ \Omega} = 0.4\ \text{A}$$

Let us consider the resistor as the system under study. Work is done on the resistor, so, by our convention, work will be negative. If voltage, ξ, and current, I, are considered to be positive, then

$$W = -\int \xi I \, dt = -\xi I \int dt = -(10 \text{ V})(0.4 \text{ A})(5 \text{ min}) \left[\frac{60 \text{ s}}{1 \text{ min}} \right]$$

The voltage and current have been removed from the integral because they do not vary with time [A1][A2]. The calculated value of work is

A1. Voltage does not vary with time.

A2. Current does not vary with time.

$$W = -1200 \text{ J}$$

Volts, amps, and seconds are all part of the consistent set of SI units; therefore, work will be in the SI unit for work, which is joules.

2.8.3 Shaft Work

In many practical situations, work is transmitted via a rotating shaft. A good example is the driveshaft of a car. Another example is the stirring of a fluid by a paddlewheel, as shown in Figure 2-21. As always, work is a force through a distance, or

$$W = \int F \, dx$$

If we let F represent the tangential force that the rotating member exerts on its environment and R represent the radius at which the force is applied (Figure 2-22), then the torque on the shaft is

$$\Im = FR$$

As the shaft rotates, the differential distance dx traveled at radius R is related to the angle θ by

$$dx = R \, d\theta$$

as long as θ is given in radians. Solving these two equations for F and dx, respectively, and substituting into the equation for work gives

$$W = \int \frac{\Im}{R} R \, d\theta = \int \Im \, d\theta$$

An alternate expression for work can be obtained using the angular velocity. The angular velocity ω is defined as

$$\omega = \frac{d\theta}{dt}$$

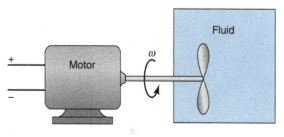

FIGURE 2-21 Stirring of a fluid by a paddlewheel.

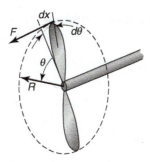

FIGURE 2-22 Tangential force exerted by a rotating blade.

This allows us to express work as

$$W = \int \Im \, d\theta = \int \Im \frac{d\theta}{dt} \, dt = \int \Im \omega \, dt$$

EXAMPLE 2-8   Shaft work with variable torque

A constant-speed motor drives a paddlewheel that is submersed in a viscous liquid. With time, the temperature of the liquid increases, the liquid viscosity decreases, and less torque is needed for the stirring action. The torque applied as a function of time is determined experimentally to be

$$\Im = A + Be^{-mt}$$

where $A = 55$ ft-lbf, $B = 20$ ft-lbf, and $m = 2.3$ hr^{-1}. If the motor rotates at a constant speed of 80 rpm, calculate the work done by the motor on the liquid in the first 10 minutes of operation.

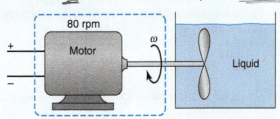

Approach:

We define the system as the motor. The work done by the motor is given by

$$W = \int \Im \omega \, dt$$

The expression for torque as a function of time is given in the problem statement. This expression is substituted into the above equation for work. Since rotation occurs at a constant speed, the angular velocity, ω, will be a constant and can be removed from the integral.

Assumptions: *Solution:*

Substituting the expression for torque into the equation for work results in

$$W = \int_0^{t_f} (A + Be^{-mt}) \omega \, dt$$

$$W = \int_0^{t_f} A\omega \, dt + \int_0^{t_f} B\omega e^{-mt} \, dt$$

A1. Motor speed is constant.

where t_f has been used to designated the final time, 10 minutes. Because motor speed is constant, ω can be removed from the integral [A1]. Performing the integration results in

$$W = A\omega t_f - \frac{B\omega}{m}\left[e^{-mt_f} - 1\right]$$

Substituting values gives

$$W = (55 \text{ ft·lbf})\left[80\frac{\text{rev}}{\text{min}}\right]\left[\frac{2\pi \text{ rad}}{1 \text{ rev}}\right](10 \text{ min})$$

$$-\frac{(20 \text{ ft·lbf})\left[80\frac{\text{rev}}{\text{min}}\right]\left[\frac{2\pi \text{ rad}}{1 \text{ rev}}\right]}{\left[2.3\frac{1}{\text{h}}\right]\left[\frac{1 \text{ h}}{60 \text{ min}}\right]}\left[e^{-\left[2.3\frac{1}{\text{h}}\right](10 \text{ min})\left[\frac{1 \text{ h}}{60 \text{ min}}\right]} - 1\right]$$

$$W = 193,000 \text{ ft·lbf}$$

Note that it was necessary to convert from revolutions to radians. Radians are essentially dimensionless units.

2.9 KINETIC ENERGY

In Section 2.1, kinetic energy was given by

$$KE = \frac{1}{2}m\mathcal{V}^2$$

It is important to understand the origin of this equation. Kinetic energy is related to the work done in changing the velocity of a body. From Newton's second law for a constant mass,

$$F = m\frac{d\mathcal{V}}{dt} \tag{2-20}$$

If we consider motion in the s direction, the velocity is given by

$$\mathcal{V} = \frac{ds}{dt} \tag{2-21}$$

Next, substitute Eq. 2-20 into Eq. 2-15 to get

$$W^* = \int_{s_1}^{s_2} F\,ds = \int_{s_1}^{s_2} m\frac{d\mathcal{V}}{dt}\,ds \tag{2-22}$$

where W^* is the work done *on* the mass to change its velocity. This is an awkward equation because the velocity is a function of t, while the integral is taken over the distance variable s. To fix this, apply the chain rule from calculus to the velocity derivative and substitute Eq. 2-21 to get

$$\frac{d\mathcal{V}}{dt} = \frac{d\mathcal{V}}{ds}\frac{ds}{dt} = \frac{d\mathcal{V}}{ds}\mathcal{V} \tag{2-23}$$

Now substitute Eq. 2-23 into Eq. 2-22 to yield

$$W^* = \int_{s_1}^{s_2} m\frac{d\mathcal{V}}{ds}\mathcal{V}\,ds = \int_{\mathcal{V}_1}^{\mathcal{V}_2} m\mathcal{V}\,d\mathcal{V} = m\int_{\mathcal{V}_1}^{\mathcal{V}_2} \mathcal{V}\,d\mathcal{V}$$

where $\mathcal{V}_1$ and $\mathcal{V}_2$ are the velocities at s_1 and s_2, and a constant mass has been assumed. Taking the integral and evaluating at the limits gives

$$W^* = m\left[\frac{\mathcal{V}_2^2}{2} - \frac{\mathcal{V}_1^2}{2}\right] \tag{2-24}$$

Eq. 2-24 is the work done in changing the velocity of a body. Although work, in general, is a function of the path, the work done to change velocity is a special case. This work is path-independent; that is, the work does not depend on how the velocity varies between the initial and the final positions, but only on the values of velocity at these positions. Thus it is possible to define a property of the system called kinetic energy, which is

$$KE = \frac{1}{2}m\mathcal{V}^2$$

With this definition the work becomes

$$W^* = KE_2 - KE_1 \tag{2-25}$$

where KE_1 and KE_2 are the kinetic energies at $\mathcal{V}_1$ and $\mathcal{V}_2$. In analyzing processes using the first law, changes in velocity will be regarded as part of the change in total energy. In fact, Eq. 2-25 can be derived from the first law, which is

$$\Delta KE + \Delta PE + \Delta U = Q - W$$

If there are no changes in potential or internal energy and no heat is transferred, this becomes

$$\Delta KE = -W$$

or

$$KE_2 - KE_1 = -W \tag{2-26}$$

The work, W^*, in Eq. 2-25 is the work done **on** a mass, m, to change its velocity. The work, W, in Eq. 2-26 is the work done **by** a system. If we select the mass as the system, then

$$W = -W^*$$

and Eq. 2-25 and Eq. 2-26 are consistent.

2.10   POTENTIAL ENERGY

Potential energy is related to the work done by gravity on a body. Imagine a body of mass m falling in a gravitational field, as shown in Figure 2-23. The amount of work done on the body as it falls from z_1 to z_2 is:

$$W^* = \int_{z_1}^{z_2} F\, dz$$

Earth   **FIGURE 2-23** A body falling under the influence of gravity.

The force, F, is

$$F = mg$$

where g is the acceleration of gravity. With this substitution, the work becomes

$$W^* = \int_{z_1}^{z_2} mg \, dz$$

Since mg is a constant, this can be integrated to give

$$W^* = mg \, (z_2 - z_1)$$

By definition, the potential energy is

$$PE = mgz$$

The work done by gravity on the body can be expressed in terms of the potential energy as

$$\boxed{W^* = (PE_2 - PE_1)} \tag{2-27}$$

This is the work done **on** the mass by gravity. As with kinetic energy, potential energy can be defined as a property of a system because the work is path-independent. The work depends only on the elevations at start and finish, not on intermediate elevations. In analyzing processes using the first law, work against gravity will be regarded as part of the change in total energy. Eq. 2-27 can be derived from the first law, which is

$$\Delta KE + \Delta PE + \Delta U = Q - W$$

If there are no changes in kinetic or internal energy and no heat is transferred, this becomes

$$\Delta PE = -W$$

or

$$PE_2 - PE_1 = -W = W^* \tag{2-28}$$

2.11 SPECIFIC HEAT OF IDEAL GASES

In the case of compressible substances, two different types of specific heat are used. To illustrate these, consider the two cases shown in Figure 2-24. In case (a), heat is added to an ideal gas in a rigid tank. We define the gas as the system. During this process, the volume

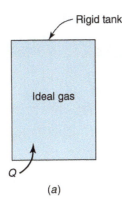

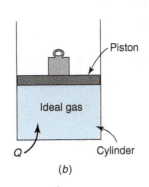

FIGURE 2-24 Addition of heat to an ideal gas (a) in a constant volume process and (b) in a constant pressure process.

of gas remains constant and no work is done; furthermore, the kinetic energy and potential energy of the system remain unchanged. From the first law,

$$\Delta U = Q - W$$

Since no work is done, the work is zero, and this becomes

$$\Delta U = Q$$

If the gas is an **ideal gas**, then it can be shown experimentally that **the internal energy is only a function of temperature**. A specific heat can be defined for gases in much the same way as the specific heat, c, was defined for solids and liquids. If the gas specific heat is constant and the gas has mass, m, then the change in internal energy can be related to temperature change by

$$\Delta U = mc_v \Delta T \qquad \text{constant specific heat, ideal gas} \qquad (2\text{-}29)$$

where c_v is called the **specific heat at constant volume**. Its name arises from the fact that it is the proportionality constant when heat is added at constant volume, as in the case just described. However, c_v has much broader application than just to this one restricted case. **Eq. 2-29 applies to all processes of an ideal gas with constant specific heat and is not restricted to constant volume processes.** Eq. 2-29 may be written in differential form as

$$dU = mc_v \, dT$$

In the general case in which specific heat varies with temperature,

$$dU = mc_v(T) \, dT \qquad (2\text{-}30)$$

and

$$\Delta U = \int dU = \int mc_v(T) \, dT$$

The internal energy is often expressed on a per unit mass basis. By definition, the specific internal energy, u, is

$$u = \frac{U}{m}$$

It follows that

$$dU = m\,du$$

Using this in Eq. 2-30 gives

$$du = c_v(T)\,dT$$

or

$$c_v(T) = \frac{du}{dT} \qquad \text{ideal gas} \tag{2-31}$$

Eq. 2-31 applies only to an ideal gas. If a gas is not ideal, a more general definition of specific heat is needed, that is,

$$c_v(T, v) \equiv \left.\frac{\partial u}{\partial T}\right|_v \tag{2-32}$$

Eq. 2-32 shows that specific heat is, in general, a function of two variables. Only in the case of an ideal gas is specific heat a function of temperature alone. A full explanation of Eq. 2-32 is beyond the scope of this text; it is shown here simply for completeness.

Now, returning to Figure 2-24, the other specific heat that is used for gases will be developed. In case (b) in Figure 2-24, heat is added to a piston–cylinder assembly, and the piston rises. A weight rests on the piston. We define the gas as the system under consideration. Some of the heat transferred into the gas is converted into internal energy and acts to raise the temperature of the gas, and the remainder is converted into expansion work. During this process, the pressure of the gas remains constant. This pressure arises from atmospheric pressure and from the force of gravity acting on the weight and the piston. Because these weights do not change during the process, neither does the pressure. The kinetic and potential energies of the system do not change; therefore, a first-law analysis of case (b) starts with

$$\Delta U = Q - W$$

Assuming that the process is slow enough to be quasi-equilibrium, the work done, from Eq. 2-17, is

$$W = \int P\,dV$$

Because pressure is constant, this becomes

$$W = P\int dV = P(V_2 - V_1) = P\Delta V$$

where V_1 is the volume at the beginning of the process and V_2 is the volume at the end of the process. Substituting this into the first law gives

$$\Delta U = Q - P\Delta V$$

This equation can be rearranged to

$$Q = \Delta U + P\Delta V = U_2 - U_1 + P(V_2 - V_1) \tag{2-33}$$

Certain simplifications will arise if we rewrite this equation in another form. Using the fact that, for this constant-pressure process, $P = P_1 = P_2$, an alternate form of Eq. 2-33 is

$$Q = (U_2 + P_2 V_2) - (U_1 + P_1 V_1) \qquad (2\text{-}34)$$

This is the first instance in which the quantity $U + PV$ appears. This quantity occurs in many different circumstances in thermodynamics and is, therefore, given a special name. It is called **enthalpy** and it is, by definition,

$$\boxed{H = U + PV} \qquad (2\text{-}35)$$

Using the definition of enthalpy, Eq. 2-33 can be rewritten as

$$\boxed{Q = H_2 - H_1 \qquad \text{constant pressure process of closed system}} \qquad (2\text{-}36)$$

Introducing the enthalpy has resulted in a simpler form for Eq. 2-33. Although we are specifically interested in ideal gases in this section, Eq. 2-36 was derived without the ideal gas law and is applicable to any constant-pressure quasi-equilibrium process of a closed system. Enthalpy is useful in many circumstances and is not limited to ideal gases.

Note that if the ideal gas law is substituted into Eq. 2-35, then

$$H = U + \frac{m\overline{R}T}{M}$$

The internal energy, U, of an ideal gas is a function only of temperature. Furthermore, m, $\overline{R}$, and M are constants. Hence, it follows that the **enthalpy, H, of an ideal gas is only a function of temperature**.

Previously, for a *constant-volume process* with constant specific heat, we had

$$\Delta U = Q = mc_v \Delta T$$

Now, for this *constant-pressure process,* a new specific heat is used to give

$$\Delta H = Q = mc_p \Delta T$$

where c_p is called the **specific heat at constant pressure**. It is the proportionality constant that determines how much the gas temperature rises when the gas is heated at constant pressure. **However, c_p is not limited in usefulness just to constant-pressure processes**. As with c_v, c_p is also useful in processes where the pressure is not constant. For all processes of an ideal gas with constant specific heat,

$$\boxed{\Delta H = mc_p \Delta T \qquad \text{constant specific heat, ideal gas}} \qquad (2\text{-}37)$$

Eq. 2-37 may be written in differential form as

$$dH = mc_p \, dT$$

In the general case in which c_p is a function of temperature,

$$dH = mc_p(T) \, dT \qquad (2\text{-}38)$$

and

$$\Delta H = \int dH = \int mc_p(T)\,dT$$

The enthalpy is often expressed on a per unit mass basis. By definition, the specific enthalpy, h, is

$$\boxed{h = \frac{H}{m}}$$

It follows that

$$dH = m\,dh$$

Using this in Eq. 2-38 gives

$$dh = c_p(T)\,dT$$

or

$$\boxed{c_p(T) = \frac{dh}{dT} \qquad \text{ideal gas}} \qquad (2\text{-}39)$$

Eq. 2-39 applies only to an ideal gas. If a gas is not ideal, a more general definition of specific heat is needed, that is,

$$c_p(T,P) \equiv \left.\frac{\partial h}{\partial T}\right|_p \qquad (2\text{-}40)$$

Eq. 2-40 shows that specific heat is, in general, a function of two variables. Only in the case of an ideal gas is specific heat a function of temperature alone. A full explanation of Eq. 2-40 is beyond the scope of this text.

In some cases, it is necessary to account for the variation of specific heat with temperature. This is especially true if there is a large temperature difference during the process or if the specific heat of the gas varies substantially with temperature. One way to account for variable specific heat is to use tables that list the values of u and h directly as a function of temperature. In this approach, no actual value of specific heat is needed, and the change in u or h during a process is determined from table values. Tables A-9 and B-9 give values of u and h for air as a function of temperature.

We can develop another important relationship for ideal gases which shows that c_v and c_p are not independent. Recall that $H = mh$, $U = mu$, and $V = mv$, by definition. Divide Eq. 2-35 by mass to obtain

$$\boxed{h = u + Pv} \qquad (2\text{-}41)$$

Eq. 2-41 is always true. In the case of an ideal gas,

$$Pv = \frac{\overline{R}T}{M}$$

Using this in Eq. 2-41 gives

$$h = u + \frac{\overline{R}T}{M}$$

Differentiate with respect to temperature to obtain

$$\frac{dh}{dT} = \frac{du}{dT} + \frac{\overline{R}}{M}$$

For an ideal gas, the term on the left-hand side is c_p (see Eq. 2-39); the first term on the right-hand side is c_v (see Eq. 2-31). Therefore,

$$c_p(T) = c_v(T) + \frac{\overline{R}}{M} \qquad (2\text{-}42)$$

Note that $c_p > c_v$ and is larger by a constant, $\overline{R}/M$, even though both c_p and c_v are functions of temperature.

EXAMPLE 2-9 Heat and work in a constant-pressure expansion

Hydrogen at 30 psia is contained in a piston–cylinder assembly. The gas has a mass of 0.009 lbm and an initial volume of 0.75 ft³. Assuming the pressure is constant during the process, how much heat must be added to double the volume? Assume specific heat is constant.

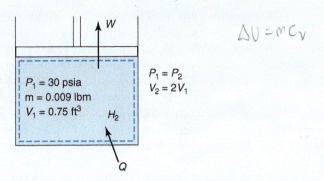

Approach:

Define the system as the hydrogen in the cylinder. The added heat is found by applying the first law:

$$\Delta E = \Delta KE + \Delta PE + \Delta U = Q - W$$

Because we have no information on potential or kinetic energy, we assume they are negligible. If we can calculate ΔU and W, then Q can be determined. The change in internal energy for an ideal gas with constant specific heat is

$$\Delta U = mc_v \Delta T$$

We are not given temperatures. But, because we know pressure and volume, the ideal gas law can be used to find the missing temperatures. Finally, the work can be determined from Eq. 2-17, which is

$$W = \int P \, dV$$

and since P is constant,

$$W = P \int dV = P (V_2 - V_1)$$

Assumptions:

A1. Kinetic energy is negligible.

A2. Potential energy is negligible.

A3. Hydrogen is an ideal gas under these conditions.

A4. Specific heat is constant.

Solution:

Define the hydrogen as the system under study. Assuming no kinetic or potential energy [A1][A2], the first law is

$$Q = \Delta U + W$$

Treating hydrogen as an ideal gas [A3],

$$\Delta U = mc_v \Delta T = mc_v (T_2 - T_1)$$

To evaluate the internal energy change, the temperatures at the initial and final states must be known. To find T_1 use

$$P_1 V_1 = \frac{m\bar{R}T_1}{M}$$

Solving for T_1

$$T_1 = \frac{MP_1 V_1}{m\bar{R}}$$

Using values for M from Table B-1 and the given information, T_1 may be calculated as

$$T_1 = \frac{\left[2.016 \, \frac{\text{lbm}}{\text{lbmol}} \right] (30 \, \text{psia}) \left(0.75 \, \text{ft}^3 \right)}{(0.009 \, \text{lbm}) \left[10.73 \, \frac{\text{psia} \cdot \text{ft}^3}{\text{lbmol} \cdot \text{R}} \right]} = 470 \, \text{R}$$

Since the volume doubles and the pressure stays constant, $V_2 = 1.5 \, \text{ft}^3$ and $P_2 = 30 \, \text{psia}$. The final temperature is

$$T_2 = \frac{MP_2 V_2}{m\bar{R}} = \frac{(2.016)(30)(1.5)}{(0.009)(10.73)} = 939 \, \text{R}$$

We could now calculate ΔU if we knew c_v. Values of c_v for hydrogen are listed in Table B-8. Note that c_v varies with temperature in this table but that the variation is very slight. Using the average temperature during the process to evaluate c_v will be a good approximation [A4]. The average temperature is

$$T_{avg} = \frac{T_1 + T_2}{2} = \frac{470 + 939}{2} = 704 \, \text{R} = 244°\text{F}$$

By interpolation in Table B-8,

$$c_v = 2.47 \frac{\text{Btu}}{\text{lbm} \cdot \text{R}}$$

Using

$$\Delta U = mc_v (T_2 - T_1)$$

$$\Delta U = (0.009 \text{ lbm}) \left(2.47 \frac{\text{Btu}}{\text{lbm·R}} \right) (939 - 470) R = 10.4 \text{ Btu}$$

Assume a quasi-equilibrium process, so that

$$W = \int P \, dV$$

Because pressure is constant

$$W = P \int dV = P (V_2 - V_1)$$

$$W = (30 \text{ psia}) (1.5 - 0.75) \text{ ft}^3 \left(\frac{1 \text{ Btu}}{5.404 \text{ psia·ft}^3} \right)$$

$$W = 4.16 \text{ Btu}$$

From the first law

$$Q = \Delta U + W = 10.4 + 4.16 = 14.6 \text{ Btu}$$

Alternative solution:

For a constant pressure process of an ideal gas in a closed system:

$$Q = \Delta H = mc_p \Delta T$$

From Table B-8, $c_p = 3.455$ Btu/(lbm·R) at $T_{avg} = 244°F$. Therefore,

$$Q = (0.009 \text{ lbm}) \left(3.455 \frac{\text{Btu}}{\text{lbm·R}} \right) (939 - 470) R = 14.6 \text{ Btu}$$

This alternative solution shows how useful enthalpy can be in a constant-pressure process of a closed system.

EXAMPLE 2-10   **First law with variable specific heat**

A rigid tank with a volume of 400 cm^3 contains air initially at 22°C and 100 kPa. A paddlewheel stirs the gas until the final temperature is 428°C. During the process, 600 J of heat are transferred from the air to the surroundings. Calculate the work done by the paddlewheel two ways:

a) Assuming constant specific heat

b) Assuming variable specific heat

Approach:

Choose the system as the gas in the tank. The work done can be determined from the first law:

$$\Delta U = Q - W$$

In part a, where specific heat is assumed to be constant, the change in internal energy is found from $\Delta U = mc_v(T_2 - T_1)$. The mass is calculated using the ideal gas law, and the specific heat

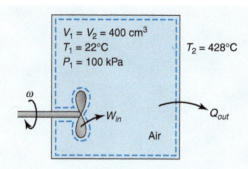

is evaluated at the average of T_1 and T_2 using data in Table A-8. Take care to note that heat is negative, since it leaves the system, and calculate the work done by the paddlewheel.

In part b, the specific heat is variable. The internal energy is calculated from $\Delta U = m(u_2 - u_1)$. Values of internal energy are obtained from Table A-9 at T_1 and T_2. As in part a, work is calculated using the first law.

Assumptions:

A1. Kinetic energy is negligible.

A2. Potential energy is negligible.

A3. Specific heat is constant.

A4. Carbon dioxide is an ideal gas under these conditions.

Solution:

a) Assuming no kinetic or potential energy [A1][A2], the first law is

$$\Delta U = Q - W$$

If specific heat is assumed constant [A3],

$$\Delta U = mc_v (T_2 - T_1)$$

The specific heat is evaluated at the average of the initial and final temperatures, which is

$$T_{avg} = \frac{22 + 428}{2} = 225°C = 498 \text{ K}$$

From Table A-8, $c_v = 0.742$ kJ/kg·K. To find mass, apply the ideal gas law [A4]

$$m = \frac{M P_1 V_1}{\bar{R} T_1}$$

From Table A-1, $M = 28.97$ for air. Substituting values

$$m = \frac{\left[28.97 \frac{\text{kg}}{\text{kmol}} \right] (100 \text{ kPa}) \left(400 \text{ cm}^3 \right) \left[\frac{1 \text{ m}}{10^6 \text{cm}^3} \right]}{\left[8.314 \frac{\text{kJ}}{\text{kmol·K}} \right] (22 + 273) \text{ K}}$$

$$m = 0.0004725 \text{ kg}$$

Rearranging the first law,

$$W = Q - \Delta U = Q - mc_v (T_2 - T_1)$$

$$W = (-600 \text{ J}) - (0.0004725 \text{ kg}) \left[0.742 \frac{\text{kJ}}{\text{kg·K}} \right] (428 - 22)°C \left[\frac{1000 \text{ J}}{1 \text{ kJ}} \right]$$

$$W = -742.3 \text{ J}$$

b) In this part, specific heat is allowed to vary with temperature. As before, the first law is

$$\Delta U = Q - W$$

$$m(u_2 - u_1) = Q - W$$

Using data for u in Table A-9 at the initial temperature of 22°C = 295 K,

$$u_1 = 210.5 \ \frac{kJ}{kg}$$

For the final state, it is necessary to interpolate. At the final temperature of $T_2 = 428°C = 701$ K,

$$\frac{u_2 - 512.3}{520.2 - 512.3} = \frac{701 - 700}{710 - 700}$$

Solving for u_2,

$$u_2 = 513.1 \frac{kJ}{kg}$$

Work is now evaluated from the first law as

$$W = (-600 \ J) - (0.0004725 \ kg)(513.1 - 210.5)\frac{kJ}{kg}\left(\frac{1000 \ J}{1 \ kJ}\right)$$

$$W = -743 \ J$$

Comments:

The work calculated using constant specific heat is very close to that calculated with variable specific heat. This is often the case as long as the constant value of specific heat is evaluated at the correct average temperature. Unless otherwise noted, we will assume constant specific heat throughout the text.

For ideal (i.e., incompressible) liquids and solids, a single specific heat, c, is used. However, real liquids and solids do change in volume if temperature and/or pressure changes. In most circumstances, the change is small enough to be neglected. For example, if a metal bar is heated, it will grow slightly longer due to thermal expansion. Just as in a gas, the added heat both increases the temperature and increases the volume. The volume change, however, is very small. If the bar is heated at constant pressure (i.e., the ends are unconstrained), the work is given by

$$W = \int P \, dV = P \Delta V$$

Because the volume change, ΔV, is very small, work is very small. For liquids and solids, the difference between constant-volume heating and constant-pressure heating is generally insignificant. This is why we were able to use the single specific heat, c, in Eq. 2-9. In effect

$$\boxed{c \approx c_v \approx c_p} \qquad \text{ideal liquids and solids}$$

In tables of specific heat of liquids or solids, values of c_p are typically reported. Although it hardly makes a difference, the use of c_p is more theoretically correct. When a solid is heated, its ends are not usually confined. The more common case is to let the solid expand freely. For such an expansion, the "correct" specific heat to use is c_p.

2.12   POLYTROPIC PROCESS OF AN IDEAL GAS

By definition, a **polytropic process** is one for which

$$PV^n = \text{constant} \quad \text{or} \quad Pv^n = \text{constant} \tag{2-43}$$

where n is a constant. Many common processes are polytropic, including constant-pressure heat addition, isothermal expansion or compression of an ideal gas, and some adiabatic processes, as shown below. The constant, n, may take any value from $-\infty$ to $+\infty$.

If a polytropic process begins at state 1 and ends at state 2, then

$$PV^n = P_1V_1^n = P_2V_2^n \tag{2-44}$$

The work done during a quasi-equilibrium, polytropic process of a closed system is

$$W = \int_1^2 P\,dV = \int_1^2 \frac{P_1V_1^n}{V^n}\,dV$$

Because P_1 and V_1 are both constants, we may write

$$W = P_1V_1^n \int_1^2 \frac{dV}{V^n} \tag{2-45}$$

If $n = 1$, this integral has one solution, and if $n \neq 1$, the integral has a different solution. Starting with the $n = 1$ case:

$$W = P_1V_1 \int_1^2 \frac{dV}{V} = P_1V_1\,(\ln V_2 - \ln V_1)$$

$$\boxed{W = P_1V_1 \ln \frac{V_2}{V_1} \qquad \text{polytropic process, } n = 1} \tag{2-46}$$

The case $n = 1$ is actually an isothermal expansion or compression. If an ideal gas is compressed or expanded isothermally from state 1 to state 2, then

$$T_1 = T_2$$

Substituting the ideal gas law in each side of this equation gives

$$\frac{P_1V_1M}{m\bar{R}} = \frac{P_2V_2M}{m\bar{R}}$$

which reduces to

$$P_1V_1 = P_2V_2$$

This is the equation for a polytropic process with $n = 1$ (see Eq. 2-44). Thus, **an isothermal process of an ideal gas is polytropic with $n = 1$**. It follows from Eq. 2-46 that the work done in an isothermal expansion or compression of an ideal gas is

$$W = P_1 V_1 \ln \frac{V_2}{V_1} \qquad \text{isothermal process, ideal gas} \qquad (2\text{-}47)$$

Substituting the ideal gas law gives an alternative form:

$$W = \frac{m\overline{R}T}{M} \ln \frac{V_2}{V_1} \qquad \text{isothermal process, ideal gas} \qquad (2\text{-}48)$$

If $n \neq 1$, then Eq. 2-45 becomes:

$$W = \left(P_1 V_1^n\right) \int_1^2 \frac{dV}{V^n} = \left(P_1 V_1^n\right) \frac{V_2^{1-n} - V_1^{1-n}}{1-n} = \frac{\left(P_1 V_1^n\right) V_2^{1-n} - \left(P_1 V_1^n\right) V_1^{1-n}}{1-n}$$

Substituting Eq. 2-44 into this expression gives

$$W = \frac{\left(P_2 V_2^n\right) V_2^{1-n} - \left(P_1 V_1^n\right) V_1^{1-n}}{1-n}$$

which simplifies to

$$W = \frac{P_2 V_2 - P_1 V_1}{1-n} \qquad \text{polytropic process, } n \neq 1 \qquad (2\text{-}49)$$

Using the ideal gas law, this equation becomes

$$W = \frac{\dfrac{m\overline{R}T_2}{M} - \dfrac{m\overline{R}T_1}{M}}{1-n} = \frac{m\overline{R}(T_2 - T_1)}{M(1-n)} \qquad \text{polytropic process, } n \neq 1 \qquad (2\text{-}50)$$

EXAMPLE 2-11   Isothermal expansion of an ideal gas

Oxygen at 300 K expands slowly and isothermally from 100 kPa to 45 kPa. The mass of oxygen is 0.052 kg. Using the ideal gas model, find the work done.

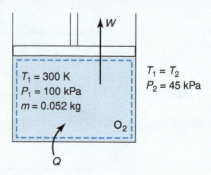

Approach:

Choose the oxygen as the system under study. Because the process is slow, we may assume it is a quasi-equilibrium process. For an isothermal expansion of an ideal gas,

$$W = \frac{m\overline{R}T}{M} \ln \frac{V_2}{V_1}$$

Use the ideal gas law to rewrite the volumes in terms of pressures and then substitute values.

Assumptions:

Solution:

A1. This is a quasi-equilibrium process.

A2. Oxygen behaves like an ideal gas under these conditions.

Define the system as the oxygen in the cylinder. For a slow, isothermal expansion of an ideal gas [A1][A2], the work done is

$$W = \frac{m\overline{R}T}{M} \ln \frac{V_2}{V_1}$$

Using the ideal gas law, this becomes

$$W = \frac{m\overline{R}T}{M} \ln \left[\frac{\frac{m\overline{R}T}{P_2 M}}{\frac{m\overline{R}T}{P_1 M}} \right] = \frac{m\overline{R}T}{M} \ln \left[\frac{P_1}{P_2} \right]$$

Substituting given values,

$$W = \frac{(0.052 \text{ kg}) \left(8.314 \frac{\text{kJ}}{\text{kmol·K}} \right) (300 \text{ K})}{32 \frac{\text{kg}}{\text{kmol}}} \ln \left(\frac{100 \text{ kPa}}{45 \text{ kPa}} \right)$$

$$W = 3.24 \text{ kJ}$$

The value of M has been taken from Table A-1.

Another polytropic process is an adiabatic, quasi-equilibrium expansion or compression of an ideal gas with constant specific heats. Although this may seem to be a very special case, in fact, it has practical importance in understanding reciprocating engines, compressors, turbines, nozzles, and many other devices. To find the value of the exponent, n, for this process, begin with the first law (not including kinetic or potential energy):

$$\Delta U = Q - W$$

For an adiabatic process between states 1 and 2, $Q = 0$, and the first law becomes

$$\Delta U = -W$$

For an ideal gas with constant specific heat, from Eq. 2-29,

$$mc_v \Delta T = mc_v (T_2 - T_1) = -W$$

Rearranging,

$$W = mc_v (T_1 - T_2)$$

Replacing temperatures using the ideal gas law gives

$$W = mc_v \left(\frac{P_1 V_1 M}{\overline{R} m} - \frac{P_2 V_2 M}{\overline{R} m} \right)$$

or

$$W = \frac{c_v M}{\overline{R}} (P_1 V_1 - P_2 V_2)$$

Solving Eq. 2-42 for $\overline{R}$ and substituting gives

$$W = \frac{c_v M}{c_p M - c_v M} (P_1 V_1 - P_2 V_2)$$

Divide the numerator and denominator by $c_v M$ to obtain

$$W = \left[\frac{1}{c_p/c_v - 1} \right] (P_1 V_1 - P_2 V_2) \tag{2-51}$$

Eq. 2-51 is identical to Eq. 2-49, which is an expression for work in a polytropic process, if $n = c_p/c_v$. Eq. 2-51 was derived assuming an adiabatic, quasi-equilibrium process of an ideal gas with constant specific heats. Therefore, such a process is polytropic with $n = c_p/c_v$. It is conventional to define k as the ratio of specific heats, that is,

$$k = \frac{c_p}{c_v}$$

From Eq. 2-43, pressure varies with volume in a polytropic process as

$$PV^n = \text{constant}$$

Therefore:

$$\boxed{PV^k = \text{constant} \qquad \text{adiabatic, quasi-equilibrium, ideal gas, constant specific heats}} \tag{2-52}$$

An alternate form is

$$Pv^k = \text{constant} \qquad \text{adiabatic, quasi-equilibrium, ideal gas,} \atop \text{constant specific heats} \qquad (2\text{-}53)$$

The quantity k is a function of temperature since the specific heats are themselves functions of temperature. Values of k are included in tables of ideal gas specific heats, such as Tables A-8 and B-8.

There are several useful equations that apply to a quasi-equilibrium, adiabatic expansion or compression of an ideal gas with constant specific heats. First, from Eq. 2-52 it follows that

$$P_1 V_1^k = P_2 V_2^k$$

An alternate way of writing this is

$$\frac{P_2}{P_1} = \left(\frac{V_1}{V_2}\right)^k \qquad \text{adiabatic, quasi-equilibrium, ideal gas,} \atop \text{constant specific heats} \qquad (2\text{-}54)$$

It is possible to eliminate pressure in this equation by inserting the ideal gas law, that is,

$$\frac{m\bar{R}T_2/MV_2}{m\bar{R}T_1/MV_1} = \left(\frac{V_1}{V_2}\right)^k$$

which simplifies to

$$\frac{T_2}{T_1} = \left(\frac{V_1}{V_2}\right)^{k-1} \qquad \text{adiabatic, quasi-equilibrium, ideal gas,} \atop \text{constant specific heats} \qquad (2\text{-}55)$$

The last expression of this series is obtained by substituting the ideal gas law into the right-hand side of Eq. 2-54 to get

$$\frac{P_2}{P_1} = \left(\frac{m\bar{R}T_1/MP_1}{m\bar{R}T_2/MP_2}\right)^k$$

After some manipulation, this equation becomes

$$\frac{T_2}{T_1} = \left(\frac{P_2}{P_1}\right)^{\frac{k-1}{k}} \qquad \text{adiabatic, quasi-equilibrium, ideal gas,} \atop \text{constant specific heats} \qquad (2\text{-}56)$$

EXAMPLE 2-12   Adiabatic compression of an ideal gas

A well-insulated piston–cylinder assembly contains 0.031 m³ of air at 40°C and 102 kPa. Find the work required to compress the air slowly to 350 kPa.

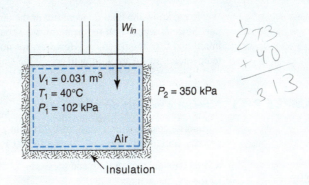

Approach:

Define the system as the air. This process is a slow, adiabatic compression of an ideal gas; therefore, the work is, from Eq. 2-51,

$$W = \left(\frac{1}{c_p/c_v - 1}\right)(P_1 V_1 - P_2 V_2)$$

The initial volume and pressure are known. To find the final volume, use

$$\frac{P_2}{P_1} = \left(\frac{V_1}{V_2}\right)^k$$

Assumptions:

A1. This is a quasi-equilibrium process.
A2. The process is adiabatic.
A3. Air may be modeled as an ideal gas.
A4. Specific heat is constant.

Solution:

The air in the cylinder is the system under study. For a slow, adiabatic compression of an ideal gas [A1][A2][A3], the work done is (see Eq. 2-51)

$$W = \left(\frac{1}{c_p/c_v - 1}\right)(P_1 V_1 - P_2 V_2)$$

To find the final volume, use [A4]

$$\frac{P_2}{P_1} = \left(\frac{V_1}{V_2}\right)^k$$

This may be rearranged to

$$V_2 = V_1 \left(\frac{P_1}{P_2}\right)^{\frac{1}{k}}$$

Substituting values:

$$V_2 = (0.031)\,\text{m}^3 \left(\frac{102}{350}\right)^{\frac{1}{1.4}}$$

$$V_2 = 0.0128\,\text{m}^3$$

Work may now be calculated as:

$$W = \frac{\left[(102\,\text{kPa})\,(0.031\,\text{m}^3) - (350\,\text{kPa})\,(0.0128\,\text{m}^3)\right]\left(\frac{1000\,\text{Pa}}{1\,\text{kPa}}\right)}{1.4 - 1}$$

$$W = -3290\,\text{J} = -3.29\,\text{kJ}$$

The value of k was taken from Table A-8 at 300 K. Work is negative because work is done on the gas during this compression process.

Comment:

We do not know the final temperature of the air, although we could calculate it from the ideal gas law. Since k depends on temperature, we should ideally use the value of k at the average of the initial and final temperatures. However, we note from Table A-8 that k varies only slightly with temperature, so we are willing to accept the small inaccuracy that results from using k at 300 K.

EXAMPLE 2-13   Adiabatic expansion of an ideal gas

Carbon monoxide with a mass of 0.221 lbm expands slowly and adiabatically in a refrigeration process. Initially, the gas is at 30 psia and 80°F. If the volume doubles during the process,

a. find the final temperature and pressure.
b. find the work done.

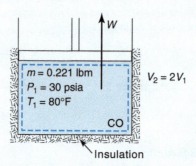

Approach:

Select the carbon monoxide as the system. This process is a slow, adiabatic expansion of an ideal gas. Assuming constant specific heats, one may write

$$\frac{T_2}{T_1} = \left(\frac{V_1}{V_2}\right)^{k-1}$$

Use this to find the final temperature. To find the final pressure, use

$$\frac{P_2}{P_1} = \left(\frac{V_1}{V_2}\right)^{k}$$

The work may calculated from the first law, assuming that no heat is transferred.

Assumptions:

A1. This is a quasi-equilibrium process.
A2. The process is adiabatic.
A3. Air may be modeled as an ideal gas.
A4. Specific heat is constant.

Solution:

The system is the carbon monoxide gas. For a slow, adiabatic expansion of an ideal gas with constant specific heats {A1}{A2}{A3}{A4},

$$\frac{T_2}{T_1} = \left(\frac{V_1}{V_2}\right)^{k-1}$$

Solving for T_2,

$$T_2 = T_1 \left(\frac{V_2}{V_1}\right)^{1-k}$$

Since the volume doubles, we may substitute $V_2/V_1 = 2$. The temperature then is

$$T_2 = (80 + 460)\, \text{R}\, (2)^{1-1.4}$$

where a value for k at 80°F from Table B-8 has been used. The result is

$$T_2 = 409\, \text{R} = -50.8°\text{F}$$

If more accuracy were desired, the calculation could be repeated with a value of k taken at the average of the initial and final temperatures; however, k does not vary significantly over this temperature range, so iteration is not necessary. To find the final pressure, use

$$\frac{P_2}{P_1} = \left(\frac{V_1}{V_2}\right)^k$$

which may be rearranged to

$$P_2 = P_1 \left(\frac{V_2}{V_1}\right)^{-k}$$

Substituting values,

$$P_2 = 30\, \text{psia}\, (2)^{-1.4} = 11.4\, \text{psia}$$

To find the work, start from the first law in the form

$$\Delta U = Q - W$$

A5. Kinetic energy is negligible.
A6. Potential energy is negligible.

where potential and kinetic energy changes have been neglected [A5][A6]. Since the process is adiabatic, this reduces to

$$\Delta U = -W$$

Assuming constant specific heat,

$$mc_v\,(T_2 - T_1) = -W$$

or

$$W = mc_v\,(T_1 - T_2)$$

Substituting values and using specific heat from Table B-8,

$$W = (0.221\, \text{lbm}) \left[0.177\, \frac{\text{Btu}}{\text{lbm·R}}\right] [80 - (-50.8)]°\text{F}$$

$$W = 5.11\, \text{Btu}$$

2.13   THE FIRST LAW IN DIFFERENTIAL FORM

The first law has been stated as

$$\Delta KE + \Delta PE + \Delta U = Q - W$$

This applies to a process in which there are finite changes in each of the constituent quantities. If, instead, the changes are differential in size, another form is needed. The differential form for work, δW, has already been discussed. Work is an inexact differential because the amount of work done depends on the process and not solely on the end states. In contrast, the change in kinetic energy, ΔKE, depends only on the end states. Specifically,

$$\Delta KE = KE_2 - KE_1 = \tfrac{1}{2}m\mathcal{V}_2^2 - \tfrac{1}{2}m\mathcal{V}_1^2$$

For closed systems, mass is constant, and the kinetic energy depends only on the velocities at states 1 and 2, not on how these velocities were attained. Therefore, change in kinetic energy is an exact differential of form dKE, and it may be integrated as follows:

$$KE_2 - KE_1 = \int_1^2 dKE$$

Change in potential energy is also an exact differential; it depends only on the end states, that is, the heights z_1 and z_2. Potential energy may be integrated as:

$$PE_2 - PE_1 = \int_1^2 dPE = mgz_2 - mgz_1$$

Change in internal energy is an exact differential as well, because internal energy change depends only on the end states. It is correct to write

$$U_2 - U_1 = \int_1^2 dU$$

where U_1 and U_2 are the internal energies at states 1 and 2.

By contrast, heat is not an exact differential. Heat is the energy transferred during a process due to a temperature difference and is not a function of the end states. It depends on the path. This can be formally demonstrated by the following argument.

Consider two different paths through $P - V$ space, as shown in Figure 2-25. For path A, the first law may be written

$$\Delta KE + \Delta PE + \Delta U = Q_A - W_A$$

$$(KE_2 - KE_1) + (PE_2 - PE_1) + (U_2 - U_1) = Q_A - W_A \tag{2-57}$$

For path B, the first law is

$$\Delta KE + \Delta PE + \Delta U = Q_B - W_B$$

$$(KE_2 - KE_1) + (PE_2 - PE_1) + (U_2 - U_1) = Q_B - W_B \tag{2-58}$$

The left-hand sides of Eq. 2-57 and Eq. 2-58 are identical. Therefore, subtracting Eq. 2-58 from Eq. 2-57 and rearranging gives

$$Q_A - W_A = Q_B - W_B \tag{2-59}$$

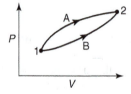

FIGURE 2-25 The paths between states 1 and 2.

Suppose now that A and B are quasi-equilibrium processes. Work is given by

$$W = \int P\,dV$$

and the work is the area under the curve in Figure 2-25. For the paths shown,

$$W_A \neq W_B$$

It follows from Eq. 2-59 that

$$Q_A \neq Q_B$$

Thus, heat depends on the path through $P - V$ space, not solely on the end states. For that reason, the differential form is written δQ, and the first law becomes

$$\boxed{dKE + dPE + dU = \delta Q - \delta W} \qquad (2\text{-}60)$$

2.14 THE "PIZZA" PROCEDURE FOR PROBLEM SOLVING

Now that you have seen a few examples of solved problems, you may have noticed some common features in the solutions. In every problem, we define the system, organize information into a schematic, apply governing principles, make assumptions, evaluate properties, and perform calculations. If problems are tackled in this systematic way, errors are avoided and correct solutions achieved. In this section, we give an overview of effective problem-solving technique and offer hints for solving tough problems.

Many students find "getting started" to be the most difficult part of the problem. As problems become more involved (see, for example, Figure 1-13, which is a schematic of a Rankine cycle power plant used to generate electricity), the tendency of many students is to immediately try to calculate something—anything. This approach can cause problems. First, many students select an equation from the text without thinking about the restrictions on that equation. For example, a student may be trying to calculate temperature and select the ideal gas law simply because it contains temperature even though the substance they are analyzing is a liquid. More subtle errors also occur, such as applying an equation that is only true for adiabatic (insulated) cases to isothermal cases.

The correct approach to solving problems involves stepping back, avoiding immediate equation grabbing, and thinking carefully about all aspects of the problem. In other words, if you sit on your hands and give some thought to the solution before trying to calculate something, you might make more and faster progress than by an immediate calculation. On the other hand, it is not always possible to see a solution procedure clearly from start to finish when beginning a problem. You may need to "play" with the equations to feel your way to a solution. Problems arise when you use equations that do not apply, so you must always be sensitive to the appropriate assumptions and restrictions on equations.

Almost all solutions to engineering problems require three tasks. If each of these tasks is handled correctly, a reasonable and/or accurate solution will be obtainable; if one task is not handled well, then the odds are poor that the answer obtained will be correct. We compare the solution procedure to a three-legged stool; remove one leg from the stool, and the stool falls. For engineering problems, the three "legs" are (1) analysis, (2) application of governing concepts, and (3) evaluation of properties.

These "legs" are described below, where we discuss a methodology to solve problems. We lay out this discussion in the format we use to solve the example problems.

Example Problem Statement

The problem statement gives much information, either explicitly or implicitly. The information may be described in terms of the value of variables at different locations in the system, a drawing of an assembly of devices working together, descriptions of how a device operates, and so on. The question to be answered is asked.

Approach:

The analysis of the problem is wrapped up in the approach to the problem solution. This can be broken into about three steps (some problems require more, others less):

- *Read the problem.* This seems to be an obvious statement. Nevertheless, many students partially read the problem and then immediately start trying to calculate something. A quick or partial reading often misses crucial pieces of information. It is useful to read the problem statement at least twice.

- *Draw a schematic and organize the inforamtion.* As noted in Chapter 1, a schematic diagram simply shows the relationship between various pieces of a system. Indicate the processes involved. Give each location or piece of information a unique symbol consistent with how that information will be used in an equation. Include units with the given information. Also, write down what is being sought—not in words copied from the problem statement, but rather with a symbol you will use in the equations.

- *Think.* This is the step some students give little attention to. Now is the time to sit on your hands for a moment or two. Consider what is occurring in the system you have drawn. Does the problem have to be solved as a steady or unsteady process? Is it a closed or an open system? Decide which governing principles (e.g., conservation of mass, conservation of energy, conservation of momentum, entropy balance, a force balance, a moment balance, etc.) are needed and how you will attack the problem.

Solution:

This phase of the solution involves application of governing concepts and evaluation of properties. Develop your equations using symbols. Do not substitute numbers until they are absolutely necessary. Do not look up properties until you know what properties are actually needed. Note that you may need to work through several equations/concepts before any calculations are possible. The following points are included in a problem solution:

- *Start the problem solution.* Where to start a problem can be a difficult decision. Begin with a governing equation (e.g., the first law) or a definition (e.g., cycle thermal efficiency) that includes the quantity you seek. This is a reasonable way to attack a problem.

- *Apply governing principles.* After you have decided how to start the problem, use whichever governing principles (e.g., conservation of mass, conservation of energy, conservation of momentum, entropy balance, a force balance, a moment balance, etc.) or definitions are needed. Leave your solution in terms of symbols/variable names. Do not substitute values of quantities until it is absolutely necessary (things may cancel and simplify). Then consider each term in the equation and whether or not sufficient information is given to evaluate it. If data are lacking, then another governing equation or definition may be required.

- *Make assumptions.* When you begin a problem solution, you may be able to immediately impose some restrictions (assumptions) on the problem (e.g., steady or unsteady). However, as you go through the solution, you may need to make additional assumptions to be able to solve the problem. Make those assumptions when required; at the very beginning of the solution, do not concern yourself about what all the assumptions must be. Let the solution procedure guide your thoughts. If you are not sure what assumptions to make, write down the governing equation with all its terms (e.g., the first law) and examine each term. Ask yourself whether you know something about the term, whether you can develop information about it from other sources, or whether you are justified in assuming that the term is zero.

- *Evaluate properties.* When you begin a problem solution, you will not know what properties are needed until you have done the analysis. Wait to evaluate properties until you need them. Likewise, make sure you evaluate the properties at the correct temperatures and/or pressures.

- *Subsitute numbers into equations.* The final step is the substitution of information (given or developed) into the equations you have developed during the problem solution. Include units; those provide a quick check on the equation. If the units do not work out, then there is an error. Use appropriate conversion factors. [Units do make a difference (see Figure 2-26.)] Likewise, the sign of the answer must make sense.

Comments:

In this part of the solution, you consider your answer and make a judgment about whether it is reasonable or not (see Figure 2-27). Is the magnitude of your answer consistent with what you might expect? Do the units match what you expect? Is the sign right? Have the assumptions you made simplified the problem too much, or can you speculate about what effect the assumptions have on the solution? This is an assessment of the overall validity and usefulness of the solution.

A10 Friday, October 1, 1999 ****   Albany, New York ■ TIMES UNION

Loss of Mars probe blamed on math conversion error

NASA scientists apparently failed to change English units of measure to metric

By MATTHEW FORDAHL
Associated Press

LOS ANGELES — The $125 million spacecraft that was destroyed on a mission to Mars last week was probably doomed by NASA scientists' embarrassing failure to convert English units of measurement to metric ones, the space agency said Thursday.

The Mars Climate Orbiter flew too close to Mars and is believed to have broken apart or burned up in the atmosphere.

NASA said the English-vs.-metric mix-up apparently caused the navigation error. The mistake was particularly embarrassing because the spacecraft had successfully flown 416 million miles over 9½ months before its disappearance Sept. 23 just as it was about to go into orbit around the Red Planet.

Agency officials said the mistake somehow escaped what is supposed to be a rigorous error-checking process. A report is expected in mid-November.

"It does not make us feel good that this happened," said Tom Gavin of NASA's Jet Propulsion Laboratory. "This mix-up has caused us to look at our entire end-to-end process. We will get to the bottom of this."

JPL said that its preliminary findings showed that Lockheed Martin Astronautics in Colorado submitted acceleration data in English units of pounds of force instead of the metric unit called newtons. At JPL, the numbers were entered into a computer that assumed metric measurements.

"In our previous Mars missions, we have always used metric," Gavin said.

The numbers were used in figuring the force of thruster firings used by the spacecraft to adjust its position. The bad numbers had been used ever since the spacecraft's launch last December, but the effect was so small that it went unnoticed. The difference 'added up over the months as the spacecraft journeyed toward Mars.

Gavin said he does not expect the error to affect NASA's relationship with Lockheed Martin Astronautics, which has built several probes for the space agency. Lockheed Martin had no immediate comment.

The orbiter's sibling spacecraft, Mars Polar Lander, is set to arrive Dec. 3. Gavin said investigators are trying to determine whether NASA made the same mistake with that spacecraft.

The Mars Climate Orbiter was on a mission to study the Red Planet's weather and look for signs of water — information key to understanding whether life ever existed or can exist there. It carried cameras along with equipment for measuring temperature, dust, water vapor and clouds.

The Mars Polar Lander will study Mars' climate history and weather with the goal of finding what happened to water on the planet. It is equipped with a robotic arm that will collect samples for testing inside the spacecraft.

FIGURE 2-26 Consequences from a major unit-conversion error. (Reprinted with permission of The Associated Press.)

FIGURE 2-27 An unreasonable solution. (Foxtrot ©1997 Bill Amend. Reprinted with permission of Universal Press Syndicate. All rights reserved.)

Because thermal systems problems can be involved, it is important to develop a systematic way to attack them. If you practice the above procedure on simpler problems, when you are faced with a more involved problem, you will have the experience and tools to attack it with confidence. Note that the above procedure is not a linear (serial) process. As you proceed through a solution and gain more knowledge, you may need to revisit earlier steps and make changes.

So how does the procedure given above correspond to the title of this section, "The 'Pizza' Procedure for Problem Solving"? Consider Figure 1-13, which is a schematic of a Rankine cycle power plant. This is a complex system. If we wanted to determine how much power could be produced for a given heat input, where would we start? We approach this problem the same way we eat a pizza. If someone told you to eat a 45-cm-diameter pizza all at once, you could not do it. Rather, you cut the pizza into manageable pieces, take bites of one piece until it is finished, and then move on to the next piece. This is identical to what we do with engineering problems. We do not solve a problem by thinking we must include everything all at once. The Rankine cycle shown in Figure 1-13 is our "pizza." Each component in the system is a piece of the "pizza." Our "bites" from the piece may be application of a governing principle or evaluation of a property at one of the locations on the device or some other action.

SUMMARY

The first law of thermodynamics states

$$\Delta E = Q - W$$

There are three types of stored energy considered in this text: internal energy, kinetic energy, and potential energy. Including these, the first law is

$$\Delta KE + \Delta PE + \Delta U = Q - W$$

The kinetic energy is due to the velocity of the system. The change in kinetic energy between two states is

$$\Delta KE = KE_2 - KE_1 = \frac{1}{2}m\mathcal{V}_2^2 - \frac{1}{2}m\mathcal{V}_1^2$$

The potential energy is due to the elevation of the system in a gravitational field. The change in potential energy between two states is

$$\Delta PE = mg\,(z_2 - z_1)$$

where z_1 and z_2 are the heights at the beginning and the end of the elevation change. In using the first law, the signs of the work and heat terms are important. The sign convention for work is:

- work done by a system is positive
- work done on a system is negative

The sign convention for heat is:

- heat transfer to a system is positive
- heat transfer from a system is negative

For an ideal (i.e., incompressible) solid or liquid, the differential change in internal energy is

$$dU = mc(T)\,dT$$

where $c(T)$ is specific heat. If specific heat is constant

$$\Delta U = mc\,\Delta T$$

If c varies with temperature, using a value evaluated at the average of the starting and ending temperatures is a reasonable approximation. By definition, the density is mass per unit volume, that is,

$$\rho = \frac{m}{V}$$

Pressure is defined as a force per unit area, or

$$P = \frac{F}{A}$$

Another important thermophysical property is temperature. Four different unit systems for temperature are in common use: Fahrenheit, Celsius, Kelvin, and Rankine. Fahrenheit and Celsius are relative scales and Kelvin and Rankine are absolute scales. The scales are related by

$$T(°F) = 1.8\,T(°C) + 32$$
$$T(°C) = \left[T(°F) - 32\right]/1.8$$
$$T(K) = T(°C) + 273.15$$
$$T(R) = T(°F) + 459.67$$
$$T(R) = 1.8\,T(K)$$

The behavior of gases can often be approximated by the ideal gas law:

$$PV = n\bar{R}T$$

The temperature in this equation must be measured using an **absolute scale**, (i.e., either Kelvin or Rankine). Likewise, the pressure must be the absolute pressure. The mass of the gas is related to the number of moles, n, by

$$m = nM$$

where M is the molecular weight. The specific volume is defined as

$$v = \frac{1}{\rho} = \frac{V}{m}$$

Alternate forms of the ideal gas law useful in engineering are

$$PV = \frac{m\bar{R}T}{M}$$

$$P = \frac{\rho \bar{R}T}{M}$$

$$Pv = \frac{\bar{R}T}{M}$$

The universal gas constant in several different units is

$$\bar{R} = 8.31434 \text{ kJ/(kmol·K)}$$
$$= 1.9858 \text{ Btu/(lbmol·R)}$$
$$= 1545.35 \text{ ft·lbf/(lbmol·R)}$$
$$= 10.73 \text{ psia·ft}^3/\text{(lbmol·R)}$$

Work is a force through a distance. In general, the work done by a force F acting in direction x is

$$W = \int F\,dx$$

A **quasi-equilibrium process** is one that passes through a series of equilibrium states. These are typically very slow processes with no abrupt changes. If a gas expands or contracts in a quasi-equilibrium process, the work done may be expressed as

$$W = \int P\,dV$$

Electrical work is given by

$$W = \int \xi i\,dt$$

Shaft work is given by

$$W = \int \Im \omega\,dt$$

Enthalpy is defined as

$$H = U + PV$$

The specific enthalpy is the enthalpy per unit mass, or

$$h = \frac{H}{m}$$

As a result

$$h = u + Pv$$

Enthalpy is useful in finding the heat transfer during a **constant-pressure**, quasi-equilibrium expansion or contraction of a closed system. This heat transfer is

$$Q = \Delta H$$

There are two different specific heats commonly used in thermodynamics: the specific heat at constant pressure, c_p, and the specific heat at constant volume, c_v. If the specific heat is not a function of temperature, then, for an ideal gas undergoing a process between two states,

$$\Delta U = mc_v \Delta T$$

$$\Delta H = mc_p \Delta T$$

If specific heat varies with temperature, then use Tables A-9 or B-9, which give values of u and h for air as a function of temperature.

The specific heats of an ideal gas are related by

$$c_p = c_v + \frac{\overline{R}}{M}$$

In a liquid or a solid, the difference between specific heat at constant pressure and specific heat at constant volume is generally negligible. Therefore,

$$c_v \approx c_p \approx c$$

A gas undergoes a polytropic process if

$$PV^n = \text{constant} \quad \text{or} \quad Pv^n = \text{constant}$$

For a polytropic process between state 1 and state 2,

$$PV^n = P_1 V_1^n = P_2 V_2^n$$

An isothermal process of an ideal gas is polytropic with $n = 1$, and the work done during such a process is

$$W = P_1 V_1 \ln \frac{V_2}{V_1} = \frac{m\overline{R}T}{M} \ln \frac{V_2}{V_1}$$

For a polytropic process in which $n \neq 1$, the work done is

$$W = \frac{P_2 V_2 - P_1 V_1}{1 - n}$$

In a polytropic process of an ideal gas with $n \neq 1$, this becomes

$$W = \frac{m\overline{R}(T_2 - T_1)}{M(1 - n)}$$

For a quasi-equilibrium, adiabatic process of an ideal gas with constant specific heats,

$$\frac{P_2}{P_1} = \left(\frac{V_1}{V_2}\right)^k \quad \frac{T_2}{T_1} = \left(\frac{V_1}{V_2}\right)^{k-1} \quad \frac{T_2}{T_1} = \left(\frac{P_2}{P_1}\right)^{\frac{k-1}{k}}$$

SELECTED REFERENCES

BLACK, W.Z., and J.G. HARTLEY, *Thermodynamics*, Harper & Row, New York, 1985.

ÇENGEL, Y.A., and M.A. BOLES, *Thermodynamics, an Engineering Approach*, 4th ed., McGraw-Hill, New York, 2002.

HOWELL, J.R., and R.O. BUCKIUS, *Fundamentals of Engineering Thermodynamics*, 2nd ed., McGraw-Hill, New York, 1992.

MORAN, M.J., and H.N. SHAPIRO, *Fundamentals of Engineering Thermodynamics*, 3rd ed., Wiley, New York, 1995.

MYERS, G., *Engineering Thermodynamics*, Prentice Hall, Englewood Cliffs, N.J., 1989.

VAN WYLEN, G.J., R.E. SONNTAG, and C. BORGNAKKE, *Fundamentals of Classical Thermodynamics*, 4th ed., Wiley, New York, 1994.

PROBLEMS

KINETIC AND POTENTIAL ENERGY

P2-1   A 2000-kg car accelerates from 20 to 60 km/h on an uphill road. The car travels 120 m and the slope of the road from the horizontal is 25°. Determine the work done by the engine.

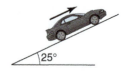

P2-2 A missile is launched vertically upward from the surface of the earth with an initial velocity of 350 m/s. If the missile mass is 1200 kg, calculate the maximum height the missile will attain. Assume no aerodynamic drag or other work during the flight and no heat transfer.

P2-3 A system of conveyor belts is used to transport a box of 30 lbm, as shown in the figure. Note that the inclined belt and the upper belt travel at a faster speed than the lower belt. Calculate the work done by the motor that drives the inclined belt. Neglect all friction.

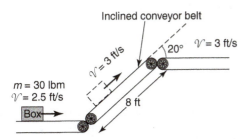

Inclined conveyor belt

$\mathcal{V}$ = 3 ft/s

20° $\mathcal{V}$ = 3 ft/s

m = 30 lbm
$\mathcal{V}$ = 2.5 ft/s

8 ft

Box

P2-4 In a front-wheel-drive car, 60% of the braking energy is dissipated in the front wheels and 40% is dissipated in the rear. If a car with a mass of 2650 lbm is decelerated from 60 mph to 15 mph on level ground by braking, calculate the energy dissipated in each front wheel (in Btu). Neglect aerodynamic drag and rolling resistance.

SPECIFIC HEAT OF SOLIDS AND LIQUIDS

P2-5 A mass of 1200 kg of fish at 20°C is to be frozen solid at –20°C. The freezing point of the fish is –2.2°C and the specific heats above and below the freezing point are 3.2 and 1.7 kJ/kg·K, respectively. The heat of fusion (the amount of heat needed to freeze 1 kg of fish) is 235 kJ/kg. Find the heat transferred.

P2-6 A steel bar initially at 1000°F is quenched by immersion in a bath of liquid water initially at 70°F. The mass of the bar is 2.5 lbm, and the volume of the water is 7 ft^3. Heat is transferred from the bath to the surroundings, which are at 70°F. After some time, the bar and water reach an equilibrium temperature of 70°F. Find the heat transferred. (For the steel, use $c = 0.106$ Btu/lbm·R.)

P2-7 A 0.14-lbm aluminum ball at 400°F is dropped into a water bath at 70°F. The bath contains 0.52 ft^3 of water and is well insulated. What is the final temperature of the ball after the ball and water reach equilibrium?

P2-8 In a new process, a thin metal film is produced when very-high-velocity particles strike a surface, melt, and adhere to the surface. Imagine an aluminum particle with a diameter of 40 μm (1 μm = 10^{-6} m) at a temperature of 20°C. The particle strikes a cold aluminum surface, also at 20°C. The particle energy is just high enough so that the particle and a portion of the surface with the same mass as the particle completely melt. What is the velocity of the particle? Assume pure aluminum with a constant specific heat of 1146 J/(kg·K). The heat of fusion (the amount of heat needed to melt 1 kg of aluminum) is 404 kJ/kg.

FORCE, MASS, UNITS

P2-9 An object weighs 40 N on a space station that has an artificial gravitational acceleration of 5 m/s^2. What is the weight of the object on earth?

P2-10 A mass of 5 lbm is acted on by an upward force of 16 lbf. The only additional force on the mass is the force of gravity. Find the acceleration in ft/s^2. Is this acceleration up or down?

P2-11 An airplane of mass 18,300 kg travels at 500 mph through the atmosphere. Calculate the kinetic energy of the plane in kJ.

PRESSURE CONCEPTS

P2-12 A gas is contained in a piston–cylinder assembly as shown in the figure. A compressed spring exerts a force of 60 N on the top of the piston. The mass of the piston is 4 kg, and the surface area is 35 cm^2. If atmospheric pressure is 95 kPa, what is the pressure of the gas in the cylinder?

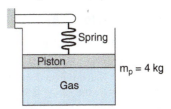

Spring

Piston

m_p = 4 kg

Gas

P2-13 A gas is contained in a piston–cylinder assembly, as shown in the figure. A compressed spring exerts a downward force on the piston. The spring is compressed 2 in., and the spring constant is 6.7 lbf/in. The piston is made of steel with a density of 490 lbm/ft^3 and a thickness of 0.5 in. The cylinder has a 7-in. diameter. Calculate the gage pressure of the gas in the tank.

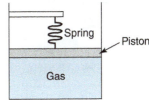

Spring

Piston

Gas

IDEAL GAS LAW

P2-14 Find the density of hydrogen at a pressure of 150 kPa and a temperature of 50°C.

P2-15 A pressurized nitrogen tank used on a paintball gun has a volume of 88 in.^{3} If the pressure of nitrogen is 4500 psia, calculate the mass of nitrogen in the tank. Assume a temperature of 70°F.

P2-16 Air is pumped from a vacuum chamber until the pressure drops to 3 torr. If the air temperature at the end of the pumping process is 5°C, calculate the air density. Eventually, the air temperature in the vacuum chamber rises to 20°C because of heat transfer with the surroundings. Assuming the volume is constant, find the final pressure, in torr.

WORK CONCEPTS

P2-17 Calculate the work, in joules, that is done in the quasi-equilibrium process from state 1 to state 2 shown in the figure.

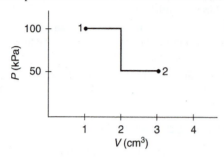

P2-18 In a certain quasi-equilibrium process, pressure increases from 200 kPa to 350 kPa. The initial gas volume is 0.2 m³. During the process, pressure varies with volume according to

$$(V - 0.1)\,10^5 = (P - 100)^2$$

where V is in m³ and P is in kPa. Calculate the work done.

P2-19 Air is contained in a piston–cylinder assembly, as shown in the figure. The piston, which is assumed massless, is held in place by a spring. Initially, the spring is not compressed and exerts no force on the piston. Then the air is heated until the volume increases by 25%. The force exerted by the spring on the piston is $F = kx$, where $k = 130$ N/cm and x is the amount by which the spring is compressed. The piston diameter is 6 cm, and the initial height of the piston is 8 cm. Calculate the amount of work done by the gas during this process. Assume atmospheric pressure is 101 kPa.

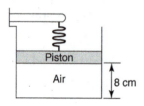

P2-20 A propeller operating at 85 rpm applies a torque of 61 N·m (Newton meters). If the propeller has been rotating for 30 minutes, find the work done in kWh (kilowatt-hours).

P2-21 A resistance heater is being used to heat a tank of nitrogen. If 3 amps are supplied to the resistor, which has a resistance of 60 Ω, how long will it take for 1200 J of work to be done?

P2-22 An electric motor operates in steady state at 1000 rpm for 45 min. The motor draws 8 amps at 110 volts and delivers a torque of 7.6 N-m. Find the total electrical energy input in kWh and the total shaft work produced in both kWh and Btu.

P2-23 Nitrogen at 28°C and 100 kPa is heated in a piston–cylinder assembly. Initially the spring shown is uncompressed and exerts no force on the piston, which is massless. If 4.5 J of work are done by the N₂,

a. how far does the piston rise?

b. what is the final temperature?

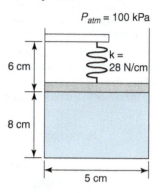

FIRST LAW

P2-24 A piston–cylinder assembly contains 0.49 g of air at a pressure of 150 kPa. The initial volume is 425 cm³. The air is then compressed while 16.4 J of work are done and 3.2 J of heat are transferred to the surroundings. Calculate the final air temperature.

P2-25 In the figure, a piston is resting on a set of stops. The cylinder contains CO_2 initially at −30°C and 45 kPa. The mass of the piston is 1.2 kg and its diameter is 0.06 m. Assuming atmospheric pressure is 101 kPa, how much heat must be added to just lift the piston off the stops?

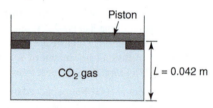

P2-26 A closed tank of volume 2.8 ft³ contains oxygen at 70°F and an absolute pressure of 14.3 lbf/in.². The gas is heated until the absolute pressure becomes 45 lbf/in.². Treating oxygen as an ideal gas,

a. find the final temperature.

b. find the total change in enthalpy, H, in Btu for this process.

P2-27 A rigid tank of volume 0.26 m³ contains hydrogen at 15°C and 101 kPa. A paddlewheel stirs the tank, adding 17.8 kJ of work. Over the same time period, the tank loses 9.3 kJ of heat to the environment. Assuming the specific heat of hydrogen does not vary with temperature, find the final temperature.

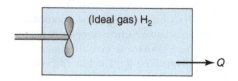

P2-28 A chamber is divided equally in two parts by a membrane. One side contains H_2 at a pressure of 130 kPa, and the other side is evacuated. The total chamber volume is 0.004 m^3. At time $t = 0$, the membrane ruptures and the hydrogen expands freely into the evacuated side. If the chamber is considered adiabatic, find the final pressure.

P2-29 Air at 20°C, 250 kPa is contained in a piston–cylinder assembly. Initially, the piston is held in place by a pin. Then the pin is removed and the gas expands rapidly. During the expansion, there is no time for any heat transfer to occur. The final air temperature and pressure are −16°C and 100 kPa. The mass of air in the cylinder is 0.4 kg. Find the work done on the atmosphere.

P2-30 Nitrogen at 50 psia and 650°F is contained in a piston-cylinder assembly. The initial volume is 25 ft^3. The nitrogen is cooled slowly while the pressure stays constant until the temperature drops to 150°F. Find the heat transferred.

P2-31 Air at 30°C is contained in a piston–cylinder assembly, as shown in the figure. The piston has a weight of 15 N and a cross-sectional area of 0.12 m^2. The initial volume of air is 3.5 m^3. Heat is added until the volume of the air becomes 6.5 m^3. Atmospheric pressure is 100 kPa.

a. Find the final air temperature.

b. Determine the work done by the air on both the piston and the atmosphere.

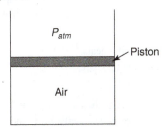

P2-32 An ideal gas with $c_p = 0.7$ kJ/kg·K and a molecular weight of 25.6 is initially at 75 kPa and 40°C. First the gas is expanded at constant pressure until its volume doubles. Then it is heated at constant volume until the pressure doubles. If the mass of gas is 4.5 kg, find

a. the total work for the entire process.

b. the heat transferred for the entire process.

ISOTHERMAL COMPRESSION OR EXPANSION OF AN IDEAL GAS

P2-33 A piston–cylinder assembly contains 0.2 kg of argon at 200 K and 50 kPa. If the argon is expanded isothermally to 30 kPa, find the work done.

P2-34 An ideal gas with a molecular weight of 37.2 is contained in a piston–cylinder assembly. The gas is initially at 130 kPa, 25°C, and has a mass of 2.34×10^{-4} kg. The gas expands slowly and isothermally until the final pressure is 100 kPa. Calculate the work done.

P2-35 An ideal gas with a volume of 0.5 ft^3 and an absolute pressure of 15 lbf/in.^{2} is contained in a piston–cylinder assembly. The gas is compressed isothermally until the pressure doubles. Calculate the heat transferred in Btu. Is the heat moving from the gas to the surroundings or vice versa?

P2-36 Air in a piston–cylinder assembly is compressed slowly and isothermally from an initial volume of 350 cm^3 to a final volume of 200 cm^3. The air is initially at 100 kPa.

a. Find the work done.

b. Find the heat transferred.

P2-37 A piston–cylinder assembly contains 0.4 kg of CO_2. The gas expands at constant temperature from an initial state of 250 kPa, 100°C to a final pressure of 100 kPa. Calculate the heat transferred during the process.

P2-38 Air at 180°F and 25 psia is compressed slowly and isothermally to 86 psia. If the initial mass of air is 0.0043 lbm, find

a. the work done.

b. the heat transferred.

P2-39 A piston–cylinder assembly of initial volume 150 cm^3 contains 0.3 g of oxygen at 120 kPa. The oxygen is then compressed slowly, isothermally, and frictionlessly, while 5.9 J of heat are removed. Find the final pressure.

P2-40 Carbon dioxide is expanded slowly and isothermally in a piston–cylinder assembly from 33.7 psia to 14.7 psia. The initial volume is 39 in.^{3} and the temperature is 100°F. Calculate the work done.

P2-41 Fifteen grams of nitrogen in a piston–cylinder assembly are compressed slowly and isothermally from 100 kPa, 25°C to 2500 kPa. Calculate the heat transferred and the work done.

POLYTROPIC PROCESS OF AN IDEAL GAS

P2-42 Air in a piston–cylinder assembly is slowly compressed from 100 kPa to 300 kPa. The mass of the air is 1.5×10^{-4} kg, and its initial temperature is 20°C. During the entire process, pressure is related to volume as

$$PV^{1.4} = \text{a constant}$$

Calculate the work done.

P2-43 Air is compressed from 150 kPa to 600 kPa while the temperature rises from 20°C to 100°C. The process is polytropic with

$$PV^n = \text{constant}$$

The initial volume of air is 1 m^3. Find

a. the value of n.

b. the work.

c. the heat transfer.

P2-44   A piston–cylinder assembly of total mass 16 lbm is free to move within a housing as shown in the figure. Initially the cylinder contains gas at an absolute pressure of 20 lbf/in.^{2} and a volume of 0.07 ft^3 and is at rest. The piston is then moved so that the entire assembly accelerates rightward and reaches a final velocity of 7.5 ft/s. During this process, the gas is compressed to a final pressure of 35 lbf/in.^{2} The process is adiabatic, and the pressure is related to the volume by $PV^{1.4} = $ constant. Calculate the change in internal energy for this process in Btu.

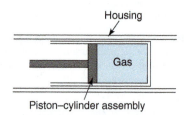

Housing

Gas

Piston–cylinder assembly

ADIABATIC COMPRESSION OR EXPANSION OF AN IDEAL GAS

P2-45   Natural gas is a mixture of methane, ethane, propane, and butane as well as other components. Composition varies by point of origin of the gas. Consider natural gas with an equivalent molecular weight of 23.6 and an equivalent specific heat, $c_p = 2.01$ kJ/kg·K. The gas is slowly compressed in a frictionless, adiabatic process from an initial volume of 212 cm^3 to a final volume of 98 cm^3. If the initial pressure is 39 kPa and the initial temperature is 15°C, find the final temperature and pressure. Assume the mixture can be modeled as an ideal gas.

P2-46   Carbon monoxide is expanded slowly in a well-insulated, frictionless piston–cylinder assembly from 300 cm^3, 25°C to 400 cm^3. Find the final temperature.

P2-47   Hydrogen with a mass of 1.1 kg is compressed slowly and adiabatically from 100 kPa, 25°C to 450 kPa in a piston–cylinder assembly. Assuming constant specific heat, calculate the final temperature and the work done.

P2-48   Air at 14.7 psia and 100°F is contained in a well-insulated piston–cylinder assembly of initial volume 0.6 ft^3.

The air is slowly expanded by applying 560 ft·lbf of work. What is the final pressure? Assume constant specific heats.

P2-49   Oxygen at 14.7 psia and 70°F is contained in a piston–cylinder assembly with an initial volume of 150 in.^{3} The oxygen is compressed slowly and adiabatically to a final volume of 50 in.^{3} Assume constant specific heat. Find

a.   the final temperature.

b.   the final pressure.

c.   the work done (in ft-lbf).

P2-50   Nitrogen at 850 K, 2 MPa expands slowly and adiabatically until the final temperature is 300 K. Assuming constant specific heat, find the final pressure and the ratio of final to initial volume.

P2-51   Air with a mass of 0.17 lbm is slowly compressed in a well-insulated, frictionless piston–cylinder assembly from 14.7 psia to 68 psia. If the air is initially at 60°F,

a.   find the final temperature.

b.   find the work done (in ft-lbf).

VARIABLE SPECIFIC HEAT

P2-52   Air is slowly expanded at constant pressure from an initial temperature of 300 K to a final temperature of 700 K in a piston–cylinder assembly. The initial volume of air is 250 cm^3, and the pressure is 150 kPa. Calculate the work done and the heat transferred

a.   using variable specific heats.

b.   using constant specific heats.

P2-53   A rigid tank of volume 4.2 ft^3 contains air initially at 100°F and 14.7 psia. Heat is added until the final pressure is 70.9 psia. Assuming variable specific heat, find the heat added.

P2-54   A rigid tank contains 0.05 kg of air at 800 K and 300 kPa. The tank is cooled while 6.35 kJ of heat are transferred. Find the final air temperature and pressure assuming variable specific heat.

CHAPTER 3

THERMAL RESISTANCES

3.1 THE FIRST LAW AS A RATE EQUATION

In the previous chapter, the first law of thermodynamics was applied to a variety of processes. In each case, we focused on the state of the system at the beginning of the process and the state at the end of the process. For example, suppose we heat a gas in a rigid tank. The gas temperature and pressure are known at the beginning of the heating process, and the first law is used to find the gas temperature and pressure at the end of the process. Intermediate conditions of the gas are not examined. Also, the time required to heat the gas was not calculated. This nonrate form of the first law is applied to processes that occur over a specific time interval and deal with fixed amounts of energy (e.g., kJ or Btu).

In this chapter, the first law is recast in the form of a *rate equation,* which is applicable at an instant of time and deals with energy rates (e.g., kW or Btu/h). This equation allows us to predict the time required for a process to occur. In addition, we examine the various ways heat can be added or removed from a system as well as the rate of heat transport.

We begin by writing the first law in differential form (see Eq. 2-60) as

$$dE = \delta Q - \delta W$$

or

$$dKE + dPE + dU = \delta Q - \delta W$$

Taking the derivative with respect to time, we obtain

$$\frac{dE}{dt} = \frac{\delta Q}{dt} - \frac{\delta W}{dt}$$

or

$$\frac{dKE}{dt} + \frac{dPE}{dt} + \frac{dU}{dt} = \frac{\delta Q}{dt} - \frac{\delta W}{dt}$$

The terms on the left-hand side represent the time rate of change of the total energy of the system. The terms on the right represent the rate at which heat and work cross the boundaries of the system and are frequently represented by the shorthand notation

$$\boxed{\begin{aligned} \dot{Q} &= \frac{\delta Q}{dt} \\ \dot{W} &= \frac{\delta W}{dt} \end{aligned}}$$

(3-1)

Incorporating this notation, the first law may be written

$$\boxed{\frac{dKE}{dt} + \frac{dPE}{dt} + \frac{dU}{dt} = \dot{Q} - \dot{W}}$$

(3-2)

The rate of work, or power, is $\dot{W}$. This can be integrated with respect to time to obtain the total work over a time interval. Likewise, the heat transfer rate, $\dot{Q}$, can be integrated with respect to time to give the total heat transferred over a time interval. Thus,

$$W = \int \dot{W}\, dt$$
$$Q = \int \dot{Q}\, dt$$

As shown in Chapter 2, for electrical work, $W = \int \xi I\, dt$. Therefore

$$\dot{W} = \xi I$$

Similarly, for shaft work, $W = \int \Im \omega\, dt$, so that

$$\dot{W} = \Im \omega \qquad\qquad (3\text{-}3)$$

EXAMPLE 3-1   Transient heating of a block

An aluminum block at 50°C is heated by an electrical resistance heater that supplies 100 W. The block has a volume of 1400 cm³. How long will it take for the block to reach 100°C? The block is covered with a very thick layer of thermal insulation.

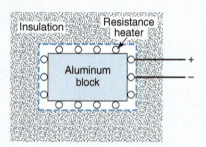

Approach:

Choose the aluminum block and the heater as the system under study. The first law in rate form, as given by Eq. 3-2, can be used to find the time to heat the block. The electrical energy added is actually the work per unit time. Because the block is well insulated, the rate of heat transfer is zero. With work and heat known, the rate of change of internal energy can be calculated from the first law. Temperature rise is related to internal energy change. Using information on the mass and specific heat of the block, you can calculate the temperature change in a given time period.

Assumptions:

A1. The block is perfectly insulated.

A2. Kinetic energy is negligible.

A3. Potential energy is negligible.

Solution:

Define the system as the aluminum block and the heater. The first law is

$$\frac{dKE}{dt} + \frac{dPE}{dt} + \frac{dU}{dt} = \dot{Q} - \dot{W}$$

Because the system is well insulated, $\dot{Q} = 0$ [A1]. There is no change in kinetic or potential energy [A2][A3], so

$$\frac{dU}{dt} = -\dot{W}$$

A4. The mass of the heater is negligible.

A5. Specific heat is constant.

We assume the mass of the heater is small compared to the mass of the aluminum [A4] and neglect the internal energy change of the heater itself. Using $dU = mc\,dT$ gives [A5]

$$mc\frac{dT}{dt} = -\dot{W}$$

For a solid $c \approx c_v \approx c_p$. Therefore

$$mc_p\frac{dT}{dt} = -\dot{W}$$

Separating variables,

$$dT = -\frac{\dot{W}}{mc_p}\,dt$$

Integrating,

$$\int dT = -\int \frac{\dot{W}}{mc_p}\,dt = -\frac{\dot{W}}{mc_p}\int dt$$

or

$$\Delta T = -\frac{\dot{W}}{mc_p}\Delta t$$

The power, $\dot{W}$, was removed from the integral because it does not vary with time. Solving for Δt,

$$\Delta t = -\frac{mc_p\,\Delta T}{\dot{W}}$$

The mass of the block is given by

$$m = \rho V$$

With values for aluminum from Table A-2,

$$m = \left(2702\,\frac{\text{kg}}{\text{m}^3}\right)(1400\,\text{cm}^3)\left(\frac{1\,\text{m}}{100\,\text{cm}}\right)^3 = 3.78\,\text{kg}$$

Electrical work is being done on the block. By our sign convention, work done on a system is negative, therefore, $\dot{W} = -100$ W. The heating time may now be calculated as

$$\Delta t = -\frac{-(3.78\,\text{kg})\left(903\,\dfrac{\text{J}}{\text{kg}\cdot{}^\circ\text{C}}\right)(100 - 50){}^\circ\text{C}}{-100\,\dfrac{\text{J}}{\text{s}}}$$

where a watt has been expressed as a joule per second and the specific heat is from Table A-2. Evaluating,

$$\Delta t = 1708\,\text{s} = 28.5\,\text{min}$$

3.2 CONDUCTION

Conduction heat transfer was first described on a mathematical basis by J. B. Fourier in 1822. He conducted an extensive series of experiments that laid the foundation for the science of heat transfer. One experiment involved the one-dimensional flow of heat in a

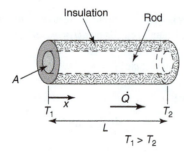

FIGURE 3-1 An insulated rod conducting heat.

rod, as shown in Figure 3-1. At the start of the experiment, the entire rod was at a uniform temperature. Then the temperature of the left end was raised to a high temperature, T_1, while the right end was lowered to a temperature, T_2. The surface of the rod was well insulated, so no heat escaped from this surface. Heat flowed from the hot end to the cold end. After a sufficiently long time, the temperatures within the rod no longer changed with time and the rate at which heat flowed reached a steady-state value.

For this perfectly insulated rod, the steady-state heat transfer rate from one end to the other can be expressed (for most common materials) as

$$\dot{Q} = \frac{kA\,(T_1 - T_2)}{L}$$
(3-4)

where $\dot{Q}$ is the rate of heat transfer, A is the cross-sectional area, L is the length, and k is a quantity called the **thermal conductivity**. In the British system, $\dot{Q}$ has units of Btu/h. In the SI system, the units of $\dot{Q}$ are joules per second or watts. The thermal conductivity has units of Btu/h·ft·°F in the British system and W/m·°C in the SI system. Thermal conductivity has a high value in electrical conductors such as copper and a low value in insulators such as glass. Table 3-1 gives some representative values of thermal conductivity for a variety of materials. Values of thermal conductivity for a variety of solids are given in Tables A-2 to A-5 and B-2 to B-5. Thermal conductivity is also included in Tables A-6 and B-6 for liquids and in Tables A-7 and B-7 for gases. In some cases, thermal conductivity varies with temperature, and then Eq. 3-4 is no longer valid.

Eq. 3-4 does not depend on the shape of the rod and can be applied to other shapes that have a constant cross-sectional area and an insulated surface. For example, it could

TABLE 3-1 **Values of Thermal Conductivity for a Variety of Materials at 300 K**

Material	k W/(m·°C)	k Btu/(h·ft·°F)
Silver	429	248
Copper	401	232
Carbon steel	60	34.7
Stainless steel	15	8.7
Plate glass	1.4	0.80
Concrete	1.4	0.80
Water	0.6	0.35
Wood (oak)	0.17	0.1
Leather	0.16	0.09
Fiberglass insulation	0.04	0.023
Air	0.026	0.015

apply to a square bar, as long as there is no heat loss from its sides. It is often applied to large flat plates, where the heat lost from the edges of the plate is negligible. In general, Eq. 3-4 applies to plane layers of material of any constant cross-sectional shape in which all the conduction occurs in only one direction.

Eq. 3-4 relates the heat transfer rate through a plane layer to the temperatures at its two faces. What happens to the temperature inside the layer? Experimentally, it has been shown that, if the thermal conductivity is constant, the temperature varies linearly between the two end values. The temperature profile within the layer is shown in Figure 3-2. The quantity $(T_2 - T_1)/L$ is, in fact, the slope of the curve of $T(x)$ versus x. Because the slope is just the derivative of a function, we may write

$$\frac{dT(x)}{dx} = \frac{T_2 - T_1}{x_2 - x_1} = \frac{T_2 - T_1}{L}$$

Substituting this into Eq. 3-4 gives

$$\dot{Q} = -kA \frac{dT}{dx}$$

This is **Fourier's law of conduction**. It is a relationship of fundamental importance in conduction heat transfer. In Figure 3-2, temperature decreases with increasing length; therefore, dT/dx is negative. It is conventional to define $\dot{Q}$ as positive in the direction of decreasing temperature (heat flows from hot to cold), so a negative sign is used in Fourier's law.

The simple one-dimensional heat conduction example described above is a convenient starting point for the discussion of conduction heat transfer. Chapter 11 discusses conduction in more detail and rigor. For now, we will use this simple model to illustrate how conduction is used in the first law.

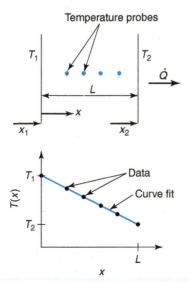

FIGURE 3-2 Temperature profile in a plane layer.

EXAMPLE 3-2 Heat loss through an oven wall

The wall of an oven measures 3 ft by 2 ft by 0.25 in. The wall is covered by a layer of insulation that is 1.5 in. thick. The thermal conductivity of the wall material is 8 Btu/h·ft·°F and the thermal conductivity of the insulation is 0.14 Btu/h·ft·°F. The inside wall temperature is 450°F, and the temperature at the outside of the insulation is 75°F. These temperatures are constant for a long time, and a steady-state temperature distribution is established in the wall. Calculate the amount of heat lost in Btu through the insulated wall in 1 h.

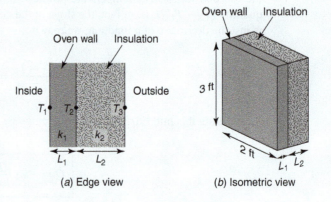

(a) Edge view (b) Isometric view

Approach:

We want to calculate the total heat lost, Q, from the heat transfer rate, $\dot{Q}$, using

$$Q = \int \dot{Q}\, dt$$

To calculate $\dot{Q}$, the insulated oven wall can be modeled as a large flat plane constructed in two layers. The formula for one-dimensional heat conduction

$$\dot{Q} = \frac{kA\,\Delta T}{L}$$

can be used for each layer. Two equations are written: one for the oven wall layer and the other for the insulation layer, with appropriate values for k, L, and ΔT for each layer. The heat flowing through the oven wall is the same as the heat flowing through the insulation, because we assume the heat travels in only one dimension. We can use this fact to find the temperature at the interface between the oven wall and the insulation. The interface temperature is needed to calculate the rate of heat transfer, $\dot{Q}$. Finally, this rate is integrated over the 1-hour time period to find the total amount of heat transferred.

Assumptions:

A1. Conduction is one-dimensional.

Solution:

The two-layer wall is sketched in the figure. Conduction through the oven wall is given by [A1],

$$\dot{Q} = \frac{k_1 A\,(T_1 - T_2)}{L_1}$$

The temperature, T_2, is unknown, as is $\dot{Q}$; therefore, we need a second equation. All the heat conducted through the first layer is also conducted through the second, because we assume none is lost at the edges of the wall. For the insulation layer, then,

$$\dot{Q} = \frac{k_2 A\,(T_2 - T_3)}{L_2}$$

which is the second equation needed. Equating these two expressions gives

$$\frac{k_1 A(T_1 - T_2)}{L_1} = \frac{k_2 A(T_2 - T_3)}{L_2}$$

Solving for the unknown temperature, T_2, yields, after some algebra,

$$T_2 = \frac{k_1 L_2 T_1 + k_2 L_1 T_3}{k_1 L_2 + k_2 L_1}$$

Substituting values,

$$T_2 = \frac{(8 \text{ Btu/h·ft·°F})(1.5 \text{ in. })(450°F) + (0.14 \text{ Btu/h·ft·°F})(0.25 \text{ in.})(75°F)}{(8 \text{ Btu/h·ft·°F})(1.5 \text{ in. }) + (0.14 \text{ Btu/h·ft·°F})(0.25 \text{ in.})}$$

$$T_2 = 448.9°F$$

To find the heat conducted through the wall in 1 hour, we can use the expression for $\dot{Q}$ for either the oven wall or the insulation. Arbitrarily selecting the oven wall:

$$\dot{Q} = \frac{k_1 A (T_1 - T_2)}{L_1}$$

By definition,

$$\dot{Q} = \frac{\delta Q}{dt}$$

therefore, we may write

$$Q = \int_0^{t_f} \dot{Q} \, dt$$

or

$$Q = \int_0^{t_f} \frac{k_1 A(T_1 - T_2)}{L_1} \, dt$$

A2. The system is in steady state.

where $t_f = 1$ hour. Because temperature does not vary with time [A2],

$$Q = \int_0^{t_f} \frac{k_1 A (T_1 - T_2)}{L_1} \, dt = \frac{k_1 A (T_1 - T_2) \, t_f}{L_1}$$

Using given values and the calculated value T_2,

$$Q = \frac{\left[8 \dfrac{\text{Btu}}{\text{h·ft·°F}} \right] (3 \text{ ft})(2 \text{ ft})(450 - 448.9)°F(1 \text{ h})}{(0.25 \text{ in. }) \left[\dfrac{12 \text{ in.}}{1 \text{ ft}} \right]}$$

$$Q = 17.45 \text{ Btu}$$

Note the very large temperature drop across the insulation ($448.9°F - 75°F = 373.9°F$) compared to the very small drop ($450°F - 448.9°F = 1.1°F$) across the metal wall of the oven.

3.3 RADIATION

Heat transfer by radiation can occur in solids, liquids, or gases. Radiation in a gas typically involves absorption and emission of photons throughout the volume of the gas. This is a very complex process and is beyond the scope of this book. In solids and liquids, photons are also emitted and absorbed throughout the volume. However, in opaque materials, the radiative behavior depends only on what happens at the surface. In an opaque solid or liquid, photons are absorbed in a very thin layer on the surface. In addition, only the photons emitted from this thin layer can escape from the opaque material. Photons emitted deep within the substance are reabsorbed in a very short distance from the point of emission and have no influence on energy transfer within the material. For opaque substances, radiation can be considered to be a surface phenomenon. Only opaque substances will be discussed in this text.

The ideal surface against which all other surfaces are compared is called a **black** surface. A black surface is defined as one that absorbs all the radiation incident upon it. As shown in Chapter 14, a black surface at temperature T emits the maximum possible radiation that can be emitted at that temperature by any surface. Imagine a black body at a uniform temperature T_s. If surrounding surfaces are far away from the body, then radiation from the body to the surroundings is governed by

$$\dot{Q} = \sigma A \left(T_s^4 - T_{surr}^4 \right) \qquad \text{black surface} \tag{3-5}$$

where T_{surr} is the absolute temperature of the surrounding surfaces (usually the ambient temperature), A is the surface area exposed to radiation, and σ is the Stefan-Boltzmann constant. The Stefan-Boltzmann constant is a fundamental physical constant, just as the acceleration of gravity, g, is a fundamental physical constant. The value of σ is

$$\sigma = 0.171 \times 10^{-8} \frac{\text{Btu}}{\text{h·ft}^2\text{·R}^4}$$

$$\sigma = 5.67 \times 10^{-8} \frac{\text{W}}{\text{m}^2\text{·K}^4}$$

Real surfaces, of course, are not black. They reflect some of the radiation incident upon them. If the fraction of incident radiation reflected by the surface is independent of the wavelength of the incident radiation, then the surface is called a **gray** surface. In addition, if the fraction reflected does not depend on the angle of incidence of the radiation, then the surface is called a **diffuse** surface. Radiation from a gray, diffuse surface to the surroundings may be written

$$\dot{Q} = \varepsilon \, \sigma A \left(T_s^4 - T_{surr}^4 \right) \qquad \text{gray, diffuse surface} \tag{3-6}$$

where ε is a quantity called emissivity. Emissivity is a property of the surface. It depends on the surface material and the surface conditions. It is a dimensionless quantity whose value varies between 0 and 1. For most electrical insulators, emissivity has a value above 0.8. For metals, the value of emissivity is a sensitive function of the surface condition. Clean and shiny metals have much lower values of emissivity than unpolished, dirty, or oxidized surfaces. If $\varepsilon = 1$, the body is a **blackbody**. Representative values of emissivity are listed in Table 3-2. In addition, Table A-17 lists emissivities for a variety of surfaces.

The temperatures in Eq. 3-5 and Eq. 3-6 must be expressed in absolute terms (i.e., degrees Kelvin or degrees Rankine). In addition, these equations only apply to bodies that are far away from other surfaces. If the surrounding surfaces are close to the body, then reflections must be taken into account and the above equations must be modified. Radiation is discussed in more detail and rigor in Chapter 14.

TABLE 3-2 Emissivity for a Variety of Materials

Material	Emissivity
Ice	0.96
Soil	0.94
Concrete, rough	0.91
Dull wrought iron	0.91
Black paint	0.90
White paint	0.90
Oak	0.88
Rubber	0.88
Coal	0.78
Oxidized brass	0.60
Oxidized cast iron	0.57
Polished wrought iron	0.29
Aluminum foil	0.05
Polished brass	0.04
Polished silver	0.02

3.4 CONVECTION

Convection heat transfer occurs whenever a gas or liquid at one temperature flows next to a surface at a different temperature. There are two basic kinds of convection, forced and natural. In forced convection, flow is induced by some external actuator, such as a pump or fan. To understand natural convection, consider, for example, the cooking element on an electric stove when there is no pot covering it. Heat conducts from the cooking element into the air just above it. As the air heats, its density decreases. The hot air rises and colder, higher-density air above the hot air flows downward under the action of gravity, creating a natural flow. We call this a buoyancy-induced flow. Natural convection also occurs on vertical surfaces and on cold surfaces exposed to a hot fluid.

Heat transferred by convection is expressed as

$$\dot{Q} = hA\Delta T \qquad (3\text{-}7)$$

where A is the surface area exposed to convection, ΔT is the temperature difference between the solid and fluid, and h is the heat transfer coefficient. The heat transfer coefficient was first introduced by Fourier. Its value depends on geometry, velocity of flow, type of gas or liquid, and sometimes temperature.

In general, h increases with velocity. For a given gas or liquid, the heat transfer coefficient is usually greater for forced convection than for natural convection. Convection in gases tends to be less effective than convection in liquids. Fortunately, one of the most effective liquids for transfering heat by convection is water. Table 3-3 shows some representative values of heat transfer coefficients.

The heat transfer coefficient is very useful in engineering calculations. Its value can theoretically be found by the solution of a set of conservation equations, but this approach is not always practical. A large body of experimental data on heat transfer coefficients for many different geometries and conditions is available in journals and books. These data are often generalized by curve fits that a designer can use to predict heat transfer coefficients as a function of velocity, fluid properties, part sizes, and other parameters. Chapter 12 presents many such curve fits, called *convective heat transfer correlations*. The theoretical basis of convective heat transfer is introduced in Chapter 12.

TABLE 3-3   **Typical Values of Heat Transfer Coefficient**

Gas or Liquid	Type of Convection	h W/(m²·°C)	h Btu/(h· ft²·°F)
Air	Natural	3–12	0.5–2
	Forced	10–150	2–30
Liquid fluorocarbon	Natural	100–300	20–50
	Forced	200–2000	35–350
Water	Natural	200–1200	35–200
	Forced	3000–7000	500–1250

It is important to recognize that the heat transfer coefficient is not a fundamental physical property of a substance, as is thermal conductivity or emissivity. Values of thermal conductivity have been tabulated for a wide variety of materials, such as copper, glass, brick, rubber, and so on. The heat transfer coefficient is determined not by picking a value from a table but rather by computing a value from a curve fit (i.e., a correlation). Table 3-3 lists values of heat transfer coefficient only to give the reader an idea of what a reasonable value might be for the condition listed. In unusual conditions, such as microchannel cooling, actual values may fall outside the ranges given.

EXAMPLE 3-3   Convection and radiation from a outdoor grill

The outside surface of a charcoal grill is at a temperature of 50°C. The grill loses heat to the surroundings by natural convection and radiation. The average heat transfer coefficient is 5.4 W/m² °C and the emissivity is 0.87. If the surface area is 0.63 m², calculate the heat transfer rate to the environment. Assume the surrounding temperature is 20°C.

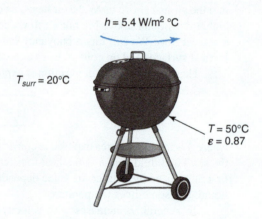

$h = 5.4$ W/m² °C

$T_{surr} = 20$°C

$T = 50$°C
$\varepsilon = 0.87$

Approach:

The heat transfer rates by convection and radiation are calculated from

$$\dot{Q}_{conv} = hA\left(T_s - T_f\right)$$

$$\dot{Q}_{rad} = \varepsilon \sigma A\left(T_s^4 - T_{surr}^4\right)$$

where T_f is the fluid (air) temperature. The total heat transfer rate to the environment is the sum of $\dot{Q}_{conv}$ and $\dot{Q}_{rad}$.

Assumption: **Solution:**

To find the total heat transfer rate, add together the convective and the radiative heat transfter contributions. For convection [A1][A2],

A1. The heat transfer coefficient is uniform over the surface of the grill.

A2. Temperature is uniform over the surface of the grill.

A3. There are no reflective surfaces near the grill.

A4. The surface of the grill is gray and diffuse

$$\dot{Q}_{conv} = hA\left(T_s - T_f\right)$$

where $\dot{Q}_{conv}$ refers to convective heat. Substituting given values yields

$$\dot{Q}_{conv} = \left[5.4\frac{W}{m^2 \cdot {}^\circ C}\right](0.63 \, m^2)\,(50 - 20)^\circ C = 102 \, W$$

For radiation [A3], [A4]

$$\dot{Q}_{rad} = \varepsilon\,\sigma A\left(T_s^4 - T_{surr}^4\right)$$

$$= (0.87)\left[5.67 \times 10^{-8}\frac{W}{m^2 \cdot K^4}\right](0.63 \, m^2)\left[(50 + 273)^4 - (20 + 273)^4\right] K^4$$

$$\dot{Q}_{rad} = 109 \, W$$

The total heat transfer rate is

$$\dot{Q}_{tot} = \dot{Q}_{conv} + \dot{Q}_{rad} = 102 \, W + 109 \, W = 211 \, W$$

3.5 THE RESISTANCE ANALOGY FOR CONDUCTION AND CONVECTION

The equations for the heat transfer rate by conduction and convection have certain features in common. These equations are

$$\dot{Q}_{cond} = \frac{kA\left(T_1 - T_2\right)}{L}$$

$$\dot{Q}_{conv} = hA\left(T_s - T_f\right)$$

In both cases, the heat transfer rate is proportional to a temperature difference. Similar types of equations are encountered in electric circuit theory. For a linear resistor, the current is related to the voltage drop by Ohm's law, which is

$$I = \frac{\Delta\xi}{R}$$

where $\Delta\xi$ is the voltage drop across the resistor, I is the current, and R is the resistance. Another way to describe this is

$$\text{flow} = \frac{\text{driving potential}}{\text{resistance to flow}}$$

We can apply this concept to heat transfer. If we compare current to heat transfer rate and voltage drop to temperature drop, we can define a so-called **thermal resistance**. Let us recast the equation for one-dimensional conduction into the form

$$\dot{Q}_{cond} = \frac{T_1 - T_2}{R_{cond}}$$

where

$$R_{cond} = \frac{L}{kA}$$

The quantity R_{cond} is the conductive thermal resistance. Likewise, the equation for convection may be rewritten in the form

$$\dot{Q}_{conv} = \frac{T_s - T_f}{\left[\frac{1}{hA}\right]} = \frac{T_s - T_f}{R_{conv}}$$

where

$$R_{conv} = \frac{1}{hA}$$

In this equation, R_{conv} is the convective thermal resistance.

The resistance analogy is especially useful in describing systems with multiple parts. For example, consider the multilayer wall in Figure 3-3. The wall consists of three layers cooled convectively on both sides and is at a steady-state condition. Each layer has a thermal resistance, and there are thermal resistances associated with convection at the boundaries as well; all these resistances are in series. Focusing for the moment on the first two resistances, R_0 and R_1:

$$R_0 = \frac{1}{h_0 A}$$

$$R_1 = \frac{L_1}{k_1 A}$$

The corresponding rate of heat transfer for each is

$$\dot{Q} = \frac{T_0 - T_1}{R_0}$$

$$\dot{Q} = \frac{T_1 - T_2}{R_1}$$

The same rate of heat transfer applies in both these equations because whatever is convected to the surface is then conducted through the wall. The heat has no place else to go. Solving

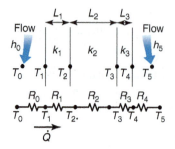

FIGURE 3-3 Heat transfer through a multilayer wall.

these two equations for T_1 gives

$$T_1 = T_0 - R_0 \dot{Q}$$

$$T_1 = T_2 + R_1 \dot{Q}$$

Eliminating T_1 yields

$$T_0 - R_0 \dot{Q} = T_2 + R_1 \dot{Q}$$

which may be rewritten as

$$\dot{Q} = \frac{T_0 - T_2}{R_0 + R_1}$$

This equation shows that the total thermal resistance between T_0 and T_2 may be regarded as the sum of the two resistances R_1 and R_0. Extending these ideas to the entire multilayer wall with convection on both surfaces results in

$$\dot{Q} = \frac{T_0 - T_5}{R_0 + R_1 + R_2 + R_3 + R_4}$$

The total thermal resistance is the sum of the individual resistances:

$$R_{tot} = R_0 + R_1 + R_2 + R_3 + R_4$$

so that

$$\dot{Q} = \frac{T_0 - T_5}{R_{tot}}$$

The idea that resistances add in series may be familiar to you from electric circuit theory.

EXAMPLE 3-4 Conduction and convection in a computer chip

A 2 cm by 2 cm chip in a small computer is cooled by forced air flow with a heat transfer coefficient of 152 W/m$^2 \cdot$°C. Electronic devices are deposited in a very thin layer on the bottom surface of the chip. If the air temperature is 20°C and the devices generate 1.6 W of heat distributed uniformly over the bottom of the chip, estimate the temperature at the device plane. Assume no heat transfer through the solder bumps.

Approach:

Heat is conducted upward through the silicon chip and then removed by convection from the top of the chip (Figure 3-4.) The problem will be solved by two methods. In the first method, the heat transfer by conduction is equated to the heat transfer by convection. Therefore,

$$\frac{kA\,(T_1 - T_2)}{L} = hA\,(T_2 - T_f)$$

This equation can be solved for T_1. The temperature at the top surface of the chip, T_2, can be found using

$$\dot{Q} = hA\,(T_2 - T_f)$$

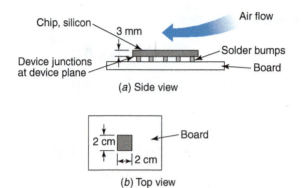

FIGURE 3-4 A chip mounted on a circuit board.

In the second method, the resistance by conduction and the resistance by convection are each calculated. The total resistance, R_{tot}, is the sum of these two resistances. Then the temperature at the device plane is calculated from

$$\dot{Q} = \frac{\Delta T}{R_{tot}} = \frac{T_1 - T_f}{R_{tot}}$$

where ΔT is the difference between the device plane temperature and the air temperature, and $\dot{Q}$ is the total heat generated.

Assumptions:

A1. All heat is conducted upward and none travels downward.

A2. Conduction is one-dimensional, and no heat leaves from the edge of the chip.

Solution:

The devices (transistors, diodes, etc.) are in a very thin layer on the solder-bump side of the chip, as shown in Figure 3-4. Heat generated in this plane is conducted through the silicon chip and then convected from the silicon surface [A1]. A sketch of the chip with temperature and thermal resistances is given in Figure 3-5. The problem will be solved in two ways.

Method 1

Conduction through the silicon chip is governed by [A2]

$$\dot{Q} = \frac{kA\,(T_1 - T_2)}{L}$$

All of the heat generated conducts through the chip and into the air. Therefore, this same heat transfer rate appears in the convection equation, which is

$$\dot{Q} = hA\,(T_2 - T_f)$$

Eliminating $\dot{Q}$ between the two equations results in

$$\frac{kA\,(T_1 - T_2)}{L} = hA\,(T_2 - T_f)$$

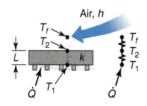

FIGURE 3-5 Thermal resistances through the chip.

Solving for T_1

$$T_1 = \frac{hL}{k}\left(T_2 - T_f\right) + T_2$$

To find T_2, rearrange the convection equation to give

$$T_2 = \frac{\dot{Q}}{hA} + T_f$$

Substituting values

$$T_2 = \frac{1.6\,\text{W}}{\left(152\dfrac{\text{W}}{\text{m}^2\cdot{}^\circ\text{C}}\right)(2\text{cm})\,(2\text{cm})\left[\dfrac{1\,\text{m}}{100\,\text{cm}}\right]^2} + 20^\circ\text{C}$$

$$T_2 = 46.32^\circ\text{C}$$

The temperature at the device plane, T_1, then becomes

$$T_1 = \frac{\left[152\dfrac{\text{W}}{\text{m}^2\cdot{}^\circ\text{C}}\right](3\,\text{mm})\left[\dfrac{1\,\text{m}}{1000\,\text{mm}}\right]}{148\dfrac{\text{W}}{\text{m K}}}(46.32 - 20)^\circ\text{C} + 46.32\,^\circ\text{C}$$

$$T_1 = 46.40^\circ\text{C}$$

where the thermal conductivity of silicon from Table A-2 has been used. Notice that the temperature difference between the top and bottom of the silicon is only 0.08°C. The chip is thin, and silicon is a good conductor.

Method 2

In this approach, the resistances for conduction and convection are calculated. For conduction,

$$R_{cond} = \frac{L}{kA} = \frac{(3\,\text{mm})\left[\dfrac{1\,\text{cm}}{10\,\text{mm}}\right]}{\left[148\dfrac{\text{W}}{\text{m}\cdot{}^\circ\text{C}}\right](2\,\text{cm})\,(2\,\text{cm})\left[\dfrac{1\,\text{m}}{100\,\text{cm}}\right]} = 0.0507\,\frac{^\circ\text{C}}{\text{W}}$$

For convection,

$$R_{conv} = \frac{1}{hA} = \frac{1}{\left[152\dfrac{\text{W}}{\text{m}^2\cdot{}^\circ\text{C}}\right](2\,\text{cm})(2\,\text{cm})\left[\dfrac{1\,\text{m}}{100\,\text{cm}}\right]^2}$$

$$R_{conv} = 16.45\,\frac{^\circ\text{C}}{\text{W}}$$

The two resistances appear in series; therefore, the total resistance is the sum of the individual resistances:

$$R_{tot} = R_{cond} + R_{conv} = 0.0507\,\frac{^\circ\text{C}}{\text{W}} + 16.45\,\frac{^\circ\text{C}}{\text{W}} = 16.5\,\frac{^\circ\text{C}}{\text{W}}$$

The temperature at the device plane is now calculated using

$$\dot{Q} = \frac{T_1 - T_f}{R_{tot}}$$

or

$$T_1 = \dot{Q}R_{tot} + T_f$$

$$T_1 = (1.6\,\text{W})(16.5)\,\frac{°\text{C}}{\text{W}} + 20 = 46.40°\text{C}$$

The convective resistance is much higher than the conductive resistance. As a result, most of the temperature drop is across the solid–fluid boundary and very little is across the silicon, as calculated in Method 1 above.

It is also possible to define a thermal resistance for conduction in cylinders. One-dimensional conduction through a cylindrical shell is depicted in Figure 3-6. Heat travels in the r-direction only. The inner surface is at temperature T_1, and the outer surface is at temperature T_2. Fourier's law may be written in cylindrical coordinates as

$$\dot{Q} = -kA\,\frac{dT}{dr}$$

In this equation, A is the area across which heat flows. This area is shown by the dotted line in Figure 3-6. Notice that the area varies with r, being smaller near the inner radius and larger near the outer radius. For a cylinder of length L, the area, A, is the circumference at radius r times the length, or

$$A(r) = 2\pi rL$$

Using this in Fourier's law gives

$$\dot{Q} = -k2\pi rL\,\frac{dT}{dr}$$

This is a first-order differential equation for T as a function of r. To solve the equation, separate the variables

$$-\frac{\dot{Q}\,dr}{2\pi rLk} = dT$$

and integrate from the inner to the outer radius

$$-\int_{r_1}^{r_2} \frac{\dot{Q}\,dr}{2\pi rLk} = \int_{T_1}^{T_2} dT$$

The heat transfer rate, $\dot{Q}$, does not vary with r. Recall that heat transfer rate is defined as heat flow per unit time. In steady state, all the heat per unit time that enters at the inner

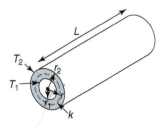

FIGURE 3-6 One-dimensional conduction through a cylindrical surface of length *L*.

radius will exit at the outer radius; thus $\dot{Q}$ is a constant. This makes the integration easy and gives the result

$$\frac{-\dot{Q}}{2\pi Lk}\ln\left(\frac{r_2}{r_1}\right) = T_2 - T_1$$

which may be rearranged to

$$\dot{Q} = 2\pi Lk \frac{(T_1 - T_2)}{\ln\left[\frac{r_2}{r_1}\right]}$$

By comparison to the standard resistance analogy,

$$\dot{Q} = \frac{T_1 - T_2}{R}$$

and the resistance through the cylindrical shell is

$$R_{cylinder} = \frac{\ln\left[\frac{r_2}{r_1}\right]}{2\pi Lk}$$

By a similar line of analysis, the resistance through a spherical shell is

$$R_{sphere} = \frac{r_2 - r_1}{4\pi r_1 r_2 k}$$

EXAMPLE 3-5 Conduction and convection from a steam pipe

An insulated steel pipe carries steam at 400°F. The heat transfer coefficient at the inner radius is 12 Btu/h·ft²·°F. On the outside of the insulation, the heat transfer coefficient is 5 Btu/h·ft²·°F. Ambient air is at 65°F. The insulation fails at 380°F. Will it survive under these conditions? See Figure 3-7 for geometry and thermal conductivities.

Approach:

To determine whether the insulation would fail, we must calculate the highest temperature reached by the insulation. Find the convection resistance on the inside, the conduction resistance of each layer, and the convection resistance on the outside. Sum these four resistances to find the total resistance. With the known gas temperatures on the inside and outside, the desired insulation temperature can be found. Because heat is flowing from inside to outside, $T_i > T_1 > T_2 > T_3 > T_f$. Therefore, the maximum insulation temperature is T_2. Use a resistance network to find T_2.

Assumption:

A1. Conduction is one-dimensional in the radial direction.

Solution:

The temperature at the pipe–insulation interface can be found from the resistance analogy assuming one-dimensional radial heat conduction [A1]. In Figure 3-7, R_1 is the resistance to convection on the inside, R_2 is conductive resistance through the steel, R_3 is conductive resistance through the insulation, and R_4 is convective resistance on the outside of the insulation. These resistances will be calculated for a 1-ft length of the pipe. The same values of temperature would be obtained if we used a 6-in. length, or a 2-ft length, or any arbitrary length.

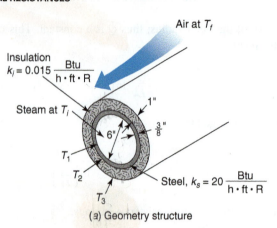

(a) Geometry structure

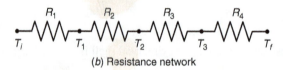

T_i T_1 T_2 T_3 T_f

(b) Resistance network

FIGURE 3-7 Heat transfer from an insulated pipe.

The resistances are

$$R_1 = \frac{1}{h_i A} = \frac{1}{\left[12\dfrac{\text{Btu}}{\text{h·ft}^2\text{·R}}\right] 2\pi \,(3\text{ in.}) \left[\dfrac{1\text{ ft}}{12\text{ in.}}\right] (1\text{ ft})}$$

$$R_1 = 0.0531 \,\frac{\text{R·h}}{\text{Btu}}$$

For the steel pipe

$$R_2 = \frac{\ln\left[\dfrac{r_2}{r_1}\right]}{2\pi L k} = \frac{\ln\left[\dfrac{3.375}{3}\right]}{2\pi \,(1\text{ ft}) \left[20\dfrac{\text{Btu}}{\text{h·ft·R}}\right]}$$

$$R_2 = 0.000937 \,\frac{\text{R·h}}{\text{Btu}}$$

For the insulation

$$R_3 = \frac{\ln\left[\dfrac{4.375}{3.375}\right]}{2\pi \,(1\text{ ft}) \left[0.015\dfrac{\text{Btu}}{\text{h·ft·R}}\right]}$$

$$R_3 = 2.75 \,\frac{\text{R·h}}{\text{Btu}}$$

For the exterior

$$R_4 = \frac{1}{h_e A} = \frac{1}{\left[5\dfrac{\text{Btu}}{\text{h·ft}^2\text{·R}}\right] 2\pi \,(4.375\text{ in.}) \left[\dfrac{1\text{ ft}}{12\text{ in.}}\right] (1\text{ ft})}$$

$$R_4 = 0.0873 \,\frac{\text{R·h}}{\text{Btu}}$$

We do not need to calculate all of the temperatures. We are only interested in whether the insulation fails. The hottest temperature in the insulation will be at the point in the insulation nearest the steam, that is, the inner radius of the insulation. The temperature at that point is T_2. Referring to Figure 3-7b, the heat flowing into T_2 from the left must equal the heat flowing out to the right. In other words,

$$Q_{i \rightarrow 2} = Q_{2 \rightarrow f}$$

or

$$\frac{T_i - T_2}{R_1 + R_2} = \frac{T_2 - T_f}{R_3 + R_4}$$

where resistances have been added in series.

Solving for T_2 gives

$$T_2 = \frac{(R_1 + R_2)\, T_f + (R_3 + R_4)\, T_i}{R_1 + R_2 + R_3 + R_4}$$

subsitituting values:

$$T_2 = \frac{(0.0531 + 0.000937)\, 65 + (2.75 + 0.0873)\, 400}{0.0531 + 0.000937 + 2.75 + 0.0873}$$

$$T_2 = 394°F$$

Too bad. The insulation fails.

3.6 THE LUMPED SYSTEM APPROXIMATION

In some circumstances involving two thermal resistances, one resistance may be much larger than the other. In that case, certain simplifications are possible. For example, consider a two-layer structure where both layers have the same thickness but different material composition, as shown in Figure 3-8. One layer is made of copper, which has a very high thermal conductivity, and the other is made of soft rubber, which has a much lower thermal conductivity. Now imagine that the left side of the copper layer is held at 40°C and the right side of the rubber layer is held at 20°C. If each layer has a thickness of 0.02 m, then the thermal resistance for a 1-m² area of the copper layer is

$$R_c = \frac{L}{k_c A} = \frac{0.02 \text{ m}}{\left[401\, \frac{\text{W}}{\text{m} \cdot \text{K}} \right] (1 \text{ m}^2)} = 4.99 \times 10^{-5}\, \frac{\text{K}}{\text{W}}$$

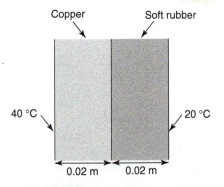

Copper

Soft rubber

40 °C

20 °C

0.02 m   0.02 m

FIGURE 3-8 A two-layer structure made of copper and rubber.

where the thermal conductivity of copper has been taken from Table A-2. For the rubber layer, the thermal resistance is

$$R_r = \frac{L}{k_r A} = \frac{0.02 \text{ m}}{\left[0.13 \frac{\text{W}}{\text{m} \cdot \text{K}}\right](1 \text{ m}^2)} = 0.154 \frac{\text{K}}{\text{W}}$$

As you can see, the resistance of the rubber layer is more than three orders of magnitude greater than the resistance of the copper layer. We can calculate the temperature at the interface between the copper and the rubber, which is T_2 in Figure 3-8, by equating the heat flux across the copper and rubber layers to obtain

$$\dot{Q} = \frac{T_1 - T_2}{R_c} = \frac{T_2 - T_3}{R_r}$$

Substituting values for all but the unknown T_2, we get

$$\frac{(40 - T_2) \text{ °C}}{4.99 \times 10^{-5} \frac{\text{K}}{\text{W}}} = \frac{(T_2 - 20) \text{ °C}}{0.154 \frac{\text{K}}{\text{W}}}$$

In this equation, the numerator contains a *difference* in temperature in °C. The denominator shows the resistance per degree Kelvin. A *difference* of 1°C is equal to a *difference* of 1 K; therefore, these units cancel and each side of the equation is in watts. Solving for T_2 gives

$$T_2 = 39.99°C$$

Recall that the temperature within a plane layer varies linearly across the layer (as long as the thermal conductivity is not a function of temperature). Figure 3-9 is a plot of the temperature variation in the copper and rubber layers. Because the thermal resistance of the copper is so low, there is virtually no temperature drop across this layer; its temperature drops from 40°C to 39.99°C. Meanwhile, almost all the temperature drop is across the rubber.

Many problems can be simplified by assuming that the thermal resistance is negligible. For example, the resistance due to conduction in a copper pipe is usually very small compared to convective resistance on the inside and/or outside of the pipe. In some transient systems, conduction resistance is much smaller than convection resistance. In that case, the so-called **lumped system approximation** can be used. The rest of this section describes and develops the lumped system approximation.

Consider an arbitrarily shaped body that is initially at a uniform high temperature T_i, as shown in Figure 3-10. The body is suddenly immersed in a cold fluid at temperature T_f.

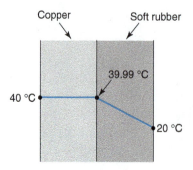

FIGURE 3-9 Temperature distribution in the copper–rubber structure.

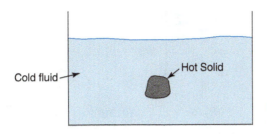

FIGURE 3-10 An arbitrarily shaped hot solid immersed in a cold fluid.

As the solid cools, the temperature will drop near the surface of the body, while the center will remain hot. With time, the effect of the cold fluid will penetrate into the solid, lowering the temperature in places nearer and nearer to the center. Eventually, the entire solid will reach the fluid temperature.

Heat transfer from the surface due to convection is calculated as

$$\dot{Q} = hA(T_s - T_f)$$

where T_s is the surface temperature. But, in this case, the surface temperature varies with time, $T_s(t)$. In addition, the solid temperature varies with location and time inside the body, so changes in internal energy will be difficult to calculate. There is, however, one condition under which the temperature will not change significantly with location in the solid. This occurs if the thermal resistance due to conduction within the solid is much smaller than the thermal resistance due to convection at the surface. Recall the copper–rubber structure example above. There was negligible variation of temperature within the copper because the copper's thermal resistance was so much lower than the rubber's thermal resistance. In the present transient situation, if the conduction resistance is much less than the convection resistance, temperature variation with location will be negligible as well. We can develop a criterion to indicate when this assumption will be valid by comparing the magnitudes of the conduction and convection resistances associated with the body.

Assume the heat transfer coefficient is uniform over the surface of the body. The convection resistance is given by

$$R_{conv} = \frac{1}{hA}$$

where A is the total surface area of the object.

The conduction resistance is more difficult to specify. For one-dimensional conduction in a plane wall, the resistance is

$$R_{cond} = \frac{L}{kA}$$

Because of the arbitrary shape of the body shown in Figure 3-10, simple one-dimensional conduction is not accurate. However, if the conduction resistance is so small that it can be ignored, then accuracy is not particularly important. Simply calculating an approximate resistance based on a one-dimensional model will be useful in deciding whether the conduction resistance is small compared to the convection resistance.

To find the conduction resistance, it is first necessary to specify a characteristic length, L_{char}, and an area, A, which characterizes heat flow inside the arbitrarily shaped solid. A natural choice for area is the total surface area of the object. The recommended choice for L_{char} is

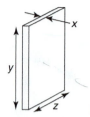

FIGURE 3-11 A thin, flat plate.

$$L_{char} = \frac{V}{A}$$ (3-8)

where V is the volume of the body and A is the surface area. To illustrate the meaning of Eq. 3-8, we will apply it to a very simple shape: a large, thin, flat plate, as shown in Figure 3-11. If the height, y, and the width, z, are very large compared to the thickness, x, then the surface area is dominated by the two large, flat sides. For all practical purposes, the surface area of the four edges is insignificant. Neglecting the edges, the ratio of volume to surface area is

$$L_{char} = \frac{V}{A} = \frac{xyz}{2yz} = \frac{x}{2}$$

If the large flat plate is immersed in a cold liquid bath, it will be cooled from both sides. Intuitively, the important distance that determines the rate at which the plate cools down is the distance between the middle of the plate and the surface, which is $x/2$. This is what the definition of L_{char} produces when applied to the flat plate. In fact, when Eq. 3-8 is applied to arbitrarily shaped objects, it produces reasonable values of L_{char}.

The relative magnitude of the conduction and the convection resistances for an arbitrarily shaped body is calculated by taking their ratio:

$$Bi = \frac{\left[\frac{L_{char}}{kA}\right]}{\left[\frac{1}{hA}\right]} = \frac{hL_{char}}{k}$$

where L_{char} is given by Eq. 3-8 and Bi is called the **Biot number**. Note that the Biot number is a ratio of resistances and is, therefore, dimensionless. If the Biot number is equal to 0.1, then the conduction resistance is 1/10 of the convection resistance. In that case, most of the temperature drop is between the surface and the fluid, and there is little temperature variation within the solid.

Therefore, the lumped system approximation is valid when the Biot number is small, typically less than about 0.1; that is,

$$Bi <\sim 0.1 \qquad \text{criterion for valid use of lumped system approximation}$$ (3-9)

All the mass within the solid body is "lumped" together and assumed to be at the same temperature. The temperature of the body varies with time but not with location. So, at any moment, the body is isothermal. With this approximation, we can now apply the first law of thermodynamics to the cooling of the arbitrary body shown in Figure 3-10.

In rate form, the first law can be written

$$\frac{dU}{dt} = \dot{Q} - \dot{W}$$

where kinetic and potential energies are ignored. No boundary work is done in this process; therefore, $\dot{W} = 0$ and

$$\frac{dU}{dt} = \dot{Q}$$

For a solid, $dU = mc\, dT \approx mc_v\, dT \approx mc_p\, dT$. We assume the entire mass of the body (including its surface) is at temperature T. Convective heat transfer is $\dot{Q} = hA\left(T - T_f\right)$, but because heat is transferred from the system, this heat is negative. The first law is then

$$mc_p \frac{dT}{dt} = -hA\left(T - T_f\right)$$

where T_f is the temperature of the fluid. To solve this differential equation, separate variables so that

$$\frac{dT}{\left(T - T_f\right)} = -\frac{hA}{mc_p}\, dt$$

Integrate from an initial temperature, T_i, at $t = 0$ to a final state T, at time t:

$$\int_{T_i}^{T} \frac{dT}{T - T_f} = -\frac{hA}{mc_p} \int_0^t dt$$

$$\ln\left(\frac{T - T_f}{T_i - T_f}\right) = -\frac{hA}{mc_p}(t - 0) = -\frac{hA}{mc_p}t$$

Taking the exponential of both sides, we obtain

$$\frac{T(t) - T_f}{T_i - T_f} = \exp\left(-\frac{hA}{mc_p}t\right) \tag{3-10}$$

Remember that $m = \rho V$ and $L_{char} = V/A$. Substituting these into Eq. 3-10 gives

$$\frac{T(t) - T_f}{T_i - T_f} = \exp\left(-\frac{hA}{\rho Vc_p}t\right) = \exp\left(-\frac{h}{\rho c_p L_{char}}t\right) \tag{3-11}$$

Note that both sides of this equation are nondimensional. It applies *only* when $Bi <\sim 0.1$ (Recall that $Bi = hL_{char}/k$). In addition, this equation applies to either a hot solid immersed in a cold fluid or to a cold solid immersed in a hot fluid. Solving for the time, t, to reach a specified temperature $T(t)$,

$$t = -\frac{\rho c_p L_{char}}{h} \ln\left[\frac{T(t) - T_f}{T_i - T_f}\right] \qquad Bi <\sim 0.1 \tag{3-12}$$

Furthermore, Eq. 3-11 can be rearranged to give the body temperature after an elapsed time, t, as

$$T(t) = \left(T_i - T_f\right) \exp\left[\frac{-h}{\rho c_p L_{char}}t\right] + T_f \qquad Bi <\sim 0.1 \tag{3-13}$$

Plots of the temperature of the solid as a function of the elapsed time are shown in Figure 3-12. In both heating and cooling, the temperature changes steeply at early times and then approaches the fluid temperature asymptotically at later times.

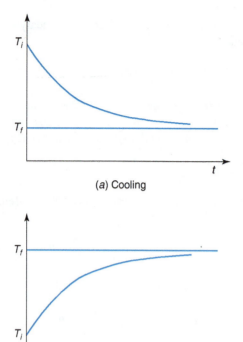

(a) Cooling

(b) Heating

FIGURE 3-12 Solid temperature versus time for a lumped system.

The exponent in Eq. 3-11 is nondimensional. The quantity $\rho c_p L_{char}/h$ is called the **time constant** of the system, and it controls the transient behavior of the body. The magnitude of the temperature difference $(T_i - T_f)$ has no effect on the speed of the transient. After about five time constants, the body temperature will have essentially reached a steady-state value.

We can rearrange the exponent to show explicitly how the Biot number affects the transient. Multiply the numerator and denominator of the exponent by (Lk) and rearrange variables to get

$$\frac{ht}{\rho c_p L_{char}}\left[\frac{L_{char}k}{L_{char}k}\right] = \left[\frac{hL_{char}}{k}\right]\left[\frac{k}{\rho c_p}\frac{t}{L_{char}^2}\right] \tag{3-14}$$

The first factor on the right-hand side, hL_{char}/k, is the nondimensional Biot number. Physically it represents a ratio of internal to external thermal resistances. (The internal resistance is conduction and the external resistance is convection.) The second factor contains the grouping of material properties $k/\rho c_p$. This group is called the **thermal diffusivity** and is given the symbol $\alpha = k/\rho c_p$. The thermal diffusivity arises naturally in the study of transient heat transfer, as will be shown in Chapter 11. The complete second factor in Eq. 3-14 is called the **Fourier number**, that is,

$$Fo = \frac{\alpha t}{L_{char}^2}$$

Physically, the Fourier number represents a ratio of the rate at which heat is conducted across a body to the rate at which heat is stored in the body, as shown in Chapter 11. Using the Fourier number, we can rewrite Eq. 3-11 as:

$$\frac{T(t) - T_f}{T_i - T_f} = \exp\left(-BiFo\right)$$

It is very common in the study of thermal and fluids engineering to encounter nondimensional numbers such as the Biot number and the Fourier number. Others will appear frequently in the chapters to come.

EXAMPLE 3-6   Annealing of a steel ball

A hot steel ball is annealed by dropping it into a cool oil bath. The ball is 0.5 in. in diameter and is initially at 400°F. If the heat transfer coefficient between the ball and oil is 16 Btu/h·ft²·°F, how long will it take for the ball to cool to 150°F? Assume the oil tank is large enough that the oil temperature does not rise during the process but remains at 70°F.

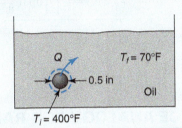

Q $T_f = 70°F$

0.5 in

Oil

$T_i = 400°F$

Approach:

First calculate the Biot number to see if the lumped system approximation can be used. If $Bi <\sim 0.1$, then the time to cool will be given by Eq. 3-12:

$$t = -\frac{\rho c_p L_{char}}{h}\ln\left[\frac{T(t) - T_f}{T_i - T_f}\right]$$

Assumptions:

Solution:

The system under consideration is the steel ball. First calculate the Biot number to see if the lumped system approach can be applied. The Biot number is

$$Bi = \frac{h L_{char}}{k}$$

A representative length for the Biot number is, from Eq. 3-8,

$$L_{char} = \frac{V}{A} = \frac{(4/3)\pi R^3}{4\pi R^2} = \frac{R}{3} = 0.0833 \text{ in.}$$

To calculate the Biot number, the thermal conductivity of steel is needed. Property values for k are given in Table B-2. Note that k varies somewhat with temperature. In this case $T_1 = 400°F \approx 860$ R and $T_2 = 150°F \approx 610$ R. The average temperature is

$$T_{avg} = \frac{T_1 + T_2}{2} = 735 \, R$$

A1. Thermal conductivity is constant.

The value of k at 720 R, which for plain carbon steel is 32.8 Btu/(h·ft·R), will be close enough [A1]. With this value,

$$Bi = \frac{\left[16 \, \frac{\text{Btu}}{\text{h·ft}^2{}^\circ\text{F}}\right](0.0833 \text{ in.})\left[\frac{1 \text{ ft}}{12 \text{ in.}}\right]}{32.8 \, \frac{\text{Btu}}{\text{h·ft·R}}} = 3.39 \times 10^{-3}$$

A2. The lumped system approximation is valid.
A3. The oil temperature is constant.

Clearly, $Bi \ll 0.1$, and the conduction resistance is so much less than the convection resistance that we can ignore conduction and use the lumped system approximation [A2]. The ball is small compared to the total mass of oil in the tank, and heat transfer from the ball will not significantly increase the oil temperature [A3]. From Eq. 3-12:

$$t = -\frac{\rho c_p L_{char}}{h} \ln\left[\frac{T(t) - T_f}{T_i - T_f}\right]$$

A4. Specific heat is constant.

Property values for ρ and c_p are given in Table B-2. Note that c_p varies somewhat with temperature. As we did above for thermal conductivity, we will use the value of c_p at 720 R, which, for plain carbon steel is 0.116 Btu/lbm·R [A4]. With these values,

$$t = \frac{-\left[490.3 \, \frac{\text{lbm}}{\text{ft}^3}\right]\left[0.116 \frac{\text{Btu}}{\text{lbm R}}\right](0.0833 \text{ in.})\left[\frac{1 \text{ ft}}{12 \text{ in.}}\right]}{16 \frac{\text{Btu}}{\text{h·ft}^2 \cdot {}^\circ\text{F}}} \ln\left[\frac{150 - 70}{400 - 70}\right]$$

$$t = 0.035 \, \text{h} = 2.1 \, \text{min}$$

3.7 THE RESISTANCE ANALOGY FOR RADIATION

If a gray, diffuse surface at temperature T_s transfers heat by radiation to surrounding surfaces at T_{surr}, and the surrounding surfaces are far away, the heat transfer rate can be written as

$$\dot{Q} = \varepsilon \sigma A \left(T_s^4 - T_{surr}^4\right)$$

This is a nonlinear equation and is not of the form

$$\dot{Q} = \frac{\Delta T}{R}$$

However, it is possible to force the equation into the desired form by factoring the term $(T_s^4 - T_{surr}^4)$. The result is

$$\dot{Q} = \varepsilon \sigma A \left(T_s - T_{surr}\right)\left(T_s + T_{surr}\right)\left(T_s^2 + T_{surr}^2\right)$$

Now define a thermal resistance for radiation in the form

$$R_{rad} = \frac{1}{\varepsilon \sigma A \left(T_s + T_{surr}\right)\left(T_s^2 + T_{surr}^2\right)}$$

This resistance is commonly written in terms of a heat transfer coefficient for radiation, which is defined as

$$\boxed{R_{rad} = \frac{1}{h_{rad} A}}$$

So that

$$\boxed{h_{rad} = \varepsilon \sigma \left(T_s + T_{surr}\right)\left(T_s^2 + T_{surr}^2\right)}$$

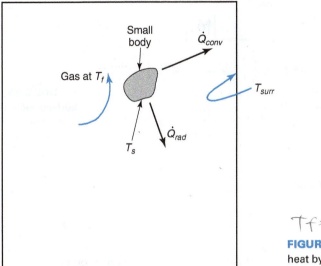

$T_f = T_{surr}$

FIGURE 3-13 A body exchanging heat by both convection and radiation.

Unfortunately, the thermal resistance for radiation depends on the temperature of the surface and the surroundings. If these are not known in advance, the values may have to be assumed. A problem of this type is given in Example 3-7. Before proceeding to the example, we consider the common case of a surface cooled by both convection and radiation.

In Figure 3-13, a small body with surface temperature T_s is placed in an oven filled with hot gas at T_f. The walls of the oven are cooler than the gas and are at T_{surr}. Heat leaving the body travels along two paths, a convective path and a radiative path. The resistance analogy may be used to represent these paths by two resistors, as shown in Figure 3-14. The convective path involves the gas temperature, and the radiative path involves the temperature of the surrounding surfaces. The total heat leaving the surface is

$$\dot{Q} = \dot{Q}_{rad} + \dot{Q}_{conv}$$

where the subscripts *rad* and *conv* refer to radiative and convective, respectively.

In many cases, the gas cooling the surface and the surrounding surfaces are both at the same temperature, T_∞. The two resistances are then connected in parallel, as shown in Figure 3-15. The total heat leaving the surface becomes

$$\dot{Q} = \dot{Q}_{rad} + \dot{Q}_{conv} = \frac{T_s - T_\infty}{R_{rad}} + \frac{T_s - T_\infty}{R_{conv}}$$

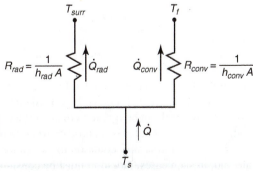

happening @ same time

can lose heat by both @ same time

FIGURE 3-14 Resistance analog for heat transfer from a surface cooled by both convection and radiation.

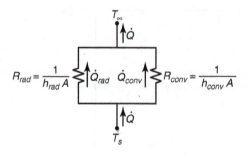

FIGURE 3-15   Resistance analog for a surface cooled by convection and radiation when the gas and the surroundings are at the same temperature, T_∞.

Combining terms gives

$$\dot{Q} = \left[\frac{1}{R_{rad}} + \frac{1}{R_{conv}} \right] (T_s - T_\infty)$$

The total resistance for the parallel combination is

$$\dot{Q} = \frac{T_s - T_\infty}{R_{tot}}$$

Comparing the last two equations reveals that

$$\boxed{\frac{1}{R_{tot}} = \frac{1}{R_{rad}} + \frac{1}{R_{conv}}} \qquad (3\text{-}15)$$

You may be familiar with this equation as the equation for two electrical resistances in parallel. Eq. 3-15 may be solved for R_{tot} to give

$$\boxed{R_{tot} = \frac{R_{rad} R_{conv}}{R_{rad} + R_{conv}}} \qquad (3\text{-}16)$$

which is an alternate form for the parallel combination of two resistors. Using $R_{rad} = 1/(h_{rad}A)$ and $R_{conv} = 1/(h_{conv}A)$ in Eq. 3-15 produces

$$h_{tot}A = h_{rad}A + h_{conv}A$$

or

$$\boxed{h_{tot} = h_{rad} + h_{conv}}$$

So for a surface exchanging heat by convection and radiation to the same temperature, the heat transfer coefficients are additive.

Radiation is important for surfaces cooled by gases. With liquid cooling, radiation is generally insignificant. If a surface is cooled by forced convection, radiation is typically small relative to convection and can be ignored unless surface temperatures are quite high. On the other hand, if a surface is cooled by natural convection in a gas, radiation is likely to be as important as convection. It is very common to encounter a surface cooled by natural convection in air, and, in such cases, radiation must be considered.

EXAMPLE 3-7   Temperature of a heating element

Someone takes a teapot off the stove and forgets to turn off the heating element, which is a coiled flat resistor with a resistance of 15 Ω. The top surface area of the element is 16 in.² Its emissivity is 0.85. The convective heat transfer coefficient from the top of the element is 3.7 Btu/h·ft²·R. If the voltage drop across the element is 30 V, how hot will it become at steady state? Assume that all the heat leaves by convection and radiation from the top of the element and that the room is at 70°F.

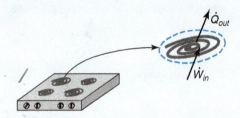

Approach:

Define the system as the heating element. The air temperature and the wall temperatures of the room are assumed to be equal at 70°F. To find the heat dissipated from the top of the element, we would use

$$\dot{Q} = h_{tot}\, A\, (T_s - T_\infty)$$

where h_{tot} is the total heat transfer coefficient, that is, the sum of the convective and radiative heat transfer coefficients. We are given the convective heat transfer coefficient. We need to calculate the radiative heat transfer coefficient with

$$h_{rad} = \varepsilon\sigma\,(T_s + T_\infty)\left(T_s^2 + T_\infty^2\right)$$

Unfortunately, h_{rad} depends on T_s, the unknown surface temperature of the heating element. To make further progress, it is necessary to *assume* a value for T_s and use this to calculate an approximate value for h_{rad}. An improved value of h_{rad} will be found later, as you will see. The next step is to use the first law in the form:

$$\frac{dU}{dt} = \dot{Q} - \dot{W}$$

Because we are interested only in steady-state temperatures, there will be no changes with time and the left-hand side of this equation becomes equal to zero. The electrical work done on the heating element is

$$\dot{W}_{in} = \xi I$$

The current, I, can be found from Ohm's law and the voltage, ξ, is given in the problem statement. After $\dot{Q}$ and $\dot{W}$ are substituted, the only unknown is the surface temperature, which can be determined.

From this point, a new value of h_{rad} is computed using the surface temperature. The calculation is repeated to find a second estimate of surface temperature, which leads to third estimate for h_{rad}, and so on. This iteration continues until the surface temperature no longer changes significantly from one iteration to the next.

Assumptions:

Solution:

We choose the heating element as the system under study. First find the rate of work, or power, in the heating element. This is

$$\dot{W}_{in} = \xi I$$

A1. The resistor is linear.
A2. The electrical resistance of the resistor does not depend on temperature.

From Ohm's law [A1][A2],

$$\xi = IR$$

where R is the *electrical resistance*. Therefore,

$$\dot{W}_{in} = \frac{\xi^2}{R}$$

Using given values,

$$\dot{W}_{in} = \frac{(30\,V)^2}{15\,\Omega} = 60\,W = -\dot{W}$$

Heat leaves the top of the element by convection and radiation. The total heat transfer coefficient is

$$h_{tot} = h_{rad} + h_{conv}$$

To evaluate the radiative heat transfer coefficient, the surface temperature must be assumed. A reasonable value to start with is 300°F. Later, we can correct it if necessary. Using our assumed value, h_{rad} becomes

$$h_{rad} = \varepsilon\sigma(T_s + T_\infty)\left(T_s^2 + T_\infty^2\right)$$

$$h_{rad} = 0.85\left[0.171 \times 10^{-8}\,\frac{\text{Btu}}{\text{h}\cdot\text{ft}^2\cdot\text{R}^4}\right](760 + 530)\left[(760)^2 + (530)^2\right]\text{R}^3$$

$$h_{rad} = 1.61\,\frac{\text{Btu}}{\text{h}\cdot\text{ft}^2\cdot\text{R}}$$

Absolute temperatures must always be used in radiative calculations, so the temperatures have been converted to Rankine. The total heat transfer coefficient is

$$h_{tot} = h_{rad} + h_{conv} = 1.61 + 3.7 = 5.96\,\frac{\text{Btu}}{\text{h}\cdot\text{ft}^2\cdot\text{R}}$$

The heat leaving the element is

$$\dot{Q} = h_{tot}A(T_s - T_\infty)$$

Applying the first law to the heating element gives

$$\frac{dU}{dt} = \dot{Q} - \dot{W}$$

A3. The system is in steady state.

At steady state [A3],

$$\frac{dU}{dt} = 0$$

Heat and work are both negative, because heat is leaving the system and work is being done on the element. The first law becomes

$$0 = -h_{tot}A(T_s - T_\infty) - (-60\,W)$$

Substituting values and converting units,

$$0 = -\left[5.96 \frac{\text{Btu}}{\text{h}\cdot\text{ft}^2\cdot\text{R}}\right](16\,\text{in.}^2)\left[\frac{1\,\text{ft}^2}{144\,\text{in.}^2}\right](T_s - 70)\,°\text{F} + (60\,\text{W})\left[\frac{3.412\frac{\text{Btu}}{\text{h}}}{1\,\text{W}}\right]$$

Evaluating and solving for T_s yields

$$T_s = 379\,°\text{F}$$

Recall that we assumed $T_s = 300°\text{F}$ in order to calculate h_{rad}. What would h_{rad} be if we use 379°F? Redoing the calculation gives $h_{rad} = 1.96$ W/m²·K and total resistance $h_{tot} = 5.66$ W/m²·K. The new value of T_s computes to 396°F. We may continue the iteration if more precision is needed. A table of the assumed and computed values of T_s is given below.

Iteration	Assumed T_s°F	Calculated T_s°F
1	300	379
2	379	396
3	396	391
4	391	392
5	392	392

This calculation has converged to three significant figures after five iterations. Notice that even the first calculated value of T_s, 379°F, is not too far from the final converged value of 392°F. This is what makes the concept of h_{rad} useful.

Comments:

Iterative solutions occur frequently in thermal and fluids engineering, although most examples will be noniterative in keeping with the introductory nature of this text. Note that it is possible to solve this problem without the use of h_{rad}. The solution would still be iterative, but it might not converge as quickly (or at all). The essential merit in h_{rad} is that it provides a very good starting guess for the subsequent iteration. If high accuracy is not needed, then using h_{rad} allows one to get a reasonable estimate in just one iteration.

3.8 COMBINED THERMAL RESISTANCES

The resistance analogy can be extended to rather complex systems. The best way to understand its scope is through examples. As a start, consider a wall in a residential building, as shown in Figure 3-16. The wall is built of wooden boards called studs. These are typically (nominally) 2 in. by 4 in. or 2 in. by 6 in. in cross-section. On the outside, a layer of foamboard is nailed to the studs. Exterior siding (wooden planks in this example) covers the foamboard. On the interior of the house, the space between the studs is filled with thermal insulation and a layer of wallboard is nailed to the studs.

In cold weather, the insulated wall prevents heat loss to the outside; in hot weather, it keeps heat from conducting into an air-conditioned room. The thermal resistance of this composite system can be estimated from the resistance analogy. The studs repeat periodically along the wall; therefore, one can define a "unit cell" of the wall. The unit cell is the section of wall between the dotted lines in Figure 3-16. The resistance through this cell will be characteristic of the wall as a whole.

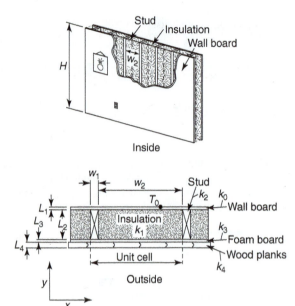

FIGURE 3-16 Cross-sectional view of a residental wall.

In the real situation, heat will be conducted in both the x- and y-directions in Figure 3-16; however, the predominant direction of heat flow will be the y-direction. It is important to correctly model heat flow in the y-direction but much less important to correctly model it in the x-direction. As a result, we may simplify the analysis by assuming that the thermal conductivity in the x-direction, k_x, is either zero or infinity. The real situation falls somewhere between these two limits.

CASE STUDY **3-1**

$k_x = 0$

In this case, no heat flows in the x-direction. The resulting resistance network is shown in Figure 3-17a. The outer surface of the wallboard is at temperature T_0. The wallboard is divided into two parts: one of width w_1 covering the stud and the other of width w_2 covering the insulation. The resistances R_0 and R_4 correspond to these two wallboard segments. The resistance through the stud is R_1 and the resistance through the insulation is R_5. The foamboard is also divided into two segments, with R_2 being the resistance through a section of width w_1 and R_6 being the resistance through a section of width w_2. Finally, R_3 and R_7 are resistances of segments of wood plank of width w_1 and w_2, respectively. Note that the left leg of the resistance network corresponds to a slice through the wall containing the stud and having a width w_1. The right leg corresponds to a slice of width w_2 containing the insulation.

The resistance R_0 can be expressed as

$$R_0 = \frac{L}{kA} = \frac{L_1}{k_0 w_1 H}$$

where H is the height of the room and the other dimensions are shown on Figure 3-16. Using values given in Table 3-4, R_0 becomes:

$$R_0 = \frac{0.375 \text{ in.}}{\left[0.098 \, \frac{\text{Btu}}{\text{h·ft·R}} \right] (1.375 \text{ in.})(8 \text{ ft})} = 0.3479 \cdot \frac{\text{h·R}}{\text{Btu}}$$

The resistance through the stud, R_1, is given by

$$R_1 = \frac{L_2}{k_2 w_1 H} = 4.87 \, \frac{\text{h·R}}{\text{Btu}}$$

(Continued)

CASE STUDY 3-1 *(Continued)*

The remaining resistances are

$$R_2 = \frac{L_3}{k_3 w_1 H} = 3.03 \frac{\text{h} \cdot \text{R}}{\text{Btu}}$$

$$R_3 = \frac{L_4}{k_4 w_1 H} = 0.696 \frac{\text{h} \cdot \text{R}}{\text{Btu}}$$

$$R_4 = \frac{L_1}{k_0 w_2 H} = 0.0211 \frac{\text{h} \cdot \text{R}}{\text{Btu}}$$

$$R_5 = \frac{L_2}{k_2 w_2 H} = 0.848 \frac{\text{h} \cdot \text{R}}{\text{Btu}}$$

$$R_6 = \frac{L_3}{k_3 w_2 H} = 0.184 \frac{\text{h} \cdot \text{R}}{\text{Btu}}$$

$$R_7 = \frac{L_4}{k_4 w_2 H} = 0.0423 \frac{\text{h} \cdot \text{R}}{\text{Btu}}$$

The total thermal resistance is found by combining resistances in series and parallel. For the left and right legs

$$R_{left} = R_0 + R_1 + R_2 + R_3 = 8.94 \frac{\text{h} \cdot \text{R}}{\text{Btu}}$$

$$R_{right} = R_4 + R_5 + R_6 + R_7 = 1.09 \frac{\text{h} \cdot \text{R}}{\text{Btu}}$$

The total resistance is

$$R_{tot} = \frac{R_{left} R_{right}}{R_{left} + R_{right}} = 0.976 \frac{\text{h} \cdot \text{R}}{\text{Btu}}$$

This is the resistance for a "unit cell," which has an area of

$$A_{cell} = H(w_1 + w_2)$$

For a wall of area A_w, the resistance would be

$$R_w = \frac{R_{tot} A_{cell}}{A_w}$$

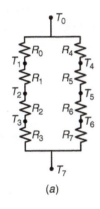

(a)

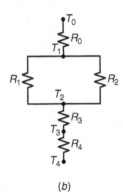

(b)

FIGURE 3-17 Alternate resistance networks.

TABLE 3-4 Parameters for the thermal analysis of the insulated wall in Figure 3-16

Dimension	Inches	Thermal Conductivity	Btu/(h ·ft ·°F)	Dimension	ft
w_1	1.375	k_0	0.098	H	8
w_2	22.625	k_1	0.022		
L_1	0.375	k_2	0.063		
L_2	3.375	k_3	0.015		
L_3	0.5	k_4	0.098		
L_4	0.75				

CASE STUDY *3-2*

$$k_x = \infty$$

In Case Study 3-1, the thermal conductivity in the x-direction, k_x, was zero. This implied that no heat could flow in the x-direction. The opposing limit is that k_x is infinity. This implies that it is extremely easy for heat to flow in the x-direction, and, as a result, there are negligibly small temperature drops in the x-direction. In Case Study 3-2, the temperature is assumed to be independent of x. The resulting resistance network is shown in Figure 3-17b.

Again, T_0 is the temperature of the outer surface of the wallboard. The resistance, R_0, in this case, applies to a segment of wallboard of width $w_1 + w_2$. T_1 is the temperature of the inner surface of the wallboard. Temperature is assumed not to vary in the x-direction; therefore, T_1 is temperature both adjacent to the stud and adjacent to the insulation. This is in contrast to Case Study 3-1 (see Figure 3-17a). In this case, T_1 is the temperature of the inner surface of the wallboard near the stud and T_4 is the temperature near the insulation.

Evaluating the resistances in Figure 3-17b gives

$$R_0 = \frac{L}{kA} = \frac{L_1}{k_0(w_1 + w_2)H} = 0.0199 \ \frac{\text{h·R}}{\text{Btu}}$$

$$R_1 = \frac{L_2}{k_0 w_1 H} = 4.87 \ \frac{\text{h·R}}{\text{Btu}}$$

$$R_2 = \frac{L_2}{k_1 w_2 H} = 0.848 \ \frac{\text{h·R}}{\text{Btu}}$$

$$R_3 = \frac{L_3}{k_3(w_1 + w_2)H} = 0.174 \ \frac{\text{h·R}}{\text{Btu}}$$

$$R_4 = \frac{L_4}{k_4(w_1 + w_2)H} = 0.0399 \ \frac{\text{h·R}}{\text{Btu}}$$

Combining resistances in series and parallel (see Figure 3-17b),

$$R_{tot} = R_0 + \frac{R_1 R_2}{R_1 + R_2} + R_3 + R_4 = 0.955 \ \frac{\text{h·R}}{\text{Btu}}$$

In Case 1, where $k_x = 0$, total resistance was 0.976, while in Case 2, where $k_x = \infty$, the resistance is 0.955. The actual resistance lies somewhere between these two extremes. The two values differ by less than 3%. For most practical circumstances, it is not necessary to know the resistance to a higher level of accuracy than 3%, thus justifying the use of a one-dimensional resistance network in this case.

SUMMARY

The first law may be written in differential form as

$$dKE + dPE + dU = \delta Q - \delta W$$

The rate equation form of the first law is

$$\frac{dKE}{dt} + \frac{dPE}{dt} + \frac{dU}{dt} = \dot{Q} - \dot{W}$$

where

$$\dot{Q} = \frac{\delta Q}{dt}$$

$$\dot{W} = \frac{\delta W}{dt}$$

or

$$Q = \int \dot{Q} \, dt$$

$$W = \int \dot{W} \, dt$$

Fourier's law for heat conduction is

$$\dot{Q} = -kA \frac{dT}{dx}$$

For one-dimensional, steady conduction through a plane layer,

$$\dot{Q} = \frac{kA(T_1 - T_2)}{L}$$

For convection on a surface of area, A,

$$\dot{Q} = hA(T_s - T_f)$$

For radiation between a diffuse, gray surface at T_s and surrounding surfaces at T_{surr}, which are large and far away from the surface,

$$\dot{Q} = \varepsilon \sigma A(T_s^4 - T_{surr}^4)$$

For electrical work,

$$\dot{W} = \xi I$$

For shaft work,

$$\dot{W} = \Im \omega$$

The Biot number is defined as

$$Bi = \frac{h L_{char}}{k}$$

where L_{char} is a characteristic length of the solid given by

$$L_{char} = \frac{V}{A}$$

If $Bi <\sim 0.1$, the lumped system approximation can be used. With this approximation, the time for a solid to heat or cool by convection is

$$t = -\frac{\rho c_p L_{char}}{h} \ln\left[\frac{T(t) - T_f}{T_i - T_f}\right]$$

where T_i is the initial temperature, $T(t)$ is the temperature at time t and T_f is the fluid temperature. The temperature of the solid after a time t is given by rearranging this equation to

$$T(t) = (T_i - T_f)\, \exp\left(\frac{-h}{\rho c_p L_{char}}\, t\right) + T_f$$

The effective thermal resistance for conduction through a plane layer is

$$R_{cond} = \frac{L}{kA}$$

For conduction through a cylindrical shell,

$$R_{cylinder} = \frac{\ln\,(r_2/r_1)}{2\pi L k}$$

For conduction through a spherical shell,

$$R_{sphere} = \frac{r_2 - r_1}{4\pi r_1 r_2 k}$$

The thermal resistance for convection is

$$R_{conv} = \frac{1}{hA}$$

The thermal resistance for radiation is

$$R_{rad} = \frac{1}{h_{rad}\, A}$$

where

$$h_{rad} = \varepsilon\sigma\,(T_s + T_{surr})\left(T_s^2 + T_{surr}^2\right)$$

The total heat transfer coefficient for a surface exchanging heat by convection and radiation to a gas and surfaces at the same temperature is:

$$h_{tot} = h_{rad} + h_{conv}$$

When two resistances are in series, the total resistance is the sum

$$R_{tot} = R_1 + R_2$$

When two resistances are in parallel, the total resistance is

$$\frac{1}{R_{tot}} = \frac{1}{R_1} + \frac{1}{R_2}$$

or

$$R_{tot} = \left\{\frac{R_1 R_2}{R_1 + R_2}\right\}$$

SELECTED REFERENCES

BECKER, M., *Heat Transfer, A Modern Approach*, Plenum, New York, 1986.

ÇENGEL, Y. A., *Introduction to Thermodynamics and Heat Transfer*, McGraw-Hill, New York, 1997.

INCROPERA, F. P., and D. P. DeWitt, *Introduction to Heat Transfer*, 4th ed., Wiley, New York, 2002.

KREITH, F., and M. S. Bohn, *Principles of Heat Transfer*, 6th ed., Brooks/Cole, Pacific Grove, CA, 2001.

MILLS, A. F., *Heat Transfer*, Irwin, Boston, 1992.

SURYANARAYANA, N. V., *Engineering Heat Transfer*, West, New York, 1995.

THOMAS, L. C., *Heat Transfer*, Prentice Hall, Englewood Cliffs, NJ, 1992.

PROBLEMS

FIRST LAW IN RATE FORM

P3-1 An arctic explorer builds a temporary shelter from wind-pack snow. The shelter is roughly hemispherical, with an inside radius of 1.5 m. After completing the shelter, the explorer crawls inside and closes off the entrance with a block of snow. Assume the shelter is now airtight and loses negligible heat by conduction through the walls. If the air temperature when the explorer completes the shelter is $-10°C$, how long will it take before the air temperature inside reaches $10°C$? Assume the explorer does not freeze to death or suffocate, but sits patiently waiting for the temperature to rise. The explorer generates body heat at a rate of 300 kJ/h.

P3-2 A well-insulated room with a volume of 60 m^3 contains air initially at 100 kPa and 25°C. A 100-W lightbulb is turned on for three hours. Assuming the room is airtight, estimate the final temperature.

P3-3   An elevator is required to carry eight people to the top of a 12-story building in less than 1 min. A counterweight is used to balance the mass of the empty elevator cage. Assume that an average person weighs 155 lbf and that each story has a height of 12 ft. What is the minimum size of motor (in hp) that can be used in this application?

P3-4   A climate-controlled room in a semiconductor factory contains a conveyor belt. Electric power is supplied to the motor of the conveyor belt at 220 V and a current that varies linearly with time as $I = 1.0\,t$, where I is in amps when t is in minutes. An air conditioner removes heat from the room at a constant rate of 2 kW. The volume of air in the room is 600 m^3. At $t = 0$, the air is at 25°C and 101 kPa. Assume the mass of air is constant during this process and assume constant specific heats.

a. Find the mass of air in the room (in kg).

b. Find the air temperature after 30 min. (in °C); ignore any temperature change of the motor or conveyor belt.

P3-5   An interplanetary probe of volume 300 ft^3 contains air at 14.7 psia and 77°F. The heaters fail and the air begins to cool. Assume heat is dissipated from the outside of the spacecraft by radiation at a steady rate of 60 Btu/h. On-board electronics generate 12 W on average. Estimate the time required for the air to cool to −30°F.

P3-6   A fan is installed in a 35-m^3 sealed box containing air at 101 kPa and 20°C. The exterior of the box is perfectly insulated. The fan does 250 W of work in stirring the air and operates for 1 h. Find the final temperature and pressure of the air. Ignore the temperature change of any fan parts.

ONE-DIMENSIONAL CONDUCTION IN RECTANGULAR COORDINATES

P3-7   A room contains four single-pane windows of size 5 ft by 2.5 ft. The thickness of the glass is 1/4 in. If the inside glass surface is at 60°F and the outside surface is at 30°F, estimate the heat loss through the windows.

P3-8   An L-shaped extrusion made of aluminum alloy 2024-T6 is well insulated on all sides, as shown in the figure. Heat flows

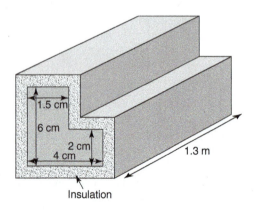

Insulation

axially in the extrusion at a rate of 35 W. If the cool end is at 25°C, find the temperature at the hot end.

P3-9   The wall of a furnace is a large surface of fire clay brick, which is 6.5 cm thick. The outer surface of the brick is measured to be at 35°C. The inner surface receives a heat flux of 2.3 W/cm^2. Estimate the temperature of the inner surface of the brick.

RADIATION HEAT TRANSFER

P3-10   A tungsten filament in a 60-W lightbulb has a diameter of 0.04 mm and an electrical resistivity of 90 $\mu\Omega$-cm. The filament loses heat to the environment, which is at 20°C, by thermal radiation. The emissivity of the filament is 0.32 and the voltage across it is 115 V. Find the length of the filament and the filament surface temperature. (Electrical resistance equals electrical resistivity times filament length divided by filament cross-sectional area.)

P3-11   On a cold winter day, the interior walls of a room are at 55°F. A man standing in the room loses heat to the walls by thermal radiation. The man's surface area is 16 ft^2, his clothing has an emissivity of 0.93 and his surface temperature is 70°F. He generates 300 Btu/h of body heat. What percentage of the man's body heat is transferred by radiation to the walls?

P3-12   The sun can be approximated as a spherical black body with a surface temperature of 5762 K. The irradiation from the sun as measured by a satellite in earth orbit is 1353 W/m^2. The distance from the earth to the sun is approximately 1.5×10^{11} m. Assuming that the sun radiates evenly in all directions, estimate the diameter of the sun.

CONVECTION AND RADIATION

P3-13   A high-torque motor has an approximately cylindrical housing 9.5 in. long and 6 in. in diameter. The motor delivers 1/8 hp in steady operation and has an efficiency of 0.72. All the heat generated by motor losses is removed by natural convection and radiation from the outer surface of the housing. The convective coefficient is 1.68 Btu/h·ft^2·°F, and the housing emissivity is 0.91. If the surroundings are at 58°F, what is the housing's outer surface temperature?

P3-14   A flat plate solar collector 6 ft by 12 ft is mounted on the roof of a house. The outer surface of the collector is at 110°F and its emissivity is 0.9. The outside air is at 70°F and the sky has an effective temperature for radiation of 45°F. The collector transfers heat by natural convection to the air with a heat transfer coefficient of 3.2 Btu/h·ft^2·°F and also transfers heat by radiation to the sky. Calculate the total heat lost from the solar collector.

P3-15   A CPU chip with a footprint of 3 cm by 2 cm is mounted on a circuit board. The chip generates 0.31 W/cm^2 and rejects heat to the environment at 28°C by convection and radiation. The outer casing of the chip has an emissivity of 0.88, and the heat transfer coefficient is 48 W/ m^2·K. Neglecting the thickness of the chip and any conduction into the circuit board, calculate the chip surface temperature.

P3-16   A metal plate 16 cm by 8 cm is placed outside on a clear night. The plate, which has an emissivity of 0.7, exchanges heat by radiation with the night sky, which is at $-40°C$. Air at $-10°C$ flows over the top of the plate, cooling it with a heat transfer coefficient of 42 W/m²·K. The plate is insulated on its underside and heated by an electric resistance heater. How much electric power must be supplied to maintain the plate at 55°C?

CONDUCTION AND CONVECTION

P3-17 A home freezer is 1.8 m wide, 1 m high, and 1.2 m deep. The interior surface of the freezer must be kept at $-10°C$. The walls of the freezer are made of polystyrene insulation sandwiched between two thin layers of steel. The combined convective/radiative heat transfer coefficient on the exterior is 8.2 W/m²·K and the ambient is at 25°C. If the power of the refrigeration unit is limited to 150 W, what thickness of polystyrene is needed? Assume the conduction resistance of the thin metal wall panels is very small and can be neglected and that the bottom of the freezer is perfectly insulated.

P3-18 The windshield of an automobile is heated on the inside by a flow of warm air. Cold air at $-15°F$ flows over the exterior of the windshield. The heat transfer coefficient on the inside is 16 Btu/h·ft²·°F, and the heat transfer coefficient on the outside is 49 Btu/h·ft²·°F. The glass of the windshield has a thickness of 0.25 in. What temperature should the inside air be so that the exterior surface temperature of the windshield is 3°F?

P3-19 A copper busbar of length 40 cm carries electricity and produces 4.8 W in joule heating. The cross-section is square, as shown in the figure, and is covered with insulation of thermal conductivity 0.036 W/m·K. All four sides are cooled by air at 20°C with an average heat transfer coefficient of 18 W/m²·K. Assuming the copper is isothermal, estimate the maximum temperature of the insulation.

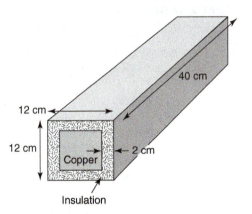

P3-20 A freezer maintains one side of a slab of ice 3 cm thick at $-10°C$. The other side exchanges heat with the ambient air and surfaces at 15°C by combined natural convection and radiation. In steady state, the ice does not melt. Find the highest possible value of the heat transfer coefficient on the ice surface exposed to the ambient air.

P3-21 The door of a kitchen oven contains a window made of a single pane of 1/4-in.-thick Pyrex glass. The interior oven temperature is 550°F and the room air is at 68°F. The combined convective/radiative heat transfer coefficient on the oven interior is 1.7 Btu/h·ft²·°F, and on the oven exterior it is 0.88 Btu/h·ft²·°F. A toddler comes by and touches the window. Calculate the temperature of the surface that the child's hand contacts.

MULTILAYER WALL

P3-22 An electronic device may be modeled as three plane layers, as shown in the figure. The entire package is cooled on both sides by air at 20°C. Heat is generated in a very thin layer between two contacting surfaces at a rate of 500 W/m², as shown. The heat transfer coefficient on both sides is 8.7 W/m²·K. Assume the layers are very large in extent in the direction not shown. Using data in the figure below, calculate the temperature T_2.

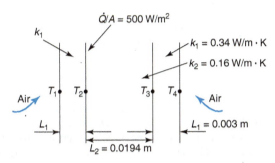

P3-23 A cardboard box is used to ship flowers on a summer day when the ambient temperature is 80°F. The air inside the box is maintained at 45°F by the use of cold packs. The box is lined with a layer of Styrofoam ($k_s = 0.015$ Btu/h·ft·R) 1/2 in. thick. The cardboard itself is 1/8 in. thick and has $k_c = 0.13$ Btu/h·ft·R. The box measures 8 in. by 8 in. by 2.5 ft. Assume h on the inside is 2.0 Btu/h·ft²·R and h on the outside is 9.3 Btu/h·ft²·R. Calculate the rate of heat transfer into the box. Neglect heat transfer on the ends.

P3-24 A living room floor 3 m by 4.5 m is constructed of a layer of oak planks 1.2 cm thick laid over plywood 2.0 cm thick. In winter, the basement air is at 15°C, while the living room air is at 20°C. The heat transfer coefficients on the living room floor and the basement ceiling are 3.6 and 6.8 W/m²·K, respectively. If the home is heated electrically and the cost of electricity is $0.08 per kWh, estimate the cost per month of the energy lost through the floor. If the room is carpeted with wall-to-wall carpeting 1.6 cm thick ($k = 0.06$ W/m·K), what would the energy cost be?

P3-25 The wall of a furnace must be designed to transmit no more than 220 Btu/h·ft². Two types of bricks are available for construction: one with a thermal conductivity of 0.38 Btu/h·ft·R and a maximum allowable temperature of 1400°F and the other with a thermal conductivity of 0.98 Btu/h·ft·R and a maximum allowable temperature of 2300°F. The inside wall of the furnace is at 2100°F and the outside wall is at 300°F. Both types of bricks have dimensions of 9 in. by 4.5 in. by 3 in., and both cost the same. If the bricks can be laid up in any manner, determine the most economical arrangement of bricks.

CONDUCTION AND CONVECTION IN CYLINDRICAL COORDINATES

P3-26 A chemical reactor is in the shape of a long cylinder, as shown in the figure. The reactor is covered with a layer of insulation 17.7 cm thick. The reactor loses heat through the insulation at a rate of 15.3 W per meter of length. The thermal conductivity of the insulation is 0.04 W/m·K. On the outside of the insulation, air at 26°C removes heat by forced convection, with a heat transfer coefficient of 32 W/m²·K. Find the maximum temperature of the insulation. Neglect radiation.

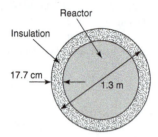

P3-27 An insulated copper wire with a length of 1.2 m carries 20 A of current. The copper is 1 mm in diameter and the insulation ($k = 0.13$ W/m·°C) has a thickness of 0.8 cm. Air at 25°C blows in crossflow over the wire to produce an external convective heat transfer coefficient of 219 W/m²·K. Assuming the copper is isothermal, find the copper temperature. Take the electrical resistivity of copper to be constant at $2.1 \times 10^{-8}\,\Omega\cdot$m.

P3-28 The wall of a submarine is 1-in-thick stainless steel (AISI 304) insulated on the interior with a 1.5-in. layer of polyurethane foam ($k = 0.017$ Btu/h·ft·°F). The heat transfer coefficient on the interior is 3.7 Btu/h·ft·°F. At full speed, the exterior heat transfer coefficient is 135 Btu/h·ft·°F. The sub is approximately cylindrical with the length 240 ft and the outer diameter 30 ft. If the seawater is at 40°F, at what rate must heat be added to the interior air to keep it at 70°F? As a first approximation, neglect heat transfer through the ends.

P3-29 An aluminum wire 2.5 m long conducts 12 A with an imposed voltage of 1.5 V. The wire, which has a diameter of 2.4 mm, is covered with a layer of insulation 2 mm thick. The thermal conductivity of the insulation is 0.15 W/m·°C. Air at 40°C flows over the exterior of the wire to give a convective heat transfer coefficient of 32 W/m²·°C. Assume the aluminum is isothermal and compute the temperature on the inside surface and also on the outside surface of the insulation.

P3-30 An insulated steel pipe carries hot water at 80°C. The outer surface of the insulation loses heat to the environment by convection and radiation. For convection, assume $h_{conv} = 5.8$ W/m²·°C. The emissivity of the insulation is 0.88. The surroundings are at 30°C. Assume the inner surface of the insulation is at the water temperature. What is the surface temperature of the insulation? Use the data shown on the figure.

P3-31 A cylinder of radius r_1 is covered with a layer of insulation of thermal conductivity, k. A fluid flows over the outside of the insulation, exchanging heat with a heat transfer coefficient, h. Let r_2 be the radius at the outer surface of the insulation. Cooling of the cylinder is controlled by the combination of conduction and convection resistances. If r_2 is small, the conduction resistance is small. As r_2 increases, the conduction resistance increases, but the surface area of exposed insulation also increases, and this results in a *decrease* in convective resistance. As a result, there is an optimal value of r_2 that produces the largest possible total resistance to heat transfer. Derive an expression for the optimum value of r_2 as a function of r_1, k, and h.

P3-32 A frozen pipe is filled with ice at 0°C. A heating tape wrapped around the pipe provides 90 W per meter of pipe length. Insulation is placed over the heating tape. The insulation has a thickness of 0.5 cm and a thermal conductivity of 0.082 W/m·°C. Convection and radiation occur from the outside of the insulation to the environment, which is at −15°C. The heat transfer coefficient is 7.7 W/m²·°C, and the emissivity is 0.94. The pipe wall remains at 0°C during the heating, and the heating tape is very thin. The pipe has an inside diameter of 3 cm and a wall thickness of 4 mm. How much time is required to completely melt the ice? (heat of fusion of ice = 3.34×10^5 J/kg, density of ice = 921 kg/m³)

CONDUCTION AND CONVECTION IN SPHERICAL COORDINATES

P3-33 Show that the conduction thermal resistance of a spherical shell of inner radius r_1 and outer radius r_2 is given by:

$$R_{sphere} = \frac{r_2 - r_1}{4\pi r_1 r_2 k}$$

P3-34 A hollow sphere made of pure aluminum has an inner radius of 3 cm and an outer radius of 18 cm. The temperature at the inner radius is maintained at 0°C. The outer surface is exposed to air at 25°C. The convective heat transfer coefficient is 65 W/m²·K, and radiation may be neglected. Calculate the rate of heat transfer and the temperature of the outer surface of the sphere.

P3-35 A bathosphere of inside diameter 3.4 m is at an ocean depth where the water temperature is 5°C. The wall of the bathosphere is made of 5-cm-thick steel. The convective heat transfer coefficient between the air and the inside wall is 9.2 W/m²·K and that between the water and the outside wall is 860 W/m²·K. After the divers return to the surface, they complain to the designer that the bathosphere was chilly. If the maximum power of the heater is 2.5 kW, estimate the air temperature inside the bathosphere.

P3-36 A high-pressure chemical reactor contains a gas mixture at 1000°F. The reactor is made of AISI 1010 carbon steel and is

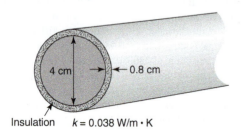

spherical, with an inner diameter of 3.2 ft and a wall thickness of 0.75 in. The outer wall of the reactor is encased in a 2.5 in. thick layer of insulation ($k = 0.03$ Btu/h·ft·R). The convective heat transfer coefficient on the inside wall of the reactor is 8.3 Btu/h·ft^2·°F, and on the outside of the insulation the combined convective/radiative heat transfer coefficient is 11.7 Btu/h·ft^2·°F. If the ambient temperature is 80°F, find the rate of heat transfer from the reactor to the surroundings.

P3-37 A novelty drink container is made of plastic in the shape of a sphere. The container has an outside diameter of 6.5 cm and a wall thickness of 2.5 mm. The container is initially filled with soda and crushed ice. The ice occupies 30% of the volume of the drink. The plastic has a conductivity of 0.07 W/m·K and an emissivity of 0.92. The inside surface of the container may be assumed to be at the freezing temperature of water. The heat transfer coefficient due to convection on the outside of the container is 9.4 W/m^2·K. Ambient temperature is 18°C. The latent heat of fusion of water is 333.7 kJ/kg, and the density of ice is 921 kg/m^3. Neglecting any transient effects, estimate the time until all the ice has just melted.

LUMPED SYSTEMS

P3-38 A small rod made of pure copper is 0.5 cm in diameter and 1.4 cm long. The rod is initially at 10°C. It is then exposed to a hot air flow at 30°C. The heat transfer coefficient between the rod and the air is 25 W/m^2·°C. What will the rod temperature be after 45 s?

P3-39 A slab of aluminum (2024-T6), which measures 16 cm by 16 cm by 1.5 cm is initially at 750 K. The slab is then annealed by a water spray at 15°C, which strikes both sides of the slab. The convective heat transfer coefficient is estimated to be 1500 W/m^2·K. How much time is required to cool the slab to 320 K? Neglect convection off the edges of the slab since almost all the surface area is on the two 16 cm by 16 cm sides.

P3-40 Buckshot initially at 450°F is quenched in an oil bath at 85°F. The buckshot is spherical with a diameter of 0.2 in. and is made of lead. The shot falls through the bath, reaching the bottom after 20 s. The convective heat transfer coefficient between buckshot and oil is 36 Btu/h·ft^2·°F. Calculate the temperature of the shot just as it reaches the bottom of the bath.

P3-41 A thermocouple is a temperature-measuring device that relies on quantum mechanical effects. Thermocouples are constructed of two thin wires of different metals welded together to form a spherical bead. Consider a thermocouple 0.1 mm in diameter that is suddenly immersed in ice water. Ideally, the thermocouple bead should immediately drop to 0°C, but in practice, there is a time delay. The heat transfer coefficient between the thermocouple and the ice water is 32 W/m^2·°C. The density, specific heat, and thermal conductivity of the bead are 8925 kg/m^3, 385 J/kg·K, and 23 W/m·K, respectively. Assuming no conduction in the thermocouple wires and an initial thermocouple temperature of 25°C, estimate the time required for the bead to reach 0.1°C.

P3-42 A long uninsulated Nichrome wire of diameter 1/16 in. is cooled convectively by air at 70°F. The heat transfer coefficient is 6.6 Btu/h·ft^2·°F. Current runs through the wire, generating heat at a rate of 1.9 W per foot.

a. Find the steady-state temperature of the wire.

b. Assume the wire is initially at 70°F. After the current is turned on, how long will it take for the wire temperature to rise to 90% of the difference between its initial temperature and its steady-state temperature?

P3-43 A copper sphere 3 cm in diameter is painted black so that it has an emissivity very close to 1. The sphere is heated to 700°C and then placed in a vacuum chamber whose walls are very cold. How long will it take for the sphere to cool to 300°C? Use the lumped system model.

COMBINED THERMAL RESISTANCES

P3-44 The roof of a house is partially covered with snow. The roof is made from plywood covered with shingles. ($k_{sh} = 0.4$ Btu/h·ft·°F). In the attic space, the heat transfer coefficient between the air and the plywood is 3.1 Btu/h·ft^2·°F. The heat transfer coefficient over the snow and the exposed shingles on the outside of the roof is 7.6 Btu/h·ft^2·°F. Assume the snow has a density of 12 lbm/ft^3 and it covers 64% of the roof area. Calculate the thermal resistance from the attic air to the outside air for the 30 ft by 60 ft roof panel shown in the figure.

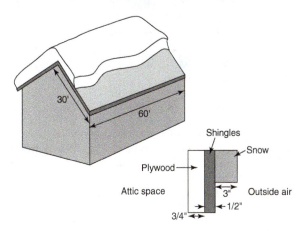

P3-45 A dining area has a glass ceiling built of square units. Each unit consists of two glass panes supported by a steel frame, as shown in the figure. The space between the panes contains a gas. The heat transfer coefficients, as shown on the figure, are

$$h_1 = 4.11 \text{ W/m}^2 \cdot °C \text{ (inside the room)}$$

$$h_2 = 3.63 \text{ W/m}^2 \cdot °C \text{ (between the panes)}$$

$$h_3 = 7.45 \text{ W/m}^2 \cdot °C \text{ (outside the room)}$$

The air in the room is at 26°C and the exterior air is at 15°C. The glass has a thermal conductivity of 1.4 W/m·K and the steel has a thermal conductivity of 37.7 W/m·K. Using dimensions on the figure, find the total heat loss through one unit.

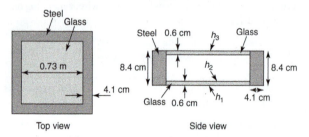

Top view   Side view

P3-46 A man is wearing a shirt and a jacket that is unzipped. His skin temperature is 70°F. The convective and radiative heat transfer coefficients on the outside of the jacket and exposed part of the shirt are estimated as 0.8 and 0.43 Btu/h·ft²·°F respectively. Model the man's torso as a cylinder of diameter 1.3 ft and height 2 ft. Assume the shirt is a layer of cloth of thickness 0.05 in. with $k_s = 0.12$ Btu/h·ft·R. The jacket is 0.4 in. thick with $k_j = 0.094$ Btu/h·ft·R. Assume the jacket covers half the man's torso. The surroundings are at 45°F. Calculate the total rate of heat loss from the man's torso. Neglect the thermal resistance due to any air layers between the shirt and the skin or the shirt and the jacket.

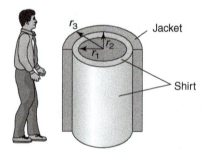

P3-47 The inside wall of a machine is covered with acoustic tile 3.5 cm thick for noise abatement. The tile increases the thermal resistance of the wall, and, as a result, the interior air temperature rises to unacceptable levels. An engineer suggests drilling holes in the tile and welding steel rods 3.5 cm long and 1.8 cm in diameter to the wall so as to increase its effective thermal conductivity, as shown in the figure. The rods are in a square array on 10-cm centers. The machine dissipates 150 W per square meter of wall area through its outer wall. The heat

transfer coefficients on the interior and exterior are 4.6 and 11.4 W/m²·K, respectively. If the exterior air temperature is 25°C, calculate the interior air temperature with and without the rods.

P3-48 In a certain localized area, the earth can be approximately represented by areas of stone, soil, and iron ore, as shown in the figure. Using data on the figure and assuming the geometry is two-dimensional, find the "effective" thermal conductivity in the vertical direction. This is the conductivity that the earth would have if it were all made of the same material.

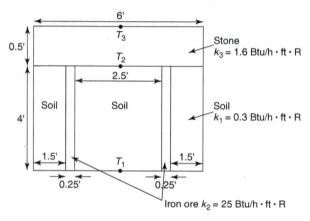

P3-49 A drinking glass with an outside diameter of 3.5 in. and a wall thickness of 0.125 in. is filled to the height of 6.2 in. with a mixture of soda and ice. The glass is placed in an insulated soft rubber sleeve 0.75 in. thick. (Cut-away view is shown in the figure.) The exposed top surface of the drink is at 32°F and gains heat by natural convection and radiation from the surroundings, which are at 85°F. On the top surface, the heat transfer coefficients for convection and radiation are 1.6 Btu/h·ft²·°F and 0.7 Btu/h·ft²·°F, respectively. The natural convective heat transfer coefficient between the soda–ice mixture and the inside wall of the glass is 57 Btu/h·ft²·°F. On the outside of the rubber sleeve, the heat transfer coefficients for convection and radiation are 2.3 Btu/h·ft²·°F and 0.85 Btu/h·ft²·°F, respectively. Assume no heat is transferred through the bottom of the glass. The initial mass of ice in the drink is 0.09 lbm. The latent heat of fusion of water is 143.5 Btu/lbm. Assuming a steady-state temperature profile in the glass wall and rubber, calculate the time required for the ice to completely melt. (Assume no one takes a sip from the glass.)

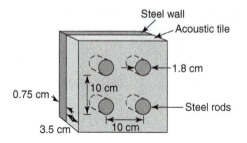

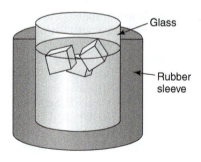

THERMODYNAMICS AND HEAT TRANSFER

P3-50 A reacting gas is contained in a cubical tank of side length 33 cm. The gas is stirred by a paddlewheel that rotates at 60 rpm under a torque of 37 J. The convective heat transfer coefficient on the interior wall of the tank is 62 W/m²·K, and, on the exterior, the combined convective/radiative heat transfer coefficient is 7 W/m²·K. The tank wall is 0.5 cm thick and is made of AISI 347 stainless steel. The bottom of the tank rests on a highly insulating surface. Due to chemical reaction, 180 W of heat are generated in the tank. The ambient temperature is 25°C. Find the steady-state temperature of the gases in the tank.

P3-51 A piston–cylinder assembly is filled with carbon dioxide gas at 250°C, 390 kPa. The piston and the curved walls of the cylinder are perfectly insulated. The bottom wall is maintained at 325°C by an external heater. Initially, the piston is 11 cm above the bottom of the cylinder, which has an inside diameter of 6 cm. As heat is transferred by convection to the CO_2 from the cylinder base, a control system lifts the piston so as to keep the average CO_2 temperature constant. If the piston rises 3 cm in 11 s, determine the convective heat transfer coefficient.

CHAPTER 4

FUNDAMENTALS OF FLUID MECHANICS

4.1 INTRODUCTION

In everyday language, the word *fluid* generally is used to mean a liquid. However, in engineering terminology, a **fluid** is a liquid or a gas. Of course, gases and liquids have significant differences. If a liquid, such as water, is poured into an open flask, it will form a layer of fluid with a distinct surface, called a free surface. Gases do not form free surfaces but rather expand to fill their containers. An "empty" glass sitting on a table is, in fact, filled with air. However, liquids and gases have similar behavior when there is no free surface present. For example, the same principles and equations can be used to analyze the flow of liquids and gases in a pipe—as long as the liquid completely fills the pipe.

The study of stationary fluids is called **fluid statics**. Water at rest in an aquarium tank exerts forces on the sides of the tank. Pressure in the atmosphere varies with height above sea level, becoming noticeably lower at high altitudes. Hydraulic brakes in a car provide mechanical advantage so that the car can be stopped with minimal force on the brake pedal. These different phenomena can all be understood using the principles of fluid statics. In addition, fluid statics deals with the buoyancy of floating objects, such as ships, buoys, swimmers, and so on.

Fluid dynamics involves moving fluids. The flow of air over a truck, the flow of oil in a pipeline, and the flow of water issuing from a fire hose are just a few of the many examples of fluid flow that can be analyzed using fluid dynamics. In this chapter, equations for conservation of mass, momentum, and energy of a moving fluid will be introduced. These equations, which are among the most basic ideas in fluid mechanics, apply to a very wide range of processes and phenomena.

4.2 FLUID STATICS

Imagine that you are standing next to a swimming pool. A pressure is exerted on you by the weight of the atmosphere above your head. You are not consciously aware of the pressure level. However, if you dive to the bottom of the pool, you would feel a significant increase in pressure on your eardrums. In both cases, the pressure in the static fluid depends on the depth of the fluid above you and on its density.

Many engineered systems must withstand pressure forces caused by a stationary fluid. As an example, consider the design of a submarine. The interior spaces in the submarine are kept at near-atmospheric pressure for the comfort of the sailors. Outside the submarine, the pressure of the water can be very large, especially at great depths. The hull must be designed to withstand the forces exerted on it by the pressure difference across the hull. Thus one of our first tasks is to develop an equation that will permit us to calculate the pressure at any depth in a fluid.

4.2.1 Pressure in a Fluid at Rest

How does pressure vary with depth and type of fluid? We will answer this question by considering how pressure varies with depth in the ocean. The result will be applicable to static fluids in general.

Figure 4-1 shows a section of the ocean. Define a control volume that encloses a portion of the ocean, as shown in Figure 4-1, and perform a force balance on the control volume. This is similar to performing a force balance on an object, except that our "object" is all the mass within the control volume.

We consider two types of forces: surface forces, which act only on the surface of the control volume, and body forces, which act on every particle throughout the control volume. There are seven forces that act on the control volume, as shown in Figure 4-1. Pressure (surface) forces act at each of the six faces of the box-shaped control volume, and a gravitational (body) force acts on the mass within the control volume. Note that the pressure forces are all normal to the control volume faces and all point inward.

The pressure forces act in the inward direction, because the water outside the control volume exerts a pressure force on the water inside the control volume. It is equally true that the water inside the control volume exerts a force on that outside, but we are not doing a force balance on the water outside the control volume and need not consider the outward forces.

For the force balance, first consider the forces in the x-direction in Figure 4-1. The water in the control volume is assumed to be at rest; therefore, the forces must be balanced. By Newton's second law of motion, the net force on the west face equals the net force on the east face, or $F_w = F_e$. By similar reasoning, a force balance in the y-direction shows that $F_n = F_s$.

In the vertical direction, there are three forces. The weight of the water due to gravity plus the force on the top of the control volume must balance the force on the bottom of the control volume. This force balance may be written

$$F_t + F_g = F_b \tag{4-1}$$

The top face of the control volume is at the ocean surface and is exposed to the atmosphere. The force there is equal to the area of the control volume face times atmospheric

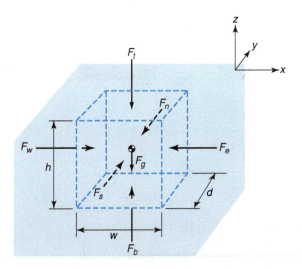

FIGURE 4-1 A control volume in the ocean.

pressure. (See the definition of pressure, Eq. 2-11.) The gravitational force is the mass times the acceleration of gravity. At depth h in Figure 4-1, the pressure will be designated as P, and F_b will equal this pressure times the area of the bottom face. With these simplifications, Eq. 4-1 becomes:

$$P_{atm} A_t + mg = PA_b \qquad (4\text{-}2)$$

At this point, we make the assumption that the ocean is a fluid of constant density (incompressible). The ocean may reasonably be assumed to be incompressible if the depth is not very large and the salinity does not vary significantly. If we write the areas in Eq. 4-2 in terms of lengths and replace the mass of water in the control volume by density times volume, Eq. 4-2 becomes

$$P_{atm}(wd) + \rho\,(wdh)\,g = P(wd)$$

which reduces to

$$\boxed{P_{atm} + \rho gh = P} \qquad (4\text{-}3)$$

Thus, for a constant density fluid (gas or liquid) at rest, the pressure is a function of the depth and density. Note that pressure is not a function of horizontal location. At any horizontal location in a stationary fluid, the pressure will be the same. The case of variable fluid density is treated in Section 4.2.2 below.

EXAMPLE 4-1 **Pressure in a stationary fluid**

A submarine dives to a depth of 4000 ft. What is the water pressure outside the hull? Assume atmospheric pressure is 14.7 psia.

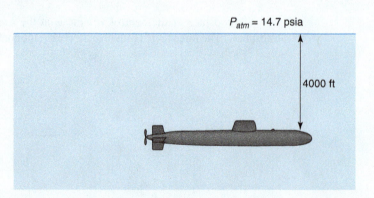

P_{atm} = 14.7 psia

4000 ft

Approach:

Apply Eq. 4-3.

Assumptions:

A1. Water density is constant with depth.

Solution:

The pressure is calculated using Eq. 4-3 [A1]

$$P = P_{atm} + \rho gh$$

Using the density of water from Table B-6, the pressure is

$$P = 14.7\frac{\text{lbf}}{\text{in.}^2} + \left(62.4\frac{\text{lbm}}{\text{ft}^3}\right)\left(32.174\frac{\text{ft}}{\text{s}^2}\right)(4000\text{ ft})\left(\frac{1\text{ lbf}}{32.174\frac{\text{lbm}\cdot\text{ft}}{\text{s}^2}}\right)\left(\frac{1\text{ ft}^2}{144\text{ in.}^2}\right)$$

$$P = 1748\frac{\text{lbf}}{\text{in.}^2}$$

Note that the relationship

$$1\text{ lbf} = 32.174\frac{\text{lbm}\cdot\text{ft}}{\text{s}^2}$$

has been used as if it were a conversion factor. This is an effective way to keep track of units when dealing with the British system.

The simple relationship between pressure and depth in a static fluid can be used as a way of measuring pressure differences. The **U-tube manometer**, shown in Figure 4-2, is a device that relies on this principle. The gas in the round container is at some pressure higher than atmospheric pressure. The bent U-tube of the manometer contains a liquid of known density. The top of the U-tube is open to the atmosphere.

The pressure difference between point A and point F is being measured. The pressure at point B is the same as that at point A because there is no vertical distance between the two and the fluids are at rest. The pressure at point C is related to that at B by

$$P_C = \rho_g g h_1 + P_B$$

where ρ_g is the density of the gas. The pressure at point D is related to the pressure at C by

$$P_D = \rho_\ell g h_2 + P_C = \rho_\ell g h_2 + \rho_g g h_1 + P_B$$

where ρ_ℓ is the density of the liquid. The pressure at point D can also be calculated starting with the pressure at point F and working downward. At F, the tube is open to the atmosphere

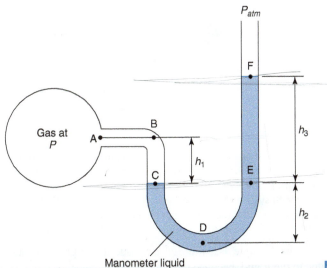

FIGURE 4-2   A manometer.

and, therefore, the pressure there is atmospheric pressure. The pressure at D can be written

$$P_D = P_F + \rho_\ell g h_3 + \rho_\ell g h_2 = P_{atm} + \rho_\ell g h_3 + \rho_\ell g h_2$$

Equating the expressions for P_D from these last two equations gives

$$\rho_\ell g h_2 + \rho_g g h_1 + P_B = P_{atm} + \rho_\ell g h_3 + \rho_\ell g h_2$$

which, with $P_A = P_B$, simplifies to

$$\boxed{P_A - P_{atm} = \rho_\ell g h_3 - \rho_g g h_1} \tag{4-4}$$

Note that the contribution from C to D, $\rho_\ell g h_2$, cancelled out with the contribution from E to D. In fact, the pressure at C is the same as that at E, since they are both the same vertical distance above D and are connected by a continuous column of the same fluid.

EXAMPLE 4-2 Manometer used to measure gas pressure

A manometer containing liquid mercury is used to measure the pressure of a mass of nitrogen gas. The density of the gas is known to be 1.6 kg/m³. If the heights h_1 and h_3 in Figure 4-2 are 1.5 cm and 4.32 cm, respectively, what is the pressure of the gas? Assume the manometer is in an environment where the temperature is 20°C and the atmospheric pressure is 101 kPa.

Approach:

Apply Eq. 4-4.

Assumptions:

A1. The density of the mercury is constant.

Solution:

Rearranging Eq. 4-4, the pressure of the nitrogen is [A1].

$$P_A = P_{atm} + \rho_\ell g h_3 - \rho_g g h_1$$

Using the density of mercury at 20°C from Table A-6, the pressure is

$$P_A = 101 \, \text{kPa} \left(\frac{1000 \, \text{Pa}}{1 \, \text{kPa}} \right) + \left(13,579 \, \frac{\text{kg}}{\text{m}^3} \right) \left(9.81 \, \frac{\text{m}}{\text{s}^2} \right) (4.32 \, \text{cm}) \left(\frac{1 \, \text{m}}{100 \, \text{cm}} \right)$$

$$- \left(1.6 \, \frac{\text{kg}}{\text{m}^3} \right) \left(9.81 \, \frac{\text{m}}{\text{s}^2} \right) (1.5 \, \text{cm}) \left(\frac{1 \, \text{m}}{100 \, \text{cm}} \right)$$

$$P_A = 101,000 \, \text{Pa} + 5,755 \, \text{Pa} - 0.24 \, \text{Pa}$$

$$P_A = 1.07 \times 10^5 \, \text{Pa} = 107 \, \text{kPa}$$

Comments:

In this calculation, each quantity in the equation was converted to the appropriate SI unit. Thus, kilopascals were converted to pascals and centimeters were converted to meters. The mass unit, kilograms, is already an SI unit, so no conversion is necessary. If this practice is followed when using the SI system, the units will always be compatible. Also note that, compared to the mercury, the gas height adds little to the pressure measurement.

In Example 4-2, the $\rho g h$ term due to the gas was much smaller than the $\rho g h$ term due to the liquid, because the density of the gas is so much smaller than that of the liquid.

In fact, this will nearly always be the case, since gases, in general, are so much less dense than liquids. As a result, Eq. 4-4 is usually approximated as

$$P_A - P_{atm} \cong \rho_\ell g h_3 \qquad (4\text{-}5)$$

When gas pressure is measured with a U-tube manometer, this approximation will be used unless otherwise noted. If a manometer is used to measure the pressure of a liquid, then Eq. 4-4 applies.

The left-hand side of Eq. 4-5 is the difference between measured pressure and atmospheric pressure. As defined in Chapter 2, this pressure is called gage pressure. In British units, the gage pressure is indicated by "psig" to distinguish it from the absolute pressure, which is called "psia."

Many types of pressure gauges, including the manometer described above, actually measure only the "gage pressure." In the manometer, the height of fluid is the measured quantity. The atmospheric pressure must be known from some other measuring device before the absolute pressure can be found. If "psi" is used, it should be clear from the context which is meant. In the SI system, pressure is typically measured in kPa. Both absolute pressures and gage pressures are expressed in kPa, and the engineer must recognize whether absolute or gage is meant.

Atmospheric pressure can be measured using a **barometer**, as shown in Figure 4-3. In a barometer, a closed tube is first filled with liquid mercury and then quickly inverted into an open container of mercury. The mercury in the tube falls under the influence of gravity. However, since the end is closed, no air can enter the tube. The space above the column of mercury contains only mercury vapor in equilibrium with the liquid mercury in the tube. The pressure of the mercury vapor is very low (much less than that of the atmosphere), so the column of mercury comes to equilibrium without emptying completely into the container.

The height of mercury in the tube is related to atmospheric pressure according to

$$P_{vapor} + \rho_\ell g h = P_{atm} \qquad (4\text{-}6)$$

where P_{vapor} is the pressure of the gaseous mercury above the liquid column and ρ_ℓ is the density of liquid mercury. P_{vapor} is the pressure at which liquid mercury will boil; that is, it is the saturation pressure of the mercury, also called vapor pressure. At a temperature of 20°C, the saturation pressure of mercury is 0.158 Pa. Atmospheric pressure is typically 101.3 kPa. Clearly P_{vapor} is much less than P_{atm}, and Eq. 4-6 can be approximated as

$$\rho_\ell g h = P_{atm}$$

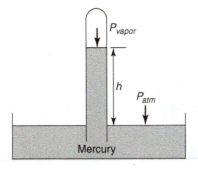

FIGURE 4-3 A simple barometer for measuring atmospheric pressure.

Thus the height of the mercury can be used to determine atmospheric pressure. The barometer was invented by Evangelista Torricelli (1608–1647). The **torr**, a unit of pressure equal to the pressure exerted by a column of mercury 1 mm in height, is named in his honor.

Manometers and barometers are simple mechanical devices for measuring pressure, but it is labor-intensive to take readings of pressure visually, especially if many measurements are needed or if measurement is continuous. They are also not suitable for making high-pressure measurements. In most modern applications, pressure transducers, which produce an electric signal in response to fluid pressure, are used. These devices can easily be incorporated into computer-controlled data acquisition systems. A wide variety are available to measure either absolute or gage pressure.

Often, in fluid statics, the density of a liquid is calculated using the **specific gravity**. By definition, specific gravity is the ratio of the density of a liquid to the density of water at 4°C:

$$SG = \frac{\rho}{\rho_{water}}$$

where SG is the specific gravity. The density of water varies slightly with temperature, reaching a maximum value of 1000 kg/m³ at 4°C, so this is chosen as the reference value. Specific gravity is a dimensionless quantity. For example, the specific gravity of kerosene is 0.817; therefore, in SI units, the density of kerosene is 817 kg/m³. In British units, the density of kerosene is 0.817 × 62.7 lbm/ft³ or 51.2 lbm/ft³.

Specific gravity applies to solids as well as to liquids. The specific gravity of a solid is the ratio of its density to that of water at 4°C. If the specific gravity of a solid is less than 1, then the solid will float in water; if specific gravity is greater than 1, the solid will sink. If the specific gravity of a liquid is less than 1 and the liquid is immiscible with water, then the liquid will form a layer on top of the water. If the specific gravity of an immiscible liquid is greater than 1, then the liquid will form droplets and sink to the bottom of the water, ending up in a layer under the water.

In some applications of fluid statics, more than one fluid is present. Consider, for example, the situation in Figure 4-4, where three fluid layers are stacked one upon the other in a tank. The fluids are assumed to be immiscible; that is, they do not mix together but stratify into distinct layers separated by fluid interfaces. Oil and water are immiscible fluids; if oil is added to a container of water, the oil will rest in a distinct layer above the water.

Pressure varies with depth in a multilayer system. If each of the three fluids has a constant density, the variation of pressure with depth is easily calculated. In Figure 4-4, the pressure at point 1 is

$$P_1 = P_{atm} + \rho_C g h_C$$

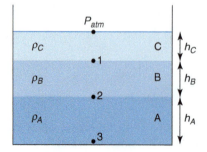

FIGURE 4-4   A multilayer fluid system.

where the subscript C denotes fluid C. The pressure at point 2 can be found once the pressure at point 1 is known:

$$P_2 = P_1 + \rho_B g h_B$$

Eliminating P_1 between the last two equations gives

$$P_2 = P_{atm} + \rho_C g h_C + \rho_B g h_B$$

The pressure at the bottom of two layers is the sum of the pressure changes across each layer. By extension, the pressure at the bottom of three layers, P_3, is

$$P_3 = P_{atm} + \rho_C g h_C + \rho_B g h_B + \rho_A g h_A$$

In terms of specific gravity, the pressure is

$$P_3 = P_{atm} + \rho_{H_2O} \left(SG_C g h_C + SG_B g h_B + SG_A g h_A \right)$$

This is a useful insight into the character of pressure in static fluid layers that can be used to simplify many analyses.

In a continuous static fluid, pressure is only a function of depth; pressure is not a function of horizontal position. When there are multiple fluids present, the pressure will be the same at two points of equal depth *if the points can be connected by an imaginary line that lies entirely within a single fluid*. For example, consider the two fluids in Figure 4-5. Points 3 and 4 can be connected by the dotted line as shown, and this dotted line in entirely within fluid A. Points 3 and 4 are at the same pressure. Points 1 and 2 are at the same horizontal location, but any line between them passes through both fluid A and fluid B. Hence, points 1 and 2 are not at the same pressure. In fact, point 1 is at atmospheric pressure and point 2 is at some pressure higher than atmospheric because of the mass of fluid B above point 2.

As another example, consider the three fluids in Figure 4-6. Points 2 and 3 both lie within fluid B and can be connected by a line that remains within fluid B; thus points 2 and 3 are at the same pressure. Point 1 is also in fluid B, and it is at the same depth as points 2 and 3; however, there is no line that joins points 1 and 2 without crossing into fluid A (or crossing into the atmosphere, which may be considered as fluid D). Points 1 and 2 are definitely not at the same pressure. A greater mass of fluid lies above point 2 than lies above point 1, so the pressure is greater at point 2.

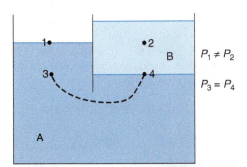

FIGURE 4-5 A divided tank with two fluid layers.

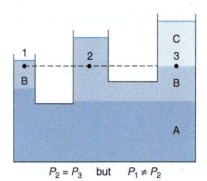

$P_2 = P_3$ but $P_1 \neq P_2$

FIGURE 4-6 Three fluids in a complex container.

EXAMPLE 4-3 Inclined manometer with multilayer fluid

A cylinder contains a layer of water covered by a layer of hydraulic fluid, as shown in the figure. The bottom of the cylinder is connected to an inclined tube that is partially filled with water.

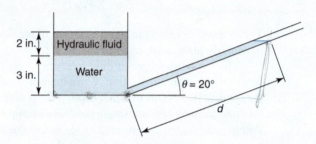

a) What is the gage pressure at the bottom of the tank?

b) What is the distance, d?

Approach:

The absolute pressure at the bottom of the tank can be found by using

$$P_{abs} = P_{atm} + \rho_{hf} g h_{hf} + \rho_w g h_w$$

where *hf* designates hydraulic fluid and *w* designates water. The gage pressure at the bottom of the tank is calculated by subtracting atmospheric pressure from this absolute pressure. The gage pressure at the bottom of the tank is equal to the gage pressure at the bottom of the inclined tube. Using this fact, the elevation of the water in the inclined tube can be determined. Trigonometry can then be used to find the distance d.

Assumptions:

Solution:

a) The pressure at the bottom of the tank will equal the pressure at the top plus the pressure due to each layer of fluid. At the top, the hydraulic fluid is exposed to the atmosphere. The absolute pressure at the tank bottom, P_{abs}, is then [A1][A2]

A1. The hydraulic fluid density is constant.
A2. The water density is constant.

$$P_{abs} = P_{atm} + \rho_{hf} g h_{hf} + \rho_w g h_w$$

and the gage pressure, P_g, would be

$$P_g = P_{abs} - P_{atm} = \rho_{hf} g h_{hf} + \rho_w g h_w$$

Values of the density of water and hydraulic fluid are given in Table B-6. Since the temperature is not specified, a reasonable assumption is to use a room temperature of 70°F [A3]. The value of density for the hydraulic fluid was approximated as the average of the values at 60° and 80°F. The gage pressure is calculated as

A3. Assume room temperature is 70°F.

$$P_g = \left(52.75 \frac{\text{lbm}}{\text{ft}^3}\right)\left(32.174 \frac{\text{ft}}{\text{s}^2}\right)(2 \text{ in.})\left(\frac{1 \text{ lbf}}{32.174 \frac{\text{lbm·ft}}{\text{s}^2}}\right)\left(\frac{1 \text{ ft}}{12 \text{ in.}}\right)\left(\frac{1 \text{ ft}^2}{144 \text{ in.}^2}\right)$$

$$+ (62.2)(32.174)(3)\left(\frac{1}{32.174}\right)\left(\frac{1}{12}\right)\left(\frac{1}{144}\right)$$

$$P_g = 0.169 \text{ psig}$$

b) In calculating the distance, d, in the inclined tube, it is important to remember that pressure in a static fluid is not a function of horizontal location but only a function of vertical location. The water fills the inclined tube to a vertical height of $d \sin \theta$. This is the equivalent height that should be used in Eq. 4-5. The upper surface of the water in the tube is exposed to the atmosphere.

The gage pressure at the bottom of the inclined tube is then

$$P_g = \rho_w g d \sin \theta$$

To find the distance, solve for d to get

$$d = \frac{P_g}{\rho_w g \sin \theta}$$

$$d = \frac{\left(0.169 \frac{\text{lbf}}{\text{in.}^2}\right)\left(\frac{144 \text{ in.}^2}{1 \text{ ft}^2}\right)\left(\frac{12 \text{ in.}}{1 \text{ ft}}\right)}{\left(62.2 \frac{\text{lbm}}{\text{ft}^3}\right)\left(32.174 \frac{\text{ft}}{\text{s}^2}\right) \sin(20)\left(\frac{1 \text{ lbf}}{32.174 \frac{\text{lbm·ft}}{\text{s}^2}}\right)}$$

$$d = 13.7 \text{ in.}$$

Comments:

Inclined manometers are often used when the pressure to be measured is small. A much larger deflection can be obtained than for a traditional U-tube manometer. The larger deflection results in a more accurate pressure reading.

EXAMPLE 4-4 **Pressure in a piston–cylinder assembly with manometer**

Water is contained in a piston–cylinder assembly, as shown in the figure. The piston is held in place by a spring with an unstretched length of 7 cm and a spring constant of 158 N/m. The piston is circular and has a diameter of 4 cm. The manometer fluid is mercury. Using data on the figure, find the mass of the piston.

Approach:

The pressure at the lowest point of the manometer, designated as P_A, can be found using

$$P_A = P_{atm} + \rho_m g h_1$$

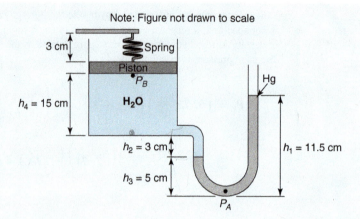

Note: Figure not drawn to scale

where ρ_m is the density of mercury. Once this pressure is known, one can deduce the pressure at the top of the water just under the piston face, P_B. Simply add the $\rho g h$ terms for each fluid layer, keeping in mind that pressure increases with depth. As a final step, draw a free-body diagram of forces on the piston. Balance the forces due to the atmosphere, the spring, and the weight of the piston against the upward pressure force of the water.

Assumptions:

A1. The mercury density is constant.

A2. The water density is constant.

Solution:

Let P_A be the pressure at the lowest point in the manometer. Then P_A may be written as [A1]

$$P_A = P_{atm} + \rho_m g h_1$$

where ρ_m is the density of mercury. Let P_B be the pressure in the water just underneath the piston. Then P_A is related to P_B by [A2]:

$$P_A = P_B + \rho_w g h_4 + \rho_w g h_2 + \rho_m g h_3$$

where ρ_w is the density of water. Eliminating P_A between these two equations gives

$$P_{atm} + \rho_m g h_1 = P_B + \rho_w g h_4 + \rho_w g h_2 + \rho_m g h_3$$

To find P_B, perform a force balance on the piston:

$$P_{atm} A_P + F_s + m_p g = P_B A_P$$

where F_s is the force of the spring, A_p is the area of the piston, and m_P is the mass of the piston. Eliminating P_B between these last two equations and combining terms gives

$$P_{atm} + \rho_m g (h_1 - h_3) = P_{atm} + \frac{F_s}{A_p} + \frac{m_p g}{A_p} + \rho_w g (h_4 + h_2)$$

Solving for the mass of the piston,

$$m_p = \rho_m (h_1 - h_3) A_p - \rho_w (h_4 + h_2) A_p - \frac{F_s}{g}$$

The spring force is given by $F_s = k \, \Delta x$, where k is the spring constant and Δx is the difference between the unstretched length and the compressed length of the spring. The area of the piston is $A_p = \pi(D_p/2)^2$. Using these relations and inserting values, the piston mass can be calculated as:

$$m_p = \left(13,579\,\frac{\text{kg}}{\text{m}^3}\right)(11.5-5)\,\text{cm}\left[\pi\left(\frac{4}{2}\right)^2\text{cm}^2\right]\left(\frac{1\,\text{m}^3}{10^6\,\text{cm}^3}\right)$$

$$-\left(998.2\,\frac{\text{kg}}{\text{m}^3}\right)(15+3)\,\text{cm}\left[\pi\left(\frac{4}{2}\right)^2\text{cm}^2\right]\left(\frac{1\,\text{m}^3}{10^6\,\text{cm}^3}\right)$$

$$-\frac{\left(158\,\frac{\text{N}}{\text{m}}\right)(7-3)\,\text{cm}\left(\frac{1\,\text{m}}{100\,\text{cm}}\right)}{9.81\,\frac{\text{m}}{\text{s}^2}}$$

$$m_p = 0.239\,\text{kg}$$

A3. Room temperature is 20°C.

The density values were taken at 20°C, which is close to room temperature [A3]. Since all quantities were converted into SI units—that is, cm to m—the units in the last term will automatically work out to be mass units, as they must to be consistent with the other terms in the equation.

The principles of fluid statics can be used to gain mechanical advantage. A hydraulic jack, as shown in Figure 4-7a , is used to lift heavy objects, such as cars, parts of buildings, packing crates, and so on. In each leg of the lift shown, the hydraulic fluid comes to the same level, h; therefore, the fluid pressure is the same at points 1 and 2. Since pressure is force per unit area,

$$\frac{F_1}{A_1} = \frac{F_2}{A_2}$$

Solving for F_1,

$$F_1 = \frac{A_1}{A_2}F_2$$

If A_1 is much smaller than A_2, then a small force F_1 can counterbalance a large force F_2. Typically, the large force is due to the weight of the object being lifted.

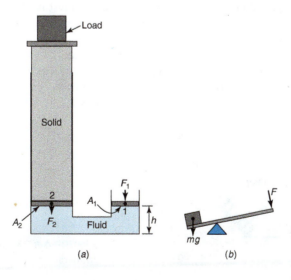

(a) (b)

FIGURE 4-7 (a) A hydraulic jack. (b) A lever–fulcrum system.

Mechanical advantage never comes without a price. If force F_1 in the small-diameter leg pushes the fluid downward a small amount, Δh_1, the level in the large tube will rise, lifting the load. However, the volume of the hydraulic fluid stays the same, so the rise in the large tube will be

$$V_1 = A_1\,\Delta h_1 = V_2 = A_2\,\Delta h_2$$

$$\Delta h_2 = \frac{A_1}{A_2}\Delta h_1$$

Since A_1 is much smaller than A_2, the load is lifted a small distance compared to the decrease in fluid level in the small tube. This is similar to what happens in a lever–fulcrum system, as shown in Figure 4-7b. The force F applied to the end of the lever is smaller than the weight of the object being lifted; however, the long arm of the lever must travel a long distance to raise the object a short distance.

In some applications, the transmission of force is all that is required. For example, in cars and trucks, the force applied by the driver to the brake pedal is magnified by the hydraulic brake system, so that a person of normal strength can stop a moving vehicle.

4.2.2   Pressure in a Static Compressible Fluid

(Go to www.wiley.com/college/kaminski)

4.2.3   Forces on Submerged Plane Surfaces

If a surface is immersed in a fluid, a force is exerted on the surface due to the pressure of the surrounding fluid. It is more difficult to remove the plug from a sink filled with water than it is to remove a plug from an empty sink. Pressure forces from the water wedge the plug into place and aid in forming a seal between the plug and the drain.

In cases where force on a submerged surface is important, the pressure is typically different on each side of the surface. For example, at the Philadelphia zoo, the polar bear exhibit includes a large pond in which the bears swim. Below the surface of the pond, a viewing area has been constructed with thick windows so visitors can watch the bears dive and navigate under water. The window is subject to hydrostatic pressure on the pond side and atmospheric pressure on the visitor side and must be designed to withstand the net force due to this pressure. Other examples of submerged surfaces where force is important occur in dams, canal locks, and submarines.

In Figure 4-8, an arbitrarily shaped flat plate is submerged in a liquid. An infinite plane that extends in the x- and y-directions and contains the plate makes an angle θ with the surface of the liquid, as shown. The x-direction is perpendicular to the y- and z-directions; it is not shown in the figure. The y-coordinate is the distance along the imaginary plane

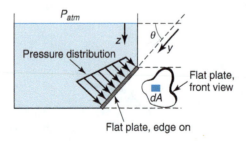

Flat plate, edge on

FIGURE 4-8   An arbitrarily shaped flat plate immersed in a static fluid.

from the liquid surface. The pressure in the liquid is a function of the depth, z, which is the coordinate perpendicular to the liquid surface. The pressure is

$$P(z) = P_{atm} + \rho g z$$

The pressure may be expressed as a function of y as

$$P(y) = P_{atm} + \rho g y \sin \theta \tag{4-7}$$

Remember that pressure is always normal to the submerged plate. The force on a differential area of the plate is the pressure at that location multiplied by the differential area. To find the total force on the plate, integrate over the plate area to obtain

$$F_R = \int_A P \, dA = \int_A \left[P_{atm} + \rho g y \sin \theta \right] dA$$

where F_R represents the resultant force on the plate due to liquid pressure. Integrating the first term and removing constants from the integral in the second term gives

$$F_R = P_{atm}A + \rho g \sin \theta \int_A y \, dA \tag{4-8}$$

The integral in this expression depends on the shape of the plate. It is possible to define an "average" y for the plate as

$$y_C A = \int_A y \, dA$$

or

$$y_C = \frac{1}{A} \int_A y \, dA \tag{4-9}$$

The reader may recognize y_C as the y-coordinate of the centroid of the area. The locations of centroids for many common shapes are listed in Table 4-1. Eliminating the integral between Eq. 4-8 and Eq. 4-9 results in

$$\boxed{F_R = P_{atm}A + \rho g \sin \theta \, y_C A} \tag{4-10}$$

This is the magnitude of the resultant force on one side of a submerged plane surface. In addition to the magnitude, it is often necessary to know the location at which the resultant force is applied. For example, in a canal lock, massive doors hold back upstream water that can reach a depth of 30 ft or more. The door hinges experience large forces and moments due to hydrostatic pressure. The moment can be calculated only if the point of application of the resultant force is known.

In Figure 4-9, point P is the location at which F_R acts. Point P always lies below point C, the centroid of the surface. The reason is simple. If a horizontal line is drawn through point C, as shown in Figure 4-9, then this line bisects the surface into two equal areas. This is one of the properties of centroids that can be demonstrated using Eq. 4-9. The area below the line is at a greater depth than that above the line, so there is more total force on the area below the line; thus the resultant force must lie in the lower half of the area.

To develop a mathematical expression for the location of P, we take moments about the centroid of the submerged area. For the resultant force to cause the same effect as the

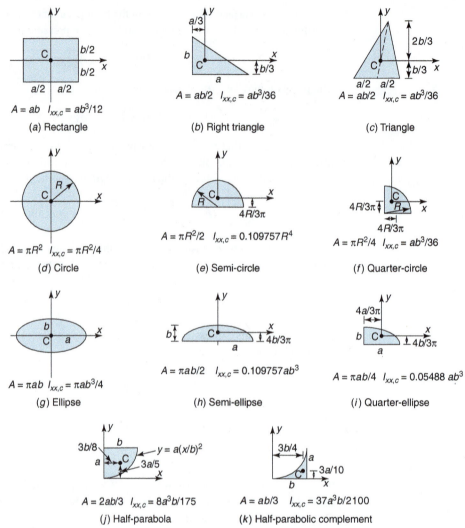

TABLE 4-1 Centroids and area moment of inertia for common surfaces.

distributed hydrostatic force, the moment produced by the resultant force must be the same as the moment produced by the distributed hydrostatic force. Mathematically,

$$(y_P - y_C) F_R = \int_A (y - y_C) P(y) \, dA \tag{4-11}$$

where y_P and y_C are shown in Figure 4-9. Note that when y lies below y_C, the pressure force tends to rotate the plate counterclockwise and the integrand on the right-hand side is positive in sign. When y lies above y_C, the pressure force tends to rotate the plate clockwise and the integrand on the right-hand side is negative. By taking the integral, the negative clockwise contributions are added to the positive counterclockwise contributions to give a net result of a counterclockwise rotation.

Substituting Eq. 4-7 into Eq. 4-11 yields

$$(y_P - y_C) F_R = \int_A (y - y_C) \left[P_{atm} + \rho g y \sin \theta \right] dA$$

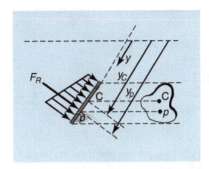

FIGURE 4-9 Point of application of the resultant force, F_R.

Substituting Eq. 4-10 into the left-hand side and expanding the right-hand side gives

$$(y_P - y_C)(P_{atm}A + \rho g y_C \sin \theta A) =$$
$$P_{atm} \int_A (y - y_C)\,dA + \rho g \sin \theta \int_A (y - y_C)y\,dA \tag{4-12}$$

The first integral on the right-hand side can be evaluated as

$$\int_A (y - y_C)\,dA = \int_A y\,dA - y_C A$$

But, using the definition of y_C given in Eq. 4-9 this reduces to

$$\int_A (y - y_C)\,dA = 0 \tag{4-13}$$

The second integral on the right-hand side of Eq. 4-12 is related to the area moment of inertia of the surface about the centroid, which is, by definition,

$$I_{xx,C} = \int_A (y - y_C)(y - y_C)\,dA$$

Expanding terms:

$$I_{xx,C} = \int_A (y - y_C)y\,dA - \int_A (y - y_C)y_C\,dA$$

The constant y_C can be removed from the second integral and, by application of Eq. 4-13, the integral is zero. Therefore,

$$I_{xx,C} = \int_A (y - y_C)y\,dA \tag{4-14}$$

Substituting Eq. 4-13 and Eq. 4-14 into Eq. 4-12 gives

$$(y_P - y_C)(P_{atm}A + \rho g y_C \sin \theta A) = \rho g \sin \theta\, I_{xx,C}$$

which can be rearranged as

$$\boxed{y_P = y_C + \frac{I_{xx,C}}{y_C A + \dfrac{P_{atm}A}{\rho g \sin \theta}}} \tag{4-15}$$

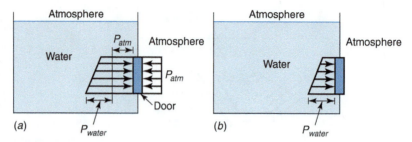

FIGURE 4-10 Pressure forces on a submerged door in a tank. (a) with forces due to atmospheric pressure and (b) without forces due to atmospheric pressure.

This equation was derived by taking into account the pressure forces on one side of a surface. In many circumstances, there are also pressure forces on the other side of the surface. Consider, for example, a tank containing water, as shown in Figure 4-10. In Figure 4-10a, the force on the left side of the door is the sum of that due to atmospheric pressure and that due to the water. The force on the right side is due solely to atmospheric pressure. This right-side pressure does not vary significantly with location on the door. The net force on the door is shown in Figure 4-10b, where atmospheric pressure has canceled out. In cases where the atmospheric pressure acts on both sides of a surface, one can set $P_{atm} = 0$ in Eq. 4-10 and Eq. 4-15, giving

$$F_R = \rho g \sin \theta \, y_C A \tag{4-16}$$

$$y_P = y_C + \frac{I_{xx,C}}{y_C A} \tag{4-17}$$

EXAMPLE 4-5 **Force and moments on a submerged flat surface**

A rectangular gate hinged at the bottom separates a tank into two compartments, one containing water and the other containing oil, as shown in Figure 4-11. The depth of the water is 6.5 ft and the height of the gate is 6 in. The oil has a specific gravity of 0.77. Initially the water and the oil are at the same depth, and the gate remains firmly closed, because the pressure due to the water is greater than that due to the oil. The oil depth is then increased, and the force on the right side of the gate becomes larger and more nearly counterbalances the force on the left. Find the height of the oil, h_2, at which the gate will just open and allow oil to bubble into the water.

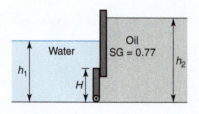

FIGURE 4-11 A divided tank containing oil and water.

Approach:

The key to the solution to this problem is to recognize that the gate will remain shut as long as the moment around the gate hinge due to the water is greater than the moment induced by the oil.

Thus we need to find the magnitude and location of the forces due to the water and the oil. The resultant forces on each side of the gate can be found from Eq. 4-16 with $\theta = 90°$. We do not need to include atmospheric pressure because both the water and the oil are exposed to the atmosphere at their upper surfaces. The gate is rectangular, so information on the location of the centroid of a rectangle, as listed in Table 4-1, is used to find the depths of the centroids on the water and oil sides. The points of application of the resultant forces for oil and water are different on each side of the gate; these are found from Eq. 4-17. The final step in the analysis is to take moments about the hinge of the gate. The moment equation contains the unknown oil height, which is the quantity requested in the problem statement.

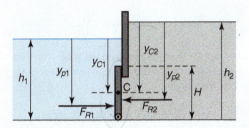

Figure 4-12   Location of resultant forces and gate centroid.

Assumptions:

Solution:

The first step is to calculate the magnitude of the resultant forces acting on the gate. We need the centroid of the area in this calculation. From Table 4-1, the centroid of a rectangle is located at its center. Therefore, the depth of the centroid on the water side of the gate is

$$y_{C1} = h_1 - \frac{H}{2} \tag{4-18}$$

Similarly on the oil side, the centroid is

$$y_{C2} = h_2 - \frac{H}{2} \tag{4-19}$$

Note that the location of the centroid on the oil side depends on the unknown depth h_2. We will keep everything in equation form for as long as is practical. This is a good practice in general, and it is especially useful in examples like this where the unknown quantity, h_2, will appear in several different equations.

A1. Water density is constant.

The resultant force on the water side of the gate is [A1]

$$F_{R1} = \rho_{H_2O} g \sin\theta \, y_{C1} A$$

In this case, $\theta = 90°$, and the area of the gate is width, W, times height, H. Therefore:

$$F_{R1} = \rho_{H_2O} g y_{C1} W H \tag{4-20}$$

A2. Oil density is constant.

By similar reasoning, the resultant force on the oil side is [A2]

$$F_{R2} = \rho_{oil} g y_{C2} W H \tag{4-21}$$

The resultant force on the water side acts at point y_{p1}, given by

$$y_{p1} = y_{C1} + \frac{I_{xx,C}}{y_{C1} A}$$

Substituting the expression for the area moment of inertia, $I_{xx,c}$, from Table 4-1 gives

$$y_{p1} = y_{C1} + \frac{WH^3}{12 y_{C1} A}$$

Since the area, A, is the width times the height, this becomes

$$y_{p1} = y_{C1} + \frac{WH^3}{12 y_{C1} WH} = y_{C1} + \frac{H^2}{12 y_{C1}} \tag{4-22}$$

The corresponding equation on the oil side is

$$y_{p2} = y_{C2} + \frac{H^2}{12 y_{C2}} \tag{4-23}$$

Finally, we take moments about the hinge on the gate. The gate will be neutrally balanced when the clockwise moment from the water pressure just balances the counterclockwise moment from the oil pressure. Referring to Figure 4-12:

$$(h_1 - y_{p1})F_{R1} = (h_2 - y_{p2})F_{R2} \tag{4-24}$$

Substituting Eq. 4-20 and Eq. 4-21 into Eq. 4-24 gives

$$(h_1 - y_{p1})(\rho_{H_2O} g y_{C1} WH) = (h_2 - y_{p2})(\rho_{oil} g y_{C2} WH)$$

Simplifying and noting that $\rho_{oil} = 0.77 \rho_{H_2O}$ produces

$$(h_1 - y_{p1}) y_{C1} = (h_2 - y_{p2})(0.77) y_{C2}$$

Substituting Eq. 4-22 and Eq. 4-23 into this expression results in

$$\left(h_1 - y_{C1} - \frac{H^2}{12 y_{C1}} \right) y_{C1} = \left(h_2 - y_{C2} - \frac{H^2}{12 y_{C2}} \right) (0.77) y_{C2}$$

The right-hand side may be simplified by substituting Eq. 4-19 into the first occurance of y_{C2} to give

$$\left(h_1 - y_{C1} - \frac{H^2}{12 y_{C1}} \right) y_{C1} = \left[h_2 - \left(h_2 - \frac{H}{2} \right) - \frac{H^2}{12 y_{C2}} \right] (0.77) y_{C2} \tag{4-25}$$

To simplify the left-hand side, first evaluate y_{C1} from Eq. 4-18 as

$$y_{C1} = h_1 - \frac{H}{2} = 6.5 - \frac{0.5}{2} = 6.25 \, \text{ft}$$

Substituting this value for y_{C1} into Eq. 4-25 and simplifying the right-hand side gives

$$\left[6.5 - 6.25 - \frac{(0.5)^2}{(12)(6.25)} \right] 6.25 = 0.77 \left[\frac{0.5}{2} y_{C2} - \frac{(0.5)^2}{12} \right]$$

Solving for y_{C2}:

$$y_{C2} = 8.09 \, \text{ft}$$

The desired oil depth, h_2, may now be found as (see Eq. 4-19)

$$h_2 = y_{C2} + \frac{H}{2} = 8.09 + \frac{0.5}{2} = 8.34 \, \text{ft}$$

Comments:

The depth on the water side is 6.5 ft. It makes sense that a greater depth of oil is needed to counterbalance the water. The oil is 77% as dense as the water, so an oil depth of 8.34 ft is an intuitively reasonable result.

4.2.4 Forces on Submerged Curved Surfaces

(Go to www.wiley.com/college/kaminski)

4.2.5 Buoyancy

Many humans love to swim. An aerial view of a typical American suburb reveals a landscape dotted with swimming pools. People are able to swim on the surface of the water because buoyant forces provide support. These forces act on all objects immersed in a fluid, including air, water, and all other gases and liquids. Buoyancy in water is critical in the design of submarines, surface vessels, buoys, offshore oil rigs, and so on. Hot-air balloons and helium-filled dirigibles depend on buoyant forces to float in the atmosphere.

The pressure in a static fluid increases with depth. An arbitrarily shaped three-dimensional object (Figure 4-13a) will experience pressure forces on all sides, but the pressure, and hence the forces, on the lower surfaces of the object will be greater than those on the upper surfaces, and the net resultant force, F_B, of all the fluid pressure forces will be upward.

It is not necessarily obvious that the net buoyant force, F_B, is vertical as shown in Figure 4-13a. In Figure 4-13b, a volume of fluid with the same shape and at the same depth as the object in Figure 4-13a is indicated with a dotted line. The buoyancy forces acting on this fluid volume are identical to those acting on the object in Figure 4-13a. The forces depend only on the pressure field outside the control volume, which is the same whether the control volume is filled with fluid or with an arbitrary object. Since the fluid in Figure 4-13b is stationary, the fluid in the control volume is in static equilibrium. Horizontal forces cancel, so there is no horizontal movement; the same is true in the vertical plane. One vertical force is the weight of the fluid in the control volume acting downward. The only other force is the buoyancy force. We conclude that the buoyancy force must be equal in magnitude and opposite in sign to the weight of the fluid in the control volume. In addition, there must be no moment on the stationary fluid in the control volume. The weight acts through the center of gravity of the object. The buoyant force F_B must, therefore, also act

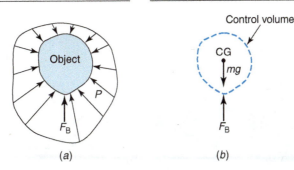

(a) (b)

FIGURE 4-13 Pressure forces on a stationary object of arbitrary shape immersed in a fluid.

through the center of gravity. For an object of uniform density, the center of gravity is the same as the centroid of the object. Summarizing:

The buoyant force on an immersed object is equal to the weight of the fluid displaced by the object. This force acts upward through the center of gravity of the object.

This principle was known to the ancient Greeks and is attributed to Archimedes (287–212 BC). In equation form, it is

$$\boxed{F_B = \rho_f V g} \tag{4-26}$$

where ρ_f is the density of the fluid in which the body is immersed and V is the immersed volume.

An object in water will experience a buoyant force that may be greater than, less than, or equal to the weight of the object. If the object is less dense than water, its weight is not sufficient to overcome the buoyant force, and it will rise. Conversely, if the object is more dense than water, it will sink. Wood rises in water and comes to rest floating on the surface, while iron sinks and comes to rest on the bottom. If the object has exactly the density of water, it will float, suspended, totally immersed, at least in theory. In practice, this is a position of unstable equilibrium, and the object would tend to drift either upward, downward, or laterally due to small currents or to slight differences between its density and that of water.

In many applications, we are interested in a body floating at the interface between two fluids, as shown in Figure 4-14. An example is a ship floating on the ocean, where the upper fluid is air and the lower fluid is water.

By an argument exactly like that given above, the buoyant force for the object in Figure 4-14 is

$$F_B = (\rho_1 V_1 + \rho_2 V_2) g$$

where ρ_1 and ρ_2 are the densities of the upper and lower fluids and V_1 and V_2 are the volumes that the body occupies in each of the two fluids. The floating body is held in static equilibrium by a balance between buoyancy forces and gravity forces. The force of gravity is

$$F = m_o g$$

Setting the buoyancy force equal to the gravity force yields

$$m_o g = (\rho_1 V_1 + \rho_2 V_2) g$$

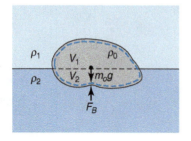

FIGURE 4-14   A body floating at the interface of two fluids.

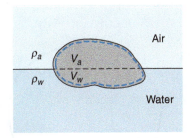

FIGURE 4-15 A body floating in water.

Rewriting this in terms of the total volume, V_{tot}, and the density ρ_o, of the object, and dividing by g gives

$$\rho_o V_{tot} = (\rho_1 V_1 + \rho_2 V_2)$$

where

$$V_{tot} = V_1 + V_2$$

A special case of great practical importance is the case of a body floating in water, as shown in Figure 4-15. For this situation, the buoyant force is

$$F_B = (\rho_a V_a + \rho_w V_w)g$$

where ρ_a is the density of air and ρ_w is the density of water. Since the density of air is so much smaller than that of water, we may write

$$F_B = \rho_w V_w g$$

The buoyant force is equal to the weight of the object, so

$$\rho_o V_{tot}\, g = \rho_w V_w g$$

where ρ_o is the density of the object, V_w is the volume submerged in the water, and V_{tot} is the total volume of the object. An alternate form is

$$\frac{\rho_o}{\rho_w} = \frac{V_w}{V_{tot}}$$

This equation can be used, for example, to calculate the submerged volume of an iceberg or a raft.

EXAMPLE 4-6 Buoyancy of a hot-air balloon

A hot-air balloon carrying a person of mass 165 lbm floats in the atmosphere at a constant altitude of 5000 ft. The balloon has a diameter of 44 ft. The mass of the balloon when uninflated, including the basket, the burner, and the fabric, is 350 lbm. The surrounding atmospheric air is at 47°F and 12.5 psia. The heat transfer coefficient between the balloon fabric and the air is 1.1 Btu/h· ft²·°F on both the inside and outside surfaces of the balloon. The fabric is 0.01 in. thick and has a thermal conductivity of 0.014 Btu/h·ft·°F. Calculate the average rate of heat generation in the burner so that the balloon maintains constant altitude.

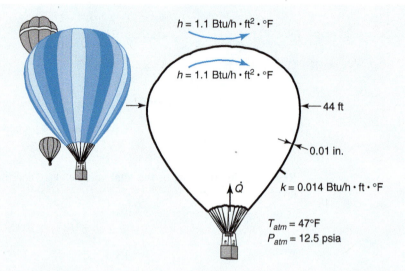

$h = 1.1\ \text{Btu/h} \cdot \text{ft}^2 \cdot °\text{F}$

$h = 1.1\ \text{Btu/h} \cdot \text{ft}^2 \cdot °\text{F}$

44 ft

0.01 in.

$\dot{Q}$

$k = 0.014\ \text{Btu/h} \cdot \text{ft} \cdot °\text{F}$

$T_{atm} = 47°\text{F}$
$P_{atm} = 12.5\ \text{psia}$

Approach:

First determine the temperature of the air inside the balloon so that it is neutrally buoyant (maintaining a constant altitude). The balloon will be neutrally buoyant when the buoyancy force on it exactly matches its weight. The weight of the air inside the balloon plus the weight of the balloon itself and the passengers must match the weight of the 47°F air that is displaced by the balloon. To find the inside air temperature from the calculated inside air mass, use the ideal gas law. Consider the balloon to be a sphere of radius 22 ft. Once the inside air temperature is known, the thermal resistances due to convection on the inside of the balloon, conduction through the fabric, and convection on the outside of the balloon are added in series to produce total thermal resistance. The required rate of heat generation is the temperature drop between inside and outside air divided by the total thermal resistance.

Assumptions:

A1. The balloon is spherical.

A2. Air behaves like an ideal gas under these conditions.

Solution:

The volume of the balloon is [A1]

$$V = \frac{4}{3}\pi R^3$$

or

$$V = \frac{4}{3}\pi (22\ \text{ft})^3 = 44,600\ \text{ft}^3$$

The mass of atmospheric air at 47°F and 12.5 psia displaced by this volume is [A2]

$$m_1 = \frac{PVM}{\overline{R}T_1}$$

Substituting values,

$$m_1 = \frac{(12.5\ \text{psia})(44,600\ \text{ft}^3)\left(28.97\ \dfrac{\text{lbm}}{\text{lbmol}}\right)}{\left(10.73\ \dfrac{\text{psia} \cdot \text{ft}^3}{\text{lbmol} \cdot \text{R}}\right)(47+460)\ \text{R}}$$

$$m_1 = 2969\ \text{lbm}$$

Define total mass as the sum of m_2, the mass of air in the balloon, m_3, the mass of the uninflated balloon, and m_4, the mass of the passenger:

$$m_{tot} = m_2 + m_3 + m_4$$

The balloon is neutrally buoyant when the buoyancy force exactly matches the weight. The buoyancy force equals the weight of the displaced air as given by

$$F_B = m_1 g$$

Setting buoyancy force equal to total weight

$$F_B = m_1 g = m_{tot} g = (m_2 + m_3 + m_4)g$$

Simplifying and solving for the unknown mass of air in the balloon, m_2,

$$m_2 = m_1 - m_3 - m_4$$
$$m_2 = 2969 - 350 - 165 = 2454 \, \text{lbm}$$

To find the temperature, use the ideal gas law. For the hot air in the balloon

$$T_2 = \frac{PVM}{\overline{R}m_2}$$

A3. Pressure is the same inside and outside of the balloon.

where we have assumed that the air inside the balloon is at the same pressure as the air outside the balloon [A3]. Substituting values,

$$T_2 = \frac{(12.5 \, \text{psia})(44,600 \, \text{ft}^3)\left(28.97\dfrac{\text{lbm}}{\text{lbmol}}\right)}{\left(10.73\dfrac{\text{psia}\cdot\text{ft}^3}{\text{lbmol}\cdot\text{R}}\right)(2454)\,\text{lbm}}$$

$$T_2 = 613\,\text{R} = 153°\text{F}$$

The heat loss from the balloon may be represented by the following resistance network:

where R_1 is the convection resistance on the inside of the balloon, R_2 is the conduction resistance through the fabric, and R_3 is the convection resistance on the outside of the balloon. The inside convection resistance is

$$R_1 = \frac{1}{hA} = \frac{1}{h4\pi R^2}$$

A4. The surface areas of the inside and the outside of the balloon are virtually the same.

where the formula for the surface area of a sphere has been used and R is the balloon radius. Because the fabric is thin, the inside and outside radii of the balloon are very nearly equal [A4]. With this simplification,

$$R_1 = \frac{1}{\left(1.1\dfrac{\text{Btu}}{\text{h}\cdot\text{ft}^2\cdot°\text{F}}\right)4\pi(22\,\text{ft})^2} = 0.0001495\,\frac{°\text{F}\cdot\text{h}}{\text{Btu}}$$

A5. Conduction through the fabric is planar.

Because the fabric is very thin compared to the radius of the balloon, the conduction through the fabric may be modeled as conduction through a plane layer [A5]. The conduction thermal resistance becomes

$$R_2 = \frac{L}{kA} = \frac{L}{k4\pi R^2} = \frac{0.01 \text{ in.}}{\left(0.014 \frac{\text{Btu}}{\text{h}\cdot\text{ft}\cdot{}^\circ\text{F}}\right)4\pi(22 \text{ ft})^2\left(\frac{12 \text{ in.}}{1 \text{ ft}}\right)} = 0.000117 \frac{{}^\circ\text{F}\cdot\text{h}}{\text{Btu}}$$

The convective resistance on the outside of the balloon is given by

$$R_3 = \frac{1}{hA} = \frac{1}{h4\pi R^2}$$

Since we have assumed A is the same on both inside and out, and since the heat transfer coefficient is also the same on both inside and out,

$$R_3 = R_1 = 0.0001495 \frac{{}^\circ\text{F}\cdot\text{h}}{\text{Btu}}$$

The total resistance is found by adding the three resistances in series to get

$$R_{tot} = R_1 + R_2 + R_3 = (0.0001495 + 0.000117 + 0.0001495)\frac{{}^\circ\text{F}\cdot\text{h}}{\text{Btu}}$$

$$R_{tot} = 0.000416 \frac{{}^\circ\text{F}\cdot\text{h}}{\text{Btu}}$$

The heat that must be added to the balloon by the burner is equal to the heat lost, so

$$\dot{Q} = \frac{T_2 - T_1}{R_{tot}} = \frac{(153 - 47){}^\circ\text{F}}{0.000416 \frac{{}^\circ\text{F}\cdot\text{h}}{\text{Btu}}} = 256,000 \frac{\text{Btu}}{\text{h}}$$

Comments:

The heat transfer coefficients used are typical of those for natural convection. If a strong wind flows over the balloon, the heat transfer coefficient would increase and more input heat would be required to maintain altitude.

EXAMPLE 4-7   Using buoyancy to determine density

Legend has it that Hiero II, king of the ancient Greek city of Syracuse, asked Archimedes in 220 BC to verify that his crown was made of pure gold. Archimedes reputedly discovered the principle of buoyancy in trying to solve this problem. Suppose that Archimedes had weighed the crown in both air and water and found it to weight 110 N in air and 103 N in water. What could Archimedes conclude? (Of course, Archimedes did not use the units of force called newtons, but the principle is the same.)

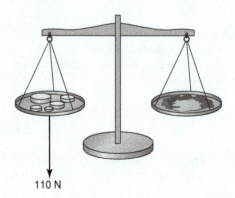

110 N

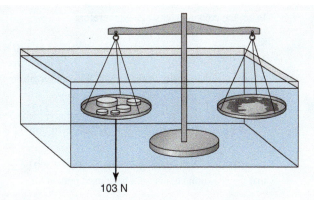

103 N

Approach:

The weight of the crown must be the same whether immersed in air or in water. The difference in the measured weights is due to buoyancy forces on the crown. Basic force balances on the crown in air and in water, including buoyancy forces, can be used to determine the volume of the crown. The weight in air can then be used to find the density of the crown. If the crown has a density less than that of gold, one can conclude that a base metal with a density lower than that of gold has been alloyed into the crown.

Assumptions:

Solution:

A force balance on the crown in air gives

$$F_{crown,air} = F_{weight} - F_{B,air}$$
$$110\,\text{N} = m_c g - \rho_a V_c g$$

where m_c is the mass of the crown, ρ_a is the density of air, and V_c is the volume of the crown. The second term on the right-hand side represents the buoyancy force of the air on the crown. This is typically very small, because the density of air is very low. Therefore, we use the common approximation [A1]

A1. Buoyancy force on the crown in air is very small.

$$110\,\text{N} = m_c g$$

Likewise, a force balance on the crown in water gives

$$F_{crown,water} = F_{weight} - F_{B,water}$$
$$103\,\text{N} = m_c g - \rho_w V_c g$$

Eliminating m_c between these two equations results in

$$103\,\text{N} = 110\,\text{N} - \rho_w V_c g$$

Solve for V_c, the volume of the crown, to get

$$V_c = \frac{(110 - 103)\text{N}}{\rho_w g}$$

Substituting values,

$$V_c = \frac{(110 - 103)\,\text{N}}{\left(997\,\dfrac{\text{kg}}{\text{m}^3}\right)\left(9.81\,\dfrac{\text{m}}{\text{s}^2}\right)} = 7.16 \times 10^{-4}\,\text{m}^3$$

In air, the crown weighs 110 N; therefore,

$$\rho_c V_c g = 110\,\text{N}$$

where ρ_c is the density of the crown. Solving for ρ_c,

$$\rho_c = \frac{110\,\text{N}}{V_c g} = \frac{110\,\text{N}}{(7.16 \times 10^{-4}\,\text{m}^3)\left(9.81\,\frac{\text{m}}{\text{s}^2}\right)} = 15{,}667\,\frac{\text{kg}}{\text{m}^3}$$

From Table A-2, the density of pure gold is 19,300 kg/m³. The crown has a lower density, implying that the crownmaker may not have used pure gold.

Comments:

One might think a little about the accuracy with which Archimedes was able to make measurements. Furthermore, a crown must have some structural strength, and pure gold is rather weak. The gold must have been alloyed in any event. The real question is how much alloying metal was used.

4.3 OPEN AND CLOSED SYSTEMS

Every system that has been considered so far has been a closed system. A **closed system** is one in which there is no mass inflow or outflow; that is, no mass crosses the system boundary. By contrast, if mass does flow across the system boundary, either entering or leaving, the system is called an **open system.** For example, consider a car tire. Define the system as the air within the tire so that the system boundary lies along the inside of the tire. When the tire is being inflated, mass (air) crosses the system boundary; therefore, this is an open system. Next consider a tire on a moving vehicle where the air within the tire is heated by frictional forces. As before, we define the system as the air within the tire, and the system boundary is coincident with the inside wall of the tire. In this case, however, no air is added or removed; therefore, we have a closed system.

In thermal–fluids analysis, it is important to carefully define the boundaries of the system under consideration. This is done by using a so-called control volume. Control volumes have been used already in this text. In this section, we use control volumes in new ways, that is, for open-system analysis. As previously stated, a control volume is a well-defined region in space that sets the boundaries of the system.

For example, if the effect of a fireplace on the room energy balance is being studied and we want to determine the air flow rate and temperature exiting the chimney, one might define the control volume to encompass all the air within the fireplace and chimney (see control volume A in Figure 4–16). For this control volume, air is drawn in from the room and exhausted through the chimney, thus crossing the boundary in two places. This is an open system. Alternatively, one might define the control volume to encompass the air within the room, the fireplace, and the chimney (see control volume B in Figure 4-16). Air enters this control volume through doors or windows and is exhausted up the chimney, crossing the boundary in several places. Control volume B is also an open system, but it is a different open system than control volume A. The details of the analysis of the effect of the fireplace on the room energy balance will change depending on the choice of control volume, but the final results (air flow rate and temperature leaving the chimney) will be the same.

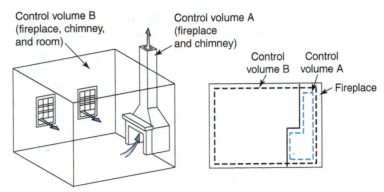

FIGURE 4-16 Two control volumes for analyzing the effect of a fireplace on a room energy balance.

Sometimes a control volume changes shape and/or size during a process. For example, suppose the filling of a washing machine with water is being analyzed. The control volume might be defined as the volume containing all the water within the tub. As the machine is filled, mass crosses the boundary, the volume of water increases, and the volume of the control volume increases. This is an open system. Now suppose that the fill cycle is finished and the machine advances to a soak cycle. The control volume is still defined as the volume containing all the water in the tub, but now no water enters or leaves. Thus the soak cycle would be a closed system.

Occasionally, the same process can be analyzed as either an open system or a closed system. For example, in Figure 4-17, two piston–cylinder assemblies are connected by a line. At the start of the process, cylinder 1 is filled with pressurized gas and cylinder 2 is evacuated. When the valve is opened, gas flows through the line and raises the piston in cylinder 2. Let control volume A be the volume containing all the gas in cylinder 1. Control volume A then defines an open system, since gas flows across its boundary. Alternatively, control volume B might be defined as all the volume in both cylinders. This is a closed system with a control volume that changes shape during the process. The choice of control volume can make the analysis easy or difficult, and it is not always obvious which choice leads to an easier analysis. Experience will help in making the choice.

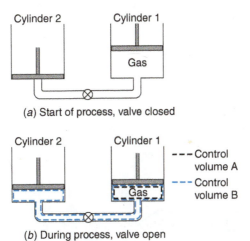

(a) Start of process, valve closed

(b) During process, valve open

FIGURE 4-17 Two alternate choices of control volume.

4.4 CONSERVATION OF MASS FOR AN OPEN SYSTEM

In an open system, we often need to keep track of how much mass is in the control volume at a given time. To do this, consider the control volume in Figure 4-18. Figure 4-18a shows the control volume at time t. Here Δm_i is a small mass that will enter the control volume during the time period Δt. Likewise, Δm_e is a small mass that will exit the control volume in the time period Δt (Δm_i is not necessarily equal to Δm_e). The mass in the control volume, m_{cv}, is all the mass within the dotted line. Figure 4-18b shows the control volume at time $t + \Delta t$. The mass at the inlet, Δm_i, has entered the control volume and the mass at the exit, Δm_e, has left the control volume. If we equate all the mass shown in Figure 4-18a with all the mass shown in Figure 4-18b, then

$$\Delta m_i + m_{cv}(t) = \Delta m_e + m_{cv}(t + \Delta t)$$

where $m_{cv}(t)$ is the mass in the control volume at time t, and $m_{cv}(t + \Delta t)$ is the mass in the control volume at time $t + \Delta t$. The quantity $m_{cv}(t)$ contains within it the differential mass Δm_e. Similarly the quantity $m_{cv}(t + \Delta t)$ includes the differential mass Δm_i. The equation may be rearranged to the form

$$m_{cv}(t + \Delta t) - m_{cv}(t) = \Delta m_i - \Delta m_e$$

Divide both sides by the time period Δt to get

$$\frac{m_{cv}(t + \Delta t) - m_{cv}(t)}{\Delta t} = \frac{\Delta m_i}{\Delta t} - \frac{\Delta m_e}{\Delta t}$$

Take the limit as Δt approaches zero so that the equation becomes

$$\frac{dm_{cv}}{dt} = \frac{dm_i}{dt} - \frac{dm_e}{dt} \tag{4-27}$$

The term dm_i/dt is called the inlet mass flow rate and is often abbreviated as

$$\frac{dm_i}{dt} = \dot{m}_i$$

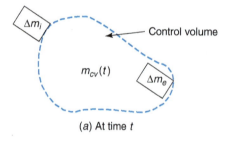

(a) At time t

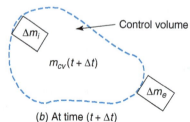

(b) At time $(t + \Delta t)$

FIGURE 4-18 Differential masses entering and exiting a control volume.

Similarly,

$$\frac{dm_e}{dt} = \dot{m}_e$$

With these substitutions, Eq. 4-27 takes the form

$$\frac{dm_{cv}}{dt} = \dot{m}_i - \dot{m}_e$$

Physically, this equation states that the time rate of change of mass within the control volume equals the mass flow rate into the control volume minus mass flow rate out of the control volume. The equation applies at an instant in time. Note that dm_{cv}/dt is *never* written as $\dot{m}_{cv}$. The different notation for $\dot{m}$ and dm_{cv}/dt is intended to signify that $\dot{m}$ is a flow rate and dm_{cv}/dt is a rate of accumulation or depletion of mass within the control volume.

For simplicity, the preceding derivation involved only one entering stream and one exiting stream. If there are multiple streams going in or out, the derivation is essentially the same except that all entering streams are added and all exiting streams are added. The resulting open system mass balance equation in rate form is

$$\frac{dm_{cv}}{dt} = \sum_{\text{in}} \dot{m}_i - \sum_{\text{out}} \dot{m}_e \qquad (4\text{-}28)$$

Open systems are often operated in steady state; that is, $dm_{cv}/dt = 0$. In this case, total mass neither increases nor decreases in the control volume, and the mass in the control volume is constant with time. If there is only one stream flowing in and one stream flowing out, then, in steady state, their flow rates are equal.

There are useful alternative expressions for the flow rates on the right-hand side of Eq. 4-28. Refer to Figure 4-19, which shows a small quantity of mass Δm, which will enter the control volume in time period Δt. This mass is contained in a differential volume that has a cross-sectional area A and a height Δx. The velocity with which the mass enters the control volume is also shown. This differential mass Δm is related to a differential volume ΔV by

$$\Delta m = \rho\,\Delta V \qquad (4\text{-}29)$$

where ρ is the density of the fluid entering the control volume. The differential volume is given by

$$\Delta V = \Delta x\,A$$

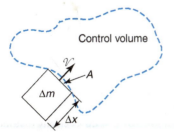

FIGURE 4-19   A differential mass entering a control volume.

The differential mass flows into the control volume with velocity of

$$\mathcal{V} = \frac{\Delta x}{\Delta t}$$

where the definition of velocity has been used, and the velocity is assumed uniform over the area A. Combining the last two equations gives

$$\Delta V = \mathcal{V} \Delta t \, A \qquad (4\text{-}30)$$

Substituting Eq. 4-30 in Eq. 4-29 produces

$$\Delta m = \rho \mathcal{V} \Delta t \, A$$

or

$$\frac{\Delta m}{\Delta t} = \rho \mathcal{V} A$$

In the limit as Δt approaches zero

$$\boxed{\dot{m} = \rho \mathcal{V} A} \qquad (4\text{-}31)$$

In this equation, velocity and density were assumed to be *uniform* over the area A. In the common situation of flow in a duct, velocity is not completely uniform but is different at each differential flow area, dA, as shown in Figure 4-20. (Likewise, the density could vary in the general case.) The molecules of fluid in a very thin layer next to the wall are at the speed of the wall, which is zero. The next layer of fluid is slowed by friction with the stagnant layer near the wall and accelerated by friction with a faster layer nearer the center of the duct. As a result, this layer takes on some velocity between that of the two bounding layers. Each successive layer of fluid is sandwiched between a slower layer and a faster layer and experiences a drag force from each. Near the center of the duct, there is often very little difference in velocity between layers, and the profile tends to flatten out. Thus, to obtain a total mass flow rate for the flow shown in Figure 4-20, we integrate over the total flow area to get

$$\dot{m} = \int_A \rho \mathcal{V} dA$$

It is useful to define an average velocity across the duct. The average velocity is defined so that the mass flow rate becomes

$$\boxed{\dot{m} = \rho \mathcal{V}_{avg} A} \qquad (4\text{-}32)$$

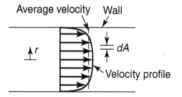

FIGURE 4-20 Velocity distribution in a duct.

Eliminating $\dot{m}$ between the last two equations (assuming density is constant) and rearranging produces

$$\mathcal{V}_{avg} = \frac{1}{A} \int \mathcal{V} dA \qquad (4\text{-}33)$$

The resulting average velocity is sketched in Figure 4-20.

It is also common to describe flow in terms of a volumetric flow rate. To derive the volumetric flow rate, start with Eq. 4-30, which is

$$\Delta V = \mathcal{V} \Delta t\, A$$

Dividing by Δt gives

$$\frac{\Delta V}{\Delta t} = \mathcal{V} A$$

In the limit as Δt approaches zero

$$\boxed{\frac{dV}{dt} = \dot{V} = \mathcal{V} A}$$

The quantity $\dot{V}$ is called the volumetric flow rate. If velocity varies over the area, A, then the volumetric flow rate may be written as

$$\boxed{\frac{dV}{dt} = \dot{V} = \mathcal{V}_{avg} A}$$

Combining this with Eq. 4-32 produces

$$\boxed{\dot{m} = \rho \dot{V}} \qquad (4\text{-}34)$$

EXAMPLE 4-8   Conservation of mass in an open system

A bathtub is being filled with water, but the drain plug does not fit properly and water leaks out. The volumetric flow rate of water down the drain is proportional to the pressure drop across the plug according to

$$\dot{V}_e = 90(P - P_{atm})$$

where flow rate is in gal/min and pressure is in psia. The flow rate from the faucet is 30 gal/min. Idealize the bathtub as a rectangular container 5 ft long, 2.5 ft wide, and 2 ft high.

a) What is the final height of water in the tub?

b) How long will it take to fill the tub to 90% of the final height?

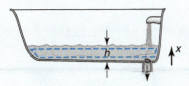

FIGURE 4-21 Filling of a bathtub with a leaky plug.

Approach:

The principle governing this problem is conservation of mass for an open system (Eq. 4-28). Defining the control volume as the water in the tub, we have an open system with one flow entering and one flow leaving. The final height of water occurs when the system is in steady state and the flow entering equals the flow exiting. To find the flow exiting, note that the pressure at the top of the drain is a function of the height, h, of liquid above the drain; that is, $P = P_{atm} + \rho gh$. Use this with the given equation for volumetric flow rate leaking by the plug to express flow rate as a function of height. Match the known entering flow rate with the exiting flow rate to determine the steady-state height.

For the second part of the problem, in which one must determine the time to fill to 90% of the final height, again start with conservation of mass (Eq. 4-28). The mass of the water is $m_{cv} = \rho V_{cv} = \rho A_{cv} x$, where x is the height of the water in the bathtub, as shown in Figure 4-21. The mass flow rate into the tub, $\dot{m}_i$, is a constant. The flow out of the tub, $\dot{m}_e$, is a function of x. Use $\dot{m}_e = \rho \dot{V}_e$ to write the rate as a volumetric flow rate and then substitute $\dot{V}_e = 90(P - P_{atm})$ and $P - P_{atm} = \rho gx$. Now you will have a differential equation in x. Separate variables and integrate to get an expression for height as a function of time. In this expression, set $x = 0.90h$ to determine the time required.

Assumptions:

Solution:

a) Define the control volume to contain all the water in the bathtub. The pressure at the bottom of the tub is given by Eq. 4-3, which is

$$P = P_{atm} + \rho gh$$

Using this, we see that the volumetric flow rate down the drain is

$$\dot{V}_e = 90(P - P_{atm}) = 90(\rho gh)$$

Conservation of mass gives

$$\frac{dm_{cv}}{dt} = \sum_{in} \dot{m}_i - \sum_{out} \dot{m}_e$$

The final height of the water will be reached when the mass in the control volume is no longer changing. At that point

$$\frac{dm_{cv}}{dt} = 0$$

so

$$\dot{m}_i = \dot{m}_e$$

where the summation, Σ, has been dropped since there is only one stream in and one stream out. Mass flow rate is related to the volumetric flow rate by

$$\dot{m} = \rho \dot{V}$$

A1. Water is incompressible.

The density of the water is constant [A1]; therefore,

$$\dot{V}_i = \dot{V}_e$$

According to the given equation for exiting volumetric flow rate, it follows that

$$\dot{V}_i = 90\rho gh$$

Solving for h,

$$h = \frac{\dot{V}_i}{90\rho g}$$

Before evaluating this equation, one must think carefully about units. From the problem statement, volumetric flow rate is in gal/min and pressure is in psia. Since $\dot{V}_e = 90(P - P_{atm})$, the units of the constant, 90, must be in gal/min·psia or gal·in.^{2}/min· lbf. Now h may be evaluated as

$$h = \frac{30\,\dfrac{\text{gal}}{\text{min}}}{\left(90\,\dfrac{\text{gal in.}^2}{\text{min·lbf}}\right)\left(62.2\,\dfrac{\text{lbm}}{\text{ft}^3}\right)\left(32.17\,\dfrac{\text{ft}}{\text{s}^2}\right)\left(\dfrac{1\,\text{lbf}}{32.17\,\dfrac{\text{lbm·ft}}{\text{s}^2}}\right)\left(\dfrac{1\,\text{ft}^2}{144\,\text{in.}^2}\right)}$$

$$h = 0.77\,\text{ft}$$

The density of water from Table B-6 has been used.

b) To determine how liquid height varies with time, start with conservation of mass in the form

$$\frac{dm_{cv}}{dt} = \dot{m}_i - \dot{m}_e$$

The control volume is all the water in the bathtub. The mass of the water is

$$m_{cv} = \rho V_{cv} = \rho A_{cv} x$$

where x, the height of the water in the bathtub, varies with time. Using this expression and

$$\dot{m} = \rho \dot{V}$$

conservation of mass becomes

$$\frac{d(\rho A_{cv} x)}{dt} = \rho(\dot{V}_i - \dot{V}_e)$$

or

$$A_{cv}\frac{dx}{dt} = \dot{V}_i - \dot{V}_e$$

Using the given expression for $\dot{V}_e$,

$$A_{cv}\frac{dx}{dt} = \dot{V}_i - 90(P - P_{atm})$$

Pressure is a function of the height of the water according to

$$P = P_{atm} + \rho g x$$

Therefore,

$$A_{cv}\frac{dx}{dt} = \dot{V}_i - 90(\rho g x)$$

Separating variables and integrating

$$\int_{x_1}^{x_2} \frac{A_{cv}\,dx}{\dot{V}_i - 90\rho g x} = \int_{t_1}^{t_2} dt$$

which becomes

$$\frac{-A_{cv}}{90\rho g} \ln \left[\dot{V}_i - 90\rho gx \right]_{x_1}^{x_2} = t_2 - t_1$$

Initially, at time $t_1 = 0$, the bathtub is empty, so $x_1 = 0$. The expression can be rearranged so that the time required to reach a water depth, x_2, is written as

$$t_2 = \frac{-A_{cv}}{90\rho g} \ln \left[\frac{\dot{V}_i - 90\rho gx_2}{\dot{V}_i} \right]$$

At the final time $(t = t_2)$, the bathtub is filled to 90% of the final height. Therefore

$$x_2 = 0.9h = (0.9)(0.77)\,\text{ft}$$

$$x_2 = 0.693\,\text{ft}$$

Substituting values

$$t_2 = \frac{-(5)(2.5)\,\text{ft}^2 \left(\dfrac{1\,\text{gal}}{0.1337\,\text{ft}^3} \right)}{90\,\dfrac{(\text{gal})(\text{in.}^2)}{(\text{min})(\text{lbf})} \left(62.2\,\dfrac{\text{lbm}}{\text{ft}^3} \right) \left(32.17\,\dfrac{\text{ft}}{\text{s}^2} \right) \left(\dfrac{1\,\text{lbf}}{32.17\,\dfrac{\text{lbm}\cdot\text{ft}}{\text{s}^2}} \right) \left(\dfrac{1\,\text{ft}^2}{144\,\text{in.}^2} \right)}$$

$$\times \ln \left(\frac{30\,\dfrac{\text{gal}}{\text{min}} - 90\,\dfrac{(\text{gal})(\text{in.}^2)}{(\text{min})(\text{lbf})} \left(62.2\,\dfrac{\text{lbm}}{\text{ft}^3} \right) \left(32.17\,\dfrac{\text{ft}}{\text{s}^2} \right) (0.693\,\text{ft}) \left(\dfrac{1\,\text{lbf}}{32.17\,\dfrac{\text{lbm}\cdot\text{ft}}{\text{s}^2}} \right) \left(\dfrac{1\,\text{ft}^2}{144\,\text{in.}^2} \right)}{30\,\dfrac{\text{gal}}{\text{min}}} \right)$$

$$t_2 = 5.5\ \text{min}$$

Comments:

After studying thermal–fluids engineering, students often begin to appreciate the merits of the metric system of units.

EXAMPLE 4-9   Conservation of mass in nozzle flow

Air enters a nozzle at 90°C, 180 kPa with an average velocity of 60 m/s. The duct has an inlet diameter of 10 cm. The exit has a diameter of 6.3 cm and is open to the atmosphere. If the velocity at the exit is 249 m/s, what is the temperature there?

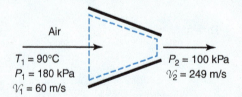

Air

$T_1 = 90°C$
$P_1 = 180\ \text{kPa}$
$\mathcal{V}_1 = 60\ \text{m/s}$

$P_2 = 100\ \text{kPa}$
$\mathcal{V}_2 = 249\ \text{m/s}$

Approach:

The control volume is chosen to follow the inside surface of the nozzle and cut across the entrance and exit. Assume the flow is steady; therefore, the inlet mass flow rate equals the outlet mass flow rate. For each mass flow rate, substitute an expression of the form $\dot{m} = \rho \mathcal{V} A$. Use the ideal gas law to eliminate density in favor of temperature and pressure. Then it will be possible to solve for the outlet temperature.

Assumptions:

A1. The nozzle operates in steady state.

Solution:

Define the control volume to follow the inside surface of the nozzle and cut across the entrance and exit. This is a steady flow with one inlet and one outlet [A1]. By conservation of mass,

$$\frac{dm_{cv}}{dt} = \dot{m}_i - \dot{m}_e$$

Because the flow is steady, the mass in the control volume does not change with time and

$$\frac{dm_{cv}}{dt} = 0$$

Therefore

$$\dot{m}_i = \dot{m}_e$$

This may also be written as

$$\rho_i \mathcal{V}_i A_i = \rho_e \mathcal{V}_e A_e$$

A2. Air is an ideal gas under these conditions.

Assuming air is an ideal gas [A2]

$$P_i v_i = \frac{\overline{R} T_i}{M} \qquad \text{and} \qquad P_e v_e = \frac{\overline{R} T_e}{M}$$

or, since

$$v = \frac{1}{\rho},$$

$$P_i = \frac{\rho_i \overline{R} T_i}{M} \qquad \text{and} \qquad P_e = \frac{\rho_e \overline{R} T_e}{M}$$

Solving for ρ and substituting gives

$$\frac{P_i M \mathcal{V}_i A_i}{\overline{R} T_i} = \frac{P_e M \mathcal{V}_e A_e}{\overline{R} T_e}$$

which simplifies to

$$\frac{P_i \mathcal{V}_i A_i}{T_i} = \frac{P_e \mathcal{V}_e A_e}{T_e}$$

Solving for the exit temperature,

$$T_e = \frac{P_e \mathcal{V}_e A_e}{P_i \mathcal{V}_i A_i} T_i$$

At the exit, the nozzle is open to the atmosphere; therefore, pressure is atmospheric. Substituting values,

$$T_e = \frac{(100\,\text{kPa}) \left(249\,\frac{\text{m}}{\text{s}}\right) \pi \left(\frac{6.3\,\text{cm}}{2}\right)^2}{(180\,\text{kPa}) \left(60\,\frac{\text{m}}{\text{s}}\right) \pi \left(\frac{10\,\text{cm}}{2}\right)^2} (90 + 273)\,\text{K} = 332\,\text{K} = 59.2°\text{C}$$

Note that absolute temperatures were necessary in this example because equations were derived using the ideal gas law.

4.5 CONSERVATION OF ENERGY FOR AN OPEN SYSTEM

The first law of thermodynamics is an expression of the principle of conservation of energy. In Chapter 2, the first law was derived for a closed system, that is, for a fixed quantity of mass with no mass crossing through the system boundary. In this section, the first law for an open system is developed.

Consider the control volume shown in Figure 4-22a. Here Δm_i is a differential quantity of mass that enters the control volume over the time period Δt. Similarly, Δm_e is a differential quantity of mass that exits the control volume over Δt; Δm_e is not necessarily equal to Δm_i. The differential quantities δQ_{cv} and δW_{cv} are the heat and work that cross the boundary during Δt. Figure 4-22b shows the control volume at time $t + \Delta t$. The mass Δm_i has entered, and the mass Δm_e has left. In the derivation that follows, several definitions are used:

$E_{cv}(t)$ = energy in the control volume at time t

$E_{cv}(t + \Delta t)$ = energy in the control volume at time $t + \Delta t$

$E(t)$ = energy of the closed system at time t

$E(t + \Delta t)$ = energy of the closed system at time $t + \Delta t$

The closed system is all the mass shown in Figure 4-22, while the control volume is all the mass inside the dotted line. Therefore, referring to Figure 4-22a, the energy of the closed system at time t is the energy within the control volume plus the energy contained in the mass Δm_i. The mass Δm_i carries energy into the control volume as it enters. This energy consists of internal energy, kinetic energy, and potential energy; the total energy in the closed system can be expressed in mathematical form as

$$E(t) = E_{cv}(t) + \Delta m_i \left(u_i + \frac{\mathcal{V}_i^2}{2} + gz_i \right)$$

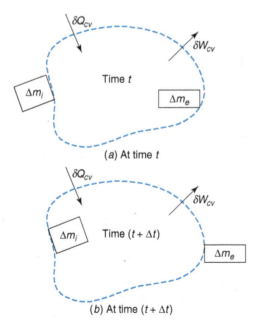

(a) At time t

(b) At time $(t + \Delta t)$

FIGURE 4-22 Control volumes for derivation of the first law.

Similarly, at time $t + \Delta t$, the total energy of the closed system is the energy within the control volume plus the energy contained in the mass Δm_e:

$$E(t + \Delta t) = E_{cv}(t + \Delta t) + \Delta m_e \left(u_e + \frac{\mathcal{V}_e^2}{2} + gz_e \right)$$

Subtracting the last two equations, we see that the change of energy of the closed system in time Δt is

$$E(t + \Delta t) - E(t) = \left[E_{cv}(t + \Delta t) + \Delta m_e \left(u_e + \frac{\mathcal{V}_e^2}{2} + gz_e \right) \right]$$

$$- \left[E_{cv}(t) + \Delta m_i \left(u_i + \frac{\mathcal{V}_i^2}{2} + gz_i \right) \right] \qquad (4\text{-}35)$$

Recall that the first law for a closed system is, from Eq. 2-1,

$$\Delta E = Q - W$$

In differential form, this is

$$dE = \delta Q - \delta W$$

which may also be written

$$dE = E(t + \Delta t) - E(t) = \delta Q - \delta W \qquad (4\text{-}36)$$

Substituting Eq. 4-35 into Eq. 4-36 gives

$$E_{cv}(t + \Delta t) - E_{cv}(t) + \Delta m_e \left(u_e + \frac{\mathcal{V}_e^2}{2} + gz_e \right)$$

$$- \Delta m_i \left(u_i + \frac{\mathcal{V}_i^2}{2} + gz_i \right) = \delta Q - \delta W \qquad (4\text{-}37)$$

The evaluation of the last term, δW, is a little tricky. This term includes expansion or contraction work and shaft work. In addition, work is required to push mass into and out of the control volume, and this form of work is called **flow work**. Using the definition of work as a force through a distance, and noting that the differential masses Δm_i and Δm_e travel a distance Δx as they enter or leave the control volume, we see that flow work is

$$\delta W_{flow} = F \, \Delta x$$

The force acting on the differential mass is due to pressure within the fluid, so

$$\delta W_{flow} = PA \, \Delta x$$

Area times the distance Δx is the volume of the differential mass; therefore,

$$\delta W_{flow} = P \, \Delta V = Pv \, \Delta m$$

As the mass Δm_i enters the control volume, work is done on the closed system. There must be some exterior force pushing the mass into the control volume. Since the mass is part of the closed system, work is being done on that system. By our sign convention, work done on the system is negative. As the mass Δm_e leaves the control volume, work is done by the closed system. Work done by the system is positive. With these considerations, flow work may be written

$$\delta W_{flow} = P_e v_e \, \Delta m_e - P_i v_i \, \Delta m_i \tag{4-38}$$

The total work for the system is the sum of the flow work and any other work on the control volume. In other words,

$$\delta W = \delta W_{cv} + \delta W_{flow} \tag{4-39}$$

The term δW_{cv} includes compression and expansion work, shaft work, and any form of work other than flow work. Eq. 4-37, Eq. 4-38, and Eq. 4-39 may be combined to give

$$E_{cv}(t + \Delta t) - E_{cv}(t) + \Delta m_e \left(u_e + \frac{V_e^2}{2} + gz_e \right) - \Delta m_i \left(u_i + \frac{V_i^2}{2} + gz_i \right)$$
$$= \delta Q_{cv} - \delta W_{cv} - P_e v_e \, \Delta m_e + P_i v_i \, \Delta m_i \tag{4-40}$$

Combining terms,

$$E_{cv}(t + \Delta t) - E_{cv}(t) + \Delta m_e \left(u_e + P_e v_e + \frac{V_e^2}{2} + gz_e \right) - \Delta m_i \left(u_i + P_i v_i + \frac{V_i^2}{2} + gz_i \right)$$
$$= \delta Q_{cv} - \delta W_{cv} \tag{4-41}$$

In this equation, δQ has been set equal to δQ_{cv}. There is no heat associated with the entering and leaving masses, so the heat for the closed system is the same as the heat for the control volume. In Eq. 4-41, the combination $u + Pv$ appears. This is enthalpy, which was introduced in Chapter 2 as a convenient definition. Now we see that enthalpy is a useful quantity when analyzing open systems. Using the definition $h = u + Pv$, Eq. 4-41 becomes

$$E_{cv}(t + \Delta t) - E_{cv}(t) + \Delta m_e \left(h_e + \frac{V_e^2}{2} + gz_e \right) - \Delta m_i \left(h_i + \frac{V_i^2}{2} + gz_i \right)$$
$$= \delta Q_{cv} - \delta W_{cv}$$

Dividing by Δt and letting Δt approach zero,

$$\boxed{\frac{dE_{cv}}{dt} = \dot{Q}_{cv} - \dot{W}_{cv} + \dot{m}_i \left(h_i + \frac{V_i^2}{2} + gz_i \right) - \dot{m}_e \left(h_e + \frac{V_e^2}{2} + gz_e \right)}$$

This equation has been derived for one inlet and one outlet. If there are multiple streams entering or leaving, the contribution of each stream is added. The resulting form of the first law is

$$\boxed{\frac{dE_{cv}}{dt} = \dot{Q}_{cv} - \dot{W}_{cv} + \sum_{in} \dot{m}_i \left(h_i + \frac{V_i^2}{2} + gz_i \right) - \sum_{out} \dot{m}_e \left(h_e + \frac{V_e^2}{2} + gz_e \right)} \tag{4-42}$$

This mathematical representation of the first law of thermodynamics, which is also called the energy equation or energy balance equation, is applicable at an instant in time and deals with rates of energy flow. It is one of the most widely used equations in thermal–fluids engineering. Physically, it states that the time rate of change of total energy within the control volume is equal to the difference between the net heat transfer to the control volume and the net power produced by the control volume plus the difference in energy flowing into and out of the control volume. Example 4-10 illustrates the use of Eq. 4-42 in one type of pipe flow. Chapter 6 is devoted entirely to applications of the energy equation in a wide range of devices and processes.

EXAMPLE 4-10 **Conservation of energy in a heated pipe**

Air flows in a pipe at a rate of 0.0064 m³/s. The air enters at 25°C and 101 kPa. A heating tape is wound around the outside of the pipe, and the tape is covered with a thick layer of thermal insulation. A voltage of 120 V is supplied to the tape, which has a resistance of 30 Ω. Assuming constant specific heat, find the exit temperature of the air.

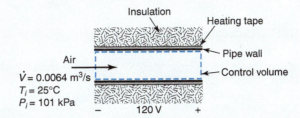

Approach:

This problem deals with heat addition in an open system. Thus we need the first law (Eq. 4-42) to determine the exit air temperature. All terms in the equation are zero except the heat input and the enthalpy terms. We can use the enthalpy of an ideal gas with constant specific heat to eliminate enthalpy from the first law in favor of temperature. The mass flow rates are related to the volumetric flow rate by

$$\dot{m} = \rho \dot{V}$$

Using Ohm's law, we calculate the heat input as voltage squared divided by current. Finally, properties are located in Table A-7 and the exit temperature is calculated.

Assumptions:

A1. Flow is steady.
A2. The pipe is perfectly insulated from the surroundings.
A3. Kinetic energy is negligible.
A4. Potential energy is negligible.

Solution:

Define the control volume to encompass the inside of the pipe and cut across the ends. The flow is steady [A1], so $dE_{cv}/dt = 0$. The electrical power supplied to the heating tape flows into the air as heat [A2]. There is no work done on or by the air. Velocity is assumed to be low, so kinetic energy effects are negligible [A3]. (Kinetic energy of flowing streams is treated in greater detail in Chapter 6.) There is no elevation change and, thus, no potential energy change [A4]. With these simplifications, the energy equation reduces to

$$0 = \dot{Q}_{cv} + \sum_{in} \dot{m}_i h_i - \sum_{out} \dot{m}_e h_e$$

The pipe has one inlet and one exit, so

$$0 = \dot{Q}_{cv} + \dot{m}_i h_i - \dot{m}_e h_e$$

At steady state, from conservation of mass, the mass flow rate entering equals the mass flow rate leaving, that is,

$$\dot{m}_i = \dot{m}_e = \dot{m}$$

Therefore,

$$0 = \dot{Q}_{cv} + \dot{m}(h_i - h_e)$$

A5. Air is an ideal gas under these conditions.
A6. Specific heat is constant.

Air may be assumed to be an ideal gas [A5], and if we further assume constant specific heat [A6], Eq. 2-37 gives

$$\Delta H = m c_p \, \Delta T$$

In our case, this becomes

$$H_i - H_e = m c_p (T_i - T_e)$$

Dividing by m and using $h = H/m$,

$$h_i - h_e = c_p (T_i - T_e)$$

Substituting this into the first law gives

$$0 = \dot{Q}_{cv} + \dot{m} c_p (T_i - T_e)$$

From Eq. 4-34, the mass flow rate may be expressed as

$$\dot{m} = \rho \dot{V}$$

which leads to

$$0 = \dot{Q}_{cv} + \rho \dot{V} c_p (T_i - T_e)$$

Rearranging and solving for exit temperature gives

$$T_e = \frac{\dot{Q}_{cv}}{\rho \dot{V} c_p} + T_i$$

Heat is generated in the heating tape at a rate of

$$\dot{Q}_{cv} = \xi I = \frac{\xi^2}{R}$$

where ξ is voltage, I is current, and R is electric resistance (by Ohm's law, $\xi = IR$). Substituting given values,

$$\dot{Q}_{cv} = \frac{(120\ \text{V})^2}{30\ \Omega} = 480\ \text{W}$$

The specific heat and density of air at atmospheric pressure are given as functions of temperature in Table A-7. The inlet air temperature is 25°C (298 K). As an approximation, we will use the table values at 300 K. One could also apply the ideal gas law to find density, but using Table A-7 is more convenient in this special case where the pressure is atmospheric. With table values, the first law becomes

$$T_e = \frac{480\ \text{W}}{\left(1.18\ \dfrac{\text{kg}}{\text{m}^3}\right)\left(0.0064\ \dfrac{\text{m}^3}{\text{s}}\right)\left(1005\ \dfrac{\text{J}}{\text{kg} \cdot \text{K}}\right)} + 25°\text{C}$$

$$T_e = 88.2°\text{C}$$

EXAMPLE 4-11   Conservation of energy in a desktop computer

Air at 17°C enters a channel between two printed circuit boards in a desktop computer. Four rows of chips are installed along one of the circuit boards, as shown in the figure. Within the chips, heat is generated at the device plane at the rate of 0.7 W per cm of row depth. The heat generated is conducted through a layer of silicon 4 mm thick and then convected to the air in the channel. The air flows at a mass flow rate of 9×10^{-4} kg/s per cm of row depth, and the resultant heat transfer coefficient in the channel is 220 W/m²·K. The components on the device plane must be kept below 85°C to ensure reliable operation. Will the design succeed or fail?

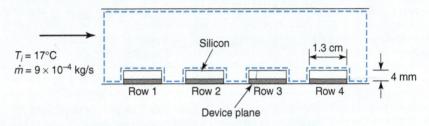

Approach:

The air will increase in temperature as it travels down the channel. The hottest chip will be the chip at the end of the chanel (in row 4) which is exposed to the highest air temperature. The exit air temperature can be found using the first law, Eq. 4-42. All terms are zero except the heat input and the enthalpy terms. The enthalpy change of an ideal gas with constant specific heat is used to eliminate enthalpy from the first law in favor of temperature. Once the air exit temperature is known, the thermal resistances for conduction through the silicon and convection from the chip surface are calculated. The total resistance, which is the sum of these two, is used to calculate the temperature at the device plane of the chip in row 4 using

$$\dot{Q}_{chip} = \frac{T_{chip} - T_e}{R_{tot}}$$

Assumptions:

A1. The flow is steady.

A2. The channel is perfectly insulated from the surroundings.

A3. No work is done.

A4. Kinetic energy change is negligible.

A5. Potential energy change is negligible.

A6. Air is an ideal gas under these conditions.

A7. Specific heat is constant.

Solution:

For the first part of the analysis, we choose the control volume as shown in the figure. The flow is steady [A1], so $dE_{cv}/dt = 0$. The electrical power generated in the chips heats the air. The channel itself is assumed to be insulated from the surroundings [A2]. There is no work done on or by the air [A3]. Velocity is assumed to be low, so kinetic energy effects are negligible [A4]. There is no elevation change and, thus, no potential energy term [A5]. With these simplifications, the energy equation reduces to

$$0 = \dot{Q}_{cv} + \sum_{in} \dot{m}_i h_i - \sum_{out} \dot{m}_e h_e$$

The channel has one inlet and one exit, so

$$0 = \dot{Q}_{cv} + \dot{m}_i h_i - \dot{m}_e h_e$$

In steady state, the mass flow rate entering equals the mass flow rate leaving, that is,

$$\dot{m}_i = \dot{m}_e = \dot{m}$$

Therefore,

$$0 = \dot{Q}_{cv} + \dot{m}(h_i - h_e)$$

Air may be assumed to be an ideal gas [A6]; therefore, for an ideal gas with constant specific heat [A7], Eq. 2-37 gives:

$$\Delta H = m c_p \, \Delta T$$

In our case, this becomes

$$H_i - H_e = mc_p \, (T_i - T_e)$$

Dividing by m and using $h = H/m$,

$$h_i - h_e = c_p \, (T_i - T_e)$$

Substituting this into the first law gives

$$0 = \dot{Q}_{cv} + \dot{m}c_p \, (T_i - T_e)$$

Solving for exit air temperature yields

$$T_e = T_i + \frac{\dot{Q}_{cv}}{\dot{m}c_p}$$

Heat is generated in each of the four rows of chips. Both the heat generated and the air mass flow rate are known for a 1-cm depth of the channel (depth is direction perpendicular to the plane of the figure.) Therefore, the exit temperature may be calculated as

$$T_e = T_i + \frac{\dot{Q}_{cv}}{\dot{m}c_p} = 17°\text{C} + \frac{4\left(0.7 \, \frac{\text{W}}{\text{cm}}\right)}{\left(9 \times 10^{-4} \, \frac{\text{kg}}{\text{s·cm}}\right)\left(1005 \, \frac{\text{J}}{\text{kg·K}}\right)} = 20.1°\text{C}$$

where the specific heat of air was taken from Table A-7. The heat generated at the device plane where transistors and other electronic components are located is conducted through the layer of silicon and then convected to the air. The resistance for conduction is

$$R_{cond} = \frac{L}{kA}$$

A8. Heat leaves only from the top of the chip (heat transfer is one-dimensional).

Here L is the thickness of the silicon and A is the surface area of the top of one row of chips per cm of depth. We neglect any heat convected off the sides of the chip[A8]. The conduction resistance is

$$R_{cond} = \frac{4 \, \text{mm}}{\left(148 \, \frac{\text{W}}{\text{m·K}}\right)(1 \, \text{cm})(1.3 \, \text{cm})\left(\frac{1 \, \text{m}^2}{10{,}000 \, \text{cm}^2}\right)} = 0.208 \, \frac{\text{K}}{\text{W}}$$

A9. The heat transfer coefficient does not change along the channel.

where the thermal conductivity of silicon has been taken from Table A-2. The convection resistance is [A9]

$$R_{conv} = \frac{1}{hA} = \frac{1}{\left(220 \, \frac{\text{W}}{\text{m}^2\text{·K}}\right)(1 \, \text{cm})(1.3 \, \text{cm})\left(\frac{1 \, \text{m}^2}{10{,}000 \, \text{cm}^2}\right)} = 35.0 \, \frac{\text{K}}{\text{W}}$$

Heat first flows through the silicon and then leaves by convection, so these two resistances are in series. The total resistance is

$$R_{tot} = R_{cond} + R_{conv} = 35.2 \, \frac{\text{K}}{\text{W}}$$

The heat generated in the last row of chips is related to the temperature drop across the chip by

$$\dot{Q}_{chip} = \frac{T_{chip} - T_e}{R_{tot}}$$

where T_{chip} is the temperature at the device plane in the last row of chips and T_e is the exit air temperature. Solving for T_{chip},

$$T_{chip} = \dot{Q}_{chip}\, R_{tot} + T_e = (0.7\ \text{W})\left(35.2\ \frac{\text{K}}{\text{W}}\right) + 20.1°\text{C} = 119°\text{C}$$

This is higher than the allowable limit of 85°C. The chip will fail and the system needs redesign.

Comments:

The controlling resistance is the convective resistance. Improving the design will require either a lower heat generation rate or a higher heat transfer coefficient.

4.6 THE BERNOULLI EQUATION

In this section, the energy equation is applied to a particular class of problems. We restrict our discussion to isothermal, incompressible flows with zero viscosity (inviscid). That may seem like a lot of restrictions, but, in fact, many important cases are in this category or may be approximated with these assumptions, including the flow of air and water through short pipelines, the draining of a sink, and the flow of water issuing from a hose.

By definition, an incompressible flow is a flow with constant density. Most common liquid flows are virtually incompressible. Liquids strongly resist changes in volume under mild pressure. But even liquids sometimes exhibit compressibility effects if the pressure is high enough.

In Chapter 2, we used the ideal gas law to calculate changes in the density of gases as a function of pressure and temperature. However, you may be surprised to learn that gases are often assumed to be incompressible. Air at standard temperature and pressure is, in fact, rather difficult to compress. Try squeezing a balloon full of air into a smaller volume. Not so easy. It is not difficult to *deform* a balloon of air, but it is difficult to reduce its volume. Air and other gases are approximately incompressible for small pressure changes and are compressible for large pressure changes. Gases, however, will flow under rather small pressure differences. As a rule of thumb for flow problems, if the gas velocity is less than about 100 m/s, the flow can be considered incompressible.

The next assumption that we discuss is the **inviscid** (zero viscosity) assumption. There are, in general, three forces that act on flowing fluids—gravity, pressure, and friction. In an inviscid flow, frictional effects are zero. This is often valid if the viscosity of the fluid is very low and the flow channel is short. Other sources of frictional losses in pipe flow include sudden contractions, serpentine passages through valves, and flow through porous media. If the flow has a smooth route with rounded corners and no major flow restrictions, it can often be approximated as inviscid.

The final assumption is the isothermal assumption. This simply means that the fluid does not change temperature.

For flow in a pipe or duct with one inlet and one outlet, the steady-state form of the energy equation (i.e., $dE_{cv}/dt = 0$) is

$$0 = \dot{Q}_{cv} - \dot{W}_{cv} + \dot{m}\left(h_1 + \frac{\mathcal{V}_1^2}{2} + gz_1\right) - \dot{m}\left(h_2 + \frac{\mathcal{V}_2^2}{2} + gz_2\right) \qquad (4\text{-}43)$$

Substitute the definition of enthalpy ($h = u + Pv$) into this equation to get

$$0 = \dot{Q}_{cv} - \dot{W}_{cv} + \dot{m}\left(u_1 + P_1v_1 + \frac{\mathcal{V}_1^2}{2} + gz_1\right) - \dot{m}\left(u_2 + P_2v_2 + \frac{\mathcal{V}_2^2}{2} + gz_2\right)$$

Assume the flow is incompressible; therefore, $v_1 = v_2 = v = 1/\rho$. Substitute this and also divide by the mass flow rate to get

$$0 = q_{cv} - w_{cv} + \left(u_1 + \frac{P_1}{\rho} + \frac{\mathcal{V}_1^2}{2} + gz_1 \right) - \left(u_2 + \frac{P_2}{\rho} + \frac{\mathcal{V}_2^2}{2} + gz_2 \right) \qquad (4\text{-}44)$$

where, by definition,

$$q_{cv} = \frac{\dot{Q}_{cv}}{\dot{m}}$$

$$w_{cv} = \frac{\dot{W}_{cv}}{\dot{m}}$$

The units of q_{cv} and w_{cv} are Btu/lbm or J/kg. They represent heat transfer per unit mass of fluid flowing and control volume work per unit mass of fluid flowing, respectively. We now consider a flow that does not exchange heat with its surroundings, (i.e., $q_{cv} = 0$). The flow is inviscid, meaning there is no friction. No heat is produced by friction, and no heat enters from outside the flow. The flow is incompressible, so there is no temperature change due to expansion or contraction. Therefore, the flow is isothermal. Recall that for an incompressible fluid, internal energy, u, depends only on temperature. So, for an isothermal, incompressible flow, $u_1 = u_2$. Applying these conditions and assumptions to Eq. 4-44 and rearranging gives

$$\frac{P_1}{\rho} + gz_1 + \frac{\mathcal{V}_1^2}{2} = \frac{P_2}{\rho} + gz_2 + \frac{\mathcal{V}_2^2}{2} + w_{cv} \qquad (4\text{-}45)$$

If, in addition, no shaft work is added to or removed from the flow via a pump, turbine, or other device (i.e., $w_{cv} = 0$),

$$\boxed{\frac{P_1}{\rho} + gz_1 + \frac{\mathcal{V}_1^2}{2} = \frac{P_2}{\rho} + gz_2 + \frac{\mathcal{V}_2^2}{2}} \qquad (4\text{-}46)$$

This is the Bernoulli equation, first presented by Daniel Bernoulli (1700–1782). It is one of the most famous and useful equations in fluid mechanics. It applies for a steady, incompressible, inviscid, and isothermal flow with no work.

EXAMPLE 4-12 Draining of a tank assuming frictionless flow

Water drains at a steady rate from a very large tank through a pipe of diameter 4 cm. Assume frictionless, incompressible flow. Because the area of the top surface of the water is large compared to the outlet pipe diameter, we also assume that the velocity of the receding top surface is negligible. Find the rate at which mass drains from the tank.

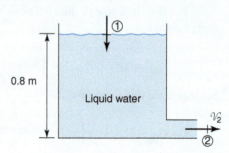

Approach:

Define station 1 at the top surface of the water in the tank and station 2 at the exit. We assume that the tank diameter is very large compared to the pipe diameter, so the velocity at station 1 is small relative to the velocity at 2 and can be approximated as zero. Apply the Bernoulli equation, noting that P_1 and P_2 are both atmospheric pressure, and solve for $\mathcal{V}_2$. Once the velocity is known, the mass flow rate is found from $\dot{m} = \rho \mathcal{V}_2 A$.

Assumptions:

A1. The flow is frictionless.

A2. The flow is incompressible.

A3. The tank is large, so the velocity of receeding water surface is very small.

Solution:

Define station 1 at the top surface of the water in the tank and station 2 at the exit. Assume the flow is frictionless and incompressible [A1][A2]. The Bernoulli equation is

$$\frac{P_1}{\rho} + gz_1 + \frac{\mathcal{V}_1^2}{2} = \frac{P_2}{\rho} + gz_2 + \frac{\mathcal{V}_2^2}{2}$$

The velocity at station 1 is assumed to be zero [A3]. With this simplification, and setting P_1 and P_2 equal to atmospheric pressure,

$$\frac{P_{atm}}{\rho} + gz_1 = \frac{P_{atm}}{\rho} + \frac{\mathcal{V}_2^2}{2} + gz_2$$

which reduces to

$$g(z_1 - z_2) = \frac{\mathcal{V}_2^2}{2}$$

or

$$\mathcal{V}_2 = \sqrt{2g(z_1 - z_2)}$$

$$= \sqrt{2\left(9.8 \, \frac{\text{m}}{\text{s}^2}\right)(0.8 \, \text{m})}$$

$$= 3.96 \, \frac{\text{m}}{\text{s}}$$

The mass flow rate is found from

$$\dot{m} = \rho \mathcal{V}_2 A_2$$

A4. The water is at room temperature.

Using the density of water at 25°C [A4] from Table A-6,

$$\dot{m} = \left(997 \, \frac{\text{kg}}{\text{m}^3}\right)\left(3.96 \, \frac{\text{m}}{\text{s}}\right)\pi(2)^2 \, \text{cm}^2 \left(\frac{1 \, \text{m}}{100 \, \text{cm}}\right)^2$$

$$= 4.96 \, \frac{\text{kg}}{\text{s}}$$

EXAMPLE 4-13   Flow in a free jet

Water issues from a pipe into the atmosphere as a vertical free jet. If the velocity at the pipe exit is 6 m/s, calculate the jet height, h. Assume fluid friction is negligible.

Approach:

Assume the flow is incompressible. For an incompressible, isothermal, frictionless flow, the Bernoulli equation applies. The pressure at both the exit of the pipe and the top of the jet is atmospheric. The velocity at the top of the jet is zero. The height is calculated from the elevation term in the Bernoulli equation.

Assumptions:

A1. The flow is incompressible.

A2. The flow is frictionless.

A3. The flow is isothermal.

Solution:

Assume the density of the water is constant [A1] and the flow is inviscid [A2] and isothermal [A3]. From the Bernoulli equation,

$$0 = \left(\frac{P_1}{\rho} + \frac{\mathcal{V}_1^2}{2} + gz_1 \right) - \left(\frac{P_2}{\rho} + \frac{\mathcal{V}_2^2}{2} + gz_2 \right)$$

where station 1 is the exit of the pipe and station 2 is the top of the jet. Within an inviscid free jet, the pressure is the same at all radial locations. At the jet exit, the water contacts the atmosphere and the pressure at station 1 is atmospheric. The pressure is also atmospheric at station 2; therefore, $P_1 = P_2 = P_{atm}$. Setting $z_1 = 0$, $z_2 = h$, and $\mathcal{V}_2 = 0$,

$$0 = \left(\frac{\mathcal{V}_1^2}{2} \right) - \left(\frac{\mathcal{V}_2^2}{2} + gh \right)$$

Solving for h,

$$h = \frac{\mathcal{V}_2^2 - \mathcal{V}_1^2}{2g}$$

Substituting values:

$$h = \frac{0 - \left(6 \, \frac{m}{s} \right)^2}{2 \left(9.81 \, \frac{m}{s^2} \right)} = 1.84 \, m$$

The Bernoulli equation has been used for finite control volumes with one inlet and one outlet. It can also be applied to infinitesimal control volumes and to control volumes with more than one inlet and/or outlet. In Figure 4-23, a flow through a nozzle is shown.

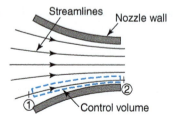

FIGURE 4-23 Streamlines in a nozzle.

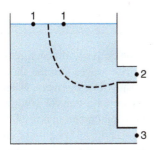

FIGURE 4-24 Water issuing from a tank through two different exits.

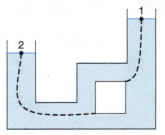

FIGURE 4-25 Water flowing between two tanks by two different routes.

The figure includes so-called streamlines. A **streamline** indicates the path that a fluid particle takes as it accelerates through the nozzle. Because a streamline is always tangent to the velocity vector, no mass flow crosses a streamline.

Figure 4-23 shows a control volume aligned with the streamline and extending from point 1 to point 2. This control volume is finite in length but infinitesimally thin in the dimension perpendicular to the streamline. Such a control volume is sometimes called a **streamtube**. Fluid enters the left side of the streamtube at point 1 and leaves through the right side at point 2. No fluid flows out the lateral sides. The flow within the streamtube meets the criteria for the Bernoulli equation; that is, it is incompressible, inviscid, and isothermal. Therefore, Bernoulli's equation applies along a streamline.

The Bernoulli equation can sometimes be used if there is more than one inlet or outlet. For example, in Figure 4-24 water drains from a tank through two different outlets. The dotted line is a streamline that divides the flow into two regions. The upper region contains all the fluid that leaves through the upper outlet, and the lower region contains all the fluid that leaves through the lower outlet. It is possible to calculate the shape and placement of the dividing streamline, but that is beyond the scope of this text. The dividing streamline is parallel to the velocity vector. No mass flows across the dividing streamline. As a result, we may write the Bernoulli equation for the upper region as

$$\frac{P_1}{\rho} + gz_1 + \frac{\mathcal{V}_1^2}{2} = \frac{P_2}{\rho} + gz_2 + \frac{\mathcal{V}_2^2}{2}$$

For the lower region, the Bernoulli equation is

$$\frac{P_1}{\rho} + gz_1 + \frac{\mathcal{V}_1^2}{2} = \frac{P_3}{\rho} + gz_3 + \frac{\mathcal{V}_3^2}{2}$$

Both of these equations apply simultaneously.

In some circumstances, the fluid can take multiple paths from one point to another. For example, in Figure 4-25 fluid flows from the right tank to the left tank by one of two paths. Again, it is possible to imagine the dividing streamline that separates the flow into two regions. The Bernoulli equation applies simultaneously to each region.

4.7 FLOW MEASUREMENT

(Go to www.wiley.com/college/kaminski)

4.8 CONSERVATION OF LINEAR MOMENTUM FOR AN OPEN SYSTEM

Fluids flowing over or against surfaces can exert significant forces on those surfaces. Surfers take advantage of this effect when they ride waves into the shore. Wind flowing over sails can drive a boat across the ocean, and water flowing over a water wheel is used to produce mechanical work. To calculate the forces on a control volume due to the flow of fluid through the control volume, conservation of momentum is used. While this is a conservation equation, similar to conservation of mass and conservation of energy, there is a significant difference between momentum and the two quantities, mass and energy. Mass and energy only have magnitude (i.e., they are scalars). Momentum involves force that has magnitude and direction (i.e., it is a vector). Hence, the momentum equation is a vector equation that can be written as three scalar equations—one for each component of momentum in the x-, y-, and z-directions—or comparable directions in cylindrical or spherical coordinates.

We now develop the momentum equation in the x-direction in a rectangular coordinate system. The equations in the y- and z-directions are exactly analogous. We use the same approach to develop the open-system momentum equation as we used to develop the open-system energy equation. Figure 4-26 shows a control volume at two different times. For interpreting this figure, a few definitions are useful:

$B_{x,cv}(t)$ = total momentum in the control volume at time t in the x-direction

$B_{x,cv}(t + \Delta t)$ = total momentum in the control volume at time $t + \Delta t$ in the x-direction

$B_x(t)$ = total momentum of the closed system at time t in the x-direction

$B_x(t + \Delta t)$ = total momentum of the closed system at time $t + \Delta t$ in the x-direction

The closed system is all the mass shown in Figure 4-26a, while the control volume is all the mass inside the dotted line. Within the time period Δt, a small quantity of mass Δm_i enters the control volume, and a small quantity of mass Δm_e exits; Δm_e is not necessarily equal to Δm_i. Momentum is mass times velocity, or

$$B = m\mathcal{V}$$

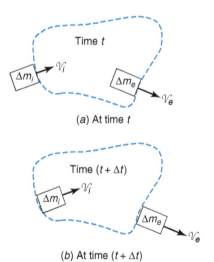

(a) At time t

(b) At time $(t + \Delta t)$

FIGURE 4-26 A control volume at two times.

In the x-direction,

$$B_x = m\mathcal{V}_x$$

where $\mathcal{V}_x$ is the component of velocity in the x-direction. The momentum of the closed system at time t is the momentum within the control volume plus the momentum carried in with the mass Δm_i. In other words,

$$B_x(t) = B_{x,cv}(t) + \Delta m_i \mathcal{V}_{x,i}$$

Figure 4-26b contains the same amount of mass as Figure 4-26a. Therefore, at time $t + \Delta t$, the momentum of the closed system is the momentum within the control volume plus the momentum carried out with the mass Δm_e. Mathematically, this is

$$B_x(t + \Delta t) = B_{x,cv}(t + \Delta t) + \Delta m_e \mathcal{V}_{x,e}$$

The change of momentum of the closed system in time Δt is

$$B_x(t + \Delta t) - B_x(t) = B_{x,cv}(t + \Delta t) - B_{x,cv}(t) + \Delta m_e \mathcal{V}_{x,e} - \Delta m_i \mathcal{V}_{x,i}$$

Divide this expression by Δt to get

$$\frac{B_x(t + \Delta t) - B_x(t)}{\Delta t} = \frac{B_{x,cv}(t + \Delta t) - B_{x,cv}(t)}{\Delta t} + \frac{\Delta m_e}{\Delta t}\mathcal{V}_{x,e} - \frac{\Delta m_i}{\Delta t}\mathcal{V}_{x,i} \qquad (4\text{-}47)$$

From Newton's second law, the sum of the forces is equal to the rate of change of momentum, or

$$\sum F_x = m\frac{d\mathcal{V}_x}{dt} = \frac{dB_x}{dt} = \lim_{\Delta t \to 0}\left(\frac{B_x(t + \Delta t) - B_x(t)}{\Delta t}\right) \qquad (4\text{-}48)$$

Substituting Eq. 4-47 into Eq. 4-48 gives

$$\sum F_x = \lim_{\Delta t \to 0}\left(\frac{B_{x,cv}(t + \Delta t) - B_{x,cv}(t)}{\Delta t} + \frac{\Delta m_e}{\Delta t}\mathcal{V}_{x,e} - \frac{\Delta m_i}{\Delta t}\mathcal{V}_{x,i}\right)$$

Taking the limit,

$$\sum F_x = \frac{d}{dt}(B_{x,cv}) + \dot{m}_e \mathcal{V}_{x,e} - \dot{m}_i \mathcal{V}_{x,i}$$

This is the momentum equation for an open system for the x-direction. It states that the sum of the forces is equal to the time rate of change of momentum within the control volume plus the momentum leaving minus the momentum entering. The equation has been written for the x-direction. The equivalent equations for the y- and z-directions are

$$\sum F_y = \frac{d}{dt}(B_{y,cv}) + \dot{m}_e \mathcal{V}_{y,e} - \dot{m}_i \mathcal{V}_{y,i}$$

$$\sum F_z = \frac{d}{dt}(B_{z,cv}) + \dot{m}_e \mathcal{V}_{z,e} - \dot{m}_i \mathcal{V}_{z,i}$$

If there is more than one stream in or out, the contribution of each stream is simply added. For example, the x-momentum equation becomes

$$\sum F_x = \frac{d}{dt}(B_{x,cv}) + \sum_{out} \dot{m}_e \mathcal{V}_{x,e} - \sum_{in} \dot{m}_i \mathcal{V}_{x,i}$$

There are similar expressions in the y- and z-directions.

EXAMPLE 4-14 Anchoring force in a pipe

Water flows at a constant rate into a pipe junction at station 1 at a velocity of 8 m/s, as shown in Figure 4-27. Water leaves at station 2 at 6 m/s. The diameter of all three pipes is 4 cm. The values of pressure are: $P_1 = 106$ kPa, $P_2 = P_3 = P_{atm} = 100$ kPa. Calculate the anchoring force needed to hold the pipe in place. Neglect the weight of the pipe and the water.

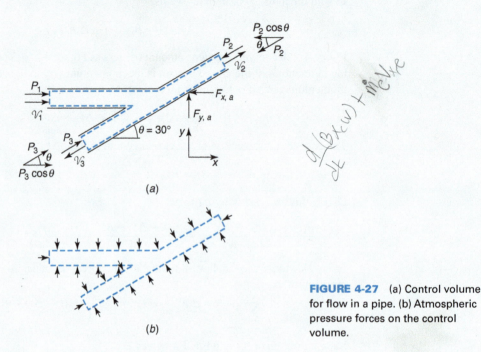

(a)

(b)

FIGURE 4-27 (a) Control volume for flow in a pipe. (b) Atmospheric pressure forces on the control volume.

Approach:

The anchoring force is the force that the pipe exerts on the water, which is balanced by the force of the water on the pipe wall. The anchoring force is found with the linear momentum equations in the x- and y-directions applied to the water. The unknown anchoring force is one of the forces on the left-hand side of the momentum equation; pressure forces are also on the left-hand side. Since the flow outlets are not normal to either the x- or y-direction, the pressure force, which *is* normal to the exit, will have both an x- and a y-component. Pressure is tricky because pressure forces act on the sides of the pipes as well as on the inlets and outlets. It is possible to subtract out the action of these side pressure forces by using gage pressure instead of absolute pressure. This is similar to what was done in computing forces on submerged surfaces. In evaluating the velocity terms, care must be taken to use negative values for velocity components in the negative x- and y-directions.

Assumptions:

Solution:

We choose as our control volume the volume inside all three branches of the pipe. The water flow exerts a force on the pipe, pushing it rightward and downward. The anchoring force counteracts the water force. To find the anchoring force, use conservation of momentum. Starting with the x-direction,

$$\sum F_x = \frac{d}{dt}\left(B_{x,cv}\right) + \sum_{out} \dot{m}_e \mathcal{V}_{x,e} - \sum_{in} \dot{m}_i \mathcal{V}_{x,i}$$

The control volume is shown by a dotted line in Figure 4-27a. The forces consist of pressure forces at the inlet and outlets and the anchoring force [A1]; therefore,

A1. The weight of the water and pipe are negligible compared to the other forces in the system.

$$\sum F_x = F_{x,a} + P_1 A_1 - P_2 \cos \theta A_2 + P_3 \cos \theta A_3$$

The x-component of the anchoring force, $\boxed{F_{x,a}}$ is assumed to be positive in the positive x-direction. (Physical intuition would lead us to conclude that the actual anchoring force must point in the negative x-direction, and it does. We will find that $F_{x,a}$ has a negative value.)

As always, force is pressure times area. All values of pressure are taken as positive. Note from the figure that the pressure forces always point inward and are normal to the control volume faces. At station 2, the component of force due to pressure in the x-direction is the projection of this force in the x-direction. It is negative, since it points in the negative x-direction. Similar reasoning is applied to the pressure force at station 3.

A2. The system is in steady state.

This is a steady flow situation [A2], and the control volume is stationary, so

$$\frac{d}{dt}(B_{x,cv}) = 0$$

The final terms in the momentum equation are

$$\sum_{\text{out}} \dot{m}_e \mathcal{V}_{x,e} - \sum_{\text{in}} \dot{m}_i \mathcal{V}_{x,i}$$

$$= \dot{m}_2 \mathcal{V}_{x,2} + \dot{m}_3 \mathcal{V}_{x,3} - \dot{m}_1 \mathcal{V}_{x,1}$$

$$= \dot{m}_2 \mathcal{V}_2 \cos \theta + \dot{m}_3 \left[-\mathcal{V}_3 \cos \theta \right] - \dot{m}_1 \mathcal{V}_1$$

Note the sign change in the second term, which occurs because the projection of $\mathcal{V}_3$ in the x-direction is in the negative x-direction.

Substituting all these terms into the momentum equation results in

$$F_{x,a} + P_1 A_1 - P_2 \cos \theta A_2 + P_3 \cos \theta A_3$$

$$= \dot{m}_2 \mathcal{V}_2 \cos \theta - \dot{m}_3 \mathcal{V}_3 \cos \theta - \dot{m}_1 \mathcal{V}_1$$

The cross-sectional area of all three pipes is the same:

$$A_1 = A_2 = A_3 = \pi (2 \text{ cm})^2 \left(\frac{1 \text{ m}}{100 \text{ cm}} \right)^2 = 0.00126 \text{ m}^2$$

At station 1,

$$\dot{m}_1 = \rho \mathcal{V}_1 A_1$$

$$= \left(997 \, \frac{\text{kg}}{\text{m}^3} \right) \left(8 \frac{\text{m}}{\text{s}} \right) (0.00126 \text{ m}^2)$$

$$= 10.05 \, \frac{\text{kg}}{\text{s}}$$

where the density of water from Table A-6 has been used. Similarly, at station 2

$$\dot{m}_2 = \rho \mathcal{V}_2 A_2$$

$$= (997)(6)(0.00126)$$

$$= 7.537 \, \frac{\text{kg}}{\text{s}}$$

By conservation of mass

$$\dot{m}_1 = \dot{m}_2 + \dot{m}_3$$

so

$$\dot{m}_3 = \dot{m}_1 - \dot{m}_2$$

$$= 10.05 - 7.537 = 2.512 \, \frac{\text{kg}}{\text{s}}$$

The velocity at station 3 can be found from

$$\mathcal{V}_3 = \frac{\dot{m}_3}{\rho A_3} = \frac{2.512 \, \frac{\text{kg}}{\text{s}}}{\left(997 \, \frac{\text{kg}}{\text{m}^3}\right)(0.00126 \, \text{m}^2)}$$

$$= 2.00 \, \frac{\text{m}}{\text{s}}$$

We might think that it is time to plug in values and get the anchoring force, but, in fact, there is just one more important thing to think about. We have not really included all the forces. We left out the forces exerted by atmospheric pressure! Figure 4-27b shows the pipe as it would look with no flow. Atmospheric pressure pushes on it from all sides. The vector sum of all these pressure forces equals the buoyancy force of the air on the pipe. Since the density of air is small, this buoyancy force is very small and will be neglected [A3]. The pressure forces at inlet and outlets include the atmospheric pressure portion, which contributes only to the buoyancy. We can remove buoyancy effects by subtracting atmospheric pressure from the pressures at the inlet and outlets of the control volume. In other words, the above momentum equation will be correct if we use *gage* pressure rather than absolute pressure at inlet and outlets. With that, the anchoring force is

A3. Buoyancy forces on the pipe are negligible.

$$F_{x,a} = \dot{m}_2 \mathcal{V}_2 \cos\theta - \dot{m}_3 \mathcal{V}_3 \cos\theta - \dot{m}_1 \mathcal{V}_1 - P_1 A_1 + P_2 \cos\theta \, A_2 - P_3 \cos\theta \, A_3$$

$$= \left(7.537 \, \frac{\text{kg}}{\text{s}}\right)\left(6 \, \frac{\text{m}}{\text{s}}\right)\cos(30) - (2.512)(2.00)\cos(30) - (10.05)(8)$$

$$- (106 - 100)\,\text{kPa} \left(\frac{1000 \, \text{Pa}}{1 \, \text{kPa}}\right)(0.00126 \, \text{m}^2)$$

$$+ (100 - 100)\cos(30)(0.00126) - (100 - 100)\cos(30)(0.00126)$$

$$F_{x,a} = -53.1 \, \text{N}$$

The force is negative, implying that a force must be applied in the negative x-direction to hold the pipe in place. This accords with intuition.

We also need the component of the anchoring force in the y-direction. This is found from

$$\sum F_y = \frac{d}{dt}(B_{y,cv}) + \sum_{\text{out}} \dot{m}_e \mathcal{V}_{y,e} - \sum_{\text{in}} \dot{m}_i \mathcal{V}_{y,i}$$

which, following the same steps as in the x-direction, becomes

$$F_{y,a} - P_2 \sin\theta \, A_2 + P_3 \sin\theta \, A_3$$

$$= \dot{m}_2 \mathcal{V}_2 \sin\theta - \dot{m}_3 \mathcal{V}_3 \sin\theta$$

or

$$F_{y,a} = \left(7.537\,\frac{kg}{s}\right)\left(6\,\frac{m}{s}\right)\sin 30 - (2.512)(2.00)\sin 30$$

$$F_{y,a} = 20.1\,N$$

As expected, this force points upward.

Comment:

In this example, we retained four significant figures in the calculated mass flow rates. If only three significant figures had been retained, the calculated velocity at station 3 would have been 1.96 m/s. Since all pipes have the same diameter and the flow is incompressible, $\mathcal{V}_1 = \mathcal{V}_2 + \mathcal{V}_3$. Given values of $\mathcal{V}_1$ and $\mathcal{V}_2$ are 8 m/s and 6 m/s; therefore, $\mathcal{V}_3$ is 2 m/s. Roundoff error often occurs during the subtraction of two large numbers that are close in size. In this example, mass flow rate at station 3 was computed by such a subtraction. If only three significant figures had been used, the final values of force in the x- and y-directions would have been -52.6 and 20.1 N, an error of about 1%.

EXAMPLE 4-15   Anchoring force in a gradual expansion

Oil with a density of 52 lbm/ft^3 flows at a steady rate through the gradual pipe expansion shown in the figure below. Using data on the figure, calculate the anchoring force needed to hold the pipe expansion in place. Assume frictionless and incompressible flow and neglect the weight of the oil and the pipe.

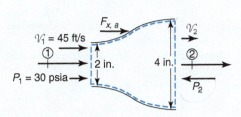

Approach:

Define the volume in the expansion as the control volume. All the forces on this control volume are in the x-direction; therefore, only conservation of momentum in the x-direction will be needed. The pressure and velocity at the outlet will have to be evaluated. To find the velocity, use conservation of mass; to find the pressure, use Bernoulli's equation. As in the previous example, gage pressure must be used.

In this problem, we need to solve conservation of mass, momentum, and energy simultaneously. (Recall that Bernoulli's equation was derived from the energy equation.) It is very common in thermal–fluids applications to use these three equations simultaneously.

Assumptions:

A1. Neglect the weight of the oil and the pipe.

A2. The flow is steady.

Solution:

The control volume is chosen to lie along the inside of the pipe wall and cut across the ends. We ignore the vertical anchoring force needed to support the weight of the oil and the pipe itself [A1]. To find the anchoring force in the x-direction, use conservation of momentum:

$$\sum F_x = \frac{d}{dt}(B_{x,cv}) + \sum_{out} \dot{m}_e \mathcal{V}_{x,e} - \sum_{in} \dot{m}_i \mathcal{V}_{x,i}$$

For steady flow [A2], this becomes

$$F_{x,a} + P_1 A_1 - P_2 A_2 = \dot{m}_2 \mathcal{V}_2 - \dot{m}_1 \mathcal{V}_1$$

The mass flow rates are given by

$$\dot{m}_1 = \dot{m}_2 = \rho \mathcal{V}_1 A_1$$

$$= \left(52\,\frac{\text{lbm}}{\text{ft}^3}\right)\left(45\,\frac{\text{ft}}{\text{s}}\right)(\pi 1^2)\,\text{in.}^2\left(\frac{1\,\text{ft}^2}{144\,\text{in.}^2}\right)$$

$$= 51.1\,\frac{\text{lbm}}{\text{s}}$$

To find $\mathcal{V}_2$, use

$$\rho \mathcal{V}_1 A_1 = \rho \mathcal{V}_2 A_2$$

$$\mathcal{V}_2 = \frac{\mathcal{V}_1 A_1}{A_2} = \left(45\,\frac{\text{ft}}{\text{s}}\right)\left(\frac{\pi(1)^2}{\pi(2)^2}\right)$$

$$= 11.25\,\frac{\text{ft}}{\text{s}}$$

A3. The flow is incompressible.
A4. The flow is inviscid.

To find the pressure at station 2, use Bernoulli's equation [A3][A4],

$$\frac{P_1}{\rho} + \frac{\mathcal{V}_1^2}{2} + gz_1 = \frac{P_2}{\rho} + \frac{\mathcal{V}_2^2}{2} + gz_2$$

For this frictionless flow with no elevation change, this becomes

$$\frac{P_1}{\rho} + \frac{\mathcal{V}_1^2}{2} = \frac{P_2}{\rho} + \frac{\mathcal{V}_2^2}{2}$$

Solving for P_2,

$$P_2 = P_1 + \frac{\rho}{2}(\mathcal{V}_1^2 - \mathcal{V}_2^2)$$

$$P_2 = 30\,\text{psia} + \frac{52\,\frac{\text{lbm}}{\text{ft}^3}}{2}\left[(45)^2 - (11.25)^2\right]\frac{\text{ft}^2}{\text{s}^2}\left(\frac{1\,\text{lbf}}{32.17\,\frac{\text{lbm}\cdot\text{ft}}{\text{s}^2}}\right)\left(\frac{1\,\text{ft}^2}{144\,\text{in.}^2}\right)$$

$$P_2 = 40.7\,\text{psia}$$

The anchoring force may now be calculated from

$$F_{x,a} = \dot{m}_2\,(\mathcal{V}_2 - \mathcal{V}_1) + P_2 A_2 - P_1 A_1$$

$$= \left(51.1\,\frac{\text{lbm}}{\text{s}}\right)(11.25 - 45)\,\frac{\text{ft}}{\text{s}}\left(\frac{1\,\text{lbf}}{32.17\,\frac{\text{lbm}\cdot\text{ft}}{\text{s}^2}}\right) + (40.7 - 14.7)\,\frac{\text{lbf}}{\text{in.}^2}\,(\pi 2^2)\,\text{in.}^2$$

$$-(30 - 14.7)\,\frac{\text{lbf}}{\text{in.}^2}\,(\pi 1^2)\,\text{in.}^2$$

$$= -53.6 + 326.7 - 48.1$$

$$F_{x,a} = 225\ \text{lbf}$$

Comments:

Note that in the momentum equation, gage pressure has been used, as explained in Example 4-14. The positive value of $F_{x,a}$ tells us that we assumed the correct direction.

SUMMARY

The **specific gravity** of a liquid or solid is the ratio of its density to the density of water at 4°C:

$$SG = \frac{\rho}{\rho_{water}}$$

Pressure is a force per unit area:

$$P = \frac{F}{A}$$

The pressure force on a surface immersed in a fluid is always normal to the surface. In a static fluid of constant density, the pressure is a function of depth, h, according to

$$P = P_{atm} + \rho g h$$

In a static fluid of variable density, the pressure variation with depth, z, is found by solving the differential equation

$$\frac{dP}{dz} = -\rho(z)g$$

A functional form for the variation of density, ρ, with depth, z, is needed before this equation can be solved.

The magnitude of the resultant force on one side of a submerged plane surface is

$$F_R = P_{atm}A + \rho g \sin \theta \, y_C A$$

where y_C is the oblique depth of the centroid of the surface (see Figure 4-9). The resultant force is applied at

$$y_P = y_C + \frac{I_{xx,C}}{y_C A + \dfrac{P_{atm} A}{\rho g \sin \theta}}$$

where $I_{xx,C}$ is the area moment of inertia of the surface. The location of the centroid and the area moments of inertia for some common shapes are given in Table 4-1. If the shape of the immersed surface is not in Table 4-1, the values of y_C and $I_{xx,C}$ can be computed by integration from Eq. 4-9 and Eq. 4-14.

If the submerged surface is exposed to the atmosphere on one side and to a fluid whose upper surface is at atmospheric pressure, the equations for F_R and y_P simplify to

$$F_R = \rho g \sin \theta \, y_C A$$

$$y_P = y_C + \frac{I_{xx,C}}{y_C A}$$

If the submerged surface is curved, then the horizontal and vertical projections of the curved surface are used to define a volume of fluid bounded by these projections and by the curved surface. A static equilibrium analysis on this fluid volume yields the force on the submerged curved surface.

The principle of buoyancy can be simply stated as: *The buoyant force on an immersed object is equal to the weight of the fluid displaced by the object. This force acts upward through the center of gravity of the object.*

If an object is floating in water exposed to the atmosphere, then the submerged volume of the object can be calculated from

$$\frac{\rho_o}{\rho_w} = \frac{V_w}{V_{tot}}$$

where ρ_o is the density of the object, ρ_w is the density of water, V_w is the volume submerged in the water, and V_{tot} is the total volume of the object.

A **closed system** is one in which no mass crosses the system boundary. By contrast, if mass flows across the system boundary (either entering or leaving), the system is called an **open system.**

In an open system, conservation of mass is given by

$$\frac{dm_{cv}}{dt} = \sum \dot{m}_i - \sum \dot{m}_e$$

The mass flow rate may be expressed in terms of the average velocity using

$$\dot{m} = \rho \mathcal{V}_{avg} A$$

The volumetric flow rate is defined as

$$\dot{V} = \mathcal{V}_{avg} A$$

The volumetric flow rate is related to the mass flow rate by

$$\dot{m} = \rho \dot{V} = \frac{\dot{V}}{v}$$

Conservation of energy for an open system may be given as:

$$\frac{dE_{cv}}{dt} = \dot{Q}_{cv} - \dot{W}_{cv} + \sum_{in} \dot{m}_i \left(h_i + \frac{\mathcal{V}_i^2}{2} + gz_i \right)$$

$$- \sum_{out} \dot{m}_e \left(h_e + \frac{\mathcal{V}_e^2}{2} + gz_e \right)$$

If the energy equation is applied to an isothermal, incompressible, frictionless flow with no heat transfer, then

$$0 = -w_{cv} + \left(\frac{P_1}{\rho} + \frac{\mathcal{V}_1^2}{2} + gz_1 \right) - \left(\frac{P_2}{\rho} + \frac{\mathcal{V}_2^2}{2} + gz_2 \right)$$

where w_{cv} is the work done per unit mass flow:

$$w_{cv} = \frac{\dot{W}_{cv}}{\dot{m}}$$

The Bernoulli equation, which applies to an isothermal, incompressible, frictionless flow with no work or heat transfer is

$$\frac{P_1}{\rho} + \frac{V_1^2}{2} + gz_1 = \frac{P_2}{\rho} + \frac{V_2^2}{2} + gz_2$$

The flow rate of a Venturi meter is given by

$$\dot{V} = CA_2 \sqrt{\frac{2(P_1 - P_2)}{\rho} \left[\frac{1}{1 - (A_2 / A_1)^2} \right]}$$

where C is the discharge coefficient.

The conservation of momentum equation in the x-direction for an open system is

$$\sum F_x = \frac{d}{dt}(B_{x,cv}) + \sum_{out} \dot{m}_e V_{x,e} - \sum_{in} \dot{m}_i V_{x,i}$$

There are similar equations in the y- and z-directions. In using this equation for a pipe flow, it is important to *use gage pressure instead of absolute pressure.*

SELECTED REFERENCES

Fox, R. W., and A. T. McDonald, *Introduction to Fluid Mechanics*, 5th ed., Wiley, New York, 1998.

Munson, B. R., D. F. Young, and T. H. Okiishi, *Fundamentals of Fluid Mechanics*, 4th ed., Wiley, New York, 2002.

Potter, M. C., and D. C. Wiggert, *Mechanics of Fluids*, 3rd ed., Brooks/Cole, Pacific Grove, CA, 2002.

Roberson, J. A., and C. T. Crowe, *Engineering Fluid Mechanics*, 6th ed., Wiley, New York, 1997.

White, F. M., *Fluid Mechanics*, McGraw-Hill, New York, 1979.

PROBLEMS

Problems designated with WEB refer to material available at www.wiley.com/college/kaminski.

PRESSURE VARIATION WITH DEPTH

P4-1 If atmospheric pressure is 14.7 lbf/in.^{2}, what is the pressure at a depth of 10 ft of water?

P4-2 Hoover Dam stands at a height of 725 ft above the Colorado river. Assuming atmospheric pressure is 14.7 lbf/in.^{2}, calculate the pressure in the reservoir at the base of the dam.

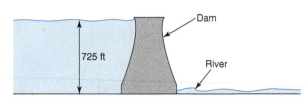

P4-3 In a manometer containing liquid mercury, the differential height is read as 6 in. If atmospheric pressure is 100 kPa, what pressure is the manometer reading (in kPa)?

P4-4 A vat in a chemical processing plant contains liquid ethylene glycol at 20°C. The air space at the top of the closed vat is maintained at 110 kPa. If the depth of the liquid is 0.8 m, what is the pressure at the bottom of the tank?

P4-5 (WEB) At great ocean depths, the hydrostatic pressure is very high. Suppose that seawater density varies with pressure according to

$$P = C_1 \ln \left(\frac{\rho}{\rho_o} \right) + C_2$$

where $C_1 = 2.24 \times 10^9$ Pa, $C_2 = 1 \times 10^5$ Pa, and $\rho_o = 1024$ kg/m^3. Assume this relation holds at any depth, z, and use it to find the pressure at a depth of 3000 m. What would the pressure be if seawater density were assumed to be constant at 1024 kg/m^3? Assume atmospheric pressure is 1×10^5 Pa.

P4-6 (WEB) A large tank contains a liquid solution whose density varies with depth as shown in the table. A gas space at the top of the tank contains air at 60 psia. Find the pressure at a depth of 30 ft using numerical integration.

depth, ft	density, lbm/ft^3
0	40.2
5	41.0
10	42.7
15	44.9
20	47.7
25	50.9
30	54.6

MANOMETERS

P4-7   A manometer is attached to a rigid tank containing gas at pressure P. The manometer fluid is mercury at 20°C. Using data on the figure, find the pressure in the tank.

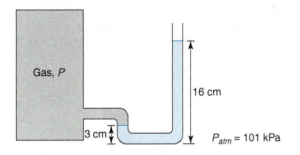

P4-8   Write an equation for the mass of the piston, m_p, in terms of ρ_{H_2O}, l, θ, and A_p. See the figure.

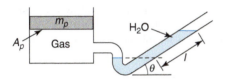

P4-9   Liquid water is contained in a piston–cylinder assembly as shown in the figure. An inclined manometer filled with water is attached to the bottom of the cylinder. Using data given on the figure, calculate the force exerted by the spring on the piston.

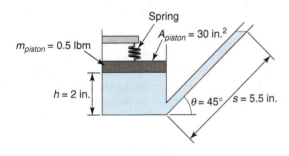

P4-10   A tank contains air at 80°F. A manometer connected to the tank contains liquid mercury, also at 80°F. Assuming atmospheric pressure is 14.2 psia and using data on the figure, calculate the density of the air in the tank.

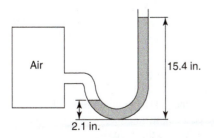

P4-11   Two piston–cylinder assemblies are connected by a tube, as shown in the figure. The diameter of each cylinder is 8 cm, and the mass of each piston is 0.4 kg. A mass rests on top of each piston. The fluid in the tube is mercury, at 20°C. Using data on the figure, calculate the unknown mass m_2.

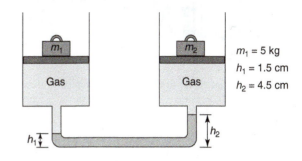

P4-12   In the device shown in the figure, calculate the gage pressure of the gas in the tank.

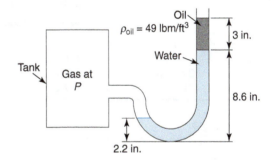

P4-13   In the manometer shown in the figure, 2 g of oil and 11 g of water are introduced. The oil has a density of 620 kg/m³. Find the length, l, to which the water rises in the inclined section.

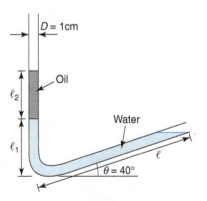

P4-14   A manometer connects a large water tank open to the atmosphere to a closed spherical tank of air. The manometer contains both oil and water. Using data on the figure, find the gage pressure of the air in tank A.

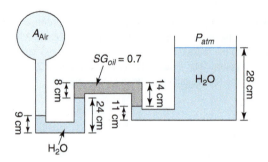

P4-15   The manometer in the figure is designed to measure small changes in pressure. Using the data in the table, answer the following questions.

a. Determine the initial gage pressure of the gas in the sphere

b. The pressure is increased so that h_2 becomes 2.0 cm. During this process, none of the liquid interfaces change in diameter. Find the final gage pressure of the gas.

Unchanged quantities				Initial value	Final value
d	1 cm	h_1	9.1 cm	
D	9 cm	h_2	2.2 cm	2.0 cm
SG_1	0.86 kg/m^3	h_3	4.6 cm	
SG_2	11.3 kg/m^3	h_4	10.7 cm	

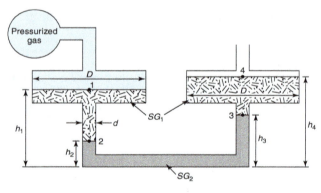

STATIC PRESSURE APPLICATIONS

P4-16   A glass tube containing oil is inserted into a tank of water, as shown in the figure. Using data on the figure, calculate the oil density. Assume the temperature is 20°C.

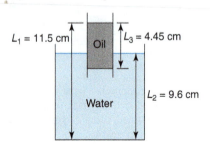

P4-17   A hydraulic lift is used to raise a crate, as shown in the figure. The large tube has a 3-ft diameter and the small tube has a 1-in. diameter. The fluid is oil with a specific gravity of 0.86. The plate under the crate has negligible mass. Initially the system is at rest with the oil heights of 1.4 ft and 6 in., as shown.

a. Find the mass of the crate in lbm.

b. Oil is poured into the small tube until the crate rises 1 in. Calculate the volume of oil added.

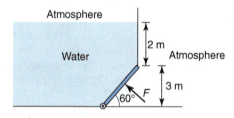

FORCES ON SUBMERGED PLANE SURFACES

P4-18   An underwater gate 8 m wide is held closed by a force, F, as shown in the figure. If the force is applied at the middle of the gate, what is the minimum value required to keep the gate closed?

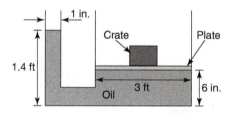

P4-19   A horizontal pipe 1.4 m in diameter is half filled with liquid oxygen ($SG = 1.18$). The gas above the liquid is at a pressure of 250 kPa. The pipe is closed on both ends by vertical, flat surfaces. Find the magnitude of the resultant force of the fluid and gas acting on one of the end surfaces.

P4-20   A gravity dam made of concrete ($\rho = 2200$ kg/m^3) holds back water that is 5.5 m deep, as shown. The bottom of the dam rests on the soil and is held in place by friction. Calculate the minimum coefficient of friction between the dam and the soil so that the dam does not slide.

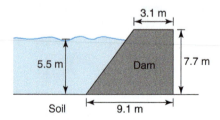

P4-21   A conduit leading from a reservoir is closed by a square gate pivoted along its midline, as shown. Calculate the force of the gate on the stop that holds it closed.

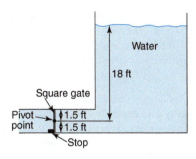

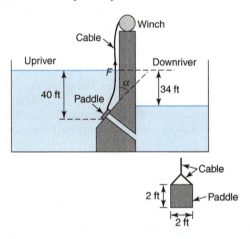

P4-22 A square plate, called a paddle, covers a passage in a canal lock, as shown. The angle, α, is 15°. Find the vertical force, F, needed to open the paddle.

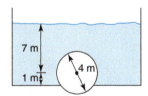

FORCES ON SUBMERGED CURVED SURFACES

P4-23 (WEB) A cylinder 5 m long and 4 m in diameter is wedged into a rectangular opening in the bottom of a tank of water. The cylinder seals the opening, which is also 5 m long. The center of the cylinder is 1 m above the floor of the tank, and the water depth is 8 m. Find the net force of the water on the cylinder.

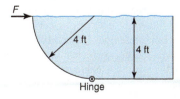

P4-24 (WEB) A semicircular gate hinged at the bottom holds back a tank of water 4 ft deep. If the gate is 15 ft wide, what force, F, is required to keep the gate closed?

P4-25 (WEB) A hemisphere filled with oil ($SG = 0.72$) is inverted on a flat surface. A narrow tube partially filled with oil protrudes from the top of the hemisphere, which has a mass of 5 kg. At what height, h, will the hemisphere lift off the surface? Neglect the weight of the tube and the oil it contains.

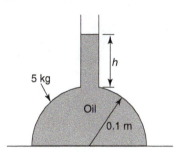

BUOYANCY

P4-26 If the "tip of the iceberg" (that is, the volume of the iceberg above the water surface) is 79 m^3, what is the volume of the submerged iceberg? For seawater density use $\rho_{seawater} = 1.027$ g/cm^3.

P4-27 A small boat has a mass of 650 lbm when empty. If the volume of the hull is 166 ft^3, determine the maximum load the boat can carry in fresh water.

P4-28 A layer of oil 6 cm thick covers a layer of water. A cylinder made of soft pine floats in this two-layer fluid, as shown. Using data on the figure, find the height, x, by which the cylinder protrudes from the fluid.

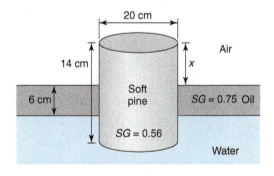

P4-29 A hot-air balloon has a mass of 250 kg and carries two passengers whose average weight is 185 lbf. The balloon, which has a diameter of 12 m, rises through atmospheric air, which is at 20°C. Find the minimum possible average temperature of the air inside the balloon. Atmospheric pressure is 100 kPa.

P4-30 A rectangular gate 12 ft high and 3 ft wide is held closed by water pressure, as shown in the figure. A counterweight of mass m is connected to the gate by a cable that runs over a pulley and attaches to the top of the gate. The counterweight, which is partially immersed in the water, is cylindrical with a diameter of 1.5 ft and a mass of 800 lbm. Air at atmospheric pressure is above the water and on the back side of the gate. Calculate the minimum water depth, h, for which the gate will stay closed.

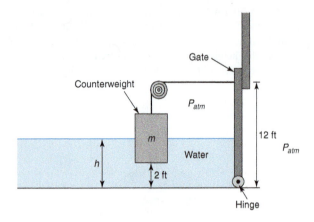

CONSERVATION OF MASS

P4-31 Rain falling on a roof flows downward over the shingles and is collected in a gutter. The gutter, which is closed at one end and open at the other, is slightly inclined so that water runs out the open end. In a heavy downpour, rain falls steadily for several hours and the flow in the gutter reaches steady state. Due to the addition of runoff from the roof, the depth of water in the gutter increases gradually along the length of the gutter in the direction of flow. Using data in the figure, calculate the inches per hour of rainfall that will cause the gutter to be completely filled with water at the open end. At rainfall rates higher than this, the gutter is inadequate to handle the water flow and excess water spills over the sides of the gutter before reaching the end. Assume that the exit velocity of the water from the gutter is 3 ft/s and that all the rain striking the roof is collected in the gutter. (Note that the homeowner has been too cheap to install downspouts.)

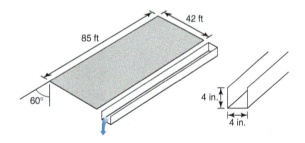

P4-32 In a home, air infiltrates from the outside through cracks around doors and windows. Consider a residence where the total length of cracks is 62 m and the total internal volume is 210 m³. Due to the wind, 9.4×10^{-5} kg/s of air enters per meter of crack and exits up a chimney. Assume that air temperature is the same inside and out and that air density is constant at 1.186 kg/m³. If windows and doors are not opened or closed, estimate the time required for one complete air change in the building.

P4-33 The hull of a vessel develops a leak and takes on water at a rate of 57.5 gal/min. When the leak is discovered, the lower deck is already submerged to a level of 7.5 in. At this time, a sailor turns on the bilge pump, which begins to remove water at a

rate of 73.8 gal/min. As an approximation, the lower deck can be modeled as a flat-bottomed container with a bottom surface area of 510 ft² and straight vertical sides. How long will it be after the pump is turned on until the deck is clear of water?

P4-34 On April 1, a reservoir has a water depth of 11 m. The reservoir is fed by a stream that becomes swollen with snowmelt as the month progresses. The volumetric flow rate of stream water entering the reservoir during the month of April is

$$\dot{V}_1 = 2.5 \times 10^7 \exp(0.026\,t)$$

where t is the time in days and the volumetric flow rate has units of m³/day ($t = 0$ at 12:01 A.M. on April 1). Water issues from the reservoir through a dam. The flow rate of the discharge at the dam is steady at a rate of 0.4×10^7 m³/day for the first 15 days of the month. At midnight on April 15, the sluice gates are adjusted to allow a higher flow rate of 6.35×10^7 m³/day. This rate remains constant until the end of the month. If the surface area of the reservoir is 2.8×10^6 m³, find the depth on April 30. Assume that the surface area remains unchanged during the month and that the effects of rainfall and evaporation are negligible.

CONSERVATION OF ENERGY IN OPEN SYSTEMS

P4-35 Hydraulic fluid enters a square conduit 2 in. on a side at a velocity of 14.2 ft/s and a temperature of 60°F. The fluid leaves the conduit at 100°F. Neglecting frictional heating and kinetic energy, find the rate of heat addition to the fluid in steady state.

P4-36 Air at 20°C and 101 kPa enters a passage between two printed circuit boards inside a desktop computer. One board contains nine chips each dissipating 2 W and five chips each dissipating 1.3 W. No heat enters the passage from the other card. If the exit temperature is 28.5°C, find the volumetric flow rate of the air. Neglect kinetic and potential energy changes.

P4-37 Water flows in a pipe 1.7 cm in diameter and 480 cm long at a velocity of 8.8 m/s. The water enters at 100°C and exits at 80°C. Calculate the rate of heat removal per square centimeter of pipe wall. Neglect frictional losses and kinetic energy.

P4-38 In a solar collector, air is heated as it flows in a rectangular channel under the collector surface, as shown in the figure. Assume the rate of heat addition on the top surface is constant and uniform at 400 W/m² and that all other sides of the channel are insulated. The air enters at 15°C and 101 kPa with a velocity of 5 m/s. The heat transfer coefficient inside the channel is 155 W/m²·K. Find the minimum and maximum temperatures of the collector surface.

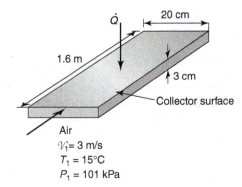

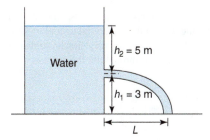

BERNOULLI EQUATION

P4-39 Water is siphoned from a waterbed into a bathtub through a 1-in.-diameter hose. The top surface of the water in the bed is 26 in. above the exit of the hose. Assume that the pressure in the waterbed is atmospheric and that the top surface recedes with negligibly small velocity. Also assume negligible frictional effects in the hose. What is the flow rate, in gal/min, of water into the bathtub?

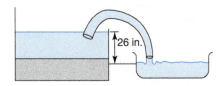

P4-40 Water flows at 25 kg/s through a gradual contraction in a pipe. The upstream diameter is 8 cm and the downstream diameter is 5.6 cm. If the exit pressure is 60 kPa, find the entrance pressure. Assume frictionless flow.

P4-41 A constriction in a water tube is used to provide suction on a submerged thin disk, as shown in the figure. Find the mass of the heaviest disk that can be supported. Assume that pressure is constant across the cross-sectional area of the constriction.

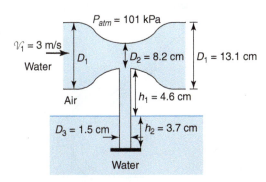

P4-42 Water issues from a hole in a large tank, as shown. Assuming frictionless flow, find L.

P4-43 A firefighter aims a jet of water at a window in a burning building, as shown in the figure. The jet is horizontal when it enters the window. If the nozzle has a diameter of 2.5 in., what is the mass flow rate of the water? (Assume there is no aerodynamic drag on the jet and the water is at 50°F)

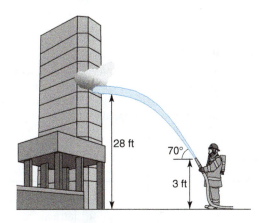

P4-44 Water flowing in a horizontal pipe branches into two pipes, as shown in the figure, and issues into the atmosphere. Neglecting all viscous effects, find the volumetric flow rate in each pipe and the diameter, D_3.

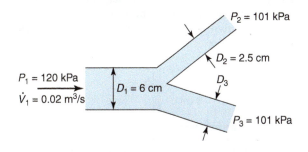

P4-45 Water at 20°C issues from the bottom of a large tank that is filled to a height of 4 m. The air pressure above the water is atmospheric. The water flows over an aluminum rod of diameter 1.1 cm and length 3 cm. The heat transfer coefficient between the water and the rod depends on velocity and is given by

$$h = 160\mathcal{V}^{0.8}$$

where h is in W/m^2·K and $\mathcal{V}$ is in m/s. The rod is initially at 65°C. How long will it take for the rod to cool to 30°C? Assume the surface of the water recedes very slowly, so that the depth of the water remains at 4 m throughout the process.

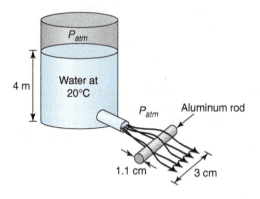

FLOW MEASUREMENT

P4-46   (WEB) Air at 17°C, 100 kPa flows in a duct. A stagnation tube connected to a U-tube manometer filled with mercury is placed in the duct. Using data on the figure, find the air velocity. Assume atmospheric pressure is 100 kPa.

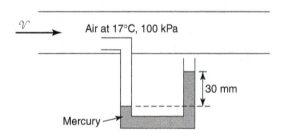

P4-47   (WEB) Oil ($SG = 0.77$) flows in a pipe with a sudden contraction, as shown in the figure. A stagnation tube open to the atmosphere is placed in the upstream section. If the oil in the stagnation tube rises to a height of $h = 22$ cm, find the velocity at the exit of the pipe.

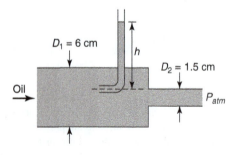

P4-48   (WEB) Hydraulic fluid at 80°F flows through a Venturi meter. The diameter at the entrance is 8.1 in., and at the throat it is 5.2 in. The pressure at the entrance is 14.7 psia. If the pressure

at the throat is measured to be 10.8 psia, find the velocity at the entrance.

CONSERVATION OF LINEAR MOMENTUM

P4-49   A block of mass 4 kg is propelled along a flat surface by a water jet, as shown in the figure. It moves to the right at a constant velocity of 2.5 m/s. The coefficient of friction between block and surface is $\mu = 0.33$. If the inlet jet area is 6.7 cm^2, find the inlet and outlet velocities of the water. Neglect wall shear and elevation changes.

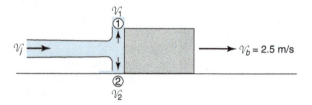

P4-50   Water at 20°C flows through a pipe, as shown in the figure. At a point 0.8 meter above the pipe bend, the velocity is 6 m/s. The exit pressure is 101 kPa. Assuming frictionless, incompressible, steady flow, find the magnitude and direction of the anchoring force.

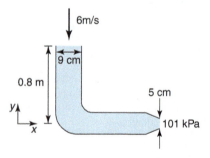

P4-51   A water jet moving at 18 ft/s strikes a vane and is turned through 120°, as shown in the figure. If the flow rate is 27 gal/min and the flow is frictionless, find the magnitude and direction of the anchoring force needed to hold the vane in place.

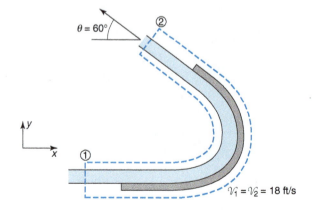

P4-52 A jet engine on a commercial aircraft exhausts combustion gases at a rate of 8 kg/s. Upon landing, a thrust reverser blocks the exhaust and redirects the flow forward, as shown in the figure. This aids in braking the plane. The exhaust gases may be assumed to flow at 200 m/s relative to the plane and to have properties very similar to those of air. Find the anchoring force needed to hold the thrust reverser on the back of the engine.

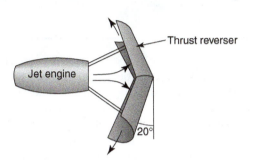

P4-53 A jet of water of area 1.6 in.^{2} and velocity 22 ft/s strikes a plate and is deflected into two symmetrical streams, as shown in the figure. If $\theta = 42°$, find the anchoring force necessary to hold the plate in place.

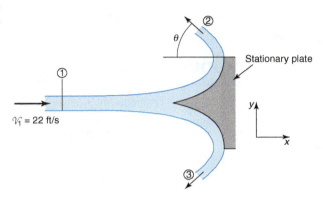

CHAPTER 5

THERMODYNAMIC PROPERTIES

5.1 INTRODUCTION

Problems in thermal–fluids engineering often involve properties such as temperature, pressure, density, internal energy, and enthalpy. These properties have been used extensively in prior chapters and will continue to be important in the chapters that follow. Until now we have focused on ideal gases because of the ease of use and utility of that model. In this chapter, however, we depart from ideal gases. The limitations of the ideal gas model are discussed and alternatives for real gas behavior are presented.

In addition, for all processes discussed thus far, substances have remained in the same phase (i.e., solid, liquid, or gas). For example, while gas temperature may have decreased in a problem, the gas remained a gas at the end of the process and did not condense into a liquid state. Likewise, liquids did not boil, and solids did not melt. In this chapter, processes involving phase change are described, and methods to evaluate the properties of substances changing phase are presented. Examples of first-law applications with boiling, condensation, sublimation, freezing, and other phase-change processes are given.

5.2 PROPERTIES OF PURE SUBSTANCES

The ideal gas law gives an excellent approximation of real gas behavior in many circumstances. It is a relationship among pressure, temperature, and density, that is,

$$P = \frac{\rho \bar{R} T}{M}$$

A more accurate way to determine the properties of a gas is from a table of experimental values. In these tables, it is typical to use specific volume instead of density. Recall that the specific volume, v, is defined as

$$v = \frac{1}{\rho}$$

Since density represents the mass per unit volume, specific volume represents the volume per unit mass. The term *specific* is frequently used to indicate a quantity defined per unit mass.

Tables of experimental values are available for a variety of pure substances. For example, Table A-12 gives the properties of the gaseous form of water (i.e., steam) as a function of pressure. For each pressure level, four properties—v, specific volume; u, internal energy; h, enthalpy; and s, entropy—are listed at different temperatures (see Table A-12). While we have already discussed v, u, and h, entropy, s, is new. Entropy is a major topic in Chapter 7 and will be presented in detail there. Table A-12 gives property values in SI units. The corresponding British unit table is Table B-12.

As long as we are dealing with water in its gaseous form (steam), the values in Table A-12 can be used instead of the ideal gas law to find property information.

In Example 5-1, we compare the results of a problem using an idealization—the ideal gas law—to that using the more accurate data from a property table; this example also illustrates how Table A-12 can be used.

EXAMPLE 5-1   Use of steam tables to determine properties

Pressure vessels are designed to withstand high pressures. The application pressure must be lower than the design rupture pressure, since rupture of the vessel could lead to loss of life or, at a minimum, extensive property damage. A pressure vessel is filled with 20.26 lbm of pressurized steam at 550°F. If the volume of the vessel is 15 ft³, determine the gage pressure of the steam using

a) the ideal gas law.

b) Table B-12.

Approach:

Part a of the problem is a straightforward application of the ideal gas law. Part b is a little tricky because of the way information is arranged in Table B-12. The table gives the specific volume for a known pressure and temperature. In this case, pressure is unknown. To use the table, first calculate the specific volume of the steam in the pressure vessel by dividing volume by mass. Then assume a value for the pressure, check the specific volume for the given temperature of 550°F, and see if the table value matches that calculated. It is unlikely that the first guess for pressure will be correct, so a second guess will be needed. Continue iterating on pressure until the table value for specific volume matches the calculated value.

Assumptions:

A1. In part a, the steam is assumed to behave like an ideal gas.

Solution:

a) The ideal gas law may be written as [A1]

$$P = \frac{\overline{R}T}{vM}$$

The specific volume, v, of the steam in the pressure vessel is

$$v = \frac{V}{m} = \frac{15 \text{ ft}^3}{20.62 \text{ lbm}} = 0.7274 \frac{\text{ft}^3}{\text{lbm}}$$

The pressure can now be calculated as

$$P = \frac{\overline{R}T}{vM} = \frac{\left(10.73 \frac{\text{psia·ft}^3}{\text{lbmol·R}}\right)(550 + 460) \text{ R}}{\left(0.7274 \frac{\text{ft}^3}{\text{lbm}}\right)\left(18.015 \frac{\text{lbm}}{\text{lbmol}}\right)}$$

$$P = 826.7 \text{psia}$$

To find the gage pressure, subtract the atmospheric pressure of 14.7 psia to get

$$P = 812 \text{ psig}$$

b) To find the pressure using Table B-12, one must look for the pressure at which the specific volume is 0.7274 ft³/lbm when the temperature is 550°F. For example, at a pressure of 180 lbf/in.² and a temperature of 550°F, the specific volume is 3.228 ft³/lbm. This specific volume is too high, so try another pressure. It does not matter if you choose a higher or a lower pressure. If you go the wrong way, you will soon discover that and realize you should reverse direction. If you pick a pressure of 250 lbf/in.², the specific volume at 550°F is 2.29 ft³/lbm. This is still too high, but at least it is closer to the desired result. At a pressure of 700 lbf/in.^{2}, the specific volume is 0.7275 ft^3/lbm at 550°F. So the final result is

$$P = 700 \, \text{psia} = 685.3 \, \text{psig}$$

Comments:

Note that the ideal gas law predicts a much higher pressure than the tables show. The tables are more accurate. Steam does not behave like an ideal gas under these conditions. Under other conditions, the steam tables and the ideal gas law will often give much closer results. How to evaluate properties correctly is the focus of the remainder of this chapter.

At low enough temperatures, all gases condense into liquids. To illustrate this process, consider the piston–cylinder assembly shown in Figure 5-1a. This closed system contains only water vapor. The weight holding the piston in place establishes the pressure in the cylinder. Now imagine that the water vapor is cooled so that it changes from state A to state B, as shown in Figure 5-1b. The temperature decreases, but the pressure remains constant because the same weight is used to compress the water vapor. The lower-temperature molecules now move less vigorously, so more collisions per unit area with the lower piston face are needed to maintain equilibrium. More collisions will occur if the molecules occupy a smaller volume; hence, state B has a lower volume than state A. The same mass fills a smaller volume, so the specific volume decreases. Process A-B can be plotted on a T-v diagram, as shown in Figure 5-2.

Now further cool the water vapor until it exists at state C. As the temperature falls, the water vapor molecules travel more and more slowly on the average and are closer together. At some point, they become close enough and are traveling slowly enough that intermolecular attractive forces become important. If more energy is removed from the gas at state C by further cooling, some of the slower molecules can no longer resist attractive forces from other molecules, and groups of molecules begin to coalesce into liquid drops. With further cooling, more molecules enter the liquid state. During this condensation process, the specific volume decreases because liquid water occupies less space than water vapor. The molecules left in the vapor phase travel at the same average speed throughout the

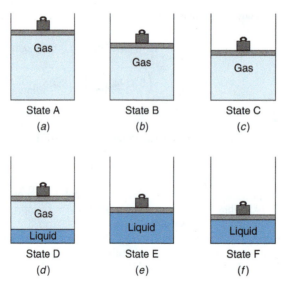

FIGURE 5-1   Condensation of a gas at constant pressure (not to scale).

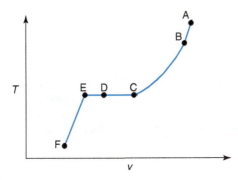

FIGURE 5-2   A plot of the condensation process at constant pressure.

condensation process. This is the mimimum speed that will keep them in the vapor phase. If energy is removed from them, they do not travel more slowly; instead, they coalesce into droplets. So the temperature, which is proportional to the average speed of the molecules, remains constant during this constant-pressure condensation.

The line CDE in Figure 5-2 represents the condensation process. At state C, only vapor exists. At state E, only liquid exists. Between these two, the cylinder contains a mixture of liquid and vapor. If the liquid in state E is cooled further, the temperature falls and the specific volume decreases slightly to point F. (The change in specific volume from E to F for the temperature change T_E to T_F is exaggerated for illustration purposes only.)

Line ABCDEF in Figure 5-2 is called an **isobar**, that is, a line of constant pressure. Certain physical states along the isobar have special names. The water vapor at point C is called a **saturated vapor**. This means that the vapor is just at the point where, if any energy is removed from it, some of the vapor will turn into liquid. Removing energy from a saturated vapor does not lower the temperature; instead, it alters the state from vapor to liquid. If energy is added to a saturated vapor, its temperature will increase and it enters the **superheated vapor** region. Points A and B in Figure 5-2 indicate superheated vapor states. The mixture of liquid and vapor that exists at any point between C and E (e.g., point D) is called a **two-phase mixture**. At point E, all the vapor has condensed. The liquid at state E is called a **saturated liquid**. The addition of any energy to a saturated liquid vaporizes some of it and the fluid enters the two-phase region, but its temperature remains constant. Removal of any energy from the liquid at point E results in a drop in temperature and the liquid enters the **subcooled (or compressed) liquid** region (e.g., point F). States E, D, and C are all at the same temperature. This special temperature is called the **saturation temperature**.

What happens at higher pressures? In Figure 5-3, two isobars at different pressures are shown. Compare points E and E′, which are both saturated liquids. At the higher pressure

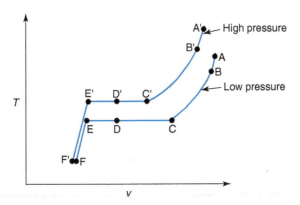

FIGURE 5-3   Two isobars at different pressures.

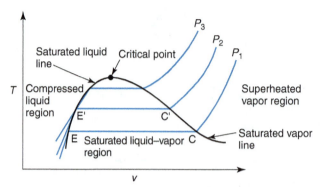

FIGURE 5-4 The liquid–vapor dome.

of point E′, water boils at a higher temperature; the higher pressure acts to keep the molecules in the liquid state. Essentially, the vapor molecules must move faster to resist the higher pressure, and faster movement implies a higher temperature.

If isobars for a variety of pressures are drawn, one can map out different regions of the T-v diagram, as shown in Figure 5-4. All the points at which water vapor just begins to condense are on the **saturated vapor line**. The points C and C′ from Figure 5-3 fall on this line. All the points at which water just begins to boil are on the **saturated liquid line**. The **critical point** separates these two lines.

The region enclosed by the saturated liquid and saturated vapor lines is the two-phase region, also called the **liquid–vapor dome**. Under the dome, the liquid phase and vapor phase exist together in equilibrium. The surface of the liquid separates the two phases. The superheated vapor region lies to the right of the saturated vapor line, while the subcooled liquid region (also called the compressed liquid region) lies to the left of the saturated liquid line.

The critical point merits special attention. In Figure 5-5, the critical isobar—the isobar that passes through the critical point—is shown. The pressure of the critical isobar is called the critical pressure, P_c. Likewise, for the critical point there is a critical temperature, T_c, and a critical specific volume, v_c. At temperatures higher than the critical temperature, liquid and vapor cannot exist together in equilibrium. At pressures higher than the critical pressure, liquid and vapor cannot exist together in equilibrium either.

For example, suppose a stoppered test tube contains only liquid water and water vapor at state 1 in Figure 5-5. There will be a meniscus separating the two phases, as shown in Figure 5-6. When the test tube is heated, the temperature rises, but the volume remains the same, as does the mass. On the T-v diagram of Figure 5-5, the state will move from

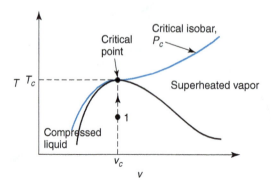

FIGURE 5-5 The critical point and the critical isobar.

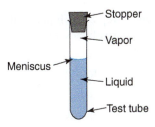

FIGURE 5-6 Test tube containing a two-phase mixture.

state 1 along the constant specific volume line shown toward the critical point. When the temperature reaches the critical temperature, the meniscus will disappear and two separate phases can no longer be identified. For substances above the critical temperature, it is not meaningful to talk of liquids and vapors. Here we have a fluid that is neither a liquid nor a vapor.

Under the liquid–vapor dome, the specific volume depends on the relative proportions of liquid and vapor present. In Figure 5-7, state A is saturated liquid with no vapor present. State B contains mostly liquid with some vapor, state C contains more vapor and less liquid, and state D contains saturated vapor with no liquid present. We can specify the mass fraction of vapor present in a two-phase mixture with a quantity called **quality**. Quality is defined as the mass of vapor divided by the total mass of the vapor–liquid mixture:

$$x = \frac{m_g}{m_f + m_g} \qquad (5\text{-}1)$$

where m_g is the mass of vapor and m_f is the mass of liquid. (The unusual subscripts designating vapor and liquid, respectively, come from the German words for vapor and liquid. These subscripts are widely used in engineering practice.) If a state has a quality of zero ($x = 0$), only saturated liquid is present. Likewise, if a state has a quality of one ($x = 1$), only saturated vapor is present. If a mixture of liquid and vapor exists together, the quality will equal some value between zero and one. The idea of quality does not apply to superheated vapors or subcooled liquids. It is not meaningful to talk about the quality of a superheated vapor or a subcooled liquid.

The specific volume is the volume per unit mass. The specific volume for a two-phase mixture, for the liquid part of the mixture, and for the vapor part of the mixture are

$$v = \frac{V_{tot}}{m_{tot}} \qquad v_f = \frac{V_f}{m_f} \qquad v_g = \frac{V_g}{m_g}$$

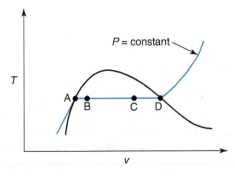

FIGURE 5-7 Points along an isobar in the liquid–vapor dome.

where *tot* designates total and f and g represent liquid and vapor, as before. The total volume is the sum of the liquid and vapor volumes, or

$$V_{tot} = V_f + V_g$$

Writing these volumes in terms of masses and specific volumes gives

$$vm_{tot} = v_f m_f + v_g m_g$$

This may be rearranged as:

$$v = \frac{v_f m_f}{m_f + m_g} + \frac{v_g m_g}{m_f + m_g}$$

where the total mass m_{tot} has been written as the sum of the masses of the vapor and the liquid. Using the definition of quality, Eq. 5-1, this equation may be rewritten as:

$$v = (1 - x)v_f + xv_g \tag{5-2}$$

The right-hand side of this equation shows that the specific volume of the mixture is the mass-weighted average of the contributions from the saturated liquid specific volume and the saturated vapor specific volume. Eq. 5-2 may be rearranged to the form

$$v = v_f + x(v_g - v_f) = v_f + xv_{fg} \tag{5-3}$$

The symbol v_{fg} is often used to designate the change in specific volume between the saturated vapor and saturated liquid states; that is, $v_{fg} = v_g - v_f$. Solving Eq. 5-3 for x gives another useful equation:

$$x = \frac{v - v_f}{v_g - v_f} = \frac{v - v_f}{v_{fg}} \tag{5-4}$$

Values for v_g, v_f, and v_{fg} are available in thermodynamic tables for many substances. For example, Table A-11 gives the properties of two-phase steam–water mixtures as a function of pressure in SI units. The saturation temperature at each pressure is indicated, along with the specific volume of the saturated liquid and saturated vapor. Other properties, such as internal energy and enthalpy for saturated liquid and vapor are also included. Table A-10 is a similar table, except that properties are given for even values of temperature instead of pressure. Tables B-11 and B-10 give the same information in British units. Tables are also available in the appendix for Refrigerant 134a (R-134a). In the following example, the use of a saturated thermodynamic table is illustrated.

EXAMPLE 5-2   Quality of a two-phase mixture

A two-phase mixture of steam and water at 100 psia occupies a volume of 0.7 ft^3. If the mass is 1.2 lbm, what is the quality?

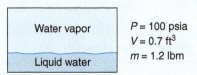

Water vapor

Liquid water

$P = 100$ psia
$V = 0.7$ ft^3
$m = 1.2$ lbm

Approach:

Divide volume by mass to find the specific volume, v, of the mixture. Determine v_f and v_g at $P = 100$ psia from Table B-11. Finally, use Eq. 5-3 to find x.

Solution:

By definition, the specific volume of the mixture is

$$v = \frac{0.7 \text{ ft}^3}{1.2 \text{ lbm}} = 0.583 \frac{\text{ft}^3}{\text{lbm}}$$

From Eq. 5-3,

$$v = v_f + x(v_g - v_f)$$

Solving for quality gives

$$x = \frac{v - v_f}{v_g - v_f}$$

The values for v_f and v_g are found in Table B-11. Inserting these and using the value for the specific volume of the mixture calculated above results in

$$x = \frac{0.583 - 0.01774}{4.434 - 0.01774}$$

$$x = 0.128$$

Comment:

This quality means that 12.8% of the total mass is vapor and 87.2% of the total mass is liquid.

EXAMPLE 5-3   Properties of a two-phase refrigerant mixture

Refrigerant 134a with a quality of 0.4 and a temperature of 12°C is contained in a rigid tank that has a volume of 0.17 m^3. Find the mass of liquid present.

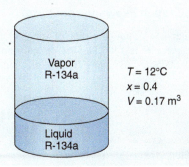

Vapor
R-134a

$T = 12$°C
$x = 0.4$
$V = 0.17$ m^3

Liquid
R-134a

Approach:

Using the known quality and temperature, you can calculate the specific volume of the mixture. Then the total mass is obtained from

$$m = \frac{V}{v}$$

From the definition of quality,

$$m_g = mx$$

Finally, the mass of liquid is just

$$m_f = m - m_g$$

Solution:

The specific volume of the two-phase mixture is given by

$$v = v_f + x(v_g - v_f)$$

Values of the specific volumes of the saturated liquid and vapor at 12°C are obtained from Table A-14. Using these property values,

$$v = 0.000797 + 0.4(0.046 - 0.000797)$$

$$v = 0.0189 \frac{\text{m}^3}{\text{kg}}$$

The total mass of the mixture is

$$m = \frac{V}{v} = \frac{0.17\,\text{m}^3}{0.0189\,\text{m}^3/\text{kg}} = 9.00\,\text{kg}$$

From the definition of quality, the mass of vapor present is

$$m_g = mx = (9)(0.4) = 3.6\,\text{kg}$$

The total mass is the sum of the vapor and liquid masses, so

$$m_f = m - m_g = 9 - 3.6 = 5.4\,\text{kg}$$

EXAMPLE 5-4   **Cavitation in a vena contracta**

Water flows through a pipe of variable area. At the entrance the water velocity is 10 m/s and the pressure is 150 kPa. The entrance area is 0.015 m². At the narrowest point, the pipe area is 0.0075 m². Assuming frictionless, isothermal flow at 20°C, find the pressure at the narrowest

$\mathcal{V}_1$ = 10 m/s

P_1 = 150 kPa

point (station 2). For the calculated pressure at station 2, will the water change phase or remain a compressed liquid?

Approach:

Assume the flow is incompressible. Use conservation of mass and the Bernoulli equation simultaneously. Apply conservation of mass between stations 1 and 2. Then use $\dot{m} = \rho \mathcal{V} A$ to determine the velocity at station 2. Apply the Bernoulli equation between 1 and 2 to find the pressure at 2. Use Table A-10 to determine the state at station 2.

Assumptions:

A1. The flow is steady.

A2. The flow is incompressible.

A3. The flow is frictionless and isothermal.

Solution:

From conservation of mass for steady flow [A1],

$$\dot{m}_1 = \dot{m}_2$$

This may be written

$$\rho \mathcal{V}_1 A_1 = \rho \mathcal{V}_2 A_2$$

The density of the water is constant [A2]; therefore,

$$\mathcal{V}_2 = \frac{\mathcal{V}_1 A_1}{A_2} = \frac{\left(10 \, \frac{m}{s}\right)\left(0.015 \, m^2\right)}{\left(0.0075 \, m^2\right)} = 20 \, \frac{m}{s}$$

From the Bernoulli equation [A3],

$$0 = \left(\frac{P_1}{\rho} + \frac{\mathcal{V}_1^2}{2} + gz_1\right) - \left(\frac{P_2}{\rho} + \frac{\mathcal{V}_2^2}{2} + gz_2\right)$$

Noting that $z_1 = z_2$ and solving for P_2,

$$P_2 = P_1 + \frac{\rho}{2}\left(\mathcal{V}_1^2 - \mathcal{V}_2^2\right)$$

$$= 150 \, kPa + \frac{997 \, \frac{kg}{m^3}}{2}\left[\left(10\frac{m}{s}\right)^2 - \left(20\frac{m}{s}\right)^2\right]\frac{1 \, kPa}{1000 \, Pa}$$

$$= 0.45 \, kPa$$

The pressure at the narrow part of the duct is very small, only 0.45 kPa. The temperature of the water is 20°C. From Table A-10, we find that the saturation pressure for water at 20°C is 2.339 kPa. This means that when the pressure is reduced to 2.339 kPa or less, the water boils and vapor bubbles form. The state at the narrow point is no longer a compressed liquid, and the density is not constant at all points in the flow. This violates the assumptions we invoked in deriving the Bernoulli equation, so the result obtained in this example is not valid and has no physical meaning.

Comments:

The formation of vapor bubbles in an isothermal liquid flow due to a localized low-pressure region is called **cavitation**. It is a serious problem in some equipment, especially pumps. The vapor bubbles that are formed are carried along with the flow. When the flow reaches a larger area, the pressure increases and the bubbles collapse suddenly. If the bubbles are near a wall, their collapse results in very high localized forces that can actually damage the solid surface. In addition, cavitation causes noise and vibration as the bubbles collapse.

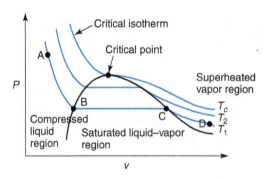

FIGURE 5-8 Evaporation of a liquid along an isotherm.

We have plotted temperature versus specific volume and examined lines of constant pressure. It is also possible to plot pressure versus specific volume and examine lines of constant temperature, as in Figure 5-8. At point A, we have a compressed liquid held at high pressure. Now let the pressure be reduced while maintaining a constant temperature through the addition of heat. At point B, the pressure is low enough that the liquid starts to boil. The reduction in pressure allows the molecules to escape from the liquid state. As more heat is added from point B to point C, all the liquid is converted to vapor. State C consists of saturated vapor. With a further reduction in pressure at constant temperature, the specific volume of the vapor increases as shown schematically by point D.

In Figure 5-9, a thermodynamic diagram that includes the solid state is shown. In addition to the liquid–vapor two-phase region, there is a solid–liquid region and a solid–vapor region. State A is a saturated solid state. This is a solid that is just about to melt. At state B, the solid and liquid exist together in equilibrium. The process from A to C is **melting**, while the process from C to A is **freezing**. State D is also a saturated solid state, but here the pressure is so low that the solid turns directly into vapor, bypassing the liquid state. The process from D to F is called **sublimation**. You may be familiar with this if you have ever seen "dry ice." Dry ice is frozen carbon dioxide that turns directly into vapor at atmosphere pressure. It is sometimes used in theatrical productions to create a dramatic fog. At state E, solid and vapor exist together in equilibrium.

Figure 5-9 also includes the so-called **triple line**. At points along this line, solid, liquid, and vapor all exist together in equilibrium. Imagine a container with a solid layer covered by a liquid layer. Above the liquid layer is a vapor region. Since we are dealing with pure substances, the solid, liquid, and vapor all have the same chemical identity.

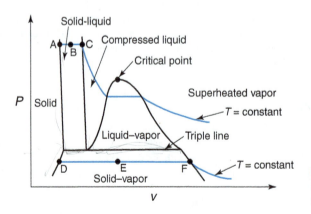

FIGURE 5-9 A diagram showing solid, liquid, and vapor states.

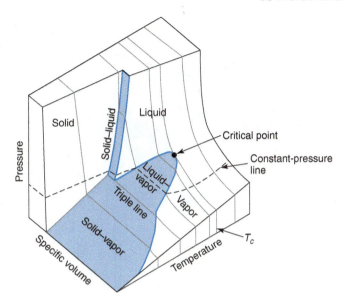

FIGURE 5-10 Isometric view of the relation among temperature, pressure, and volume.

The temperature and pressure are the same at all points along the triple line. These are called the triple-point temperature and triple-point pressure.

We have now seen temperature versus specific volume and pressure versus specific volume. Another way to visualize the behavior of pure substances is with a three-dimensional plot of pressure versus temperature and specific volume, as shown in Figure 5-10. The two-phase regions are highlighted in blue. When we project the three-dimensional plot onto the *P-v* plane, we obtain Figure 5-9. When we project the three-dimensional plot onto the *T-v* plane, we obtain a figure similar to Figure 5-4 but with a solid region included. It is also useful to consider a projection onto the *P-T* plane, as shown in Figure 5-11. Here each of the two-phase regions has collapsed onto a line. These lines separate the single-phase regions where solid, liquid, and vapor exist. The triple line projects onto a triple point. At point A, for example, liquid and vapor exist in equilibrium with a range of possible values of specific volume.

The diagrams shown so far all deal with pure substances that contract upon freezing. In Figure 5-9, the solid at state A has a lower specific volume than the liquid at state C.

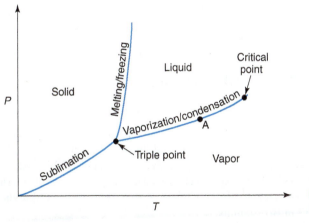

FIGURE 5-11 Projection of Figure 5-10 onto the *P-T* plane.

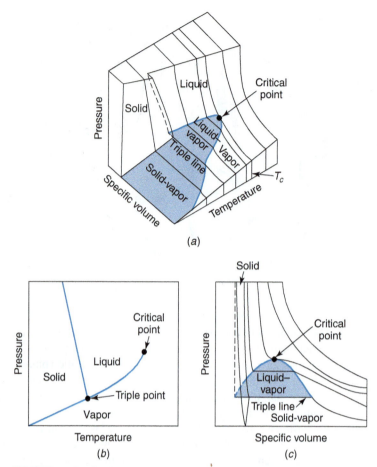

FIGURE 5-12 *P-v-T* relationship for a substance that expands on freezing (e.g., water). (a) *P-v-T* relationship. (b) Projection onto the *P-T* plane. (c) Projection onto the *P-v* plane.

It is reasonable to expect that the solid would occupy less space than the liquid. However, there is one very important substance that is an exception to this rule—water. Ice takes up more space than the corresponding volume of liquid water. The fact that water expands on freezing has had important implications for the geological history of our planet. Water that has collected in cracks in rock expands on freezing and fractures the rock, accelerating erosion. In addition, if water did not expand upon freezing, ice would not float. The thermodynamic diagrams for water are shown in Figure 5-12.

5.3 INTERNAL ENERGY AND ENTHALPY IN TWO-PHASE SYSTEMS

Thermodynamic tables generally contain more than pressure, temperature, and specific volume data. Values for internal energy and enthalpy are also available. The procedure for arriving at these values is beyond the scope of this text; however, we can say that the data result from experimental measurements. Values of internal energy and enthalpy for steam–water are given in Tables A-10 through A-13 in SI units and in Tables B-10 through B-13 in British units.

For an ideal gas, internal energy, u, and enthalpy, h, are only functions of temperature. However, in general in the single-phase regions, u and h depend on temperature and pressure, though the variation with pressure is often small. The internal energy per unit mass is called the specific internal energy. You may recall that the tables also contain the specific volume, v, which is the volume per unit mass. Similarly, h is the enthalpy per unit mass, that is,

$$ v = \frac{V}{m} \qquad u = \frac{U}{m} \qquad h = \frac{H}{m} $$

where m is the total mass of the system.

In a two-phase region, specific volume results from the mass-weighted average of the saturated liquid and saturated vapor specific volumes. In like manner, the internal energy of a mixture is also a mass-weighted average. For example, in the liquid–vapor two-phase region, the specific internal energy of the liquid phase is denoted as u_f, while the specific internal energy of the vapor phase is u_g. The specific internal energy of the mixture is

$$
\begin{aligned}
u &= (1-x)\,u_f + x u_g \\
&= u_f + x\left(u_g - u_f\right) \\
&= u_f + x u_{fg}
\end{aligned}
\tag{5-5}
$$

where $u_{fg} = u_g - u_f$. This equation is derived in exactly the same way that Eq. 5-2 and Eq. 5-3 for specific volume were derived. Enthalpy in the two-phase region is given by a similar set of equations as

$$
\begin{aligned}
h &= (1-x)\,h_f + x h_g \\
&= h_f + x\left(h_g - h_f\right) \\
&= h_f + x h_{fg}
\end{aligned}
\tag{5-6}
$$

where $h_{fg} = h_g - h_f$ is called the **enthalpy of vaporization** or **heat of vaporization**. Finally, we can write

$$ x = \frac{m_g}{m_f + m_g} = \frac{v - v_f}{v_g - v_f} = \frac{u - u_f}{u_g - u_f} = \frac{h - h_f}{h_g - h_f} = \frac{s - s_f}{s_g - s_f} \tag{5-7} $$

Entropy, s, is included here for completeness, though it will not be discussed or used until Chapter 7.

We have now introduced seven **thermodynamic properties**—T, P, v, u, h, s, and x. A property is a quantity that characterizes a system in equilibrium. We say that a system is at a given state due to the values of its properties. For example, if a gas is at temperature T_1 and pressure P_1, then T_1 and P_1 are properties of the gas at state 1. Quantities such as heat and work are not properties. They characterize the **process** that takes place when a system moves from one state to another. If a gas at state 1, characterized by T_1 and P_1, is heated while volume remains constant until it reaches T_2 and P_2, then we say that the system has moved from state 1 to state 2 while heat is transferred. The value of a property at a state is not dependent on the process used to arrive at that state.

These definitions of property and state are helpful in applying the first law of thermodynamics. A particularly powerful concept that we will use often throughout the rest of the text is the **state principle**. A simple statement of this principle is:

Two independent thermodynamic properties are needed to completely specify the state of a pure substance.

This statement implies that if we know two independent properties of a substance, then the state of the substance is fixed, and we can evaluate *all* other properties of that substance at that state. Later in this chapter, a more rigorous statement of the state principle is given. We have seen an example of the state principle in the ideal gas law. If we know T and P, then v can be found from the ideal gas law. Likewise, if v and P are known, T is easily determined. Less obvious is the fact that u and x are properties. If we know P and x, for example, it will be possible to find T and v using thermodynamic tables. That is very useful, as is demonstrated in the next example.

EXAMPLE 5-5   **Heating of a two-phase mixture in a rigid tank**

A rigid tank, which has a volume of 1.5 ft^3, contains H_2O at a quality of 0.87 and a pressure of 30 psia. Heat is added until only saturated vapor remains. How much heat is added?

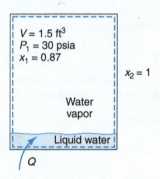

Approach:

Choose the system to be both the water vapor and the liquid water. To determine the heat transferred, apply the first law. In this process, the volume remains constant, no work is done, and there are no changes in kinetic or potential energy; therefore, the first law reduces to $\Delta U = Q$. Furthermore, the *specific* volume remains constant. The specific volume at the initial state can be calculated from the given information and the stream tables used to find the initial value of internal energy. The final state is saturated vapor with the same value of specific volume as the initial state. Knowing that, you can determine the final internal energy from the saturated steam tables.

Assumptions:

A1. Volume is constant during the process.

Solution:

The system under study is the mixture of vapor and liquid. The heating process can be visualized on a $P\text{-}v$ diagram. At state 1, the initial state, a two-phase mixture exists. Since the tank is rigid, the volume does not change [A1]. Mass is not added or removed, so the mass does not change either. It follows that the specific volume $v = V/m$ does not change during the process. At the final state, state 2, the tank is filled with saturated vapor. The process is plotted in the figure.

To find the heat added, apply the first law for a closed system:

$$\Delta KE + \Delta PE + \Delta U = Q - W$$

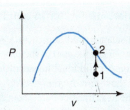

A2. Kinetic energy does not change.
A3. Potential energy does not change.

In this process, there is no change in the kinetic or potential energy [A2][A3]. There is also no work done, so the first law reduces to

$$\Delta U = Q$$

or

$$U_2 - U_1 = Q$$

The heat, Q, is the quantity we want to determine. The total internal energy, U_1, is given by

$$U_1 = mu_1$$

We can find the mass, m, if we know the volume and specific volume. At state 1, the pressure and quality are known, as given in the problem statement. Since these are two independent properties, by the state principle, we can find all other thermodynamic properties of state 1. From Table B-11 at $P_1 = 30$ psia, $v_f = 0.017$ ft³/lbm and $v_g = 13.75$ ft³/lbm. At state 1, the specific volume is

$$v_1 = v_f + x(v_g - v_f)$$

$$v_1 = 0.017 + 0.87(13.75 - 0.017)$$

$$v_1 = 11.97 \frac{ft^3}{lbm}$$

The mass can now be found as

$$m = \frac{V}{v_1} = \frac{1.5\,ft^3}{11.97\,\frac{ft^3}{lbm}} = 0.125\,lbm$$

The specific internal energy at state 1 is

$$u_1 = u_{f1} + x\left(u_{g1} - u_{f1}\right) = \left[218.84 + 0.87\left(1088 - 218.84\right)\right]\frac{Btu}{lbm} = 975.0\,\frac{Btu}{lbm}$$

where values of u_{f1} and u_{g1} were obtained from Table B-11 at $P_1 = 30$ psia. To find u_2, we need to make use of the fact that $v_1 = v_2 = 11.97$ ft³/lbm. We know that state 2 is a saturated vapor with $v_g = v_2 = 11.97$ ft³/lbm. Therefore, we know two properties at state 2—specific volume and quality ($x_2 = 1$). By the state principle, we could find all other properties at state 2 if we needed them. In Table B-11, if $v_g = 11.97$ ft³/lbm, then $P_2 = P_{sat} = 35$ psia and $u_2 = u_g = 1090.3$ Btu/lbm. We now have all the pieces we need to calculate heat as

$$Q = U_2 - U_1 = m\left(u_2 - u_1\right)$$

$$Q = 0.125\,lbm\,(1090.3 - 975.0)\,\frac{Btu}{lbm}$$

$$Q = 14.41\,Btu$$

EXAMPLE 5-6 An immersion heater in a rigid tank

A well-insulated rigid tank with a volume of 360 in.³ initially contains a two-phase mixture of R-134a at 5°F. At the start of the process, the tank is half-filled by volume with liquid and half-filled with vapor. A cylindrical heater of length 2.5 in. and diameter 0.5 in. is inserted in the liquid, as shown in the figure. A voltage drop of 60 V is imposed across the heater, which has a resistance of 35 Ω. The heat transfer coefficient on the outside of the heater is 365 Btu/ h·ft²·°F. The heater is operated until the final pressure is 100 psia. The heater remains covered by liquid during the entire process. Calculate

a) the time required for the process.

b) the maximum surface temperature of the heater.

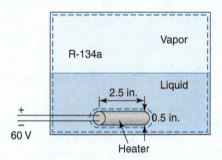

Approach:

Choose the system to be the vapor and liquid R-134a. Use the first law to find the total amount of heat transferred. Assume that all the electrical work done on the heater is transferred as heat to the R-134a and that none is diverted to raising the temperature of the heater itself (the heater mass is small). To calculate internal energy, the initial quality is needed, which can be determined from the known initial volumes of vapor and liquid. To fix the final state, use the fact that the tank is rigid and, therefore, specific volume is constant. From the given value of final pressure and the final specific volume, the final quality can be calculated. This is used to find the final internal energy and, hence, the total heat, Q.

The power generated by the heater, $\dot{Q}$, is the voltage times the current. The time for the process is found from $t = Q/\dot{Q}$. During the process, the temperature of the two-phase system increases. Therefore, the maximum surface temperature of the heater will occur at the end of the process when the R-134a is hottest. The final R-134a temperature is the saturation temperature, which corresponds to the given final pressure. To determine the surface temperature of the heater, use $\dot{Q} = hA (T_s - T_2)$.

Assumptions:

A1. The tank is rigid.

A2. The heater is submerged throughout the process.

Solution:

a) Define the closed system to be all the R-134a present. The heating process can be visualized on a P-v diagram, as shown in the figure. State 1, the initial state, is in the two-phase region. Since the tank is rigid and the total mass is constant, the specific volume does not change [A1] and the process follows a vertical line on the P-v diagram. State 2, the final state, is still in the two-phase region since the heater remains submerged in liquid at the final state [A2].

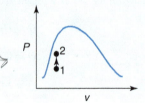

To find the heat added, apply the first law for a closed system:

$$\Delta KE + \Delta PE + \Delta U = Q - W$$

A3. Kinetic energy does not change.
A4. Potential energy does not change.

In this process, there is no change in the kinetic or potential energy, and no work is done [A3][A4]. The first law reduces to

$$\Delta U = Q$$

or

$$U_2 - U_1 = Q = m\,(u_2 - u_1)$$

To calculate the internal energy, u_1, we will need the initial quality. The mass of vapor present in the tank at the initial state is

$$m_{g1} = \frac{V_{g1}}{v_{g1}} = \frac{V/2}{v_{g1}}$$

where V is the total volume of the tank (the vapor occupies half the tank). The specific volume, v_{g1}, is found from Table B-14 at 5°F. Substituting values,

$$m_{g1} = \frac{\left(\dfrac{360\,\text{in.}^2}{2}\right)\left(\dfrac{1\,\text{ft}}{12\,\text{in.}}\right)^3}{\left(1.92\,\dfrac{\text{ft}^3}{\text{lbm}}\right)} = 0.0542\,\text{lbm}$$

Similarly, the mass of liquid present is

$$m_{f1} = \frac{V_{f1}}{v_{f1}} = \frac{V/2}{v_{f1}} = \frac{\left(\dfrac{360\,\text{in.}^2}{2}\right)\left(\dfrac{1\,\text{ft}}{12\,\text{in.}}\right)^3}{\left(0.0119\,\dfrac{\text{ft}^3}{\text{lbm}}\right)} = 8.75\,\text{lbm}$$

The total mass is

$$m = m_{g1} + m_{f1} = 0.0542 + 8.75 = 8.81\,\text{lbm}$$

By definition, the initial quality is

$$x_1 = \frac{m_{g1}}{m} = \frac{0.0542\,\text{lbm}}{8.81\,\text{lbm}} = 0.00616$$

The initial internal energy may now be determined using

$$u_1 = u_{f1} + x_1\,(u_{g1} - u_{f1})$$

With values for specific internal energy from Table B-14 at 5°F,

$$u_1 = 13.09 + 0.00616\,(94.01 - 13.09) = 13.6\,\frac{\text{Btu}}{\text{lbm}}$$

To find the final internal energy, we need to fix the final state. We know that the specific volume is constant during this process because the tank is rigid. Therefore

$$v_1 = v_2 = v_{f1} + x_1\,(v_{g1} - v_{f1})$$

Substituting values from Table B-14 at 5°F,

$$v_1 = 0.0119 + 0.00616\,(1.92 - 0.0119) = 0.0236\,\frac{\text{ft}^3}{\text{lbm}}$$

The final quality can be determined from

$$x_2 = \frac{v_2 - v_{f2}}{v_{g2} - v_{f2}}$$

At the final state, the pressure is given as 100 psia. Using values of v_{f2} and v_{g2} at 100 psia from Table B-15,

$$x_2 = \frac{0.0236 - 0.0133}{0.475 - 0.0133} = 0.0224$$

The final internal energy may now be calculated as

$$u_2 = u_{f2} + x_2\,(u_{g2} - u_{f2})$$

Using values in Table B-15,

$$u_2 = 36.75 + 0.0224\,(103.7 - 36.75) = 38.25\,\frac{\text{Btu}}{\text{lbm}}$$

The total heat transferred to the R-134a during this process is

$$Q = m\,(u_2 - u_1) = (8.81\,\text{lbm})\,(38.25 - 13.6)\,\frac{\text{Btu}}{\text{lbm}} = 217\,\text{Btu}$$

A5. The mass of the heater is small.

We assume all the electrical work in the heater is converted to heat that enters the R-134a mixture. In actuality, some heat is needed to raise the temperature of the heater itself; however, we assume the heater has a small mass and little heat is required to raise its temperature [A5].

The rate of heat generated in the heater is voltage times current, or

$$\dot{Q} = \xi I = \frac{\xi^2}{R} = \frac{(60\,\text{V})^2}{35\,\Omega} = 103\,\text{W}$$

A6. Heat generation rate is constant.

Assuming the rate of heat generation is constant with time [A6],

$$Q = \int \dot{Q}\,dt = \dot{Q}t$$

Solving for elapsed time gives

$$t = \frac{Q}{\dot{Q}} = \frac{217\,\text{Btu}}{(103\,\text{W})\left(3.412\,\dfrac{\text{Btu/h}}{1\,\text{W}}\right)} = 0.619\,\text{h}$$

b) We now need the maximum surface temperature of the heater. As heat is added to the refrigerant, the pressure and temperature of the two-phase mixture both increase. The heater surface will be hottest when the two-phase mixture is hottest, that is, at the final state. From Table B-15, the saturation temperature at the final pressure of 100 psia is $T_2 = 79.2°F$. The heat generated by the heater is related to surface temperature through

$$\dot{Q} = hA\,(T_s - T_2)$$

A7. The heat transfer coefficient is uniform over the cylinder.

where T_s is surface temperature and the heat transfer coefficient, h, is assumed to be uniform over the surface of the heater [A7]. The area of the cylindrical heater, including both ends and the curved portion, is

$$A = 2\pi r^2 + \pi rL$$

where r is radius and L is length. Substituting values,

$$A = 2\pi \,(0.25\,\text{in.})^2 + \pi \,(0.25\,\text{in.})\,(2.5\,\text{in.}) = 2.36\,\text{in.}^2$$

Solving for surface temperature and substituting values gives

$$T_s = \frac{\dot{Q}}{hA} + T_2 = \frac{351\,\dfrac{\text{Btu}}{\text{h}}}{\left(365\,\dfrac{\text{Btu}}{\text{h·ft}^2\text{·°F}}\right)(2.36\,\text{in.}^2)\left(\dfrac{1\,\text{ft}^2}{144\,\text{in}^2}\right)} + 79.2\,°\text{F} = 138\,°\text{F}$$

EXAMPLE 5-7   Heating at constant pressure

Two kilograms of saturated liquid water at 50 kPa are heated slowly at constant pressure. During this process, 5876 kJ of heat are added. Find the final water temperature.

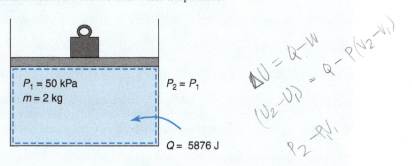

$P_1 = 50\ \text{kPa}$
$m = 2\ \text{kg}$

$P_2 = P_1$

$Q = 5876\ \text{J}$

$\Delta U = Q - W$
$(U_2 - U_1) = Q - P(V_2 - V_1)$
$P_2 - PV_1$

Approach:

Select the saturated water as the closed system. Because the process occurs at constant pressure, the first law reduces to $Q = \Delta H$. State 1 is a saturated liquid. The enthalpy of saturated liquid water at the given pressure, h_1, can be found in Table A-11. Using h_1 and the given values of mass and Q, the enthalpy of the final state, h_2, can be calculated. The final pressure is the same as the initial pressure; hence, two independent properties of the final state are known, enthalpy and pressure. Therefore, by the state principle, all other properties, including temperature, can be determined.

The final state could be either a two-phase mixture or a superheated vapor. If h_2 is greater than the enthalpy of saturated liquid but less than that of saturated vapor, the final state will be two-phase. Otherwise, it will be superheated vapor. The final state is located in the appropriate table, and the temperature is determined.

Assumptions:

Solution:

Define the system to be the water in the piston–cylinder assembly. For a closed system, the first law is

$$\Delta KE + \Delta PE + \Delta U = Q - W$$

A1. The process is slow.

Assuming that the process is slow enough to be considered quasi-equilibrium, work is given by [A1]

$$W = \int P\,dV$$

A2. Kinetic energy changes are negligible.
A3. Potential energy changes are negligible.

Substituting this into the first law and neglecting kinetic and potential energy changes [A2][A3],

$$\Delta U = Q - \int P \, dV$$

This is a constant-pressure process; therefore,

$$\Delta U = Q - P \int dV = Q - P \Delta V$$

which may be rearranged to

$$Q = \Delta U + P \, \Delta V$$

By definition, $H = U + PV$; therefore,

$$Q = \Delta H = H_2 - H_1 = m (h_2 - h_1)$$

This equation was previously derived in Chapter 2 (see Eq. 2-36). Solving for h_2 produces

$$h_2 = \frac{Q}{m} + h_1$$

The enthalpy of saturated liquid water at state 1, h_1, is found in Table A-11. Using this and the given values, h_2 becomes

$$h_2 = \frac{5876 \, \text{kJ}}{2 \, \text{kg}} + 340.5 \, \text{kJ/kg}$$

$$h_2 = 3278 \, \text{kJ/kg}$$

To find the final temperature, T_2, we need to apply the state principle. Since h_2 and P_2 are known, it should be possible to find T_2. The problem is that it is not obvious which table to look in. Is the final state a two-phase mixture, a saturated vapor, or a superheated vapor?

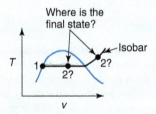

Let us assume for the moment that the final state is a two-phase mixture. In Table A-11 at 50 kPa, the enthalpies of saturated liquid and saturated vapor are

$$h_f = 340.5 \, \text{kJ/kg}$$

$$h_g = 2305.4 \, \text{kJ/kg}$$

All the values of enthalpy for a mixture of liquid and vapor fall between these two values. This is because the enthalpy of the mixture is a weighted average of the enthalpies at the liquid and vapor states.

The value of h_2 calculated above ($h_2 = 3278$ kJ/kg) is higher than h_g. This indicates that the final state is not a two-phase mixture, but a superheated vapor. Therefore, from Table A-12, at $P_2 = 50$ kPa (0.05 MPa) and $h_2 = 3278$ kJ/kg,

$$T_2 = 400°C.$$

EXAMPLE 5-8 **Power production and heat transfer in a reciprocating steam engine**

In a reciprocating steam engine, a piston–cylinder assembly is 30 cm in diameter. Initially, the piston is 7.5 cm from the end of the cylinder, and the enclosed volume contains saturated water vapor at 500 kPa. The steam expands to 100 kPa and four times the initial volume. The engine operates at 120 rpm. Outside the assembly, air at 25°C flows over the cylinder with a convective heat transfer coefficient of 100 W/m²·K. From previous experience you can estimate the outside surface temperature of the cylinder as the average of the steam's initial and final temperatures and the area for heat transfer as the area of the cylinder when the steam is fully expanded. Because of the danger of someone getting burned on the hot cylinder, the cylinder is covered with 3 cm of insulation ($k = 0.05$ W/m·K). Determine the following:

a) The work produced during one expansion before the insulation is added (in kJ)

b) The work produced during one expansion after the insulation is added (in kJ)

c) The temperature on the outside surface of the insulation (in °C)

Approach:

We want to determine work produced, so we select the steam as the system and apply the closed-system energy equation. In part a, where there is no insulation, the heat transferred is determined from $\dot{Q} = hA\left(T_{avg} - T_f\right)$, where T_{avg} is the average of the initial and final steam temperature and T_f is the air temperature. In parts b and c, where there is insulation, the thermal resistance is the series combination of the conduction resistance through the insulation and the convection resistance on the outside of the insulation.

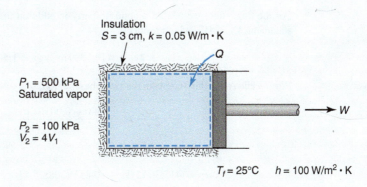

Insulation
$S = 3$ cm, $k = 0.05$ W/m · K

$P_1 = 500$ kPa
Saturated vapor

$P_2 = 100$ kPa
$V_2 = 4V_1$

Q

W

$T_f = 25°C$ $h = 100$ W/m² · K

Assumptions:

A1. Potential and kinetic energy effects are negligible.

A2. Heat transfer is one-dimensional.

A3. The heat transfer rate is constant.

Solution:

a) We define the system as the steam contained in the piston–cylinder assembly. For this system the closed system energy equation is

$$\Delta KE + \Delta PE + \Delta U = Q - W$$

Assuming negligible change in kinetic and potential energy [A1], and solving for work,

$$W = Q - \Delta U = Q - m\left(u_2 - u_1\right)$$

The heat transfer is evaluated from the basic rate equation [A2]:

$$\dot{Q} = \frac{\Delta T}{R_{tot}}$$

Integrating this equation with respect to time, and assuming the heat transfer rate is constant over the time, t, of the expansion [A3],

$$Q = \frac{\Delta T}{R_{tot}} t$$

A4. There is no heat transfer through the piston.

Because heat transfer is out of the assembly, it must be negative; and assuming there is no heat transfer through the piston [A4],

$$W = -\dot{Q}_{end} t - \dot{Q}_{side} t - m(u_2 - u_1)$$

Using the convective resistance, and assuming the average temperature of the assembly is

$$T_{avg} = \left(\frac{T_1 + T_2}{2} \right)$$

$$\dot{Q}_{end} = \frac{T_{avg} - T_f}{1/hA_{end}} \qquad \text{and} \qquad \dot{Q}_{side} = \frac{T_{avg} - T_f}{1/hA_{side}}$$

We now can begin evaluating all the parameters in the equations. The properties of the steam at the initial state (saturated vapor, $P_1 = 500$ kPa) are obtained from Table A-11: $v_1 = v_{g1} = 0.3749 \, \text{m}^3/\text{kg}; u_1 = u_{g1} = 2561.2 \, \text{kJ/kg}; T_1 = 151.86°\text{C}$. The mass is

$$m = \frac{V_1}{v_1} = \frac{(\pi/4)D^2 L_1}{v_1} = \frac{(\pi/4)(0.3 \, \text{m})^2 (0.075 \, \text{m})}{0.3749 \, \text{m}^3/\text{kg}} = 0.0141 \, \text{kg}$$

For the final state, we need a second property in addition to $P_2 = 100$ kPa. From the problem statement, $V_2 = 4V_1$, which results in

$$v_2 = 4v_1 = 4 \left(0.3749 \, \text{m}^3/\text{kg} \right) = 1.500 \, \text{m}^3/\text{kg}$$

From Table A-11, this state is in the two-phase region, so $T_2 = 99.63°\text{C}$ and

$$x_2 = \frac{v_2 - v_{f2}}{v_{g2} - v_{f2}} = \frac{(1.500 - 0.001043)}{(1.6940 - 0.001043)} = 0.885$$

$$u_2 = u_2 + x_2 u_{fg}$$

$$u_2 = 417.36 \, \text{kJ/kg} + (0.885)(2088.7 \, \text{kJ/kg}) = 2266.7 \, \text{kJ/kg}$$

The average surface temperature is:

$$T_{avg} = \frac{151.86°\text{C} + 99.63°\text{C}}{2} = 125.7°\text{C}$$

A5. The surface of the cylinder at maximum piston extension is at the average of the initial and final steam temperature.

The area of the cylinder wall exposed to the steam changes as the piston moves in and out. After many cycles, the wall reaches the approximately constant wall temperature, T_{avg}. We use the surface area of the cylinder at its maximum volume rather than at its minimum volume to calculate heat transfer because the entire wall is exposed to the steam for at least part of the cycle. The areas of the end and sides are [A5]

$$A_{end} = \frac{\pi}{4} D^2 = \frac{\pi}{4} (0.30 \, \text{m})^2 = 0.0707 \, \text{m}^2$$

$$A_{side} = \pi D 4 L_1 = \pi (0.30 \, \text{m})(4)(0.075 \, \text{m}) = 0.283 \, \text{m}^2$$

The time for one-half revolution (for the expansion process) is

$$t = \frac{(0.5 \, \text{rev})(60 \, \text{s}/1 \, \text{min})}{120 \, \text{rev/min}} = 0.25 \, \text{s}$$

Therefore,

$$\dot{Q}_{end} = \frac{(125.7 - 25)°C(1\,kJ/1000\,J)}{\dfrac{1}{(100\,W/m^2 \cdot K)(0.0707\,m^2)}} = 0.712\frac{kJ}{s}$$

$$\dot{Q}_{side} = \frac{(125.7 - 25)°C(1\,kJ/1000\,J)}{\dfrac{1}{(100\,W/m^2 \cdot K)(0.283\,m^2)}} = 2.850\frac{kJ}{s}$$

$$W = -\left(0.712\frac{kJ}{s}\right)(0.25\,s) - \left(2.850\frac{kJ}{s}\right)(0.25\,s) - (0.0141\,kg)(2266.7 - 2561.2)\frac{kJ}{kg}$$

$$= 3.62\,kJ$$

b) When insulation is added, we must take into account both the convective and conductive resistances. Note also that the area of the side increases:

$$A_{side} = \pi\,(D + 2S)\,4L_1 = \pi\,[0.30\,m + 2\,(0.03\,m)]\,(4)\,(0.075\,m) = 0.339\,m^2$$

$$\dot{Q}_{side} = \frac{T_{avg} - T_f}{\dfrac{1}{hA_{side}} + \dfrac{\ln(r_2/r_1)}{2\pi kL_1}}$$

$$\dot{Q}_{side} = \frac{(125.7 - 25)°C(1\,kJ/1000\,J)}{\dfrac{1}{(100\,W/m^2 \cdot K)(0.339\,m^2)} + \dfrac{\ln(0.36/0.30)}{2\pi(0.05\,W/m \cdot K)(4)(0.075\,m)}} = 0.051\frac{kJ}{s}$$

For the end:

$$\dot{Q}_{end} = \frac{T_{avg} - T_f}{\dfrac{1}{hA_{end}} + \dfrac{S}{kA_{end}}}$$

where S is the thickness of the insulation. Substituting values

$$\dot{Q}_{end} = \frac{(125.7 - 25)°C\,(1\,kJ/1000\,J)}{\dfrac{1}{(100\,W/m^2 \cdot K)\,(0.0707\,m^2)} + \dfrac{0.03\,m}{(0.05\,W/m \cdot K)\,(0.0707\,m^2)}} = 0.012\frac{kJ}{s}$$

$$W = -\left(0.012\frac{kJ}{s}\right)(0.25\,s) - \left(0.051\frac{kJ}{s}\right)(0.25\,s) - (0.0141\,kg)(2266.7 - 2561.2)\frac{kJ}{kg}$$

$$= 4.15\,kJ$$

c) The surface temperature of the insulation can be obtained from the rate equation and the heat transfer rate found above. For the end surface:

$$T_{end} = T_f + \frac{\dot{Q}_{end}}{hA_{end}} = 25°C + \frac{0.012\,kW\left(\dfrac{1000\,W}{1\,kW}\right)}{(100\,W/m^2 \cdot K)\,(0.0707\,m^2)} = 26.7°C$$

For the side surface:

$$T_{side} = T_f + \frac{\dot{Q}_{side}}{hA_{side}} = 25°C + \frac{0.051\,kW\left(\dfrac{1000\,W}{1\,kW}\right)}{(100\,W/m^2 \cdot K)\,(0.339\,m^2)} = 26.5°C$$

Comments:

In part a without the insulation, the heat loss was equivalent to $100(0.891/3.62) = 24.6\%$ of the work produced. In part b with the insulation, the heat loss was reduced to only 0.38% of the work produced and the surface temperature was reduced from 125.7°C to near 25°C. The addition of the insulation has a very positive impact and would pay for itself quickly.

5.4 PROPERTIES OF REAL LIQUIDS AND SOLIDS

Ideal liquids and solids are, by definition, incompressible. Their densities are constant under all conditions. Ordinarily this is a very good assumption; however, there are some important exceptions. The expansion of liquid mercury as a function of temperature is the principle used in making thermometers. Bridges are constructed with expansion joints to allow the roadway to increase in size without buckling as temperature increases. The density of seawater is elevated at great depths. In this section, we present the use of thermodynamic tables to deal with real compressible liquids and solids.

As in ideal gases, the internal energy of **ideal solids and liquids** depends only on temperature. In differential form

$$du = c_v \, dT$$

If c_v is not a function of temperature,

$$\Delta u = c_v \, \Delta T$$

What about enthalpy? The enthalpy of an ideal gas depends only on temperature. Is this true for ideal solids and liquids? By definition,

$$h = u + Pv$$

For a process that starts at state 1 and ends at state 2,

$$h_1 = u_1 + P_1 v_1$$

$$h_2 = u_2 + P_2 v_2$$

If we assume an ideal solid or liquid, the volume does not change and

$$v_1 = v_2 = v$$

The difference in enthalpy is then

$$h_2 - h_1 = u_2 - u_1 + v(P_2 - P_1) \tag{5-8}$$

The internal energy, u, is a function only of temperature for an ideal solid or liquid; however, the enthalpy depends on pressure as well. For an isothermal process, $u_1 = u_2$ and Eq. 5-8 reduces to

$$h_2 - h_1 = v \, (P_2 - P_1) \tag{5-9}$$

For real solids and liquids, internal energy and enthalpy are typically strong functions of temperature and weak functions of pressure. Because the variation with pressure is so slight, it is rare to find a table that gives properties of compressed liquids, especially for low values of pressure. Instead, practitioners typically approximate v, u, and h for the compressed liquid by using data available in the saturated liquid–vapor tables.

A word about the meaning of a "compressed liquid" is appropriate here. This means that a liquid at a given temperature, T, is at a pressure higher than the saturation pressure corresponding to T. For example, consider a glass of water sitting on a dining room table at room temperature ($\approx 20°C$). The pressure in the room is 100 kPa. At $20°C$, the water's saturation pressure is 2.38 kPa. Because the room pressure is greater than the saturation pressure, the water is in the compressed liquid region. A compressed liquid and a subcooled liquid are the same thing, and the following approximations are often used:

$$v(T,P) \approx v_f(T)$$

$$u(T,P) \approx u_f(T)$$

In words, the specific volume or internal energy of a compressed liquid at T and P is approximately equal to the specific volume or internal energy, respectively, of the saturated liquid at temperature T. Note that it is the *temperature* of the saturated liquid, not the pressure, that is the important parameter. At very high pressures, these approximations are not accurate, but at ordinary pressures, they are excellent approximations.

To approximate enthalpy for a compressed liquid, apply Eq. 5-9 to an isothermal compression from a saturated liquid to a compressed liquid. If state 1 is a saturated liquid at temperature T and state 2 is a compressed liquid at temperature T and pressure P, then Eq. 5-9 becomes

$$h(T,P) - h_f(T) \approx v[P - P_{sat}(T)]$$

In this equation specific volume is constant. The specific volume of the saturated liquid at temperature T is an excellent approximate value for the specific volume during this process. Using this and solving for enthalpy results in

$$h(T,P) \approx h_f(T) + v_f(T)[P - P_{sat}(T)] \qquad \text{compressed liquid} \qquad (5\text{-}10)$$

Consider the common process in which a solid or liquid is heated at constant pressure. For example, if an empty frying pan is heated on a range top, heat is added at constant pressure. In this case, the metal expands as it heats. The atmosphere presses on the frying pan and keeps it at constant pressure. From Eq. 2-36, the heat added in a constant-pressure process of a closed system is

$$Q = H_2 - H_1$$

The most accurate way to find the enthalpy in this equation is to use thermodynamic tables. However, it is often more convenient to use specific heat data. By definition, the specific heat at constant pressure is (see Eq. 2-40)

$$c_p(T,P) \equiv \left. \frac{\partial h}{\partial T} \right|_p$$

When enthalpy is a function only of temperature, this may be written as

$$c_p(T,P) = \frac{dh}{dT}$$

Separating variables and integrating from state 1 to state 2 gives

$$\int_1^2 dh = \int_1^2 c_p\,(T,P)\,dT$$

If we assume that c_p is not a function of temperature or pressure, then

$$\boxed{h_2 - h_1 = c_p\,(T_2 - T_1)} \tag{5-11}$$

Substituting $h = H/m$, this becomes

$$H_2 - H_1 = mc_p\,(T_2 - T_1)$$

As mentioned above, for a constant-pressure heating process, $Q = H_2 - H_1$; therefore,

$$\boxed{Q = mc_p\,(T_2 - T_1)} \tag{5-12}$$

This equation applies as long as c_p is not a function of temperature. It can also be used as an approximation when c_p is a function of temperature. In that case, the value of c_p at the average temperature of the process is used.

EXAMPLE 5-9 Heating of a subcooled liquid at constant pressure

Estimate the heat required to raise 3 lbm of liquid water at atmospheric pressure from 40°F to 160°F (in Btu).

a) Use specific heat data.

b) Use the steam tables.

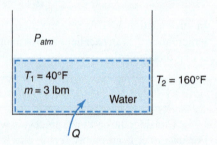

Approach:

Choose the water as the closed system. If specific heat data are used, the heat is given by

$$Q = mc_p\,\Delta T$$

The specific heat is calculated at the average temperature for the process. Values are available in Table B-6. As an alternate approach, the steam tables can be consulted. To find heat transfer, use

$$Q = m\,(h_2 - h_1)$$

To find the enthalpy of the compressed liquid at h_1, use the following approximation:

$$h_1 = h_f\,(T_1) + v_f\,(T_1)\,[P_1 - P_{sat}\,(T_1)]$$

Values of properties to insert in the right-hand side of this equation may be found in Table B-10. The final enthalpy, h_2, is calculated similarly.

Assumptions:

A1. Specific heat is constant.

Solution:

a) Define the liquid water as the system. This is a constant pressure process, so

$$Q = mc_p \, \Delta T$$

Use the value of c_p at the average water temperature of [A1]:

$$T_{avg} = \frac{40 + 160}{2} = 100°\text{F}$$

With the value of c_p at 100°F from Table B-6,

$$Q = (3 \text{ lbm})(0.998 \, \frac{\text{Btu}}{\text{lbm·°F}})(160 - 40)°\text{F}$$

$$Q = 359 \text{ Btu}$$

b) Alternatively

$$Q = m \, (h_2 - h_1)$$

Using Eq. 5-10, h_1 is approximated as

$$h_1 = h_f \, (T_1) + v_f \, (T_1) \, [P_1 - P_{sat} \, (T_1)]$$

With values from Table B-10 at 40°F,

$$h_1 = 8.02 \, \frac{\text{Btu}}{\text{lbm}} + 0.016 \, \frac{\text{ft}^3}{\text{lbm}} \, [14.7 - 0.122] \, \frac{\text{lbf}}{\text{in.}^2} \left(\frac{1 \text{ Btu}}{778 \text{ ft·lbf}} \right) \left(\frac{144 \text{ in.}^2}{1 \text{ft}^2} \right)$$

$$h_1 = 8.06 \, \frac{\text{Btu}}{\text{lbm}}$$

Similarly for h_2,

$$h_2 = 128 \, \frac{\text{Btu}}{\text{lbm}} + 0.0164 \, \frac{\text{ft}^3}{\text{lbm}} \, [14.7 - 4.75] \, \frac{\text{lbf}}{\text{in.}^2} \left(\frac{1 \text{ Btu}}{778 \text{ ft·lbf}} \right) \left(\frac{144 \text{ in.}^2}{\text{ft}^2} \right)$$

$$h_2 = 128 \, \frac{\text{Btu}}{\text{lbm}} + 0.0302 \, \frac{\text{Btu}}{\text{lbm}} \approx 128 \, \frac{\text{Btu}}{\text{lbm}}$$

Notice that the pressure-correction term was so small that it makes no difference to three significant figures. Finally, the heat is calculated as

$$Q = 3 \text{ lbm} \, (128 - 8.06) \, \frac{\text{Btu}}{\text{lbm}}$$

$$Q = 360 \text{ Btu}$$

As you can see, the differences are slight. Both of these methods are approximate.

Comment:

If the pressure is moderate, as in this example, the enthalpy of the compressed liquid is nearly equal to the enthalpy of the saturated liquid at the same temperature.

5.5 THE STATE PRINCIPLE

Several thermodynamic properties have been used to characterize the state of a system. These properties fall into two categories: those that depend on the size of the system and those that do not. Pressure and temperature, for example, do not depend on the size of the system. If a system at temperature T is divided in half, each half will have temperature T. Volume, on the other hand, does depend on the size of the system. Cutting a system in half halves the volume. Specific volume, which is the reciprocal of density, does not depend on the size of the system.

We call those properties that depend on the size, or extent, of the system **extensive properties**. Those that are independent of size are called **intensive properties**. The extensive properties include V, U, and H. Some intensive properties are T, P, u, v, and h. The total internal energy, U, depends on the mass of the system, while the specific internal energy, u, which is internal energy per unit mass, does not depend on the total mass of the system.

The intensive properties of a system are not all independent. The state principle, which addresses this point, was introduced above in simplified form. We are now in a position to develop a more rigorous statement of the state principle.

If the pressure and specific volume of an ideal gas are known, then the temperature can be determined from the ideal gas law. Furthermore, this temperature can be used to determine u and h. In this case, specifying the two thermodynamic properties P and v is enough to uniquely determine all the other thermodynamic properties. We could also have specified P and T and calculated v, u, and h. As a further example, if v and u were known, it would be possible to find P, T, and h. Note, however, that we run into trouble if we only specify T and u. Since u depends only on T for an ideal gas, there is no way to determine what P and v are. These observations are embodied in the **state principle**:

> *For a pure substance consisting of a single chemical species, specifying any two independent intensive thermodynamic properties uniquely determines all the remaining intensive thermodynamic properties.*

In the case of an ideal gas, P and T are independent, but T and u are not. The state principle applies to real gases and to solids and liquids as well. For example, in a two-phase mixture of liquid and vapor, specifying T and v allows calculation of quality, x. The quality can then be used to find u and h. As soon as the temperature of a two-phase mixture is known, the pressure can be determined. So for a two-phase mixture, T and v are independent properties. Note that T and P are not independent. In the two-phase region, quality, x, is also an intensive thermodynamic property. Pressure and quality are independent parameters, as are temperature and quality.

In a compressed liquid, pressure and temperature can be used to find v, u, and h. However, since v, u, and h are weakly dependent on pressure, it is not practical to choose T and u or T and h as independent properties.

The state principle given above applies to a simple compressible substance, that is, one that is subject only to expansion or compression work. If other forms of work are present, then additional parameters are needed to specify the state of the system. For example, if an object is falling in a gravitational field, then the earth does work on the object. In this case, work is related to the change in potential energy, and the elevation is needed to specify the state. In other systems, work may be done by magnetic fields or surface tension or other effects. In those cases, additional parameters will be needed to specify the state.

$Pv = \dfrac{MRT}{M}$

$Pv = \dfrac{RT}{M}$

$u = mC(T_2 - T_1)$

5.6 USE OF TABLES TO EVALUATE PROPERTIES

We have been using various tables to find values of thermodynamic properties. It is not always obvious which table applies in a given situation, especially if the phase of the substance is unknown. Furthermore, extracting data from the tables can present challenges to the novice user. This section gives an overview of the use of thermodynamic tables. A systematic approach to evaluating properties is described.

Thermodynamic property tables may be classified according to the type of data they contain: saturation properties (liquid–vapor and sometimes solid–vapor), superheated vapor properties, and compressed liquid properties. The two main questions we have to answer when evaluating properties at a given state are: What table do we use? How do we obtain the necessary data from the table?

Consider the T-v diagram in Figure 5-4. For a given T and v, we could fix the region in which the state lies; that is, we could determine whether the substance were two-phase, superheated vapor, saturated liquid, saturated vapor, or subcooled liquid. But without these two independent properties, we could not fix the location on this diagram.

The procedure we use to evaluate properties from tables is similar to using a roadmap. For example, suppose a classmate invites you home for semester break and tells you only the name of the town. To go there, do you just jump in your car and drive blindly? Eventually, you might arrive at your destination, but chances are you would get lost. Instead, you might get a map and, using the map's index, find two coordinates associated with the town. These two coordinates are sufficient to locate the town on the map. At that point you can plan a logical route to your friend's home. The "index" to use when evaluating thermodynamic properties is the table containing the saturation properties. As described below, comparison of given information with saturated liquid or saturated vapor properties will indicate in which region the state lies.

Evaluating properties at a given state requires us to do the following:

1. Identify two independent properties.
2. Determine the region in which the state lies (saturated, superheated, or subcooled/compressed liquid) using these two properties.
3. Use the table for that region to find all additional properties of interest.

This task is easy when using a graphical representation of the property data. It is slightly more difficult when we use tables, but the resulting data are more accurate.

The key to determining which table to use lies in the saturated liquid–vapor table. Below is a step-by-step procedure to obtain correct data. (We will deal primarily with liquids and vapors. Solid–vapor saturation tables will be ignored.) For completeness in the discussion that follows, we will include the common property entropy, s, described in Chapter 7.

Identify two independent properties   In theory, any two of the properties listed below for a particular region are independent:

Compressed liquid region:

P, T, v, u, h, s

Superheated vapor region:

P, T, v, u, h, s

Two-phase region: (P and T are not independent)

P, v, u, h, s, x

T, v, u, h, s, x

However, some combinations (e.g., u and h) would be very difficult to use in practice. Most often, one of the independent properties is either P or T. The second independent property then can be any of the others.

Determine the region   Given two properties, we compare one of the properties to the saturated liquid and/or vapor property.

1. Assume that we are given P and T. Two different approaches can be taken.

    A. Evaluate the saturation temperature, T_{sat}, of the fluid at P. Compare T_{sat} to T.

- If $T > T_{sat}$, then the substance is a superheated vapor.
- If $T < T_{sat}$, then the substance is a subcooled/compressed liquid.
- If $T = T_{sat}$, then the state is indeterminate. The substance could be a saturated liquid, a saturated vapor, or a two-phase mixture.

    B. Evaluate the saturation pressure, P_{sat}, of the fluid at T. Compare P_{sat} to P.

- If $P < P_{sat}$, then the substance is a superheated vapor.
- If $P > P_{sat}$, then the substance is a subcooled/compressed liquid.
- If $P = P_{sat}$, then the state is indeterminate. The substance could be a saturated liquid, a saturated vapor, or a two-phase mixture.

2. Assume that we are given either P or T and one of v, u, h, or s (entropy). At either P or T, evaluate the corresponding saturated liquid (subscript f) and/or saturated vapor (subscript g) property.

    A. If v, u, h, or s is greater than the corresponding saturated vapor property (v_g, u_g, h_g, or s_g), then the substance is a superheated vapor.

    B. If v, u, h, or s is less than the corresponding saturated liquid property (v_f, u_f, h_f, or s_f), then the substance is a subcooled/compressed liquid.

    C. If v, u, h, or s has a value between the corresponding saturated liquid property (v_f, u_f, h_f, or s_f) and the corresponding saturated vapor property (v_g, u_g, h_g, or s_g), then the substance is in the two-phase mixture region.

Evaluate the property   Once the region has been identified, then the property can be evaluated by the following methods:

    1. Saturated liquid, saturated vapor, superheated vapor, subcooled/compressed liquid: For these regions, the value of the property can be read directly from a table. If the desired value falls between table values, then the property sought is evaluated by interpolation. Suppose we have a table, as shown below, where Z is an independent property and Y is the property whose value we need to determine.

$$
\begin{array}{cc}
Z_1 & Y_1 \\
Z_2 & Y_2 \\
Z_3 & Y_3
\end{array}
$$

If we know Z_1, then we can read Y_1 directly. On the other hand, if our given Z is between Z_1 and Z_2, then we must interpolate. Many different interpolation schemes are available.

We will rely only on *linear* interpolation. To evaluate the property Y at the given value of Z, use the following formula:

$$\frac{Z - Z_1}{Z_2 - Z_1} = \frac{Y - Y_1}{Y_2 - Y_1}$$

2. Two-phase mixture: The liquid and the vapor are in thermal equilibrium in this region. Both fluids are at their saturation values. To obtain a property (other than P or T) in this region, the superposition principle is used. The mixture property is the mass-weighted average of the contributions from the saturated liquid and the saturated vapor properties.

Recall that x is the mass fraction of vapor in the liquid–vapor mixture; x is called the *quality*. Note that $(1 - x)$ is the mass fraction of liquid in the mixture. When $x = 0$, we have all saturated liquid. When $x = 1$ (or 100%), we have all saturated vapor. Letting m_g equal the mass of vapor and m_f equal the mass of liquid in a mixture, then

$$x = \frac{m_g}{m_g + m_f} = \frac{v - v_f}{v_g - v_f} = \frac{u - u_f}{u_g - u_f} = \frac{h - h_f}{h_g - h_f} = \frac{s - s_f}{s_g - s_f}$$

3. Subcooled/compressed liquid approximation: If subcooled/compressed liquid tables are not available for a particular fluid (and this is typically the norm), then we can use an approximation to evaluate the properties of a subcooled liquid. In general, states are a function of two properties; for example, $Y(P, T)$ where Y is v, u, or s. However, in the sub-cooled/compressed liquid region, the effect of changing pressure on properties is small. The properties are mostly a function of temperature. Hence, to evaluate subcooled/compressed liquid properties, the saturated liquid values at the given temperature are used.

Let Y equal one of the properties v, u, or s. For a given P and T, if T is less than the saturation temperature at P, then we are in the subcooled liquid region and the approximation is

$$Y(P, T) \cong Y_f(T)$$

That is, we use the saturated liquid value evaluated at the given temperature to approximate the subcooled/compressed liquid property of interest. This approximation can also be used for enthalpy. However, noting that $h = u + Pv$, we can obtain a better approximation with

$$h(P, T) \cong h_f(T) + v_f(T)[P - P_{sat}(T)]$$

This equation was derived in Section 5.4. The second term often is quite small and can be neglected if improved accuracy is not needed.

5.7 REAL GASES AND COMPRESSIBILITY

(Go to www.wiley.com/college/kaminski)

SUMMARY

The ideal gas law is not the most accurate way to relate P, T, and v. Thermodynamic tables, which give the most accurate relationship among the three variables, are available for a variety of substances. These tables give values for liquids as well as for vapors. When liquid and vapor exist together in equilibrium, the quality of the mixture is defined as

$$x = \frac{m_g}{m_f + m_g}$$

If the quality is zero ($x = 0$), only saturated liquid is present. If the quality is one ($x = 1$), only saturated vapor is present. The specific volume of a two-phase mixture of saturated liquid and vapor is

$$v = v_f + x\left(v_g - v_f\right) = v_f + xv_{fg}$$

The specific internal energy of a two-phase mixture is given by

$$u = u_f + x\left(u_g - u_f\right) = u_f + xu_{fg}$$

The specific enthalpy is

$$h = h_f + x\left(h_g - h_f\right) = h_f + xh_{fg}$$

The specific entropy is

$$s = s_f + x\left(s_g - s_f\right) = s_f + xs_{fg}$$

Furthermore,

$$x = \frac{m_g}{m_f + m_g} = \frac{v - v_f}{v_g - v_f} = \frac{u - u_f}{u_g - u_f} = \frac{h - h_f}{h_g - h_f} = \frac{s - s_f}{s_g - s_f}$$

The specific volume, internal energy, enthalpy, and entropy of **compressed or subcooled liquids** may be approximated using values for the saturated liquid at the same temperature, that is,

$$v(T,P) \approx v_f(T)$$
$$u(T,P) \approx u_f(T)$$

$$s(T,P) \approx s_f(T)$$
$$h(T,P) \approx h_f(T) + v_f(T)\left[P - P_{sat}(T)\right]$$

Thermodynamic properties may be classified as intensive (independent of the size of the system, such as T, P, or u) or extensive (dependent of the size of the system, such as m, V, or H).

The **state principle** is

> For a pure substance consisting of a single chemical species, specifying any two independent intensive thermodynamic properties uniquely determines all the remaining intensive thermodynamic properties.

Therefore, knowing any two independent properties allows evaluation of all the other properties of a system.

The compressibility factor, which can be used to determine whether a gas is ideal, is given by

$$Z = \frac{MPv_{act}}{\bar{R}T}$$

If $Z = 1$, the gas is perfectly ideal. Gases are typically ideal at high temperatures and low pressures. The reduced temperature and pressure are defined as

$$T_R = \frac{T}{T_C} \qquad P_R = \frac{P}{P_C}$$

where all temperatures and pressures are absolute. The compressibility chart in Section 5.7 gives Z as a function of P_R and T_R.

SELECTED REFERENCES

BLACK, W. Z., and J. G. HARTLEY, *Thermodynamics*, Harper & Row, New York, 1985.

CENGEL, Y. A., and M. A. BOLES, *Thermodynamics, an Engineering Approach*, 4th ed., McGraw-Hill, New York, 2002.

HOWELL, J. R., and R. O. BUCKIUS, *Fundamentals of Engineering Thermodynamics*, 2nd ed., McGraw-Hill, New York, 1992.

MORAN, M. J., and H. N. SHAPIRO, *Fundamentals of Engineering Thermodynamics*, 3rd ed., Wiley, New York, 1995.

MYERS, G., *Engineering Thermodynamics*, Prentice Hall, Englewood Cliffs, NJ, 1989.

VAN WYLEN, G. J., R. E. SONNTAG, and C. BORGNAKKE, *Fundamentals of Classical Thermodynamics*, 4th ed., Wiley, New York, 1994.

PROBLEMS

Problems designated with WEB refer to material available at www.wiley.com/college/kaminski.

VAPOR–LIQUID EQUILIBRIUM TABLES

P5-1 At what pressure (in kPa) does water boil if $T = 170°C$?

P5-2 What is the specific volume of saturated water vapor at 600 kPa?

P5-3 What is the temperature of saturated water vapor with $v = 0.3468 \text{m}^3/\text{kg}$?

P5-4 Find the temperature in °F at which

a. water boils if P = 35 psia.

b. the specific volume of saturated water vapor is 1207 ft^3/lbm.

P5-5 Find the pressure in kPa at which

a. water condenses if $T = 195°C$.

b. the specific volume of saturated water vapor is 0.05 m^3/kg.

P5-6 A rigid can contains 0.90 g of saturated water vapor at 450 kPa. Calculate the volume of the can in cubic centimeters.

P5-7 A piston–cylinder assembly contains 0.12 ft³ of saturated water vapor at 350°F. What is the mass of vapor in the tank?

SUPERHEATED VAPOR TABLES

P5-8 Find the specific volume of gaseous R-134a at 40°C for $P = 100$ kPa, 400 kPa, and 800 kPa. Use both the ideal gas law and tabulated values.

P5-9 Find the density of steam at 3.5 MPa and 415°C

a. using the steam tables.

b. using the ideal gas law.

P5-10 Refrigerant 134a at a pressure of 20 psia and a temperature of 40°F occupies a volume of 0.5 ft³. Find the mass

a. from table values.

b. from the ideal gas law.

P5-11 A container is filled with 0.026 kg of R-134a at a temperature of 40°C. What is the pressure if the volume is

a. 364 cm³.

b. 1560 cm³.

P5-12 A tank contains 0.05 lbm of water vapor at 20 psia and 500°F. Find the volume of the tank (in ft³).

P5-13 A container of volume 0.047 m³ is filled with 6.7 kg of steam at 600°C. Calculate the system pressure.

QUALITY

P5-14 A piston–cylinder assembly with a volume of 400 in.³ contains a steam–water mixture at 80 psia. If the total mass of the mixture is 0.066 lbm, find the volume of liquid present (in in.³).

P5-15 A two-phase mixture of steam and water has a quality of 0.79 and occupies a space of 0.51 ft³. If the total mass is 0.087 lbm,

a. find the temperature.

b. find the volume of liquid present (in in.³).

P5-16 A tank contains a two-phase mixture of steam and water at 40 psia. If the volume of the vapor is 10 times that of the liquid, what is the quality?

P5-17 A tank of volume 530 cm³ contains a two-phase mixture of R-134a at −12°C. The mass of liquid present is four times the mass of vapor.

a. Find the total mass of R-134a in the tank.

b. Find the volume of liquid present.

P5-18 A tank with a volume of 4.8 ft³ contains 6 lbm of liquid water. The tank also contains water vapor in equilibrium with the liquid. If the pressure in the tank is 30 psia, calculate the quality.

P5-19 A vial of volume 280 cc contains a two-phase mixture of steam and water at 30°C. The quality is 0.45. Find the mass in grams.

SATURATED, SUPERHEATED, AND COMPRESSED LIQUID TABLES

P5-20 A tank of volume 0.04 m³ contains 0.6 kg of R-134a at a pressure of 0.2 MPa.

a. Find the temperature.

b. If the volume is 0.068 m³, then what is the temperature?

P5-21 Find the specific volume of H_2O in each of the following states:

a. Saturated liquid at 160°F

b. Superheated vapor at 80 psia and 440°F

c. Two-phase mixture at a quality of 0.7 and a pressure of 40 psia

d. Subcooled liquid at 120°F, 14.7 psia

P5-22 Determine the volume, in m³, of 0.23 kg of H_2O at a temperature of 150°C and

a. a pressure of 0.2 MPa.

b. a quality of 0.6.

c. a pressure of 5 MPa.

P5-23 Find the specific volume of

a. compressed liquid water at 100°F, 1000 psia.

b. saturated liquid water at 100°F.

c. saturated liquid water at 1000 psia.

P5-24 Fill in the values of the specific volume of compressed liquid water at the conditions shown in the table. Use scientific notation with four significant figures, for example, 0.6216×10^{-3}.

	P_{sat} 2.339 kPa	P_{sat} 0.3613 MPa	5 MPa	10 MPa
20°C	$\times 10^{-3}$		$\times 10^{-3}$	$\times 10^{-3}$
140°C		$\times 10^{-3}$	$\times 10^{-3}$	$\times 10^{-3}$

Does v depend more on temperature or pressure?

P5-25 Calculate the enthalpy of compressed liquid water at 40°C two ways: using the approximate relationship for enthalpy of a compressed liquid and using the compressed liquid tables. Perform the calculation at these pressures:

a. 10 MPa

b. 20 MPa

c. 50 MPa

FIRST-LAW APPLICATIONS

P5-26   A rigid tank of volume 0.6 m^3 contains saturated R-134a vapor at 24°C. The contents are cooled until the temperature is 0°C. How much heat is removed? Show the process on a P-v diagram.

P5-27   A mixture of steam and water is contained in a rigid tank of volume 3050 cm^3. The mixture has a quality of 0.55 and a temperature of 120°C. Heat is added until the temperature is 140°C. Find

a. the final quality.

b. the amount of heat added.

P5-28   A rigid tank contains a two-phase mixture of water and steam at a quality of 0.65 and a pressure of 20 psia. The mass of the mixture is 0.26 lbm. The mixture is heated until the final quality is 0.95. Compute the final pressure and the heat added.

P5-29   A two-phase mixture of steam and water at 800 kPa, $x = 0.85$, is contained in a rigid, well-insulated tank. An electric resistance heater supplies 50 W to the mixture, which has a total mass of 1.3 kg. How long must the heater operate for the steam to reach a final temperature of 190°C?

P5-30   R-134a at 40°F, 15 psia, is contained in a rigid tank of volume 228 in.^3. The tank is cooled at a rate of 6 Btu/h. How much time is needed to cool the R-134a to the point where it just begins to condense?

P5-31   A rigid tank filled with 0.7 lbm of saturated water vapor at 400°F is cooled at constant volume. If the final temperature is 260°F,

a. find the final mass of liquid.

b. find the heat transferred.

P5-32   A well-insulated piston–cylinder assembly of volume 0.006 m^3 contains 6.25 g of steam at 150°C. The steam expands and, during this process, 0.759 kJ of work is done. If the final temperature is 95°C, what is the final volume?

P5-33   A two-phase mixture of steam and water with a temperature of 160°C and a quality of 0.6 is contained in a piston–cylinder assembly. The two-phase mixture, which has a total mass of 0.9 kg, is compressed slowly and isothermally until only saturated liquid is present. What is the work done on the system?

P5-34   Refrigerant 134a is contained in a perfectly insulated piston–cylinder assembly. The refrigerant is initially a saturated vapor at 10°F with a volume of 0.32 ft^3. It is then compressed to a superheated vapor at 120°F and 80 psia. Find the work done.

P5-35   A two-phase mixture of water and steam with a quality of 0.63 and $T = 300$°F expands isothermally until only saturated vapor remains. The initial volume is 0.114 ft^3. During the process, 16.2 Btu of heat are added. Find the work done.

P5-36   A piston–cylinder assembly contains 0.25 kg of saturated Refrigerant 134a vapor at 16°C. The refrigerant is cooled at constant pressure until the volume is one-half of its original value. Calculate the heat transferred.

P5-37   A piston–cylinder assembly contains 0.15 kg of saturated steam at 130°C. The piston is held in place by a weight. To reach the final state, 8300 J of heat are added. Find the final temperature.

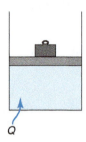

P5-38   Twelve kilograms of H_2O at 800 kPa and 400°C are cooled in a constant-pressure process until 2 kg of liquid water are present. Find the heat transferred.

P5-39   R-134a at −20°C and 200 kPa is heated at constant pressure. If the mass of refrigerant present is 6.2 kg and the heat added is 380 kJ, determine the final state.

P5-40   Saturated liquid water at 70 psia is cooled at constant pressure to 80°F. If the volume of water present is 0.71 ft^3, find the heat transferred.

P5-41   A piston–cylinder assembly of initial volume 0.6 m^3 contains H_2O at 500 kPa and 280°C. The system is cooled in a two-step process:

| 1–2 | Constant volume cooling until only saturated vapor remains |
| 2–3 | Constant temperature cooling until only saturated liquid remains |

a. Sketch the process on a T-v diagram.

b. For process 1–2 , calculate the work done and the heat transferred.

c. For process 2–3, calculate the work done.

P5-42   A two-phase mixture of water and steam at 190°F is contained in a piston–cylinder assembly. Initially the piston rests on stops. The combined mass of the water and steam is 0.06 lbm, and the initial quality is 0.3. The piston has a diameter of 6 in. and a mass of 12 lbm. How much heat must be added to triple the volume? Assume $P_{atm} = 14.7$ psia. Sketch the process on a P-v diagram.

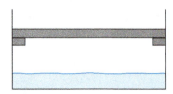

FIRST-LAW APPLICATIONS WITH THERMAL RESISTANCES

P5-43 A runner whose surface area is 1.8 m^2 generates 650 W of body heat. On a hot and cloudy day, the air is at 85°F. The heat transfer coefficient between runner and air is 14 Btu/ h·ft^2·°F.

a. If the runner is wearing only shorts and does not sweat, what would the skin temperature be (in°F)?

b. What volume of sweat (in fluid oz) must be evaporated per hour to keep the skin temperature at 70°F? Assume sweat has the properties of water.

P5-44 R-134a with an initial quality of 0.73 is contained in a piston–cylinder assembly, as shown in the figure. The curved walls of the cylinder are perfectly insulated. Initially, the R-134a occupies a volume of height 20 cm and diameter 7.5 cm. The piston–cylinder assembly is placed on a surface at 10°C, and heat conducts upward through the bottom wall and boils the liquid R-134a. The piston may be assumed to be massless and frictionless. The cylinder is constructed of stainless steel (AISI 304) with a wall thickness of 0.8 cm. The heat transfer coefficient between the liquid and the bottom of the cylinder is 268 W/m^2·K. How long will it take for the piston to rise 5 cm?

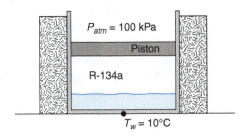

P_{atm} = 100 kPa

Piston

R-134a

T_w = 10°C

P5-45 A rigid box made of aluminum with a wall thickness of 0.25 in. contains saturated steam at a pressure of 60 psia. The convective heat transfer coefficient on the interior is 1.9 Btu/ h·ft^2·°F, and on the exterior, it is 3.6 Btu/ h·ft^2·°F. The box is a cube with a side length of 1.4 ft. A heater inside the box maintains the steam at a steady-state temperature. The exterior air temperature is 60°F. Find the power input to the heater.

P5-46 A piston–cylinder assembly contains water and steam at a quality of 0.7. The piston, which is made of carbon steel, is 1.5 cm thick and 6 cm in diameter. Initially the piston rests on the steam 9 cm above the bottom of the cylinder, compressing the two-phase system by its weight. The sides and bottom of the cylinder are well insulated, but heat is lost off the top of the piston. The convective heat transfer coefficient on the top of the piston is 9 W/m^2. The convective heat transfer coefficient on the bottom of the piston is 6.2 W/m^2·K. How long will it be before the piston sinks to half its initial height? Assume the surroundings are at 20°C.

COMPRESSIBILITY

P5-47 (WEB) Estimate the specific volume of carbon monoxide at 150 K and 10 MPa using

a. the compressibility chart.

b. the ideal gas law.

P5-48 (WEB) Steam at 800°F and 5000 psia has a mass of 25 lbm. Calculate the volume using

a. the steam tables.

b. the compressibility chart.

P5-49 (WEB) Does sulfer dioxide at a pressure of 2000 psia behave like an ideal gas at these temperatures?

a. 1500°F

b. 850°F

c. 350°F

P5-50 (WEB) A rigid container with a volume of 0.77 m^3 contains 110 kg of gaseous propane at 208°C. Using the compressibility chart, estimate the pressure of the gas.

CHAPTER 6

APPLICATIONS OF THE ENERGY EQUATION TO OPEN SYSTEMS

6.1 INTRODUCTION

An open system is one in which mass enters and/or leaves the control volume during a process that can be either steady or transient. In a steady flow process, nothing changes as a function of time. Although mass flows in and out, no mass accumulates in the control volume. The thermodynamic state of the masses entering and leaving are also constant with time. Any heat or work crossing the boundary enters or leaves at a constant rate. If one took a snapshot of the control volume at a given time and noted the temperature, pressure, amount of mass, and so on of the material in the control volume, and then later took another snapshot, none of the properties would have changed, even though mass flowed in and out.

In a transient system, properties do change with time. Consider the filling of a partially full bathtub with hotter water. Initially, the water in the tub is at one temperature. Adding hot water changes both the mass and total internal energy of the water in the tub and results in a higher final temperature. The system's properties have changed with time.

There are many devices that operate for long periods of time in steady state. Some of the most important ones include nozzles, diffusers, turbines, compressors, pumps, heat exchangers, mixing chambers, and throttles. These devices will be introduced in the first part of this chapter. The last part of the chapter will deal with transient processes.

6.2 NOZZLES AND DIFFUSERS

A **nozzle** is a duct with a smoothly varying cross-sectional area in which the fluid velocity increases from the entrance to the exit, as shown in Figure 6-1a. Nozzles are used in a wide variety of applications ranging from the ordinary garden hose to jet engines and rockets. A **diffuser**, shown in Figure 6-1b, also has a smoothly varying cross-sectional area similar to a nozzle, but in a diffuser the velocity *decreases* from entrance to exit. Diffusers are used, for example, downstream of steam turbines and at the entrance to jet engines. In both nozzles and diffusers, there is a trade-off between pressure and velocity.

In subsonic flow, the area of a nozzle decreases in the flow direction and the area of a diffuser increases. In supersonic flow, the opposite is true: at the exit a nozzle flares out and a diffuser necks down. To increase the velocity in supersonic flow, it is necessary first to decrease the flow area and then to increase the area, counter to intuition (Figure 6-2). In such a nozzle, the inlet flow is subsonic and the exit flow is supersonic.

Both subsonic and supersonic nozzles and diffusers can be analyzed using the first law. The first law for an open system is, from Eq. 4-42:

$$\frac{dE_{cv}}{dt} = \dot{Q}_{cv} - \dot{W}_{cv} + \sum_{in} \dot{m}_i \left(h_i + \frac{\mathcal{V}_i^2}{2} + gz_i \right) - \sum_{out} \dot{m}_e \left(h_e + \frac{\mathcal{V}_e^2}{2} + gz_e \right) \quad (6\text{-}1)$$

FIGURE 6-1 Example nozzle and diffuser: (a) a subsonic nozzle; (b) a subsonic diffuser.

FIGURE 6-2 A converging–diverging nozzle for supersonic flow at the exit.

Nozzles and diffusers are typically operated as steady flow devices, so the left-hand side of Eq. 6-1 is zero. The rate of heat transfer is usually very small and is neglected. There is no control volume work in a nozzle or diffuser, since there are no turning shafts, expanding boundaries, or electrical circuits. The elevation change between inlet and outlet is usually small or zero, so the potential energy change is rarely important. On the other hand, the conversion between pressure and velocity is the purpose of nozzles and diffusers, so kinetic energy is very important. Neglecting transient terms, heat, work, and potential energy, Eq. 6-1 becomes

$$0 = \dot{m}_1 \left(h_1 + \frac{\mathcal{V}_1^2}{2} \right) - \dot{m}_2 \left(h_2 + \frac{\mathcal{V}_2^2}{2} \right)$$

The subscript 1 designates the stream entering and 2 designates the stream exiting. Because nozzles and diffusers have only one inlet and one outlet, from conservation of mass,

$$\dot{m}_1 = \dot{m}_2$$

Incorporating this result, the first law for a nozzle or diffuser can be written as

$$\boxed{h_1 + \frac{\mathcal{V}_1^2}{2} = h_2 + \frac{\mathcal{V}_2^2}{2} \qquad \text{nozzle or diffuser}} \qquad (6\text{-}2)$$

EXAMPLE 6-1 Flow in a nozzle

Water vapor enters a well-insulated nozzle at 300 kPa and 500°C, with a velocity of 75 m/s. The entrance area is 0.5 m^2. The water vapor exits at 100 kPa and 200°C.

a) Find the exit velocity.

b) Find the exit area.

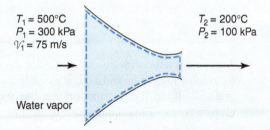

$T_1 = 500°C$
$P_1 = 300$ kPa
$\mathcal{V}_1 = 75$ m/s

$T_2 = 200°C$
$P_2 = 100$ kPa

Water vapor

Approach:

Solve the energy equation for a nozzle (Eq. 6-2) for exit velocity. Use values for enthalpy of water vapor from Table A-12. To find the exit area, apply conservation of mass. Substitute $\dot{m} = \rho \mathcal{V} A$ and $\rho = 1/v$ into the conservation of mass equation. Values for specific volume, v, are in the steam tables.

Assumptions:

A1. The nozzle is adiabatic.

A2. The flow is steady.

Solution:

a) Since the nozzle is well insulated and the flow is steady [A1][A2],

$$h_1 + \frac{\mathcal{V}_1^2}{2} = h_2 + \frac{\mathcal{V}_2^2}{2}$$

Solving for exit velocity

$$\mathcal{V}_2 = \sqrt{2\,(h_1 - h_2) + \mathcal{V}_1^2}$$

Values for the enthalpy of water vapor can be found in Table A-12. With these values,

$$\mathcal{V}_2 = \sqrt{2\,(3486 - 2875.3)\,\frac{\text{kJ}}{\text{kg}}\left(\frac{1000\,\text{J}}{1\,\text{kJ}}\right) + \left(75\,\frac{\text{m}}{\text{s}}\right)^2} = 1108\,\frac{\text{m}}{\text{s}}$$

The flow is supersonic at the exit. Note that all the units were converted to consistent SI units. Both kilograms and joules are consistent SI units, but kJ are not. If this is done carefully, the result will be in the appropriate SI units, in this case m /s.

b) From conservation of mass,

$$\dot{m}_1 = \dot{m}_2$$

This may also be written

$$\rho_1 \mathcal{V}_1 A_1 = \rho_2 \mathcal{V}_2 A_2$$

Substituting $\rho = \frac{1}{v}$,

$$\frac{\mathcal{V}_1 A_1}{v_1} = \frac{\mathcal{V}_2 A_2}{v_2}$$

Solving for A_2,

$$A_2 = \left(\frac{v_2}{v_1}\right)\left(\frac{\mathcal{V}_1}{\mathcal{V}_2}\right) A_1$$

With given values and values from Table A-12,

$$A_2 = \left(\frac{2.172}{1.1867}\right)\left(\frac{75}{1108}\right)(0.5)\,\text{m}^2$$

$$A_2 = 0.0619\,\text{m}^2$$

6.3 TURBINES

A turbine is a modern-day descendant of the windmill. It is a device for extracting energy from a flowing stream and using it to rotate a shaft. In the past, windmills were used to grind grain into flour. Today, wind turbines like those shown in Figure 6-3 are used to produce electric power. The rotating shaft of the turbine drives an electrical generator. The power

FIGURE 6-3 Three wind turbines near Madison, New York. Each turbine is 67 m high and can produce 1.65 MW. (Photo by author.)

induced within the coils of the generator is provided to homes and businesses through a distribution network of electric cables.

High-pressure steam can also be used to drive turbines. In a power plant, a boiler fueled by oil, coal, or nuclear energy produces high-temperature, high-pressure steam. The steam is fed into a turbine that is then used to drive an electrical generator. Steam turbines typically have several sets of blades on the same shaft, as shown schematically in Figure 6-4.

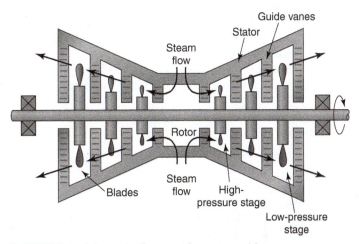

FIGURE 6-4 Schematic diagram of a steam turbine.

FIGURE 6-5 GE's MS7001FB gas turbine on the half shell during assembly. This unit is rated at 280 MW in combined cycle (Photo courtesy of Maximilian Stock Ltd./Getty Images, Inc.)

Figure 6-5 shows an actual turbine installed in a test bed. Each set of blades on the shaft is called a stage. At the inlet to the turbine, high-pressure fluid strikes the stage with the smallest diameter. The fluid imparts some of its energy to the turbine as it flows over the blades and exits the first stage at a lower pressure. The second stage is larger in diameter so that it can extract energy from a fluid that is now somewhat depleted. Each successive stage has a larger diameter and higher surface area to draw energy from an increasingly low-pressure and low-density fluid. The characteristic symbol for a turbine, shown in Figure 6-6, emphasizes its expanding shape.

The turbine rotor shown in Figure 6-4 is supported by bearings and encased within a stator, which does not rotate. The stator has several stages of stationary blades that are

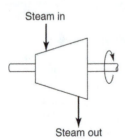

Steam in

Steam out

FIGURE 6-6 Typical symbol used to designate a turbine.

FIGURE 6-7   GE's A Series reheat steam turbine rated from 85 to 150 MW—final assembly.
(Photo courtesy of Monty Rakusen/Getty Images, Inc.)

designed to direct flow over the rotating blades. A photograph of a steam turbine being lowered into a stator is reproduced as Figure 6-7.

Another type of turbine, the hydroturbine, is used to extract energy from flowing water. Hydroturbines are descendants of the water wheel. As shown in Figure 6-8a, water is delivered to the top of the wheel and flows over the paddles, impelled by gravitational force. The water wheel was used extensively up until about 1850, when hydroturbines began to appear. The hydroturbine occupies less space, operates at higher speeds, works submerged, and is not limited by ice formation. The blades on a hydroturbine rotor are shown in Figure 6-8b.

Turbines are used not only in generating electric power but also in many other applications. Gas turbines are used in jet engines, in turbo-charged vehicles, in flow meters, and even in dental drills.

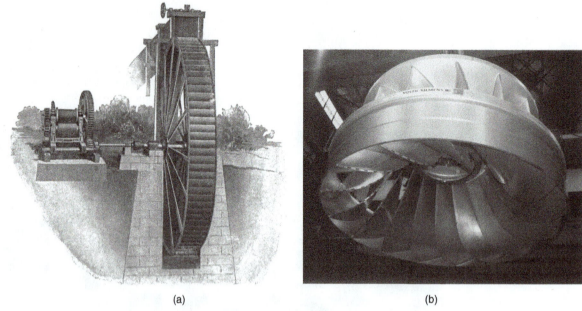

(a)   (b)

FIGURE 6-8 (a) A steel-overshot water wheel used extensively in the 19th century. (*Source:* Daugherty, R.L., *Hydraulic Turbines*, McGraw-Hill, New York, 1913) (b) A 4.5-m diameter hydrotrubine runner that delivers 54 MW. (Photograph courtesy of Voith Siemens Hydropower Generation, Inc.)

To determine the power output of a turbine, the first law for an open system, Eq. 6-1, is used. In some circumstances, terms can be dropped from the equation because they are relatively small. For example, the heat loss from the turbine can often be neglected. Any heat loss to the environment reduces the amount of work that can be produced, so this heat loss is minimized by design. Recall that a system with no heat transfer in or out is said to be adiabatic. Turbines are frequently idealized as adiabatic.

In a turbine, velocities of entering and leaving streams are often low, so kinetic energy can be neglected. A useful rule of thumb is that kinetic energy is probably unimportant if air velocities are less than 30 m/s. In addition, for all turbines except hydroturbines, potential energy changes are unimportant. In a hydroturbine, the elevation change is the major source of energy for the turbine, and it certainly must be included.

For an adiabatic turbine with a single inlet at state 1 and a single outlet at state 2 operating in steady state with no kinetic or potential energy effects, Eq. 6-1 simplifies to

$$0 = -\dot{W}_{cv} + \dot{m}\left(h_1 - h_2\right) \qquad \text{adiabatic turbine, no } KE \text{ or } PE \qquad (6\text{-}3)$$

This is the simplest meaningful equation for analyzing turbines. If heat transfer, kinetic energy, or potential energy must be considered, then it is best to start from the first law, Eq. 6-1, and eliminate small terms, as illustrated in the examples that follow. The work and the enthalpy terms are never eliminated from the first law for a turbine.

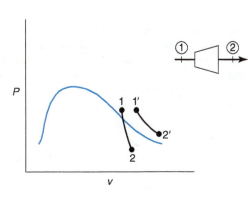

FIGURE 6-9 *P-v* diagram for a steam turbine showing two possible operating conditions starting from two different states, 1 and 1′. The exit state can be either single- or two-phase. The inlet state is typically superheated vapor or saturated vapor. (In the two examples shown, both 1 and 1′ are superheated.)

Figure 6-9 shows a *P-v* diagram for a steam turbine. High-pressure steam enters at state 1 and low-pressure steam exits at state 2. Sometimes state 2 is in the two-phase region. Although steam turbines can tolerate some liquid droplets, a quality less than about 0.9 can lead to unacceptable erosion of the turbine blades. As the steam flows through the turbine, temperature decreases and specific volume increases.

EXAMPLE 6-2 **Expansion in a steam turbine**

Steam enters an adiabatic turbine at 1.2 MPa and 500°C. The inlet pipe is 0.5 m in diameter, and the steam flows at 18 m/s. The exit pressure is 20 kPa, and the exit quality is 0.98.

a) Find the exit temperature.

b) Find the power produced.

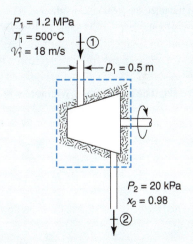

Approach:

To solve part a, one must recognize that the exit state has a quality between 0 and 1; therefore, it is a two-phase mixture of liquid and vapor. Knowing the exit pressure, you can read the exit temperature directly from Table A-11.

For part b, the energy equation for an open system, Eq. 6-1 is used. The turbine is assumed to be adiabatic with no change in potential energy. Inlet kinetic energy is calculated from the known inlet velocity, while exit kinetic energy is assumed to be negligible. The mass flow rate through the turbine is calculated from $\dot{m}_1 = \rho_1 \mathcal{V}_1 A_1$. The specific volume and enthalpy for the inlet state are available in Table A-12, while properties at the exit are found using Table A-11. After careful consideration of units, the power can be calculated from the energy equation.

Assumptions:

Solution:

a) For the exit state, the pressure and quality are given. Because the quality is between 0 and 1, the exit condition must be two-phase. Table A-11 gives the properties of a saturated steam–water mixture with entries in even units of pressure. At 20 kPa, the saturation temperature is 60.06°C. This is the exit temperature.

b) Define a control volume enclosing the turbine. The first law for this open system is

$$\frac{dE_{cv}}{dt} = \dot{Q}_{cv} - \dot{W}_{cv} + \sum_{in} \dot{m}_i \left(h_i + \frac{\mathcal{V}_i^2}{2} + gz_i \right) - \sum_{out} \dot{m}_e \left(h_e + \frac{\mathcal{V}_e^2}{2} + gz_e \right)$$

A1. The flow is steady.
A2. The turbine is adiabatic.
A3. Potential energy change is zero.
A4. Kinetic energy at the exit is negligible.

Assume the turbine is operating in a steady state, [A1] so the left-hand side is zero. There is no heat transferred for an adiabatic turbine [A2]. In addition, there is no elevation change, so the potential energy terms may be dropped [A3]. As mentioned previously, if the velocity is less then about 30 m/sec, the kinetic energy terms may be neglected. But, since we happen to know the velocity at the inlet, we include the kinetic energy term just to see how big it is. There is not enough information given to calculate the velocity at the exit, but it is probably negligible [A4]. With these considerations, the first law becomes

$$0 = -\dot{W}_{cv} + \dot{m} \left(h_1 + \frac{\mathcal{V}_1^2}{2} \right) - \dot{m} h_2$$

At the inlet, the steam is superheated. This is evident from Table A-12, because there is an entry at 1.2 MPa and 500°C. The enthalpy from the table is

$$h_1 = 3476 \, \frac{kJ}{kg}$$

The exit state is two-phase. With values of h_f and h_g from Table A-11,

$$h_2 = h_f + x_2 \left(h_g - h_f \right)$$

$$h_2 = 251.4 + 0.98 \, (2609.7 - 251.4)$$

$$h_2 = 2563.5 \, \frac{kJ}{kg}$$

To find the power produced, the mass flow rate is needed. This may be found from

$$\dot{m}_1 = \rho_1 \mathcal{V}_1 A_1 = \frac{\mathcal{V}_1 A_1}{v_1}$$

Using values for v from Table A-12,

$$\dot{m}_1 = \frac{\left(18 \, \frac{m}{s} \right) \left(\frac{0.5}{2} \right)^2 \pi \, m^2}{0.2946 \, \frac{m^3}{kg}} = 12 \, \frac{kg}{s}$$

Because there is only one stream in and one stream out and the system is steady, conservation of mass gives

$$\dot{m}_1 = \dot{m}_2 = \dot{m}$$

Power is calculated from the first law as

$$\dot{W}_{cv} = \dot{m}\left(h_1 + \frac{\mathcal{V}_1^2}{2}\right) - \dot{m}h_2$$

$$\dot{W}_{cv} = \left(12\frac{\text{kg}}{\text{s}}\right)\left[3476\frac{\text{kJ}}{\text{kg}} - 2563.5\frac{\text{kJ}}{\text{kg}} + \frac{(18)^2\frac{\text{m}^2}{\text{s}^2}}{2}\frac{1\,\text{kJ}}{1000\,\text{J}}\right]$$

$$\dot{W}_{cv} = 41,712 - 30,762 + 1.944 = 10,952\frac{\text{kJ}}{\text{s}} = 10,952\,\text{kW}$$

Notice that the last term arose from the kinetic energy, which has a value of only 1.944 kW. It is negligible compared to the enthalpy terms, which are 41,712 and 30,762 kW. The velocity is only 18 m/s. Recall the rule of thumb that the kinetic energy is important for velocities above about 30 m/s. It would have been appropriate to drop the kinetic energy term in this case.

EXAMPLE 6-3   Expansion in a gas turbine

A gas turbine is designed to operate with a mass flow rate of 5.4 kg/s. The turbine drives both a compressor and a generator, providing 881 kW to the compressor and 1.4 MW to the generator. A total of 22 kW of heat are lost to the environment from the turbine's outer casing. If the exit temperature is 110°C, find the inlet temperature. For gas properties, use the properties of air at 110°C.

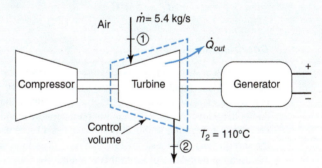

Approach:

Select the turbine as the control volume. Start with the first law, because it includes power ($\dot{W}$) and fluid properties from which the inlet temperature may be determined. Assume kinetic and potential energy are negligible. To find the enthalpy change, assume an ideal gas and use $\Delta H = mc_p\,\Delta T$. The total work is the sum of the work to the compressor and the work to the generator. With this information and the values of specific heat from Table A-7, the inlet temperature can be calculated.

Assumptions: *Solution:*

Define the turbine as the control volume under study. The first law for an open system is

$$\frac{dE_{cv}}{dt} = \dot{Q}_{cv} - \dot{W}_{cv} + \sum \dot{m}_i\left(h_i + \frac{\mathcal{V}_i^2}{2} + gz_i\right) - \sum \dot{m}_e\left(h_e + \frac{\mathcal{V}_e^2}{2} + gz_e\right)$$

A1. The flow is steady.
A2. Kinetic energy is negligible.
A3. Potential energy is negligible.
A4. Combustion gases have the properties of air.
A5. The combustion gases behave like an ideal gas.
A6. Specific heat is constant.

The turbine is assumed to operate in steady state with one inlet and one outlet [A1]. Kinetic and potential energy effects are assumed to be negligible [A2][A3]. The first law reduces to

$$0 = \dot{Q}_{cv} - \dot{W}_{cv} + \dot{m}(h_1 - h_2)$$

The gas is assumed to be air with constant specific heat [A4][A5][A6]. The enthalpy change is then

$$\Delta H = mc_p\,\Delta T$$

Substituting $H = mh$ and expanding the Δ terms gives

$$h_1 - h_2 = c_p(T_1 - T_2)$$

Substituting this into the first law results in

$$0 = \dot{Q}_{cv} - \dot{W}_{cv} + \dot{m}c_p(T_1 - T_2)$$

Solving for the inlet temperature,

$$T_1 = \frac{\dot{W}_{cv} - \dot{Q}_{cv}}{\dot{m}c_p} + T_2$$

The specific heat of air at 110°C (383 K) is found by interpolation in Table A-7. The two work terms are added to give the total control volume work, $\dot{W}_{cv}$. By the sign convention, $\dot{Q}_{cv} = -\dot{Q}_{out}$. Then the inlet temperature is

$$T_1 = \frac{[881 + 1400 - (-22)]\text{kW}\left(\dfrac{1000\,\text{W}}{1\,\text{kW}}\right)}{\left(5.4\,\dfrac{\text{kg}}{\text{s}}\right)\left(1.012\,\dfrac{\text{kJ}}{\text{kg·°C}}\right)\left(\dfrac{1000\,\text{J}}{1\,\text{kJ}}\right)} + 110\text{°C}$$

$$T_1 = 531\text{°C}$$

Comments:

In the problem statement, it was suggested that the specific heat be evaluated at the exit temperature of 110°C. At that point, the inlet temperature was unknown. Now that a value for the inlet temperature is available, it would be more appropriate to estimate the specific heat at the average of the inlet and outlet temperatures. Then the inlet temperature could be recalculated using this new value of specific heat. Iterating in this fashion produces a better estimate of the inlet temperature. Because the specific heat is not a strong function of temperature, only a few iterations would be needed.

We used the properties of air in this example. Actual gas flowing through a gas turbine typically results from combustion of natural gas and contains nitrogen, carbon dioxide, carbon monoxide, water vapor, nitrous oxides, excess oxygen, and other gases. Nevertheless, nitrogen dominates, since it is the major constituent of air and the major constituent in combustion gas. Therefore, air properties are reasonable to use in this example.

6.4 COMPRESSORS, BLOWERS, FANS, AND PUMPS

A **compressor** is a device used to raise the pressure of a gas. There are many different types of compressors, including axial flow, reciprocating, and centrifugal. Figure 6-10 shows an axial flow compressor, which resembles a gas turbine. Flow enters the stage with the largest diameter at low pressure and exits the stage with the smallest diameter at high

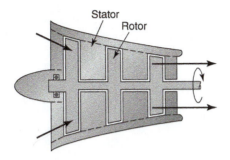

FIGURE 6-10   An axial flow compressor.

pressure. Hence, this compressor is similar to a turbine with the gas flowing in the opposite direction. Figure 6-11 shows two other types of compressors, one reciprocating and the other centrifugal. Additional compressor configurations are also available.

Although these various compressors differ greatly in geometry, they can all be analyzed in the same manner. To drive every compressor, work is added by some external agent. This agent may be a motor or a turbine. Figure 6-12 shows an axial compressor driven by a turbine in a jet engine. Heat transfer is typically small in a compressor and is often neglected. Other terms that are usually insignificant are kinetic and potential energy. As a result, the equation for a compressor is identical to that for a turbine, and Eq. 6-3 may be used.

Blowers and **fans** also raise the pressure of a gas, but to a lower level than a compressor. If the gas density increases more than 7% from inlet to exit, then the device is called a compressor. For density increases less than 7%, the gas is assumed to be incompressible and the device is called a blower or fan. Note that a fan has the lowest pressure rise, and often the goal is high gas velocity rather than pressure rise. Whether the device is a compressor, blower, or fan, the same form of the first law, Eq. 6-3, can be used in its analysis.

A **pump** is used to increase the pressure of a liquid. Eq. 6-3 applies to pumps as well as to compressors and turbines. Eq. 6-3 is

$$0 = -\dot{W}_{cv} + \dot{m}\,(h_1 - h_2) \tag{6-4}$$

Typically, a pump does not significantly increase the temperature of a liquid. There are some frictional losses, and these result in the exiting liquid being slightly hotter than the entering liquid; however, this effect is not usually significant. Therefore, the pumping process is

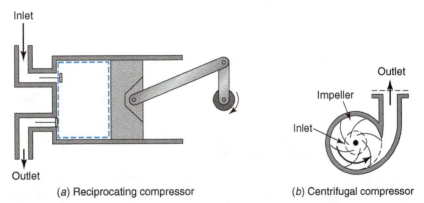

(a) Reciprocating compressor   (b) Centrifugal compressor

FIGURE 6-11   Two types of compressors: (a) reciprocating compressor; (b) centrifugal compressor.

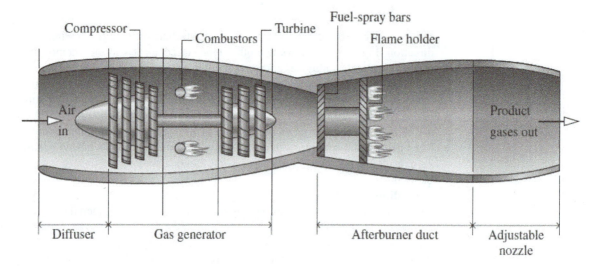

FIGURE 6-12 In this jet engine, air flows through the compressor, where its pressure and temperature increase. Fuel is injected into the air stream and burned in the combustion chamber. The high-temperature combustion gases flow through the turbine, which then drives the compressor. The gases leave the turbine and are expanded in the nozzle, providing thrust. (*Source*: Moran and Shapiro, *Fundamentals of Engineering Thermodynamics*, 6th Ed., John Wiley and Sons. Used with permission.)

often considered isothermal. The enthalpy change of an ideal (incompressible) liquid in an isothermal process is given by Eq. 5-9, which is

$$h_2 - h_1 = v(P_2 - P_1) \tag{6-5}$$

The rise in enthalpy across a pump becomes

$$\boxed{h_2 = h_1 + v\,(P_2 - P_1) \qquad \text{pump}} \tag{6-6}$$

Substituting Eq. 6-5 into Eq. 6-4 and solving for power gives an expression for the work done by a pump:

$$\boxed{\dot{W}_{cv} = \dot{m}v\,(P_1 - P_2) \qquad \text{pump}} \tag{6-7}$$

This is a useful equation for determining pumping power.

EXAMPLE 6-4   **Compression of saturated steam**

Saturated steam at 230°F enters a compressor and is compressed to 80 psia. The mass flow rate is 50 lbm/h. Heat loss from the compressor to the surroundings occurs at a rate of 112 Btu/h. The power input is 1.5 hp. Find the exit steam temperature.

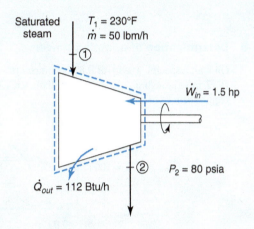

Approach:

Choose a control volume around the compressor. Start from the first law, eliminating kinetic energy, potential energy, and the transient term to get

$$0 = \dot{Q}_{cv} - \dot{W}_{cv} + \dot{m}\,(h_1 - h_2)$$

The enthalpy at the inlet can be determined from the given information. The first law can then be solved for the exit enthalpy. Using this exit enthalpy and the given outlet pressure, you can determine the exit temperature from property tables.

Assumptions:

A1. Kinetic energy is negligible.

A2. Potential energy is negligible.

A3. Flow is steady.

Solution:

Define a control volume to enclose the compressor. If you neglect kinetic and potential energy and assume steady flow [A1][A2][A3], the first law is

$$0 = \dot{Q}_{cv} - \dot{W}_{cv} + \dot{m}\,(h_1 - h_2)$$

State 1 is saturated steam at 230°F. From Table B-10, $h_1 = 1157.1$ Btu/lbm. The first law may be solved for h_2 to get

$$h_2 = \frac{\dot{Q}_{cv} - \dot{W}_{cv}}{\dot{m}} + h_1$$

Because of the sign convention for heat and work, $\dot{Q}_{cv} = -\dot{Q}_{out}$ and $\dot{W}_{cv} = -\dot{W}_{in}$. Substituting values,

$$h_2 = \frac{-112\,\frac{\text{Btu}}{\text{h}} - (-1.5\,\text{hp})\left(\dfrac{2544\,\frac{\text{Btu}}{\text{h}}}{1\,\text{hp}}\right)}{50\,\frac{\text{lbm}}{\text{h}}} + 1157.1\,\frac{\text{Btu}}{\text{lbm}}$$

$$h_2 = 1231.2\,\frac{\text{Btu}}{\text{h}}$$

The final state is determined by two independent properties. State 2 has an enthalpy, h_2, of 1231.2 Btu/h and a pressure, P_2, of 80 psia. The expected outlet state of a compressor is superheated vapor. From Table B-12,

$$T_2 \approx 400°F$$

EXAMPLE 6-5 Determination of pumping power

Oil with a specific gravity of 0.82 is pumped from a pressure of 100 kPa to 550 kPa. The oil flows at 12 m/s through a pipe 2.5 cm in diameter. Neglecting frictional losses, determine the power requirement of the pump.

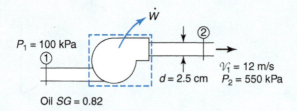

$P_1 = 100$ kPa
①
$\dot{W}$
②
$d = 2.5$ cm
$\mathcal{V}_1 = 12$ m/s
$P_2 = 550$ kPa

Oil $SG = 0.82$

Approach:

Choose a control volume to enclose the pump. The power of the pump may be found from Eq. 6-7, which is $\dot{W}_{cv} = \dot{m}v\,(P_1 - P_2)$. To find the mass flow rate, use $\dot{m} = \rho \mathcal{V}A$. The specific volume of the oil is the inverse of the density. The density is found from the definition of specific gravity, that is, $\rho = 0.82\rho_{water}$.

Assumptions:

Solution:

The control volume encloses the pump. The mass flow rate of the oil is

$$\dot{m} = \rho \mathcal{V}A = \rho \mathcal{V}\pi \left(\frac{d}{2}\right)^2$$

Since the specific gravity of the oil is 0.82, this becomes

$$\dot{m} = 0.82\rho_{water}\mathcal{V}\pi \left(\frac{d}{2}\right)^2 = 0.82\left(1000\,\frac{\text{kg}}{\text{m}^3}\right)\left(12\,\frac{\text{m}}{\text{s}}\right)\pi\left(\frac{0.025\text{m}}{2}\right)^2 = 4.83\,\frac{\text{kg}}{\text{s}}$$

where the definition of specific gravity has been used. The pumping power is given by Eq. 6-7, which is [A1][A2][A3]

A1. The oil is incompressible.
A2. The flow is steady.
A3. The flow is frictionless.

$$\dot{W}_{cv} = \dot{m}v\,(P_1 - P_2)$$

Substituting $v = 1/\rho$ gives

$$\dot{W}_{cv} = \dot{m}\,(P_1 - P_2)\,/\rho$$

Inserting values,

$$\dot{W}_{cv} = \frac{4.83\,\frac{\text{kg}}{\text{s}}\,(100 - 550)\,\text{kPa}\left(\dfrac{1000\,\text{Pa}}{1\,\text{kPa}}\right)}{0.82\left(1000\,\frac{\text{kg}}{\text{m}^3}\right)}$$

$$\dot{W}_{cv} = -2651\,\text{W} = -2.65\,\text{kW}$$

Comments:

The work is negative because work is being done on the fluid in the control volume. By our sign convention, work done by the control volume is positive and work done on the control volume is negative.

6.5 THROTTLING VALVES

A throttling valve (also called a flow restriction) is a device used to reduce the pressure of a flowing fluid. The flow restriction produces a large reduction in pressure over a short distance. The water faucet in a shower is a common example. Other examples of throttles are shown in Figure 6-13.

The first law for a throttling valve takes on a particularly simple form. Recall that the first law is

$$\frac{dE_{cv}}{dt} = \dot{Q}_{cv} - \dot{W}_{cv} + \sum_{\text{in}} \dot{m}_i \left(h_i + \frac{\mathcal{V}_i^2}{2} + gz_i \right) - \sum_{\text{out}} \dot{m}_e \left(h_e + \frac{\mathcal{V}_e^2}{2} + gz_e \right)$$

No work is done by throttling valves, and they typically operate in steady state with one inlet and one outlet. Kinetic and potential energy changes are usually negligible. Over the short distance of the flow restriction, there is insignificant heat transfer. With these assumptions, the first law for a throttling valve reduces to

$$h_1 = h_2 \qquad \text{throttling valve} \tag{6-8}$$

A throttling valve is said to be an isenthalpic device.

Throttling devices are often used in refrigerators and air conditioners. In this application, the inlet is usually a saturated liquid and the outlet is a two-phase mixture.

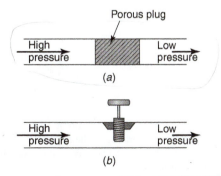

FIGURE 6-13 Throttling devices.

EXAMPLE 6-6 Throttling of a refrigerant

R-134a at 140 kPa with a quality of 0.6 is throttled to 100 kPa. Find the exit state.

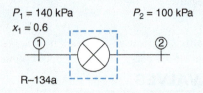

$P_1 = 140$ kPa
$x_1 = 0.6$

$P_2 = 100$ kPa

R–134a

Approach:

The control volume is drawn around the throttling valve. The first law for a throttling valve is $h_1 = h_2$. State 1 is two-phase, so enthalpy is given by $h_1 = h_{f1} + x_1 \left(h_{g1} - h_{f1} \right)$. Since $h_1 = h_2$, the exit enthalpy, h_2, may be calculated. The exit state has an enthalpy of h_2 and a pressure of 100 kPa. However, the exit state could be either two-phase or superheated vapor. To determine which, check Table A-15. If the outlet enthalpy is between that of the saturated liquid and the saturated vapor, then it is two-phase. Otherwise, it is superheated vapor.

Assumptions:

A1. Kinetic energy is negligible.
A2. Potential energy is negligible.
A3. The flow is steady.
A4. The process is adiabatic.
A5. No work is done on or by the device.

Solution:

Define a control volume around the trottling valve. For a throttling valve [A1][A2][A3][A4][A5],

$$h_1 = h_2$$

State 1 is in a two-phase region. Enthalpy is given by

$$h_1 = h_{f1} + x_1 \left(h_{g1} - h_{f1} \right)$$

Using values from Table A-15,

$$h_1 = 25.77 + 0.6 \left(236.04 - 25.77 \right)$$

$$h_1 = 151.93 \ \frac{\text{kJ}}{\text{kg}}$$

The exit state will have the same enthalpy as state 1 and a pressure of 100 kPa. From Table A-15, h_2, which equals 151.93, falls between h_f and h_g at 100 kPa. Therefore, state 2 is also in the two-phase region, and is given by

$$h_2 = h_{f2} + x_2 \left(h_{g2} - h_{f2} \right)$$

Solving for x_2,

$$x_2 = \frac{h_2 - h_{f2}}{h_{g2} - h_{f2}} = \frac{151.93 - 16.29}{231.35 - 16.29} = 0.63$$

Comments:

As the pressure is lowered, the quality rises. Less pressure means a higher proportion of vapor in the two-phase mixture. Note that the temperature of the refrigerant decreases from $-18.8°C$ to $-26.43°C$. In refrigeration systems, large changes in temperature across throttling valves are used to obtain the cooling effect. This is discussed in Chapter 8.

6.6 MIXING CHAMBERS

In a mixing chamber, two or more streams of fluid are mixed and a single stream of fluid exits, as illustrated in Figure 6-14. A simple example of a mixing chamber is a faucet that is fed by both hot and cold water lines. Adjusting the valve of this mixing chamber produces water at a variety of temperatures. We restrict attention to mixing chambers in which all entering streams have the same chemical composition. Typically, kinetic and potential energy effects are unimportant for mixing chambers. There may or may not be any control volume work or heat transfer. When dealing with mixing chambers, it is usually best to start with the first law and eliminate terms as appropriate.

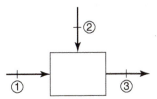

FIGURE 6-14   Fluids at different states enter at stations 1 and 2, are intimately mixed within the mixing chamber, and exit at station 3.

EXAMPLE 6-7   Heating of a room with infiltration from the exterior

A room is heated by a hot-air system, which supplies 8.5 lbm/min of air at 85°F. Air infiltrates into the room at a rate of 3.1 lbm/min from the outside, which is at 0°F. Air leaves through the open damper in the fireplace. Two people are sitting in the room watching TV. Each person generates 72 Cal/h and the TV generates 330 W. What is the steady-state temperature of the air in the room assuming perfect mixing?

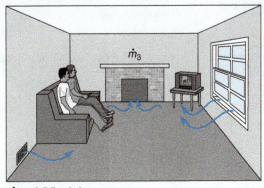

$\dot{m}_2 = 3.1$ lbm/min
$T_2 = 0°$ F

$\dot{m}_1 = 8.5$ lbm/min
$T_1 = 85°$ F

Approach:

This problem is solved by using both conservation of mass and conservation of energy. The flow rates of air into the room are known. The mass flow rate out must equal the sum of the two entering mass flow rates. (We assume steady state.) Use conservation of energy assuming no work and negligible kinetic and potential energy. The heat generated by the two people and the TV set must be taken into account. To find the enthalpies, assume air is an ideal gas with constant specific heat, so that $\Delta h = c_p \, \Delta T$. Combining the conservation of mass and energy equation, we can solve for the unknown exit temperature.

Assumptions:

A1. The flow is steady. The mass of air within the room does not change with time.

A2. The air is perfectly mixed before it leaves the room.

A3. Kinetic energy is negligible.

A4. Potential energy is negligible.

A5. Conduction through the walls is insignificant.

A6. Air may be treated as an ideal gas.

A7. Specific heat is constant.

Solution:

We define the control volume to include the volume inside the room, excluding the people, the TV set, and any other furnishings. The room is a mixing chamber with two streams entering, one at 85°F and the other at 0°F. There is also air leaving the room through the fireplace [A1]. Conservation of mass requires

$$\dot{m}_1 + \dot{m}_2 = \dot{m}_3$$

where stream 1 is the hot-air supply, stream 2 is the infiltration of cold air that leaks into the room, and stream 3 is the mixed air that leaves the room [A2]. Using given values for the entering streams,

$$\dot{m}_3 = 8.5 + 3.1 = 11.6 \frac{\text{lbm}}{\text{min}}$$

The first law for a steady process is

$$0 = \dot{Q}_{cv} - \dot{W}_{cv} + \sum_{\text{in}} \dot{m}_i \left(h_i + \frac{V_i^2}{2} + gz_i \right) - \sum_{\text{out}} \dot{m}_e \left(h_i + \frac{V_e^2}{2} + gz_e \right)$$

There is no work done on or by the air in the control volume. Kinetic and potential energy may be neglected [A3][A4]. There is heat added to the control volume by the two people and the TV. We assume no heat is conducted through the walls [A5]. With these considerations, the first law reduces to

$$0 = \dot{Q}_{cv} + \dot{m}_1 h_1 + \dot{m}_2 h_2 - \dot{m}_3 h_3$$

Using conservation of mass, this may be rewritten as

$$0 = \dot{Q}_{cv} + \dot{m}_1 h_1 + \dot{m}_2 h_2 - (\dot{m}_1 + \dot{m}_2) h_3$$

Rearranging terms,

$$0 = \dot{Q}_{cv} + \dot{m}_1 (h_1 - h_3) + \dot{m}_2 (h_2 - h_3)$$

If we make the reasonable assumption that air is an ideal gas with constant specific heat, then [A6][A7]

$$0 = \dot{Q}_{cv} + \dot{m}_1 c_p (T_1 - T_3) + \dot{m}_2 c_p (T_2 - T_3)$$

Solving for T_3,

$$T_3 = \frac{\dot{Q}_{cv} + \dot{m}_1 c_p T_1 + \dot{m}_2 c_p T_2}{\dot{m}_1 c_p + \dot{m}_2 c_p}$$

or

$$T_3 = \frac{\frac{\dot{Q}_{cv}}{c_p} + \dot{m}_1 T_1 + \dot{m}_2 T_2}{\dot{m}_1 + \dot{m}_2}$$

The heat generated by the people and the TV set is

$$\dot{Q}_{cv} = (2) \left(72 \frac{\text{Cal}}{\text{h}} \right) \left(\frac{1\,\text{Btu}}{0.252\,\text{Cal}} \right) \left(\frac{1\,\text{h}}{60\,\text{min}} \right) + (330\,\text{W}) \left(\frac{1\,\text{Btu/s}}{1.055\,\text{kW}} \right) \left(\frac{1\,\text{kW}}{1000\,\text{W}} \right) \left(\frac{60\,\text{s}}{1\,\text{min}} \right)$$

$$\dot{Q}_{cv} = 28.3 \frac{\text{Btu}}{\text{min}}$$

The specific heat of air may be found in Table B-8. Note that the variation with temperature is insignificant in this case. With given values, the exit temperature is

$$T_3 = \frac{\dfrac{\left(28.3 \dfrac{\text{Btu}}{\text{min}}\right)}{\left(0.24 \dfrac{\text{Btu}}{\text{lbm} \cdot \text{R}}\right)} + \left(8.5 \dfrac{\text{lbm}}{\text{min}}\right)(85°\text{F}) + \left(3.1 \dfrac{\text{lbm}}{\text{min}}\right)(0°\text{F})}{\left(8.5 \dfrac{\text{lbm}}{\text{min}}\right) + \left(3.1 \dfrac{\text{lbm}}{\text{min}}\right)}$$

$$T_3 = 72.5°\text{F}$$

Comments:

In reality, there will be some heat loss by conduction through the walls of the room, but that effect is not considered here. Note that it is not necessary to use absolute temperature because only temperature *differences* are involved in the energy equation. If absolute temperature had been used, the final result would have been the same. If the power generated by the TV and the people had not been included in the analysis, the final temperature would have been 62.3°F.

6.7 HEAT EXCHANGERS

In a heat exchanger, two different fluid streams exchange heat across a wall. An example is shown in Figure 6-15. Here a cold fluid enters the inner tube of the heat exchanger. A hot fluid enters the surrounding jacket and flows over the outside of the tube. The cold fluid temperature increases, while the hot fluid temperature decreases.

Heat exchangers are built in a wide variety of configurations for many different purposes. Chapter 13 discusses heat exchangers in depth, and Figure 13-1 illustrates several

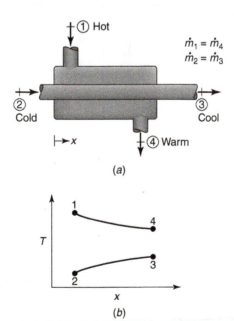

FIGURE 6-15 A one-pass shell and tube heat exchanger: (a) geometry; (b) temperature as a function of *x*.

types of heat exchangers. In this section, we focus attention on one aspect of heat exchanger design: the application of the energy equation to heat exchangers.

You may be familiar with the heat exchanger in your automobile, which, for historical reasons, is called a radiator. Air flows over the radiator as the vehicle moves and cools the water–antifreeze mixture inside the radiator. The water–antifreeze mixture, in turn, is circulated through the engine block to cool it. Heat exchangers are widely used in the chemical process industry, in power plants, in refrigeration equipment, and in numerous other applications. Sometimes the fluid within the heat exchanger changes phase. In that case, the heat exchanger is usually called a condenser or an evaporator.

All heat exchangers share certain features in common. Usually kinetic and potential energy changes are negligible. In addition, no control volume work is done within a heat exchanger. While there may be significant energy flow from one fluid to another within the heat exchanger, very little heat usually leaves or enters the heat exchanger from the surroundings. So, if the entire heat exchanger is selected as the control volume, it is reasonable to assume adiabatic conditions. On the other hand, if a control volume is defined as just the cold or just the hot fluid, the heat transfer into or out of the control volume is a major term in the energy equation. The selection of control volume depends on the nature of the analysis, on what is known, on what is sought, and so on. Choosing an appropriate control volume is a matter of experience.

The pressure changes within a heat exchanger are typically very small. There must be some pressure change so that the fluids will flow; however, this pressure change is small enough to have negligible influence on the thermodynamic states. If, for example, a fluid condenses within a heat exchanger, the condensation may be approximated as a constant-pressure process even though a small pressure change is necessary to drive the flow.

EXAMPLE 6-8   Evaporation of refrigerant in a heat exchanger

Refrigerant R-134a enters the evaporator of an air-conditioning system with a quality of 0.42 and a temperature of $-12°C$. It exits as saturated vapor. The flow rate of refrigerant is 8 kg/min. The other fluid, air, enters at 25°C and atmospheric pressure and exits at 18°C. Find the volumetric flow rate of air at the exit.

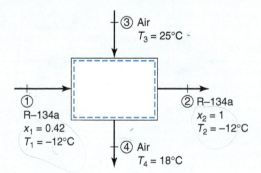

Approach:

Select the entire heat exchanger as the control volume. All terms in the energy equation are dropped except the enthalpy terms. The air and the R-134a do not mix, so $\dot{m}_1 = \dot{m}_2$ and $\dot{m}_3 = \dot{m}_4$. For the refrigerant, use Table A-14 to find enthalpy. For the air, use ideal gas relations to find enthalpy.

Assumptions:

Solution:

Define the control volume to enclose the entire heat exchanger. The first law is

$$\frac{dE_{cv}}{dt} = \dot{Q}_{cv} - \dot{W}_{cv} + \sum_{in} \dot{m}_i \left(h_i + \frac{V_i^2}{2} + gz_i \right) - \sum_{out} \dot{m}_e \left(h_e + \frac{V_e^2}{2} + gz_e \right)$$

A1. The flow is steady.
A2. Kinetic energy is negligible.
A3. Potential energy is negligible.
A4. The exterior casing of the heat exchanger is perfectly insulated.

Assume that the heat exchanger operates in steady state [A1]. Kinetic and potential energy changes are negligible, and no work is done [A2][A3]. Also assume that there is no heat transfer between the outside of the heat exchanger and the environment [A4]. The first law then becomes

$$0 = \dot{m}_1 h_1 + \dot{m}_3 h_3 - \dot{m}_2 h_2 - \dot{m}_4 h_4$$

The air and the refrigerant do not mix in the heat exchanger. Therefore, by conservation of mass,

$$\dot{m}_1 = \dot{m}_2 = \dot{m}_r$$
$$\dot{m}_3 = \dot{m}_4 = \dot{m}_a$$

Using these in the first law,

$$0 = \dot{m}_r (h_1 - h_2) + \dot{m}_a (h_3 - h_4)$$

Solving for the mass flow rate of the air,

$$\dot{m}_a = -\dot{m}_r \frac{(h_1 - h_2)}{(h_3 - h_4)}$$

A5. Specific heat is constant.

Assuming a constant specific heat for the air [A5],

$$\dot{m}_a = -\dot{m}_r \frac{(h_1 - h_2)}{c_p (T_3 - T_4)}$$

The enthalpy at state 1 is

$$h_1 = h_f + x_1 \left(h_g - h_f \right)$$

Using Table A-14,

$$h_1 = 34.39 \, \frac{kJ}{kg} + (0.42)(240 - 34.39) \, \frac{kJ}{kg}$$

$$h_1 = 121 \, \frac{kJ}{kg}$$

With values from Tables A-14 and A-8,

$$\dot{m}_a = -8 \, \frac{kg}{min} \frac{(121 - 240) \, \frac{kJ}{kg}}{\left(1 \, \frac{kJ}{kg \cdot K} \right)(25 - 18)°C} = 136 \, \frac{kg}{min}$$

The volumetric flow rate at the exit is

$$\dot{V}_4 = \frac{\dot{m}_a}{\rho_4}$$

The density of air could be evaluated from the ideal gas law using atmosphere pressure and the exit temperature of the air stream. However, it is easier to use Table A-7, which give the density of air at atmospheric pressure for a range of temperatures. From this table, $\rho = 1.22$ kg/m^3 and the volumetric flow rate becomes

$$\dot{V}_4 = \frac{136 \; \frac{\text{kg}}{\text{min}}}{1.22 \; \frac{\text{kg}}{\text{m}^3}}$$

$$\dot{V}_4 = 111 \; \frac{\text{m}^3}{\text{min}}$$

This is the amount of cold air that can be supplied by the air conditioner.

EXAMPLE 6-9 Design of an air conditioner

The cooling coil of an air conditioner is to be designed for a cooling capacity of 25 kW. The cooling coil is a heat exchanger that has an array of inline copper tubes (as shown in the schematic) with inside diameter of 4.8 mm and outside diameter of 6.0 mm. Refrigerant 134a, which flows inside the tubes, enters the heat exchanger at a pressure of 0.20 MPa and 15% quality and leaves as a saturated vapor. The outlet velocity of the refrigerant is set at 40 m/s, and the resulting refrigerant-side heat transfer coefficient is 350 W/m$^2 \cdot$K. Air at 40°C flows perpendicular to the outside of the tubes, with a heat transfer coefficient of 175 W/m$^2 \cdot$K; because of the high air mass flow rate, we can assume the air temperature remains relatively constant (as a first approximation). Determine the following:

a) The mass flow rate of the refrigerant (in kg/s)

b) The number of tubes required, and

c) The length of the tubes (in m)

$D_i = 4.8$ mm
$D_o = 6.0$ mm

Air
40°C
$h_{air} = 175$ W/m$^2 \cdot$ K

R–134a
$P_1 = 0.2$ MPa
$x_1 = 0.15$
$h_{ref} = 350$ W/m$^2 \cdot$ K

Approach:

Define the control volume to include only the volume inside the tubes where refrigerant flows (exclude the air). Because we are given the heat transfer rate, we can use the open-system energy equation to calculate the required total mass flow rate. Once the mass flow rate is known, the given diameter and velocity in each tube can be used to calculate the required number of tubes. The heat transfer rate equation then is used to calculate the required tube length.

Assumptions:

A1. The system is steady.
A2. No work is done on or by the control volume.

Solution:

a) We define the control volume to contain only the refrigerant. For this control volume, assuming [A1], [A2], and [A3], the open system energy equation is

$$0 = \dot{Q} + \dot{m}_1 h_1 - \dot{m}_2 h_2$$

A3. Potential and kinetic energy effects are negligible.

From the conservation of mass equation, using the same assumptions,

$$0 = \dot{m}_1 - \dot{m}_2 \quad \rightarrow \quad \dot{m}_1 = \dot{m}_2 = \dot{m}$$

Combining the two equations, and solving for mass flow rate,

$$\dot{m} = \frac{\dot{Q}}{h_2 - h_1}$$

From Table A-15 for R-134a at 0.20 MPa, $T_{sat} = -10.09°C$, $h_f = 36.84$ kJ/kg, $h_g = 241.30$ kJ/kg, and $v_g = 0.0993$ m³/kg. For the entering refrigerant,

$$h_1 = h_f + x_1 \left(h_g - h_f\right)$$

$$h_1 = 36.84 + (0.15)(241.30 - 36.84) = 67.51 \text{ kJ/kg}$$

A4. There is no pressure drop across the heat exchanger.

For the leaving refrigerant, assuming [A4], $h_2 = h_g$. Because heat transfer is to the control volume, the heat transfer rate is positive:

$$\dot{m} = \frac{(25 \text{ kW}) \left(1 \text{ kJ/kW·s}\right)}{(241.3 - 67.51) \text{ kJ/kg}} = 0.144 \frac{\text{kg}}{\text{s}}$$

b) The definition of mass flow rate is

$$\dot{m} = \rho \mathcal{V} A = \frac{\mathcal{V}}{v} A$$

For multiple tubes, N, in parallel, $A = N \pi D_i^2 / 4$. Substituting this into the mass flow rate expression and solving for N:

$$N = \frac{4 \dot{m} v}{\pi D_i^2 \mathcal{V}} = \frac{4 \left(0.144 \text{ kg/s}\right) \left(0.0993 \text{ m}^3/\text{kg}\right)}{\pi \left(0.0048 \text{ m}\right)^2 \left(40 \text{ m/s}\right)} = 19.73 \text{ tubes}$$

We need an integral number of tubes, so we use $N = 20$.

A5. The heat transfer is one-dimensional.

c) The air and the refrigerant have constant temperatures; so assuming [A5], the heat transfer rate equation is

$$\dot{Q} = \frac{\Delta T}{R_{tot}}$$

The total thermal resistance is composed two convective resistances and one conduction resistance:

$$R_{tot} = R_{ref} + R_{cond} + R_{air}$$

where

$$R_{ref} = \frac{1}{h_{ref} \pi D_i L N} = \frac{1}{\left(350 \text{ W/m}^2 \cdot \text{K}\right) \pi \left(0.0048 \text{ m}\right) L \left(20\right)} = \frac{0.00947}{L} \frac{\text{m} \cdot \text{K}}{\text{W}}$$

$$R_{air} = \frac{1}{h_{air} \pi D_o L N} = \frac{1}{\left(175 \text{ W/m}^2 \cdot \text{K}\right) \pi \left(0.006 \text{ m}\right) L \left(20\right)} = \frac{0.0152}{L} \frac{\text{m} \cdot \text{K}}{\text{W}}$$

From Table A-2 for copper, $k = 401$ W/m·K.

$$R_{cond} = \frac{\ln \left(D_o/D_i\right)}{2 \pi k L N} = \frac{\ln \left(0.006/0.0048\right)}{2 \pi \left(401 \text{ W/m} \cdot \text{K}\right) L \left(21\right)} = \frac{4.43 \times 10^{-6}}{L} \frac{\text{m} \cdot \text{K}}{\text{W}}$$

At 0.2 MPa, the saturation temperature of R-134a is $T_{sat} = -10.09°C$. (See Table A-15.) Combining the resistances with the heat transfer rate equation, and solving for length, L,

$$L = \frac{(25000\ \text{W})\,(0.00947 + 4.43 \times 10^{-6} + 0.0152)\,(\text{m·K/W})}{[40 - (-10.09)]\ \text{K}} = 12.3\ \text{m}$$

Comments:

In this example, we needed to use conservation of mass, conservation of energy, and the heat transfer rate equation, with appropriate resistances, in a sequential order. To accommodate this length in a reasonably sized package, the tubes would be laid out in a serpentine pattern. Note that the thermal resistance for conduction through the metal tube wall is very small compared to the convective resistances and could have been neglected.

6.8 TRANSIENT PROCESSES

(Go to www.wiley.com/college/kaminski)

SUMMARY

The first law for an open system is

$$\frac{dE_{cv}}{dt} = \dot{Q}_{cv} - \dot{W}_{cv} + \sum_{in} \dot{m}_i \left(h_i + \frac{\mathcal{V}_i^2}{2} + gz_i \right)$$

$$- \sum_{out} \dot{m}_e \left(h_e + \frac{\mathcal{V}_e^2}{2} + gz_e \right)$$

In an adiabatic nozzle or diffuser, the first law often reduces to

$$h_1 + \frac{\mathcal{V}_1^2}{2} = h_2 + \frac{\mathcal{V}_2^2}{2}$$

In an adiabatic turbine or compressor where kinetic and potential energy effects are negligible, the first law becomes

$$W_{cv} = \dot{m}\,(h_1 - h_2)$$

This equation must be modified if there is more than one inlet and outlet to a turbine. The work required to pump an incompressible liquid is

$$\dot{W}_{cv} = \dot{m}\,v\,(P_1 - P_2)$$

In a throttling device, the first law becomes

$$h_i = h_e$$

SELECTED REFERENCES

BLACK, W. Z., and J. G. HARTLEY, *Thermodynamics*, Harper & Row, New York, 1985.

CENGEL, Y. A., and M. A. BOLES, *Thermodynamics, an Engineering Approach*, 4th ed., McGraw-Hill, New York, 2002.

DAUGHERTY, R. L., *Hydraulic Turbines*, McGraw-Hill, New York, 1913.

HOWELL, J. R., and R. O. BUCKIUS, *Fundamentals of Engineering Thermodynamics*, 2nd ed., McGraw-Hill, New York, 1992.

MORAN, M. J., and H. N. SHAPIRO, *Fundamentals of Engineering Thermodynamics*, 3rd ed., Wiley, New York, 1995.

MYERS, G., *Engineering Thermodynamics*, Prentice Hall, Englewood Cliffs, NJ, 1989.

VAN WYLEN, G. J., R. E. SONNTAG, and C. BORGNAKKE, *Fundamentals of Classical Thermodynamics*, 4th ed., Wiley, New York, 1994.

PROBLEMS

Problems designated with WEB refer to material available at www.wiley.com/college/kaminski.

NOZZLES AND DIFFUSERS

P6-1 Steam at 160 psia and 400°F enters a nozzle with a volumetric flow rate of 6,615 cfm (cubic feet per minute). The inlet area is 14.5 in.^{2}. If the steam leaves at 1500 ft/s at a pressure of 40 psia, find the exit temperature.

P6-2 Oxygen at 220°F enters a well-insulated nozzle of inlet diameter 0.6 ft. The inlet velocity is 60 ft/sec. The oxygen leaves at 75°F, 10 psia. The exit area is 0.01767 ft^2. Calculate the pressure at the inlet.

P6-3 A well-insulated nozzle has an entrance area of 0.28 m^2 and an exit area of 0.157 m^2. Air enters at a velocity of 65 m/s and leaves at 274 m/s. The exit pressure is 101 kPa, and the exit temperature is 12°C. What is the entrance pressure?

P6-4 Carbon monoxide enters a nozzle at 520 kPa, 100°C, with a velocity of 10 m/s. The gas exits at 120 kPa and 500 m/s. Assuming no heat transfer and ideal gas behavior, find the exit temperature.

P6-5 Low-velocity steam with negligible kinetic energy enters a nozzle at 320°C, 3 MPa. The steam leaves the nozzle at 2 MPa with a velocity of 410 m/s. The mass flow rate is 0.37 kg/s.

a. Determine the exit state.

b. Determine the exit area.

P6-6 Steam enters a diffuser at 250°C and 50 kPa and exits at 300°C and 150 kPa. The diameter at the entrance is 0.25 m and the diameter at the exit is 0.5 m. If the mass flow rate is 9.4 kg/s, find the heat transfer to the surroundings.

P6-7 Air enters a diffuser at 50 kPa, 85°C with a velocity of 250 m/s. The exit pressure is atmospheric at 101 kPa. The exit temperature is 110°C. The diameter at the inlet is 8 cm.

a. Find the exit velocity.

b. Find the diameter at the exit.

Assume constant specific heats.

P6-8 Superheated steam enters a well-insulated diffuser at 14.7 psia, 320°F, and 400 ft/s. The steam exits as saturated vapor at a very low speed. Find the exit pressure and temperature.

TURBINES

P6-9 Steam enters an adiabatic turbine at 0.8 MPa and 500°C. It exits at 0.05 MPa and 150°C. If the turbine develops 24.5 MW of power, what is the mass flow rate?

P6-10 Air enters an adiabatic turbine at 900 K and 1000 kPa. The air exits at 400 K and 100 kPa with a velocity of 30 m/s. Kinetic and potential energy changes are negligible. If the power delivered by the turbine is 1000 kW,

a. find the mass flow rate.

b. find the diameter of the duct at the exit.

P6-11 Saturated steam at 320°C enters a well-insulated turbine. The mass flow rate is 2 kg/s and the exit pressure is 50 kPa. Determine the final state if the power produced is

a. 100 kW.

b. 400 kW.

P6-12 Superheated steam at 1.6 MPa, 600°C enters a well-insulated turbine. The exit pressure is 50 kPa. The turbine produces 10 MW of power. If the exit pipe is 1.6 m in diameter and carries 11 kg/s of flow, find the velocity at the exit. Neglect kinetic energy.

P6-13 Air at 550°C and 900 kPa is expanded through an adiabatic gas turbine to final conditions of 100 kPa and 300°C. The total power output desired is 1 MW. If the inlet velocity is 30 m/s, what should the inlet pipe diameter be? Neglect kinetic and potential energy.

P6-14 Air at 510°C and 450 kPa enters an ideal, adiabatic turbine. The exit pressure is 101 kPa. In steady state, the turbine produces 50 kW of power.

a. Find the exit temperature. (Hint: use Eg. 2-56)

b. Find the mass flow rate.

P6-15 Saturated steam at 3 MPa enters a well-insulated turbine operating in steady state. The turbine produces 600 kW of power. The mass flow rate through the turbine is 84 kg/min and the exit quality is 0.93. Find the exit temperature.

COMPRESSORS

P6-16 In a 3-hp compressor, carbon dioxide flowing at 0.023 lbm/s is compressed to 120 psia. The gas enters at 60°F and 14.7 psia. The inlet and outlet pipes have the same diameter. Find the final temperature and the volumetric flow rate at the exit (in ft^3/min). Assume constant specific heat at 100°F.

P6-17 A well-insulated compressor is used to raise saturated R-134a vapor at a pressure of 360 kPa to a final pressure of 900 kPa. The compressor operates in steady state with a power input of 850 W. If the flow rate is 0.038 kg/s, what is the final temperature?

P6-18 Air flowing at 0.5 m^3/min enters a compressor at 101 kPa and 25°C. The air exits at 600 kPa and 300°C. During this process, 250 W of heat are lost to the environment. What is the required power input?

P6-19 Refrigerant 134a enters a compressor at 0°F and 10 psia with a volumetric flow rate of 15 ft^3/min. The refrigerant exits at 70 psia and 140°F. If the power input is 2 hp, find the rate of heat transfer in Btu/h.

PUMPS

P6-20   A pump is used to raise the pressure of a stream of water from 10 kPa to 0.7 MPa. The temperature of the water is the same at the inlet and outlet and equal to 20°C. The velocity also does not change across the pump. If the mass flow rate is 14 kg/s, what power is needed to drive the pump? Assume frictionless flow and no significant elevation change.

P6-21   A 2-hp pump is used to raise the pressure of saturated liquid water at 5 psia to a higher value. Assume the velocity is constant, the water is incompressible, and the flow is frictionless. If the mass flow rate is 6 lbm/s, find the final pressure.

P6-22   Water is pumped at 12 m/s through a pipe of diameter 1.2 cm. The inlet pressure is 30 kPa. If the pump delivers 6 kW, find the final pressure. Assume frictionless, incompressible flow with no elevation or velocity changes.

P6-23   A 1-hp pump delivers oil at a rate of 10 lbm/s through a pipe 0.75 in. in diameter. There is no elevation change between inlet and exit, no velocity change, and no oil temperature change. The oil density is 56 lbm/ft^3. Find the pressure rise across the pump.

P6-24   An architect needs to pump 2.3 lbm/s of water to the top of the Empire State Building, which is about 1000 ft high. Assume water at 45 psia is available at the base of the building. What is the power of the pump needed, in hp, if the flow is assumed to be frictionless? The velocity of the water is constant.

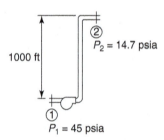

THROTTLING DEVICES

P6-25   Air at 150°C, 40 kPa is throttled to 100 kPa. The inlet velocity is 3.6 m/s. Find the exit velocity.

P6-26   Saturated liquid R-134a at 24°C is throttled until the final quality is 0.116. Find the final temperature and pressure.

P6-27   Saturated liquid R-134a at 80°F undergoes a throttling process. The pressure decreases to one-fourth of its original value. Find the exit quality.

P6-28   A supply line, contains a two-phase mixture of steam and water at 240°C. To determine the quality of the mixture, a throttling calorimeter is used. In this device, a small sample of the two-phase mixture is bled off from the line and expanded through a throttling valve to atmospheric pressure. If the temperature on the downstream side of the throttling valve is measured to be 125°C, what is the quality of the mixture in the main steam line?

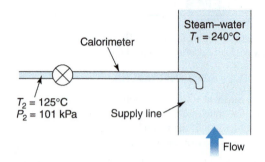

P6-29   In a heat pump, R-134a is throttled through an expansion coil, which is a long copper tube of small diameter. The tube is bent in a coil both to fit in a compact space and to provide a large pressure drop. The refrigerant enters as saturated liquid at 5°C with a flow rate of 0.025 kg/s and exits as a two-phase mixture at a pressure of 200 kPa. The wall of the coil may be assumed to be at the average temperature of the inlet and outlet. Heat is exchanged by natural convection and radiation from the outer surface of the coil with a combined heat transfer coefficient of 6 W/m^2·°C to the surroundings at 20°C. The expansion coil has an outside diameter of 8 mm and a length of 2.2 m. Calculate the quality at the exit state.

MIXING CHAMBERS

P6-30   One way to produce saturated liquid water is to mix subcooled liquid water with steam. In the tank shown in the figure, 40 kg/s of subcooled liquid water enter at 15°C and 50 kPa. Superheated steam enters at 200 °C and 50 kPa. What mass flow rate of steam is required so that the exit stream is saturated liquid water at 50 kPa? Assume the tank is well insulated.

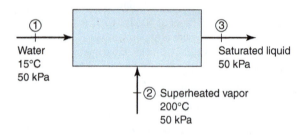

P6-31   In a desuperheater, superheated steam is converted to saturated steam by spraying liquid water into the steam. Using data on the figure, calculate the mass flow rate of liquid water.

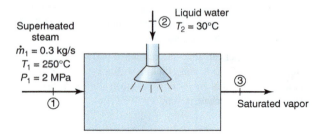

P6-32 A laundry requires a stream of 8 kg/s of hot water at 40°C. To obtain this supply, liquid water at 20°C is mixed in an adiabatic chamber with saturated steam. All three process streams are at 100 kPa. What are the required mass flow rates of the two inlet streams?

PIPE HEAT TRANSFER

P6-33 Steam with a quality of 0.88 and a pressure of 20 kPa enters a condenser. The steam flow is divided equally among 20 tubes 2.1 cm in diameter that run in parallel through the condenser. The same amount of heat is removed from each tube. Liquid water exits each tube with a velocity of 1.5 m/s and a temperature of 55°C. Find the total amount of heat removed from the entire condenser.

P6-34 Saturated steam at 120°F is condensed in a tube, as shown in the figure. Cooling water at 50°F flows in crossflow over the exterior of the pipe, giving a heat transfer coefficient of 200 Btu/h·ft²·°F. Find the exit quality.

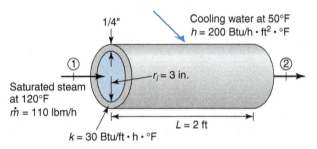

HEAT EXCHANGERS

P6-35 Superheated R-134a enters a well-insulated heat exchanger at 0.7 MPa, 70°C. It exits as saturated liquid at 0.7 MPa with a volumetric flow rate of 6000 cm³/min. The R-134a exchanges heat with an air flow, which enters at 18°C at a mass flow rate of 195 kg/min. Find the exit air temperature.

P6-36 R-134a flows through the evaporator of a refrigeration cycle at a rate of 5 kg/s. The R-134a enters as saturated liquid and leaves as saturated vapor at 12°C. Air at 25°C enters the shell side of the heat exchanger. If the air leaves at 15°C, what mass flow rate of air is required?

P6-37 Superheated steam at 5 psia and 200°F is condensed in a heat exchanger. The steam flows at 39 lbm/s and exits as saturated

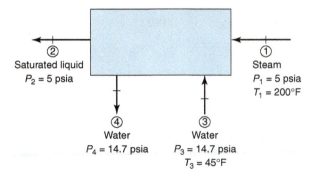

liquid. Cooling water at 45°F is used to condense the steam. The water and steam are not mixed in the heat exchanger, but enter and leave as separate streams. If the maximum allowable rise in water temperature is 15°F and the maximum allowable water velocity is 11 ft/s, what is the diameter of the pipe that carries water to the heat exchanger?

P6-38 A two-phase mixture of steam and water with a quality of 0.93 and a pressure of 5 psia enters a condenser at 14.3 lbm/s. The mixture exits as saturated liquid. River water at 45°F is fed to the condenser through a large pipe. The exit temperature of the river water is 70°F less than the exit temperature of the other stream. If the maximum allowable average velocity in the pipe carrying river water is 15 ft/s, calculate the pipe diameter.

P6-39 A heat exchanger is used to cool engine oil. The specific heat of the oil is 0.6 Btu/lbm·°F. Using data on the figure, find the exit temperature of the air.

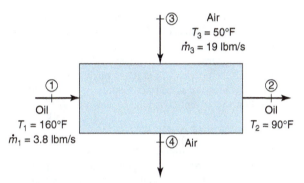

TWO-COMPONENT SYSTEMS

P6-40 Saturated liquid R-134a at 36°C is throttled to −8°C. The refrigerant then enters an evaporator and exits as saturated vapor. The evaporator is used to cool liquid water from 20°C to 10°C. If the mass flow rate of refrigerant is 0.013 kg/s, what is the mass flow rate of the water?

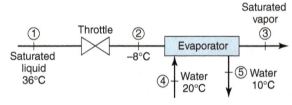

P6-41 In a flash chamber, a pressurized liquid is throttled to a lower pressure, where it becomes a two-phase mixture. The saturated liquid and vapor streams are removed in separate lines. In the figure, liquid R-134a at 10°F and 30 psia is throttled to 5 psia. If 21.6 lbm/h of saturated vapor exits the flash chamber, what is the inlet flow rate? Assume the flash chamber is adiabatic.

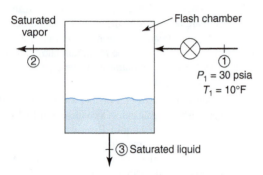

Saturated vapor

Flash chamber

②

③ Saturated liquid

$P_1 = 30$ psia
$T_1 = 10°F$

①

P6-42 Saturated liquid water at 40 kPa enters a 140-kW pump. The output of the pump is fed into a boiler, where heat is added at a rate of 302 MW. There is negligible pressure drop across the boiler. If the mass flow rate of water is 70 kg/s, determine the boiler pressure and the state at the exit of the boiler.

P6-43 Air at 2000 R enters the turbine of a turbojet engine. The turbine is well insulated and produces 100 Btu of work per pound mass of air flowing through the engine. Upon exiting the turbine, the air enters the inlet of an insulated nozzle at 20 ft/s. The air leaves the nozzle at 2800 ft/s through an exit flow area of 0.6 ft^2. The pressure at the nozzle exit is 10 lbf/in.^{2}. What is the mass flow rate of air through the engine in lbm/s?

OPEN-SYSTEM TRANSIENTS

P6-44 (WEB) A well-insulated, rigid tank of volume 0.7 m^3 is initially evacuated. The tank develops a leak and atmospheric air at 20°C, 100 kPa enters. Eventually the air in the tank reaches a pressure of 100 kPa. Find the final temperature.

P6-45 (WEB) Helium at 150°F and 40 psia is contained in a rigid, well-insulated tank of volume 5 ft^3. A valve is cracked open and the helium slowly flows from the tank until the pressure drops to 20 psia. During this process, the helium in the tank is maintained at 150°F with an electric resistance heater.

a. Find the mass of helium withdrawn.

b. Find the energy input to the heater.

P6-46 (WEB) A residential hot-water heater initially contains water at 140°F. Someone turns on a shower and draws water from the tank at a rate of 0.2 lbm/s. Cold makeup water at 50°F is added to the tank at the same rate. A burner supplies 5472 Btu/h of heat. The water tank, which is a cylinder of diameter 1.8 ft, is filled to a height of 4 ft. How long will it be before the exiting water reaches 100°F? Assume a well-mixed tank.

P6-47 (WEB) In an industrial process, two streams are mixed in a tank and a single stream exits. Both streams may be assumed to have the properties of water. The volume of fluid in the tank is constant. A paddlewheel stirs the tank contents, doing work $\dot{W}$. Initially, the water in the tank is at temperature T_1. At time $t = 0$, stream A at temperature T_A enters with mass flow rate $\dot{m}_A$, and stream B enters at T_B with rate $\dot{m}_B$. The quantities T_A, T_B,

$\dot{m}_A$, and $\dot{m}_B$ are all constant with time. Assuming a well-mixed tank, derive a formula for the time, t_2, at which the tank water temperature is T_2. The tank is well insulated.

P6-48 (WEB) A well-insulated tank of volume 0.035 m^3 is initially evacuated. A valve is opened, and the tank is charged with superheated steam from a supply line at 600 kPa, 500°C. The valve is closed when the pressure reaches 300 kPa. How much mass enters?

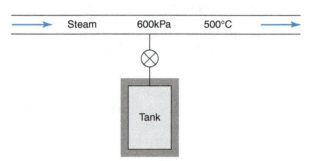

Steam   600kPa   500°C

Tank

P6-49 (WEB) A well-insulated piston–cylinder assembly contains 0.06 kg of R-134a at −15°C with a quality of 0.92. A supply line introduces superheated R-134a at 10°C, 200 kPa into the cylinder. Assuming the pressure in the cylinder is constant, calculate the volume just when all the liquid has evaporated.

P6-50 (WEB) The pressure inside a pot is maintained at an elevated level by a steel bob that rests on an open tube of inside diameter 0.5 cm. The bob, which has a mass of 0.401 kg, jiggles whenever the pressure in the pot is high enough to displace it, and steam is released. Heat is added to the bottom of the pot at a rate of 900 W. Heat is lost from the sides and the top of the pot by natural convection with a heat transfer coefficient of 3.9 W/m^2·K. The pot has a height of 0.154 m and a diameter of 0.256 m. The ambient temperature is at 20°C. Assume that conduction resistance through the pot sides and top is very small and that there is no air in the pot (only water and steam). The pot is half filled with water when the bob first lifts.

a. Find the temperature inside the pot.

b. Find the *net* rate of heat addition to the pot.

c. Find the initial mass of the two-phase mixture in the pot.

d. Find the fraction of the pot that is filled with liquid after 1 h.

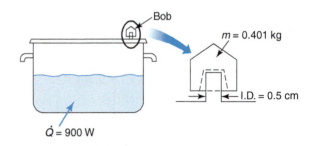

Bob

$m = 0.401$ kg

I.D. = 0.5 cm

$\dot{Q} = 900$ W

CHAPTER 7

THERMODYNAMIC CYCLES AND THE SECOND LAW

7.1 INTRODUCTION

We have studied conservation of mass and conservation of energy. Both are valid for every situation we can imagine. (Remember, we are ignoring nuclear reactions.) However, while these two laws are necessary to solve many problems, for other problems they are not sufficient to obtain a valid answer. This does not mean conservation of mass or conservation of energy is ever invalid. It means that conservation of energy cannot do certain things; it cannot determine:

1. Whether a process is possible or not
2. Whether there is any limitation to the conversion of heat into work
3. In what direction a process must proceed

For example, none of the following processes violate the first law:

1. Warm a bicycle inside a house. Then take it outside in cold weather. Convert the stored thermal energy to kinetic energy so that the bicycle moves without pedaling.
2. Burn fuel in a car engine and convert all the heat generated into work. The engine never gets hot but rather remains at ambient temperature.
3. Drop an ice cube into a cup of warm coffee. Part of the coffee boils and the ice cube becomes larger.

Of course, none of these processes ever occur in reality. Such limitations to conservation of energy necessitate another fundamental law of thermodynamics—the second law. The basis of the second law is experimental evidence; the second law has never been disproved. Fundamentally, the second law states that heat does not move spontaneously from cold to hot bodies. The implications of this statement are wide-reaching and include:

1. Every process has losses.
2. We cannot build a perpetual motion machine.
3. Heat cannot be converted into work with 100% efficiency.

For engineering use, the second law must be expressed on a mathematical basis. In the course of developing the second law, a new thermodynamic property, entropy, is needed. To derive the second law, we begin with an examination of thermodynamic cycles.

7.2 THERMODYNAMIC CYCLES

The prosperity of our society depends on the availability of inexpensive electric power, and the thermal-fluids sciences are a critically important technology in the production of this power. In power plants that use fossil fuels or nuclear power as the energy source,

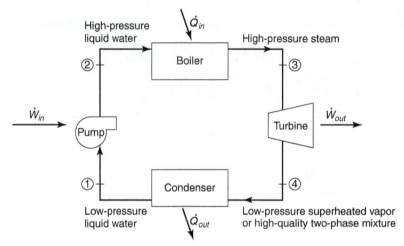

FIGURE 7-1   Example of a power cycle.

electricity is generated when heat is converted into work using a turbine, and the turbine is used to drive an electric generator. This chapter introduces the thermodynamic cycles used for generating power, as well as those used in refrigeration systems.

Figure 7-1 shows a simple power cycle using water as the working fluid. The cycle begins at station 1, where low-pressure liquid water is fed into a pump. The pump does work on the water to raise its pressure. At station 2, the high-pressure water exiting from the pump enters the boiler, where heat is added, and the liquid water is vaporized at constant pressure. The high-pressure steam that exits the boiler at station 3 is passed through a turbine. The turbine extracts energy from the flowing fluid and produces work during this process. Typically, the turbine is connected to an electric power generator, which is not shown in the figure. The fluid leaving the turbine could be either low-pressure superheated vapor or a high-quality two-phase mixture. The turbine exhaust enters a condenser at station 4, where heat is removed, and the steam condenses to the liquid state.

This system is called a cycle because the fluid, which starts at station 1, returns to its initial state after passing through several different states at stations 2, 3, and 4. The heat added to the cycle at the boiler is partially converted into work by the turbine. Because the fluid at the turbine exhaust must be returned to its initial state before it takes another pass through the cycle, some of the heat added in the boiler must be rejected in the condenser. This rejected heat is unavailable for conversion into work by the turbine.

A P-v diagram of the cycle in Figure 7-1 is given in Figure 7-2. At station 1, water exists as a saturated liquid. The pump raises it to a high-pressure subcooled liquid at

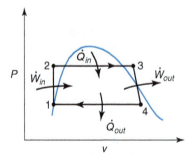

FIGURE 7-2 P-v diagram for the power cycle in Figure 7-1.

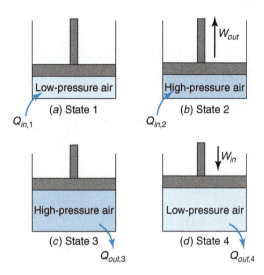

FIGURE 7-3   An air power cycle.

station 2. Within the boiler, the subcooled water is first heated to the saturated liquid state, then heated to the saturated vapor state, and finally superheated to state 3. The superheated steam at station 3 then expands in the turbine to a higher specific volume and a lower pressure. As shown in the figure, the fluid at station 4 is a two-phase mixture. Heat is removed from the mixture until it reaches a saturated liquid at station 1. This is only an example. State 3, for example, might be saturated steam in some practical systems and state 4 might fall in the superheated vapor region.

In the above example, four open systems were linked together to form a cycle. A closed system can also execute a cycle by going through a series of processes. An example is shown in Figure 7-3. State 1 is low-pressure air contained in a piston–cylinder assembly. Heat is added at constant volume until the air reaches state 2, where its pressure has increased. From state 2 to state 3, heat is added at constant pressure; during this process, the air expands, and work is done on the surroundings. From state 3 to state 4, the air is cooled at constant volume so that its pressure decreases. Finally, the air is compressed at constant pressure from state 4 back to state 1. The air must be cooled during the compression to maintain constant pressure. The cycle then can be repeated.

Figure 7-4 shows the P-v diagram for this cycle. State 1 to state 2 involves heating at constant volume. The heat added is designated on the figure as $Q_{in,1}$. Heat, $Q_{in,2}$, is also added in going from state 2 to state 3, but here the piston is allowed to rise and work, W_{out}, is also done. From state 3 to state 4, the air is cooled at constant volume, and the heat removed is designated as $Q_{out,3}$. Finally, as the system returns to its initial state, work, W_{in},

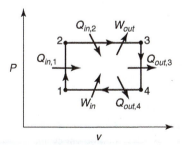

FIGURE 7-4 P-v diagram for the cycle in Figure 7-3.

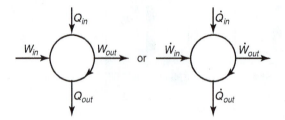

FIGURE 7-5   Schematic representations of power cycles.

is done on the air to compress it, and heat $Q_{out,4}$ is removed to maintain the air at constant pressure.

Both of these cycles have certain features in common. Heat is added over part of the cycle and removed over another part of the cycle. Work is done *by* the cycle, and work is done *on* the cycle. In addition, the *P-v* diagram is a set of lines segments forming a closed figure. Note that in both *P-v* diagrams, the direction of the cycle is clockwise. This is characteristic of thermal power cycles, which are also called heat engines.

Power cycles, in general, can be represented by the diagrams in Figure 7-5. These indicate the heat and work interactions without specifying the details of the cycle. The arrow on the circle is meant to indicate that this is a power cycle whose *P-v* diagram executes a clockwise path. A further shorthand is typically used in which the net work, given by

$$W_{net} = W_{out} - W_{in}$$

is used to replace the separate W_{in} and W_{out}. The diagram for a closed-system power cycle in terms of net work is shown in Figure 7-6. Applying the first law to the cycle shown in Figure 7-6 produces

$$Q_{in} = W_{net} + Q_{out}$$

As noted above, not all of the heat added to a power cycle is converted to useful work; some heat is always rejected (Q_{out}). We can characterize how well a cycle converts heat to work by comparing the useful work obtained to the heat added. Define a cycle thermal efficiency as

$$\eta_{cycle} = \frac{\text{energy we want to use}}{\text{energy we purchase}}$$

Applying this to the power cycle, we obtain

$$\eta_{cycle} = \frac{W_{net}}{Q_{in}} = \frac{\dot{W}_{net}}{\dot{Q}_{in}}$$

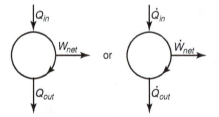

FIGURE 7-6   Alternate schematic representations of a power cycle.

The cycle thermal efficiency is a measure of the success of the conversion of heat into work. Ideally, one would like to convert all the heat into work, to achieve an efficiency of 1 (or 100%), but that is not possible, as will be discussed below. The worst case occurs when heat is added and no work at all is done. Such a cycle has an efficiency of zero. Real cycles vary in efficiency between zero and something less than 1.

The net heat for a cycle is

$$Q_{net} = Q_{in} - Q_{out}$$

As can be seen by applying the first law to the cycle in Figure 7-6,

$$Q_{net} = W_{net} \quad \text{or} \quad \dot{Q}_{net} = \dot{W}_{net}$$

For a cycle, the net heat added equals the net work done.

EXAMPLE 7-1   Energy balances in a power cycle

A power cycle operates with a cycle thermal efficiency of 36%. Fuel is burned in the boiler at a rate of 12,000 kg/h. If the heating value of the fuel is 35,000 kJ/kg, find the heat rejected in the condenser.

Approach:

The heat rejected in the condenser can be obtained from the first law, once the heat added in the boiler and the net work produced are known. Work is related to heat added in the boiler through the cycle thermal efficiency.

Solution:

The power cycle is similar to the one depicted in Figure 7-1. The rate of heat added in the boiler is

$$\dot{Q}_{in} = \left(12,000\ \frac{\text{kg}}{\text{h}} \right) \left(35,000\ \frac{\text{kJ}}{\text{kg}} \right) \left(\frac{1\,\text{h}}{3600\,\text{s}} \right)$$

$$= 1.17 \times 10^5\ \text{kW}$$

From the definition of efficiency

$$\dot{W}_{net} = \eta \dot{Q}_{in} = (0.36)\left(1.17 \times 10^5\ \text{kW}\right)$$

$$= 4.12 \times 10^4\ \text{kW}$$

From the first law

$$\dot{Q}_{out} = \dot{Q}_{in} - \dot{W}_{net}$$

$$= 1.17 \times 10^5 - 4.12 \times 10^4$$

$$= 7.49 \times 10^4\ \text{kW} = 74.9\,\text{MW}$$

Thermodynamic cycles are also used for refrigeration. An example of a refrigeration cycle is given in Figure 7-7. This cycle moves energy from a low-temperature environment to a high-temperature environment by taking advantage of the fact that the saturation temperature of a fluid depends on its pressure. It may seem almost paradoxical at first, but in this cycle, the fluid evaporates at low temperature and condenses at high temperature. This

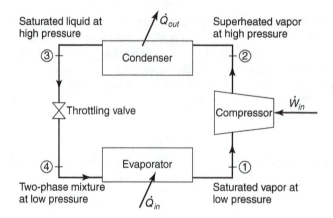

FIGURE 7-7 A refrigeration cycle.

occurs because the pressure is set at different levels in different sections of the cycle. The evaporator is used to absorb heat from a space at low temperature, such as the inside of a refrigerator. In the condenser, heat is rejected at high temperature. In a home refrigerator, this heat is typically rejected to the room air and acts to heat the room.

The details of the cycle are shown in Figure 7-7. At station 1, a saturated vapor at low pressure is fed into a compressor, which adds energy in the form of work. Superheated vapor exits the compressor at high pressure. In the condenser, heat is removed to condense the flowing fluid as it moves from station 2 to station 3. The saturated liquid at station 3 is expanded through a throttling device to return it to low pressure. As the pressure is reduced, some of the liquid vaporizes, so the fluid at station 4 is a two-phase mixture at low pressure and temperature. As the fluid flows through the evaporator, heat is added to vaporize fluid from state 4 to state 1. The cycle is completed as the fluid returns to station 1.

A P-v diagram of this refrigeration cycle is shown in Figure 7-8. Station 1 is saturated vapor. In the compressor, the specific volume is reduced and the pressure is raised. It is important to recognize that as the pressure increases, so does the temperature. The high-temperature, high-pressure, superheated vapor at station 2 is cooled until it begins to condense and then is further cooled until no vapor remains at station 3. The saturated liquid at station 3 expands across the throttling valve to station 4, where the specific volume is greater, and the pressure and temperature are lower than at station 3. Finally, the two-phase mixture at station 4 is vaporized in the evaporator at low temperature until only saturated vapor remains at station 1.

Figure 7-9 gives a schematic representation of a refrigeration cycle. Note that the cycle is represented as moving in the counterclockwise direction. (See the P-v diagram in Figure 7-8.) Refrigeration cycles always travel counterclockwise, while power cycles always travels clockwise.

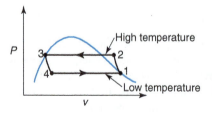

FIGURE 7-8 P-v diagram for the refrigeration cycle in Figure 7-7.

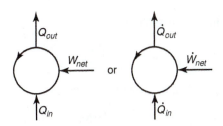

FIGURE 7-9 Schematic representation of a refrigeration cycle.

The objective of a refrigeration cycle is to remove as much heat as possible from a cold environment for a given amount of work input. The performance is characterized by the **coefficient of performance**, defined as

$$COP = \frac{\text{energy we want to use}}{\text{energy we purchase}}$$

In equation form:

$$COP = \frac{Q_{in}}{W_{in}} = \frac{\dot{Q}_{in}}{\dot{W}_{in}}$$

The COP is the ratio of the heat removed by the cycle to the work that must be done to remove this heat. The COP for a refrigeration cycle is a measure of performance analogous to the thermal efficiency for a power cycle. The value of COP is typically greater than 1; however, this does not imply that energy is created. Rather, a refrigeration cycle uses work input (W_{in}) to move energy (Q_{in}) from a low temperature to a high temperature. Energy is conserved, as expressed by $Q_{out} = Q_{in} + W_{in}$.

EXAMPLE 7-2   **Performance of a refrigeration cycle**

A 1-gal bottle of water at 70°F is placed in a refrigerator which has a COP of 1.5. The refrigerator requires 500 W to operate. How long would it take for the water to cool to 40°F if all the refrigeration power were used to cool the water?

Approach:

The rate of heat entering the refrigeration cycle is equal to the rate of heat leaving the gallon of water. Use the COP and the given power input to the refrigerator to determine the cycle input heat rate. Write the first law for the gallon of water in rate form. The time to cool the water is then found by assuming that the rate is constant with time.

Assumptions:

Solution:

Define the system as the bottle of water. We need a time, so use the closed-system energy equation in rate form:

$$\dot{Q} - \dot{W} = \frac{dE}{dt}$$

A1. Kinetic energy is negligible.
A2. Potential energy is negligible.

Ignoring potential and kinetic energy effects [A1][A2], and recognizing that no work is done on the water in the bottle,

$$\dot{Q} = \frac{dU}{dt}$$

Separating variables and integrating with respect to time

$$dU = \int \dot{Q}\, dt$$

A3. Rate of heat transfer is constant with time.

Assume that the heat transfer rate is constant [A3], so that

$$\Delta U = \dot{Q}\Delta t$$

or

$$\Delta t = \frac{\Delta U}{\dot{Q}} = \frac{m\,\Delta u}{\dot{Q}}$$

The heat transfer rate is the heat entering the refrigeration cycle, $\dot{Q}_{in}$. Using the definition of COP, we obtain

$$\dot{Q}_{in} = (COP)\left(\dot{W}_{in}\right) = (1.5)\,(500\ \text{W}) = 750\ \text{W}$$

A4. Specific heat is constant.

Since the water is liquid and assuming a constant specific heat [A4],

$$\Delta u = c_p\,\Delta T$$

Mass can be expressed in terms of density and volume:

$$m = \rho V$$

With c_p and ρ from Table A-6 evaluated at the average temperature of 55°F ($\approx$ 13°C),

$$\Delta t = \frac{\rho V c_p\,\Delta T}{\dot{Q}} = \frac{\left(999.2\,\frac{\text{kg}}{\text{m}^3}\right)(1\ \text{gal})\left(\frac{3.79\times10^{-3}\text{m}^3}{1\ \text{gal}}\right)\left(4190\,\frac{\text{J}}{\text{kg}\cdot\text{K}}\right)(70-40)°\,\text{F}\left(\frac{1°\text{C}}{1.8°\text{F}}\right)}{750\ \text{W}}$$

$$\Delta t = 351\ \text{s} = 5.86\ \text{min}$$

7.3 THE CARNOT CYCLE AND THE SECOND LAW OF THERMODYNAMICS

As mentioned above, no thermal power cycle can convert 100% of the heat input into useful work. In this section, we formally demonstrate this fact. In the early part of the 19th century, a young French engineer, Sadi Carnot, was studying the efficiency of heat engines. He wanted to identify the factors that were most important in improving cycle efficiency. He arrived at an unexpected conclusion—that there was a natural limit to the performance of heat engines governed by the temperatures at which heat was added to and removed from the engine. Other factors, such as type of working fluid, frictional losses, pressures, and so on, certainly influenced cycle efficiency, but it was temperature alone that determined the *maximum possible* efficiency. To prove his point, Carnot devised an ingenious "thought experiment."

His goal was to isolate the effect of temperature on the maximum cycle efficiency of a power cycle. To do this, he began by defining a thermal reservoir. A **thermal reservoir** is a body so large in extent that its temperature does not change when heat is added or removed. The ocean and the atmosphere are good examples of thermal reservoirs. Adding a small amount of heat to the ocean does not raise its temperature.

Carnot then imagined a cycle that received heat from a hot reservoir and rejected it to a cold reservoir. Thus all heat addition and rejection were at unique constant temperatures. He used an adiabatic expansion to reduce the fluid temperature from a high value to a low value. To return the fluid from the low temperature to the high temperature, an adiabatic compression was used, and this completed the cycle. In summary, the Carnot cycle consisted of four steps:

1–2	Isothermal heat addition
2–3	Adiabatic expansion
3–4	Isothermal cooling
4–1	Adiabatic compression

Here is one way a Carnot cycle could be built:

1. A piston–cylinder assembly containing a gas is brought into contact with a high-temperature reservoir at temperature T_H. Heat is transferred into the gas while the gas expands. The rate of expansion is very slow, and the gas remains isothermal during the process.

2. The piston–cylinder assembly is removed from the thermal reservoir and perfectly insulated. The gas is allowed to expand further while its temperature drops to T_L.

3. The assembly is brought into contact with a low-temperature reservoir at temperature T_L. The gas is cooled at constant temperature while it contracts.

4. The assembly is removed from the cold reservoir and insulated. The gas is then compressed until it reaches temperature T_H. At this point, the cycle is complete. The assembly is again brought into contact with the high-temperature reservoir, and step 1 is repeated.

This imagined cycle is not practical, but it does have several important features. All heat is added at one particular temperature, T_H, rather than over a range of temperatures, and all heat is rejected at a second, lower temperature, T_L. Furthermore, the cycle is composed of processes that are easily reversed. For example, adding heat at constant temperature while allowing the gas to expand is the opposite of removing heat at constant temperature while the gas is compressed. If the effects of friction are negligible, and the temperature difference between the gas and reservoir is infinitesimally small, the amount of work done and heat transferred will have the same magnitude but opposite directions in the two processes. The same is true for a slow, adiabatic compression of a gas, which can be reversed to a slow, adiabatic expansion. However, in a real expansion process followed by a compression, friction and other nonideal processes are present, and more work must be done to compress the system back to its initial state than the work produced during the expansion.

Reversible processes are very important in understanding the limits of real processes. Therefore, we formally define a reversible process as follows:

> *A system undergoes a process from state 1 to state 2 while heat, Q, is added to the system and work, W, is done by the system. The process is **reversible** if it is possible for the system to return from state 2 to state 1 while the same amount of heat, Q, is removed and the same amount of work, W, is done on the system.*

Some common processes can be idealized as reversible, at least in the limit of zero friction. For example, flow through a nozzle is the opposite of flow through a diffuser. If these processes occur under ideal conditions—that is, no heat transfer and no viscous

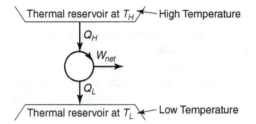

FIGURE 7-10   A power cycle operated between two thermal reservoirs.

effects in the fluid—then they are reversible. Other processes are never reversible, such as sliding a block on a table (adding heat to the block will not cause it to move) or a sudden expansion of a fluid through a throttling valve (reversing the flow through the valve will not raise the fluid pressure). More discussion of the differences between reversible and irreversible processes is included later in this chapter.

Carnot's thought experiment was aimed at defining the best possible power cycle. Therefore, he imagined all the processes within his cycle as frictionless. He also introduced the idea of a reversible process and built the Carnot cycle from four reversible processes. A schematic of a reversible power cycle operating between two thermal reservoirs at T_H and T_L is shown in Figure 7-10. In Figure 7-11, this cycle is reversed to produce a refrigerator operating between the same two reservoirs. Because the cycle is reversible, the magnitudes of Q_H, Q_L, and W_{net} are the same in both cases, but their directions are opposite.

The cycle thermal efficiency of a power cycle is given by

$$\eta_{cycle} = \frac{W_{net}}{Q_H}$$

By the first law,

$$W_{net} = Q_H - Q_L$$

so the cycle efficiency becomes

$$\eta_{cycle} = \frac{Q_H - Q_L}{Q_H} = 1 - \frac{Q_L}{Q_H}$$

This equation implies that one way to increase the cycle efficiency is to reduce the magnitude of Q_L. The closer Q_L is to zero, the closer the efficiency will be to unity.

Is it possible to construct a cycle in which $Q_L = 0$? Let us assume that it is possible. Such a cycle is pictured in Figure 7-12. This hypothetical cycle exchanges heat with a single reservoir and turns all the heat into work. Now we use this hypothetical power cycle to drive a refrigeration cycle, as shown in Figure 7-13. The net work from the power cycle would supply the work needed for the refrigeration cycle. Let Q_H be the heat transferred from the hot reservoir at T_H to the power cycle. The quantity Q'_H is the heat rejected by the

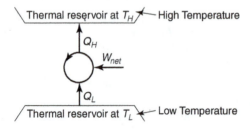

FIGURE 7-11   A refrigeration cycle operated between the same two thermal reservoirs as the power cycle in Figure 7-10.

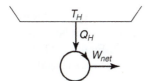

FIGURE 7-12 A power cycle exchanging heat with a single reservoir.

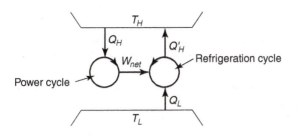

Power cycle

Refrigeration cycle

FIGURE 7-13 A power cycle driving a refrigeration cycle.

refrigeration cycle to the hot reservoir, while Q_L is the heat removed by the refrigeration cycle from the cold reservoir at temperature T_L. An energy balance on the power cycle gives

$$W_{net} = Q_H$$

An energy balance on the refrigeration cycle yields

$$W_{net} + Q_L = Q'_H$$

Eliminating net work between these two equations

$$Q_H + Q_L = Q'_H$$

Now consider an energy balance on the combination of the two cycles, as indicated by the dotted line in Figure 7-14. The heat added to the high-temperature reservoir is just $Q'_H - Q_H$. From the previous equation, this heat is equal to Q_L, the heat removed from the low-temperature reservoir:

$$Q_L = Q'_H - Q_H$$

Note that no work crosses the dotted line. In this hypothetical device, the heat Q_L moves spontaneously from the low-temperature reservoir to the high-temperature reservoir.

Experience tell us that heat never moves *spontaneously* from cold to hot, but always from hot to cold. For example, a glass of cold water left in a hot room will spontaneously heat up as heat travels from the hot room air to the cold water. The water will never get

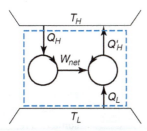

FIGURE 7-14 A cycle formed by the combination of two cycles.

colder spontaneously due to heat transfer from the cold water to the hot air. Note that it is possible to move heat from a cold body to a hot body, but only if work is added. This occurs in a refrigeration cycle, as shown in Figure 7-11. It is just that heat never moves *spontaneously* from cold to hot. Work must always be added.

This seemingly innocent and self-evident statement has unexpectedly far-reaching consequences. It is called the Clausius statement of the **second law of thermodynamics**:

> *Heat cannot move spontaneously from cold to hot bodies.*

The second law is a fundamental premise on which the science of thermodynamics rests.

Returning to the hypothetical power cycle that exchanges heat with a single reservoir, we see that such a cycle violates the Clausius statement of the second law of thermodynamics. Since the existence of a power cycle that exchanges heat with a single reservoir and converts it all into work leads to a contradiction, such a cycle cannot exist. In other words:

> *A power cycle cannot receive heat from a single thermal reservoir and convert all the heat into work*

Ideally, one would prefer to convert all heat added to a power cycle into useful work, but the second law implies that this can never be the case. Since it is not possible to convert all the heat to work, then what is the maximum work that can be produced for a given heat input? To answer this question, compare the performance of a reversible power cycle to the performance of an irreversible cycle operated between the same two reservoirs (Figure 7-15). Let the same amount of heat, Q_H, be added from the high-temperature reservoir to each cycle. In general, a different amount of work will be done by each cycle and, therefore, a different amount of heat will be rejected to the low-temperature reservoir. Let us assume the irreversible cycle is better, that is,

$$W_I > W_R$$

Now take the reversible cycle in Figure 7-15 and run it in the opposite direction. The result is shown in Figure 7-16. Since the cycle is reversible, all the magnitudes of heat and work shown in Figure 7-15 carry over to Figure 7-16, except that now they have the opposite directions. The reversible power cycle has become a reversible refrigeration cycle. We use the heat rejected from this reversible refrigeration cycle as the high-temperature heat input to the irreversible cycle, thus bypassing the high-temperature reservoir altogether. The combined system of the two cycles indicated by the dotted line in Figure 7-16 is a cycle that exchanges heat with a single reservoir. By the second law, such a cycle can never produce a net positive amount of work. Therefore W_I cannot be greater than W_R. This means that

$$W_I \leq W_R$$

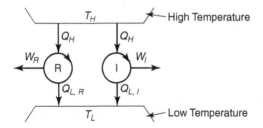

FIGURE 7-15 A reversible (R) and an irreversible (I) cycle operated between the same two reservoirs.

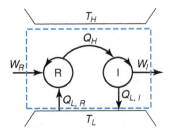

FIGURE 7-16 The reversible cycle is used to drive the irreversible cycle.

and our original assumption was incorrect. The important conclusion here is:

The work produced by an irreversible cycle will never be greater than the work produced by a reversible cycle operating between the same two temperatures.

In this example, Q_H was the same for both cycles. When the reversible cycle is operated as a power cycle, its efficiency is

$$\eta_R = \frac{W_R}{Q_H}$$

The efficiency of the irreversible cycle is

$$\eta_I = \frac{W_I}{Q_H}$$

Since W_I is less than or equal to W_R,

$$\eta_I \leq \eta_R$$

This conclusion is embodied in the first Carnot principle, which may be stated as:

The efficiency of an irreversible heat engine is never greater than the efficiency of a reversible one operating between the same two temperatures.

Some further thought experiments will produce surprising results. For example, suppose two reversible cycles, a refrigeration cycle, R_1, and a power cycle, R_2, are operated between the same two reservoirs. Let the heat rejected by cycle 1 be used to drive cycle 2, as shown in Figure 7-17. By the same reasoning used above, the combined cycle exchanges heat with a single reservoir and, therefore, cannot produce a net positive amount of work. This implies that

$$W_{R2} \leq W_{R1}$$

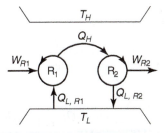

FIGURE 7-17 Two reversible cycles, one driving the other.

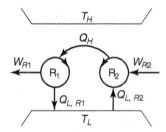

FIGURE 7-18 Both cycles in Figure 7-17 reversed.

But suppose both cycles are now reversed, as shown in Figure 7-18. The same line of reasoning leads to the conclusion that

$$W_{R1} \leq W_{R2}$$

The last two equations taken together imply that

$$W_{R1} = W_{R2}$$

No other conclusion is possible. Both reversible cycles must produce the same amount of work. The efficiencies of these two cycles are

$$\eta_1 = \frac{W_{R1}}{Q_H} = \frac{W_{R2}}{Q_H} = \eta_2$$

This conclusion is called the second Carnot principle:

> *All reversible cycles operating between the same two temperatures have the same efficiency.*

This is a remarkable statement. Note that nowhere in the derivation have we specified the details of the cycle. In fact, the cycle thermal efficiency does not depend on the processes that make up the cycle. It also does not depend on the working fluid used. The efficiency depends only on the temperatures of the two reservoirs.

So what is this efficiency? One way to calculate it is to consider the Carnot cycle described above. The Carnot cycle is an example of a reversible cycle in which all heat is added at a given single temperature and all heat is rejected at a second, lower temperature. Since any reversible cycle will have the same efficiency as the Carnot cycle, it will suffice to find the efficiency of the Carnot cycle. This, then, represents the upper limit on what is possible for all cycles, either reversible or irreversible.

The Carnot cycle may seem like a special case of a reversible cycle, but, in fact, we need only one case. If we can calculate the efficiency for this one case, we will know the cycle thermal efficiency for all reversible cycles. The easiest calculation to perform will be the one in which the working fluid is an ideal gas. Therefore, below we analyze the efficiency of a Carnot cycle operated with an ideal gas.

To recap, the Carnot cycle is

1–2	isothermal heat addition
2–3	adiabatic expansion
3–4	isothermal cooling
4–1	adiabatic compression

The efficiency of a power cycle is

$$\eta_{cycle} = \frac{W}{Q_H} = \frac{Q_H - Q_L}{Q_H} = 1 - \frac{Q_L}{Q_H} \tag{7-1}$$

All the heat is added isothermally during process 1–2. Neglecting potential and kinetic energy effects, the first law gives

$$Q_{1-2} = U_2 - U_1 + W_{1-2}$$

where W_{1-2} is the work done as the gas expands from 1 to 2. For an ideal gas with a constant specific heat, this becomes

$$Q_{1-2} = mc_v(T_2 - T_1) + W_{1-2}$$

Since process 1–2 is isothermal, $T_1 = T_2 = T_H$. The heat transferred is then

$$Q_{1-2} = W_{1-2}$$

In Chapter 2, we developed an expression for the work done in an isothermal expansion of an ideal gas (see Eq. 2-48):

$$Q_{1-2} = \frac{m\bar{R}T_H}{M} \ln\left(\frac{V_2}{V_1}\right)$$

By a similar line of reasoning, the heat rejected in the low temperature reservoir is

$$Q_{3-4} = \frac{m\bar{R}T_L}{M} \ln\left(\frac{V_4}{V_3}\right)$$

The heat Q_{1-2} is positive, because heat is added to the system. The heat Q_{3-4} is negative, because heat is removed. In dealing with cycles, however, we have used only positive values of heat and work, for example, Q_H and Q_L have both been assumed to be positive. Therefore,

$$Q_{1-2} = Q_H = \frac{m\bar{R}T_H}{M} \ln\left(\frac{V_2}{V_1}\right) \tag{7-2}$$

$$Q_{3-4} = -Q_L = \frac{m\bar{R}T_L}{M} \ln\left(\frac{V_4}{V_3}\right) \tag{7-3}$$

Substituting Eq. 7-2 and Eq. 7-3 into Eq. 7-1 gives

$$\eta = 1 - \frac{-\left\{\dfrac{m\bar{R}T_L}{M} \ln\left(\dfrac{V_4}{V_3}\right)\right\}}{\dfrac{m\bar{R}T_H}{M} \ln\left(\dfrac{V_2}{V_1}\right)}$$

which may also be written

$$\eta = 1 - \frac{T_L \ln\left(\dfrac{V_4}{V_3}\right)^{-1}}{T_H \ln\left(\dfrac{V_2}{V_1}\right)} = 1 - \frac{T_L \ln\left(\dfrac{V_3}{V_4}\right)}{T_H \ln\left(\dfrac{V_2}{V_1}\right)} \tag{7-4}$$

Process 2–3 is an adiabatic expansion of an ideal gas with constant specific heat. From Eq. 2-55,

$$\frac{T_2}{T_3} = \left(\frac{V_3}{V_2}\right)^{k-1} = \frac{T_H}{T_L}$$

Similarly, for the adiabatic compression from 4 to 1,

$$\frac{T_4}{T_1} = \left(\frac{V_1}{V_4}\right)^{k-1} = \frac{T_L}{T_H}$$

Combining the last two equations gives

$$\left(\frac{V_3}{V_2}\right)^{k-1} = \left(\frac{V_4}{V_1}\right)^{k-1}$$

which leads to

$$\frac{V_3}{V_4} = \frac{V_2}{V_1} \qquad (7\text{-}5)$$

Using Eq. 7-5 in Eq. 7-4 results in

$$\boxed{\eta_{Carnot} = 1 - \frac{T_L}{T_H} \qquad \text{reversible}} \qquad (7\text{-}6)$$

In developing Eq. 7-6, we assumed c_v does not vary with temperature. This does not limit our final conclusion as long as we can show that an ideal gas with constant specific heat can exist; in fact, experimental data show this. As long as we can calculate the Carnot efficiency for *any* case, (e.g., an ideal gas with constant c_v), we can calculate the Carnot efficiency for *all* cases. The temperatures in Eq. 7-6 must be absolute temperatures, that is, either Rankine or Kelvin. Note that since, by definition, T_L is always less than T_H,

$$0 < \eta_{Carnot} < 1$$

Eq. 7-6 gives the efficiency for *any* reversible cycle in which all heat is added at the constant temperature T_H and all the heat is removed at the constant temperature T_L. Furthermore, reversible cycles are more efficient than irreversible cycles operated between the same two temperature limits. Therefore, Eq. 7-6 represents the upper limit on what is possible. This is very useful in engineering, as shown by the next example.

EXAMPLE 7-3 **A Carnot power cycle**

The Niagara Mohawk power company wishes to construct a power plant in East Greenbush, New York. River water is available for cooling at 25°C, and the maximum temperature that the turbine blades can withstand is 600°C. If the plant is to produce 130 MW, what is the minimum possible heat transfer that must be added in the boiler?

Approach:

The goal is to get the required power output adding as little heat as possible, which occurs when the cycle has the maximum possible efficiency. The most efficient cycle is a Carnot cycle operated between the hot temperature (the turbine blade temperature) and the cold temperature (the river water).

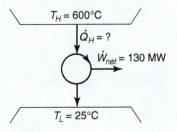

Solution:

The minimum heat transfer would be the heat transfer for a reversible cycle, since such a cycle has the maximum possible efficiency. The cycle thermal efficiency is

$$\eta_{Carnot} = 1 - \frac{T_L}{T_H}$$

$$= 1 - \frac{(25 + 273)}{(600 + 273)}$$

$$= 0.659$$

where we use absolute temperatures, as required. The efficiency is related to the heat input by

$$\eta_{Carnot} = \frac{\dot{W}}{\dot{Q}_H}$$

Solving for the heat input

$$\dot{Q}_H = \frac{\dot{W}}{\eta_{Carnot}} = \frac{130 \, \text{MW}}{0.659} = 197 \, \text{MW}$$

Comments:

Note that the heat rejected to the river water is $197 - 130 = 67 \, \text{MW}$. This large amount of energy could adversely affect the ecology of the river, changing habitats for flora and fauna and causing deaths. To protect the river species, power plants are designed with cooling towers in which the water from the condenser is cooled before it is returned to the river. In the cycle analyzed in this example, a Carnot efficiency was assumed. Real cycles have lower efficiency and consequently greater amounts of rejected heat; therefore, the thermal pollution problem is a major concern in power plant design.

7.4 THE THERMODYNAMIC TEMPERATURE SCALE

Recall that in Chapter 2, we defined temperature in terms of the behavior of an ideal gas. The problem with such a definition is that it depends on the properties of a given substance, that is, an ideal gas. It would be better to have a definition that is independent of the properties of any particular substance. In fact, Eq. 7-6 allows us to develop such a definition.

The efficiency of a power cycle is

$$\eta = \frac{W_{net}}{Q_H}$$

Since $W_{net} = Q_H - Q_L$,

$$\eta = \frac{Q_H - Q_L}{Q_H} = 1 - \frac{Q_L}{Q_H}$$

This applies to any cycle; therefore, it also applies to a reversible cycle:

$$\eta_R = 1 - \frac{Q_L}{Q_H}$$

But, for a reversible cycle, from Eq. 7-6,

$$\eta_R = 1 - \frac{T_L}{T_H}$$

This leads to the conclusion that

$$\boxed{\frac{T_L}{T_H} = \frac{Q_L}{Q_H}} \qquad \text{reversible cycle} \qquad (7\text{-}7)$$

In deriving Eq. 7-7, we used Eq. 7-6. To find Eq. 7-6, we used the ideal gas law. Now suppose we look at things from a different perspective. Assume we have not yet assigned a temperature scale, so we do not yet know the form of the ideal gas law. Then, we might decide to regard Eq. 7-7 as the primary definition of temperature. That is, temperature could be defined in terms of the heat added and rejected in a reversible power cycle. This definition is theoretically satisfying, because it is independent of the properties of any particular substance.

If Eq. 7-7 is used as the primary definition of temperature, then, by working backwards, we can recover the ideal gas law, and it will have its same familiar form. Note that this is not the only possible definition of temperature—it is just a very simple and useful one. It was first suggested by Lord Kelvin and is called the **Kelvin temperature scale.**

Although it is reassuring to have a firm theoretical definition for temperature, it is, in fact, impractical to make standard temperature measurements by using reversible cycles. Generally, temperature standards are still set by measuring the behavior of ideal gases.

7.5 REVERSIBLE REFRIGERATION CYCLES

We have introduced two Carnot principles for power cycles. A similar analysis can be performed for refrigeration cycles. This analysis leads to the Carnot principles for refrigerators:

> *The coefficient of performance of an irreversible refrigeration cycle is never greater than the coefficient of performance of a reversible cycle when both operate between the same two temperatures.*

> *All reversible refrigeration cycles operating between the same two temperatures have the same coefficient of performance.*

The coefficient of performance is

$$COP = \frac{Q_L}{W_{in}}$$

This may also be written

$$COP = \frac{Q_L}{Q_H - Q_L} = \frac{1}{\dfrac{Q_H}{Q_L} - 1}$$

Since, for any reversible cycle,

$$\frac{T_L}{T_H} = \frac{Q_L}{Q_H}$$

The coefficient of performance for a reversible refrigerator may be written

$$COP_{Carnot} = \frac{1}{\dfrac{T_H}{T_L} - 1} \qquad \text{reversible refrigeration cycle}$$

EXAMPLE 7-4 Comparison of a real refrigeration cycle with a Carnot cycle

An inventor claims to have a very efficient air conditioner installed in her home. When the outside air is at 95°F, this air conditioner keeps the interior at 70°F while consuming only 450 W of power. A thermal analysis of the building shows that under these conditions, the heat transfer rate through the walls and roof is 28,000 Btu/h. Is the inventor's claim possible or impossible?

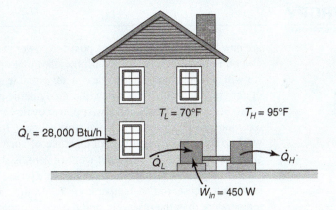

$T_L = 70°F \qquad T_H = 95°F$

$\dot{Q}_L = 28{,}000 \text{ Btu/h}$

$\dot{Q}_L$ \qquad $\dot{Q}_H$

$\dot{W}_{in} = 450 \text{ W}$

Approach:

The heat entering the building through the walls and roof is equal to the heat that must be removed by the air conditioner ($\dot{Q}_L$). Calculate the *COP* of a reversible air conditioner operating between the given interior and exterior temperatures and find the power input to this best possible air conditioner. If the claimed air conditioner requires less power, it is impossible.

Assumptions:

A1. Air in the house is at a constant temperature, and the outside air temperature is also constant.

Solution:

The coefficient of performance of a reversible air conditioner operating between the given interior and exterior temperatures is [A1]

$$COP_{Carnot} = \frac{1}{\dfrac{T_H}{T_L} - 1} = \frac{1}{\left(\dfrac{95 + 460}{70 + 460}\right) - 1} = 21.2$$

By definition

$$COP_{Carnot} = \frac{\dot{Q}_L}{\dot{W}_{in}}$$

The heat removed by the air conditioner equals the heat entering the house through the walls and roof. This heat is removed from the low-temperature space (the interior of the house), so $\dot{Q}_L = 28,000$ Btu/h and

$$\dot{W}_{in} = \frac{\dot{Q}_L}{COP_{Carnot}} = \frac{\left(28,000\ \frac{\text{Btu}}{\text{h}}\right)\left(\dfrac{1\ \text{W}}{3.412\ \frac{\text{Btu}}{\text{hr}}}\right)}{21.2} = 387\ \text{W}$$

This is less than the claimed value of 450 W; therefore, the device is possible.

Comments:

Although the calculation shows that this air conditioner is theoretically possible, do not buy one. Real refrigeration cycles typically have *COP* values under 5. Furthermore, an air conditioner as efficient as the one claimed here would have to be very large, perhaps larger than the house itself, in order to transfer heat in the evaporator and condenser across small temperature differences. (Reversible heat transfer occurs only across infinitesimally small temperature differences.)

7.6 ENTROPY

We now have criteria for the best possible power and refrigeration cycles. We can use these criteria to make judgments about processes that operate between two states. Specifically, we will demonstrate which processes are possible and which are impossible. In the course of deriving these relations, a new thermodynamic property—entropy—will be developed.

Because of the second law, a power cycle cannot receive heat from a single thermal reservoir and produce net positive work (Figure 7-19). It is possible, however, for the net work to be zero or negative. An example of such a cycle, which exchanges heat with a single temperature and has a net work of zero, is shown in Figure 7-20.

In Figure 7-20, a gas is contained in a piston–cylinder assembly. The gas exchanges heat with a reservoir at temperature T. At first, the gas is at an infinitesimally higher temperature than the reservoir, and the gas is cooled slowly and isothermally. An amount of heat, Q, conducts to the reservoir during this process. The pressure and volume of the gas decrease, and the mass rises. Work, W, is done on the gas during this compression

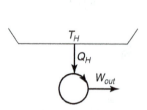

FIGURE 7-19 An impossible power cycle.

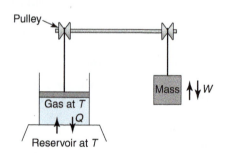

FIGURE 7-20 A cycle that produces zero net work.

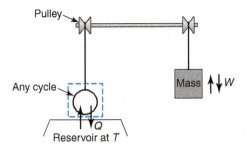

FIGURE 7-21 Another cycle that produces zero net work.

process. In the second half of the cycle, the gas is at an infinitesimally lower temperature than the reservoir, and the gas is heated slowly and isothermally. The pressure and volume of the gas increase, and the mass sinks. If this process occurs without friction and if only differential-sized temperature differences are involved, then the amount of heat, Q, added to the gas during the expansion process will equal that removed from the gas during the compression process. The first law then implies that the work, W, done by the gas during the expansion process will equal the work done on the gas during the compression process, and, hence, the net work for the cycle is zero.

In a practical sense, this is a useless cycle. However, it is important for use in a subsequent thought experiment. We now replace the piston–cylinder assembly in Figure 7-20 with any cycle, as shown in Figure 7-21. If net work for this cycle is zero, then the total work done on the cycle must equal the total work done by the cycle. This implies that the heat added to the reservoir must equal the heat removed. If that is the case, then the cycle is reversible, since the magnitudes of Q and W are the same when the cycle is run backwards. This conclusion may be stated as:

For a reversible cycle exchanging heat with a single temperature, net work is zero.

Now suppose a reversible cycle receives a differential quantity of heat, δQ_H, from a thermal reservoir at constant temperature T_H, as shown in Figure 7-22. The cycle rejects a differential quantity of heat, δQ, to system 1, which is at temperature T. During the cycle, the reservoir temperature is maintained at T_H but T, the temperature of system 1, may vary.

From the first law,

$$Q = \Delta E + W$$

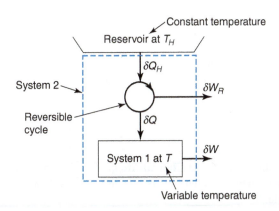

FIGURE 7-22 A system interacting with a reversible cycle.

Applying the first law to closed system 2, indicated by the dotted line in Figure 7-22, gives

$$\delta Q_H = \delta E + \delta W_R + \delta W$$

Since the cycle is reversible, Eq. 7-7 applies, and

$$\frac{\delta Q_H}{T_H} = \frac{\delta Q}{T}$$

Solving for δQ_H and substituting in the previous equation gives

$$T_H \frac{\delta Q}{T} = \delta E + \delta W_R + \delta W$$

Let δW_2 be the total work done by system 2. Then

$$T_H \frac{\delta Q}{T} = \delta E + \delta W_2 \tag{7-8}$$

Now let system 2 undergo a cycle. To find the total work done during the cycle, integrate Eq. 7-8 over the complete cycle to get

$$\oint T_H \frac{\delta Q}{T} = \oint \delta E + \oint \delta W_2 \tag{7-9}$$

The integral sign with the superimposed circle indicates that the limits of the integral are for one complete cycle. Since, during a cycle, the system returns to its initial state, the energy E at the beginning of the cycle equals the energy at the end of the cycle, and the integral over this energy is zero. Letting W_2 be the net work done by system 2 during the cycle, Eq. 7-9 becomes

$$W_2 = \oint T_H \frac{\delta Q}{T}$$

The temperature of the reservoir is constant, so it can be removed from under the integral:

$$W_2 = T_H \oint \frac{\delta Q}{T}$$

System 2 is a cycle that exchanges heat with a single reservoir. Therefore,

$$W_2 \le 0$$

This implies that

$$T_H \oint \frac{\delta Q}{T} \le 0$$

Temperature is always greater than zero, so we may conclude that

$$\boxed{\oint \frac{\delta Q}{T} \le 0}$$

This relationship is called the **Clausius inequality**. It applies for any system. The "equals" sign applies when the system is reversible, and the "less than" sign applies when the system is irreversible.

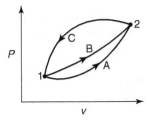

FIGURE 7-23 The paths of three possible reversible processes.

The Clausius inequality can be used to derive a new thermodynamic property. Consider three reversible processes as shown in Figure 7-23. For a reversible cycle,

$$\oint \frac{\delta Q}{T} = 0 \tag{7-10}$$

Let a system begin at point 1, follow path A to point 2, and then complete the cycle by returning along path C. Applying Eq. 7-10 to this cycle gives

$$\left(\int_1^2 \frac{\delta Q}{T} \right)_A + \left(\int_2^1 \frac{\delta Q}{T} \right)_C = 0 \tag{7-11}$$

where the subscripts A and C on the integrals indicate the path. An alternative route is to start at point 1 and take path B to point 2 before returning to point 1 along C. For this cycle,

$$\left(\int_1^2 \frac{\delta Q}{T} \right)_B + \left(\int_2^1 \frac{\delta Q}{T} \right)_C = 0 \tag{7-12}$$

Subtracting Eq. 7-12 from Eq. 7-11 and rearranging:

$$\left(\int_1^2 \frac{\delta Q}{T} \right)_A = \left(\int_1^2 \frac{\delta Q}{T} \right)_B$$

This states that the integral of $\delta Q/T$ for a reversible process does not depend on the path. It depends only on the end states. This allows us to define a new thermodynamic property called **entropy**. By definition, the entropy change for a reversible process is given by

$$S_2 - S_1 = \int_1^2 \frac{\delta Q}{T} \qquad \text{reversible path} \tag{7-13}$$

Entropy is a property like temperature, pressure, or enthalpy. Values of entropy are tabulated for many substances. We can theoretically construct a table of entropy values in the following manner: First a reference state is defined. For example, we might choose the triple point of water as the state where entropy will be defined as zero and call it state 1. To find the entropy at another state, which we call state 2, construct a reversible process from state 1 to state 2. Then measure the integral of $\delta Q/T$ for this process and calculate S_2 from Eq. 7-13. (S_1 is zero by definition). It does not matter which reversible process is chosen. We always obtain the same value of S_2. This process could be repeated with any desired number of states to produce an entire table of entropy values. In practice, there are more accurate ways to construct a property table, but that topic is beyond the scope of this text. Note that if an irreversible path is taken from point 1 to point 2, entropy change is still $S_2 - S_1$; however, the heat is not given by the right-hand side of Eq. 7-13.

The units of entropy are J/K or Btu/R. It is also useful to define **specific entropy**, which is the entropy per unit mass. Entropy is related to specific entropy through

$$S = ms$$

where s is specific entropy. In the two-phase region, specific entropy is given by the following equation:

$$s = s_f + x \left(s_g - s_f \right)$$

which has the same form as the equations already presented for other properties in the two-phase region.

Recall that for a compressed liquid, the specific volume can be approximated by the specific volume of the saturated liquid at the same temperature, as developed in Chapter 5. This subcooled liquid approximation is also valid for entropy, and the entropy of a compressed liquid can be approximated as

$$s \left(T, P \right) \approx s_f \left(T \right)$$

The verification of this equation is left for a more advanced treatment of the subject.

EXAMPLE 7-5   Isothermal compression of steam

Three kilograms of steam in a piston–cylinder assembly are compressed slowly, isothermally, and reversibly from 100 kPa, 500°C to 300 kPa. Find the heat transfer.

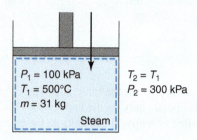

$P_1 = 100$ kPa
$T_1 = 500$°C
$m = 31$ kg

Steam

$T_2 = T_1$
$P_2 = 300$ kPa

Approach:

Since the process is reversible, Eq. 7-13 can be used to find the heat transfer. The temperature is constant, and T may be removed from the integral. Values of entropy at the initial and final states are available in Table A-12.

Assumptions:

A1. The process is reversible.

Solution:

For a reversible process [A1],

$$S_2 - S_1 = \int_1^2 \frac{\delta Q}{T}$$

A2. The steam is isothermal.

Since the process is isothermal [A2],

$$S_2 - S_1 = \frac{1}{T} \int_1^2 \delta Q$$

$$S_2 - S_1 = \frac{Q}{T}$$

Solving for Q and using $S = ms$ gives

$$Q = Tm\,(s_2 - s_1)$$

Using values of entropy from Table A-12,

$$Q = (500 + 273)\,\text{K}\,(3\,\text{kg})(8.3251 - 8.8342)\frac{\text{kJ}}{\text{kg} \cdot \text{K}}$$

$$= -1,180\,\text{kJ}$$

Comments:

The heat is negative because the steam must be cooled during compression in order to keep it at a constant temperature. Note that you must use absolute temperature in this calculation.

7.7 COMPARISON OF ENTROPY AND INTERNAL ENERGY

In the previous section, the integral of $\delta Q/T$ around a reversible cycle was shown to be equal to zero, that is,

$$\oint \frac{\delta Q}{T} = 0$$

This fact was then used to derive the property entropy, so that, for a reversible process,

$$S_2 - S_1 = \int_1^2 \frac{\delta Q}{T}$$

Now consider the first law in the form

$$dU = \delta Q - \delta W$$

Intergrated from state 1 to state 2, this becomes

$$U_2 - U_1 = \int_1^2 (\delta Q - \delta W)$$

When the first law is integrated around a cycle, the result is

$$0 = \oint (\delta Q - \delta W)$$

because the internal energy is the same at the beginning and the end of the cycle. Scanning the last five equations, one can see that U is a state variable associated with the first law and S is a similar state variable associated with the second law. Physically, U can be thought of as the *quantity* of stored heat and S can be thought of as the *quality* of stored heat. Not all the heat that is stored can be converted into useful work. Entropy gives a measure of the limits on practical work production.

7.8 REVERSIBLE AND IRREVERSIBLE PROCESSES

A system undergoes a reversible process if it can be returned to its initial state with no net change to the surroundings. For example, consider a rubber ball dropped from a height of 1 m above a smooth concrete floor. The ball accelerates as it falls. When it hits the floor, it deforms, comes to rest, and then springs back upward. If the collision with the floor were reversible, the ball would return to its original starting height of 1 m. The work done on the ball by the floor as the ball deforms would exactly equal the work done by the ball on the floor as the ball springs back to its spherical shape. The ball has returned to its initial state—that is, having the potential energy of a ball 1 m above the floor—and the net work to the surroundings is zero.

Of course, real balls do not return to exactly the same height from which they were dropped. There is always some internal friction within the ball as it deforms at the moment of contact. There is also some air resistance, which tends to slow the flight of the ball by exerting a drag force. Nevertheless, the idea of a reversible process is useful for understanding real processes. A reversible process occurs in the limit when all irreversible effects, such as friction, are eliminated and sets an upper bound on the behavior of real processes. For example, we know intuitively that the ball mentioned above will never return to a higher position than its starting position. The reversible process produces the greatest possible final height for the ball.

Some thermodynamic processes can be idealized as reversible. For example, a slow, isothermal compression of a gas can be reversed to a slow, isothermal expansion. The amount of work done and heat transferred will have the same magnitude but opposite directions in these two processes. The same is true for a slow, adiabatic compression of a gas, which can be reversed to a slow, adiabatic expansion. In a real expansion process followed by a compression, friction introduces irreversibilities, and more work must be done to compress the system back to its initial state than the work produced during the expansion.

Some processes, however, are inherently irreversible. If two fluids are mixed together, there is no way to unmix them by simply reversing the process. A mass sliding along a surface is opposed by friction, which heats the surface and the mass. Trying to force this process to reverse itself is hopeless; adding heat to a moving mass will not cause it to speed up. Hence, friction is inherently irreversible.

Some engineering devices described in earlier chapters can be idealized as reversible. Ideal turbines and compressors, for example, are reversible devices that act in opposite directions. Ideal nozzles and diffusers are also a reversible pair; fluid flows in one direction in a nozzle and in the opposite direction in a diffuser. Pumps may also be idealized as reversible, with their alter ego being the hydroturbine. In addition, heat transfer across an infinitesimal temperature difference is considered reversible. However, there are irreversibilities in the real devices that must be taken into account.

Some devices are never reversible. A throttling valve, for example, is inherently irreversible. A fast expansion or contraction is irreversible. In addition, a heat exchanger in which heat is transferred across a finite temperature difference is irreversible. This is because heat can never travel from cold to hot spontaneously.

One way to distinguish between reversible and irreversible processes is to imagine a movie of the process. When the movie is run backwards, a reversible process will seem physically possible, but an irreversible process will stand out as absurd. For example, a movie of a windmill rotating in one direction will not look absurd when it is run backwards and the windmill rotates in the opposite direction. This is a reversible process. A movie of a diver jumping off a diving board can be quite comical when run backwards, with the diver rising feet first out of the water into the air, arcing upward, and landing on the diving board. This is an irreversible process. However, there are more subtle situations that are difficult to distinguish as reversible or irreversible. For these cases, we use another fundamental equation that is developed in Section 7.11.

7.9 THE TEMPERATURE–ENTROPY DIAGRAM

In Chapter 5, we introduced several useful thermodynamic diagrams, such as P-v plots and T-v plots. In discussions of the second law, a plot of temperature versus entropy, as shown in Figure 7-24, yields important physical insight. In this section, we show that the area under the curve of a reversible process on a T-s diagram equals the heat transferred per unit mass during the process.

Consider a reversible process between states 1 and 2, as shown in Figure 7-24. From Eq. 7-13

$$S_2 - S_1 = \int_1^2 \frac{\delta Q}{T}$$

This may be written in differential form as

$$dS = \frac{\delta Q}{T}$$

or

$$\delta Q = T\,dS = mT\,ds$$

Integrating both sides of this equation yields

$$\frac{Q}{m} = \int_1^2 T\,ds \tag{7-14}$$

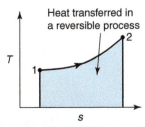

FIGURE 7-24 A reversible process plotted in T-s space. The system is being heated.

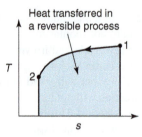

FIGURE 7-25 A reversible process in which the system is being cooled.

In Figure 7-24, temperature is plotted versus specific entropy, and a reversible process from state 1 to state 2 is shown. By Eq. 7-14, the area under the curve of temperature versus entropy is the heat transferred per unit mass. This area is shown in blue in Figure 7-24.

Another example of a process plotted on a T-s diagram is given in Figure 7-25. In Figure 7-24, the entropy at state 2 is higher than that at state 1. Therefore, the heat per unit mass calculated by Eq. 7-14 will be positive. This implies that heat is added to the system during this process. On the other hand, for the process in Figure 7-25, the entropy decreases and the value of heat per unit mass, Q/m, is negative; that is, the system is cooled during this process.

Certain processes have very simple plots in T-s space. In Figure 7-26, line A represents a reversible, isothermal heat addition. During this process, temperature is constant and entropy increases. Line B is a reversible, adiabatic compression. During an adiabatic process, the heat transferred is zero by definition. Since heat transferred per unit mass is the area under the curve on a T-s diagram, an adiabatic process must be represented by a vertical line. During an adiabatic compression, temperature increases. If process B were in the opposite direction, it would be an adiabatic expansion.

A process that is both adiabatic and reversible is called **isentropic**, because entropy is unchanged during the process. Line B in Figure 7-26 represents an isentropic compression. Many devices can be idealized as isentropic, including turbines, compressors, pumps, nozzles, and diffusers. For any of these devices, the maximum possible performance is obtained when the process is isentropic.

A T-s diagram can be useful for visualizing the amount of work done per unit mass during a cycle. In Figure 7-27, a reversible power cycle composed of three reversible steps is shown. Heat is added from state 1 to state 2. The amount of heat per unit mass is equal to the area under the curve from state 1 to state 2. The process from state 2 to state 3 is adiabatic. In the final process, from state 3 to state 1, heat is rejected by the cycle. Note that heat is rejected even though the temperature increases. In this process, a gas is compressed and cooled at the same time. The temperature rises because of the compression, but not as high as it would without the cooling.

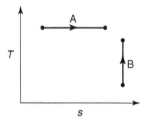

FIGURE 7-26 An isothermal heat addition and an adiabatic compression.

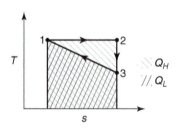

FIGURE 7-27 A reversible power cycle.

Using the first law, the net work for the cycle is

$$W_{net} = Q_H - Q_L = Q_{net}$$

Dividing by m,

$$\frac{W_{net}}{m} = \frac{Q_H}{m} - \frac{Q_L}{m} = \frac{Q_{net}}{m}$$

or

$$w_{net} = q_H - q_L = q_{net}$$

where w_{net} is the net work per unit mass, and q_H, q_L, and q_{net} are the heat added per unit mass, the heat rejected per unit mass, and the net heat of the cycle per unit mass, respectively.

Since q_H is the area under the curve from 1 to 2 and q_L is the area under the curve from 3 to 1, graphically, the net work of the cycle per unit mass is the area enclosed by the curve 1–2–3–1, as shown in Figure 7-28. The enclosed area also represents the net heat per unit mass of the cycle. Another example of a reversible power cycle is illustrated in Figure 7-29. Here, heat is added in each of the three processes from 1 to 2, from 2 to 3, and from 3 to 4. Heat is removed in the process from 4 to 1. The net work per unit mass of the cycle is represented by the area enclosed by the curve 1–2–3–4–1.

A T-s diagram can be used to visualize the efficiency of a power cycle. Cycle thermal efficiency is defined as

$$\eta = \frac{W_{net}}{Q_H} = \frac{w_{net}}{q_H}$$

In Figure 7-29, q_H is the area under the curve 1–2–3–4. As before, the net work per unit mass is the area inside the curve. Therefore, the efficiency is the ratio of the area inside the curve 1–2–3–4–1 to the area under curve 1–2–3–4.

This leads us to the interesting question of what shape cycle would produce the highest efficiency. Cycles are always limited by practical considerations. There is usually a maximum temperature that the working fluid may be allowed to attain. Higher temperatures may damage the materials in the turbine or other components. There is also a limit on the low temperature. This is typically the temperature at which cooling water is available or the temperature of the ambient air.

Assume the practical limits for T_H and T_L are known. Then the challenge becomes to design the optimum cycle that fits between T_H and T_L. One possibility is shown in Figure 7-30. As discussed above, the efficiency is the ratio of the area inside the curve

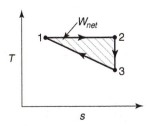

FIGURE 7-28 Net work for a reversible power cycle.

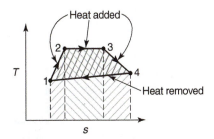

FIGURE 7-29 Another example of a reversible power cycle.

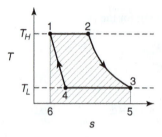

FIGURE 7-30 A power cycle operating between temperatures T_H and T_L.

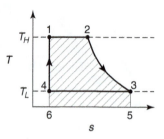

FIGURE 7-31 A more efficient cycle.

1–2–3–4–1 to the shaded area 1–2–3–5–6–1. However, this is not the optimum cycle that can operate between the two temperature limits. The cycle in Figure 7-31 has a higher efficiency. It has the same amount of heat per unit mass, q_H, and more net work per unit mass, w_{net}. As a matter of fact, the cycle that will have the highest ratio of net work to heat added will be one in the shape of a rectangle, as shown in Figure 7-32. This is none other than the Carnot cycle. As you may recall, it consists of four reversible steps:

1–2	Isothermal heat addition
2–3	Adiabatic expansion
3–4	Isothermal cooling
4–1	Adiabatic compression

The efficiency of a Carnot cycle can be deduced from its T-s diagram. The only process where heat is added to the cycle is from state 1 to state 2. The heat per unit mass is the area under the horizontal line that joins state 1 and state 2 (area 1–2–3–5–6–1) and is given by

$$q_H = T_H (s_2 - s_1)$$

The net work per unit mass of the cycle is the area inside the rectangle 1–2–3–4–1 on the T-s diagram. This is

$$w_{net} = (T_H - T_L) (s_2 - s_1)$$

The efficiency is the ratio of these two, or

$$\eta = \frac{(T_H - T_L)(s_2 - s_1)}{T_H (s_2 - s_1)} = \frac{T_H - T_L}{T_H}$$

$$\boxed{\eta_{Carnot} = 1 - \frac{T_L}{T_H} \qquad \text{Carnot cycle}}$$

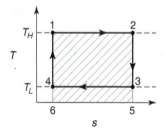

FIGURE 7-32 The Carnot cycle.

This is exactly the equation that was derived earlier for the Carnot cycle, showing the internal self-consistency of the arguments that have been introduced in this chapter.

EXAMPLE 7-6   Isentropic compression of a refrigerant

R-134a at 20 psia and 40°F is compressed slowly, adiabatically, and without friction to a final pressure of 200 psia. Ignore potential and kinetic energy effects. If the cylinder contains 4.6 lbm of refrigerant,

a) Find the final temperature (in °F).

b) Find the work done (in Btu).

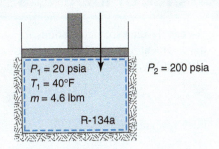

Approach:

Define the R-134a as the closed system under study. This process is adiabatic and frictionless; therefore, it is reversible. An adiabatic, reversible process is isentropic. We use this fact to determine the final temperature using data in Table B-16 and then calculate the work from the first law.

Assumptions:

A1. The process is reversible.

A2. The process is adiabatic.

Solution:

a) Choose the R-134a as the closed system. The final temperature can be determined by recognizing that this is a reversible and adiabatic process [A1][A2], that is, an isentropic process and

$$s_1 = s_2$$

From Table B-16, the entropy of superheated R-134a at 20 psia and 40°F is

$$s_1(P_1, T_1) = 0.2406 \frac{Btu}{lbm \cdot R} = s_2(P_2, T_2)$$

At state 2, the pressure is $P_2 = 200$ psia. In Table B-16, search the section where $P = 200$ psia and find the line with an entropy equal to 0.2406 Btu/(lbm·R). The corresponding temperature is about 180°F, so $T_2 = 180°F$.

A3. Potential and kinetic energy effects are negligible.

b) To find the work done, use the first law for a closed system [A3], which is

$$Q = \Delta U + W$$

or

$$Q = U_2 - U_1 + W$$

Since the process is adiabatic, $Q = 0$, and

$$W = U_1 - U_2 = m(u_1 - u_2)$$

Values of internal energy are found in Table B-16. These are easy to find now that the temperature, T_2, is known, so u_1 (20 psia, 40°F) = 100.59 Btu/lbm and u_2 (200 psia, 180° F) = 122.88 Btu/lbm

$$W = (4.6\,\text{lbm})\,(100.59 - 122.88)\,\frac{\text{Btu}}{\text{lbm}} = -102.5\,\text{Btu}$$

Comment:

Since this is a compression, work must be done on the system, and work is negative as calculated.

7.10 ENTROPY CHANGE OF IDEAL GASES

We can develop special relations for entropy change of an ideal gas. The first step in the development is the analysis of a quasi-equilibrium, reversible process of a closed system. The first law, neglecting kinetic and potential energy, is

$$\delta Q = dU + \delta W$$

For a reversible process,

$$\delta Q = T\,dS$$

From Eq. 2-18, the work done in a quasi-equilibrium process is

$$\delta W = P\,dV$$

Combining these three relations,

$$T\,dS = dU + P\,dV$$

If this equation is written in terms of specific properties, it becomes

$$mT\,ds = m\,du + mP\,dv$$

or

$$\boxed{T\,ds = du + P\,dv} \tag{7-15}$$

This equation is a relationship among thermodynamic properties. Property values depend only on the end states, not on the process. So, although the equation was derived specifically for a quasi-equilibrium, reversible process, it is not, in fact, restricted to this process; it is generally true for all processes. Eq. 7-15 is one of the most powerful and important equations in thermodynamics.

There is another equation similar to Eq. 7-15 that involves enthalpy instead of internal energy. This equation may be derived by starting from the definition of enthalpy, which is

$$h = u + Pv$$

Taking the derivative

$$dh = du + d\,(Pv)$$

Applying the chain rule,

$$dh = du + P\,dv + v\,dP$$

The first two terms on the right-hand side may be replaced with Eq. 7-15 to yield

$$dh = T\,ds + v\,dP$$

Rearranging,

$$\boxed{T\,ds = dh - v\,dP} \tag{7-16}$$

This is the second so-called $T\,ds$ equation. Like its companion, Eq. 7-15, it applies to all processes without any restriction.

The $T\,ds$ equations can be used to find the entropy change of an ideal gas undergoing a process. The ideal gas law is

$$P = \frac{\overline{R}T}{Mv}$$

Furthermore, for an ideal gas

$$du = c_v\,dT$$

Substituting these last two relations into Eq. 7-15 gives

$$T\,ds = c_v\,dT + \frac{\overline{R}T}{Mv}\,dv$$

Dividing by T,

$$ds = c_v\frac{dT}{T} + \frac{\overline{R}\,dv}{Mv}$$

Integrating this equation from state 1 to state 2,

$$\int_1^2 ds = \int_1^2 c_v\frac{dT}{T} + \int_1^2 \frac{\overline{R}\,dv}{Mv}$$

which simplifies to

$$s_2 - s_1 = \int_1^2 c_v\frac{dT}{T} + \frac{\overline{R}}{M}\,\ln\left(\frac{v_2}{v_1}\right) \tag{7-17}$$

If the specific heat is not a function of temperature, then

$$\boxed{s_2 - s_1 = c_v\,\ln\left(\frac{T_2}{T_1}\right) + \frac{\overline{R}}{M}\,\ln\left(\frac{v_2}{v_1}\right)} \qquad \text{ideal gas, constant specific heat} \tag{7-18}$$

An alternative expression for the entropy change of an ideal gas can be found by starting with the second $T\,ds$ equation. This equation is

$$T\,ds = dh - v\,dP$$

For an ideal gas,

$$dh = c_p \, dT$$

and

$$v = \frac{\overline{R}T}{MP}$$

Substituting these gives

$$T \, ds = c_p \, dT - \frac{\overline{R}T}{MP} \, dP$$

Dividing by T,

$$ds = c_p \frac{dT}{T} - \frac{\overline{R} \, dP}{MP}$$

Integrating from state 1 to state 2,

$$\int_1^2 ds = \int_1^2 c_p \frac{dT}{T} - \int_1^2 \frac{\overline{R} \, dP}{MP}$$

or

$$s_2 - s_1 = \int_1^2 c_p \frac{dT}{T} - \frac{\overline{R}}{M} \ln\left(\frac{P_2}{P_1}\right) \tag{7-19}$$

If specific heat is assumed to be constant, then

$$\boxed{s_2 - s_1 = c_p \ln\left(\frac{T_2}{T_1}\right) - \frac{\overline{R}}{M} \ln\left(\frac{P_2}{P_1}\right) \qquad \text{ideal gas, constant specific heat}} \tag{7-20}$$

If the temperature change of the ideal gas is large, the variation in specific heat could be significant. To account for variable specific heat, we first define a reference state. Only *differences* in entropy have physical meaning, so we are free to choose the value of entropy at one reference state and then specify values for other states based on the reference value. Arbitrarily, we choose entropy to be zero at a temperature of 0 K and a pressure of 1 atm. We then define $s°(T)$ as the entropy of the ideal gas at temperature, T, and at 1 atm pressure. From Eq. 7-19,

$$s°(T) - 0 = \int_0^T c_p \frac{dT}{T} - \frac{\overline{R}}{M} \ln \frac{(1 \text{ atm})}{(1 \text{ atm})}$$

which reduces to

$$s°(T) = \int_0^T c_p \frac{dT}{T} \tag{7-21}$$

Because $s°(T)$ depends only on temperature, the integration can be performed once using specific heat data and then tabulated as a function of temperature. Values of $s°(T)$ for air are given in Tables A-9 and B-9.

The integral in Eq. 7-19 may be expressed in terms of $s°(T)$ using

$$\int_{T_1}^{T_2} c_p \frac{dT}{T} = \int_{0}^{T_2} c_p \frac{dT}{T} - \int_{0}^{T_1} c_p \frac{dT}{T}$$

$$\int_{T_1}^{T_2} c_p \frac{dT}{T} = s^o(T_2) - s^o(T_1)$$

With this relation, Eq. 7-19 becomes

$$s_2 - s_1 = s^o(T_2) - s^o(T_1) - \frac{\overline{R}}{M} \ln\left(\frac{P_2}{P_1}\right) \qquad \text{ideal gas, variable specific heat} \qquad (7\text{-}22)$$

which applies when specific heat varies with temperature.

For the special case of an isentropic process, Eq. 7-22 reduces to

$$0 = s^o(T_2) - s^o(T_1) - \frac{\overline{R}}{M} \ln\left(\frac{P_2}{P_1}\right)$$

This equation can be manipulated into a more convenient form by solving for P_2:

$$P_2 = P_1 \exp\left\{\frac{M[s^o(T_2) - s^o(T_1)]}{\overline{R}}\right\}$$

The exponential of the difference of two functions can be expressed as a quotient; therefore,

$$\frac{P_2}{P_1} = \frac{\exp\left[Ms^o(T_2)/\overline{R}\right]}{\exp\left[Ms^o(T_1)/\overline{R}\right]} \qquad (7\text{-}23)$$

This equation has one important characteristic. The numerator on the right-hand side is a function only of the temperature T_2 and the denominator is a function only of T_1. We can use this fact to create a table of values that are functions of temperature. To produce the table, we define the **relative pressure**, P_r, as

$$P_r(T) = \exp\left[\frac{Ms^o(T)}{\overline{R}}\right]$$

Using relative pressure, Eq. 7-23 becomes

$$\frac{P_2}{P_1} = \frac{P_{r2}}{P_{r1}} \qquad \begin{array}{c} \text{isentropic process, ideal gas,} \\ \text{variable specific heats} \end{array} \qquad (7\text{-}24)$$

Eq. 7-24 is useful in evaluating the pressure change during an isentropic process of an ideal gas with variable specific heats. The values of relative pressure for air are listed in Tables A-9 and B-9.

There is an equation similar to Eq. 7-24 that involves volume change during an isentropic process. Using the ideal gas law, we can express the ratio of the specific volume at any two states as

$$\frac{v_2}{v_1} = \frac{\dfrac{\overline{R}T_2}{MP_2}}{\dfrac{\overline{R}T_1}{MP_1}}$$

We now assume an isentropic process and substitute Eq. 7-24 to get

$$\frac{v_2}{v_1} = \frac{\dfrac{\overline{R}T_2}{MP_{r2}}}{\dfrac{\overline{R}T_1}{MP_{r1}}}$$

Define **relative volume** as

$$v_r = \frac{\overline{R}T}{MP_r}$$

so that

$$\boxed{\frac{v_2}{v_1} = \frac{v_{r2}}{v_{r1}} \qquad \begin{array}{c} \text{isentropic process, ideal gas,} \\ \text{variable specific heat} \end{array}} \qquad (7\text{-}25)$$

Like relative pressure, relative volume is a function only of temperature and is easily tabulated. Tables A-9 and B-9 contain values of relative pressure for air. Eq. 7-25 is used to find the volume change for an isentropic process of an ideal gas with variable specific heat.

EXAMPLE 7-7   Isothermal expansion of air

Two kilograms of air at 300 K are expanded isothermally and reversibly to twice the initial volume. Find the heat transferred.

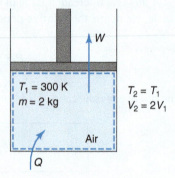

Approach:

Select the air in the cylinder as the system. For this reversible process, $\delta Q = T\, dS$. Integrate this expression to find the heat transferred in terms of the entropy change of an ideal gas.

Assumptions:

A1. The process is reversible.

A2. The process is isothermal.

A3. Air behaves like an ideal gas under these conditions.
A4. Specific heat is constant.

Solution:

Define the air in the cylinder as the system. We could begin with either the first law or the second law. Choosing the latter, we know that for a reversible process [A1]

$$\delta Q = T\, dS$$

Integrating from state 1 to state 2,

$$\int_1^2 \delta Q = \int_1^2 T\, dS$$

For a constant temperature process [A2],

$$Q = T\,(S_2 - S_1)$$

or

$$Q = Tm\,(s_2 - s_1)$$

The entropy change for an ideal gas with constant specific heat is [A3][A4]

$$s_2 - s_1 = c_v \ln\left(\frac{T_2}{T_1}\right) + \frac{\overline{R}}{M}\ln\left(\frac{v_2}{v_1}\right)$$

In this process, the temperature does not change and the volume doubles; therefore,

$$s_2 - s_1 = c_v \ln\,(1) + \frac{\overline{R}}{M}\ln\,(2)$$

Note that $\ln(1) = 0$, so with values of $\overline{R}/M$ from Table A-1,

$$s_2 - s_1 = \left(0.287\frac{kJ}{kg \cdot K}\right)\ln\,(2) = 0.199\frac{kJ}{kg \cdot K}$$

Substituting values,

$$Q = (300\,K)\,(2\,kg)\left(0.199\,\frac{kJ}{kg \cdot K}\right)$$
$$Q = 119\,kJ$$

Comments:

Expanding a gas will cause its temperature to drop. Thus, to maintain a constant temperature, we must add heat to the gas. The positive value of heat we calculated demonstrates this.

EXAMPLE 7-8 **Adiabatic compression with variable specific heat**

Air at 250 K and 100 kPa is slowly compressed to 1575 kPa in a well-insulated piston–cylinder assembly. Calculate the final temperature two ways:

a) Assuming constant specific heat

b) Assuming variable specific heat

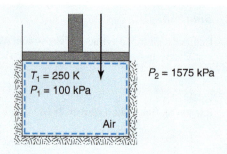

$T_1 = 250$ K
$P_1 = 100$ kPa

$P_2 = 1575$ kPa

Air

Approach:

Since the process is slow, we assume it is quasi-equilibrium, that is, reversible. Define the air in the cylinder as the closed system. The well-insulated piston–cylinder assembly is assumed to be adiabatic. For part a, in which constant specific heat is used, the final temperature is found (see Eq. 2-56) from

$$\frac{T_2}{T_1} = \left(\frac{P_2}{P_1}\right)^{\frac{k-1}{k}}$$

For part b, in which specific heat is variable, use

$$\frac{P_2}{P_1} = \frac{P_{r2}}{P_{r1}}$$

This equation applies because the process is isentropic. (All adiabatic, reversible processes are isentropic.) Values of P_r are available in Table A-9 as a function of temperature. The known initial temperature is used to find P_{r1}. Then the value of P_{r2} is calculated from the above equation, and the final temperature corresponding to that value of P_{r2} is found by interpolation in Table A-9.

Assumptions:

A1. The process is quasi-equilibrium.
A2. The process is adiabatic.
A3. Air behaves like an ideal gas under these conditions.
A4. Specific heat is constant.

Solution:

a) The process is slow, and therefore it is reasonable to assume that it is a quasi-equilibrium process [A1]. The piston–cylinder assembly is well-insulated and, hence, adiabatic [A2]. With the further assumption of ideal gas behavior and constant specific heat [A3] [A4], we may apply Eq. 2-56, which is

$$\frac{T_2}{T_1} = \left(\frac{P_2}{P_1}\right)^{\frac{k-1}{k}}$$

To evaluate k, we need the average of the initial and final temperatures; however, the final temperature is unknown. We will need an iterative approach. First, use the value of k at the given initial temperature to find an estimate of the final temperature, T_2. Then calculate the average temperature and recalculate T_2.

Using the value of k at 250 K from Table A-8,

$$T_2 = T_1 \left(\frac{P_2}{P_1}\right)^{\frac{k-1}{k}} = 250 \text{ K} \left(\frac{1575 \text{ kPa}}{100 \text{ kPa}}\right)^{\frac{1.401-1}{1.401}} = 550 \text{ K}$$

The temperature change is substantial, so we will redo the calculation with an estimate of the average temperature:

$$T_{avg} = \frac{250 + 550}{2} = 400 \text{ K}$$

Using k at 400 K,

$$T_2 = 250 \text{ K} \left(\frac{1575 \text{ kPa}}{100 \text{ kPa}} \right)^{\frac{1.395-1}{1.395}} = 545.7 \text{ K}$$

This is close to the previous value of 550 K, so no further iteration is needed.

b) To take variable specific heat into account, we first recognize that the process is reversible and adiabatic, and, hence, isentropic. Therefore, from Eq. 7-24,

$$P_{r2} = P_{r1} \frac{P_2}{P_1}$$

Using the value of P_{r1} in Table A-9 at 250 K,

$$P_{r2} = 0.7329 \left(\frac{1575}{100} \right) = 11.54$$

By interpolation in Table A-9,

$$T_2 = 545.8 \text{ K}$$

Comments:

The results for constant and variable specific heat are very close. Since the final temperature is unknown in this case, the variable specific heat approach is more convenient and direct. In this case using constant specific heat would give an error of 5 K if there is no iteration.

7.11 ENTROPY BALANCES FOR OPEN AND CLOSED SYSTEMS

The conservation equations for mass and energy may be viewed as mass balances and energy balances. We need a similar mathematical expression for use with the second law, that is, an entropy balance. Note that entropy is not a conserved quantity, so there is no "conservation of entropy equation." This is discussed in detail below.

To develop the entropy balance equation, start with the Clausius inequality:

$$\oint \frac{\delta Q}{T} \leq 0$$

We apply this inequality to the irreversible cycle shown in Figure 7-33. The cycle begins at state 1 and follows an irreversible path to state 2. Then the cycle is completed by

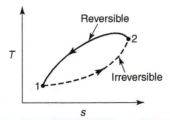

FIGURE 7-33 A cycle composed of a reversible and an irreversible step.

returning from 2 to 1 along a different but reversible path. The Clausius inequality for this cycle is

$$\int_1^2 \frac{\delta Q}{T} + \left(\int_2^1 \frac{\delta Q}{T} \right)_R \leq 0$$

For the reversible path, the integral may be expressed in terms of the entropy, so that

$$\int_1^2 \frac{\delta Q}{T} + (S_1 - S_2) \leq 0$$

or

$$S_2 - S_1 \geq \int_1^2 \frac{\delta Q}{T}$$

We remove the inequality from this equation by introducing the concept of **entropy generation**. This results in

$$S_2 - S_1 = \int_1^2 \frac{\delta Q}{T} + S_{gen} \qquad (7\text{-}26)$$

where S_{gen} is entropy generated. The magnitude of S_{gen} tells us whether the process is real, reversible, or impossible:

Real process	$S_{gen} > 0$
Reversible process	$S_{gen} = 0$
Impossible process	$S_{gen} < 0$

To use Eq. 7-26, we either assume $S_{gen} = 0$ (a reversible process), or we calculate the magnitude of S_{gen} to determine whether a process is possible or impossible.

The integral in Eq. 7-26 can be difficult to evaluate if the system does not have the same temperature everywhere. For example, suppose a closed system exchanges heat with its surroundings through a finite number of surfaces, N, as shown in Figure 7-34. Each of the surfaces is at a different temperature. Real systems with continuous temperature variations can be approximated by such discrete systems if the number of surfaces, N, is chosen to be large. For the system shown in Figure 7-34, the integral in Eq. 7-26 can be

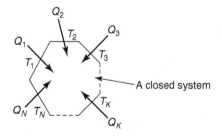

FIGURE 7-34   A system with non-uniform temperature exchanging heat with the surroundings.

replaced by a summation to give

$$S_2 - S_1 = \sum_{k=1}^{N} \frac{Q_k}{T_k} + S_{gen}$$

In differential form,

$$dS = \sum_{k=1}^{N} \frac{\delta Q_k}{T_k} + \delta S_{gen}$$

If each term is taken per unit time,

$$\frac{dS}{dt} = \sum_{k=1}^{N} \frac{\dot{Q}_k}{T_k} + \frac{\delta S_{gen}}{dt}$$

or

$$\frac{dS}{dt} = \sum_{k=1}^{N} \frac{\dot{Q}_k}{T_k} + \dot{S}_{gen} \tag{7-27}$$

Eq. 7-27 can be used to demonstrate that the entropy of the universe always increases. To prove this, consider a closed system at uniform temperature T_H that exchanges heat with surroundings at T_L, as shown in Figure 7-35. We define the surroundings to include everything in the universe except the closed system. In the vicinity of the closed system, the surroundings are at a uniform temperature T_L, which is less than the system temperature, T_H. The rate of heat transfer, $\dot{Q}_{sys}$, between system and surroundings is defined to be positive. Applying Eq. 7-27 to the closed system gives

$$\frac{dS_{sys}}{dt} = -\frac{\dot{Q}_{sys}}{T_H} + \dot{S}_{gen,sys} \tag{7-28}$$

The first term on the right-hand side is negative because heat leaves the system.
Applying Eq. 7-27 to the surroundings gives

$$\frac{dS_{surr}}{dt} = \frac{\dot{Q}_{sys}}{T_L} + \dot{S}_{gen,surr}$$

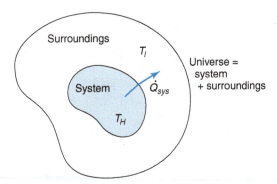

FIGURE 7-35 The universe is divided into a system and its surroundings. The system exchanges heat with the surroundings.

In this equation, the first term on the right-hand side is positive because heat enters the surroundings. The entropy change of the universe is the sum of the entropy change of the system and the entropy change of the surroundings. Adding the last two equations produces

$$\frac{dS_{universe}}{dt} = \frac{dS_{sys}}{dt} + \frac{dS_{surr}}{dt} = \left[-\frac{\dot{Q}_{sys}}{T_H} + \dot{S}_{gen,sys} \right] + \left[\frac{\dot{Q}_{sys}}{T_L} + \dot{S}_{gen,surr} \right]$$

which simplifies to

$$\frac{dS_{universe}}{dt} = \dot{Q}_{sys} \left(\frac{1}{T_L} - \frac{1}{T_H} \right) + \dot{S}_{gen,sys} + \dot{S}_{gen,surr}$$

All quantities on the right-hand side ($T_L, T_H, \dot{Q}_{sys}, \dot{S}_{gen,sys}$, and $\dot{S}_{gen,surr}$) are positive, and $T_H > T_L$; therefore,

$$\frac{dS_{universe}}{dt} > 0$$

The time rate of change of entropy of the universe is always positive; hence, the entropy of the universe is always increasing. Note that the entropy of a closed system can either increase or decrease by the addition or removal of energy. In Eq. 7-28, if $\dot{S}_{gen,sys} > \dot{Q}_{sys}/T_H$, the entropy increases, and if $\dot{S}_{gen,sys} < \dot{Q}_{sys}/T_H$, the entropy decreases.

Because entropy is generated in all real processes, entropy is *not* conserved, as are energy and mass. Hence, Eq. 7-27 is an *entropy balance equation, not* a conservation of entropy equation.

Eq. 7-27 applies to a closed system. To find the entropy balance for an open system, a derivation similar to those that were used to find the first law for an open system, the mass balance for an open system, and the momentum balance for an open system can be applied. The derivation is left as a homework exercise. The resulting entropy balance equation for an open system is

$$\boxed{\frac{dS_{cv}}{dt} = \sum_{k=1}^{N} \frac{\dot{Q}_k}{T_k} + \sum_{in} \dot{m}_i s_i - \sum_{out} \dot{m}_e s_e + \dot{S}_{gen}} \qquad (7\text{-}29)$$

EXAMPLE 7-9 **Entropy generated in expansion of a refrigerant**

R-134a is quickly expanded in a piston–cylinder assembly from a superheated vapor at 200 psia, 140°F to a two-phase mixture at 5 psia with a quality of 0.98. If the process is adiabatic, find the entropy generated per unit mass.

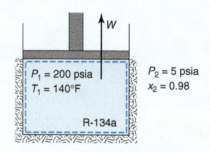

Approach:

The system is the R-134a within the cylinder. The process is fast and, therefore, irreversible. Because this is a closed system, use Eq. 7-26 to find entropy generated. The process is adiabatic, so heat transfer is zero. Values for entropy may be obtained in Table B-16.

Assumptions:

Solution:

Define the R-134a as the system under study. The entropy generated is given by a rearrangement of Eq. 7-26,

$$S_{gen} = S_2 - S_1 - \int_1^2 \frac{\delta Q}{T}$$

A1. The process is adiabatic.

For an adiabatic process, $\delta Q = 0$, so [A1]

$$S_{gen} = S_2 - S_1 = m(s_2 - s_1)$$

and the entropy generated per unit mass is

$$\frac{S_{gen}}{m} = s_2 - s_1$$

For state 1, using Table B-16, $s_1(200\,\text{psia}, 140°\text{F}) = 0.2226\,\text{Btu/lbm} \cdot \text{R}$. At state 2, the entropy may be evaluated from

$$s_2 = s_f + x_2\left(s_g - s_f\right)$$

With values from Table B-15, $s_f(5\,\text{psia}) = -0.009\,\text{Btu/lbm·R}$ and $s_g(5\,\text{psia}) = 0.2311\,\text{Btu/lbm·R}$; therefore,

$$s_2 = -0.009 + 0.98\,[0.2311 - (-0.009)] = 0.226\,\frac{\text{Btu}}{\text{lbm} \cdot \text{R}}$$

$$\frac{S_{gen}}{m} = 0.226 - 0.2226 = 0.0034\,\frac{\text{Btu}}{\text{lbm} \cdot \text{R}}$$

Comments:

The entropy generation is positive, which means that this process is possible. Note that just because a process is *possible*, that does not mean it could take place in reality. Practical limitations might prevent actual achievement of a thermodynamically possible process.

7.12 SECOND-LAW ANALYSIS OF TURBINES, PUMPS, AND COMPRESSORS

An isentropic process represents the limit of the possible; real processes fall short of this limit. To evaluate the actual behavior of a real component, such as a turbine, pump, or compressor, the real process is compared to an idealized, isentropic process using a quantity called the isentropic efficiency.

To develop isentropic efficiency, start with the entropy balance for an open system, that is,

$$\frac{dS_{cv}}{dt} = \sum_{k=1}^{N} \frac{\dot{Q}_k}{T_k} + \sum_{in} \dot{m}_i s_i - \sum_{out} \dot{m}_e s_e + \dot{S}_{gen}$$

We now apply this equation to an ideal turbine, that is, one that operates adiabatically and reversibly. Restricting the discussion to steady processes gives $dS_{cv}/dt = 0$. Because the turbine is adiabatic $\dot{Q}_k = 0$. Because it is reversible, $\dot{S}_{gen} = 0$. If there is only one stream in and one stream out, $\dot{m}_i = \dot{m}_e = \dot{m}$. With all these simplifications, we are left with

$$0 = \dot{m}s_e - \dot{m}s_i$$

or

$$\boxed{s_e = s_i \qquad \text{isentropic process}}$$

A similar analysis applies to an ideal compressor or an ideal pump; that is, turbines, compressors, and pumps are isentropic devices in the limit of no friction or heat transfer.

Figure 7-36 shows an expansion through an ideal turbine on a T-s diagram. Since entropy is the same at the inlet and the outlet, the expansion through the turbine is a vertical line. Pressure and temperature both decrease in the expansion. The input to the turbine may be either saturated vapor (state 1) or superheated vapor (state 3 or 3′). If the input is saturated vapor, the output of an ideal turbine will always fall in the two-phase region (state 2), as shown in Figure 7-36. If the input is superheated, the output may be either a two-phase mixture (state 4) or a superheated vapor (state 4′).

Next consider a real, adiabatic turbine with friction. If the turbine is assumed to operate in steady state with one inlet and one outlet, then Eq. 7-29 reduces to

$$0 = \dot{m}s_i - \dot{m}s_e + \dot{S}_{gen}$$

which may also be written

$$\frac{\dot{S}_{gen}}{\dot{m}} = s_e - s_i$$

Because $\dot{S}_{gen}$ is always positive for real processes, s_e is always greater than s_i. An expansion through an ideal turbine (process 1–2s) is compared to an expansion through a real turbine (process 1–2) in Figure 7-37. In this comparison, both the ideal and real turbines are expanded *from the identical inlet state to the same outlet pressure*. State 2s has the same entropy as state 1 and the same pressure as state 2. The outlet states for the ideal

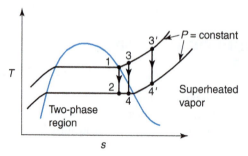

FIGURE 7-36 An expansion through an ideal turbine.

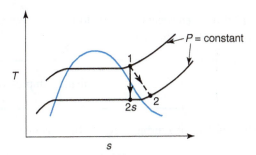

FIGURE 7-37 Expansion through an ideal (1–2s) and a real (1–2) turbine.

and real processes may be both two-phase, one two-phase and one superheated (as shown in Figure 7-37), or both superheated, depending on the problem at hand.

The **isentropic efficiency** of a turbine is, by definition

$$\eta_T = \frac{\dot{W}_{act}}{\dot{W}_{ideal}} \qquad \text{adiabatic turbine} \qquad (7\text{-}30)$$

where $\dot{W}_{act}$ is the actual work done and $\dot{W}_{ideal}$ is the isentropic work. The power produced by an adiabatic turbine (neglecting kinetic and potential energy effects) is

$$\dot{W} = \dot{m}\,(h_i - h_e)$$

Therefore, the isentropic efficiency for a turbine may be written as

$$\eta_T = \frac{h_1 - h_2}{h_1 - h_{2s}} \qquad \text{adiabatic turbine} \qquad (7\text{-}31)$$

The isentropic efficiency is always between 0 and 1. Typical values range from 0.7 to 0.9. For a turbine, the actual work delivered is always *less than* the ideal, isentropic work.

Compressors can also be characterized by an isentropic efficiency. A *T-s* diagram for both an ideal and a real compression process is given in Figure 7-38. The input to a compressor may be saturated vapor or superheated vapor, and the output is always super-heated vapor. The ideal and real compressors *have identical inlet states and the same outlet pressure*. State 2s in Figure 7-38 has the same entropy as state 1 and the same pressure as state 2. For a turbine, the actual work delivered is always *less than* the ideal, isentropic work; for a compressor, the actual work required is always *greater than* the ideal, isentropic

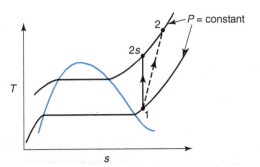

FIGURE 7-38 An ideal (1–2s) and a real (1–2) compression process.

work. Since we want the isentropic efficiency to be in the range of 0 to 1, the isentropic efficiency of a compressor is defined as

$$\eta_C = \frac{\dot{W}_{ideal}}{\dot{W}_{act}} \qquad \text{adiabatic compressor}$$

which is the inverse of that for a turbine (Eq. 7-30). This may also be written as

$$\eta_C = \frac{h_{2s} - h_1}{h_2 - h_1} \qquad \text{adiabatic compressor} \qquad (7\text{-}32)$$

Pumps are analogous to compressors. The work required by a real pump is always greater than the work for an ideal pump. Thus the isentropic efficiency of a pump is defined as

$$\eta_P = \frac{\dot{W}_{ideal}}{\dot{W}_{act}} = \frac{h_{2s} - h_1}{h_2 - h_1} \qquad \text{adiabatic pump} \qquad (7\text{-}33)$$

where h_1 is the enthalpy at the inlet state, h_{2s} is the enthalpy at the state with the same entropy as state 1 and the same pressure as state 2, and h_2 is the enthalpy at the actual exit state of the pump. The ideal and real pumping processes are shown diagrammatically in Figure 7-39.

Pumps are designed to handle only liquids. The ideal work for a pump can be found from Eq. 4-45, which is

$$\frac{P_1}{\rho} + gz_1 + \frac{\mathcal{V}_1^2}{2} = \frac{P_2}{\rho} + gz_2 + \frac{\mathcal{V}_2^2}{2} + w_{cv} \qquad (7\text{-}34)$$

This equation applies in the special case of an adiabatic, incompressible flow with no friction. Such a flow is isentropic. In a pump, there is no significant change in potential or kinetic energy. By definition, $w_{cv} = \dot{W}/\dot{m}$. With these considerations, Eq. 7-34 becomes

$$\frac{P_1}{\rho} = \frac{P_2}{\rho} + \frac{\dot{W}}{\dot{m}}$$

Rearranging and using $\rho = 1/v$,

$$\dot{W} = \dot{W}_{ideal} = \dot{m}v\left(P_1 - P_2\right) \qquad (7\text{-}35)$$

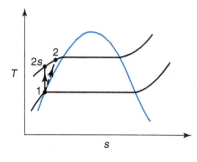

FIGURE 7-39   An ideal (1–2s) and a real (1–2) pumping process.

The work done by an ideal adiabatic pump may be found from the first law (see Eq. 6-4):

$$\dot{W}_{ideal} = \dot{m}\,(h_1 - h_{2s})$$

Combining this with Eq. 7-35 and solving for h_{2s} gives the enthalpy rise across an ideal pump as

$$\boxed{h_{2s} = h_1 + v(P_2 - P_1) \qquad \text{ideal, adiabatic pump}} \qquad (7\text{-}36)$$

For an actual pump, the enthalpy rise may be found by solving Eq. 7-33 for h_2 to get

$$\boxed{h_2 = h_1 + \frac{h_{2s} - h_1}{\eta_P} \qquad \text{adiabatic pump}} \qquad (7\text{-}37)$$

EXAMPLE 7-10 Isentropic efficiency of a turbine

Steam at 160 psia and 600°F expands through a well-insulated turbine to an exhaust pressure of 5 psia. If the mass flow rate is 15 lbm/s, and the isentropic efficiency is 0.82, calculate the power produced (in hp), the temperature of the exit state, and the quality at the exit state. Show the process on a T-s diagram.

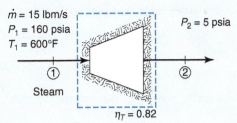

$\dot{m} = 15$ lbm/s
$P_1 = 160$ psia
$T_1 = 600°F$

$P_2 = 5$ psia

① Steam ②

$\eta_T = 0.82$

Approach:

We define the control volume to enclose the turbine. Begin with the energy equation for an open system, and ignore potential and kinetic energy effects. Assuming steady flow, adiabatic operation, and one inlet and one outlet, the energy equation reduces to $\dot{W} = \dot{m}\,(h_1 - h_2) = \dot{W}_{act}$. Use Eq. 7-30 to relate the actual work to the ideal work:

$$\eta_T = \frac{\dot{W}_{act}}{\dot{W}_{ideal}} = \frac{h_1 - h_2}{h_1 - h_{2s}}$$

State 2s is found by setting $s_1 = s_{2s}$, which applies to an ideal, adiabatic turbine. The enthalpy at state 1, h_1, is evaluated at P_1 and T_1. State 2s is specified because two independent properties, s_{2s} and P_2, are known. Therefore, h_2 can be evaluated. To find the exit temperature, use the energy equation again. Because we determined $\dot{W}_{act}$ above, the only unknown is h_2. With both h_2 and P_2 known, T_2 can be determined.

Assumptions:

Solution:

The energy equation for an open system is

$$\frac{dE_{cv}}{dt} = \dot{Q}_{cv} - \dot{W}_{cv} + \sum_{in} \dot{m}_i \left(h_i + \frac{\mathcal{V}_i^2}{2} + g z_i\right) - \sum_{out} \dot{m}_e \left(h_e + \frac{\mathcal{V}_e^2}{2} + g z_e\right)$$

A1. Kinetic energy is negligible.
A2. Potential energy is negligible.
A3. The flow is steady.
A4. The turbine is perfectly insulated.

Ignoring potential and kinetic energy effects [A1][A2] and assuming steady flow [A3], adiabatic operation [A4], and one inlet and one outlet,

$$\dot{W} = \dot{m}\left(h_1 - h_2\right) = \dot{W}_{act}$$

Using Eq. 7-30,

$$\eta_T = \frac{\dot{W}_{act}}{\dot{W}_{ideal}} = \frac{h_1 - h_2}{h_1 - h_{2s}}$$

and solving for actual work

$$\dot{W}_{act} = \eta_T \dot{m}\left(h_1 - h_{2s}\right)$$

State 2s is found by recognizing, for an ideal, adiabatic turbine,

$$s_1 = s_{2s}$$

where state 2s has the same pressure as state 2. Using values for the entropy of steam at 160 psia and 600°F from Table B-12,

$$s_1 = 1.70 \frac{\text{Btu}}{\text{lbm} \cdot \text{R}} = s_{2s}$$

State 2s could be either superheated vapor or two-phase mixture. To determine which it is, consult the saturated steam table, Table B-11. At the exhaust pressure, P_2, of 5 psia, the entropies of saturated liquid and vapor are

$$s_f = 0.235 \frac{\text{Btu}}{\text{lbm} \cdot \text{R}}$$

$$s_g = 1.84 \frac{\text{Btu}}{\text{lbm} \cdot \text{R}}$$

Since the value of s_{2s} falls between these two values, state 2s must be in the two-phase region. The quality is determined from

$$x_{2s} = \frac{s_{2s} - s_f}{s_g - s_f}$$

Substituting values,

$$x_{2s} = \frac{1.70 - 0.235}{1.84 - 0.235} = 0.913$$

We are now ready to calculate the enthalpy at state 2s. It is

$$h_{2s} = h_f + x\left(h_g - h_f\right)$$

$$h_{2s} = 130 \frac{\text{Btu}}{\text{lbm}} + 0.913\left(1131 - 130\right)\frac{\text{Btu}}{\text{lbm}} = 1040 \frac{\text{Btu}}{\text{lbm}}$$

where enthalpy values have been taken from Table B-11 at 5 psia. The power produced by the turbine, using the value of h_1 from Table B-12, is

$$\dot{W}_{act} = \eta_T \dot{m}\left(h_1 - h_{2s}\right) = 0.82\left(15 \frac{\text{lbm}}{\text{s}}\right)\left(1325 - 1044\right)\frac{\text{Btu}}{\text{lbm}} = 3460 \frac{\text{Btu}}{\text{s}}$$

$$\dot{W}_{act} = 3460 \frac{\text{Btu}}{\text{s}}\left(\frac{1.055\,\text{kW}}{1 \frac{\text{Btu}}{\text{s}}}\right)\left(\frac{1\,\text{hp}}{0.746\,\text{kW}}\right)$$

$$\dot{W}_{act} = 4893\,\text{hp}$$

To find the actual exit state, note that the actual work is given by

$$\dot{W}_{act} = \dot{m}\,(h_1 - h_2)$$

Solving for h_2,

$$h_2 = h_1 - \frac{\dot{W}_{act}}{\dot{m}}$$

$$h_2 = 1325\,\frac{\text{Btu}}{\text{lbm}} - \frac{3460\,\dfrac{\text{Btu}}{\text{s}}}{15\,\dfrac{\text{Btu}}{\text{lbm}}} = 1094\,\frac{\text{Btu}}{\text{lbm}}$$

The value of h_2 falls between the enthalpy for a saturated liquid and the enthalpy for a saturated vapor at $P_2 = 5$ psia (see Table B-11). Therefore, the actual exit state is a two-phase liquid with a saturation temperature of 162°F. The quality at state 2 is

$$x_2 = \frac{h_2 - h_f}{h_g - h_f} = \frac{1094 - 130}{1131 - 130} = 0.963$$

The *T-s* diagram for this process is shown below.

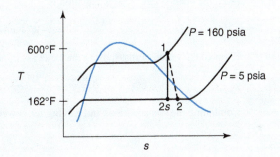

Comments:

The isentropic work is greater than the actual work. Because more energy is extracted from the steam during the isentropic process, $x_{2s} < x_2$, as calculated.

EXAMPLE 7-11   **Expansion in a gas turbine**

A small gas turbine is designed for a flow rate of 0.2 kg/s with an inlet pressure of 1750 kPa and an inlet temperature of 1125°C. The exit pressure is 100 kPa. The surface area of the turbine is 2.1 m², and the air surrounding the turbine is at 40°C. The uninsulated turbine case has an emissivity of 0.6, and the convective heat transfer coefficient is 10 W/m²· K. Assume the combustion gases in the turbine have the properties of air and use variable specific heats.

a) Determine the power output for an isentropic turbine.

b) Estimate the power output for the uninsulated turbine. From previous experience, the temperature of the outside surface of the turbine can be estimated from the average of the gas inlet temperature and the exit temperature.

c) Estimate the thickness of insulation ($k = 0.03$ W/m·K, $\varepsilon = 0.35$) required to cut the heat losses by 99.5%.

Approach:

With the given information we can use the open-system energy equation to calculate the turbine power for isentropic operation; we can also obtain the gas exit temperature from this. For the uninsulated turbine, we will calculate the convective and radiative heat transfer from the turbine and subtract this total from the isentropic power. Finally, we can determine the insulation thickness required by assuming one-dimensional heat transfer through the insulation and using the heat loss information from part b.

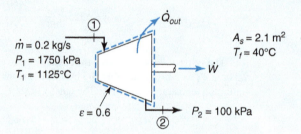

Assumptions:

A1. The flow is steady.
A2. Neglect potential and kinetic energy effects.

Solution:

a) We define the control volume around the turbine. For this control volume, assuming [A1] and [A2], the open-system energy equation is

$$0 = -\dot{W} + \dot{m}_1 h_1 - \dot{m}_2 h_2$$

From conservation of mass, using the same assumptions,

$$0 = \dot{m}_1 - \dot{m}_2 \quad \rightarrow \quad \dot{m}_1 = \dot{m}_2 = \dot{m}$$

Combining the two equations and solving for power,

$$\dot{W} = \dot{m}\left(h_1 - h_2\right)$$

A3. The flow is isentropic.
A4. Air is the working fluid.

For part a, we assume isentropic operation [A3] and use relative pressures to determine the exit state. Assuming the gas properties are the same as those of air [A4], by interpolation in Table A-9 at $T_1 = 1125 + 273 = 1398$ K, $h_1 = 1513.0$ kJ/kg and $P_{r1} = 447.9$. Using the relative pressure for the exit condition,

$$P_{r2} = P_{r1}\left(P_2/P_1\right) = 447.9\left(100/1750\right) = 25.6$$

By interpolation at P_{r2}, $h_2 = 689.8$ kJ/kg, and $T_2 = 678$ K. Therefore,

$$\dot{W} = \left(0.2\ \frac{\text{kg}}{\text{s}}\right)(1513.0 - 689.8)\ \frac{\text{kJ}}{\text{kg}}\left(\frac{1\ \text{kW}}{1\ \text{kJ/s}}\right) = 164.6\ \text{kW}$$

b) If we include heat losses, the energy equation is rewritten as

$$\dot{W} = \dot{m}\left(h_1 - h_2\right) - \dot{Q}_{out}$$

A5. Radiation from a gray diffuse surface to distant surroundings is assumed.

where $\dot{Q}_{out}$ is defined as a positive number so that the direction of the heat transfer is taken into account. Assuming the turbine is small compared to its surroundings [A5] and the surroundings are at the air temperature,

$$\dot{Q}_{out} = h_{conv}A\left(T_s - T_f\right) + \varepsilon\sigma A\left(T_s^4 - T_f^4\right) = \left(h_{conv} + h_{rad}\right)A\left(T_s - T_f\right)$$

where the radiative heat transfer coefficient is

$$h_{rad} = \varepsilon\sigma\left(T_s + T_f\right)\left(T_s^2 + T_f^2\right)$$

A6. The surface temperature of the turbine is the average of gas inlet and outlet temperatures.

We assume the surface temperature is the average of the gas inlet and outlet temperatures for isentropic operation [A6], $T_s = (1398 + 678)/2 = 1038$ K. Therefore,

$$h_{rad} = (0.6)\left(5.67 \times 10^{-8}\,\frac{\text{W}}{\text{m}^2 \cdot \text{K}^4}\right)(1038\,\text{K} + 313\,\text{K})\left[(1038\,\text{K})^2 + (313\,\text{K})^2\right] = 54.0\,\frac{\text{W}}{\text{m}^2 \cdot \text{K}}$$

$$\dot{Q}_{out} = (10 + 54.0)\,\frac{\text{W}}{\text{m}^2 \cdot \text{K}}\,(2.1\,\text{m}^2)\,(1038\,\text{K} - 313\,\text{K})\left(\frac{1\,\text{kW}}{1000\,\text{W}}\right) = 97.5\,\text{kW}$$

$$\dot{W} = 164.5 - 97.5 = 67.0\,\text{kW}$$

c) We want to reduce the heat losses by 99.5% by adding insulation. The heat transfer rate then is $(1 - 0.995)(97.5\,\text{kW}) = 0.488\,\text{kW}$. We need to rewrite the heat transfer equation to include the conduction resistance. Assume one-dimensional conduction [A7] across the insulation and treat it as a plain wall. The resistance network is

A7. Conduction is one-dimensional.

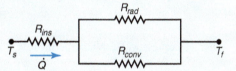

The rate of heat transfer may be calculated as

$$\dot{Q} = \frac{T_s - T_f}{R_{ins} + \left(\dfrac{1}{R_{conv}} + \dfrac{1}{R_{rad}}\right)^{-1}} = \frac{T_s - T_f}{\dfrac{t}{kA} + \left(\dfrac{1}{1/h_{conv}A} + \dfrac{1}{1/h_{rad}A}\right)^{-1}} = \frac{T_s - T_f}{\dfrac{t}{kA} + (h_{conv}A + h_{rad}A)^{-1}}$$

Solving for the thickness, t,

$$t = kA\left[\frac{T_s - T_f}{\dot{Q}} - (hA + h_rA)^{-1}\right]$$

Note that the radiative heat transfer coefficient is different for this calculation because the outside surface temperature of the insulation and its emissivity must be used. An iterative solution is required. We can calculate the outside surface temperature, T_{ins}, using the heat transfer rate equation applied only to the insulation:

$$\dot{Q} = \frac{kA}{t}(T_s - T_{ins})$$

Iteratively solving the thickness equation, the heat transfer equation across the insulation, and the radiative heat transfer coefficient, we obtain

$$T_{ins} = 331.4\,\text{K} = 58.4\,°\text{C}$$

$$t = 9.1\,\text{cm}$$

$$h_{rad} = 2.66\,\text{W/m}^2 \cdot \text{K}$$

Comments:

This is an approximate analysis. Heat loss from the uninsulated turbine will affect the outlet air temperature, so subtracting the heat loss from the isentropic power is not exact, but it is a logical approximation. For the insulated turbine, the effect of heat loss is minor, so that analysis is reasonable.

EXAMPLE 7-12 Isentropic efficiency of a compressor

Air enters a well-insulated compressor at 25°C and 101 kPa and exits at 300°C and 650 kPa. Assuming constant specific heat, calculate the isentropic efficiency of the compressor.

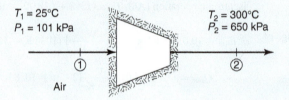

$T_1 = 25°C$
$P_1 = 101$ kPa

$T_2 = 300°C$
$P_2 = 650$ kPa

①

②

Air

Approach:

Let the control volume enclose the compressor. The isentropic efficiency of a compressor is defined as

$$\eta_C = \frac{h_1 - h_{2s}}{h_1 - h_2}$$

For an ideal gas with constant specific heat,

$$\Delta h = c_p \, \Delta T$$

Therefore,

$$\eta_C = \frac{c_p \, (T_1 - T_{2s})}{c_p \, (T_1 - T_2)}$$

We can determine the final temperature for an isentropic expansion of an ideal gas using Eq. 2-56:

$$T_{2s} = T_1 \left(\frac{P_2}{P_1} \right)^{\frac{k-1}{k}}$$

Now everything is known, and η_C can be determined.

Assumptions:

A1. Air may be considered an ideal gas under these conditions.
A2. Specific heat is constant.

Solution:

The control volume is drawn around the compressor. For an isentropic expansion of an ideal gas with constant specific heat [A1][A2],

$$T_{2s} = T_1 \left(\frac{P_2}{P_1} \right)^{\frac{k-1}{k}}$$

The ratio of specific heats, k, is a function of temperature, as shown in Table A-8. Using a value of k at the average of T_1 and T_2, we have

$$T_{2s} = (25 + 273) \, \text{K} \left(\frac{650}{101} \right)^{\frac{1.4-1}{1.4}} = 507 \, \text{K} = 234°C$$

The isentropic efficiency of a compressor is

$$\eta_C = \frac{h_1 - h_{2s}}{h_1 - h_2}$$

Assume constant specific heats, so that $\Delta h = c_p \Delta T$. Because the outlet temperature for the isentropic compression is close to that for the actual compression, the specific heats used to calculate enthalpy change for each process will be essentially the same. Therefore,

$$\eta_C = \frac{c_p(T_1 - T_{2s})}{c_p(T_1 - T_2)} = \frac{(T_1 - T_{2s})}{(T_1 - T_2)} = \frac{(25 - 234)}{(25 - 300)} = 0.76$$

7.13 MAXIMUM POWER CYCLE

The Carnot cycle represents one limit on power cycles operating between two temperature differences. The Carnot cycle assumes infinite reservoirs of heat that do not change temperature when heat is added or removed. In practical cycles, the fluid used to supply heat to and remove heat from the cycle changes temperature. To account for this effect, consider the cycle shown in Figure 7-40. This cycle is reversible and receives heat $\dot{Q}_H$ from a high-temperature ideal gas that enters at T_1 and exits at T_2. The cycle rejects heat to a second, low-temperature ideal gas that enters at T_4 and exits at T_3. Both streams are assumed to have the same mass flow rate, $\dot{m}$, and no heat is transferred to the surroundings. We will develop an expression for the maximum possible power that such a cycle can produce.

The heat added from the high-temperature fluid is

$$\dot{Q}_H = \dot{m}c_p\,(T_1 - T_2) \tag{7-38}$$

and the heat removed by the low-temperature fluid is

$$\dot{Q}_L = \dot{m}c_p\,(T_3 - T_4)$$

From the first law, the work produced by this cycle is

$$\dot{W} = \dot{Q}_H - \dot{Q}_L = \dot{m}c_p\,[(T_1 - T_2) - (T_3 - T_4)] \tag{7-39}$$

For the control volume shown in Figure 7-40, the second law may be written (see Eq. 7-29):

$$\frac{dS_{cv}}{dt} = \sum_{k=1}^{N} \frac{\dot{Q}_k}{T_k} + \sum_{in} \dot{m}_i s_i - \sum_{out} \dot{m}_e s_e + \dot{S}_{gen}$$

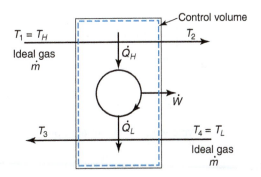

FIGURE 7-40 A reversible cycle with finite heat capacity fluid streams.

The cycle is assumed to operate in steady state, and the control volume is adiabatic; therefore,

$$\dot{S}_{gen} = \sum_{out} \dot{m}_e s_e - \sum_{in} \dot{m}_i s_i = \dot{m}\,(s_2 - s_1) + \dot{m}\,(s_3 - s_4)$$

Each fluid stream enters and exits with negligible change in pressure; therefore, using Eq. 7-20,

$$\dot{S}_{gen} = \dot{m}c_p \, \ln\left(\frac{T_2}{T_1}\right) + \dot{m}c_p \, \ln\left(\frac{T_3}{T_4}\right) \tag{7-40}$$

We assume all processes in the control volume are reversible, so that $\dot{S}_{gen} = 0$. Under these circumstances, Eq. 7-40 reduces to

$$\ln\left(\frac{T_2}{T_1}\right) = -\ln\left(\frac{T_3}{T_4}\right) = \ln\left(\frac{T_4}{T_3}\right)$$

or

$$T_3 = \frac{T_1 T_4}{T_2} \tag{7-41}$$

Substituting Eq. 7-41 into Eq. 7-39,

$$\dot{W} = \dot{m}c_p \left[(T_1 - T_2) - \left(\frac{T_1 T_4}{T_2} - T_4\right) \right] \tag{7-42}$$

We now adjust T_2, the exit temperature of the hot ideal gas, to produce the maximum possible work. To maximize work, take the derivative of $\dot{W}$ with respect to T_2,

$$\frac{d\dot{W}}{dT_2} = \dot{m}c_p \left[-1 + \frac{T_1 T_4}{T_2^2} \right]$$

At the optimum value of exit temperature, $T_{2,opt}$, the derivative is zero, therefore

$$0 = \left[-1 + \frac{T_1 T_4}{T_{2,opt}^2} \right]$$

Solving for $T_{2,opt}$ gives

$$T_{2,opt} = (T_1 T_4)^{1/2} \tag{7-43}$$

Substituting this into Eq. 7-42 gives the maximum work of the cycle as

$$\dot{W}_{max} = \dot{m}c_p \left\{ \left[T_1 - (T_1 T_4)^{1/2} \right] - \left[\frac{T_1 T_4}{(T_1 T_4)^{1/2}} - T_4 \right] \right\}$$

After some algebraic simplification,

$$\dot{W}_{max} = \dot{m}c_p \left(T_1^{1/2} - T_4^{1/2} \right)^2 \tag{7-44}$$

The efficiency of the cycle with the maximum power output is

$$\eta = \frac{\dot{W}_{max}}{\dot{Q}_{H,max}} \tag{7-45}$$

To find the heat added for the maximum condition, evaluate Eq. 7-38 at the optimum exit temperature, $T_{2,opt}$ (see Eq. 7-43) to get

$$\dot{Q}_{H,max} = \dot{m}c_p \left(T_1 - T_{2,opt}\right) = \dot{m}c_p \left[T_1 - (T_1 T_4)^{1/2}\right]$$

Substituting this and Eq. 7-44 into Eq. 7-45 gives

$$\eta = \frac{\dot{m}c_p \left(T_1^{1/2} - T_4^{1/2}\right)^2}{\dot{m}c_p \left[T_1 - (T_1 T_4)^{1/2}\right]}$$

After considerable simplification,

$$\eta = 1 - \left(\frac{T_4}{T_1}\right)^{1/2}$$

Referring to Figure 7-40, T_1 is the temperature at which hot gas enters the control volume and T_4 is the temperature at which cold gas enters. Therefore, the efficiency becomes

$$\boxed{\eta = 1 - \left(\frac{T_L}{T_H}\right)^{1/2} \qquad \text{maximum power cycle}} \qquad (7\text{-}46)$$

This is the efficiency of a cycle that generates maximum work with minimum entropy production while accounting for the finite heat capacity of hot and cold temperature sources. The efficiency is lower than that of a Carnot cycle. The cycle in Figure 7-40 can be reduced to a Carnot cycle if T_2 approaches T_1, so there is no temperature change of the hot source. In that case, however, $\dot{Q}_H$ becomes zero and the cycle produces zero net work. While the Carnot cycle has the highest possible efficiency, it may not be the best yardstick against which to compare real cycles. High efficiency is not important if there is no work output.

SUMMARY

A **thermodynamic cycle** is a series of processes in which a working fluid begins at an initial state, moves through one or more intermediate states, and then returns to its initial state. If net work is produced by the cycle, then it is labeled a **power cycle**. If work is added to the cycle and heat is pumped from a cold space to a hot space, then the cycle is a **refrigeration cycle**. On a P-v diagram, power cycles proceed clockwise and refrigeration cycles proceed counterclockwise.

The **cycle thermal efficiency** of a power cycle is defined as:

$$\eta_{cycle} = \frac{W_{net}}{Q_{in}} = \frac{\dot{W}_{net}}{\dot{Q}_{in}}$$

The efficiency is the fraction of the heat input that is converted into work. The value of efficiency varies between 0 and 1.

For a refrigeration cycle, the **coefficient of performance** is defined as

$$COP = \frac{Q_{in}}{W_{in}} = \frac{\dot{Q}_{in}}{\dot{W}_{in}}$$

The COP is the ratio of the heat removed by the cycle to the work that must be done to remove this heat. The value of COP is typically greater than 1. The COP for a refrigeration cycle is a measure of performance analogous to the thermal efficiency for a power cycle. In both cases, high values are good.

A **thermal reservoir** is a body so large in extent that its temperature does not change appreciably when moderate amounts of heat are added or removed. The ocean and the atmosphere are good examples of thermal reservoirs.

A **Carnot cycle** is a special cycle operated between two constant temperatures. In a Carnot cycle, all heat is added at one particular temperature and all heat is removed at a different, single temperature. All processes in a Carnot cycle are reversible. This means that the cycle can be operated in the opposite direction, and the values of heat and work will remain the same in magnitude but will change sign. The four reversible steps in a Carnot cycle are:

1–2   Isothermal heat addition
2–3   Adiabatic expansion
3–4   Isothermal cooling
4–1   Adiabatic compression

The efficiency of a Carnot power cycle is

$$\eta_{Carnot} = 1 - \frac{T_L}{T_H}$$

Furthermore, for any power cycle,

$$\eta_{cycle} = 1 - \frac{Q_L}{Q_H}$$

Therefore, for a Carnot power cycle,

$$\frac{T_L}{T_H} = \frac{Q_L}{Q_H}$$

This last equation is also true for a Carnot refrigeration cycle, which is a power cycle run in the opposite direction. The *COP* of a Carnot refrigerator is

$$COP_R = \frac{1}{\frac{T_H}{T_L} - 1}$$

Investigating the maximum performance of cycles leads, after a long chain of reasoning, to the need for the **second law of thermodynamics**. The Clausius statement of the second law is:

Heat cannot move spontaneously from cold to hot bodies.

The second law is a fundamental axiom, like the first law. It cannot be proven, but it is assumed to be true based on all our experience. These two laws together form the theoretical basis for thermodynamics. The second law can be used to prove that:

A power cycle cannot receive heat from a single thermal reservoir and convert it all into work.

Further analysis leads to:

The work produced by an irreversible cycle will never be greater than the work produced by a reversible cycle operating between the same two reservoirs.

The first Carnot principle is a natural consequence of this statement. It reads:

The efficiency of an irreversible heat engine is never greater than the efficiency of a reversible one operating between the same two reservoirs.

The second Carnot principle is:

All reversible cycles operating between the same two reservoirs have the same efficiency.

The efficiency does not depend on what the processes are that make up the cycle. It also does not depend on the working fluid used.

There are also two Carnot principles for refrigerators:

The coefficient of performance of an irreversible refrigeration cycle is never greater than the coefficient of performance of a reversible cycle when both operate between the same two thermal reservoirs.

All reversible refrigeration cycles operating between the same two reservoirs have the same coefficient of performance.

The second law leads to a new thermodynamic property, called **entropy**, which is a state variable. Entropy depends only on the end states of a process, not on the path. Values of entropy for many substances are included in thermodynamic tables.

The units of entropy are J/K or Btu/R. It is also useful to define a specific entropy, which is the entropy per unit mass. Entropy is related to specific entropy through

$$S = ms$$

where s is specific entropy. In the two-phase region, specific entropy is given by the following equation:

$$s = s_f + x\left(s_g - s_f\right)$$

For a compressed liquid, entropy can be approximated by the value of the saturated liquid at the same *temperature*, that is,

$$s\left(T, P\right) \approx s_f\left(T\right)$$

For a reversible process,

$$S_2 - S_1 = \int_1^2 \frac{\delta Q}{T}$$

An alternative statement of this equation is

$$\delta Q = T\, dS$$

or

$$Q = \int_1^2 T\, dS$$

The last three equations all apply only to a reversible process. If a reversible process is plotted on a *T-s* diagram, then the area under the curve is the heat transferred.

Some thermodynamic processes that can be made **reversible** in the limit include the following:

- Slow, isothermal compression of a gas
- Slow, isothermal expansion of a gas
- Slow, adiabatic compression of a gas
- Slow, adiabatic expansion of a gas
- Turbines and compressors
- Nozzles and diffusers
- Pumps
- Heat transfer across an infinitesimal temperature difference

Some processes which are **inherently irreversible** include the following:

- Mixing of two fluids
- Throttling processes
- Expansions of a gas into vacuum
- Heat transfer across a finite temperature difference (as in most heat exchangers)
- Any process involving friction

There are two equations that relate entropy to other thermophysical properties:

$$T\,ds = du + P\,dv$$

$$T\,ds = dh - v\,dP$$

For an ideal gas with constant specific heats, the entropy change is

$$s_2 - s_1 = c_v \ln\left(\frac{T_2}{T_1}\right) + \frac{\overline{R}}{M} \ln\left(\frac{v_2}{v_1}\right)$$

$$s_2 - s_1 = c_p \ln\left(\frac{T_2}{T_1}\right) - \frac{\overline{R}}{M} \ln\left(\frac{P_2}{P_1}\right)$$

These apply for any process of an ideal gas with constant specific heats, either reversible or irreversible. If specific heat varies with temperature, then use

$$s_2 - s_1 = s^o(T_2) - s^o(T_1) - \frac{\overline{R}}{M} \ln\left(\frac{P_2}{P_1}\right)$$

where values of $s^o(T)$ for air are available in Tables A-9 and B-9.

If a process is reversible and adiabatic, then

$$s_1 = s_2$$

In such a process, the entropy is constant and the process is called **isentropic**.

For an isentropic process of an ideal gas with variable specific heat,

$$\frac{P_2}{P_1} = \frac{P_{r2}}{P_{r1}} \quad \text{and} \quad \frac{v_2}{v_1} = \frac{v_{r2}}{v_{r1}}$$

The values of relative pressure and relative volume for air are listed in Tables A-9 and B-9.

Using the concept of entropy generation, the entropy balance equation for a closed system is

$$S_2 - S_1 = \int_1^2 \frac{\delta Q}{T} + S_{gen}$$

where

$S_{gen} > 0$	Real process
$S_{gen} = 0$	Reversible process
$S_{gen} < 0$	Impossible process

If the system is not isothermal, it can be difficult to evaluate the entropy generated. In that case, the system boundary can be divided into a finite number of isothermal zones, and the entropy balance can be written as

$$S_2 - S_1 = \sum_{k=1}^{N} \frac{Q_k}{T_k} + S_{gen}$$

The entropy balance can also be written in rate form as

$$\frac{dS}{dt} = \sum_{k=1}^{N} \frac{\dot{Q}_k}{T_k} + \dot{S}_{gen}$$

The entropy balance for an open system is

$$\frac{dS_{cv}}{dt} = \sum_{k=1}^{N} \frac{\dot{Q}_k}{T_k} + \sum_{in} \dot{m}_i s_i - \sum_{out} \dot{m}_e s_e + \dot{S}_{gen}$$

The isentropic efficiency of a turbine is, by definition

$$\eta_T = \frac{\dot{W}_{act}}{\dot{W}_{ideal}} = \frac{h_1 - h_2}{h_1 - h_{2s}}$$

State $2s$ has the same entropy as state 1 and the same pressure as state 2. The isentropic efficiency of a compressor is

$$\eta_C = \frac{\dot{W}_{ideal}}{\dot{W}_{act}} = \frac{h_1 - h_{2s}}{h_1 - h_2}$$

and for a pump is

$$\eta_P = \frac{\dot{W}_{ideal}}{\dot{W}_{act}} = \frac{h_1 - h_{2s}}{h_1 - h_2}$$

The efficiency of a cycle that generates maximum work with minimum entropy production while accounting for the finite heat capacity of hot and cold temperature sources is

$$\eta = 1 - \left(\frac{T_L}{T_H}\right)^{1/2}$$

SELECTED REFERENCES

BEJAN, A. "Models of Power Plants That Generate Minimum Entropy While Operating at Maximum Power," *Am. J. Phys.*, Vol. 64, No. 8, 1996.

BLACK, W. Z., and J. G. HARTLEY, *Thermodynamics*, Harper & Row, New York, 1985.

CARNOT, S., *Reflections on the Motive Power of Heat*, Translation by R.H. Thurston, ASME, New York, 1943. (Originally published in 1824 by Chez Bachelier, Paris, under the title *Reflexions sur la Puissance Motrice du Feu*)

CENGEL, Y. A., and M. A. BOLES, *Thermodynamics, an Engineering Approach*, 4th ed., McGraw-Hill, New York, 2002.

FENN, J. B., *Engines, Energy, and Entropy*, Freeman, New York, 1982.

HOWELL, J. R., and R. O. BUCKIUS, *Fundamentals of Engineering Thermodynamics*, 2nd ed., McGraw-Hill, New York, 1992.

MORAN, M. J., and H. N. SHAPIRO, *Fundamentals of Engineering Thermodynamics*, 3rd ed., Wiley, New York, 1995.

MYERS, G., *Engineering Thermodynamics*, Prentice Hall, Englewood Cliffs, NJ, 1989.

VAN WYLEN, G. J., R. E. SONNTAG, and C. Borgnakke, *Fundamentals of Classical Thermodynamics*, 4th ed., Wiley, New York, 1994.

PROBLEMS

CYCLE EFFICIENCY

P7-1 Heat is removed from a power cycle by water flowing at 6.5 kg/s. The water enters at 10°C and leaves at 26°C. If the cycle is 64% efficient, what is the net power produced?

P7-2 A simple power cycle consists of a boiler, turbine, condenser, and pump. All the heat is removed from the condenser by cooling water, which enters at 60°F and exits at 75°F. The mass flow rate of the water is 18.8 lbm/sec. If the heat added to the boiler is 396 Btu/s, find the cycle efficiency and the net power produced in hp.

P7-3 A new coal-fired power plant, which will produce 400 MW of power, is planned. The efficiency of the cycle is 37%. The average heating value of coal is 28,000,000 kJ per ton, and the price of coal is $50 per ton. What is the estimated fuel bill for this plant in the first five years of operation?

P7-4 A room air conditioner is used to cool a 10 ft by 12 ft by 8 ft high room from 95°F to 68°F in 15 min. If the *COP* is 2.8, estimate the power (in W) required. For simplicity, assume the mass of air and pressure in the room remain constant.

P7-5 A refrigerator is capable of delivering 3 kW of cooling while requiring 950 W of electric power to operate. If the *COP* of the refrigerator is improved by 20%, how much electric power would be required to deliver 3 kW of cooling?

P7-6 An ice machine produces ice at a rate of 15 lbm/h. Water is fed to the machine at 50°F, and the temperature of the ice is 25°F. If the *COP* of the ice machine is 1.6, determine the required power input, in kW. The latent heat of fusion of ice is 143.5 Btu/lbm.

CARNOT CYCLE

P7-7 An air conditioner is used to cool a classroom on a day when the exterior temperature is 95°F. Room air is maintained at 68°F, and there are 35 students in the class. The average heat generated per student is 90 W. The room is lit by 12 fluorescent lightbulbs, each generating 40 W. If the *COP* of the air conditioner is 1.2, what power input is required, in kW? If the air conditioner operated on a Carnot cycle, what power input would be required?

P7-8 A reversible power cycle produces 150 hp of work while operating between temperature limits of 800°F and 60°F. If the maximum temperature is increased to 850°F and the same amount of heat is added from the high-temperature reservoir, find the new power output of the cycle.

P7-9 Heat leaks into a refrigerator at a rate of 215 Btu/min. The interior is kept at 38°F while the room is at 68°F. If electricity costs $0.03 per kWh, what is the minimum cost of operating the refrigerator for one month (30 days)?

P7-10 In a proposed ocean thermal power plant, warm water near the surface is used to provide heat input to a power cycle. The cycle rejects heat to cold water at the ocean floor. If the temperatures of the water at the surface and at the ocean floor are 24°C and 3°C, respectively, what is the maximum possible efficiency of this cycle?

P7-11 A refrigerated compartment is maintained at 2°C. Heat from the surroundings, at 25°C, leaks into the compartment at a rate of 1.2 kW. The refrigeration cycle used to maintain the temperature difference has a coefficient of performance that is half the Carnot value. How large a compressor, in hp, is needed to drive this cycle?

P7-12 A power cycle delivers 7.5 hp of work while rejecting 8300 Btu/h of heat to a reservoir at 45°F. If the high-temperature reservoir is at 360°F, is this cycle irreversible, reversible, or impossible? Support your answer with calculations.

P7-13 A power cycle operates between temperature limits of 400°C and 15°C. The cycle requires a heat input of 10.1 kW and rejects 6 kW to the low-temperature reservoir. Is the cycle irreversible, reversible, or impossible? Support your answer with calculations.

P7-14 A power cycle consumes fuel at a rate of 8 gal/h. The maximum temperature in the cycle is 600°F, and heat is rejected to the atmosphere at 70°F. If the fuel has an energy content of 22,000 Btu/lbm and a density of 45 lbm/ft³, what is the maximum possible power that can be produced by the cycle?

P7-15 A freezer rejects 210 W of heat into a room at 22°C. The freezer temperature is –3°C. An ice-cube tray containing 0.5 kg of liquid water at 20°C is placed in the freezer and is completely solidified in 18 minutes. Is this freezer irreversible, reversible, or impossible? Ignore the energy required to cool the plastic tray and consider only the water. The latent heat of fusion of water is 333.7 kJ/kg.

P7-16 In a Carnot vapor power cycle, the turbine produces 2000 hp. The maximum temperature that the turbine blades will withstand is 1250°F. Cooling water is available at 50°F. If fuel can be burned to produce heat at the rate of 1700 Btu/lbm and the fuel costs $1.41 per lbm, what is the fuel bill for this plant for one month of operation (30 days)?

P7-17 The reversible cycle in Figure 7-41a receives heat Q_1 from a reservoir at T_H and does work W_3. In Figure 7-41b, two reversible cycles are shown. The one that operates between T_H and T_M receives the same amount of heat, Q_1, as is shown in Figure 7-41a, and does work W_1. This top cycles rejects heat Q_2 to a reservoir at T_M and supplies the same heat Q_2 to the bottom cycle. $T_L < T_M < T_H$.

Show that *one* of the following three statements is true:

$$W_3 > W_1 + W_2$$
$$W_3 < W_1 + W_2$$
$$W_3 = W_1 + W_2$$

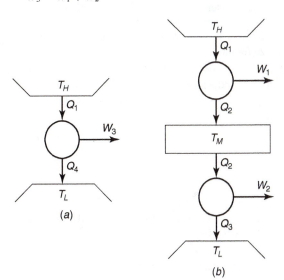

FIGURE 7-41 Two reversible cycles.

P7-18 A reversible power cycle operates according to the *T-s* diagram shown. Using data in the diagram, calculate the cycle efficiency.

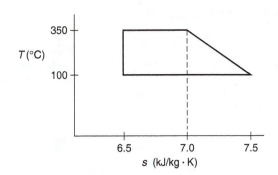

SECOND LAW, CLOSED SYSTEM

P7-19 A cylinder of volume 300 cm³ contains saturated steam at 0.6 MPa. The steam is then allowed to expand adiabatically and reversibly to a final pressure of 0.2 MPa.

a. Find the final quality.

b. Find the work done.

P7-20 R-134a at 10 psia and 20°F is compressed reversibly and isothermally in a piston–cylinder assembly. During the compression, 0.0945 Btu of heat is removed. The initial volume is 69 in.³. Find the final pressure and the work done, in ft-lbf.

P7-21 Saturated water vapor at 200 kPa is expanded slowly and without friction (i.e., reversibly) and isothermally in a piston–cylinder device until the pressure of the H_2O is 50 kPa.

a. What is the entropy change of the H_2O (in kJ/kg·K)?

b. Determine a value for the work done per unit mass of the H_2O (in kJ/kg).

P7-22 Steam at 400 kPa, 200°C is contained in a well-insulated piston–cylinder assembly of initial volume 0.13 m³. The steam expands while 32 kJ of work is done. It is claimed that the final state is a two-phase mixture. Is this possible?

P7-23 A piston–cylinder assembly contains 0.2 m³ of air initially at 3.5 MPa and 330°C. The air expands in a slow, frictionless, isothermal process to 150 kPa. Find the heat transferred.

P7-24 Argon with a mass of 0.9 lbm is initially at 14.7 psia and 75°F. The gas is compressed reversibly to 100 psia. Find the work required if the process is isothermal.

P7-25 Air is compressed from 100 kPa and 20°C to 850 kPa in a well-insulated piston–cylinder assembly. The mass of air present is 0.04 kg. If the process is reversible, determine the final temperature and work required assuming

a. constant specific heat.

b. variable specific heat.

P7-26 Determine the entropy change if 3 kg of air at 30°C and 95 kPa are compressed to 425 kPa and 200°C. Do the calculation two ways:

a. with constant specific heat.

b. with variable specific heat.

ENTROPY GENERATION

P7-27 A well-insulated chamber is divided in two parts by a thin membrane, as shown in the figure. The left side is initially filled with air at 30°C and 100 kPa. The right side is initially evacuated. The membrane ruptures, and air expands freely into the evacuated section. Calculate the entropy generated during this process.

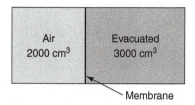

P7-28 A steam–water mixture with an initial quality of 0.05 is contained in a piston–cylinder assembly. The initial pressure and volume are 125 kPa and 175 cm³. Heat is added at constant pressure until only vapor remains. Calculate the entropy generated (in kJ/K).

P7-29 In a vortex tube, air enters perpendicular to the tube axis and flows in a swirling motion around the inside periphery of the tube. Because of centrifugal forces, the air is separated into a hot and a cold stream, with one stream exiting at the top and the other exiting at the bottom. It is proposed to construct a vortex tube to use as a refrigeration system. Air enters at 20°C and 300 kPa and exits from the top at 60°C and 280 kPa and from the bottom at 0°C and 280 kPa. The tube is perfectly insulated. Is it possible for such a device to operate as described?

P7-30 A salesman is promoting a well-insulated steam turbine that produces 3000 kW. Steam enters the turbine at 700 kPa, 250°C and exits at a pressure of 10 kPa. The mass flow rate is 3.7 kg/s. Is such a turbine possible?

P7-31 Saturated vapor R-134a at 0°F enters the tube side of a well-insulated heat exchanger and exits at 20°F. Compressed liquid R-134a at 100°F and 180 psia enters the shell side of the heat exchanger. Both streams flow at 8 lbm/s. Calculate the total entropy generated.

P7-32 Saturated R-134a vapor at 5°F enters the compressor of a refrigeration cycle. Superheated vapor at 120°F and 180 psia exits. Is the compressor losing heat to the surroundings or gaining heat?

SECOND LAW IN OPEN SYSTEMS

P7-33 Air expands through an ideal, insulated nozzle from an inlet pressure of 1500 kPa to an exit pressure of 500 kPa. The inlet temperature is 100°C and the inlet area is 0.04 m². If the mass flow rate is 22 kg/s, find the exit velocity. Assume constant specific heats.

P7-34 Air flowing at 200 ft/s enters an ideal, well-insulated nozzle. The inlet conditions are 50 psia and 500°F, and the exit

pressure is 10 psia. Using variable specific heat, determine the final velocity.

P7-35 Air at 0.5 kPa and 0°C expands through an ideal, well-insulated diffuser. The air enters at 250 m/s and leaves at 120 m/s. Assuming constant specific heat, find the exit air temperature and pressure.

P7-36 Steam at 200 kPa and 250°C is expanded in an ideal, adiabatic diffuser. The steam enters at 705 m/s and leaves at 500 kPa. Find the exit velocity and temperature of the steam.

P7-37 Steam at 1.4 MPa, 600°C expands through an ideal, adiabatic turbine to a final temperature of 50°C. The mass flow rate is 6 kg/s. Calculate the final pressure and the power produced.

P7-38 Steam expands in an adiabatic, reversible turbine from 3.0 MPa, 700°C to a final pressure of 10 kPa. If the flow rate is 1.7 kg/s,

a. calculate the power output.

b. show the process on a *T-s* diagram.

P7-39 Saturated vapor R-134a at 0.24 MPa is compressed adiabatically and reversibly in a compressor to 1.6 MPa. If the flow rate is 48 kg/h, calculate the work input to the compressor and the final refrigerant temperature.

P7-40 Saturated R-134a vapor at –10°F enters a frictionless, well-insulated compressor. If the pressure at the exit is 400 psia, what is the temperature at the exit?

P7-41 A tank of volume 400 ft³ is initially filled with air at 70°F and 14.7 lbf/in.². A vacuum pump slowly removes air from the tank until a very low pressure is achieved. The tank is uninsulated and may exchange heat with the surroundings at 70°F. During the process, the contents of the tank remain at a constant temperature of 70°F. The pump exhausts to the surroundings, which are at 70°F and 14.7 lbf/in.². Determine the minimum possible work required for this process.

P7-42 A rigid, well-insulated tank with a volume of 12 ft³ is initially filled with water vapor at 800°F and 40 psia. A leak develops in the tank and steam escapes until the pressure reaches 14.7 psia.

a. Determine the final temperature of the tank contents.

b. Determine the amount of mass that escapes.

ISENTROPIC EFFICIENCY

P7-43 Steam at 500°F enters a well-insulated turbine and exits at 5 psia. The isentropic efficiency is 89%, and the quality at the exit state is 0.9. Find the inlet pressure.

P7-44 Steam enters a turbine at 600 kPa and 300°C and exits at 5 kPa. The turbine efficiency is 75%.

a. Calculate the state at the exit.

b. Calculate the work produced per kg of steam flow.

P7-45 A turbine and a throttling valve are operated in parallel, as shown in the figure. Steam enters the system at 550 psia, 600°F and leaves at 130 psia, 500°F. If the mass flow rate is 30 lbm/s and the turbine is well insulated with an isentropic efficiency of 88%, find the power produced by the turbine (in hp).

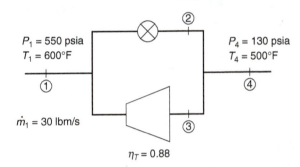

$P_1 = 550$ psia
$T_1 = 600°F$

$P_4 = 130$ psia
$T_4 = 500°F$

$\dot{m}_1 = 30$ lbm/s

$\eta_T = 0.88$

P7-46 Air with a flow rate of 20 m³/s enters a compressor at 100 kPa and 20°C and is compressed to 400 kPa. Assume the process is adiabatic and the compressor has an efficiency of 84%.

a. Calculate the exit temperature.

b. Calculate the power input to the compressor.

P7-47 Air at 450 kPa and 500 K expands through a well-insulated turbine to an exit pressure of 150 kPa. The mass flow rate is 40 kg/s, and the isentropic efficiency of the turbine is 82%. Calculate the work done and the exit temperature.

MAXIMUM POWER CYCLE

P7-48 Combustion gases at 850°F provide heat to a power cycle. The cycle operates reversibly at maximum power and produces 40 MW, while rejecting heat to air at 50°F. Assume that the combustion gases and cooling gases have the same mass flow rate and that the combustion gases have the properties of air. Find the mass flow rate.

P7-49 A hot air stream at 310°C with a flow rate of 3.2 kg/s is available to power a cycle. If the cycle is reversible and rejects heat to air at 20°C, find the maximum possible power output. Assume the cold stream has the same mass flow rate as the hot stream.

P7-50 An actual coal-fired power plant operating between temperature limits of 350°C and 20°C produces 150 MW while burning coal at a rate of 62 metric tons per hour. The heating value of the coal is 30,000 kJ/kg. Compare the efficiency of this cycle to the efficiency of a theoretical maximum power cycle operating between these temperature limits. Also compute the efficiency of the Carnot cycle for the same temperature limits.

CHAPTER 8

REFRIGERATION, HEAT PUMP, AND POWER CYCLES

8.1 INTRODUCTION

The Carnot heat power cycle and the Carnot refrigeration cycle discussed in Chapter 7 are idealized cycles whose performances cannot be attained in reality. However, their characteristics can be used to guide the development of practical cycles. In everyday life we use refrigeration and heat pump cycles for conditioning the interior of buildings and vehicles. Whenever we drive a car or fly in an airplane, a heat power cycle is used to provide the motive force. The electricity we use for so many different tasks is generated by a heat power cycle (unless we live in an area where hydroelectric dams or wind power are used extensively). In this chapter, when we discuss power production, we focus only on heat engines, so hereafter we will use *power cycle* to mean *heat power cycle*.

In each of these applications, the job of the engineer is to devise the cycle that will accomplish the goals most effectively at the lowest cost. Many trade-offs are required when designing the cycle. For example, over the projected life of an electric power plant, implementing modifications that improve the cycle thermal efficiency (and thus lower the fuel operating costs) have to be balanced against the increased capital costs needed to modify the plant.

In the sections that follow, several common cycles are examined. We start with the simplest cycle—the *vapor-compression refrigeration cycle*—to illustrate how an analysis requires us to view the cycle from two vantage points: at times, we must stand back and look at the cycle in aggregate; at other times, we must get closer and focus on a specific component or several components of the cycle. After the refrigeration cycle, we present power cycles: first, a *vapor power cycle* in which the working fluid is alternately vaporized and condensed and, second, *a gas power cycle* in which the working fluid remains a gas during all processes. Each of these power cycles can be modified in a variety of ways to improve its performance. Some of the modifications possible for a power cycle are discussed along with the trade-offs that might result.

We have studied all the components that are used to construct cycles (turbines, compressors, pumps, heat exchangers, valves). Likewise, we have used the open-system conservation of mass and energy equations, and we have studied the use of entropy. Now we put all of these devices and processes together in the analysis of refrigeration and power cycles.

8.2 VAPOR-COMPRESSION REFRIGERATION CYCLES

The Clausius statement of the second law of thermodynamics is: "Heat does not move spontaneously from cold to hot bodies." There are devices, such as refrigerators and air conditioners, that do move heat from cold to hot spaces. For example, a refrigerator absorbs heat from its interior compartment and rejects this heat to the surrounding room. An air

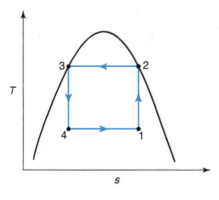

FIGURE 8-1   Carnot refrigeration cycle.

conditioner removes heat from a living space and rejects it outdoors. Neither the refrigerator nor the air conditioner operates spontaneously. Work (in the form of electricity) must be added to drive each cycle.

There are several types of refrigeration cycles, but we focus only on the vapor-compression refrigeration cycle because it is commonly used in a wide range of applications (automobiles, homes, buildings). This is considered an *energy-consuming cycle* because we use energy in the form of work to move energy in the form of heat from a low-temperature reservoir to a high-temperature reservoir. We showed in Chapter 7 that the Carnot refrigeration cycle has four processes: (1) isentropic compression, (2) reversible, isothermal heat rejection, (3) isentropic expansion, and (4) reversible, isothermal heat addition. Figure 8-1 shows an implementation of the Carnot cycle using phase change. There are two practical problems with building this cycle as shown. In process 1–2, a two-phase mixture is compressed isentropically to a saturated vapor. No device is known to exist that can accomplish this task. Real compressors require vapor at both inlet and exit. In process 3–4, a saturated liquid is expanded isentropically to a two-phase mixture. Using a turbine to expand the liquid is not economically feasible because the power that could be extracted is very small compared to the power input to the compressor.

Because of these practical considerations, the Carnot cycle is modified in two ways, as shown in Figure 8-2: state 1 is moved to the saturated vapor line so that the compressor receives only vapor at its inlet, not a two-phase mixture. In addition, the expansion from state 3 to state 4 is accomplished through an inexpensive, but irreversible, throttling valve.

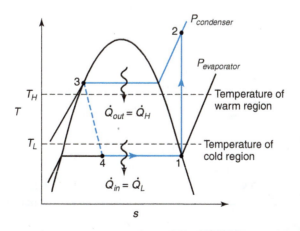

FIGURE 8-2   *T-s* diagram for ideal vapor-compression refrigeration cycle.

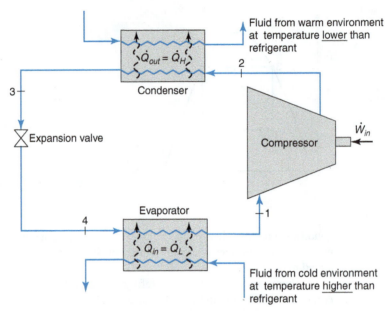

FIGURE 8-3 Schematic of vapor-compression refrigeration cycle.

The cycle in Figure 8-2 is called an *ideal vapor-compression refrigeration cycle*. It is composed of four processes that are caused by four separate devices: a compressor, a condenser, an expansion (throttling) valve, and an evaporator, as shown in Figure 8-3. The four processes are:

1–2	Compression (work *into* cycle)
2–3	Heat rejection *from* cycle *to* environment
3–4	Expansion
4–1	Heat addition *to* cycle *from* cooled space

For the ideal cycle, we assume that the compressor is reversible and adiabatic (i.e., isentropic), that no pressure drop occurs across the evaporator or condenser, that the fluid exits the condenser as a saturated liquid, and that the fluid exits the evaporator as a saturated vapor. The *T-s* diagram in Figure 8-4 is for a *real vapor-compression cycle*. Note that the

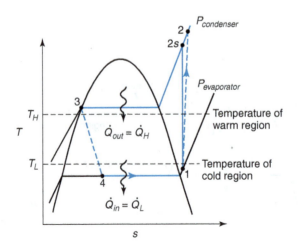

FIGURE 8-4 *T-s* diagram of a real vapor-compression refrigeration cycle.

compressor is nonisentropic, that slightly subcooled liquid exits the condenser, and that slightly superheated vapor exits the evaporator. Pressure drop across the evaporator and condenser may or may not be neglected.

The working fluid in a vapor-compression refrigeration cycle is called a refrigerant; typical refrigerants are R-12, R-134a, ammonia, and carbon dioxide. In process 1–2 the working fluid enters the compressor as a saturated (or slightly superheated) vapor, is compressed to a higher pressure and temperature using $\dot{W}_{in}$, and exits as a superheated vapor. The hot vapor flows through a condenser (process 2–3), where it is cooled and condensed by a lower-temperature fluid from the environment; this heat transfer rate out of the cycle is $\dot{Q}_{out} = \dot{Q}_H$. The fluid exits the heat exchanger as a saturated (or a slightly subcooled) liquid and flows through the throttling valve (process 3–4); the fluid flashes to a low-quality vapor–liquid mixture at a much lower pressure and temperature. Finally, the two-phase mixture flows through the evaporator (process 4–1), where the fluid vaporizes (and possibly superheats) with heat transfer from the higher temperature in the conditioned space; this heat transfer rate into the cycle is $\dot{Q}_{in} = \dot{Q}_L$. Hence, this process uses $\dot{W}_{in}$ to extract energy $(\dot{Q}_L)$ from a low-temperature environment and to reject energy $(\dot{Q}_H)$ to a high-temperature environment. Remember from our discussion of cycles in Chapter 7 that energy conservation for a complete cycle yields $\dot{Q}_H = \dot{Q}_L + \dot{W}_{in}$.

The performance of a refrigeration cycle is evaluated through the use of two quantities that describe the overall effect of the components working together: the *cooling* or *refrigerating capacity* and the coefficient of performance (*COP*). The energy we "use" in a refrigeration cycle—the refrigerating capacity—is equal to $\dot{Q}_L$. This is the heat that is removed from the refrigerated space. It is a heat transfer rate with units of W, kW, or Btu/h. Often $\dot{Q}_L$ is expressed in terms of *"tons of refrigeration"*. This unit of measure appeared early in the use of mechanical refrigeration systems, when their primary function was to supply blocks of ice for use in iceboxes in homes. It refers to the heat transfer rate required to freeze 1 ton (2000 lbm) of 0°C water into ice at 0°C in 24 hours. A "ton of refrigeration" is equivalent to 211 kJ/min or 200 Btu/min.

The *COP* of a refrigeration cycle was defined in Chapter 7:

$$COP_{Ref} = \frac{\text{energy we want to use}}{\text{energy we purchase}} = \frac{\dot{Q}_L}{\dot{W}_{in}} = \frac{\dot{Q}_L}{\dot{Q}_H - \dot{Q}_L} \qquad (8\text{-}1)$$

The magnitude of the *COP* of a refrigeration cycle can theoretically range from a value much less than 1 to a number much greater than 1. Typical values are from 1.5 to 5. This means that 1.5 to 5 units of useful cooling capacity are obtained from each unit of input power.

To obtain values for either the heat transfer rates or the input power, we must analyze the individual components by applying conservation of energy to each component. For the compressor we assume steady, adiabatic operation and negligible potential and kinetic energy effects to obtain:

$$\dot{W} = -\dot{m}\,(h_2 - h_1)$$

With $h_2 > h_1$ and $\dot{m}$ always positive, strict application of the energy equation shows that $\dot{W}$ is negative; power is input to the compressor. However, when we deal with cycles, we prefer to use positive values of work, power, and heat transfer, so we use a subscript to indicate the direction of energy flow. Therefore,

$$-\dot{W} = \dot{W}_{in} = \dot{m}(h_2 - h_1) \qquad (8\text{-}2)$$

For the heat exchangers, we assume steady operation with no work and negligible potential and kinetic energy effects (and again use subscripts to indicate direction) to obtain:

$$\dot{Q}_H = \dot{m}(h_2 - h_3) \quad \text{condenser} \tag{8-3}$$

$$\dot{Q}_L = \dot{m}(h_1 - h_4) \quad \text{evaporator} \tag{8-4}$$

Note that $\dot{Q}_H$ occurs at the hottest temperature in the cycle and is also referred to as $\dot{Q}_{out}$; $\dot{Q}_L$ occurs at the lowest temperature in the cycle and is also referred to as $\dot{Q}_{in}$.

Across the expansion (throttling) valve, we have an isenthalpic process, so that

$$h_4 = h_3$$

Substituting Eq. 8-2 and Eq. 8-4 into Eq. 8-1 results in

$$COP_{Ref} = \frac{\dot{Q}_L}{\dot{W}_{in}} = \frac{h_1 - h_4}{h_2 - h_1} \tag{8-5}$$

For a Carnot refrigeration cycle $COP_{Carnot\ Ref} = T_L/(T_H - T_L)$, so if the difference between the high and low temperatures decreases, then the COP increases. For a real vapor-compression cycle, if the difference between the high and low pressures is reduced, then the difference between the two saturation temperatures also is reduced, and COP increases.

EXAMPLE 8-1   Ideal vapor-compression refrigeration cycle

An ideal vapor-compression refrigeration cycle operates as a refrigerator at steady state with refrigerant R-134a as the working fluid. Saturated vapor enters the compressor at 30 lbf/in.², and saturated liquid leaves the condenser at 140 lbf/in.². The refrigeration capacity is 3 tons.

a) Determine the compressor power (in hp).

b) Determine the heat transfer rate from the condenser (in Btu/min).

c) Determine the coefficient of performance.

Approach:

Referring to the figure, we define a control volume around the compressor and develop an expression for the power consumed by it using conservation of energy. From previous use of the energy equation,

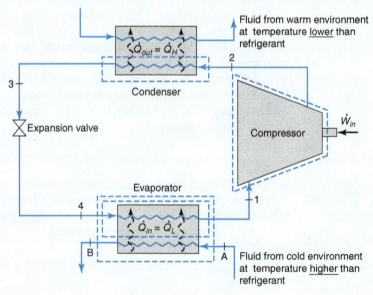

we know we will need the mass flow rate. Since it is not given, we evaluate the mass flow rate by analyzing another part of the cycle for which enough information is given. Because refrigeration capacity is given, we apply conservation of energy to the evaporator. There is sufficient information available to evaluate properties, so the only unknown in the equation is the mass flow rate. Finally, the heat transfer rate from the condenser is obtained by applying conservation of energy to the refrigerant, and the coefficient of performance is then given by Eq. 8-5.

Assumptions:

A1. The overall system and individual components are steady.

A2. The compressor is adiabatic.

A3. Potential and kinetic energy effects are negligible.

A4. Compressor is reversible.

A5. No work occurs in the heat exchangers.

Solution:

a) For the control volume around the compressor, we assume steady, adiabatic, and negligible potential and kinetic energy effects [A1], [A2], [A3]. Applying these assumptions to conservation of energy results in an expression for the compressor power:

$$\dot{W}_C = \dot{m}\,(h_2 - h_1)$$

We also apply the second law to the compressor in the form [A4]

$$s_1 = s_2$$

We have enough information to evaluate the two enthalpies, $h_1(P_1, \text{sat. vapor})$ and $h_2(P_2, s_2 = s_1)$, but the mass flow rate is unknown.

The mass flow rate is obtained by analyzing the evaporator with conservation of energy because refrigeration capacity is given ($\dot{Q}_{in} = 3$ tons). For the control volume drawn only around the refrigerant in the evaporator, we assume steady, negligible potential and kinetic energy effects, and no work [A1], [A3], [A5], thus giving:

$$\dot{Q}_{in} = \dot{m}\,(h_1 - h_4)$$

The enthalpies at the entrance and exit of the evaporator can be determined from the given information, and the refrigeration capacity is given in the problem statement. Therefore, the only unknown in this equation is the mass flow rate. Solving for $\dot{m}$,

$$\dot{m} = \frac{\dot{Q}_{in}}{h_1 - h_4}$$

b) The heat transfer rate from the condenser is obtained by applying the energy equation to the refrigerant and assuming steady, negligible potential and kinetic energy effects, and no work [A1], [A3], [A4], thus giving

$$\dot{Q}_{out} = \dot{m}\,(h_2 - h_3)$$

c) The coefficient of performance is then given by the ratio $\dot{Q}_{in}/\dot{W}_C$. We can now evaluate all the unknown quantities. The enthalpies are evaluated using Tables B-15 and B-16.

State 1: $P_1 = 30$ lbf/in.^{2}, saturated vapor $\rightarrow h_1 = h_{f1} = 103.96$ Btu/lbm, $s_1 = 0.2209$ Btu/lbm·R

State 2: $P_2 = 140$ lbf/in.^{2}, $s_2 = s_1 \rightarrow$ by interpolation $h_2 = 117.7$ Btu/lbm

State 3: $P_3 = 140$ lbf/in.^{2}, saturated liquid $\rightarrow h_3 = h_{f3} = 44.43$ Btu/lbm

State 4: Throttling process $\rightarrow h_4 = h_3 = 44.43$ Btu/lbm

Solving the refrigeration capacity expression given above for the mass flow rate:

$$\dot{m} = \frac{\dot{Q}_{in}}{h_1 - h_4} = \frac{3\ \text{tons}}{(103.96 - 44.43)\ \frac{\text{Btu}}{\text{lbm}}} \left(\frac{200\ \text{Btu}/\text{min}}{1\ \text{ton}} \right) = 10.1\ \text{lbm/min}$$

Now we can calculate the compressor power:

$$\dot{W}_C = \left(10.1\frac{\text{lbm}}{\text{min}}\right)(117.7 - 103.96)\frac{\text{Btu}}{\text{lbm}}\left(\frac{60\ \text{min}}{1\ \text{h}}\right)\left(\frac{1\ \text{hp}}{2545\ \text{Btu/h}}\right) = 3.27\ \text{hp}$$

The heat transfer rejected from the condenser is

$$\dot{Q}_{out} = \left(10.1\frac{\text{lbm}}{\text{min}}\right)(117.7 - 44.43)\frac{\text{Btu}}{\text{lbm}} = 740\frac{\text{Btu}}{\text{min}}$$

The coefficient of performance is

$$COP_{Ref} = \frac{\dot{Q}_{in}}{\dot{W}_C} = \frac{3\ \text{ton}}{3.27\ \text{hp}}\left(\frac{200\ \text{Btu/min}}{1\ \text{ton}}\right)\left(\frac{1\ \text{hp}}{2545\ \text{Btu/h}}\right)\left(\frac{60\ \text{min}}{1\ \text{h}}\right) = 4.33$$

Comments:

For the use of 3.27 hp of input power, we obtain $4.33 \times 3.27\ \text{hp} = 14.2\ \text{hp}$ of useful cooling.

EXAMPLE 8-2   Vapor-compression refrigeration cycle with irreversibilities

A water chiller is to be designed using a R-134a vapor-compression refrigeration cycle. Existing equipment is to be used to cool the water from 38°C to 10°C. The refrigeration cycle compressor has an isentropic efficiency of 82% and has a rated volumetric flow rate of 5.5 m³/min at the compressor inlet. The high-pressure side is at 1.4 MPa, and the low-pressure side is at 0.2 MPa. To protect the compressor from damage, the evaporator is designed to ensure 5°C of superheat at its outlet. The condenser is designed to ensure 5°C of subcooling at its outlet.

a) Determine the water flow rate that can be cooled by the chiller (in kg/s).

b) Determine the cycle coefficient of performance.

Approach:

To find the water flow rate, we begin the solution by applying conservation of energy to the evaporator as shown below in the figure, because that is the only piece of equipment that has water flowing through it. We need information on the enthalpies and flow rates of the water and R-134a in the

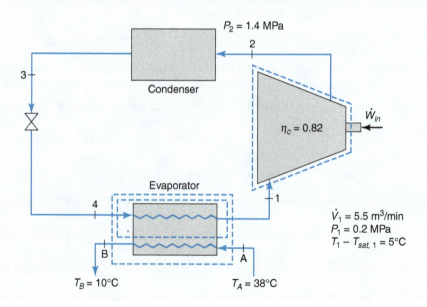

evaporator. The refrigerant flow and all enthalpies can be determined from the given information; the only unknown is the water flow rate. For the coefficient of performance, the input heat transfer rate can be determined from an analysis of the evaporator. Compressor power can be determined from conservation of energy applied to a control volume drawn around the compressor.

Assumptions:

A1. The complete system and individual components are steady.
A2. The heat exchangers are adiabatic.
A3. There is no work in the heat exchangers.
A4. Potential and kinetic energy effects are negligible.
A5. Water is an ideal liquid with constant specific heat.

Solution:

a) To determine the water flow rate, we begin with an energy balance around the evaporator, since that is the piece of equipment through which the water flows. We define a control volume to include the whole evaporator. Assuming steady state, adiabatic conditions, no work, and negligible potential and kinetic energy effects [A1], [A2], [A3], [A4], conservation of mass gives us:

$$\dot{m}_4 = \dot{m}_1 \quad \text{and} \quad \dot{m}_A = \dot{m}_B$$

The energy equation gives us

$$\dot{m}_4 h_4 + \dot{m}_A h_A - \dot{m}_1 h_1 - \dot{m}_B h_B = 0$$

Combining terms and using $\Delta h = c_p \, \Delta T$ [A5], we obtain

$$\dot{m}_1 (h_4 - h_1) + \dot{m}_A c_p (T_A - T_B) = 0$$

Solving for the water flow rate:

$$\dot{m}_A = \frac{\dot{m}_1 (h_4 - h_1)}{c_p (T_B - T_A)}$$

The refrigerant flow rate, $\dot{m}_1$, is determined from given information and the definition of mass flow rate, $\dot{m}_1 = \rho_1 \mathcal{V}_1 A_1 = \rho_1 \dot{V}_1 = \dot{V}_1 / v_1$. Enough information is given in the problem statement to evaluate all the properties in these two equations.

b) The cycle coefficient of performance is defined as

$$COP_{Ref} = \frac{\dot{Q}_L}{\dot{W}_{in}}$$

From part a. above, if we perform an energy balance around only the refrigerant flowing through the evaporator (see figure for control volume), then application of the energy equation assuming steady conditions, no work, and negligible potential and kinetic energy effects [A1], [A3], [A4] gives us

$$\dot{Q}_L = \dot{m}_1 (h_1 - h_4)$$

A6. The compressor is adiabatic.

To evaluate $\dot{W}_{in}$, we define a control volume around the compressor and assume steady conditions, negligible potential and kinetic energy effects, and adiabatic conditions [A1], [A4], [A6]. From the energy equation we obtain

$$\dot{W} = -\dot{W}_{in} = \dot{m}_1 (h_1 - h_2)$$

Now we use the definition of isentropic efficiency:

$$\eta_C = \frac{\dot{W}_{ideal}}{\dot{W}_{act}} = \frac{\dot{W}_{ideal}}{\dot{W}_{in}} = \frac{h_{2s} - h_1}{h_2 - h_1}$$

so that $\dot{W}_{in} = \dot{m}_1 (h_{2s} - h_1) / \eta_C$. We have enough information to evaluate the properties and, hence, the *COP*.

The R-134a properties are evaluated using Tables A-14 and A-16.

> **State 1:** $P_1 = 0.20$ MPa, $T_1 = T_{sat}(P_1) + 5°C = -10.09°C + 5°C = -5.09°C →$ by interpolation $h_1 = 245.74$ kJ/kg, $s_1 = 0.9419$ kJ/kg·K, $v_1 = 0.1019$ m³/kg
>
> **State 2:** $P_2 = 1.40$ MPa, $s_{2s} = s_1 →$ by interpolation $h_{2s} = 287.2$ kJ/kg·K
>
> **State 3:** $P_3 = P_2$, $T_3 = T_{sat}(P_3) - 5°C = 52.43°C - 5°C = 47.43°C →$ $h_3 \approx h_{f3}(T_3) = 117.5$ kJ/kg
>
> **State 4:** Throttling process $→ h_4 = h_3 = 117.5$ kJ/kg

The specific heat of water, evaluated at $T_{avg} = (38 + 10)/2 = 24°C$, is obtained from Table A-6; $c_p = 4179$ J/kg·K. We now can calculate all the unknowns.

The refrigerant flow rate is

$$\dot{m}_1 = \frac{\dot{V}_1}{v_1} = \frac{(5.5 \, \text{m}^3/\text{min})}{(0.1019 \, \text{m}^3/\text{kg})} \left(\frac{1 \, \text{min}}{60 \, \text{s}} \right) = 0.90 \, \frac{\text{kg}}{\text{s}}$$

The water mass flow rate is

$$\dot{m}_A = \frac{\dot{m}_1 (h_4 - h_1)}{c_p (T_B - T_A)} = \frac{\left(0.90 \, \frac{\text{kg}}{\text{s}} \right)(117.5 - 245.74) \, \frac{\text{kJ}}{\text{kg}} \left(\frac{1000 \, \text{J}}{1 \, \text{kJ}} \right)}{\left(4179 \, \frac{\text{J}}{\text{kg} \cdot \text{K}} \right)(10 - 38)°C} = 0.99 \, \frac{\text{kg}}{\text{s}}$$

The cooling capacity is

$$\dot{Q}_L = \dot{m}_1 (h_1 - h_4) = \left(0.90 \frac{\text{kg}}{\text{s}} \right)(245.74 - 117.5) \, \frac{\text{kJ}}{\text{kg}} \left(\frac{1 \, \text{kW}}{1 \, \text{kJ/s}} \right) = 115.4 \, \text{kW}$$

The compressor power input is

$$\dot{W}_{in} = \dot{m}_1 (h_{2s} - h_1)/\eta_C = \frac{\left(0.90 \, \frac{\text{kg}}{\text{s}} \right)}{0.82} (287.2 - 245.74) \, \frac{\text{kJ}}{\text{kg}} \left(\frac{1 \, \text{kW}}{1 \, \text{kJ/s}} \right) = 45.5 \, \text{kW}$$

The cycle coefficient of performance is

$$COP_{Ref} = \frac{\dot{Q}_L}{\dot{W}_{in}} = \frac{115.4 \, \text{kW}}{45.5 \, \text{kW}} = 2.54$$

Comments:

Note that if an ideal compressor were used, the compressor power would decrease to $\dot{W}_{in} = 37.3$ kW, and we would obtain an increase in the coefficient of performance to 3.09.

8.3 HEAT PUMPS

Consider Figure 8-2 once again. In a refrigeration cycle, the heat transfer rate $\dot{Q}_L$ is removed from the cool space; this is the heat transfer rate that is used. The heat rejected, $\dot{Q}_H$, is discarded and not put to any practical use. Note, however, $\dot{Q}_H$ is at a temperature higher than the environment to which it is rejected. Consider the cycle operation if we view if from a different perspective. Suppose $\dot{Q}_H$ is the energy we want to use, and we use this energy for heating. Now, $\dot{Q}_L$ is the unimportant heat transfer rate. When a refrigeration cycle is used for heating, it is called a *heat pump*.

As with the refrigeration cycle, the overall performance of a heat pump usually is described in terms of its capacity and *COP*. The latter quantity is defined as

$$COP_{HP} = \frac{\text{energy we want to use}}{\text{energy we purchase}} = \frac{\dot{Q}_H}{\dot{W}_{in}} = \frac{\dot{Q}_H}{\dot{Q}_H - \dot{Q}_L} = \frac{h_2 - h_3}{h_2 - h_1} \qquad (8\text{-}6)$$

In Eq. 8-6, the enthalpy differences are obtained by using Eq. 8-2 and Eq. 8-3.

We can show that the relationship between the *COP* of the refrigeration cycle and heat pump cycle is

$$COP_{HP} = COP_{Ref} + 1 \qquad (8\text{-}7)$$

Because the *COP* of a refrigeration cycle, COP_{Ref}, is always positive, $COP_{HP} > 1$. Typical values range from 3 to 6, which means that 3 to 6 units of useful heating are obtained for each unit of power input into the system.

Heat pumps are more expensive and complex than traditional gas- or oil-fired furnaces or electric resistance heaters. Nevertheless, because the COP_{HP} is always greater than unity, over the life of the heat pump total energy costs (electricity to drive the compressor) typically are lower than the cost of oil or natural gas. Hence, total life-cycle costs (capital plus operating costs) are also lower for the heat pump than for traditional heating methods. In the limit, the worst $COP_{HP} = 1$, which implies that the heat pump works as a resistance heater. However, there is another benefit associated with heat pumps. The same cycle used to heat can be used to cool, thus eliminating the need to have both a furnace and an air-conditioning system. The only needed modification to the cycle is to add a reversing valve, as shown in Figure 8-5; Figure 8-6 shows a typical arrangement for a heat pump/air-conditioning system in a house. Depending on the mode of operation (cooling or heating), either heat exchanger can serve as the evaporator or condenser.

In most heat pumps, the evaporator is placed outside the building, and atmospheric air is used to vaporize the refrigerant (which boils at a very low temperature). Because the *COP* and heating capacity decrease as the source temperature becomes low, facilities

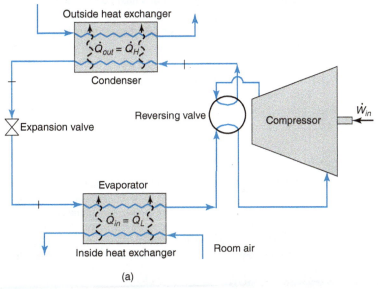

(a)

FIGURE 8-5 Continued

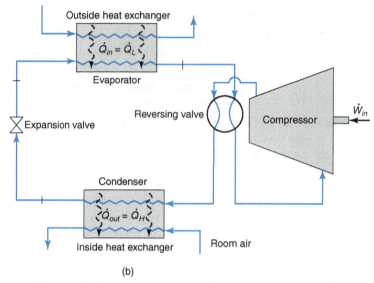

(b)

FIGURE 8-5 Heat pump with reversing valve showing summer and winter operation: (a) cooling mode; (b) heating mode.

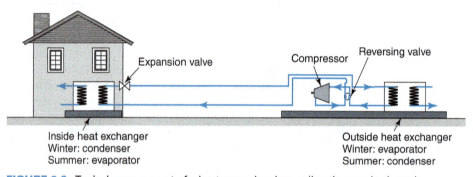

FIGURE 8-6 Typical arrangement of a heat pump in a house (heating mode shown).

that have air-source heat pumps often require a back up source of heat (e.g., gas or electric heating). To avoid this problem, ground-source (burying evaporator tubing in the ground, where the temperature is relatively constant) and water-source (submerging the evaporator in a large body of water whose temperature is relatively constant) heat pumps have become more popular in recent years. Although these systems are more expensive than air-source heat pumps, they have the advantage of a relatively constant-temperature energy source and do not need a backup system.

EXAMPLE 8-3 Ideal vapor-compression heat pump cycle

In the winter a large house requires 35 kW to maintain an indoor temperature of 21.3°C while the outdoor temperature is –6°C. An ideal vapor-compression heat pump cycle is to be designed with R-134a as the working fluid. To ensure the evaporator and condenser sizes (heat transfer areas) are not too large, a 10°C temperature difference is specified between the saturation temperature of the working fluid in each heat exchanger and the air flowing into it. Assume the refrigerant is saturated vapor at the evaporator exit and saturated liquid at the condenser exit.

a) Determine the compressor power (in kW).

b) Determine the coefficient of performance.

Approach:

Compressor power is determined by applying conservation of energy to the compressor, as shown in the figure. Sufficient information is given to evaluate all properties. Refrigerant flow, needed for the compressor power calculation, is unknown, so we use other given information. The heat output from the cycle is given ($\dot{Q}_H = 35$ kW), so application of conservation of energy to the condenser is used, and the only unknown is the refrigerant flow rate. Because both compressor power and input heat transfer rate are known, the coefficient of performance can be calculated.

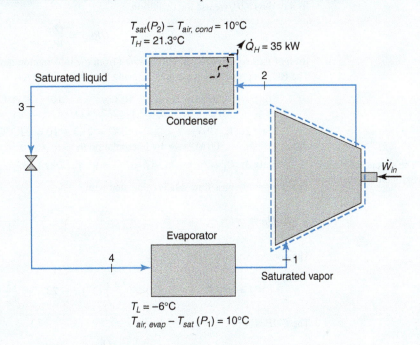

Assumptions:

A1. The complete system and individual components are steady.

A2. The compressor is adiabatic.
A3. Potential and kinetic energy effects are negligible.

A4. The compressor is isentropic.

Solution:

a) We determine the compressor power by using its control volume, as shown in the figure. Performing a mass balance and assuming steady conditions [A1] we obtain

$$\dot{m}_1 = \dot{m}_2 = \dot{m}$$

From an energy balance and assuming adiabatic conditions with negligible potential and kinetic energy effects [A2], [A3], we obtain

$$\dot{W} = -\dot{W}_{in} = \dot{m}\,(h_1 - h_2)$$

To evaluate the enthalpies, we need to determine the system pressures.

The difference between the saturation temperatures in the evaporator and the condenser and the air entering those two heat exchangers is 10°C. Hence,

$$\Delta T = 10°C = T_{air,evap} - T_{sat}(P_1) \quad \text{and} \quad T_{air,evap} = -6°C \quad \text{evaporator}$$

$$\Delta T = 10°C = T_{sat}(P_2) - T_{air,cond} \quad \text{and} \quad T_{air,cond} = 21.3°C \quad \text{condenser}$$

We have enough information to evaluate P_1 and P_2, as well as $h_1 = h_{g1}(P_1)$. Because we have an ideal system [A4], the second law gives: $s_2 = s_1$. We evaluate the enthalpy at s_2 and P_2; $h_2 = h_2(P_1, s_2 = s_1)$.

The mass flow rate is evaluated from the remaining piece of information given in the problem statement. The 35 kW is the power used to heat the house, so $\dot{Q}_H = 35$ kW. Applying the

A5. No work is done in the heat exchangers.

energy equation to the control volume around the refrigerant in the condenser [A1], [A3], [A5], we obtain:

$$\dot{Q} = -\dot{Q}_H = \dot{m}\,(h_3 - h_2)$$

We have enough information to evaluate the two enthalpies, and the only unknown in the equation is the refrigerant flow rate. Finally, combining all the information we can calculate the compressor power.

b) For the *COP*, we use the definition

$$COP_{Ref} = \frac{\dot{Q}_H}{\dot{W}_{in}}$$

Both of these quantities can be evaluated from the information developed above. The R-134a properties are evaluated using Tables A-14 and A-16.

> **State 1:** $T_{sat}(P_1) = T_{air,evap} - 10°C = -6 - 10 = -16°C \rightarrow P_1(T_{sat} = -16°C) = 0.157$ MPa $\rightarrow h_1 = h_{g1}(P_1) = 237.74$ kJ/kg, $s_1 = s_{g1} = 0.9298$ kJ/kg·K

> **State 2:** $T_{sat}(P_2) = T_{air,cond} + 10°C = 21.3 + 10 = 31.3°C \rightarrow P_2(T_{sat} = 31.3°C) = 0.800$ MPa $\rightarrow$ by interpolation $h_2 = h_2(P_2, s_2 = s_1) = 271.3$ kJ/kg·K

A6. The subcooled liquid approximation is valid.

> **State 3:** $P_3 = P_2, \rightarrow h_3 \approx h_{f3}(P_3) = 93.42$ kJ/kg [A6]

The refrigerant mass flow rate is calculated with

$$\dot{Q}_H = \dot{m}\,(h_2 - h_3) \rightarrow \dot{m} = \frac{\dot{Q}_H}{h_2 - h_3} = \frac{35\,\text{kW}}{(271.3 - 93.42)\,\dfrac{\text{kJ}}{\text{kg}}}\left(\frac{1\,\text{kJ/s}}{1\,\text{kW}}\right) = 0.197\,\frac{\text{kg}}{\text{s}}$$

The compressor power is

$$\dot{W}_{in} = \dot{m}\,(h_2 - h_1) = \left(0.197\,\frac{\text{kg}}{\text{s}}\right)(271.3 - 237.74)\,\frac{\text{kJ}}{\text{kg}}\left(\frac{1\,\text{kW}}{1\,\text{kJ/s}}\right) = 6.61\,\text{kW}$$

The *COP* is

$$COP_{Ref} = \frac{\dot{Q}_H}{\dot{W}_{in}} = \frac{35\,\text{kW}}{6.61\,\text{kW}} = 5.29$$

Comments:

Note that if this cycle were used for cooling in the summer, its coefficient of performance would be $COP_{Ref} = COP_{HP} - 1 = 5.29 - 1 = 4.29$. With the same compressor power input, the cooling capacity would be $\dot{Q}_L = \dot{W}_{in}COP_{Ref} = 6.61\,\text{kW} \times 4.29 = 28.36\,\text{kW}$.

8.4 THE RANKINE CYCLE

Power cycles are used to convert one type of energy into another, more usable form. In heat power cycles, the chemical energy stored in fossil fuels (i.e., coal, natural gas, oil), wood, or other substances is released through a combustion process. Alternatively, the energy contained in uranium is released through a nuclear reaction and used to drive a heat power cycle. In both of these types of processes, thermal energy is produced. Whatever the source of the energy, the engineer's objective is to convert this thermal energy into mechanical energy or electricity though the use of appropriate devices. In transportation (e.g., automobiles, trucks, trains, airplanes, ships), the objective of the power cycle is the production of mechanical energy to propel the vehicle. In electricity production, the thermal energy is first converted to mechanical energy and then the mechanical energy is converted to electrical energy in a generator.

Different power cycles are used for different purposes. For transportation, a lightweight power plant with a high power density (net power output/mass of engine) is needed, and the power plant should be able to change operating conditions quickly. For an electrical power plant, the goal is to produce large amounts of electricity for a long time without shutting down; so called *base-loaded power plants* handle the majority of the electric power consumed, and reliability takes precedence over compactness. However, at certain times during the day, demand for electric power surges quickly. Additional power is needed, and so-called *peaking power plants* are used to respond rapidly to the changing demand. In this and following sections, we discuss several different power cycles and point out their applications.

Vapor power plants use a working fluid (most often water, but other fluids have been used) that is alternately vaporized and condensed. The most common vapor power plant (a base-loaded power plant) is the *Rankine cycle*. A schematic of the basic components of a simple ideal Rankine cycle is shown in Figure 8-7. The Carnot heat power cycle has four ideal processes, and the simple Rankine cycle has four processes:

1–2	Compression of the working fluid with work input
2–3	Heat addition to the working fluid, which is vaporized
3–4	Expansion of the working fluid with work output
4–1	Heat rejection from the working fluid, which is condensed

For a real Rankine cycle, none of these processes would be reversible. However, for an *ideal Rankine cycle*, we assume that the turbine and pumps are reversible and adiabatic (i.e., isentropic), that no pressure drop occurs across the boiler or condenser, that the fluid exits the condenser as a saturated liquid, and that the fluid exits the boiler as a saturated vapor.

Compare the T-s diagram of the ideal Rankine cycle in Figure 8-8 with that of a Carnot cycle shown in Figure 8-9. In the Carnot cycle, all heat is rejected at a constant temperature (process 4–1); the same is true of the ideal Rankine cycle. However, because it is difficult, if not impossible, to pump a two-phase mixture isentropically from state 1 to state 2, as would be required in the Carnot cycle, the mixture is condensed completely in the ideal Rankine cycle (state 1). The heat addition in the Carnot cycle is isothermal

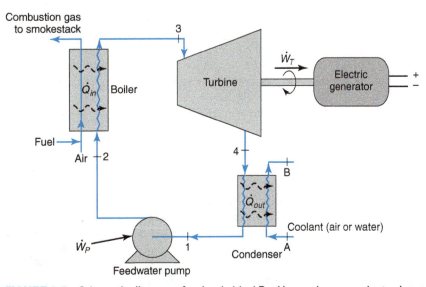

FIGURE 8-7 Schematic diagram of a simple ideal Rankine cycle power plant using a fossil fuel.

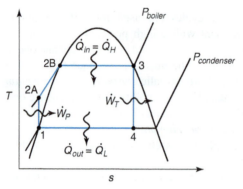
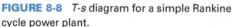

FIGURE 8-8 *T-s* diagram for a simple Rankine cycle power plant.

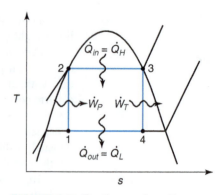

FIGURE 8-9 *T-s* diagram for a Carnot cycle on a vapor dome.

(process 2–3), but the heat addition in the ideal Rankine cycle is not. In the ideal Rankine cycle, the temperature of the liquid working fluid increases as heat is added (nonisothermal heat addition) until the saturation temperature (state 2B) is reached; after that point, the liquid vaporizes, and heat addition is isothermal. Because of the nonisothermal heat addition (process 2A–2B), the cycle thermal efficiency of the Rankine cycle is less than that of the Carnot cycle operating at the same high and low temperatures. If we now use turbines and pumps that have isentropic efficiencies less than 100%, then the real Rankine cycle thermal efficiency will be even lower than the ideal Rankine cycle.

The analysis of power cycles generally focuses on two cycle-aggregate quantities: the *cycle thermal efficiency*, η_{cycle}, and the *net power output*, $\dot{W}_{net}$. As introduced in Chapter 7, cycle thermal efficiency is defined as

$$\eta_{cycle} = \frac{\text{energy we want to use}}{\text{energy we purchase}} = \frac{\dot{W}_{net}}{\dot{Q}_H} = 1 - \frac{\dot{Q}_L}{\dot{Q}_H} \qquad (8\text{-}8)$$

and η_{cycle} is always less than unity. The net power is defined as:

$$\dot{W}_{net} = \dot{W}_{turbine} - \dot{W}_{pump} = \dot{W}_T - \dot{W}_P \qquad (8\text{-}9)$$

To evaluate cycle thermal efficiency and net power output, we must analyze individual components of the cycle. First, we apply conservation of energy to a control volume around the boiler. Assuming steady state, neglecting changes in kinetic and potential energy, and with no work done on or by the boiler (recognizing from conservation of mass that mass flow in is equal to mass flow out), we obtain

$$\dot{Q}_H = \dot{m}\,(h_3 - h_{2A}) \qquad \text{boiler} \qquad (8\text{-}10)$$

(As in the vapor-compression refrigeration cycle, we use a subscript to indicate the direction of the energy flow, so that all power and heat transfer terms are positive.) A similar analysis can be performed on each of the other three components in the system. In each case, kinetic and potential energy are ignored. The turbine and pump are assumed to be adiabatic, and no work is done on or by the condenser. These assumptions result in the following expressions:

$$\dot{W}_T = \dot{m}\,(h_3 - h_4) \qquad \text{turbine} \qquad (8\text{-}11)$$

$$\dot{Q}_L = \dot{m}\,(h_4 - h_1) \qquad \text{condenser} \qquad (8\text{-}12)$$

$$\dot{W}_P = \dot{m}\,(h_{2A} - h_1) \qquad \text{pump} \qquad (8\text{-}13)$$

Substituting Eq. 8-10 and Eq. 8-12 into Eq. 8-8 and noting that all the mass flow rates cancel out, we can express the cycle thermal efficiency as

$$\eta_{cycle} = \frac{(h_3 - h_4) - (h_{2A} - h_1)}{(h_3 - h_{2A})} = 1 - \frac{h_4 - h_1}{h_3 - h_{2A}} \qquad (8\text{-}14)$$

In addition to the net power output and the thermal efficiency, the *back work ratio*, *BWR*, is used to characterize a power cycle and is defined as the ratio of the pump work (Eq. 8-13) to the turbine work (Eq. 8-11):

$$BWR = \frac{\dot{W}_P}{\dot{W}_T} = \frac{h_2 - h_1}{h_3 - h_4} \qquad (8\text{-}15)$$

For vapor power cycles, such as the Rankine cycle, the power needed to compress a liquid is quite small compared to the power extracted from the expanding steam. Hence, the *BWR* is low, in the order of 1–3%. However, when we study gas power cycles (beginning in Section 8.6 with the Brayton cycle), we will see that the *BWR* can be quite high (on the order of 40–80%).

EXAMPLE 8-4   Ideal Rankine cycle

An ideal Rankine cycle uses water as the working fluid. The boiler operates at 4 MPa and produces saturated vapor. Saturated liquid exits the condenser at 20 kPa. Net power produced by the cycle is 50 MW.

a) Determine the cycle thermal efficiency.

b) Determine the mass flow rate of the steam.

c) Determine the heat transfer rate into the boiler.

d) Determine the back work ratio.

Approach:

Cycle thermal efficiency is $\eta_{cycle} = \dot{W}_{net}/\dot{Q}_{in}$. Net power is given. The input heat transfer rate is determined from an energy balance on the boiler, but the mass flow is required. Because we know that $\dot{W}_{net} = \dot{W}_T - \dot{W}_P$, and the work terms can be evaluated from conservation of energy applied to the turbine and to the pump, we can calculate mass flow rate.

Assumptions:

Solution:

Given information is shown in the diagram.

a) Cycle thermal efficiency is defined as

$$\eta_{cycle} = \frac{\dot{W}_{net}}{\dot{Q}_{in}}$$

A1. The complete system and individual components are steady.

Net power, $\dot{W}_{net}$, is given in the problem statement. The input heat transfer rate is determined by evaluating the energy equation using a control volume around the water flowing through the boiler, assuming steady-state conditions, no work, and negligible potential and kinetic energy effects [A1], [A2], [A3]:

$$\dot{Q} = \dot{Q}_{in} = \dot{m}(h_3 - h_2)$$

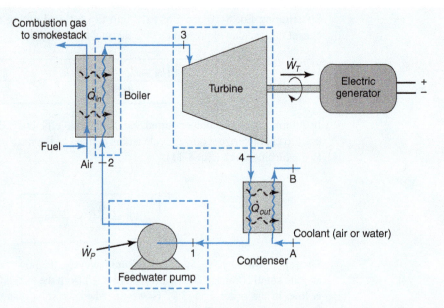

Using the net power specified in the problem statement and knowing that $\dot{W}_{net} = \dot{W}_T - \dot{W}_P$, we can obtain the mass flow rate. Applying the mass and energy balance equations to the control volumes around the turbine and pump [A1], [A3], [A4], we obtain

A2. No work is done in the boiler or condenser.

A3. Potential and kinetic energy effects are negligible.

A4. The pump and turbine are adiabatic.

$$\dot{m}_1 = \dot{m}_2 = \dot{m}_3 = \dot{m}_4 = \dot{m}$$

$$\dot{W} = \dot{W}_T = \dot{m}(h_3 - h_4) \quad \text{and} \quad \dot{W} = -\dot{W}_P = \dot{m}(h_1 - h_2)$$

We have enough given information to evaluate all four enthalpies, and mass flow rate cancels from the cycle efficiency expression.

For the back work ratio, $BWR = \dot{W}_P / \dot{W}_T$. We have developed expressions for both of these quantities, so we can begin the evaluation of all parameters. The water properties are evaluated using Table A-11.

State 1: $P_1 = 20$ kPa $\rightarrow h_1 = h_{f1}(P_1) = 251.4$ kJ/kg, $v_1 = v_{f1}(P_1) = 0.001017$ m³/kg

A5. The pump and turbine are isentropic.

A6. The subcooled liquid approximation is valid.

State 2: $P_2 = 4$ MPa $\rightarrow$ Because the pump is ideal [A5], $s_2 = s_1$, and the outlet enthalpy is given by (see Eq. 7-36) [A6]

$$h_2 \approx h_1 + v_1(P_2 - P_1)$$

$$= 251.4 \frac{\text{kJ}}{\text{kg}} + \left(0.001017 \frac{\text{m}^3}{\text{kg}}\right)(4000 - 20)\text{kPa}\left(\frac{1 \text{ kN/m}^2}{1 \text{ kPa}}\right)\left(\frac{1 \text{ kJ}}{1 \text{ kN} \cdot \text{m}}\right)$$

$$h_2 = 251.4 + 4.04 = 255.45 \frac{\text{kJ}}{\text{kg}}$$

State 3: $P_3 = 4$ MPa $\rightarrow h_3 = h_{g3}(P_3) = 2801.4$ kJ/kg, $s_3 = s_{g3}(P_3) = 6.0701$ kJ/kg·K

State 4: $P_4 = 20$ kPa $\rightarrow$ Because the turbine is ideal [A5], $s_4 = s_3$, so that $h_4 = h_4(P_4, s_4 = s_3)$

Note that

$$x_4 = \frac{h_4 - h_{f4}}{h_{g4} - h_{f4}} = \frac{s_4 - s_{f4}}{s_{g4} - s_{f4}} = \frac{6.0701 - 0.832}{7.9085 - 0.832} = 0.740$$

Solving for h_4,

$$h_4 = h_{f4} + x_4(h_{g4} - h_{f4}) = 251.4 + 0.740(2609.7 - 251.4) = 1996.5 \frac{\text{kJ}}{\text{kg}}$$

b) Mass flow rate is obtained from

$$\dot{W}_{net} = \dot{W}_T - \dot{W}_P = \dot{m}\,(h_3 - h_4) - \dot{m}\,(h_2 - h_1) \rightarrow \dot{m} = \frac{\dot{W}_{net}}{(h_3 - h_4) - (h_2 - h_1)}$$

$$\dot{m} = \frac{50,000\,\text{kW}}{\left[(2801.4 - 1996.5) - (255.45 - 251.4)\right]\frac{\text{kJ}}{\text{kg}}\left(\frac{1\,\text{kW}}{1\,\text{kJ/s}}\right)} = 62.4\,\frac{\text{kg}}{\text{s}}$$

The turbine power output is

$$\dot{W}_T = \dot{m}\,(h_3 - h_4) = \left(62.4\,\frac{\text{kg}}{\text{s}}\right)(2801.4 - 1996.5)\frac{\text{kJ}}{\text{kg}}\left(\frac{1\,\text{kW}}{1\,\text{kJ/s}}\right) = 50,226\,\text{kW}$$

The pump work input is

$$\dot{W}_P = \dot{m}\,(h_2 - h_1) = \left(62.4\,\frac{\text{kg}}{\text{s}}\right)(255.45 - 251.4)\frac{\text{kJ}}{\text{kg}}\left(\frac{1\,\text{kW}}{1\,\text{kJ/s}}\right) = 252.7\,\text{kW}$$

c) The input heat transfer rate is

$$\dot{Q}_{in} = \dot{m}(h_3 - h_2) = \left(62.4\,\frac{\text{kg}}{\text{s}}\right)(2801.4 - 255.45)\frac{\text{kJ}}{\text{kg}}\left(\frac{1\,\text{kW}}{1\,\text{kJ/s}}\right) = 158,870\,\text{kW}$$

So, finally, the cycle thermal efficiency is

$$\eta_{cycle} = \frac{\dot{W}_{net}}{\dot{Q}_{in}} = \frac{(50,226 - 252.7)\,\text{kW}}{158,870\,\text{kW}} = 0.315$$

d) and the back work ratio is

$$BWR = \frac{\dot{W}_P}{\dot{W}_T} = \frac{252.7\,\text{kW}}{50,226\,\text{kW}} = 0.005 = 0.5\%$$

Comments:

The cycle thermal efficiency of 0.315 means that for each unit of input energy, 31.5% of it is converted to useful work and 68.5% of it is discarded. Note also that the power required by the pump is nearly negligible compared to the power output by the turbine.

The Carnot cycle efficiency ($\eta_{Carnot} = 1 - T_L/T_H$) increases when the temperature at which heat is added is increased or when the temperature at which heat is rejected is decreased. The same is true for the Rankine cycle if we consider the average temperature at which heat is added or rejected from the cycle. In addition, the net work per unit mass increases as the separation between the high and low temperatures increases. Figure 8-10 shows two T-s diagrams of an ideal Rankine cycle with different condensing and boiling pressures. From Chapter 7, we know that the area enclosed by curves on a T-s diagram is equal to the net work per unit mass produced by the cycle; the area under a curve is equal to heat transfer per unit mass. As shown in Figure 8-10a, a lower condenser pressure results in an increase in cycle efficiency because of a lower heat-rejection temperature. The net work per unit mass increases; compare area 1–2–3–4–1 to the area 1'–2'–3–4'–1'. The increase in work per unit mass is equal to the area 1'–2'–2–1–4–4'–1'. The heat transfer per unit mass added to the original cycle is equal to the area under the curve 2–3, and for

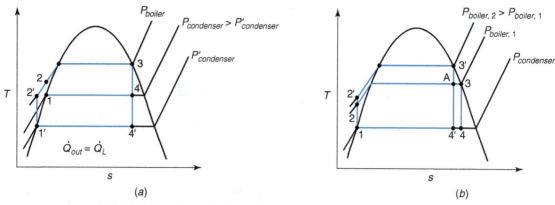

FIGURE 8-10 Effect of boiler and condenser pressure on Rankine cycle.

the cycle with the lower condenser pressure, the heat transfer per unit mass added is the area under the curve 2′–2–3. Note that the heat added must increase slightly (area under curve 2′–2); however, the increase in net work is greater than the increase in heat added, so the efficiency increases. The temperature of the ultimate heat sink, the ambient air or water surrounding the condenser, dictates the lowest feasible condenser pressure. For condensing water, this results in a condenser operating pressure below atmospheric.

For an increased boiler pressure (Figure 8-10b), the cycle efficiency increases because of the higher average temperature at which heat is added. The work per unit mass may increase or decrease; compare the relative magnitudes of the areas A–3–4–4′–A and 2–2′–3′–A–2. The first area (A–3–4–4′–A) is the *decrease* in work per unit mass when the boiler pressure is increased; the latter area (2–2′–3′–A–2) is the *increase* in work per unit mass associated with the same boiler pressure increase. The heat transfer per unit mass added to the lower boiler pressure cycle is the area under the curve 2–A–3, and for the higher pressure cycle, the heat transfer added is the area 2′–3′. The combination of change in net work and heat added results in a higher cycle efficiency.

The ideal Rankine cycle assumes an isentropic turbine and pump; no pressure drops in the boiler, condenser, or piping; and no extraneous heat losses anywhere in the system. A real Rankine cycle does have irreversibilities associated with it. Nonisentropic expansion through the turbine is the greatest source of loss in the cycle, followed by the nonisentropic compression in the pump. Pressure losses in piping, heat exchangers, valves, and so on are of secondary importance and would be taken into account in a detailed thermodynamic analysis of a real system. As shown on Figure 8-11, the process 3–4s represents an isentropic

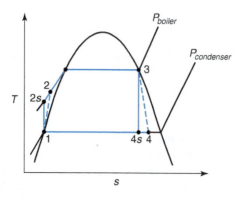

FIGURE 8-11 *T-s* diagram for Rankine cycle with nonisentropic turbine and pump.

expansion through the turbine of a Rankine cycle. Process 3–4 represents the actual process when the turbine isentropic efficiency is less than 100%. Likewise, process 1–2s represents the isentropic compression in the pump, and 1–2 is the actual process with a pump isentropic efficiency less than 100%.

In Chapter 7, we defined isentropic efficiencies of turbines and pumps, respectively, as:

$$\eta_T = \frac{h_3 - h_4}{h_3 - h_{4s}} \qquad \text{turbine} \qquad (8\text{-}16)$$

$$\eta_P = \frac{h_{2s} - h_1}{h_2 - h_1} \qquad \text{pump} \qquad (8\text{-}17)$$

Solving Eq. 8-16 and Eq. 8-17 for the actual enthalpy changes across each device and substituting into Eq. 8-14 results in

$$\eta_{cycle} = \frac{(h_3 - h_4) - (h_2 - h_1)}{(h_3 - h_2)} = \frac{\eta_T (h_3 - h_{4s}) - (h_{2s} - h_1)/\eta_P}{(h_3 - h_2)} = \frac{\dot{W}_{net}/\dot{m}}{\dot{Q}_H/\dot{m}} \qquad (8\text{-}18)$$

Real turbine and pump isentropic efficiencies (η_T and η_P, respectively) are less than 100%. For a given cycle, h_1, h_{2s}, h_3, and h_{4s} are fixed irrespective of the magnitude of the turbine and pump efficiencies. From Eq. 8-18 we can see that the work output from the turbine and the work input to the pump are less than those for the ideal Rankine cycle, and the heat transfer input to the cycle does not change. Thus the cycle thermal efficiency with nonideal turbines and pumps is less than the ideal cycle efficiency.

Another loss is associated with the combustion of the fossil fuel used to vaporize the water in the boiler. When the coal, natural gas, or oil is burned, products of combustion, such as water vapor, carbon dioxide, NOx, and SOx, are formed, and a minor amount of the hydrocarbons escape unburned. We want to extract as much energy as possible from these hot gases to boil the water. However, if we cool the exhaust gas too much in the boiler, the water vapor can combine with other constituents in the products of combustion and form acids, which will attack and corrode the metal it comes into contact with. Hence, we define a boiler efficiency to take into account that portion of the energy supplied by the fuel that is not transferred to the water in the boiler:

$$\eta_{boiler} = \frac{\dot{Q}_H}{\dot{m}_F HV} = \frac{\dot{m}(h_3 - h_2)}{\dot{m}_F HV} \qquad (8\text{-}19)$$

where $\dot{m}_F$ is the fuel flow rate and HV, the heating value of the fuel, is the energy released per unit mass during complete combustion of the fuel. Thus the overall conversion efficiency from fuel input to net power output is

$$\eta_{overall} = \frac{\dot{W}_{net}}{\dot{m}_F HV} = \eta_{boiler} \eta_{cycle} \qquad (8\text{-}20)$$

While the cycle thermal efficiency is useful for evaluation and comparison of the overall performance of a cycle, in practice operators of thermal power plants are more interested in the energy input to the plant for each kilowatt-hour of output electricity.

One measure most often used to describe plant performance is *net station heat rate*, which has units of Btu/kWh. This quantity can be determined by measuring the net electrical output in kilowatt-hours over a specified time period, t, and the total fossil fuel input during the same time period. Thus, we can calculate this quantity with

$$\text{net station heat rate} = \frac{\text{total fossil fuel input}}{\text{net electrical power produced}} = \frac{\dot{m}_F t HV}{\dot{W}_{net} t} = \frac{1}{\eta_{overall}} = \frac{1}{\eta_{boiler} \eta_{cycle}}$$

As given, this expression is dimensionless, so to put this quantity into correct units, we need to use the conversion from Btu/h to kW:

$$\boxed{\text{net station heat rate} = \frac{3413 \, \text{Btu}/\text{kWh}}{\eta_{boiler} \eta_{cycle}}} \tag{8-21}$$

Note that we want the cycle thermal efficiency to be as large as possible. However, we want the net station heat rate to be as low as possible.

EXAMPLE 8-5   Rankine cycle with inefficiencies

A Rankine cycle uses water as the working fluid. The boiler operates at 8 MPa and produces saturated vapor. Saturated liquid exits the condenser at 7.5 kPa. The pump has an isentropic efficiency of 82%, and the turbine has an isentropic efficiency of 88%. The steam flow rate is 2.8×10^4 kg/h. Cooling water enters the condenser at 20°C and leaves at 38°C.

a) Determine the net power produced (in kW).

b) Determine the heat transfer rate into the boiler (in kW).

c) Determine the cycle thermal efficiency.

d) Determine the net station heat rate if the boiler efficiency is 92%.

e) Determine the cooling water flow rate (in kg/h).

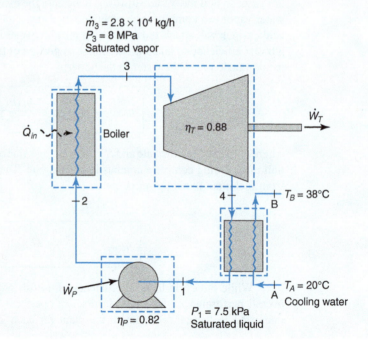

Approach:

The approach to this solution is very similar to that used in Example 8-4; the primary difference is in the evaluation of the fluid properties, which now requires the use of the isentropic efficiency. The net power produced can be determined from energy balances around the turbine and pump, as can the heat transfer rate into the boiler. With both of these known, the cycle thermal efficiency can be calculated. To obtain the cooling water flow rate, we apply conservation of energy to the condenser because that is the only location where cooling water is present. Sufficient information is given to evaluate all properties.

Assumptions:

Solution:

a) The net power produced by the cycle is $\dot{W}_{net} = \dot{W}_T - \dot{W}_P$. Using the control volumes defined in the figure around the turbine and pump, and applying conservation of mass and energy assuming steady-state conditions, adiabatic conditions, and negligible potential and kinetic energy effects [A1], [A2], [A3], we obtain

A1. The complete system and individual components are steady.
A2. Potential and kinetic energy effects are negligble.
A3. The pump, turbine, and condenser are adiabatic.

$$\dot{m}_1 = \dot{m}_2 = \dot{m}_3 = \dot{m}_4 = \dot{m}$$
$$\dot{W} = \dot{W}_T = \dot{m}\,(h_3 - h_4)$$
$$\dot{W} = -\dot{W}_P = \dot{m}\,(h_1 - h_2)$$

Mass flow rate is given, and the inlet enthalpies (h_1 and h_3) to both devices can be evaluated with the given information. The outlet enthalpies (h_2 and h_4) can be determined, because using the definitions of isentropic efficiency we can obtain

$$h_4 = h_3 - \eta_T\,(h_3 - h_{4s}) \quad \text{and} \quad h_2 = h_1 + \frac{h_{2s} - h_1}{\eta_P}$$

b) The input heat transfer rate is obtained by applying conservation of energy to the water flowing through the boiler assuming steady state, negligible potential and kinetic energy effects, and no work [A1], [A2], [A4]:

A4. No work occurs in the boiler or condenser.

$$\dot{Q} = \dot{Q}_{in} = \dot{m}\,(h_3 - h_2)$$

All these quantities are known or can be evaluated.

c) With net power and input heat transfer rate known, the cycle thermal efficiency is calculated with $\eta_{cycle} = \dot{W}_{net}/\dot{Q}_{in}$.

d) The net station heat rate is then determined from Eq. 8-21.

e) The cooling water flow rate is obtained by applying conservation of mass and energy to the control volume drawn around the condenser as shown above, and assuming steady state, negligible potential and kinetic energy effects, adiabatic conditions, and no work [A1], [A2], [A3], [A4], we obtain:

$$\dot{m}_A = \dot{m}_B = \dot{m}_{\substack{cooling \\ water}} \quad \text{and} \quad \dot{m}_4 h_4 + \dot{m}_A h_A - \dot{m}_1 h_1 - \dot{m}_B h_B = 0$$

Combining these two equations and assuming the cooling water is incompressible with constant specific heat [A5], [A6] (from Table A-6 evaluated at the average temperature), so that $\Delta h = c_p \Delta T$, we obtain

A5. The liquid water is incompressible.
A6. The water has a constant specific heat.

$$\dot{m}_{\substack{cooling \\ water}} = \frac{\dot{m}\,(h_4 - h_1)}{c_p\,(T_B - T_A)}$$

Now all quantities can be evaluated. The water properties are determined from Table A-11.

State 1: $P_1 = 7.5$ kPa, saturated liquid → $h_1 = h_{f1}(P_1) = 168.79$ kJ/kg, $v_1 = v_{f1}(P_1) = 0.001008$ m^3/kg

A7. The pump and turbine are isentropic.
A8. The subcooled liquid approximation is valid.

State 2: $P_2 = 8$ MPa $\rightarrow$ First, we evaluate the pump as if it were ideal [A7], $s_2 = s_1$, and the exit enthalpy can be determined with [A8]:

$$h_{2s} \approx h_1 + v_1 (P_2 - P_1)$$

$$= 168.79 \frac{kJ}{kg} + \left(0.001008 \frac{m^3}{kg}\right) (8000 - 7.5) \text{ kPa} \left(\frac{1 \text{ kN/m}^2}{1 \text{ kPa}}\right) \left(\frac{1 \text{ kJ}}{1 \text{ kN·m}}\right)$$

$$h_{2s} = 168.79 + 8.06 = 176.85 \frac{kJ}{kg}$$

Second, we use the definition of isentropic efficiency

$$h_2 = h_1 + \frac{h_{2s} - h_1}{\eta_P} = 168.79 + \frac{176.85 - 168.79}{0.82} = 178.6 \frac{kJ}{kg}$$

State 3: $P_3 = 8$ MPa, saturated vapor $\rightarrow$ $h_3 = h_{g3}(P_3) = 2758.0$ kJ/kg, $s_3 = s_{g3}(P_3) = 5.7432$ kJ/kg·K

State 4: $P_4 = 7.5$ kPa $\rightarrow$ First, we evaluate the turbine as if it were ideal ([A7]), $s_4 = s_3$, so that $h_{4s} = h_{4s}(P_4, s_4 = s_3)$

Note that

$$x_{4s} = \frac{h_{4s} - h_{f4}}{h_{g4} - h_{f4}} = \frac{s_{4s} - s_{f4}}{s_{g4} - s_{f4}} = \frac{5.7432 - 0.5764}{8.7515 - 0.5764} = 0.673$$

Solving for h_{4s}

$$h_{4s} = h_{f4} + x_{4s} \left(h_{g4} - h_{f4}\right) = 168.75 + 0.673 (2574.8 - 168.75) = 1788.0 \frac{kJ}{kg}$$

Second, we use the definition of isentropic efficiency

$$h_4 = h_3 - \eta_T (h_3 - h_{4s}) = 2758.0 - 0.88 (2758.0 - 1788.0) = 1904.4 \frac{kJ}{kg}$$

Note that

$$x_4 = \frac{h_4 - h_{f4}}{h_{g4} - h_{f4}} = \frac{1904.4 - 168.75}{2574.8 - 168.75} = 0.721$$

Turbine power is obtained from

$$\dot{W}_T = \dot{m} (h_3 - h_4)$$

$$= \left(2.8 \times 10^4 \frac{kg}{h}\right) \left(\frac{1 \text{ h}}{3600 \text{ s}}\right) (2758.0 - 1904.4) \frac{kJ}{kg} \left(\frac{1 \text{ kW}}{1 \text{ kJ/s}}\right) = 6639 \text{ kW}$$

Pump power is obtained from

$$\dot{W}_P = \dot{m} (h_2 - h_1) = \left(2.8 \times 10^4 \frac{kg}{h}\right) \left(\frac{1 \text{ h}}{3600 \text{ s}}\right) (178.6 - 168.79) \frac{kJ}{kg} \left(\frac{1 \text{ kW}}{1 \text{ kJ/s}}\right) = 76.3 \text{ kW}$$

Net cycle power is

$$\dot{W}_{net} = \dot{W}_T - \dot{W}_P = 6639 - 76.3 = 6562.8 \text{ kW}$$

Heat transfer rate into the boiler is obtained from

$$\dot{Q}_{in} = \dot{m}(h_3 - h_2)$$

$$= \left(2.8 \times 10^4 \, \frac{kg}{h}\right)\left(\frac{1\,h}{3600\,s}\right)(2758.0 - 178.6) \, \frac{kJ}{kg}\left(\frac{1\,kW}{1\,kJ/s}\right) = 20{,}062 \, kW$$

The cycle thermal efficiency is

$$\eta_{cycle} = \frac{\dot{W}_{net}}{\dot{Q}_{in}} = \frac{6562.8 \, kW}{20062 \, kW} = 0.327$$

The net station heat rate is

$$\text{net station heat rate} = \frac{3413 \, Btu/kWh}{\eta_{boiler}\,\eta_{cycle}} = \frac{3413 \, Btu/kWh}{0.92 \times 0.327} = 11{,}345 \, Btu/kWh$$

The cooling water flow rate is

$$\dot{m}_{\substack{cooling \\ water}} = \frac{\dot{m}(h_4 - h_1)}{c_p(T_B - T_A)} = \frac{\left(2.8 \times 10^4 \, \frac{kg}{h}\right)(1904.4 - 168.79) \, \frac{kJ}{kg}\left(\frac{1000\,J}{1\,kJ}\right)}{\left(4176 \, \frac{J}{kg \cdot K}\right)(38 - 20)\,K} = 646{,}500 \, \frac{kg}{h}$$

Comments:

If the pump and turbine were isentropic ($\eta_s = 100\%$), the cycle thermal efficiency would be

$$\dot{W}_T = \dot{m}(h_3 - h_{4s}) = \left(2.8 \times 10^4 \, \frac{kg}{h}\right)\left(\frac{1\,h}{3600\,s}\right)(2758.0 - 1788.0) \, \frac{kJ}{kg}\left(\frac{1\,kW}{1\,kJ/s}\right) = 7544 \, kW$$

$$\dot{W}_P = \dot{m}(h_{2s} - h_1) = \left(2.8 \times 10^4 \, \frac{kg}{h}\right)\left(\frac{1\,h}{3600\,s}\right)(176.85 - 168.79) \, \frac{kJ}{kg}\left(\frac{1\,kW}{1\,kJ/s}\right) = 62.7 \, kW$$

$$\dot{Q}_{in} = \dot{m}(h_3 - h_{2s}) = \left(2.8 \times 10^4 \, \frac{kg}{h}\right)\left(\frac{1\,h}{3600\,s}\right)(2758.0 - 176.85) \, \frac{kJ}{kg}\left(\frac{1\,kW}{1\,kJ/s}\right) = 20{,}076 \, kW$$

$$\eta_{cycle} = \frac{\dot{W}_{net}}{\dot{Q}_{in}} = \frac{(7544 - 62.7) \, kW}{20076 \, kW} = 0.373$$

Note that the inefficiencies of the pump and turbine reduced the cycle efficiency from 0.373 to 0.327.

Also, compare the present cycle efficiency (using the isentropic pump and turbine) with that given in Example 8-4. Raising the boiler pressure and reducing the condenser pressure results in an increase in the cycle efficiency from 0.315 to 0.373.

Practical considerations affect the operating conditions of a Rankine cycle. For example, we have assumed that saturated liquid exits the condenser. In practice, the flow is probably slightly subcooled. While this requires more heat to be added in the boiler to raise the water to saturation, the subcooling is needed to protect the pump. Water at a given temperature, T, enters a pump; the water accelerates, and its pressure decreases. The pressure it reaches depends on the pump design. If the pressure decreases far enough,

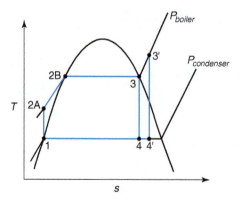

FIGURE 8-12 Effect of superheating on Rankine cycle.

then the saturation pressure of the water, $P_{sat}(T)$, can be reached. Should that happen or if the operating pressure in the pump falls below $P_{sat}(T)$, then the water will boil and vapor bubbles will form. The liquid water and vapor bubbles pass from the low-pressure region of the pump to the high-pressure region, the vapor bubbles rapidly collapse, and a pressure wave is transmitted through the liquid. Thus, when pumping a liquid, we want to avoid *cavitation* (the very rapid formation and collapse of vapor bubbles in the liquid), which can quickly and severely damage a pump impeller. Cavitation and its prevention are discussed in Section 9.11 (available on the web).

A second practical consideration is related to the outlet condition from the turbine. As shown on Figure 8-10b, an increase in the boiler pressure causes a decrease in the turbine outlet quality. A steam turbine can withstand some level of liquid drops flowing through it, but if the quality is too low, water drops can damage the turbine blades by erosion. Coatings and special metals are used to alleviate this problem. In addition, damage can be minimized if we *superheat* the steam before it enters the turbine. When saturated steam enters a turbine and expands (Figure 8-12, process 3–4), the exit state may be well under the vapor dome; the exit quality is relatively low. We superheat the steam (process 3–3') so that after the expansion (process 3'–4'), the turbine outlet quality increases. Added benefits of superheating are an increase in cycle thermal efficiency because of a higher average temperature at which heat is added to the cycle and an increase in work per unit mass. Metallurgical capabilities of the turbine blades limit the maximum temperature (in the range of 600–650°C) that can be used at the turbine inlet.

EXAMPLE 8-6 Rankine cycle with superheating

For the Rankine cycle described in Example 8-5, the boiler exit temperature is raised to 550°C while all other conditions are kept the same.

a) Determine the net power produced (in kW).

b) Determine the heat transfer rate into the boiler (in kW).

c) Determine the cycle thermal efficiency.

d) Determine the net station heat rate (in Btu/kWh).

e) Determine the cooling water flow rate (in kg/h).

Approach:

The analysis of this problem is identical to that in Example 8-5. The only difference is in the evaluation of the turbine inlet and outlet enthalpies.

Assumptions:

Same as in Example 8–5

Solution:

Following Example 8-5, we have:

State 1: Same as before $\rightarrow h_1 = 168.79$ kJ/kg

State 2: Same as before $\rightarrow h_2 = 178.6 \dfrac{\text{kJ}}{\text{kg}}$

State 3: $P_3 = 8$ MPa, $T_3 = 550°C \rightarrow$ superheated vapor $\rightarrow h_3 = h_3(P_3, T_3) = 3521.0$ kJ/kg, $s_3 = s_3(P_3, T_3) = 6.8778$ kJ/kg·K

State 4: $P_4 = 7.5$ kPa $\rightarrow$ First, we evaluate the turbine as if it were ideal (isentropic), $s_4 = s_3$, so that $h_{4s} = h_{4s}(P_4, s_4 = s_3)$

Note that,

$$x_{4s} = \frac{h_{4s} - h_{f4}}{h_{g4} - h_{f4}} = \frac{s_{4s} - s_{f4}}{s_{g4} - s_{f4}} = \frac{6.8778 - 0.5764}{8.2515 - 0.5764} = 0.821$$

Solving for h_{4s},

$$h_{4s} = h_{f4} + x_{4s}\left(h_{g4} - h_{f4}\right) = 168.75 + 0.821\,(2574.8 - 168.75) = 2144.1\,\frac{\text{kJ}}{\text{kg}}$$

Second, we use the definition of isentropic efficiency:

$$h_4 = h_3 - \eta_T\,(h_3 - h_{4s}) = 3521.0 - 0.88\,(3521.0 - 2144.1) = 2309.3\,\frac{\text{kJ}}{\text{kg}}$$

Note that

$$x_4 = \frac{h_4 - h_{f4}}{h_{g4} - h_{f4}} = \frac{2309.3 - 168.75}{2574.8 - 168.75} = 0.890$$

a) Therefore, turbine power output is

$$\dot{W}_T = \dot{m}\,(h_3 - h_4) = \left(2.8 \times 10^4\,\frac{\text{kg}}{\text{h}}\right)\left(\frac{1\,\text{h}}{3600\,\text{s}}\right)(3521.0 - 2309.4)\,\frac{\text{kJ}}{\text{kg}}\left(\frac{1\,\text{kW}}{1\,\text{kJ/s}}\right) = 9424\,\text{kW}$$

Pump power input is unchanged at

$$\dot{W}_P = 76.6\,\text{kW}$$

Therefore, net power is $\dot{W}_{net} = 9424 - 76.6 = 9347$ kW

b) Input heat transfer rate is

$$\dot{Q}_{in} = \dot{m}\,(h_3 - h_2) = \left(2.8 \times 10^4\,\frac{\text{kg}}{\text{h}}\right)\left(\frac{1\,\text{h}}{3600\,\text{s}}\right)(3521.0 - 178.6)\,\frac{\text{kJ}}{\text{kg}}\left(\frac{1\,\text{kW}}{1\,\text{kJ/s}}\right) = 26,000\,\text{kW}$$

c) The cycle thermal efficiency is

$$\eta_{cycle} = \frac{\dot{W}_{net}}{\dot{Q}_{in}} = \frac{(9424 - 76.3)\ \text{kW}}{26,000\,\text{kW}} = 0.360$$

d) The net station heat rate is

$$\text{net station heat rate} = \frac{3413\,\text{Btu}/\text{kWh}}{\eta_{boiler}\eta_{cycle}} = \frac{3413\,\text{Btu}/\text{kWh}}{0.92 \times 0.360} = 10,305\,\text{Btu}/\text{kWh}$$

e) The cooling water flow rate is

$$\dot{m}_{cooling\ water} = \frac{\dot{m}\,(h_4 - h_1)}{c_p\,(T_B - T_A)} = \frac{\left(2.8 \times 10^4\,\frac{kg}{h}\right)(2309.3 - 168.79)\,\frac{kJ}{kg}\left(\frac{1000\,J}{1\,kJ}\right)}{\left(4176\,\frac{J}{kg \cdot K}\right)(38 - 20)\,K} = 797{,}340\,\frac{kg}{h}$$

Comments:

Compared to the results of Example 8-5, superheating increased the turbine outlet quality from 0.721 to 0.890 and the cycle thermal efficiency from 0.327 to 0.360.

8.5 THE RANKINE CYCLE WITH REHEAT AND REGENERATION

In the above discussion of the Rankine cycle, we used only four basic components: pump, boiler, turbine, and condenser. We can introduce two cycle modifications to increase cycle efficiency. These modifications require additional equipment and additional complexity, but the increase in the complexity of the analysis is due solely to the need to keep track of additional fluid states and the need to account for the interactions among the various components of the system.

If we increase the boiler pressure in a simple Rankine cycle and use the same maximum superheat temperature, we would obtain a higher cycle efficiency, but the outlet steam quality would drop and this could damage the turbine. In the *reheat* cycle shown on Figure 8-13, the steam enters the turbine from the boiler, but the steam is not allowed

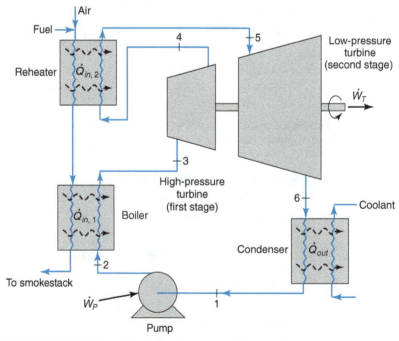

FIGURE 8-13 Schematic of a Rankine cycle with reheat.

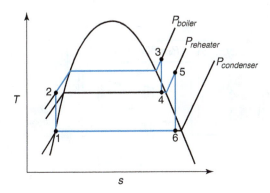

FIGURE 8-14 *T-s* diagram of a reheat Rankine cycle.

to expand to the condenser pressure. (See *T-s* diagram on Figure 8-14.) Instead, after expansion through the high-pressure turbine (process 3–4), the steam is extracted from the turbine at an intermediate pressure and is routed to another heat exchanger, where it is reheated to a higher temperature (process 4–5). The steam then completes its expansion in the low-pressure turbine to the condenser pressure (process 5–6).

Reheat permits the use of higher boiler pressures, which results in higher cycle efficiencies, higher turbine outlet qualities, and increased work per unit mass. The disadvantages include more complexity in the cycle and additional capital cost. Because of this latter consideration, either one or two reheat processes are typically used; additional reheat stages cannot be justified economically.

The main differences in the analysis of a reheat Rankine cycle compared to the simple Rankine cycle are associated with the work produced by the turbines and the heat addition to the cycle. In the reheat cycle, each turbine is addressed separately. The inlet and outlet conditions are different for the high- and low-pressure turbines; in addition, the isentropic efficiencies also could be different. For the heat addition, two processes need to be taken into account, the main vaporization/superheating process that sets the high-pressure turbine inlet steam conditions and the reheating process(es) that set(s) the inlet condition(s) for the other turbine(s).

EXAMPLE 8-7   Rankine cycle with reheat

One reheat stage is used in a Rankine cycle. The steam enters the high-pressure turbine at 1250 psia and 1000°F, is extracted at 600 psia, is reheated to 1000°F, and then expands in the low-pressure turbine to a pressure of 1.0 psia. The water leaves the condenser as a saturated liquid. Both turbine stages have an isentropic efficiency of 89%, and the pump isentropic efficiency is 83%. The net power output is 50 MW.

a) Determine the required mass flow rate (in lbm/s).

b) Determine the cycle thermal efficiency.

Approach:

From our examination of previous Rankine cycles, the application of conservation of energy to each component resulted in either a work (power) term or a heat transfer rate that was calculated by multiplying together a mass flow rate and an enthalpy difference. Because net power is given, we can use energy balances around the turbines and pump to determine the mass flow rate. For the cycle thermal efficiency, in addition to the net work, the heat inputs at the boiler and reheater are required. Energy balances around those two heat exchangers are used to find the heat transfer rates.

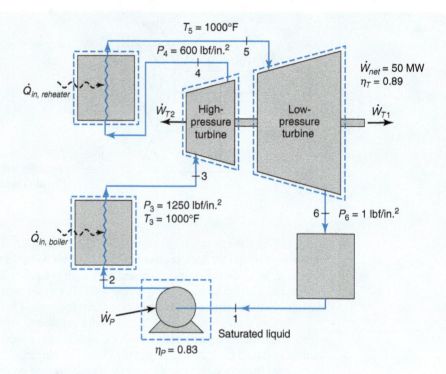

Assumptions:

Solution:

a) We begin the present analysis for the mass flow rate by using the given net power. Referring to the schematic, two turbines must be taken into account:

$$\dot{W}_{net} = 50\,\text{MW} = \dot{W}_{T1} + \dot{W}_{T2} - \dot{W}_P$$

A1. The complete system and individual components are steady.

Defining control volumes around each turbine and pump, and assuming [A1], [A2], [A3], we obtain

$$\dot{W}_{T1} = \dot{m}\,(h_3 - h_4) \quad \text{and} \quad \dot{W}_{T2} = \dot{m}\,(h_5 - h_6)$$

A2. The pump and turbines are adiabatic.

$$\dot{W} = -\dot{W}_P = \dot{m}\,(h_1 - h_2)$$

A3. Potential and kinetic energy effects are negligible.

The inlet enthalpies (h_1, h_3, and h_5) can be evaluated with the given information. The outlet enthalpies (h_2, h_4, and h_6) can be evaluated with the given information and the definition of isentropic efficiencies. Thus the only unknown in the above four equations is the mass flow rate.

b) The cycle thermal efficiency is defined by

$$\eta_{cycle} = \frac{\dot{W}_{net}}{\dot{Q}_{in}}$$

A4. No work occurs in either the boiler or reheater.

We obtain the input heat transfer rate, $\dot{Q}_{in} = \dot{Q}_{in,boiler} + \dot{Q}_{in,reheater}$, by analyzing control volumes around the boiler and reheater. Assuming [A1], [A3], [A4], the energy equation gives us:

$$\dot{Q} = \dot{Q}_{in,boiler} = \dot{m}\,(h_3 - h_2) \quad \text{boiler}$$

$$\dot{Q} = \dot{Q}_{in,reheater} = \dot{m}\,(h_5 - h_4) \quad \text{reheater}$$

We have already seen that the enthalpies can be evaluated, and the mass flow rate can be determined using the equations given above.

The water properties can be evaluated with Table B-11 and B-12.

State 1: $P_1 = 1.0$ psia, saturated liquid $\rightarrow h_1 = h_{f1}(P_1) = 69.74$ Btu/lbm, $v_1 = v_{f1}(P_1) = 0.016136$ ft^3/lbm

A5. Initially, assume the pump and turbines are isentropic to evaluate ideal performance.

A6. The subcooled liquid approximation is valid.

State 2: $P_2 = 1250$ psia $\rightarrow$ First, we evaluate the pump as if it were ideal [A5], $s_2 = s_1$, and the outlet enthalpy is given by [A6]:

$$h_{2s} \approx h_1 + v_1 (P_2 - P_1)$$

$$= 69.74 \frac{Btu}{lbm} + \left(0.016136 \frac{ft^3}{lbm}\right)(1250 - 1)\frac{lbf}{in.^2}\left(\frac{144\,in.^2}{1\,ft^2}\right)\left(\frac{Btu}{778.2\,ft\cdot lbf}\right)$$

$$h_{2s} = 69.74 + 3.73 = 73.47 \frac{Btu}{lbm}$$

Second, we use the isentropic efficiency

$$h_2 = h_1 + \frac{h_{2s} - h_1}{\eta_P} = 69.74 + \frac{73.47 - 69.74}{0.82} = 74.29 \frac{Btu}{lbm}$$

State 3: $P_3 = 1250$ psia, $T_3 = 1000°F \rightarrow$ superheated vapor $\rightarrow h_3 = 1498.2$ Btu/lbm, $s_3 = 1.6244$ Btu/lbm·R

State 4: $P_4 = 600$ psia $\rightarrow$ First, we evaluate the turbine as if it were ideal [A5], $s_4 = s_3$, so that $h_{4s} = h_{4s}(P_4, s_4 = s_3) \rightarrow$ superheated vapor $\rightarrow$ by interpolation $h_{4s} = 1395.6$ Btu/lbm

Second, we use the isentropic efficiency

$$h_4 = h_3 - \eta_T (h_3 - h_{4s}) = 1498.2 - 0.89(1498.2 - 1395.6) = 1406.9 \frac{Btu}{lbm}$$

State 5: $P_5 = 600$ psia, $T_5 = 1000°F \rightarrow$ superheated vapor $\rightarrow h_5 = 1517.8$ Btu/lbm, $s_5 = 1.7155$ Btu/lbm·R

State 6: $P_6 = 1$ psia $\rightarrow$ First, we evaluate the turbine as if it were ideal (isentropic), $s_6 = s_5$, so that $h_{6s} = h_{6s}(P_6, s_6 = s_5)$

Note that

$$x_{6s} = \frac{h_{6s} - h_{f6}}{h_{g6} - h_{f6}} = \frac{s_{6s} - s_{f6}}{s_{g6} - s_{f6}} = \frac{1.7155 - 0.13266}{1.9779 - 0.13266} = 0.858$$

Solving for h_{6s}:

$$h_{6s} = h_{f6} + x_{6s}\left(h_{g6} - h_{f6}\right) = 69.74 + 0.858(1105.8 - 69.74) = 958.7 \frac{Btu}{lbm}$$

Second, we use the isentropic efficiency

$$h_6 = h_5 - \eta_T (h_5 - h_{6s}) = 1517.8 - 0.89(1517.8 - 958.7) = 1020.2 \frac{Btu}{lbm}$$

Note that

$$x_6 = \frac{h_6 - h_{f6}}{h_{g6} - h_{f6}} = \frac{1020.2 - 69.74}{1105.8 - 69.74} = 0.917$$

Now, to calculate the mass flow rate, we combine the expressions for turbine and pump power:

$$\dot{W}_{net} = 50\,MW = \dot{W}_{T1} + \dot{W}_{T2} - \dot{W}_P = \dot{m}(h_3 - h_4) + \dot{m}(h_5 - h_6) - \dot{m}(h_2 - h_1)$$

and solve for mass flow rate:

$$\dot{m} = \frac{\dot{W}_{net}}{(h_3 - h_4) + (h_5 - h_6) - (h_2 - h_1)}$$

$$= \frac{50\,\text{MW}(1000\,\text{kW}/1\,\text{MW})}{[(1498.2 - 1395.6) + (1517.8 - 1020.2) - (74.29 - 69.74)]\frac{\text{Btu}}{\text{lbm}}} \left(\frac{1\,\text{Btu/s}}{1.055\,\text{kW}}\right)$$

$$= 79.5\,\frac{\text{lbm}}{\text{s}}$$

The input heat transfer rate is

$$\dot{Q}_{in} = \dot{Q}_{in,boiler} + \dot{Q}_{in,reheater} = \dot{m}\left[(h_3 - h_2) + (h_5 - h_4)\right]$$

$$= \left(79.5\,\frac{\text{lbm}}{\text{s}}\right)[(1498.2 - 74.29) + (1517.8 - 1406.9)]\frac{\text{Btu}}{\text{lbm}}\left(\frac{1.055\,\text{kW}}{\text{Btu/s}}\right)$$

$$= 128,800\,\text{kW}$$

The cycle thermal efficiency is

$$\eta_{cycle} = \frac{\dot{W}_{net}}{\dot{Q}_{in}} = \frac{50000\,\text{kW}}{128800\,\text{kW}} = 0.388$$

Comments:

If this cycle did not have a reheat process, there would be only one turbine and one heat input. Using the same outlet conditions from the boiler and the same isentropic efficiencies, the resulting turbine exit enthalpy would be 974.5 Btu/lbm, the turbine exit quality would be 0.873, the mass flow rate would be 91.2 lbm/s, and the cycle thermal efficiency would be 0.365, which is lower than with reheat (0.388).

The second modification to the basic Rankine cycle is to add *regeneration*, which involves rerouting a portion of the steam flow from the turbine and passing this steam through additional heat exchangers that raise the boiler feedwater inlet temperature. As shown on Figure 8-12, heat is added *from an external source* to the Rankine cycle during three processes: nonisothermal heating of the feedwater until it reaches the saturation temperature (process 2A–2B), isothermal vaporization until saturated vapor is attained (process 2B–3), and nonisothermal superheating to the desired peak temperature (process 3–3′). The Carnot cycle achieves its high cycle efficiency with all the heat addition *from an external source* occurring isothermally. *Regeneration* is the process in which energy *internal* to the system (contained in the working fluid) is used to raise the feedwater temperature before it reaches the boiler; this process takes place in *feedwater heaters*. When feedwater heaters are used, less energy is added to the working fluid in the boiler *from an external source* to raise the feedwater temperature to the saturation temperature. This results in a higher cycle efficiency.

Figure 8-15a shows a regenerative Rankine cycle with a *closed feedwater heater*, which is a heat exchanger in which condensing steam is used to heat liquid water and the two streams remain separate. The process is illustrated on a *T-s* diagram on Figure 8-15b. The energy used to heat the feedwater comes from steam extracted from the turbine. A fraction (y) of the total steam flow that enters the high-pressure turbine is extracted and flows to the closed feedwater heater. The extracted steam condenses (process 5–7), and the

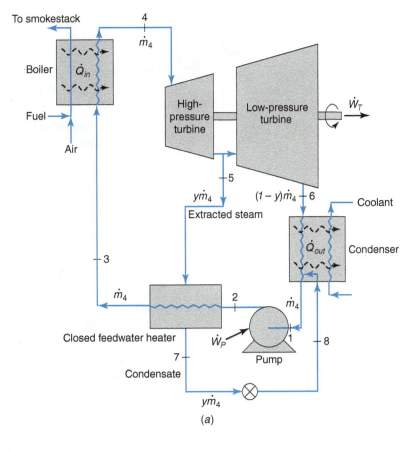

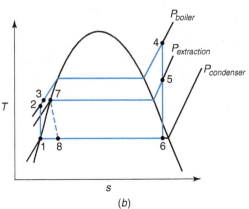

FIGURE 8-15 Regenerative Rankine cycle with a closed feedwater heater.

condensate (often assumed to be saturated liquid) is routed through a throttling valve/steam trap to a lower-pressure location (process 7–8) in the cycle (as shown on Figure 8-15a) or the condensate could be pumped to a higher pressure and injected into the feedwater downstream of the closed feedwater heater (as shown on Figure 8-16).

Figure 8-17 shows an *open feedwater heater* (which is basically a mixing chamber). The process is illustrated on a *T-s* diagram on Figure 8-17b. A fraction (y) of the total steam flow entering the turbine is extracted from the turbine and is mixed directly with the incoming feedwater. Note that because of the low pressure in the open feedwater heater, a

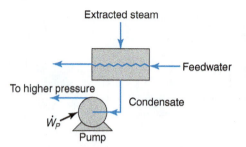

FIGURE 8-16 Condensate from closed feedwater heater pumped to higher pressure.

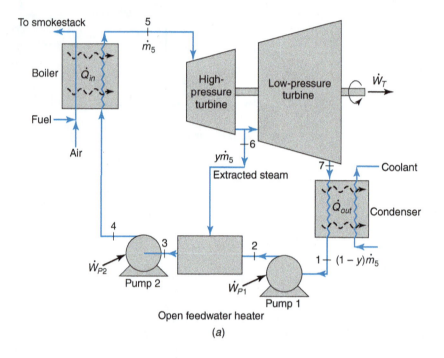

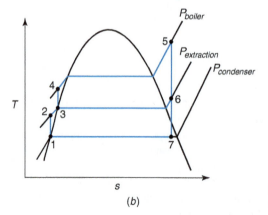

FIGURE 8-17 Regenerative Rankine cycle with an open feedwater heater.

second pump must be added to the system to raise the feedwater to the boiler pressure. The open feedwater heater is also used to drive dissolved gases out of the liquid water (called *deaeration*), because dissolved gases can cause corrosion in the equipment and pipes and can significantly degrade the operation of the condenser. Example 8-8 demonstrates how to calculate the extraction steam flow fraction, y, for an open feedwater heater, and its effect on cycle thermal efficiency; Example 8-9 does the same for a closed feedwater heater.

Because of the flow extracted from the turbine, the flow rate is different through the various turbines, and less power is produced than if the total flow were allowed to expand from the inlet to the exit. Thus, in setting the operating conditions (number of turbines, extraction pressures, extraction flow rates, etc.), the reduction in power produced and added capital cost to install feedwater heaters must be balanced against the improved cycle thermal efficiency (reduced fuel cost) to make the addition of the feedwater heaters economically feasible.

Modern power plants include all of the devices and modifications described above. Figure 8-18 shows a typical schematic of an actual Rankine cycle power plant. Superheat, reheat, and regeneration (both open and closed feedwater heaters) are used to raise the cycle efficiency. The number of stages of each process (reheat and regeneration), the maximum pressure level and extraction pressure levels, the net power output, and so on are determined by the requirements of the company that will own the plant and by an economic analysis that balances capital costs against operating and maintenance costs. With the tools we have developed in this chapter, an analysis of such an involved plant is relatively straightforward.

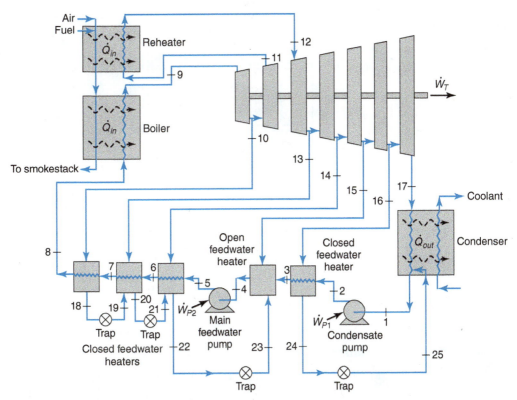

FIGURE 8-18 Modern Rankine cycle.

The main difficulty is keeping track of all the flows and fluid properties at all the locations in the system.

EXAMPLE 8-8 Rankine cycle with regeneration — open feedwater heater

One open feedwater heater is used in an ideal Rankine cycle. Steam leaves the boiler at 40 bar, 400°C and expands in the high-pressure turbine to a pressure of 8 bar, at which point a fraction y of the total flow is extracted from the turbine and the remainder of the flow expands in the low-pressure turbine to a pressure of 20 kPa. The flow from the condenser is saturated liquid water, and it is pumped to the open feedwater heater, where the extracted steam flow is mixed with the feedwater to produce saturated liquid water at 8 bar. At the outlet of the open feedwater heater, a second pump raises the pressure to 40 bar and pumps the water to the boiler.

a) Determine the fraction of the total flow extracted from the turbine.

b) Determine the cycle thermal efficiency.

Approach:

Using the figure, the fraction of the total flow extracted from the turbine can be determined through an energy and mass balance on a control volume around the open feedwater heater. Insufficient information is given to evaluate the mass flow rate entering the high-pressure turbine. However, sufficient information is given to evaluate the mass flow rates at all locations on the control volume in terms of the total mass flow entering the high-pressure turbine and fractions of that flow. Once that fraction is determined, mass flow through every device is known, so to obtain the cycle thermal efficiency, we apply conservation of energy to turbines, pumps, and boiler to obtain the net power.

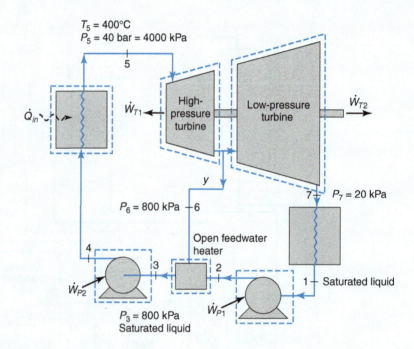

Assumptions:

Solution:

a) First we define $\dot{m}_6$ in terms of the total flow entering the high-pressure turbine:

$$y = \frac{\dot{m}_6}{\dot{m}_5}$$

so that the extracted flow equals $\dot{m}_6 = y\dot{m}_5$. The remainder of the total flow, $(1 - y)\,\dot{m}_5$, continues through the second stage of the turbine, the condenser, and the pump until it enters the open feedwater heater and mixes with the extracted steam.

We define a control volume around the feedwater heater and assume steady-state conditions, negligible potential and kinetic energy effects, and no work [A1], [A2], [A3]. Conservation of mass gives us:

A1. The complete system and individual components are steady.
A2. Potential and kinetic energy effects are negligible.
A3. No work occurs in the feedwater heater or boiler.

$$\dot{m}_6 + \dot{m}_2 - \dot{m}_3 = 0$$

Rewriting all the flows in terms of $\dot{m}_5$, we obtain

$$y\dot{m}_5 + (1 - y)\,\dot{m}_5 - \dot{m}_5 = 0$$

This does not help us.

Conservation of energy gives us:

$$y\dot{m}_5 h_6 + \dot{m}_2 h_2 - \dot{m}_3 h_3 = 0$$

Again, expressing the mass flows at 2 and 3 in terms of $\dot{m}_5$ we, obtain

$$y\dot{m}_5 h_6 + (1 - y)\,\dot{m}_5 h_2 - \dot{m}_5 h_3 = 0$$

Solving for y:

$$y = \frac{h_3 - h_2}{h_6 - h_2}$$

These three enthalpies can be determined from the given information.

b) The cycle thermal efficiency is defined as

$$\eta_{cycle} = \frac{\dot{W}_{net}}{\dot{Q}_{in}}$$

From the schematic, we can see that

$$\dot{W}_{net} = \dot{W}_{T1} + \dot{W}_{T2} - \dot{W}_{P1} - \dot{W}_{P2}$$

For the control volumes defined around the pumps and turbines, applying conservation of mass and energy, and assuming [A1], [A2], [A4], we have:

A4. The pump and turbine are adiabatic.

$$\dot{W}_{T1} = \dot{m}_5\,(h_5 - h_6)$$

$$\dot{W}_{T2} = (1 - y)\,\dot{m}_5\,(h_6 - h_7)$$

$$\dot{W}_{P1} = (1 - y)\,\dot{m}_5\,(h_2 - h_1)$$

$$\dot{W}_{P2} = \dot{m}_5\,(h_4 - h_3)$$

For the control volume defined around the water flowing through the boiler, applying conservation of mass and energy, and assuming steady-state conditions, negligible potential and kinetic energy effects, and no work [A1], [A2], [A3]:

$$\dot{Q}_{in} = \dot{m}_5\,(h_5 - h_4)$$

All the enthalpies can be determined from the given information. The water properties can be evaluated with Tables A-11 and A-12.

State 1: $P_1 = 20$ kPa, saturated liquid $\rightarrow h_1 = h_{f1}(P_1) = 251.4$ kJ/kg,
$v_1 = v_{f1}(P_1) = 0.001017$ m^3/kg

A5. We assume the pump and turbine are isentropic initially to determine ideal performance.

A6. The subcooled liquid approximation is valid.

State 2: $P_2 = 8\,\text{bar} = 800\,\text{kPa}$, subcooled liquid $\rightarrow$ Because the pump is ideal [A5], $s_2 = s_1$, and the outlet enthalpy is given by [A6]:

$$h_2 \approx h_1 + v_1\,(P_2 - P_1)$$

$$= 251.4\,\frac{\text{kJ}}{\text{kg}} + \left(0.001017\,\frac{\text{m}^3}{\text{kg}}\right)(800 - 20)\,\text{kPa}\left(\frac{1\,\text{kN/m}^2}{1\,\text{kPa}}\right)\left(\frac{1\,\text{kJ}}{1\,\text{kN}\cdot\text{m}}\right)$$

$$h_2 = 251.4 + 0.8 = 252.2\,\frac{\text{kJ}}{\text{kg}}$$

State 3: $P_3 = 8\,\text{bar} = 800\,\text{kPa} \rightarrow$ saturated liquid $\rightarrow h_3 = h_{f3}(P_3) = 721.11\,\text{kJ/kg}$, $v_3 = v_{f3}(P_3) = 0.001115\,\text{m}^3/\text{kg}$

State 4: $P_2 = 40\,\text{bar} = 4000\,\text{kPa}$, subcooled liquid $\rightarrow$ The same method as used for state 2 is used:

$$h_4 \approx h_3 + v_3\,(P_4 - P_3)$$

$$= 721.11\,\frac{\text{kJ}}{\text{kg}} + \left(0.001115\,\frac{\text{m}^3}{\text{kg}}\right)(4000 - 800)\,\text{kPa}\left(\frac{1\,\text{kN/m}^2}{1\,\text{kPa}}\right)\left(\frac{1\,\text{kJ}}{1\,\text{kN}\cdot\text{m}}\right)$$

$$h_4 = 721.11 + 3.6 = 724.7\,\frac{\text{kJ}}{\text{kg}}$$

State 5: $P_5 = 40\,\text{bar} = 4000\,\text{kPa}$, $T_5 = 400°\text{C} \rightarrow$ superheated vapor $\rightarrow h_5 = 3213.6\,\text{kJ/kg}$, $s_3 = 6.7690\,\text{kJ/kg}\cdot\text{K}$

State 6: $P_6 = 8\,\text{bar} = 800\,\text{kPa}$, $s_6 = s_5 \rightarrow$ superheated vapor $\rightarrow$ by interpolation $h_6 = 2817.8\,\text{kJ/kg}$

State 7: $P_7 = 20\,\text{kPa}$, $s_7 = s_6 = s_5$, two-phase

Note that

$$x_7 = \frac{h_7 - h_{f7}}{h_{g7} - h_{f7}} = \frac{s_7 - s_{f7}}{s_{g7} - s_{f7}} = \frac{6.7690 - 0.832}{7.9085 - 0.832} = 0.839.$$

Solving for h_7,

$$h_7 = h_{f7} + x_7\,(h_{g7} - h_{f7}) = 251.4 + 0.839\,(2609.7 - 251.4) = 2230.0\,\frac{\text{kJ}}{\text{kg}}$$

The fraction extracted from the turbine is:

$$y = \frac{h_3 - h_2}{h_6 - h_2} = \frac{721.11 - 252.2}{2817.8 - 252.2} = 0.183$$

Because the mass flow rate is unknown and cannot be determined, we will work with the turbine power outputs per unit mass:

$$\frac{\dot{W}_{T1}}{\dot{m}_5} = h_5 - h_6 = 3213.6 - 2817.8 = 395.8\,\frac{\text{kJ}}{\text{kg}}$$

$$\frac{\dot{W}_{T2}}{\dot{m}_5} = (1 - y)\,(h_6 - h_7) = (1 - 0.183)\,(2817.8 - 2230.0) = 480.2\,\frac{\text{kJ}}{\text{kg}}$$

The pump power inputs per unit mass are:

$$\frac{\dot{W}_{P1}}{\dot{m}_5} = (1 - y)\,(h_2 - h_1) = (1 - 0.183)\,(252.2 - 251.8) = 0.33\,\frac{\text{kJ}}{\text{kg}}$$

$$\frac{\dot{W}_{P2}}{\dot{m}_5} = h_4 - h_3 = 724.7 - 721.11 = 3.6\,\frac{\text{kJ}}{\text{kg}}$$

The input heat transfer rate per unit mass is

$$\frac{\dot{Q}_{in}}{\dot{m}_5} = h_5 - h_4 = 3213.6 - 724.7 = 2488.9 \; \frac{kJ}{kg}$$

The cycle thermal efficiency is

$$\eta_{cycle} = \frac{\dot{W}_{net}}{\dot{Q}_{in}} = \frac{\dot{W}_{net}/\dot{m}_5}{\dot{Q}_{in}/\dot{m}_5} = \frac{(395.8 + 480.2 - 0.33 - 3.6) \; kJ/kg}{2388.9 \, kJ/kg} = 0.350$$

Comments:

If the open feedwater heater were removed from the cycle, then the total steam flow would expand from 40 bar, 400°C to 20 kPa in one turbine, and there would be only one pump. With all other conditions remaining the same, an analysis of such a cycle would give:

$$\frac{\dot{W}_T}{\dot{m}_5} = h_5 - h_6 = 3213.6 - 2230 = 983.6 \; \frac{kJ}{kg}$$

$$\frac{\dot{W}_P}{\dot{m}_5} = v_1 (P_2 - P_1) = \left(0.001017 \; \frac{m^3}{kg} \right) (4000 - 20) \; \frac{kN}{m^2} = 4.1 \; \frac{kJ}{kg}$$

$$\frac{\dot{Q}_{in}}{\dot{m}_5} = h_5 - h_4 = 3213.6 - (251.4 + 4.1) = 2958.2 \; \frac{kJ}{kg}$$

$$\eta_{cycle} = 0.331$$

Note the decrease in cycle efficiency.

EXAMPLE 8-9   Rankine cycle with regeneration—closed feedwater heater

An ideal Rankine cycle has one stage of reheat and one closed feedwater heater, as shown in the figure. The desired power output is 100 MW. The steam conditions at the inlet to the high-pressure turbine are 10 MPa, 550°C. A fraction y of the steam is extracted at 1.0 MPa and is used to heat the feedwater in the closed feedwater heater; the outlet temperature of the feedwater equals the saturation temperature of the extracted steam. The remainder of the steam is reheated to 550°C and then expanded in the low-pressure turbine to a pressure of 10 kPa. The condensate from the condenser is saturated liquid.

a) Determine the fraction of the total flow extracted from the turbine.

b) Determine the mass flow rate entering the high-pressure turbine.

c) Determine the cycle thermal efficiency.

Approach:

The approach to this problem is very similar to that followed in Example 8-8. We define all the mass flow rates into and out of a control volume around the closed feedwater heater in terms of the mass flow entering the high-pressure turbine and apply conservation of mass and energy. The mass flow rate can be canceled out, leaving the fraction we seek. The total mass flow rate can be determined through the use of the given net power output and energy balances around the turbines and pump. For cycle thermal efficiency, we are given the net power output; the total heat input can be determined with an energy balance around the boiler and reheater.

Assumptions:

Solution:

a) Referring to the system schematic, the fraction of the total flow extracted from the turbine can be determined through a mass and energy balance around the closed feedwater heater.

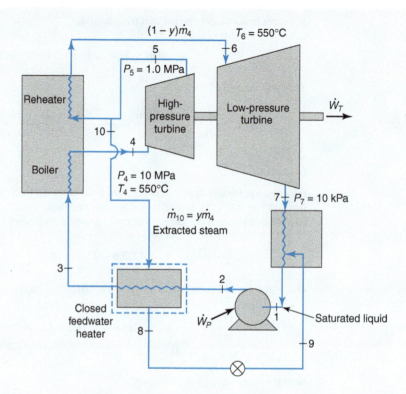

We define $\dot{m}_{10}$ in terms of the total flow entering the high-pressure turbine:

$$y = \frac{\dot{m}_{10}}{\dot{m}_4}$$

so that the extracted flow equals $\dot{m}_{10} = y\dot{m}_4$. The extracted flow does not mix with the liquid feedwater as it did in the open feedwater heater; the heat exchanger tubes in the closed feedwater heater separate the two flows.

Applying conservation of mass separately to the feedwater flow and the extracted steam flow and assuming steady flow [A1], we obtain:

A1. The complete system and individual components are steady

$$\dot{m}_{10} - \dot{m}_8 = y\dot{m}_4 - \dot{m}_8 = 0 \rightarrow \dot{m}_8 = y\dot{m}_4 \qquad \text{extracted flow}$$

$$\dot{m}_2 - \dot{m}_3 = 0 \rightarrow \dot{m}_2 = \dot{m}_3 = \dot{m}_4 \qquad \text{feedwater}$$

Applying conservation of energy to the overall control volume around the feedwater heater, assuming steady-state conditions, negligible potential and kinetic energy effects, adiabatic conditions, and no work [A1], [A2], [A3], [A4], we obtain:

A2. These are negligible potential and kinetic energy effects.
A3. The feedwater heater, pump, and turbine are adiabatic.
A4. No work occurs in the feedwater heater, boiler, or reheater.

$$y\dot{m}_4 h_{10} - \dot{m}_8 h_8 + \dot{m}_2 h_2 - \dot{m}_3 h_3 = 0$$

Combining the mass and energy equations,

$$y\dot{m}_4 h_{10} - y\dot{m}_4 h_2 + \dot{m}_4 h_2 - \dot{m}_4 h_3 = 0$$

and solving for y,

$$y = \frac{h_3 - h_2}{h_{10} - h_8}$$

where $h_{10} = h_5$. The enthalpies can be determined from the given information.

b) The mass flow rate entering the high-pressure turbine is calculated with the given net power output (100 MW) and application of the energy equation to the two turbine stages and the pump:

$$\dot{W}_{net} = 100 \text{ MW} = \dot{W}_{T1} + \dot{W}_{T2} - \dot{W}_{P}$$

Applying conservation of energy to the two turbines and the pump assuming steady-state conditions, negligible potential and kinetic energy effects, and adiabatic conditions [A1], [A2], [A3] gives:

$$\dot{W}_{T1} = \dot{m}_4 \, (h_4 - h_5)$$

$$\dot{W}_{T2} = (1 - y) \, \dot{m}_4 \, (h_6 - h_7)$$

$$\dot{W}_{P} = \dot{m}_4 \, (h_2 - h_1)$$

The enthalpies in these equations can be determined from the given information. When the last four equations are combined, the only unknown is $\dot{m}_4$, which is one of the quantities we seek.

c) The cycle thermal efficiency is

$$\eta_{cycle} = \frac{\dot{W}_{net}}{\dot{Q}_{in}}$$

We need the input heat transfer rate, and that can be determined from conservation of energy around the water flowing through the boiler and reheater assuming steady-state conditions, negligible potential and kinetic energy effects, and no work [A1], [A2], [A4]. Therefore, we obtain

$$\dot{Q}_{in} = \dot{m}_4 \, (h_4 - h_3) + (1 - y) \, \dot{m}_4 \, (h_6 - h_5)$$

The enthalpies can be determined from the given information.

The water properties can be evaluated with Tables A-11 and A-12.

State 1: $P_1 = 10$ kPa, saturated liquid $\rightarrow h_1 = h_{f1}(P_1) = 45.81$ kJ/kg, $v_1 = v_{f1}(P_1) = 0.001010$ m^3/kg

A5. The pump and turbines are isentropic.
A6. The subcooled liquid approximation is valid.

State 2: $P_2 = 10$ MPa, subcooled liquid $\rightarrow$ The pump is ideal [A5], so the enthalpy change across the pump is [A6]

$$h_2 \approx h_1 + v_1 \, (P_2 - P_1)$$

$$= 45.81 \, \frac{\text{kJ}}{\text{kg}} + \left(0.001010 \, \frac{\text{m}^3}{\text{kg}}\right) (10000 - 10) \, \text{kPa} \left(\frac{1 \, \text{kN/m}^2}{1 \, \text{kPa}}\right) \left(\frac{1 \, \text{kJ}}{1 \, \text{kN} \cdot \text{m}}\right)$$

$$h_2 = 45.81 + 10.1 = 55.9 \, \frac{\text{kJ}}{\text{kg}}$$

State 3: $P_3 = 10$ MPa, $T_3 = T_{sat}(P_2) \rightarrow$ using [A6]:

$$h_3 \approx h_{f2} + v_{f2} \, (P_3 - P_2)$$

$$= 762.81 \, \frac{\text{kJ}}{\text{kg}} + \left(0.001127 \, \frac{\text{m}^3}{\text{kg}}\right) (10000 - 1000) \, \text{kPa} \left(\frac{1 \, \text{kN/m}^2}{1 \, \text{kPa}}\right) \left(\frac{1 \, \text{kJ}}{1 \, \text{kN} \cdot \text{m}}\right)$$

$$h_3 = 762.81 + 10.1 = 773.0 \, \frac{\text{kJ}}{\text{kg}}$$

State 4: $P_4 = 10$ MPa, $T_4 = 550°C \rightarrow$ superheated vapor $\rightarrow h_4 = 3500.9$ kJ/kg, $s_4 = 6.7561$ kJ/kg·K

State 5: $P_5 = 1$ MPa, $s_5 = s_4 \rightarrow$ superheated vapor $\rightarrow$ by interpolation, $h_5 = 2860.2$ kJ/kg

State 6: $P_6 = 1$ MPa, $T_6 = 550°C \rightarrow$ superheated vapor $\rightarrow$ by interpolation, $h_6 = 3588.2$ kJ/kg, $s_6 = 7.8956$ kJ/kg·K

State 7: $P_7 = 10$ kPa, $s_7 = s_6 \rightarrow$ two-phase

Note that

$$x_7 = \frac{h_7 - h_{f7}}{h_{g7} - h_{f7}} = \frac{s_7 - s_{f7}}{s_{g7} - s_{f7}} = \frac{7.8956 - 0.6943}{8.1502 - 0.6943} = 0.972$$

Solving for h_7:

$$h_7 = h_{f7} + x_7 \left(h_{g7} - h_{f7}\right) = 191.83 + 0.972 \,(2584.7 - 191.83) = 2517.4 \,\frac{\text{kJ}}{\text{kg}}$$

State 8: $P_1 = 1$ MPa, saturated liquid $\rightarrow h_8 = h_{f8}(P_8) = 762.81$ kJ/kg

Using these enthalpies, we obtain the fraction extracted:

$$y = \frac{h_3 - h_2}{h_{10} - h_8} = \frac{773.0 - 55.9}{2860.2 - 762.81} = 0.342$$

We combine the expressions for the individual power terms to obtain:

$$\dot{W}_{net} = \dot{W}_{T1} + \dot{W}_{T2} - \dot{W}_P = \dot{m}_4 \,(h_4 - h_5) + (1 - y)\, \dot{m}_4 \,(h_6 - h_7) - \dot{m}_4 \,(h_2 - h_1)$$

Solving for $\dot{m}_4$,

$$\dot{m}_4 = \frac{\dot{W}_{net}}{(h_4 - h_5) + (1 - y)(h_6 - h_7) - (h_2 - h_1)}$$

$$= \frac{100{,}000\,\text{kW} \left(\frac{1\,\text{kJ}}{\text{s}} \big/ 1\,\text{kW}\right)}{[(3500.9 - 2860.2) + (1 - 0.342)(3588.2 - 2517.4) - (55.9 - 45.8)]\,\frac{\text{kJ}}{\text{kg}}} = 74.9 \,\frac{\text{kg}}{\text{s}}$$

The input heat transfer rate is

$$\dot{Q}_{in} = \dot{m}_4 \,(h_4 - h_3) + (1 - y)\, \dot{m}_4 \,(h_6 - h_5)$$

$$= \left(74.9 \,\frac{\text{kg}}{\text{s}}\right)[(3500.9 - 773.0) + (1 - 0.342)(3588.2 - 2860.2)]\,\frac{\text{kJ}}{\text{kg}} \left(\frac{1\,\text{kW}}{1\,\text{kJ}/\text{s}}\right)$$

$$= 240{,}200\,\text{kW}$$

The cycle thermal efficiency is

$$\eta_{cycle} = \frac{\dot{W}_{net}}{\dot{Q}_{in}} = \frac{100{,}000\,\text{kW}}{240{,}200\,\text{kW}} = 0.416$$

Comments:

As noted above, the method used to evaluate the fraction extracted from the closed feedwater heater is the same as that used to evaluate the fraction extracted from the open feedwater heater. However, because one is simply a mixing device (open feedwater heater) and the other one is a heat exchanger (closed feedwater heater), the expressions we developed for the fraction extracted are different.

8.6 THE BRAYTON CYCLE

A power plant used for transportation needs to be lightweight with a high power density. A Rankine cycle power plant is heavy because of the piping, pumps, turbines, and water. The *Brayton cycle*, however, is a *gas power cycle* that uses a rotating gas compressor and turbine for the compression and expansion processes, respectively. Because the equipment is relatively compact and light, and because the working fluid is a gas (air and combustion gases), the Brayton cycle is used for jet engines, helicopters, and ships. In addition, because it can be built in a shorter time than a Rankine cycle power plant and its characteristics are such that it can change operating conditions rapidly, the Brayton cycle is also used for peaking power plants. More recently, the Brayton cycle has been used in base-loaded power plants in conjunction with Rankine cycles, which is discussed on the web (Section 8.8).

Shown on Figure 8-19 are two schematics of *ideal Brayton cycles*. In an *open Brayton cycle* (Figure 8-19a) air is continuously drawn into a compressor. The high-pressure air is mixed with fuel and is burned in the combustion chamber (assumed to be at constant pressure). The high-temperature and -pressure combustion products then expand through the turbine and are exhausted to the atmosphere and not used again. (See Figure 8-20 for the *T-s* and *P-v* diagrams for this cycle.) As with any heat power cycle, the Brayton cycle has four processes:

1–2 Compression of the working fluid with work input
2–3 Heat addition to the working fluid in the combustor
3–4 Expansion of the working fluid with work output
4–1 Heat rejection from the working fluid

The compression, heat addition, and expansion processes are comparable to those in the Rankine cycle. The heat rejection in this open system requires an explanation. The combustion products exit the turbine at a high temperature and ambient pressure. We can imagine a fictitious heat exchanger that uses ambient air to cool the turbine exhaust gases, which

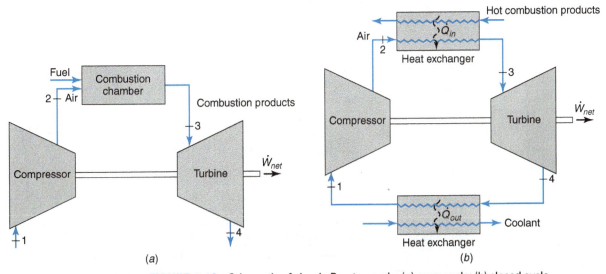

FIGURE 8-19 Schematic of simple Brayton cycle: (a) open cycle; (b) closed cycle.

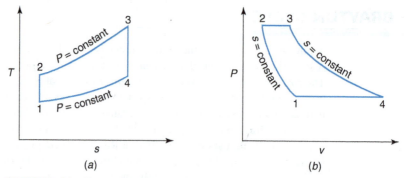

FIGURE 8-20   *T-s* and *P-v* diagrams for the ideal Brayton cycle.

then would return to the compressor inlet at ambient temperature and pressure, thus completing the cycle. We consider the cycle heat rejection to be associated with the difference in enthalpies between the turbine exhaust and the ambient air entering the compressor.

The open cycle is the most commonly used form of the gas power cycle. To simplify its analysis, we make several assumptions. First, we model the cycle as a *closed cycle* by replacing the heat addition and heat rejection processes with constant pressure heat exchangers (Figure 8-19b). Second, we assume that the combustion takes place external to the cycle and that the energy from the combustion process is transferred to the working fluid in a heat exchanger. Likewise, the working fluid rejects heat to the ambient air in the second heat exchanger. Third, because the working fluid is continuously recycled, we can choose any gas that gives the desired output. Air, helium, and hydrogen all have favorable properties. The evaluation of the properties does not change the analysis of the cycle. Therefore, for simplicity we assume the working fluid is only air and ignore the products of combustion. This common assumption is called the *air-standard analysis*.

The cycle thermal efficiency for the Brayton cycle is expressed in the same manner as for the Rankine cycle:

$$\eta_{cycle} = \frac{\text{energy we want to use}}{\text{energy we purchase}} = \frac{\dot{W}_{net}}{\dot{Q}_H} = 1 - \frac{\dot{Q}_L}{\dot{Q}_H} \tag{8-22}$$

The net power is defined as:

$$\dot{W}_{net} = \dot{W}_{Turbine} - \dot{W}_{Compressor} = \dot{W}_T - \dot{W}_C \tag{8-23}$$

The same assumptions used to analyze the Rankine cycle components (steady-state, adiabatic turbine and compressor, negligible potential and kinetic energy effects, no work in the heat exchangers) are also applied to the Brayton cycle components; thus

$$\dot{W}_T = \dot{m}\,(h_3 - h_4) \qquad \text{turbine} \tag{8-24}$$

$$\dot{Q}_L = \dot{m}\,(h_4 - h_1) \qquad \text{heat rejection} \tag{8-25}$$

$$\dot{W}_C = \dot{m}\,(h_2 - h_1) \qquad \text{compressor} \tag{8-26}$$

$$\dot{Q}_H = \dot{m}\,(h_3 - h_2) \qquad \text{heat addition} \tag{8-27}$$

Incorporating Eq. 8-23 through Eq. 8-27 into Eq. 8-22 and noting that all the mass flow rates cancel, we express the Brayton cycle efficiency as

$$\eta_{cycle} = \frac{(h_3 - h_4) - (h_2 - h_1)}{(h_3 - h_2)} = 1 - \frac{(h_4 - h_1)}{(h_3 - h_2)} \tag{8-28}$$

Because air is an ideal gas, we can express these enthalpy differences in terms of specific heat: $\Delta h = \int c_p\, dT$. If we assume a constant specific heat, then $\Delta h = c_p\, \Delta T$. None of the four processes in the Brayton cycle is at constant temperature (Figure 8-20a). Thus the temperature dependence of the specific heat must be taken into account by evaluating the specific heat at the average process temperature. The analysis can be simplified by using a *cold-air-standard analysis*, which ignores the temperature dependence of specific heat and assumes all air properties are evaluated at room temperature (25°C or 77°F). The value of this assumption is that we can develop equations to show clearly the effects of different operating parameters on the performance of a Brayton cycle.

Using $\Delta h = c_p\, \Delta T$ and applying the cold-air-standard assumptions, we rewrite Eq. 8-28 as

$$\eta_{cycle} = 1 - \frac{c_p\,(T_4 - T_1)}{c_p\,(T_3 - T_2)} = 1 - \frac{(T_4 - T_1)}{(T_3 - T_2)} = 1 - \frac{T_1}{T_2}\frac{(T_4/T_1 - 1)}{(T_3/T_2 - 1)} \tag{8-29}$$

For the ideal Brayton cycle (isentropic compressor and turbine), the isentropic relation for an ideal gas is (see Chapter 2)

$$\frac{T_2}{T_1} = \left(\frac{P_2}{P_1}\right)^{(k-1)/k} = \frac{T_3}{T_4} = \left(\frac{P_3}{P_4}\right)^{(k-1)/k} \tag{8-30}$$

Because the pressure at the compressor exit is the same as the pressure at the entrance to the turbine, and the pressure at the compressor inlet is the same as the pressure at the turbine outlet, $P_2/P_1 = P_3/P_4$. Therefore, from Eq. 8-30 we can show that $T_4/T_1 = T_3/T_2$. Substituting this into Eq. 8-29:

$$\eta_{cycle} = 1 - \frac{T_1}{T_2} \tag{8-31}$$

Define a pressure ratio, r_p, as

$$\boxed{r_p = \frac{P_2}{P_1}} \tag{8-32}$$

Substituting Eq. 8-30 and Eq. 8-32 into Eq. 8-31, we can express the thermal efficiency of an ideal Brayton cycle ($\eta_T = \eta_C = 100\%$) as

$$\boxed{\eta_{cycle} = 1 - \frac{1}{r_p^{(k-1)/k}}} \tag{8-33}$$

Figure 8-21 shows the ideal Brayton cycle thermal efficiency as a function of pressure ratio. From this figure, it would appear that the only determinant of the cycle efficiency is pressure ratio. That is true theoretically. However, very high pressure ratios are not practical because they require stronger (thicker, heavier, and, hence, more expensive) components. In addition, another practical consideration is a constraint on the materials used to construct the turbine. Turbine blades must withstand both very high temperatures and high centrifugal forces. While many advances in material development and blade design have raised the maximum operating temperature of gas tubines, the metallurgy associated with the blade material, design, and construction is the main limiting factor. Once the turbine blade designers specify the maximum operating temperature, the maximum air

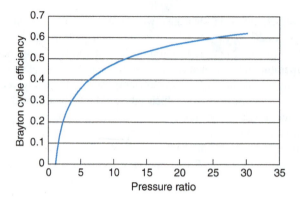

FIGURE 8-21 Ideal Brayton cycle thermal efficiency ($k = 1.4$).

temperature is controlled by using significantly more air than the minimum required to burn the fuel completely.

Depending on the combination of maximum cycle temperature (point 3 on Figure 8-20a) and the pressure ratio, the cycle performance can be affected significantly. Figure 8-22 demonstrates the trade-off between pressure ratio and net work produced by the cycle. Three Brayton cycles are illustrated for different pressure ratios. The area enclosed by the curves represents the net work output per unit mass. For a low pressure ratio (cycle 1), the temperature rise across the compressor is small and much heat must be added to raise the air to T_{max}; this cycle would have the lowest thermal efficiency of the three cycles shown. For a moderate pressure ratio (cycle 2), less heat must be added compared to cycle 1 to raise the air to T_{max}, and the net work per unit mass and cycle efficiency are greater than in cycle 1. For a high pressure ratio (cycle 3), compression raises the air temperature to a very high value and only a small amount of heat must be added to reach T_{max}; the net work per unit mass is less than in cycle 2, but the thermal efficiency of cycle 3 is greater than that in cycle 2. With an increasing pressure ratio, net work increases, reaches a maximum, then decreases again. The pressure ratio at which this maximum occurs can be determined for a given ratio T_3/T_1 by differentiating an expression for net work with respect to the pressure ratio, setting the resulting equation to zero, and solving for r_p to get

$$r_{p,opt} = \left(\frac{T_3}{T_1}\right)^{k/2(k-1)} \tag{8-34}$$

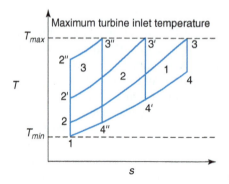

FIGURE 8-22 Combined effects of turbine inlet temperature and pressure ratio on ideal Brayton cycle work output ($k = 1.4$).

Similarly, the maximum net work out at this optimum value of $r_{p,opt}$ is

$$\left.\frac{\dot{W}}{\dot{m}c_p T_1}\right|_{max,opt} = \left(\sqrt{\frac{T_3}{T_1}} - 1\right)^2 \tag{8-35}$$

and the cycle optimum efficiency is

$$\eta_{cycle,max,opt} = 1 - \sqrt{\frac{T_1}{T_3}} \tag{8-36}$$

The trends in net work per unit mass for given temperature and pressure ratios are illustrated in Figure 8-23.

For a given power plant net output, a larger net work per unit mass requires smaller equipment (compressor, turbine, piping, etc.). This reduced capital cost must to be balanced against higher fuel costs (lower efficiency) over the life of the plant to determine the operating conditions that would be most economically feasible.

The Brayton cycle back work ratio, $BWR = \dot{W}_C/\dot{W}_T$, ranges from 40% to 80% compared to 1% to 3% for the Rankine cycle. Thus, an enormous fraction of the power produced by the turbine in the Brayton cycle is needed for the compression process. For peaking power, it would be useful to have all the turbine power available. One way to achieve this is to store compressed air in large underground caverns. During off hours, the Brayton cycle is used to compress and store air. When electrical power demand peaks during the day, the stored compressed air is then used in the turbine, thus allowing more power to be available when needed.

A real Brayton cycle has inefficiencies associated with the compression and expansion processes. Similar to what was done with the Rankine cycle, the compressor and turbine isentropic efficiencies can be incorporated into the analysis to obtain the cycle thermal efficiency and the net power produced. In the limit with $\eta_T = \eta_C = 100\%$ and ignoring the effect of temperature level on the specific heat, the cycle thermal efficiency will equal that calculated with Eq. 8-33. For any compressor or turbine whose isentropic efficiency is

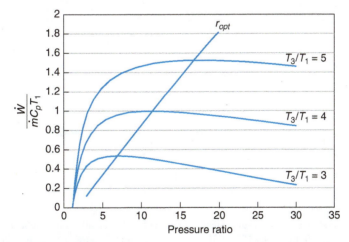

FIGURE 8-23 Net power as function of pressure ratio and T_{max}/T_{min} for ideal Brayton cycle ($k = 1.4$).

less than 100%, the cycle thermal efficiency must be calculated with Eq. 8-28 and its value will be less than that calculated with Eq. 8-33.

EXAMPLE 8-10   Ideal Brayton cycle—cold-air-standard analysis

Air at 96 kPa and 17°C enters the compressor of a simple ideal Brayton cycle that has a pressure ratio of 9. The turbine is limited to a temperature of 927°C. The mass flow rate is 3.5 kg/s.

a) Determine compressor power (in kW).

b) Determine turbine power (in kW).

c) Determine net power output (in kW).

d) Determine cycle thermal efficiency.

e) Determine volumetric flow rate at the compressor inlet (in m³/min).

f) Determine back work ratio.

Approach:

The solution approach to the Brayton cycle is similar to that used for the Rankine cycle. The main difference is in the evaluation of fluid properties. Compressor and turbine power are evaluated with energy balances around those two devices. Cycle thermal efficiency requires the heat input in addition to the net work output, so an energy balance around the combustor is used. Because we know the mass flow rate, the inlet volumetric flow rate is obtained by direct application of the mass equation and the ideal gas equation to evaluate the inlet specific volume of the air.

Assumptions:

Solution:

a) In this Brayton cycle the compressor work can be determined by applying the mass and energy balance equations to a control volume around the compressor, as shown in the schematic. For the

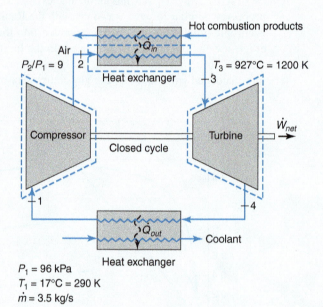

$P_1 = 96$ kPa
$T_1 = 17°C = 290$ K
$\dot{m} = 3.5$ kg/s

A1. The complete system and individual components are steady.
A2. Potential and kinetic energy effects are negligible.
A3. The compressor and turbine are adiabatic.

compressor, we assume steady state, negligible potential and kinetic energy effects, and adiabatic conditions [A1], [A2], [A3]. From conservation of mass, the steady-state assumption gives us:

$$\dot{m}_1 = \dot{m}_2 = \dot{m}$$

From conservation of energy we obtain

$$\dot{W} = -\dot{W}_C = \dot{m}(h_1 - h_2)$$

The cold-air-standard analysis assumes air, an ideal gas, is the working fluid with constant specific heat evaluated at room temperature [A4], [A5], [A6], so $\Delta h = c_p \Delta T$ and

A4. Air is the working fluid.

A5. Air is an ideal gas.

A6. Constant specific heats are evaluated at room temperature.

A7. The compressor and turbine are isentropic.

$$\dot{W}_C = \dot{m} c_p (T_2 - T_1)$$

The compressor outlet temperature [A7] is obtained with the isentropic relationship for an ideal gas:

$$\frac{T_2}{T_1} = \left(\frac{P_2}{P_1}\right)^{(k-1)/k}$$

Both c_p and $k = c_p/c_v$ are found in the ideal gas Table A-8.

b) The turbine work is evaluated in a similar manner. For a control volume around the turbine and assuming steady state, negligible potential and kinetic energy effects, adiabatic conditions, and constant specific heat [A1], [A2], [A3], [A6], the energy and mass equations give:

$$\dot{W} = \dot{W}_T = \dot{m}(h_3 - h_4) = \dot{m} c_p (T_3 - T_4)$$

and the outlet temperature is found from

$$\frac{T_4}{T_3} = \left(\frac{P_4}{P_3}\right)^{(k-1)/k}$$

c) The net power is $\dot{W}_{net} = \dot{W}_T - \dot{W}_C$

d) Cycle thermal efficiency is defined as:

$$\eta_{cycle} = \frac{\dot{W}_{net}}{\dot{Q}_{in}}$$

The input heat transfer rate is evaluated with conservation of energy applied to a control volume around the air flowing through the heat exchanger assuming steady state, negligible potential and kinetic energy effects, constant specific heat, and no work [A1], [A2], [A6], [A8], which gives us

A8. No work occurs in the heat exchangers.

$$\dot{Q} = \dot{Q}_{in} = \dot{m}(h_3 - h_2) = \dot{m} c_p (T_3 - T_2)$$

where, again, $\Delta h = c_p \Delta T$.

e) The volumetric flow rate at the compressor inlet is obtained from

$$\dot{m} = \rho_1 \dot{V}_1 = \frac{\dot{V}_1}{v_1} \rightarrow \dot{V}_1 = \dot{m} v_1$$

Because we have an ideal gas, $Pv = (\bar{R}/M) T$ or $v = \bar{R}T/PM$.

For the cold-air-standard analysis, we use Table A-8 to evaluate c_p and k at room temperature:

$$c_p = 1.005 \, \text{kJ/kg} \cdot \text{K} \quad \text{and} \quad k = 1.40$$

The compressor and turbine outlet temperatures, respectively, are

$$T_2 = T_1 \left(\frac{P_2}{P_1}\right)^{(k-1)/k} = (17 + 273) \, \text{K}(9)^{(1.4-1)/1.4} = 543.3 \, \text{K}$$

$$T_4 = T_3 \left(\frac{P_4}{P_3} \right)^{(k-1)/k} = (927 + 273) \, \text{K} \left(\frac{1}{9} \right)^{(1.4-1)/1.4} = 640.5 \, \text{K}$$

Compressor power is

$$\dot{W}_C = \dot{m} c_p (T_2 - T_1) = \left(3.5 \, \frac{\text{kg}}{\text{s}} \right) \left(1.005 \, \frac{\text{kJ}}{\text{kg} \cdot \text{K}} \right) (543.3 - 290) \, \text{K} \left(\frac{1 \, \text{kW}}{1 \, \text{kJ/s}} \right) = 891 \, \text{kW}$$

Turbine power is

$$\dot{W}_T = \dot{m} c_p (T_3 - T_4) = \left(3.5 \, \frac{\text{kg}}{\text{s}} \right) \left(1.005 \, \frac{\text{kJ}}{\text{kg} \cdot \text{K}} \right) (1200 - 640.5) \, \text{K} \left(\frac{1 \, \text{kW}}{1 \, \text{kJ/s}} \right) = 1968 \, \text{kW}$$

Net power is

$$\dot{W}_{net} = \dot{W}_T - \dot{W}_C = 1968 - 891 = 1077 \, \text{kW}$$

Input heat transfer rate is

$$\dot{Q}_{in} = \dot{m} c_p (T_3 - T_2) = \left(3.5 \, \frac{\text{kg}}{\text{s}} \right) \left(1.005 \, \frac{\text{kJ}}{\text{kg} \cdot \text{K}} \right) (1200 - 543.3) \, \text{K} \left(\frac{1 \, \text{kW}}{1 \, \text{kJ/s}} \right) = 2310 \, \text{kW}$$

Cycle thermal efficiency is

$$\eta_{cycle} = \frac{\dot{W}_{net}}{\dot{Q}_{in}} = \frac{1077 \, \text{kW}}{2310 \, \text{kW}} = 0.466$$

The specific volume at the inlet is

$$v = \frac{\overline{R} T}{P M} = \frac{\left(8.314 \, \frac{\text{kJ}}{\text{kmol} \cdot \text{K}} \right) (290 \, \text{K})}{(96 \, \text{kPa}) \left(28.97 \, \frac{\text{kg}}{\text{kmol}} \right) \left(\frac{1 \, \text{kJ}}{1 \, \text{kN} \cdot \text{m}} \right)} = 0.867 \, \frac{\text{m}^3}{\text{kg}}$$

The volumetric flow rate at the inlet is

$$\dot{V}_1 = \dot{m} v_1 = \left(3.5 \, \frac{\text{kg}}{\text{s}} \right) \left(0.867 \, \frac{\text{m}^3}{\text{kg}} \right) = 3.03 \, \frac{\text{m}^3}{\text{s}}$$

f) The back work ratio is:

$$BWR = \frac{\dot{W}_C}{\dot{W}_T} = \frac{891 \, \text{kW}}{1968 \, \text{kW}} = 0.453$$

Comments:

Eq. 8-33 gives the Brayton cycle thermal efficiency for a cycle with an isentropic turbine and compressor assuming a cold-air-standard analysis.

Thus

$$\eta_{cycle} = 1 - \frac{1}{r_p^{(k-1)/k}} = 1 - \frac{1}{9^{(1.4-1)/1.4}} = 0.466$$

as we would expect. Also, compare the *BWR* for this cycle (0.453) with that calculated for the Rankine cycle in Example 8-4 (0.005). This illustrates quite clearly how much energy is needed to compress a gas compared to a liquid.

EXAMPLE 8-11   Ideal Brayton cycle, air-standard analysis

Rework Example 8-10 but use an air-standard analysis instead of a cold-air-standard analysis.

Approach:

The approach is the same as in Example 8-10. The only difference is in the evaluation of the fluid properties.

Assumptions:

Same as in Example 8-10, except we do not assume constant specific heats at room temperature.

Solution:

The Brayton cycle in Example 8-10 can be evaluated in two ways when we use the air-standard analysis rather than the cold-air-standard analysis. The governing equations, before invoking the constant specific heat assumption, are the same. Thus,

$$\dot{W}_C = \dot{m}\,(h_2 - h_1)$$

$$\dot{W}_T = \dot{m}\,(h_3 - h_4)$$

$$\dot{Q}_{in} = \dot{m}\,(h_3 - h_2)$$

The difference between this solution and the solution given in Example 8-10 is in the evaluation of the enthalpy differences. In this solution we take into account the effect of temperature level on the specific heat. Below are the two approaches to evaluating these enthalpy differences.

Method 1: Assuming variable specific heat

Because T_1 and T_3 are given and we are treating the air as an ideal gas for which enthalpy is a function of temperature only, we can look up the enthalpies for states 1 and 3 directly in Table A-9. To find the enthalpy at state 2, we use the given pressure ratio and the relative pressure functions:

$$P_2/P_1 = P_{r2}/P_{r1} \quad \rightarrow \quad P_2 = P_1\left(P_{r2}/P_{r1}\right)$$

The relative pressure, P_{r1}, is evaluated at T_1. From the calculated P_{r2}, we again use Table A-9 to find h_2. A similar approach is used to evaluate h_4.

Method 2: Assuming constant specific heat at the average process temperature

In this approach, we use specific heat (as we did for the cold-air-standard analysis), but now we evaluate the specific heats at the average temperature of each process. Hence,

$$\dot{W}_T = \dot{m}c_{p34}\,(T_3 - T_4)$$

$$\dot{W}_C = \dot{m}c_{p12}\,(T_2 - T_1)$$

$$\dot{Q}_{in} = \dot{m}c_{p23}\,(T_3 - T_2)$$

The subscripts given on each specific heat are used to indicate the specific heat evaluated at the average process temperature, for example, c_{p34} is evaluated at $(T_3 + T_4)/2$; the other specific heats are found in a like manner. Note that we need T_2 and T_4 for these averages, so we use the isentropic relations for an ideal gas:

$$\frac{T_2}{T_1} = \left(\frac{P_2}{P_1}\right)^{(k-1)/k} \quad \text{and} \quad \frac{T_4}{T_3} = \left(\frac{P_4}{P_3}\right)^{(k-1)/k}$$

The problem is that the property k is a function of temperature and must be evaluated at the average process temperature, but we cannot get that until we have the outlet temperature.

This circular argument is handled by an iterative solution. We first guess an outlet temperature, calculate the average process temperature, evaluate k, and then calculate the outlet temperature using the isentropic relations. If the calculated and guessed temperatures are "close enough," then we can continue with the remainder of the problem solution. If they are significantly different, then we use the newly calculated temperature as our new guess, then step through the process again.

Method 1:

Using Table A-9, we evaluate the two inlet enthalpies:

State 1: $T_1 = 290$ K $\quad h_1 = 290.16$ kJ/kg $\quad P_{r1} = 1.2311$

State 3: $T_3 = 1200$ K $\quad h_3 = 1277.79$ kJ/kg $\quad P_{r3} = 238.0$

Using the definition of relative pressure and the fact that $P_2 = P_3$ and $P_1 = P_4$

$$P_{r2} = P_{r1}\left(\frac{P_2}{P_1}\right) = 1.2311 \times 9 = 11.08 \quad \text{and} \quad P_{r4} = P_{r3}\left(\frac{P_4}{P_3}\right) = 238.0 \times \frac{1}{9} = 26.44$$

From Table A-9 at this value of P_{r2}, by interpolation, $h_2 = 544.35$ kJ/kg and $T_2 = 540$ K; likewise, $h_4 = 696.2$ kJ/kg and $T_4 = 684.1$ K.

Using these values of enthalpy we obtain:

$$\dot{W}_C = \dot{m}\,(h_2 - h_1) = \left(3.5\,\frac{\text{kg}}{\text{s}}\right)(544.35 - 290.16)\,\frac{\text{kJ}}{\text{kg}}\left(\frac{1\,\text{kW}}{1\,\text{kJ/s}}\right) = 890\,\text{kW}$$

$$\dot{W}_T = \dot{m}\,(h_3 - h_4) = \left(3.5\,\frac{\text{kg}}{\text{s}}\right)(1277.79 - 696.2)\,\frac{\text{kJ}}{\text{kg}}\left(\frac{1\,\text{kW}}{1\,\text{kJ/s}}\right) = 2036\,\text{kW}$$

$$\dot{Q}_{in} = \dot{m}\,(h_3 - h_2) = \left(3.5\,\frac{\text{kg}}{\text{s}}\right)(1277.79 - 544.35)\,\frac{\text{kJ}}{\text{kg}}\left(\frac{1\,\text{kW}}{1\,\text{kJ/s}}\right) = 2567\,\text{kW}$$

$$\dot{W}_{net} = \dot{W}_T - \dot{W}_C = 2036 - 890 = 1146\text{ kW}$$

$$\eta_{cycle} = \frac{\dot{W}_{net}}{\dot{Q}_{in}} = \frac{1146\,\text{kW}}{2567\,\text{kW}} = 0.446$$

Method 2:

First evaluate the compressor outlet temperature (T_2) as described above using data from Table A-8.

Guess T_2 (K)	$T_{avg} = (T_1 + T_2)/2$ (K)	c_{p12} (kJ/kg· K)	k_{12}	$T_2 = T_1\,(P_2/P_1)^{(k-1)/k}$
400	345	1.008	1.398	542
542	416	1.015	1.394	540

The calculated T_2 is close to the guessed T_2, so we will end this iteration and now proceed to calculate the turbine outlet temperature.

Guess T_4 (K)	$T_{avg} = (T_3 + T_4)/2$ (K)	c_{p34} (kJ/kg·K)	k_{34}	$T_4 = T_3\,(P_4/P_3)^{(k-1)/k}$
800	1000	1.142	1.366	666
666	933	1.128	1.341	686
686	943	1.130	1.341	686

We can stop the iteration and calculate the rest of the quantities. For the heat addition, the average temperature of that process is $T_{avg} = (T_2 + T_3)/2 = (542 + 1200)/2 = 871$ K, and the corresponding specific heat is 1.115 kJ/kg·K. Therefore,

$$\dot{W}_C = \dot{m}c_{p12}\,(T_2 - T_1) = \left(3.5\,\frac{\text{kg}}{\text{s}}\right)\left(1.015\,\frac{\text{kJ}}{\text{kg·K}}\right)(540 - 290)\,\text{K}\left(\frac{1\,\text{kW}}{1\,\text{kJ/s}}\right) = 888\,\text{kW}$$

$$\dot{W}_T = \dot{m}c_{p34}\,(T_3 - T_4) = \left(3.5\,\frac{\text{kg}}{\text{s}}\right)\left(1.130\,\frac{\text{kJ}}{\text{kg·K}}\right)(1200 - 686)\,\text{K}\left(\frac{1\,\text{kW}}{1\,\text{kJ/s}}\right) = 2033\,\text{kW}$$

$$\dot{Q}_{in} = \dot{m}c_{p23}\,(T_3 - T_2) = \left(3.5\,\frac{\text{kg}}{\text{s}}\right)\left(1.115\,\frac{\text{kJ}}{\text{kg·K}}\right)(1200 - 540)\,\text{K}\left(\frac{1\,\text{kW}}{1\,\text{kJ/s}}\right) = 2576\,\text{kW}$$

$$\dot{W}_{net} = \dot{W}_T - \dot{W}_C = 2033 - 888 = 1145\,\text{kW}$$

$$\eta_{cycle} = \frac{\dot{W}_{net}}{\dot{Q}_{in}} = \frac{1145\,\text{kW}}{2576\,\text{kW}} = 0.444$$

Comments:

The cycle thermal efficiencies calculated by the two methods are very close to each other, as might have been expected. Both methods take into account the effect of temperature on the specific heat. Comparing the cycle thermal efficiency from the cold-air-standard analysis ($\eta_{cycle} = 0.466$) versus the air-standard analysis ($\eta_{cycle} = 0.444$), it is clear that the fluid properties do have a measureable and significant effect on the results.

EXAMPLE 8-12   Brayton cycle, air-standard analysis with isentropic efficiencies

Air at 14 lbf/in.², 60°F and a volumetric flow rate of 10,000 ft³/min enters the compressor of a Brayton cycle that has a pressure ratio of 14. The turbine is limited to a temperature of 2040°F. The isentropic efficiencies of the compressor and turbine are 83% and 87%, respectively. On the basis of an air-standard analysis with variable specific heats, determine:

a) the cycle thermal efficiency.

b) net power output (in kW).

Approach:

The analysis of this Brayton cycle is similar to that in Example 8-10 or Example 8-11. Again, the only difference is in the evaluation of the properties, which now need to incorporate the effects of the isentropic efficiencies.

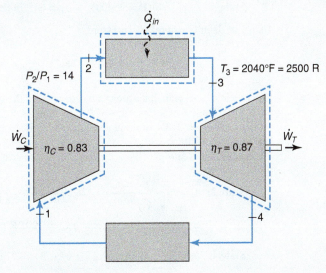

$P_1 = 14$ lbf/in.²
$T_1 = 60°\text{F} = 520$ R
$\dot{V}_1 = 10{,}000$ ft³/min

Assumptions:

A1. The complete cycle and individual components are steady.
A2. Potential and kinetic energy effects are negligible.
A3. Compressor and turbine are adiabatic.
A4. No work occurs in the heat exchangers.

Solution:

a) To determine the cycle thermal efficiency, we need the net power output and the heat input. Applying conservation of energy and mass to the control volumes around the compressor and turbine, and assuming steady state, negligible potential and kinetic energy effects, and adiabatic conditions [A1], [A2], [A3], we obtain:

$$\dot{W}_C = \dot{m}\,(h_2 - h_1)$$
$$\dot{W}_T = \dot{m}\,(h_3 - h_4)$$

Applying conservation of energy and mass to the control volume around the air flowing through the combustor, and assuming steady state, negligible potential and kinetic energy effects, and no work [A1], [A2], [A4], we obtain

$$\dot{Q}_{in} = \dot{m}\,(h_3 - h_2)$$

To evaluate the actual enthalpy differences $(h_2 - h_1)$ and $(h_3 - h_4)$, we use the isentropic efficiency:

$$\eta_T = \frac{h_3 - h_4}{h_3 - h_{4s}} \quad \text{and} \quad \eta_C = \frac{h_{2s} - h_1}{h_2 - h_1}$$

Using these definitions in the turbine and compressor power terms, we get

$$\dot{W}_T = \dot{m}\eta_T\,(h_3 - h_{4s}) \quad \text{and} \quad \dot{W}_C = \dot{m}(h_{2s} - h_1)\big/\eta_C$$

The mass flow rate is obtained from

$$\dot{m} = \rho_1 \dot{V}_1 = \frac{\dot{V}_1}{v_1}$$

Because we have an ideal gas, $Pv = (\overline{R}/M)\,T$ or $v = \overline{R}T/PM$.
Using Table A-9, we evaluate the two inlet enthalpies:

State 1: $T_1 = 60°F = 520\ R \rightarrow h_1 = 124.27\ \text{Btu/lbm} \quad P_{r1} = 1.2147$

State 3: $T_3 = 2040°F = 2500\ R \rightarrow h_3 = 645.78\ \text{Btu/lbm} \quad P_{r3} = 435.7$

Using the definition of relative pressure,

$$P_{r2} = P_{r1}\left(\frac{P_2}{P_1}\right) = 1.2147 \times 14 = 17.0 \quad \text{and} \quad P_{r4} = P_{r3}\left(\frac{P_4}{P_3}\right) = 435.7 \times \frac{1}{14} = 31.1$$

The two ideal exit enthalpies are evaluated, as were the exit enthalpies in Method 1 in Example 8-11. From Table A-9 at this value of P_{r2}, by interpolation, $h_{2s} = 264.1$ Btu/lbm and $T_{2s} = 1092$ R; likewise, $h_{4s} = 313.4$ Btu/lbm and $T_{4s} = 1286$ R. Note that the subscript s is used to indicate that this is for an isentropic process.
We now use the isentropic efficiency definitions to evaluate the actual exit enthalpies:

State 2: $h_2 = h_1 + \dfrac{h_{2s} - h_1}{\eta_C} = 124.27 + \dfrac{264.1 - 124.27}{0.83} = 292.7\ \dfrac{\text{Btu}}{\text{lbm}}$

By interpolation, we obtain $T_2 = 1205\ R = 745°F$.

State 4: $h_4 = h_3 - \eta_T\,(h_3 - h_{4s}) = 645.78 - 0.87\,(645.78 - 313.4) = 356.6\ \dfrac{\text{Btu}}{\text{lbm}}$

By interpolation, we obtain $T_4 = 1452\ R = 992°F$.

The specific volume at the compressor inlet is:

$$v_1 = \frac{\overline{R}T_1}{P_1 M} = \frac{\left(10.73 \, \frac{\text{psia} \cdot \text{ft}^3}{\text{lbmol} \cdot \text{R}}\right)(520 \, \text{R})}{(14 \, \text{psia})\left(28.97 \, \frac{\text{lbm}}{\text{lbmol}}\right)} = 13.8 \, \frac{\text{ft}^3}{\text{lbm}}$$

The mass flow rate is calculated from

$$\dot{m} = \frac{\dot{V}_1}{v_1} = \frac{10{,}000 \, \frac{\text{ft}^3}{\text{min}}}{13.8 \, \frac{\text{ft}^3}{\text{lbm}}} = 724.6 \, \frac{\text{lbm}}{\text{min}} = 12.1 \, \frac{\text{lbm}}{\text{s}}$$

Using these values of enthalpy, we obtain

$$\dot{W}_C = \dot{m}\,(h_2 - h_1) = \left(12.1 \, \frac{\text{lbm}}{\text{s}}\right)(292.7 - 124.27)\frac{\text{Btu}}{\text{lbm}}\left(\frac{1.055 \, \text{kW}}{1 \, \text{Btu/s}}\right) = 2150 \, \text{kW}$$

$$\dot{W}_T = \dot{m}\,(h_3 - h_4) = \left(12.1 \, \frac{\text{lbm}}{\text{s}}\right)(645.78 - 356.6)\frac{\text{Btu}}{\text{lbm}}\left(\frac{1.055 \, \text{kW}}{1 \, \text{Btu/s}}\right) = 3685 \, \text{kW}$$

Thus, the net power output is

$$\dot{W}_{net} = \dot{W}_T - \dot{W}_P = 3685 - 2150 = 1535 \, \text{kW}$$

The input heat transfer rate is

$$\dot{Q}_{in} = \dot{m}\,(h_3 - h_2) = \left(12.1 \, \frac{\text{lbm}}{\text{s}}\right)(645.78 - 292.7)\frac{\text{Btu}}{\text{lbm}}\left(\frac{1.055 \, \text{kW}}{1 \, \text{Btu/s}}\right) = 4507 \, \text{kW}$$

b) Finally, the cycle thermal efficiency is

$$\eta_{cycle} = \frac{\dot{W}_{net}}{\dot{Q}_{in}} = \frac{1535 \, \text{kW}}{4507 \, \text{kW}} = 0.340$$

Comments:

If we let the turbine and compressor be isentropic, the turbine power would be 4243 kW, the compressor power would be 1784 kW, the net power would be 2458 kW (which is 60% more than when the isentropic efficiencies were used), and the cycle thermal efficiency would be 0.504 instead of 0.340. Finally, if we used a cold-air-standard analysis with an ideal compressor and turbine, the cycle thermal efficiency would be

$$\eta_{cycle} = 1 - \frac{1}{r_p^{(k-1)/k}} = 1 - \frac{1}{14^{(1.4-1)/1.4}} = 0.530$$

As can be seen, the Brayton cycle performance is very sensitive to irreversibilities in the turbine and compressor.

8.7 THE BRAYTON CYCLE WITH REGENERATION

(Go to www.wiley.com/college/kaminski)

8.8 COMBINED CYCLES AND COGENERATION

(Go to www.wiley.com/college/kaminski)

8.9 OTTO AND DIESEL CYCLES

The Rankine cycle is used for large-scale power generation. The Brayton cycle is used for large-scale power generation and in transportation (jets, helicopters, and ships). Both of these cycles use rotary machines to produce the power and are considered *external* combustion cycles. (*External* means that the fuel is burned outside the system boundary, and heat is transferred into the working fluid across the system boundary.) While the Brayton cycle also can be an *internal* combustion cycle (fuel is burned inside the system boundary), most often the term *internal combustion engine* is used to refer to cycles that use a reciprocating engine (a piston–cylinder assembly), such as the engines used in most automobiles, trucks, lawnmowers, light planes, motor boats, snowmobiles, portable generators, chainsaws, etc. The range of applications of the reciprocating internal combustion engine is enormous because of its relatively high power-to-weight ratio, scalability of size (from small to large power output), low cost, reasonable efficiency, and reliability.

The two most common types of internal combustion engines are based on the *Otto cycle* and the *Diesel cycle*. In the Otto cycle, air and fuel are mixed before the mixture is introduced into the piston–cylinder assembly; then an electric spark is used to ignite the air–fuel mixture. Hence, the Otto cycle is called a *spark-ignition engine*. The Diesel cycle, on the other hand, is a *compression-ignition engine*. The air in the piston–cylinder assembly is compressed to a high pressure. Fuel is then injected into the piston–cylinder assembly, which now contains high-temperature air (due to the compression). This air temperature is greater than the ignition temperature of the fuel, so the fuel ignites.

Shown in Figure 8-24a is an idealized schematic of the piston–cylinder assembly of a reciprocating engine. Depending on whether the engine is a *two-stroke engine* or a *four-stroke engine*, the four processes that comprise the thermodynamic cycle may be completed in either one complete revolution of the crankshaft or two, respectively. The analysis of a real Otto or Diesel cycle is quite complex. With simplifying assumptions, an ideal cycle can be formulated that is easily analyzed while capturing the essential aspects of the real cycle. We begin with an analysis of the *ideal Otto cycle*. Figure 8-24b shows the four processes (real and ideal) in a four-stroke engine on a *P-v* diagram, while Figure 8-24c shows the thermodynamic processes of the ideal Otto cycle on a *T-s* diagram. In addition to the air-standard assumptions used previously, the ideal Otto cycle is analyzed assuming it is a

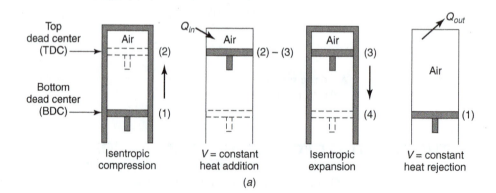

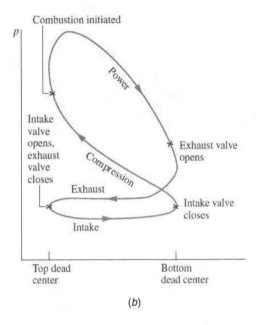

(b)

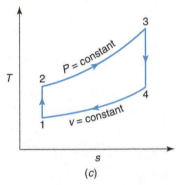

FIGURE 8-24 (a) Schematic of piston–cylinder assembly. (b) Ideal and real Otto cycle *P-v* diagram. (c) Ideal Otto cycle *T-s* diagram. (Part b adapted from Moran and Shapiro, *Fundamentals of Engineering Thermodynamics*, 6th Ed., John Wiley and Sons. Used with permission.)

closed cycle. Referring to Figure 8-24c for the *T-s* behavior, the four internally reversible processes are:

1–2	Isentropic compression
2–3	Constant-volume heat addition
3–4	Isentropic expansion
4–1	Constant-volume heat rejection

These processes are shown on Figure 8-24b with solid lines indicating the ideal *P-v* behavior. The dotted lines represent the actual *P-v* behavior in a real Otto cycle. Note that the area enclosed by the curves can be interpreted as the net work output per unit mass.

For a four-stroke engine operating on an Otto cycle (Figure 8-24a and Figure 8-25), starting with the piston at *bottom dead center* (BDC) and the intake and exhaust valves closed, the *compression stroke* moves the piston from BDC to *top dead center* (TDC). When the piston approaches TDC, the combustion process is initiated by igniting the air–fuel mixture with a spark. The *power stroke* now takes place; the expanding gas does work on the piston as it returns to BDC. After the piston reaches BDC, the exhaust valve opens

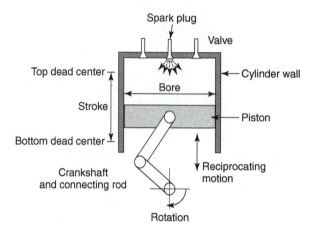

FIGURE 8-25 Schematic of one-cylinder four-stroke reciprocating piston–cylinder engine.

and the piston moves upward through the *exhaust stroke*, during which oxygen, nitrogen, unburned fuel, and products of combustion are expelled from the cylinder. When the piston reaches TDC, the exhaust valve closes, the intake valve opens, and the *intake stroke* occurs, drawing in fresh air (or the air–fuel mixture).

In two-stroke engines (Figure 8-26), all four processes occur in one revolution. During the downward and near the bottom of the power stroke, the exhaust port opens and exhaust gases are released; in addition, the air–fuel mixture is compressed slightly in the engine crankcase. Further movement toward the bottom of the stroke opens the intake port, thus admitting the air–fuel mixture into the piston–cylinder assembly. At BDC the piston reverses direction and, during the upward stroke, first the air–fuel port is covered. Further movement upward continues to expel exhaust gases until the exhaust port, too, is covered. The piston continues toward TDC, compressing the air–fuel mixture. At TDC, the air–fuel mixture is ignited, and the process repeats.

Two-stroke engines are less efficient than four-stroke engines because of poorer intake and exhaust characteristics. However, because two-stroke engines produce power on every revolution, have simpler construction, and have higher power-to-weight and volume-to-weight ratios than four-stroke engines, they are popular for small-scale applications such as motorcycles, motorboats, lawnmowers, snowmobiles, chainsaws, and so on. One approach to increase the power-to-weight ratio in four-stroke engines is to use *turbocharging*. In a

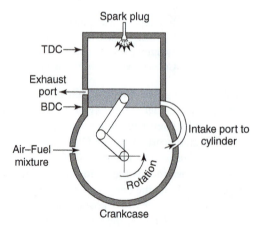

FIGURE 8-26 Schematic of one-cylinder two-stroke reciprocating piston–cylinder engine.

turbocharger, engine exhaust gases are used to drive a turbine connected to a compressor. The compressor forces more air into the piston–cylinder assembly than would flow in a *naturally aspirated* engine. The overall effect of the increased air (and fuel) charge is to increase the power output for an engine of a given size.

The cycle efficiency calculation is similar to that performed for the Rankine and Brayton cycles, but the Otto cycle differs from the Rankine and Brayton cycles in that the processes of the Otto cycle are closed-system processes rather than open-system processes. Instead of using heat transfer rates and power to find efficiency, we use heat transfer and work:

$$\eta_{cycle} = \frac{W_{net}}{Q_{in}} = \frac{W_{34} - W_{12}}{Q_{23}} = \frac{Q_{net}}{Q_{in}} = \frac{Q_{in} - Q_{out}}{Q_{in}} = 1 - \frac{Q_{out}}{Q_{in}} = 1 - \frac{Q_{41}}{Q_{23}} \qquad (8\text{-}37)$$

Each of the four processes can be evaluated with a closed-system energy balance, that is;

$$\Delta E = \Delta PE + \Delta KE + \Delta U = Q - W$$

For example, for process 1–2, we assume adiabatic compression with negligible potential and kinetic energy effects, so that

$$W_{12} = m\,(u_2 - u_1) \qquad (8\text{-}38)$$

For process 2–3, we assume constant-volume heat addition, so there is no work. Again, neglecting potential and kinetic energy effects, the energy equation gives

$$Q_{23} = m\,(u_3 - u_2) \qquad (8\text{-}39)$$

With similar analyses, for processes 3–4 and 4–1, respectively, we obtain

$$W_{34} = m\,(u_3 - u_4) \qquad (8\text{-}40)$$

$$Q_{41} = m\,(u_4 - u_1) \qquad (8\text{-}41)$$

Substituting Eq. 8-39 and Eq. 8-41 into Eq. 8-37, we obtain the cycle thermal efficiency of the ideal Otto cycle:

$$\eta_{cycle} = 1 - \frac{u_4 - u_1}{u_3 - u_2} \qquad (8\text{-}42)$$

The internal energies can be determined using the tabulated data in the air tables, Tables A-9 or B-9, which take into account the temperature dependence of specific heat and are the most accurate way to determine efficiency.

If we use $\Delta u = \int c_v\,dT = c_v\,\Delta T$ and invoke the cold-air-standard analysis assumptions, the cycle efficiency becomes:

$$\eta_{cycle} = 1 - \frac{c_v\,(T_4 - T_1)}{c_v\,(T_3 - T_2)} = 1 - \frac{T_4 - T_1}{T_3 - T_2} = 1 - \frac{T_1}{T_2}\left(\frac{T_4/T_1 - 1}{T_3/T_2 - 1}\right) \qquad (8\text{-}43)$$

Because the compression (1–2) and expansion (3–4) processes are isentropic, we can use the isentropic relations for an ideal gas to show:

$$\frac{T_1}{T_2} = \left(\frac{v_2}{v_1}\right)^{k-1} = \frac{T_4}{T_3} = \left(\frac{v_3}{v_4}\right)^{k-1} \qquad (8\text{-}44)$$

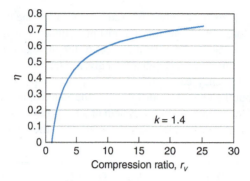

FIGURE 8-27 Thermal efficiency of the ideal Otto cycle ($k = 1.4$).

Recognizing that $v_2 = v_3$ and $v_4 = v_1$, it is clear that $T_4/T_1 = T_3/T_2$. Thus, Eq. 8-43 reduces to

$$\eta_{cycle} = 1 - \frac{T_1}{T_2} \tag{8-45}$$

From Eq. 8-44 we have this temperature ratio in terms of specific volumes and specific heat ratio, k. We define a volume compression ratio, r_v, as

$$\boxed{r_v = \frac{V_1}{V_2} = \frac{v_1}{v_2}} \tag{8-46}$$

where V_1 is the maximum volume and V_2 is the minimum volume in the piston–cylinder assembly. Substituting Eq. 8-44 and Eq. 8-46 into Eq. 8-45, we obtain

$$\boxed{\eta_{cycle} = 1 - \frac{1}{r_v^{k-1}}} \tag{8-47}$$

Figure 8-27 illustrates the efficiency given by Eq. 8-47. For a given gas (value of k), the thermal efficiency of the ideal Otto cycle is a function only of the compression ratio. There is a diminishing effect on efficiency with increasing compression ratio. In addition, there are practical limitations to the maximum usable compression ratio. Volatile fuel is mixed with air before the compression process. With increasing compression ratio, the combustible air–fuel mixture may reach a temperature greater than the self-ignition temperature of the fuel before the fuel can be ignited with a spark. This preignition results in "knock," the noise an engine makes when it is running poorly; it can damage an engine. Knock can be suppressed with additives to the fuel that raises its octane rating.

EXAMPLE 8-13 Otto cycle

Following an air-standard Otto cycle, air at 14.7 psia and 60°F enters a piston–cylinder assembly (compression ratio of 11.5) with an initial volume of 40 in.[3]. Heat is added until the maximum temperature is 3500°F.

a) Determine the heat addition (in Btu).

b) Determine the net work (in Btu).

c) Determine the cycle thermal efficiency.

d) If the engine has a four-stroke process, eight cylinders, and runs at 3000 revolutions per minute (rpm), what is the net power output (in kW and hp)?

Approach:

The heat addition (Q_{23}) is determined from application of the closed-system energy equation to process 2–3. The net work is determined from $W_{net} = Q_{net} = Q_{23} - Q_{41}$. To evaluate the heat output (Q_{41}), use the closed-system energy balance on process 4–1. Then, both Q_{23} and Q_{41} will be known and net work can be calculated. With both the net work and heat input known, the cycle thermal efficiency is easily calculated. The net work output is for one cylinder during one power stroke. For a four-stroke process, there is only one power stroke per two revolutions. We need to take into account the number of power strokes per minute to obtain the net power output from the engine.

$$P_1 = 14.7 \text{ lbf/in.}^2$$
$$T_1 = 60°F = 520 \text{ R}$$
$$T_3 = 3500°F = 3960 \text{ R}$$
$$V_1 = 40 \text{ in.}^3$$
$$P_2/P_1 = 11.5$$
$$N = 3000 \text{ rpm}$$

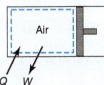

Assumptions:

A1. No work occurs during the heat addition process.

A2. Potential and kinetic energy effects are negligible.

A3. The compression and expansion processes are adiabatic.

Solution:

a) The heat addition, Q_{23}, is obtained by applying conservation of energy to the closed system defined in the figure. For the closed system assuming no work and negligible potential and kinetic energy effects [A1] [A2], we obtain for process 2–3 (volume is constant):

$$Q = \Delta U = m\,(u_3 - u_2) = Q_{23}$$

The internal energies can be evaluated from the given information. The mass is calculated with the ideal gas equation:

$$m = \frac{P_1 V_1 M}{\bar{R} T_1}$$

b) Recognizing that processes 1–2 and 3–4 are adiabatic [A3], we can determine the cycle net work by

$$W_{net} = Q_{net} = Q_{23} - Q_{41}$$

The heat rejection, Q_{41}, is determined in the same manner as the heat addition, so for process 4–1:

$$Q = \Delta U = m\,(u_1 - u_4) = -Q_{41}$$

so that

$$W_{net} = m\big[\,(u_3 - u_2) - (u_4 - u_1)\,\big]$$

The internal energies can be determined from the given information.

c) The cycle thermal efficiency then is:

$$\eta_{cycle} = \frac{W_{net}}{Q_{in}} = \frac{W_{net}}{Q_{23}}$$

d) The net work calculated above is produced by one cylinder during each power stroke, which occurs every other revolution of the crankshaft. Let n = number of cylinders, N = engine speed in revolutions per minute, and X = the number of crankshaft revolutions per power stroke ($X = 2$ for a four-stroke engine, and $X = 1$ for a two-stroke engine). Hence, power produced is

$$\dot{W}_{net} = \frac{W_{net}nN}{X}$$

Using Table B-9, we evaluate the two initial internal energies:

State 1: $T_1 = 60°F = 520\,R$ $\quad u_1 = 88.62\,Btu/lbm$ $\quad v_{r1} = 158.58$

State 3: $T_3 = 3500°F = 3960\,R$ $\quad$ by interpolation $u_3 = 804.8\,Btu/lbm$ $\quad v_{r3} = 0.468$

A4. The compression and expansion processes are isentropic.

Assuming the compression and expansion processes are isentropic [A4] and using the definition of relative volume,

$$v_{r2} = v_{r1}\left(\frac{v_2}{v_1}\right) = 158.58\left(\frac{1}{11.5}\right) = 13.79$$

$$v_{r4} = v_{r3}\left(\frac{v_4}{v_3}\right) = 0.468\,(11.5) = 5.38$$

From Table B-9 at this value of v_{r2}, by interpolation, $u_2 = 234.7\,Btu/lbm$; likewise, $u_4 = 336.5\,Btu/lbm$.

The mass of air in the piston–cylinder assembly is

$$m = \frac{P_1 V_1 M}{\overline{R} T_1} = \frac{\left(14.7\,\frac{lbf}{in.^2}\right)(40\,in.^3)\left(28.97\,\frac{lbm}{lbmol}\right)}{\left(10.73\,\frac{psia \cdot ft^3}{lbmol \cdot R}\right)(520\,R)\,(12in./1ft)^3} = 0.00177\,lbm$$

The net work is:

$$W_{net} = m\left[(u_3 - u_2) - (u_4 - u_1)\right]$$

$$= (0.00177\,lbm)\,[(804.8 - 234.7) - (336.5 - 88.62)]\,\frac{Btu}{lbm} = 0.569\,Btu$$

The input heat transfer is:

$$Q_{23} = m\,(u_3 - u_2) = (0.00177\,lbm)\,(804.8 - 234.8)\,\frac{Btu}{lbm} = 1.01\,Btu$$

Therefore, the cycle thermal efficiency is

$$\eta_{cycle} = \frac{\dot{W}_{net}}{\dot{Q}_{23}} = \frac{0.569\,Btu}{1.01\,Btu} = 0.564$$

The power produced at 3000 rpm with eight cylinders in this four-stroke engine is

$$\dot{W}_{net} = \frac{W_{net}nN}{X}$$

$$= \frac{\left(0.569\,\frac{Btu}{cylinder\text{-}powerstroke}\right)(8\,cylinders)\left(3000\,\frac{rev}{min}\right)\left(\frac{1\,min}{60\,s}\right)\left(\frac{1.055\,kW}{1\,Btu/s}\right)}{2\,\frac{rev}{powerstroke}}$$

$$= 120\,kW = 161\,hp$$

Comments:

For a cold-air-standard analysis, the cycle thermal efficiency using Eq. 8-47 is 0.624.

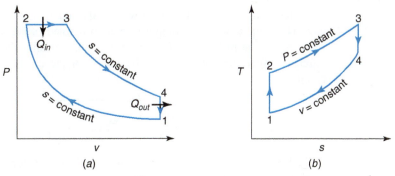

FIGURE 8-28 The *P-v* and *T-s* diagrams for the ideal Diesel Cycle.

The *ideal Diesel cycle* is analyzed in much the same way as the Otto cycle. Again, we assume it is a closed cycle. The four internally reversible processes illustrated in Figure 8-28 are:

1–2	Isentropic compression (from BDC to TDC)
2–3	Constant-pressure heat addition (the volume expands from TDC to the *cutoff point*, the location in the stroke where the heat addition process stops)
3–4	Isentropic expansion (additional expansion from cutoff point to BDC)
4–1	Constant-volume heat rejection (occurs at BDC)

The area enclosed by the curves can be interpreted as the net work output per unit mass. The primary difference between the Diesel cycle and Otto cycle is the process from 2–3. In the Diesel cycle, heat is transferred at constant pressure (with increasing volume) through part of the power stroke until point 3 (the cutoff point) is reached as compared to the constant-volume (and increasing pressure) heat addition in the Otto cycle; then, from 3–4, the remainder of the power stroke (expansion) is performed isentropically.

The thermal efficiency is calculated with Eq. 8-37. However, the heat in, Q_{23}, for the Diesel cycle is different than that for the Otto cycle because the volume changes during the constant-pressure heat addition. Applying conservation of energy (ignoring potential and kinetic energy), we obtain

$$
\begin{aligned}
Q_{23} &= \Delta U_{23} + W_{23} \\
&= m\,(u_3 - u_2) + mP\,(v_3 - v_2) \\
&= m\,(h_3 - h_2)
\end{aligned}
\tag{8-48}
$$

The heat out, Q_{41}, is the same as in the Otto cycle (see Eq. 8-38). Inserting the expressions for heat transfer into the efficiency equation (Eq. 8-37) results in

$$
\eta_{cycle} = 1 - \frac{u_4 - u_1}{h_3 - h_2}
\tag{8-49}
$$

These enthalpies and internal energies can be obtained from the air Tables A-9 or B-9. We can simplify this expression by using the cold-air-standard assumptions to get $\Delta u = \int c_v\, dT = c_v\, \Delta T$ and $\Delta h = \int c_p\, dT = c_p\, \Delta T$, so that efficiency becomes

$$
\eta_{cycle} = 1 - \frac{c_v\,(T_4 - T_1)}{c_p\,(T_3 - T_2)} = 1 - \frac{(T_4 - T_1)}{k\,(T_3 - T_2)}
\tag{8-50}
$$

where $k = c_p/c_v$ by definition.

To further simplify Eq. 8-50 we use the ideal gas isentropic process expressions (similar to Eq. 8-44), the compression ratio defined in Eq. 8-46, and the *cutoff ratio*, r_c, which is the ratio of the volume in the cylinder after the heat addition (point 3 on Figure 8-28) to the volume at the beginning of the combustion process or end of the compression process (point 2 on Figure 8-28):

$$r_c = \frac{V_3}{V_2} = \frac{v_3}{v_2} \tag{8-51}$$

Thus the thermal efficiency of the ideal Diesel cycle when using the cold-air-standard analysis is:

$$\eta_{cycle} = 1 - \frac{1}{r_v^{k-1}} \left[\frac{r_c^k - 1}{k(r_c - 1)} \right] \tag{8-52}$$

Figure 8-29 illustrates Eq. 8-52. Note that $r_c = 1$ implies that all the heat would be added at a constant volume; in the Otto cycle, all the heat is added at constant volume while the pressure rises. Thus, if $r_c = 1$, the Diesel cycle essentially becomes exactly like the Otto cycle. Comparison of the efficiency expressions for the Otto cycle (Eq. 8-47) and the Diesel cycle (Eq. 8-52) shows that they differ only by the term in brackets; with $r_c = 1$, the bracketed term equals unity. Hence, for the same compression ratio, the Otto cycle efficiency is always greater than that of the Diesel cycle.

Why are Diesel engines used in so many applications, from locomotives, heavy trucks, and buses to ships and stand-by generator power given the apparent advantage of the Otto cycle over the Diesel cycle? Because Diesel engines are compression ignition, the compression ratio for the Diesel engine (typically in the range of 12 to 20) is higher than that of the Otto cycle (typically 7 to 10). This results in a higher cycle thermal efficiency. In addition, because fuel is not injected into the piston–cylinder assembly until compression is complete, preignition or "knock" cannot occur. Therefore, less refined fuel can be used in a Diesel engine compared to the more volatile (and expensive) fuel used in the Otto cycle.

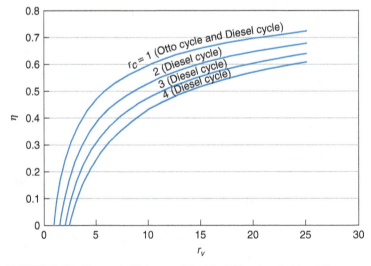

FIGURE 8-29   Thermal efficiency of the ideal Diesel cycle ($k = 1.4$).

Finally, because the Diesel cycle must withstand higher pressures, they are more rugged and require less maintenance. Nevertheless, because Otto cycle engines tend to be lighter, less expensive to manufacture, and operate over a wider range of engine speeds (which results in less stringent transmission requirements), Otto cycle engines are more widely used than Diesel cycles in typical passenger cars.

EXAMPLE 8-14 **Diesel cycle**

Air at 100 kPa and 300 K undergoes an air-standard Diesel cycle. The air is compressed to 4000 kPa, and heat is added until the maximum temperature is 2100 K.

a) Determine the compression ratio.

b) Determine the cutoff ratio.

c) Determine the cycle thermal efficiency.

Approach:

The analysis of the Diesel cycle is similar in some ways to that of the Otto cycle. For this particular problem, the given information on states is used with property evaluation techniques and definitions to determine the compression and cutoff ratios. For the cycle thermal efficiency, we use Eq. 8-58 and evaluate properties at the four states in the cycle.

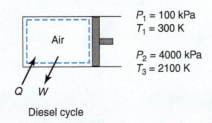

$P_1 = 100$ kPa
$T_1 = 300$ K

$P_2 = 4000$ kPa
$T_3 = 2100$ K

Diesel cycle

Assumptions:

A1. The compression and expansion processes are isentropic.
A2. Air is an ideal gas.

Solution:

a) The compression ratio is defined as:

$$r_v = \frac{V_1}{V_2}$$

We have neither of these volumes, so we need to use what we do have. As shown on the T-s diagram for the Diesel cycle, we see that the process 1–2 is isentropic [A1], and assuming an ideal gas [A2] the relative volume is:

$$r_v = \frac{V_1}{V_2} = \frac{v_{r1}}{v_{r2}}$$

Relative volumes are a function of temperature only. We have T_1, so v_{r1} can be evaluated. State 2 and, hence, v_{r2}, can be determined with the properties known at 2: $P_2 = 4000$ kPa and $s_2 = s_1$.

b) The cutoff ratio is:

$$r_c = \frac{V_3}{V_2} = \frac{v_3}{v_2}$$

From the P-v diagram, $P_3 = P_2$, T_2 is evaluated above and T_3 is given. Using the ideal gas equation,

$$r_c = \frac{v_3}{v_2} = \frac{\overline{R}T_3/P_3M}{\overline{R}T_2/P_2M} = \frac{T_3}{T_2}$$

c) The cycle thermal efficiency is

$$\eta_{cycle} = 1 - \frac{u_4 - u_1}{h_3 - h_2}$$

These two internal energies and two enthalpies can be evaluated from the given information.

Using Table A-9, we evaluate the two initial internal energies:

State 1: $T_1 = 300$ K $\rightarrow u_1 = 214.07$ kJ/kg , $v_{r1} = 621.2$, $P_{r1} = 1.386$

State 2: $P_2 = 4000$ kPa, $s_2 = s_1 \rightarrow$ Using the definition of relative pressure

$$P_{r2} = P_{r1}\left(\frac{P_2}{P_1}\right) = 1.386\left(\frac{4000}{100}\right) = 55.44$$

By interpolation, $h_2 = 856.6$ kJ/kg, $v_{r2} = 43.14$, and $T_2 = 831.4$ K.

State 3: $T_3 = 2100$ K $\rightarrow h_3 = 2377.4$ kJ/kg, $v_{r3} = 2.356$

State 4: $s_4 = s_3 \rightarrow \dfrac{v_{r4}}{v_{r3}} = \dfrac{V_4}{V_3} = \dfrac{V_4}{V_2}\dfrac{V_2}{V_3} = \dfrac{V_1}{V_2}\dfrac{V_2}{V_3} = \dfrac{r_v}{r_c}$

Therefore, $v_{r4} = v_{r3}\dfrac{r_v}{r_c}$. From above, $r_v = \dfrac{v_{r1}}{v_{r2}} = \dfrac{621.2}{43.14} = 14.4$,

$$r_c = \frac{T_3}{T_2} = \frac{2100K}{831.4K} = 2.53, \text{ and}$$

$$v_{r4} = v_{r3}\frac{r_v}{r_c} = 2.356\frac{14.4}{2.53} = 13.43. \text{ By interpolation, } u_4 = 959.4 \text{ kJ/kg}$$

and $T_4 = 1229.3$ K.

The cycle thermal efficiency is:

$$\eta_{cycle} = 1 - \frac{u_4 - u_1}{h_3 - h_2} = 1 - \frac{(959.4 - 214.07)\,kJ/kg}{(2377.4 - 856.6)\,kJ/kg} = 0.510$$

Comments:

If we perform a cold-air-standard analysis using the compression and the cutoff ratios developed above, the cycle thermal efficiency using Eq. 8-52 is 0.572. Variable specific heat effects are significant.

SUMMARY

Refrigeration cycles use input mechanical energy to move low-temperature energy to a higher temperature, and vapor-compression refrigeration cycles have four processes: compression (work input), heat rejection, expansion, and heat addition. Refrigeration cycles can be used for two purposes. If we use the heat added to the cycle, then we are using the cycle for refrigeration/cooling, and the performance of the cycle is described by the coefficient of performance:

$$COP_{Ref} = \frac{\text{energy we want to use}}{\text{energy we purchase}} = \frac{\dot{Q}_L}{\dot{W}_{in}} = \frac{\dot{Q}_L}{\dot{Q}_H - \dot{Q}_L}$$

If we use the heat rejected by the cycle, then we are using the cycle for heating, and the performance of this *heat pump cycle* is described by the coefficient of performance:

$$COP_{HP} = \frac{\text{energy we want to use}}{\text{energy we purchase}} = \frac{\dot{Q}_H}{\dot{W}_{in}} = \frac{\dot{Q}_H}{\dot{Q}_H - \dot{Q}_L}$$

These two *COP*s are related by:

$$COP_{HP} = COP_{Ref} + 1$$

The *Rankine cycle* is a *heat power cycle* and is used to convert chemical energy stored in coal, natural gas, oil, wood, or other substances into useful mechanical energy through a combustion process; energy contained in uranium and released through a nuclear reaction can also be used. The cycle has a minimum of four processes: compression (work input), heat addition and vaporization of the working fluid, expansion (work output), and heat rejection and condensation of the working fluid. The performance of a power cycle is described by the *cycle thermal efficiency*:

$$\eta_{cycle} = \frac{\text{energy we want to use}}{\text{energy we purchase}} = \frac{\dot{W}_{net}}{\dot{Q}_H} = 1 - \frac{\dot{Q}_L}{\dot{Q}_H}$$

Another way to describe the performance of a power cycle is to use the *net station heat rate*:

$$\text{net station heat rate} = \frac{3413 \text{ Btu}/\text{kWh}}{\eta_{boiler}\,\eta_{cycle}}$$

which incorporates the inefficiency of the boiler in addition to the performance of the cycle itself.

To increase the cycle thermal efficiency, we modify the basic Rankine cycle by including additional devices and processes. *Reheat* involves the partial expansion of the working fluid through a high-pressure turbine, extraction of the fluid and addition of more heat to the fluid (*reheating*), then completion of the expansion in a low-pressure turbine; several stages of reheat can be used. *Regeneration* involves extracting a portion of the vapor flow from a turbine and rerouting this fluid through a *feedwater heater* that raises the boiler feedwater inlet temperature. Several o*pen* (a mixer) and *closed* (a heat exchanger) feedwater heaters can be used in a cycle.

The *Brayton cycle* is a heat power cycle that must have a minimum of four processes, as described above. In this *gas power cycle,* the working fluid remains a gas through all processes and uses a rotating gas compressor and turbine for the compression and expansion, respectively. An *open cycle* incorporates combustion of the fuel into the process and rejects heat to the atmosphere directly; a *closed cycle* replaces the heat addition and heat rejection processes with constant-pressure heat exchangers. A common assumption is called the *air-standard analysis;* for simplicity, we assume the working fluid is only air and ignore the products of combustion. The analysis can be simplified further by using a *cold-air-standard analysis,* which ignores the temperature dependence of specific heat and assumes all air properties are evaluated at room temperature (25°C or 77°F). For an ideal Brayton cycle (turbine and compressor are isentropic) using a cold-air-standard analysis, the cycle thermal efficiency is:

$$\eta_{cycle} = 1 - \frac{1}{r_p^{(k-1)/k}}$$

where r_p is the pressure ratio across the compressor.

To increase the cycle thermal efficiency, we modify the basic Brayton cycle by including additional devices and processes. *Reheat* is similar to the reheat described above for the Rankine cycle. *Regeneration* involves the addition of a heat exchanger to the cycle; the hot exhaust gases from the turbine are used to heat the compressed air exiting from the compressor. *Intercooling* involves the partial compression of the incoming air to an intermediate pressure, routing the air to a heat exchanger where the air is cooled, then completion of the compression process in a second compressor; several stages of intercooling can be used.

Combined cycles use two cycles (e.g., a Brayton cycle and a Rankine cycle) operating in tandem so that the overall plant thermal efficiency is greater than that of either of the individual cycles. A Brayton cycle has a higher high temperature than a Rankine cycle, and the Rankine cycle has a lower low temperature than the Brayton cycle. The exhaust gases from the Brayton cycle are used as the heat input to the Rankine cycle. The combined cycle efficiency is defined as

$$\eta_{cycle} = \frac{\dot{W}_{net,Brayton} + \dot{W}_{net,Rankine}}{\dot{Q}_{in}}$$

Cogeneration is a system in which electricity and steam are produced in the same plant, and the cost to do this is often much less than if two separate plants (one strictly for electricity and one strictly for steam) were used. For industrial use, cogeneration is called *combined heat and power (CHP)* while for community use it is called *district heating.* The overall plant efficiency is defined as

$$\eta_{plant} = \frac{\dot{W}_{net} + \dot{Q}_{delivered}}{\dot{Q}_{in}} = 1 - \frac{\dot{Q}_{out}}{\dot{Q}_{in}}$$

The term *internal combustion engine* is used to refer to heat power cycles that use a reciprocating engine (a piston–cylinder assembly). In an *Otto cycle,* air and fuel are mixed before the mixture is introduced into the piston–cylinder assembly; then an electric spark is used to ignite the air–fuel mixture. Hence, the Otto cycle is called a *spark-ignition engine.* The cycle has four processes (compression, constant-volume heat addition, expansion, constant-volume heat rejection). For an ideal Otto cycle (compression and expansion processes are isentropic) using a cold-air-standard analysis, the cycle thermal efficiency is

$$\eta_{cycle} = 1 - \frac{1}{r_v^{k-1}}$$

where r_v is the compresson ratio based on initial and final volumes.

The *Diesel cycle* is a *compression-ignition engine.* The air in the piston–cylinder assembly is compressed to a high pressure. Fuel is then injected into the piston–cylinder assembly, which now contains high-temperature air (due to the compression). This air temperature is greater than the ignition temperature of the fuel, so the fuel ignites. The cycle has four processes (compression, constant-pressure heat addition, expansion, constant-volume heat rejection). For an ideal Diesel cycle (compression and expansion processes are isentropic) using a cold-air-standard analysis, the cycle thermal efficiency is

$$\eta_{cycle} = 1 - \frac{1}{r_v^{k-1}}\left[\frac{r_c^k - 1}{k\,(r_c - 1)}\right]$$

where r_v is the compresson ratio based on initial and final volumes; r_c is the cutoff ratio and represents the fraction of the stroke during which heat is added.

SELECTED REFERENCES

ALTHOUSE, A. D., C. H. TURNQUIST, and A. F. BRACCIANO, *Modern Refrigeration and Air Conditioning*, 2nd ed., Goodheart & Willcox, Tinley Park, Il. 2001.

ÇENGEL, Y. A., and M. A. BOLES, *Thermodynamics: An Engineering Approach*, 3rd ed., McGraw-Hill, New York, 1998.

HEYWOOD, J. B., *Internal Combustion Engine Fundamentals*, McGraw-Hill Science, New York, 1988.

KEHLHOFER, R., et al., ed., *Combined-Cycle Gas & Steam Turbine Power Plants*, 2nd ed., PENNWELL Publishing, Tulsa, 1999.

KITTO, J. B., ed., *Steam: Its Generation and Use*, 40th ed., Babcock & Wilcox, New York, 1992.

MORAN, M. J., and H. N. SHAPIRO, *Fundamentals of Engineering Thermodynamics*, 5th ed., Wiley, New York, 2003.

Parsons, R., ed., *Fundamentals: 2001 ASHRAE Handbook*, American Society of Heating, Refrigerating, and Air-Conditioning Engineers, Atlanta, GA, 2001.

SONNTAG, R. E., C. BORGNAKKE, and G. J. VAN WYLEN, *Fundamentals of Thermodynamics*, 5th ed., Wiley, New York, 1998.

TAYLOR, C. F., *The Internal Combustion Engine in Theory and Practice: Vol. 1. Thermodynamics, Fluid Flow, and Performance*, 2nd ed., MIT Press, Cambridge, MA, 1985.

PROBLEMS

Problems designated with WEB refer to material available at www.wiley.com/college/kaminski

REFRIGERATION CYCLE

P8-1 In an ideal vapor-compression refrigeration cycle using refrigerant R-134a, saturated vapor enters the compressor at a temperature of –20°C with a volumetric flow rate of 1.5 m^3/min. The refrigerant leaves the condenser at 35°C, 10 bar.

a. Determine the compressor power (in kW).

b. Determine the refrigerating capacity (in tons).

c. Determine the coefficient of performance (*COP*).

P8-2 R-134a is used in a vapor-compression refrigeration cycle. Saturated vapor at 20 psia enters the compressor, which has an isentropic efficiency of 80%, and leaves at 120 psia. Saturated liquid exits the condenser, and saturated vapor exits the evaporator. The mass flow rate is 15 lbm/min.

a. Determine the compressor power (in hp).

b. Determine the refrigerating capacity (in tons).

c. Determine the coefficient of performance (*COP*).

P8-3 A vapor-compression refrigeration cycle uses R-134a. Liquid at 1200 kPa exits the condenser at 40°C. The evaporator operates at a pressure of 240 kPa. The compressor isentropic efficiency is 75%. Determine the cycle coefficient of performance (*COP*) if the refrigerant leaves the evaporator as superheated vapor at 0°, 5°, 10°, 15°, and 20°C above the saturation temperature.

P8-4 An ice-making plant is designed to produce 10,000 lbm of ice each day. Liquid water enters the plant at 50°F and solid ice leaves it at 20°F; the enthalpy change for the water when it goes from the liquid to the solid is 167.4 Btu/lbm. R-134a enters the compressor as saturated vapor at 20 lbf/in.^{2} and leaves the condenser as a saturated liquid at 100 lbf/in.^{2}. The compressor isentropic efficiency is 85%.

a. Determine the refrigerant flow rate (in lbm/s).

b. Determine the compressor input power (in hp).

c. Determine the cycle coefficient of performance.

P8-5 A large frozen food storage building is to be maintained at –10°C. The cooling load is 243 kW. An ideal R-134a vapor-compression refrigeration cycle is to be used for the cooling. Saturated vapor enters the compressor at 100 kPa, and saturated liquid leaves the condenser at 800 kPa. Water used to cool the condenser experiences a 10°C temperature rise.

a. Determine the mass flow rate of the refrigerant (in kg/s).

b. Determine the power input to the compressor (in kW).

c. Determine the cycle coefficient of performance.

d. Determine the water mass flow rate (in kg/s).

P8-6 Data from an experiment on a new R-134a vapor-compression refrigeration cycle were obtained. The motor driving the compressor consumed 2.14 hp. The refrigerant entered the compressor at 20°F and 20 lbf/in.^{2} and exited at 170°F and 160 lbf/in.^{2}. Refrigerant exited the condenser at 155 lbf/in.^{2} as a saturated liquid, and the pressure just downstream of the expansion valve was 22 lbf/in^2.

a. Determine the compressor isentropic efficiency.

b. Determine the cooling capacity (in tons).

c. Determine the cycle coefficient of performance (*COP*).

HEAT PUMP CYCLE

P8-7 A home is heated with a groundwater heat pump, which uses subterranean water at 50°F as the low-temperature reservoir. The heat pump is designed to blow air in the residential space at 30°F above the thermostat set point. Heat loss from the building to the outside air is 358 W per degree Fahrenheit of temperature difference between the inside of the house and the outside air. On a winter day when the outside temperature is 20°F, the thermostat is set at 65°F. Determine the *minimum possible* electric power that must be supplied to the heat pump under these conditions.

P8-8 In winter, a building requires 94,000 kJ/h of heat, and an ideal vapor-compression heat pump is used. R-134a enters the isentropic compressor of the heat pump at 0.4 MPa, 10°C and exits at 1 MPa. Saturated liquid leaves the condenser.

a. Determine the mass flow rate of the refrigerant (in kg/s).

b. Determine the power input to the compressor (in kW).

c. Determine the cycle coefficient of perfomance (*COP*).

P8-9 A vapor-compression refrigeration system with a cooling capacity of 6 tons is to be used as a heat pump to warm liquid water. The working fluid is R-134a. The water enters the condenser at 55°F and leaves at 80°F. Saturated vapor enters the compressor at 40 lbf/in.^{2}, and superheated vapor leaves at 120 lbf/in.^{2}, 110°F. Heat transfer between the compressor and the surroundings occurs at a rate of 1.0 Btu/lbm of refrigerant flowing through the compressor. Liquid refrigerant leaves the condenser at 85°F, 120 lbf/in^2.

a. Determine the compressor power input (in Btu/min).

b. Determine the water flow rate through the condenser (in lbm/min).

c. Determine the coefficient of performance.

P8-10 An ideal vapor-compression heat pump cycle using R-134a is used to heat a house. The inside temperature is 22°C; the outside temperature is 0°C. Saturated vapor at 2.2 bar enters the compressor, and saturated liquid leaves the condenser at 8 bar. The mass flow rate is 0.2 kg/s.

a. Determine the power input to the compressor (in kW).

b. Determine the coefficient of performance.

c. Determine the coefficient of performance if the system were used as a refrigeration cycle.

d. Determine the maximum theoretical coefficient of performance working between thermal reservoirs at 22°C and 0°C.

P8-11 R-134a is used in a vapor-compression heat pump cycle in which the refrigerant enters the adiabatic compressor at 2.4 bar, 0°C, with a volumetric flow rate of 0.8 m^3/min, and leaves at 10 bar, 55°C. Liquid leaves the condenser at 34°C.

a. Determine the power input to the compressor (in kW).

b. Determine heating capacity of the system (in kW).

c. Determine the coefficient of performance.

d. Determine the isentropic compressor efficiency.

SIMPLE RANKINE CYCLE

P8-12 In an ideal Rankine cycle, saturated water vapor enters the turbine at 20 MPa and exits at 10 kPa. Saturated liquid exits the condenser.

a. Determine the net work per unit mass of steam flow (in kJ/kg).

b. Determine heat input per unit mass of steam flow (in kJ/kg).

c. Determine the cycle thermal efficiency.

d. Determine the heat rejection per unit mass of steam flow (in kJ/kg).

P8-13 Data from a simple Rankine cycle power plant were measured to determine actual performance. The measured steam flow rate was 6.8 kg/s. The measured conditions of the water are shown in the table.

Device	Inlet conditions	Outlet conditions
Pump	$P_1 = 10$ kPa; $T_1 = 45°C$	$P_2 = 5.2$ MPa; $T_2 = 46°C$
Boiler	$P_3 = 5.1$ MPa; $T_3 = 45°C$	$P_4 = 5.0$ MPa; $T_4 = 500°C$
Turbine	$P_5 = 4.5$ MPa; $T_5 = 500°C$	$P_6 = 15$ kPa; $x_6 = 0.97$
Condenser	$P_6 = 15$ kPa; $x_6 = 0.97$	$P_7 = 12$ kPa; $T_7 = 45°C$

a. Determine the heat addition (in kW).

b. Determine the net power produced (in kW).

c. Determine the heat rejection (in kW).

d. Determine the turbine isentropic efficiency.

e. Determine the pump isentropic efficiency.

f. Determine the cycle thermal efficiency.

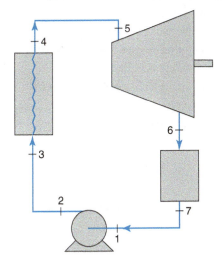

P8-14 A Rankine cycle power plant has a flow control valve (for use during partial load conditions) located between the boiler and the turbine. At one partial load condition, steam leaves the boiler at 6.0 MPa, 300°C; flows through the valve, which drops the pressure to 4.5 MPa; then enters the turbine, in which it expands to the condenser pressure of 10 kPa. Net power output is 500 MW. The turbine isentropic efficiency is 82% and that of the pump is 68%. The condensate from the condenser leaves at 40°C.

a. Determine the steam flow rate (in kg/s).

b. Determine the cycle thermal efficiency.

c. Determine the steam flow rate and cycle thermal efficiency if no valve is present between the boiler and turbine.

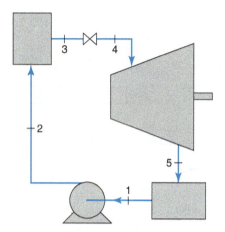

P8-15   In a 2-MW Rankine cycle, saturated vapor leaves the boiler at 2 MPa and expands in the turbine to an outlet condition of 15 kPa, 94% quality. Saturated liquid leaves the condenser. The pump is ideal. The temperature rise of the cooling water in the condenser is 10°C.

a. Determine the mass flow rate of steam (in kg/s).

b. Determine the input heat transfer rate (in MW).

c. Determine the cycle thermal efficiency.

d. Determine the cooling water flowrate (in kg/s).

P8-16   A 1000-MW coal-fired Rankine cycle power plant uses coal that has a heating value of 13,390 Btu/lbm. The inlet

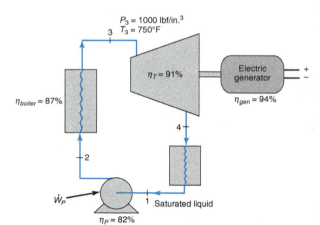

conditions to the steam turbine are 1000 psia, 750°F. Saturated liquid exits the condenser at a pressure of 1 psia. The boiler has a combustion efficiency of 87%, and the electric generator has an efficiency of 94%. The turbine isentropic efficiency is 91%, and the pump isentropic efficiency is 82%.

a. Determine the overall plant efficiency.

b. Determine the coal flow rate (in tons/day).

P8-17   For Problem P 8-12, assume that the turbine has an isentropic efficiency of 91% and the pump has an isentropic efficiency of 78%.

a. Determine the net work per unit mass of steam flow (in kJ/kg).

b. Determine the heat input per unit mass of steam flow (in kJ/kg).

c. Determine the cycle thermal efficiency.

d. Determine the heat rejection per unit mass of steam flow (in kJ/kg).

P8-18   Steam flowing at 15.9 lbm/s enters the turbine of a simple Rankine cycle power plant at 1000 psia, 800°F and exits at 2 psia. Saturated liquid exits the condenser. The turbine and pump are isentropic.

a. Determine the power output of the turbine (in hp and kW).

b. Determine the power input to the pump (in hp and kW).

c. Determine the heat input (in hp and kW).

d. Determine the cycle thermal efficiency.

P8-19   For Problem P 8-18, assume the turbine has an isentropic efficiency of 87% and the pump has an isentropic efficiency of 75%. All other conditions remain the same.

a. Determine the power output of the turbine (in hp and kW).

b. Determine the power input to the pump (in hp and kW).

c. Determine the heat input (in hp and kW).

d. Determine the cycle thermal efficiency.

RANKINE CYCLE WITH REHEAT

P8-20   An ideal Rankine cycle uses one stage of reheat. Steam enters the high-pressure turbine at 10 MPa, 550°C; expands to 1 MPa, where it is extracted and routed to a reheater, where the steam temperature is raised to 500°C. The steam is then expanded in the low-pressure turbine to the condenser pressure of 20 kPa. Saturated liquid exits the condenser. The net power produced by the plant is 100 MW. Both turbine stages and the pump are isentropic.

a. Determine the mass flow rate of the steam (in kg/s).

b. Determine the cycle thermal efficiency.

P8-21   For Problem P 8-20, both stages of the turbine have isentropic efficiencies of 85%, and the pump isentropic efficiency is 78%. All other conditions remain the same.

a. Determine the mass flow rate of the steam (in kg/s).

b. Determine the cycle thermal efficiency, and compare the efficiency to that calculated in Problem P8-20.

P8-22 A Rankine cycle has three turbine stages with two reheats between the stages. Superheated vapor leaves the boiler at 30 MPa, 550°C; the vapor leaves the first reheater at 5 MPa, 500°C and the second reheater at 0.5 MPa, 400°C. The condenser pressure is 0.05 bar, and saturated liquid exits the condenser. Total mass flow is 2.5×10^6 kg/h.

a. Determine the net power (in kW).

b. Determine the cycle thermal efficiency.

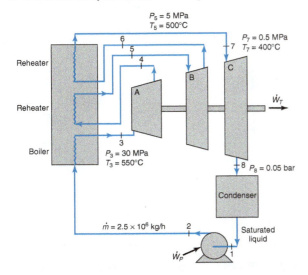

P8-23 For Problem P8-22, assume that the three turbine stages have isentropic efficiencies of 91%, 87%, and 83%, respectively, and that the pump isentropic efficiency is 80%.

a. Determine the net power (in kW).

b. Determine the cycle thermal efficiency.

P8-24 For the Rankine cycle power plant shown in the figure, determine the following:

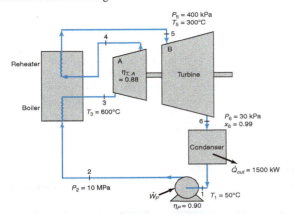

a. The mass flow rate (in kg/s).

b. The total heat addition (in kW).

c. The net power output (in kW).

d. The cycle thermal efficiency.

P8-25 In a Rankine cycle power plant, superheated steam leaves the boiler at 1250 psia, 1000°F. Saturated liquid exits the condenser, which operates at a pressure of 2 psia. Pump isentropic efficiency is 90%.

a. Determine the cycle thermal efficiency if expansion is through a single turbine with an isentropic efficiency of 90%.

b. Determine the cycle thermal efficiency if reheat is used in which the steam is extracted at 100 psia, is reheated to 800°F, and is expanded to the same condenser pressure; both low- and high-pressure turbines have isentropic efficiencies of 90%.

P8-26 In an ideal reheat Rankine cycle, steam enters the high-pressure turbine at 9 MPa, 500°C and is extracted at a lower pressure, reheated to 500°C, and then, in the low-pressure turbine, expanded to 10 kPa. To minimize possible damage in the low-pressure turbine, the minimum quality at the turbine outlet is specified to be 90%.

a. Determine the pressure at which reheating takes place (in kPa).

b. Determine the total heat addition per unit mass (in kJ/kg).

c. Determine the total heat rejection per unit mass (in kJ/kg).

d. Determine the cycle thermal efficiency.

P8-27 In an ideal reheat Rankine cycle, steam enters the high-pressure turbine at 800 psia, 900°F and exits as a saturated vapor. The steam then enters the reheater, where its temperature is raised to 800°F. The steam then expands in the low-pressure turbine to 1 psia. Total heat addition is 2.2×10^8 Btu/h.

a. Determine the pressure at which reheat takes place (in psia).

b. Determine the mass flow rate (in lbm/s).

c. Determine the heat rejection (in Btu/h).

d. Determine the cycle thermal efficiency.

e. Determine the net power output (in Btu/h, hp, and kW).

P8-28 An ideal Rankine cycle uses one stage of reheat. Vapor leaves the boiler at 2000 psia, 1000°F, expands in the high-pressure turbine to 500 psia, at which point the steam is extracted and routed back through the reheater, where the steam temperature is raised to 1000°F. The steam then expands through the low-pressure turbine to the condenser pressure of 5 psia. The condensate leaves the condenser as a saturated liquid. The steam mass flow rate is 5 lbm/s.

a. Determine the net power produced (in kW).

b. Determine the heat input to the boiler (without reheat) (in kW).

c. Determine the heat input in the reheater (in kW).

d. Determine the cycle thermal efficiency.

RANKINE CYCLE WITH REGENERATION

P8-29 A closed feedwater heater is used in a Rankine cycle. Steam leaves the boiler at 20 MPa, 600°C. Between the high- and low-pressure turbines, steam at 1 MPa is extracted and delivered to the closed feedwater heater. Feedwater exits the feedwater heater at 20 MPa and the saturation temperature of the 1-MPa steam; saturated liquid condensate is fed through a steam trap back to the condenser. Steam from the second-stage turbine enters the condenser at 10 kPa, and saturated liquid exits the condenser. Both stages of the turbine have isentropic efficiencies of 90%, and the pump isentropic efficiency is 85%.

a. Determine the fraction of steam entering the turbine that must be extracted.

b. Determine the net work per unit mass of steam entering the first turbine stage (in kJ/kg).

c. Determine the heat added per unit mass of steam entering the first turbine stage (in kJ/kg).

d. Determine the cycle thermal efficiency.

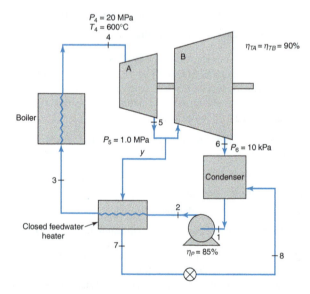

P8-30 For Problem P8-29 the closed feedwater heater is changed to an open feedwater heater, from which saturated liquid at 1 MPa exits. With this change, a second pump must be added to the system.

a. Determine the fraction of steam entering the turbine that must be extracted.

b. Determine the net work per unit mass of steam entering the high-pressure turbine (in kJ/kg).

c. Determine the heat added per unit mass of steam entering the high-pressure turbine (in kJ/kg).

d. Determine the cycle thermal efficiency.

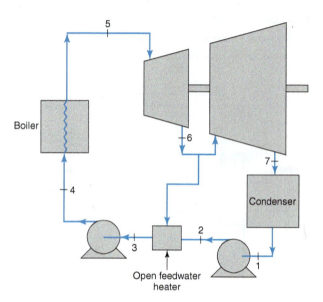

P8-31 A Rankine cycle power plant uses one open feedwater heater. Steam leaves the boiler at 1000 psia, 800°F and enters the turbine. At 100 psia, steam is extracted and routed to the open feedwater heater; the feedwater exits the feedwater heater as a saturated liquid. The condenser pressure is 2 psia. The turbine and pump are isentropic.

a. Determine the fraction of steam entering the turbine that is extracted.

b. Determine the cycle thermal efficiency.

c. Determine the cycle thermal efficiency if there were no feedwater heater.

P8-32 A Rankine cycle power plant uses one closed feedwater heater. Steam leaves the boiler at 6 MPa, 400°C. At 400 kPa, 15% of the steam entering the turbine is extracted and routed to the closed feedwater heater; the condensate exits the feedwater heater as a saturated liquid and is routed back to the condenser. The condenser pressure is 7.5 kPa. The turbine and pump are isentropic.

a. Determine the cycle thermal efficiency.

b. Determine the cycle thermal efficiency if there were no feedwater heater.

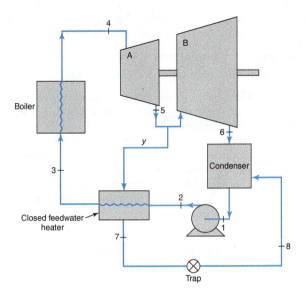

a. Determine the fraction of steam entering the high-pressure turbine that is extracted for each of the feedwater heaters.

b. Determine the power required to operate each pump (in kW).

c. Determine the power produced by each turbine stage (in kW).

d. Determine the heat input in the boiler (in kW).

e. Determine the cycle thermal efficiency.

P8-34 A Rankine cycle uses one stage of reheat and one open feedwater heater. Steam leaves the boiler at 5 MPa, 550°C and expands in the high-pressure turbine to a pressure of 1 MPa. The steam is extracted, some of which flows through the reheater, where the steam temperature is raised to 500°C and the remainder flows to the open feedwater heater. The steam from the reheater enters the low-pressure turbine and expands to the condenser pressure of 20 kPa; saturated liquid exits the condenser. Feedwater from the open feedwater heater leaves as saturated liquid at 1 MPa. Each pump has an isentropic efficiency of 88%, and each turbine stage has an isentropic efficiency of 92%.

a. Determine the fraction of the mass flow entering the high-pressure turbine that is extracted to flow to the open feedwater heater.

b. Determine the work output of the high- and low-pressure turbines per unit of mass flowing into the high-pressure turbine (in kJ/kg).

c. Determine the work input of the low-pressure and high-pressure pumps per unit of mass flowing into the high-pressure turbine (in kJ/kg).

P8-33 An ideal Rankine cycle uses two open feedwater heaters and three pumps. Steam at a flow rate of 12 lbm/s leaves the boiler at 1500 psia, 1600°F, enters the high-pressure turbine and expands to 250 psia, where steam is extracted for the first open feedwater heater. The steam expands in the intermediate-pressure turbine to a pressure of 100 psia, where steam is extracted for the second open feedwater heater. The steam expands in the low-pressure turbine to the condenser pressure of 4 psia. Water leaves the condenser and the feedwater heaters as saturated liquid. All pumps and turbines are isentropic.

(Continued)

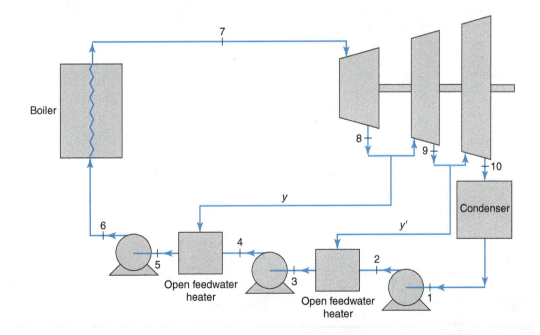

d. Determine heat input in the boiler and reheater per unit of mass entering the high-pressure turbine.

e. Determine the cycle thermal efficiency.

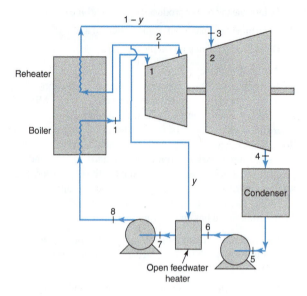

P8-35 A regenerative Rankine cycle that produces 500 MW uses one closed and one open feedwater heater. Steam exits the boiler at 10 MPa, 550°C. Steam at 1 MPa is extracted between the high- and intermediate-pressure turbines and is sent to the closed feedwater heater, from which the feedwater leaves at 10 MPa, 150°C; saturated liquid condensate is pumped forward and injected into the boiler feedwater line. Steam at 0.15 MPa is extracted between the intermediate- and low-pressure turbines and is sent to an open feedwater heater, from which saturated liquid water leaves at 0.15 MPa. Steam leaves the low-pressure turbine at 6 kPa. The turbines and pumps are isentropic.

a. Determine the cycle thermal efficiency.

b. Determine the required mass flow rate (in kg/s).

P8-36 A Rankine cycle power plant uses reheat and one closed feedwater heater. Steam leaves the boiler at 6 MPa, 400°C and enters the high-pressure turbine. At 400 kPa all the steam is extracted; 85% of the flow is routed through a reheater, where the steam temperature is raised to 400°C and then is returned to the low-pressure turbine. The other 15% of the extracted steam is routed to the closed feedwater heater; the condensate exits the feedwater heater as a saturated liquid and is routed back to the condenser. The condenser pressure is 7.5 kPa. The turbine and pump are isentropic.

a. Determine the cycle thermal efficiency.

b. Compare this result to that obtained in Problem p 8-32.

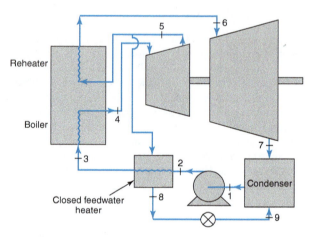

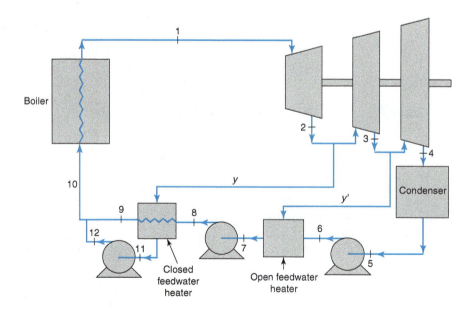

BRAYTON CYCLE

P8-37 Air at 1 atm, 40°F enters the compressor of an ideal Brayton cycle. The pressure ratio is 10. The maximum temperature in the cycle is 1500°F. Using an air-standard analysis, determine the following:

a. The cycle thermal efficiency

b. The back work ratio

c. The temperature of the air at the turbine exit (in°C)

P8-38 For Problem P8-37, rework the problem using an air-standard analysis with a turbine efficiency of 89% and a compressor efficiency of 83%.

a. Determine the cycle thermal efficiency.

b. Determine the back work ratio.

c. Determine the temperature of the air at the turbine exit (in°C).

Compare with the results from Problem P8-37.

P8-39 Air at 100 kPa, 27°C at a volumetric flow rate of 10 m³/s enters an ideal Brayton cycle and is compressed to 1750 kPa. The air temperature at the entrance to the turbine is 1073°C. Using a cold-air-standard analysis, determine the following:

a. The net power (in kW)

b. The heat addition (in kW)

c. The cycle thermal efficiency

d. Back work ratio

P8-40 Re-work Problem P8-39 with an air-standard analysis.

P8-41 The compressor and turbine of a simple Brayton cycle each have an isentropic efficiency of 82%. The compressor pressure ratio is 12. The minimum and maximum temperatures are 290 K and 1400 K. On the basis of a cold-air-standard analysis, determine the following quantities for both an ideal cycle (isentropic compressor and turbine) and a nonideal cycle (compressor and turbine isentropic efficiencies of 82%):

a. The net work per unit mass of air flowing (in kJ/kg)

b. The heat rejected per unit mass of air flowing (in kJ/kg)

c. The cycle thermal efficiency

P8-42 An ideal air-standard Brayton cycle has a compressor pressure ratio of 10. Air enters the compressor at $P_1 = 14.7$ lbf/in.², $T_1 = 70°F$, with a mass flow rate of 90,000 lbm/h. The turbine inlet temperature is 1740°F. Use an air-standard analysis with constant specific heats.

a. Determine the net power developed (in hp and kW).

b. Determine the cycle thermal efficiency.

P8-43 For the cycle in Problem P8-42, include in the analysis turbine and compressor isentropic efficiencies of 88% and 84%, respectively. For the modified cycle, determine the following:

a. The net power developed (in hp and kW)

b. The cycle thermal efficiency

P8-44 (WEB) For the cycle in Problem P8-43, add a regenerator with an effectiveness of 80% to the cycle. For the modified cycle, determine the following:

a. The net power developed (in hp and kW)

b. The thermal efficiency

P8-45 An ideal Brayton cycle is to be operated at full load and at part load. Air enters the compressor at 14.5 psia, 77°F. At full load, the air leaves the compressor at 116 psia, and the turbine inlet temperature is 1800°F; the air flow rate is 17.9 lbm/s. At part load, air leaves the compressor at 58 psia, and the turbine inlet temperature is 1340°F; the air flow rate is 12.7 lbm/s.

a. For full load, determine the net work (in hp), the heat addition (in hp), and the cycle thermal efficiency.

b. For part load, determine the net work (in hp), the heat addition (in hp), and the cycle thermal efficiency.

P8-46 The back work ratio in a Brayton cycle is defined as the ratio of the compressor work divided by the turbine work.

a. Using a cold-air-standard analysis, show that this ratio in an ideal Brayton cycle is equal to the absolute temperature at the compressor inlet divided by the absolute temperature at the turbine outlet.

b. Develop an expression for the back work ratio, again using a cold-air-standard analysis, but with turbine and compressor efficiencies less than 1.

P8-47 For an ideal Brayton cycle with given low temperature T_1 and high temperature T_3, derive the expression (Eq. 8-35) for the maximum work at the optimum pressure ratio using a cold-air-standard analysis.

P8-48 For an ideal Brayton cycle with given low temperature T_1 and high temperature T_2, derive the expression (Eq. 8-34) for the optimum pressure ratio to produce the maximum work.

BRAYTON CYCLE WITH REGENERATION, REHEAT, AND/OR INTERCOOLING

P8-49 (WEB) Experimental data are obtained from a regenerative Brayton cycle. Air enters the compressor, which has an isentropic efficiency of 75%, at 100 kPa, 27°C, with a flow rate of 12.4 kg/s, and exits the compressor at 1050 kPa, 400°C. The air passes through the regenerator to the combustor, where 15.21 MW of heat are added, and then the air expands in the turbine to a pressure of 100 kPa, 967°C. At the exit of the regenerator, the air temperature is 727°C.

a. Determine the turbine isentropic efficiency.

b. Determine the net power output (in kW).

c. Determine the cycle thermal efficiency.

d. Determine the regenerator effectiveness.

P8-50 (WEB) The following temperatures were measured on a test of a regenerative Brayton cycle with a pressure ratio of 5.41.

$T_1 = 290.2$ K $\quad T_2 = 505.0$ K
$T_x = 629.4$ K $\quad T_3 = 1046.7$ K
$T_4 = 713.7$ K $\quad T_y = 590.1$ K

Use the numbering system shown in the figure and a cold-air-standard analysis.

a. Determine the cycle thermal efficiency.

b. Determine the regenerator effectiveness.

c. Assuming the regenerator is replaced with one with an effectiveness of 85%, what is the new cycle efficiency?

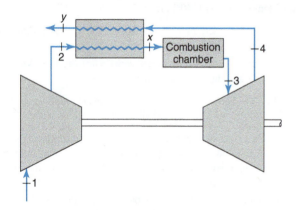

P8-51 (WEB) A regenerative Brayton cycle power plant was designed and built. Before releasing the plant to the owners, the construction company had to verify the actual cycle performance. The test data shown in the table were obtained.

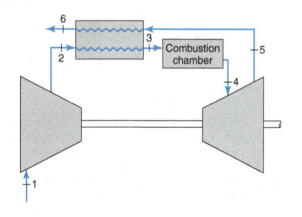

	Entering	Leaving
Compressor	$P_1 = 97$ kPa, $T_1 = 17°C$	$P_2 = 525$ kPa, $T_2 = 229°C$
Regenerator	$P_2 = 525$ kPa, $T_2 = 229°C$	$P_3 = 510$ kPa, $T_3 = 348°C$
Combustor	$P_3 = 510$ kPa, $T_3 = 348°C$	$P_4 = 502$ kPa, $T_4 = 727°C$
Turbine	$P_4 = 502$ kPa, $T_4 = 727°C$	$P_5 = 104$ kPa, $T_5 = 427°C$
Regenerator	$P_5 = 104$ kPa, $T_5 = 427°C$	$P_6 = 97$ kPa, $T_6 = 311°C$

The velocity at the entrance to the compressor was measured to be 135 m/s in a pipe with a diameter of 1.6 m.

a. Determine the compressor isentropic efficiency.

b. Determine the turbine isentropic efficiency.

c. Determine the regenerator effectiveness.

d. Determine the net power output (in kW).

e. Determine the cycle thermal efficiency.

P8-52 (WEB) A regenerative Brayton cycle develops a net power output of 10^7 Btu/h. Air enters the compressor at 14 lbf/in.^{2} and 80°F and is compressed to 70 lbf/in.^{2}. The air then passes through the regenerator and exits at 500°F. The temperature at the turbine inlet is 1080°F. The compressor and turbine are ideal. Using an air-standard analysis, determine the following:

a. The cycle thermal efficiency

b. The regenerator effectiveness

c. The volumetric flow rate of air entering the compressor (in ft^3/min)

P8-53 (WEB) Air enters the compressor of an ideal regenerative Brayton cycle at 14.2 psia, 60°F and exits at a pressure of 250 psia. The air flows through the regenerator, which has an effectiveness of 78%, to the combustor, where the air temperature is raised to 2500°F. A net power output of 10-MW electric power is needed. The compressor and turbine are isentropic. The electric generator has an efficiency of 94%.

a. Determine the cycle thermal efficiency.

b. Determine the overall plant efficiency.

c. Determine the required mass flow rate of air (in lbm/s).

P8-54 (WEB) A regenerative Brayton cycle power plant is shown in the figure. Air enters the compressor at 1 bar, 27°C and is compressed to 4 bar. The isentropic efficiency of the compressor is 80%, and the regenerator effectiveness is 90%. All the power developed by the high-pressure turbine is used to run the compressor, and the low-pressure turbine provides the net power output of 97 kW. Each turbine has an isentropic efficiency of 87%. The temperature at the inlet to the high-pressure

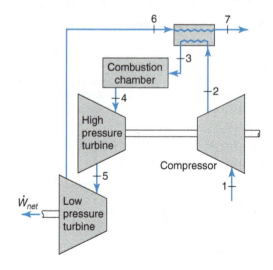

turbine is 1200 K. Using a cold-air-standard analysis, determine the following:

a. The mass flow of air into the compressor (in kg/s);

b. The thermal efficiency of the cycle;

c. The temperature of the air at the exit of the regenerator (in K).

P8-55 Air enters the compressor of an ideal reheat Brayton cycle at 100 kPa, 300 K and leaves at 1600 kPa. It is heated to 1300 K before it enters the high-pressure turbine. The air expands to 400 kPa, is extracted, and sent to a reheater, from which it exits at 1300 K. It expands in the low-pressure turbine to a pressure of 100 kPa. The compressor and both turbines are ideal.

a. Determine the net work per unit mass of air flowing (in kJ/kg).

b. Determine the heat transfer per unit mass of air flowing in each heat transfer process (in kJ/kg).

c. Determine the cycle thermal efficiency.

d. Determine the net work per unit mass of air flowing (in kJ/kg) and the cycle efficiency if the expansion occurs in one stage with no reheat.

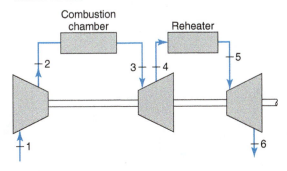

P8-56 (WEB) Data from a regenerative Brayton cycle power plant with one stage of reheat were measured to determine the cycle's actual performance. The measured air flow rate was 1.4 kg/s. The measured conditions of the air are shown in the table.

Device	Inlet conditions	Outlet conditions
Compressor	$P_1 = 0.10$ MPa, $T_1 = 27°C$	$P_2 = 1.72$ MPa, $T_2 = 427°C$
Regenerator	$P_3 = 1.69$ MPa, $T_3 = 417°C$	$P_4 = 1.67$ MPa, $T_4 = 614°C$
Combustor	$P_5 = 1.66$ MPa, $T_5 = 607°C$	$P_6 = 1.65$ MPa, $T_6 = 1021°C$
High-pressure turbine	$P_7 = 1.64$ MPa, $T_7 = 1000°C$	$P_8 = 0.42$ MPa, $T_8 = 657°C$
Reheater	$P_9 = 0.40$ MPa, $T_9 = 642°C$	$P_{10} = 0.38$ MPa, $T_{10} = 950°C$
Low-pressure turbine	$P_{11} = 0.37$ MPa, $T_{11} = 941°C$	$P_{12} = 0.11$ MPa, $T_{12} = 667°C$
Regenerator	$P_{13} = 0.11$ MPa, $T_{13} = 662°C$	$P_{14} = 0.10$ MPa, $T_{14} = 467°C$

a. Determine the heat addition (in kW).

b. Determine the net power produced (in kW).

c. Determine the heat rejection (in kW).

d. Determine the isentropic efficiencies of the high- and low-pressure turbines and the compressor.

e. Determine the regenerator effectiveness.

f. Determine the cycle thermal efficiency.

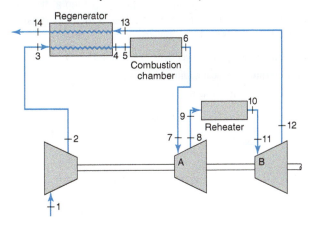

P8-57 For an ideal Brayton cycle with two turbines, assume that the reheat process raises the temperature of the air entering the low-pressure turbine to the same temperature as the air entering the high-pressure turbine. Using a cold-air-standard analysis, show that the maximum work is developed when the pressure ratio is the same across each turbine. Assume that the inlet state (pressure and temperature) to the high-pressure turbine is known and that the inlet pressure to the low-pressure turbine is the same as the outlet pressure from the first turbine.

P8-58 (WEB) Air is compressed in two stages from 95 kPa, 27°C to 1350 kPa. After the air is compressed to 400 kPa in the first compressor, it is routed to an intercooler, where the temperature of the air is lowered to 27°C. In the second compressor, the air is further compressed to 1350 kPa. Both compressors are isentropic. For a flow rate of 0.5 kg/s, determine the following:

a. The power required for the compression process (in kW)

b. The power required for the compression process if it occurs in one stage and there is no intercooling (in kW)

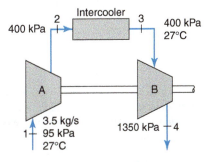

P8-59 (WEB) A Brayton cycle has both intercooling and reheat. Air enters compressor A at 105 kPa, 300 K with a volumetric flow rate of 15 m³/s, where it is compressed to 400 kPa. The intercooler cools the air to 300 K. Compressor B compresses the air to 1500 kPa.

In the combustion chamber, the air is heated to 1200 K before it enters the turbine A, where it expands to 400 kPa, and then is routed to the reheater, where it is reheated to 1200 K. The air finally expands back to 105 kPa in turbine B. Both compressor and turbine stages are isentropic.

a. Determine the net power developed (in kW).

b. Determine the heat addition (in kW).

c. Determine the cycle thermal efficiency.

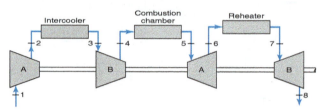

P8-60 (WEB) As shown in the figure, a Brayton cycle is equipped with a regenerator whose effectiveness is 70% (ε_{reg} = 0.70). There are two stages of compression and two stages of expansion, with a pressure ratio of 2.5 across each stage. The air enters the first compressor at 100 kPa and 22°C. The intercooler reduces the compressed air temperature back to 22°C before entering the second compressor. Both compressors have isentropic efficiencies of 78%. At the entrance to the first turbine, the temperature is 827°C. The temperature of the air exiting the reheater is 827°C. Both turbines have isentropic efficiencies of 84%. Using a cold-air-standard analysis, determine the following:

a. The compressor work per unit mass of flowing air (in kJ/kg)

b. The turbine work per unit mass of flowing air (in kJ/kg)

c. The heat addition per unit mass of flowing air (in kJ/kg)

d. The cycle thermal efficiency.

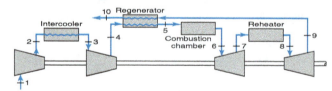

P8-61 (WEB) For an ideal Brayton cycle with an ideal regenerator and given low temperature T_1 and high temperature T_2, derive the expression for cycle thermal efficiency given in section 8.7 using a cold-air-standard analysis.

COMBINED CYCLES AND COGENERATION

P8-62 (WEB) Mercury and steam are used in a combined cycle, as shown on the figure. The mercury leaves the mercury–steam heat exchanger as a saturated liquid at 0.04 MPa, flows through a pump where the pressure is raised to 1.6 MPa, leaves the boiler as a saturated vapor, expands in an isentropic turbine cycle, and condenses at 0.04 MPa. Use the mercury properties given in the table; the specific volume of saturated liquid mercury at 0.04 MPa is 7.35×10^{-5} m³/kg. In the steam cycle, water leaves the condenser as a saturated liquid at 10 kPa, flows through a pump where the pressure is raised to 5 MPa, leaves the mercury–steam heat exchanger as a saturated vapor, leaves the steam superheater at 500°C, then expands in the turbine to a pressure of 10 kPa.

a. Determine the ratio of mercury flow rate to steam flow rate.

b. Determine the total heat addition per unit mass of steam flowing (in kJ/kg water).

c. Determine the total heat rejection per unit mass of steam flowing (in kJ/kg steam).

d. Determine the cycle thermal efficiency.

P(MPa)	T(°C)	h_f (kJ/kg)	h_g (kJ/kg)	s_f (kJ/kg·K)	s_g (kJ/kg·K)
0.04	309	42.21	335.64	0.1034	0.6073
1.60	562	75.37	364.04	0.1498	0.4954

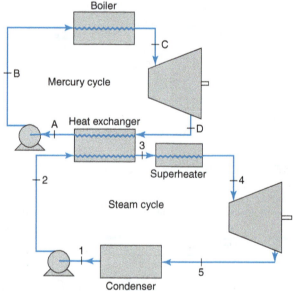

P8-63 (WEB) In a combined cycle, hot air leaving the exit of the turbine in a simple air-standard Brayton cycle is used in the boiler of a simple Rankine cycle. Air enters the compressor of the Brayton cycle at 14.7 psia, 60°F and enters the turbine at 115 psia, 1840°F. In the air–steam heat exchanger, the steam leaves the boiler in the Rankine cycle at 500 psia, 650°F and expands in the turbine to a pressure of 1 psia. Saturated liquid leaves the condenser in the Rankine cycle. Air leaves the air–steam heat exchanger at 14.7 psia, 620°F. Turbines, pump, and compressor are isentropic. For a net power output of 250 MW, determine the following:

a. The air flow rate (in lbm/s)

b. The steam flow rate (in lbm/s)

c. The combined cycle thermal efficiency

P8-64 (WEB) In a combined cycle, hot air leaving the exit of the turbine in a simple air-standard Brayton cycle is used in the boiler of a regenerative Rankine cycle. Air enters the compressor ($\eta_C = 85\%$) of the Brayton cycle at 100 kPa, 27°C and is compressed to 1500 kPa. At the inlet to the turbine ($\eta_T = 88\%$), the air temperature is 1227°C. The exhaust gas from the turbine is used in an air–steam heat exchanger, from which the steam exits as a superheated vapor at 10 MPa, 500°C and the air exits at 227°C. The steam flows to the turbine ($\eta_T = 90\%$) and expands to 0.5 MPa, at which point some of the steam is extracted and routed to the open feedwater heater. The remainder of the steam continues its expansion to the condenser pressure of 15 kPa. Liquid water exits the condenser and the open feedwater heater as saturated liquids at the appropriate pressures. The two water pumps in the Rankine cycle are isentropic. The net power output of the combined cycle is 1000 MW.

a. Determine the ratio of steam mass flow rate to air mass flow rate.

b. Determine the air and steam flow rates (in kg/s).

c. Determine the heat input to the combined cycle (in MW).

d. Determine the combined cycle thermal efficiency.

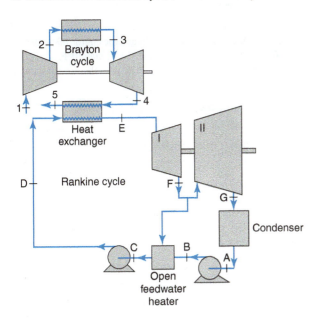

P8-65 (WEB) A cogeneration steam (Rankine) cycle is used to produce electric power and process steam for heating in an oil refinery. The boiler generates 10 kg/s of steam at 8 MPa, 500°C. This steam is expanded in a turbine to a pressure of 0.5 MPa, at which point 4 kg/s of steam are extracted for process heating; the

remainder of the steam expands to the condenser pressure of 10 kPa. Saturated liquid leaves the condenser and is pumped to a mixing tank in which the saturated liquid condensate from the process heaters is drained. The mixture is pumped to the boiler pressure. Both pumps and turbines are isentropic.

a. Determine the heat input to the boiler (in kW).

b. Determine the net power output (in kW).

c. Determine the process heat supplied (in kW).

d. Determine the cogeneration system efficiency.

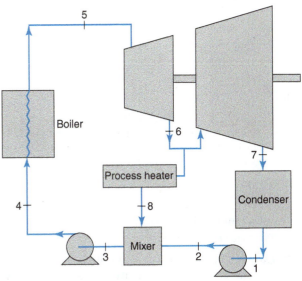

P8-66 (WEB) A cogeneration plant is designed to provide 50 MW of electric power and 100 MW of process heating. A reheat Rankine cycle is used. Steam exits the boiler at 10 MPa, 450°C

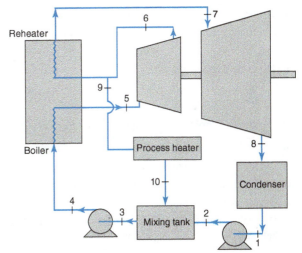

and expands in a high-pressure turbine ($\eta_T = 85\%$) to a pressure of 0.8 MPa. Part of the steam is extracted and routed to the heating load, while the remainder is routed to a reheater in which the steam temperature is raised to 400°C. The reheated steam is fed to the low-pressure turbine ($\eta_T = 85\%$) and is expanded to the condenser pressure of 50 kPa; saturated liquid exits the condenser and is pumped to a mixing tank. Saturated liquid at 0.5 MPa from the process heating load is drained into the mixing tank. The mixture exiting the mixing tank is pumped to the boiler. Both pumps have an isentropic efficiency of 75%.

a. Determine the mass flow rates through the high-pressure turbine, the process heating load, and the low-pressure turbine (in kg/s).

b. Determine the total heat input (in MW).

c. Determine the cogeneration system efficiency.

OTTO CYCLE

P8-67 Air at 100 kPa, 300 K enters an ideal Otto cycle. The initial volume is 500 cm³. The compression ratio is 8.5, and the maximum temperature in the cycle is 2100 K. Using a cold-air-standard analysis, determine the following:

a. The heat addition (in kJ)

b. The heat rejection (in kJ)

c. The net work (in kJ)

d. The cycle thermal efficiency

P8-68 For Problem P8-67, rework the problem with an air-standard analysis. Compare the results of these two problems.

P8-69 An ideal Otto cycle uses air as the working fluid. The minimum temperature in the cycle is 70°F, and the maximum temperature is 2000°F. The compression ratio is 10. Using a cold-air-standard analysis, determine the following:

a. The heat addition per unit mass (in Btu/lbm)

b. The heat rejection per unit mass (in Btu/lbm)

c. The net work per unit mass (in Btu/lbm)

d. The cycle thermal efficiency

e. The Carnot cycle efficiency when operating between the same two temperatures

P8-70 Rework Problem P8-69 using an air-standard analysis.

a. Determine the heat addition per unit mass (in Btu/lbm).

b. Determine the heat rejection per unit mass (in Btu/lbm).

c. Determine the net work per unit mass (in Btu/lbm).

d. Determine the cycle thermal efficiency.

e. Determine the Carnot cycle efficiency when operating between the same two temperatures.

Compare to the previous results.

P8-71 Air at 100 kPa, 33°C enters an ideal Otto cycle. The peak pressure and temperature in the cycle are 4.6 MPa and 1950°C, respectively. Using an air-standard analysis, determine the following:

a. The net work per unit mass (in kJ/kg)

b. The cycle thermal efficiency

P8-72 Otto cycles are used in spark-ignition engines. Too high a compression ratio will cause gasoline to autoignite (that is, ignite without the use of a spark) and the engine will "knock." If the autoignition temperature of gasoline is 700°F and the air–fuel mixture is compressed isentropically, determine the maximum compression ratio that will prevent autoignition at the end of the compression stroke. Perform the calculations for initial air temperatures of 0°, 40°, 80°, and 120°F.

P8-73 Air at 96 kPa, 27°C enters an eight-cylinder, four-stroke Otto cycle that operates at 3000 rpm. Each cylinder has a bore of 9 cm and a stroke of 8.5 cm. At top dead center, the volume is 15% of the cylinder volume at bottom dead center. The maximum temperature in the cycle is 2200 K. Using an air-standard analysis, determine the following:

a. The net work per cycle per cylinder (in kJ)

b. The cycle thermal efficiency

c. The power developed (in kW and hp)

P8-74 An eight-cylinder, four-stroke Otto cycle runs at 5000 rpm. The compression ratio is 12.2, and the engine displacement is 396 in.³. Air enters the engine at 14.4 psia, 70°F. The maximum temperature in the engine is 4000°F. Using an air-standard analysis, determine the following:

a. The net work per cycle per cylinder (in Btu)

b. The cycle thermal efficiency

c. The net power output (in kW and hp)

d. The air flow through the engine (in ft³/min)

P8-75 Many motorcycles have two-stroke engines. Assume we have a two-cylinder, 250 cm³ displacement engine that operates on an ideal Otto cycle at 4500 rpm. The compression ratio is 9. Air enters the engine at 14.7 lbf/in.², 77°F and peak temperature is 3690°F. Use an air-standard analysis.

a. Determine the net work per cylinder (in Btu).

b. Determine the cycle thermal efficiency.

c. Determine the net power output (in kW and hp).

d. What is the cycle efficiency if a cold-air-standard analysis is used?

DIESEL CYCLE

P8-76 Air at 14.7 lbf/in.², 80°F enters a Diesel cycle, which has a compression ratio of 15 and a cutoff ratio of 2.2. The volume at the beginning of the compression process is 0.45 ft^3. Using an air-standard analysis, determine the following:

a. The heat addition (in Btu)

b. The maximum temperature in the cycle (in R)

c. The heat rejection (in Btu)

d. The cycle thermal efficiency

P8-77   Air at 100 kPa, 300 K enters a Diesel cycle, which has a compression ratio of 18 and a volume at the beginning of the compression process of 0.05 m^3. The maximum temperature of the cycle is 2200 K. Using an air-standard analysis, determine the following:

a. The net work per cycle (in kJ)

b. The cycle thermal efficiency

c. The cutoff ratio

P8-78   For Problem P8-77, rework the problem using a cold-air-standard analysis and compare the results of the two problems.

P8-79   A Diesel cycle produces net work of 617 kJ/kg of air flowing through the cycle with a heat addition of 947 kJ/kg. Air enters the compressor at 27°C and leaves at 727°C. Using a cold-air-standard analysis, determine the following:

a. The compression ratio

b. The maximum temperature in the cycle

c. The cycle thermal efficiency

d. The cutoff ratio

P8-80   Air at 102 kPa, 7°C enters a six-cylinder, four-stroke Diesel engine whose total displacement is 5.0 L, compression ratio is 20, and cutoff ratio is 2. The engine runs at 2000 rpm. Using an air-standard analysis, determine the following:

a. The power produced (in kW and hp)

b. The cycle thermal efficiency

P8-81   Rework Problem P8-80 with cutoff ratios of 1.5 and 2.5, and discuss the results.

P8-82   Air at 14.7 psia, 40°F enters an ideal four-stroke Diesel cycle that has a compression ratio of 20 and a displacement of 300 in.^{3}. It runs at 2100 rpm, and the maximum temperature is 3300°F. The fuel has an energy content of 19,360 Btu/lbm.

a. Determine the power output of the engine (in hp and kW).

b. Determine the cycle thermal efficiency.

c. Determine the fuel consumption rate (in lbm/h).

P8-83   For an ideal Diesel cycle, assuming the compression and cutoff ratios are known, use a cold-air-standard analysis to develop the Diesel cycle thermal efficiency expression (Eq. 8-52).

P8-84   The Stirling cycle is similar to the Otto and Diesel cycles in that it has a sequence of processes occurring in a reciprocating piston–cylinder assembly. However, unlike the Otto or Diesel cycles, the Stirling cycle is an *external* combustion engine. The four processes are shown in the *P-v* and *T-s* diagrams. There is an isothermal compression from 1–2, constant volume heat addition from 2–3, isothermal expansion from 3–4, and constant-volume heat rejection from 4–1. The heat transfer from process 4–1 is used for the heat transfer required by process 2–3, so that all the cycle heat addition occurs in process 3–4 and all the cycle heat rejection occurs in process 1–2. If all processes are reversible, show that the cycle thermal efficiency is given by

$$\eta_{Stirling} = 1 - T_1/T_3$$

and compare this expression to the Carnot cycle thermal efficiency.

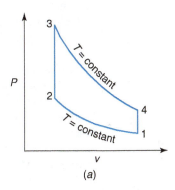

(a)

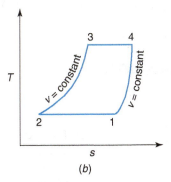

(b)

CHAPTER 9

INTERNAL FLOWS

9.1 INTRODUCTION

Whenever we turn on a water faucet or open the valve on a propane-fired grill or fill a car with gasoline, we use a system in which a fluid (liquid or gas) flows inside a closed conduit. The proper design of that system requires a means to move the fluid from one place to another (pump or compressor) and a determination of pressure differences, flow rates, and velocities. The power required by the pump or compressor also must be determined. The needed information is calculated from an analysis of fluid flow through the pipes that comprise the system.

In earlier chapters, we dealt with one-dimensional flows. We did not take into account the effects of fluid friction in pipe flows, and the average velocity in the pipe was used to describe the flow. In real flows, we still use the average velocity to describe the flow but recognize that the velocity is not the same at every distance from the wall. Fluid near the wall moves more slowly than fluid near the center of the pipe, and at the wall the fluid velocity is zero. The variation in velocity is a direct result of fluid friction. Fluid friction also causes pressure losses as fluid flows through a conduit. In this chapter, we use the principles of conservation of mass and momentum to develop equations for the velocity as a function of position and apply that information to the prediction of pressure changes in internal flows. Fluid friction can be an important contributor to pressure losses. Important frictional losses also occur in many common pipeline elements, such as valves, elbows, junctions, and so on. In addition, pressure and gravity forces must be taken into account.

To deal with frictional effects, the fundamental fluid property called *viscosity* is introduced. Most people have some experience with viscous fluids and would describe honey or oil as more viscous than water. Less obvious is the magnitude of the viscosity of gases. Is air more or less viscous than water? In this chapter, viscosity is defined on a mathematical basis and used in the development of equations for internal flow.

9.2 VISCOSITY

When a solid is placed under a shear stress, it deforms. For example, if you hold a cube of gelatin between your hands and move one hand slightly, the cube will change shape (see Figure 9-1); you must maintain the force to keep the cube deformed. If too much force is applied, the cube will eventually tear. Solids typically behave this way, but most solids are so strong that large forces are needed to produce noticeable deformation.

Fluids behave differently than solids and, under a shear stress, deform continuously. One way to visualize this behavior is by analogy to a stack of papers on a horizontal surface subjected to a force. If one presses downward on the stack, the papers do not move. However, if one slides the top paper horizontally, then it will move forward and may drag along the papers beneath it. As long as a shearing force is applied, the sheets of paper will continue to slide over each other. Fluid in low-speed flow can be thought of as sliding in "laminae," or sheets that rub against each other. The formal definition of **a fluid is a substance that**

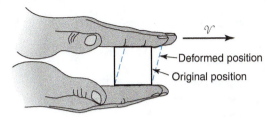

FIGURE 9-1 Deformation of a gelatin cube under the action of a shear stress.

deforms continuously under the application of a shear stress. Both liquids and gases behave this way, and both are classified as fluids.

Consider a fluid flowing between parallel plates, as shown in Figure 9-2. The bottom plate is stationary, and the top plate moves to the right with constant velocity. The fluid resists motion because of its internal friction, and a force must be applied to the top plate to move it. The velocity of the fluid is zero where it touches the stationary bottom plate. The velocity of the fluid is equal to the plate velocity where it touches the moving top plate. Setting the fluid velocity at a solid boundary equal to the velocity of that boundary is called the **no-slip condition**. The no-slip condition applies for most common flows and for all flows considered in this text. It is not applicable in rarified gas flows such as found high in the atmosphere.

The tangential force on the top plate, F_t (also called a shear force), is due to internal friction as layers of fluid slide over each other. The magnitude of the shear force per unit area is called the **shear stress**, τ. The shear stress is

$$\tau = \frac{F_t}{A}$$

where A is the area on which the force acts. Figure 9-2 shows the deformation of a fluid element under the action of a shear stress. The fluid begins in position ABCD and moves to position $A'B'CD$ in the differential time interval δt. The differential angle of deformation is $\delta\alpha$, and the top of the fluid element translates a distance $\delta\ell$ in this time interval. We define the rate of deformation of the fluid as

$$\lim_{\delta t \to 0} \frac{\delta\alpha}{\delta t} = \frac{d\alpha}{dt}$$

The distance traveled by the top of the element is related to the velocity of the top plate, $\delta\mathcal{V}$, through

$$\delta\ell = \delta\mathcal{V}\delta t \tag{9-1}$$

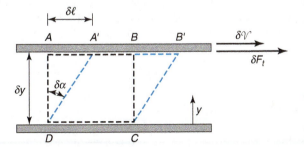

FIGURE 9-2 A fluid flowing between parallel plates. The top plate moves to the right, and the bottom is stationary.

From trigonometry,

$$\tan(\delta\alpha) = \frac{\delta\ell}{\delta y}$$

where δy is the plate spacing, as shown in Figure 9-2. In the limit, $\delta\alpha$ is a differentially small angle; therefore, we may use the approximation that $\tan(\delta\alpha) \approx \delta\alpha$, which applies for small angles. Hence

$$\delta\ell = \delta\alpha\,\delta y$$

Substituting Eq. 9-1 and rearranging gives

$$\frac{\delta\alpha}{\delta t} = \frac{\delta\mathcal{V}}{\delta y}$$

Taking the limit of both sides

$$\frac{d\alpha}{dt} = \frac{d\mathcal{V}}{dy}$$

where y is the position coordinate normal to a wall, as shown in Figure 9-2. This equation shows that the rate of deformation of the fluid element is equal to the derivative of the velocity. For so-called **Newtonian fluids**, the rate of deformation is directly proportional to the applied shear stress, that is,

$$\tau \propto \frac{d\mathcal{V}}{dy}$$

To make the equation an equality, a proportionality factor, μ, is used, yielding

$$\boxed{\tau = \mu\frac{d\mathcal{V}}{dy}} \qquad (9\text{-}2)$$

The quantity, μ, is called the **viscosity**, which is a property of the fluid. It is an indication of how much internal friction is present. Some fluids, such as oils, have high viscosity, and a substantial applied stress is required to cause these fluids to flow. Other fluids, such as water, have lower viscosity and flow more easily for the same applied stress. In general, liquid viscosity decreases exponentially with increasing temperature, as shown in Figure 9-3, but gas viscosity increases with temperature. Viscosity often varies considerably with temperature, and that effect must be considered in calculations. For example, motorists in New England use different oils in their engines in summer and winter because the seasonal temperature variation alters the viscosity of the oil.

Returning to the case of a Newtonian fluid flowing between two parallel plates with one plate stationary and the other moving, we find from experiment that the velocity profile between the plates is linear, as shown in Figure 9-4. The shear stress required to move the plate at velocity $\mathcal{V}_P$ is determined by application of Eq. 9-2:

$$\tau = \mu\frac{d\mathcal{V}}{dy} = \mu\frac{(\mathcal{V}_P - 0)}{(b - 0)}$$

or

$$\boxed{\tau = \mu\frac{\mathcal{V}_P}{b}} \qquad (9\text{-}3)$$

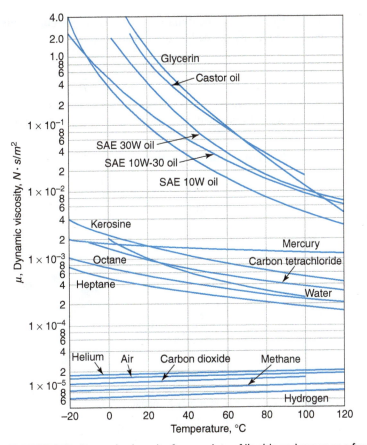

FIGURE 9-3 Dynamic viscosity for a variety of liquids and gases as a function of temperature. (*Source*: Munson, B.R., D.F. Young, and T.H. Okiishi, *Fundamentals of Fluid Mechanics*, 4th ed., Wiley, New York, 2002, p.829. Used with permission.)

For this linear velocity profile, the derivative is just equal to the slope of the profile.

Fluids that behave according to Eq. 9-2 are called Newtonian fluids. Most common fluids are Newtonian, including air and water, although many important non-Newtonian fluids are used. For example, paint, molten plastic, ketchup, and toothpaste behave as non-Newtonian fluids. This text focuses on Newtonian fluids.

As can be seen from Eq. 9-2, the units of viscosity are $(N \cdot s)/m^2$ or $(lbf \cdot s)/ft^2$. Other units are sometimes used and will be introduced as needed. Values of viscosity for many fluids as a function of temperature are given in Tables A-6, A-7, B-6, and B-7. The viscosity

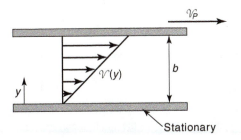

FIGURE 9-4 Velocity profile between two plates. The top plate moves to the right and the bottom is stationary.

defined by Eq. 9-2 is the **dynamic viscosity** μ. **Kinematic viscosity** is, by definition, the dynamic viscosity divided by the density, that is,

$$\nu = \frac{\mu}{\rho}$$

To avoid confusion, the dynamic viscosity, μ, is emphasized in this text.

EXAMPLE 9-1   Shear stress in a journal bearing

In the journal bearing shown, the inner cylinder rotates while the outer cylinder is stationary. The gap between the cylinders contains light oil at 100°F. Using the dimensions on the figure, calculate the power required (in hp) to rotate the bearing at 3600 rpm.

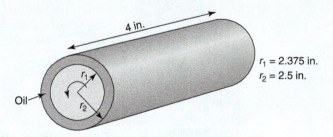

Approach:

For a rotating system, power may be obtained from the torque, $\Im$, and rotational speed, ω, using

$$\dot{W} = \Im\omega$$

Torque is $\Im = Fr$. Our task is to determine the force acting in the journal bearing from knowledge of the viscous forces in the oil.

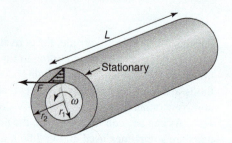

Assumptions: *Solution:*

The power required to rotate a shaft is, from Eq. 3-3,

$$\dot{W} = \Im\omega$$

The torque, $\Im$, is the force multiplied by the moment arm, that is,

$$\Im = Fr_1$$

where F is the force that the rotating inner cylinder applies to the oil. The space between the two cylinders is small compared to the radii; therefore, the oil layer may be treated as

A1. The oil layer is virtually planar.

A2. The flow is Newtonian.

planar [A1]. Assuming the oil is Newtonian [A2], the shear stress at the inner cylinder is (see Eq. 9-3)

$$\tau = \frac{F}{A} = \mu \frac{\mathcal{V}}{r_2 - r_1}$$

where A is the area of the outer surface of the inner cylinder and $\mathcal{V}$ is the velocity of the outer surface of the inner cylinder. Solving for F and substituting into the torque relationship gives

$$\Im = \frac{\mu A \mathcal{V} r_1}{r_2 - r_1}$$

The velocity is the angular velocity multiplied by the radius, that is,

$$\mathcal{V} = \omega r_1$$

Using this and an expression for the area of the inner cylinder, the torque becomes

$$\Im = \frac{\mu \left(2\pi r_1 L\right) \omega r_1^2}{r_2 - r_1} = \frac{2\pi \mu L \omega r_1^3}{r_2 - r_1}$$

The power required to rotate the shaft is

$$\dot{W} = \Im \omega = \frac{2\pi \mu L \omega^2 r_1^3}{r_2 - r_1}$$

The viscosity of light oil may be obtained from Table B-6. Substituting values,

$$\dot{W} = \frac{2\pi \left(1530 \times 10^{-5} \, \frac{\text{lbm}}{\text{ft} \cdot \text{s}}\right)\left(\frac{4}{12} \, \text{ft}\right)\left(3600 \, \frac{\text{rev}}{\text{min}}\right)^2 \left(\frac{2\pi \, \text{rad}}{1 \, \text{rev}}\right)^2 \left(\frac{2.375}{12} \, \text{ft}\right)^3}{\left(\frac{2.5 - 2.375}{12} \, \text{ft}\right)}$$

$$\times \left(\frac{1 \, \text{lbf}}{32.2 \, \frac{\text{lbm} \cdot \text{ft}}{\text{s}^2}}\right)\left(\frac{1 \, \text{min}}{60 \, \text{s}}\right)^2 \left(\frac{1 \, \text{hp}}{550 \, \frac{\text{lbf} \cdot \text{ft}}{\text{s}}}\right)$$

$$\dot{W} = 0.191 \, \text{hp}$$

9.3 FULLY DEVELOPED LAMINAR FLOW IN PIPES

Flow in various types of round tubes is a topic of great practical importance. Water is distributed to homes via pipes, blood flows in the body through blood vessels, gas is piped from remote sites for thousands of miles, and oil is pumped from wells through circular pipes. These are only a few of the numerous applications of pipe flow in industrial, commercial, residential, military, and natural systems.

In electric circuits, current, I, is driven by an applied voltage difference, ξ, and is limited by the electrical resistance, $R_{electric}$:

$$I = \frac{\xi}{R_{electric}}$$

In heat transfer, the heat transfer rate ($\dot{Q}$) is proportional to the driving temperature difference, ΔT, and is inversely proportional to the thermal resistance:

$$\dot{Q} = \frac{\Delta T}{R_{thermal}}$$

For internal fluid flows, a driving pressure difference, ΔP, causes flow ($\dot{V}$) and is resisted by the hydraulic resistance:

$$\dot{V} = \frac{\Delta P}{R_{hydraulic}}$$

In all three cases, the flow (electrical current, heat transfer, or fluid flow) is proportional to a driving potential (voltage drop, temperature drop, or pressure drop, respectively) divided by a resistance to the flow.

For a given hydraulic resistance, the higher the pressure drop, the higher the flow rate. To develop a relationship between flow rate and pressure drop, we need information about the hydraulic resistance. Hydraulic resistance depends on the velocity profile in the pipe, so the first step is to examine the velocity field.

Figure 9-5 is a sketch of the velocity field in a round tube. Flow enters at the left with uniform velocity across the cross-sectional area of the pipe, as indicated by the arrows of equal length. Once fluid comes in contact with the stationary wall of the pipe, it decelerates to zero velocity and a layer of slow-moving fluid called the **boundary layer** forms near the wall of the pipe. In addition, conservation of mass shows that if the fluid near the wall slows down, the fluid near the center must accelerate. This is caused by convection of mass from the low-velocity near-wall region into the tube core. Slow fluid near the wall exerts a frictional force on fluid farther away from the wall. The accumulating effects of friction and convection cause the boundary layer to increase in thickness as the fluid moves downstream.

In the entrance region, fluid is forced away from the wall and into the center region of the pipe. Figure 9-5 shows the velocity profile in the axial direction; however, there is also a component of velocity in the radial direction in the entrance region. Eventually, the boundary layer extends to the center of the pipe, and the flow profile takes on a rounded shape. At a point farther downstream, the flow profile stops changing, and no further flow occurs in the radial direction. All flow is in the axial direction. This is called the **fully developed flow** region, where the velocity profile is independent of the distance from the pipe entrance.

To analyze flow in a pipe, we begin with a fully developed, steady, incompressible, laminar flow. The term *laminar* implies that the fluid moves in sheets, or "laminae," that slip relative to each other. Under other conditions, flow becomes unstable and turbulent, and the velocity at a given location fluctuates. The topic of laminar versus turbulent flow is

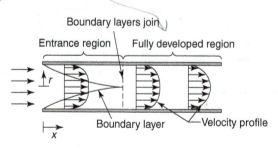

Boundary layers join

Entrance region   Fully developed region

Boundary layer   Velocity profile

FIGURE 9-5   Development of the internal flow field in a pipe.

discussed in detail in the following section. The equations developed in this section apply to a laminar flow.

For generality, we consider an inclined pipe, as shown in Figure 9-6. Define the open system as the control volume of fluid contained within the dotted line in Figure 9-6. This control volume is cylindrical with length, L, radius, r, and axis coincident with the axis of the pipe. We apply conservation of momentum in the x-direction to obtain:

$$\sum F_x = \frac{d}{dt}\left(B_{x,cv}\right) + \dot{m}_e \mathcal{V}_{x,e} - \dot{m}_i \mathcal{V}_{x,i} \tag{9-4}$$

where $\mathcal{V}_{x,e}$ is the average velocity at the exit of the control volume, $\mathcal{V}_{x,i}$ is the average velocity at the entrance, and $B_{x,cv}$ is the x-component of momentum. We are analyzing steady flow, so the first term on the right-hand side of Eq. 9-4 is identically zero. From conservation of mass,

$$\dot{m}_e = \dot{m}_i = \rho \mathcal{V}_{x,e} A_e = \rho \mathcal{V}_{x,i} A_i$$

Note that in the fully developed region, there is no flow in the radial direction, so no fluid enters or exits across the cylindrical surface of the control volume. The flow is incompressible; therefore, density is constant. In addition, the areas of the control volume inlet and exit are the same. Thus,

$$\mathcal{V}_{x,e} = \mathcal{V}_{x,i}$$

Using these simplifications, Eq. 9-4 reduces to

$$\sum F_x = 0$$

Three types of forces act on the control volume: forces due to pressure, gravity, and friction. Pressure forces are always normal to a control volume face, so the only pressure forces in the x-direction act on the two ends of the control volume. Viscous forces are tangential to control volume faces, so all viscous forces in the x-direction occur on the curved surface of the control volume. Finally, the gravity force is mg, where m is the

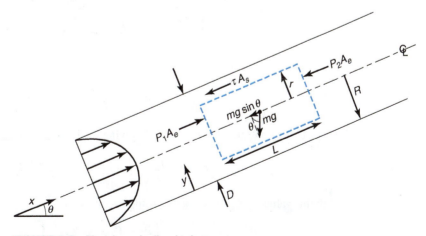

FIGURE 9-6   Flow in an inclined tube.

mass of fluid in the control volume. With these considerations, the force balance becomes (see Figure 9-6)

$$P_1 A_e = P_2 A_e + \tau A_s + mg \sin \theta$$

Substituting appropriate expressions for the end area A_e, and the curved side area A_s, and expressing mass in terms of density and volume, we obtain

$$P_1 \pi r^2 = P_2 \pi r^2 + \tau (2\pi rL) + \rho \left(\pi r^2 L\right) g \sin \theta \tag{9-5}$$

The shear stress is related to the velocity at a solid wall by

$$\tau = \mu \frac{d\mathcal{V}_x}{dy} \tag{9-6}$$

We need to describe this in terms of r. From Figure 9-6, $r = R - y$ and $dr = d(R - y) = -dy$. Thus,

$$\tau = -\mu \frac{d\mathcal{V}_x}{dr}$$

The viscous forces on the fluid in the control volume oppose the flow and, therefore, point in the negative x direction.

Substituting Eq. 9-6 into Eq. 9-5 and simplifying yields

$$P_1 - P_2 = -\mu \frac{d\mathcal{V}_x}{dr} \frac{2L}{r} + \rho Lg \sin \theta$$

which may be rewritten as

$$\boxed{\frac{P_1 - P_2}{L} - \rho g \sin \theta = -\frac{2\mu}{r} \frac{d\mathcal{V}_x}{dr}} \tag{9-7}$$

This is a differential equation for x-direction velocity as a function of radial distance, r. At the wall, the velocity is zero; therefore, the boundary condition for Eq. 9-7 is

$$\text{at} \quad r = R, \qquad \mathcal{V}_x = 0 \tag{9-8}$$

where R is the radius of the pipe. Separating variables, Eq. 9-7 becomes

$$\left\{ -\frac{P_1 - P_2}{2L\mu} + \frac{\rho g \sin \theta}{2\mu} \right\} r dr = d\mathcal{V}_x$$

Integrating both sides gives

$$\frac{1}{2\mu} \left\{ \rho g \sin \theta - \frac{P_1 - P_2}{L} \right\} \frac{r^2}{2} + C = \mathcal{V}_x$$

where C is a constant of integration. Applying the boundary condition, Eq. 9-8,

$$C = -\frac{1}{2\mu} \left\{ \rho g \sin \theta - \frac{P_1 - P_2}{L} \right\} \frac{R^2}{2}$$

With this value for C, the velocity profile becomes

$$\mathcal{V}_x(r) = \frac{1}{4\mu} \left\{ \frac{P_1 - P_2}{L} - \rho g \sin \theta \right\} \left(R^2 - r^2\right) \tag{9-9}$$

At this point, we define the pressure drop as

$$\Delta P = P_1 - P_2$$

Note that ΔP is defined in this way so that it has a positive value. Multiplying and dividing the right-hand side of Eq. 9-9 by R^2 and using the definition of pressure drop, we get

$$\mathcal{V}_x(r) = \frac{R^2}{4\mu}\left\{\frac{\Delta P}{L} - \rho g \sin\theta\right\}\left[1 - \left(\frac{r}{R}\right)^2\right]$$ (9-10)

This expression shows that in the fully developed flow region, the velocity profile is quadratic in r, with a parabolic shape, as shown in Figure 9-6.

Our goal is to relate pressure drop to the *average* velocity. To find the average velocity in the pipe, consider a differential cross-sectional pipe area, dA, as shown in Figure 9-7. The velocity at location r is multiplied by the differential cross-sectional area. To get the average, all such velocity–area products are summed (i.e., integrated) and then divided by the total pipe area, A. This results in the following expression:

$$\mathcal{V}_m = \frac{\int_A \mathcal{V}_x(r)\,dA}{A}$$ (9-11)

where $\mathcal{V}_m$ is the mean (average) velocity of flow in the pipe. It is easy to see that if the velocity were constant with r, $\mathcal{V}_m$ would be equal to this constant velocity, as it should be.

To develop an expression for the mean velocity, substitute Eq. 9-10 into Eq. 9-11 to get

$$\mathcal{V}_m = \frac{\int_0^R \frac{R^2}{4\mu}\left\{\frac{\Delta P}{L} - \rho g \sin\theta\right\}\left[1 - \left(\frac{r}{R}\right)^2\right]2\pi r\,dr}{\pi R^2}$$

where the differential area has been expressed as $dA = 2\pi\,r\,dr$ and R is the radius of the tube. Simplifying yields

$$\mathcal{V}_m = \frac{2R^2\left\{\frac{\Delta P}{L} - \rho g \sin\theta\right\}}{4\mu R^2}\int_0^R\left[r - \frac{r^3}{R^2}\right]dr$$

Integrating and evaluating at the limits gives

$$\mathcal{V}_m = \frac{(\Delta P - \rho g L \sin\theta)}{2\mu L}\left[\frac{R^2}{2} - \frac{R^4}{4R^2}\right]$$

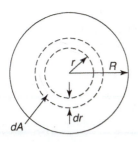

dA

FIGURE 9-7   Differential cross-sectional area of a pipe.

Simplifying,

$$\mathcal{V}_m = \frac{(\Delta P - \rho g L \sin \theta) R^2}{8 \mu L} \qquad (9\text{-}12)$$

Solving for pressure drop gives

$$\Delta P = \frac{8 \mu L \mathcal{V}_m}{R^2} + \rho g L \sin \theta \qquad \text{laminar, fully developed flow, inclined pipe} \qquad (9\text{-}13)$$

In the common case of a horizontal pipe, $\theta = 0$ (see Figure 9-6) and Eq. 9-13 becomes

$$\Delta P = \frac{8 \mu L \mathcal{V}_m}{R^2} \qquad \text{laminar, fully developed flow, horizontal pipe} \qquad (9\text{-}14)$$

Solving for mean velocity,

$$\mathcal{V}_m = \frac{R^2 \, \Delta P}{8 \mu L} \qquad (9\text{-}15)$$

If we multiply both sides of this equation by the cross-sectional area of the pipe and rearrange it, we obtain the volumetric flow rate in terms of the driving pressure drop and a hydraulic resistance:

$$\dot{V} = \mathcal{V}_m \pi R^2 = \frac{\Delta P}{\left(\dfrac{8 \mu L}{\pi R^4} \right)}$$

Note that from Eq. 9-10, at $r = R$, the velocity is zero, which is consistent with the no-slip condition. Moving from one wall through the centerline of the tube to the other wall, we know that a maximum velocity must occur. Because this flow is axisymmetric, it makes intuitive sense for the maximum to occur at the center of the pipe. We may also demonstrate this mathematically by calculating the r location at which the derivative of Eq. 9-10 equals zero. For a horizontal pipe, the maximum velocity is found by evaluating Eq. 9-10 at $r = 0$ and $\theta = 0$ to give

$$\mathcal{V}_{max} = \frac{R^2 \, \Delta P}{4 \mu L}$$

Comparing this with Eq. 9-15 shows that, for fully developed laminar pipe flow in a horizontal circular pipe, the mean velocity is half the maximum velocity, that is,

$$\mathcal{V}_m = \frac{\mathcal{V}_{max}}{2}$$

The velocity profile in a horizontal pipe (Eq. 9-10 with $\theta = 0$) may be expressed in terms of the mean velocity (Eq. 9-15). Combining these two expressions and simplifying, we obtain

$$\mathcal{V}_x (r) = 2 \mathcal{V}_m \left[1 - \left(\frac{r}{R} \right)^2 \right] \qquad \text{laminar, fully developed flow} \qquad (9\text{-}16)$$

Laminar fully developed flow in a pipe is called Hagen-Poiseuille flow in honor of the investigators who first derived the equations. Other laminar flow situations are discussed in subsequent sections.

EXAMPLE 9-2   Pressure change and shear stress in a pipe

Water flows at an average velocity of 0.3 ft/s in a horizontal pipe of diameter 0.5 in. and length 6 ft. Assuming a fully developed laminar flow, calculate the shear stress at the wall and the pressure drop from inlet to outlet. Water temperature is 70°F.

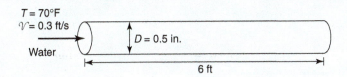

Approach:

The definition of shear stress is

$$\tau = -\mu \frac{d\mathcal{V}}{dr}$$

We need to evaluate this at the wall. The velocity gradient is obtained by differentiating the velocity profile:

$$\mathcal{V}(r) = \mathcal{V}_{max}\left(1 - \frac{r^2}{R^2}\right)$$

To obtain the pressure drop, use

$$\Delta P = \frac{8\mu L \mathcal{V}_m}{R^2}$$

Assumptions:

A1. The flow is laminar.
A2. The flow is fully developed.

A3. The fluid is Newtonian.

Solution:

The velocity profile in a fully developed laminar flow [A1][A2] is

$$\mathcal{V}(r) = 2\mathcal{V}_m\left(1 - \frac{r^2}{R^2}\right)$$

By definition, the shear stress is [A3]

$$\tau = -\mu \frac{d\mathcal{V}}{dr}$$

Substituting the velocity distribution into this expression and taking the derivative yields

$$\tau = -\mu\left(-4\mathcal{V}_m \frac{r}{R^2}\right)$$

Evaluating this expression at the wall ($r = R$) gives

$$\tau_w = \frac{4\mu \mathcal{V}_m}{R}$$

Substituting given values and using the viscosity from Table B-6,

$$\tau_w = \frac{4\left(65.8 \times 10^{-5} \frac{\text{lbm}}{\text{ft} \cdot \text{s}}\right)\left(0.3 \frac{\text{ft}}{\text{s}}\right)\left(\dfrac{1 \text{ lbf}}{32.2 \dfrac{\text{lbm} \cdot \text{ft}}{\text{s}^2}}\right)}{\left(\dfrac{0.25}{12}\right) \text{ft}} = 0.00118 \frac{\text{lbf}}{\text{ft}^2}$$

This is the stress that the wall exerts on the moving fluid due to friction. To find the pressure drop, use

$$\Delta P = \frac{8\mu L \mathcal{V}_m}{R^2}$$

Inserting values,

$$\Delta P = \frac{8\left(65.8 \times 10^{-5} \frac{\text{lbm}}{\text{ft} \cdot \text{s}}\right)(6 \text{ ft})\left(0.3 \frac{\text{ft}}{\text{s}}\right)\left(\dfrac{1 \text{ lbf}}{32.2 \dfrac{\text{lbm} \cdot \text{ft}}{\text{s}^2}}\right)}{(0.25 \text{ in.}^2)} = 0.00471 \frac{\text{lbf}}{\text{in.}^2}$$

Comments:

Water flows under very small pressure drops. The flow velocity is very slow in this example, and little force is required to move the fluid.

9.4 LAMINAR AND TURBULENT FLOW

In the previous section, we assumed a steady, laminar flow. Under certain conditions, flow is no longer laminar but becomes unstable and chaotic. Such a flow is termed **turbulent**. Laminar and turbulent flow patterns can be seen in the smoke plume shown in Figure 9-8. The smoke initially rises in a straight, stable column indicative of laminar flow. At some height above the cigarette, the neat, orderly column gives way to a swirling, complex flow field indicative of turbulence.

Figure 9-9 shows the trajectories of fluid "particles" in laminar and turbulent flow. In laminar flow, fluid particles follow a smooth trajectory, with no oscillations about the average velocity. By contrast, in turbulent flow, the particles follow a rough trajectory, with many random oscillations about the average velocity. You may have felt the effects of turbulence during an airplane flight. Normally, the flight is very smooth; however, if the plane hits a pocket of turbulence, the plane bounces around.

The regime of the flow, either laminar or turbulent, depends, in part, on the velocity. In low-speed flow, layers of fluid, or "laminae," slide smoothly relative to each other. As the velocity increases, small disturbances may appear in the flow. If the velocity remains low, these disturbances are damped out. However, as the velocity increases, the flow becomes unstable, and the disturbances grow and become random. These random fluctuations in velocity result in the characteristic jagged oscillations of turbulent flow.

Laminar flow is analogous to an army marching off to war in neat, straight rows and columns. Turbulent flow is like an army in retreat—disorganized, somewhat random, and moving quickly. The faster people move, the more difficult it is to maintain a marching formation. Fluid particles behave just this way.

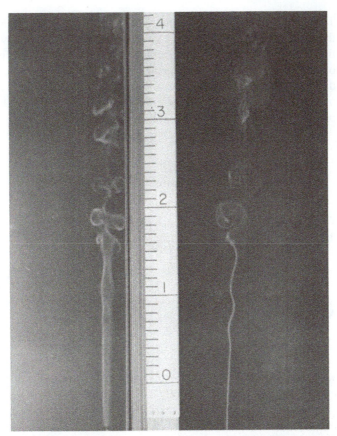

FIGURE 9-8 Smoke plume showing laminar, transition, and turbulent regions. (*Source*: Bejan, A., *Convection Heat Transfer*, 2nd ed., Wiley, New York, 1995. Used with permission.)

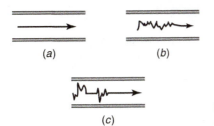

FIGURE 9-9 Trajectory of a particle in different flow regimes: (a) laminar, (b) turbulent, (c) transition.

The analysis of turbulent flow is difficult. No closed-form analytic solutions to the momentum equation exist. It is possible to use numerical analysis to predict turbulent flow behavior, but even in these cases, simplifying assumptions are necessary to solve practical problems. The results are often quite good, but sometimes they differ substantially from experimental data. Unfortunately, turbulent flow is more the norm than the exception in cases of practical engineering interest. Because of its great importance, we must deal in some way with turbulent flow, typically through a combination of theory and experiment.

To begin our study, we take a second look at the solution for fully developed laminar flow in a horizontal pipe, which is, from Eq. 9-14

$$\Delta P = \frac{8\mu L \mathcal{V}_m}{R^2}$$

The pressure drop depends on four parameters: $\mu, L, \mathcal{V}_m$, and R. Experiments show that these same four parameters come into play in turbulent flow; however, the functional relationship is more complex. It is possible to reduce the number of parameters by the process of nondimensionalization. Reducing parameters minimizes the amount of experimental data that must be gathered to describe turbulent flow. Virtually all published information on turbulent flow is in terms of nondimensional parameters; therefore, to make further progress, we must determine what the nondimensional parameters are.

First, we rewrite the preceding equation in terms of diameter instead of radius:

$$\Delta P = \frac{8\mu L \mathcal{V}_m}{\left(\dfrac{D}{2}\right)^2} = \frac{32\mu L \mathcal{V}_m}{D^2} \tag{9-17}$$

We now define a nondimensional pressure, which is the pressure drop divided by the dynamic pressure (see Section 4.7):

$$\Delta P^* = \frac{\Delta P}{\frac{1}{2}\rho \mathcal{V}_m^2} \tag{9-18}$$

A nondimensional pipe length may be defined as

$$L^* = \frac{L}{D} \tag{9-19}$$

We have a free choice of length scale and could, for example, have selected the pipe radius instead of the diameter. We merely need some parameter present in the system to scale the variable. The diameter is convenient because it is easily measured. Substituting these last two expressions into Eq. 9-17 gives

$$\Delta P^* \left(\frac{1}{2}\rho \mathcal{V}_m^2\right) = \frac{32\mu \, (DL^*) \, \mathcal{V}_m}{D^2}$$

Rearranging,

$$\frac{\Delta P^*}{L^*} = \frac{64\mu}{\rho D \mathcal{V}_m} \tag{9-20}$$

We now define the **Reynolds number** as

$$\boxed{Re = \frac{\rho L_{char} \mathcal{V}}{\mu}} \tag{9-21}$$

where L_{char} is a length characteristic of the geometry and $\mathcal{V}$ is a velocity appropriate for the flow. Osborn Reynolds identified this parameter in 1883 as being important in fluid mechanics, and it is named in his honor. For the case of pipe flow, the characteristic length is the pipe diameter and the characteristic velocity is the mean velocity in the pipe. Using the Reynolds number, Eq. 9-20 becomes

$$\frac{\Delta P^*}{L^*} = \frac{64}{Re_D} \tag{9-22}$$

where the subscript on the Reynolds number indicates that it is based on pipe diameter. The left-hand side of this equation is dimensionless; therefore, the right-hand side is dimensionless as well. Hence, *the Reynolds number is a nondimensional parameter*, as can be easily verified. The left-hand side of Eq. 9-22 is a nondimensional pressure drop per unit length of pipe. We define the **Darcy friction factor,** f, as

$$f = \frac{\Delta P^*}{L^*} \tag{9-23}$$

With this definition, Eq. 9-22 becomes

$$\boxed{f = \frac{64}{Re_D} \qquad \text{fully developed laminar flow}} \tag{9-24}$$

Substituting the definitions of ΔP^* and L^* (Eq. 9-18 and Eq. 9-19) and rearranging yields

$$\boxed{\Delta P = f \frac{L}{D} \frac{\rho V_m^2}{2} \qquad \text{laminar or turbulent}} \tag{9-25}$$

This last equation is not restricted to laminar flow. It comes directly from the definition of the friction factor and is used in turbulent as well as laminar flow.

Eq. 9-14 expresses pressure drop in laminar pipe flow as a function of four variables. Eq. 9-24 is an equivalent nondimensional equation that expresses friction factor as a function of only one variable, the Reynolds number. The same idea is applicable to turbulent flow. That is, in turbulent flow, the friction factor for flow in a pipe with a smooth wall is only a function of Reynolds number, that is,

$$f = g(Re)$$

where g is a function determined by experiment. Reduction in the number of variables has the great advantage of minimizing the amount of experimental data that must be collected.

The Reynolds number is a dimensionless quantity whose importance in fluid mechanics cannot be overemphasized. We use it extensively throughout the rest of the book. Although we derived it for one particular case of internal flow in a pipe, in fact, it appears in many other flow problems, both internal and external. It is always based on a characteristic length and a characteristic velocity appropriate for the situation. In the present example, the characteristic length is the pipe diameter and the mean velocity; in other cases, the Reynolds number might be expressed in terms of the height of a plate or the speed of a blade.

The Reynolds number is important because flows with the same Reynolds number behave similarly. At lower Reynolds numbers, flows remain laminar. At high Reynolds numbers, the flow becomes turbulent. Between these two regimes the flow is transitional; that is, the flow becomes more and more unstable with increasing Reynolds number until, finally, the flow is fully turbulent.

It is possible to perform a formal, mathematical, stability analysis of the flow to show that Reynolds number determines stability, but that analysis is beyond the scope of this text. We must be satisfied with recognizing that since Re is the only parameter on which the solution depends, Re must determine the character of the flow. Experiments show that the flow regime in a typical pipe is a function of Reynolds number according to

$$
\begin{array}{ll}
Re < 2100 & \text{Laminar} \\
2100 < Re < 4000 & \text{Transitional} \\
4000 < Re & \text{Turbulent}
\end{array}
$$

These ranges are approximate and depend on various factors such as the roughness of the pipe wall and the nature of the inlet flow. Transitional flow is characterized by bursts of turbulence, which are eventually damped out, as illustrated in Figure 9-9c. The disturbance typically starts near the wall and is carried into the interior of the flow, where it is smoothed out into laminar flow. If the Reynolds number is high enough, however, the disturbances are not damped out. The critical Reynolds number at which transitional flow occurs depends on the geometry and flow situation.

For a given fluid (ρ and μ) and characteristic length, increasing velocity tends to destabilize the flow. Higher velocity implies a higher Reynolds number, since velocity appears in the numerator of the Reynolds number. A high Reynolds number is characteristic of turbulent flow.

The characteristic length, L_{char}, appears in the numerator of the Reynolds number. For a given flow rate and fluid, small diameters tend to be stabilizing and lead to lower Reynolds numbers. Disturbances that start at the wall can be damped by the presence of the other wall. If there is ample separation between the walls, a packet of perturbed flow has time to develop into full-blown turbulence before getting near the other wall, where it can be stabilized by viscous forces.

The final parameters in the Reynolds number are the fluid properties, density and viscosity. These are not independent but are a function of the type of fluid. With a given velocity and characteristic length, high viscosity results in lower Reynolds numbers and pushes the flow toward the laminar regime. Viscosity is a stabilizing influence in the flow. It is very difficult to perturb a viscous fluid out of its flow pattern. Conversely, higher densities tend to destabilize the flow. Once a high-density packet of fluid is perturbed, it is difficult to force it back into a smooth, laminar flow pattern.

The Reynolds number can be viewed as the ratio of inertial forces to viscous forces. To understand this concept, imagine a small cube of fluid where each side has a length h. To accelerate this cube against inertia, a force equal to the mass times the acceleration is applied, that is,

$$
F_{in} = ma = m\frac{d\mathcal{V}}{dt}
$$

where the acceleration has been written as the derivative of the velocity. Now suppose the cube is accelerated from rest and the derivative is approximated by its equivalent difference formula, so that

$$
F_{in} \approx m\frac{\Delta\mathcal{V}}{\Delta t} = m\frac{\mathcal{V}_2 - \mathcal{V}_1}{t_2 - t_1} = m\frac{\mathcal{V}_2}{t_2}
$$

or, dropping the subscript 2,

$$
F_{in} \approx m\frac{\mathcal{V}}{t}
$$

The velocity can be written as the distance per unit time, or

$$
\mathcal{V} = \frac{h}{t}
$$

Eliminating t between these two equations gives

$$F_{in} = m\frac{V^2}{h}$$

The mass is density times volume, so

$$F_{in} = (\rho h^3)\left(\frac{V^2}{h}\right) = \rho h^2 V^2$$

For the viscous forces, the shear stress is

$$\tau = \frac{F_t}{A} = \mu\frac{V}{h}$$

Solving for the tangential force gives

$$F_t = \mu A\frac{V}{h}$$

where area, A, of the side of the cube is h^2, so

$$F_t = \mu h^2\frac{V}{h} = \mu h V$$

Dividing the inertial force by the viscous (tangential) force gives

$$\frac{F_{in}}{F_t} = \frac{\rho h^2 V^2}{\mu h V} = \frac{\rho h V}{\mu} = Re$$

This dimensionless ratio is the Reynolds number based on the characteristic length, h. As mentioned earlier, flows at high Reynolds number are turbulent while flows at low Reynolds number are laminar. A high Reynolds number means that inertial forces are large compared to viscous forces, and any perturbations that occur in the flow will not be damped out by fluid friction. Conversely, if the Reynolds number is low, then the viscous forces are high and stabilizing, and laminar flow is likely.

9.5 HEAD LOSS

In the last two sections, we investigated pipeline flow using the momentum equation. In this section, we apply the energy equation to viscous flow in a pipe. In Figure 9-10, an incompressible, viscous fluid flows through an inclined pipe of length L. We define a

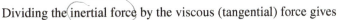

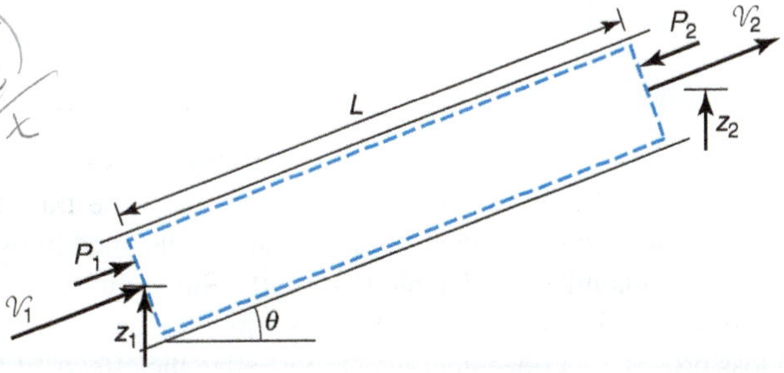

FIGURE 9-10 Flow in an inclined pipe.

control volume as enclosing all the fluid within the pipe and assume the flow is steady and incompressible. Under these circumstances, the energy equation is given by Eq. 4-44, which is

$$0 = q_{cv} - w_{cv} + \left(u_1 + \frac{P_1}{\rho} + \frac{V_1^2}{2} + gz_1 \right) - \left(u_2 + \frac{P_2}{\rho} + \frac{V_2^2}{2} + gz_2 \right) \qquad (9\text{-}26)$$

where

$$q_{cv} = \frac{\dot{Q}_{cv}}{\dot{m}}$$

$$w_{cv} = \frac{\dot{W}_{cv}}{\dot{m}}$$

and we have written enthalpy in terms of internal energy and flow work, $h = u + Pv$.

There is no work done on or by the fluid, so $w_{cv} = 0$. The fluid enters at the same temperature as the surroundings. Energy is dissipated within the fluid due to friction, and this causes a rise in internal energy of the fluid. The rise in internal energy implies a rise in temperature, although the temperature increase is usually quite small. In addition, some heat transfer between the pipe and the surroundings may occur. The flow is incompressible and the entrance and exit areas are equal, so velocity is constant. With these simplifications, Eq. 9-26 becomes

$$0 = (q_{cv} + u_1 - u_2) + \left(\frac{P_1}{\rho} + gz_1 \right) - \left(\frac{P_2}{\rho} + gz_2 \right) \qquad (9\text{-}27)$$

The first term in this equation arises from the effects of friction. It occurs in all real fluid flows and represents the irreversible conversion of mechanical energy to thermal energy. (Rub your hands together. They get hot from friction. Stored chemical energy in your muscles has been converted to mechanical energy to move your hands, then converted again to thermal energy.) To visualize the first term in Eq. 9-27, divide the equation by the acceleration of gravity, g, and rearrange to get

$$0 = \frac{(q_{cv} + u_1 - u_2)}{g} + \left(\frac{P_1 - P_2}{\rho g} \right) + (z_1 - z_2) \qquad (9\text{-}28)$$

Every term in Eq. 9-28 has the units of length. We define **head loss** caused by fluid frictional effects as

$$h_L = \frac{(u_2 - u_1) - q_{cv}}{g} \qquad (9\text{-}29)$$

so that Eq. 9-28 becomes

$$\boxed{h_L = \left(\frac{P_1 - P_2}{\rho g} \right) + (z_1 - z_2) \qquad \text{inclined pipe}} \qquad (9\text{-}30)$$

The quantity $\Delta P / \rho g$ is called the **pressure head**. In the absence of friction (i.e., when $h_L = 0$), the pressure head would raise the fluid a distance $z_1 - z_2$. When friction is present, some of the pressure head is consumed in overcoming viscous friction and only the remainder is available to raise the fluid to a higher elevation. If, for example, the pressure head is 15 m but the fluid rise ($z_2 - z_1$) is only 12 m, then head loss is 3 m. Head loss provides a convenient way to visualize the effects of friction.

In a horizontal pipe, $z_1 = z_2$, and head loss is

$$\boxed{h_L = \left(\frac{P_1 - P_2}{\rho g}\right) = \frac{\Delta P}{\rho g} \qquad \text{horizontal pipe}} \qquad (9\text{-}31)$$

Head loss is related to friction factor. Substituting Eq. 9-25 into Eq. 9-31 gives

$$\boxed{h_L = f\frac{L}{D}\frac{V_m^2}{2g} \qquad \text{laminar or turbulent, any pipe orientation}} \qquad (9\text{-}32)$$

Eq. 9-32 is called the Darcy-Weisbach formula. Although the derivation above has been for a horizontal pipe, this equation also applies for inclined pipes and is used for both laminar and turbulent flows. In many laminar flows, an analytic expression for friction factor can be obtained (in turbulent flow, one must rely on experiments). For fully developed laminar flow in a straight circular tube, the friction factor is, from Eq. 9-24,

$$f = \frac{64}{Re_D}$$

Analyses can be performed for pipes with other cross-sectional shapes, such as rectangular, triangular, annular, and so on. The length scale for the circular pipe is simply the diameter of the pipe. For other shapes, the commonly used length scale is the so-called **hydraulic diameter** defined as

$$\boxed{D_h = \frac{4A_c}{P_{wetted}}} \qquad (9\text{-}33)$$

where A_c is the cross-sectional area of the conduit and P_{wetted} is the wetted perimeter, that is, any portion of the conduit perimeter touched by the fluid. The hydraulic diameter has the benefit of being applicable to any shape, though it tends to be more accurate in turbulent rather than laminar flows. For a circular cross-section, it reduces to the ordinary diameter. Table 9-1 contains expressions for friction factors for fully developed laminar flows in conduits of many common shapes. These expressions were obtained by a combination of theory and experiment. In this table, the Reynolds number is based on the hydraulic diameter. Once the friction factor has been obtained from Table 9-1, head loss and pressure drop may be found from

$$\boxed{h_L = f\frac{L}{D_h}\frac{V_m^2}{2g} \qquad \text{laminar or turbulent, any pipe orientation}}$$

$$\boxed{\Delta P = f\frac{L}{D_h}\frac{\rho V^2}{2} \qquad \text{horizontal conduit}}$$

These expressions are identical to Eq. 9-32 and Eq. 9-25 except that the hydraulic diameter is used in place of the ordinary diameter.

TABLE 9-1 **Friction factors for fully developed laminar flow***

Shape	a/b	Friction factor
Rectangle	0	96.0/Re
	0.05	89.9/Re
	0.10	84.7/Re
	0.25	72.9/Re
	0.5	62.2/Re
	0.75	57.9/Re
	1.00	56.9/Re
	θ	
Isoceles triangle	10°	50.8/Re
	30°	52.3/Re
	60°	53.3/Re
	90°	52.6/Re
	120°	51.0/Re
	D_1/D_2	
Concentric annulus	0.0001	71.8/Re
	0.01	80.1/Re
	0.1	89.4/Re
	0.6	95.6/Re
	1.0	96.0/Re
	a/b	
Ellipse	1	64.0/Re
	2	67.3/Re
	4	73.0/Re
	8	76.6/Re
	16	78.2/Re

Source: Adapted from Munson, B.L., D.F. Young, and T.H. Okiishi, *Fundamentals of Fluid Mechanics,* 4th ed., Wiley, New York, 2002, Table 8.3, and Çengel, Y.A., and R.H. Turner, *Fundamentals of Thermal-Fluid Sciences,* McGraw-Hill, New York, 2001, Table 12.1. * Reynolds number is based on hydraulic diameter.

EXAMPLE 9-3 **Laminar flow in a triangular duct**

A vertical conduit with a triangular cross-section carries liquid ethylene glycol at 0°C. All three sides of the conduit are of length 0.82 cm, and the conduit has a height of 2 m. If the fluid flows

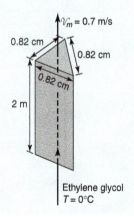

upward at 0.7 m/s, find the pressure change between the inlet and the outlet. Assume fully developed flow.

Approach:

First check the Reynolds number using the hydraulic diameter to be sure the flow is laminar. Then obtain the friction factor for a triangular duct from Table 9-1. Use the friction factor to find head loss via

$$h_L = f \frac{L}{D_h} \frac{V_m^2}{2g}$$

Note that hydraulic diameter, D_h, should be used here as well. Finally, to include hydrostatic pressure loss, use Eq. 9-30 to find the total pressure change.

Assumptions:

Solution:

Since the conduit is not round, the analysis must be based on the hydraulic diameter, which is

$$D_h = \frac{4A}{P}$$

The area of the triangular cross-section is

$$A = \frac{bh}{2} = \frac{b^2 \sin \alpha}{2} = \frac{(0.82)^2 \sin (60°)}{2} = 0.291 \, \text{cm}^2$$

Using this area in the hydraulic diameter gives

$$D_h = \frac{4 \, (0.29 \, \text{cm}^2)}{3 \, (0.82 \, \text{cm})} = 0.473 \, \text{cm}$$

At this point, we check to see if the flow is laminar. Using properties of ethylene glycol from Table A-6, we see that the Reynolds number is

$$Re = \frac{\rho D_h V}{\mu} = \frac{\left(1131 \, \frac{\text{kg}}{\text{m}^3}\right) \left(\frac{0.473}{100} \, \text{m}\right) \left(0.7 \, \frac{\text{m}}{\text{s}}\right)}{65.1 \times 10^{-3} \, \frac{\text{kg}}{\text{m} \cdot \text{s}}} = 57.6$$

The Reynolds number is less than 2100; therefore, the flow is laminar. This is an isosceles triangle for which all three internal angles are 60°. From Table 9-1, the friction factor for fully developed [A1] laminar flow in such a triangle is

A1. The flow is fully developed.

$$f = \frac{53.3}{Re} = \frac{53.3}{57.6} = 0.926$$

The head loss, is, from Eq. 9-32,

$$h_L = f \frac{L}{D_h} \frac{V_m^2}{2g}$$

$$h_L = 0.926 \left(\frac{200 \text{ cm}}{0.473 \text{ cm}} \right) \left[\frac{\left(0.7 \frac{\text{m}}{\text{s}} \right)^2}{2 \left(9.81 \frac{\text{m}}{\text{s}^2} \right)} \right] = 9.78 \text{ m}$$

A2. The flow is incompressible.

From Eq. 9-30 [A2],

$$h_L = \left(\frac{P_1 - P_2}{\rho g} \right) + (z_1 - z_2)$$

Solving for $\Delta P = P_1 - P_2$,

$$\Delta P = \rho g \left(h_L + z_2 - z_1 \right)$$

$$\Delta P = \left(1131 \frac{\text{kg}}{\text{m}^3} \right) \left(9.81 \frac{\text{m}}{\text{s}^2} \right) (9.78 + 2) \text{ m} = 130,685 \text{ Pa} = 131 \text{ kPa}$$

9.6 FULLY DEVELOPED TURBULENT FLOW IN PIPES

Analytical solutions for turbulent flows are not possible, even for the simplest geometries. Rather, turbulent flow is studied using a combination of simplifying assumptions, numerical methods, and experiments. In this section, the results of experimental study are presented.

The nondimensional Reynolds number and friction factor are especially helpful in guiding experimental work in turbulent flow. For a given geometry, the friction factor depends only on the Reynolds number and the relative wall roughness, which is the roughness height divided by the pipe diameter. If we compare the flow in two different pipes with the same relative roughness, a flow that is double that in a tube which has half the diameter will be similar to the original flow since both have the same Reynolds number $(Re = \rho D V_m / \mu)$. One need only measure pressure drop, velocity, and so on for the original case, not for both cases, to determine the friction factor at the given Reynolds number.

Experimenters use the Reynolds number to reduce the number of cases that they must measure. If it were necessary to measure pressure drop for every combination of possible densities, viscosities, velocities, and geometries, the number of experiments would be prohibitively large. By measuring frictional effects only for given values of Reynolds number in geometrically similar conduits, the amount of experimental effort is greatly reduced. This is the value of using dimensionless parameters in thermal–fluids engineering. Many additional nondimensional parameters will be introduced in the chapters that follow.

The friction factors for circular tubes with fully developed turbulent flow have been determined by many experiments. Those data were curve-fit using a regression analysis to produce the following correlation:

$$\frac{1}{\sqrt{f}} = -2.0 \log \left(\frac{\varepsilon/D}{3.7} + \frac{2.51}{Re \sqrt{f}} \right) \qquad \text{turbulent flow} \qquad (9\text{-}34)$$

This implicit relation for f is known as the **Colebrook equation**. The parameter, ε, is the roughness of the pipe wall. It is the average height (typically small) of naturally occurring protrusions on the wall. Table 9-2 gives values of ε/D for some common pipe surfaces. Note that the relative roughness, ε/D, is an additional nondimensional parameter.

The Colebrook equation is awkward to use. To simplify calculations, an approximate explicit relation for f was published by **Haaland** in 1983:

$$\frac{1}{\sqrt{f}} \approx -1.8 \log \left[\left(\frac{\varepsilon/D}{3.7} \right)^{1.11} + \frac{6.9}{Re} \right] \qquad \text{turbulent flow} \qquad (9\text{-}35)$$

This relation gives values within 2% of the Colebrook equation.

The Colebrook equation is presented in graphical form in Figure 9-11, the so-called Moody chart, which includes both laminar and turbulent flow regimes. In the turbulent regime, the chart contains a family of curves corresponding to different values of relative roughness, ε/D. The higher the relative roughness, the greater the friction factor becomes, as expected. In the laminar regime, surface roughness is unimportant. Friction factor decreases

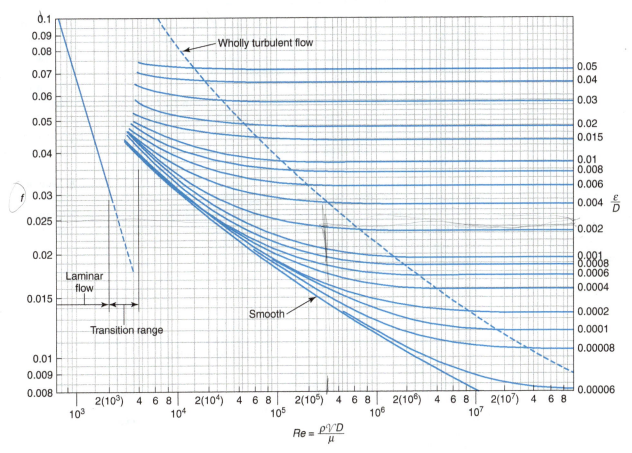

FIGURE 9-11 Friction factor as a function of Reynolds number and relative roughness for round pipes — the Moody chart.
(*Source*: Munson, B. R., D. F. Young, and T. H. Okiishi, *Fundamentals of Fluid Mechanics*, 4th ed., Wiley, New York, 2002, p. 477. Used with permission.)

with increasing Reynolds number. For a given relative roughness in the turbulent regime, friction-factor curves eventually flatten out at high Reynolds numbers; for this situation, friction factor depends almost exclusively on the relative roughness.

For the case of smooth pipes, where relative roughness is zero, the friction factor is given by **Petukhov** as

$$f = (0.79 \ln Re - 1.64)^{-2} \qquad \text{turbulent flow, smooth wall } 3000 < Re_D < 5 \times 10^6$$

(9-36)

Two other simple, but less accurate, correlations for smooth tubes that apply over a more limited range of Reynolds number are

$$
\begin{aligned}
&\text{turbulent flow, smooth wall}\\
f = 0.316/Re^{0.25} \quad &3000 < Re_D < 2 \times 10^4\\
f = 0.184/Re^{0.2} \quad &Re_D > 2 \times 10^4
\end{aligned}
$$

(9-37)

The Colebrook equation is implicit in friction factor. To actually obtain a value for f, an iterative approach is needed, and equation-solving software tools are commonly available for this purpose. In an iterative calculation of friction factor, the Moody chart or the Haaland equation can be used to give a good initial guess for f. On the other hand, it may not be necessary to use the Colebrook equation at all. In many cases, the condition of the pipe wall is uncertain and the relative roughness values in Table 9-2 may not apply. Using the Colebrook equation and/or Moody chart with Table 9-2 for new, clean pipes typically gives values within 15% of experimental results. If significant corrosion or deposits have formed on the pipe wall during use, the relative roughness can increase by as much as an order of magnitude, and it may be difficult to predict pressure drop with accuracy. Given the uncertainties in pipe wall condition, the Haaland relation is often of sufficient accuracy and may be used instead of the Colebrook equation.

The Colebook equation, Haaland equation, and Moody chart are especially useful when the pipe diameter and velocity are known and the pressure drop is to be calculated. However, in many design situations, the appropriate diameter for a given pressure drop and length is wanted. This calculation is difficult because the Reynolds number, which depends on diameter, is unknown. It is possible to assume a value for the Reynolds number and iterate. A similar situation arises when the diameter, length, and pressure drop are known,

TABLE 9-2 Equivalent roughness for clean pipes

Pipe	Equivalent Roughness, ε	
	Feet	Millimeters
Riveted steel	0.003–0.03	0.9–9.0
Concrete	0.001–0.01	0.3–3.0
Wood stave	0.0006–0.003	0.18–0.9
Cast iron	0.00085	0.26
Galvanized iron	0.0005	0.15
Commercial steel	0.00015	0.045
Wrought iron	0.00015	0.045
Drawn tubing	0.000005	0.0015
Plastic, glass	0.0	0.0

and the velocity is to be calculated. To avoid iteration in cases like these, **Swamee and Jain** in 1976 published three explicit relations based on the Colebrook equation:

$$h_L = 1.07 \frac{\dot{V}^2 L}{g D^5} \left\{ \ln \left[\frac{\varepsilon}{3.7D} + 4.62 \left(\frac{\mu D}{\rho \dot{V}} \right)^{0.9} \right] \right\}^{-2} \qquad \begin{array}{l} 10^{-6} < \varepsilon/D < 10^{-2} \\ 3000 < Re < 3 \times 10^8 \end{array}$$

$$\dot{V} = -0.965 \left(\frac{g D^5 h_L}{L} \right)^{0.5} \ln \left[\frac{\varepsilon}{3.7D} + \left(\frac{3.17 \mu^2 L}{g \rho^2 D^3 h_L} \right)^{0.5} \right] \qquad Re > 2000$$

$$D = 0.66 \left[\varepsilon^{1.25} \left(\frac{L \dot{V}^2}{g h_L} \right)^{4.75} + \mu \dot{V}^{9.4} \left(\frac{L}{g h_L} \right)^{5.2} \right]^{0.04} \qquad \begin{array}{l} 10^{-6} < \varepsilon/D < 10^{-2} \\ 5000 < Re < 3 \times 10^8 \end{array}$$

$$(9\text{-}38)$$

These expressions are accurate to within 2% of the Colebrook equation. **Note that the equations are not dimensionless.** In the SI system, express variables in terms of m, s, and kg. For example, use $N \cdot s/m^2$ for dynamic viscosity, m^3/s for volumetric flow rate, meters for head loss, and meters for diameter. In the British system, express variables in terms of ft, s, and lbm. For example, use $lbm/ft \cdot s$ for dynamic viscosity, ft^3/s for volumetric flow rate, ft for diameter, and ft for head loss.

All of the relations for turbulent flow in this section can be used for noncircular pipes. The hydraulic diameter given by Eq. 9-33 is used in place of the actual diameter in the Reynolds number and head loss relations. Such a practice will give a reasonable approximation of actual behavior.

EXAMPLE 9-4   Friction factor in a pipe

A horizontal cast-iron pipe of diameter 4 in. carries 30,000 gal/h of water. The length of the pipe is 50 ft. Calculate the pressure drop using the Moody chart. Assume the water is at room temperature.

Approach:

The first step is to determine whether the flow is laminar or turbulent, so we need to calculate the Reynolds number. Once the flow regime is known, the appropriate friction factor can be evaluated and the pressure drop calculated with $\Delta P = \rho g h_L$, where

$$h_L = f \frac{L}{D} \frac{\mathcal{V}^2}{2g}$$

Assumptions:

Solution:

The Reynolds number is defined as

$$Re = \frac{\rho D \mathcal{V}}{\mu}$$

The velocity can be found from

$$\mathcal{V} = \frac{\dot{V}}{A} = \frac{\left(30{,}000 \frac{gal}{h} \right) \left(\frac{231 \, in.^3}{gal} \right) \left(\frac{1 \, h}{3600 \, s} \right) \left(\frac{1 \, ft}{12 \, in.} \right)}{2^2 \pi \, in.^2} = 12.8 \frac{ft}{s}$$

A1. The water is at room temperature.

With water properties from Table B-6 [A1], the Reynolds number is

$$Re = \frac{\left(62.2 \, \frac{\text{lbm}}{\text{ft}^3}\right)\left(\frac{4}{12}\right)\text{ft}\left(12.76 \, \frac{\text{ft}}{\text{s}}\right)}{\left(65.8 \times 10^{-5} \, \frac{\text{lbm}}{\text{ft} \cdot \text{sec}}\right)} = 3.51 \times 10^5$$

The Reynolds number is greater than 4000; therefore, the flow is turbulent. Now check the relative roughness. From Table 9-2, the surface roughness of a cast-iron pipe is

$$\varepsilon = 0.00085 \, \text{ft}$$

The relative roughness is then

$$\frac{\varepsilon}{D} = \frac{0.00085 \, \text{ft}}{4 \, \text{in.} \left(\frac{1 \, \text{ft}}{12 \, \text{in.}}\right)} = 0.00255$$

A2. The flow is incompressible.

From the Moody chart [A2] (Figure 9-11) at $Re = 3.5 \times 10^5$ and $\varepsilon/D = 0.00255$,

$$f \approx 0.025$$

The head loss may now be calculated as

$$h_L = f \frac{L}{D} \frac{\mathcal{V}^2}{2g}$$

$$= (0.025)\left[\left(\frac{50 \, \text{ft}}{4 \, \text{in.}}\right)\left(\frac{12 \, \text{in.}}{1 \, \text{ft}}\right)\right] \frac{\left(12.8 \, \frac{\text{ft}}{\text{s}}\right)^2}{2\left(32.17 \, \frac{\text{ft}}{\text{s}^2}\right)} = 9.49 \, \text{ft}$$

A3. The flow is fully developed.

Pressure drop in a horizontal pipe is [A3]

$$\Delta P = \rho g h_L$$

$$= \left(62.2 \, \frac{\text{lbm}}{\text{ft}^3}\right)\left(32.17 \, \frac{\text{ft}}{\text{s}^2}\right)(9.49 \, \text{ft})\left(\frac{1 \, \text{lbf}}{32.17 \, \frac{\text{lbm} \cdot \text{ft}}{\text{s}^2}}\right)\left(\frac{1 \text{ft}^2}{144 \, \text{in.}^2}\right)$$

$$= 4.09 \, \frac{\text{lbf}}{\text{in.}^2} = 4.09 \, \text{psi}$$

9.7 ENTRANCE EFFECTS

The previous sections dealt with *fully developed* flow in internal passages. But how does one know that the flow is actually fully developed? By definition, in a fully developed flow, the velocity profile does not change with downstream position and the only component of velocity is in the axial direction. Recall that the friction factor is a nondimensional pressure drop per unit length. While the friction factor is constant in the fully developed region, it varies in the entrance region, where the velocity profile is changing. At the pipe entrance, large frictional effects result from the difference in velocity between the wall and the core of the flow, and the friction factor is high. The friction factor diminishes throughout the

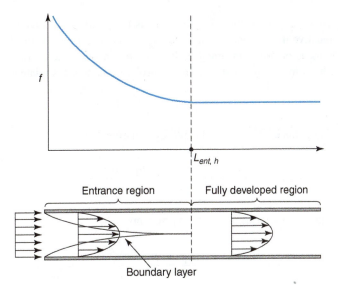

FIGURE 9-12 Friction factor in the entrance region of a pipe.

entrance region as the flow develops and reaches an asymptotic value in the fully developed region, as illustrated in Figure 9-12. Consequently, **the pressure drop per unit length is greater in the entrance region than in the fully developed region**.

The entrance length can be determined by advanced analyses that are beyond the scope of this text. Some useful relations for entrance length are

$$
\begin{array}{ll}
L_{ent,h} \approx 0.065\, Re\, D & \text{laminar}, Re < 2100 \\
L_{ent,h} \approx 4.4\, (Re)^{1/6} D & \text{turbulent}, Re > 4000
\end{array}
\tag{9-39}
$$

The subscript h refers to *hydrodynamic* entrance length. (Later, we will also introduce a thermal entrance length.) Entrance effects tend to be important in laminar flow for short pipes. For example, if $Re = 1000$, then $L_{ent,h} = 65D$. If the pipe diameter were 1 cm and the total pipe length 1 m (100 cm), then the first 65 cm would be in the entrance region and only the last 35 cm would have a fully developed flow. Using relationships for fully developed flow for the whole pipe would result in an underestimate of the pressure drop. Turbulent flow is less sensitive to entrance effects. For example, if $Re = 10,000$, then $L_{ent,h} = 20.4D$. Nevertheless, one should always check to detect the presence of entrance effects for both laminar and turbulent flow.

9.8 STEADY-FLOW ENERGY EQUATION

Pipeline systems frequently include components such as pumps, fans, turbines, and other devices that add or remove energy from a fluid flow. We refer to all these energy flows as shaft work. In this section, we consider systems that involve both shaft work and frictional losses in pipes. To begin, we assume a steady, incompressible flow and apply the energy equation in the form (see Eq. 9-26)

$$
0 = q_{cv} - w_{cv} + \left(u_1 + \frac{P_1}{\rho} + \frac{V_1^2}{2} + gz_1 \right) - \left(u_2 + \frac{P_2}{\rho} + \frac{V_2^2}{2} + gz_2 \right)
$$

Our sign convention for work is that work is positive if it is done by the fluid and negative if it is done on the fluid. By such a definition, pump work is negative and turbine work is positive. In practice, it is often convenient to define new variables, w_P and w_T, which are always positive. The control volume work is related to these new variables by

$$w_{cv} = w_T - w_P$$

Using this expression in the energy equation results in

$$0 = q_{cv} + w_P - w_T + \left(u_1 + \frac{P_1}{\rho} + \frac{V_1^2}{2} + gz_1\right) - \left(u_2 + \frac{P_2}{\rho} + \frac{V_2^2}{2} + gz_2\right)$$

Rearranging terms and dividing by g produces

$$\frac{P_1}{\rho g} + \frac{V_1^2}{2g} + z_1 + \frac{w_P}{g} = \frac{P_2}{\rho g} + \frac{V_2^2}{2g} + z_2 + \frac{w_T}{g} + \frac{(u_2 - u_1) - q_{cv}}{g} \tag{9-40}$$

We assume the flow enters at the same temperature as the surroundings. The fluid temperature increases (usually slightly) due to frictional heating, and, as a result, there may be some heat transfer to the surroundings. These effects are incorporated into the last term of the equation, which has previously been defined as the head loss. We now define pump head as

$$h_P = \frac{w_P}{g} = \frac{\dot{W}_P}{\dot{m}g}$$

and the head available to drive a turbine as

$$h_T = \frac{w_T}{g} = \frac{\dot{W}_T}{\dot{m}g}$$

Using these last two expressions and the definition of head loss from Eq. 9-29 in Eq. 9-40:

$$\boxed{\frac{P_1}{\rho g} + \frac{V_1^2}{2g} + z_1 + h_P = \frac{P_2}{\rho g} + \frac{V_2^2}{2g} + z_2 + h_T + h_L} \tag{9-41}$$

Each term in this equation has the units of length and can be visualized as a "head." By definition:

$$\frac{P}{\rho g} = \text{Pressure head}$$

$$\frac{V^2}{2g} = \text{Velocity head}$$

$$h_L = \text{Head loss}$$

$$h_P = \text{Pump head}$$

$$h_T = \text{Turbine head}$$

Head is a sort of "equivalent elevation." For example, if the pump head supplied is 6 m, then the pump can lift a frictionless fluid a net elevation of 6 m provided there is no change in pressure or velocity.

The quantity $\dot{W}_P$ is the power that the pump actually supplies to the fluid. If the pump operation were ideal, less work would be needed to raise the pressure of the fluid. The ratio of the ideal work, $\dot{W}_{ideal}$, to the actual work was defined in Chapter 7 as the isentropic efficiency (see Eq. 7-33):

$$\eta_{s,P} = \frac{\dot{W}_{ideal}}{\dot{W}_P}$$

The two power terms in this ratio are related to what happens to the fluid as the pump works on it. Inefficiencies arise from frictional effects within the fluid.

There are also mechanical inefficiencies in pumps. Mechanical losses due to friction occur in bearings, gears, and couplings to the motor or turbine driving the pump. This friction results in heating of the mechanical components and the losses are to the surroundings, not to the fluid flowing through the pump. The enthalpy change of the fluid is not affected by these losses. We define a pump *mechanical efficiency* as

$$\eta_{m,p} = \frac{\dot{W}_P}{\dot{W}_{in}}$$

where $\dot{W}_{in}$ is the input power required at the pump shaft. Mechanical efficiency is always between 0 and 1. The actual power input required for a given pressure rise is

$$\dot{W}_{in} = \frac{\dot{W}_P}{\eta_{m,P}} = \frac{\dot{W}_{ideal}}{\eta_{s,P}\eta_{m,P}}$$

In many cases, the pump is driven by an electric motor. There are losses in the motor that arise from eddy currents, bearing friction, aerodynamic windage, joule heating, and so on. To determine the electric power needed as input to the motor, $\dot{W}_{electric}$, we define the motor efficiency as

$$\eta_{motor} = \frac{\dot{W}_{out}}{\dot{W}_{electric}}$$

The mechanical power output, $\dot{W}_{out}$, from the electric motor is equal to the power input to the pump, $\dot{W}_{in}$. Hence, the electric power input required to drive the pump–motor combination is

$$\dot{W}_{electric} = \frac{\dot{W}_{out}}{\eta_{motor}} = \frac{\dot{W}_P}{\eta_{motor}\eta_{m,P}} = \frac{\dot{W}_{ideal}}{\eta_{motor}\eta_{m,P}\eta_{s,P}}$$

An efficiency for hydroturbines can be defined similarly. In a hydroturbine, the power output is always less than the power, $\dot{W}_T$, supplied to the turbine from the flowing fluid. The mechanical efficiency of a hydroturbine is defined as

$$\eta_{m,T} = \frac{\dot{W}_{out}}{\dot{W}_T}$$

where $\dot{W}_{out}$ is the power available at the output shaft of the turbine. This power is typically used to drive an electric generator that produces electricity for homes and businesses. The power supplied to the turbine by the fluid is related to the electric power produced as

$$\dot{W}_T = \frac{\dot{W}_{out}}{\eta_{m,T}} = \frac{\dot{W}_{electric}}{\eta_{generator}\,\eta_{m,T}} = \frac{\dot{W}_{ideal}}{\eta_{generator}\,\eta_{m,T}\,\eta_{s,T}}$$

Like pump efficiency, turbine efficiency is between 0 and 1.

EXAMPLE 9-5 A pump with inefficiencies

A pump with a mechanical efficiency of 0.88 is used to pump water through a horizontal commercial steel pipe of length 55 m and inner diameter 1.5 cm. Water enters the pump at 100 kPa and exits the pipe at 140 kPa with a velocity of 2.2 m/s. Assume the water is at 25°C. Find the power input to the pump.

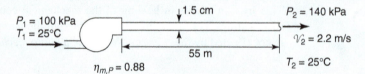

$P_1 = 100$ kPa
$T_1 = 25°C$

1.5 cm

$P_2 = 140$ kPa
$\mathcal{V}_2 = 2.2$ m/s
$T_2 = 25°C$

55 m

$\eta_{m,P} = 0.88$

Approach:

We can use Eq. 9-4 to calculate the pump head required for the given conditions. The head loss is determined with Eq. 9-32, which requires a friction factor. Calculate the Reynolds number to determine whether the flow is laminar or turbulent. Then evaluate the appropriate friction factor. Finally, use $\eta_{m,P} = \dot{W}_P/\dot{W}_{in}$ to find the input power to the pump.

Assumptions:

Solution:

We begin by finding the friction factor in the pipe. The Reynolds number is

$$Re = \frac{\rho D \mathcal{V}}{\mu}$$

A1. The water is at room temperature.

Using properties of water at 25°C from Table A-6 [A1],

$$Re = \frac{\left(997\,\frac{kg}{m^3}\right)\left(\frac{1.5}{100}\,m\right)\left(2.2\,\frac{m}{s}\right)}{8.72 \times 10^{-4}\,\frac{N \cdot s}{m^2}} = 37{,}730$$

A2. Friction factor may be approximated with the Haaland relationship.

Because the Reynolds number is greater than 4000, the flow is turbulent. We use the Haaland relationship to find the friction factor [A2]:

$$\frac{1}{\sqrt{f}} \approx -1.8 \log\left[\left(\frac{\varepsilon/D}{3.7}\right)^{1.11} + \frac{6.9}{Re}\right]$$

For commercial steel pipe, from Table 9-2, $\varepsilon = 0.045$ mm; therefore,

$$\frac{1}{\sqrt{f}} \approx -1.8 \log\left[\left(\frac{\left(\frac{0.045\,mm}{1.5\,cm}\right)\left(\frac{1\,cm}{10\,mm}\right)}{3.7}\right)^{1.11} + \frac{6.9}{37{,}730}\right]$$

$$f = 0.0291$$

The head loss is related to the friction factor by

$$h_L = f \frac{L}{D} \frac{V^2}{2g}$$

Substituting values,

$$h_L = 0.0291 \left(\frac{55 \, \text{m}}{0.015 \, \text{m}} \right) \left[\frac{\left(2.2 \, \frac{\text{m}}{\text{s}} \right)^2}{2 \left(9.81 \, \frac{\text{m}}{\text{s}^2} \right)} \right] = 26.3 \, \text{m}$$

A3. The flow is steady.
A4. The flow is incompressible.

To get the pump head, use the steady-flow energy equation, which is [A3][A4]

$$\frac{P_1}{\rho g} + \frac{V_1^2}{2g} + z_1 + h_P = \frac{P_2}{\rho g} + \frac{V_2^2}{2g} + z_2 + h_T + h_L$$

In this case, there is no change in pipe area and the flow is incompressible; therefore, $V_1 = V_2$. The pipe is horizontal, so $z_1 = z_2$. There is no turbine, so turbine head is zero. With these simplifications, the energy equation becomes

$$\frac{P_1}{\rho g} + h_P = \frac{P_2}{\rho g} + h_L$$

Solving for pump head gives

$$h_P = \frac{P_2 - P_1}{\rho g} + h_L$$

$$h_P = \frac{(140 - 100) \, \text{kPa} \left(\frac{1000 \, \text{Pa}}{1 \, \text{kPa}} \right)}{\left(997 \, \frac{\text{kg}}{\text{m}^3} \right) \left(9.81 \, \frac{\text{m}}{\text{s}^2} \right)} + 26.3 \, \text{m} = 30.4 \, \text{m}$$

The pump head is related to the pump power by

$$h_P = \frac{\dot{W}_P}{\dot{m} g}$$

To find the mass flow rate, use

$$\dot{m} = \rho V A = \left(997 \, \frac{\text{kg}}{\text{m}^3} \right) \left(2.2 \, \frac{\text{m}}{\text{s}} \right) \left[\pi \left(\frac{0.015}{2} \right)^2 \, \text{m}^2 \right] = 0.388 \, \frac{\text{kg}}{\text{s}}$$

The useful work supplied by the pump to the flow is

$$\dot{W}_P = \dot{m} g h_P$$

$$\dot{W}_P = \left(0.388 \, \frac{\text{kg}}{\text{s}} \right) \left(9.81 \, \frac{\text{m}}{\text{s}^2} \right) (30.4 \, \text{m}) = 116 \, \text{W}$$

The input shaft power to the pump is

$$\dot{W}_{in} = \frac{\dot{W}_P}{\eta_{m,P}} = \frac{116 \, \text{W}}{0.88} = 132 \, \text{W}$$

9.9 MINOR LOSSES

In addition to viscous losses due to friction in straight sections of pipe, there are often other sources of losses in pipeline flow. For example, if the flow turns through an elbow or passes through a valve, there is an added frictional loss. All pipeline losses from sources other than wall friction are traditionally called **minor losses.** Do not be fooled by the name. Sometimes the so-called minor losses are greater than the head loss in straight sections of pipe, and these "minor" losses then dominate the flow situation.

The head loss due to minor losses can be determined experimentally and correlated as

$$h_L = K_L \frac{V^2}{2g}$$

where K_L is called the loss coefficient. The loss coefficient is dimensionless. The head loss due to minor losses is added to the head loss in straight pipe sections so that the steady-flow energy equation becomes

$$\frac{P_1}{\rho g} + \frac{V_1^2}{2g} + z_1 + h_P = \frac{P_2}{\rho g} + \frac{V_2^2}{2g} + z_2 + h_T + \sum h_L \qquad (9\text{-}42)$$

where $\sum h_L$ represents the sum of all frictional losses, whatever the source.

Table 9-3 lists many pipeline components and their associated loss coefficients. In fact, loss coefficients depend to some extent on Reynolds number and on pipe diameter. The numbers listed should be used only as guidelines. Figure 9-13 and Figure 9-14 give loss coefficients for pipeline entrances and exits. Figure 9-15 plots the loss coefficients in sudden expansions and contractions.

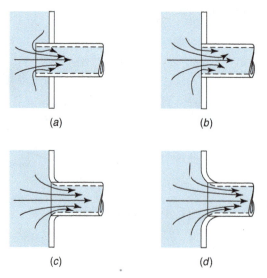

(a)

(b)

(c)

(d)

FIGURE 9-13 Entrance flow conditions and loss coefficient: (a) reentrant, $K_L = 0.8$, (b) sharp-edged, $K_L = 0.5$, (c) slightly rounded, $K_L = 0.2$, (d) well-rounded, $K_L = 0.04$.

TABLE 9-3 Loss Coefficients for Pipe Components (*Source*: Munson, B. R., D. F. Young, and T. H. Okiishi, *Fundamentals of Fluid Mechanics*, 4th ed., Wiley, New York, 2002. p.489 Used with permission.)

Component	K_L	
a. Elbows		
Regular 90°, flanged	0.3	
Regular 90°, threaded	1.5	
Long radius 90°, flanged	0.2	
Long radius 90°, threaded	0.7	
Long radius 45°, flanged	0.2	
Regular 45°, threaded	0.4	Regular 90°, flanged
		Long radius 45°, flanged
b. 180° return bends		
180° return bend, flanged	0.2	
180° return bend, threaded	1.5	
c. Tees		
Line flow, flanged	0.2	
Line flow, threaded	0.9	
Branch flow, flanged	1.0	180° return bend, flanged
Branch flow, threaded	2.0	
d. Union, threaded	0.08	
		Line flow, flanged
e. Valves		
Globe, fully open	10	
Angle, fully open	2	
Gate, fully open	0.15	
Gate, ¼ closed	0.26	
Gate, ½ closed	2.1	
Gate, ¾ closed	17	Branch flow, flanged
Swing check, forward flow	2	
Swing check, backward flow	∞	
Ball valve, fully open	0.05	
Ball valve, ⅓ closed	5.5	
Ball valve, ⅔ closed	210	Union, threaded

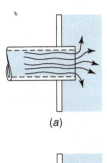

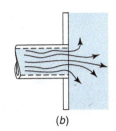

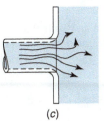

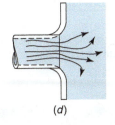

FIGURE 9-14 Exit flow conditions and loss coefficient: (a) reentrant, $K_L = 1.0$, (b) sharp-edged, $K_L = 1.0$, (c) slightly rounded, $K_L = 1.0$, (d) well-rounded, $K_L = 1.0$.

(a)

(b)

(c)

(d)

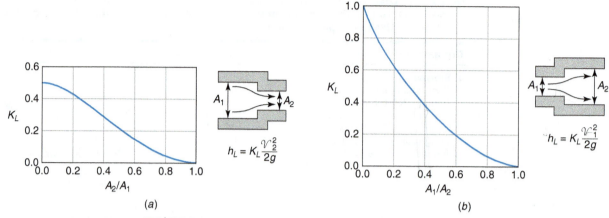

FIGURE 9-15 (a) Loss coefficient in a sudden contraction. (b) Loss coefficient in a sudden expansion.

EXAMPLE 9-6 Pumping water to a higher elevation

A pump is needed to remove water from a mine shaft. How much pump power (in kW) is needed to remove water at a rate of 65 kg/s? Assume the pump is ideal and use data on the figure.

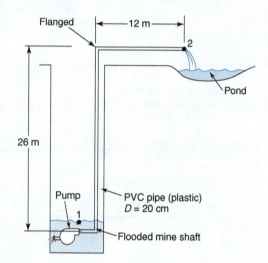

Approach:

Use the steady-flow energy equation (Eq. 9-42) to relate pump head to elevation changes and head losses. The losses consist of three components: the loss at the entrance, the loss at the elbows, and the loss due to wall friction in the lines. Since velocity is easily calculated from flow rate, it is possible to find the minor losses using

$$h_{L,in} = K_L \frac{V^2}{2g}$$

where values of K_L are available in Figure 9-13 and Table 9-3. To determine the head loss due to wall friction, use the Reynolds number to determine whether the flow is laminar or turbulent, and then evaluate the appropriate friction factor. Once the pump head is known, the power may be calculated from

$$h_p = \frac{\dot{W}_p}{\dot{m}g}$$

Assumptions:

A1. The flow is steady.
A2. The flow is incompressible.

Solution:

Define station 1 on the surface of the water in the mine shaft and station 2 at the exit of the pipe. The steady flow energy equation is [A1][A2]

$$\frac{P_1}{\rho g} + \frac{\mathcal{V}_1^2}{2g} + z_1 + h_P = \frac{P_2}{\rho g} + \frac{\mathcal{V}_2^2}{2g} + z_2 + h_T + \sum h_L$$

This is an incompressible flow and the pipe diameter does not change; therefore,

$$\dot{m}_1 = \dot{m}_2 = \rho \mathcal{V}_1 A_1 = \rho \mathcal{V}_2 A_2$$

which reduces to

$$\mathcal{V}_1 = \mathcal{V}_2$$

In addition, the pressure is atmospheric at both inlet and outlet, so

$$P_1 = P_2$$

There is no turbine, so $h_T = 0$. The energy equation then reduces to

$$z_1 + h_P = z_2 + \sum h_L$$

To find head loss, we need the flow velocity:

$$\mathcal{V} = \frac{\dot{m}}{\rho A} = \frac{65 \, \frac{\text{kg}}{\text{s}}}{\left(997 \, \frac{\text{kg}}{\text{m}^3}\right) \pi \, (10 \, \text{cm})^2 \left(\frac{1 \, \text{m}}{100 \, \text{cm}}\right)^2} = 2.08 \, \frac{\text{m}}{\text{s}}$$

where Table A-6 has been used to find the density and viscosity.

The head loss has three components: the loss at the entrance, the loss at the elbows, and the loss due to wall friction in the lines. The head loss at the entrance is

$$h_{L,in} = K_L \frac{\mathcal{V}^2}{2g}$$

The loss coefficient in Figure 9-13 is used, so that

$$h_{L,in} = 0.8 \, \frac{\left(2.08 \, \frac{\text{m}}{\text{s}}\right)^2}{2 \left(9.81 \, \frac{\text{m}}{\text{s}^2}\right)} = 0.176 \, \text{m}$$

At the 90° elbow, the loss is

$$h_{L,90} = K_L \frac{\mathcal{V}^2}{2g} = (0.3) \, \frac{(2.08)^2}{2 \, (9.81)} = 0.066 \, \text{m}$$

where Table 9-3 has been used for the loss coefficient. To find the head loss in the long runs of the pipe, the Reynolds number is needed. It is

$$Re = \frac{\rho D \mathcal{V}}{\mu} = \frac{\left(997 \, \frac{\text{kg}}{\text{m}^3}\right) (20 \, \text{cm}) \left(\frac{1 \, \text{m}}{100 \, \text{cm}}\right) \left(2.08 \, \frac{\text{m}}{\text{s}}\right)}{1.12 \times 10^{-3} \, \frac{\text{N} \cdot \text{s}}{\text{m}^2}} = 370,314$$

Because $Re > 4000$, the flow is turbulent. According to Table 9-2, a plastic pipe is very smooth, and $\varepsilon = 0$. The friction factor can be determined by either the Colebrook equation (Eq. 9-34), the Haaland equation (Eq. 9-35), the Petukhov equation (Eq. 9-36), or the Moody chart (Figure 9-11). For illustration, we choose the Haaland relationship [A3]. (The other approaches would give similar results.) The Haaland equation is

A3. Approximate friction factor with the Haaland relationship.

$$\frac{1}{\sqrt{f}} \approx -1.8 \log \left[\left(\frac{\varepsilon/D}{3.7} \right)^{1.11} + \frac{6.9}{Re} \right] \approx -1.8 \log \left[\frac{6.9}{370,314} \right]$$

$$f = 0.0138$$

Therefore, the head loss due to wall friction is

$$h_{L,w} = f \frac{L}{D} \frac{V^2}{2g} = 0.0138 \frac{(26+12)\,\text{m}}{(20\,\text{cm})\left(\frac{1\,\text{m}}{100\,\text{cm}}\right)} \frac{\left(2.08\,\frac{\text{m}}{\text{s}}\right)^2}{2\left(9.81\,\frac{\text{m}}{\text{s}^2}\right)} = 0.578\,\text{m}$$

Solving the energy equation for the pump head,

$$h_P = z_2 - z_1 + \sum h_L$$

Adding together the three components of loss and using the definition of h_p,

$$\frac{\dot{W}_P}{\dot{m}g} = z_2 - z_1 + h_{L,in} + 2h_{L,90} + h_{L,w}$$

Solving for work,

$$\dot{W}_P = \dot{m}g \left[(z_2 - z_1) + h_{L,in} + 2h_{L,90} + h_{L,w} \right]$$

$$\dot{W}_{cv} = \left(65\,\frac{\text{kg}}{\text{s}} \right) \left(9.81\,\frac{\text{m}}{\text{s}^2} \right) [(26 - 0) + 0.176 + (2)(0.066) + 0.578]\,\text{m}$$

$$= 17,140\,\text{W} = 17.1\,\text{kW}$$

EXAMPLE 9-7 Loss coefficient for a valve

An experiment is designed to measure the loss coefficient of a gate valve. The valve is installed in a pipe with a diameter of 1 in. When the valve is half closed, hydraulic fluid flows through the valve at a mass flow rate of 1.9 lbm/s. If the pressure drop across the valve is 0.5 psi, what is the loss coefficient?

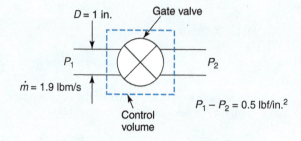

Approach:

The loss coefficient is related to head loss through

$$h_v = K_L \frac{V^2}{2g}$$

The velocity may be found from the given mass flow rate and pipe diameter. Use the steady-flow energy equation (Eq. 9-42) to relate pressure drop to head loss in the valve.

Assumptions:

A1. The flow is steady.

A2. The flow is incompressible.

A3. Wall friction is negligible in the short pipe segments.

Solution:

The control volume includes the valve and a short section of pipe upstream and downstream of the valve. The steady-flow energy equation [A1][A2] applied to this control volume is

$$\frac{P_1}{\rho g} + \frac{\mathcal{V}_1^2}{2g} + z_1 + h_P = \frac{P_2}{\rho g} + \frac{\mathcal{V}_2^2}{2g} + z_2 + h_T + \sum h_L$$

Since there are no area changes and the flow is incompressible,

$$\mathcal{V}_1 = \mathcal{V}_2 = \mathcal{V}$$

There are no elevation changes and no pumps or turbines within the control volume. The straight pipe sections are short and, therefore, frictional head loss may be neglected [A3]. With these simplifications, the energy equation becomes

$$\frac{P_1}{\rho g} = \frac{P_2}{\rho g} + h_v$$

where h_v is the head loss in the valve. This may also be written as

$$h_v = \frac{P_1 - P_2}{\rho g}$$

Using the density of hydraulic fluid from Table B-6, the head loss is

$$h_L = \frac{\left(0.5 \, \frac{\text{lbf}}{\text{in.}^2}\right)\left(\frac{144 \, \text{in.}^2}{1 \, \text{ft}^2}\right)}{\left(53.0 \, \frac{\text{lbm}}{\text{ft}^3}\right)\left(32.2 \, \frac{\text{ft}}{\text{s}^2}\right)\left(\frac{1 \, \text{lbf}}{32.2 \, \frac{\text{lbm} \cdot \text{ft}}{\text{s}^2}}\right)} = 1.36 \, \text{ft}$$

To find the velocity, use conservation of mass in the form

$$\mathcal{V} = \frac{\dot{m}}{\rho A} = \frac{1.9 \, \frac{\text{lbm}}{\text{s}}}{\left(53.0 \, \frac{\text{lbm}}{\text{ft}^3}\right) \pi \left[\frac{\left(\frac{1}{12} \, \text{ft}\right)}{2}\right]^2} = 6.57 \, \frac{\text{ft}}{\text{s}}$$

The loss coefficient is found from the head loss using

$$h_v = K_L \frac{\mathcal{V}^2}{2g}$$

Rearranging,

$$K_L = \frac{2 g h_v}{\mathcal{V}^2} = \frac{2\left(32.2 \, \frac{\text{ft}}{\text{s}^2}\right)(1.36 \, \text{ft})}{\left(6.57 \, \frac{\text{ft}}{\text{s}}\right)^2} = 2.02$$

Comments:

The loss coefficient for a gate valve from Table 9-3 is 2.1. The difference between this value and the experimental result of 2.02 could be due to experimental error, the imprecision of the term *half closed*, and/or differences in valve construction.

9.10 PIPELINE NETWORKS

(Go to www.wiley.com/college/kaminski)

9.11 PUMP SELECTION

(Go to www.wiley.com/college/kaminski)

SUMMARY

The **shear stress** in a fluid is related to the **dynamic viscosity** by

$$\tau = \mu \frac{d\mathcal{V}}{dy}$$

In flow between two parallel plates with one plate moving at velocity $\mathcal{V}_P$ and the other stationary, the shear stress is

$$\tau = \mu \frac{\mathcal{V}_P}{b}$$

where b is the plate spacing. In **fully developed laminar flow in a circular pipe**, the velocity profile is

$$\mathcal{V}(r) = 2\mathcal{V}_m \left(1 - \frac{r^2}{R^2}\right)$$

The pressure drop in fully developed laminar flow in an inclined circular pipe is

$$\Delta P = \frac{8\mu L \mathcal{V}_m}{R^2} + \rho g L \sin\theta$$

For a **horizontal pipe**, the pressure drop becomes

$$\Delta P = \frac{8\mu L \mathcal{V}_m}{R^2}$$

The flow regime is controlled by the **Reynolds number**, defined as

$$Re = \frac{\rho D \mathcal{V}_m}{\mu}$$

The Reynolds number can be viewed as the ratio of inertial forces to viscous forces. In internal flow in a pipe,

$Re < 2100$	Laminar flow
$2100 < Re < 4000$	Transitional flow
$4000 < Re$	Turbulent flow

The head loss in a pipe is given in terms of a friction factor using

$$h_L = f \frac{L}{D} \frac{\mathcal{V}_m^2}{2g}$$

This applies to laminar or turbulent flow with any pipe orientation. Frictional pressure drop is related to the friction factor by

$$\Delta P = f \frac{L}{D} \frac{\rho \mathcal{V}_m^2}{2}$$

This applies to a simple pipe with no minor losses, no pumps or turbines, and so on—just the pipe itself. In **fully developed laminar flow in a circular pipe**, the friction factor is

$$f = \frac{64}{Re}$$

For conduits with noncircular cross-section, the hydraulic diameter, given by

$$D_h = \frac{4A_c}{P_{wetted}}$$

is used in place of the diameter in the Reynolds number, h_L, and ΔP expressions. For a circular pipe, the hydraulic diameter reduces to the ordinary diameter. Friction factors for laminar flow in some noncircular geometries are given in Table 9-1. The hydraulic diameter is also used in turbulent flow equations.

In turbulent flow, the friction factor is given by the **Colebrook equation**:

$$\frac{1}{\sqrt{f}} = -2.0 \log \left(\frac{\varepsilon/D}{3.7} + \frac{2.51}{Re\sqrt{f}} \right)$$

which is a curve fit to experimental results. The Colebrook equation is implicit in f. A more convenient expression, which gives results within 2% of the Colebrook equation, is the explicit **Haaland equation**:

$$\frac{1}{\sqrt{f}} \approx -1.8 \log \left[\left(\frac{\varepsilon/D}{3.7} \right)^{1.11} + \frac{6.9}{Re} \right]$$

Friction factors may also be read graphically from the **Moody** chart (Figure 9-11). For smooth pipes, the friction factor can be found from the **Petukhov** formula, which is

$$f = (0.79 \ln Re - 1.64)^{-2} \qquad \text{turbulent flow, smooth wall,} \\ 3000 < Re_D < 5 \times 10^6$$

Flow in pipes may be calculated using the **Swamee-Jain formulae**, which are

$$h_L = 1.07 \frac{\dot{V}^2 L}{gD^5} \left\{ \ln \left[\frac{\varepsilon}{3.7D} + 4.62 \left(\frac{\mu D}{\rho \dot{V}} \right)^{0.9} \right] \right\}^{-2}$$

$$10^{-6} < \varepsilon/D < 10^{-2} \\ 3000 < Re < 3 \times 10^8$$

$$\dot{V} = -0.965 \left(\frac{gD^5 h_L}{L} \right)^{0.5} \ln \left[\frac{\varepsilon}{3.7D} + \left(\frac{3.17\mu^2 L}{g\rho^2 D^3 h_L} \right)^{0.5} \right]$$

$$Re > 2000$$

$$D = 0.66 \left[\varepsilon^{1.25} \left(\frac{L\dot{V}^2}{gh_L} \right)^{4.75} + \mu\dot{V}^{9.4} \left(\frac{L}{gh_L} \right)^{5.2} \right]^{0.04}$$

$$10^{-6} < \varepsilon/D < 10^{-2} \\ 5000 < Re < 3 \times 10^8$$

These are curve fits to the results obtained from combining the Colebrook relation with the energy equation for flow in a single pipe. If SI units are used, all quantities must be in m, s, and kg. If British units are used, all quantities must be in ft, s, and lbm.

Pressure drop per unit length is higher in the entrance region of a pipe than in the fully developed region. The entrance length is given by

$$L_{ent,h} \approx 0.065 \, Re \, D \qquad \text{laminar, Re} < 2100 \\ L_{ent,h} \approx 4.4 \, (Re)^{1/6} \, D \qquad \text{turbulent, Re} > 4000$$

The steady-flow energy equation is

$$\frac{P_1}{\rho g} + \frac{V_1^2}{2g} + z_1 + h_P = \frac{P_2}{\rho g} + \frac{V_2^2}{2g} + z_2 + h_T + \sum h_L$$

where

$$\frac{P}{\rho g} = \text{pressure head}$$

$$\frac{V^2}{2g} = \text{velocity head}$$

$$h_L = \text{head loss}$$

$$h_P = \text{pump head}$$

$$h_T = \text{turbine head}$$

The pump head is related to pump power by

$$h_P = \frac{w_P}{g} = \frac{\dot{W}_P}{\dot{m}g}$$

and the head available to drive the turbine as

$$h_T = \frac{w_T}{g} = \frac{\dot{W}_T}{\dot{m}g}$$

The mechanical efficiency of a pump is

$$\eta_{m,P} = \frac{\dot{W}_P}{\dot{W}_{in}}$$

The mechanical efficiency of a hydroturbine is

$$\eta_{m,T} = \frac{\dot{W}_{out}}{\dot{W}_T}$$

The head loss due to minor losses can be determined experimentally and correlated as

$$h_L = K_L \frac{V^2}{2g}$$

Values for loss coefficient, K_L, can be found in Table 9-3 and in Figure 9-12 through Figure 9-14.

SELECTED REFERENCES

Fox, R. W., and A. T. McDonald, *Introduction to Fluid Mechanics*, 5th ed., Wiley, New York, 1998.

Munson, B. R., D. F. Young, and T. H. Okiishi, *Fundamentals of Fluid Mechanics*, 4th ed., Wiley, New York, 2002.

Potter, M. C., and D. C. Wiggert, *Mechanics of Fluids*, 3rd ed., Brooks/Cole, Pacific Grove, CA, 2002.

Roberson, J. A., and C. T. Crowe, *Engineering Fluid Mechanics*, 6th ed., Wiley, New York, 1997.

White, F. M., *Fluid Mechanics*, McGraw-Hill, New York, 1979.

PROBLEMS

Problems designated with WEB refer to material available at www.wiley.com/college/kaminski.

LAMINAR FLOWS

P9-1 Fully developed, laminar flow of a viscous fluid ($\mu = 2.17\,\mathrm{N \cdot s/m^2}$) flows between horizontal parallel plates 1 m long that are spaced 3.0 mm apart. The pressure drop is 1.25 kPa. Determine the volumetric flow rate (per unit width) through the channel (in $\mathrm{m^3/s \cdot m}$).

P9-2 Journal bearings are constructed with concentric cylinders with a very small gap between the two cylinders; the gap is filled with oil. Because of the very small gap, the flow in the gap is laminar. Consider a sealed journal bearing with inner and outer diameters of 50 and 51 mm, respectively, and a length of 75 mm. The shaft (inner cylinder) rotates at 3000 rpm. At startup the torque needed to turn the shaft is 0.25 N-m. Determine the viscosity of the oil (in $\mathrm{N \cdot s/m^2}$). After an hour of operation, will the torque have increased or decreased? Explain.

P9-3 Consider laminar water flow at 20°C between two very large horizontal plates. The lower plate is stationary, and the upper plate moves to the right at a velocity of 0.25 m/s. For a plate spacing of 2 mm, determine the pressure gradient and its direction required to produce zero net flow at a cross-section.

P9-4 In the 3/4-in. pipe shown in the figure, oil flows downward at 6 gal/min. The oil has a specific gravity of 0.87 and a dynamic viscosity of 0.4 lbm/ft·s. The specific gravity of the manometer fluid is 2.9. Determine the manometer deflection, h (in ft).

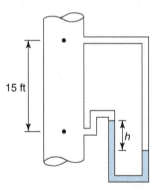

P9-5 In Problem P9-4, if the flow is upward instead of downward, determine the manometer deflection, h (in ft).

P9-6 Data are read from and written to spinning computer disks (3600 rpm) by small read–write heads that float above the disk on a thin (0.5-μm) film of air. Consider a 10 mm by 10 mm head located 55 mm from the disk centerline. For air at 25°C, and assuming the flow is similar to that between infinite parallel plates, determine

a. the Reynolds number based on the gap dimension.

b. the power required to overcome the viscous shear (in W).

P9-7 Skimmers are used to remove viscous fluids, such as oil, from the surface of water. As shown in the diagram, a continuous belt moves upward at velocity $\mathcal{V}_0$ through the fluid, and the more viscous liquid (with density ρ and viscosity μ) adheres to the belt. A film with thickness h forms on the belt. Gravity tends to drain the liquid, but the upward belt velocity is such that net liquid is transported upward. Assume the flow is fully developed, laminar, with zero pressure gradient, and zero shear stress at the outer film surface where air contacts it. Determine an expression for the velocity profile and flow rate. Use a differential analysis similar to that used for fully developed laminar flow through an inclined pipe. Clearly state the velocity boundary conditions at the belt surface and at the free surface.

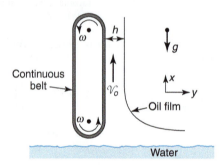

P9-8 Consider a fully developed laminar flow of 20°C water down an inclined plane that is 20° to the horizontal. The water thickness is 1 mm. The water is exposed to atmosphere everywhere, and the air exerts zero shear on the water. Using a differential analysis similar to that used for fully developed laminar flow through an inclined pipe, determine the volumetric flow rate per unit width (in $\mathrm{m^3/s \cdot m}$).

P9-9 A biomedical device start-up company is developing a liquid drug injection device. The device uses compressed air to drive the plunger in a piston–cylinder assembly that will push the drug (viscosity and density similar to water at 10°C) through the hypodermic needle (inside diameter 0.25 mm and length 50 mm). If the flow must remain laminar in the hypodermic needle, determine

a. the maximum flow possible (in $\mathrm{cm^3/s}$).

b. the required air pressure for the maximum flow if the pressure at the end of the needle must be 105 kPa (in kPa). (Assume fully developed flow.)

P9-10 The viscosity of liquids is measured with a capillary viscometer, in which a laminar flow is maintained in a small-diameter tube and the pressure drop and flow rate are measured.

If the flow is fully developed, then Eq. 9-13 can be used to calculate the liquid viscosity. However, entrance effects often are present. Consider the flow of a liquid ($SG = 0.92$) through a tube 450 mm long and 0.75 mm in diameter. A flow of 1 cm³/s is obtained when the pressure drop is 65 kPa.

a. Determine the viscosity if the flow is fully developed (in N·s/m²).

b. Determine the viscosity if the pressure drop in the entrance length is twice that for the same length of fully developed flow (in N·s/m²).

P9-11 A machine tool manufacturer is considering using gravity flow to supply cutting oil ($SG = 0.87$, $\mu = 0.003$ N·s/m²) to the tool and workpiece. The vertical 5-mm-diameter tube connecting the oil reservoir to the workpiece is very long, so the flow can be assumed to be fully developed; in addition, the depth of oil in the reservoir is negligible compared to the tube length. The pressure is atmospheric at the exit of the tube and at the surface of the reservoir. Determine the volumetric flow rate of the oil (in cm³/s).

P9-12 A manometer, with pressure taps 25 ft apart, is used to measure the pressure drop of oil ($SG = 0.82$) flowing in a 1.5-in. pipe with a volumetric flow rate of 4 ft³/min. The manometer fluid is mercury ($SG = 13.6$). The distance from the lower-pressure tap to the surface of the mercury highest in the manometer is 2 ft, and the distance from the upper pressure tap to the same height in the mercury is 4 ft. For a manometer deflection of 4 in., determine

a. the flow direction.

b. the friction factor.

c. whether the flow is laminar or turbulent.

d. the oil viscosity (in lbm/ft·s).

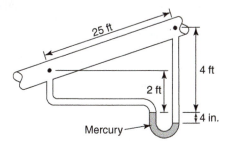

P9-13 Develop an expression for the velocity profile for fully developed laminar flow between stationary infinite parallel plates. Use an approach similar to that applied in Section 9.3 for a circular tube.

P9-14 In an inclined 50-mm-diameter pipe, a fluid ($SG = 0.88$) flows with a volumetric flow rate of 0.003 m³/s. The gage pressure at the pipe inlet is 720 kPa. The pipe outlet is at atmospheric pressure and is 15 m above the inlet. Determine the head loss between the inlet and outlet (in m).

P9-15 The pipe exit in Problem P9-14 is lowered to the same elevation as the inlet. Determine the inlet pressure for this new condition (in kPa).

TURBULENT FLOW

P9-16 An air-conditioning duct is 25 cm square and must convey 25 m³/min of air at 100 kPa, 25°C. The duct is made of sheet metal that has a roughness of approximately 0.05 mm. Determine the pressure drop for 25 m of horizontal duct run (in kPa and mm of water).

P9-17 A manufacturer develops a new type of flow control valve. Before it can be advertised and sold, its loss coefficient must be determined. The valve is installed in a 6-in. pipe, and 2 ft³/s of water flows through it. The pressure drop is measured with a manometer whose fluid has a specific gravity of 1.3. The manometer deflection is 7.5 in. Determine the loss coefficient for the valve.

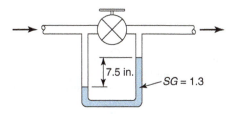

P9-18 When pumping a fluid, the pressure at the entrance to the pump must never drop below the saturation pressure of the fluid. If the pressure does drop below the saturation pressure, cavitation (the forming of vapor bubbles) occurs, which can damage the pump impeller. Consider the system shown in the figure, which is constructed of commercial steel pipe and threaded connections. For water at 10°C, determine the maximum possible flow rate without cavitation occurring (in m³/s).

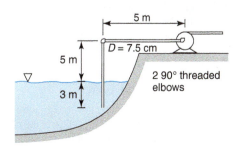

P9-19 Fire codes mandate that the pressure drop in horizontal runs of commercial steel pipe must not exceed 1.0 lbf/in.² per 150 ft of pipe for flows up to 500 gal/min. For a water temperature of 50°F, determine the minimum pipe diameter required (in in.). Is the number you calculated feasible?

P9-20 The owners of a luxurious mountain resort want to install a fancy water fountain. The artist's initial design uses 75 m of 7.5-cm-diameter commercial steel pipe ending in a nozzle with a diameter of 3.75 cm with a 40-kW pump to pull water from a lake above the resort at a flow rate of 0.05 m³/s. To save operating costs, the owners want to remove the pump and rely only on a gravity head to power the fountain. Assuming the friction factor is 0.016 for both cases and neglecting minor losses, determine

a. the flow rate if the pump is removed from the system (in m³/s).

b. the height of the water jet with and without the pump if the nozzle is pointed vertically upward (in m).

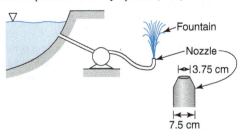

P9-21   At an oil tank farm, a vandal opens a valve at the end of a 5-cm-diameter, 50-m-long horizontal pipe from the bottom of a large-diameter oil tank. The oil tank is open to the atmosphere, and the oil depth is 6.5 m. The oil has $SG = 0.85$ and a kinematic viscosity of 6.8×10^{-4} m²/s. Neglecting minor losses, determine the initial flow rate from the tank (in m³/s).

P9-22   In a large convention center, heated air at 85°F must be conveyed from the furnace room to the display rooms through a 500-ft smooth duct. The required flow rate is 7500 ft³/min. If the pressure loss must not exceed 2.5 in. of water,

a. Determine the minimum diameter required (in in.).

b. Determine the pumping power required (in hp).

P9-23   Consider a heat exchanger that has 1000 2.5-cm-diameter smooth tubes in parallel, each 6 m long. The total water flow of 1 m³/s at 10°C flows through the tubes. Neglecting entrance and exit losses, determine

a. the pressure drop (in kPa).

b. the pumping power required (in kW).

c. the pumping power for the same flow rate if solid deposits from the water build up on the inner surface of the pipes with a thickness of 1 mm and an equivalent roughness of 0.4 mm.

P9-24   The piping system that connects one reservoir to a second reservoir consists of 150 ft of 3-in. cast-iron pipe that has four flanged elbows, a well-rounded entrance, sharp-edged exit, and a fully open gate valve. For 75 gal/min of water at 50°F, determine the elevation difference between the two reservoirs (in ft).

P9-25   Vandals open the drain valve on a water tower that is 10 m in diameter with a water depth of 8 m. The water flows out a sharp-edged opening into a horizontal 30-m-long pipe that is 10 cm in diameter; the gate valve in the pipe is half opened. Assuming the friction factor is 0.016, determine

a. the time required for the tank to drain (in min).

b. the time required for the tank to drain if only the sharp-edged opening and the valve are present (in min).

c. the appropriateness of the friction factor value used.

P9-26   An oil transporter truck is filled from the top with 15 m³ of fuel oil ($SG = 0.86$, $\mu = 5.3 \times 10^{-2}$ N·s/m²) from a reservoir that is 4 m below the truck top. A 10-m-long flexible hose 6 cm in diameter whose surface roughness is equivalent to that of galvanized iron connects the truck to the reservoir. A one-third-closed ball valve and two bends that are equivalent to 90° threaded elbows are in the hose. For a filling time of 15 min and a pump mechanical efficiency of 75%, determine the required pump power (in kW).

P9-27   Large office buildings use circulating hot water systems to ensure that hot water is available instantly in all restrooms. Consider a system that consists of 200 m of 2.5-cm commercial steel pipe. It has 15 90° regular threaded elbows, two fully open gate valves, three half-open gate valves, and one three-quarter-closed gate valve. For water at 50°C and a pump with a mechanical efficiency of 75%, determine

a. the power required if the water velocity is 2 m/s (in kW).

b. the power required if the water velocity is 1 m/s (in kW).

P9-28   To ensure adequate water supplies to a town, a municipal water department developed a second reservoir and wants to connect the new reservoir to the old one using a concrete pipe. The reservoirs are 1.5 miles apart with a difference in surface elevations of 25 ft. Determine the minimum pipe diameter needed to carry 10 ft³/s of water at 50°F.

P9-29   The reservoir behind a dam is connected to a hydroelectric power plant with a penstock (a large pipe to convey the water). At a particular plant, the elevation difference between the reservoir surface and the hydroturbine is 50 m, and the penstock is constructed of 150 m of 1-m-diameter cast-iron pipe. The turbine has a mechanical efficiency of 78%, and the electric generator has an efficiency of 94%. For a 1 m³/s flow of 10°C water, determine

a. the power output from the plant (in kW).

b. the power output if a fully open gate valve and two long-radius 45° flanged elbows also are in the pipe (in kW).

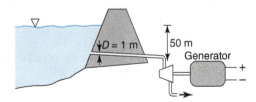

P9-30 The drain at the bottom of a swimming pool (10 m in diameter and 2 m deep) is well rounded and is connected to a 5-cm-diameter, 20-m-long plastic pipe. The water is at 20°C. For a friction factor of 0.021, determine

a. the time required to drain the pool (in min).

b. whether this value of friction factor is appropriate.

P9-31 If the pool in Problem P9-30 has a sharp-edged entrance and two 90° regular threaded elbows, determine the time required to drain the pool (in min).

P9-32 A pipe connects two reservoirs at different elevations. The pipe is constructed of 12-in.-diameter commercial steel with flanged fittings. The gate valve is one-fourth closed. The water temperature is 50°F. Determine the required elevation difference between the two reservoirs to produce a water flow rate of 10 ft³/s (in ft).

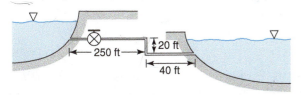

P9-33 A liquid ($SG = 0.93, \mu = 0.00068$ N·s/m²) is contained in a vertical 2-cm-diameter pipe. At one elevation the fluid pressure is 230 kPa; at an elevation 10 m higher, the pressure is 110 kPa.

a. Determine whether the flow is moving and in what direction.

b. Determine the flow velocity if it is flowing (in m/s).

P9-34 The designers of a large shopping mall install 18-in.-diameter smooth concrete storm sewers to channel away runoff after heavy rainstorms. Each storm sewer will need to carry a flow of 10 ft³/s. The pressures at the entrance and exit of the sewer are atmospheric. If the sewers are 200 ft long before they join with larger pipes, determine the required elevation change per 100 ft of pipe (in ft).

P9-35 In mountainous regions, tunnels are often used for cars, trucks, and trains. If the tunnel is too long, ventilation air must be supplied to dilute and purge vehicle exhaust gases from the tunnel. Consider a 3-ft-diameter, 2500-ft-long duct constructed of commercial steel pipe that carries air at 45°F, 14.1 psia with a flow rate of 10,000 ft³/min.

a. Determine the pressure drop (in in. of water).

b. Determine the power required (in hp).

P9-36 Water at 10°C flows from a lake at a flow rate of 0.1 m³/s. A 15-cm-diameter, 100-m-long galvanized iron pipe connects the lake to a building in which either a pump or a turbine is located. The elevation difference between the lake surface and the building is 10 m.

a. Determine whether the device in the building is a pump or a turbine.

b. Determine the power of the device (in W).

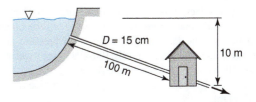

P9-37 Ski resorts pump water to make snow when the weather does not cooperate. Consider a resort that uses 100 gal/min of 35°F water. It is pumped from the water holding pond through a 4-in.-diameter, 3000-ft steel pipe to the top of the mountain. The elevation difference is 950 ft. The gage pressure required at the nozzle at the end of the pipe is 150 lbf/in.² Determine the required pumping power (in hp).

P9-38 In the western United States, many crops are irrigated, and water must be pumped long distances. Consider a system that consists of a 1-m-diameter, 2-km-long steel pipe, which connects a river to an irrigation canal. The canal's elevation is 50 m higher than that of the river. For water at 15°C and a pump with a mechanical efficiency of 80%, and neglecting minor losses, determine the power required to pump 2.5 m³/s of water (in kW).

P9-39 Air at 105 kPa and 25°C flows from a 7.5-cm circular duct into a 22.5-cm circular duct. The downstream pressure is 6.5 mm of water higher than the upstream pressure.

a. Determine the average air velocity approaching the expansion (in m/s).

b. Determine the volumetric flow rate (in m³/s).

c. Determine the mass flow rate (in kg/s).

P9-40 In some high-rise buildings, water is stored in an elevated tank on the roof to minimize pressure fluctuations in the system. Consider water that is pumped through a 10-cm steel pipe to the roof of a 200-m-tall building; the pump is on the ground floor. For a water temperature of 10°C and a flow of 0.02 m³/s, what is the pressure at the pump discharge (in kPa)?

P9-41 A fluid flows by gravity down an 8-cm galvanized iron pipe. The pressures at the higher and lower locations are 120 kPa and 140 kPa, respectively. The horizontal distance between the two locations is 30 m, and the pipe has a slope of 1-m rise per 10 m of run (horizontal distance). For a fluid with a kinematic viscosity of 10^{-6} m²/s and a density of 900 kg/m³, determine the flow rate (in m³/s).

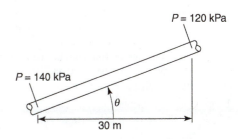

P9-42 A new factory is to be built that requires 0.03 m³/s of water. The water main from which the water will be obtained is 150 m from the factory. The water main pressure is 400 kPa (gage), and the factory needs 100-kPa (gage) water at a location 10 m above the water main. Assuming that galvanized steel pipe will be used, determine the minimum pipe diameter needed (in m).

P9-43 For the system shown in the figure, a water flow rate of 3 m³/s is to be pumped from the lower to the upper reservoir through a 1-m-diameter commercial steel pipe. The pump has a mechanical efficiency of 80%. Neglecting minor losses, determine the power required (in kW).

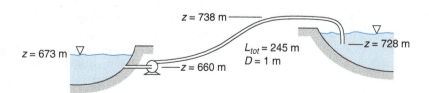

P9-44 Fresh air is distributed in a factory through a 250-ft-long rectangular galvanized duct, which is 36 in. by 6 in. For a flow rate of 5000 ft³/min of 60°F air at 14.2 lbf/in.², determine the fan pumping power required if the fan has a mechanical efficiency of 65% (in hp).

P9-45 Water at 20°C is to be siphoned from a large tank, as shown in the figure. The siphon is a 2.5-cm-diameter smooth tube and has a reentrant inlet.

a. Determine the volumetric flow rate if only the minor losses are taken into account (in m³/s).

b. Determine the volumetric flow rate if both the minor and line losses are taken into account (in m³/s).

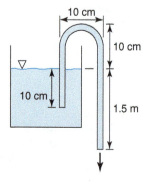

P9-46 A pump draws 40°F water from a lake through 20 ft of commercial steel pipe; the line has a reentrant inlet and a 90° regular flanged elbow. The pump elevation is 12 ft above the lake surface. For a design flow rate of 100 gal/min, the head at the suction side of the pump must not be less than −20 ft of water. Determine the minimum pipe diameter (in in.).

P9-47 Large farm implements and road construction equipment use hydraulically actuated cylinders to position scoops,

cutting blades, and other tools. High-pressure pumps are used to circulate the hydraulic fluid ($\rho = 880$ kg/m³ and $\mu = 0.033$ N·s/m²). Consider a hydraulic system that has a pump outlet pressure of 20 MPa and that requires a minimum pressure at the hydraulic cylinder of 18 MPa at a flow rate of 0.0005 m³/s. If the hydraulic fluid flows through 25 m of smooth, drawn steel tubing, determine the minimum tubing diameter required (in cm).

P9-48 Water at 70°F with a flow rate of 30 gal/min flows from a 1-in.-diameter tube into a 2-in.-diameter tube through a sudden expansion. Determine the pressure rise across the expansion (in lbf/in.²).

P9-49 A Class 100 clean room is to be supplied with 15 m³/min of air, which enters the duct (shown in the figure) at 100 kPa, 25°C. All entrances and exits are sharp edged.

a. Determine the pressure in the clean room (in kPa).

b. Determine the fan power required (in W).

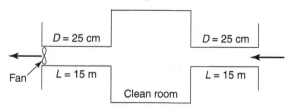

P9-50 In Problem P9-49, the sharp-edged entrances and exits are replaced with well-rounded entrances and exits. For the same fan power as in the original installation, determine the new volumetric flow rate (in m³/min).

P9-51 Frictional pressure loss in fluid flow is converted to unwanted thermal energy. Consider an 18-gal/min flow of 70°F water through a 1.25-in.-diameter smooth tube. The tube is sloped so that the pressure remains constant throughout the tube.

a. Determine the slope (in ft/100 ft).

b. Determine the heat transfer per 100 ft of tube if the temperature remains constant (Btu/hr).

c. Determine the temperature rise if the tube is perfectly insulated (in °F).

P9-52 A gas turbine power plant consists of a compressor, a combustor in which the fuel and air are mixed and combusted, and a turbine that drives an electrical generator. The air compression process takes from 40–80% of the turbine output power, leaving only 20–60% to drive the electric generator. Some gas turbine plants store compressed air in salt domes or caverns for use during times when additional electric power is needed; the compressed air to be supplied to the power plant is taken from the stored air instead of just using the air compressor. Consider the system shown in the figure. The air reservoir, which fills with 10°C water when the air has been used, is connected to the outside by a 30-cm cast-iron pipe. During charging of the reservoir with air, the air pressure, P, increases. Determine the gage pressure P required to produce a water flow rate of 0.15 m³/s (in kPa).

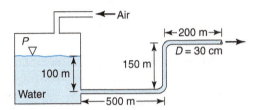

P9-53 Water at 20°C is pumped from a reservoir through a 20-cm commercial steel pipe for 5 km from the pump outlet to a reservoir whose surface is 150 m above the pump. The flow rate is 0.10 m³/s.

a. Determine the pressure at the pump outlet (in kPa).

b. Determine the pumping power required (in W).

P9-54 Two reservoirs are connected by three galvanized iron pipes in series. The first pipe is 600 m long, 20 cm in diameter; the second pipe is 800 m long, 30 cm in diameter; and the third pipe is 1200 m long, 40 cm in diameter. For a flow of 0.15 m³/s of water at 10°C, determine the elevation difference between the reservoirs (in m).

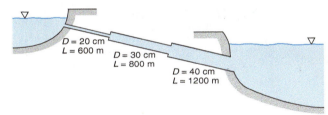

P9-55 In a water system, a reservoir is connected to a canal with a 8-in. cast-iron pipe. The system has three regular 90° threaded elbows and a half-closed gate valve, and the exit from the reservoir is sharp edged. With an elevation difference of 55 ft between the reservoir surface and the pipe outlet, 3 ft³/s of water at 50°F flows through the pipe. Determine the total length of straight pipe in the system (in ft).

P9-56 On your land high in the Rocky Mountains, you decide to produce your own electric power for your vacation home using a hydroturbine. The surface of the small lake from which you will get the 50°F water is 500 ft above where you will locate the turbine. You connect the lake and turbine with 1000 ft of 6-in.

cast-iron pipe. The turbine discharge is the same diameter as the inlet and is open to the atmosphere. Determine the maximum power that can be produced (in W).

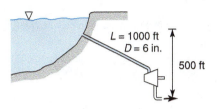

P9-57 Fire trucks have pumps to boost the pressure of the water supplied by a fire hydrant. Consider a fire truck that has a 250-ft-long, 2-in.-diameter smooth fire hose. Water must reach the nozzle at the hose exit at 100 lbf/in.² (gage). Water from the hydrant reaches the pump inlet at 60 lbf/in.² (gage). If the design pressure drop specification for the hose is 25 psi/100 ft of length, determine

a. the design flow rate (in gal/min).

b. the nozzle exit velocity (in ft/s).

c. the pump power required if the pump has a mechanical efficiency of 75% (in hp).

P9-58 Water is pumped from a lake to a pond that is 50 m above the lake. A suction pipe runs from the lake to a pump, and a connecting pipe runs from the pump to the pond. The suction pipe is constructed of 10-cm-diameter cast-iron pipe (assume no minor losses). The connecting pipe is also 10-cm-diameter cast iron and has five long-radius 90° threaded elbows. The pump can be located in one of three places: (1) *level* with the lake surface, and the suction pipe would be 6 m long and the connecting pipe would total 150 m long; (2) 10 m *below* the lake surface, and the suction pipe would be 11 m long and the connecting pipe would total 160 m long; or (3) 5 m *above* the lake surface, and the suction pipe would be 8 m long and the connecting pipe would total 145 m long. For a flow of 0.025 m³/s, determine which installation requires the smallest required pumping power (in W).

P9-59 Many universities have a central facility that produces chilled water for use in cooling all the buildings on campus. The water at 10°C is continuously circulated through a closed-flow loop and used as needed. Consider a system that consists of 5 km of 30-cm commercial steel pipe with a flow rate of 0.15 m³/s. The pump has a mechanical efficiency of 75% and is driven by a motor that has an efficiency of 92%.

a. Determine the pressure drop (in kPa).

b. Determine the pumping power required (in kW).

c. Determine the annual cost if electricity costs $0.10/kWh and the system runs 7,500 h/yr.

P9-60 The Alaskan oil pipeline is 48 in. in diameter, with a wall roughness of approximately 0.0005 ft. The design flow rate is 1.6×10^6 barrels per day (1 barrel = 42 gal). To limit the required pipe wall thickness, the maximum allowable oil pressure is 1200 psig. To keep dissolved gases in solution in the crude oil, the minimum oil pressure is 50 psig. The oil has $\rho = 58$ lbm/ft^3 and $\mu = 0.0113$ lbm/ft·s.

a. Determine the maximum spacing between pumping stations (in km).

b. Determine the pumping power at each station if the pump mechanical efficiency is 85% (in kW).

P9-61 (WEB) For air at 300 K and 1 atm, a fan performance curve can be approximated with $h = 70 - 3 \times 10^{-4} \dot{V}^2$, where h is the pressure rise across the fan in cm of water and $\dot{V}$ is the air flow rate in m^3/min. The fan discharges into a smooth rectangular duct 20 cm by 40 cm.

a. Determine the flow rate if the duct is 30 m long (in m^3/min).

b. Determine the flow rate if the duct is 75 m long (in m^3/min).

P9-62 A town water system is constructed to supply water at a flow rate of 0.04 m^3/s, as shown in the figure. Available cast-iron pipe is to be used, and the gate valve is fully open. The water is at 20°C. Determine the height to which the upper reservoir dam (reservoir surface elevation) must be built (in m).

P9-64 (WEB) In drier regions, large central pivot sprinkler systems are used to irrigate large areas. Consider the simplified schematic of a portion of such a sprinkler (shown in the figure). Water at 10°C is pumped through the spray arm, which is constructed of 2.5-cm-diameter galvanized iron. The flow area of each nozzle is 1.5 cm^2. The pressure at the first nozzle is 250 kPa (gage). Ignoring friction in each nozzle but not in the connecting lengths of pipe, determine the flow rate through the sprinkler (in m^3/s).

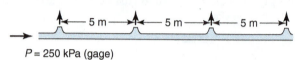

$P = 250$ kPa (gage)

P9-65 (WEB) Two pipes are connected in parallel. The first pipe is 2 cm in diameter and 100 m long with a friction factor of 0.012. The second pipe is 5 cm in diameter and 50 m long with a friction factor of 0.010. Determine the ratio of the flow rates in the two pipes.

P9-66 (WEB) For a storm sewer modification project, a 24-in. pipe and a 30-in. pipe both open at their ends to the atmosphere are to be joined using three existing (but underutilized pipes), as shown in the figure. All the pipes are concrete, and the friction factors are in the figure. The branches are horizontal. For a total flow rate of 20 ft^3/s of 60°F water, determine the flow rate

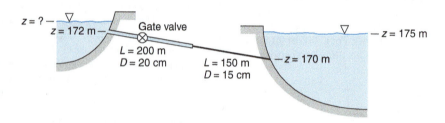

P9-63 (WEB) The pump in an existing water system (shown in the figure) fails and must be replaced. A duplicate is not available, so a manufacturer proposes a pump with a pump curve: $h_P = -4 \times 10^{-6} \dot{V}^2 + 0.0038\dot{V} + 86$, where $\dot{V}$ is in gal/min and h_p is in ft. The gate valve is fully open, and a friction factor of 0.018 can be used for both pipes. Determine the flow rate in the system (in gal/min).

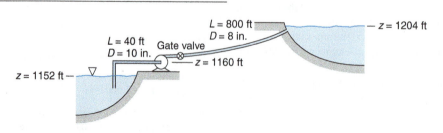

in each of the three connecting pipes (in ft³/s) and the elevation difference from the entrance to the exit.

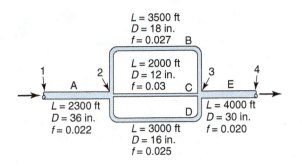

$L = 3500$ ft
$D = 18$ in.
$f = 0.027$ B

$L = 2000$ ft
$D = 12$ in.
$f = 0.03$ C

$L = 2300$ ft
$D = 36$ in.
$f = 0.022$

$L = 3000$ ft
$D = 16$ in.
$f = 0.025$

$L = 4000$ ft
$D = 30$ in.
$f = 0.020$

P9-67 (WEB) For the town water supply system described in Problem P9-62, adequate flow has been obtained initially, but with the town expanding in population, the city planners want to increase the flow to 0.08 m³/s. Two approaches have been suggested. One is to raise the water level of the upper reservoir by increasing the height of the dam. For the second approach, because the water department has 350 m of 15-cm cast-iron pipe and a gate valve available, the managers suggest running a second pipe parallel to the original 15-cm pipe.

a. Determine the upper reservoir elevation required if the first approach is used (in m).

b. Determine the flow rate if the second approach is used (in m³/s).

CHAPTER 10

EXTERNAL FLOWS

10.1 INTRODUCTION

When an object moves through a stationary fluid (e.g., an airplane, a golf ball, or a hailstone moving through the air or a submarine moving through the ocean) or a fluid flows past a stationary object (e.g., river water sweeping past the pylons of a bridge or wind blowing past a building, a communications tower, or a person), forces are generated because of the *relative motion* between the fluid and the object. The flow around an object may be *unbounded*, in which case all surfaces are far away from the object and have no effect on the flow around the object (such as when a plane flies through the air); or the flow may be *bounded on one side*, in which case the flow field extends a very long distance in one direction from the object and nothing in that direction affects the flow over the object (such as when air flows over the roof of a house). In either case we call these *external flows*. Our study of external flows focuses on the determination of the forces that result whenever there is relative motion between a fluid and a solid object.

A simple demonstration can illustrate the two forces we study in this chapter. Suppose you drive in a car with your arm held out the window and orient your hand parallel to the wind, as shown in Figure 10-1. If you did not resist the force acting on it, your hand would be pushed backward. There is no force to push the hand up or down. Now you tilt your hand upward. Because of the new orientation of your hand relative to the wind, a force pushes it upward; a force also continues to push it backward. If you tilt your hand downward, the air pushes it downward and backward. If you place your hand perpendicular to the airflow, the force pushing your hand backward is greater than with your hand parallel to the wind, but, again, there is no up or down force acting on the hand.

The force that acts *parallel* to the direction of the fluid flow is called a *drag force*. It is composed of the combined effects of viscous forces (shear stresses) and pressure forces. The force that acts *perpendicular* to the direction of the fluid flow is called a *lift force* and also results from viscous and pressure forces. The magnitude of these forces depends on several quantities. You experience one force if you insert your hand into air moving at 50 km/h. If you attempted to do the same thing in water that was flowing at 50 km/h, the force would be much greater; both the density and viscosity of the fluid affect the forces. Tilting the hand up or down demonstrates that the forces are affected by the relative orientation (*angle of attack*) of the object relative to the fluid velocity direction. If you perform the hand-out-the-window experiment at 50 km/h and 100 km/h, you would experience a much greater force at the higher velocity. Finally, we can compare the power required to drive a sleek sports car versus a minivan at 100 km/h; the effects of *streamlining* a body (e.g., compare an automobile from the 1950s to one built in 2004 or a military jet to a small private plane) show that both size and shape affect the forces. Likewise, we can change an object's surface geometry to influence the drag forces as well. (Why does a golf ball have dimples on its surface rather than being smooth?)

In all these examples, the drag and lift forces result from an interaction between the object and the fluid. Changes in the flow field around an object are what generate the forces. The flow field is divided into two parts. A thin layer called the *boundary layer* is next to

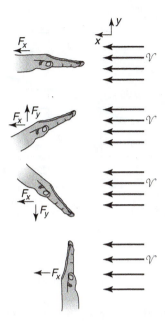

FIGURE 10-1 Forces acting parallel and perpendicular to velocity direction.

the body, and effects of viscosity are felt there. The "thinness" of the boundary layer is a relative quantity. For example, the motion of treetops illustrates the size of a boundary layer for wind blowing over the ground; for a 20-m-tall tree, the top may be strongly affected by the wind, but at the tree base, a person might feel only a light wind. For a plane flying at 300 km/h, the boundary-layer thickness on the plane's wing may be on the order of 30 mm. Outside the boundary layer, viscous effects are negligible, and the velocity (called the *free-stream velocity*) is the same as the fluid velocity upstream of the object; the upstream velocity is also called the *approach velocity*.

In this chapter, we discuss drag and lift forces and how the external flow field and boundary layer affect them. The effects of size, shape, orientation, fluid (density and viscosity), and velocity on drag and lift forces are quantified through the use of information obtained from experimental, theoretical, and numerical investigations.

10.2 BOUNDARY-LAYER CONCEPTS

In Chapter 9, the flow was either laminar or turbulent depending on the magnitude of the Reynolds number, which provides information about the fluid state. The Reynolds number plays an important role in external flows, too, but for a given flow, the state of the flow is also dependent on the location on the body. This concept can be illustrated well by considering a smoke plume rising into a quiescent volume, as shown in Figure 9-8. Initially, the plume is well organized and mixes little with the surrounding air. In this laminar flow region, diffusion governs the interaction between the smoke and clean air, and the process of mixing is very slow. Farther up the plume, some waviness starts and grows in magnitude. This instability marks the transition region between laminar and turbulent flows, but there still is some coherence in the plume. At some point in the flow, the instability grows too large and turbulence initiates. This is reflected by rapid large-scale bulk mixing of the smoke plume and the surrounding air, and all regular structure in the plume disappears. The smoke plume has approximately a constant rise velocity but, depending on how far

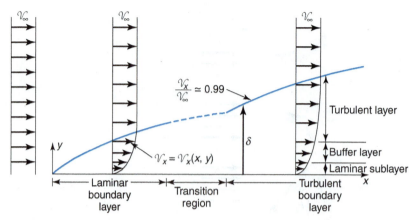

FIGURE 10-2 Steady, viscous flow over a flat plate.

from the smoke source, the character of the flow changes. This is also true for flow over a solid surface.

While the flow of a plume or jet into a quiescent volume or into a moving flow is one manifestation of an external flow, we are more concerned with how flows interact with solid surfaces. Consider the flow over the infinitesimally thin flat plate shown in Figure 10-2. The upstream uniform velocity is aligned with the plate. At the plate surface, the fluid has the same velocity as the plate. This so-called *no-slip condition* is present in all flows of real fluids for which the continuum assumption is valid (e.g., it is not valid on high altitude airplanes). Fluid viscosity provides internal friction that tends to retard the flow, and these viscous forces cause each layer of the fluid to exert a small force on the layer above it, reducing its velocity slightly. Because conservation of mass must be satisfied, when the velocity in the *x*-direction decreases, a velocity in the *y*-direction is produced that is much smaller than that in the *x*-direction. Near the start of the plate in the laminar flow region, the fluid layers remain well defined, and the track of a particle injected into the flow can be traced with ease. (The glacier shown in Figure 10-3 is a dramatic example of laminar flow. The dark lines in the glacier are medial moraines, which can be traced back to their origins.) Farther down the flat plate, an instability initiates the transition region, which lasts over an uncertain length. At the end of the transition region, turbulent flow continues until the end of the plate; in this region rapid and chaotic bulk mixing of the fluid occurs.

In the laminar region, viscous forces retard the velocity of each layer slightly. However, viscous effects eventually die out some distance perpendicular from the plate, and at all distances past this location the velocity is again uniform, as it was upstream of the plate; in this region, there are no velocity gradients ($d\mathcal{V}_x/dy = 0$, where $\mathcal{V}_x$ is the local velocity). From the definition of local shear stress for a Newtonian fluid, $\tau = \mu \left(d\mathcal{V}_x/dy \right)$, we can see that outside of the viscous-influenced region, there are no viscous stresses. We call this *inviscid flow*. The location at which we separate the viscous and inviscid regions is arbitrarily defined at that distance—the *boundary-layer thickness*, δ—from the plate to where the boundary layer velocity, $\mathcal{V}_x$, reaches 99% of the free-stream velocity, $\mathcal{V}_\infty$. This thickness increases the farther the flow is from the leading edge of the plate.

In the turbulent region, bulk mixing of the flow is more efficient than viscous forces in smoothing out velocity gradients, and the boundary-layer thickness grows at a faster rate than in the laminar region. Unlike laminar flow, we can identify three distinct flow behaviors in the turbulent boundary layer. Near the wall, where fluid velocities are low, viscous forces dominate, the flow is laminar, and the region is called the *laminar sublayer*. Far from the

FIGURE 10-3 Glacier with medial moraines tracing the flow. (Copyright by Marli Miller. Used with permission.)

wall, turbulence dominates, and the region is called the *turbulent layer*. Between these two layers, the flow has characteristics of both the laminar sublayer and the turbulent layer and is called the *buffer layer*. We still define the boundary-layer thickness as that location where the boundary-layer velocity reaches $0.99\mathcal{V}_\infty$.

Figure 10-4 shows nondimensional laminar and turbulent velocity profiles. The laminar profile is smooth and uniformly changing. The turbulent one has a higher gradient at the wall and changes shape abruptly near the wall. These two profiles reflect the different mechanisms—viscous effects versus turbulent mixing—that govern the flow in the laminar and turbulent regions.

Transition begins at a distance from the leading edge of a body and depends on many variables, including flow velocity; type of fluid; surface roughness; free-stream turbulence level; whether the surface is heated, cooled, or adiabatic; surface shape (e.g.,

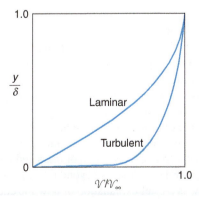

FIGURE 10-4 Nondimensional velocity profiles in laminar and turbulent flows.
(*Source*: Adapted from R. W. Fox and A. T. McDonald, *Introduction to Fluid Mechanics*, 3rd ed., Wiley, New York, 1985. Used with permission.)

flat plate, cylinder, airfoil, etc.); orientation of the surface to the free stream; and so on. In addition, the boundary layer does not become fully turbulent instantaneously at some location but transitions over some finite length of the body. The transition location is best characterized by the *length Reynolds number, Re = $\rho \mathcal{V}_\infty L / \mu$*, where L is the distance from the leading edge of the body. *Transition* or *critical Reynolds numbers*—those length Reynolds numbers at which transition occurs—for different surfaces/bodies have been established for "typical" or "average" conditions. For flow over a flat plate, the value typically used is $Re_{crit} = 500,000$. Nontypical conditions can be created so that the transition Reynolds number is smaller or larger than this value. For example, a small-diameter "trip wire" can be placed near the leading edge of a plate to induce the transition to turbulence at a lower Reynolds number, or a plate of finite thickness can be used that can also induce early transition.

10.3 DRAG ON A FLAT PLATE

Experimental investigations are used extensively to produce drag and lift data for use in a variety of applications. However, we can also solve the governing mass, momentum, and energy differential equations for flow over an object to obtain drag and lift information. While analytic solutions are not possible, for a few simple external flows we can solve the equations through a simple numerical analysis, and for more complex flows and bodies, we resort to using large computational fluid dynamics (CFD) codes to develop data on the pressure and shear distributions over the complete surface of the body. From those data, we can calculate the drag and lift forces.

Drag force on a body is the sum of the pressure and shear forces acting *parallel* to the flow velocity. The magnitude of a pressure force (relative to the free-stream pressure) can be calculated as *PA*, where *P* is the pressure and *A* is the area perpendicular to the pressure. A shear force can be calculated using τA, where τ is the shear stress and *A* is an area parallel to the surface along which the shear force acts. To illustrate the effects of pressure and shear forces, let us consider the evaluation of the drag force for two special situations involving a flat plate.

Consider the simplest body shape, an infinitesimally thin flat plate aligned with a flow past it. Because the plate is infinitesimally thin (area perpendicular to flow director is zero, $A = 0$) and parallel to the flow, no pressure force ($F = PA$) can exist in the *x*-direction. However, because of the no-slip condition, the flow past the flat plate will exert a force, and that force is due solely to shearing stresses (viscous effects) caused by the relative motion of the fluid past the plate. Thus, in this configuration, the drag force (force parallel to the flow velocity) is caused only by shear stress. Before we consider the second special situation with a flat plate, we focus on the development of quantitative information for drag force on a flat plate aligned with the flow.

In Chapter 9 we developed a differential equation that described the flow in a circular tube. We solved that equation for laminar flow and obtained the velocity profile. Once the velocity profile was available, we developed an expression for the friction factor. For flow over a flat plate, if we can obtain information about the velocity profile, then we can derive expressions for the drag. Therefore, we want to solve the governing equations for the velocity field on one side of this flat plate. Rather than developing a different equation for each specific situation, we would like to use equations that are general; that is, we want differential equations that describe conservation of mass and momentum for any situation. The development of the general equations is beyond the scope of this text but can be found in many references. Below is a brief explanation of how the equations are obtained.

In the development in Chapter 9 of the equation governing laminar pipe flow, we defined a control volume that encompassed a cross-section of the tube, and we examined the forces (pressure, shear, and gravity) acting on the boundaries. For the general situation, consider the boundary layer shown in Figure 10-2. We define a differential control volume (Figure 10-5) in the boundary layer and evaluate the control volume with conservation of mass and momentum. As with any application of the conservation laws, we examine what happens at and across the boundaries. Because of the velocity profile, the velocities, and, hence, the mass flows $(\dot{m}_x, \dot{m}_y)$ across each of the sides of the control volume are slightly different. Likewise, pressure forces (PA) and shear forces (τA) acting on each boundary are different; gravity force $(g\rho V)$ is evaluated assuming density is uniform across the volume. Once we have assessed each term in the mass and momentum equations, we take the limit as the control volume is shrunk to zero size. The general equations then are applied to specific applications. For flow over a flat plate, we assume steady, two-dimensional laminar flow of a Newtonian fluid with constant properties and no gravitational effects; we impose several other simplifying assumptions.

The resulting conservation of mass and momentum equations for a two-dimensional flow in rectangular coordinates are

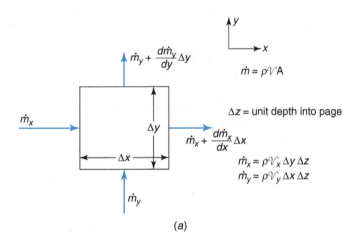

(a)

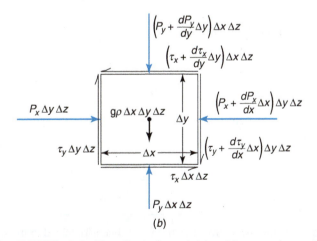

(b)

FIGURE 10-5 Mass flows through (a) and forces acting on (b) differential control volumes.

Conservation of mass

$$\frac{\partial \mathcal{V}_x}{\partial x} + \frac{\partial \mathcal{V}_y}{\partial y} = 0 \tag{10-1}$$

Conservation of momentum in the x-direction

$$\mathcal{V}_x \frac{\partial \mathcal{V}_x}{\partial x} + \mathcal{V}_y \frac{\partial \mathcal{V}_x}{\partial y} = g_x - \frac{1}{\rho}\frac{\partial P}{\partial x} + \frac{\mu}{\rho}\left(\frac{\partial^2 \mathcal{V}_x}{\partial x^2} + \frac{\partial^2 \mathcal{V}_x}{\partial y^2}\right) \tag{10-2}$$

Conservation of momentum in y-direction

$$\mathcal{V}_x \frac{\partial \mathcal{V}_y}{\partial x} + \mathcal{V}_y \frac{\partial \mathcal{V}_y}{\partial y} = g_y - \frac{1}{\rho}\frac{\partial P}{\partial y} + \frac{\mu}{\rho}\left(\frac{\partial^2 \mathcal{V}_y}{\partial x^2} + \frac{\partial^2 \mathcal{V}_y}{\partial y^2}\right) \tag{10-3}$$

These equations are called the *Navier-Stokes equations* in honor of the individuals who first developed them. The coordinate along the plate is the x-axis, and $x = 0$ is at the leading edge of the plate; the y-axis is perpendicular to the plate. The boundary conditions for this flow are: at $y = 0, \mathcal{V}_x = 0, \mathcal{V}_y = 0$; at $y \to \infty, \mathcal{V}_x = \mathcal{V}_\infty$ (the free-stream velocity), $\mathcal{V}_y = 0$.

Blasius, who obtained the laminar velocity profile shown on Figure 10-4, was the first person to solve this system of equations. The laminar velocity profile changes smoothly from the wall to the free-stream velocity. The velocity profile in Figure 10-4 has been nondimensionalized so that it applies to any position along the plate (in the laminar region). From the velocity profile, the boundary-layer thickness, δ, can be determined as a function of the length Reynolds number:

$$\frac{\delta}{x} = \frac{5}{\sqrt{\rho \mathcal{V}_\infty x/\mu}} = \frac{5}{\sqrt{Re_x}} \qquad \text{laminar flow} \tag{10-4}$$

At the distance, δ, from the wall, the velocity is 99% of the free-stream value. This definition is somewhat arbitrary and could have been based on 99.9% or 97% of the free-stream velocity. Conventional practice is to use 99%. The Reynolds number falls out naturally when the governing equations are nondimensionalized. Note that as we move along the plate away from the leading edge, the boundary-layer thickness, δ, grows thicker as $x^{1/2}$.

The laminar shear stress is evaluated from the velocity profile:

$$\tau_w = \mu \left.\frac{\partial \mathcal{V}_x}{\partial y}\right|_{y=0} = 0.332 \mathcal{V}_\infty^{3/2}\sqrt{\rho\mu/x}$$

To create a nondimensional shear stress, we recast this expression in terms of the *local skin friction coefficient*:

$$C_{f,x} = \frac{\tau_w}{\frac{1}{2}\rho \mathcal{V}_\infty^2} = \frac{0.664}{\sqrt{Re_x}} \tag{10-5}$$

In Eq. 10-5, the shear stress has been nondimensionalized in the same way as the pressure drop was nondimensionalized in internal flow to create the friction factor. The skin friction coefficient decreases as $x^{-1/2}$ as we move away from the leading edge. We also need an average drag coefficient over a plate of length L. For a flat plate the drag force is

$$F_D = \int_A \tau_w \, dA = \overline{\tau}_w A = \overline{\tau}_w WL$$

where W is the plate width and $\overline{\tau}_w$ is the average wall shear stress.

We define the average drag coefficient as

$$C_D = \frac{F_D}{\frac{1}{2}\rho V_\infty^2 A} = \frac{1}{A}\int_A C_{f,x}\, dA \tag{10-6}$$

Substituting Eq. 10-5 into this expression and integrating gives

$$C_D = \frac{1}{WL}\int_0^L \frac{0.664}{\sqrt{\rho V_\infty x/\mu}} W\, dx = \frac{1.328}{\sqrt{Re_L}} \qquad \text{for } Re_L < 10^5 \tag{10-7}$$

where the Reynolds number is defined in terms of the total length of the plate, L. The area A is the planform area of the plate, $A = WL$. Note that these equations are for only *one* side of a flat plate.

A similar solution is not possible for turbulent flow over a flat plate. Instead, we rely on experimental data. Shown in Figure 10-4 is the nondimensionalized turbulent velocity profile. It changes much more rapidly near the wall than does the laminar velocity profile and is blunter over a large region as well. This change in the velocity profile affects the drag coefficient. Thus, for turbulent flow that begins at the leading edge of the plate, best fits with experimental data are

$$\frac{\delta}{x} = \frac{0.37}{Re_x^{1/5}} \qquad \text{turbulent flow} \tag{10-8}$$

$$\tau_w = 0.0225\rho V_\infty^2 \left(\frac{\mu}{\rho V_\infty \delta}\right)^{1/4} \qquad \text{turbulent flow}$$

$$C_D = \frac{0.074}{Re_L^{1/5}} \qquad \text{for } 10^5 < Re_L < 10^7 \tag{10-9}$$

$$C_D = \frac{0.455}{(\log Re_L)^{2.58}} \qquad \text{for } 10^7 < Re_L < 10^9 \tag{10-10}$$

Note that from Eq. 10-8 the turbulent boundary-layer thickness grows as $x^{4/5}$ as compared to $x^{1/2}$ for laminar flow.

Turbulent flow most often does not begin at the leading edge of a flat plate. We can induce turbulent flow by placing a disturbance, such as a "trip wire," at the leading edge, but in the usual case laminar flow at the leading edge is followed by a transition region of indefinite length until fully turbulent flow is reached. For those situations in which a plate is long enough to have both laminar and turbulent regions (such as shown on Figure 10-2), we can combine Eq. 10-6 and Eq. 10-9 to obtain a correlation for the drag coefficient for a plate that has combined laminar and turbulent flow. Ignoring the transition region and assuming that turbulent flow commences immediately upon attaining the critical (or transition) Reynolds number, the composite drag coefficient can be determined from

$$C_D = \frac{1}{L} \left(\int_0^{x_{crit}} C_{D,lam}\, dx + \int_{x_{crit}}^{L} C_{D,turb}\, dx \right) \tag{10-11}$$

Carrying out the integration and assuming a critical Reynolds number of 500,000, we obtain

$$C_D = \frac{0.074}{Re_L^{1/5}} - \frac{1740}{Re_L} \tag{10-12}$$

When the Reynolds number is 10^6, the laminar contribution to the total drag coefficient is 27%; at $Re_L = 10^7$, the contribution decreases to 5.6%. As the Reynolds number becomes larger, the laminar contribution quickly becomes negligible and can be ignored.

The flow described above was for a situation (flow parallel to an infinitesimally thin plate) where there was no pressure drag. Shearing forces caused all the drag force. Now consider the other extreme configuration: an infinitesimally thin plate in which the flat plate is set perpendicular to the flow field (Figure 10-6). Again, we want the forces acting parallel to the flow direction. Because the plate is infinitesimally thin ($A = 0$), no shear force, τA, can exist. However, we know that the flow does exert a force on the plate. (Consider the hand oriented perpendicular to the flow in Figure 10-1.) This force is due solely to a pressure distribution imbalance around the plate. The pressure on the upstream face of the plate is greater than that on the downstream face. We cannot solve the governing equations analytically for this situation. Instead, the drag coefficient is obtained experimentally, and the variation in C_D with Reynolds number is shown on Figure 10-7a. Notice that the drag coefficient has little dependence on the Reynolds number. This is typical of very blunt bodies (such as flat plates perpendicular to the flow). However, above a Reynolds number of about 1000, the shape of the blunt body will affect the drag coefficient (Figure 10-7b).

In neither of the above configurations is there an imbalance in the forces in the direction perpendicular to the flow field. Because of this, no lift force is generated.

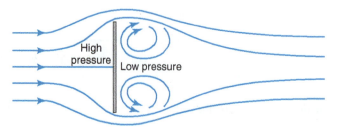

FIGURE 10-6 Flow perpendicular to thin flat plate.

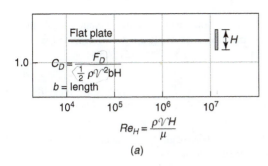

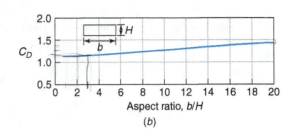

FIGURE 10-7 Drag coefficient for flow perpendicular to flat plate.
(*Source*: Adapted from Young, D. F., B. R. Munson, and T. H. Okiishi, *A Brief Introduction to Fluid Mechanics*, Wiley, 1997, and Fox, R. W., and A. T. McDonald, *Introduction to Fluid Mechanics*, 3rd ed., Wiley, 1985. Used with permission.)

EXAMPLE 10-1   Drag on a flat plate

An advertising banner for an ice cream shop is towed behind a small plane flying low over an ocean beach on a hot summer day. The plane speed is 90 km/h. The banner is $H = 1$ m tall and $L = 20$ m long. The air temperature is 32°C.

a) Determine the power (in kW and hp) required to tow the banner assuming the banner acts as a flat plate.

b) Determine the power (in kW and hp) if experiments have shown that a banner drag coefficient based on the area HL can be approximated with $C_D = 0.05L/H$.

c) Explain why there is a difference between these two results.

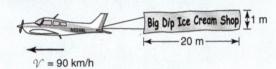

Approach:

Power is calculated by multiplying force times the velocity. We know the plane's velocity. The force is determined from a drag calculation $F_D = C_D \rho \mathcal{V}_\infty^2 A/2$. Thus, we need to determine the drag coefficient.

Assumptions:

Solution:

a) Power is $\dot{W} = F_D \mathcal{V}_\infty$, where $\mathcal{V}_\infty = 90$ km/h = 25 m/s. Force is calculated from

$$F_D = C_D \left(\frac{1}{2} \rho \mathcal{V}_\infty^2 A \right)$$

The drag coefficient must be evaluated for the total length of the banner, so we need to calculate the length Reynolds number, $Re_L = \rho \mathcal{V}_\infty L/\mu$. We assume the atmospheric pressure is 101 kPa

A1. Air is at 1 atm pressure.

[A1]. By interpolation in Table A-7 at 32°C, $\mu = 1.875 \times 10^{-5}$ N·s/m² and $\rho = 1.161$ kg/m³. Therefore,

$$Re_L = \frac{\rho \mathcal{V}_\infty L}{\mu} = \frac{\left(1.161 \text{ kg/m}^3 \right) (25 \text{ m/s}) (20 \text{ m})}{\left(1.875 \times 10^{-5} \text{ N·s/m}^2 \right) \left(1 \text{ kg·m/1 N·s}^2 \right)} = 3.096 \times 10^7$$

This Reynolds number indicates that turbulent flow covers most of the banner. While this is a large Reynolds number and the laminar contribution is probably small, we assume the flow behaves similar to that over a thin flat plate and that Eq. 10-12 applies [A2] [A3]:

A2. Laminar flow begins from the leading edge.
A3. The transition Reynolds number is 500,000.

$$C_D = \frac{0.074}{Re_L^{1/5}} - \frac{1740}{Re_L} = \frac{0.074}{(3.096 \times 10^7)^{1/5}} - \frac{1740}{3.096 \times 10^7} = 0.00235 + 0.000056 = 0.00241$$

The laminar contribution is small and could have been neglected. Remember that this drag coefficient is for *one* side of the banner, and we must take into account *both* sides. The drag force is

$$F_D = 2C_D \left(\frac{1}{2}\rho V_\infty^2 A \right)$$

$$= 2\,(0.00241)\left(\frac{1}{2}\right)\left(1.161\,\frac{kg}{m^3}\right)\left(25\,\frac{m}{s}\right)^2 (1\,m)\,(20\,m)\left(\frac{1\,N \cdot s^2}{1\,kg \cdot m}\right) = 35.0\,N$$

The power required to drag the banner is

$$\dot{W} = F_D V_\infty = (35.0\,N)\left(25\,\frac{m}{s}\right)\left(\frac{1\,J}{1\,N \cdot m}\right)\left(\frac{1\,W}{1\,J/s}\right) = 875\,W = 1.17\,hp$$

b) We now calculate the power using the expression for the experimentally determined drag coefficient, which is the total for both sides of the banner:

$$C_D = 0.05L/H = (0.05)(20\,m)/(1\,m) = 1.00$$

$$F_D = 2C_D \left(\frac{1}{2}\rho V_\infty^2 A \right)$$

$$= 2(1)\left(\frac{1}{2}\right)\left(1.161\,\frac{kg}{m^3}\right)\left(25\,\frac{m}{s}\right)^2 (1\,m)(20\,m)\left(\frac{1\,N \cdot s^2}{1\,kg \cdot m}\right) = 14,500\,N$$

$$\dot{W} = F_D V_\infty = (14,500\,N)\left(25\,\frac{m}{s}\right)\left(\frac{1\,J}{1\,N \cdot m}\right)\left(\frac{1\,W}{1\,J/s}\right) = 362,800\,W = 362.8\,kW = 487\,hp$$

Comments:

c) The large difference between the power assuming the banner acts as a flat plate and that using the empirical drag coefficient is due to the inappropriateness of assuming that the banner acts as a rigid flat plate. In reality, the banner flutters and waves, causing it to have a thickness in the direction of flow. This thickness (which varies with time) causes significant pressure drag, unlike the flat plate, on which only viscous drag exists.

EXAMPLE 10-2   Drag on plate perpendicular to flow

Your start-up company has little money to buy a chemical mixer it needs, so you decide to design and build a rotary mixer, as shown on the schematic. You need a motor to drive the mixer. The mixer paddles are $L_2 = 6$ in. square and rotate at 60 rpm. As a first approximation, you ignore drag from the connecting ($L_1 = 18$ in.) and drive rods. The fluid has a density $\rho = 79.3$ lbm/ft^3 and a viscosity $\mu = 2.56$ lbm/ft · s. Determine the required motor power (in hp).

why can't you use figure or py.to... Fasting chart or by c Re > 1000? before do b/c Re > 1000? confused when you use this chart

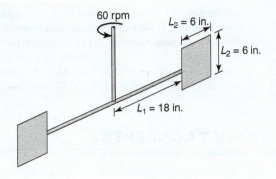

Approach:

For a rotating device, power is torque times rotational speed. Torque is calculated as force times the radius of rotation. The force is the drag force exerted on the two paddles as they rotate in the fluid. For the drag force, we need to estimate the drag coefficient for this situation.

Assumptions:

A1. Ignore drag on the connecting rods.

A2. Ignore velocity variation across the paddle.

A3. The fluid is stationary.

A4. Ignore interactions between the paddles and the side of the tank.

Solution:

We begin with the expression for power to drive a rotating device, $\dot{W} = \Im\omega$, where $\Im$ is the torque and ω is the rotational speed.

   Ignoring drag forces on the connecting rods [A1], torque is determined from $\Im = 2RF_D$, where the factor 2 is needed because of the two paddles. We assume the force acts through the centerline of the paddles, so the radius is $R = L_1 + L_2/2$.

   Rearranging Eq. 10-6, We see that the drag force is found from $F_D = C_D(1/2\rho\mathcal{V}^2A)$, where $A = L_2 L_2$ is the cross-sectional area of a paddle. The velocity varies from the inner to the outer edge of the paddle. We assume that using the velocity at the centerline of the paddle will account for the variation [A2]. In addition, we assume the fluid is stationary [A3]. Thus, $\mathcal{V} = R\omega$.

   Because the flow is perpendicular to the paddles and for very blunt bodies the drag coefficient is nearly independent of the Reynolds number, we assume the drag coefficient from Figure 10-7b is reasonable for this situation [A4]. Thus, from the figure at an aspect ratio of 1, $C_D \approx 1.1$. Therefore,

$$R = L_1 + L_2/2 = 18\,\text{in.} + 6\,\text{in.}/2 = 21\,\text{in.} = 1.75\,\text{ft}$$

$$\mathcal{V} = R\omega = (1.75\,\text{ft})(60\,\text{rev}/1\,\text{min})\,(2\pi\,\text{rad}/1\,\text{rev})\,(1\,\text{min}/60\,\text{s}) = 11\,\text{ft/s}$$

$$F_D = C_D\left(\frac{1}{2}\rho\mathcal{V}^2 A\right)$$

$$= 1.1\left(\frac{1}{2}\right)\left(79.3\,\frac{\text{lbm}}{\text{ft}^3}\right)\left(11\,\frac{\text{ft}}{\text{s}}\right)^2\left(\frac{6}{12}\,\text{ft}\right)\left(\frac{6}{12}\,\text{ft}\right)\left(\frac{1\,\text{lbf}\cdot\text{s}^2}{32.17\,\text{ft}\cdot\text{lbm}}\right) = 41.0\,\text{lbf}$$

$$\Im = 2RF_D = 2(1.75\,\text{ft})(41\,\text{lbf}) = 143.5\,\text{ft}\cdot\text{lbf}$$

$$\dot{W} = \Im\omega$$

$$= (143.5\,\text{ft}\cdot\text{lbf})\,(60\,\text{rev}/1\,\text{min})\,(2\pi\,\text{rad}/1\,\text{rev})(1\,\text{min}/60\text{s})\left(\frac{1\,\text{hp}}{550\,\text{ft}\cdot\text{lbf/s}}\right) = 1.64\,\text{hp}$$

Comments:

We made several assumptions that can have an effect on the calculated power. The first assumption was that drag forces on the connecting and drive rods were negligible compared to the drag force on the two paddles. This is probably a reasonable assumption because the rods have small diameters and lengths. The third assumption—the fluid is stationary—is weaker. The calculated numbers give a good estimate of the required torque and power at start-up (when the fluid velocity is zero). Once the mixer has been operating, the fluid in the mixer will have a rotational speed less than the paddle speed. Likewise, interactions between the outer edge of the paddles and the tank containing the liquid also will affect the drag and, hence, the required motor power. Nevertheless, we can get an *estimate* of the required power from this procedure.

10.4 DRAG AND LIFT CONCEPTS

Most bodies are blunt; that is, they are not infinitesimally thin but have finite dimensions in the direction transverse to the flow and shapes dictated by the functions of the objects. As a consequence, their drag and lift characteristics are set by some combination of pressure and shear forces. *Drag force* on a body is the sum of the *pressure* and *shear* forces acting *parallel* to the flow velocity. *Lift force* on a body is the net pressure and shear forces acting *perpendicular* to the flow velocity. Consider the forces acting on the infinitely long (two-dimensional) cylinder in crossflow shown on Figure 10-8. Pressure (relative to the free-stream pressure) is shown as an inward-pointing normal force when it is positive; the length of the arrow is proportional to the magnitude of the local pressure force, $P\,dA$. Local shear forces, $\tau\,dA$, are shown as arrows parallel to the surface of the cylinder. The *streamlines* indicate the paths of particles and show the flow field around the cylinder. The flow is *symmetric* around the horizontal centerline of the cylinder. Upstream of the cylinder the streamlines are evenly spaced, and between any two streamlines the flow rate is constant.

Fluid that strikes the cylinder exactly at its centerline is stopped completely; that is, the velocity is directly perpendicular to the surface at that location and looses all forward motion. This point is called the *forward stagnation point*, and the flow divides here, with half flowing over and half flowing under the cylinder. As the flow progresses around the cylinder, fluid away from the surface is not pushed out of the way completely, and the streamlines near the cylinder have a decreased spacing, which indicates a decreased flow area. From the Bernoulli equation, decreased flow area means increased velocity, which causes a decrease in the pressure. On the back half of the cylinder, the flow decelerates and the pressure rises but does not return to its original value. (Details of the processes on the back side of the cylinder are discussed below.)

The magnitude and direction of the pressure (normal) and viscous (shear) forces acting on a differential area of the cylinder are shown on Figure 10-8. To determine the

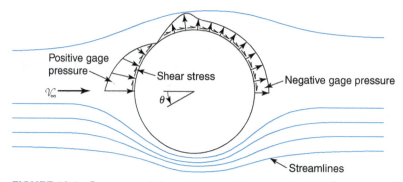

FIGURE 10-8 Pressure and shear stresses acting on cylinder ($10^3 < Re_D < 10^5$).

drag force, F_D, and the lift force, F_L, we need to separate pressure and viscous forces into components that are parallel and perpendicular to the free-stream velocity and then integrate over the complete surface area:

$$F_D = -\int_A P \cos \theta \, dA + \int_A \tau \sin \theta \, dA \qquad (10\text{-}13)$$

$$F_L = -\int_A P \sin \theta \, dA + \int_A \tau \cos \theta \, dA \qquad (10\text{-}14)$$

where θ is the angular position on the body. The first term on the right-hand side of Eq. 10-13 represents the pressure or *form drag* because the shape or form of the object has such a strong effect on the pressure distribution; the second term represents the friction drag. Note that because of the front-to-back asymmetrical pressure and shear distributions on the cylinder, both terms in Eq. 10-13 are nonzero, resulting in a drag force. Because of symmetry around the horizontal centerline of the cylinder, both terms in Eq. 10-14 are identically zero, resulting in zero lift force. For *blunt bodies* (bodies with finite transverse dimensions), friction drag does not contribute much to the total drag at higher Reynolds numbers. For *thin bodies* pressure drag can be much smaller than friction drag.

Now consider an *asymmetrical* body, such as the infinitely long (two-dimensional) airfoil shown on Figure 10-9. The angle between the airfoil profile and the velocity vector is called the *angle of attack*, α. Because of the shape of the surface and its orientation to the airflow, air accelerates more as it flows over the top of the airfoil than it does as it goes under the airfoil. From the Bernoulli equation, increased velocity implies decreased pressure, and, therefore, pressure is lower on the top surface than on the bottom surface. Application of Eq. 10-14 would show a nonzero lift force; the drag force also would be nonzero. (This is discussed in more detail in Section 10.6.)

For any body shape, the actual drag and lift forces can be determined from these two equations if the body shape is known, along with the pressure and shear stress distributions. However, except for a few special situations, pressure and shear stress distributions typically are not available, so we resort to the use of overall drag and lift force measurements. To make these measurements most useful, we must generalize the data so that they can be used for applications different from the experiments used to generate the data. We do this through the use of similitude and dimensionless quantities.

Let us consider a simple experiment designed to measure the drag force on a sphere. We have already established that the drag (or lift) force is dependent on the velocity, density,

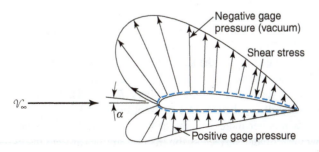

FIGURE 10-9 Pressure and shear acting on an airfoil. (*Source*: J. A. Robertson and C. T. Crowe, *Engineering Fluid Mechanics*, 6th ed., Wiley, New York, 1997. Used with permission.)

viscosity, and size of a body: $F_D = f(\mathcal{V}, \rho, \mu, D)$. We want to cover a wide range of conditions, so we choose 10 velocities $\mathcal{V}$, 10 fluid densities ρ, 10 viscosities μ, and 10 diameters D (which is representative of the size of the body). To cover all combinations of these variables, we would need to run 10^4 tests and would need 100 figures on which to plot all the data. A tremendous amount of time would be required to obtain the data, and these figures would be cumbersome and difficult to use.

Through the use of dimensional analysis and similitude concepts, we can develop nondimensional parameters that describe the drag force behavior much more effectively than if we used the individual variables. The drag force is nondimensionalized with the dynamic pressure $\rho \mathcal{V}^2/2$ and an area A. (Recall that the dynamic pressure was used to nondimensionalize pressure drop in internal flow calculations as well.) The Reynolds number, as has been used previously, describes the fluid state. Thus

$$\frac{F_D}{\frac{1}{2}\rho \mathcal{V}^2 A} = func\left(\frac{\rho \mathcal{V} L_{char}}{\mu}\right) \tag{10-15}$$

The quantity on the left-hand side of Eq. 10-15 is the drag coefficient, C_D:

$$C_D = \frac{F_D}{\frac{1}{2}\rho \mathcal{V}^2 A} \tag{10-16}$$

which was introduced earlier in Eq. 10-6, and the quantity on the right-hand side of Eq. 10-15 is the Reynolds number, $Re = \rho \mathcal{V} L_{char}/\mu$, so that $C_D = func(Re)$. The functional relationship is determined experimentally or by solution of the equations governing the flow around the body (if possible). A length characteristic of the body, L_{char}, is used for the nondimensionalization in the Reynolds number. Likewise, the *lift coefficient*, C_L, is defined as

$$C_L = \frac{F_L}{\frac{1}{2}\rho \mathcal{V}^2 A} \tag{10-17}$$

Depending on the body and application, the area to use in these equations is different. For most bodies except flat plates and airfoils (from which lift is desired), the area to use in Eq. 10-15 and Eq. 10-17 is the *projected frontal area* of the body, A, which is the area an observer would see if the object were viewed head-on (i.e., parallel to the direction of flow). However, for flat plates and airfoils, the *projected planform area* is used; this is the maximum area an observer of the body would see if the body were viewed from above it (i.e., perpendicular to the direction of flow).

Shown in Figure 10-10 are drag coefficients for an infinitely long smooth cylinder in a flow perpendicular to the axis of the cylinder and for a smooth sphere. The general trends in the curves are the same, although the cylinder's drag coefficient is greater than that of the sphere for most of the Reynolds number range. Consider the flow around the cylinder. At very low Reynolds numbers ($Re < \sim 1$), the flow around the cylinder is symmetric and follows the cylinder surface; Figure 10-11a illustrates the flow pattern. As the Reynolds number increases, the character of the flow changes. Instead of continuing to follow the cylinder surface as the flow progresses around the cylinder, the boundary layer

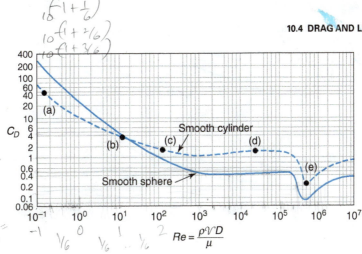

FIGURE 10-10 Drag coefficients for a smooth cylinder and sphere.
(*Source*: Adapted from D. F. Young, B. R. Munson, and T. H. Okiishi, *A Brief Introduction to Fluid Mechanics*, Wiley, New York, 1997. Used with permission.)

separates from the surface, and two symmetric stable vortices form behind the cylinder in the *separated flow region* (Figure 10-11b). (Sometimes behind the corner of a building, vortices similar to these are created by the passing wind and entrained dust or papers show the wind pattern.) With a further increase in Reynolds number ($\sim 90 < Re < \sim 1000$), vortices begin to shed off the back of the cylinder, alternating sides at a regular frequency that is dependent on the Reynolds number (Figure 10-11c). This is called an oscillating von Karman vortex street wake; the frequency of shedding is well known and predictable. (A von Karman vortex street can form behind an object as small as a thin telephone wire or as large as an island, as shown on Figure 10-12.) Each vortex that is shed causes a small transverse force to be imparted to the cylinder; this can result in *flow-induced vibration* of the solid. The failure of the Tacoma Narrows Bridge, as shown in classic movies of the event, has been attributed to vortex shedding; "singing" from wires is also credited to vortex shedding. An increase in fluid velocity increases the vortex frequency, the wake becomes more turbulent, and the drag coefficient attains almost a constant value (Figure 10-11d). At a Reynolds number of around 200,000, a sudden drop in the drag coefficient occurs that is caused by a narrowing of the turbulent wake on the backside of the cylinder (Figure 10-11e).

At very low Reynolds numbers over a sphere, no flow separation occurs. Stokes obtained an analytic solution of the governing momentum equation for this special situation:

$$C_D = \frac{24}{Re} \qquad Re \leq 1 \tag{10-18}$$

Care must be taken when using *Stokes' law*, because there are significant deviations between it and data for $Re > 1$.

The changing characteristics of the drag coefficient can be explained by examining the forces acting on the flow as it passes by the cylinder. First, though, consider the flow through the converging–diverging channel shown in Figure 10-13. This flow has many similarities to flow around a cylinder or sphere, but flow behavior is easier to visualize in a channel. Using the Bernoulli equation, we can show how the pressure varies with length. All along and near the solid wall, a shear force acts against the flow. In the converging portion of the channel, the pressure decreases; this is favorable to the flow because pressure forces

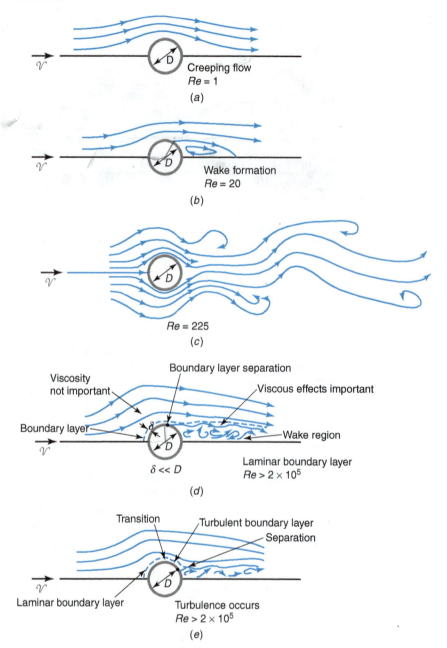

FIGURE 10-11   Flow patterns around a cylinder in crossflow at different Reynolds numbers.

act in the same direction as the flow. In the diverging portion of the channel, the pressure increases, and this is unfavorable to the flow (*an adverse pressure gradient*). As the flow moves from the minimum flow area of the channel into the diffusing section, the pressure increases, so a fluid particle experiences a net pressure force in the direction opposite to that of the flow. If the sum of the forces due to the adverse pressure gradient and the wall shear stress is large enough to overcome the momentum of the flow, then the fluid particles near the wall can be brought to rest and forced back upstream. When that occurs, fluid is deflected away from the wall into the mainstream, and *flow separation* occurs. This is

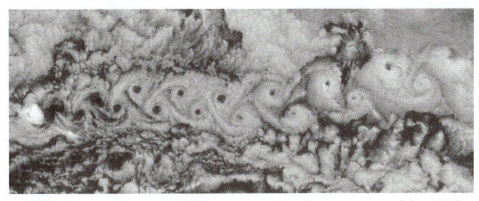

FIGURE 10-12   Von Karman vortex street behind an island (view covers area of about 365 km by 158 km).
(*Source*: NASA GSFC/LaRC/JPL, 2001. Used with permission.)

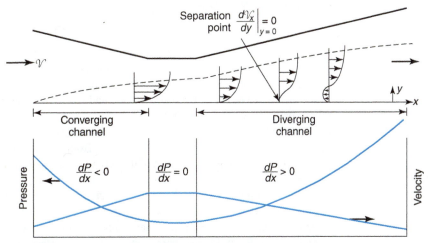

FIGURE 10-13   Flow through converging–diverging channel and pressure distribution.
(*Source*: Adapted from R. W. Fox and A. T. McDonald, *Introduction to Fluid Mechanics*, 3rd ed., Wiley, New York, 1985. Used with permission.)

what happens on flow over a cylinder at relatively low Reynolds numbers (Figure 10-11b). The flow accelerates from the forward stagnation point on the cylinder and the pressure decreases. At some point around the cylinder, the velocity reaches a maximum (and, hence, the minimum pressure). With further progress around the cylinder, the flow decelerates, pressure rises, and an adverse pressure gradient exists, which causes flow separation.

   The sudden drop in the drag coefficient that begins at a Reynolds number of about 200,000 is related to a change in the character of the flow and boundary-layer separation. For a Reynolds number less than about 200,000, the flow remains laminar on the front side of the cylinder, and boundary-layer separation occurs at about 81° from the forward stagnation point. For a Reynolds number greater than about 200,000, transition to turbulence occurs ahead of the vertical centerline of the cylinder, and separation occurs on the order of 120° from the forward stagnation point. A turbulent velocity profile is much blunter than a laminar flow profile (see Figure 10-4). A consequence of this is that, for a given free-stream velocity, turbulent flow has greater momentum nearer the wall and thus can withstand a greater adverse pressure gradient than a laminar flow. This permits a turbulent flow to reach

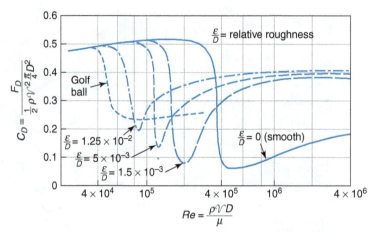

FIGURE 10-14 Roughness effects on the drag coefficient of a sphere.
(*Source*: Adapted from D. F. Young, B. R. Munson, and T. H. Okiishi, *A Brief Introduction to Fluid Mechanics*, Wiley, New York, 1997. Used with permission.)

farther around a cylinder than a laminar flow, the low-pressure wake region decreases in size, the asymmetry in the pressure distribution decreases, and the drag coefficient drops. Note the differences between the cylinder and sphere drag coefficients. This is reflective of the difference between two-dimensional and three-dimensional flow around each body, respectively. When the boundary layer on the cylinder becomes turbulent, the separation point can only shift toward the rear from the top and bottom in a two-dimensional way. For the sphere, the separation point shifts not only from the top and bottom but also from both sides. Thus the decrease in the area of the separation zone for the sphere is greater than that for the cylinder.

This sharp decrease in the drag coefficient on spheres when the flow changes from laminar to turbulent (on the order of a factor of 5) is the reason why golf balls have dimples. Dimples or surface roughness cause a boundary layer to transition to turbulence at much lower Reynolds numbers than for smooth surfaces (Figure 10-14). With a lower drag coefficient, the ball goes farther for the same effort.

EXAMPLE 10-3 Drag on a circular cylinder

Fishing trawlers drag large nets to catch large numbers of fish at one time. Consider a net that is 120 m long, 6 m high and is woven into a 10-cm square mesh using 5-mm-diameter string. Four cables connect the net to the trawler. If the engine power is 170 kW and the sea temperature is 5°C, determine:

a) the maximum speed (in km/h) the trawler can achieve when dragging the nets.

b) the force (in N) each connecting cable must withstand.

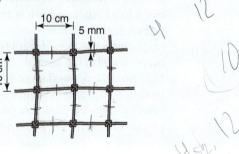

Approach:

Power is calculated from force times the velocity. The force is determined from application of the drag coefficient. We assume the flow through the strings in the net behaves as crossflow over circular cylinders; this ignores the effect of all the locations where two perpendicular strings are knotted. We need to determine the total length of the string in the net and the drag coefficient. Because the drag coefficient is dependent on the Reynolds number and, hence, velocity, we may need to iterate to find a solution.

Assumptions:

Solution:

a) Power is

$$\dot{W} = F_D \mathcal{V}$$

where $\mathcal{V}$ is the trawling speed we want to determine. Force is calculated from

$$F_D = C_D \left(\frac{1}{2} \rho \mathcal{V}^2 A \right)$$

Combining these two equations and solving for the velocity, we obtain

$$\mathcal{V} = \left[\frac{2\dot{W}}{C_D \rho A} \right]^{1/3}$$

A1. The string acts as a circular cylinder.
A2. The flow is perpendicular to all strings in the net.
A3. The knots are ignored.

The drag coefficient is dependent on the Reynolds number. We assume the situation can be approximated as flow perpendicular to a circular cylinder [A1] [A2] [A3]. For a first estimate, we assume that the Reynolds number is in the range in Figure 10-10, where the drag coefficient is approximately constant at $C_D \approx 1$. After we calculate the velocity, we will check this assumption.

The area is thickness of the string times the total length of the strings in the net. From the schematic, we estimate the total length of string in $1\,m^2$ to be $10 \times 10 \times 1\,m = 100\,m/m^2$. (Do not double count the edges in the 1-m^2 section.) Hence, the total length is $(100\,m/m^2)(120\,m)(6\,m) = 72,000\,m$.

A4. Freshwater properties are a good approximation to seawater properties.

Therefore, assuming we can use freshwater fluid properties [A4],

$$\mathcal{V} = \left[\frac{2\dot{W}}{C_D \rho A} \right]^{1/3} = \left[\frac{2\,(170,000\,W) \left(\frac{1\,J/s}{1\,W} \right) \left(\frac{1\,N\cdot m}{1\,J} \right) \left(\frac{1\,kg\cdot m}{1\,N\cdot s^2} \right)}{1\,(1000\,kg/m^3)\,(0.005\,m)\,(72,000\,m)} \right]^{1/3} = 0.98\,\frac{m}{s} = 3.53\,\frac{km}{h}$$

Now check the Reynolds number. At 5°C the viscosity is $\mu = 1.519 \times 10^{-3}\,kg/m\cdot s$, so that

$$Re = \frac{\rho \mathcal{V} D}{\mu} = \frac{(1000\,kg/m^3)\,(0.98\,m/s)\,(0.005\,m)}{1.50 \times 10^{-3}\,N\cdot s/m^2} = 3267$$

From Figure 10-10 at this Reynolds number, the drag coefficient is about constant   the value we assumed, so no iteration is required.

b) The force each of the four connecting cables must withstand is

$$F_{cable} = \frac{1}{4} F_D = \frac{1}{4} C_D \left(\frac{1}{2} \rho \mathcal{V}^2 A \right)$$

$$= \left(\frac{1}{4} \right) (1) \left(\frac{1}{2} \right) \left(1000\,\frac{1kg}{m^3} \right) \left(0.98\,\frac{m}{s} \right)^2 (0.005\,m)\,(72,000\,m) \left(\frac{N\cdot s^2}{kg\cdot m} \right) = 43,320\,N$$

Comments:

The assumption that the flow is perpendicular to all the strings in the net may or may not be good. Likewise, the locations where two strings cross and connect to each other (to form the mesh) will affect the drag, but we are not sure by how much. Nevertheless, we need to obtain an estimate from the information we have available, and what we have is reasonable. Also, seawater has different properties than fresh water. Note that this example used an *empty* net. Once fish are gathered, the drag force will increase significantly.

EXAMPLE 10-4   Drag on sphere

A group of meteorological scientists want to use a 14-ft-diameter hydrogen-filled balloon to make weather measurements from a (relatively) stationary location at 1000 ft above the ground. At that height, the wind speed is 25 mph, the pressure is 14.2 psia, and the temperature is 55.4°F. The instruments weigh 20 lbf. The weight of the balloon material is 5 lbf. The hydrogen has a density of 0.0054 lbm/ft^3. Determine the length of cable required (in ft).

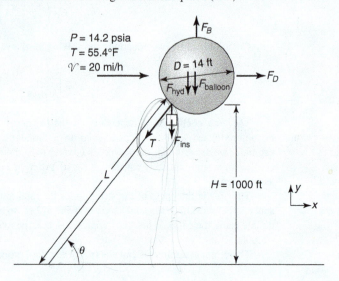

Approach:

The wind will push the balloon to one side; for the balloon to remain stationary, the net force in the vertical direction and the net force in the horizontal direction both must be zero. The forces acting on the balloon are the buoyancy force; the weight of the instruments, balloon material, and hydrogen; the drag force; and the tension in the cable. Force balances in the two directions are needed.

Assumptions:

Solution:

Consider the diagram shown above. The length, L, we seek is obtained with trigonometry: $L = H / \sin \theta$, where H, the vertical distance above the ground, is known.

The angle θ is determined from force balances on the balloon. All the forces are shown, with F_B the buoyancy force, F_D the drag force, F_{ins} the weight of the instruments, F_{hyd} the weight of the hydrogen, $F_{balloon}$ the weight of the balloon material, and T the tension force in the cable. We assume the cable acts as a rigid body [A1], so that the tension force is directed along the cable to the point of connection with the ground. Force balances in the two coordinate directions are

A1. The cable can be considered a rigid link.

$$\sum F_x = 0 \rightarrow F_D - T \cos \theta = 0$$

$$\sum F_y = 0 \rightarrow F_B - F_{hyd} - T \sin \theta - F_{ins} - F_{balloon} = 0$$

The x-direction equation is solved for T and substituted into the y-direction equation:

$$F_B - F_{hyd} - F_D \frac{\sin \theta}{\cos \theta} - F_{ins} - F_{balloon} = 0$$

where $\sin \theta / \cos \theta = \tan \theta$. Solving for the angle θ,

$$\theta = \tan^{-1} \left(\frac{F_B - F_{hyd} - F_{ins} - F_{balloon}}{F_D} \right)$$

A2. Air and hydrogen are ideal gases.

The buoyancy force is $F_B = \rho_{air} Vg$. At the given pressure and temperature, the ideal gas equation [A2] is

$$\rho_{air} = \frac{PM}{RT} = \frac{\left(14.2 \frac{lbf}{in.^2} \right) \left(28.97 \frac{lbm}{lbmol} \right) \left(\frac{144 \, in.^2}{1 \, ft^2} \right)}{\left(1545 \frac{ft \cdot lbf}{lbmol \cdot R} \right) (515.4 \, R)} = 0.0745 \frac{lbm}{ft^3}$$

so that the buoyancy force is

$$F_B = \left(0.0745 \frac{lbm}{ft^3} \right) \left[\frac{4}{3} \pi \, (7 \, ft)^3 \right] \left(32.17 \frac{ft}{s^2} \right) \left(\frac{lbf \cdot s^2}{32.17 \, ft \cdot lbm} \right) = 107 \, lbf$$

In a similar manner, the weight of the hydrogen is

$$F_{hyd} = \left(0.0054 \frac{lbm}{ft^3} \right) \left[\frac{4}{3} \pi \, (7ft)^3 \right] \left(32.17 \frac{ft}{s^2} \right) \left(\frac{lbf \cdot s^2}{32.17ft \cdot lbm} \right) = 7.8 \, lbf$$

The drag force is calculated with the drag coefficient, which requires the Reynolds number. The air viscosity at 55.4°F is obtained from Table B-7 by interpolation, $\mu = 1.196 \times 10^{-5}$ lbm/ft·s, so that the Reynolds number is:

$$Re = \frac{\rho V D}{\mu} = \frac{\left(0.0745 \frac{lbm}{ft^3} \right) \left(25 \frac{mi}{h} \right) \left(\frac{5280 \, ft}{1 \, mi} \right) \left(\frac{1 \, h}{3600 \, s} \right) (14 \, ft)}{1.196 \times 10^{-5} \, lbm/ft \cdot s} = 3.2 \times 10^6$$

A3. The balloon acts as a smooth sphere.
A4. Ignore drag on the cable and instrument package.

Assuming that the balloon acts as a smooth sphere and ignoring the drag on the instrument package and connecting cable [A3] [A4], we obtain the drag coefficient from Figure 10-10, $C_D \approx 0.15$, and the drag force from

$$F_{drag} = C_D \left(\frac{1}{2} \rho V^2 A \right)$$

$$= 0.15 \left(\frac{1}{2} \right) \left(0.0745 \frac{lbm}{ft^3} \right) \left[\left(25 \frac{mi}{h} \right) \left(\frac{5280 \, ft}{1 \, mi} \right) \left(\frac{1 \, h}{3600 \, s} \right) \right]^2 \frac{\pi}{4} (14 \, ft)^2 \left(\frac{lbf \cdot s^2}{32.17 \, ft \cdot lbm} \right)$$

$$= 35.9 \, lbf$$

Substituting these values into the equation for θ,

$$\theta = \tan^{-1} \left(\frac{107 - 7.8 - 20 - 5}{35.9} \right) = 64°$$

The length is

$$L = (1000 \, ft) / \sin (64°) = 1111 \, ft$$

Comments:

The drag force is underestimated because of the presence of the instrument package and its attachment wires, the tethering cable, and so on. Likewise, we treated this balloon as a smooth sphere, which is probably not the case. Finally, the cable will not act as a rigid link. If the angle were 90°, the analysis is fine. However, as the angle becomes smaller the cable will sag more and more, and the tension force will act not toward the ground attachment point but more toward the ground under the balloon. As a result, a longer cable would be needed to reach the desired elevation.

10.5 DRAG ON TWO- AND THREE-DIMENSIONAL BODIES

Drag coefficients for many two-dimensional bodies are given in Table 10-1 and for three-dimensional bodies in Table 10-2. These data are only representative of the vast body of knowledge available on this topic.

TABLE 10-1 **Drag coefficients for two-dimensional bodies***

Object	Description	Drag coefficient $C_D = F_D/(\rho \mathcal{V}^2 A/2)$	Area	Reynolds number $Re = \rho \mathcal{V} L/\mu$
Rectangular rod, sharp corners	$\begin{array}{cc} D/L & C_D \\ <0.1 & 1.9 \\ 0.5 & 2.5 \\ 1.0 & 2.2 \\ 2.0 & 1.7 \\ 3.0 & 1.3 \end{array}$	$A = LW$	$>10^4$
	Round front edge	$\begin{array}{cc} D/L & C_D \\ 0.5 & 1.2 \\ 1.0 & 0.9 \\ 2.0 & 0.7 \\ 4.0 & 0.7 \end{array}$	$A = LW$	$>10^4$
	Square rod, round corners	$\begin{array}{cc} R/L & C_D \\ 0 & 2.2 \\ 0.02 & 2.0 \\ 0.17 & 1.2 \\ 0.33 & 1.0 \end{array}$	$A = LW$	10^5
	Equilateral triangle, round corners	$\begin{array}{ccc} & \multicolumn{2}{c}{C_D} \\ R/L & \rightarrow & \leftarrow \\ 0 & 1.4 & 2.1 \\ 0.02 & 1.2 & 2.0 \\ 0.08 & 1.3 & 1.9 \\ 0.25 & 1.1 & 1.3 \end{array}$	$A = LW$	10^5
	Circular rod	1.2	$A = LW$	$6 \times 10^3 < Re < 2 \times 10^5$
	Semicircular rod	$\begin{array}{cc} \rightarrow & \leftarrow \\ 2.2 & 1.2 \end{array}$	$A = LW$	$>10^4$
	Semicircular shell	$\begin{array}{cc} \rightarrow & \leftarrow \\ 2.3 & 1.1 \end{array}$	$A = LW$	10^4

*W is length out of plane of the page

TABLE 10-2 Drag coefficients for three-dimensional bodies

Object	Description	Drag coefficient $C_D = F_D/(\rho \mathcal{V}^2 A/2)$	Area	Reynolds number $Re = \rho \mathcal{V}L/\mu$
	Cube Square to flow	1.05	$A = L^2$	$>10^4$
	45° to flow	0.80	$A = L^2$	
	Cone	$\begin{array}{cc} \alpha & C_D \\ 30° & 0.6 \\ 60° & 0.8 \\ 90° & 1.2 \end{array}$	$A = \pi L^2/4$	$>10^4$
	Short horizontal cylinder, axis parallel to flow	$\begin{array}{cc} D/L & C_D \\ <0.5 & 1.1 \\ 1 & 0.93 \\ 2 & 0.83 \\ 4 & 0.85 \\ 8 & 1.0 \end{array}$	$A = \pi L^2/4$	$>10^4$
	Short horizontal cylinder, axis perpendicular to flow	$\begin{array}{cc} L/D & C_D \\ 1 & 0.6 \\ 2 & 0.7 \\ 5 & 0.8 \\ 10 & 0.9 \end{array}$	$A = \pi L^2/4$	$>10^3$
	Sphere	0.4 24/Re	$A = \pi L^2/4$	$2 \times 10^3 < Re < 2 \times 10^5$ $Re < \sim 1$
	Hemisphere Solid	$\xrightarrow{\;}$ 1.2 $\quad \xleftarrow{\;}$ 0.4	$A = \pi L^2/4$	$>10^4$
	Hollow	1.4 $\quad$ 0.4		
	Streamlined body	0.04	$A = \pi L^2/4$	$>10^5$
	Parachute	1.3	$A = \pi L^2/4$	
	Tree	$\begin{array}{cc} \text{Velocity (m/s)} & C_D \\ 10 & 0.4\text{–}1.2 \\ 20 & 0.3\text{–}1.0 \\ 30 & 0.2\text{–}0.7 \end{array}$	Frontal Area	

(Continued)

TABLE 10-2 *(Continued)*

Object	Description	Drag coefficient $C_D = F_D/(\rho \mathcal{V}^2 A/2)$	Area	Reynolds number $Re = \rho \mathcal{V} L/\mu$
	Tall office building	1.3–1.5	Frontal area	
	Person (average size)	standing C_D = 1.0–1.3 standing $C_D A$ = 0.84 m^2 sitting $\quad C_D A$ = 0.56 m^2 crouching $C_D A$ = 0.23 m^2	Frontal area	

Effect of Reynolds Number   As seen on Figure 10-7, the drag coefficient for flow perpendicular to a flat plate is nearly independent of the Reynolds number of the flow. Cylinders and spheres (see Figure 10-10) have relatively constant drag coefficients for Reynolds numbers from 1000 to 100,000. Indeed, for many blunt bodies over practical ranges of Reynolds numbers, the effect of Reynolds number on the drag coefficient is small and can be ignored. This is reflected in the drag coefficients for two-dimensional bodies given in Table 10-1 and for three-dimensional bodies in Table 10-2.

Effect of Shape Including Streamlining   The projected frontal area and the shape of an object affect the magnitude of the drag force. The definition of drag coefficient shows that the drag force is directly proportional to the area: decrease the projected area, and the drag force is reduced.

The effect of other changes in the object shape is slightly more complicated. From the discussion above about the changing drag characteristics around an infinite cylinder with increasing Reynolds number, we can see that if we can minimize the separated flow region, then the drag coefficient can be reduced. We can accomplish this by *streamlining* or modifying the shape of the body so that the adverse pressure gradient is reduced. Consider the streamlined shape in Figure 10-15 compared to the two circular cylinders. The streamlined shape has more surface area than either cylinder, and the total drag on the objects is the sum of the pressure and skin friction forces. For many blunt bodies when streamlining techniques are implemented, the increase in skin friction is much less than the decrease in pressure drag. In such cases, changes in pressure drag have the greatest effect on the drag coefficient. The streamlined body has a separated flow region that is much smaller than that on the two cylinders. At the same velocity and air density, the drag force on the large cylinder is on the order of 20–30 times that of the streamlined shape. The small cylinder has a drag force approximately the same as the much larger streamlined body.

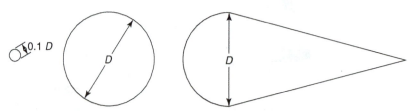

FIGURE 10-15   Streamlined shape.

TABLE 10-3 **Drag coefficients of vehicles***

Object	Description	Drag coefficient $C_D = F_D/(\rho \mathcal{V}^2 A/2)$
	Plymouth Voyager minivan	0.40
	Volkswagen "bug"	0.46
	1932 Fiat Balillo	0.60
	GM Sunraycer (experimental solar vehicle)	0.12
	Mercedes Benz E320	0.29
	Ford Taurus	0.30
	Tractor trailer Standard	0.96
Deflector	With deflector	0.76
Gap seal	With deflector and gap seal	0.70
	Pickup truck	0.5
	Bicycle, upright	1.1
	Bicycle, racing	0.9
	Bicycle, tandem drafting	0.5
	Bicycle, single with fairing	0.12

* Area is projected frontal area.

Sources: Adapted from J. A. Robertson and C. T. Crowe, *Engineering Fluid Mechanics*, 6th ed., Wiley, 1997; D. F. Young, B. R. Munson, and T. H. Okiishi, *A Brief Introduction to Fluid Mechanics*, Wiley, New York, 1997. Used with permission.

Automobiles and airplanes are the most obvious examples of where streamlining is used to great effect. Examples of the drag coefficients of some vehicles are given in Table 10-3. One of the techniques used to reduce the drag coefficient is to round off all the sharp corners, eliminate rain gutters (which used to be common on cars), and generally help make the flow smoother over the car. Consider the outside rearview mirror. On older

TABLE 10-4 **Relative magnitudes of friction and pressure drag on objects**

Object	Friction drag	Pressure drag
Thin flat plates aligned with flow velocity	Large	Zero to negligible
Thin flat plates perpendicular to flow velocity	Zero	Large
Blunt bodies (cyclinders, spheres, etc.)	Negligible to small	Large
Streamlined bodies	Negligible to small	Large

vehicles, the mirror was simply a thin disk with a drag coefficient of about 1.1. Newer cars have streamlined mirrors; the side of the mirror facing the direction of the car movement has a fairing that allows the air to pass over the surface smoothly. This shape has a drag coefficient on the order of 0.3 to 0.4. By itself, the change in the mirror shape will decrease only a little the overall drag coefficient of the car. However, the combination of changes in every aspect of the vehicle configuration can result in significant drag coefficient reductions. (Compare a minivan with a passenger car on Table 10-3.)

The total drag force on a body is some combination of friction (shear) drag and pressure drag. Depending on the body shape, the magnitudes of these two contributions are significantly different. Qualitative assessments of the relative contributions from each are given in Table 10-4. Note that seldom are the two drag forces large simultaneously.

EXAMPLE 10-5   Effect of streamlining

Wind deflectors and fairings have been demonstrated to reduce the drag coefficient on tractor-trailer rigs. (See Table 10-3.) The cost of the streamlining attachments must be less than the savings in fuel costs to make them cost-effective.

Consider a 25,000-kg tractor-trailer with a frontal area of 8.5 m^2 and a rolling resistance of $C_{rolling} = 2\%$ of the vehicle weight. The engine's specific fuel consumption is $SFC = \dot{m}_F / \dot{W} = 0.21$ kg fuel/kWh, and the drivetrain efficiency is $\eta_{drive} = 85\%$. Diesel fuel specific gravity is $SG = 0.83$. With a cost of diesel fuel of \$1.50/gal and for annual usage of 150,000 miles per year at 70 mph, determine:

a) yearly fuel expense (in \$) for a tractor-trailer without any streamlining.

b) yearly fuel expense savings (in \$) if a top wind deflector is used.

c) yearly fuel expense savings (in \$) if a top wind deflector and gap seal are used.

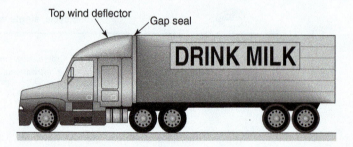

Top wind deflector
Gap seal
DRINK MILK

Approach:

The fuel expense is determined by knowing how much fuel is used each year with and without the use of streamlining devices. The engine's specific fuel consumption is used to calculate the total fuel used. The power required by the tractor-trailer can be calculated from the drag force plus the rolling resistance and tractor-trailer speed; this power must be increased by use of the drivetrain efficiency to give the required engine power.

Assumptions:

Solution:

a) We let m_F equal the total mass of fuel used in a year, so that fuel expense is determined from:

$$\text{fuel expense} = \frac{(\text{fuel cost})(\text{total fuel used})}{\text{fuel density}} = \frac{(\$/\text{gal})(m_F \text{ kg})}{\rho \text{ kg/m}^3}$$

Fuel cost is given. Total fuel used, m_F, is determined from the definition of SFC:

$$m_F = SFC \times \dot{W} \times t$$

where t is the total time in a year that the tractor-trailer is used, $t = (150,000 \text{ mi})/(70 \text{ mi/h}) = 2143$ h.

The required engine power (taking into account the drivetrain efficiency) is

$$\dot{W} = F_{tot}\mathcal{V}/\eta_{drive}$$

where $\mathcal{V}$ is the tractor-trailer speed 70 mi/h = 31.3 m/s, and F_{tot} is the total drag and rolling forces [A1].

A1. The relative speed between the truck and air is 70 mph (i.e., the air is calm).

The total force is

$$F_{tot} = F_{rolling} + F_{drag}$$

where the drag force is calculated from $F_D = C_D\left(0.5\rho\mathcal{V}^2 A\right)$, and the rolling force is $F_{rolling} = C_{rolling}W_{tractor}$. We have enough information to evaluate the rolling force:

$$F_{rolling} = C_{rolling}W_{tractor} = (0.02)(25,000 \text{ kg})(9.81 \text{ m/s}^2)(1 \text{ N·s}^2/1 \text{ kg·m}) = 4905 \text{ N}$$

A2. Air is at 1 atm and 27°C.

From Table 10-3, the drag coefficient for an unstreamlined tractor-trailer is 0.96. Assuming [A2], the air density is $\rho = 1.177 \text{ kg/m}^3$, and the drag force is

$$F_D = C_D\left(\frac{1}{2}\rho\mathcal{V}^2 A\right)$$

$$= 0.96\left(\frac{1}{2}\right)(1.177 \text{ kg/m}^3)(31.3 \text{ m/s})^2 (8.5 \text{ m}^2)(1 \text{ N·s}^2/1 \text{ kg·m}) = 4705 \text{ N}$$

The total force is $F_{tot} = 4905 \text{ N} + 4705 \text{ N} = 9610 \text{ N}$. The engine power is:

$$\dot{W} = [(9610 \text{ N})(31.3 \text{ m/s})/(0.85)](1 \text{ W·s}/1 \text{ N·m}) = 354,000 \text{ W} = 354 \text{ kW}$$

The total mass of fuel used is

$$m_F = (0.21 \text{ kg/kWh})(354 \text{ kW})(2143 \text{ h}) = 159,300 \text{ kg}$$

The fuel expense is

$$(\$1.50/\text{gal})(159,300 \text{ kg})(1 \text{ gal}/3.78 \times 10^{-3} \text{ m}^3)(0.83)(1 \text{ m}^3/1000 \text{ kg}) = \$52,470$$

b and **c**. The same calculations are done for parts b and c as was done for part a. The only difference is in magnitude of the drag coefficient. From Table 10-3 we obtain the other drag coefficients, and the table below gives the calculated quantities.

	C_D	F_{tot} (N)	$\dot{W}$ (kW)	m_F (kg)	Annual fuel expense ($)	Annual savings ($)
Without streamlining	0.96	9610	354	159,300	52,470	NA
With air deflector only	0.76	8630	318	143,000	47,100	5,370
With air deflector and gap seal	0.70	8336	307	138,150	45,500	6,970

Comments:

Note that compared to the tractor-trailer without any streamlining, just the air deflector saves 10.2% of the fuel expense; using both the air deflector and gap seals saves 13.2%. Small changes can have large effects. These devices are cost-effective, as demonstrated by how many are seen on tractor-trailers traveling the highways.

Drag on Composite Bodies   Drag coefficients of many bodies have been given in Table 10-1 and Table 10-2 . However, data may not be available for some objects of interest. For example, consider the highway sign shown in Figure 10-16, which is composed of a flat plate located on top of two circular cylinders. This combination of objects is not generally available in tables, but if we need an *estimate* of the possible wind loading to design the structure to withstand a 100-mph wind, then we can use an approximation technique called *superposition*. With superposition, the individual parts of the composite body are analyzed as if the other parts were not present, and the contributions from each are added to get the total force.

Note that several assumptions and approximations are involved in this procedure. Consider the highway sign. We know that a boundary layer forms on the ground, so the wind velocity is probably not uniform over the cylinder or the flat plate. There are probably interactions between the cylinder and the flat plate that affect/modify their individual drag characteristics. If a drag coefficient for an infinite cylinder (Figure 10-10) is used, then this is only an approximation for the finite cylinders, one end of which is attached to the

FIGURE 10-16   Highway sign subject to wind loading.

earth and the other to the flat plate. Hence, the estimated overall results must be interpreted carefully, even though they are often accurate.

EXAMPLE 10-6   Drag on composite body

A design engineering firm has been asked to develop a portable sign (see schematic for dimensions) for use in warmer climates and in the summer. The design air temperature is 25°C. Water, which can be easily drained or filled, will be used as the weighting material for ease of movement. What should the base volume be so that the sign can withstand a wind of 100 km/h? (That is, because we know the base is 1 m by 2 m, what should the height, H, be?) Ignore the weight of the sign and poles.

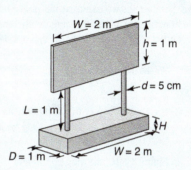

Approach:

We recognize that the sign will tip along the long edge of the sign base rather than the short edge. The problem to be solved is one involving the force required to tip the sign being counterbalanced by the weight in the base of the sign. The sum of the moments on the sign must be zero, so that the sign remains upright. Moments are caused by the drag force from the wind and the weight (volume) of the water in the base.

Assumptions:

Solution:

We begin with a moment balance taken about point O on the back lower edge of the sign base, as shown on the schematic:

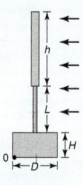

$$\sum M = 0$$

The forces that cause moments are the drag force due to the wind on the sign and on its supporting poles and the water weight [A1].

A1. Ignore drag forces on the base.

$$R_{sign}F_{D,sign} + 2R_{pole}F_{D,pole} - R_{water}F_{water} = 0$$

The factor of 2 on the second term is to account for both poles.

The moment caused by the base is the water weight times its moment arm. The weight of the water acts through the center of the base, so the moment arm for the water-filled base is $R_{water} = D/2$. The water weight is $F_{water} = mg = \rho Vg = \rho HDWg$, where H is the quantity we want to determine. At 25°C water has a density of 997 kg/m³.

The drag force on each pole acts through its center, so the moment arm of each pole is $R_{pole} = H + L/2$, and the drag force on one pole is

$$F_{D,pole} = C_{D,pole}\left(\frac{1}{2}\rho \mathcal{V}^2 A\right)$$

A2. Total drag on the composite body can be approximated by adding the drag on the individual (isolated) parts.

We assume that we can treat the poles and sign as a composite body [A2]. Hence, for the poles we assume we can use the drag coefficient on an infinite cylinder (Figure 10-10). At 25°C, the air properties are: $\rho = 1.186$ kg/m³, $\mu = 1.84 \times 10^{-5}$ N·s/m. Hence, the Reynolds number is

$$Re = \frac{\rho \mathcal{V} d}{\mu}$$

$$= \frac{(1.186 \text{ kg/m}^3)(100 \text{ km/h})(0.05 \text{ m})(1000 \text{ m/1 km})(1 \text{ h/3600 s})}{1.84 \times 10^{-5} \text{ N·s/m}^2} = 89,522$$

From Figure 10-10, the drag coefficient is $C_{D,pole} \approx 1.5$, so drag force on one pole is

$$F_{D,pole} = 1.5\left(\frac{1}{2}\right)\left(1.186\frac{\text{kg}}{\text{m}^3}\right)\left(\frac{100 \times 1000}{3600}\frac{\text{m}}{\text{s}}\right)^2 (0.05 \text{ m})(1 \text{ m})(1 \text{ N·s}^2/1 \text{ kg·m}) = 34.3 \text{ N}$$

The drag force on the sign acts through its center, so its moment arm is $R_{sign} = H + L + h/2$. The drag coefficient for flow perpendicular to a rectangular plate with an aspect ratio of 2 is determined from Figure 10-7; the value is $C_{D,sign} \approx 1.1$.

$$F_{D,sign} = 1.1\left(\frac{1}{2}\right)\left(1.186\frac{\text{kg}}{\text{m}^3}\right)\left(\frac{100 \times 1000}{3600}\frac{\text{m}}{\text{s}}\right)^2 (2 \text{ m})(1 \text{ m})(1 \text{ N·s}^2/1 \text{ kg·m}) = 1007 \text{ N}$$

From the moment equation above, the only unknown is the height of the base volume, so solving for it:

$$\left(H + L + h/2\right) F_{D,sign} + 2\left(H + L/2\right) F_{D,pole} - \left(D/2\right)\left(\rho HDWg\right) = 0$$

$$H = \frac{-\left(L + h/2\right) F_{D,sign} - 2\left(L/2\right) F_{D,pole}}{F_{D,sign} + 2F_{D,pole} - \left(D^2/2\right)\left(\rho Wg\right)}$$

$$= \frac{-\left(1 \text{ m} + 1 \text{ m}/2\right)(1007 \text{ N}) - 2\left(1 \text{ m}/2\right)(34.3 \text{ N})}{1007 \text{ N} + 2(34.3 \text{ N}) - \left(1 \text{ m}^2/2\right)\left(997 \text{ kg/m}^3\right)(2 \text{ m})\left(9.81 \text{ m/s}^2\right)\left(1 \text{ N·s}^2/1 \text{ kg·m}\right)}$$

$$= 0.177 \text{ m}$$

Comment:

The drag force due to the poles is insignificant compared to that of the sign.

10.6 LIFT

A symmetric flow around a symmetric object causes a drag force parallel to the direction of the free-stream velocity, but there will be no force perpendicular to the direction of the flow. Only if the object is asymmetrical or if the flow field is asymmetrical around an object will a *lift force*—a force *perpendicular* to the fluid motion—be created. Airfoils, such as used on wings of airplanes (see Figure 10-9), are the most common examples of asymmetric bodies employed for their ability to create lift. (The "spoiler" on the rear of sports and racing cars is an example of an inverted airfoil; it creates a downward force—negative lift—for improved car handling and stability to counteract lift forces produced by the car body.) Shown on Figure 10-17 are some of the terms used to describe the physical characteristics of an airfoil.

From Eq. 10-14, if the pressure distribution around an airfoil is integrated over the airfoil's complete perimeter, the total lift force can be calculated. So how is lift created? Lift is created by the unequal flow of fluid over the top and bottom surfaces of the airfoil. The flow reaching the leading edge of an airfoil splits into two streams. Because of the airfoil's curved geometry, the upper stream accelerates, while the lower one decelerates. Boundary layers on airfoils are thin; viscous effects are significant only near the wing and in the wake trailing the wing. As a result, the flow field outside the boundary layer may be treated as an inviscid flow. Inviscid flow is a topic of considerable importance in vehicle design and analysis. However, it is beyond the scope of this text. The pressure distribution in the flow field around the airfoil can be obtained from the solution of an inviscid flow around an object. The boundary layer is very thin, and there is negligible pressure change across it. Therefore, the pressure distribution from the inviscid flow solution is used to describe the pressure distribution on the surface of the airfoil, and these pressures determine the lift forces. How does the pressure vary? Consider the Bernoulli equation applied to the airfoil's flow field. Just before the flow splits at the leading edge, the velocity and pressure are known. Use this as the reference point for the Bernoulli equation. The second location can be taken along the surface of the airfoil. The velocities over the top surface are greater than those over the bottom surface. Therefore, from the Bernoulli equation, the pressures on the top surface will be smaller than those on the bottom. (See Figure 10-9 for a schematic representation of the pressure distribution.) Integrating the pressure distribution results in a net upward force, which we call lift.

Although an airfoil is designed for lift, it also experiences drag, so airfoils are streamlined to minimize drag forces. Note that the lift force on the wings maintains a plane in the air; a plane's engines provide power to overcome the drag force induced with forward motion. Figure 10-18 demonstrates how the lift and drag coefficients vary with a changing angle of attack. Three characteristics should be noted. First, typical airfoils have a positive lift coefficient at zero angle of attack, because airfoils are not symmetric. Second, the lift coefficient is approximately linear with angle of attack; in addition, the drag coefficient also

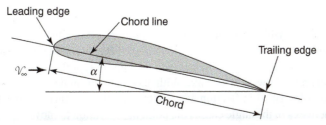

FIGURE 10-17
Terminology for an airfoil.

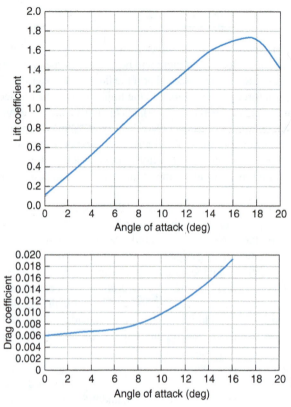

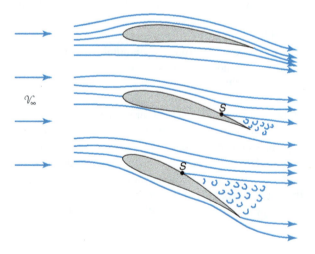

FIGURE 10-18 Effect of angle of attack on lift and drag coefficients.

FIGURE 10-19 Separation due to increased angle of attack.

increases with angle of attack, but not as dramatically as the lift coefficient. Third, if the angle of attack becomes large enough, the lift coefficient reaches a peak, then decreases precipitously; the drag coefficient increases just as rapidly. The airfoil then *stalls*. Figure 10-19 illustrates how stall occurs. With increases in angle of attack, boundary-layer separation (described in Section 10.4) occurs on the top surface near the trailing edge. Further increases in the angle causes the point of separation to move forward, but the lift coefficient still increases. At some critical angle of attack characteristic of the particular airfoil, the forward movement of the separation point no longer produces an increase in lift, and above this critical angle the lift decreases significantly. Stall is an extremely dangerous condition, and many inexperienced pilots have crashed because they did not have the skill or time and altitude to recover from a stall.

The definition of the lift coefficient (Eq. 10-17) indicates that the lifting force increases with the square of the velocity. At takeoff, a plane has its maximum weight and its slowest speed. The minimum speed needed to obtain sufficient lift can be determined by rearranging Eq. 10-17. In steady flight, the lift force (F_L) must equal the airplane's weight (W):

$$F_L = W = C_L \frac{1}{2} \rho \mathcal{V}_\infty^2 A$$

Thus the minimum velocity required for steady-state flight would be

$$\mathcal{V}_{min} = \left[\frac{2W}{\rho C_{L,max} A} \right]^{1/2} \tag{10-19}$$

At low takeoff and landing speeds, how can sufficient lift be generated to accommodate the plane's weight? Or, conversely, how can we design the wing so that the takeoff (and landing) speed is low? From Eq. 10-19, we see that we need either a high lift coefficient and/or a large wing area. Both of these can be obtained with movable flaps that extend from the trailing edge (and sometimes from the leading edge, too). These flaps increase the lift, the planform area, and drag, as shown on Figure 10-20. However, because flaps are used only at low speeds, the increased drag is not as important as the decrease in landing and takeoff speeds.

Minimum takeoff speed is inversely proportional to air density; lift force is directly proportional to density. Airplanes land and take off at airports that are at various elevations above sea level, and cruising heights range from only a few thousand meters for small private planes to over 12,000 m for commercial jets. Air density and temperature vary dramatically from sea level to cruising elevation. Table 10-5 shows that, for example, at

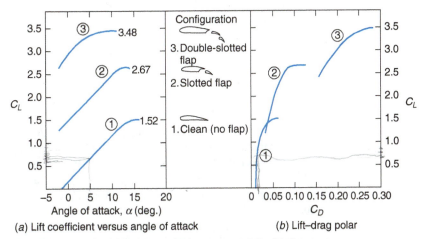

(a) Lift coefficient versus angle of attack   (b) Lift–drag polar

FIGURE 10-20 Lift and drag coefficients on airfoil with flaps.
(*Source:* R. W. Fox and A. T. McDonald, *Introduction to Fluid Mechanics*, 3rd ed., Wiley, New York, 1985. Used with permission.)

TABLE 10-5 **Properties of U.S. Standard Atmosphere**

Altitude (m)	Temperature (K)	P/P_o	ρ/ρ_o
0	288.2	1.0000	1.0000
500	284.9	0.9421	0.9529
1,000	281.7	0.8870	0.9075
1,500	278.4	0.8345	0.8638
2,000	275.2	0.7846	0.8217
3,000	268.7	0.6920	0.7423
4,000	262.2	0.6085	0.6689
5,000	255.7	0.5334	0.6012
6,000	249.2	0.4660	0.5389
8,000	236.2	0.3519	0.4292
10,000	223.3	0.2615	0.3376
12,000	216.7	0.1915	0.2546
15,000	216.7	0.1195	0.1590
20,000	216.7	0.05457	0.07258

$P_0 = 101.325$ kPa, $\rho_0 = 1.2250$ kg/m³; data up to 90,000 m are available

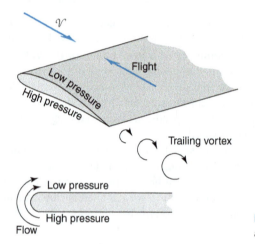

FIGURE 10-21 Trailing vortices from a wing tip.

12,000 m air density is only about 25% that at sea level (0 m). This variation must be taken into account in any calculation.

Most of the discussion above focused on infinite-length airfoils. The flow around finite-length airfoils is different because of what occurs at the airfoil tip. As discussed above, the pressure on the bottom surface is greater than that on the top. This pressure imbalance causes fluid to flow from the bottom side around the wing tip to the top side, and the forward motion of the plane sweeps the fluid downstream. The resulting swirl flow from each wing tip is called a *trailing vortex*. Figure 10-21 schematically shows what happens, and Figure 10-22 is a photograph of vortices trailing from a cropduster. Tip vortices add drag (*induced drag*), and methods are available to calculate their effect. Note that trailing vortices occur with small and large planes, and the strength of the trailing vortices is proportional to the lift generated by the wing. At airports, specified separation distances are required between planes landing or taking off. This distance is to allow the vortices to lessen in strength, because when these vortices are formed, they are strong enough to affect a plane flying through them.

FIGURE 10-22 Trailing vortex from a cropduster.
(*Source:* NASA Langley, 1990. Used with permission.)

Note that not all aspects of drag or lift are discussed in this section; only the basic ideas are presented. Additional information can be found in advanced books on fluid mechanics.

EXAMPLE 10-7   Power required for a large commercial jet

A large commercial jet has a wing planform area of 450 m² and a mass of 300,000 kg at takeoff, where it is 27°C and 101 kPa. At 12,000 m, where it is –56°C and 19.2 kPa, the plane's cruising speed is 900 km/h. Assume the lift and drag characteristics of the wing can be represented by the data on Figure 10-20. At takeoff, double-slotted flaps are used that increase the wing surface area by 10%, while at cruising speed, the flaps are retracted. The engine efficiency is 35%, and the fuel has a heating value (energy content) of 40,000 kJ/kg.

a) Determine the angle of attack required at cruising altitude.

b) Determine the power (in kW) required to cruise.

c) Determine the minimum speed (in km/h) for takeoff.

d) Determine the amount of fuel required for a flight of 8000 km (in kg).

Approach:

At the steady cruising altitude, there is no net vertical force on the plane. The lift force exactly balances the weight of the plane, so we can use the definition of lift coefficient (Eq. 10-17) and the data in Figure 10-20 to determine the required angle of attack. At that angle, the drag coefficient can be evaluated, too, and power is the drag force times the cruising velocity. For the minimum takeoff speed, we can use Eq. 10-19. The total fuel used is determined by integrating conservation of energy.

Assumptions:

Solution:

a) The lift coefficient is defined as:

$$C_L = \frac{F_L}{\frac{1}{2}\rho \mathcal{V}^2 A}$$

A1. Only the wings, not the fuselage, tail, and so on, cause lift and drag.
A2. Air is an ideal gas.

At steady cruising altitude, $F_L = mg$, and the planform area [A1] and velocity are known. The density is determined from the ideal gas equation [A2]:

$$\rho = \frac{PM}{\overline{R}T} = \frac{(19.2\,\text{kPa})\,(28.97\,\text{kg/kmol})\,(1\,\text{kN}/1\,\text{m}^2\cdot\text{kPa})}{(8.314\,\text{kJ/kmol}\cdot\text{K})\,(-56+273)\,\text{K}\,(1\,\text{kN}\cdot\text{m}/1\,\text{kJ})} = 0.308\,\frac{\text{kg}}{\text{m}^3}$$

Therefore,

$$C_L = \frac{(300{,}000\,\text{kg})\,(9.81\,\text{m/s}^2)}{\frac{1}{2}\left(0.308\,\dfrac{\text{kg}}{\text{m}^3}\right)\left(\dfrac{900\times1000}{3600}\,\dfrac{\text{m}}{\text{s}}\right)^2(450\,\text{m}^2)} = 0.68$$

From Figure 10-20, the angle of attack at this lift coefficient is about 5°.

b) Cruising power is obtained from

$$\dot{W} = F_D \mathcal{V}$$

where the drag force is obtained from

$$F_D = C_D\left(\frac{1}{2}\rho\mathcal{V}^2 A\right)$$

At the angle of attack determined above, the drag coefficient is approximately 0.02. Hence,

$$F_D = 0.02\left(\frac{1}{2}\right)\left(0.308\,\frac{\text{kg}}{\text{m}^3}\right)\left(\frac{900\times1000}{3600}\,\frac{\text{m}}{\text{s}}\right)^2(450\,\text{m}^2)\,(1\,\text{N}\cdot\text{s}^2/1\,\text{kg}\cdot\text{m}) = 86{,}625\,\text{N}$$

and the cruising power is

$$\dot{W} = (86{,}625\,\text{N})\left(\frac{900\times1000}{3600}\,\frac{\text{m}}{\text{s}}\right)\left(\frac{1\,\text{J}}{1\,\text{N}\cdot\text{m}}\right)\left(\frac{1\,\text{W}}{1\,\text{J/s}}\right) = 2.17\times10^7\,\text{W} = 21{,}700\,\text{kW}$$

c) The minimum takeoff speed is obtained from Eq. 10-19. From Figure 10-20, the maximum lift coefficient with flaps fully extended is about 3.4. Remember that the problem statement indicates that the wing area is increased by 10% with the flaps extended. At takeoff the air density is 1.177 kg/m³, so that

$$\mathcal{V}_{min} = \left[\frac{2W}{\rho C_{L,max}A}\right]^{1/2}$$

$$= \left[\frac{2\,(300{,}000\,\text{kg})\,(9.81\,\text{m/s}^2)}{(1.177\,\text{kg/m}^3)\,(3.4)\,(450\,\text{m}^2\times1.10)}\right]^{1/2} = 54.5\,\frac{\text{m}}{\text{s}} = 196\,\frac{\text{km}}{\text{h}} = 122\,\frac{\text{mi}}{\text{h}}$$

d) Using the definition of thermal cycle efficiency, we know that

$$\eta = \frac{\dot{W}}{\dot{Q}_{in}} \qquad \text{or} \qquad \dot{Q}_{in} = \frac{\dot{W}}{\eta}$$

A3. The flight is steady.
A4. No work occurs in the combustor.

Defining a control volume around a combustion chamber with assumptions [A3] and [A4] and applying conservation of mass and energy,

$$\dot{Q} + \dot{m}_F\,(h_{in} - h_{out}) = 0$$

Solving for the heat transfer rate and substituting into the previous equation:

$$\dot{m}_F \Delta h = \frac{\dot{W}}{\eta}$$

A5. Power, mass flow rate, and efficiency are constant with respect to time.

We integrate this with respect to time [A5], where the time period is $t = L/\mathcal{V}$, and L is the length of the flight:

$$m = \dot{m}_F t = \frac{\dot{W}}{\eta \Delta h} t = \frac{\dot{W} L}{\eta \Delta h \mathcal{V}}$$

$$= \frac{(21,700\,\text{kW})\,(8,000,000\,\text{m})\left(\dfrac{3600\,\text{s}}{1\,\text{h}}\right)}{(0.35)\left(40,000\,\dfrac{\text{kJ}}{\text{kg}}\right)\left(900,000\,\dfrac{\text{m}}{\text{h}}\right)\left(\dfrac{1\,\text{kW·s}}{1\,\text{kJ}}\right)} = 49,600\,\text{kg}$$

Comments:

Only the lift and drag on the main wings were taken into account in this solution. Lift and drag contributed by the fuselage and tail section were ignored.

10.7 MOMENTUM-INTEGRAL BOUNDARY LAYER ANALYSIS

(Go to www.wiley.com/college/kaminski)

SUMMARY

The *boundary layer* for flow past a solid surface is a thin layer in which viscous effects are important. Velocity varies from zero at the wall (*no-slip condition*) to the free-stream velocity at the edge of the boundary layer. The *boundary-layer thickness*, δ, is defined (arbitrarily) as the distance from the surface to where the boundary-layer velocity, $\mathcal{V}_x$, reaches 99% of the free-stream velocity, $\mathcal{V}_\infty$. Outside the boundary layer, viscous effects are negligibly small and are ignored, and we treat the flow as *inviscid*.

Depending on the flow velocity, surface geometry, and fluid, the flow near a surface may be *laminar* or *turbulent*. In laminar flows, viscous forces dominate, and the velocity profile is smooth and uniformly changing. In turbulent flows, bulk mixing of the flow dominates, and the velocity profile has a higher gradient at the wall and changes shape abruptly near the wall. *Transition* from laminar to turbulent flow begins at a distance from the leading edge of a body and depends on many flow, fluid, and geometric variables. In addition, the boundary-layer flow does not become fully turbulent instantaneously at some location but transitions over some finite length of the body. The transition location is best characterized by the length Reynolds number, $Re_x = \rho \mathcal{V} x / \mu$, where x is the distance from the leading edge of the body.

Drag force on a body is the sum of the *pressure* and *shear* forces acting *parallel* to the flow velocity. *Lift force* on a body is the net *pressure* and *shear* force acting *perpendicular* to the flow velocity. Drag and lift characteristics are described with nondimensional parameters. The *local skin friction coefficient* is defined as

$$C_{f,x} = \frac{\tau_w}{\frac{1}{2}\rho \mathcal{V}_\infty^2}$$

For laminar flow over an infinitesimally thin flat plate aligned with the velocity field,

$$C_{f,x} = \frac{0.664}{\sqrt{Re_x}}$$

where the characteristic length used in the Reynolds number is defined as the distance from the leading edge.

The *drag coefficient* is defined as

$$C_D = \frac{F_D}{\frac{1}{2}\rho \mathcal{V}_\infty^2 A}$$

where A is the projected frontal area for blunt objects or the projected planform area for flat plates and airfoils.

For laminar flow over an infinitesimally thin flat plate aligned with the velocity field, we can obtain the drag coefficient by integrating the local skin friction coefficient over the plate length:

$$C_D = \frac{1}{A} \int_A C_{f,x} \, dA = \frac{1.328}{\sqrt{Re_L}}$$

Similar expressions for turbulent flow on a flat plate have been obtained from experimental data:

$$C_D = \frac{0.074}{Re_L^{1/5}} \qquad for \; 10^5 < Re_L < 10^7$$

$$C_D = \frac{0.455}{(\log Re_L)^{2.58}} \qquad for \; 10^7 < Re_L < 10^9$$

For other geometries, drag coefficients are listed in figures and tables (e.g., Figure 10-7, Figure 10-10, Figure 10-15, Table 10-1, Table 10-2, Table 10-3).

The *lift coefficient*, C_L, is defined as:

$$C_L = \frac{F_L}{\frac{1}{2}\rho \mathcal{V}^2 A}$$

Typical data for lift coefficients for airfoils are given in Figure 10-18 and Figure 10-20.

For some geometries and flow conditions, the conservation of mass and momentum equations can be solved for the velocity profile; from the velocity profile, the drag coefficient can be determined. Blasius obtained the first exact solution to the governing partial differential equations (Eq. 10-1, Eq. 10-2, Eq. 10-3) for flow over a flat plate. An approximate approach, which focuses on the integrated effects of flow in the near-wall region, is called a *momentum integral analysis*.

SELECTED REFERENCES

ANDERSON, J. D., *Computational Fluid Dynamics*, 6th ed., McGraw-Hill, New York, 1995.

ANDERSON, J. D., *Fundamentals of Aerodynamics*, 3rd ed., McGraw-Hill, New York, 2001.

BATCHELOR, G. K., *An Introduction to Fluid Mechanics*, Cambridge University Press, New York, 2000.

DOUGLAS, J. F., J. M. GASIOREK, and J. A. SWAFFIELD, *Fluid Mechanics*, 4th ed., Prentice Hall, Harlow, England 2001.

FOX, R. W., and A. T. McDONALD, *Introduction to Fluid Mechanics*, 5th ed., Wiley, New York, 1999.

MUNSON, B. R., D. F. YOUNG, and T. H. OKIISHI, *Fundamentals of Fluid Mechanics*, 4th ed., Wiley, New York, 2002.

POTTER, M. C., and D. C. WIGGERT, *Mechanics of Fluids*, 3rd ed., Brooks/Cole, Pacific Grove, CA, 2002.

SCHLICHTING, H., *Boundary Layer Theory*, 8th ed., Springer-Verlag, Berlin, New York, 1999.

VAN DYKE, M., *Album of Fluid Motion*, Parabolic Press, Stanford, CA, 1982.

WHITE, F. M., *Viscous Fluid Flow*, 2nd edition, McGraw-Hill, New York, 1991.

PROBLEMS

Problems designated with WEB refer to material available at www.wiley.com/college/kaminski

FLOW OVER FLAT PLATES

P10-1 Air at 23°C, 100 kPa with a free-stream velocity of 100 km/h flows along a flat plate. How long does the plate have to be to obtain a boundary layer thickness of 8 mm?

P10-2 A large cruise ship has length $L = 250$ m, beam (width) $W = 65$ m, and draft (depth) $D = 20$ m. It cruises at 25 km/h. Assume the flow over the hull can be approximated as that over a flat plate.

a. Estimate the total skin friction drag (in N) and the power (in kW) to propel the ship on a voyage in the Caribbean, where the water temperature is 28°C.

b. Estimate the total skin friction drag (in N) and the power (in kW) to propel the ship on a voyage to Alaska, where the water temperature is 4°C.

P10-3 A 5 mm × 1.5 m × 4 m plastic panel ($SG = 1.75$) is lowered from a ship to a construction site on a lake floor at

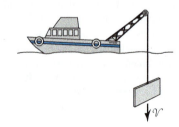

a rate of 1.5 m/s. Determine the tension in the cable lowering the panel

a. assuming the panel descends vertically with its wide end down (in N).

b. assuming the panel descends vertically with its narrow end down (in N).

P10-4 Your car has broken down on an interstate highway, and you are on the median strip between the lanes. By the time you decide you need to cross to the other side of the empty road, a stream of cars moving bumper to bumper at 120 km/h passes only 1 m away from you. For an air temperature of 25°C, determine the velocity (in km/h) of the wind that will hit you 10 seconds after the first car has passed.

Top view of cars

P10-5 A flat-bottomed river barge 60 m long and 12 m wide is towed through still water (at 25°C) at 10 km/h.

a. Determine the force required to overcome the drag (in N).

b. Determine the power required by the towboat (in kW).

c. Determine the boundary-layer thickness at the end of the barge (in mm).

P10-6 Outboard racing boats are designed for part of the hull to rise completely out of the water when a high speed is reached. Then the boat "planes" on the remainder of the hull. At 50 mi/h on 60°F water the area of the hull planing is 6 ft long and 5 ft wide. Determine the power required to overcome the friction drag (in hp).

P10-7 A ceiling fan has five thin blades, each 55 cm long and 15 cm wide. Assume the blades can be approximated as flat plates. Air temperature is 27°C. For rotational speeds of 50 rpm, 100 rpm, and 150 rpm, determine the power (in W) needed to overcome the drag force. (*Hint*: Because velocity varies with distance from the center of rotation, you must integrate the drag coefficient.)

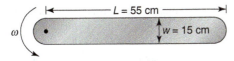

DRAG ON BLUFF BODIES

P10-8 The high-speed trains in France and Japan are streamlined to reduce drag forces. Consider a 120-m-long train whose outer surface can be approximated by a flat plate with a width of 10 m.

a. At 101 kPa, 20°C, determine the drag (in N) due to skin friction only and the power required (in kW) to overcome this drag at 100 km/h, 200 km/h, and 300 km/h.

b. The front of the train can be approximated by a hemisphere facing forward with a circumference of 15 m. Estimate the drag force caused by the front of the train and the power required to overcome it.

P10-9 In large electric power plants, cool water flows through condensers downstream of the steam turbines that drive the electric generators. This water is recirculated, so it is often cooled in cooling towers, such as shown in the figure. The design specification is that the tower must withstand a 100 mi/h wind at 70°F. Approximating the drag coefficient from information given in one of the tables, determine

a. the drag force on the tower (in lbf).

b. the moment that must be resisted by the foundation of the tower (in ft · lbf).

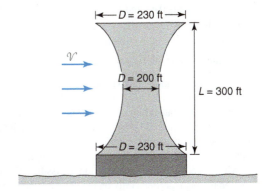

P10-10 A child releases a helium-filled balloon that is spherical in shape into 80°F, 14.7-psia air. If the balloon weighs 0.01 lbf and has a diameter of 1 ft, determine its terminal velocity.

P10-11 You hike to the top of a mountain and climb the fire tower. The wind is blowing at 80 km/h. The air temperature is 17°C, and the pressure is 94 kPa. Estimate the wind force (in N) that would act on you.

P10-12 In the western United States, empty boxcars are sometimes blown over by strong crosswinds. Shown in the figure are the dimensions of one type of a 20,000-kg boxcar. Determine the minimum wind velocity (in m/s and in mi/h) normal to the side of the boxcar needed to blow it over. Evaluate the air at 22°C, 101 kPa.

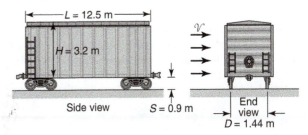

P10-13 In a bicycle race, a bicyclist coasts down a hill with a 7% grade to save energy. The mass of the bicycle and rider is 85 kg, the projected area is 0.22 m², and the drag coefficient is 0.9. Air temperature is 17°C. Neglecting rolling friction and bearing friction, determine

a. the maximum velocity if the air is still (in m/s).

b. the maximum speed if there is a headwind of 5 m/s (in m/s).

c. the maximum speed if there is a tailwind of 5 m/s (in m/s).

P10-14 A 2.5-cm sphere with a specific gravity of 0.25 is released into a fluid with a specific gravity of 0.71. The sphere rises at a terminal velocity of 0.5 cm/s. Determine the dynamic viscosity of the fluid (in N·s/m²).

P10-15 Many sports cars are convertibles. The airflow over such a car is significantly different depending on whether the convertible top is up or down. The engine of the 1000-kg car delivers 135 kW to the wheels, the car frontal area is 1.9 m², and rolling resistance is 2.5% of the car weight. The drag coefficient when the top is down is 0.43 and 0.31 when it is up. For 20°C air at 1 atm, determine

a. the maximum speed with the top up (in m/s).

b. the maximum speed with the top down (in m/s).

P10-16 Wind speed is measured with an anemometer. A home-made anemometer can be constructed from a thin plate hinged on one end; when the plate is hung from the hinge, wind imping-ing on the plate will cause the plate to rotate around the hinge. The angular deflection is a measure of the wind speed. For a brass plate 20 mm wide and 50 mm long, derive a relationship between wind speed and angular deflection θ. Assume that the drag force on the plate depends only on the velocity component normal to the surface for angles less than about 40° and that the air temperature is 25°C.

a. Determine the relationship between wind speed and angular deflection.

b. Determine the thickness of brass needed for $\theta = 30°$ at a wind speed of 60 km/h (in mm).

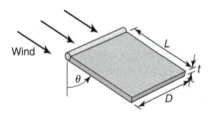

P10-17 A 70-kg bicycle racer in the Tour de France can main-tain about 40 km/h on a calm day over level ground. The bike has a mass of about 10 kg and has a rolling resistance of 1% of the weight of bicycle and rider. The drag coefficient of the bike and rider is 1.1, and their frontal area is 0.24 m². The air temperature is 25°C.

a. Determine the power output by the rider (in kW) on level ground.

b. Determine the velocity the rider could attain going up a hill that has a slope of 6° (in km/h).

P10-18 Assume the bicycle rider in Problem P10-17 adds a fairing to streamline his bike and body. The drag coefficient is reduced to 0.24, but the frontal area is increased to 0.29 m². From the power the rider can produce, estimate the new speed (in km/h) the rider can maintain on level ground.

P10-19 A parachutist controls her free-fall speed by falling spread-eagle ($C_D \approx 1.2$) to slow down or head down ($C_D \approx 0.4$) to speed up. The frontal areas in the two positions are about 0.70 m² and 0.25 m², respectively. For a 55-kg skydiver at 3000 m (assume the density and temperature are approximately constant at this elevation), determine

a. the terminal speed in each position (in km/h).

b. the time (in s) and distance (in m) to reach 95% of the terminal speed.

P10-20 In the United States, the Bonneville Salt Flats in Utah are used by individuals trying to set land-speed records in various classes of vehicles. One challenger has developed a 1750-lbf car that has a 675-hp engine, a streamlined body with a drag coefficient $C_D \approx 0.29$, a frontal area of 13.5 ft², and rolling resistance of only 3% of the body weight. The car's transmission has an efficiency of 88% (that is, 88% of the engine power is transferred to the tires). On a day when the air temperature is 95°F, determine the maximum speed of the car (in mi/h).

P10-21 A BMW 520 has a drag coefficient of 0.31 and a frontal area of 22.5 ft². It weighs 3,500 lbf. If rolling resistance is 1.5% of the weight, determine

a. the speed at which drag resistance becomes larger than the rolling resistance (in mi/h).

b. the power (in kW and hp) required to cruise at 45 mi/h and 75 mi/h.

P10-22 Some military jets deploy parachutes when they land to reduce the distance required to stop. Suppose a 14,500-kg jet uses two 6-m-diameter parachutes and lands at 300 km/h in 20°C air.

a. Determine the total force the cables connecting the parachutes to the plane must withstand (in N).

b. Determine the time (in s) and distance (m) required to deceler-ate the plane to 150 km/h (without using brakes and ignoring drag from the plane).

P10-23 In some automobiles, gas mileage (km/L) is calculated and displayed on the instrument panel. One day on a long drive, a bored engineering student realizes that his gas mileage is 20% lower traveling into a headwind than when there was no head-wind. The road is level, the temperature is 7°C, and his speed is 120 km/h. The driver (a car enthusiast) knows that the drag coefficient of his car is 0.35, frontal area is 2.1 m², mass is 950 kg, and rolling resistance is 3% of the body weight. To pass the time, he uses this information to calculate the headwind velocity. What is it (in km/h)?

P10-24 If you have ever been hit by a hailstone, you know it can hurt because of its high speed. Consider a 4-cm hailstone falling in 17°C, 96-kPa air. Assume the hailstone has a specific gravity of 0.84. Determine its terminal velocity (in m/s and mi/h)

a. for a smooth hailstone.

b. for a hailstone with a surface roughness similar to that of a golf ball.

P10-25 A beginning bicyclist can produce 84 W for short periods of time. On a hot day (32°C), how fast can the bicyclist travel if the projected area of the bike and cyclist is 0.5 m² and the drag coefficient is 1.1?

P10-26 A copper sphere 10 mm in diameter is dropped into a 1-m-deep drum of asphalt. The asphalt has a density of 1150 kg/m³ and a viscosity of 10^5 N·s/m². Estimate the time (in hours) it takes for the sphere to reach the bottom of the drum.

P10-27 A meteorological balloon is to be filled with helium at 0°C, 100 kPa. The surrounding air is at the same pressure and temperature. The instrument package the balloon must lift has a mass of 30 kg, and the balloon material has a mass of 0.15 kg/m². If an upward vertical velocity of 3 m/s is desired, what diameter (in m) balloon is required?

P10-28 In dry regions, wind storms can entrain much dust into the air. For a particle 0.05 mm in diameter with a density of 1.8 g/cm³ raised to a height of 100 m in such a storm, estimate how long it will take the particle to settle back to earth. Assume that the air is at 27°C, 100 kPa and that the time required to reach the terminal velocity is negligible.

P10-29 A 5-mm iron sphere is dropped into a tank of 17°C unused engine oil. Determine the sphere's terminal velocity (in cm/s).

P10-30 The military sometimes needs to move large equipment into remote areas where there are no landing strips, so the equipment is parachuted to the ground. To prevent damage, a bulldozer weighing 45 kN cannot strike the ground at a velocity greater than 10 m/s. Determine how many 20-m-diameter parachutes are required when the air is at 17°C, 95 kPa?

P10-31 A helium-filled spherical balloon is released into air at 40°F, 14.0 psia. The combined weight of the balloon and its payload is 300 lbf. If a vertical velocity of 10 ft/s is desired, what diameter balloon is required (in ft)? Assume the helium is at the same temperature and pressure as the air. If this balloon is tethered to the ground in a 10 mi/h wind, what angle does the restraining cable make with the ground?

P10-32 A 40-mm Ping Pong ball weighing 0.025 N is released from the bottom of a 4-m-deep swimming pool whose temperature is 20°C. Ignoring the time to reach terminal velocity, how long does the ball take to reach the pool surface (in s)?

P10-33 A 50 mi/h, 60°F wind blows perpendicular to an outdoor movie screen that is 70 ft wide and 35 ft tall; the screen is supported on 10-ft-tall pilings.

a. Estimate the drag force on the screen (in lbf).

b. Estimate the moment at the base of the pilings (in ft·lbf).

P10-34 A hotdog company decides to create a giant helium-filled balloon of a hotdog to float in parades for advertising purposes. It will float 75 ft above the street and will be controlled by people holding onto tethering lines. The balloon is 50 ft long and 10 ft in diameter and can be approximated as a cylinder. Air at 70°F, 14.7 psia is funneled down the street between the buildings at a velocity of 25 mi/h. Determine the drag force (in lbf).

P10-35 A telephone wire 5 mm in diameter is suspended between telephone poles spaced 50 m apart. If the wind velocity is 100 km/h and the air is at 2°C, 1 atm, determine the horizontal force (in N) the wire exerts on the poles.

P10-36 When parachuting, an Army Ranger and his gear may weigh as much as 250 lbf. To prevent injury, the Ranger's vertical landing speed must be less than 15 ft/s. If the parachute can be approximated as an open hemisphere and the air is at 70°F, 14.7 psia, what diameter (in ft) parachute is required?

P10-37 An office building, approximately 90 m wide and 150 m tall, is to be built in a new development far from any other building. Its drag coefficient is 1.4.

a. Determine the drag force (in N) if the wind at 17°C is uniform at 15 m/s.

b. Determine the drag force (in N) if the velocity profile can be approximated with the one-seventh power law $\mathcal{V}_x / \mathcal{V}_\infty = (y/\delta)^{1/7}$ with a boundary-layer thickness of 100 m and a free-stream velocity of 15 m/s (*Hint*: Integrate to obtain the total drag force.)

DRAG ON COMPOSITE BODIES

P10-38 The superintendent of a national cemetery wants to erect a larger than usual flagpole and flag. The flagpole is 125 ft tall. The flag has a height $H = 20$ ft and a length $L = 38$ ft. Assume the flagpole must withstand a wind of 60 mi/h at 32°F when the flag is flying. The drag coefficient of the flag based on area (LH), can be estimated by $C_D = 0.05L/H$. If the pole has a diameter of 9 in., determine

a. the total force exerted on the pole (in lbf).

b. the moment at the base of the pole (in ft·lbf).

P10-39 Antennas on old cars are vertical circular cylinders 0.25 in. in diameter and 4 ft long. Some people attach objects to the top of their antenna so that their car is more easily found in crowded parking lots. If the car is driven at 65 mi/h, and the air is at 80°F, 14.7 lbf/in.², determine

a. the bending moment (in ft·lbf) at the base of the antenna without the object attached.

b. the bending moment (in ft·lbf) at the base of the antenna if an object shaped like a sphere 3 in. in diameter is attached to the top of the antenna.

P10-40 The external rearview mirrors (two each) on old cars were circular disks 10 cm in diameter. New cars use streamlined rearview mirrors (two each) to reduce drag losses; these mirrors can be approximated as hemispheres facing upstream. A car without mirrors has a drag coefficient of 0.36, a frontal area of 1.5 m^2, and rolling resistance can be ignored. For a car speed of 125 km/h in air at 23°C, 100 kPa, what percent increase in gas mileage could be obtained by replacing the old mirrors with two new ones of the same diameter?

P10-41 Taxis carry advertising signs on their roofs to generate extra income for the operator. If the sign is a rectangular box 30 cm high, 1.2 m wide, and 1.2 m long, estimate the increased fuel cost caused by the addition of the sign. Assume the taxi is driven 100,000 km annually at an average speed of 50 km/h. Its engine cycle thermal efficiency is 25%. A reasonable average air condition is 10°C, 100 kPa. The fuel costs $0.40/L, its specific gravity is 0.82, and its energy content is 40,000 kJ/kg.

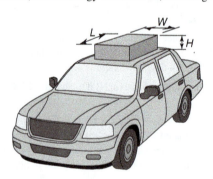

P10-42 A thin flat plate 10 ft long and 2 ft wide is mounted horizontally on a 10-ft-long, 3-in.-diameter pole. Air flows at 60°F, 14.7 psia along the 10-ft length of the plate. The velocity profile of the air flow varies from 0 at the base of the pole to 50 ft/s at 10 ft (along the top of the plate). Taking into account the variation in velocity, determine the total drag force (in lbf) acting on the composite body.

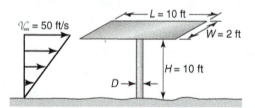

P10-43 A large family is going on a vacation in their minivan, which has a drag coefficient of 0.44 and a frontal area of 3.5 m^2. Because they need more room for their luggage, they will use a rectangular car-top carrier that is 1.5 m wide, 30 cm high, and

2 m long. Estimate the increase in power required to drive at 100 km/h with the car-top carrier compared to without it.

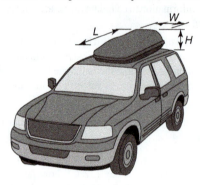

LIFT

P10-44 A small aircraft has a wing area of 27 m^2, a take-off mass of 2500 kg, a lift coefficient at takeoff of 0.49, and a drag coefficient at takeoff of 0.0074. For standard atmospheric conditions, determine

a. the takeoff speed at sea level (in km/h).

b. the power required at takeoff (in kW).

c. the maximum mass (in kg) possible at takeoff speed using the power from part b if the airport is at 2500 m.

P10-45 For a small plane, the lift coefficient at the landing speed is 1.15 and the maximum lift coefficient (at the stall speed) is 1.42. The landing speed of the airplane is 8 m/s faster than its stall speed. Determine both the landing and stalling speeds (in m/s).

P10-46 A 250-kg glider with a wing area of 22 m^2 has a minimum glide angle of 1.7°. Its lift coefficient is 1.1. For a still day at 15°C, 100 kPa, determine

a. the total horizontal distance for the glider to descend from 1500 m to sea level (in km).

b. the time required (in min).

P10-47 When a plane glides at its shallowest angle, lift, drag, and weight forces are all in equilibrium. Show that the glide slope angle, θ, is given by $\theta = \tan^{-1}(C_D/C_L)$.

P10-48 A hydrofoil is a watercraft that rides above the surface of the water on foils, which are essentially wings attached to the bottom of struts connecting the foils to the hull of the boat. Suppose the area of the foils in contact with the water on a 2000-kg hydrofoil is 1.1 m^2. Their lift and drag coefficients are 1.72 and 0.45, respectively.

a. Determine the minimum speed required for the foils to support the hydrofoil (in km/h).

b. Determine the power required to propel the hydrofoil at the speed calculated in part a (in kW).

c. the top speed if the boat has a 175-kW engine (in km/h). (Note that at higher speeds the hydrofoil rises farther out of the water and the lifting area is decreased.)

P10-49 Consider a U-2 reconnaissance plane that loses power at 35,000 ft over hostile territory. If its lift and drag characteristics are similar to those given in Figure 10-18, determine

a. the optimum angle of attack for maximum glide distance.

b. whether it can make it to an airport 425 miles away (Ignore initial velocity).

P10-50 If you have ever flown a kite in a strong wind, you know that the pull on the string can be quite strong. Consider a 1.2 m by 0.8 m kite that has a mass of 0.5 kg. Its lift coefficient can be approximated by $C_L = 2\pi \sin \alpha$ where α is the angle of attack. In a 45 km/h wind, the kite's string has an angle of 50° to the horizontal and the kite has an angle of attack of 5°. Determine the force on the string (in N).

P10-51 A small experimental plane has a mass of 750 kg, a drag coefficient of 0.063, and a lift coefficient of 0.4. In level flight, it is flown at 175 km/h. For standard conditions (1 atm, 25°C), determine the effective lift area of the plane (in m^2).

P10-52 Because of the decrease in density and temperature with increasing elevation in the atmosphere, lift and drag forces change. Consider a plane flying at velocity $\mathcal{V}$ at sea level. For the same lift and drag coefficients, determine

a. the speed required at 10,000 m to generate the same lift force.

b. the change in drag force.

P10-53 When planes take off and land at airports at higher elevations, the lower-density air (due to reduced atmospheric pressure) must be taken into account because of its effect on lift and drag. If a plane requires 15 s to reach its takeoff speed of 220 km/h at sea level in a distance of 500 m, estimate for an airport at 2000 m

a. the takeoff speed (in km/h).

b. the takeoff time (in s).

c. the additional runway length required (in m). Assume the same constant acceleration for both cases.

P10-54 An airplane is to be designed that will fly at 650 km/h at an altitude where the density is 0.655 kg/m^3 and the kinematic viscosity is 2×10^{-5} m^2/s. A one-fifteenth scale model (that is, the model is geometrically similar to the prototype but is one-fifteenth its size) is built for use in a wind tunnel whose velocity is 650 km/h. The air in the wind tunnel is at 55°C, and viscosity is independent of pressure.

a. Determine the test section pressure (in kPa) so that the model data are useful in designing the prototype. (*Hint:* Match Reynolds numbers.)

b. Determine the relation between the drag on the prototype and that on the model.

P10-55 A 1 m by 1.5 m plate moves through still water at 3 m/s and is at an angle of 12° to the velocity vector. For this situation the drag coefficient is 0.17 and the lift coefficient is 0.72.

a. Determine the resultant force on the plate (in N).

b. Determine the angle at which this force acts on the plate.

c. Determine the power required to move the plate (in kW).

d. Determine the drag force (in N) and power (in kW) if the plate moves through 20°C air instead of water.

MOMENTUM INTEGRAL EQUATION

P10-56 (WEB) An expression for the laminar velocity profile on a flat plate is

$$\mathcal{V}_x = C_1 \sin(C_2 y) + C_3$$

where the argument of the sine function is in radians. Using the three common physical conditions that the velocity profile should satisfy, determine

a. the constants C_1, C_2, and C_3.

b. the nondimensional velocity profile.

c. the boundary thickness (δ/x) as a function of length.

d. the skin friction coefficient, C_f, as a function of length.

P10-57 (WEB) The velocity distribution in a laminar boundary layer on a smooth flat plate is given by $\mathcal{V}_x/\mathcal{V}_\infty = 3(y/\delta) - 2(y/\delta)^2$. Develop an expression for the drag coefficient.

P10-58 (WEB) Measurements from flow over a flat plate result in the turbulent velocity profile $\mathcal{V}_x/\mathcal{V}_\infty = (y/\delta)^{1/9}$ and skin friction coefficient $C_f = 0.046/Re_x^{1/5}$. Develop a relationship that describes the growth of the boundary-layer thickness. (Note that Eq. 10-8 cannot be used.)

P10-59 (WEB) Assume a cubic velocity profile for flow over a flat plate of the form $\mathcal{V}_x/\mathcal{V}_\infty = C_0 + C_1\eta + C_2\eta^2 + C_3\eta^3$ where $\eta = y/\delta$. The profile should satisfy the three common physical conditions given in the momentum integral material located on the web. A fourth condition can be determined from the differential momentum equation and is: at $y = 0$, $d^2\mathcal{V}_x/dy^2 = 0$.

a. Determine the constants C_0, C_1, C_2, and C_3.

b. Determine the boundary thickness (δ/x) as a function of length.

c. Determine the skin friction coefficient, C_f, as a function of length.

CHAPTER 11

CONDUCTION HEAT TRANSFER

11.1 INTRODUCTION

Conduction heat transfer occurs in solids, liquids, and gases. A brief discussion of the mechanism governing this phenomenon is given in Chapter 3, where we addressed simple, one-dimensional situations. In this chapter, we develop a more general view of conduction in solids. Consider several examples:

- In hot climates, we cool the inside of buildings; in cold climates, we require heating. In both cases, continuous cooling or heating is needed because of heat gains or losses through the walls and roofs of the buildings. The heat transfer rate through the walls and roofs depends on conduction.

- To cool an engine block, conduction through the solid material is needed to transfer heat from the cylinder (where combustion takes place) through the engine block to the coolant in the cooling channels.

- Whenever a dam is constructed, because the curing of the concrete is exothermic (heat generating), attention must be paid to the transient temperature distribution through the concrete so that thermally induced stresses do not crack the dam.

- Computer chips are constructed of many different materials and must be connected to the rest of the computer through interconnects on the boards. Thermal expansion effects must be considered because component materials have different coefficients of thermal expansion. To determine the thermal stresses, the temperature distribution throughout the chip must be determined.

- Favorable metal properties (such as a very hard surface and more resilient core) can be obtained by the rapid immersion of a hot metal into a cooler fluid. The rate of cooling and its penetration depth affect the growth of a favorable grain structure in the metal, which dictates the strength, hardness, and so on of the metal.

Part of the analysis of each of these examples involves conduction heat transfer. The objective of the analysis may be to determine the steady-state heat transfer rate or temperature distribution through the solid, or we may want the transient (time-dependent) temperature distribution or heat transfer rate in the solid. We can accomplish all these tasks with the appropriate application of a single equation—the heat conduction equation—which is developed in the next section. Before we develop the conduction equation, we want to reconsider the heat transfer rate when it occurs in more than one direction.

In Chapter 3, we briefly examined Fourier's law for steady one-dimensional conduction:

$$\dot{Q} = -kA\frac{dT}{dx} \tag{11-1}$$

When we divide the heat transfer rate, $\dot{Q}$, by the surface area, A, we obtain the *heat flux*, q'':

$$q'' = \frac{\dot{Q}}{A} = -k\frac{dT}{dx} \tag{11-2}$$

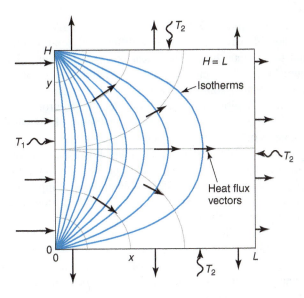

FIGURE 11-1 Lines of constant temperature in a solid body. (*Source*: Adapted from A. Bejan, *Heat Transfer*, 2nd ed., Wiley, New York, 1993. Used with permission.)

Heat flux is often used when we need the heat transfer rate per unit area at a specific location on a surface, particularly if the heat flux varies over a surface. In Eq. 11-2, temperature varies in only one direction, so an ordinary differential is used to describe the temperature gradient, dT/dx.

However, there are applications in which temperature varies in more than one direction. Consider the simple two-dimensional situation shown in Figure 11-1. One side of this rectangle is held at one temperature, T_1, and the three other sides are held at a second temperature, T_2, $T_1 > T_2$. The lines shown in the figure indicate lines of constant temperature (*isotherms*). From Eq. 11-2, the heat flux direction is aligned with the temperature gradient, so in Figure 11-1 the arrows drawn perpendicular to the isotherms represent the heat fluxes at those locations. Where the isotherms are more closely spaced, the heat flux is greater. Hence, heat flux has both magnitude and direction; it is a vector quantity. For a three-dimensional temperature field, we can write the heat flux vector ($\bar{q}''$) in terms of its x-, y-, and z-components, (q_x'', q_y'', q_z''), respectively:

$$\bar{q}'' = q_x''\hat{i} + q_y''\hat{j} + q_z''\hat{k} \tag{11-3}$$

where $\hat{i}$, $\hat{j}$, and $\hat{k}$ are unit direction vectors. Now the one-dimensional expression for heat flux (Eq. 11-2) is no longer valid. Instead, we must use partial differentials to describe the heat flux in the three directions to obtain Fourier's law in multidimensions:

$$q_x'' = -k\frac{\partial T}{\partial x} \qquad q_y'' = -k\frac{\partial T}{\partial y} \qquad q_z'' = -k\frac{\partial T}{\partial z} \tag{11-4}$$

11.2 THE HEAT CONDUCTION EQUATION

The conservation of energy equation we have used is an integral expression that applies to a control volume. We treat the volume as a blackbox and assume its properties are uniform throughout the volume. We infer what occurs in the box from what crosses its boundaries. In the above examples, temperature is not uniform. Hence, we need a new approach to tackle such problems.

The development of the *heat conduction equation* begins with application of the conservation of energy equation to a differential control volume. For example, perhaps we want to evaluate the transient heat loss from a steam pipe to the ground, as shown on Figure 11-2a. Heat spreads in the x-, y-, and z-directions and changes with time. We define a differential control volume ($\Delta x \times \Delta y \times \Delta z$) that is representative of all such volumes in the region of interest. Consider this control volume centered about a point $(x + \Delta x/2, y + \Delta y/2, z + \Delta z/2)$ as shown in Figure 11-2b. We analyze the heat transfer on the boundaries of this differential control volume, as well as energy storage and (if present) energy generation in the volume. Then we shrink the volume to zero size, which results in a partial differential equation describing the conservation of energy at a point. When we apply this equation to a specific problem, invoke simplifying assumptions, and impose appropriate boundary conditions, the solution gives the temperature field (distribution with space and/or time) in the material. Once this distribution is known, then the heat transfer rate at any point in the material or on its surface can be computed from Fourier's law.

The differential volume shown in Figure 11-2b contains a homogenous material; that is, all the material properties (density, thermal conductivity, specific heat) are uniform throughout the volume. In addition, the temperature is uniform. No mass flows through any of the faces of the volume. In the most general situation, there might be energy generation (e.g., joulean heating from the flow of electricity through the material, nuclear reaction,

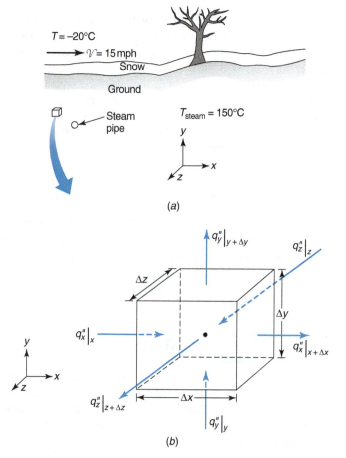

(a)

(b)

FIGURE 11-2
Differential control volume for conduction heat transfer.

chemical reaction). Likewise, if the material temperature changes with time, there is energy storage or depletion. We now apply conservation of energy to the volume:

$$\frac{dE}{dt} = \dot{Q} - \dot{W} \tag{11-5}$$

Our task is to evaluate each of these three terms as they apply to this differential control volume.

For the energy storage term, dE/dt, we neglect potential and kinetic energy and assume we have an ideal solid ($c \approx c_v \approx c_p$), so that

$$\frac{dE}{dt} = \frac{dU}{dt} = m\frac{du}{dt} = mc_p\frac{\partial T}{\partial t} = \rho V c_p \frac{\partial T}{\partial t} = \rho\,(\Delta x\,\Delta y\,\Delta z)\,c_p \frac{\partial T}{\partial t} \tag{11-6}$$

Note that the ordinary derivative on the left-hand side of Eq. 11-6 is changed to a partial derivative on the right-hand side, because now temperature can vary with position and time.

The heat transfer rate, $\dot{Q}$, is the net heat transfer by conduction into and out of the six faces of the volume. Consider the net heat conduction rate in the x-direction. For heat conduction into the left face of the volume, we multiply the heat flux at that face, $q_x''|_x$, by the area of that face, $\Delta y\,\Delta z$. The symbol $|_x$ indicates that the quantity q_x'' is evaluated at location x. In a similar manner, we can obtain the heat conduction out of the right face, so that the net heat conduction in the x-direction is

$$\dot{Q}_x = q_x''|_x\,(\Delta y\,\Delta z) - q_x''|_{x+\Delta x}\,(\Delta y\,\Delta z) \tag{11-7}$$

The net heat conduction rates in the y- and z-directions use the appropriate heat fluxes and areas on the remaining four faces of the volume, so that the net heat conduction rate in all directions is

$$\begin{aligned}
\dot{Q}_x &+ \dot{Q}_y + \dot{Q}_z \\
&= \left[q_x''|_x\,(\Delta y\,\Delta z) - q_x''|_{x+\Delta x}\,(\Delta y\,\Delta z) \right] + \left[q_y''|_y\,(\Delta x\,\Delta z) - q_y''|_{y+\Delta y}\,(\Delta x\,\Delta z) \right] \\
&+ \left[q_z''|_z\,(\Delta x\,\Delta y) - q_z''|_{z+\Delta z}\,(\Delta x\,\Delta y) \right]
\end{aligned} \tag{11-8}$$

The energy generation term is equivalent to the power term, $\dot{W}$. (Remember, for example, electric power is considered to be included in this term.) We let q''' represent a volumetric heat generation rate (W/m³), so that the total energy generation in the volume is given by

$$\dot{W} = -q'''\Delta x\,\Delta y\,\Delta z \tag{11-9}$$

The negative sign is needed because, by definition, $\dot{W}$ is positive when the system does work.

Substituting Eq. 11-6, Eq. 11-8, and Eq. 11-9 into Eq. 11-5, we obtain

$$\begin{aligned}
\rho(\Delta x\,&\Delta y\,\Delta z)\,c_p \frac{\partial T}{\partial t} \\
&= \left[q_x''|_x\,(\Delta y\Delta z) - q_x''|_{x+\Delta x}\,(\Delta y\,\Delta z) \right] + \left[q_y''|_y\,(\Delta x\,\Delta z) - q_y''|_{y+\Delta y}\,(\Delta x\,\Delta z) \right] \\
&+ \left[q_z''|_z\,(\Delta x\,\Delta y) - q_z''|_{z+\Delta z}\,(\Delta x\,\Delta y) \right] - (-q'''dx\,dy\,dz)
\end{aligned} \tag{11-10}$$

We divide Eq. 11-10 by the volume, $\Delta x \, \Delta y \, \Delta z$, gather like terms, and rearrange to obtain

$$\left[\frac{q''_x|_x - q''_x|_{x+\Delta x}}{\Delta x}\right] + \left[\frac{q''_y|_y - q''_y|_{y+\Delta y}}{\Delta y}\right] + \left[\frac{q''_z|_z - q''_z|_{z+\Delta z}}{\Delta z}\right] + q''' = \rho c_p \frac{\partial T}{\partial t} \quad (11\text{-}11)$$

Note that the first three bracketed terms on the left-hand side of Eq. 11-11 are definitions of derivatives.

We take the limit as the volume shrinks uniformly to zero size, so that $\Delta x \to 0$, $\Delta y \to 0$, and $\Delta z \to 0$ simultaneously and obtain

$$-\frac{\partial q''_x}{\partial x} - \frac{\partial q''_y}{\partial y} - \frac{\partial q''_z}{\partial z} + q''' = \rho c_p \frac{\partial T}{\partial t} \quad (11\text{-}12)$$

The negative signs arise due to consistent use of the positive direction in the derivatives.

The heat fluxes in the three directions are expressed with Fourier's law (Eq. 11-4), so that

$$\frac{\partial}{\partial x}\left(k\frac{\partial T}{\partial x}\right) + \frac{\partial}{\partial y}\left(k\frac{\partial T}{\partial y}\right) + \frac{\partial}{\partial z}\left(k\frac{\partial T}{\partial z}\right) + q''' = \rho c_p \frac{\partial T}{\partial t} \quad (11\text{-}13)$$

which is the general form, in Cartesian coordinates, of the heat conduction equation. The thermal conductivity, k, cannot be removed from the brackets, because in the most general situation, the thermal conductivity is a function of temperature (i.e., a function of position).

This equation is the basic tool for heat conduction analysis. Both steady and nonsteady (transient) problems can be solved with the equation. From its solution, we obtain a temperature distribution. If the problem is one-dimensional and steady, we would obtain $T = T(x)$. For a three-dimensional and transient problem, we would obtain $T = T(x, y, z, t)$. Using the temperature distribution (or temperature field) and Fourier's law, we can determine the heat flux at any location.

It is important to remember the physical significance of this partial differential equation (Eq. 11-13). It is simply a statement of the conservation of energy at a point. What it says is: at any point in a medium the net conduction heat transfer (per unit volume) into the volume plus the volumetric heat generation rate must equal the time rate of change of energy storage (per unit volume) within the volume. The first three terms on the left-hand side of the equation represent the net conduction (per unit volume) in the x-, y-, and z-directions, respectively. The fourth term is analogous to the work term (per unit volume) in Eq. 11-5. The right-hand side is (on a per unit volume basis) identical to the right-hand side of Eq. 11-5 (ignoring potential and kinetic energy effects). Similar analyses can be performed for conduction in cylindrical and spherical coordinate systems. Below are the general equations in those coordinates:

$$\boxed{\frac{1}{r}\frac{\partial}{\partial r}\left(kr\frac{\partial T}{\partial r}\right) + \frac{1}{r^2}\frac{\partial}{\partial \theta}\left(k\frac{\partial T}{\partial \theta}\right) + \frac{\partial}{\partial z}\left(k\frac{\partial T}{\partial z}\right) + q''' = \rho c_p \frac{\partial T}{\partial t} \quad \text{cylindrical}} \quad (11\text{-}14)$$

where r is in the radial direction, θ is in the circumferential direction, and z is in the axial direction.

$$\boxed{\frac{1}{r^2}\frac{\partial}{\partial r}\left(kr^2\frac{\partial T}{\partial r}\right) + \frac{1}{r^2 \sin^2\theta}\frac{\partial}{\partial \phi}\left(k\frac{\partial T}{\partial \phi}\right) + \frac{1}{r^2 \sin\theta}\frac{\partial}{\partial \theta}\left(k\sin\theta\frac{\partial T}{\partial \theta}\right) + q''' = \rho c_p \frac{\partial T}{\partial t}}$$

$$\text{spherical}$$

$$(11\text{-}15)$$

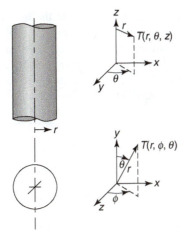

FIGURE 11-3 Directions in cylindrical and spherical coordinates.

where r is in the radial direction, θ is in the polar direction, and ϕ is in the azimuthal direction. The relationship of the cylindrical and spherical coordinate systems to the rectangular coordinate system is shown on Figure 11-3.

As with any differential equation, an appropriate number of boundary and initial conditions are needed to complete the solution. A *boundary condition* is a known physical condition imposed on the boundaries of the volume or area being analyzed and is a restriction that must be satisfied by the solution. An *initial condition* is a known physical condition imposed at a fixed time, after which the transient behavior of the system is desired. The number of boundary or initial conditions required is determined from the differential equation. For example, if the equation is second-order in x (i.e., $\partial^2 T / \partial x^2$), then two x boundary conditions are needed; if the equation is first-order in time (i.e., $\partial T / \partial t$), then one initial condition is required. Three common boundary conditions used with the heat conduction equation are given below.

The boundary condition of the *first kind* (or a *Dirichlet condition*) occurs when the temperature is prescribed on a boundary surface. One of the more common conditions is a surface with a constant temperature. For example, if the boundary is at $x = 0$:

$$T(0,t) = T_s \qquad \text{constant surface temperature} \qquad (11\text{-}16)$$

The temperature T_s must be a known quantity, and it must be fixed at T_s for all times, t. Physically, this mathematical boundary condition is approximated when a constant-pressure boiling or condensing fluid with a very large heat transfer coefficient touches the solid surface of a body. Consider the equation $q'' = h\,\Delta T$, where ΔT is the difference between the fluid and the surface. For a finite heat transfer rate, the temperature difference must approach zero if the heat transfer coefficient is very large, so the surface temperature remains constant. In general, the prescribed surface temperature could be constant, a known function of location, or a known function of time.

A boundary condition of the *second kind* (or a *Neumann condition*) occurs when heat flux is prescribed on the boundary. For example, if a constant heat flux, q''_s, is applied at a boundary at $x = 0$, then

$$-k\left.\frac{\partial T}{\partial x}\right|_{x=0} = q''_s \qquad \text{constant surface heat flux} \qquad (11\text{-}17)$$

Fourier's law is used on the left-hand side to denote the heat conducted into the solid at the boundary. Heat conducted into the solid is balanced by the heat applied to the boundary, q_s''. This boundary condition is approximated when an electric resistance heater is pressed against a body. A special case of this boundary condition is the perfectly insulated (or adiabatic) surface. In Eq. 11-17, when the heat flux is set equal to zero, the thermal conductivity can be removed from the right-hand side of the equation, so that

$$\left.\frac{\partial T}{\partial x}\right|_{x=0} = 0 \qquad \text{adiabatic surface} \tag{11-18}$$

In general, the prescribed surface heat flux could be constant, a known function of location, or a known function of time.

A boundary condition of the *third kind* (or a *convection boundary condition*) occurs when a fluid at T_f (which could be constant, a known function of location, or a known function of time) flows past a body whose surface is at a different temperature, $T(0,t)$:

$$-k \left.\frac{\partial T}{\partial x}\right|_{x=0} = h\left[T_f - T(0,t)\right] \qquad \text{convection at surface} \tag{11-19}$$

We obtain Eq. 11-19 by performing an energy balance at the surface of the body. Heat conducted out of or into the wall (the left-hand side of the equation) is balanced by the heat convected into or out of the fluid (the right-hand side of the equation).

If we make simplifying assumptions in the specification of the geometry, boundary conditions, material properties, and so on, the partial differential heat conduction equation can often be reduced in order and solved analytically. Sometimes the resulting equation is still a partial differential equation, and an infinite series solution can be determined. In other situations, the equation can be reduced such that the temperature becomes a function of only one variable, and the partial differential equation is changed into an ordinary differential equation that can be solved by simple integration. In the next sections, solutions to the governing heat conduction equation and the use of the results are demonstrated.

11.3 STEADY ONE-DIMENSIONAL CONDUCTION

Consider heat transfer through the plane wall shown in Figure 11-4. This may represent a wall in a house, a window, or some other large flat plane with uniform known and constant temperatures on each face. The material properties are all specified. We want to determine the temperature distribution in the wall and the heat transfer through the wall.

To begin, we assume the wall is very large and the temperature at every point inside the wall is constant (i.e., steady state). Furthermore, we assume that the thermal conductivity of the material is constant and that there is no internal heat generation. We use this information and our assumptions to modify the general conduction equation. We work in Cartesian coordinates, so the applicable form of the equation is Eq. 11-13. Because the wall is very large and the temperatures are uniform in the y- and z-directions, there cannot be any conduction in the y- or z-directions, so we eliminate the second and third terms on the left-hand side of the equation. The fourth term is eliminated because there is no internal heat generation, and the steady-state assumption sets the right-hand side equal to zero, so

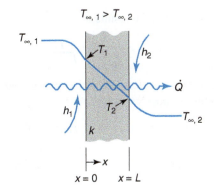

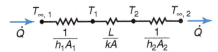

FIGURE 11-4   Heat transfer through a plane wall.

Eq. 11-13 reduces to

$$\frac{\partial}{\partial x}\left(k\frac{\partial T}{\partial x}\right) = 0 \tag{11-20}$$

Thermal conductivity is constant and is removed from inside the brackets. Now temperature is only a function of x, so we can change the partial differentials to ordinary differentials to obtain

$$\frac{d^2T}{dx^2} = 0 \tag{11-21}$$

Because this equation is second-order in x, we need two boundary conditions to solve the problem. Using the two specified surface temperatures,

$$\begin{aligned} \text{at} \quad x = 0, \quad T = T_1 \\ \text{at} \quad x = L, \quad T = T_2 \end{aligned} \tag{11-22}$$

Integrating Eq. 11-21 once gives us

$$\frac{dT}{dx} = C_1 \tag{11-23}$$

Integrating Eq. 11-23 gives us

$$T = C_1 x + C_2 \tag{11-24}$$

Applying the boundary conditions to Eq. 11-24, for $x = 0$ we obtain

$$T_1 = C_2 \tag{11-25}$$

and for $x = L$ we obtain

$$T_2 = C_1 L + C_2 = C_1 L + T_1 \tag{11-26}$$

Combining Eq. 11-25 and Eq. 11-26 results in

$$C_1 = \frac{T_2 - T_1}{L} \tag{11-27}$$

Substituting the two constants into Eq. 11-24 gives the temperature distribution in this wall subject to the restrictions/assumptions we imposed:

$$T(x) = (T_2 - T_1)\frac{x}{L} + T_1 \tag{11-28}$$

From this linear temperature distribution, using Fourier's law we obtain the heat transfer rate:

$$\dot{Q} = -kA\frac{dT}{dx} = -kA\frac{T_2 - T_1}{L} = \frac{T_1 - T_2}{\left[\dfrac{L}{kA}\right]} \tag{11-29}$$

Note that we obtain the heat transfer rate in terms of the thermal resistance, as used in Chapter 3.

Another important geometry involves tubes, as shown in Figure 11-5, which can be heat exchanger tubes, steam pipes, and other common applications. The inside and outside wall temperatures are known, as are the geometry and tube properties. The approach to

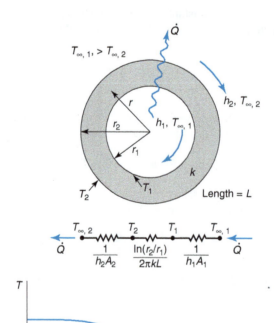

FIGURE 11-5 Heat transfer through the wall of a tube.

this problem is similar to that for the plane wall, because T is only a function of r. The steady-state temperature distribution and heat transfer rate are sought. We assume the tube is very long (so that the wall temperature varies only in the radial direction), has constant thermal conductivity, and has no internal heat generation. Applying this information to Eq. 11-14, we obtain

$$\frac{1}{r}\frac{\partial}{\partial r}\left(kr\frac{\partial T}{\partial r}\right) = 0 \qquad (11\text{-}30)$$

The partial derivative can be converted to an ordinary derivative because T depends only on the radial position. Thermal conductivity can be removed from inside the brackets but r cannot because it is a variable. Hence, the equation becomes

$$\frac{d}{dr}\left(r\frac{dT}{dr}\right) = 0 \qquad (11\text{-}31)$$

subject to the two boundary conditions:

$$\begin{aligned} \text{at} \quad r = r_1, \quad T = T_1 \\ \text{at} \quad r = r_2, \quad T = T_2 \end{aligned} \qquad (11\text{-}32)$$

Integration of Eq. 11-31 once gives us

$$r\frac{dT}{dr} = C_1 \qquad (11\text{-}33)$$

Separating variables, we get

$$dT = \frac{C_1}{r}\,dr \qquad (11\text{-}34)$$

Integrating Eq. 11-34, we obtain the general solution

$$T(r) = C_1 \ln r + C_2 \qquad (11\text{-}35)$$

Following a procedure similar to that used for the plane wall, we can evaluate the two constants in Eq. 11-35 by application of the two boundary conditions (Eq. 11-32), and the temperature distribution in the tube is

$$T(r) = \frac{T_1 - T_2}{\ln\left(r_1/r_2\right)} \ln\left(\frac{r}{r_2}\right) + T_2 \qquad (11\text{-}36)$$

Using Fourier's law, we can determine the heat transfer rate (for a tube of length L):

$$\dot{Q} = -kA_1 \left.\frac{dT}{dr}\right|_1 = -kA_2 \left.\frac{dT}{dr}\right|_2 \qquad (11\text{-}37)$$

Evaluating the area and the temperature gradient at the inner surface 1 or at the outer surface 2, we substitute the expressions into Eq. 11-37 to obtain

$$\boxed{\dot{Q} = \frac{2\pi kL\,(T_1 - T_2)}{\ln\left(r_2/r_1\right)} = \frac{(T_1 - T_2)}{\left[\dfrac{\ln\left(r_2/r_1\right)}{2\pi kL}\right]}} \qquad (11\text{-}38)$$

Again, the heat transfer rate is described by a driving temperature potential and a thermal resistance.

A similar analysis can be performed for heat conduction through a spherical shell. With the same assumptions as used for the previous two analyses, the heat transfer rate is

$$\dot{Q} = \frac{4r_1r_2\pi k\,(T_1 - T_2)}{r_2 - r_1} = \frac{(T_1 - T_2)}{\left[\dfrac{r_2 - r_1}{4r_1r_2\pi k}\right]} \qquad (11\text{-}39)$$

EXAMPLE 11-1   One-dimensional conduction

A nuclear fuel rod assembly, consisting of an outer cladding and the inner nuclear material, has an outside diameter of 75 mm. The outer cladding is 10 mm thick and is made of a material with a thermal conductivity of 3.2 W/m·K. The nuclear reaction generates 60,000 W/m³ uniformly in the inner nuclear material. The outside of the assembly is surrounded by water at 300°C, and the convection coefficient is 100 W/m²·K.

a) Determine the temperature at the assembly surface (in °C).

b) Determine the temperature at the interface between the inner nuclear material and the outer cladding (in °C).

Approach:

The given information is shown on the schematic.

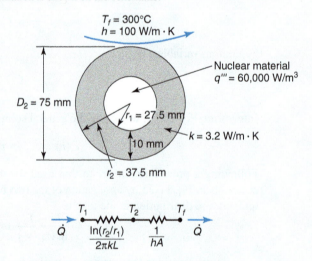

From the thermal circuit, we can see that the outside surface temperature, T_2, can be calculated from the basic rate equation applied across the convective resistance at the outer surface of the cladding, $\dot{Q} = \Delta T/R$. The temperature difference is $\Delta T = T_2 - T_f$, and the thermal resistance is $R = 1/(hA) = 1/(h\pi 2r_2 L)$. Total power produced by the nuclear fuel is calculated using the volume of the fuel and the volumetric heat generation rate. Finally, the interface temperature is calculated from the basic rate equation, too, but now applying it across the cladding material.

Assumptions:

A1. The system is steady.

Solution:

a) The total power generated by the nuclear reaction is $\dot{Q} = q'''V = q'''A_x L = q'''\pi r_1^2 L$. Assuming a steady system [A1], this heat transfer rate equals the heat leaving by convection at the outer surface of the assembly:

$$\dot{Q} = \frac{T_2 - T_f}{(1/hA)} = \frac{T_2 - T_f}{(1/h\pi 2r_2 L)} = q'''\pi r_1^2 L$$

This expression can be solved for T_2:

$$T_2 = T_f + \frac{q''' r_1^2}{h 2 r_2}$$

Note that the length cancels, and the variables in the above equation are given in the problem statement, so that

$$T_2 = T_f + \frac{q''' r_1^2}{h 2 r_2} = 300°C + \frac{(60,000 \, \text{W/m}^3)(0.0275 \, \text{m})^2}{(100 \, \text{W/m}^2 \cdot \text{K}) \, 2 \, (0.0375 \, \text{m})} = 306.1°C$$

b) The temperature at the interface between the fuel and the cladding can also be calculated with the basic rate equation if we now apply it across the cladding material. Assuming steady, one-dimensional heat transfer with constant properties and no internal heat generation [A1][A2][A3][A4] and referring to the thermal circuit:

A2. Heat transfer is one-dimensional.
A3. All properties are constant.
A4. There is no internal heat generation in the outer cladding.

$$\dot{Q} = \frac{\Delta T}{R} = \frac{T_1 - T_2}{\left[\dfrac{\ln(r_2/r_1)}{2\pi k L}\right]}$$

where the thermal resistance of the cladding is for a circular tube. Incorporating the power expression into the last equation and solving for the interface temperature, T_1:

$$T_1 = T_2 + \frac{q''' r_1^2 \ln(r_2/r_1)}{2k}$$

Again, the length cancels and all the needed information is given, so that

$$T_1 = T_2 + \frac{q''' r_1^2 \ln(r_2/r_1)}{2k}$$

$$= 306.1°C + \frac{(60,000 \, \text{W/m}^3)(0.0275 \, \text{m})^2 \ln(0.0375/0.0275)}{2 \, (3.2 \, \text{W/m} \cdot \text{K})} = 308.3°C$$

Comments:

This is a straightforward application of basic conduction and thermal resistance concepts.

EXAMPLE 11-2 **Conduction with variable thermal conductivity**

A wafer of silicon 3 mm thick and 2 cm square is used in an electronic device. One side of the device is held at 85°C and the other is held at 25°C. The thermal conductivity of silicon varies with temperature as $k = k_0(1 + BT)$, where $k_0 = 175 \, \text{W/m} \cdot \text{K}$, $B = 0.00556°C^{-1}$, and T is in °C.

a) Determine the heat transfer rate (in W) if the thermal conductivity is evaluated at its average temperature.

b) Determine the heat transfer rate (in W) if the temperature dependence of thermal conductivity is formally taken into account in the governing differential equation.

Approach:

A schematic of the problem is provided here. Because the thickness of the silicon wafer is small relative to the width and length, we assume the heat transfer is one-dimensional through the wafer (i.e., edge effects are negligible). When thermal conductivity is assumed constant, as in part a, the rate equation across a plane wall is $\dot{Q} = \Delta T/R$, where $\Delta T = T_1 - T_2$ and $R = t/(kA) = t/(k_{avg}LW)$. The thermal conductivity, k_{avg}, is evaluated at the average temperature of the silicon,

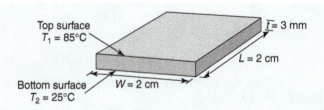

Top surface
$T_1 = 85°C$

$t = 3$ mm

$L = 2$ cm

Bottom surface
$T_2 = 25°C$

$W = 2$ cm

$T_{avg} = (T_1 + T_2)/2$, where T_{avg} is substituted into the expression for thermal conductivity, $k = k_0(1 + BT)$. Everything is known, so the heat transfer rate can be calculated.

When thermal conductivity is not constant, then we cannot use the expression for the wall resistance given above. Rather, we must return to the governing differential equation and incorporate the expression for thermal conductivity as a function of temperature.

Assumptions:

Solution:

a) To determine the heat transfer rate using the thermal conductivity evaluated at the average temperature, we first calculate the average temperature:

$$T_{avg} = \frac{T_1 + T_2}{2} = \frac{85 + 25}{2} = 55°C$$

The thermal conductivity at the average temperature is

$$k_{avg} = k_0 \left(1 + BT_{avg}\right) = 175 \, \frac{W}{m \cdot K} \left(1 + \left[0.00556 \, \frac{1}{°C}\right] 55°C\right) = 228.5 \, \frac{W}{m \cdot K}$$

A1. The system is steady.
A2. Heat transfer is one dimensional.
A3. All properties are constant in part (a).
A4. There is no internal heat generation.

Assuming steady one-dimensional heat transfer with constant properties and no internal heat generation [A1][A2][A3][A4], the heat transfer rate is

$$\dot{Q} = \frac{\Delta T}{R} = \frac{T_1 - T_2}{\left(\dfrac{t}{k_{avg}LW}\right)} = \frac{(85 - 25)°C}{\left[\dfrac{0.003 \text{ m}}{(228.5 \text{ W/m·K}) (0.02 \text{ m}) (0.02 \text{ m})}\right]} = 1,828.12 \, W$$

Two decimal places are shown for comparison to the results in part b.

b) To incorporate the temperature dependence of thermal conductivity for conduction in a plane wall, we begin with Eq. 11-20. We assume steady one-dimensional heat transfer with no internal heat generation [A1], [A2], [A4]. Recognizing that temperature varies only in the x-direction,

$$\frac{d}{dx}\left[k\frac{dT}{dx}\right] = \frac{d}{dx}\left[k_0 \left(1 + BT\right) \frac{dT}{dx}\right] = 0$$

The boundary conditions are the same as before: at $x = 0$, $T = T_1$, and at $x = t$, $T = T_2$. Multiply through by dx and integrate once to obtain

$$\int d\left[k_0 \left(1 + BT\right) \frac{dT}{dx}\right] = 0 \quad \rightarrow \quad k_0 \left(1 + BT\right) \frac{dT}{dx} = C_1$$

Separate variables and integrate a second time to obtain

$$\int k_0 \left(1 + BT\right) dT = \int C_1 \, dx \quad \rightarrow \quad k_0 \left(T + \frac{BT^2}{2}\right) = C_1 x + C_2$$

We apply the first boundary condition and obtain $C_2 = k_0 \left(T_1 + BT_1^2/2 \right)$. We apply the second boundary condition (also using the expression for C_2) and obtain

$$C_1 = \frac{k_0}{t} \left[(T_2 - T_1) + \frac{B}{2} \left(T_2^2 - T_1^2 \right) \right]$$

Thus, we have an expression for the temperature profile through the plane wall with variable thermal conductivity. Note that it is a nonlinear equation (a quadratic).

We use Fourier's law to determine the heat transfer rate:

$$\dot{Q} = q''A = -kA\frac{dT}{dx}$$

From the second differential equation above, we can see that

$$\frac{dT}{dx} = \frac{C_1}{k_0 \left(1 + BT \right)}$$

so that

$$q'' = -k\frac{dT}{dx} = -\left[k_0 \left(1 + BT \right) \right] \left[\frac{C_1}{k_0 \left(1 + BT \right)} \right] = -C_1 = -\frac{k_0}{t} \left[(T_2 - T_1) + \frac{B}{2} \left(T_2^2 - T_1^2 \right) \right]$$

The total heat transfer rate is

$$\dot{Q} = q''LW = \frac{k_0}{t} \left[(T_1 - T_2) + \frac{B}{2} \left(T_1^2 - T_2^2 \right) \right] LW$$

Now everything can be calculated:

$$\dot{Q} = \frac{(175\,\text{W/m} \cdot \text{K})}{0.003\,\text{m}} \left[(85 - 25)\,°\text{C} + \frac{(0.00556/°\text{C})}{2} \left(85^2 - 25^2 \right) °\text{C}^2 \right] (0.02\,\text{m})\,(0.02\,\text{m})$$

$$= 1828.12\,\text{W}$$

Comments:

There is no difference between the heat transfer rates when we evaluate the thermal conductivity at the average temperature versus formally taking into account its temperature variation in the governing differential equation. Note that if we had evaluated the thermal conductivity at the incorrect temperature (for example, at 25°C, $k = 199.3$ W/m·K or at 85°C, $k = 257.7$ W/m·K), we would have obtained a significantly different heat transfer rate. Also note that if we simply integrate the expression for thermal conductivity to obtain an average thermal conductivity over a temperature range, we obtain

$$k_{avg} = \frac{1}{T_2 - T_1} \int_{T_1}^{T_2} [k_0 \left(1 + BT \right)] dT = \frac{k_0}{T_2 - T_1} \left[(T_2 - T_1) + \frac{B}{2} \left(T_2^2 - T_1^2 \right) \right]$$

$$= k_0 \left[\frac{(T_2 - T_1)}{T_2 - T_1} + \frac{B}{2} \frac{\left(T_2^2 - T_1^2 \right)}{T_2 - T_1} \right] = k_0 \left[1 + B\frac{T_1 + T_2}{2} \right] = k_0 \left[1 + BT_{avg} \right]$$

which is the same result we found above in part a. This expression is valid only if thermal conductivity has a linear temperature variation.

EXAMPLE 11-3   Conduction with internal heat generation

We want to determine experimentally the heat transfer coefficient of a single-phase fluid flowing inside a straight circular tube. The tube is 3 m long, has a 12.4-mm inner diameter and a 15.4-mm outer diameter, and is well insulated. The tube's thermal conductivity is 14.3 W/m·K. The fluid ($c_p = 2.4$ kJ/kg·K) enters the tube at 21°C at a flow rate of 0.7 kg/s. Electric current heats the tube; the current is $I = 473$ A, and the voltage drop across the length of the tube is $\xi = 5.6$ V. At a distance of 2.4 m from the inlet, the measured temperature on the outside surface of the tube is 41.2°C. Determine the heat transfer coefficient at that location (in W/m²·K).

Approach:

The schematic of this problem is shown.

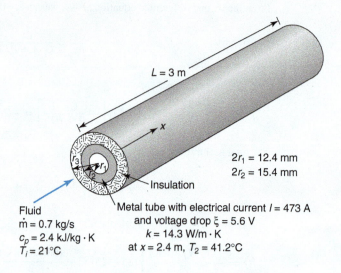

The definition of the heat transfer coefficient is $h = q''/\Delta T$. From the experimental data, we can determine the heat flux. The appropriate temperature difference is defined as $\Delta T = T_w - T_f$. The fluid temperature is the average or bulk temperature of the fluid at the location in the tube and is determined from the application of the energy equation to the fluid. The wall temperature, T_w, is the *inside* wall temperature, T_1. From the experiment we measure the *outside* wall temperature, T_2. Because we have heat generation within the tube wall, we must develop the temperature profile starting from the governing differential conduction equation, Eq. 11-14.

Assumptions:

A1. Heat flux is uniform over the length and circumference of the tube.

A2. The system is steady.

A3. Potential and kinetic energy effects are negligible.

A4. There is no work.

Solution:

Heat transfer coefficient is defined as $h = q''/\Delta T$. Heat flux is defined as $q'' = \dot{Q}/A$, which assumes a uniform heat flux [A1]. The total heat transfer rate, $\dot{Q}$, is determined from the joulean heating due to the current flow through the tube, $\dot{Q} = \xi I$, and the area is the inside surface area of the tube, $A = \pi 2 r_i L$.

The local fluid temperature $T_f = T_x$ is evaluated at $x = 2.4$ m. With a control volume drawn around the fluid, we assume [A2], [A3], and [A4] with one inlet and one outlet. Using conservation of mass and energy and eliminating terms, we obtain

$$\dot{m}_i = \dot{m}_x = \dot{m} \quad \text{and} \quad \dot{Q}_x + \dot{m}(h_i - h_x) = 0$$

where x is the location of interest, h represents enthalpy, and $\dot{Q}_x = q'' 2\pi r_1 x$ represents the heat added to the fluid between the inlet and location x. We assume the fluid is ideal with a constant specific heat [A5], [A6], so that we can use $\Delta h = c_p (T_i - T_x)$. Substituting this expression into the energy equation and solving for T_x, we obtain:

A5. The fluid is incompressible.
A6. Specific heat is constant.

$$T_x = T_i + \frac{\dot{Q}_x}{\dot{m} c_p} = T_i + \frac{q'' \pi 2 r_1 x}{\dot{m} c_p}$$

The inside wall temperature, T_w, is obtained by solving the governing differential conduction equation, Eq. 11-14. Assuming [A2], [A7], [A8], and [A9] (there are no circumferential temperature variations or axial conduction because axial conduction would be small compared to the radial conduction), we see that the temperature is only a function of radial position, and Eq. 11-14 reduces to

A7. Heat transfer is one-dimensional.
A8. Thermal conductivity is constant.
A9. The volumetric heat generation rate is uniform.

$$\frac{1}{r} \frac{d}{dr} \left(r \frac{dT}{dr} \right) + \frac{q'''}{k} = 0$$

The first boundary condition for this problem is at $r = r_2, T = T_2$. The second boundary condition needs to be a known value, too. At the outer surface, we assume the tube is well insulated [A10], so the second boundary condition is at $r = r_2$, $dT/dr = 0$.

A10. The outer surface is insulated.

Separating variables and integrating once gives us

$$r \frac{dT}{dr} = -\frac{q'''}{k} \frac{r^2}{2} + C_1$$

Again, separating variables and integrating, we obtain

$$T = -\frac{q'''}{k} \frac{r^2}{4} + C_1 \ln r + C_2$$

Applying the first boundary condition given above results in:

$$T_2 = -\frac{q'''}{k} \frac{r_2^2}{4} + C_1 \ln r_2 + C_2$$

For the second boundary condition, we use the differential equation that was obtained after the first integration, so that

$$0 = -\frac{q'''}{k} \frac{r_2^2}{2} + C_1$$

Solving for C_1 and C_2, we obtain

$$C_1 = \frac{q''' r_2^2}{2k} \quad \text{and} \quad C_2 = T_2 + \frac{q''' r_2^2}{4k} - \frac{q''' r_2^2}{2k} \ln r_2$$

Using these constants in the general expression for the temperature distribution and simplifying terms:

$$T(r) = T_2 + \frac{q'''}{4k} \left(r_2^2 - r^2 \right) - \frac{q''' r_2^2}{2k} \ln \left(\frac{r_2}{r} \right)$$

This equation can be evaluated at the inside surface to determine T_1.

The volumetric heat generation rate, q''', can be calculated from the total electric power and the volume of the tube wall, $q''' = \dot{Q}/V = \dot{Q}/\left[\pi \left(r_2^2 - r_1^2 \right) L \right]$.

The total electric power dissipated in the tube wall is

$$\dot{Q} = \xi I = (473 \, \text{A}) (5.6 \, \text{V}) = 2649 \, \text{W}$$

The heat flux is

$$q'' = \frac{\dot{Q}}{A} = \frac{\dot{Q}}{\pi 2 r_1 L} = \frac{2649\,\text{W}}{\pi 2\,(0.0062\,\text{m})\,(3\,\text{m})} = 22{,}670\,\frac{\text{W}}{\text{m}^2}$$

The volumetric heat generation rate is

$$q''' = \frac{\dot{Q}}{V} = \frac{\dot{Q}}{\pi \left(r_2^2 - r_1^2 \right) L} = \frac{2649\,\text{W}}{\pi \left[(0.0077\,\text{m})^2 - (0.0062\,\text{m})^2 \right] (3\,\text{m})} = 13{,}480{,}000\,\frac{\text{W}}{\text{m}^3}$$

The fluid temperature at location x is

$$T_x = T_i + \frac{q'' \pi 2 r_1 x}{\dot{m} c_p} = 21°\text{C} + \frac{(22{,}670\,\text{W}/\text{m}^2)\,\pi 2\,(0.0062\,\text{m})\,(2.4\,\text{m})}{(0.7\,\text{kg}/\text{s})\,(2.4\,\text{kJ}/\text{kg·K})\,(1000\,\text{J}/1\,\text{kJ})}$$

$$= 21 + 1.26 = 22.26°\text{C}$$

The inside wall temperature at location x is

$$T\,(r_1) = T_2 + \frac{q'''}{4k}\left(r_2^2 - r_1^2 \right) - \frac{q''' r_2^2}{2k} \ln\left(\frac{r_2}{r_1} \right)$$

$$= 41.2°\text{C} + \frac{13{,}480{,}000\,\text{W}/\text{m}^3}{4\,(14.3\,\text{W}/\text{m·K})} \left([0.0077\,\text{m}]^2 - [0.0062\,\text{m}]^2 \right)$$

$$- \frac{(13{,}480{,}000\,\text{W}/\text{m}^3)\,(0.0077\,\text{m})^2}{2\,(14.3\,\text{W}/\text{m·K})} \ln\left(\frac{0.0077}{0.0062} \right) = 41.2 - 1.14 = 40.06°\text{C}$$

The heat transfer coefficient is

$$h = \frac{q''}{T_1 - T_x} = \frac{22{,}670\,\text{W}/\text{m}^2}{(40.06 - 22.26)\,\text{K}} = 1{,}273\,\frac{\text{W}}{\text{m}^2\text{·K}}$$

Comments:

Once a solution to an application is developed, it can be applied to other similar situations. Note that the temperature drop across the wall is only 1.26°C. For other operating conditions, the temperature drop across the wall can be much greater.

11.4 STEADY MULTIDIMENSIONAL CONDUCTION

There are many situations in which temperature varies only in one direction in a solid, as shown in previous section. There are other applications (e.g., Figure 11-1 and Figure 11-2) for which the assumption of one-dimensional heat conduction may be too much of a simplification or inappropriate. In that case, a multidimensional conduction problem must be solved.

A classical example of such a problem is shown in Figure 11-1. This rectangular system is often used in differential equation courses to introduce students to the solution of partial differential equations. The steady-state temperature distribution and the heat transfer rate (per unit depth) through the left face of the domain are sought. With no internal heat generation, constant thermal conductivity, and ignoring the z direction, Eq. 11-13 reduces to

$$\frac{\partial^2 T}{\partial x^2} + \frac{\partial^2 T}{\partial y^2} = 0 \tag{11-40}$$

The boundary conditions needed to complete the problem formulation are

$$
\begin{aligned}
T(0, y) &= T_1 & T(L, y) &= T_2 \\
T(x, 0) &= T_2 & T(x, H) &= T_2
\end{aligned}
\tag{11-41}
$$

An analytic solution to this equation can be developed in terms of an infinite series:

$$\frac{T(x, y) - T_2}{T_1 - T_2} = \frac{4}{\pi} \sum_{n=1}^{\infty} \frac{\sinh\left[(2n+1)\pi(L-x)/H\right]}{\sinh\left[(2n+1)(\pi L/H)\right]} \frac{\sin\left[(2n+1)(\pi y/H)\right]}{2n+1}$$

$$\tag{11-42}$$

Eq. 11-42 can be used to determine the temperature at any x and y location in the domain. The heat transfer rate can also be determined by combining this equation with Fourier's law.

Many other exact solutions to multidimensional heat conduction problems have been derived, and examples of these can be found in the literature. However, the range of these solutions is limited; only simple geometries and boundary conditions can be handled. For complex or realistic applications, a numerical solution of the governing equation (and its boundary conditions) is the best approach.

Numerical approaches (*finite difference*, *finite volume*, and *finite element*) involve the development of an approximate solution to the governing partial differential equation. The equation is reformulated into a system of simultaneous algebraic equations, each of which is applicable to a very small area or volume in the body under study. The solution of this system results in discrete values of temperature at the center of each volume. Heat fluxes then can be obtained by applying Fourier's law.

Complex geometries with steady or transient conditions can be handled with numerical analysis. Different equations are formulated depending on the location of the volume (e.g., interior or on a boundary) and the physics involved (e.g., a convective or constant temperature boundary condition, transient heat storage, etc.). A complete discussion of this topic is beyond the scope of this book. However, many references in the literature give good explanations of numerical analysis and guidance about development of computer codes, limitations, cautions, more accurate approximations, and so on.

Once solutions are obtained, then the results must be presented in a manner that, ideally, makes them easy to use. For many situations, the temperature profile is not what is sought; the steady-state heat transfer rate is the goal of the analysis, and a simplified equation that generalizes the results is presented so that the governing equations do not need to be solved again. The main concept behind this simplification is to express the results of a multidimensional conduction problem as a one-dimensional problem.

Consider the heat transfer between two isothermal surfaces for the three standard geometries:

$$\dot{Q} = kA\frac{\Delta T}{\Delta x} = \frac{\Delta T}{\Delta x/kA} = \frac{\Delta T}{1/kS} \qquad \text{plane wall}$$

$$\dot{Q} = \frac{2\pi L k}{\ln\left(r_2/r_1\right)}\Delta T = \frac{\Delta T}{\ln\left(r_2/r_1\right)/2\pi L k} = \frac{\Delta T}{1/kS} \qquad \text{hollow cylinder} \qquad \text{(11-43)}$$

$$\dot{Q} = \frac{4\pi\, r_1 r_2 k}{r_2 - r_1}\Delta T = \frac{\Delta T}{\left(r_2 - r_1\right)/4\pi\, r_1 r_2 k} = \frac{\Delta T}{1/kS} \qquad \text{hollow sphere}$$

The thermal resistance in each of these expressions is described with the thermal conductivity, k, and a *conduction shape factor, S,* that is characteristic of the specific geometry. The shape factor, S, has units of length and is used in the defining equation:

$$\dot{Q} = kS\Delta T = \frac{\Delta T}{1/kS} \qquad \text{(11-44)}$$

The shape factor is related to the thermal resistance by

$$R = \frac{1}{kS} \qquad \text{(11-45)}$$

For multidimensional geometries (geometries more complex than the one-dimensional ones described above), the governing equations can be solved for the heat transfer rate. The results are then manipulated so that a conduction shape factor can be calculated and generalized (correlated) for the specific geometry. This shape factor is used in Eq. 11-44 whenever that geometry is encountered in the future. Table 11-1 lists some of these shape factors.

EXAMPLE 11-4 Multidimensional conduction with shape factor

The buildings at a university are steam-heated in winter and are connected to the boiler plant by a network of approximately 1 mile of 6-in.-outside-diameter steel pipe (wall thickness is 0.3 in.) that is buried 6 ft below ground level. A layer of insulation ($k = 0.25$ Btu/h·ft·°F) 2 in. thick covers the pipe. The ground thermal conductivity is estimated to be 1.1 Btu/h·ft·°F. If the steam is at 350°F and the surface temperature of the ground is 15°F, determine the heat loss from the pipes (in Btu/h).

Approach:

The schematic of the system is shown. The thermal circuit is also shown.

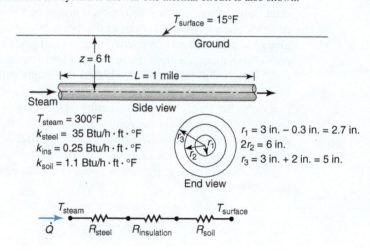

This is a steady, multidimensional heat conduction problem. However, assuming that it can be treated as a one-dimensional problem, it can be solved by application of the basic rate equation, $\dot{Q} = \Delta T / R_{tot}$. Conduction between the circular pipe and the earth surface is taken into account with a shape factor. We need to evaluate that resistance and the other resistances in the circuit.

Assumptions:

A1. The system is steady.
A2. Heat transfer is one-dimensional.
A3. All properties are constant.
A4. There is no internal heat generation.
A5. The heat transfer coefficient inside the pipe is negligible.

Solution:

Begin with the heat transfer rate equation assuming steady, one-dimensional, constant properties and no internal heat generation [A1][A2][A3][A4]: $\dot{Q} = \Delta T / R_{tot}$. The overall driving temperature difference is $\Delta T = T_{steam} - T_{surface}$.

If the convective heat transfer resistance between the steam and the pipe is negligible [A5], the total thermal resistance is $R_{tot} = R_{steel} + R_{insulation} + R_{soil}$. For constant properties, the individual thermal resistances are determined with

$$R_{steel} = \frac{\ln\left(r_2/r_1\right)}{2\pi k_{steel}L} \qquad R_{insulation} = \frac{\ln\left(r_3/r_2\right)}{2\pi k_{ins}L} \qquad R_{soil} = \frac{1}{k_{soil}S}$$

The soil thermal resistance requires a shape factor. For a circular tube enclosed in a semi-infinite medium with $z > D$, we obtain from Table 11-1,

$$S = \frac{2\pi L}{\ln\left[(2z/r_3) + \sqrt{(2z/r_3)^2 - 1}\right]}$$

With everything known, we can calculate the three resistances. From Appendix B-2, the steel thermal conductivity $k_{steel} = 35$ Btu/h·ft·°F. Therefore,

$$R_{steel} = \frac{\ln\left(r_2/r_1\right)}{2\pi k_{steel}L} = \frac{\ln\left(3/2.7\right)}{2\pi\left(35\,\text{Btu/h·ft·°F}\right)\left(1\,\text{mi}\right)\left(5280\,\text{ft}/1\,\text{mi}\right)} = 9.07 \times 10^{-8}\,\frac{\text{h·°F}}{\text{Btu}}$$

$$R_{insulation} = \frac{\ln\left(5/3\right)}{2\pi\left(0.25\,\dfrac{\text{Btu}}{\text{h·ft·°F}}\right)\left(1\,\text{mi}\right)\left(5280\,\text{ft}/1\,\text{mi}\right)} = 6.16 \times 10^{-5}\,\frac{\text{h·°F}}{\text{Btu}}$$

$$S = \frac{2\pi\left(1\,\text{mi}\right)\left(5280\,\text{ft}/1\,\text{mi}\right)}{\ln\left[\left(\dfrac{(2)(6\,\text{ft})}{(10/12)\,\text{ft}}\right) + \sqrt{\left(\dfrac{(2)(6\,\text{ft})}{(10/12)\,\text{ft}}\right)^2 - 1}\,\right]} = 9876\,\text{ft}$$

$$R_{soil} = \frac{1}{k_{soil}S} = \frac{1}{\left(1.1\,\text{Btu/h·ft·°F}\right)\left(9876\,\text{ft}\right)} = 9.21 \times 10^{-5}\,\frac{\text{h·°F}}{\text{Btu}}$$

Therefore the total resistance is

$$R_{tot} = R_{steel} + R_{insulation} + R_{soil} = 9.07 \times 10^{-8} + 6.16 \times 10^{-5} + 9.21 \times 10^{-5} = 1.538 \times 10^{-4}\,\frac{\text{h·°F}}{\text{Btu}}$$

Finally, the total heat transfer rate is

$$\dot{Q} = \frac{\Delta T}{R_{tot}} = \frac{(350 - 15)°\,\text{F}}{1.538 \times 10^{-4}\,\dfrac{\text{h·°F}}{\text{Btu}}} = 2.18 \times 10^6\,\frac{\text{Btu}}{\text{h}}$$

Comments:

The pipe wall adds little to the total thermal resistance in this problem. In other situations, wall resistance can be significant.

TABLE 11-1 Conduction shape factors for selected isothermal configurations

Physical Configuration	Schematic	Shape Factor S $\dot{Q} = kS\,(T_1 - T_2)$	Restrictions
Horizontal cylinder in a semi-infinite solid		$2\pi L/\cosh^{-1}(2z/D)$ $2\pi L/\ln\left[(2z/D)+\sqrt{(2z/D)^2-1}\right]$ $2\pi L/\ln(4z/D)$	$z \approx D$ $z > D$ $z \gg D$
Row of horizontal cylinders in a semi-infinite solid	 L = length	$2\pi L/\ln\left[(2x/\pi D)\sinh(2\pi z/x)\right]$	$z > D$
Horizontal cylinder at midplane of an infinite wall	 L = length	$2\pi L/\ln(8t/\pi D)$	$t > D/2$
Row of horizontal cylinders at midplane of an infinite wall	 L = length	$2\pi L/\ln\left[(2x/\pi D)\sinh(\pi t/x)\right]$	$t > D$
Circular hole centered in a square solid	 L = length	$\approx 2\pi L/\ln(1.08t/D)$	$t > D$
Two cylinders in an infinite solid	 L = length	$2\pi L/\cosh^{-1}\left[(4x^2-D^2-d^2)/2Dd\right]$	$D > d$
Vertical cylinder in a semi-infinite solid		$2\pi L/\ln(4L/D)$	$L \gg D$

(Continued)

TABLE 11-1 *(Continued)*

Physical Configuration	Schematic	Shape Factor S $\dot{Q} = kS(T_1 - T_2)$	Restrictions
Sphere in a semi-infinite solid		$2\pi D / \left[1 - (D/4z)\right]$	$z > D/2$
Thin rectangular plate in a semi-infinite solid parallel to surface		$\pi a / \ln(4a/b)$ $2\pi a / \ln(2\pi z/b)$ $2\pi a / \ln(4a/b)$	$z = 0$ $a \gg b, z > 2b$ $z \gg a$
Rectangular hole in a semi-infinite solid		$(a/2b + 5.7)L / \ln(3.5z/a^{1/4}b^{3/4})$	$a > b$
Horizontal rectangular block in a semi-infinite solid		$2.756a \left[\ln(1 + z/L)\right]^{-0.59}(b/z)^{0.078}$	$a > L$
Thin circular disk in a semi-infinite solid		$2D$ $4D$	$z = 0$ $z \gg D$
Edge section of two intersecting walls		$aL/t + bL/t + 0.54L$	$L > t/5$
Corner section of three intersecting walls		$0.15t$	$t \ll L_1, L_2, L_3$

11.5 LUMPED SYSTEM ANALYSIS FOR TRANSIENT CONDUCTION

As discussed in Chapter 3 and Section 11.1, transient heat conduction occurs in many different applications. In the general case, temperature varies with time and location. Under some circumstances, the variation with location can be ignored, and the entire solid is assumed to be at a uniform temperature at a given time. This is the so-called *lumped system analysis* (also called *lumped parameter* and *lumped heat capacity analysis*). As given in Chapter 3, the lumped system analysis is valid (that is, when internal conduction may be ignored) when the nondimensional *Biot number*, $Bi = hL_{char}/k <\sim 0.1$. In that case, we assume there is negligible temperature variation within the body, and all the resistance to heat transfer is contained in the convective heat transfer process. The solution to this zero-dimensional heat transfer problem is given in Chapter 3:

$$\boxed{\frac{T(t) - T_f}{T_i - T_f} = \exp\left(-\frac{hA}{mc_p}t\right) = \exp\left(-BiFo\right) \qquad Bi <\sim 0.1} \qquad (11\text{-}46)$$

where the Fourier number, $Fo = \alpha t / L_{char}^2$, and Biot number use a characteristic length, $L_{char} = V/A$.

The total heat transfer that occurs between time $t = 0$ and $t = t$ is determined by integrating the above expression with respect to time. The instantaneous heat transfer rate is given by

$$\dot{Q} = hA\left[T_f - T(t)\right] = hA\left[T_f - T_i\right]\exp\left(-BiFo\right) = hA\left[T_f - T_i\right]\exp\left(-\frac{hA}{mc_p}t\right) \quad (11\text{-}47)$$

and the total heat transfer is

$$Q = \int_0^t \dot{Q}\,dt = \int_0^t hA\left[T_f - T_i\right]\exp\left(-\frac{hA}{mc_p}t\right)\,dt \qquad (11\text{-}48)$$

With constant h, A, m, c_p, T_i, and T_f, we can carry out the integration to obtain

$$Q = mc_p\left[T_f - T_i\right]\left[1 - \exp\left(-BiFo\right)\right]$$

When $T_f > T_i$, the heat transfer, Q, is positive, which is consistent with heat transfer being defined as positive into a control volume.

We nondimensionalize the total heat transfer, Q, with the maximum possible energy, $Q_{max} = mc_p\left(T_f - T_i\right)$, that could be absorbed or lost from the body after an infinite period of time. This occurs when the thermal capacity of the body, $mc_p = \rho V c_p$, experiences the maximum possible temperature change from the initial temperature of the body, T_i, to the temperature of the surrounding fluid, T_f:

$$\boxed{\frac{Q}{Q_{max}} = \frac{Q}{mc_p\left[T_f - T_i\right]} = 1 - \exp\left(-BiFo\right)} \qquad (11\text{-}49)$$

EXAMPLE 11-5   Lumped system with internal heat generation

The Hot Stuff Clothes Iron Company is working on a new design for an iron that reaches its operating temperature of 110°C in 1 min. The baseplate of the iron is 1.02 kg, and the exposed surface area is 258 cm². The design room temperature is 20°C, and the heat transfer coefficient is estimated to be 10 W/m²·K. The baseplate is made of steel with a density of 7,800 kg/m³, specific heat of 444 J/kg·K, and a thermal conductivity of 37.7 W/m·K. Determine the power (in W) needed for the baseplate to go from room temperature to operating temperature in 1 min.

Approach:

The given information is shown on the schematic. The lumped system approach discussed in Section 11.5 is not valid because the present problem has electric power addition as well as heat losses. We first need to determine if a lumped system approach is valid at all. If it is, then we need to redo the lumped system analysis, but this time including power addition.

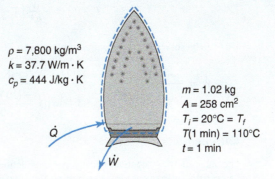

$\rho = 7,800 \text{ kg/m}^3$
$k = 37.7 \text{ W/m} \cdot \text{K}$
$c_p = 444 \text{ J/kg} \cdot \text{K}$

$m = 1.02 \text{ kg}$
$A = 258 \text{ cm}^2$
$T_i = 20°\text{C} = T_f$
$T(1 \text{ min}) = 110°\text{C}$
$t = 1 \text{ min}$

$\dot{Q}$

$\dot{W}$

Assumptions:

Solution:

The Biot number for the lumped system approach is defined as $Bi = h\left(V/A\right)/k$. Substituting $V = m/\rho$ into this expression, we obtain $Bi = h\left(m/\rho A\right)/k$. All of these quantities are given in the problem statement, so that

$$Bi = \frac{hm}{kA\rho} = \frac{\left(10\,\text{W/m}^2 \cdot \text{K}\right)\left(1.02\,\text{kg}\right)}{\left(37.7\,\text{W/m} \cdot \text{K}\right)\left(0.0258\,\text{m}^2\right)\left(7800\,\text{kg/m}^3\right)} = 0.00134$$

Because $Bi < 0.1$, we can analyze the iron baseplate as a lumped system problem.

We apply conservation of energy to the closed system defined in the schematic and assume

A1. Potential and kinetic energy effects are negligible.

negligible potential and kinetic energy effects [A1], giving us

$$\frac{dU}{dt} = m\frac{du}{dt} = \dot{Q} - \dot{W}$$

A2. Specific heat is constant.

Assuming the iron is an ideal solid with a constant specific heat [A2] ($c \approx c_v \approx c_p$), so that $du = c_p\,dT$, and describing the heat transfer in terms of the rate equation for convective heat transfer, we obtain

$$mc_p\frac{dT}{dt} = -hA\left(T - T_f\right) - \dot{W}$$

Rearranging this equation to separate the variables,

$$\frac{dT}{dt} = -\frac{hA}{mc_p}\left(T - T_f + \frac{\dot{W}}{mc_p}\right)$$

Separating the variables,

$$\int_{T_i}^{T} \frac{dT}{T - T_f + \dfrac{\dot{W}}{mc_p}} = -\int_{0}^{t} \frac{hA}{mc_p}\,dt$$

and integrating

$$\ln\left[\frac{T - T_f + \dot{W}/mc_p}{T_i - T_f + \dot{W}/mc_p}\right] = -\frac{hA}{mc_p}t$$

Exponentiating both sides,

$$\frac{T - T_f + \dot{W}/mc_p}{T_i - T_f + \dot{W}/mc_p} = \exp\left(-\frac{hA}{mc_p}t\right)$$

Letting $T_i = T_f$ and rearranging once again,

$$T - T_f = -\frac{\dot{W}}{hA}\left[1 - \exp\left(-\frac{hA}{mc_p}t\right)\right]$$

This can be solved for the required input power, $\dot{W}$, to give:

$$\dot{W} = \frac{-hA\,(T - T_f)}{1 - \exp\left(\frac{-hA}{mc_p}t\right)} = \frac{-\left(10\,\text{W/m}^2\cdot\text{K}\right)\left(0.0258\,\text{m}^2\right)(110 - 20)\,\text{K}}{1 - \exp\left(\dfrac{-\left(10\,\text{W/m}^2\cdot\text{K}\right)\left(0.0258\,\text{m}^2\right)}{(1.02\,\text{kg})\left(444\,\text{J/kg}\cdot\text{K}\right)}60\,\text{s}\right)} = -691\,\text{W}$$

Comments:

The value of $\dot{W}$ is negative, because work is defined as positive out and negative in, and we add power to this iron to raise its temperature. Note that the electric power can be treated as either power input ($\dot{W}$) as done above, or it could be treated as a second heat transfer term ($\dot{Q}$), in which case the control boundary would include the baseplate but not the electric resistance heater wires. If the heat transfer approach is used, then the heat transfer term would be positive because of the sign convention (heat transfer positive when an input). The magnitude of the final answer remains the same.

Because the lumped system equation (Eq. 11-46) is simple, we may be tempted to use it in inappropriate applications. Consider the example of cooking a baked potato or thawing a frozen piece of meat. If a potato is not cooked long enough, the outside layer of the potato may be at a high temperature and soft but the center may still be cooler and hard. Likewise, for the thawing meat, the outside may thaw while the inside remains frozen. It is an easy step to imagine engineering applications (e.g., heat treatment of metals) in which a non-negligible temperature variation exists in a body. In those situations, the lumped system method is not suitable and a solution that incorporates the spatial variations in temperature must be used. Thus, a one-, two-, or three-dimensional transient partial differential equation must be solved subject to applicable boundary and initial conditions. The results of such analyses are given in the next two sections for simple geometries.

11.6 ONE-DIMENSIONAL TRANSIENT CONDUCTION

The temperature in zero-dimensional transient heat conduction in a solid (i.e., lumped system analysis) is dependent only on the Biot and Fourier numbers. With a uniform temperature in the body, the location within the body is not part of the solution. However, often we need to know the temperature at various locations in a body and, intuitively, the solution to the governing partial differential equation should depend on the Biot number, Fourier number, and something that represents the location.

Consider one-dimensional transient conduction in a plane wall with thickness $2L$, as shown on Figure 11-6. Initially, the wall is at a uniform temperature, T_i, as is the surrounding fluid. At time equal zero, the surrounding fluid temperature instantaneously changes from T_i to T_f and the convective heat transfer coefficient, h, is the same and uniform on both sides of the wall. We want to determine the temperature variation with time at every location within this solid. In Cartesian coordinates, assuming constant properties and no internal heat generation, the governing equation is, from Eq. 11-13,

$$\frac{\partial^2 T}{\partial x^2} = \frac{1}{\alpha}\frac{\partial T}{\partial t} \tag{11-50}$$

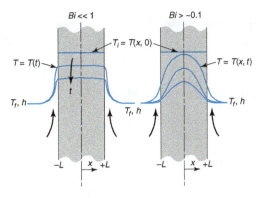

FIGURE 11-6   Infinite plane wall in which heat transfer is one dimensional.

where $\alpha = k/\rho c_p$ is the thermal diffusivity.

The initial condition and boundary conditions are:

$$T(x,0) = T_i \qquad\qquad \text{at} \qquad t = 0$$

$$-kA\frac{\partial T}{\partial x} = hA\left(T - T_f\right) \qquad \text{at} \qquad x = +L \qquad\qquad (11\text{-}51)$$

$$-kA\frac{\partial T}{\partial x} = hA\left(T_f - T\right) \qquad \text{at} \qquad x = -L$$

The boundary conditions are obtained from energy balances on the two faces of the plane wall. The different signs on the boundary conditions result from consistent use of the positive direction in defining the direction of heat flow.

This problem can be solved analytically in terms of $T_i, T_f, t, h, k, L, x,$ and α. We will not do this here. However, it is instructive to recast the equation and the boundary conditions in terms of nondimensional groups so that the results of the analysis can be generalized. Previously, we obtained the Biot and Fourier numbers by manipulating the resulting equation from the lumped system analysis. Now we nondimensionalize the governing differential equation for one-dimensional conduction (Eq. 11-50) to see how nondimensional groups fall out naturally from the equations.

We nondimensionalize the temperature with

$$\theta(x,t) = \frac{T(x,t) - T_f}{T_i - T_f} \qquad\qquad (11\text{-}52)$$

Solving for $T(x,t)$,

$$T(x,t) = T_f + (T_i - T_f)\,\theta(x,t)$$

The following derivatives are obtained through use of the chain rule and other manipulations:

$$\frac{\partial T}{\partial t} = \frac{\partial T}{\partial \theta}\frac{\partial \theta}{\partial t} = \left(T_i - T_f\right)\frac{\partial \theta}{\partial t}$$

$$\frac{\partial T}{\partial x} = \frac{\partial T}{\partial \theta}\frac{\partial \theta}{\partial x} = \left(T_i - T_f\right)\frac{\partial \theta}{\partial x} \qquad\qquad (11\text{-}53)$$

$$\frac{\partial^2 T}{\partial x^2} = \frac{\partial}{\partial x}\left(\frac{\partial T}{\partial x}\right) = \frac{\partial}{\partial x}\left[\left(T_i - T_f\right)\frac{\partial \theta}{\partial x}\right] = \left(T_i - T_f\right)\frac{\partial^2 \theta}{\partial x^2}$$

Substitute these expressions into Eq. 11-50 and simplify to obtain

$$\frac{\partial^2 \theta}{\partial x^2} = \frac{1}{\alpha}\frac{\partial \theta}{\partial t} \qquad\qquad (11\text{-}54)$$

Define a nondimensional variable for space as $X = x/L$. Again using the chain rule,

$$\frac{\partial^2 \theta}{\partial x^2} = \frac{\partial}{\partial x}\left(\frac{\partial \theta}{\partial x}\right) = \frac{\partial}{\partial X}\left(\frac{\partial \theta}{\partial X}\frac{\partial X}{\partial x}\right)\left(\frac{\partial X}{\partial x}\right) = \frac{\partial}{\partial X}\left(\frac{\partial \theta}{\partial X}\frac{1}{L}\right)\left(\frac{1}{L}\right) = \frac{1}{L^2}\frac{\partial^2 \theta}{\partial X^2} \quad (11\text{-}55)$$

Incorporating Eq. 11-55 into Eq. 11-54 and rearranging, we get

$$\frac{\partial^2 \theta}{\partial X^2} = \frac{L^2}{\alpha}\frac{\partial \theta}{\partial t} \quad (11\text{-}56)$$

Define a nondimensional time,

$$\boxed{\tau = \alpha t / L^2 = Fo,}$$

which is the Fourier number, and incorporate this into Eq. 11-56 to obtain the nondimensionalized conduction equation:

$$\frac{\partial^2 \theta}{\partial X^2} = \frac{\partial \theta}{\partial \tau} \quad (11\text{-}57)$$

Now we nondimensionalize the boundary and initial conditions. At $t = 0$, $T(x, t) = T_i$. Hence, using the definitions of θ and τ, we can show that at $\tau = 0$, $\theta = 1$. Likewise, for boundary condition 2, where at $x = L$, $-kA\left(\partial T/\partial x\right) = hA\left(T - T_f\right)$, we can show that at $X = 1$, $\partial \theta/\partial X = -\left(hL/k\right)\theta = -Bi\theta$.

The nondimensional temperature, θ, depends on X, $\tau = Fo$, and Bi. Compared to the lumped system analysis, the only additional variable needed is a parameter to account for spatial variations in temperature, as expected. For a given wall and applicable boundary conditions, if the Biot number is very small, then the exact solution and the lumped system solution give the same result.

In a like manner, we can nondimensionalize the conduction equation in cylindrical and spherical coordinates, and the Biot number, Fourier number, and nondimensional length are obtained. Note that the above example of transient conduction in a plane wall is symmetric around its centerline ($x = 0$). Instead of analyzing the complete wall, we could have just as easily analyzed only half the wall. In that case, a boundary condition would be needed for the centerline. Because of symmetry, no heat crosses the wall's midplane. Thus the appropriate boundary condition would be an adiabatic surface ($q'' = 0 = -k\partial T/\partial x$) at $x = 0$, $\partial T/\partial x = 0$. Lines of symmetry are always treated as adiabatic lines or surfaces.

The solution to Eq. 11-57 involves an infinite series. From the solution, two items of engineering interest are obtained: (1) the temperature profile at any location in the body as a function of time, and (2) the heat transfer rate at any instant in time or the total heat added or removed from the body over a time interval (discussed below). This information is presented for a plane wall in Figure 11-7 (called a *Heisler chart* after the individual who first presented data in this manner). For an infinite plane wall, the initial condition is uniform temperature throughout the body and fluid, and the boundary condition is the instantaneous application of a convective heat transfer coefficient with a step change in the fluid temperature. Note that the characteristic length used in the Biot and Fourier numbers is the wall half-width. The transient solutions for an infinite cylinder and a sphere are given in Figure 11-8 and Figure 11-9, respectively, and the characteristic length used for both is the radius.

While these charts were developed for the sudden application of a convective boundary condition, they also can be used for the situation when a sudden change in the surface temperature is imposed. From the convective rate equation, $q'' = h(T_s - T_f)$, if the temperature difference is allowed to go to zero, the convective heat transfer coefficient must

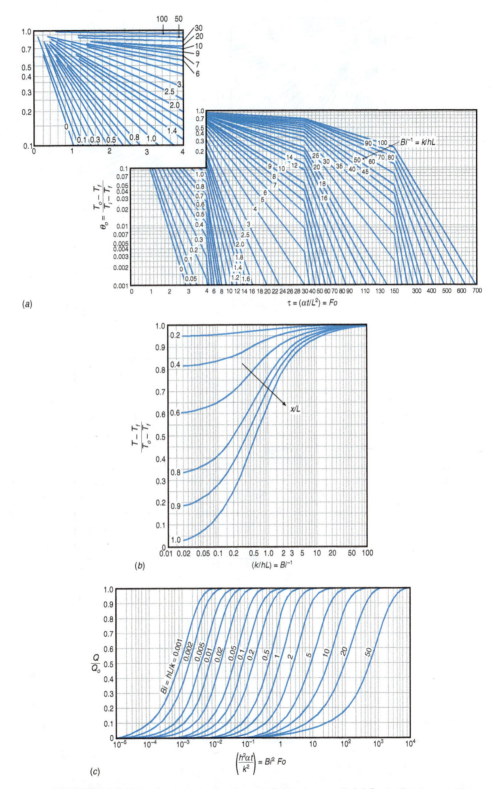

FIGURE 11-7 Transient conduction in an infinite plane wall. (a) Centerline temperature.
(b) Other locations. (c) Total heat transferred. T_o is the centerline temperature.
(*Source*: F. P. Incropera and D. P. DeWitt, *Introduction to Heat Transfer*, 3rd ed., Wiley, New York, 1996. Used
with permission.)

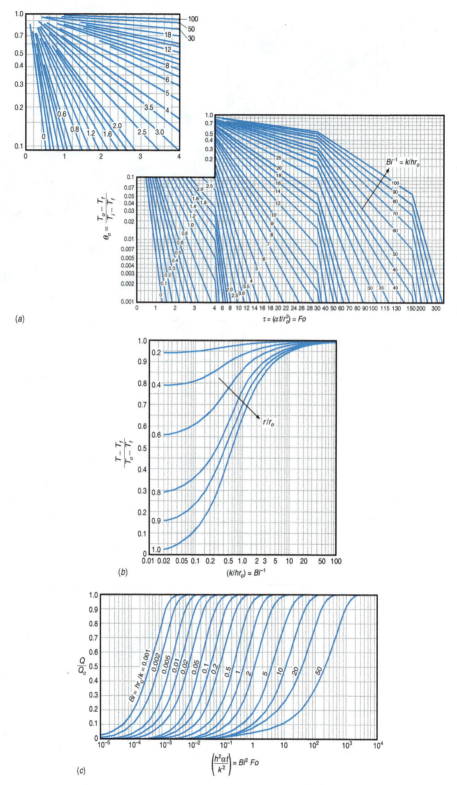

FIGURE 11-8 Transient conduction in an infinite cylinder. (a) Centerline temperature. (b) Other locations. (c) Total heat transferred. T_o is the centerline temperature.

(*Source*: F. P. Incropera and D. P. DeWitt, *Introduction to Heat Transfer*, 3rd ed., Wiley, New York, 1996. Used with permission.)

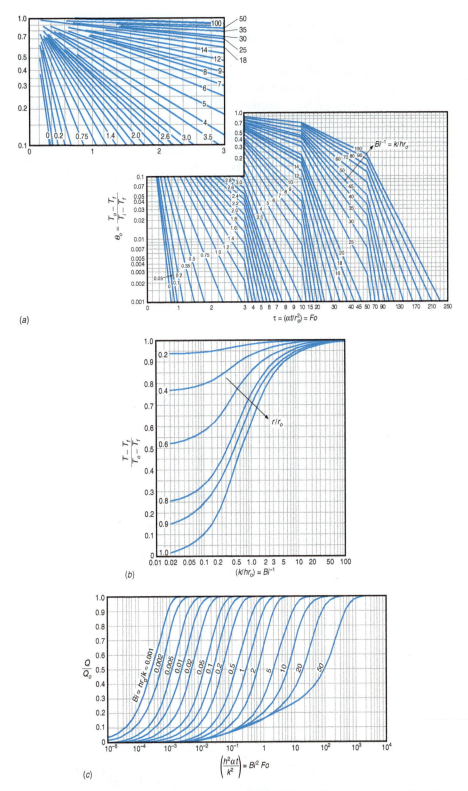

FIGURE 11-9 Transient conduction in a sphere. (a) Centerline temperature. (b) Other locations. (c) Total heat transferred. T_o is the centerline temperature.

(*Source*: F. P. Incropera and D. P. DeWitt, *Introduction to Heat Transfer*, 3rd ed., J. Wiley, New York, 1996. Used with permission.)

approach infinity in order for the heat flux to remain finite. This is equivalent to setting $1/Bi = k/hL_{char} = 0$.

Reasonable solutions can be obtained through the use of these charts. Nevertheless, they are difficult to read (e.g., the upper left-hand corner of Figure 11-7a) and accuracy can be compromised. To avoid this problem, we note that the higher-order terms in the infinite-series solution decrease in size quickly after the first few terms. For $\tau = Fo > \sim 0.2$, the first term in the series provides a solution that is within 2% of the exact solution. Hence, the *one-term approximations* are given by

$$\theta\,(x,t)_{\substack{plane \\ wall}} = \frac{T\,(x,t) - T_f}{T_i - T_f} = C_1 \exp\left(-\lambda_1^2 \tau\right) \cos\left(\lambda_1 x / L\right) \qquad \tau > \sim 0.2 \tag{11-58}$$

$$\theta\,(r,t)_{\substack{infinite \\ cylinder}} = \frac{T\,(r,t) - T_f}{T_i - T_f} = C_1 \exp\left(-\lambda_1^2 \tau\right) J_0\left(\lambda_1 r / r_0\right) \qquad \tau > \sim 0.2 \tag{11-59}$$

$$\theta\,(r,t)_{sphere} = \frac{T\,(r,t) - T_f}{T_i - T_f} = C_1 \exp\left(-\lambda_1^2 \tau\right) \frac{\sin\left(\lambda_1 r / r_0\right)}{\lambda_1 r / r_0} \qquad \tau > \sim 0.2 \tag{11-60}$$

Values for the constants C_1 and λ_1 are given in Table 11-2. Values of the Bessel function of the first kind, J_0 and J_1, are well known and are listed in Table 11-3. Note that at the centerline where $x = r = 0$, the cosine term and the Bessel function equal 1, and in the limit as $\lambda_1 r / r_0$ approaches zero, the term $\left[\sin\left(\lambda_1 r / r_0\right)\right] / \left(\lambda_1 r / r_0\right)$ goes to 1. It is also important to note that the form of these equations—an exponential variation of temperature with respect to time—is similar to that obtained with the lumped system analysis.

At a given time after the transient has begun, there is a temperature variation in the body. The total heat transfer that has occurred up to that time can be determined by integrating the temperature profile over the volume, similar to the approach taken with the lumped system analysis. However, in that analysis, the integration was over time. In the present analysis, the integration is over the volume. Performing the integration with the temperature distributions described by the one-term approximations and nondimensionalizing the result with $Q_{max} = mc_p\left(T_f - T_i\right)$ as we did for the lumped system analysis, we obtain

$$\left(\frac{Q}{Q_{max}}\right)_{\substack{plane \\ wall}} = 1 - \theta_{0,\,\substack{plane \\ wall}} \frac{\sin \lambda_1}{\lambda_1} \qquad \tau > \sim 0.2 \tag{11-61}$$

$$\left(\frac{Q}{Q_{max}}\right)_{\substack{infinite \\ cylinder}} = 1 - 2\theta_{0,\,\substack{infinite \\ cylinder}} \frac{J_1\left(\lambda_1\right)}{\lambda_1} \qquad \tau > \sim 0.2 \tag{11-62}$$

$$\left(\frac{Q}{Q_{max}}\right)_{sphere} = 1 - 3\theta_{0,\,sphere} \frac{\sin \lambda_1 - \lambda_1 \cos \lambda_1}{\lambda_1^3} \qquad \tau > \sim 0.2 \tag{11-63}$$

where θ_0 is the nondimensional centerline temperature.

TABLE 11-2 Constants used in the one-term approximation for one-dimensional transient conduction

Bi = hL_{char}/k	Infinite plane wall with thickness 2L ($L_{char} = L$)		Infinite cylinder ($L_{char} = r_0$)		Sphere ($L_{char} = r_0$)	
	λ_1 (rad)	C_1	λ_1 (rad)	C_1	λ_1 (rad)	C_1
0.01	0.0998	1.0017	0.1412	1.0025	0.1730	1.0030
0.02	0.1410	1.0033	0.1995	1.0050	0.2445	1.0060
0.03	0.1732	1.0049	0.2439	1.0075	0.2989	1.0090
0.04	0.1987	1.0066	0.2814	1.0099	0.3450	1.0120
0.05	0.2217	1.0082	0.3142	1.0124	0.3852	1.0149
0.06	0.2425	1.0098	0.3438	1.0148	0.4217	1.0179
0.07	0.2615	1.0114	0.3708	1.0173	0.4550	1.0209
0.08	0.2791	1.0130	0.3960	1.0197	0.4860	1.0239
0.09	0.2956	1.0145	0.4195	1.0222	0.5150	1.0268
0.10	0.3111	1.0160	0.4417	1.0246	0.5423	1.0298
0.15	0.3779	1.0237	0.5376	1.0365	0.6608	1.0445
0.20	0.4328	1.0311	0.6170	1.0483	0.7593	1.0592
0.25	0.4801	1.0382	0.6856	1.0598	0.8448	1.0737
0.3	0.5218	1.0450	0.7465	1.0712	0.9208	1.0880
0.4	0.5932	1.0580	0.8516	1.0932	1.0528	1.1164
0.5	0.6533	1.0701	0.9408	1.1143	1.1656	1.1441
0.6	0.7051	1.0814	1.0185	1.1346	1.2644	1.1713
0.7	0.7506	1.0919	1.0873	1.1539	1.3525	1.1978
0.8	0.7910	1.1016	1.1490	1.1725	1.4320	1.2236
0.9	0.8274	1.1107	1.2048	1.1902	1.5044	1.2488
1.0	0.8603	1.1191	1.2558	1.2071	1.5708	1.2732
2.0	1.0769	1.1795	1.5995	1.3384	2.0288	1.4793
3.0	1.1925	1.2102	1.7887	1.4191	2.2889	1.6227
4.0	1.2646	1.2287	1.9081	1.4698	2.4556	1.7201
5.0	1.3138	1.2402	1.9898	1.5029	2.5704	1.7870
6.0	1.3496	1.2479	2.0490	1.5253	2.6537	1.8338
7.0	1.3766	1.2532	2.0937	1.5411	2.7165	1.8674
8.0	1.3978	1.2570	2.1286	1.5526	2.7654	1.8921
9.0	1.4149	1.2598	2.1566	1.5611	2.8044	1.9106
10.0	1.4289	1.2620	2.1795	1.5677	2.8363	1.9249
20.0	1.4961	1.2699	2.2881	1.5919	2.9857	1.9781
30.0	1.5202	1.2717	2.3261	1.5973	3.0372	1.9898
40.0	1.5325	1.2723	2.3455	1.5993	3.0632	1.9942
50.0	1.5400	1.2727	2.3572	1.6002	3.0788	1.9962
100.0	1.5552	1.2731	2.3809	1.6015	3.1102	1.9990
∞	1.5707	1.2733	2.4050	1.6018	3.1415	2.0000

TABLE 11-3 Bessel functions of the first kind

Z	$J_0(Z)$	$J_1(Z)$
0.0	1.0000	0.0000
0.1	0.9975	0.0499
0.2	0.9900	0.0995
0.3	0.9776	0.1483
0.4	0.9604	0.1960
0.5	0.9385	0.2423

(*Continued*)

TABLE 11-3 (Continued)

z	$J_0(Z)$	$J_1(Z)$
0.6	0.9120	0.2867
0.7	0.8812	0.3290
0.8	0.8463	0.3688
0.9	0.8075	0.4059
1.0	0.7652	0.4400
1.1	0.7196	0.4709
1.2	0.6711	0.4983
1.3	0.6201	0.5220
1.4	0.5669	0.5419
1.5	0.5118	0.5579
1.6	0.4554	0.5699
1.7	0.3980	0.5778
1.8	0.3400	0.5815
1.9	0.2818	0.5812
2.0	0.2239	0.5767
2.1	0.1666	0.5683
2.2	0.1104	0.5560
2.3	0.0555	0.5399
2.4	0.0025	0.5202
2.5	−0.0484	0.4971
2.6	−0.0968	0.4708
2.7	−0.1424	0.4416
2.8	−0.1850	0.4097
2.9	−0.2243	0.3754
3.0	−0.2601	0.3391

EXAMPLE 11-6 Transient conduction in a sphere

To obtain a hard surface on a metal object with a somewhat softer core for toughness, heat treatment is required. The hot metal is plunged into a much colder bath so that the surface temperature drops very rapidly. The rapid temperature drop results in a favorable grain structure that causes a very hard material. The slower cooling of the core results in a softer material.

At Ball Bearings 'R Us, 316 stainless-steel ball bearings 6 mm in diameter are quenched in an oil bath to harden the outside surface. The balls are uniformly heated to 925°C, then plunged into a 25°C oil bath. Because of agitation in the bath, the convective heat transfer coefficient is 3600 W/m²·K. The surface temperature of the ball must be lowered to 200°C for adequate hardening to occur.

a) Determine the time required for the ball surface to reach 200°C (in s).

b) Determine the centerline temperature when the surface temperature reaches 200°C.

c) Determine the energy removed from each ball bearing during the cooling process (in kJ).

Approach:

This is a transient conduction heat transfer problem. The first task is to check whether we can solve it with the lumped system approach or if we need to use a one-dimensional analysis. If the one-dimensional analysis is needed, then we can use either the Heisler charts or the one-term approximation to get the centerline and surface temperatures and the heat transfer.

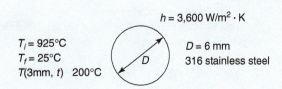

$T_i = 925°C$
$T_f = 25°C$
$T(3mm, t)$ $200°C$

$h = 3,600$ W/m$^2 \cdot$ K

$D = 6$ mm
316 stainless steel

Assumptions:

Solution:

a) The Biot number is $Bi = hL_{char}/k$, where $L_{char} = V/A = (4/3\pi r_o^3)/(4\pi r_o^2) = r_o/3$, so that for this sphere $Bi = h(r_o/3)/k$. From Table A-2, the properties of 316 stainless steel are: $k = 13.4$ W/m·K; $c_p = 468$ J/kg·K; $\rho = 8,238$ kg/m^3; and $\alpha = 3.48 \times 10^{-6}$m^2/s. Therefore,

$$Bi = \frac{h(r_0/3)}{k} = \frac{(3600\,\text{W}/\text{m}^2\cdot\text{K})(0.003\,\text{m}/3)}{(13.4\,\text{W}/\text{m}^2\cdot\text{K})} = 0.27$$

Because $Bi > 0.1$, the lumped system approach is not acceptable.

Assuming one-dimensional conduction with constant properties and no internal heat generation [A1][A2][A3], we choose to use the one-term approximation for the one-dimensional transient heat transfer in the ball bearing, Eq. 11-60:

A1. Conduction is one dimensional in r-direction.
A2. Properties are constant.
A3. There is no internal heat generation.

$$\frac{T(r,t) - T_f}{T_i - T_f} = C_1 \exp\left(-\lambda_1^2 \tau\right) \frac{\sin(\lambda_1 r/r_0)}{\lambda_1 r/r_0}$$

After solving for time, we will check to be sure that $\tau > 0.2$ as required, using Eq. 11-60. The three temperatures on the left-hand side of the equation are known. The radii to use in the sine term are all known ($r = r_0$), and we can obtain the time from the Fourier number, $\tau = \alpha t/r_0^2$. We can obtain the constants λ_1 and C_1 from Table 11-2 if we have the Biot number.

We first calculate the Biot number using the characteristic length required for the one-term approximation:

$$Bi = \frac{hr_0}{k} = \frac{(3600\,\text{W}/\text{m}^2\cdot\text{K})(0.003\,\text{m})}{(13.4\,\text{W}/\text{m}^2\cdot\text{K})} = 0.806$$

From Table 11-2 we obtain $\lambda_1 \approx 1.4320$ and $C_1 \approx 1.2236$. Solving Eq. 11-60 and letting $r = r_0$, we obtain:

$$\tau = -\frac{1}{\lambda_1^2}\ln\left\{\left[\frac{\lambda_1 r_0/r_0}{C_1 \sin(\lambda_1 r_0/r_0)}\right]\left[\frac{T(r_0,t) - T_\infty}{T_i - T_\infty}\right]\right\}$$

$$= -\frac{1}{1.4320^2}\ln\left\{\left[\frac{1.4320}{1.2236\sin(1.4320)}\right]\left[\frac{200 - 25}{925 - 25}\right]\right\} = 0.717$$

Note that the argument of the sine function is in radians.

Because $\tau > 0.2$, the one-term approximation is valid, and since $\tau = \alpha t/r_0^2$,

$$t = \frac{\tau r_0^2}{\alpha} = \frac{0.717\,(0.003\,\text{m})^2}{3.48 \times 10^{-6}\text{m}^2/\text{s}} = 1.85\,\text{s}$$

b) To obtain the centerline temperature when the surface temperature reaches 200°C, we use the same equation but recognize that at the centerline, where $r = 0$, in the limit as $\lambda_1 r / r_0 \to 0$, then $[\sin(\lambda_1 r / r_0)] / (\lambda_1 r / r_0) \to 1$. Thus,

$$T(0, t) = T_f + (T_i - T_f) C_1 \exp\left(-\lambda_1^2 \tau\right)$$

$$= 25°C + (925°C - 25°C)(1.2236) \exp\left[-(1.4320)^2 (0.717)\right] = 278.1°C$$

c) The energy removed from each ball bearing is determined using Eq. 11-63:

$$\left(\frac{Q}{Q_{max}}\right)_{sphere} = 1 - 3\theta_{0,\,sphere} \frac{\sin \lambda_1 - \lambda_1 \cos \lambda_1}{\lambda_1^3}$$

$$Q_{max} = mc_p (T_i - T_f) = \rho V c_p (T_i - T_f) = \rho \left(4\pi r_0^3 / 3\right) c_p (T_i - T_f)$$

$$= \left(8238 \, \frac{kg}{m^3}\right) \left(\frac{4\pi}{3}\right) (0.003 \, m)^3 \left(468 \, \frac{J}{kg \cdot K}\right) (925 - 25) \, K = 392 \, J$$

The nondimensional centerline temperature is

$$\theta_{0,\,sphere} = \frac{T(0, t) - T_f}{T_i - T_f} = \frac{278.1 - 25}{925 - 25} = 0.281$$

Thus,

$$Q = \left[1 - 3\theta_{0,\,sphere} \frac{\sin \lambda_1 - \lambda_1 \cos \lambda_1}{\lambda_1^3}\right] Q_{max}$$

$$= \left[1 - 3(0.281) \frac{\sin(1.4320) - 1.4320 \cos(1.4320)}{1.4320^3}\right] (392 \, J) = 303 \, J$$

Comments:

If the lumped system approach had been used and the ball bearing cooled (uniformly) to 200°C, the time required would have been 1.75 s. The difference in time is not large, because the Biot number for the lumped system approach is not much larger than 0.1. Note that the Heisler charts could have been used to solve this problem. In an actual factory, with a production rate of ball bearings, the cooling rate of the oil would need to be determined. Let us assume we produce 5,000 balls/h. Then the heat transfer rate would be $\dot{Q} = (303 \, J/ball)(5000 \, ball/h)(1 \, h/3600 \, s) = 421 \, W$.

EXAMPLE 11-7   Transient conduction in a plane wall

Arctic Motor Company specializes in vehicles for use in cold climates. They equip the engine with an electric block heater to keep the engine and coolant warm during frigid nights. In some locations, by the time morning arrives, an insulating ice layer (at −30°C) has formed on the 4.5-mm thick glass windshield. The company design specification for the window defroster is that when the engine is started and the defroster turned on, the outside surface (which is touching the ice) of the windshield reaches 0°C in 1 min. The air from the defroster is at 40 °C. Determine the convective heat transfer coefficient required on the inside surface of the windshield. Assume the glass has a density of 2,300 kg/m³, a thermal conductivity of 1.4 W/m·K, and a specific heat of 800 J/kg·K.

Approach:

The schematic of the problem is shown.

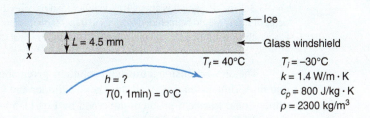

We assume the windshield is large and edge effects can be ignored, so that we can analyze it as a one-dimensional transient conduction problem in a plane. With the ice acting as an insulating layer, the windshield is half of a plane wall with a thickness of 2×4.5 mm $= 9$ mm. Using the Heisler chart approach (Figure 11-7), we can calculate the nondimensional temperature and the Fourier number, and then pick off the corresponding Biot number, which contains the heat transfer coefficient.

Assumptions:

A1. Heat transfer is one dimensional in the x-direction.

A2. Properties are constant.

A3. There is no internal heat generation.

Solution:

Assuming a one-dimensional conduction problem with constant properties and no internal heat generation [A1][A2][A3], we calculate the non-dimensional centerline temperature at 60 s into the transient for use with the Heisler charts as (see schematic):

$$\frac{T(0, 60\,s) - T_f}{T_i - T_f} = \frac{(0 - 40)°C}{(-30 - 40)°C} = 0.57$$

The Fourier number is

$$\tau = \frac{\alpha t}{L^2} = \frac{kt}{\rho c_p L^2} = \frac{\left(1.4\,\frac{W}{m \cdot K}\right)(60\,s)\left(\frac{1\,J/s}{1\,W}\right)}{\left(2300\,\frac{kg}{m^3}\right)\left(800\,\frac{J}{kg \cdot K}\right)(0.0045\,m)^2} = 2.25$$

From the Heisler chart, Figure 11-7a, for the midplane temperature we find

$$Bi^{-1} \approx 3 = \frac{k}{hL} \rightarrow h = \frac{k}{Bi^{-1}L} = \frac{1.4\,W/m \cdot K}{3\,(0.0045\,m)} = 104\,\frac{W}{m^2 \cdot K}$$

Note that the chart reading accuracy may be within 5%, though in some regions of the chart (such as toward the upper left hand corner of Figure 11-7a), the accuracy is poorer.

Comments:

Once the required heat transfer coefficient is determined, through the use of an appropriate heat transfer coefficient correlation for this geometry (as discussed in Chapter 12), then the needed air velocity past the window could be calculated. With the velocity known, and estimating the size of the flow area, the defroster fan characteristics could be determined and a fan specified.

If we had used the one-term approximation, Eq. 11-58 is applicable:

$$\frac{T(x, t) - T_f}{T_i - T_f} = C_1 \exp\left(-\lambda_1^2 \tau\right) \cos\left(\lambda_1 x/L\right)$$

Note that $\tau = \alpha t/L^2 >\sim 0.2$, so the one-term approximation is applicable. To evaluate λ_1 and C_1, we require the heat transfer. However, we do not have an explicit equation we can solve for the heat transfer coefficient. Rather, an iterative solution to the equation is needed. Doing so, we get $h = 93.3$ W/m^2·K, which is about 10% smaller than the previos answer. In an actual application, a factor of safety would be included in the heat transfer coefficient to ensure that the target temperature was reached in 1 min or less. Thus a higher value of the heat transfer coefficient would be used.

The one-dimensional transient solutions given above are for bodies that have at least one finite dimension (i.e., wall thickness, cylinder and sphere radii). There is another one-dimensional transient problem governed by Eq. 11-50—the case of a semi-infinite solid. Consider Figure 11-10, which depicts the transient temperature distribution that occurs, for example, at the earth's surface. Between day and night, the air temperature changes and the sun's heat flux goes from a maximum to zero. The effect of this variation penetrates only a very short distance into the ground within a period of 24 hours ($T_1(t)$ on Figure 11-10). On a longer time scale, from the hottest day of summer to the coldest day of winter, the effect of the air temperature variation, $T_2(t)$, will penetrate into the ground for only 1 or 2 m, depending on the soil composition, water content, and other factors; the time period is one year. The daily cycle is superimposed on the annual cycle. Below the penetration depth, the ground temperature remains constant (T_3), so it is irrelevant whether the depth of the solid below the penetration depth is 10 m, 1000 km, or infinity. Hence, we call this situation a *semi-infinite solid*. A much smaller-scale example is the heat treatment of a thick slab of metal. The hot slab is plunged into a water or oil bath that is at a much colder temperature than the metal. The resulting temperature penetration depth may be millimeters or less and the time period may be tens of seconds. While the slab surface and nearby regions are affected by the sudden change in the boundary condition, the interior of the slab remains at its initial temperature, and we also analyze this heat-treating situation as a semi-infinite conduction problem.

Temperature in a semi-infinite solid is governed by $\partial^2 T/\partial x^2 = (1/\alpha)\,\partial T/\partial t$. Three common sets of boundary conditions (Figure 11-11) have been used to obtain analytic solutions to this equation. The initial condition is the same $T(x \to \infty, t) = T_i$. The solutions use the *Gaussian error function*, erf (Z), and the *complementary error function*,

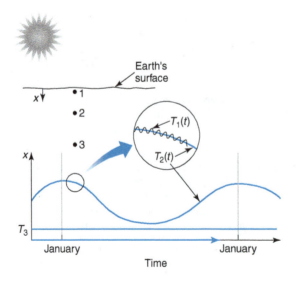

FIGURE 11-10 Temperature variation at the earth's surface through the year.

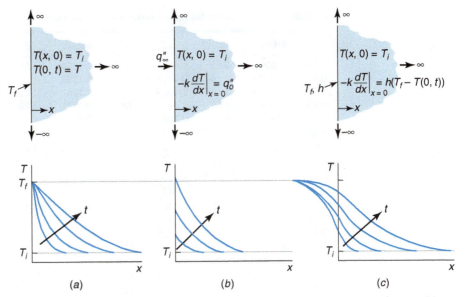

FIGURE 11-11   Examples of transient temperature distributions in a semi-infinite solid.
(a) Case 1. (b) Case 2. (3) Case 3.

erfc$(Z) = 1 - erf(Z)$, where Z is any positive number. This standard mathematical function is tabulated in Table 11-4 for different values of Z. The transient temperature solutions are:

Case 1: Step change in constant surface temperature from T_i to T_∞

$$\theta(x,t)_{semi\text{-}infinite} = \frac{T(x,t) - T_f}{T_i - T_f} = 1 - \text{erfc}\left(\frac{x}{2\sqrt{\alpha t}}\right)$$
$$q''(0,t) = \frac{k(T_f - T_i)}{\sqrt{\pi \alpha t}}$$

(11-64)

Case 2: Step change in constant surface heat flux from 0 to q_0''

$$\frac{T(x,t) - T_i}{(q_0''/k)\sqrt{\alpha t}} = \frac{2}{\sqrt{\pi}} \exp\left(\frac{-x^2}{4\alpha t}\right) - \sqrt{\frac{x^2}{\alpha t}}\,\text{erfc}\left(\frac{x}{2\sqrt{\alpha t}}\right)$$

(11-65)

Case 3: Step change in surface convection from 0 to h

$$\theta(x,t)_{semi\text{-}infinite} = \frac{T(x,t) - T_f}{T_i - T_f}$$
$$= 1 - \text{erfc}\left(\frac{x}{2\sqrt{\alpha t}}\right) + \left[\exp\left(\frac{hx}{k} + \frac{h^2\alpha t}{k^2}\right)\right]\left[\text{erfc}\left(\frac{x}{2\sqrt{\alpha t}} + \frac{h\sqrt{\alpha t}}{k}\right)\right]$$
$$q''(0,t) = h(T_f - T_i)\exp\left(\frac{h^2\alpha t}{k^2}\right)\text{erfc}\left(\frac{h\sqrt{\alpha t}}{k}\right)$$

(11-66)

In practice, Case 1 is approximated when a boiling or condensing fluid with a very high heat transfer coefficient is brought in contact with the surface. Note that this case is a

TABLE 11-4 **Gaussian complementary error function**

Z	erfc(Z)	Z	erfc(Z)	Z	erfc(Z)	Z	erfc(Z)	Z	erfc(Z)
0.00	1.0000	0.52	0.4621	1.04	0.1414	1.56	0.0274	2.08	0.00327
0.02	0.9774	0.54	0.4451	1.06	0.1339	1.58	0.0255	2.10	0.00298
0.04	0.9549	0.56	0.4284	1.08	0.1267	1.60	0.0237	2.12	0.00272
0.06	0.9324	0.58	0.4121	1.10	0.1198	1.62	0.0220	2.14	0.00247
0.08	0.9099	0.60	0.3961	1.12	0.1132	1.64	0.0204	2.16	0.00225
0.10	0.8875	0.62	0.3806	1.14	0.1069	1.66	0.0189	2.18	0.00205
0.12	0.8652	0.64	0.3654	1.16	0.1009	1.68	0.0175	2.22	0.00169
0.14	0.8431	0.66	0.3506	1.18	0.0952	1.70	0.0162	2.26	0.00139
0.16	0.8210	0.68	0.3362	1.20	0.0897	1.72	0.0150	2.30	0.00114
0.18	0.7991	0.70	0.3222	1.22	0.0845	1.74	0.0139	2.34	0.00094
0.20	0.7773	0.72	0.3086	1.24	0.0795	1.76	0.0128	2.38	0.00076
0.22	0.7557	0.74	0.2953	1.26	0.0748	1.78	0.0118	2.42	0.00062
0.24	0.7343	0.76	0.2825	1.28	0.0703	1.80	0.0109	2.46	0.00050
0.26	0.7131	0.78	0.2700	1.30	0.0660	1.82	0.0101	2.50	0.00041
0.28	0.6921	0.80	0.2579	1.32	0.0619	1.84	0.0093	2.55	0.00031
0.30	0.6714	0.82	0.2462	1.34	0.0581	1.86	0.0085	2.60	0.00024
0.32	0.6509	0.84	0.2349	1.36	0.0544	1.88	0.0078	2.65	0.00018
0.34	0.6306	0.86	0.2239	1.38	0.0510	1.90	0.0072	2.70	0.00013
0.36	0.6107	0.88	0.2133	1.40	0.0477	1.92	0.0066	2.75	0.00010
0.38	0.5910	0.90	0.2031	1.42	0.0446	1.94	0.0061	2.80	0.00008
0.40	0.5716	0.92	0.1932	1.44	0.0417	1.96	0.0056	2.85	0.00006
0.42	0.5525	0.94	0.1837	1.46	0.0389	1.98	0.0051	2.90	0.00004
0.44	0.5338	0.96	0.1746	1.48	0.0363	2.00	0.0047	2.95	0.00003
0.46	0.5153	0.98	0.1658	1.50	0.0339	2.02	0.0043	3.00	0.00002
0.48	0.4973	1.00	0.1573	1.52	0.0316	2.04	0.0039	3.20	0.00001
0.50	0.4795	1.02	0.1492	1.54	0.0294	2.06	0.0036	3.40	0.00000

special situation for Case 3 (convective boundary condition) with $h \rightarrow \infty$, which results in the surface temperature equaling the fluid temperature. Case 2 occurs when, for example, an electric resistance heater is pressed against a surface such that a known heat flux is imposed on the surface. It could also be an approximation of when a radiant heat source with a very high source temperature is directed toward a surface with a much lower temperature.

EXAMPLE 11-8 Semi-infinite Transient Conduction

During a fire investigation, an insurance company wants to estimate how long it would take for a large sheet of yellow pine 3 cm thick to reach ignition temperature if it is exposed to a fire on one side. The temperature of the fire is 625°C, and the heat transfer coefficient is 50 W/m²·K. The initial temperature of the material is 25°C, and the ignition temperature is 275°C. Estimate the time required (in s) for the material to start burning when suddenly exposed to this operating condition. Ignore radiation. The properties of the pine are: $\rho = 640$ kg/m³, $c_p = 2805$ J/kg·K, and $k = 0.15$ W/m·K.

Approach:

This is a semi-infinite conduction with a step change in surface convection (Eq. 11-66):

$$\frac{T(x,t) - T_f}{T_i - T_f} = 1 - \mathrm{erfc}\left(\frac{x}{2\sqrt{\alpha t}}\right) + \left[\exp\left(\frac{hx}{k} + \frac{h^2 \alpha t}{k^2}\right)\right]\left[\mathrm{erfc}\left(\frac{x}{2\sqrt{\alpha t}} + \frac{h\sqrt{\alpha t}}{k}\right)\right]$$

We want to determine the time, t, at the surface of the wood, $x = 0$, when $T(0, t) = 275$°C. The initial and fluid temperatures ($T_i = 25$°C and $T_f = 625$°C, respectively) are given, as are the heat transfer coefficient, h, and thermal conductivity, k. Everything is known except time, which is the quantity sought.

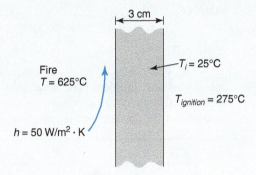

Assumptions:

A1. Semi-infinite conduction occurs.
A2. Properties are constant.
A3. There is no internal heat generation.

Solution:

Assuming semi-infinite conduction with constant properties and no internal heat generation [A1][A2][A3], we start with Eq. 11-66. At the surface, $x = 0$, so we can eliminate several terms from the equation. First, as shown in Table 11-4, erfc(0) = 1. Second, the two other terms involving x go to zero. We can calculate the thermal diffusivity from the given material properties, $\alpha = k/\rho c_p = 8.36 \times 10^{-8}\ \mathrm{m^2/s}$.

Substituting what is known (and leaving 0's and 1's to show where the terms were eliminated and/or evaluated), we obtain

$$\frac{(275 - 625)°C}{(25 - 625)°C} = 1 - 1 + \left[\exp\left(0 + \frac{h^2 \alpha t}{k^2}\right)\right]\left[\mathrm{erfc}\left(0 + \frac{h\sqrt{\alpha t}}{k}\right)\right]$$

On the right-hand side of the equation, h, α, and k are known. Only time is unknown. We cannot solve this equation explicitly for time. Hence, an iterative solution is required. This is a relatively straightforward iterative solution using common software. Without giving details of the iteration, we determine the time to be 34.5 s.

Comments:

To ensure that the semi-infinite approach is valid, we calculate the temperature of the side of the panel not exposed to the fire. If that temperature is not different from the initial temperature, then the semi-infinite assumption is reasonable. Using Eq. 11-66 in its entirety, now with $t = 34.5$ s and $x = 0.03$ m (the thickness of the wood), we calculate $T(0.03\ \mathrm{m}, 34.5\ \mathrm{s}) = 25°C$, which is the initial temperature.

Because the temperatures are high and from experience we know that fires radiate much energy, the assumption that radiation can be ignored should also be evaluated. One method is to compare the heat flux caused by convection to an estimated heat flux caused by radiation. Such a radiation analysis is beyond the scope of this text but can be found in textbooks on radiation heat transfer.

11.7 MULTI-DIMENSIONAL TRANSIENT CONDUCTION

There are applications in which a finite-size body can be treated as infinite in one or more dimensions, and the above one-dimensional transient solutions would be used. Consider heat treatment of a large metal slab whose length and width are much greater than its thickness. Far away from its edges, where edge effects would influence the cooling, the slab could be treated as one-dimensional. The same is true for a cylinder whose length is much longer than its radius. However, when all the dimensions of a body are of comparable size (e.g., a short cylinder), then the one-dimensional conduction analysis breaks down, and heat transfer occurs in two dimensions. Where multi-dimensional conduction occurs, we must solve the appropriate equation. For the short cylinder example, with no circumferential variations in conditions, constant properties, and no internal heat generation, Eq. 11-14 reduces to

$$\frac{1}{r}\frac{\partial}{\partial r}\left(r\frac{\partial T}{\partial r}\right) + \frac{\partial}{\partial z}\left(\frac{\partial T}{\partial z}\right) = \frac{1}{\alpha}\frac{\partial T}{\partial t} \tag{11-67}$$

The analytic solution to this equation involves an infinite series, as did the one-dimensional problem. The separation of variables approach is used, and the two-dimensional solution is expressed as the product of two one-dimensional solutions. Mathematically, this is expressed as:

$$\left[\frac{T(r,z,t)-T_f}{T_i-T_f}\right]_{\substack{short \\ cylinder}} = \left[\frac{T(z,t)-T_f}{T_i-T_f}\right]_{\substack{plane \\ wall}} \times \left[\frac{T(r,t)-T_f}{T_i-T_f}\right]_{\substack{infinite \\ cylinder}} \tag{11-68}$$

$$\theta(r,z,t)_{\substack{short \\ cylinder}} = \theta(z,t)_{\substack{plane \\ wall}} \times \theta(r,t)_{\substack{infinite \\ cylinder}}$$

Graphically, this solution (Figure 11-12a) is the intersection of an infinite plane wall and an infinite cylinder. A three-dimensional rectangular solid is formed by the intersection of three infinite plane walls (Figure 11-12b). Other two- and three-dimensional transient solutions for a variety of semi-infinite and finite geometries can be constructed from the

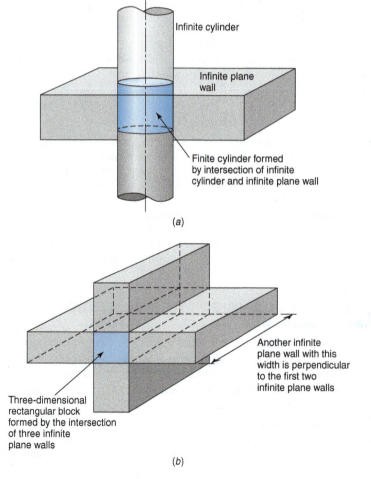

Infinite cylinder

Infinite plane wall

Finite cylinder formed by intersection of infinite cylinder and infinite plane wall

(a)

Another infinite plane wall with this width is perpendicular to the first two infinite plane walls

Three-dimensional rectangular block formed by the intersection of three infinite plane walls

(b)

FIGURE 11-12 (a) A short cylinder formed by the intersection of a plane wall and infinite cylinder. (b) A three-dimensional rectangular solid formed by the intersection of three infinite plane walls.

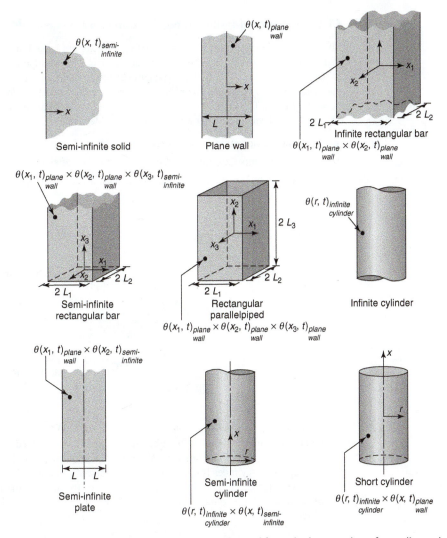

FIGURE 11-13 Multidimensional shapes formed from the intersection of one-dimensional solutions.

one-dimensional solutions shown above (Figure 11-13). Note that the same convective heat transfer coefficient must be used in all the one-dimensional solutions.

The total heat transfer to or from a multidimensional solid after a period of time can be determined using the one-dimensional solutions. For a two-dimensional body,

$$\left[\frac{Q}{Q_{max}} \right]_{2D} = \left[\frac{Q}{Q_{max}} \right]_1 + \left[\frac{Q}{Q_{max}} \right]_2 \left(1 - \left[\frac{Q}{Q_{max}} \right]_1 \right) \tag{11-69}$$

where the subscripts 1 and 2 indicate the two geometries. For a three-dimensional body,

$$\left[\frac{Q}{Q_{max}} \right]_{3D} = \left[\frac{Q}{Q_{max}} \right]_1 + \left[\frac{Q}{Q_{max}} \right]_2 \left(1 - \left[\frac{Q}{Q_{max}} \right]_1 \right)$$
$$+ \left[\frac{Q}{Q_{max}} \right]_3 \left(1 - \left[\frac{Q}{Q_{max}} \right]_1 \right) \left(1 - \left[\frac{Q}{Q_{max}} \right]_2 \right) \tag{11-70}$$

EXAMPLE 11-9   Multidimensional transient conduction in a short cylinder

Short plastic cylinders 6 cm long and 3 cm in diameter are heated in an oven prior to additional manufacturing steps. Initially, the cylinders are at 25°C. The manufacturing requires that no portion of the plastic be below 125°C. The oven temperature is 175°C, and the heat transfer coefficient on the cylinders is 10 W/m²·K. The plastic properties are: $\rho = 1215$ kg/m³, $k = 0.42$ W/m·K, and $c_p = 800$ J/kg·K. The supplier of the oven states that the plastic cylinders should be in the oven for 15 min. To confirm the supplier's recommendation, determine

a)   the center temperature of the cylinder (in °C) after 15 min.

b)   the surface temperature at the edge of the end (in °C) after 15 min.

c)   the rate at which heat must be added to the oven if 100 cylinders per minute are heated (in W).

Approach:

The schematic of the problem is shown.

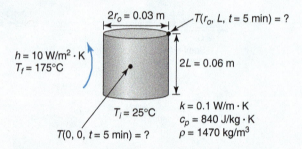

This is a transient conduction problem in a finite body. We first check the Biot number to see if the lumped system approach can be used. If it cannot, then we need to use a multidimensional approach. Note that if a multidimensional approach is required, from Figure 11-13 or Eq. 11-68, we see that a short cylinder is created from the intersection of an infinite plane wall and an infinite cylinder. Hence, the solution procedure would be to calculate the temperature response of the plane wall and the infinite cylinder, and then combine those results to obtain the desired quantities.

Assumptions:

A1. Properties are constant.

A2. Heat transfer is two-dimensional conduction.

A3. There is no internal heat generation.

Solution:

a) We begin by assuming constant properties [A1] and checking the lumped system Biot number. The characteristic length is $L_{char} = V/A$. The volume is $V = \pi D^2 2L/4 = \pi (0.03\,\text{m})^2 2(0.03\,\text{m})/4 = 4.24 \times 10^{-5}\,\text{m}^3$, and the surface area is $A = \pi D 2L + 2\pi D^2/4 = \pi (0.03\,\text{m})2(0.03\,\text{m}) + 2\pi (0.03\,\text{m})^2/4 = 0.00707\,\text{m}^2$, which gives a characteristic length of 0.006 m. Therefore,

$$ Bi = \frac{hL_{char}}{k} = \frac{(10\,\text{W}/\text{m}^2\cdot\text{K})\,(0.006\,\text{m})}{(0.42\,\text{W}/\text{m}^2\cdot\text{K})} = 0.143 $$

Although this is larger than 0.1, it is not significantly larger, and we could assume the lumped system approach is valid. However, we will use a multidimensional transient conduction calculation.

Assuming two-dimensional conduction, constant properties, and no internal heat generation [A1][A2][A3], and using Eq. 11-68, we see that a short cylinder is created from the intersection of an infinite plane wall and an infinite cylinder:

$$ \left[\frac{T\,(r,z,t) - T_f}{T_i - T_f}\right]_{\substack{short \\ cylinder}} = \left[\frac{T\,(z,t) - T_f}{T_i - T_f}\right]_{\substack{plane \\ wall}} \times \left[\frac{T\,(r,t) - T_f}{T_i - T_f}\right]_{\substack{infinite \\ cylinder}} $$

Thus, we need to evaluate the dimensionless temperatures for the plane wall and the infinite cylinder at both the centerline and at the edge of the end of the cylinder.

We assume two-dimensional heat transfer with constant properties, no internal heat generation, uniform heat transfer coefficient, and negligible radiation [A1][A2][A3][A4][A5]. We begin with the plane wall, whose thickness is $2L = 0.06$ m. The Fourier number is:

A4. The heat transfer coefficient is uniform over the surface.

A5. Radiation is ignored.

$$\tau_w = \frac{\alpha t}{L^2} = \frac{kt}{\rho c_p L^2} = \frac{(0.42 \text{ W/m·K}) (900 \text{ s})}{(1215 \text{ kg/m}^3) (800 \text{ J/kg·K}) (0.03 \text{ m})^2} = 0.432$$

Because this is greater than 0.2, we can proceed with the one-term approximation.

For the plane wall we recalculate the Biot number using the appropriate characteristic length, so that

$$Bi_w = \frac{hL}{k} = \frac{(10 \text{ W/m}^2\text{·K}) (0.03 \text{ m})}{0.42 \text{ W/m·}K} = 0.714$$

At this Biot number from Table 11-2, we obtain (by interpolation) $\lambda_1 = 0.7563$ and $C_1 = 1.0933$. Now using Eq. 11-58, we evaluate the nondimensional temperature at the centerline of the infinite plane wall ($x = 0$):

$$\theta_{0, \text{ plane} \atop \text{wall}} = \left[\frac{T(0, 900 \text{ s}) - T_f}{T_i - T_f} \right]_{\text{plane} \atop \text{wall}} = C_1 \exp \left(-\lambda_1^2 \tau \right)$$

$$= 1.0933 \exp \left[- (0.7563)^2 (0.432) \right] = 0.854$$

For the infinite cylinder, its Biot and Fourier numbers are

$$Bi_{\text{cylinder}} = \frac{hr_0}{k} = \frac{(10 \text{ W/m}^2\text{·K}) (0.015 \text{ m})}{0.42 \text{ W/m·K}} = 0.357$$

$$\tau_{\text{cylinder}} = \frac{\alpha t}{r_0^2} = \frac{kt}{\rho c_p r_0^2} = \frac{(0.42 \text{ W/m·K}) (900 \text{ s})}{(1215 \text{ kg/m}^3) (800 \text{ J/kg·K}) (0.015 \text{ m})^2} = 1.728$$

At this Biot number from Table 11-2, we obtain (by interpolation) $\lambda_1 = 0.8064$ and $C_1 = 1.0837$. Now using Eq. 11-59, we evaluate the nondimensional temperature at the centerline of the infinite cylinder ($r = 0$, $t = 900$ s):

$$\theta_{0, \text{ infinite} \atop \text{cylinder}} = \left[\frac{T(0, 900 \text{ s}) - T_f}{T_i - T_f} \right]_{\text{infinite} \atop \text{cylinder}}$$

$$= C_1 \exp \left(-\lambda_1^2 \tau \right) = 1.0837 \exp \left[- (0.8064)^2 (1.728) \right] = 0.352$$

Finally, we solve Eq. 11-68 for the center temperature:

$$T(0, 0, 900 \text{ s}) = T_f + (T_i - T_f) \left[\frac{T(0, 900 \text{ s}) - T_f}{T_i - T_f} \right]_{\text{plane} \atop \text{wall}} \times \left[\frac{T(0, 900 \text{ s}) - T_f}{T_i - T_f} \right]_{\text{infinite} \atop \text{cylinder}}$$

$$= 175°C + (25°C - 175°C) (0.854) (0.352) = 130°C$$

b) The same approach is used to find the temperature at the edge of the end of the cylinder. We determine the surface temperature of the plane wall ($x = L = 0.03$ m) and that of the infinite cylinder ($r = r_0 = 0.015$ m). For the plane wall,

$$\left[\frac{T\,(0.03\,\text{m}, 900\,\text{s}) - T_f}{T_i - T_f} \right]_{\substack{plane \\ wall}} = \left[C_1 \exp\left(-\lambda_1^2 \tau\right) \right] \cos\left(\lambda_1 x/L\right)$$

$$= [0.854] \cos\,(0.7563) = 0.621$$

For the infinite cylinder,

$$\left[\frac{T\,(0.015\,\text{m}, 900\,\text{s}) - T_f}{T_i - T_f} \right]_{\substack{infinite \\ cylinder}} = \left[C_1 \exp\left(-\lambda_1^2 \tau\right) \right] J_0\left(\lambda_1 r/r_0\right)$$

$$= [0.352]\, J_0\,(0.8064) = 0.297$$

Finally, for the edge temperature,

$$T\,(0.03\,\text{m}, 0.015\,\text{m}, 900\,\text{s})$$

$$= T_f + \left(T_i - T_f\right) \left[\frac{T\,(0.03\,\text{m}, 900\,\text{s}) - T_f}{T_i - T_f} \right]_{\substack{plane \\ wall}} \times \left[\frac{T\,(0.015\,\text{m}, 900\,\text{s}) - T_f}{T_i - T_f} \right]_{\substack{infinite \\ cylinder}}$$

$$= 175°\text{C} + (25°\text{C} - 175°\text{C})\,(0.621)\,(0.297) = 147°\text{C}$$

c) For the heat transfer into one of the short cylinders, we use Eq. 11-69:

$$\left[\frac{Q}{Q_{max}} \right]_{2D} = \left[\frac{Q}{Q_{max}} \right]_1 + \left[\frac{Q}{Q_{max}} \right]_2 \left(1 - \left[\frac{Q}{Q_{max}} \right]_1 \right)$$

along with the appropriate expressions for an infinite plane wall and an infinite cylinder. For the heat transfer into the plane wall, use Eq. 11-61:

$$\left(\frac{Q}{Q_{max}} \right)_{\substack{plane \\ wall}} = 1 - \theta_{0,\,plane \atop wall} \frac{\sin \lambda_1}{\lambda_1} = 1 - 0.854 \frac{\sin\,(0.7563)}{0.7563} = 0.225$$

For the heat transfer into the infinite cylinder, use Eq. 11-62 (evaluating the Bessel function with Table 11-3):

$$\left(\frac{Q}{Q_{max}} \right)_{\substack{infinite \\ cylinder}} = 1 - 2\theta_{0,\,infinite \atop cylinder} \frac{J_1\,(\lambda_1)}{\lambda_1} = 1 - 2\,(0.352) \frac{0.3712}{0.8064} = 0.676$$

We solve Eq. 11-69 for the maximum possible heat transfer into one cylinder using

$$Q_{max} = m c_p \left(T_f - T_i\right) = \rho V c_p \left(T_f - T_i\right)$$

$$= \left(1215\,\text{kg/m}^3\right) \left(4.24 \times 10^{-5} \text{m}^3\right) \left(800\,\text{J/kg·K}\right) (175 - 25)\,\text{K} = 6182\,\text{J}$$

$$Q = Q_{max} \left\{ \left[\frac{Q}{Q_{max}} \right]_1 + \left[\frac{Q}{Q_{max}} \right]_2 \left(1 - \left[\frac{Q}{Q_{max}} \right]_1 \right) \right\}$$

$$= (6182\,\text{J})\,\{0.225 + 0.676\,(1 - 0.225)\} = 4630\,\text{J}$$

For the heat transfer rate to process 100 cylinders per minute:

$$\dot{Q} = (4630\,\text{J/cylinder})\,(100\,\text{cylinder/min})\,(1\,\text{min}/60\,\text{s}) = 7717\,\text{W}$$

Comments:

For better processing, the plastic temperature probably should be more uniform; that is, there should be a smaller temperature difference between the center and the surface of the cylinder. One way to accomplish this would be to reduce the heating rate. Reduction in the heat transfer coefficient could achieve this. Because the Biot number for the lumped system approach was not significantly greater than 0.1, using that approach probably would not have had much effect on the final answers.

11.8 EXTENDED SURFACES

Consider the basic equation governing convective heat transfer:

$$\dot{Q} = hA_s\left(T_s - T_f\right) \tag{11-71}$$

If we wanted to increase the heat transfer rate in a given application, how could we accomplish that task? The heat transfer coefficient, h, could be increased, for example, by switching from single-phase natural convection to boiling. However, in most applications, the fluids are fixed, and such a drastic change in the mode of heat transfer is not feasible. For a single-phase forced convection situation, a larger pump or fan to increase flow would result in an increased h, but the cost may be too much or the additional flow may affect the process so much as to make the increased flow impractical. Likewise, we could increase the temperature difference between the surface and the fluid. Again, though, often the operating conditions are such that large changes in either temperature may not be realistic. Hence we are left with only the heat transfer area, A_s, as the parameter with which we have some flexibility.

The heat transfer area can be increased by adding *fins* or *extended surfaces*—thin or slender pieces of metal—to the primary heat transfer surface (Figure 11-14). Applications of fins include car radiators, finned evaporators and condensers in air conditioners, single-finned tubes used for convectors in home-heating situations, and a myriad of heat exchanger types in industrial and commercial installations. Fins are used with all modes of heat transfer (natural convection, forced convection, boiling, and condensation) and many types of fluids. Note that the benefit of increased heat transfer with added fins should be balanced against increased pressure drop/pumping power that may accompany the added surface area, as well as increased cost compared to an unfinned tube or surface.

The fin may be integral to the surface (e.g., the cooling fins on an air-cooled motorcycle engine are cast as part of the cylinder head, or the fins on a tube are formed during the swaging operations), or the fins may be manufactured during an operation separate from that used to form the base tube. The fins then are either brazed or press-fit onto the tube. Many different fin shapes are possible. When considering the use of fins, the choice of the number of fins, spacing, length, thickness, shape, material, and so on will depend on both heat transfer and fluid flow considerations. Manufacturing, maintenance, and operating costs must also be considered in fin design.

Shown in Figure 11-15 is a schematic of the heat transfer processes that occur in a fin. We assume the base is hotter than the fluid, although the same processes apply when the fluid is hotter than the base. Heat conducts from the base into the fin and is removed

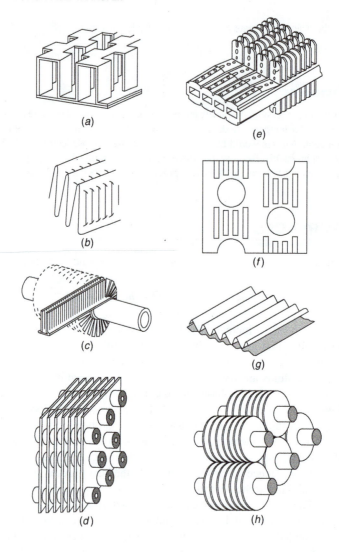

FIGURE 11-14 Examples of finned tubes and finned heat exchangers. (a) Offset strip fins used in plate-fin heat exchangers. (b) Louvered fins used in automotive heat exchangers. (c) Segmented fins for circular tubes. (d) Plate-fin and tube heat exchangers. (e) Integral aluminum strip finned tube. (f) Louvered tube-and-plate fin. (g) Corrugated plates used in rotary regenerators. (h) Individually finned tubes. (*Source*: R. L. Webb, *Principles of Enhanced Heat Transfer*, Wiley, New York, 1994. Used by permission.)

by convection at the outer surface of the fin. The fin is usually made of a high-thermal-conductivity material to facilitate the flow of heat from the base to the tip. With heat being convected away from the surface, the temperature of the fin decreases from base to tip. Often much of the fin is at a temperature not significantly different from the base temperature.

To calculate the heat transfer rate from a simple fin, we invoke several assumptions:

1. The heat transfer coefficient is uniform over the fin surface.
2. Temperature varies only along the length of the fin, and temperature does not vary across the fin.
3. The fin thermal conductivity is constant.
4. The fin shape is constant over the length of the fin.
5. There is no internal heat generation.
6. Conduction is steady.
7. There is no thermal resistance between the fin and the base material.

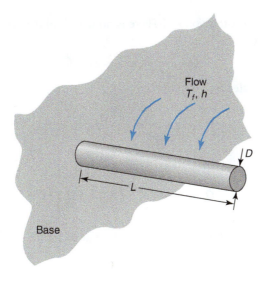

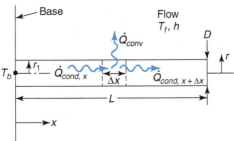

FIGURE 11-15   Schematic of heat transfer in a pin fin.

Although any constant-shape fin could be used, we apply these assumptions to the pin fin shown in Figure 11-15. According to assumption 2, temperature, T, varies only with x. In reality, temperature will vary slightly with r. Because the pin fin is always long compared to its diameter and has a high thermal conductivity, this transverse conductive resistance is usually negligible. Hence, temperature varies with distance x along the fin but not with r.

A criterion can be developed to indicate when assumption 2 is suspect. In Section 11.5, we developed the Biot number, which compared conduction resistance to convection resistance in transient systems; when the internal resistance (conduction) was small compared to the external resistance (convection), then temperature variations in the solid were negligible. The same approach can be used with fins to assess assumption 2. That is, if we compare the fin conduction resistance in the transverse direction to the convection resistance, we can ignore temperature variations in the transverse direction if

$$Bi = \frac{hL_{char}}{k} <\sim 0.2 \tag{11-72}$$

where the characteristic length, L_{char}, is equal to fin thickness in a rectangular fin and diameter in a pin fin.

To begin the analysis, we define a control volume of differential width Δx and finite radius r_1, as shown in Figure 11-15. Note that heat crosses the boundary in three places. Heat flows in by conduction at the left face, out by conduction at the right face, and out by convection over the curved outer surface of the fin. We apply conservation of energy to the closed-system control volume. There is no work and the system is steady, so for this control volume

$$\dot{Q}_{cond,\, x} - \dot{Q}_{cond,\, x+\Delta x} - \dot{Q}_{conv} = 0 \tag{11-73}$$

The heat transfer rates are given by:

$$\dot{Q}_{cond,\, x} = -kA_x \left.\frac{dT}{dx}\right|_x$$

$$\dot{Q}_{cond,\, x+\Delta x} = -kA_x \left.\frac{dT}{dx}\right|_{x+\Delta x} \tag{11-74}$$

$$\dot{Q}_{conv} = hA_f \left(T - T_f\right) = hp\Delta x \left(T - T_f\right)$$

where A_x is the cross-sectional area of the fin, A_f is the fin surface area, and p is the perimeter of the fin. Substituting Eq. 11-74 into Eq. 11-73 and simplifying results in

$$\left[\frac{\left.\dfrac{dT}{dx}\right|_{x+\Delta x} - \left.\dfrac{dT}{dx}\right|_x}{\Delta x} \right] - \frac{hp}{kA_x}\left(T - T_f\right) = 0 \tag{11-75}$$

Taking the limit as $\Delta x \to 0$ (the first bracketed term in Eq. 11-75 is the definition of a second derivative),

$$\frac{d^2T}{dx^2} - \frac{hp}{kA_x}\left(T - T_f\right) = 0 \tag{11-76}$$

This fin equation must be solved for $T(x)$. It can be applied to any fin with a cross-sectional area that does not vary with x, with constant thermal conductivity, and a heat transfer coefficient that does not vary along the fin. Once the temperature distribution is determined, the heat transfer rate can be calculated.

To solve the fin equation, it is necessary to specify boundary conditions. Typically, the temperature is known at the base of the fin (i.e., at $x = 0$). At the tip of the fin (i.e., at $x = L$), several different boundary conditions are possible. We can specify the temperature or the heat transfer coefficient, or the tip might be treated as an insulated surface. Each of these is an approximation to the real behavior. Usually, the predicted overall performance of the fin does not depend critically on the choice of fin-tip boundary condition because the tip has a small area compared to the rest of the fin. As an illustration, the fin equation is solved with an adiabatic boundary condition at the tip. Thus, the boundary conditions are

$$T = T_b \qquad \text{at} \quad x = 0$$
$$dT/dx = 0 \qquad \text{at} \quad x = L \tag{11-77}$$

We introduce a change in variable to convert Eq. 11-76 into a homogeneous differential equation. Let $\theta(x) = T(x) - T_f$ and for convenience define

$$m^2 = \frac{hp}{kA_x}$$

Thus, Eq. 11-76 is converted to

$$\frac{d^2\theta}{dx^2} - m^2\theta = 0 \tag{11-78}$$

with boundary conditions

$$\theta = \theta_b = T_b - T_f \qquad \text{at} \quad x = 0$$

$$d\theta/dx = 0 \qquad \text{at} \quad x = L \tag{11-79}$$

The general solution to Eq. 11-78 is

$$\theta(x) = C_1 \exp(mx) + C_2 \exp(-mx) \tag{11-80}$$

where C_1 and C_2 are constants to be determined from the boundary conditions (Eq. 11-79). Solving for the constants and substituting back into the general solution results in the temperature distribution in the fin:

$$\theta(x) = (T_b - T_f) \left[\frac{e^{mx}}{1 + e^{2mL}} + \frac{e^{-mx}}{1 + e^{-2mL}} \right] \tag{11-81}$$

Using the definition of the hyperbolic cosine, this equation can be rewritten in the form

$$\theta(x) = (T_b - T_f) \frac{\cosh[m(L - x)]}{\cosh(mL)} \tag{11-82}$$

The heat transfer rate from this fin can be determined in two ways, but the same result is obtained. All heat convected from the surface must conduct through the base of the fin first. Hence,

$$\dot{Q} = -kA_x \frac{dT}{dx}\bigg|_{x=0} = (T_b - T_f)\sqrt{hpkA_x}\tanh(mL) \tag{11-83}$$

or we can integrate along the fin surface to obtain all the heat transfer convected from the fin:

$$\dot{Q} = \int_{A_s} h\,\Delta T\,dA_s = \int_0^L h(T - T_f)p\,dx = (T_b - T_f)\sqrt{hpkA_x}\tanh(mL) \tag{11-84}$$

Because there are many parameters that can be changed during the design of a fin, we need a method for comparing the effects of different geometries. In addition, we would like to have a method that permits us to easily utilize these results. When we assess turbines and compressors, we compare their actual performance to their ideal performance. For fins we do likewise. We define *fin efficiency* as

$$\eta_f = \frac{\dot{Q}_{act,fin}}{\dot{Q}_{ideal,fin}} \tag{11-85}$$

The actual heat transfer rate, $\dot{Q}_{act,\,fin}$, is calculated from equations such as Eq. 11-83. The *ideal heat transfer rate from a fin*, $\dot{Q}_{ideal,\,fin}$, is the maximum possible heat transfer rate that would occur when the entire fin is at the fin base temperature. This is obtained from

$$\dot{Q}_{ideal,\,fin} = hA_f(T_b - T_f) \tag{11-86}$$

When Eq. 11-83 and Eq. 11-86 are combined in Eq. 11-85 and simplified, the fin efficiency is

$$\boxed{\eta_f = \frac{\tanh(mL)}{mL}} \tag{11-87}$$

TABLE 11-5 **Thermal performance of uniform cross-section fins ($m^2 = hp/kA_x$)**

Case	Tip boundary condition	Thermal performance
A	Adiabatic tip $dT/dx\big|_{x=L} = 0$	$T(x) - T_f = (T_b - T_f)\dfrac{\cosh\left[m(L-x)\right]}{\cosh(mL)}$ $\dot{Q} = (T_b - T_f)\sqrt{hpkA_x}\tanh(mL)$ $\eta_f = \dfrac{\tanh(mL)}{mL}$
B	Convection from tip $h\left[T(L) - T_f\right] = -k\dfrac{dT}{dx}\bigg|_{x=L}$	$T(x) - T_f = (T_b - T_f)\dfrac{\cosh\left[m(L-x)\right] + (h/mk)\sinh\left[m(L-x)\right]}{\cosh(mL) + (h/mk)\sinh mL}$ $\dot{Q} = (T_b - T_f)\sqrt{hpkA_x}\dfrac{\sinh mL + (h/mk)\cosh mL}{\cosh(mL) + (h/mk)\sinh mL}$ $\eta_f = \dfrac{\sinh mL + (h/mk)\cosh mL}{\cosh(mL) + (h/mk)\sinh mL}\left(\dfrac{1}{mL}\right)$
C	Fixed tip temperature $T(L) = T_L\text{(known)}$	$T(x) - T_f = (T_b - T_f)\dfrac{\left(\dfrac{T_L - T_f}{T_b - T_f}\right)\sinh(mx) + \sinh\left[m(L-x)\right]}{\sinh(mL)}$ $\dot{Q} = (T_b - T_f)\sqrt{hpkA_x}\dfrac{\cosh(mL) - \left(\dfrac{T_L - T_f}{T_b - T_f}\right)}{\sinh(mL)}$ $\eta_f = \dfrac{\cosh(mL) - \left(\dfrac{T_L - T_f}{T_b - T_f}\right)}{mL\sinh(mL)}$
D	Very long fin ($L \to \infty$) $T(L) = T_f$	$T(x) - T_f = (T_b - T_f)\exp(-mx)$ $\dot{Q} = (T_b - T_f)\sqrt{hpkA_x}$ $\eta_f = \dfrac{1}{mL}$

This expression is valid for any fin with a constant cross-sectional area, a uniform heat transfer coefficient, and an insulated fin tip.

Solutions for fins with other tip boundary conditions are given in Table 11-5. Note that in Case B, the terms preceded by (h/mk) are associated with the convection from the tip. If h is set to zero (insulated tip) in these terms, then the expressions for the convective tip fin (Case B) collapse to the insulated tip fin expressions (Case A). Likewise, if a fin is very long ($mL >\sim 3$), then Case D is the limit of Case A.

Fin efficiencies for several other fin shapes, including circumferential fins and fins with non-uniform cross-sectional areas, are given in Figure 11-16 and Figure 11-17.

The expressions for the fin with convection from its tip are cumbersome compared to those of the insulated tip fin. We can obtain a reasonable approximation to the convective

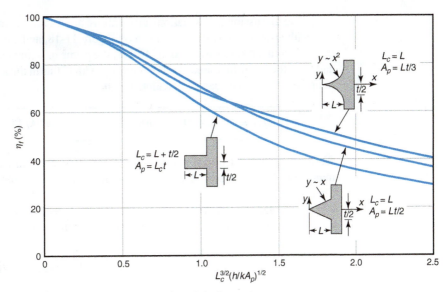

FIGURE 11-16 Fin efficiency of straight fins.
(*Source*: F. P. Incropera and D. P. DeWitt, *Introduction to Heat Transfer*, 3rd ed., Wiley, New York, 1996. Used with permission.)

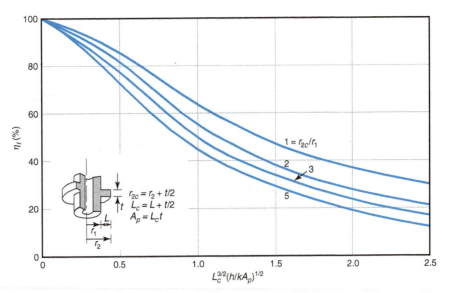

FIGURE 11-17 Fin efficiency of circular fins.
(*Source*: F. P. Incropera and D. P. DeWitt, *Introduction to Heat Transfer*, 3rd ed., Wiley, New York, 1996. Used with permission.)

tip fin if in the insulated tip fin expressions we use a "corrected" length, L^*, to account for the additional area of the fin tip.

$$L^* = L + A_x/p \tag{11-88}$$

For the pin fin, $L^* = L + D/4$, and for a rectangular fin, $L^* = L + t/2$, where t is the fin thickness.

The discussion and equations given above are for single fins. In practice, we usually deal with arrays of fins, such as shown on Figure 11-18, and we must account for the heat transfer from the base surface area, A_b (i.e., the prime or unfinned area on the base surface) and from the fin surface area, A_f. For a surface with N fins in the array, the total heat transfer to or from the surface can be calculated with

$$\dot{Q}_{tot} = \dot{Q}_{base} + \dot{Q}_{fins} = hA_b \left(T_b - T_f\right) + \eta_f hNA_f \left(T_b - T_f\right) \tag{11-89}$$

We assume the heat transfer coefficient is the same on both the base and fin areas. We manipulate Eq. 11-89 so that we can express the heat transfer in terms of a thermal resistance. This is done as follows.

The total heat transfer area is equal to $A_{tot} = A_b + NA_f$. Solve this for the base surface area and substitute into Eq. 11-89. The resulting equation is rearranged to give

$$\dot{Q}_{tot} = \left[1 - \frac{NA_f}{A_{tot}} \left(1 - \eta_f\right)\right] hA_{tot} \left(T_b - T_f\right) \tag{11-90}$$

We now define a total or *overall surface efficiency*, η_o, as

$$\eta_o = \frac{\dot{Q}_{tot}}{\dot{Q}_{max}} \tag{11-91}$$

where

$$\dot{Q}_{max} = hA_{tot} \left(T_b - T_f\right) \tag{11-92}$$

is the maximum possible heat transfer from the total surface. That occurs when the total surface area (base plus fins) is at the base temperature. Substituting Eq. 11-90 and Eq. 11-92 into Eq. 11-91 we obtain

$$\eta_o = 1 - \frac{NA_f}{A_{tot}} \left(1 - \eta_f\right) \tag{11-93}$$

To calculate the total heat transfer from an array of fins, use Eq. 11-91, Eq. 11-92, and Eq. 11-93 along with the appropriate expression for fin efficiency.

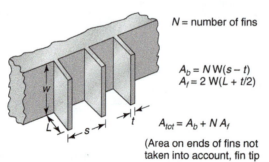

N = number of fins

$A_b = NW(s - t)$
$A_f = 2W(L + t/2)$

$A_{tot} = A_b + NA_f$

(Area on ends of fins not taken into account, fin tip area is.)

FIGURE 11-18   Array of fins on a plane wall.

We can also recast the equation to be consistent with the thermal resistance concept introduced earlier:

$$R_{fin,o} = \frac{T_b - T_f}{\dot{Q}_{tot}} = \frac{1}{\eta_o h A_{tot}}$$ (11-94)

A second indicator of fin performance is called the *fin effectiveness*, ε_f, which is defined as the ratio of the heat transfer from a fin with base area A_x to the heat transfer rate from the same base area that would exist without the fin:

$$\varepsilon_f = \frac{\eta_f h A_f \left(T_b - T_f\right)}{h A_x \left(T_b - T_f\right)} = \frac{A_f}{A_x} \eta_f$$ (11-95)

If $\varepsilon_f = 1$, then the addition of the fin does not help the heat transfer; the added material is wasted. If $\varepsilon_f < 1$, then the fin insulates the surface. Hence, the value of ε_f should be as large as possible, taking into account practical considerations, and fins with $\varepsilon_f < \sim 2$ may not be justified. A fin is often specified as the result of optimization, taking into account cost, weight, manufacturability, pressure drop, and so on.

While it is possible to have very high values of fin effectiveness, practicality again suggests there are limitations. Consider a very long fin, whose efficiency is given by $\eta_f = 1/mL$. If we substitute this into the expression for fin effectiveness and simplify, we obtain

$$\varepsilon_{f, long\, fin} = \left(\frac{kp}{h A_x} \right)^{1/2}$$

To obtain a large value of fin effectiveness, we would want to use the fin when the convective heat transfer coefficient, h, is low. For example, consider the radiator used in a car; fins are used on the air side (low heat transfer coefficient) while water (high heat transfer coefficient) flows inside unfinned tubes. The fin thermal conductivity, k, should be high; aluminum and copper are often used, though steel may be used in some applications. The ratio of fin perimeter to cross-sectional area (p/A_x) should be large, so this suggests that slender or thin fins be used. Again, examination of a car radiator will show very thin fins with very close spacing. This combination is typical because it ensures a large surface area without impeding the flow so much that the heat transfer coefficient is reduced to an unacceptable level.

We can make a fin as long as we want, but the law of diminishing returns comes into play. Consider the ratio the heat transfer rate for a fin with an adiabatic tip ($\dot{Q}_{finite}$) with the heat transfer rate for an infinitely long fin, ($\dot{Q}_\infty$), $\dot{Q}_{finite}/\dot{Q}_\infty = \tanh(mL)$. Figure 11-19

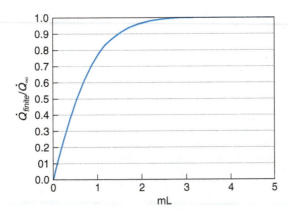

FIGURE 11-19 Comparison of heat transfer from a finite fin versus an infinite fin.

shows the comparative magnitude of the heat transfer rates as the value of mL is increased. When the ratio approaches 1, the fin can be considered infinitely long. Notice that after a value of $mL \approx 2.5$ or so, very little increase in heat transfer is obtained by lengthening the fin. Hence, the additional (minor) increase in heat transfer probably cannot be justified for the additional cost of the longer fin.

EXAMPLE 11-10 Extended surface (fin) heat transfer

An electronic device 48 mm by 48 mm dissipates 25 W and is cooled by air flowing over its surface. The air temperature is 25°C. Reliability and life of the device can be improved if the surface temperature is lowered. This is often accomplished by adding a finned heat sink to the device surface to increase the surface area available for convective heat transfer.

a) Determine the surface temperature for the unfinned configuration (in °C) if the convective heat transfer coefficient is 200 W/m²·K.

b) Determine the surface temperature (in °C) if a copper finned heat sink with 8 equally spaced fins—2 mm thick, 20 mm high, 48 mm long—are attached to the device, and the heat transfer coefficient is reduced to 150 W/m²·K because of the addition of the fins.

Approach:

A schematic of the device is given.

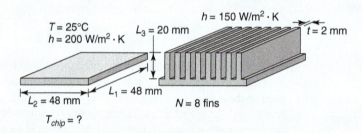

We can approach this problem in a straightforward manner. The governing rate equation is $\dot{Q} = \Delta T / R_{tot} = \Delta T / R_{fin,o}$. We assume that there is no thermal resistance between the chip surface and the fins. The temperature difference is $\Delta T = T_{chip} - T_f$. The heat transfer rate and the air temperature are known, so we solve the rate equation for the surface temperature. We can use Eq. 11-94 to determine the total resistance with fins, $R_{tot} = R_{fin,o} = 1/\eta_o h A_{tot}$; the overall surface efficiency, η_o, is from Eq. 11-93. Without fins, $\eta_o = 1$, and the total surface area, A_{tot}, is simply the plan area of the chip without fins.

Assumptions:

A1. The system is steady.

A2. Properties are constant.

A3. Heat transfer coefficient is uniform over the surface.

Solution:

a) Without fins, we obtain

$$T_{chip} = T_f + \dot{Q}R_{tot} = T_f + \frac{\dot{Q}}{hA}$$

$$= 25°C + \frac{25\,W}{\left(200\,W/m^2\cdot K\right)(0.048\,m)\,(0.048\,m)} = 79.3°C$$

b) Following the approach outlined above, for the fin thermal resistance, we need the fin efficiency, the total surface area, and the area of a single fin. We assume [A1], [A2], [A3], [A4], and [A5]. From the schematic, we see that this is a rectangular fin on a plane surface, so we use the fin efficiency for Case B in Table 11-5 to evaluate the fin efficiency:

$$\eta_f = \frac{\sinh mL + (h/mk)\cosh mL}{\cosh (mL) + (h/mk)\sinh mL}\left(\frac{1}{mL}\right)$$

A4. Conduction is one-dimensional along the fin.
A5. The fin tip is adiabatic.

The parameter m is obtained from:

$$m = \left[\frac{hp}{kA_x}\right]^{1/2} = \left[\frac{(150\,\text{W/m}^2\cdot\text{K})\,2\,(0.048\,\text{m})}{(400\,\text{W/m}\cdot\text{K})\,(0.048\,\text{m})\,(0.002\,\text{m})}\right]^{1/2} = 19.36\,\text{m}^{-1}$$

$$\rightarrow mL_3 = \left(19.36\,\text{m}^{-1}\right)(0.02\,\text{m}) = 0.387$$

where the copper thermal conductivity ($k = 400$ W/m·K) was obtained from Table A-2. Note that for the perimeter, $p = 2L_1$, we ignored the contributions from the fin ends because they add little to the total area. Therefore,

$$\eta_f = \frac{\sinh(mL_3) + (h/mk)\cosh(mL_3)}{\cosh(mL_3) + (h/mk)\sinh(mL_3)}\left(\frac{1}{mL_3}\right)$$

$$= \frac{\sinh(0.387) + \left(\dfrac{150\,\text{W/m}^2\cdot\text{K}}{\left(19.36\,\text{m}^{-1}\right)\left(400\,\text{W/m}\cdot\text{K}\right)}\right)\cosh(0.387)}{\cosh(0.387) + \left(\dfrac{150\,\text{W/m}^2\cdot\text{K}}{\left(19.36\,\text{m}^{-1}\right)\left(400\,\text{W/m}\cdot\text{K}\right)}\right)\sinh(0.387)}\left(\frac{1}{0.387}\right) = 0.996$$

Ignoring the ends of each fin, the surface area of one fin is

$$A_f = 2L_1L_3 = 2\,(0.048\,\text{m})\,(0.02\,\text{m}) = 0.00192\,\text{m}^2$$

The area of the base not covered by the fins is

$$A_b = L_1L_2 - NL_1t = (0.048\,\text{m})\,(0.048\,\text{m}) - 8\,(0.048\,\text{m})\,(0.002\,\text{m}) = 0.00154\,\text{m}^2$$

Hence, the total area with the fins is

$$A_{tot} = A_b + NA_f = 0.00154 + 8\,(0.00192) = 0.0169\,\text{m}^2$$

Now using Eq. 11-93 to calculate the overall surface efficiency,

$$\eta_o = 1 - \frac{NA_f}{A_{tot}}\left(1 - \eta_f\right) = 1 - \frac{8\,(0.00192\,\text{m}^2)}{0.0169\,\text{m}^2}\,(1 - 0.996) = 0.996$$

Using Eq. 11-94, we find that the total fin thermal resistance is

$$R_{tot} = \frac{1}{\eta_o h A_{tot}} = \frac{1}{0.996\,(150\,\text{W/m}^2\cdot\text{K})\,(0.0169\,\text{m}^2)} = 0.396\,\frac{\text{K}}{\text{W}}$$

Solving the rate equation for the surface temperature with fins,

$$\dot{Q} = \frac{T_{chip} - T_f}{R_{tot}} \rightarrow T_{chip} = T_f + \dot{Q}R_{tot} = 25°\text{C} + (25\,\text{W})\,(0.396\,\text{K/W}) = 34.9°\text{C}$$

Comments:

The addition of fins decreases the chip temperature from 79.3°C to 34.9°C, which shows that the fins are very effective. Because the efficiency equation used above is cumbersome, an alternative approach is to assume an insulated tip fin and use the appropriate fin efficiency equation, Eq. 11-87, with a corrected length. Hence, the fin corrected length is $L^* = L_3 + t/2 = 0.020 + 0.002/2 = 0.021$ m, $mL^* = \left(19.36\,\text{m}^{-1}\right)0.021$ m $= 0.407$, $\eta_f = \tanh(mL^*)/mL^* = 0.948$, $\eta_o = 0.953$, $R_{tot} = 0.414$ K/W, and $T_{chip} = 35.3°$C. As can be seen, this approximation is very good.

11.9 CONTACT RESISTANCE

We have considered conduction heat transfer in layers of several materials (see Chapter 3). Several explicit assumptions were made to solve for the temperature drop or heat transfer across the layers of the wall. One implicit assumption was also made: we assumed the two materials were in *perfect contact*. However in practice, except in special situations, there is imperfect contact between the two materials, which results in an additional thermal resistance. A crude example of the effect of the surface roughness can be demonstrated by considering the heat flow from a warm surface to your hand. If you gently lay your hand on the surface, heat will conduct to your skin everywhere it is in contact with the surface, but because of the nonplanar geometry of your hand, not all portions of it will contact the surface. If you press hard, your skin will deflect, more of your hand will contact the surface, and more heat will flow to your hand.

Shown in Figure 11-20 is a representation of the effects of *contact resistance*. Rather than a simple change in slope of the temperature–position curve where the two materials are in contact, there is a finite drop in temperature over a short distance (on the order of 0.5 to 50 μm) due to microroughness and trapped fluids between the surfaces. Where the peaks of the roughness on the two surfaces meet, there is good thermal contact. Where valleys coincide, the gap causes an insulating layer to exist. The fluids may be a gas (e.g., air) or a liquid (e.g., oil or lubricant used in the manufacturing process).

The analysis of heat transfer across an interface is difficult. All modes of heat transfer (conduction, convection, and radiation) contribute to the heat transfer across the interface. Quantities that can affect the contact resistance include surface roughness, type of fluid in the gap, pressure holding the two materials together, the solids' properties, and the temperature at the gap. Contact resistance can be decreased by decreasing the surface roughness, increasing the pressure holding the two materials together, inserting a liquid that has a thermal conductivity greater than air or oil into the gap, or by inserting a soft metal foil or a thermal grease at the interface between the two materials.

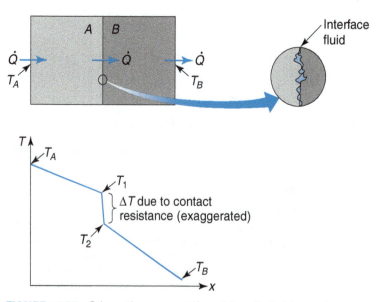

FIGURE 11-20 Schematic representation of the effect of thermal contact resistance.

TABLE 11-6   Representative values of thermal contact resistance

Material pair	Interface material	Surface roughness (μm)	Contact pressure (MPa)	Contact resistance (m$^2 \cdot$ K/W $\times 10^4$)
Stainless/Aluminum	Air	20 – 30	7 – 30	4 – 2.4
Stainless/Aluminum	Air	1 – 2	5 – 20	0.9 – 0.48
Steel Ct-30/ Steel Ct-30	Air	7.2 – 5.1	5 – 35	3.0 – 1.5
Steel Ct-30/Aluminum	Air	7.2 – 4.5	3 – 35	3.9 – 1.2
Steel Ct-30/Copper	Air	7.2 – 4.4	5 – 35	2.0 – 0.95
Steel Ct-30/Copper	Air	2 – 1.4	10 – 35	0.05
Brass/Copper	Air	5.1 – 4.4	3 – 35	3.0 – 0.22
Aluminum/Copper	Air	1.4	2 – 35	0.26 – 0.095
Aluminum/Aluminum	Air	10	0.1	2.75
Aluminum/Aluminum	Helium	10	0.1	1.05
Aluminum/Aluminum	Silicone oil	10	0.1	0.53
Aluminum/Aluminum	Glycerin	10	0.1	0.27
Stainless/Stainless	Vacuum	10	0.1	6 – 25
Stainless/Stainless	Vacuum	10	10	0.4 – 4.0
Aluminum/Aluminum	Vacuum	10	0.1	1.5 – 5.0
Aluminum/Aluminum	Vacuum	10	10	0.2 – 0.4
Silicon chip/Aluminum	Air		0.227 – 0.5	0.3 – 0.6
Silicon chip/Aluminum	Epoxy (0.02 mm)			0.2 – 0.9
Aluminum/Aluminum	Thermal grease		0.1	0.07
Stainless/Stainless	Thermal grease		3.5	0.04
Brass/Brass	Tin solder (15μm)			0.025 – 0.14

Source: Adapted from F. P. Incropera and D. P. DeWitt, *Introduction to Heat Transfer*, 3rd ed., Wiley, 1996; F. Kreith and M. S. Bohn, *Principles of Heat Transfer*, 5th ed., West Publishing Co., 1993. Used with permission.

The **thermal contact resistance**, $R''_{contact}$, is defined by

$$\boxed{R''_{contact} = \frac{T_1 - T_2}{q''}}$$ (11-96)

where T_1 and T_2 are the surface temperatures on either side of the interface and q'' is the heat flux through the interface. Note that $R''_{contact}$ is based on heat transfer per unit area. The thermal resistance that appears in a thermal circuit and has units of °C/W is:

$$R_{contact} = \frac{R''_{contact}}{A}$$

Representative values of contact resistance for a variety of materials and surface roughness are given in Table 11-6. Note that there are large uncertainties associated with both the evaluation and use of contact resistance data, so care must be exercised when estimating its effect in a heat transfer problem.

EXAMPLE 11-11   Contact resistance

Your company makes baseboard convectors for hot water heating systems. Heat from the hot water inside the aluminum tubes (25-mm outer diameter, 22-mm inner diameter) is convected to room air through the outside finned surface. Annular aluminum fins (1 mm thick and 15 mm long) are spaced 1 cm apart and are press-fit to the tube; contact resistance between the fins and the tube is about 2.75×10^{-4} $m^2 \cdot K/W$. To eliminate the contact resistance, you want to replace the press-fit fins with brazed fins. The water is 85°C, the water heat transfer coefficient 1000 $W/m^2 \cdot K$, the room temperature is 22°C, and the air convective heat transfer coefficient is 25 $W/m^2 \cdot K$. From previous calculations, total outside area (fins and bare tube) is 0.448 m^2, and overall surface efficiency is 0.958.

a) Determine the heat transfer rate per unit length with the press-fit fins (in W/m).

b) Determine the heat transfer rate per unit length with the brazed fins (in W/m).

Approach:

A schematic and a thermal circuit of the problem are given. The basic one-dimensional heat transfer rate equation, $\dot{Q} = \Delta T / R_{tot}$, is used. We use the fin equations we developed above to obtain the fin resistance. All other thermal resistances can be evaluated from information given in the problem statement. Note that for part b, we set $R_{contact} = 0$. The temperature difference is $\Delta T = T_{water} - T_f$.

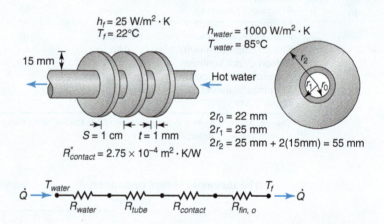

Assumptions:

A1. The system is steady.
A2. Heat transfer is one dimensional.
A3. The heat transfer coefficient is uniform.
A4. Properties are constant.

Solution:

We assume [A1], [A2], [A3], and [A4] for the system, so that the heat transfer rate is obtained from:

$$\dot{Q} = \frac{\Delta T}{R_{tot}} = \frac{\Delta T}{R_{water} + R_{tube} + R_{contact} + R_{fin,\,o}}$$

We use a tube length, $Z = 1$ m, to determine the heat transfer rate per unit length. The thermal resistance due to the water inside the tube is

$$R_{water} = \frac{1}{h_{water} A_{tube}} = \frac{1}{h_{water} 2\pi r_0 Z}$$

$$= \frac{1}{(1000\,W/m^2 \cdot K)\,\pi\,(0.022\,m)\,(1\,m)} = 0.0145\,K/W$$

The tube wall thermal resistance is:

$$R_{tube} = \frac{\ln(r_1/r_0)}{2\pi k_{al} Z} = \frac{\ln(25/22)}{2\pi\,(237\,W/m \cdot K)\,(1\,m)} = 0.000086\,K/W$$

The aluminum thermal conductivity ($k_{al} = 237$ $W/m \cdot K$) was obtained from Table A-2.

Now we calculate the number of fins in 1 meter of the tube:

$$N = \frac{Z}{s} = \frac{1\,\text{m}}{0.01\,\text{m}} = 100$$

Contact resistance is present under each fin and is calculated with $R_{contact} = R''_{contact}/A_{contact}$. For the fins, we need to use the total area under the fins $A_{contact} = N2\pi r_1 t$, so that

$$R_{contact} = \frac{R''_{contact}}{A_{contact}} = \frac{R''_{contact}}{N2\pi r_1 t} = \frac{2.75 \times 10^{-4}\,\text{m}^2\cdot\text{K}/\text{W}}{100\pi\,(0.025\,\text{m})\,(0.001\,\text{m})} = 0.0350\,\text{K}/\text{W}$$

Using the overall surface efficiency in Eq. 11-94, we obtain the total fin thermal resistance

$$R_{fin,\,o} = \frac{1}{\eta_o h A_{tot}} = \frac{1}{0.958\,(25\,\text{W}/\text{m}^2\cdot\text{K})\,(0.448\,\text{m}^2)} = 0.0932\,\frac{\text{K}}{\text{W}}$$

Finally, the heat transfer rate per meter of length *without* the contact resistance is

$$\dot{Q} = \frac{\Delta T}{R_{water} + R_{tube} + R_{fin,\,o}} = \frac{(85 - 22)\,\text{K}}{(0.0145 + 0.000086 + 0.0932)\,\text{K}/\text{W}} = 584\,\text{W}$$

The heat transfer rate per meter of length *with* the contact resistance is:

$$\dot{Q} = \frac{\Delta T}{R_{water} + R_{tube} + R_{contact} + R_{fin,\,o}}$$

$$= \frac{(85 - 22)\,\text{K}}{(0.0145 + 0.000085 + 0.0350 + 0.0932)\,\text{K}/\text{W}} = 441\,\text{W}$$

Comments:

As can be seen, contact resistance can have a significant negative effect on heat transfer. In this example, the heat transfer rate increased by 32% when contact resistance was eliminated. Thus much effort should be made to minimize this thermal resistance so that system performance is improved.

SUMMARY

Heat flux has both magnitude and direction, so it is a vector quantity. Thus, for a three-dimensional temperature field, we write the heat flux vector ($\bar{q}''$) in terms of its x-, y-, and z-components, (q_x'', q_y'', q_z''), respectively, as $\bar{q}'' = q_x''\hat{i} + q_y''\hat{j} + q_z''\hat{k}$. We must use partial differentials to describe the heat flux in the three directions to obtain Fourier's law in multidimensions:

$$q_x'' = -k\frac{\partial T}{\partial x}, \qquad q_y'' = -k\frac{\partial T}{\partial y}, \qquad q_z'' = -k\frac{\partial T}{\partial z}$$

The general heat conduction equation is the basic tool for heat conduction analysis. In the three coordinate systems the equations are:

$$\frac{\partial}{\partial x}\left(k\frac{\partial T}{\partial x}\right) + \frac{\partial}{\partial y}\left(k\frac{\partial T}{\partial y}\right) + \frac{\partial}{\partial z}\left(k\frac{\partial T}{\partial z}\right) + q'''$$

$$= \rho c_p \frac{\partial T}{\partial t} \qquad \text{rectangular} \qquad (11\text{-}13)$$

$$\frac{1}{r}\frac{\partial}{\partial r}\left(kr\frac{\partial T}{\partial r}\right) + \frac{1}{r^2}\frac{\partial}{\partial \theta}\left(k\frac{\partial T}{\partial \theta}\right) + \frac{\partial}{\partial z}\left(k\frac{\partial T}{\partial z}\right) + q'''$$

$$= \rho c_p \frac{\partial T}{\partial t} \qquad \text{cylindrical} \qquad (11\text{-}14)$$

$$\frac{1}{r^2}\frac{\partial}{\partial r}\left(kr^2\frac{\partial T}{\partial r}\right) + \frac{1}{r^2\sin^2\theta}\frac{\partial}{\partial \phi}\left(k\frac{\partial T}{\partial \phi}\right)$$

$$+\frac{1}{r^2\sin\theta}\frac{\partial}{\partial \theta}\left(k\sin\theta\frac{\partial T}{\partial \theta}\right) + q'''$$

$$= \rho c_p \frac{\partial T}{\partial t} \qquad \text{spherical} \qquad (11\text{-}15)$$

Both steady and nonsteady (transient) problems can be solved with these equation. From their solution, we obtain a temperature distribution. Using the temperature distribution (or temperature field) and Fourier's law, we can determine the heat flux at any location.

Three common one-dimensional geometries can be evaluated in a straightforward manner with the conduction equation. For steady state and constant thermophysical properties, the heat transfer rate in those situations is described by $\dot{Q} = \Delta T / R$, where the resistance is:

$$R_{\substack{plane \\ wall}} = \frac{L}{kA} \qquad \text{plane wall}$$

$$R_{cylinder} = \frac{\ln(r_2/r_1)}{2\pi kL} \qquad \text{cylindrical shell}$$

$$R_{sphere} = \frac{r_2 - r_1}{4 r_1 r_2 \pi k} \qquad \text{spherical shell}$$

In situations where temperature varies in more than one direction, a multidimensional analysis is needed. If the temperature profile is not sought, but the steady-state heat transfer rate is, then the results can be simplified through the use of a *conduction shape factor*, S, and the results of the multidimensional solution are expressed as a one-dimensional problem. The conduction shape factor is defined as

$$\dot{Q} = kS\,\Delta T = \frac{\Delta T}{1/kS}$$

For transient heat transfer in objects with negligible internal conduction resistance and significant convection thermal resistance at the surface, the lumped system analysis is used if the *Biot number* ($Bi = hL_{char}/k$, $L_{char} = V/A$) is small enough. The time-varying uniform temperature is described by:

$$\frac{T(t) - T_f}{T_i - T_f} = \exp\left(-\frac{hA}{mc_p}t\right) = \exp\left(-BiFo\right) \qquad Bi <\sim 0.1$$

Transient conduction in one dimension with non-negligible conduction resistance can be analyzed with the general conduction equation. Solutions to some simple geometries and boundary conditions consist of infinite series. Because higher-order terms decrease in size quickly after the first few terms, an *approximate solution* often is used, which involves using only the first term in the infinite series. The *one-term approximations* for three common geometries are

$$\theta(x,t)_{\substack{plane \\ wall}} = \frac{T(x,t) - T_f}{T_i - T_f} = C_1 \exp\left(-\lambda_1^2 \tau\right) \cos\left(\lambda_1 x/L\right)$$
$$\tau >\sim 0.2 \qquad \text{plane wall}$$

$$\theta(r,t)_{\substack{infinite \\ cylinder}} = \frac{T(r,t) - T_f}{T_i - T_f} = C_1 \exp\left(-\lambda_1^2 \tau\right) J_0\left(\lambda_1 r/r_0\right)$$
$$\tau >\sim 0.2 \qquad \text{cylinder}$$

$$\theta(r,t)_{sphere} = \frac{T(r,t) - T_f}{T_i - T_f} = C_1 \exp\left(-\lambda_1^2 \tau\right) \frac{\sin\left(\lambda_1 r/r_0\right)}{\lambda_1 r/r_0}$$
$$\tau >\sim 0.2 \qquad \text{sphere}$$

The constants used in the solutions are given in Table 11-2 and Table 11-3. Complementary equations are given for the total heat transfer that occurs over a given time period.

Another one-dimensional transient problem is that which occurs in a *semi-infinite solid*. In such a situation, the surface and nearby regions in the solid are affected by a sudden change in the surface boundary condition, but the interior of the solid remains at its initial temperature. Solutions for three different boundary conditions are given in the text.

For multidimensional transient conduction, the analytic solution to the general conduction equation involves an infinite series. Because the equation is linear, for some multidimensional geometries, one-dimensional solutions can be combined to give two- and three-dimensional solutions. For example, the product of the one-dimensional plane wall solution and the infinite cylinder solution can describe the transient heat transfer in a short cylinder:

$$\theta(r,x,t)_{\substack{short \\ cylinder}} = \theta(x,t)_{\substack{plane \\ wall}} \times \theta(r,t)_{\substack{infinite \\ cylinder}}$$

Examples of other multidimensional bodies are given in Figure 11-13.

Adding fins or extended surfaces to the primary heat transfer surface can increase heat transfer area. This is an effective way to increase the heat transfer from a surface for a given temperature difference or to decrease the driving temperature difference if the heat transfer rate is fixed. Fin performance is often expressed in terms of the *fin efficiency*:

$$\eta_f = \frac{\dot{Q}_{act,fin}}{\dot{Q}_{ideal,fin}}$$

where the fin efficiency depends on the fin geometry, fin material, and heat transfer coefficient. The ideal heat transfer rate, $\dot{Q}_{ideal,fin} = hA_f\left(T_b - T_f\right)$, is the maximum possible heat transfer rate that would occur when the complete fin is at the fin base temperature. Fin efficiency expressions for several fin geometries are given in Table 11-5, and graphical representations are given in Figure 11-16 and Figure 11-17. For an array of fins, the total or *overall surface efficiency*, η_o (similar to the fin efficiency), is defined as:

$$\eta_o = \frac{\dot{Q}_{tot}}{\dot{Q}_{max}}$$

where $\dot{Q}_{max} = hA_{tot}\left(T_b - T_f\right)$ is the maximum possible heat transfer from the total surface. In terms of the fin efficiency, we can develop an expression for the overall surface efficiency:

$$\eta_o = 1 - \frac{NA_f}{A_{tot}}\left(1 - \eta_f\right)$$

When two materials are brought into contact, unless very exacting preparations and care are used, there is imperfect contact, which results in a *thermal contact resistance*, defined by

$$R''_{contact} = \frac{T_1 - T_2}{q''} \qquad \text{and} \qquad R_{contact} = \frac{R''_{contact}}{A}$$

where T_1 and T_2 are the surface temperatures in the two materials on either side of the interface and q'' is the heat flux through the interface. Representative values of contact resistance are given in Table 11-6.

SELECTED REFERENCES

ARPACI, V. S., *Conduction Heat Transfer*, Addison-Wesley Reading, MA, 1999.

HUGHES, T. J. R., *The Finite Element Method: Linear Static and Dynamic Finite Element Analysis* Dover, Mineola, NY, 2000.

INCROPERA, F. P., and D. P., DEWITT, *Fundamentals of Heat and Mass Transfer*, 5th ed., Wiley, New York, 2001.

JAEGER, J. C., and H. S. CARSLAW, *Conduction of Heat in Solids*, 2nd ed., Oxford University Press, New York, 1986.

KRAUS, A. D., J. WELTY, and A. AZIZ, *Extended Surface Heat Transfer*, Wiley, New York, 2000.

OZISIK, M. N., *Boundary Value Problems of Heat Conduction*, Dover, New York, 2002.

PATANKAR, S. V., *Numerical Heat Transfer and Fluid Flow*, Hemisphere Publishing, New York, 1980.

ROHSENOW, W. M., J. P. HARTNETT, and Y. I. CHO, ed., *Handbook of Heat Transfer*, 3rd ed., McGraw-Hill Professional, New York, 1998.

MINKOWICZ, W. J., E. M., SPARROW, G. E. SCHNEIDER and R. H. PLETCHER, *Handbook of Numerical Heat Transfer*, Wiley Interscience, New York, 1988.

TANNEHILL, J. C., D. A. ANDERSON, and R. H. PLETCHER, *Computational Fluid Mechanics and Heat Transfer*, 2nd ed., Taylor & Francis, Washington, D. C., 1997.

PROBLEMS

ONE-DIMENSIONAL STEADY

P11-1 One approach used to determine the thermal conductivity of metals is to sandwich an electric heater between two identical plates. Consider two pieces of a metal; each piece is 1 cm thick, 10 cm wide, and 10 cm long. All edges are heavily insulated, and the exposed faces have the same convective boundary conditions. For an applied power input of 173 W to the heater, the temperatures of the inner and outer faces of the metal plates are 42.3°C and 38.7°C, respectively. Determine the thermal conductivity of the metal.

P11-2 In an experiment, the boiling heat transfer coefficient is to be measured using the apparatus shown in the figure. Condensing steam at 120°C is used to heat the end of the 304 stainless-steel rod with a 25-mm diameter. The outside perimeter of the rod is heavily insulated, and the temperature in the rod is measured in two places, $T_4 = 91.19$°C and $T_3 = 100.20$°C. The boiling fluid is at 10°C. The condensing heat transfer coefficient is 7500 W/ m²·K.

a. Determine the heat transfer rate (in W).

b. Determine the temperatures (T_2 and T_5) on the two ends of the rod (in °C).

c. Determine the heat transfer coefficient on the test specimen end (in W/m²·K).

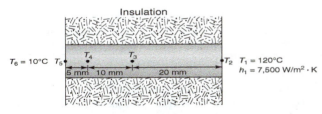

P11-3 Large electrical currents are often carried in aluminum conductors. Consider a long, 2-cm-diameter cable covered by insulation 2 mm thick. For a particular application, the outside insulation temperature is limited to 35°C, and none of the insulation can exceed 50°C. The cable is in an environment in which the convective heat transfer coefficient is 25 W/m²·K, the air temperature is 24°C, the insulation thermal conductivity is 0.10 W/m·K, and the electrical resistance per unit length of the wire is $3.9 \times 10^{-4}\,\Omega$/m. Determine the maximum current allowed (in A).

P11-4 For the design of a chemical processing plant in Hawaii, you need to determine the thickness of insulation on several steam lines that would make the most sense economically. The steel steam lines have a total length of 350 m, have an outside diameter of 2.5 cm, and carry saturated steam at 200 kPa. The design air temperature is 27°C. We assume we can obtain insulation with thickness in 1.0-cm increments. The insulation thermal conductivity is 0.05 W/m·K and costs $0.00015/cm³. From previous studies when you took into account the prevailing wind, you developed a simple correlation for the heat transfer coefficient to be $h = 75(D/10)^{-0.38}$, where D is the outside diameter in cm and h is in W/m²·K. Natural gas is used in the boiler ($\eta_{boiler} = 87\%$) and costs $0.50/10^5$ kJ. For a 20-year life, assuming the plant runs 8200 h/yr, determine:

a. the recommended insulation thickness.

b. the maximum savings compared to no insulation.

P11-5 Consider a solid cylindrical rod with radius R_0, a uniform volumetric heat generation q''', a known outside wall temperature T_w, and constant thermophysical properties. Using Example 11-3 as a guide and the general conduction equation in cylindrical coordinates, develop the following expression for the steady temperature profile in the rod:

$$T = T_w + \frac{q''' R_o^2}{4k}\left[1 - \left(\frac{r}{R_o}\right)^2\right]$$

P11-6 Gargantuan Motors has developed a new rear window defogging system. The electric heating element is a thin and transparent film applied to the entire inner surface of the rear window. The glass is 4 mm thick with a thermal conductivity of 0.94 W/m·K. The design operating condition for the defogger is for an outside condition of $-10°C$ with a convective heat transfer coefficient of 65 W/m²·K and an inside condition of $-10°C$ with a convective heat transfer coefficient of 10 W/m²·K. The inner surface of the window is to be maintained at 10°C.

a. Determine the heat flux that must be supplied to the heater (in W/m²).

b. Determine the temperature on the outside surface of the window (in °C).

P11-7 An approach to determining the thermal conductivity of a material is to put a known material in series with the unknown material, as shown in the figure. A heat input is applied to one end of the assembly, and the other end is cooled. Temperatures are measured at specific locations in both materials. Material A is stainless steel with a thermal conductivity of 15.2 W/m·K. The specimens are rods 2 cm in diameter. The rods are heavily insulated. In one test the following temperatures were measured:

$T_1 = 93.00°C \quad T_2 = 82.57°C \quad T_3 = 69.21°C \quad T_4 = 66.28°C$

Determine the thermal conductivity of the unknown material (in W/m·K).

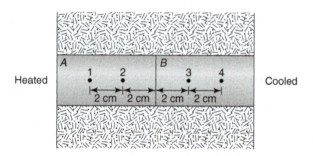

Heated | A | B | Cooled
2 cm | 2 cm | 2 cm | 2 cm

P11-8 When natural gas is burned, water vapor and other products of combustion are produced. These products can mix with the water vapor to produce a dilute acid. To prevent acid attack on the chimney, the gases should be kept above a minimum temperature. Consider a 304 stainless-steel chimney of 200-mm inside diameter with walls 1 mm thick. Insulation ($k = 0.075$ W/m·K) 10 mm thick covers the outside of the metal. The outside air is $-5°C$ with a convective heat transfer coefficient of 25 W/m²·K. The inside convective heat transfer coefficient is 15 W/m²·K. To avoid condensation forming on the inside surface of the chimney, that surface temperature must be greater than 100°C. Determine the minimum required gas temperature (in °C).

P11-9 Consider an infinite plane wall $2L$ thick, in which there is uniform volumetric heat generation rate, q'''. The wall

surfaces are maintained at $T = T_1$ at $x = -L$ and $T = T_2$ at $x = +L$. For constant thermal conductivity k, steady-state operating conditions, and defining the origin of the x-coordinate from the centerline of the plane, show that the solution of the general conduction equation for the temperature distribution in the wall is

$$T(x) = \frac{q'''L^2}{2k}\left[1 - \left(\frac{x}{L}\right)^2\right] + \frac{T_2 - T_1}{2}\left(\frac{x}{L}\right) + \frac{T_1 + T_2}{2}$$

P11-10 A large 2-kW electric heater (30 cm by 30 cm square by 0.1 cm thick) can be approximated as an infinite plane wall of thickness $2L = 0.1$ cm. The heater element is exposed on both sides to air at $T_\infty = 25°C$ and a heat transfer coefficient $h = 75$ W/m²·K. The heater material has a thermal conductivity of $k = 0.5$ W/m·K. We need to determine the maximum steady-state temperature in this heater. With the equation given in P11-9, determine

a. the maximum temperature inside the heater (in °C).

b. the location at which it occurs.

P11-11 Water heaters are insulated to minimize heat losses. Consider an electric hot water heater made from 316 stainless steel that has a 60-cm inside diameter, is 175 cm tall, has a wall thickness of 4 mm, and is covered with a 5-cm layer of fiberglass. The basement where the tank is kept is at 15°C. The air-side convective heat transfer coefficient is 10 W/m²·K and that on the water side is 100 W/m²·K. If electricity is $0.05/kWh, determine:

a. the cost required to maintain the water at 60°C for 24 h when no water is removed or added to the tank.

b. the cost required to maintain the water at 60°C for 24 h if 100 L of water is removed and replaced with 100 L of 10°C water.

P11-12 A plane wall has a thickness of 0.1 m and a thermal conductivity of 25 W/m·K. One side is insulated and the other side is exposed to a fluid at 92°C with a convective heat transfer coefficient of 500 W/m²·K. The wall has a uniform volumetric heat generation rate of 0.3 MW/m³. The wall is at steady state. Using the result given in Problem P11-9, determine

a. the maximum temperature in the wall (in °C).

b. the location where it occurs.

P11-13 The curing of concrete is an exothermic reaction; that is, the curing process produces heat. If a concrete slab is large enough, the temperature can rise to the point where the magnitude of thermal stresses may cause cracking. Consider a large slab of concrete 1 m thick. Both sides are maintained at 20°C. The curing process produces a uniform internal heat generation rate of 60 W/m³. The thermal conductivity of the concrete is 1.1 W/m·K. Using the results from Problem P11-9, determine the steady temperature at the centerline of the slab.

P11-14 A large steel plate is exposed to 800°C combustion gases on one side; the heat transfer coefficient is 300 W/m²·K, and the plate is 25 mm thick with a thermal conductivity of 40 W/m·K. The other side of the plate is to be insulated so that the insulation's outside temperature does not exceed 35°C. To save money, two layers of insulation are to be used. An expensive, high-temperature insulation ($k = 0.055$ W/m·K) is to be applied to the steel, and the second layer is a less expensive insulation ($k = 0.071$ W/m·K). The maximum allowable temperature for the less expensive insulation is 350°C. The heat transfer coefficient on the insulation surface is 10 W/m²·K, and the ambient air temperature is 30°C. Determine the thickness of each of the layers of insulation.

FINS/EXTENDED SURFACES

P11-15 A brass rod 0.2-in. in diameter and 3 in. long with $k = 64$ Btu/h·ft·°F connects two plates, each of which is at 150°F. Air at 75°F flows over the rod with $h = 40$ Btu/h·ft²·°F.

a. Determine the temperature of the rod midway between the two plates (in °F).

b. Determine the total heat transfer rate from the rod (in Btu/h).

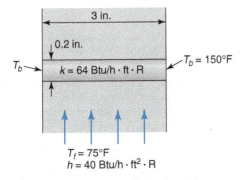

P11-16 A solar collector is constructed as shown in the figure. The plate that absorbs the incident solar heat flux is copper and 2 mm thick, and the space between the absorber plate and the glass cover plate is evacuated so there are no convective heat transfer losses. The tubes for the water flow are spaced 20 cm apart, and the water flowing in the tubes is at 50°C. Because of excellent conduction, the temperature of the absorber plate directly above the tubes is at the same temperature as the water. For a net steady-state radiation heat flux of 900 W/m² absorbed by the plate, what is the maximum temperature on the plate? Develop the differential equation for this geometry similar to what was done for a convecting fin, and then solve the equation.

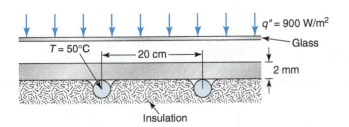

P11-17 Baseboard convectors using hot water are used to heat rooms in houses. A common design is to attach annular fins to a horizontal tube and use natural convection to add heat to the room. Consider a design that has 75 mm-diameter fins attached to a 25-mm-outside-diameter tube. The fins are 1 mm thick and are spaced 5 mm apart; the tube is 3 m long with a wall thickness of 2 mm. The water is at 55°C and has a convective heat transfer coefficient of 1250 W/m²·K. Assume the temperature decrease of the water is small. The room air is at 20°C and has a convective heat transfer coefficient of 10 W/m²·K. Determine the heat transfer rate to the air (in W).

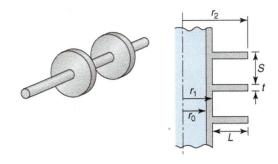

P11-18 In a heat transfer experiment, the objective is to determine the convective heat transfer coefficients on three fins. Each of the three fins (one of which is shown here) is a solid rod 15 cm long with a diameter of 1 cm. The first fin is a pure copper, the second is 2024-T6 aluminum, and the third is 304 stainless steel. Small electric heaters are attached to the bases of the fins, and the power is adjusted such that the base temperature is 100°C when the room temperature is 25°C. The measured powers are: copper rod, 4.1 W; aluminum rod, 4.0 W; 304 stainless steel, 2.7 W.

a. Estimate the convective heat transfer coefficient on each of the three fins (in W/m²·K).

b. Estimate the tip temperature on each fin (in °C).

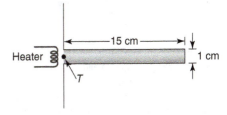

P11-19 Electronic equipment is to be encased in an aluminum box whose temperature must be limited to 60°C. Vertical rectangular pure aluminum fins are to be attached to the box top to aid

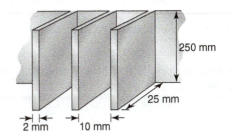

heat removal. Air at 25°C will flow over the fins with a convective heat transfer coefficient of 50 W/m²·K. If 10 fins spaced 1 cm apart are used that are 25 mm long, 250 mm high, and 2 mm thick, what is the heat transfer from the top of the box (in W)?

P11-20 The failure rate of integrated chips increases rapidly with higher operating temperatures. The ability to pack more components into smaller areas results in higher power and, therefore, more severe cooling requirements. One way to improve cooling is to add fins to chips to increase surface area. Consider a 12.5 mm by 12.5 mm chip. A 4×4 array of pure copper circular pin fins, 1.5 mm in diameter and 10 mm long, is attached to the outer surface of the chip. The surface has a maximum operating temperature of 75°C, and the heat transfer coefficient (with air cooling) around the fins is 200 W/m²·K. The design air temperature is 35°C.

a. Determine the maximum power dissipation from the surface if no fins were used (in W).

b. Determine the maximum power dissipation from the surface if the pin fins are used (in W).

P11-21 Air flows over a plane wall with a convective heat transfer coefficient of 40 W/m²·K. Insufficient heat transfer is obtained from this situation, so aluminum alloy fins (alloy 2024-T6) of rectangular profile are attached to the plane wall. The fins are 50 mm long, 0.5 mm thick, and are equally spaced at a distance of 4 mm (250 fins/m). With the fins, the convective heat transfer coefficient is reduced to 30 W/m²·K. What percentage increase in heat transfer is obtained with the fins compared to the plane wall arrangement without fins?

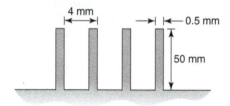

P11-22 Motorcycle engines often are air-cooled with annular fins attached to the cylinder head. Consider a cylinder ($k = 55$ W/m·K) with an inside diameter of 100 mm and a wall thickness of 6 mm. Over the cylinder a fin assembly is interference-fit so that there is negligible contact resistance between the cylinder and the fin assembly. The fin assembly ($k = 230$ W/m·K) has a base thickness of 4 mm; the six fins are 25 mm long, 2 mm thick, and spaced 2 mm apart. A heat flux (assumed constant) of 10^5 W/m² is imposed at the inside surface of the cylinder. Air outside the engine is at 30°C with a convective heat transfer coefficient of 100 W/m²·K.

a. Determine cylinder wall temperature (in °C).

b. Determine the interface temperature between the cylinder and the base of the fin assembly (in °C).

c. Determine the fin base temperature (in °C).

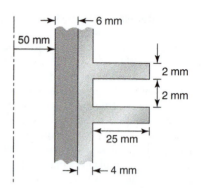

P11-23 A densely populated circuit board has 113 electronic devices attached to it. Forty of the devices dissipate 0.3 W each, 30 dissipate 0.2 W each, and the rest dissipate 0.15 W each. Because this is a critical installation, the circuit board, 3 mm thick, is constructed of a high-thermal-conductivity material ($k = 15.6$ W/m·K) so that the heat from the devices is spread out evenly over the 10 cm by 20 cm area of the board. The backside of the board is cooled with air at 25°C with a heat transfer coefficient of 25 W/m²·K. To obtain a worst-case estimate, no heat transfer credit is taken for convection from the device side of the circuit board.

a. Determine the temperature of the surface of the circuit board cooled by air (in °C).

b. Determine the temperature of the surface of the circuit board cooled by air if 500 pin fins, 0.2 cm in diameter and 2 cm long, are attached to an aluminum plate (2 mm thick, $k = 177$ W/m·K) that is epoxied to the circuit board. The epoxy ($k = 2$ W/m·K) is 0.1 mm thick.

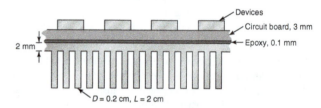

MULTIDIMENSIONAL STEADY CONDUCTION

P11-24 In many new homes, hot water pipes are encased in the floors to provide uniform heating in the winter. Consider 2.5-cm hot water pipes located at the midplane of a 10-cm-thick concrete floor ($k = 1.1$ W/m·K). The pipes are spaced 20 cm apart. The air temperature is 23°C, and its convective heat transfer coefficient

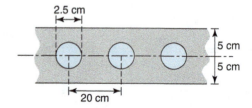

is 10 W/m²·K. The water is at 60°C. Assume the heat transfer coefficient inside the pipes is very high.

a. Determine the heat transfer rate per unit length of the pipe (in W/m).

b. Determine the surface temperature of the concrete (in °C).

P11-25 Liquid nitrogen is stored in a 3-m sphere buried in the earth ($k = 0.17$ W/m·K), with its center 4 m below the surface. A 10-cm-thick layer of insulation ($k = 0.05$ W/m·K) covers the sphere. The nitrogen is at -180°C, and the surface of the earth is at 10°C. Nitrogen vaporizes (and the vapor is vented to the surface) because of heat transfer from the earth to the tank, thus maintaining a constant temperature and pressure in the nitrogen. The enthalpy of vaporization of the nitrogen is 198.6 kJ/kg.

a. Determine the heat transfer rate to the nitrogen (in W).

b. Determine the nitrogen vaporization rate (in kg/h).

P11-26 Hot water is pumped between two buildings in an office complex. The 2-in.-inside-diameter pipe is buried 1.5 ft below the earth's surface ($k = 0.10$ Btu/h·ft·°F). Insulation 1 in. thick ($k = 0.02$ Btu/h·ft·°F) covers the carbon steel pipe, whose wall thickness is 0.25 in. The water has a convective heat transfer coefficient of 2350 Btu/h·ft²·°F. If the soil surface temperature is 10°F and the water enters the pipe at 170°F, determine the initial heat transfer per unit length (in Btu/h·ft) when the water enters the pipe.

P11-27 A very long electrical conductor is buried in a large trench filled with sand ($k = 0.03$ W/m·K) to a centerline depth of 0.5 m. The conductor has an outer diameter of 25 mm, and the current flow and resistance of the cable cause a dissipation of 1 W per meter of length. The conductor is covered with an insulating sleeve of thickness 3 mm with $k = 0.01$ W/m·K. At the surface a 25°C wind blows such that the heat transfer coefficient is 75 W/m²·K. Determine the temperature at the interface between the conductor and the insulating sleeve. (Note: you will need to make an assumption about the convective resistance, and you will need to justify it.)

P11-28 Your next-door neighbor decides to construct a small underground room in his backyard. The room will be 8 ft tall and 12 ft square. He will construct it of concrete and bury it under 2 ft of earth ($k = 0.62$ Btu/h·ft·°F). He wants to buy a heater that will maintain the room at 65°F when the outdoor temperature is 0°F, and he asks you to tell him how big the heater should be.

a. Determine the steady-state heat transfer rate from the room (in Btu/h).

b. Determine the steady-state heat transfer rate if 5 in. of insulation ($k = 0.02$ Btu/h·ft·°F) is added to the outside of the room (in Btu/h).

P11-29 Molds for plastics and other materials are sometimes heated by the insertion of electrical resistance heaters at appropriate locations on the body. Consider a 150-mm-long, 12.5-mm-diameter, 100-W electric heater inserted into a hole

drilled perpendicular to the surface of a large mold whose thermal conductivity is 10 W/m·K. If the surface temperature of the mold is maintained at 25°C a long way from the heater, estimate the steady-state temperature of the heater.

P11-30 A thin electronic component 20 mm square is epoxied to a large 2024-T6 aluminum heat sink. The thermal resistance of the epoxy is 0.35×10^{-4} m²·K/W. The temperature of the aluminum block far away from the electronic component is 25°C. The top of the component is swept by air at 25°C with a convective heat transfer coefficient of 50 W/m²·K. If the maximum component temperature is 75°C, what is its maximum allowable operating power (in W)?

LUMPED SYSTEMS ANALYSIS

P11-31 You are late preparing for a party; the soft drinks should have been placed into the refrigerator much sooner. It is only $2^{1}/_{2}$ hours until the party. When the party has started, will the drinks have reached 5°C? Each can is 6 cm in diameter and 12 cm high and is initially at 20°C. The air in the refrigerator is at 0°C, and you estimate $h \approx 4$ W/m²·K. Assume the properties of the drink can be approximated as those of water and the can's contribution to the transient is negligible.

a. Determine the estimated time required for a drink to reach 5°C.

b. Determine a method to speed up the cooling process.

P11-32 You are designing a radiant energy test facility and need to determine how long it will take a test specimen to reach a steady temperature. Initially, a 3-cm-thick brass plate is at a uniform temperature of 50°C. At time zero, one side of the plate is exposed to a radiant heat flux of 6000 W/m² and the other side is exposed to air at 20°C with $h = 75$ W/m²·K.

a. Determine the steady-state temperature of the plate (in °C).

b. Determine the temperature of the plate 15 min after the start of heating (in °C).

c. Determine the time to reach steady state if steady state is when 99% of the difference between the final plate temperature and the air temperature is achieved (in s).

P11-33 In some lumped systems, different parts of the surface may be exposed to different conditions. Consider the exhaust pipe of an automobile engine. Just before the engine starts, the exhaust pipe is at a uniform temperature T_i. When the engine starts at $t = 0$, exhaust gases at T_g (assumed constant with time) flow through the tube with a convective heat transfer coefficient of h_g. Outside the pipe, the air is at T_a with a convective heat transfer coefficient of h_a. Using a lumped system analysis, develop an expression for the pipe temperature as a function of time.

P11-34 Steel balls with a diameter of 2 cm are annealed by heating them uniformly to 950°C and then cooling them to 125°C in air at 35°C. The convective heat transfer coefficient is 25 W/m²·K. For 347 stainless-steel balls, determine the time required for the cooling process (in s).

P11-35 During the start-up of a new natural gas furnace, the test engineers want to monitor the exhaust gas temperature to assess the unit's performance. A 1-mm-diameter copper constantan thermocouple is used to make the temperature measurement. Before they use the thermocouple, they want to determine its response characteristics and estimate the heat transfer coefficient on the thermocouple. They develop the following experiment. Initially, the thermocouple is at 25°C. They insert it quickly into a gas stream that is at 200°C and has the same velocity as the stream they want to monitor. In 8.3 s, the thermocouple reads 199°C. Determine the heat transfer coefficient (in $W/m^2 \cdot K$). (Assume the properties of the thermocouple are the same as copper.)

P11-36 An electronic device that dissipates 30 W is used infrequently. Its maximum allowable operating temperature is limited to 65°C; as soon as 65°C is reached, the device must be shut off. The device and an attached heat sink have a combined mass of 0.25 kg, a surface area of 56.3 cm^2, and an effective specific heat of 800 J/kg·K. The device is initially at a uniform temperature of 25°C in air at 25°C with a heat transfer coefficient of 10 $W/m^2 \cdot K$.

a. Determine the steady-state operating temperature (in °C).

b. Determine the time required to reach the maximum operating temperature (in s).

c. If a heat sink is to be added to the device so that the operating time is to be doubled, what additional mass and area are needed? Assume the mass-to-area ratio of the added material is the same as that of the original device.

ONE-DIMENSIONAL TRANSIENT CONDUCTION

P11-37 To obtain a hard surface on a metal plate with a somewhat softer core for toughness, heat treatment is required. Often the hot metal plate is plunged into a much colder bath so that the surface temperature drops very rapidly. The rapid temperature drop results in a favorable grain structure that causes a very hard surface material. The slower cooling of the core of the plate results in a softer core material. Consider a large stainless-steel slab, initially at a uniform temperature of 1250°F. It is to be plunged into a heat-treatment bath which is at 150°F. The research engineers state that from their studies, the correct surface hardness will be reached if the temperature at a depth of 0.2 in. in the slab reaches 700°F in less than 35 s. The steel has a thermal diffusivity of $\alpha = 0.135$ ft²/h and a thermal conductivity of $k = 7$ Btu/h·ft·°F.

a. Determine the time required for the temperature to reach 700°F at a depth of 0.2 in. if you assume a very large heat transfer coefficient.

b. Determine the time required for the temperature to reach 700°F at a depth of 0.2 in. if the heat transfer coefficient is 250 Btu/h·ft²·F.

P11-38 Heat-treating furnaces are heavily insulated on their outer surfaces and are lined with fire clay bricks that are 10 cm thick. Initially, the bricks are at a uniform temperature of 25 °C.

After start-up, the furnace walls are exposed to 1300°C gases with a combined convective/radiative heat transfer coefficient of 50 $W/m^2 \cdot K$. Determine the temperature (in°C) of

a. the outer surface of the bricks after 30 min.

b. the outer surface of the bricks after 4 h.

c. the temperature next to the insulation after 30 min and 4 h.

P11-39 Hot metals are quenched in cold fluids to change the material properties. Consider a long 7.5-cm-diameter cylinder of 316 stainless steel that is taken out of a furnace at 500°C and plunged into a cold bath at 25°C. The convective heat transfer coefficient is 1000 $W/m^2 \cdot K$.

a. Determine the centerline temperature of the cylinder 90 s after it is quenched (in °C).

b. Determine the surface temperature of the cylinder 5 min after it is quenched (in °C).

c. Determine the time required for the centerline temperature to reach 50°C (in s).

P11-40 An 8-cm-diameter potato ($\rho = 1100$ kg/m³, $c_p = 3900$ J/kg·K, $k = 0.6$ W/m·K), initially at a uniform temperature of 25°C, is baked in an oven at 170°C until a temperature sensor inserted to the center of the potato indicates a temperature of 70°C. The potato is then taken out of the oven and is wrapped in thick towels so that no heat is lost. Assume the heat transfer coefficient in the oven to be 25 $W/m^2 \cdot K$.

a. Determine how long the potato is baked (in min).

b. Determine the surface temperature of the potato before it is wrapped in the towel (in °C).

c. Determine the final equilibrium temperature of the potato (in °C).

P11-41 In the manufacturing of laminated wood tabletops, a more expensive and attractive wood surface layer (1.5 mm thick) is glued to a less expensive structural wood. Heat is applied to the surface of the table to speed the curing of the glue. A heater consists of a massive plate maintained at 150°C by an embedded electrical heater. The glue will cure sufficiently if heated above 50°C for at least 2 min, but its temperature should not exceed 120°C to avoid deterioration of the glue. Assume that the laminate and structural wood have an initial temperature of 25°C and that they have equivalent thermophysical properties of $k = 0.15$ W/m·K and $\rho c_p = 1.5 \times 10^6$ J/m³·K.

a. Determine how long it will take to heat the glue.

b. Determine the glue temperature at the end of the 2-min curing time.

c. Determine the energy removed from the heater during the time it takes to cure the glue if the heater has a square surface area of 250 mm to the side (in kJ).

1.5 mm — Laminate — Epoxy — Structural wood

P11-42 You have built your dream cabin and now need to lay a water line from the well to the cabin. You need to estimate how deep the water line should be placed. From historical records, you discover that there has never been a cold spell of $-10°C$ weather for longer than 4 weeks, but often the temperature is as low as $0°C$ for 12 weeks. Throughout the fall, you note that the ground temperature for a reasonable depth is approximately uniform at about $10°C$. You assume that if the ground temperature at depth does not reach $0°C$ in 12 weeks of $-10°C$ weather (this includes a factor of safety), then the water in the pipes will not freeze. The soil on your property has $k = 2.1$ W/m·K and $\alpha = 7 \times 10^{-7}$ m²/s. How deep should you lay the water pipe?

P11-43 When cold weather reaches Florida, if the air temperature remains below freezing ($0°C$) for an extended period of time, the orange, lime, and other citrus fruit crops can be severely damaged. Consider an 8-cm orange, whose properties can be approximated as those of water. If the orange is initially at $10°C$ and the air temperature drops suddenly to $-5°C$, determine how long it will take for any part of the orange to begin to freeze (ignore radiation)

a. if it is a relatively still night and the heat transfer coefficient is 10 W/m²·K.

b. if it is a windy night and the heat transfer coefficient is 40 W/m²·K.

P11-44 Stainless-steel (AISI 304) ball bearings, which have been uniformly heated in an oven to $850°C$, are hardened by quenching them in an oil bath that is maintained at about $40°C$ by removing warm oil from the top of the bath and adding cool oil at the bottom. The ball diameter is 20 mm. The balls move through the bath on a conveyor belt at a velocity of 0.15 m/s. Assume the oil has $c_p = 1960$ J/kg·K and its convective heat transfer coefficient is 1230 W/m²·K.

a. Determine how long the balls must be in the bath until their surface temperature reaches $100°C$.

b. Determine the center temperature at the conclusion of the cooling.

c. Determine what oil flow is required if the oil temperature cannot rise more than $5°C$ and 10,000 balls per hour are to be quenched.

P11-45 For a large party celebrating your graduation from college, your parents buy a case of frozen steaks. Before you can throw the steaks onto the grill, they need to be thawed at room temperature of $77°F$ with an estimated convective heat transfer coefficient of 5 Btu/h·ft²·°F on both sides of the steak. The very large 1-in.-thick steaks are initially at $10°F$, and they are thawed when the centerline temperature is at $32°F$. Assume the steaks' properties can be approximated as that of water. Neglect the energy associated with the melting-phase change.

a. Determine the time required for the steaks to thaw (in min).

b. Determine the time required if you try to speed up the thawing by hanging the steaks on hooks and blowing air over them with a fan such that the heat transfer coefficient is increased to 50 Btu/h·ft²·°F (in min).

P11-46 One method for experimentally determining thermal conductivity is to measure the temperature response of a thick slab when it is subjected to a step change in surface temperature. Consider a solid with $\rho = 2500$ kg/m³ and $c_p = 630$ J/kg·K. The solid's temperature, initially a uniform $25°C$, is measured with a thermocouple embedded 6 mm from the surface. Boiling water at $100°C$ is brought into contact with the surface, and the heat transfer coefficient is very large. After 90 s, the thermocouple reads $73°C$. Determine the thermal conductivity of the solid (in W/m·K).

P11-47 Curing of the epoxy in a laminated material can be accelerated by the application of heat. Consider an electric heater that is pressed tightly against a thick slab of a laminated material whose properties are estimated to be $\rho = 1200$ kg/m³, $c_p = 1350$ J/kg·K, and $k = 1.3$ W/m·K. The laminate initially is at a uniform $25°C$. The heater has a heat flux of 350 W/m².

a. Determine the temperature at the surface of the laminate 3 min after the heat is applied (in °C).

b. Determine the temperature 5 mm into the laminate 3 min after the heat is applied (in °C).

P11-48 A consulting engineer is asked to investigate a suspicious fire. A room paneled with thick oak planks burned very quickly. The insurance company wants an estimate of the time required for the surface of the oak planks to reach their ignition temperature of $400°C$. Initially, the wood was at $25°C$. The temperature of the hot gases from the fire was estimated to be $850°C$ with a heat transfer coefficient of 25 W/m²·K.

a. Determine the time required to reach the ignition temperature (in s).

b. Determine the temperature 1 cm inside the wall at this time.

c. Comment on the influence of radiation.

P11-49 For proper heat treatment of metals, the temperature distribution in the metal must be carefully controlled. Consider the annealing of a large slab of 304 stainless steel. Initially, the slab is at $150°C$. It is placed in an oven in which the air temperature is $1040°C$ and the heat transfer coefficient is 450 W/m²·K. The *average* temperature of the slab must be raised to $800°C$, but the surface temperature should not rise above $900°C$.

a. Determine the maximum slab thickness that can be processed (in cm).

b. Determine the time required for the annealing process (in s and min).

P11-50 Quenching of a metal slab in an oil bath requires that the bath temperature does not rise significantly. The temperature rise in the bath can be determined if the heat transferred to the oil from the slab is known. Consider a plain carbon steel slab 2.5 cm thick and 2 m square. Initially at $1000°C$, it is quenched in an oil bath at $100°C$. The heat transfer coefficient is 450 W/m²·K. Ignoring edge effects, determine:

a. the time required for the centerline temperature to reach $425°C$.

b. the temperature at a depth of 0.5 cm from the surface at the same time (in °C).

c. the heat transfer from the slab to the oil (in kJ).

MULTIDIMENSIONAL TRANSIENT CONDUCTION

P11-51 For the cylinder in Problem P11-39, assume it now is only 15 cm long. For the same conditions as given in that problem, determine

a. the center temperature of the cylinder 1 min after it is quenched (in °C).

b. the center temperature of the cylinder 5 min after it is quenched (in °C).

c. the time required for the center temperature to reach 50°C (in s).

P11-52 Raw clay molded into bricks is fired in a kiln at 1300°C and cooled in air at 25°C with a convective heat transfer coefficient of 50 W/m²·K. The 5.7 cm by 10 cm by 20 cm brick has the following properties: $\rho = 2050$ kg/m³; $k = 1.0$ W/m·K; $c_p = 960$ J/kg·K; and $\alpha = 0.51 \times 10^{-6}$ m²/s. After 50 min of cooling, determine

a. the temperature at the center (in °C).

b. the temperature of the corners of the brick (in °C).

P11-53 Exposed ceiling beams are popular in many modern houses. However, because they are exposed, their ignition in the event of a fire is of concern. Consider a 15 cm by 15 cm yellow pine beam, initially at 25°C, attached to a ceiling, thus insulating the base. The three other sides are exposed to fire at 600°C with a combined convective/radiative heat transfer coefficient of 50 W/m²·K. Determine the beam's maximum temperature 5 min after fire starts.

P11-54 For a fire investigation, the insurance company for whom you work wants to know how long it would take for an oak beam 2 in. by 4 in. to ignite under certain conditions. The air temperature is 1000°F with a convective heat transfer coefficient of 1.8 Btu/h·ft²·°F. The initial temperature of the wood is 75°F, and its ignition temperature is 900°F. The wood has $\rho = 45$ lbm/ft³, $c_p = 0.30$ Btu/lbm·°F, and $k = 0.10$ Btu/h·ft·°F. Ignore radiation. Determine the time required for any of the wood to start burning when suddenly exposed to these operating conditions (in s).

P11-55 A hotdog 20 mm in diameter and 15 cm long at 5°C is placed into boiling 100°C water; because of the vigorous boiling, the heat transfer coefficient on the hotdog is 250 W/m²·K. To be fully cooked, its center should be at 80°C. The hotdog's properties are: $\rho = 890$ kg/m³, $c_p = 3350$ J/kg·K, and $k = 0.5$ W/m·K. Determine how long the hot dog should be in the water (in s).

P11-56 The same heat-treating oven used in Problem P11-44 is used to prepare the same geometry balls for a different application. If the hot balls are cooled by natural convection in air at 20°C such that the heat transfer coefficient is 15 W/m²·K, determine the time it takes for the surface temperature to reach 100°C (in s).

CONTACT RESISTANCE

P11-57 In the thermal conductivity measurement device described in Problem P11-7, you ask that a new sample of the material be tested. The temperatures have changed to

$$T_1 = 93.00°C \quad T_2 = 84.97°C \quad T_3 = 71.39°C \quad T_4 = 69.13°C$$

The same thermal conductivity for the unknown material is obtained. However, you are puzzled by the different temperatures and a decreased heat transfer rate. When you disassemble the test piece, you discover that the new operator did not assemble the pieces carefully enough, and the two ends have some roughness that should not be there. As a result of the roughness, there is thermal contact resistance. Determine the magnitude of the thermal contact resistance (in m²·K/W).

P11-58 A device is to be constructed with a 3-mm 304 stainless-steel plate and a 9-mm layer of 2024-T6 aluminum. The temperature drop across the composite wall will be 150°C. Two different manufacturing methods can be used. The first method will result in a surface roughness of the contacting parts of 25 μm and a contact pressure of 25 MPa. The second method will result in a surface roughness of the contacting parts of 1.5 μm and a contact pressure of 7 MPa.

a. Estimate the heat flux for these two manufacturing methods (in W/m²).

b. Estimate the percent decrease in the heat flux compared to no contact resistance.

P11-59 Significant manufacturing advances have been made to improve the speed of computer chips. The technique is to place more discrete electronic components closer together. However, this increases the electric power dissipation to such high levels that cooling is becoming a problem. To accommodate the high-power densities, direct cooling with boiling has been investigated. Consider the 10 mm by 10 mm thin chip shown in the figure, which is cooled by liquid at 25°C with a boiling heat transfer coefficient of 750 W/m²·K. The chip is attached to the circuit board, and the contact resistance between the chip and board is estimated to be 1.4×10^{-4} m²·K/W. The circuit board is 4 mm thick with a thermal conductivity of 2 W/m·K. The backside of the board is exposed to air at 25°C with a heat transfer coefficient of 35 W/m²·K.

a. Determine the thermal circuit of the chip, board, and cooling fluid combinations.

b. Determine the chip temperature if the chip heat dissipation rate is 4 W (in °C).

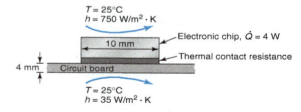

$T = 25°C$
$h = 750$ W/m²·K

10 mm

Electronic chip, $\dot{Q} = 4$ W

Thermal contact resistance

4 mm   Circuit board

$T = 25°C$
$h = 35$ W/m²·K

P11-60 Annular fins 1.5 mm thick and 15 mm long are attached to a 30-mm-diameter tube. Tube and fins are 2024-T6 aluminum. The thermal contact resistance between the fins and the tube is 2.5×10^{-4} m$^2 \cdot$K/W. The tube wall is at 75°C, the surrounding air temperature is 25°C, and the convective heat transfer coefficient is 100 W/m$^2 \cdot$K. For a single fin, determine:

a. the heat transfer rate without contact resistance (in W).

b. the heat transfer rate with contact resistance (in W).

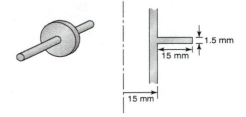

CHAPTER *12*

CONVECTION HEAT TRANSFER

12.1 INTRODUCTION

Whenever a fluid at one temperature flows past a solid surface at a different temperature, heat transfer occurs. Shown in Figure 12-1 is a schematic of a computer chip on a circuit board cooled by the flow of air over it. Heat from the chip first conducts to the air, and the air movement convects heat away from the surface. Similar processes with heat transfer to or from the fluid take place in car engines, to or from the roofs of houses, in heat exchangers in oil refineries, and in numerous other applications.

Fluid movement occurs due to two fundamentally different mechanisms: natural and forced convection. The resulting heat transfer characteristics are significantly different. Consider convective heat transfer from a hot surface to a cooler fluid. In **natural convection** (also called free convection), heat conducts into the fluid near the hot surface, thereby raising the temperature of the fluid and decreasing its density. Surrounding cooler fluid at higher density then flows under gravity to displace the hot fluid. (This is what the phrase *hot air rises* means.) Whenever hot fluid rises, cooler fluid falls to fill the void. An example of a circuit board cooled by natural convection is shown in Figure 12-2. Natural convection occurs with either a hot surface/cold fluid or a cold surface/hot fluid arrangement.

In **forced convection**, fluid is moved by mechanical means, such as a fan or pump. However, it also can occur without fans being involved. For example, an ice skater racing across a lake experiences considerable forced convection as the air rushes past. As with natural convection, forced convection can involve either heating or cooling of a solid surface.

In previous chapters, the convective heat transfer coefficient, h, was introduced. Its definition is (for a surface hotter than a fluid)

$$h \equiv \frac{q''}{\Delta T} = \frac{q''}{T_s - T_f}$$

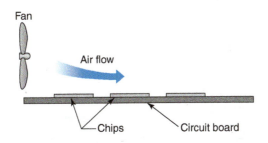

FIGURE 12-1 Forced convection cooling of electronic components.

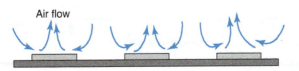

FIGURE 12-2 Natural convection cooling of electronic components.

where T_s is surface temperature and T_f is fluid temperature. The heat flux q'' and the temperature difference ΔT are such that h is always positive. In this chapter, common relationships for calculating or estimating the value of h will be examined.

12.2 FORCED CONVECTION IN EXTERNAL FLOWS

When fluid flows next to a stationary surface, a velocity boundary layer forms, as shown in Figure 12-3. The mechanisms underlying the formation and growth of the velocity boundary layer have been discussed in previous chapters. The edge of the velocity boundary layer is arbitrarily defined as the point where the velocity is 99% of the free-stream value. If the fluid is at a different temperature than the plate, a thermal boundary layer also forms, as shown in Figure 12-4. The edge of the thermal boundary layer is arbitrarily defined as the point where the fluid temperature minus the surface temperature is 99% of the difference between the free-stream temperature and the surface temperature. For a surface being cooled, heat conducts from the surface into the fluid and then is swept downstream with the flow. Heat is added to the fluid all along the plate, so the thermal boundary layer grows in thickness with distance along the plate, as shown in Figure 12-4.

The thermal boundary layer acts as a barrier to heat transfer. The heat transfer coefficient is defined as

$$h \equiv \frac{q''}{T_s - T_f} \tag{12-1}$$

where T_s is surface temperature and T_f is fluid temperature. Heat conducts from the surface into the fluid according to Fourier's law:

$$q'' = -k\frac{dT}{dy}$$

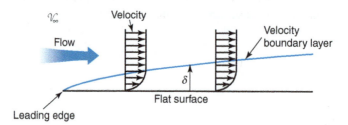

FIGURE 12-3 Growth of the velocity boundary layer on a flat surface.

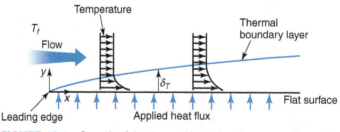

FIGURE 12-4 Growth of the thermal boundary layer on a flat surface.

where y is distance from the surface, as shown in Figure 12-4. To get an idea of how boundary-layer thickness affects heat transfer, we will approximate the temperature gradient as

$$\frac{dT}{dy} \approx \frac{\Delta T}{\Delta y}$$

Using the temperature difference across the boundary layer for ΔT and the thickness of the thermal boundary layer, δ_T, for Δy,

$$\frac{\Delta T}{\Delta y} \approx \frac{T_f - T_s}{\delta_T}$$

The heat flux now may be approximated as

$$q'' \approx -k\frac{\Delta T}{\Delta y} \approx -k\left(\frac{T_f - T_s}{\delta_T}\right)$$

Substituting this into Eq. 12-1 gives

$$h \approx \frac{\left[-k\left(\dfrac{T_f - T_s}{\delta_T}\right)\right]}{T_s - T_f} \approx \frac{k}{\delta_T}$$

From this equation, we see that a larger boundary layer implies a smaller heat transfer coefficient. Because the boundary-layer thickness grows with distance along the plate, the local heat transfer coefficient decreases along the plate. Heat transfer is substantially better at the leading edge than it is downstream, as shown in Figure 12-5.

The relative sizes of the velocity and thermal boundary layers depend on three important physical properties: the thermal conductivity, k; specific heat, c_p; and viscosity, μ. Thermal conductivity has an effect because it controls how easily heat is conducted in the fluid. Specific heat determines the temperature rise in the fluid as a result of conduction. Finally, viscosity affects the velocity field and thus the rate at which heat is convected. More advanced analysis shows that these important physical parameters appear in the following combination:

$$Pr = \frac{c_p\mu}{k}$$

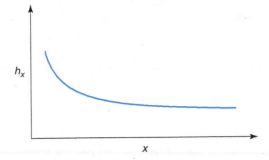

FIGURE 12-5 Variation of heat transfer coefficient along the surface.

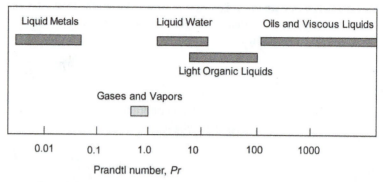

FIGURE 12-6 Typical Prandtl numbers for common fluids.

where Pr is called the **Prandtl number**. The Prandtl number is dimensionless, and, since it is composed of thermophysical properties, it is itself a thermophysical property. For example, the Prandtl number is included in Table A-6 for the properties of liquids.

The Prandtl number depends only on the fluid properties, not on the flow velocity or geometry. We can rearrange the Prandtl number as

$$Pr = \frac{\left(\frac{\mu}{\rho}\right)}{\left(\frac{k}{\rho c_p}\right)} = \frac{\nu}{\alpha}$$

where $\nu = \mu/\rho$ is the kinematic viscosity and $\alpha = k/\rho c_p$ is the thermal diffusivity. Physically, the Prandtl number is the ratio of momentum transport (ν) to thermal transport (α) in the boundary layer. As such, the Prandtl number determines the relative size of the velocity and thermal boundary layers. If the Prandtl number is 1, the velocity and thermal boundary layers have the same thickness at a given location on the plate. If the Prandtl number is less than 1, then the velocity boundary layer is thinner than the thermal boundary layer. Conversely, if the Prandtl number is greater than 1, then the velocity boundary layer is thicker than the thermal boundary layer.

Figure 12-6 shows Prandtl number ranges for typical fluids. Liquid metals have very low Prandtl numbers. For liquid metals, thermal transport is much more effective than momentum transport. Gases have Prandtl numbers ranging between 0.5 and 1, while liquids typically have Prandtl numbers greater than 1. Note that there are no common fluids with Prandtl numbers on the order of 0.1.

Another important dimensionless parameter in convective heat transfer is found by comparing convection across a thermal boundary layer of thickness δ_T to conduction across a stagnant layer of the same thickness. Convection is given by

$$\dot{Q}_{conv} = hA\left(T_s - T_f\right)$$

where T_s is the temperature of the surface and T_f is the temperature of the fluid. Conduction is given by

$$\dot{Q}_{cond} = \frac{kA}{\delta_T}\left(T_s - T_f\right)$$

The ratio of convection to conduction would then be

$$\frac{\dot{Q}_{conv}}{\dot{Q}_{cond}} = \frac{hA\left(T_s - T_f\right)}{\frac{kA}{\delta_T}\left(T_s - T_f\right)} = \frac{h\delta_T}{k}$$

This dimensionless ratio is called the **Nusselt number**, the nondimensional heat transfer coefficient. The boundary-layer thickness is not easily determined for geometries other than flow over a flat plate. It is an inconvenient length scale to use in the Nusselt number, so a dimension characteristic of the geometry is used instead. The Nusselt number then becomes

$$\boxed{Nu = \frac{hL_{char}}{k}}$$

where L_{char} is a characteristic length that will be specified as needed. As we will see later in this chapter, the Nusselt number arises naturally from a solution of the governing equations for flow with convection.

The Nusselt number is to the thermal boundary layer what the friction coefficient is to the velocity boundary layer. Just as the friction coefficient can be calculated as a function of the Reynolds number, so the Nusselt number can also be calculated as a function of the Reynolds number. In addition, the Nusselt number depends on the Prandtl number. For forced convection, most experimental data can be correlated by an equation of the form

$$Nu = f(Re, Pr)$$

The first case we consider is flow over an isothermal flat plate, as shown in Figure 12-7. The velocity boundary layer at the leading edge of the plate is laminar. If the plate is sufficiently long, the boundary layer transitions to turbulence at a distance x_{crit} along the plate. The location of the transition point is found by experiment and depends on the turbulence of the free-stream flow, the roughness of the plate, the shape of the leading edge, and other factors. A reasonable estimate of the transition point is given by

$$Re_{x,crit} = \frac{\rho \mathcal{V} x_{crit}}{\mu} = 5 \times 10^5$$

where $Re_{x,crit}$ is the so-called critical Reynolds number.

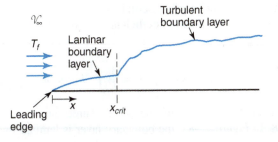

FIGURE 12-7 Velocity boundary layer on a flat plate.

It is possible to use conservation of mass, momentum, and energy to solve for the heat transfer coefficient along a flat plate in laminar flow; however, the analysis is beyond the scope of this text. The resulting equation for the local Nusselt number is

$$Nu_x = \frac{h_x x}{k} = 0.332\,Re_x^{1/2}\,Pr^{1/3} \qquad \begin{array}{l} Re_x < 5 \times 10^5 \\[4pt] Pr > 0.6 \\[4pt] \text{isothermal plate} \end{array} \qquad (12\text{-}2)$$

where h_x is the local heat transfer coefficient, that is, the heat transfer coefficient at a distance x from the leading edge of the plate. The local heat transfer coefficient, which is shown in Figure 12-5, depends on the local Reynolds number, Re_x, defined as

$$Re_x = \frac{\rho \mathcal{V} x}{\mu}$$

The fluid properties in Eq. 12-2 are evaluated at the film temperature, which is defined as the average of the surface and fluid temperatures, that is,

$$T_{film} = \frac{T_s + T_f}{2}$$

If $x > x_{crit}$, the boundary layer is turbulent. The local Nusselt number in a turbulent boundary layer is given by

$$Nu_x = \frac{h_x x}{k} = 0.0296\,Re_x^{4/5}\,Pr^{1/3} \qquad \begin{array}{l} Re_x > 5 \times 10^5 \\[4pt] 0.6 < Pr < 60 \\[4pt] \text{isothermal plate} \end{array} \qquad (12\text{-}3)$$

Fluid properties in this equation are also evaluated at the film temperature.

Heat flux is related to heat transfer coefficient by

$$q''(x) = h_x \left(T_s - T_f \right)$$

where $q''(x)$ is the local heat flux at location x. For the isothermal plate under consideration, heat flux is large near the leading edge, where the heat transfer coefficient is high, and smaller downstream, where the heat transfer coefficient is lower. In most practical applications, detailed information on the local variation of heat flux is not needed and the average heat flux over the entire plate is the quantity of interest. To find average heat flux, we need an average heat transfer coefficient.

The average heat transfer coefficient can be calculated from the local heat transfer coefficient using

$$h = \frac{1}{L} \int_0^L h_x \, dx \qquad (12\text{-}4)$$

where L is the total length of the plate. Three cases are of interest, as shown in Figure 12-8. In Figure 12-8a, the boundary layer is laminar over the entire length of the

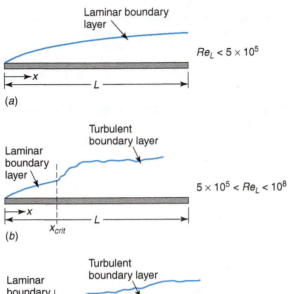

FIGURE 12-8 Three cases of boundary-layer development on an isothermal flat plate. (a) The boundary layer is laminar over the entire plate. (b) The laminar and turbulent boundary layers are of comparable extent. (c) The turbulent boundary layer extends over almost the entire plate.

plate. In Figure 12-8b, the boundary layer is laminar on the first part of the plate and turbulent on the rest of the plate. In Figure 12-8c the boundary layer is turbulent over a very large portion of the plate.

To find the average heat transfer coefficient in Figure 12-8a, where the boundary layer is laminar over the whole plate, solve Eq. 12-2 for h_x and substitute into Eq. 12-4 to get

$$h = \frac{1}{L} \int_0^L \frac{0.332k\, Re_x^{1/2} Pr^{1/3}}{x} \, dx = \frac{0.332k\, Pr^{1/3}}{L} \int_0^L \left(\frac{\rho \mathcal{V} x}{\mu} \right)^{1/2} \frac{dx}{x}$$

$$= \frac{0.332k\, (\rho \mathcal{V})^{1/2} Pr^{1/3}}{L\mu^{1/2}} \int_0^L \frac{dx}{x^{1/2}}$$

Performing the integration and writing the result in terms of nondimensional parameters yields

$$Nu_L = \frac{hL}{k} = 0.664 Re_L^{1/2} Pr^{1/3} \qquad \begin{array}{l} Re_L < 5 \times 10^5 \\ Pr > 0.6 \\ \text{isothermal plate} \end{array} \qquad (12\text{-}5)$$

This equation gives the average heat transfer coefficient on an isothermal flat plate when the boundary layer is laminar over the entire plate. As before, fluid properties are evaluated at the film temperature, which is the average of the surface and fluid temperatures.

In Figure 12-8b, the boundary layer is laminar up to x_{crit} and turbulent beyond that. To find the average heat transfer coefficient in this case, we use

$$h = \frac{1}{L} \left(\int_0^{x_{crit}} h_{x, \, laminar} \, dx + \int_{x_{crit}}^{L} h_{x, \, turbulent} \, dx \right)$$

Solving Eq. 12-2 and Eq. 12-3 for the local heat transfer coefficients, substituting into this equation, and performing the integration results in

$$Nu_L = \frac{hL}{k} = \left(0.037 Re_L^{4/5} - 871 \right) Pr^{1/3} \qquad \begin{array}{l} 5 \times 10^5 < Re_L \leq 10^8 \\ 0.6 < Pr < 60 \\ \text{isothermal plate} \end{array} \qquad (12\text{-}6)$$

where the transition Reynolds number is assumed to be 5×10^5. As before, properties are evaluated at the film temperature. If the Reynolds number is very high, the boundary layer is turbulent over a great portion of the plate, as illustrated in Figure 12-8c. In this case we may neglect the small part of the plate covered by a laminar boundary layer and assume that the boundary layer is turbulent over the entire plate. To find the average heat transfer coefficient, we proceed as before. Solving Eq. 12-3 for h_x, substituting into Eq. 12-4, and performing the integration leads to

$$Nu_L = \frac{hL}{k} = 0.037 Re_L^{4/5} Pr^{1/3} \qquad \begin{array}{l} 10^8 < Re_L \\ 0.6 < Pr < 60 \\ \text{isothermal plate} \end{array} \qquad (12\text{-}7)$$

In many practical cases, the flow is actually turbulent starting from the leading edge. This can occur if the boundary layer is disturbed at the leading edge. For example, if the plate has a finite thickness, the corner of the leading edge can trip the boundary layer into turbulence. In that case, Eq. 12-7 can be used to find the average heat transfer coefficient over the whole plate.

Up to this point, we have assumed that the plate is isothermal. If, instead, heat flux is constant on the plate, the local Nusselt numbers in laminar and turbulent flows are given by

$$Nu_x = \frac{h_x x}{k} = 0.453 Re_x^{1/2} Pr^{1/3} \qquad \begin{array}{l} Re_x < 5 \times 10^5 \\ Pr > 0.6 \\ \text{constant heat flux plate} \end{array} \qquad (12\text{-}8)$$

$$Nu_x = \frac{h_x x}{k} = 0.0308 Re_x^{4/5} Pr^{1/3} \qquad \begin{array}{l} Re_x > 5 \times 10^5 \\ 0.6 < Pr < 60 \\ \text{constant heat flux plate} \end{array} \qquad (12\text{-}9)$$

For the laminar case (Eq. 12-8), the local heat transfer coefficient is 36% higher on a constant heat flux plate than on an isothermal plate. The difference is much smaller in turbulent flow, amounting to only 4%. The local heat transfer coefficient on a constant heat flux plate is used to find the variation of surface temperature along the plate. Correlations for the average heat transfer coefficient have been obtained from these local correlations, and they can be used to find the average temperature of the plate. However, the results are very close (within 2%) to those obtained by using the correlations given above for constant surface temperature. Therefore, if average values of the heat transfer coefficient are needed, the isothermal correlations may be used.

Many empirical equations have been developed to estimate convective heat transfer coefficients for a very wide range of geometries, flow conditions, and fluids. Because of the difficulty of making heat transfer measurements and the inherent uncertainty in their results, there is typically a fairly large uncertainty in predictions from such correlations, often ranging from 5% to 25%. Therefore, one must account for such uncertainty when using correlations for thermal engineering design.

EXAMPLE 12-1   Cooling of a computer chip

Computer chips generate heat during operation, and their failure rate increases with temperature. In the figure, a chip is cooled by a flow of air at 20°C and atmospheric pressure. If the maximum allowable chip surface temperature is 65°C and the air velocity is 3.4 m/s, how much heat can be removed by forced convection (in W)? Use data given in the figure.

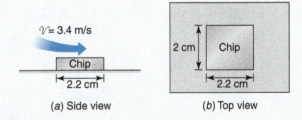

(a) Side view   (b) Top view

Approach:

The heat removed may be calculated using $\dot{Q} = hA\left(T_s - T_f\right)$. The temperature and areas are given, so we only need to find the heat transfer coefficient. Computer chips are thin, and it is reasonable to assume that the chip can be approximated as a flat plate. The lip at the leading edge of the chip might improve heat transfer, but we assume that no data or correlations are available to quantify that effect. Using flat plate correlations will give conservative results, that is, lower values of heat transfer coefficient. We also assume that no heat is conducted into the substrate under the chip.

We seek an average heat transfer coefficient over the whole chip. To determine whether the boundary layer is laminar, turbulent, or a mix of laminar and turbulent, calculate the length Reynolds number. Then select the appropriate correlation for Nusselt number and use it to find h.

Assumptions:

A1. All heat leaves by convection from the top of the chip and none is conducted into the substrate.

Solution:

The heat transfer rate is [A1]

$$\dot{Q} = hA\left(T_s - T_f\right)$$

We need to evaluate the heat transfer coefficient. To do so, first determine the flow regime by calculating the Reynolds number at the end of the plate:

$$Re = \frac{\rho \mathcal{V} L}{\mu}$$

To evaluate the properties, use the film temperature, which is

$$T_{film} = \frac{T_f + T_s}{2} = \frac{20 + 65}{2} = 42.5°C = 315\,K$$

Using data from Table A-6 and the given information,

$$Re = \frac{\left(1.126 \frac{kg}{m^3}\right)\left(3.4 \frac{m}{s}\right)(2.2 \text{ cm})\left(\frac{1 \text{ m}}{100 \text{ cm}}\right)}{1.92 \times 10^{-5} \frac{kg}{m \cdot s}} = 4387$$

A2. Flat plate correlations apply.

Since the Reynolds number is less than 5×10^5, the flow is laminar. We assume that the flow over the top of the chip can be approximated as flow over a flat plate [A2], so that from Eq. 12-5,

$$Nu_L = 0.664 \, Re_L^{1/2} \, Pr^{1/3}$$

The Prandtl number, found in Table A-6, is

$$Pr = 0.71$$

The above equation for the Nusselt number applies if $Pr > 0.6$; therefore, it is valid in this case. Substituting Re and Pr, the Nusselt number becomes

$$Nu_L = 0.664 \, (4387)^{1/2} \, (0.71)^{1/3} = 39.2$$

Using the definition of the Nusselt number,

$$Nu_L = \frac{hL}{k} = 39.2$$

Solving for h,

$$h = \frac{39.2 \, k}{L} = \frac{(39.2)\left(0.0271 \frac{W}{m \cdot °C}\right)}{(2.2 \text{ cm})\left(\frac{1 \text{ m}}{100 \text{ cm}}\right)} = 48.3 \frac{W}{m^2 \cdot °C}$$

Therefore, the heat transferred is

$$\dot{Q} = hA \, (T_s - T_f) = \left(48.3 \frac{W}{m^2 \cdot °C}\right) [(2.0 \text{ cm})(2.2 \text{ cm})] \left(\frac{1 \text{ m}}{100 \text{ cm}}\right)^2 (65 - 20) \, °C = 0.957 \, W$$

EXAMPLE 12-2   Heat loss from a residential building

An exterior wall of a building is exposed to wind blowing at 20 ft/s parallel to the wall. The wall is insulated with 6 in. of fiberglass that has a thermal conductivity of 0.026 Btu/h·ft·R. The air inside the building exchanges heat by natural convection with the inside of the wall. The inside heat transfer coefficient is 1 Btu/h·ft²·R. If the outside air is at 20°F and the inside air is at 70°F, estimate

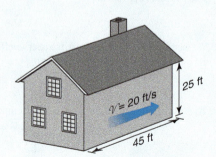

the heat loss through the wall. Assume the conduction resistance is dominated by the insulation and neglect the resistance of the wallboard, wall studs, siding material, and so on.

Approach:

The rate of heat loss through the wall is $\dot{Q} = \Delta T / R_{tot}$. The total resistance, R_{tot}, is the sum of three resistances: convective resistance on the inside of the house, conduction resistance through the fiberglass, and convective resistance on the outside of the house. We know the natural convection heat transfer coefficient on the inside and so can calculate the inside convective resistance. To find the heat transfer coefficient on the outside, assume the wall of the house may be approximated as a flat plate. Determine the Reynolds number and use the appropriate correlation.

Assumptions:

A1. Conduction resistance in the siding, wallboard, and so on is small compared to conduction resistance of the fiberglass.
A2. Heat transfer is one-dimensional.
A3. The side of the building may be modeled as an infinitely wide flat plate.

Solution:

The heat transfer can be modeled as three resistances in series [A1][A2]:

$$R_1 = \frac{1}{h_1 A} \qquad R_2 = \frac{L_2}{k_2 A} \qquad R_3 = \frac{1}{h_3 A}$$

where R_1 is convection resistance on the inside of the house, R_2 is conduction resistance through the fiberglass, R_3 is convection resistance on the outside of the house, L_2 is the insulation thickness, and k_2 is the insulation thermal conductivity. Everything is known except the outside heat transfer coefficient, h_3, so approximate the flow as flow over a flat plate [A3]. First find the Reynolds number:

$$Re = \frac{\rho L \mathcal{V}}{\mu}$$

For this external flow, we need to evaluate the properties at the film temperature, but we do not know the temperature of the outside surface of the wall. We could assume a value and then iterate, but, for simplicity, we will just use a guessed value and tolerate the small error that this introduces. With properties of air at 32°F from Table B-7, the Reynolds number is

$$Re_L = \frac{\left(0.081 \,\frac{\text{lbm}}{\text{ft}^3}\right)(45 \text{ ft})\left(20 \,\frac{\text{ft}}{\text{s}}\right)}{1.165 \times 10^{-5} \,\frac{\text{lbm}}{\text{ft·s}}} = 6.26 \times 10^6$$

A4. The critical Reynolds number is 5×10^5.

Referring to Eq. 12-6, this Reynolds number falls in the range in which both laminar and turbulent flow exist on the plate. Assuming the critical Reynolds number is 5×10^5 [A4],

$$Nu_L = \frac{h_3 L}{k} = \left(0.037 Re_L^{0.8} - 871\right) Pr^{1/3}$$

Solving for h_3,

$$h_3 = \frac{k}{L} \left(0.037 Re_L^{0.8} - 871\right) pr^{1/3}$$

Using property values from Table B-7,

$$h_3 = \left(\frac{0.014 \,\frac{\text{Btu}}{\text{h·ft·°F}}}{45 \text{ ft}}\right) \left[0.037 \left(6.26 \times 10^6\right)^{0.8} - 871\right] (0.72)^{1/3}$$

$$h_3 = 2.58 \,\frac{\text{Btu}}{\text{h·ft}^2\text{·°F}}$$

The external thermal resistance, R_3, is then

$$R_3 = \frac{1}{h_3 A} = \frac{1}{\left(2.58 \, \dfrac{\text{Btu}}{\text{h·ft}^2\text{·°F}}\right)\left[(45)\,(25)\,\text{ft}^2\right]} = 3.45 \times 10^{-4} \, \frac{\text{°F·h}}{\text{Btu}}$$

The thermal resistance across the insulation is

$$R_2 = \frac{L_2}{k_2 A} = \frac{0.5 \, \text{ft}}{\left(0.026 \, \dfrac{\text{Btu}}{\text{h·ft·R}}\right)\left[(45)\,(25)\,\text{ft}^2\right]} = 0.0171 \, \frac{\text{°F·h}}{\text{Btu}}$$

Finally, the internal thermal resistance, R_1, is

$$R_1 = \frac{1}{h_1 A} = \frac{1}{\left(1 \, \dfrac{\text{Btu}}{\text{h·ft}^2\text{·°F}}\right)\left[(45)\,(25)\,\text{ft}^2\right]} = 8.89 \times 10^{-4} \, \frac{\text{°F·h}}{\text{Btu}}$$

The total resistance is

$$R_{tot} = R_1 + R_2 + R_3 = 8.89 \times 10^{-4} + 0.0171 + 3.45 \times 10^{-4} = 0.0183 \, \frac{\text{°F·h}}{\text{Btu}}$$

The rate of heat loss is

$$\dot{Q} = \frac{\Delta T}{R_{tot}} = \frac{(70 - 20) \, \text{°F}}{0.0183 \, \dfrac{\text{°F·h}}{\text{Btu}}} = 2732 \, \frac{\text{Btu}}{\text{h}}$$

Comment:

Clearly, the insulation conduction resistance is the dominant resistance in this case.

Flow over a flat plate is the simplest case of external flow. Another geometry often encountered in applications is the cylinder in crossflow. Examples include flow over tube banks, pipes, wires, extrusions, and filaments.

When fluid flows perpendicular to an infinite cylinder, complex flow patterns arise. As discussed in Chapter 10, a velocity boundary layer forms on the windward (upstream) side of the cylinder. This boundary layer separates from the cylinder at some location, and a wake forms on the leeward (downstream) side of the cylinder. Because of the velocity variations around the circumference, the convective heat transfer coefficient also varies, as shown in Figure 12-9. However, in most circumstances, only the circumferentially averaged heat transfer coefficient is needed.

The flow over a cylinder is strongly dependent on the Reynolds number. Different flow patterns occur at different Reynolds numbers (see Figure 10-11). As a result, it is difficult to find a simple equation for the convective heat transfer coefficient that applies to all Reynolds number ranges. However, the following equation for the average Nusselt number for convection over a right circular cylinder in crossflow is both useful and simple:

$$Nu_D = \frac{h\,D}{k} = C\,Re_D^m\,Pr^{1/3} \tag{12-10}$$

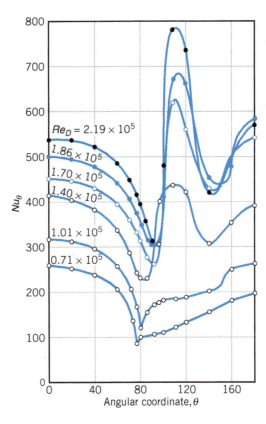

FIGURE 12-9 Local Nusselt number for air in crossflow over a right circular cylinder. Air impinges at $\theta = 0$.
(*Source*: From F. P. Incropera and D. P. Dewitt, *Introduction to Heat Transfer*, 4th ed., Wiley, New York, 2002, p. 384. Used with permission.)

The values of C and m for different Reynolds number ranges are given in Table 12-1. The same equation can be used for certain other bodies in crossflow, such as squares and ellipses, and these are also included in the table. All fluid properties in Eq. 12-10 are evaluated at the film temperature, which is the average of the fluid temperature and the surface temperature.

For convection over a sphere, the following correlation applies:

$$Nu_D = \frac{hD}{k} = 2 + \left[0.4\,Re_D^{1/2} + 0.06\,Re_D^{2/3}\right] Pr^{0.4} \left(\frac{\mu_f}{\mu_s}\right)^{1/4} \tag{12-11}$$

which is valid for $3.5 < Re_D < 80{,}000$ and $0.7 < Pr < 380$. In all correlations presented so far, properties have been evaluated at the film temperature. This correlation, however, is different. All properties are evaluated at the fluid temperature except for μ_s, which is evaluated at the surface temperature. The choice of temperature at which to evaluate properties is set by the individual researcher who developed the correlation and is generally selected to produce the best fit to experimental data. In more advanced treatments, the heat transfer coefficient may be found from the solution to a set of partial differential equations. In such a theoretical treatment, there is no ambiguity about the correct temperature to use for evaluating properties.

TABLE 12-1 Correlations of average Nusselt number for various bodies in crossflow

Cross-section of the cylinder	Fluid	Range of Re	Nusselt number
Circle	Gas or liquid	0.4–4 4–40 40–4000 4000–40,000 40,000–400,000	$Nu = 0.989Re^{0.330} Pr^{1/3}$ $Nu = 0.911Re^{0.385} Pr^{1/3}$ $Nu = 0.683Re^{0.466} Pr^{1/3}$ $Nu = 0.193Re^{0.618} Pr^{1/3}$ $Nu = 0.027Re^{0.805} Pr^{1/3}$
Ellipse	Gas	2500–15,000	$Nu = 0.248Re^{0.612} Pr^{1/3}$
Ellipse (tilted 90°)	Gas	3000–15,000	$Nu = 0.094Re^{0.804} Pr^{1/3}$
Square	Gas	2500–8000 5000–100,000	$Nu = 0.177Re^{0.699} Pr^{1/3}$ $Nu = 0.102Re^{0.675} Pr^{1/3}$
Square (tilted 45°)	Gas	2500–7500 5000–100,000	$Nu = 0.289Re^{0.624} Pr^{1/3}$ $Nu = 0.246Re^{0.588} Pr^{1/3}$
Hexagon	Gas	5000–100,000	$Nu = 0.153Re^{0.638} Pr^{1/3}$
Hexagon (tilted 45°)	Gas	5000–19,500 19,500–100,000	$Nu = 0.160Re^{0.638} Pr^{1/3}$ $Nu = 0.0385Re^{0.782} Pr^{1/3}$
Vertical Plate	Gas	4000–15,000	$Nu = 0.228Re^{0.731} Pr^{1/3}$

EXAMPLE 12-3   Combined conduction and convection in a wire

A copper wire of diameter 1/8 in. is covered with insulation 1/16 in. thick. Air blows in crossflow over the wire at 19 ft/s. The wire carries a current of 300 A and the air is at 80°F. The insulation has a thermal conductivity of 0.21 Btu/h·ft·°F, and the copper has an electrical resistivity of $1.72 \times 10^{-6} \, \Omega\cdot$cm. Find the maximum temperature of the insulation.

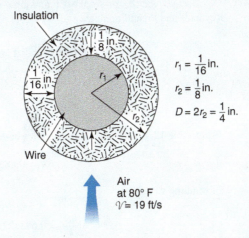

$r_1 = \frac{1}{16}$ in.

$r_2 = \frac{1}{8}$ in.

$D = 2r_2 = \frac{1}{4}$ in.

Air at 80° F
$\mathscr{V} = 19$ ft/s

Approach:

The heat generated in the wire is conducted through the insulation and convected to the air. The maximum temperature of the insulation, T_{max}, occurs at the inner radius of the insulation. The heat generated is related to temperature drop by

$$\dot{Q} = \frac{T_{max} - T_{air}}{R_{tot}}$$

where R_{tot} is the sum of the resistance across the insulation and the convective resistance on the outside of the wire. To find convective resistance, the heat transfer coefficient must be known. Correlations for forced convection over an infinite cylinder available in Table 12-1 can be used to obtain the heat transfer coefficient. The value of the Reynolds number is used to determine which correlation in this table applies.

Perform all calculations per unit length of wire. The heat generated in a 1-ft segment of wire is a function of current and electrical resistance. Electrical resistance is related to electrical resistivity by $R_{electric} = \rho_{electric} L/A_c$, where $\rho_{electric}$ is the electrical resistivity, L is the length of the wire, and A_c is the cross-sectional area of the copper.

Assumptions:

A1. Conduction is one-dimensional.

Solution:

To find the maximum insulation temperature, the thermal resistances for conduction across the insulation and convection to the air must be known [A1]. The thermal circuit is

$$T_{air} \quad R_1 = \frac{1}{hA} \quad R_2 = \frac{\ln\left(\frac{r_2}{r_1}\right)}{2\pi L k} \quad T_{max}$$

where T_{max} is the temperature at the inner radius of the insulation, R_1 is convection resistance, and R_2 is conduction resistance across the insulation. All calculations are performed for a unit length of wire of 1 ft.

The convective resistance, R_1, depends on the heat transfer coefficient. Correlations for heat transfer from a cylinder in crossflow depend on fluid properties evaluated at the film temperature. The surface temperature of the insulation, needed to calculate the film temperature, is unknown. For simplicity, we evaluate properties at the given air temperature of 80°F as an approximation. After calculating the surface temperature, we may recalculate the properties at the correct film temperature and iterate, if necessary.

Using properties of air at 80°F and atmospheric pressure from Table B-7, we find that the Reynolds and Prandtl numbers are

$$Re = \frac{\rho D \mathcal{V}}{\mu} = \frac{\left(0.074 \, \frac{\text{lbm}}{\text{ft}^3}\right)\left[0.25 \, \text{in.} \left(\frac{1 \, \text{ft}}{12 \, \text{in.}}\right)\left(19 \, \frac{\text{ft}}{\text{s}}\right)\right]}{1.25 \times 10^{-5} \, \frac{\text{lbm}}{\text{ft·s}}} = 2343$$

$$Pr = 0.72$$

From Table 12-1, a correlation that applies in this range of Reynolds and Prandtl number is

$$Nu = 0.683 \, Re^{0.466} Pr^{1/3}$$

Substituting values,

$$Nu = \frac{hD}{k} = 0.683 \, (2343)^{0.466} \, (0.72)^{1/3} = 22.8$$

It follows that, with air properties from Table B-7, the heat transfer coefficient is

$$h = \frac{(22.8)\left[0.25 \, \text{in.} \left(\frac{1 \, \text{ft}}{12 \, \text{in.}}\right)\right]}{\left(0.015 \, \frac{\text{Btu}}{\text{h·ft·°F}}\right)} = 16.4 \, \frac{\text{Btu}}{\text{h·ft}^2 \text{·°F}}$$

The convective resistance on the outside of the wire is, therefore,

$$R_1 = \frac{1}{hA} = \frac{1}{h\,(2\pi r_2)\,L} = \frac{1}{\left(16.4 \, \frac{\text{Btu}}{\text{h·ft}^2 \text{·°F}}\right)\left[2\pi \left(\frac{1}{8} \text{in.}\right)\left(\frac{1 \, \text{ft}}{12 \, \text{in}}\right)\right](1 \, \text{ft})} = 0.932 \, \frac{\text{°F·h}}{\text{Btu}}$$

The conduction resistance through the insulation is

$$R_2 = \frac{\ln\left(\frac{r_2}{r_1}\right)}{2\pi L k} = \frac{\ln\,(2)}{2\pi \, (1 \, \text{ft}) \left(0.21 \, \frac{\text{Btu}}{\text{h·ft}^2 \text{·°F}}\right)} = 0.525 \, \frac{\text{°F·h}}{\text{Btu}}$$

Therefore, the total resistance is

$$R_{tot} = R_1 + R_2 = 1.46 \, \frac{\text{°F·h}}{\text{Btu}}$$

Next, find the total heat generated per unit foot in the wire. The electrical resistance per unit foot is

$$R_{electric} = \frac{\rho_{electric} L}{A_c} = \frac{\left(1.72 \times 10^{-6} \Omega \text{·cm}\right)(1 \, \text{ft})}{\pi \left[\left(\frac{1}{16} \, \text{in.}\right)\left(\frac{1 \, \text{ft}}{12 \, \text{in.}}\right)\right]^2 (1 \, \text{ft})} = 6.62 \times 10^{-4} \Omega$$

Therefore, the power generated is

$$\dot{Q} = I^2 R_{electric} = (300\,\text{A})^2 \left(6.62 \times 10^{-4}\,\Omega\right) = 59.6\,\text{W}$$

Heat transferred is related to temperature drop by

$$\dot{Q} = \frac{T_{max} - T_{air}}{R_{tot}}$$

Therefore, the maximum temperature of the insulation is

$$T_{max} = \dot{Q}R_{tot} + T_{air} = (59.6\,\text{W}) \left(1.46\,\frac{°\text{F·h}}{\text{Btu}}\right) \left(\frac{3.412\,\frac{\text{Btu}}{\text{h}}}{1\,\text{W}}\right) + 80°\text{F} = 376°\text{F}$$

Comment:

At this elevated temperature, the insulation is likely to fail.

12.3 LAMINAR CONVECTION IN PIPES

When convection occurs inside a pipe or channel, boundary layers form as they do in external flow. Figure 12-10a shows the development of the velocity boundary layer in a channel. At the entrance to the channel, the fluid at the center is at the free-stream velocity, but the fluid at the wall has zero velocity (no-slip condition). A velocity boundary layer starts to grow on each wall, just as it does for external flow over a flat plate. With lower velocities near the wall, the centerline velocity must increase to satisfy conservation of mass. At some distance from the entrance, the boundary layers become so thick that they extend to the center of the channel and meet. At that point, the velocity profile no longer changes shape with downstream location, and the flow is considered fully developed.

If the channel wall is heated or cooled, a thermal boundary layer also forms, as shown in Figure 12-10b. It may be thicker, thinner, or the same size as the velocity boundary layer, depending on the Prandtl number. Like the velocity boundary layer, the thermal boundary layer grows from the wall until it reaches the center of the channel. At that point, the temperature profile becomes fully developed and its shape no longer changes with downstream location, although its temperature level does change.

The convective heat transfer coefficient varies with location along the channel, as shown in Figure 12-10c. In the entrance region, where both velocity and thermal boundary layers are thin, the heat transfer coefficient is high. Farther downstream, as the boundary layers thicken, the heat transfer coefficient decreases. Finally, at some distance down the channel, the heat transfer coefficient levels off to an asymptotic value called the **fully developed heat transfer coefficient**.

We will explore laminar convective heat transfer with two different boundary conditions. One condition is a constant and uniform heat flux on the wall and the other condition is a constant wall temperature. The fluid temperature develops quite differently in these two cases. Furthermore, because of its importance, we will focus on flow in a circular pipe. Note that these two boundary conditions are two limiting cases. The results from the third common boundary condition—convection to another fluid—fall between the constant wall temperature and the constant wall heat flux results.

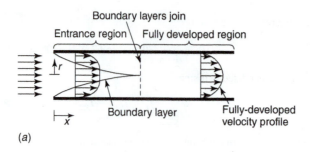

(a)

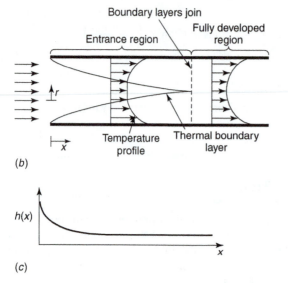

(b)

(c)

FIGURE 12-10 Boundary layer development and local heat transfer coefficient. (a) Velocity boundary layer in the entrance region of a channel. (b) Thermal boundary layer in the entrance region of a channel. (c) Local heat transfer coefficient variation with length.

The first case of internal flow we consider is heating of a constant property fluid in a pipe by application of a constant heat flux at the wall. In this case, the temperature field varies throughout the cross-section of the pipe, as shown in Figure 12-11. The flow field and the temperature field are both *fully developed*; that is, the profile shape does not vary with downstream location. There is one important difference between the fully developed velocity and fully developed temperature fields. The average velocity does not change with downstream location; however, heat is added, so the average temperature *does* increase. Although the average temperature changes, the *shape* of the temperature profile does not. This condition will be stated on a mathematical basis in the derivation that follows.

The temperature profile in the pipe can be determined from an energy balance on the differential control volume shown in Figure 12-12. The control volume is an annular ring of thickness Δr and length Δx and is concentric with the centerline of the pipe. Fluid flows

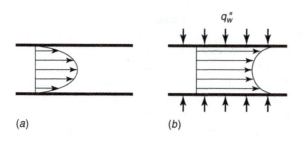

(a)   *(b)*

FIGURE 12-11 Fully developed velocity and temperature profiles in a pipe. (a) Velocity profile. (b) Temperature profile.

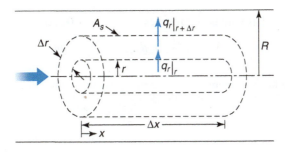

FIGURE 12-12 Control volume for fully developed laminar convection in a pipe.

into the left face and out the right face of the control volume. The first law for such an open system flow is

$$\dot{Q} - \dot{W} = \sum_{\text{out}} \dot{m}_e h_e - \sum_{\text{in}} \dot{m}_i h_i \tag{12-12}$$

Recall that in this equation, h is enthalpy, not heat transfer coefficient. There is no boundary work on the control volume. Furthermore, the work done by friction is usually very small. Such work is called **viscous dissipation**, and it can be significant for high-speed flows or flows with very high viscosity. For flow of common fluids, such as air or water at common velocities, viscous dissipation is negligible, and the term $\dot{W}$ may be set equal to zero.

Heat enters the control volume by conduction in both the axial and radial directions. Fluid flows in the axial direction and carries enthalpy into and out of the control volume. This enthalpy (energy) flow is often much greater than simple conduction in the axial direction, so we ignore conduction in the flow direction. (Exceptions occur in the case of liquid metal flow, where thermal conductivity is high, and very-low-speed flow, where convection is very weak.)

Assuming no viscous dissipation ($\dot{W} = 0$) and conduction only in the radial direction, Eq. 12-12 becomes

$$\left(q_r'' A_s \right)\big|_r - \left(q_r'' A_s \right)\big|_{r+\Delta r} = \dot{m}_e h_e - \dot{m}_i h_i$$

where q_r'' is the heat flux due to conduction in the radial direction and A_s is the surface area of the curved side of the control volume. Flow enters only in the axial direction, and because the flow is incompressible, density and velocity do not change. Under these circumstances, we may write

$$\left(q_r'' 2\pi r\, \Delta x \right)\big|_r - \left(q_r'' 2\pi r\, \Delta x \right)\big|_{r+\Delta r} = \left(\rho \mathcal{V}_x \Delta A_x \right)\left(h_{x+\Delta x} - h_x \right) \tag{12-13}$$

where ΔA_x, which is the differential cross-sectional area of the control volume in the axial direction, is given by

$$\Delta A_x = \pi\,(r + \Delta r)^2 - \pi r^2 = \pi\,(r^2 + 2r\Delta r + \Delta r^2) - \pi r^2$$

We may drop the term $\pi \Delta r^2$ because, in the limit as $\Delta r \to 0$, $\pi \Delta r^2$ will approach zero more quickly than the term in Δr and will be negligibly small compared to it. Therefore,

$$\Delta A_x = 2\pi r\, \Delta r$$

Substituting this into Eq. 12-13, dividing by $\Delta r \Delta x$, and rearranging gives

$$-\frac{2\pi \left[(rq_r'')\big|_{r+\Delta r} - (rq_r'')\big|_r \right]}{\Delta r} = \frac{2\pi \rho \mathcal{V}_x r\,(h_{x+\Delta x} - h_x)}{\Delta x}$$

Taking the limit as volume approaches zero ($\Delta r \to 0$ and $\Delta x \to 0$) gives

$$-\frac{1}{r}\frac{d}{dr}(rq_r) = \rho \mathcal{V}_x \frac{dh}{dx} \tag{12-14}$$

Because the flow is incompressible, enthalpy is related to temperature through

$$dh = c_p \, dT$$

The heat flux by conduction in the radial direction is given by Fourier's law as

$$q_r = -k\frac{dT}{dr}$$

Substituting the last two expressions into Eq. 12-14 and noting that T is a function of both r and x produces

$$\frac{1}{r}\frac{\partial}{\partial r}\left(rk\frac{\partial T}{\partial r}\right) = \rho \mathcal{V}_x c_p \frac{\partial T}{\partial x} \tag{12-15}$$

This partial differential equation arises from applying conservation of energy principles. The left-hand side represents conduction in the radial direction, while the right-hand side accounts for convected energy. Assuming a constant thermal conductivity, k, and rearranging Eq. 12-15,

$$\frac{1}{r}\frac{\partial}{\partial r}\left(r\frac{\partial T}{\partial r}\right) = \frac{\mathcal{V}_x}{\alpha}\frac{\partial T}{\partial x} \tag{12-16}$$

where α is the thermal diffusivity previously defined as $\alpha = k/\rho c_p$. In Chapter 9, the fully developed velocity profile in a circular tube was developed. Using the results of that analysis as given by Eq. 9-16 in Eq. 12-16 produces

$$\frac{1}{r}\frac{\partial}{\partial r}\left(r\frac{\partial T}{\partial r}\right) = \frac{2\mathcal{V}_m}{\alpha}\left[1-\left(\frac{r}{R}\right)^2\right]\frac{\partial T}{\partial x} \tag{12-17}$$

where $\mathcal{V}_m$ is the mean velocity in the pipe and R is the pipe radius. To make further progress in solving this equation, we focus attention on the derivative $\partial T/\partial x$. This is the rate of change in temperature with axial position. Figure 12-13 shows the temperature profile at two axial locations. The mean fluid temperature, $T_m(z)$, increases with downstream location, since heat is added at the wall. The wall temperature, $T_w(z)$, also increases, but the *difference* between the mean temperature and wall temperature remains the same. This is because the shape of the temperature profile does not change in the fully developed region. More generally, for any point on the temperature profile,

$$T(r,x) - T_m(x) = f(r)$$

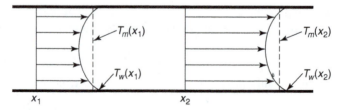

FIGURE 12-13 Temperature profiles in fully developed laminar convection.

where $f(r)$ is some (as yet unknown) function of r. The difference between the temperature at point (r, x) and the mean temperature at that value of x is not changing as a function of downstream location and thus is only a function of r. We are now in a position to evaluate the axial derivative as

$$\frac{\partial T}{\partial x} = \frac{\partial T(r,x)}{\partial x} = \frac{\partial f(r)}{\partial x} + \frac{\partial T_m(x)}{\partial x} = \frac{dT_m(x)}{dx} \tag{12-18}$$

where the ordinary derivative has been used because T_m is a function only of x and $\partial f(r)/\partial x = 0$.

We can find an expression for $T_m(x)$ by choosing an appropriate control volume and applying an energy balance. In Figure 12-14, a fluid at temperature T_i enters a pipe whose walls are heated with a constant heat flux. Imagine a control volume that starts at the entrance to the pipe and extends a distance x downstream, as shown in the figure. The first law for this control volume, neglecting kinetic and potential energy, is

$$\frac{dE_{cv}}{dt} = \dot{Q}_{cv} - \dot{W}_{cv} + \dot{m}_i h_i - \dot{m}_e h_e$$

Since the flow is steady and no work is done, this equation reduces to

$$\dot{Q} = \dot{m}(h_e - h_i)$$

where the subscript on the control volume heat transfer has been dropped for convenience. If the flow is considered incompressible with constant specific heat, then the enthalpy may be replaced by

$$\dot{Q} = \dot{m}c_p(T_e - T_i) \tag{12-19}$$

The fluid temperature at the inlet is a constant; however, since heat is being added at the wall, the fluid temperature increases with distance x. We write the exit temperature in terms of the mean temperature and Eq. 12-19 becomes

$$\dot{Q}(x) = \dot{m}c_p[T_m(x) - T_i] \tag{12-20}$$

The heat added has been written as a function of x, because the total amount of heat added depends on how far downstream the control volume extends. In terms of the heat flux, the heat added to the control volume is

$$\dot{Q}(x) = q_w'' A(x)$$

where $A(x)$ is the pipe surface area for a control volume of length x, and q_w'' is the heat flux at the wall. Since the inside pipe surface area is cylindrical, the heat added may be written

$$\dot{Q}(x) = q_w'' \pi D x$$

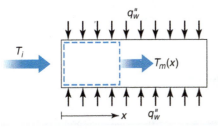

FIGURE 12-14 A control volume that extends a distance x along the inside of a pipe with constant heat flux on the walls.

where D is the pipe diameter. Substituting this into Eq. 12-20,

$$q_w'' \pi D x = \dot{m} c_p \left[T_m(x) - T_i \right]$$

Solving for $T_m(x)$ produces

$$T_m(x) = \frac{q_w'' \pi D}{\dot{m} c_p} x + T_i \qquad (12\text{-}21)$$

Taking the derivative with respect to x,

$$\frac{dT_m(x)}{dx} = \frac{q_w'' \pi D}{\dot{m} c_p}$$

Using $\dot{m} = \rho \mathcal{V} A_x$, where A_x is the cross-sectional area of the pipe, this becomes

$$\frac{dT_m(x)}{dx} = \frac{q_w'' \pi D}{\rho \mathcal{V}_m A_x c_p} = \frac{4 q_w'' \pi D}{\rho \mathcal{V}_m \pi D^2 c_p} = \frac{4 q_w''}{\rho \mathcal{V}_m D c_p} = \frac{2 q_w''}{\rho \mathcal{V}_m R c_p} \qquad (12\text{-}22)$$

where R is the radius of the pipe. We are now in a position to return to the differential equation for the temperature profile. Combining Eq. 12-22, Eq. 12-18, and Eq. 12-17 gives

$$\frac{1}{r} \frac{\partial}{\partial r} \left(r \frac{\partial T}{\partial r} \right) = \frac{4 q_w''}{\alpha \rho R c_p} \left[1 - \left(\frac{r}{R} \right)^2 \right]$$

Noting that $\partial T / \partial r$ is only a function of r and using $\alpha = k / \rho c_p$, we may rewrite this as the following ordinary differential equation:

$$\frac{1}{r} \frac{d}{dr} \left(r \frac{dT}{dr} \right) = \frac{4 q_w''}{kR} \left[1 - \left(\frac{r}{R} \right)^2 \right]$$

Separating variables and integrating twice gives

$$T(r) = \frac{4 q_w''}{kR} \left[\frac{r^2}{4} - \frac{r^4}{16 R^2} \right] + C_1 \ln r + C_2$$

The first boundary condition is that the temperature must be finite at the pipe center where $r = 0$; therefore, $C_1 = 0$. The second boundary condition is that the temperature at the wall is T_w. Applying the second boundary condition gives

$$C_2 = T_w - \frac{4 q_w''}{kR} \left(\frac{3 R^2}{16} \right)$$

Thus the temperature profile in a fully developed laminar flow with constant wall heat flux and constant fluid properties is

$$T(r) = T_w - \frac{4 q_w'' R}{k} \left[\frac{3}{16} - \frac{1}{4} \left(\frac{r}{R} \right)^2 + \frac{1}{16} \left(\frac{r}{R} \right)^4 \right] \qquad (12\text{-}23)$$

We would like to relate this profile to the convective heat transfer coefficient, h. Convection is given by

$$q_w'' = \frac{\dot{Q}_w}{A} = h \left(T_w - T_m \right) \qquad (12\text{-}24)$$

where T_m is the so-called **bulk mean temperature**. The bulk mean temperature is the temperature that would be achieved if heat addition were stopped and the fluid allowed to come to equilibrium at a uniform temperature. The bulk mean temperature is found by an energy balance on the control volume, shown in Figure 12-15. Fluid enters with the velocity and temperature profiles characteristic of fully developed convection and leaves with a fully developed velocity profile and a uniform value of T_m. The purpose of the control volume is to provide an equivalent (average) temperature field. Applying the first law for an open steady system to this control volume gives

$$0 = \dot{Q}_{cv} - \dot{W}_{cv} + \sum_{in} \dot{m}_i h_i - \sum_{out} \dot{m}_e h_e$$

No heat enters the control volume and no work is done. The velocity is not uniform over the cross-section of the pipe, so the mass flow must be computed by adding contributions at each differential area, dA. In the limit, the summations become integrals and the first law may be written

$$0 = \left[\int_A \rho \mathcal{V}_x h_i \, dA \right] - \left[\int_A \rho \mathcal{V}_x h_e \, dA \right]$$

Combining the integrals and noting that this is an incompressible flow,

$$0 = \rho \int_A \mathcal{V}_x (h_i - h_e) \, dA$$

The enthalpy change may be rewritten in terms of temperature by $\Delta h = c_p \, \Delta T$, so that

$$0 = \rho c_p \int_A \mathcal{V}_x [T(r) - T_m] \, dA$$

where we have assumed a constant specific heat. Integrating each term produces

$$0 = \int_A \mathcal{V}_x T(r) \, dA - \int_A \mathcal{V}_x T_m \, dA$$

The mean velocity is, by definition (see Eq. 9-11),

$$\mathcal{V}_m = \frac{\int_A \mathcal{V}_x \, dA}{A}$$

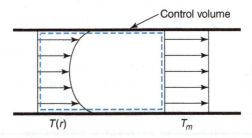

FIGURE 12-15 In the control volume shown, fluid enters with the actual temperature profile and leaves with a flat profile corresponding to the bulk mean temperature.

Using this to simplify the second term and noting that T_m is constant across the area and can be removed from the integral, we get

$$0 = \int_A \mathcal{V}_x T(r) \, dA - \mathcal{V}_m A T_m$$

Solving for T_m,

$$T_m = \frac{\int_A \mathcal{V}_x T(r) \, dA}{\mathcal{V}_m A}$$

As developed previously, the differential area $dA = 2\pi r \, dr$, so the bulk mean temperature may be written

$$T_m = \frac{\int_0^R \mathcal{V}_x(r) T(r) 2\pi r \, dr}{\mathcal{V}_m \pi R^2} \tag{12-25}$$

where the dependence of the velocity field on r has been explicitly shown. At this point, the velocity field for a fully developed flow, as given by Eq. 9-16, and the temperature field for a fully developed flow, as given by Eq. 12-23, are substituted into Eq. 12-25 to get

$$T_m = \frac{\int_0^R 4\mathcal{V}_m \left(1 - \frac{r^2}{R^2}\right) \left\{ T_w - \frac{4q_w'' R}{k} \left[\frac{3}{16} - \frac{1}{4}\left(\frac{r}{R}\right)^2 + \frac{1}{16}\left(\frac{r}{R}\right)^4 \right] \right\} r \, dr}{\mathcal{V}_m R^2}$$

Carrying out the integration yields

$$T_m = T_w - \frac{Dq_w''}{k}\left(\frac{11}{48}\right)$$

where $D = 2R$ is the diameter of the pipe. Combining this with Eq. 12-24,

$$h = \frac{k}{D}\left(\frac{48}{11}\right)$$

Rearranging this equation

$$\boxed{Nu_D = \frac{hD}{k} = 4.36 \qquad Re_D < 2100 \quad \text{laminar, constant heat flux}} \tag{12-26}$$

where the nondimensional Nusselt number, Nu_D, has been used.

The Nusselt number developed naturally from the analysis. Indeed, all the nondimensional parameters used in fluid mechanics and heat transfer result from solution of the appropriate differential equations. In Chapter 9, the Reynolds number appeared when the momentum balance was nondimensionalized. It is possible to derive the Prandtl number as well from solution of the conservation equations, but that demonstration is beyond the scope of this text.

In fully developed laminar flow, Nusselt number is a constant and the heat transfer coefficient, h, is not a function of x. However, both the wall temperature and the mean

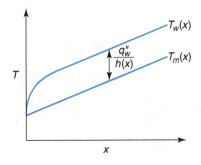

FIGURE 12-16   Fluid and surface temperature in a pipe with constant heat flux on the walls.

temperature vary with x. The bulk mean fluid temperature (as given by Eq. 12-21) increases linearly along the pipe starting at the initial temperature as plotted in Figure 12-16. The wall temperature is related to the bulk mean temperature through

$$q''_w = \frac{\dot{Q}_w}{A} = h\,(T_w - T_m)$$

Solving for the wall temperature gives

$$T_w = \frac{q''_w}{h} + T_m$$

Referring to Figure 12-10c, the heat transfer coefficient is high at the entrance and decreases to an asymptotic value in the fully developed region. Using this information and the linear variation of $T_m(x)$, the behavior of $T_w(x)$ can be inferred. Figure 12-16 shows this behavior graphically. In the entrance, where $h(x)$ is high, the surface temperature is close to the fluid temperature. As $h(x)$ approaches a constant value, the surface temperature tracks the fluid temperature, maintaining a near constant distance above it.

Up to this point, we have considered laminar flow with a constant wall heat flux. Another common boundary condition encountered in practice is a constant wall temperature. It is possible to find the temperature profile in fully developed laminar flow with a constant wall temperature analytically. The procedure is similar to that used in the case of constant heat flux, but additional complications arise. The derivation is beyond the scope of this text, and here we merely state the final result for the Nusselt number. To find the heat transfer coefficient in a pipe assuming constant wall temperature, fully developed laminar flow, and constant properties, use

$$\boxed{Nu_D = \frac{hD}{k} = 3.66 \qquad Re_D < 2100 \quad \text{laminar, constant wall temperature}} \qquad (12\text{-}27)$$

The fluid properties in these equations are based on the bulk mean temperature, T_m, which is the average of the inlet and exit temperatures.

EXAMPLE 12-4   Convection in a pipe with a constant heat flux at the wall

Water flows in an insulated pipe with an inside diameter of 0.8 cm and a length of 6.7 m. A constant heat flux of 0.7 W/cm^2 is applied on the outside wall of the pipe under the insulation. The inlet water temperature is 15°C. If the inside wall temperature must stay below 85°C everywhere along the pipe, what minimum flow velocity is needed? Neglect entrance effects.

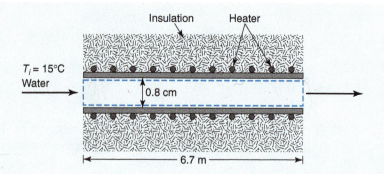

Approach:

Because heat is added to the water flow and its temperature increases all along the pipe, the highest wall temperature will occur at the pipe exit where the fluid temperature is highest. The exit temperature of the water can be determined from the given wall temperature at the exit using the heat transfer coefficient calculated with Eq. 12-26 and

$$T_w = \frac{q_w''}{h} + T_m$$

Applying conservation of energy to the water flow, the mass flow rate and, hence, the velocity can be determined.

Assumptions:

A1. The pipe is perfectly insulated from the environment and all the heat enters the water.
A2. The flow is steady.
A3. Potential and kinetic energy changes are negligible.
A4. No work is done on or by the control volume.
A5. Water is an ideal liquid with a constant specific heat.

Solution:

From conservation of energy (Eq. 12-19), with T_i as the inlet water temperature and T_e as the exit water temperature, the heat added is [A1][A2][A3][A4][A5]

$$\dot{Q} = \dot{m} c_p (T_e - T_i)$$

Because $\dot{Q} = q'' A_s$ and $\dot{m} = \rho \mathcal{V} A_x$,

$$q'' A_s = \rho \mathcal{V} A_x c_p (T_e - T_i)$$

where we have been careful to distinguish between A_s, the surface area of the pipe, and A_x, the cross-sectional area of the pipe. In terms of the diameter, D, and length, L, of the pipe,

$$q'' \pi D L = \rho \mathcal{V} \frac{\pi D^2}{4} c_p (T_e - T_i)$$

Solving for velocity,

$$\mathcal{V} = \frac{4 q'' L}{\rho D c_p (T_e - T_i)}$$

To find the exit temperature, use

$$\dot{Q} = q_w'' A_s = h A_s (T_s - T_e)$$

Solving for T_e,

$$T_e = T_s - \frac{q_w''}{h}$$

To evaluate h, we need to use the appropriate heat transfer coefficient correlation. However, we do not know the velocity. Therefore, we cannot determine whether the flow is laminar or turbulent.

A6. The flow is laminar.

A7. The flow is fully developed.

To make further progress, we assume the flow is laminar [A6]. Once we calculate a velocity, we will check the Reynolds number. We could just as easily assume the flow is turbulent, but the calculation would be a little more complex. Assuming fully developed laminar flow with a constant wall heat flux [A7], from Eq. 12-26:

$$Nu_D = \frac{hD}{k} = 4.36$$

Recall that this correlation is based on fluid properties evaluated at the bulk mean temperature T_m, which is $T_m = (T_i + T_e)/2$. We do not know the exit temperature, so we will need to estimate it. The wall temperature at the exit is 85°C, and T_e is less than the wall temperature. As a first approximation, the bulk mean temperature for the whole pipe is

$$T_m \approx \tfrac{1}{2}(T_i + T_s) \approx \tfrac{1}{2}(15 + 85)\,°C = 50°C$$

We evaluate the thermal conductivity at 50°C. After the water exit temperature is calculated, we will check this approximation. Using data from Table A-6, we see that the heat transfer coefficient is

$$h = \frac{4.36k}{D} = \frac{4.36\left(0.643\,\dfrac{W}{m\cdot K}\right)}{(0.8\ cm)\left(\dfrac{1\ m}{100\ cm}\right)} = 350\,\frac{W}{m^2\cdot K}$$

The exit temperature may now be calculated as

$$T_e = T_s - \frac{q_w''}{h} = 85°C - \frac{\left(0.7\,\dfrac{W}{cm^2}\right)\left(\dfrac{100\ cm}{1\ m}\right)^2}{350\,\dfrac{W}{m^2\cdot K}} = 65°C$$

With this value of exit temperature, the bulk mean temperature is $T_m = (15 + 65)/2 = 40°C$. The thermal conductivity of water at 40°C is, from Table A-6, $k = 0.631$ W/m·K, which is close to the assumed value of 0.643 W/ m·K. Therefore, we do not need to iterate, and we accept the exit temperature as 65°C. Using values of density and specific heat for water at 40°C, we find that the mean velocity is

$$\mathcal{V}_m = \frac{4\left(0.7\,\dfrac{W}{cm^2}\right)(670\ cm)\left(\dfrac{100\ cm}{1\ m}\right)^2}{\left(992.2\,\dfrac{kg}{m^3}\right)(0.8\ cm)\left(4175\,\dfrac{J}{kg\cdot K}\right)(65-15)\,°C} = 0.113\,\frac{m}{s}$$

Now check the Reynolds number using

$$Re = \frac{\rho D \mathcal{V}_m}{\mu} = \frac{\left(992.2\,\dfrac{kg}{m^3}\right)(0.8\ cm)\left(\dfrac{1\ m}{100\ cm}\right)\left(0.113\,\dfrac{m}{s}\right)}{6.34\times10^{-4}\,\dfrac{N\cdot s}{m^2}} = 1417$$

Since this is less than the Reynolds number of 2100 given in Eq. 12-26, the flow is laminar, and it was appropriate to use the laminar correlation.

Comments:

If the Reynolds number had been greater than 2100, we would have had to repeat the calculation assuming the flow is turbulent. Correlations for Nusselt number in turbulent flow are given in the next section.

12.4 TURBULENT CONVECTION IN PIPES

There are no completely analytical solutions for turbulent flow in pipes. Instead, one must rely on experimental and numerical investigations to generate data that are then correlated with regression analyses. A wide variety of convective heat transfer coefficient correlations are available, and this section summarizes some of the more common ones for pipe flows. Unlike laminar flow, in which the boundary condition (constant wall temperature or constant wall heat flux) changes the heat transfer coefficient, turbulent correlations can generally be used for either situation. (The one exception is liquid metal flow, which we will not address.)

Perhaps the most widely used correlation for turbulent flow in pipes is an adaptation of a correlation originally published by **Dittus and Boelter**, that is,

$$Nu_D = 0.023 Re_D^{0.8} Pr^n \qquad \begin{array}{l} Re_D > 10,000 \\[4pt] 0.7 < Pr \le 160 \qquad \text{turbulent} \\[4pt] \dfrac{L}{D} \ge 10 \\[12pt] n = 0.3 \quad \text{cooling} \quad T_w < T_m \\[4pt] n = 0.4 \quad \text{heating} \quad T_m < T_w \end{array} \tag{12-28}$$

In this equation all properties are evaluated at the mean temperature, which is the average of the inlet and outlet temperatures. Note that the exponent on the Prandtl number depends on whether the fluid is being heated or cooled. Fluid properties vary with temperature, and the variation has different implications in heating and cooling.

If temperature differences are large, so that property variations become significant, then the following correlation by **Seider and Tate** is recommended:

$$Nu_D = 0.027 Re_D^{0.8} Pr^{1/3} \left(\frac{\mu}{\mu_w}\right)^{0.14} \qquad \begin{array}{l} Re_D > 10,000 \\[4pt] 0.7 < Pr \le 16,700 \qquad \text{turbulent} \\[4pt] \dfrac{L}{D} \ge 10 \\[12pt] n = 0.3 \quad \text{cooling} \quad T_w < T_m \\[4pt] n = 0.4 \quad \text{heating} \quad T_m < T_w \end{array}$$

$$\tag{12-29}$$

All properties are evaluated at the mean temperature except μ_w, which is the dynamic viscosity evaluated at the wall temperature. This equation may be used for either constant surface temperature or constant heat flux.

In turbulent flow, the roughness of the pipe wall augments heat transfer. Turbulent velocity profiles are characterized by a thin laminar sublayer near the wall, as shown in Figure 12-17a. Small protrusions on the wall disturb the laminar sublayer, causing mixing and improving convection. Laminar velocity profiles are not as steep at the wall as turbulent profiles and are unaffected by wall roughness. In general for turbulent flow, an increase in roughness improves heat transfer; however, at some point the peaks of the rough surface extend beyond the laminar sublayer and no further improvement in heat transfer occurs. Since roughness improves heat transfer, it is sometimes artificially added for that purpose. The improvement in heat transfer must be balanced against the increase in pressure drop, which may lead to larger pumps or fans.

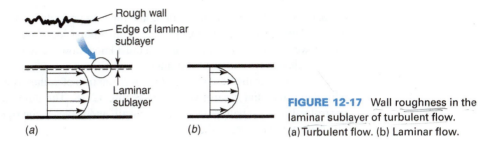

FIGURE 12-17 Wall roughness in the laminar sublayer of turbulent flow. (a) Turbulent flow. (b) Laminar flow.

Forced convection heat transfer coefficients in smooth or rough pipes are given approximately by Petukhov as

$$Nu_D = \frac{(f/8)\, Re_D Pr}{1.07 + 12.7\,(f/8)^{1/2}\,(Pr^{2/3} - 1)} \left(\frac{\mu}{\mu_w}\right)^n \qquad
\begin{aligned}
& 10^4 < Re_D < 5 \times 10^6 \\[4pt]
& 0.5 < Pr < 2000 \\[4pt]
& 0.08 < \mu/\mu_w < 40 \\[10pt]
& n = 0 \qquad \text{constant heat flux} \\
& n = 0.11 \quad T_w > T_m \\
& n = 0.25 \quad T_m > T_w
\end{aligned}$$

(12-30)

This equation gives excellent estimates of the heat transfer coefficient for smooth pipes and reasonable estimates for rough pipes. For the friction factor, use Petukhov's correlation for smooth pipes, which is given in Chapter 9 as

$$f = (0.79\ \ln\ Re - 1.64)^{-2} \qquad
\begin{aligned}
& \text{turbulent flow, smooth wall} \\
& 3000 < Re_D < 5 \times 10^6
\end{aligned}$$

The above correlations apply in the fully turbulent regime, where $Re > 10{,}000$. When $2100 < Re < 10{,}000$, the flow may be in transition between laminar and turbulent. A useful correlation by Gnielinski for low Reynolds number is

$$Nu_D = \frac{(f/8)\,(Re_D - 1000)\, Pr}{1 + 12.7\,(f/8)^{1/2}\,(Pr^{2/3} - 1)} \qquad
\begin{aligned}
& 3000 < Re_D < 5 \times 10^6 \\
& 0.5 < Pr < 2000
\end{aligned} \qquad (12\text{-}31)$$

This equation applies for constant surface temperature or constant heat flux. Properties are evaluated at the average temperature of the inlet and outlet. The pipe wall is assumed to be smooth, and Eq. 9-36 for the friction factor applies.

12.5   INTERNAL FLOW WITH CONSTANT WALL TEMPERATURE

In some cases, a pipe wall is held at a nearly constant temperature. For example, if a fluid is boiling or condensing at constant pressure on the outside of a pipe, then the fluid is at the saturation temperature all along the pipe. Because of the high heat transfer coefficients typical of boiling and condensation, the wall temperature is approximately equal to the fluid temperature and is also uniform along the pipe. In previous sections, correlations for heat transfer coefficient with a constant wall temperature boundary condition were presented. Before we apply those correlations, we must know how the temperature of the fluid inside the pipe changes as it flows from inlet to exit.

To calculate the fluid temperature as a function of x, the distance along the pipe, start with the differential control volume in Figure 12-18. The fluid enters the left face of the control volume, exchanges heat with the wall, and leaves at the right face. The first law for this control volume is

$$\frac{dE_{cv}}{dt} = \dot{Q} - \dot{W} + \dot{m}_i h_i - \dot{m}_e h_e$$

where h is *enthalpy*, not heat transfer coefficient. Since the flow is steady and no work is done, the equation reduces to

$$\dot{Q} = \dot{m}\,(h_e - h_i)$$

For an incompressible fluid with a constant specific heat, $\Delta h = c_p\,\Delta T$ and

$$\dot{Q} = \dot{m} c_p\,(T_e - T_i)$$

Using the mean temperatures at the inlet and exit of the control volume, as shown in Figure 12-18, this equation may be written

$$\dot{Q} = \dot{m} c_p \left(T_m\big|_{x+\Delta x} - T_m\big|_x\right)$$

Substituting $\dot{Q} = h\,\Delta A\,(T_w - T_m)$ into the left-hand side gives

$$h \Delta A \left(T_w - T_m\right) = \dot{m} c_p \left(T_m\big|_{x+\Delta x} - T_m\big|_x\right)$$

If we assume fully developed conditions, the only quantity that is a function of x in this equation is T_m. The differential area, ΔA, may be written in terms of the perimeter, P, and the length of the control volume, Δx, to yield

$$hP \left(T_w - T_m\right) \Delta x = \dot{m} c_p \left(T_m\big|_{x+\Delta x} - T_m\big|_x\right)$$

Dividing by $hP\,\Delta x$ and taking the limit as Δx approaches zero,

$$T_w - T_m = \frac{\dot{m} c_p}{hP} \lim_{\Delta x \to 0} \left[\frac{\left(T_m\big|_{x+\Delta x} - T_m\big|_x\right)}{\Delta x}\right] = \frac{\dot{m} c_p}{hP}\frac{dT_m}{dx}$$

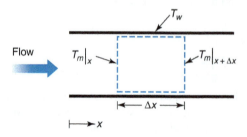

FIGURE 12-18 Differential control volume in a pipe with a constant temperature wall.

Separating variables in preparation for integration produces

$$dx = \frac{\dot{m}c_p}{hP} \left(\frac{dT_m}{T_w - T_m(x)} \right)$$

where the dependence of T_m on x has been explicitly shown. Integrating this equation for a pipe of length L with inlet temperature T_i and exit temperature T_e gives

$$\int_0^L dx = \frac{-\dot{m}c_p}{hP} \int_{T_i}^{T_e} \frac{dT_m}{T_m(x) - T_w}$$

Performing the integration,

$$L = \frac{-\dot{m}c_p}{hP} [\ln(T_e - T_w) - \ln(T_i - T_w)] = \frac{-\dot{m}c_p}{hP} \ln\left(\frac{T_e - T_w}{T_i - T_w}\right)$$

Since the surface area of the pipe is length times perimeter, an alternate form is

$$\ln\left(\frac{T_e - T_w}{T_i - T_w}\right) = -\frac{hA}{\dot{m}c_p} \tag{12-32}$$

Exponentiating both sides,

$$\frac{T_e - T_w}{T_i - T_w} = \exp\left(-\frac{hA}{\dot{m}c_p}\right)$$

Solving for exit temperature gives

$$\boxed{T_e = (T_i - T_w)\exp\left(-\frac{hA}{\dot{m}c_p}\right) + T_w} \tag{12-33}$$

This result is plotted in Figure 12-19. At the inlet, the fluid temperature rises sharply toward the surface temperature, then approaches the surface temperature asymptotically. Near the exit of the pipe, the difference between wall and mean fluid temperature is smaller and less heat is transferred. From the first law, the total heat transferred between the wall and fluid is

$$\dot{Q} = \dot{m}c_p(T_e - T_i)$$

Rearranging,

$$\dot{m}c_p = \frac{\dot{Q}}{T_e - T_i} \tag{12-34}$$

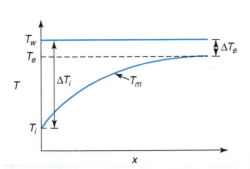

FIGURE 12-19 Fluid and surface temperature in a pipe with constant wall temperature.

We may rewrite this in terms of the difference between wall and fluid temperature at inlet and exit. By definition,

$$\Delta T_i = T_w - T_i$$

$$\Delta T_e = T_w - T_e$$

Using these expressions, Eq. 12-34 becomes

$$\dot{m} c_p = \frac{\dot{Q}}{T_e - T_i} = \frac{\dot{Q}}{(T_w - T_i) - (T_w - T_e)} = \frac{\dot{Q}}{\Delta T_i - \Delta T_e}$$

Substituting this into Eq. 12-32 and expressing the left-hand side in terms of ΔT_i and ΔT_e gives

$$\ln\left(\frac{\Delta T_e}{\Delta T_i}\right) = \frac{-hA\,(\Delta T_i - \Delta T_e)}{\dot{Q}}$$

We now define an "equivalent" temperature, ΔT_{LM}, as

$$\Delta T_{LM} = \frac{(\Delta T_e - \Delta T_i)}{\ln\left(\dfrac{\Delta T_e}{\Delta T_i}\right)} \qquad (12\text{-}35)$$

so that the heat transferred becomes

$$\dot{Q} = hA\,\Delta T_{LM} \qquad (12\text{-}36)$$

The quantity ΔT_{LM} is called the **log mean temperature difference**. What does this represent? Consider the temperatures in Figure 12-19. The difference between the wall and the fluid temperature varies along the pipe, being large at the inlet and small at the outlet. The *average* difference in temperature is not easily determined. Clearly, from the figure, something such as $\Delta T_{avg} = T_w - (T_i + T_e)/2$ is not accurate. Eq. 12-36 shows that the appropriate average temperature difference to use is, in fact, ΔT_{LM}. The log mean temperature difference has a value between ΔT_i and ΔT_e.

In some circumstances, convection occurs on both the inside and the outside of a pipe. For example, water pipes in residential basements are often exposed to the air. If the water is at a different temperature than the air, natural convection occurs on the outside of the pipe and forced convection occurs on the inside. If the fluid on the outside of the pipe is at a constant temperature, T_∞, along the pipe length (Figure 12-20), then Eq. 12-35 and

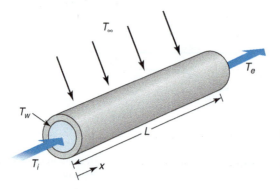

FIGURE 12-20   Convection on both the inside and outside of a pipe. The exterior fluid has a uniform temperature along the pipe length.

Eq. 12-36 can be adapted to compute heat transfer. To demonstrate this, consider an insulated pipe of differential length Δx with convection on both inside and outside, as shown in Figure 12-21. The thermal circuit for this geometry consists of four resistances in series: the convection resistance on the inside, the conduction resistance through the wall, the conduction resistance through the insulation, and the convection resistance on the outside. Using the notation in Figure 12-21, the total thermal resistance is

$$R_{tot} = \frac{1}{h_i A_1} + \frac{\ln(r_2/r_1)}{2\pi k_1 \Delta x} + \frac{\ln(r_3/r_2)}{2\pi k_2 \Delta x} + \frac{1}{h_\infty A_3}$$

The areas for convection on the inside and outside are $A_1 = 2\pi r_1 \Delta x$ and $A_3 = 2\pi r_3 \Delta x$, respectively; therefore, the total resistance becomes

$$R_{tot} = \frac{1}{2\pi r_1 \Delta x\, h_i} + \frac{\ln(r_2/r_1)}{2\pi k_1 \Delta x} + \frac{\ln(r_3/r_2)}{2\pi k_2 \Delta x} + \frac{1}{2\pi r_3 \Delta x\, h_\infty} \tag{12-37}$$

For this differential length of pipe, Δx, the fluid temperature is T_m and the heat transfer rate is

$$\dot{Q} = \frac{T_m - T_\infty}{R_{tot}} \tag{12-38}$$

We now define an **overall heat transfer coefficient**, U, using the expression

$$\dot{Q} = UA\,(T_m - T_\infty)$$

The overall heat transfer coefficient may be based on either the inside area or the outside area, that is,

$$\dot{Q} = U_1 A_1\,(T_m - T_\infty) = U_3 A_3\,(T_m - T_\infty) \tag{12-39}$$

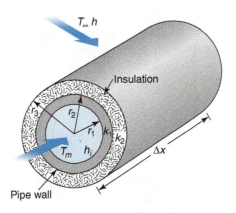

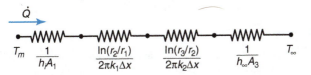

FIGURE 12-21 Thermal circuit for an insulated pipe with convection on both the inside and outside.

Arbitrarily using the inside area as an example, eliminate $\dot{Q}$ between Eq. 12-38 and Eq. 12-39 to get

$$U_1 A_1 = \frac{1}{R_{tot}}$$

Substituting R_{tot} from Eq. 12-37 and using $A_1 = 2\pi r_1 \Delta x$ gives

$$U_1 (2\pi r_1 \Delta x) = \frac{1}{\dfrac{1}{2\pi r_1 \Delta x\, h_i} + \dfrac{\ln (r_2/r_1)}{2\pi k_1 \Delta x} + \dfrac{\ln (r_3/r_2)}{2\pi k_2 \Delta x} + \dfrac{1}{2\pi r_3 \Delta x\, h_\infty}}$$

which simplifies to

$$U_1 = \frac{1}{\dfrac{1}{h_i} + \dfrac{r_1}{k_1} \ln \left(\dfrac{r_2}{r_1}\right) + \dfrac{r_1}{k_2} \ln \left(\dfrac{r_3}{r_2}\right) + \dfrac{r_1}{r_3 h_\infty}} \qquad (12\text{-}40)$$

We have solved for the overall heat transfer coefficient for the case of an insulated pipe. Similar expressions may be written for uninsulated pipes or for pipes with three or more layers.

The advantage of defining an overall heat transfer coefficient is that the preceding equations for flow in a pipe with an isothermal wall can be applied. In Eq. 12-32, simply replace T_w by T_∞ and hA by UA. The total heat transferred is, from Eq. 12-36, $\dot{Q} = UA\, \Delta T_{LM}$. This is applicable only if T_∞ does not vary along the pipe.

EXAMPLE 12-5   Heating of water in a solar collector

A solar collector is used to supply hot water to a home. A copper tube of diameter 1.2 cm is soldered to the back of the collector plate, which is maintained at a uniform temperature of 75°C by incident sunlight, as shown in the figure. Water enters the tube at 25°C with a mass flow rate of 0.0122 kg/s. Assume that the tube wall is at the same temperature as the plate, and neglect entrance effects and the effects of bends in the tube. Determine the total length of tube needed so that the exit temperature is 55°C.

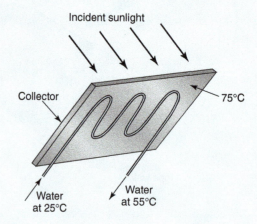

Approach:

The wall of the tube is at a constant temperature, so Eq. 12-32 may be solved for the required tube surface area. Once the area is known, the length can be determined. The heat transfer coefficient in Eq. 12-32 is found by applying an appropriate correlation. Calculate the Reynolds number to determine whether the flow is laminar or turbulent, and then choose a correlation. Check to be sure that the Prandtl number of the fluid is within the range for which the correlation applies.

Assumptions:

A1. The flow is fully developed.
A2. The tube wall is at a constant temperature.
A3. The effect of the bends in the tube is negligible.

Solution:

The heat transfer coefficient inside the tube depends on the Reynolds number, which is given by

$$Re = \frac{\rho D \mathcal{V}}{\mu}$$

Using $\dot{m} = \rho \mathcal{V} A_x$, where A_x is cross-sectional area, this becomes

$$Re = \frac{D\dot{m}}{\mu A_x}$$

With water properties at the bulk mean temperature of $T_m = (25 + 55)/2 = 40°C$ from Table A-6,

$$Re = \frac{D\dot{m}}{\mu A_x} = \frac{(0.012\,\text{m})\left(0.0122\,\frac{\text{kg}}{\text{s}}\right)}{\left(6.34 \times 10^{-4}\,\frac{\text{N·s}}{\text{m}^2}\right)\pi\left(\frac{0.012}{2}\text{m}\right)^2} = 2042$$

$$Pr = 4.19$$

This Reynolds number is less than 2100; therefore, the flow is laminar. Eq. 12-27 applies for this range of Reynolds and Prandtl numbers when the pipe wall temperature is constant. Therefore [A1][A2][A3],

$$Nu_D = \frac{hD}{k} = 3.66$$

Again using properties at 40°C from Table A-6,

$$h = \frac{3.66k}{D} = \frac{3.66\left(0.631\,\frac{\text{W}}{\text{m·K}}\right)}{0.012\,\text{m}} = 193\,\frac{\text{W}}{\text{m}^2\cdot\text{K}}$$

Solving Eq. 12-32 for the unknown total tube surface area, A_s, and substituting values gives

$$A_s = \frac{-\dot{m}c_p}{h}\ln\left(\frac{T_e - T_w}{T_i - T_w}\right) = \frac{-\left(0.0122\,\frac{\text{kg}}{\text{s}}\right)\left(4175\,\frac{\text{J}}{\text{kg·K}}\right)}{193\,\frac{\text{W}}{\text{m}^2\cdot\text{K}}}\ln\left(\frac{55 - 75}{25 - 75}\right) = 0.242\,\text{m}$$

The required tube length is now found from

$$L = \frac{A_s}{\pi D} = \frac{0.242\,\text{m}^2}{\pi\,(0.012\,\text{m})} = 6.41\,\text{m}$$

EXAMPLE 12-6 **Wall temperature in an exhaust stack**

A factory discharges hot exhaust gases into the atmosphere through a vertical stack 9 m high and 0.7 m in diameter. The gases enter at 5 m/s, 502°C, and near-atmospheric pressure and may be assumed to have the properties of air. Wind at 10°C blows over the outside of the stack, giving an exterior heat transfer coefficient of 17 W/m²·K. If the wall temperature at the exit is too low, condensation of some exhaust gas species will occur. These species form acids that corrode the metal stack wall. Assume the stack wall is thin and has a high thermal conductivity. Find the wall temperature at the exit.

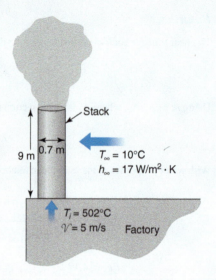

Approach:

The wall temperature depends on the exhaust gas exit temperature. To find exit temperature, use Eq. 12-33. In this case, the exterior air is at a constant temperature, so the overall heat transfer coefficient, U, should be used in place of h in this equation. To determine U, the heat transfer coefficient inside the duct is needed. After calculating the Reynolds and Prandtl numbers, an appropriate correlation can be selected based on whether the flow is laminar or turbulent. Finally, once exit temperature is known, a thermal circuit similar to that shown in Figure 12-21 can be applied to find the wall temperature.

Assumptions:

A1. The stack gases have the properties of air.

A2. The wind flows at a constant velocity.

A3. The flow is fully developed.

Solution:

We need the heat transfer coefficient for flow inside the stack. The properties for internal flow correlations are based on the average of inlet and exit temperatures, however, the exit temperature is unknown. To make further progress, we use properties at the inlet temperature and correct them later, if needed. Using air properties from Table A-7 at 502°C (775 K), the Reynolds number is [A1][A2]

$$Re = \frac{\rho D \mathcal{V}}{\mu} = \frac{\left(0.456 \, \frac{\text{kg}}{\text{m}^3}\right)(0.7\,\text{m})\left(5 \, \frac{\text{m}}{\text{s}}\right)}{3.55 \times 10^{-5} \, \frac{\text{kg}}{\text{m·s}}} = 44{,}890$$

At this Reynolds number, the flow is turbulent and the Dittus-Boelter relation given by Eq. 12-28 may be used. We must check to be sure that the Prandtl number and the L/D ratio are in range for the correlation. The Prandtl number is given in Table A-7 as $Pr = 0.687$. This is close enough to the lower limit of 0.7 specified in the correlation. The ratio $L/D = 9/0.7 = 12.9$ is higher than the lower limit of 10. Therefore, from the Dittus-Boelter equation [A3],

$$Nu_D = 0.023 Re_D^{0.8} Pr^n = 0.023 \, (44{,}890)^{0.8} \, (0.687)^{0.3} = 108$$

Note that the stack gases are cooling, so the exponent on the Prandtl number, n, is 0.3. The interior heat transfer coefficient is now calculated as

$$h_i = \frac{108k}{D} = \frac{108\left(0.056\,\frac{W}{m \cdot K}\right)}{0.7\,m} = 8.73\,\frac{W}{m^2 \cdot K}$$

A4. The wall is thin and highly conducting.

To calculate the exit temperature, we need the overall heat transfer coefficient given by Eq. 12-40. Rewriting for the case of negligible conduction resistance in the wall and $r_1 \approx r_3$ [A4],

$$U = \frac{1}{\dfrac{1}{h_i} + \dfrac{1}{h_\infty}} = \frac{1}{\dfrac{1}{8.73} + \dfrac{1}{17}} = 5.77\,\frac{W}{m^2 \cdot K}$$

The exit temperature of the gas is, from Eq. 12-33,

$$T_e = (T_i - T_\infty) \exp\left(\frac{-UA_s}{\dot{m}c_p}\right) + T_\infty$$

where T_w is replaced by the exterior air temperature, T_∞, and the heat transfer coefficient is replaced by the overall heat transfer coefficient. The mass flow rate is

$$\dot{m} = \rho \mathcal{V} A_x = \left(0.456\,\frac{kg}{m^3}\right)\left(5\,\frac{m}{s}\right)\left[\pi\left(\frac{0.7}{2}\,m\right)^2\right] = 0.877\,\frac{kg}{s}$$

Substituting values in the expression for exit temperature gives

$$T_e = \left[(502 - 10)\,°C\right] \exp\left(\frac{-\left(5.77\,\dfrac{W}{m^2 \cdot K}\right)\pi\,(0.7\,m)\,(9\,m)}{\left(0.877\,\dfrac{kg}{s}\right)\left(1092\,\dfrac{J}{kg \cdot °C}\right)}\right) + 10°C = 447°C$$

To find the wall temperature at the exit, consider the thermal circuit shown.

The heat transfer is given by

$$\dot{Q} = \frac{T_e - T_\infty}{\dfrac{1}{h_i A} + \dfrac{1}{h_\infty A}} = \frac{T_e - T_w}{\dfrac{1}{h_i A}}$$

Solving for T_w,

$$T_w = T_e + (T_\infty - T_e)\left[\frac{\dfrac{1}{h_i}}{\dfrac{1}{h_i} + \dfrac{1}{h_\infty}}\right] = 447°C + (10 - 447)\,°C \left[\frac{\dfrac{1}{8.73}}{\dfrac{1}{8.73} + \dfrac{1}{17}}\right] = 158°C$$

Comments:

Properties were evaluated, as a preliminary estimate, at the inlet temperature. The correlation requires properties evaluated at the average of the inlet and exit temperatures, which is $T_{avg} = (502 + 447)/2 = 475°C$. This is not far different than the inlet temperature of 502°C, so we do not need to iterate.

12.6 NONCIRCULAR CONDUITS

The preceding sections on internal flow have dealt with round tubes. To correlate heat transfer for other shapes, the hydraulic diameter is used. The hydraulic diameter has previously been defined as

$$D_h = \frac{4A}{P_{wetted}}$$

where A is the cross-sectional flow area and P_{wetted} is the wetted perimeter of the tube. For a circular pipe, the hydraulic diameter reduces to the ordinary diameter.

For laminar flow in various noncircular shapes, Nusselt number relations are given in Table 12-2 for both constant heat flux and constant-temperature boundary conditions. The hydraulic diameter is used in the Nusselt number. For turbulent flows, any of the turbulent correlations presented above give reasonable results as long as the diameter in both the Nusselt and Reynolds numbers is replaced by the hydraulic diameter.

12.7 ENTRANCE EFFECTS IN FORCED CONVECTION

In previous sections, entrance effects were described qualitatively. Here we add more detail on the development of velocity and thermal boundary layers and present correlations useful in the entrance region. We consider two limiting cases: either the velocity profile is developing at the same time as the temperature profile or the velocity profile is already fully developed before heat transfer begins. In Figure 12-22a, flow enters a pipe with constant velocity and temperature. Heating begins immediately at the entrance of the pipe, and the thermal and velocity profiles develop simultaneously. In Figure 12-22b, flow enters a pipe, and the velocity profile develops fully within the unheated starting length.

The rate of boundary-layer development depends on the Prandtl number. For the flow shown in Figure 12-22a, the Prandtl number is greater than unity. For fluids with high Prandtl numbers, such as oils, heat conductance is poor and the thermal boundary layer grows slowly compared to the velocity boundary layer. For the opposite situation, when Prandtl number is low, the thermal boundary layer grows more quickly than the velocity boundary layer. Low Prandtl numbers are characteristic of liquid metal flows. If the Prandtl number equals unity, the boundary layers grow at the same rate. Gases typically have Prandtl numbers close to unity.

The heat transfer coefficient in the entrance region is higher than in the fully developed region. It is possible to solve for velocity and temperature fields in the entrance region using conservation equations and to predict heat transfer coefficients; however, that analysis is beyond the scope of this text. As a result of such analysis, a new nondimensional parameter called the **Graetz number** appears. By definition,

$$Gz = RePr\frac{D}{L}$$

TABLE 12-2 Nusselt numbers and friction factors for fully developed laminar flow in tubes of differing cross section

$$Nu_D = \frac{hD_h}{k}$$

Cross Section	$\dfrac{b}{a}$	(Uniform q_s'')	(Uniform T_s)	$f\,Re_{D_h}$
◯	—	4.36	3.66	64
△	—	3.11	2.49	53
a ☐ b	1.0	3.61	2.98	57
a ▭ b	1.43	3.73	3.08	59
a ▭ b	2.0	4.12	3.39	62
a ▭ b	3.0	4.79	3.96	69
a ▭ b	4.0	5.33	4.44	73
▭ b	8.0	6.49	5.60	82
═══	∞	8.23	7.54	96
Heated / Insulated	∞	5.39	4.86	96

A correlation due to Hausen for average heat transfer coefficient in a circular pipe with constant surface temperature can be written in terms of the Graetz number as:

$$Nu = 3.66 + \frac{0.0668Gz}{1 + 0.04Gz^{2/3}} \quad \text{entrance region, constant temperature wall, unheated starting length, } Re < 2100$$

(12-41)

This equation applies when the velocity profile is fully developed in an unheated starting length. Properties are evaluated at the bulk temperature, which is the average of the inlet and outlet temperatures. If the pipe is long, the Graetz number becomes very small and the correlation approaches $Nu = 3.66$, which is the result for fully developed flow.

When the velocity and temperature fields are both developing simultaneously, an appropriate equation for the average heat transfer coefficient published by Seider and

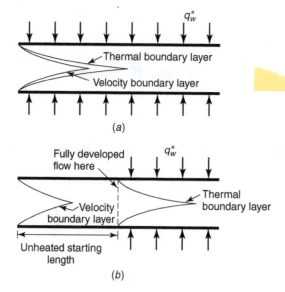

FIGURE 12-22 (a) Simultaneously developing velocity and thermal profiles. (b) Development of boundary layers with an unheated starting length.

Tate is

$$Nu = 1.86Gz^{1/3}\left(\frac{\mu}{\mu_s}\right)^{0.14}$$

entrance region, constant temperature wall, simultaneously developing

$Re < 2100$

$0.48 < Pr < 16{,}700$

$0.0044 < \left(\frac{\mu}{\mu_s}\right) < 9.75$

(12-42)

All properties in this equation are evaluated at the bulk temperature except μ_s, which is the viscosity evaluated at the surface temperature. This equation does not reduce to the correct limit for long pipes. It should be used only when it gives Nusselt numbers larger than 3.66. If the Nusselt number predicted by the equation falls below 3.66, then the flow may be presumed to be fully developed and a constant value of 3.66 should be applied.

In Chapter 9, a criterion for the entrance length for development of the velocity boundary layer was given as (see Eq. 9-39):

$$L_{ent,\,h} \approx 0.065ReD \qquad \text{laminar,}\, Re < 2100$$
$$L_{ent,\,h} \approx 4.4\,(Re)^{1/6}\,D \qquad \text{turbulent,}\, Re > 4000$$

Similar relationships are available for the thermal entry length. In the case of simultaneously developing velocity and temperature profiles, the thermal entry length may be approximated as

$$L_{ent,\,t} \approx 0.037Re_D PrD \qquad \text{laminar,}\, Re < 2100, \text{constant temperature wall}$$
$$L_{ent,\,t} \approx 0.053Re_D PrD \qquad \text{laminar,}\, Re < 2100, \text{constant heat flux}$$

(12-43)

In turbulent flow, the thermal entrance length is similar in size to the hydrodynamic entrance length and both may be approximated as

$$L_{ent,h} \approx L_{ent,t} \approx 4.4\,(Re)^{1/6}D \qquad \text{turbulent}, Re > 4000 \qquad (12\text{-}44)$$

Entrance effects may be significant in laminar flows and should always be checked.

12.8 NATURAL CONVECTION OVER SURFACES

In forced convection, the heat transfer coefficient depends on the imposed flow velocity. In natural convection, the velocity is also important. However, this velocity is unknown. Consider Figure 12-23, where forced and natural convection over a cylinder are depicted. In the forced convective situation, the flow is driven by some external agent and the velocity is known. This velocity is used to compute the Reynolds number for the flow. In the natural convective situation, the flow is induced by changes in fluid density. There is no single velocity analogous to the free-stream velocity that can be used to characterize the flow, so it is not possible to compute a Reynolds number for natural convection.

In natural convection, velocity and temperature boundary layers form along the surface, just as they do in forced convection. For example, the boundary layers for natural convection on a vertical flat plate are illustrated in Figure 12-24. The thermal boundary layer is similar to that for forced convection. The velocity boundary layer is different because velocity far from the plate is zero. The velocity boundary layer is zero at both extremes, at the surface and at the free stream.

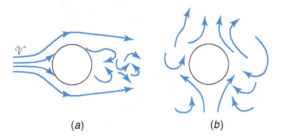

(a)   (b)

FIGURE 12-23 (a) Forced and (b) natural convection over a cylinder.

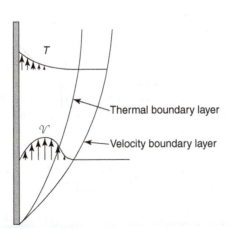

Thermal boundary layer

Velocity boundary layer

FIGURE 12-24 Development of boundary layers on a vertical flat plate in natural convection.

Because the boundary layers grow along the surface, the heat transfer coefficient in natural convection varies with vertical position. As in forced convection, the heat transfer coefficient is higher where the boundary layers are thinner. For most purposes, it is not necessary to take the variation of heat transfer coefficient with position into account, and an average value for the entire surface can be used.

Natural convection is governed by how density changes with temperature. This process takes place while the pressure of the surrounding fluid is constant. Therefore, an important factor that describes the fluid behavior is

$$\left(\frac{\partial \rho}{\partial T} \right)_p$$

In practice, we are interested in the relative change in density, so we define the **volume expansivity** as

$$\boxed{\beta = -\frac{1}{\rho} \left(\frac{\partial \rho}{\partial T} \right)_p}$$

The volume expansivity is a function of thermodynamic properties, so it, too, is a thermodynamic property.

The volume expansivity of an ideal gas can be calculated exactly. From the ideal gas law,

$$\rho = \frac{PM}{\overline{R}T}$$

Taking the partial derivative of density with respect to temperature while holding pressure constant yields

$$\left(\frac{\partial \rho}{\partial T} \right)_p = -\frac{PM}{\overline{R}T^2}$$

Substituting this into the definition of volume expansivity produces

$$\beta = -\frac{1}{\rho} \left(\frac{\partial \rho}{\partial T} \right)_p = \frac{PM}{\rho \overline{R}T^2}$$

Note that, from the ideal gas law, $PM = \rho \overline{R}T$, so that

$$\boxed{\beta = \frac{\rho \overline{R}T}{\rho \overline{R}T^2} = \frac{1}{T} \quad \text{ideal gas}}$$

It is easy to calculate the volume expansivity for an ideal gas. For other fluids, experimental measurements must be used (see Tables A-6 and B-6). This derivation reveals that the units of volume expansivity are inverse temperature, that is, 1/K or 1/R.

Just as the Reynolds number is used to correlate forced convective flows, a nondimensional parameter called the **Grashof number** is used to correlate natural convective flows. The Grashof number is defined as

$$Gr = \frac{g\beta\rho^2 \left(T_s - T_f \right) L_{char}^3}{\mu^2}$$

where g is the acceleration of gravity, T_s is surface temperature, T_f is fluid temperature, and L_{char} is an appropriate length scale. Grashof number increases with increasing temperature difference between wall and fluid and with increasing size of the surface exchanging heat. Convective heat transfer is better at higher Grashof numbers. The Grashof number results naturally from a more advanced analysis of convection, which is beyond the scope of this text.

Correlations for the average heat transfer coefficient in natural convection are frequently written in the form

$$Nu = \frac{hL}{k} = C\,Ra_L^n = C(Gr_L Pr)^n$$

where Ra, the Rayleigh number, is the product of the Grashof and Prandtl numbers, that is,

$$Ra = \frac{g\beta\rho^2\left(T_s - T_f\right)L_{char}^3}{\mu^2}Pr$$

The characteristic length, L_{char}, depends on the geometry. The values of C and n also depend on the geometry and on the flow regime. Correlations for several geometries are presented in Table 12-3. Fluid properties should be evaluated at the film temperature.

TABLE 12-3 **Correlations for natural convection over various surfaces**

Geometry	Characteristic Length, L_{char}	Ra Range	Correlation
Vertical plate	L	$Ra_L \le 10^9$	$Nu_L = 0.68 + \dfrac{0.670\,Ra_L^{1/4}}{[1+(0.492/Pr)^{9/16}]^{4/9}}$
		all Ra_L	$Nu_L = \left\{0.825 + \dfrac{0.387\,Ra_L^{1/6}}{[1+(0.492/Pr)^{9/16}]^{8/27}}\right\}^2$
		$10^4 < Ra_L < 10^9$	$Nu_L = 0.59\,Ra^{1/4}$
		$10^9 < Ra_L < 10^{13}$	$Nu_L = 0.1\,Ra^{1/3}$
Inclined plate	L	$Ra_L < 10^9$	Replace g in Ra_L by $g\cos\theta$, as long as $0 < \theta \le 60°$
Hot plate facing up or cold plate facing down	$\dfrac{area}{perimeter}$	$10^4 < Ra_L < 10^7$ $10^7 < Ra_L < 10^{11}$	$Nu_L = 0.54\,Ra_L^{1/4}$ $Nu_L = 0.15\,Ra_L^{1/3}$
Hot plate facing down or cold plate facing up	$\dfrac{area}{perimeter}$	$10^5 < Ra_L < 10^{10}$	$Nu_L = 0.27\,Ra_L^{1/4}$

TABLE 12-3   (Continued)

Geometry	Characteristic Length, L_{char}	Ra Range	Correlation
Horizontal cylinder 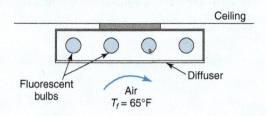	D	$10^{-10} < Ra_D < 10^{-2}$ $10^{-2} < Ra_D < 10^{2}$ $10^{2} < Ra_D < 10^{4}$ $10^{4} < Ra_D < 10^{7}$ $10^{7} < Ra_D < 10^{12}$ $Ra_D < 10^{12}$	$Nu_D = 0.675 Ra_D^{0.058}$ $Nu_D = 1.02 Ra_D^{0.148}$ $Nu_D = 0.85 Ra_D^{0.188}$ $Nu_D = 0.48 Ra_D^{0.25}$ $Nu_D = 0.125 Ra_D^{0.333}$ $Nu_D = \left\{ 0.60 + \dfrac{0.387 Ra_D^{1/6}}{[1+(0.559/Pr)^{9/16}]^{8/27}} \right\}^2$
Vertical cylinder	L		Use vertical plate correlations as long as $D \geq \dfrac{35L}{Gr_L^{1/4}}$
Sphere ↓D	D	$Ra_D < 10^{11}$	$Nu_D = 2 + \dfrac{0.589 Ra_D^{1/4}}{[1+(0.469/Pr)^{9/16}]^{4/9}}$ $Pr > 0.7$

EXAMPLE 12-7   Temperature of a diffuser in a fluorescent light

A fluorescent light is covered with a diffuser, which is a sheet of translucent plastic of size 4 ft by 2 ft. The electronics controlling the light are temperature-sensitive and must be kept cool. If 65 W of heat are dissipated by the light and removed by natural convection from the bottom surface of the diffuser to room air at 65°F, estimate the surface temperature of the diffuser.

Approach:

The heat transferred is given by $\dot{Q} = hA (T_s - T_f)$. The heat transfer coefficient can be found using a correlation for a hot plate facing downward from Table 12-3. Note that the heat transfer coefficient depends on the Grashof number, which is a function of surface temperature. After solving the correlation for the heat transfer coefficient, substitute into $\dot{Q} = hA (T_s - T_f)$ and solve for surface temperature.

Assumptions:

A1. All the generated heat flows through the diffuser.

A2. Radiation is negligible.

Solution:

Because of heat dissipation within the light, the diffuser is hotter than the surrounding air. It can be modeled as a hot plate facing downward, for which the Nusselt number is, from Table 12-3,

$$Nu = \frac{hL}{k} = 0.27 Ra^{1/4} = 0.27 \left[\frac{g \beta \rho^2 \left(T_s - T_f\right) L^3}{\mu^2} Pr \right]^{1/4}$$

We should check to be sure that the Rayleigh number falls in the range of the correlation. However, the surface temperature is unknown and the Rayleigh number cannot be calculated. Instead, we will assume that the correlation applies and check the Rayleigh number range at the end of the calculation.

The rate of heat transfer is related to the temperature drop via [A1][A2]

$$\dot{Q} = hA \left(T_s - T_f\right)$$

Solving for h and substituting into the correlation above gives

$$\frac{\dot{Q}L}{kA \left(T_s - T_f\right)} = 0.27 \left[\frac{g \beta \rho^2 \left(T_s - T_f\right) L^3 Pr}{\mu^2} \right]^{1/4}$$

Raising both sides to the fourth power,

$$\frac{\dot{Q}^4 L^4}{k^4 A^4 \left(T_s - T_f\right)^4} = (0.27)^4 \left[\frac{g \beta \rho^2 \left(T_s - T_f\right) L^3 Pr}{\mu^2} \right]$$

which reduces to

$$\left[\frac{\dot{Q}}{(0.27) kA} \right]^4 \frac{L \mu^2}{\rho^2 g \beta Pr} = \left(T_s - T_f\right)^5$$

Taking both sides to the one-fifth power,

$$\left[\frac{\dot{Q}}{(0.27) kA} \right]^{\frac{4}{5}} \left(\frac{L \mu^2}{\rho^2 g \beta\, Pr} \right)^{\frac{1}{5}} = T_s - T_f$$

Note that $L = \dfrac{A}{P} = \dfrac{(4)(2)\ \text{ft}^2}{2(4+2)\ \text{ft}} = 0.667\ \text{ft}$

The fluid properties should be evaluated at the film temperature, that is, at the average of the surface and fluid temperatures. However, in this case, the surface temperature is unknown, and there is no obvious value to use as a guess. So, to start somewhere, assume a film temperature of 80°F and correct later, if necessary. With properties at 80°F, the temperature difference becomes

$$T_s - T_f = \left[\frac{(65\ \text{W}) \left(\dfrac{3.412 \frac{\text{Btu}}{\text{h}}}{1\ \text{W}} \right)}{(0.27) \left(0.015 \frac{\text{Btu}}{\text{h} \cdot \text{ft} \cdot °\text{F}} \right) \left[(4)(2)\ \text{ft}^2 \right]} \right]^{4/5}$$

$$\times \left[\frac{(1.25 \times 10^{-5})^2 \left(\dfrac{\text{lbm}}{\text{ft} \cdot \text{s}} \right)^2 (0.667\ \text{ft})}{\left(32.17 \frac{\text{ft}}{\text{s}^2} \right) \left(0.074 \frac{\text{lbm}}{\text{ft}^3} \right)^2 \left(1.86 \times 10^{-3} \frac{1}{°\text{F}} \right) (0.72)} \right]^{1/5} = 62.9°\text{F}$$

The surface temperature is

$$T_s = 62.9 + T_f = 62.9 + 65 = 128°F$$

So the film temperature predicted by this result would be

$$T_{film} = \frac{T_s + T_f}{2} = \frac{128 + 65}{2} = 96.5°F$$

We assumed 80°F. If we want a precise result, we would repeat the calculation with properties evaluated at 96°F. But, for our purposes, we will accept the inaccuracy. It is also necessary to check whether the Rayleigh number falls in the correct range. The Rayleigh number is

$$Ra = \frac{g\beta\rho^2 \left(T_s - T_f\right) L^3 Pr}{\mu^2}$$

Using data from Table B-7,

$$Ra = \left(2.097 \times 10^6\right) \frac{1}{°F \cdot ft^3} (128 - 65)\,°F\,(0.667\,ft)^3\,(0.72) = 2.82 \times 10^7$$

From Table 12-3, the correlation is valid if $10^5 < Ra < 10^{11}$. Since the Rayleigh number does fall in this range, it was appropriate to use the correlation.

Comments:

The calculated surface temperature is higher than what occurs in practice. We have not included the effect of thermal radiation, which is important in this case. Furthermore, some of the heat generated by the lamp is conducted into the ceiling. Incorporating these factors in the analysis would result in a lower surface temperature.

12.9 NATURAL CONVECTION IN VERTICAL CHANNELS

(Go to www.wiley.com/college/kaminski)

12.10 NATURAL CONVECTION IN ENCLOSURES

(Go to www.wiley.com/college/kaminski)

12.11 MIXED FORCED AND NATURAL CONVECTION

(Go to www.wiley.com/college/kaminski)

12.12 DIMENSIONAL SIMILITUDE

In this chapter many nondimensional parameters have been introduced. These include the Nusselt, Prandtl, Graetz, Grashof, and Rayleigh numbers. Convection also depends on Reynolds number, which was introduced earlier. Although various arguments have been given to justify the reasonableness of using these parameters from a physical point of view, no proof has been given that these are indeed the correct parameters and that they are the only parameters needed. While it is possible to derive all these nondimensional parameters mathematically, the analysis involves a system of coupled nonlinear partial differential equations and is beyond the scope of this text.

Many students, when first introduced to nondimensional groups, find them difficult to understand and awkward to use. It may be helpful to explain why they are so valuable to thermal analysis. Consider an engineer who requires heat transfer coefficients for forced convection over a circular cylinder and who has decided to run an experiment to generate the data. Heat transfer coefficients depend on six quantities: velocity, diameter, density, viscosity, specific heat, and thermal conductivity. If the engineer did not use the nondimensional groups, he or she might decide to make measurements varying each of these quantities independently to determine their effect. If 5 values of each parameter were tested for all combinations of the 6 parameters, the number of tests required would be $6^5 = 7776$. On the other hand, if the engineer used the fact that the Nusselt number depends on only two quantities—the Reynolds number and Prandtl number—then the amount of experimental data necessary is vastly reduced. This, then, is a very practical reason that correlations are written in terms of nondimensional groups.

A second reason is that if we match nondimensional parameters for two geometrically similar situations, we can expect them to behave similarly. If the actual situation of interest is difficult to measure for some reason, the nondimensional groups can indicate an equivalent system that is easier to handle. For example, if forced convection over a submarine needs to be determined, difficulties might arise because the submarine is so large. On the other hand, since the convection depends on the Reynolds and Prandtl numbers, it might be possible to construct a small-scale model of the ship and test it in a fluid other than water. As long as the Reynolds and Prandtl numbers are the same in both cases, the heat transfer coefficient can be measured for the small system and the results used to infer behavior in the large system.

12.13 GENERAL PROCEDURE FOR EVALUATING HEAT TRANSFER COEFFICIENTS

In this chapter, we have presented many different convective heat transfer coefficient correlations. These are only a small sampling of the large number of correlations available in the literature. The correlations vary widely in format and are subject to various constraints. To avoid errors, it is useful to have a general procedure for determining which correlation to use. A suggested procedure is the following:

1. **Determine whether convection is forced or natural**. Do not apply a forced convection correlation for a natural convection situation or vice versa. In some cases of forced convection, velocities are very low and the flow is actually a combination of forced and natural effects. Such a case is called mixed convection, and correlations are available in the literature for some geometries.

2. **Choose the correct geometry**. The flow is either internal or external, and many geometries have been tested.

3. **Evaluate the fluid properties at the correct temperature**. In internal flow, properties are usually evaluated at the average of the inlet and exit temperatures. In external flow, properties are usually evaluated at the film temperature, which is the average of the surface temperature and the fluid temperature. It is always necessary to check, since some researchers use other conventions.

    4a. Calculate the Reynolds number if forced convection. Confirm that the correlation applies to the calculated Reynolds number.

    4b. Calculate the Grashof number if natural convection. Confirm that the correlation applies. In some cases, natural convection correlations are written in terms of the Rayleigh number instead of the Grashof number. The Rayleigh number is simply the product of the Grashof and Prandtl numbers.

    5. Check the Prandtl number. All correlations have restricted ranges of Prandtl numbers. Some apply only to gases or only to liquids. Generally speaking, special correlations are needed for liquid metal flows.

    6. Calculate any other parameters in the correlation. For example, sometimes the length-to-diameter ratio appears in a correlation.

    7. Note the wall boundary condition. Some correlations, especially in laminar flow, apply only for a constant wall temperature boundary condition while others apply only for a constant heat flux boundary condition. In turbulent flow, the wall boundary condition is not important.

    8. Determine whether the correlation applies for local or average heat transfer coefficient. Local coefficients vary with location on the body and give temperatures or heat fluxes at that location only. In most analyses, the average heat transfer coefficient is needed.

    9. Calculate the Nusselt number. Based on the process of elimination using steps 1 through 8, a reasonable convective heat transfer coefficient correlation can usually be identified. Calculate the Nusselt number. The heat transfer coefficient can be determined using $Nu = hL_{char}/k$, where L_{char} is the appropriate characteristic length.

SUMMARY

Convective heat transfer is related to temperature difference by

$$\dot{Q}_{conv} = hA\left(T_s - T_f\right)$$

Convection correlations for the heat transfer coefficient, h, are written in terms of the nondimensional **Nusselt** and **Prandtl** numbers, defined as

$$Nu = \frac{hL_{char}}{k}$$

$$Pr = \frac{c_p\mu}{k}$$

where L_{char} is a characteristic length for the geometry. The Prandtl number is a fluid property that varies with temperature.

    For forced convection over an isothermal flat plate, the **local** heat transfer coefficient is

$$Nu_x = \frac{h_x x}{k} = 0.332Re_x^{1/2}Pr^{1/3} \qquad \begin{array}{l} Re_x < 5 \times 10^5 \\[4pt] Pr > 0.6 \\[4pt] \text{isothermal plate} \end{array}$$

$$Nu_x = \frac{h_x x}{k} = 0.0296Re_x^{4/5}Pr^{1/3} \qquad \begin{array}{l} Re_x > 5 \times 10^5 \\[4pt] 0.6 < Pr < 60 \\[4pt] \text{isothermal plate} \end{array}$$

with properties evaluated at the film temperature $T_{film} = \left(T_s + T_f\right)/2$. For forced convection over an isothermal flat plate with a laminar boundary layer, the **average** heat transfer coefficient is

$$Nu_L = \frac{hL}{k} = 0.664Re_L^{1/2}Pr^{1/3} \qquad \begin{array}{l} Re_L < 5 \times 10^5 \\[4pt] Pr > 0.6 \\[4pt] \text{isothermal plate} \end{array}$$

For a plate with both laminar and turbulent boundary layers,

$$Nu_L = \frac{hL}{k} = \left(0.037Re_L^{4/5} - 871\right)Pr^{1/3} \qquad \begin{array}{l} 5 \times 10^5 < Re_L \leq 10^8 \\[4pt] 0.6 < Pr < 60 \\[4pt] \text{isothermal plate} \end{array}$$

For a plate with only a turbulent boundary layer,

$$Nu_L = \frac{hL}{k} = 0.037Re_L^{4/5}Pr^{1/3} \qquad \begin{array}{l} 10^8 < Re_L \\ 0.6 < Pr < 60 \\ \text{isothermal plate} \end{array}$$

The critical Reynolds number for transition from laminar to turbulent flow is 5×10^5.

For forced convection over a flat plate with constant heat flux, the **local** heat transfer coefficient is

$$Nu_x = \frac{h_x x}{k} = 0.453Re_x^{1/2}Pr^{1/3} \qquad \begin{array}{l} Re_x < 5 \times 10^5 \\ Pr > 0.6 \\ \text{constant heat flux plate} \end{array}$$

$$Nu_x = \frac{h_x x}{k} = 0.0308Re_x^{4/5}Pr^{1/3} \qquad \begin{array}{l} Re_x > 5 \times 10^5 \\ 0.6 < Pr < 60 \\ \text{constant heat flux plate} \end{array}$$

To find the average heat transfer coefficients for a flat plate with constant heat flux, use the correlations for the average heat transfer coefficient with constant surface temperature given above.

For crossflow over a cylinder, the average heat transfer coefficient is

$$Nu_D = \frac{hD}{k} = C\,Re_D^m Pr^{1/3}$$

Values of C and m, which depend on Reynolds number range, are given in Table 12-1. Table 12-1 also includes correlations for other bodies in crossflow, such as rectangular bars and flat plates perpendicular to the flow direction. For a sphere in crossflow, the average Nusselt number is

$$Nu_D = \frac{hD}{k} = 2 + \left[0.4Re_D^{1/2} + 0.06Re_D^{2/3}\right]Pr^{0.4}\left(\frac{\mu_f}{\mu_s}\right)^{1/4}$$

In the preceeding correlations for external flow, properties are evaluated at the film temperature. In this correlation, μ_f is evaluated at the film temperature and μ_s is evaluated at the surface temperature.

For internal convection in a single-phase fluid in a conduit, the total heat transferred to or from the wall is related to temperatures at the exit and inlet by

$$\dot{Q} = \dot{m}c_p\,(T_e - T_i)$$

The heat flux equation for internal flow in a conduit is

$$q_w'' = \frac{\dot{Q}_w}{A} = h\,(T_w - T_m)$$

where T_m is the bulk mean temperature of the fluid and T_w is the wall temperature. For fully developed laminar convection in a circular pipe with constant heat flux,

$$Nu_D = \frac{hD}{k} = 4.36 \qquad Re_D < 2100 \quad \begin{array}{l}\text{laminar, constant} \\ \text{heat flux}\end{array}$$

For fully developed laminar convection in a circular pipe with constant wall temperature,

$$Nu_D = \frac{hD}{k} = 3.66 \qquad Re_D < 2100 \quad \begin{array}{l}\text{laminar, constant} \\ \text{wall temperature}\end{array}$$

The fluid properties in these equations are based on the average of the inlet and exit temperatures.

For turbulent flow in a pipe, the **Dittus-Boelter** correlation is

$$Nu_D = 0.023Re_D^{0.8}Pr^n \qquad \begin{array}{l} Re_D > 10{,}000 \\ 0.7 < Pr \leq 160 \\ \dfrac{L}{D} \geq 10 \end{array}$$

$$\begin{array}{ll} n = 0.3 & \text{cooling} \quad T_w < T_m \\ n = 0.4 & \text{heating} \quad T_m < T_w \end{array}$$

If temperature differences are large, so that property variations become significant, then the following correlation by **Seider and Tate** is recommended:

$$Nu_D = 0.027Re_D^{0.8}Pr^{1/3}\left(\frac{\mu}{\mu_w}\right)^{0.14} \qquad \begin{array}{l} Re_D > 10{,}000 \\ 0.7 < Pr \leq 16{,}700 \\ \dfrac{L}{D} \geq 10 \end{array}$$

$$\begin{array}{ll} n = 0.3 & \text{cooling} \quad T_w < T_m \\ n = 0.4 & \text{heating} \quad T_m < T_w \end{array}$$

Forced convection heat transfer coefficients in smooth or rough pipes are also given approximately by Petukhov as

$$Nu_D = \frac{(f/8)\,Re_D Pr}{1.07 + 12.7\,(f/8)^{1/2}\,(Pr^{2/3} - 1)}\left(\frac{\mu}{\mu_w}\right)^n$$

$$\begin{array}{l} 10^4 < Re_D < 5 \times 10^6 \\ 0.5 < Pr < 2000 \\ 0.08 < \mu/\mu_w < 40 \\[4pt] n = 0 \quad \text{constant heat flux} \\ n = 0.11 \quad T_w > T_m \\ n = 0.25 \quad T_m > T_w \end{array}$$

For the friction factor, use Petukhov's correlation for smooth pipes, which is

$$f = (0.79\ln Re - 1.64)^{-2} \qquad \begin{array}{l}\text{turbulent flow, smooth wall} \\ 3000 < Re_D < 5 \times 10^6\end{array}$$

The above correlations apply in the fully turbulent regime, where $Re > 10{,}000$. When $2100 < Re < 10{,}000$, the flow may be in transition between laminar and turbulent. A useful correlation by Gnielinski for low Reynolds number is

$$Nu_D = \frac{(f/8)\,(Re_D - 1000)\,Pr}{1 + 12.7\,(f/8)^{1/2}\left(Pr^{2/3} - 1\right)} \qquad \begin{array}{l} 3000 < Re_D < 5 \times 10^6 \\[4pt] 0.5 < Pr < 2000 \end{array}$$

This equation applies for constant surface temperature or constant heat flux.

The total heat transferred from a single-phase fluid to or from the wall of a pipe with constant wall temperature is

$$\dot{Q} = hA\,\Delta T_{LM}$$

where the log mean temperature difference is

$$\Delta T_{LM} = \frac{(\Delta T_e - \Delta T_i)}{\ln\left(\dfrac{\Delta T_e}{\Delta T_i}\right)}$$

The exit temperature of the fluid flowing inside the pipe with constant wall temperature may be found from

$$\ln\left(\frac{T_e - T_w}{T_i - T_w}\right) = -\frac{hA}{\dot{m}c_P}$$

If heat transfer occurs both inside and outside of a pipe, and the exterior flow is at a uniform temperature along the pipe, then the heat transfer rate may be expressed in terms of an overall heat transfer coefficient, U, where

$$\dot{Q} = UA\,(T_m - T_\infty)$$

and U is given by

$$U_1 = \frac{1}{\dfrac{1}{h_i} + \dfrac{r_1}{k_1}\ln\left(\dfrac{r_2}{r_1}\right) + \dfrac{r_1}{k_2}\ln\left(\dfrac{r_3}{r_2}\right) + \dfrac{r_1}{r_3 h_\infty}}$$

In this equation, U_1 is based on the inside surface area of the pipe, A_1. A similar equation can be written if U_2, based on the outside area, is used.

In noncircular conduits, the preceeding correlations for circular pipes apply if the hydraulic diameter is used in both the Reynolds and Nusselt numbers. Hydraulic diameter is

$$D_h = \frac{4A}{P_{wetted}}$$

where A is the cross-sectional flow area and P_{wetted} is the wetted perimeter of the tube. Nusselt number relations are given in Table 12-2 for various noncircular conduits.

For internal flow, the thermal entry length is given by

$$L_{ent,t} \approx 0.037 Re_D Pr D \qquad \begin{array}{l}\text{laminar}, Re < 2100, \\ \text{constant temperature wall}\end{array}$$

$$L_{ent,t} \approx 0.053 Re_D Pr D \qquad \begin{array}{l}\text{laminar}, Re < 2100, \\ \text{constant heat flux}\end{array}$$

$$L_{ent,h} \approx L_{ent,t} \approx 4.4\,(Re)^{1/6} D \qquad \text{turbulent}, Re > 4000$$

Entrance effects in laminar flow may be significant and should always be checked.

In natural convection, the Nusselt number depends on the Grashof number, defined as

$$Gr = \frac{g\beta\rho^2\,(T_s - T_f)\,L_{char}^3}{\mu^2}$$

Correlations for the average heat transfer coefficient in natural convection over surfaces are frequently written in the form

$$Nu = \frac{hL}{k} = C\,Ra_L^n = C(Gr_L\,Pr)^n$$

where Ra, the Rayleigh number, is the product of the Grashof and Prandtl numbers. Values of C and n, which depend on the geometry and on the flow regime, are given in Table 12-3. Fluid properties should be evaluated at the film temperature.

Correlations for natural convection in vertical parallel plate channels are given in section 12.9.

In mixed forced and natural convection, the Richardson number is defined as

$$Ri = \frac{Gr}{Re^2}$$

The range in which mixed forced and natural convective effects are important is

$$0.1 < Ri < 10$$

The Nusselt number for mixed convection can be approximated by a relation of the form

$$Nu^n = Nu_{forced}^n \pm Nu_{natural}^n$$

where the Nusselt numbers for pure forced and pure natural convection are found from known correlations. The plus sign is used when forced flow is aiding or perpendicular to the natural flow and the minus sign is used when forced flow opposes natural flow. For vertical geometries, $n = 3$, while for horizontal geometries, $n = 3.5$. For cylinders or spheres, use $n = 4$.

SELECTED REFERENCES

BECKER, M., *Heat Transfer, A Modern Approach*, Plenum Press, New York, 1986.

CENGEL, Y. A., and R. H. TURNER, *Fundamentals of Thermal–Fluid Sciences*, McGraw-Hill, New York, 2001.

INCROPERA, F. P., and D. P. DeWitt, *Introduction to Heat Transfer*, 4th ed., Wiley, New York, 2002.

KREITH, F., and M. S. Bohn, *Principles of Heat Transfer*, 6th ed., Brooks/Cole, Pacific Grove, CA, 2001.

MILLS, A. F., *Heat Transfer*, Irwin, Boston, 1992.

SURYANARAYANA, N. V., *Engineering Heat Transfer*, West, New York, 1995.

THOMAS, L. C., *Heat Transfer*, Prentice Hall, Englewood Cliffs, NJ, 1992.

PROBLEMS

Problems designated with WEB refer to material available at www. wiley.com/college/kaminski.

EXTERNAL FORCED FLOW—FLAT PLATES

P12-1   An undergraduate heat transfer lab has an experiment to illustrate the effects of different boundary conditions on heat transfer from a flat plate. A test section is installed into a wind tunnel. The test section consists of 100 thin strip heaters placed on a 2-m-long flat plate. Each heater, 20 mm long and 250 mm wide, is located so that there is no space between adjoining heaters and is electrically and thermally insulated from the adjacent heaters; the backside of the plate is heavily insulated. The power to each heater can be individually controlled. The free-stream air temperature is 25°C and has a velocity of 4 m/s. By controlling the power to each strip, two different boundary conditions can be modeled. Ignoring radiation, determine for strips 1, 5, 25, 100, and 200

a. the heat transfer rate when the power is adjusted in each heater to maintain a uniform plate temperature of 50°C (in W).

b. the wall temperature on strip number 25 when the power is adjusted in each heater to maintain a uniform heat flux (equal to that on strip 25 from the first part) over the entire plate (in °C).

P12-2   Many schemes have been proposed to supply arid regions with fresh water. One plan involves towing icebergs from the polar regions to dry regions that need fresh water. Consider an iceberg, 1000 m long and 500 m wide, that is towed through 10°C water at a velocity of 1 km/h. The density of ice is 917 kg/m³ and the heat of fusion is 333.4 kJ/kg.

a. Determine the average rate at which the flat bottom of the iceberg will melt (in mm/h).

b. Determine how much ice will melt if the voyage is 1500 km long (in kg).

P12-3   Two brothers have rooms side by side in a flat-roofed mobile home. The older brother continually complains that his room is colder than that of the younger brother. The older brother

decides to add more heating to his room. As shown in the figure, the first room is 4 m long and the second one is 3 m long; each is 4 m deep. The roof thickness is 0.25 m with a thermal conductivity of 1.2 W/m·K. The outside wind is parallel to the roof at a velocity of 20 km/h at −10°C the inside temperature is to be maintained at 21°C and the inside heat transfer coefficient is 7.5 W/m²·K. Determine the heat loss from the roof of each of the two rooms (in W).

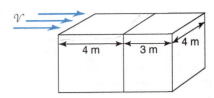

P12-4   Rolling mills are used to reduce the thickness of steel plates to create thin steel strips. The metal must be at a high temperature so that the power (force) required to reduce the metal thickness is not excessive and so that the desired material properties are obtained. Consider a 304 stainless-steel strip 3 mm thick leaving a rolling mill at 1000°C at a speed of 20 m/s. A length of 50 m is exposed to air at 35°C. Convective heat transfer occurs on both the top and bottom surfaces of the strip. Ignoring radiation and axial conduction in the steel, estimate the temperature of the strip when it reaches 50 m from the roller (in °C).

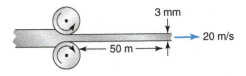

P12-5   The failure rate of computer chips increases with increasing operating temperature. Consider a 15 mm by 15 mm chip that is cooled on its top surface by a 5-m/s flow of 25°C air. Any heat transfer from its bottom surface to the circuit board is ignored. Because of the chip construction, the electrical power dissipated in the chip results in a uniform heat flux over the surface of the chip. The maximum temperature that any part of the chip can experience is 80°C.

a. Determine the maximum allowable chip power (in W).

b. Determine the maximum allowable chip power if this chip is the fifth in a column of identical chips all mounted flush to the surface with no space between the chips (in W).

P12-6 The walls of a house are constructed of an exterior sheathing, insulation, framing timber, and drywall; their composite resistance is estimated to be 4.15 $m^2 \cdot K/W$. A winter wind blows parallel to the 3-m-high, 18-m-long wall. The wind velocity is 30 km/h, and its temperature is −5°C. The heat transfer coefficient at the interior of the house is 5 $W/m^2 \cdot K$. For an inside air temperature of 21°C, determine the heat transfer rate through the wall (in W).

P12-7 One wall of an older office building (6 m high and 30 m long) is all glass 7 mm thick. Wind blows parallel to it at 20 km/h and 5°C. The inside surface temperature of the glass is 20°C.

a. Determine the heat transfer rate from the glass (in W).

b. Determine the heat transfer rate if the wind velocity is tripled (in W).

P12-8 The roof of a minivan can be approximated as a flat plate, 2 m wide and 3.5 m long. The sun beats down on the roof such that the net solar radiation absorbed is 350 W/m^2. If the ambient air is 32°C, the car is moving at 100 km/h, and the inside surface of the roof is heavily insulated, determine the steady-state temperature of the roof (in °C).

P12-9 Power transformers change the voltage of electricity, but the devices are not 100% efficient. Dissipated heat must be removed from transformers so that they do not reach a temperature that could damage them. Consider a transformer that dissipates 30 W. It is 10 cm wide and 20 cm long, with eight fins 2 cm tall, 2 mm thick, and 20 cm long evenly distributed across the surface of the transformer. Air at 25°C is blown parallel along the length of the fins. Assume the fins have a fin efficiency of 100%. Ignoring radiation, and for a base temperature less than 65°C, determine the minimum air velocity required (in m/s).

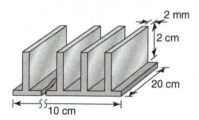

P12-10 Solar-powered planes have been designed to be able to stay aloft for very long times. Proposed uses include meteorology and surveillance. Photovoltaic (PV) cells are mounted on the top surface of the wing. The PV panel is 1.5 m wide and 8 m long.

The solar energy absorbed but not converted to electricity equals about 850 W/m^2, and the PV cell conversion efficiency decreases with increasing temperature. Determine the temperature of the trailing edge of the panel when the plane flies at 110 km/h at 5000 m, where the pressure is 54 kPa and temperature is 256 K.

EXTERNAL FORCED FLOW—CYLINDERS AND SPHERES

P12-11 For a quick solution to an overheating problem, brass rods, 6 mm in diameter and 5 cm long, are attached to a surface of a power supply. Air at 20°C and 5 m/s is blown perpendicular to the rods. If the base temperature must not exceed 75°C, how much power can be dissipated by one rod (in W)?

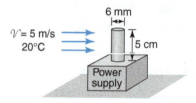

P12-12 If the brass rods in Problem P 12-11 are replaced by rectangular aluminum alloy fins 10 cm wide (in the direction of air flow), 2 mm thick, and 5 cm long, determine how much power can be dissipated by one fin (in W).

P12-13 In an electric hair drier, air at 25°C flows with a velocity of 5 m/s perpendicular to a Nichrome heating element. The heating element is 1 mm in diameter and 40 cm long, with a resistance of 1.38 Ω/m. The wire temperature cannot exceed 430°C so that the wire will not lose strength and sag.

a. Determine the total power dissipated (in W).

b. Determine the electric current in the wire (in A).

P12-14 An existing electric power line is being examined to determine whether a higher current can be used. You are asked to calculate the maximum power dissipation per meter of length that is permissible by joulean heating such that the inside surface of the cable insulation does not exceed 77°C. The copper wire in the cable is 2 cm in diameter, and the insulation is 0.1 cm thick with a thermal conductivity of 0.08 $W/m \cdot K$. The wind velocity perpendicular to the wire is 5 km/h and the air temperature is 27°C. Neglecting radiation, determine the allowable power (heat generation rate) per unit length (in W/m).

P12-15 The insulation on a 15-cm steam pipe deteriorates over time and is to be removed and replaced. The outer surface of the steam pipe is at 110°C. Air at −6°C blows perpendicular to the pipe at 40 km/h.

a. Determine the heat transfer rate per unit length of pipe if it is left bare (in W/m).

b. Determine the heat transfer rate per unit length of pipe if 4-cm insulation ($k = 0.04$ $W/m \cdot K$) is applied to the pipe (in W/m).

P12-16 A very long cylinder 25 mm in diameter is placed in a large oven whose walls are maintained at 400°C. Air at 77°C flows perpendicular to the cylinder at a velocity of 2.5 m/s. The emissivity of the cylinder is 0.65. Determine the steady-state temperature of the cylinder (in °C).

P12-17 After the extrusion of a long solid plastic rod, it is cooled by a crossflow of 30°C air at a velocity of 20 m/s. The rod, whose diameter is 3.5 cm, initially has a uniform temperature of 200°C. The plastic's properties are: $\rho = 2300$ kg/m^3, $c_p = 850$ J/kg·K, and $k = 1.2$ W/m·K.

a. Determine the time required for the surface temperature of the rod to drop to 100°C (in s).

b. Determine the centerline temperature at the same time (in °C).

P12-18 Lead shot is made by dropping molten lead ($\rho = 10{,}600$ kg/m^3) from a drop tower. Each pellet, a sphere 2 mm in diameter, is cooled as it passes through air at 10°C. Assume the shot falls at its terminal velocity. The lead must be solidified from its molten state at 327°C to a solid state before it reaches the pool of water at the bottom of the drop tower. The enthalpy of fusion for lead is 24.5 kJ/kg. Ignoring radiation, determine the minimum required height of the drop tower (in m).

P12-19 False temperature readings can be obtained from temperature sensors if they are used incorrectly or if the effects of radiation are not taken into account. (For example, if a thermometer were used in direct sunlight or in shade to measure air temperature, significantly different readings would be obtained.) Consider a thermocouple that is a 1.5-mm sphere, which is used to measure the temperature of an air stream in a large duct. The air velocity is 4 m/s. The walls of the duct are at 150°C. The thermocouple indicates an air temperature of 300°C and has an emissivity of 0.5. Determine the actual air temperature (in °C).

P12-20 Fluid velocities are often measured with hot wire anemometers. In such a device, the temperature of a small-diameter cylinder is maintained constant by varying the electric current through it in response to varying fluid velocity; a wheatstone bridge is used to control the current. A typical hot wire is constructed of a 0.2-mm-diameter polished platinum wire 10 mm long. Air at 23°C flows over the hot wire maintained at 200°C. The electrical resistivity of platinum is 17 $\mu\Omega$-cm. Determine the electric current required for a velocity of

a. 1 m/s.

b. 10 m/s.

P12-21 After a heat-treating process, a 2024-T6 aluminum sphere 20 mm in diameter is removed from an oven that is at 85°C. The sphere is placed in an air stream at 27°C that has a velocity of 10 m/s. Determine the time required for the sphere's temperature to cool to 40°C (in s).

P12-22 In an oil refinery, a steam pipe ($k = 15$ W/m·K) with an inside diameter of 10 cm and an outside diameter of 11 cm is covered with 3.5 cm of insulation ($k = 0.03$ W/m·K). The steam

is at 300°C. Air and surrounding surfaces are at 17°C, and the air flows perpendicular to the pipe at a velocity of 2 m/s. The heat transfer coefficient of the steam is 100 W/m^2·K. Determine the heat transfer rate per unit length of pipe (in W/m).

P12-23 If you have ever changed a hot incandescent lightbulb, then you know that much of the power going into the bulb is converted to heat (about 90%) rather than to light (about 10%). (Fluorescent light bulbs are much more efficient.) All the heat is dissipated from the glass bulb. Consider an 8-cm-diameter 100-W lightbulb cooled by air at 30°C. Both convection and radiation ($\varepsilon = 0.85$) cool the glass. Assuming the surroundings are at 30°C for radiation purposes, determine the temperature of the glass bulb if air at 2 m/s flows across it (in °C).

P12-24 After long 316 stainless-steel rods 50 mm in diameter are preheated to a uniform temperature of 1000°C, they must be conveyed to another location in the plant for additional processing. The conveyor moves the rods perpendicular to the direction of travel at a velocity of 3 m/s. The rod emissivity is 0.5, and the air and surrounding temperatures are at 27°C. The centerline temperature of the rod must be greater than 900°C for the next processing step.

a. Determine the convective heat transfer coefficient at the start of the travel (in W/m^2·K).

b. Determine the radiation heat transfer coefficient at the start of the travel between one processing station and another (in W/m^2·K).

c. Determine the allowable time for transit between the two stations assuming the total heat transfer coefficient (radiation and convection) is the sum of those calculated in parts a and b (in s).

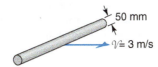

50 mm

$\mathcal{V} = 3$ m/s

INTERNAL FORCED CONVECTION

P12-25 Ventilation ducts are often uninsulated when they run through attics and other uninhabited spaces. Consider an airflow at 70°C and 15 m^3/min that enters a 20-m-long, 30-cm-square duct. The duct runs through a space that is at 10°C. Ignoring the temperature drop across the metal duct, determine:

a. the outlet temperature of the air (in °C).

b. the heat transfer rate from the hot air (in W).

P12-26 Because of a shallow ocean floor and deep draft, an oil tanker must use an offshore oil depot to unload. The depot is connected to a shore installation by a 1100-m-long, 45-cm pipe. In the winter, the ocean water temperature is 5°C. The oil (properties equivalent to unused engine oil), initially at a temperature of 20°C, is pumped from the tanker at a flow rate of

0.08 m³/min. Ignoring the water and pipe thermal resistances, determine

a. the outlet temperature of the oil (in °C).

b. the heat transfer rate (in W).

P12-27 A steam condenser downstream of a turbine in a Rankine cycle power plant has 5,000 tubes, each with an internal diameter of 0.75 in. The steam condenses at 120°F on the outside of the tubes. The total cooling water flow rate, at 3500 lbm/s, enters the tubes at 54°F and leaves at 85°F. Because the condensation heat transfer coefficient is very high, ignore the steam (and tube wall) thermal resistances.

a. Determine the heat transfer rate (in Btu/h).

b. Determine the tube length required (in ft).

P12-28 An air compressor used in a large car body shop is located in an inside equipment room. Fresh air at 5°C, 96 kPa is conveyed to the compressor from outside through a 30-cm circular duct that is 15 m long. The duct runs along the ceiling of the facility, where the temperature is 34°C. If the air flow rate is 0.35 m³/s, determine

a. the temperature of the air when it reaches the compressor (in °C).

b. the heat transfer rate (in W).

P12-29 A proposed cooling technique for high-power computer chips is to machine microchannels into the backside of the silicon chip. Because the Nusselt number is defined as $Nu = hk/L_{char}$, for a given value of Nu, as the characteristic length decreases, the heat transfer coefficient increases. Consider a 1.5 cm by 1.5 cm computer chip that dissipates 50 W. Water at 20°C is used as the coolant, and its outlet temperature is limited to 25°C. Heat transfer is primarily from the base of the channel (that is, ignore heat transfer to the water from the channel sides). Assume the distance between the sides of two channels is 0.1 mm.

a. Determine the number of 0.25-mm-deep and 0.25-mm-wide microchannels that are on a chip and the average surface temperature of the base of the microchannels (in °C).

b. Determine the number of 1-mm-deep and 1-mm-wide microchannels that are on a chip and the average surface temperature of the base of the microchannels (in °C).

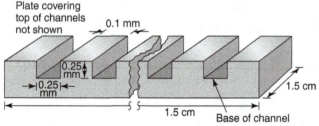

P12-30 Consider turbulent flow of a fluid through a tube maintained at constant temperature. The mass flow rate is 0.32 kg/s, and the heat transfer coefficient is 250 W/m²·K. Now the free-stream velocity of the fluid is doubled. Assume the flow regime remains unchanged.

a. Estimate the percent change in the pressure drop of the fluid between the old and new flow rates.

b. Estimate the percent change in the *local* heat flux between the fluid and the walls of the channel.

c. Estimate the percent change in the *total* heat transfer rate over the length of the channel if the heat transfer area is 3.7 m², specific heat is 2200 J/kg·K, inlet temperature is 25°C, and wall temperature is 75°C.

P12-31 Glycerin is pumped through a 1.5-cm-diameter tube that is 5 m long. The inlet temperature is 32°C, the required outlet temperature is 22°C, and the flow rate is 100 kg/h. Determine the wall temperature required to obtain this outlet temperature (in °C).

P12-32 In a pharmaceutical application, the product is subjected to a final sterilization by heating it from 32°C to 80°C. A flow of 60 cm³/s is passed through a 10-mm tube that is heated with a uniform heat flux produced by wrapping the tube with an electric resistance heater. If product properties can be approximated by those of ethylene glycol and the tube is 25 m long, determine

a. the required power (in W).

b. the wall temperature at the tube exit (in °C).

P12-33 Air enters a compressor operating at steady state with a volumetric flow rate of 37 m³/min at 105 kPa and 30°C and exits with a pressure of 690 kPa and temperature of 240°C. The compressor is cooled with 40 kg/min of water that circulates in a water jacket enclosing the compressor. The water jacket can be approximated as 25 3-cm-diameter, 2-m-long pipes. The water enters at 20°C, and the wall temperature of the pipes can be approximated as being constant at 135°C. Assume fully developed flow.

a. Determine the heat transfer rate from the compressor to the water (in kW).

b. Determine the mass flow rate of air (in kg/s).

c. Determine the power input to the compressor (in kW).

P12-34 The condenser downstream of the turbine in a large Rankine cycle power plant is constructed of 30,000 25-mm tubes. The steam condenses at 50°C with a heat transfer coefficient of 9000 W/m²·K on the outside of the tubes. The cooling water enters the tube side of the condenser at 20°C at a flow rate of 17,000 kg/s. For a 1000-MW (net) power output and a cycle thermal efficiency of 42%, determine

a. the cooling rate required (in MW).

b. the outlet temperature of the cooling water (in °C).

c. the length of tubing required (in m).

P12-35 Air at a mass flow rate of 0.0015 lbm/s and an inlet temperature of 80°F enters a rectangular duct 3.5 ft long, 0.15 in. high, and 0.60 in. wide. A uniform heat flux of 50 Btu/h·ft^2 is imposed on the duct surface.

a. Determine the outlet temperature of the air (in °F).

b. Determine the highest wall temperature and its location (in °F and ft).

P12-36 Water enters at 40°F and flows at a rate of 0.25 ft^3/s inside a 20-ft-long annulus whose inner and outer radii are 1 in. and 2 in., respectively. The inner surface is maintained at 150°F, and the outer surface is heavily insulated.

a. Determine the outlet temperature (in °F).

b. Determine the heat transfer rate (in Btu/h).

P12-37 Unused engine oil is to be heated from 20°C to 65°C using condensing steam at 100°C. The oil flows inside a 1-cm-diameter tube at a flow rate of 0.1 kg/s. The resistance of the condensing steam and the tube wall can be ignored. Determine the length of tube required (in m).

P12-38 In a small ship with limited space, water must be heated from 10°C to 50°C with condensing steam at 100°C. The water flow rate is 1.5 kg/s. Either one 4-cm-diameter tube, two 3-cm tubes, or three 2-cm tubes can be used in parallel. Determine which configuration will yield the shortest length.

P12-39 A thick, stainless-steel (AISI 316) pipe with inside and outside diameters of 20 mm and 40 mm, respectively, is heated electrically to provide a uniform heat generation rate of 10^7 W/m^3. This pipe is encased within a larger concentric tube with an inside diameter of 50 mm whose outer surface is heavily insulated. Pressurized water flows through the annular region between the two tubes with a flow rate of 0.6 kg/s. The water inlet temperature is 20°C.

a. Determine the required pipe length if the desired outlet temperature is 40°C (in m).

b. Determine the highest surface temperature and its location (in °C and m).

P12-40 Pasteurization is the sterilization of milk to ensure no diseases are transmitted with the milk. Consider a flow of 1.5 kg/s of milk whose temperature must be raised from 35°C to 75°C in a 2-cm-diameter tube. The wall temperature is 100°C. Milk properties are: $\rho = 1030$ kg/m^3, $\mu = 2.12 \times 10^{-3}$ N·s/m^2, $c_p = 3850$ J/kg·K, and $k = 0.6$ W/m·K. Determine the required tube length (in m).

P12-41 Pressurized liquid water enters a 2-cm-diameter, 6-m-long tube at 20°C at a flow rate of 0.5 kg/s. The tube surface temperature is constant, and the total power transferred to the water is 150 kW. Determine the surface temperature (in °C).

P12-42 Parts of the Alaskan oil pipeline (1 m in diameter) are buried 3 m below the surface of the earth ($k = 0.65$ W/m·K) and covered with 20 cm of insulation ($k = 0.05$ W/m·K). Pumping stations are 60 km apart. To decrease pumping power, the oil is heated to about 100°C (to reduce its viscosity) before it enters the pipeline at a pumping station. In the winter the surface temperature of the earth is −30°C. Assume the oil properties can be approximated with those of unused oil given in Table A-6. For a flow rate of 0.5 m^3/s, and using properties evaluated at the average temperature of the oil, determine

a. the oil temperature when it reaches the next pumping station (in °C).

b. the heat transfer required at the pumping station to raise the oil temperature back to 100°C (in W).

c. the pumping power required (in W).

d. the pumping power if the inlet oil temperature is 50°C instead of 100°C (in W).

P12-43 Pressurized liquid water flowing inside a tube at a rate of 1 kg/s is to be heated from 25°C to 90°C using condensing steam. The 304 stainless-steel tube has an inside diameter of 25 mm, a wall thickness of 1 mm, and a length of 6 m. The condensing coefficient on the outside of the tube is 6,500 W/m^2·K.

a. Determine the steam temperature and pressure required (in °C and kPa).

b. Determine the condensation rate of the steam assuming the steam enters as a saturated vapor and exits as a saturated liquid (in kg/s).

P12-44 The oil from a large diesel engine flows through an oil cooler before it is returned to the engine. Consider a flow rate of 0.1 kg/s that must be cooled from 90°C to 40°C by passing through a thin-walled tube with a diameter of 12.7 mm. Air at 30°C is in crossflow outside the tubes with a velocity of 10 m/s. Determine the required tube length (in m).

P12-45 For some applications, enhanced cooling capabilities are obtained by attaching a heat-generating system to a *cold plate*, which is maintained at a cold temperature by passing water through it. Consider the copper cold plate (shown in the figure) that has heat-generating equipment attached to its top and bottom surfaces. Each of the six channels is 6 mm square and 100 mm long, and the walls of each channel are 4 mm thick. If chilled water at 10°C is pumped through the channels at a velocity of 0.5 m/s and the surfaces of the cold plate must stay below 45°C, determine:

a. the maximum allowable power (top and bottom surfaces) to the cold plate (in W).

b. the water outlet temperature (in °C).

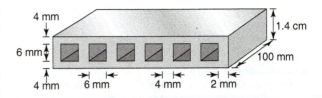

NATURAL CONVECTION

P12-46 The total thermal resistance between the outside and inside of a home consists of the external convective resistance, the wall resistance, and the internal convective resistance. Adding additional insulation reduces the heat transfer but also changes the inside-wall temperature. Compare the average natural convection heat transfer coefficient on a 2.5-m-tall wall for two situations:

a. Inside air temperature of 22°C and wall temperature of 10°C

b. Inside air temperature of 22°C and wall temperature of 17°C

P12-47 Consider again the lightbulb in Problem P 12-23. For all the same conditions, determine the temperature of the glass bulb if it is cooled by natural convection (in °C).

P12-48 In any design process, decisions have to be made about placement of components. Cost, performance, and maintainability are some of the criteria used. Consider the placement of a 1.2-W, 60 mm by 60 mm electric component in a larger device. The component's surface temperature must not exceed 85°C. The air is quiescent at 25°C. Neglecting radiation, determine whether the component can be located facing downward or facing upward. (In other words, what is the component's surface temperature if it is facing upward or downward?)

P12-49 To lower the viscosity of an oil before it is used in a process, an electric resistance heater 1.5 mm in diameter and 30 mm long is immersed horizontally in a vat of unused engine oil which is at 20°C. If the heater surface should not rise above 150°C so that the oil does not smoke, determine the maximum power that can be dissipated in the heater (in W).

P12-50 An electric resistance heater, 10 mm in diameter and 300 mm long, is rated at 550 W. If the heater is horizontally positioned in a large tank of water that is at 20°C, estimate the surface temperature of the heater (in °C).

P12-51 A passive solar-heating technique is to use a massive masonry wall (a Trombe wall) to absorb solar energy and then to release it slowly when the air temperature surrounding the wall is lower than that of the wall. Consider a long 3-m-tall wall, well insulated on its backside, that has a net radiant solar energy flux into the wall of 150 W/m². The air temperature is 21°C. Assuming that the temperature of the wall changes very slowly and the wall operation can be approximated as quasi-steady, determine the average surface temperature of the wall (in °C).

P12-52 A steam-heated cooking vat in a food-processing plant has a bottom that is 1.5 m by 1.5 m. The vat is filled with water, initially at 25°C, and the bottom is heated with condensing steam at 105°C.

a. Determine the initial heat transfer rate from the bottom of the vat to the water (in W).

b. Determine how long it would take for the water temperature to rise to 30°C if the water depth is 60 cm (in min).

P12-53 You devise a transient heat transfer experiment to measure the natural convection heat transfer coefficient on a 2024-T6 aluminum sphere. Initially, the 3-cm sphere is at a uniform temperature of 90°C as measured by a thermocouple inserted into the sphere's center. You plunge the sphere into 10°C water and record the center temperature as it decreases with time. The center temperature reaches 80°C after 0.83 s, 50°C after 5.66 s, and 40°C after 9.78 s.

a. Determine the average heat transfer coefficient as the sphere changes temperature from 90° to 80°C (in W/m²·K).

b. Determine the average heat transfer coefficient as the sphere changes temperature from 50° to 40°C (in W/m²·K).

c. Determine whether or not this approach is valid.

P12-54 Farmer Brown installs an electric resistance heater in the watering trough for his cows so that the water will not freeze during the long cold winter. He does not want a cow to burn its tongue if it accidentally touches the heater. He places a 25-mm-diameter, 30-cm-long, 100-W heater horizontally in the water, which is maintained at 5°C.

a. Determine the surface temperature of the heater (in °C).

b. Determine the surface temperature of the heater if the water trough develops a leak, all the water drains out, and the air temperature is −15°C (in °C).

P12-55 The manufacturer of the electric resistance heater described in Problem P 12-54 wants to expand her sales and considers using the heater for fuel oil tanks, too. Concern about a possible fire hazard if the oil anywhere in an oil tank reaches too high a temperature makes her contact a consulting engineer for an analysis. If the oil has properties of unused engine oil and the oil is at 0°C, determine the surface temperature of the heater (in °C).

P12-56 The coils in electric power transformers mounted on telephone poles in every neighborhood are cooled by oil. If the coils reach too high a temperature, the transformer can fail. To prevent this problem, the transformer is externally cooled by air. The worst-case scenario occurs on hot, still summer days. Consider a transformer that is 55 cm in diameter and 1.5 m tall on a day when the temperature is 40°C. Assume the heat transfer coefficient on the ends is the same as on the cylinder sides. If 250 W must be dissipated, determine

a. the surface temperature of the transformer (in °C) when ignoring radiation.

b. the surface temperature if radiation is included with an emissivity of 0.6 (in °C).

P12-57 To improve the heat transfer from the transformer described in Problem P 12-56, 16 longitudinal fins made of plain carbon steel are attached. Each fin has the same length as the transformer, is 4 mm thick, and extends from the surface 100 mm. Using only natural convection, determine the surface temperature of the transformer (in °C).

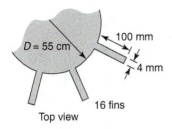

D = 55 cm

100 mm

4 mm

16 fins

Top view

P12-58 Two 1.5-m-diameter, 2-m-tall tanks connected to a common piping header are used to store propane for use in an isolated cabin in the mountains. Unknown to the owner, the spring in the pressure relief valve on the system weakens and allows the pressure in the two tanks to drop to atmospheric pressure. The temperature of the propane falls to $-42°C$ when the pressure inside the two tanks reaches 1 atm. The still ambient air is at $15°C$. Heat transfer from the air to the propane causes it to vaporize, and the vapor is vented from the tank. Properties of propane are $v_f = 0.001755$ m^3/kg, $v_g = 0.4127$ m^3/kg, $h_{fg} = 425$ kJ/kg, and $h_g = 493$ kJ/kg. Ignoring the wall resistance of the tank and radiation, determine how long it will take for the tank to empty (in days).

P12-59 An experiment is performed to determine the heat transfer coefficient on a horizontal circular cylinder. Radiation effects are minimized by polishing the cylinder's surface. The 30-cm-long, 2.5-cm-diameter cylinder has well-insulated ends. Measurements show that 30 W are dissipated when the cylinder surface temperature is $95°C$ and the surrounding air and surfaces are at $20°C$.

a. Determine the natural convection heat transfer coefficient from the data.

b. Determine the natural convection heat transfer coefficient if radiation is taken into account and the surface emissivity is estimated to be 0.07 (in W/m^2·K).

c. Compare the calculated heat transfer coefficient to one calculated with the appropriate correlation (in W/m^2·K). Comment on the accuracy of the experimental results.

P12-60 Arrays of vertical fins are often attached to equipment to aid passive (i.e., natural convection) cooling of the device. Consider the assembly shown in the figure, which is located in air at $20°C$. Each fin has a length of 25 mm, a thickness of 1.5 mm, and a height of 100 mm. Assume the fin has a fin efficiency of

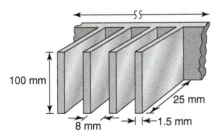

100 mm

25 mm

8 mm

1.5 mm

100%, the base temperature is $75°C$, and each fin operates as if it were independent of all other surfaces nearby. For a fin spacing of 8 mm, determine the heat transfer rate from an array of fins that covers 150 mm of wall (in W).

P12-61 A window 30 cm tall and 45 cm wide is centered in an oven door that is 50 cm tall and 75 cm wide. During operation when the room temperature is $24°C$, the window reaches a temperature of $45°C$ and the door surface reaches $33°C$. Assume that both the door and window have an emissivity of 1.0 and the surroundings also are at $24°C$.

a. Estimate the heat transfer from the door and window (in W).

b. Estimate the heat transfer if the door did not have a window (in W).

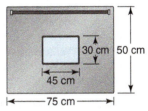

30 cm   50 cm

45 cm

75 cm

P12-62 The heat-loss situation described in Problem P 12-15 changes when the wind stops and the air is calm. For this new condition, determine

a. the heat transfer rate per unit length of pipe (in W/m) with no insulation.

b. the heat transfer rate per unit length of pipe if 4-cm insulation ($k = 0.04$ W/m·K) is applied to the pipe (in W/m).

P12-63 A power amplifier is mounted vertically in air that is at $27°C$. The case is made of anodized aluminum with a surface area of 3800 mm^2 and a height of 40 mm. If the amplifier operates at $127°C$, estimate the total power dissipation (natural convection and radiation) from the unit (in W). Assume a surface emissivity of 0.76.

P12-64 In car paint shops and other drying applications, radiant heaters are often used because the radiant thermal energy heats the surface directly with minimal heating of the surrounding air. Consider a vertical flat panel 1 m tall and 4 m long with an emissivity of 0.85 mounted on the wall of a large room. The panel is maintained at a uniform temperature of $330°C$, and the walls and air in the room are maintained at $25°C$. Determine the heat transfer rate from the panel to the room (in W).

P12-65 A home hobbyist builds a kiln to fire her ceramic pots. Plans obtained from the Internet state that because of the thick fire clay bricks used to construct the kiln, insulation on the outside surface is not required. When she uses the 1 m by 1 m by 1 m kiln for the first time in a room at $30°C$, the kiln's outside wall temperature is $90°C$. Assuming that there is heat loss from the four sides and the top only and that these surfaces have an emissivity of 0.8, determine

a. the total heat loss from the kiln (in W).

b. the total heat loss from the kiln if 4-cm-thick insulation with $k = 0.04$ W/m·K and $\varepsilon = 0.1$ is used (assume the outside brick temperature remains at 90°C).

c. the simple payback time for the insulation if the insulation costs $400, the cost of natural gas is $0.50/$10^5$ kJ, the furnace has an efficiency of 84%, and the furnace operates 2000 h/yr.

P12-66 In oil refineries and chemical processing plants, insulation on pipes is wrapped in thin aluminum metal sheaths to protect the insulation from the weather. After weathering and exposure to harsh air-borne chemicals around the plants, the surface of the metal sheath corrodes and the emissivity is about 0.4. Consider a 30-cm-I.D., 40-cm-O.D. carbon steel pipe ($k = 60$ W/m·K) carrying saturated steam at 350°C covered with 7.5 cm of fiberglass insulation ($k = 0.036$ W/m·K). The steam convective heat transfer coefficient is 600 W/m²·K. With an air temperature of 0°C, determine the heat transfer rate per unit meter of pipe length when

a. a crossflow at 30 km/h is on the outside of the pipe (in W/m).

b. natural convection is on the outside of the pipe (in W/m) (use an approximation based on the results of part a and justify).

P12-67 The power cables for an electric welding rig are suspended above the floor of a factory so that a tripping hazard is not created. The cables are 25 m long; the copper is 10 mm in diameter and is covered by a 2-mm-thick rubberized cover ($k = 0.26$ W/m·K), which is black, has an emissivity of 0.9, and cannot have a temperature greater than 65°C. The cable resistance is $4 \times 10^{-4} \Omega$. If the cable is suspended in calm air at 27°C, determine the maximum allowable current (in A).

P12-68 (WEB) A manufacturer of prefabricated buildings is considering using the same structure (shown in the figure) for walls and roofs for quickly and cheaply assembled buildings. The inner and outer surfaces are 1.27-cm-thick plywood ($k = 0.115$ W/m·K), and the air-filled gap is 10 cm wide; the panels are 2.5 m long and 1.25 m deep out of the plane of the page. For temperatures on the outside of the two plywood sheets of $-10°C$ and 15°C, determine

a. the heat transfer rate for both horizontal and vertical orientations (in W).

b. the effect of inserting a baffle at midheight for when the panel would be used in a vertical orientation (in W).

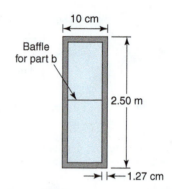

P12-69 (WEB) A thermal pane window is often constructed of two panes of plate glass separated by a short distance; this arrangement provides increased thermal resistance compared to a single pane of glass. Consider a 4-ft-wide and 6-ft-high window whose inside-wall temperature is 75°F and outside-wall temperature is 45°F. It is constructed of glass that is 0.25 in. thick. Determine the heat transfer rate

a. for a single pane of glass (in Btu/h).

b. for two panes separated by a 1-in. air gap (in Btu/h).

c. for two panes of glass if a thin (0.1-in.) sheet of glass is inserted between the other two panes that are still separated by 1 in. (in Btu/h).

P12-70 (WEB) Flat plate solar collectors have their best efficiency if they are tilted toward the sun at an angle that equals the latitude of the location of the collector. Consider a 3-m-wide and 2-m-high solar collector. The solar absorber plate, maintained at 65°C, is separated from the glass cover plate, which is at 30°C, by a distance of 5 cm.

a. Determine the heat loss from the collector if it is horizontal (in W).

b. Determine the heat loss from the collector if it is tilted at an angle of 33° from the horizontal (in W).

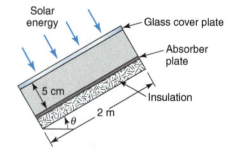

CHAPTER *13*

HEAT EXCHANGERS

13.1 INTRODUCTION

In earlier chapters we analyzed heat transfer between two fluids separated by a solid wall. The temperatures of the hot and cold fluids were constant, and the governing equation was

$$\dot{Q} = \frac{T_H - T_C}{R_{tot}} \tag{13-1}$$

Now we consider the situation where the fluid temperatures are not constant along the separating wall, and Eq. 13-1 is not valid. Different analysis methods must be developed that take into account the varying temperature difference between the two fluids. One special situation—a single-phase fluid in a constant wall temperature tube—was addressed in Section 12.5. Now we will look at more general situations.

The device used to transfer thermal energy from a hotter fluid to a cooler fluid is called a *heat exchanger*. In automobiles, hot liquid coolant from the engine is pumped through small tubes inside the radiator; on the outside of the radiator, cooler air is blown over fins attached to the tubes. In evaporators of vapor-compression air conditioners, boiling refrigerant inside tubes absorbs heat from air passing over the cooling coils; in condensers, condensing refrigerant gives up heat to air passing through the condenser. In the common radiator used to heat older buildings or convectors in newer homes and buildings, either hot water or steam is pumped through the device, and natural convection to air on the outside removes heat from the water or steam. In these heat exchangers, a solid surface separates the two fluids, and all operate in a steady-state mode. Figure 13-1 shows a variety of heat exchanger configurations.

Industrial heat exchangers have a large variety of construction types, but the heat transfer mechanisms can be the same. Shell-and-tube heat exchangers are commonly used in power plants, oil refineries, and chemical processing plants. Plate-and-frame heat exchangers are also used in these applications, as well as in the food-processing industry. Cooling towers are heat exchangers in which direct contact occurs between the hot fluid (water) and the cold fluid (air); both heat and mass transfer occur in this arrangement. These heat exchangers operate in steady state; others operate in a transient mode. For example, in Chapter 8 we discussed the regenerator used in the Brayton cycle. Typically, this heat exchanger is rotary. When a portion of the heat exchanger is in the hot stream, the solid wall of the regenerator absorbs heat. This section then rotates into the cold stream, where the hot wall releases heat to the cold air. Thus each section of the heat exchanger operates in a periodic (transient) mode as it rotates continuously; however, we can view the overall operation as steady.

Each of the heat exchanger types has advantages and disadvantages. The choice of which configuration to use will depend on the application, types of fluids, pressure and temperature levels, modes of heat transfer, pressure-drop restrictions, maintenance and cleaning requirements, cost, size, weight, construction materials needed, and so on. However, the primary consideration is whether or not the chosen heat exchanger will handle the heat transfer rate required in the specific application.

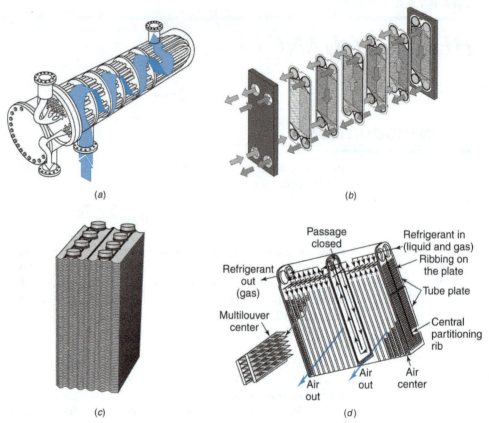

FIGURE 13-1 Several heat exchanger types: (a) shell-and-tube, (b) plate-and-frame, (c) plate-fin, (d) automotive evaporator.
(*Sources:* Adapted from F. Kreith, ed., *The CRC Handbook of Thermal Engineering*, CRC Press, 2000; Used with permission of Alfa Laval; A.P. Fraas, *Heat Exchanger Design*, 2nd Ed. Wiley, New York, 1989. Used with permission.)

The simplest heat exchanger geometry is the *tube-in-tube* (also known as the *concentric tube* and the *double-pipe*) heat exchanger (Figure 13-2a and b). In *counter-flow*, two fluids at different temperatures enter the heat exchanger at opposite ends, and heat is transferred continuously from the hotter to the cooler fluid along the length of the heat exchanger. In *parallel* flow, two fluids enter the heat exchanger at the same end and flow in the same direction. An increase in complexity occurs if *crossflow* is used; that is, the two fluids follow flow paths that are perpendicular to each other (Figure 13-2c). A further increase in complexity occurs if the fluids make multiple *passes* through the heat exchanger; that is, the fluid traverses the length of the heat exchanger several times before exiting (Figure 13-2d).

Figure 13-3 shows examples of how fluid temperatures vary with different heat exchanger configurations and flow conditions. At different locations in the heat exchanger, the temperature differences between the two fluids are different. We must account for this variation in our analysis. The general analysis of all these heat exchanger configurations is the same; the geometry and operating conditions are taken into account when we examine a specific heat exchanger.

Many parameters affect the overall performance of a heat exchanger. Consider the unfinned multitube heat exchanger in Figure 13-4a. For the fluid flowing outside the tubes, *mixing* of the fluid in the direction transverse to the main flow occurs as the fluid proceeds from the entrance to the exit of the heat exchanger. *Mixed flow* reduces temperature variations in the transverse direction. However, the fluid inside one tube does not mix with the fluid in any of the other tubes until the fluid exits all the tubes. This is called *unmixed flow*. To visualize these two processes, consider what happens when food coloring is continuously dropped into a flowing stream of water. As the food coloring moves downstream, the width of the colored water increases perpendicular to the flow direction, and the color becomes less intense; the color intensity is analogous to temperature. Now consider food coloring continuously dropped into water entering one tube in a tube array. Once the colored water fills the single tube, that water does not mix with water in the other tubes. For the finned tube heat exchanger in Figure 13-4b, unmixed flow is present on both sides of the heat exchanger; the fins prevent mixing in the same manner as flow inside a tube.

While numerous configurations, heat transfer mechanisms, and modes of operation are used, the most common type of heat exchanger operates at steady state and has a solid surface separating the two fluid streams. The analysis of these types of heat exchangers involves the use of conservation of energy, the heat transfer rate equation, and the evaluation of the total thermal resistance between the two fluid streams. The objective of the analysis is either (1) to determine the heat transfer rate (or *heat duty*) possible with a given heat

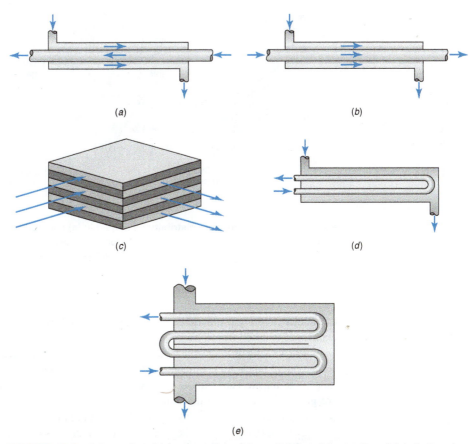

FIGURE 13-2 Schematics of (a) counterflow, (b) parallel flow, (c) crossflow, (d) 1 shell and 2 tube passes, and (e) 2 shell and 4 tube passes.

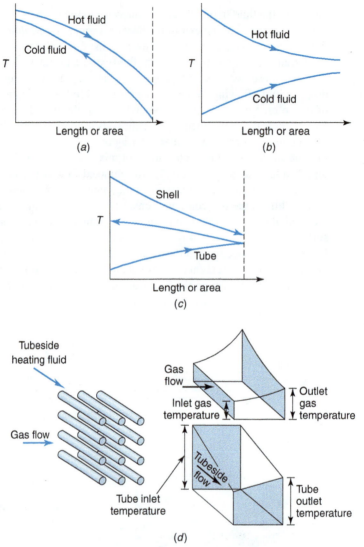

FIGURE 13-3 Temperature distributions of fluid in (a) counterflow, (b) parallel flow, (c) 1 shell pass and 2 tubes passes, and (d) crossflow heat exchanger.
(*Source for d:* G. Walker, *Industrial Heat Exchangers: A Basic Guide,* 2nd ed., Hemisphere Publishing, New York, 1990. Used with permission.)

exchanger (a *rating* problem) or (2) to develop the information needed to build a new heat exchanger that will transfer a given heat transfer rate (a *design* or *sizing* problem).

For a *rating problem,* the heat exchanger exists. The geometry (number, size, spacing, and layout of tubes, fin geometry, shell geometry, etc.) and heat exchanger type (shell-and-tube, plate-fin, etc.) are known, and the two fluids, the flow rates, and the inlet temperatures are given. The task is to determine the overall heat transfer rate. (This is equivalent to determining the outlet temperatures of the two fluids.)

For a *design problem,* the type of heat exchanger is known, along with the two fluids and their flow rates. In addition, the inlet temperatures of the two fluids and the required heat duty are known. (This is equivalent to specifying the outlet temperatures of the two

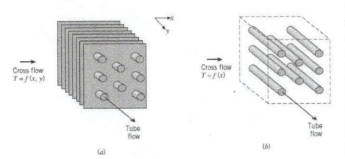

fluids.) The task is to determine the heat exchanger size or area needed to provide a specified heat transfer rate.

Two analyses of heat transfer in a heat exchanger are commonly used: the *log mean temperature difference (LMTD)* method and the *effectiveness-NTU (ε-NTU)* method. While either approach can be used for either a design- or a rating-type problem, typically the *LMTD* method is used for design problems and the *ε-NTU* approach is used for rating. These two approaches are discussed below in Sections 13.3 and 13.4, respectively. In both analysis methods, the *overall heat transfer coefficient* is needed; this quantity was discussed briefly in Chapter 12 and is discussed in more detail in the next section.

13.2 THE OVERALL HEAT TRANSFER COEFFICIENT

The total thermal resistance between the two fluids in a heat exchanger depends on resistances due to convective heat transfer of both fluids, conduction through the solid wall, and (possibly) conduction through additional material called *fouling* (e.g., dirt, crud, algae, mineral deposits, corrosion, etc.) that can adhere to the wall surfaces. For a particular application, some or all of these resistances may be negligible. In general, there are five contributors to the overall thermal resistance: two convective resistances, two fouling resistances, and a wall resistance, as shown in Eq. 13-2 and illustrated in Figure 13-5.

$$R_{tot} = R_{conv,i} + R_{fouling,i} + R_w + R_{fouling,o} + R_{conv,o}$$

$$= \frac{1}{\eta_{o,i} h_i A_i} + \frac{R_i''}{\eta_{o,i} A_i} + R_w + \frac{R_o''}{\eta_{o,o} A_o} + \frac{1}{\eta_{o,o} h_o A_o} \tag{13-2}$$

Because fins could be used, we have included the overall fin efficiencies $\eta_{o,o}$ and $\eta_{o,i}$ for fins on the exterior and interior surfaces, respectively. When no fins are present, these efficiencies become unity.

An *overall heat transfer coefficient, U,* is typically used to describe the total thermal resistance in a heat exchanger and is defined in Eq. 13-3:

$$R_{tot} = \frac{1}{UA} = \frac{1}{\eta_{o,i} h_i A_i} + \frac{R_i''}{\eta_{o,i} A_i} + R_w + \frac{R_o''}{\eta_{o,o} A_o} + \frac{1}{\eta_{o,o} h_o A_o} \tag{13-3}$$

The area used in the UA product can be either the inside area, A_i, or the outside area, A_o, of the tubes or channels in a heat exchanger. However, the product $UA = U_i A_i = U_o A_o$ is a constant because a heat exchanger has only one total thermal resistance:

$$R_{tot} = \frac{1}{UA} = \frac{1}{U_i A_i} = \frac{1}{U_o A_o} \tag{13-4}$$

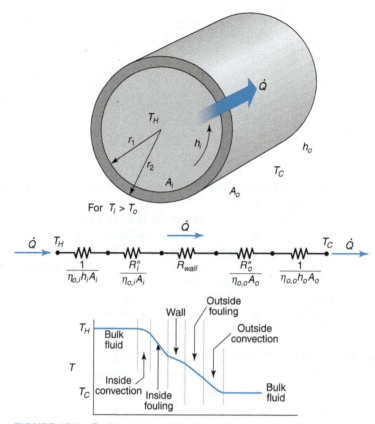

FIGURE 13-5 Resistances contributing to the overall heat transfer coefficient.

The value of U generally changes depending on whether the inside area is used to define it (U_i) or the outside area is used (U_o). For example, consider a circular tube with fins on its outside; the outside heat transfer area is significantly different from that on the inside of the tube, so U_i would be different from U_o. The magnitude of U can be determined by evaluating each term in Eq. 13-3 using information about the fluids, flow rates, and geometry of a particular heat exchanger. The convective resistances, $R_{conv} = 1/\eta_o hA$, are calculated using correlations for the convective heat transfer coefficients, h, found, for example, in Chapter 12. The overall surface efficiency, η_o, is unity if the surface has no fins or is evaluated with information about the efficiency of fins/extended surfaces given in Chapter 11.

The wall resistance, R_w, is due to conduction through the solid wall; for example,

$$R_w = \frac{\ln(r_o/r_i)}{2\pi kL} \qquad \text{circular tube}$$

$$R_w = \frac{\Delta x}{kA} \qquad \text{plane wall}$$

In the wall thermal resistance, the *wall* thermal conductivity (not the fluid thermal conductivity) is used.

Fouling thermal resistance, $R_{fouling} = R''/\eta_o A$, is present whenever an unwanted substance coats a heat transfer surface. The effects of fouling are usually expressed in terms of a *fouling factor*, R'', which has units of $m^2 \cdot K/W$ or $h \cdot ft^2 \cdot {}^\circ F/Btu$. Typical fouling materials are mineral deposits, corrosion products, dirt, biological growths, deposits caused by chemical reactions in the fluid, and sedimentation. In addition, if the fouling thickness becomes great enough, it can increase fluid flow resistance (pressure drop). Other consequences of fouling include oversized or redundant equipment to accommodate the increased thermal resistance; use of special materials or construction to minimize the effects of fouling; and increased cleaning requirements of heat exchangers with their attendant increased downtime and loss of production.

Fouling factors (see Table 13-1) must be used with caution because the degree of fouling in a heat exchanger is often unknown. Fouling occurs over time, often at an uneven rate. By analogy, consider an initially clean window in a dusty environment. Over time the glass will become dirty as the thickness of the material fogging the glass increases, thus decreasing the amount of light that comes through the window. In an analogous manner, a heat exchanger surface can also be coated with an unwanted layer of material. Hence, the thermal and hydraulic performance of a heat exchanger can decrease with time. If fouling is taken into account when a heat exchanger is designed, then the heat exchanger will be "oversized" when it is first put into service (clean state) and may be "undersized" after a long period of time (fouled state) before it is cleaned. Parameters affecting fouling include type of fluid, flow velocity, temperature of the surface, particle concentration in the fluid, and surface conditions. Therefore, fouling factors should be considered to be estimates with large uncertainties, and a designer must consider the effect of the factor on a final design.

Occasionally, one of the resistances in Eq. 13-2 may be significantly larger than the others, and that resistance is the *dominant* or *controlling* resistance. This might occur with a gas flow (low heat transfer coefficient) on one side of a thin-walled heat exchanger and boiling (high heat transfer coefficient) on the other side; if the areas on the two sides were comparable with negligible wall conduction resistance, then the gas-side thermal resistance would dominate the overall thermal resistance. When designing a heat exchanger with a dominant resistance, a better estimate of the area can be obtained if more effort is expended to quantify the dominant resistance rather than focusing on an estimate of a minor resistance.

Finally, the relative magnitudes of the thermal resistances dictate the temperature of the wall, which is needed, for example, if a thermal stress calculation is to be done, if a

TABLE 13-1   Typical Fouling Factors

Type of Fluid	Fouling Factor, R'', $m^2 \cdot K/W$
Water	
Seawater	0.000275–0.00035
Treated cooling tower water	0.000175–0.00035
River water	0.00035–0.00053
Treated boiler feedwater	0.00009
Liquids	
No. 6 fuel oil	0.0009
Engine lube oil	0.000175
Refrigerants	0.000175
Ethylene glycol solutions	0.00035
Kerosene	0.00035–0.00053
Heavy fuel oil	0.00053–0.00123
Gas or Vapor	
Steam (non-oil-bearing)	0.0009
Exhaust steam (oil-bearing)	0.00026–0.00035
Compressed air	0.000175
Natural gas flue gas	0.0009

temperature sensitive fluid is to be heated/cooled, if an estimate is needed about where fouling might preferentially occur, or if a condensing or a boiling heat transfer coefficient must be calculated. Assuming negligible wall resistance, when the hot-side thermal resistance is low, the wall temperature will be close to the hot fluid temperature, and with a low cold-side thermal resistance, wall temperature tracks the cold fluid temperature. If the two resistances are about the same, the wall temperature will be about the average of the two fluid temperatures.

EXAMPLE 13-1 Evaluation of overall heat transfer coefficient

The inner tube of a 10-ft-long double-pipe heat exchanger has an inner diameter of $D_i = 1$ in., an outer diameter of $D_{o,i} = 1.5$ in., and is made of brass. The outer tube has an inner diameter of $D_{i,o} = 2$ in. Water enters the annulus at 70°F at a flow rate of 3.5 lbm/s. Hydraulic fluid enters the inner tube at 100°F at a flow rate of 0.5 lbm/s.

a) Determine the overall heat transfer coefficient based on the inside surface area (in Btu/h·ft²·°F).

b) Determine the fraction that each individual resistance contributes to the overall resistance.

Approach:

A schematic of the double pipe heat exchanger is shown here.

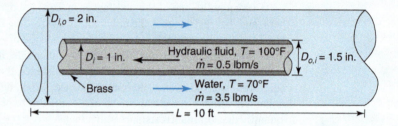

The overall heat transfer coefficient can be determined using Eq. 13-3 and Eq. 13-4. The two convective heat transfer coefficients must be calculated with the given information and appropriate convective heat transfer coefficient correlations. Fouling resistance will be ignored since no information is given. The wall resistance can be calculated for a cylindrical tube once the thermal conductivity of the metal is found.

Assumptions:

Solution:

a) To evaluate the overall heat transfer coefficient based on the inside area of this thin-walled tube, we begin with

$$\frac{1}{U_i A_i} = \frac{1}{\eta_{o,i} h_i A_i} + \frac{R_i''}{\eta_{o,i} A_i} + \frac{\ln\left(r_o/r_i\right)}{2\pi k L} + \frac{R_o''}{\eta_{o,o} A_o} + \frac{1}{\eta_{o,o} h_o A_o}$$

Multiplying through by A_i,

$$\frac{1}{U_i} = \frac{A_i}{\eta_{o,i} h_i A_i} + \frac{A_i R_i''}{\eta_{o,i} A_i} + \frac{A_i \ln\left(r_o/r_i\right)}{2\pi k L} + \frac{A_i R_o''}{\eta_{o,o} A_o} + \frac{A_i}{\eta_{o,o} h_o A_o}$$

A1. Ignore fouling.

With no fins ($\eta_{o,i} = \eta_{o,o} = 1$), ignoring the fouling resistances [A1], and defining $A_i = \pi D_i L$ and $A_o = \pi D_o L$, where L is the tube length, we obtain

$$\frac{1}{U_i} = \frac{1}{h_i} + \frac{D_i \ln\left(r_o/r_i\right)}{2k} + \frac{D_i}{h_o D_{o,i}}$$

The tube length L cancels out of every term.

To evaluate the heat transfer coefficient for the hydraulic fluid, its Reynolds number ($Re = \rho \mathcal{V} D_i / \mu$) is needed. Using an expression for mass flow rate ($\dot{m} = \rho \mathcal{V} A = \rho \mathcal{V} (\pi D_i^2 / 4) \rightarrow \mathcal{V} = 4\dot{m} / \rho \pi D_i^2$), we rewrite the Reynolds number as $Re = 4\dot{m} / \pi \mu D_i$. We should evaluate the hydraulic fluid properties at its average temperature. However, we do not have the outlet temperature, so we will evaluate the properties at its inlet temperature, 100°F: $\mu = 0.00556$ lbm/ft·s, $c_p = 0.467$ Btu/lbm·°F, $k = 0.069$ Btu/h·ft·°F, $Pr = 136$. Therefore,

$$Re = \frac{4\dot{m}}{\pi \mu D_i} = \frac{4\,(0.5\ \text{lbm/s})}{\pi\,(0.00556\ \text{lbm/ft·s})\,(1/12\ \text{ft})} = 1374$$

Because this is laminar flow, we check whether entrance effects are important:

$$L_{ent,t,lam} \approx 0.05 Re Pr D_i = 0.05\,(1374)\,(136)\left(\frac{1}{12}\ \text{ft}\right) = 779\ \text{ft}$$

This is much longer than the 10-ft-long tubes, so entrance effects must be included. An appropriate equation for laminar flow with entrance effects (see Eq. 12-42) is

$$Nu = 1.86 \left(\frac{Re Pr D}{L}\right)^{1/3} \left(\frac{\mu_b}{\mu_w}\right)^{0.14}$$

A2. Assume the wall temperature is close to the water inlet temperature.

The Prandtl and Reynolds numbers fall within the applicable range for this correlation. We have the viscosity at the bulk temperature. For the viscosity at the wall temperature, we estimate the wall temperature as being close to the water inlet temperature, 70°F [A2], so by interpolation in Table B-6, $\mu_w = 0.00903$ lbm/ft·s. Therefore,

$$Nu = 1.86 \left[\frac{1374\,(136)\,(1\ \text{in.})}{(10\ \text{ft})\,(12\ \text{in.}/1\ \text{ft})}\right]^{1/3} \left(\frac{0.00556}{0.00903}\right)^{0.14} = 20.1$$

$$h_i = \frac{Nu\,k}{D_i} = \frac{20.1\,(0.069\ \text{Btu/h·ft·°F})}{1/12\ \text{ft}} = 16.6\ \frac{\text{Btu}}{\text{h·ft}^2\text{·°F}}$$

For the water-side heat transfer coefficient, we begin with its Reynolds number, $Re = \rho \mathcal{V} D_{h,o} / \mu$. We must calculate the annulus hydraulic diameter:

$$D_{h,o} = \frac{4A_x}{P_{wetted}} = \frac{4\left[\pi D_{i,o}^2/4 - \pi D_{o,i}^2/4\right]}{\pi D_{i,o} + \pi D_{o,i}} = D_{i,o} - D_{o,i} = 2\ \text{in.} - 1.5\ \text{in.} = 0.5\ \text{in.}$$

The water properties are evaluated with Table B-6 at the water inlet temperature, 70°F: $\mu = 0.000658$ lbm/ft·s, $c_p = 0.998$ Btu/lbm·°F, $k = 0.347$ Btu/h·ft·°F, $\rho = 62.2$ lbm/ft^3, $Pr = 6.82$. The water velocity is determined from

$$\dot{m} = \rho \mathcal{V} A \qquad \text{or} \qquad \mathcal{V} = \frac{\dot{m}}{\rho A}$$

$$\mathcal{V} = \frac{\dot{m}}{\rho\left[\pi D_{i,o}^2/4 - \pi D_{o,i}^2/4\right]} = \frac{3.5\ \dfrac{\text{lbm}}{\text{s}}}{62.2\ \dfrac{\text{lbm}}{\text{ft}^3}\left[\pi\left(\dfrac{2}{12}\ \text{ft}\right)^2/4 - \pi\left(\dfrac{1.5}{12}\ \text{ft}\right)^2/4\right]} = 5.9\ \frac{\text{ft}}{\text{s}}$$

$$Re = \frac{\rho \mathcal{V} D_{h,o}}{\mu} = \frac{\left(62.2\ \dfrac{\text{lbm}}{\text{ft}^3}\right)\left(5.9\ \dfrac{\text{ft}}{\text{s}}\right)\left(\dfrac{0.5}{12}\ \text{ft}\right)}{0.000658\ \dfrac{\text{lbm}}{\text{ft·s}}} = 23,238$$

This is turbulent flow. The Dittus-Boelter equation is applicable for these Reynolds and Prandtl numbers. Because the water is heated, the Prandtl number exponent is 0.4:

$$Nu = 0.023 Re^{0.8} Pr^{0.4} = 0.023 (23,238)^{0.8} (6.82)^{0.4} = 154$$

$$h_o = \frac{Nu\,k}{D_{h,o}} = \frac{154\,(0.347\,\text{Btu/h·ft·°F})}{0.5/12\,\text{ft}} = 1283\,\frac{\text{Btu}}{\text{h·ft}^2\text{·°F}}$$

To evaluate the overall heat transfer coefficient, we need the thermal conductivity of the brass wall, which is 63.6 Btu/h·ft·°F from Table B-2, so that

$$\frac{1}{U_i} = \frac{1}{h_i} + \frac{D_i \ln\left(r_o/r_i\right)}{2k} + \frac{D_i}{h_o D_{o,i}}$$

$$= \frac{1}{16.6\,\dfrac{\text{Btu}}{\text{h·ft}^2\text{·°F}}} + \frac{\left(\frac{1}{12}\,\text{ft}\right)\ln\left(1.5/1\right)}{2\,(63.6\,\text{Btu/h·ft·°F})} + \frac{1\,\text{in.}}{\left(1283\,\dfrac{\text{Btu}}{\text{h·ft}^2\text{·°F}}\right)(1.5\,\text{in.})}$$

$$= 0.0602\,\frac{\text{h·ft}^2\text{·°F}}{\text{Btu}} + 0.000266\,\frac{\text{h·ft}^2\text{·°F}}{\text{Btu}} + 0.000520\,\frac{\text{h·ft}^2\text{·°F}}{\text{Btu}} = 0.0610\,\frac{\text{h·ft}^2\text{·°F}}{\text{Btu}}$$

or

$$U_i = 16.4\,\frac{\text{Btu}}{\text{h·ft}^2\text{·°F}}$$

b) The fractions that each individual resistance contributes to the overall resistance are

$$\frac{0.0602}{0.0610} = 0.987 \qquad \frac{0.000266}{0.0610} = 0.00435 \qquad \frac{0.000520}{0.0610} = 0.00852$$

Comments:

The oil-side thermal resistance dominates the total resistance. Ignoring the water convective resistance and the wall conduction resistance would change the final number by only about 1%. Also, because the water and wall resistances are so small, the wall temperature will track the water temperature, as assumed. If we had assumed fully developed flow with a constant wall heat flux boundary condition, $Nu = 4.36$, $h_i = 3.6$ Btu/h·ft²·°F, and $U_i = 2.96$ Btu/h·ft²·°F—which is only 18% of the correct value. This example illustrates the need to carefully evaluate the heat transfer coefficients.

13.3 THE LMTD METHOD

The governing equation for the *log mean temperature difference* (*LMTD*) heat exchanger thermal analysis method is the heat transfer rate equation:

$$\dot{Q} = \frac{\Delta T_{mean}}{R_{tot}} \tag{13-5}$$

In Figure 13-3, the temperature distributions of the two fluid streams are shown for a variety of heat exchangers. What is the appropriate temperature difference to use in Eq. 13-5 for these situations? That is, what true or effective mean temperature difference (ΔT_{mean}) should be used that is consistent with the total resistance? In this section we develop a relationship between the effective mean temperature difference and the heat exchanger configuration and operating conditions.

Shown in Figure 13-6 are a counterflow heat exchanger and the hot and cold fluid temperatures. The appropriate temperature difference to use in Eq. 13-5 is obtained by applying conservation of energy and the heat transfer rate equation to the differential segment shown in the figure. We define the positive x direction from the left end to the right end of the heat exchanger. We assume the following:

1. Steady state exists.

2. There are constant specific heats if the flow is single phase. If there is phase change (boiling or condensation), it occurs at a constant temperature (constant pressure).

3. The constant overall heat transfer coefficient applies over the complete heat exchanger.

4. If there are multiple tubes, each tube has the same flow rate. Likewise, the flow outside the tubes is evenly distributed across the heat exchanger.

5. Temperatures and velocities are uniform over all cross-sectional flow areas.

6. The two fluids exchange heat only with each other, and there is no shaft work or heat generation. Potential and kinetic energy effects are ignored.

7. Axial conduction along the solid surfaces is ignored.

At the differential element in the heat exchanger, the rate equation gives

$$\delta \dot{Q} = U(T_H - T_C)\, dA \tag{13-6}$$

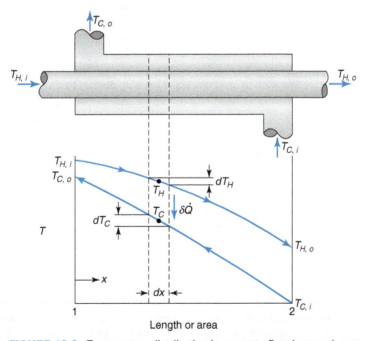

FIGURE 13-6   Temperature distribution in a counterflow heat exchanger.

where $\delta\dot{Q}$ is the differential heat transfer rate in the differential area dA, which stretches from x to $x + dx$. The temperature difference $T_H - T_C$ is the *local* temperature difference, which varies all along the heat exchanger, and U is the overall heat transfer coefficient. Eq. 13-6 must be integrated over the heat exchanger area to obtain the total heat transfer rate. To do this, we must express each variable in an appropriate manner.

Conservation of energy applied to the hot fluid gives

$$\dot{Q}_H = \dot{m}_H \left(h_{H,o} - h_{H,i} \right) = \dot{m}_H c_{p,H} \left(T_{H,o} - T_{H,i} \right) = C_H (T_{H,o} - T_{H,i}) \qquad (13\text{-}7)$$

where we define $C = \dot{m}c_p$, the *heat capacity rate*, and h signifies enthalpy rather than the heat transfer coefficient in this equation. (The two quantities h—enthalpy and heat transfer coefficient—should never be confused because the context in which they are used is always different.) Likewise, for the cold fluid

$$\dot{Q}_C = \dot{m}_C \left(h_{C,o} - h_{C,i} \right) = \dot{m}_C c_{p,C} \left(T_{C,o} - T_{C,i} \right) = C_C (T_{C,o} - T_{C,i}) \qquad (13\text{-}8)$$

From conservation of energy, $\dot{Q}_H = -\dot{Q}_C$.

We want the energy balance for a differential segment, and we use differential heat transfer rates and temperature changes of the fluids. Both the hot and cold fluid temperatures decrease in the positive x-direction. Thus, by analogy to Eq. 13-7, the differential heat transfer rate between x and $x + dx$ are:

$$\delta\dot{Q}_H = -\dot{m}_H c_{p,H}\, dT_H = -C_H\, dT_H \qquad (13\text{-}9)$$

$$\delta\dot{Q}_C = -\dot{m}_C c_{p,C}\, dT_C = -C_C\, dT_C \qquad (13\text{-}10)$$

To drop the subscripts C and H on the heat transfer rate, we let $\delta\dot{Q} = \delta\dot{Q}_H = \delta\dot{Q}_C$, so that:

$$\delta\dot{Q} = -C_H\, dT_H = -C_C\, dT_C \qquad (13\text{-}11)$$

We want to integrate Eq. 13-6 with respect to $T_H - T_C$, so we need

$$d\left(T_H - T_C \right) = dT_H - dT_C \qquad (13\text{-}12)$$

From Eq. 13-11 we obtain

$$dT_H = \frac{-\delta\dot{Q}}{C_H} \qquad \text{and} \qquad dT_C = \frac{-\delta\dot{Q}}{C_C}$$

Incorporate these into Eq. 13-12 to obtain

$$d\left(T_H - T_C \right) = \frac{-\delta\dot{Q}}{C_H} + \frac{\delta\dot{Q}}{C_C} = -\delta\dot{Q}\left(\frac{1}{C_H} - \frac{1}{C_C} \right) \qquad (13\text{-}13)$$

Solve Eq. 13-13 for $\delta\dot{Q}$ and substitute into Eq. 13-6. We designate a subscript 1 to indicate the fluids entering and leaving on the left-hand side of the exchanger and a subscript 2 to indicate those fluids entering and leaving on the right-hand side. Rearranging the equation, we obtain an expression that is integrated over the heat exchanger length from end 1 to end 2 (see Figure 13-6).

$$\int_1^2 \frac{d\,(T_H - T_C)}{T_H - T_C} = -\left(\frac{1}{C_H} - \frac{1}{C_C}\right) U \int_1^2 dA \tag{13-14}$$

Performing the integration:

$$\ln\left(\frac{\Delta T_2}{\Delta T_1}\right) = -\left(\frac{1}{C_H} - \frac{1}{C_C}\right) UA \tag{13-15}$$

where for this counterflow heat exchanger

$$
\begin{aligned}
\Delta T_1 &= T_{H,1} - T_{C,1} = T_{H,i} - T_{C,o} \\
\Delta T_2 &= T_{H,2} - T_{C,2} = T_{H,o} - T_{C,i}
\end{aligned}
\qquad \text{counterflow} \tag{13-16}
$$

We solve Eq. 13-7 for the hot heat capacity rate $C_H = (T_{H,o} - T_{H,i})/\dot{Q}_H$ and Eq. 13-8 for the cold heat capacity rate $C_C = (T_{C,o} - T_{C,i})/\dot{Q}_C$; we again use $\dot{Q}_H = -\dot{Q}_C = \dot{Q}$. Substitute these three expressions into Eq. 13-15 to obtain

$$\ln\left(\frac{\Delta T_2}{\Delta T_1}\right) = \ln\left(\frac{T_{H,2} - T_{C,2}}{T_{H,1} - T_{C,1}}\right) = -UA\left[\left(\frac{T_{H,2} - T_{H,1}}{\dot{Q}}\right) - \left(\frac{T_{C,2} - T_{C,1}}{\dot{Q}}\right)\right] \tag{13-17}$$

Finally, Eq. 13-17 is solved for $\dot{Q}$:

$$\dot{Q} = UA\frac{\Delta T_1 - \Delta T_2}{\ln\left(\frac{\Delta T_1}{\Delta T_2}\right)} = UA\frac{\Delta T_2 - \Delta T_1}{\ln\left(\frac{\Delta T_2}{\Delta T_1}\right)} = UA\Delta T_{LM} \tag{13-18}$$

where the quantity ΔT_{LM} is called the *log mean temperature difference* or *LMTD* and is defined as:

$$\Delta T_{LM} = \frac{\Delta T_1 - \Delta T_2}{\ln\left(\frac{\Delta T_1}{\Delta T_2}\right)} = \frac{\Delta T_2 - \Delta T_1}{\ln\left(\frac{\Delta T_2}{\Delta T_1}\right)} \tag{13-19}$$

A similar analysis can be applied to a *parallel flow heat exchanger*. An expression identical to Eq. 13-19 is obtained. The only difference is in the evaluation of the temperature differences at the two ends of the heat exchanger. As shown on Figure 13-3b, if we define the left end of the heat exchanger as 1 and the right end as 2, then for the parallel flow heat exchanger the temperature differences are:

$$
\begin{aligned}
\Delta T_1 &= T_{H,1} - T_{C,1} = T_{H,i} - T_{C,i} \\
\Delta T_2 &= T_{H,2} - T_{C,2} = T_{H,o} - T_{C,o}
\end{aligned}
\qquad \text{parallel flow} \tag{13-20}
$$

The choice of end 1 and end 2 is arbitrary and, as shown in Eq. 13-19, has no effect on the ΔT_{LM}.

Figure 13-3 shows the temperature profiles of other heat exchanger configurations involving crossflow and/or multiple passes of one or both fluids. For these situations and others, expressions for the appropriate mean temperature difference can be determined. The results of these analyses are presented in terms of a correction factor, F, to the *LMTD* calculated as if the flow arrangement is counterflow. That is,

$$F = \frac{\Delta T_{mean}}{\Delta T_{LM,cf}} \quad \text{and} \quad \dot{Q} = UAF \, \Delta T_{LM,cf} \tag{13-21}$$

The log mean temperature difference for counterflow $\left(\Delta T_{LM,cf}\right)$ is used as the reference temperature difference, because a counterflow heat exchanger provides the greatest mean temperature difference between two fluids with specified inlet and outlet temperatures. All other heat exchangers have a ΔT_{mean} smaller than that of a counterflow heat exchanger. Hence, for a given heat duty and overall heat transfer coefficient, the counterflow heat exchanger will require the smallest surface area. Note that for a counterflow heat exchanger, $F = 1$, and the *LMTD* is calculated with Eq. 13-19 with the ΔT's given in Eq. 13-16. For a parallel flow heat exchanger, $F = 1$ and the *LMTD* also is calculated with Eq. 13-19, but with the ΔT's given by Eq. 13-20. For all other heat exchangers, $F < 1$.

The correction factor F depends on three pieces of information:

$$F = \text{func}(P, R, \text{heat exchanger geometry and flow arrangement})$$

where P, a relative measure of the tube-side temperature change compared to the inlet temperature difference, is defined by

$$P = \frac{T_{tube,o} - T_{tube,i}}{T_{shell,i} - T_{tube,i}} \tag{13-22}$$

The quantity R is a heat capacity ratio, expressed as the tube-side fluid heat capacity rate divided by the shell-side heat capacity rate. It is related to the temperatures through the use of conservation of energy:

$$R = \frac{C_{tube}}{C_{shell}} = \frac{T_{shell,i} - T_{shell,o}}{T_{tube,o} - T_{tube,i}} \tag{13-23}$$

Figure 13-7 shows curves of F for several common heat exchanger configurations.

Consider the shape of the curves in Figure 13-7 when R becomes very small, that is, when $R \to 0$. In this situation for all heat exchanger types, flow arrangements, and values of P, the value of $F \to 1$. This can occur in two situations:

1. When there is a phase change (boiling or condensing) on one side of a heat exchanger, the enthalpy changes even though the temperature does not change. The definition of specific heat is $c_p = \partial h / \partial T|_p$, where h is enthalpy. Hence, $c_p \to \infty$ during a constant-pressure phase change, and the heat capacity rate $(\dot{m}c_p)$ goes to infinity. This causes $R = C_{shell}/C_{tube} \to 0$.

2. When the heat capacity rate on one side of a heat exchanger is very large relative to the other side, perhaps due to a large difference in mass flow rates, the heat capacity ratio again approaches zero, $R \to 0$.

In both of these situations, the temperature of one of the fluids remains constant and the heat exchanger configuration becomes irrelevant in determining the value of F.

Table 13-2 describes the steps needed to design or rate a heat exchanger using the *LMTD* method. Design is a straightforward calculation using the *LMTD* method. Rating requires an iterative solution.

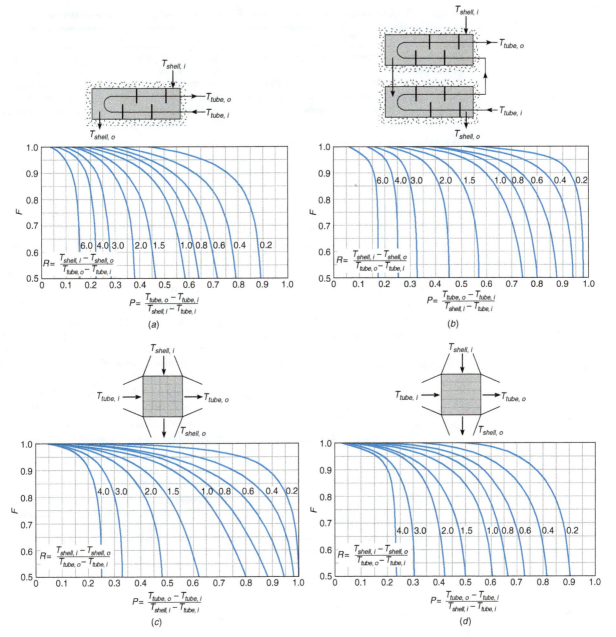

FIGURE 13-7 *F* correction factors for several heat exchanger types. (a) Shell-and-tube with one shell pass and any multiple of two tube passes (two, four, etc. tubes passes). (b) Shell-and-tube with two shell passes and any multiple of four tube passes (four, eight, etc. tube passes). (c) Single-pass, crossflow with both fluids unmixed. (d) Single-pass, crossflow with one fluid mixed and the other unmixed.

(*Source*: F. P. Incropera and D. P. DeWitt, *Introduction to Heat Transfer*, 3rd ed., Wiley New York, 1996. Used with permission.)

TABLE 13-2 Designing or Rating a Heat Exchanger using the *LMTD* Method

Design	Rating
Known information: the type of heat exchanger and basic configuration (e.g., diameter and wall thickness of tubes); the two fluids and their flow rates; the inlet temperatures of the two fluids; the required heat duty or the two outlet temperatures	**Known information:** the geometry (number, size, spacing, and layout of tubes, fin geometry, shell geometry, etc.) and type of heat exchanger (shell-and-tube, plate-fin, fin-tube, etc.); the two fluids and their flow rates; the two inlet temperatures
Objective: determine the area needed	**Objective:** determine the overall heat transfer rate or the two outlet temperatures

Steps to follow	**Steps to follow**
1. Calculate the heat transfer coefficients on each side of the heat exchanger using the given geometry, fluid, and flow rates.	**1.** Calculate the heat transfer coefficients on each side of the heat exchanger using the given geometry, fluid, and flow rates.
2. Calculate wall resistance and estimate fouling resistances if required.	**2.** Calculate wall resistance and estimate fouling resistances if required.
3. Calculate the overall heat transfer coefficient using Eq. 13-3.	**3.** Calculate the overall heat transfer coefficient using Eq. 13-3.
4. Calculate R and P.	**4.** Calculate R.
5. Evaluate F for the heat exchanger geometry using an appropriate figure or equation.	**5.** Assume a value of one of the exit temperatures, calculate the other exit temperature using Eq. 13-7 or Eq. 13-8, and calculate P, or assume a value of P and calculate the exit temperatures.
6. Use the given heat duty, or calculate the heat duty using Eq. 13-7 or Eq. 13-8.	**6.** Evaluate F for the heat exchanger geometry using an appropriate figure or equation.
7. Calculate $\Delta T_{LM,cf}$ using Eq. 13-16 and Eq. 13-19.	**7.** Calculate $\Delta T_{LM,cf}$ using Eq. 13-19.
8. Calculate area using Eq. 13-21, $$A = \dot{Q}/UF\Delta T_{LM,cf}$$	**8.** Calculate the heat duty using Eq. 13-21.
	9. Calculate outlet temperatures using Eq. 13-7 and Eq. 13-8, and compare to those assumed in step 5.
	10. Repeat steps 5 through 9 until solution converges. (Use the temperatures calculated in step 9 as the next assumed temperature.)

EXAMPLE 13-2 *LMTD* design problem

We must design a heat exchanger to heat 2.5 kg/s of water from 15°C to 85°C using hot oil ($c_p = 2.35$ kJ/kg·K), which enters the shell-side of the heat exchanger at 160°C. The oil must leave the heat exchanger at 100°C. From previous experience, we decide to use a one-shell-pass, four-tube-pass heat exchanger that contains 20 thin-walled 25-mm tubes. The oil heat transfer coefficient is 400 W/m²·K. Determine the length of the shell required to accomplish the desired heating.

Approach:

The given information is shown on the schematic below.

We want to determine the required length of the shell of this heat exchanger, which is one-fourth the tube length since we have four tube passes. The water flow rate and temperature change are given, which is sufficient information to evaluate the heat transfer rate. Because we know the number of tubes and their diameters, determination of the tube length is equivalent to determining the heat exchange surface area. Hence, this is a design problem, and the *LMTD* method is the preferred approach.

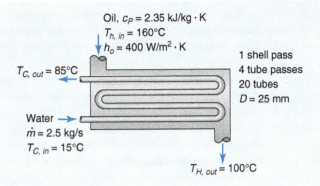

Oil, $c_P = 2.35$ kJ/kg·K
$T_{h,\,in} = 160°C$
$h_o = 400$ W/m²·K

1 shell pass
4 tube passes
20 tubes
$D = 25$ mm

$T_{C,\,out} = 85°C$

Water →
$\dot{m} = 2.5$ kg/s
$T_{C,\,in} = 15°C$

$T_{H,\,out} = 100°C$

Assumptions:

A1. Ignore wall thermal resistance.

A2. Ignore fouling.

A3. The system is steady.

A4. Potential and kinetic energy effects are negligible.

A5. No work occurs in the heat exchanger.

A6. Water is incompressible.

A7. Specific heat is constant.

Solution:

We begin by writing the governing rate equation:

$$\dot{Q} = UAF\,\Delta T_{LM,\,cf}$$

The tubes are thin (no dimension is given), which implies that we can ignore wall thermal resistance [A1], so that $U_i = U_o$. We ignore fouling since no information is given [A2]. The surface area is $A = N_t \pi D L_t = N_t \pi D N_p L_p$, where N_t is the number of tubes, L_t is the tube length, N_p is the number of tube passes, and L_p is the tube length per pass or the shell length. Combining these expressions and solving for the length of each tube pass, we obtain

$$L_p = \frac{\dot{Q}}{UF\,\Delta T_{LM,\,cf} N_t \pi D N_p}$$

The heat transfer rate is found with conservation of energy on the water. Assuming [A3], [A4], and [A5], the energy balance gives us

$$\dot{Q} = \dot{m}\left(h_{C,\,out} - h_{C,\,in}\right)$$

We assume the water is an ideal liquid [A6] with a constant specific heat [A7], so that $\Delta h = c_p\,\Delta T$. The water enters at 15°C and leaves at 85°C, and we evaluate all the properties at the average temperature $(15 + 85)/2 = 50°C$, so from Table A-6 we obtain $c_p = 4.18$ kJ/kg·K and the heat transfer rate is

$$\dot{Q} = \dot{m}\left(h_{C,out} - h_{C,in}\right) = \dot{m}c_p\left(T_{C,\,out} - T_{C,\,in}\right) = \left(2.5\,\frac{\text{kg}}{\text{s}}\right)\left(4.18\,\frac{\text{kJ}}{\text{kg·K}}\right)(85 - 15)\,\text{K} = 731.5\,\text{kW}$$

The overall heat transfer coefficient is determined from

$$\frac{1}{UA} = \frac{1}{U_i A_i} = \frac{1}{U_o A_o} = \frac{1}{\eta_{o,i} h_i A_i} + \frac{R_i''}{\eta_{o,i} A_i} + R_w + \frac{R_o''}{\eta_{o,o} A_o} + \frac{1}{\eta_{o,o} h_o A_o}$$

Because the tube wall is thin, $A_i = A_o$. There are no fins, so $\eta_{o,i} = \eta_{o,o} = 1$. We ignore the wall and fouling resistances. Applying all these conditions to the above equation, we obtain:

$$\frac{1}{U} = \frac{1}{U_i} = \frac{1}{U_o} = \frac{1}{h_i} + \frac{1}{h_o}$$

The outside (or oil) heat transfer coefficient is given. To evaluate the water convective heat transfer coefficient, we calculate its Reynolds number. From Table A-6 at the average water temperature: $\mu = 5.29 \times 10^{-4}$ N·s/m², $Pr = 3.44$, and $k = 0.643$ W/m·K. The Reynolds number is $Re = \rho V D/\mu$. For a straight circular tube (as was done in Example 13-1), we rewrite the Reynolds number as $Re = 4\dot{m}_{tube}/\pi \mu D$. The Reynolds number is for a single tube, and the given mass flow rate is the total flow through all the tubes. Therefore,

$$Re = \frac{4\left(\dot{m}_{tot}/N_t\right)}{\pi \mu D} = \frac{4\left(\frac{2.5}{20}\text{kg/s}\right)}{\pi\left(5.29 \times 10^{-4}\,\frac{\text{kg}}{\text{m}\cdot\text{s}}\right)(0.025\,\text{m})} = 12{,}034$$

This is turbulent flow. To cover the Reynolds and Prandtl numbers in the problem, we choose the Gnielinski correlation to calculate the water heat transfer coefficient:

$$Nu = \frac{hD}{k} = \frac{(f/8)(Re_D - 1000)\,Pr}{1 + 12.7\,(f/8)^{1/2}\left(Pr^{2/3} - 1\right)}$$

The friction factor is

$$f = (0.79\ln Re - 1.64)^{-2} = (0.79\ln(12034) - 1.64)^{-2} = 0.030$$

The Nusselt number is:

$$Nu = \frac{(0.03/8)(12034 - 1000)(3.44)}{1 + 12.7\,(0.03/8)^{1/2}\left(3.44^{2/3} - 1\right)} = 71.4$$

$$h = \frac{71.4\left(0.643\,\frac{\text{W}}{\text{m}\cdot\text{K}}\right)}{0.025\,\text{m}} = 1836\,\frac{\text{W}}{\text{m}^2\cdot\text{K}}$$

The overall heat transfer coefficient is:

$$U = \left[\frac{1}{h_i} + \frac{1}{h_o}\right]^{-1} = \left[\frac{1}{1836} + \frac{1}{400}\right]^{-1}\frac{\text{W}}{\text{m}^2\cdot\text{K}} = 328\,\frac{\text{W}}{\text{m}^2\cdot\text{K}}$$

The *LMTD* is calculated as if the heat exchanger were pure counterflow:

$$\Delta T_{LM,cf} = \frac{(T_{H,i} - T_{C,o}) - (T_{H,o} - T_{C,i})}{\ln\left(\frac{T_{H,i} - T_{C,o}}{T_{H,o} - T_{C,i}}\right)} = \frac{(160 - 85) - (100 - 15)}{\ln\left(\frac{160 - 85}{100 - 15}\right)} = 79.9°C$$

The F factor is evaluated with Figure 13-7. We need the P and R factors:

$$P = \frac{T_{tube,o} - T_{tube,i}}{T_{shell,i} - T_{tube,i}} = \frac{85 - 15}{160 - 15} = 0.483$$

$$R = \frac{T_{shell,i} - T_{shell,o}}{T_{tube,o} - T_{tube,i}} = \frac{160 - 100}{85 - 15} = 0.857$$

With these numbers, we determine $F \approx 0.87$.
    Finally, we can calculate the shell length:

$$L_p = \frac{\dot{Q}}{UF\Delta T_{LM,cf}N_t\pi D N_p}$$

$$= \frac{(731.5\,\text{kW})(1000\,\text{W/kW})}{\left(328\,\frac{\text{W}}{\text{m}^2\cdot\text{K}}\right)(0.87)(79.9°C)(20)\pi(0.025\,\text{m})(4)} = 5.10\,\text{m}$$

Comments:

If we had chosen the Dittus-Boelter equation (with the *Pr* exponent of 0.4) to calculate the water-side heat transfer coefficient, we would have obtained $Nu = 69.3, h = 1782$ W/m²·K, $U = 327$ W/m²·K, and $L_p = 5.12$ m. For this problem, the effect of the different heat transfer coefficient correlations is minor.

EXAMPLE 13-3 **Effect of fouling on area**

We decide that we should take into account fouling in the design of the heat exchanger in Example 13-2. Our company's policy is to use a fouling factor of 0.00035 m²·K/W for water in this application and 0.0009 m²·K/W for the oil. For all the same conditions given in Example 13-2, determine the required shell length and compare to the result with clean tubes.

Approach:

The solution to this problem is identical to that given in Example 13-2. The only change is the inclusion of fouling factors in the evaluation of the overall heat transfer coefficient.

Assumptions:

Same as in Example 13-2.

Solution:

With the same assumptions and given information from Example 13-2, the overall heat transfer coefficient is calculated with:

$$\frac{1}{U} = \frac{1}{h_i} + R_i'' + R_0'' + \frac{1}{h_o}$$

$$U = \left[\frac{1}{h_i} + R_i'' + R_0'' + \frac{1}{h_o}\right]^{-1} = \left[\frac{1}{1836} + 0.00035 + 0.0009 + \frac{1}{400}\right]^{-1} \frac{W}{m^2 \cdot K} = 232 \frac{W}{m^2 \cdot K}$$

The shell length is:

$$L_p = \frac{\dot{Q}}{UF\,\Delta T_{LM,cf}\,N_t\pi DN_p}$$

$$= \frac{(731.5\,\text{kW})\,(1000\,\text{W}/\text{kW})}{\left(232\,\dfrac{W}{m^2 \cdot K}\right)(0.87)\,(79.9°\text{C})\,(20)\,\pi\,(0.025\,\text{m})\,(4)} = 7.22\,\text{m}$$

Comments:

With a fouling factor, the shell length increased from 5.10 m to 7.22 m, and the total surface area went from 32.0 m² to 45.4 m². Both the cost of the tubes and the shell would increase. Shells are often constructed of heavy, thick steel plates that are expensive to construct, so anything that makes them larger can have a significant impact on the overall cost of the heat exchanger.

13.4 THE EFFECTIVENESS-*NTU* METHOD

The main equation for the *effectiveness-NTU (ε-NTU)* method of analyzing heat exchangers is

$$\dot{Q}_{act} = \varepsilon\dot{Q}_{max} = \varepsilon C_{min}\left(T_{H,i} - T_{C,i}\right) \tag{13-24}$$

We have previously used this equation in the discussion of the Brayton cycle in Chapter 8. The parameter needed to characterize the performance of the regenerator was the *effectiveness*, ε. Below is an analysis that shows that the effectiveness is dependent on the heat exchanger geometry, the *number of transfer units, NTU,* and the ratio of *heat capacity rates,* C_{min}/C_{max}, where:

$$NTU = \frac{UA}{C_{min}} \quad \text{and} \quad \frac{C_{min}}{C_{max}} = \frac{\dot{m}c_p|_{min}}{\dot{m}c_p|_{max}}$$

where C_{max} is the fluid with the larger value of the product of mass flow rate and specific heat.

Just as we evaluate the thermal efficiency of cycles to assess their performance and use the isentropic efficiency to evaluate the thermodynamic performance of devices, one way we can assess the performance of heat exchangers is to evaluate their effectiveness, ε. As explained in Chapter 8, effectiveness is a dimensionless parameter defined as:

$$\varepsilon = \frac{\text{actual heat transfer rate}}{\text{maximum possible heat transfer rate}} = \frac{\dot{Q}_{act}}{\dot{Q}_{max}} \tag{13-25}$$

The magnitude of the effectiveness can range from 0 (no heat transfer at all) to 1 (maximum possible heat transfer for the given fluid inlet temperatures, the flow rates, and specific heats). The actual heat transfer rate can be calculated with conservation of energy using the same assumptions invoked in Section 13.3.

$$\dot{Q}_{act} = \dot{m}_H \left(h_{H,i} - h_{H,o} \right) = \dot{m}_H c_{p,H} \left(T_{H,i} - T_{H,o} \right)$$

or

$$\dot{Q}_{act} = \dot{m}_C \left(h_{C,o} - h_{C,i} \right) = \dot{m}_C c_{p,C} \left(T_{C,o} - T_{C,i} \right) \tag{13-26}$$

where h is enthalpy in this equation.

To evaluate the maximum possible heat transfer rate (without violating the second law of thermodynamics), we need to examine how the temperatures of the two single-phase fluids behave. The maximum possible heat transfer rate would occur when one of the fluids undergoes the maximum possible temperature rise or fall. That is, the highest possible outlet temperature for the cold fluid would equal $T_{H,i}$ and the lowest possible outlet temperature for the hot fluid would equal $T_{C,i}$. Hence, the maximum possible temperature change of either fluid would be $(T_{H,i} - T_{C,i})$.

Which fluid could undergo this maximum temperature change? We have defined the *heat capacity rate* as $C = \dot{m}c_p$. In general, this product is different for the hot (C_H) and cold (C_C) fluids flowing through a heat exchanger; that is, C_H does not have to equal C_C. For example, if $C_H > C_C$, then we designate the larger heat capacity rate $C_{max} = C_H$ and the smaller heat capacity rate $C_{min} = C_C$. Because the energy balance must be satisfied,

$$\dot{Q}_{act} = C_H \, \Delta T_H = C_C \, \Delta T_C$$

Consequently, the cold fluid would undergo a larger actual temperature change (ΔT_C) than the hot fluid (ΔT_H); that is, $\Delta T_C > \Delta T_H$.

As can be seen, the fluid with the minimum heat capacity rate will undergo a greater temperature change than the fluid with the maximum heat capacity rate. Thus only the fluid with the minimum heat capacity rate, C_{min}, could experience the maximum possible temperature change $T_{H,i} - T_{C,i}$ which then leads to the maximum possible heat transfer rate:

$$\dot{Q}_{max} = C_{min}\left(T_{H,i} - T_{C,i}\right) \tag{13-27}$$

The same result is obtained if we had assumed $C_C > C_H$.

For an *existing* heat exchanger, with measured flow rates and inlet and outlet temperatures on both sides, the effectiveness can be calculated with Eq. 13-25. However, when *designing* a new heat exchanger, the heat exchanger geometry, flows, and fluid properties must be used to evaluate the effectiveness.

The ε-*NTU* method is developed with the same energy and rate equations used in the *LMTD* method, but the equations are manipulated differently to obtain a different but analogous result. The development is illustrated for a counterflow heat exchanger. We arbitrarily assume $C_{min} = C_H$, and begin by combining Eq. 13-6 and Eq. 13-13 to eliminate $\delta\dot{Q}$. After rearrangement, we obtain

$$\frac{d\left(T_H - T_C\right)}{T_H - T_C} = -\left(1 - \frac{C_H}{C_C}\right)\frac{U}{C_H}\,dA = -\left(1 - \frac{C_{min}}{C_{max}}\right)\frac{U}{C_{min}}\,dA \tag{13-28}$$

Integrating this equation from the hot fluid inlet (end 1) to the hot fluid outlet (end 2) results in

$$\frac{T_{H,2} - T_{C,2}}{T_{H,1} - T_{C,1}} = \exp\left[-\left(1 - \frac{C_{min}}{C_{max}}\right)\frac{UA}{C_{min}}\right] \tag{13-29}$$

Add and subtract $T_{H,1}$ to the numerator of the left-hand side of this equation, and also add and subtract $T_{H,2}$ to the denominator. Rearrange the temperatures to obtain the difference $\left(T_{H,1} - T_{C,2}\right) = \left(T_{H,i} - T_{C,i}\right)$. Divide the numerator and the denominator by this difference, and incorporate $\varepsilon = C_H\left(T_{H,1} - T_{H,2}\right)/C_{min}\left(T_{H,1} - T_{C,2}\right)$. From the energy balance, $\dot{Q}_H = -\dot{Q}_C$. Also input an expression for $\left(T_{C,1} - T_{C,2}\right)$ in terms of temperatures and the heat capacity ratio to obtain

$$\frac{1 - \varepsilon}{1 - \dfrac{C_{min}}{C_{max}}\varepsilon} = \exp\left[-\left(1 - \frac{C_{min}}{C_{max}}\right)\frac{UA}{C_{min}}\right] \tag{13-30}$$

This is rearranged to give

$$\varepsilon = \frac{1 - \exp\left[-\left(1 - \dfrac{C_{min}}{C_{max}}\right)\dfrac{UA}{C_{min}}\right]}{1 - \dfrac{C_{min}}{C_{max}}\exp\left[-\left(1 - \dfrac{C_{min}}{C_{max}}\right)\dfrac{UA}{C_{min}}\right]} \tag{13-31}$$

We define a dimensionless parameter, the *number of transfer units (NTU)*, as

$$NTU = \frac{UA}{C_{min}} \tag{13-32}$$

This represents the nondimensional thermal size of the heat exchanger (but does not necessarily imply the physical size). A second nondimensional parameter is the *heat capacity ratio, C**

$$C^* = \frac{C_{min}}{C_{max}} \qquad (13\text{-}33)$$

Incorporating these two definitions into Eq. 13-31 results in

$$\varepsilon = \frac{1 - \exp\left[-NTU\left(1 - C^*\right)\right]}{1 - C^* \exp\left[-NTU\left(1 - C^*\right)\right]} \qquad (13\text{-}34)$$

An identical expression would have been developed if we had started by assuming $C_{min} = C_C$.

Similar expressions for effectiveness have been developed theoretically for many other heat exchanger configurations and flow arrangements. In all cases, these analyses show that heat exchanger effectiveness depends on three pieces of information:

$$\varepsilon = \text{func}(C^*, NTU, \text{heat exchanger geometry and flow arrangement})$$

The heat exchanger geometry considerations include the type of construction (counterflow, parallel flow, etc.), number of fluid passes, and mixed or unmixed fluids. These considerations are identical to those taken into account when evaluating the F factor in the *LMTD* method.

Some common expressions for effectiveness are given in Table 13-3. Figure 13-8 shows the effect of varying parameters on the effectiveness. Note the exponential behavior of the curves. When *NTU* is large, obtaining a small increase in the effectiveness may require a significant increase in area. For example, consider a simple counterflow heat exchanger with $C^* = 0.5$ and $NTU = 3.5$, with an effectiveness of 91.3%. If we wanted to increase the effectiveness to 94% (assuming U remains constant), we would need an $NTU = 4.36$. Thus, a 2.7 percentage point increase in effectiveness would require a 25% increase in surface area. Hence, for a heat exchanger that may need to operate over a range of conditions, it would not be wise to design with an *NTU* near where the curve begins to flatten out. By comparison, a counterflow heat exchanger with $C^* = 0.5$ and $NTU = 1.5$ has an effectiveness of 69.1%. To increase this by 2.7 percentage points (to 71.8%), only a 9.5% increase in area is needed.

Note that for a given C^* and *NTU*, a counterflow heat exchanger has the highest effectiveness of any heat exchanger or flow arrangement, and a parallel flow heat exchanger has the lowest. All others fall between these two (Figure 13-9). In addition, with single-phase flow on one side of a heat exchanger and a constant wall or fluid temperature on the other side of the heat exchanger, $C_{max} \to \infty$, $C^* \to 0$, the geometry becomes irrelevant, and all heat exchangers have the same expression for effectiveness:

$$\varepsilon = 1 - \exp\left(-NTU\right) \qquad \text{for } C^* = 0 \qquad (13\text{-}35)$$

Table 13-4 gives steps used in the ε-*NTU* method. Regardless of whether the ε-*NTU* or the *LMTD* method is used for either a rating or a design problem, identical results will be obtained (within roundoff error).

TABLE 13-3 **Common equations for effectivenes and *NTU***

Type of Heat Exchanger	Effectiveness relations	NTU relations
All exchangers with $C^* = 0$	$\varepsilon = 1 - \exp(-NTU)$	$NTU = -\ln(1 - \varepsilon)$
Double pipe Counter flow	$\varepsilon = \dfrac{1 - \exp[-NTU(1 - C^*)]}{1 - C^* \exp[-NTU(1 - C^*)]}$	$NTU = \dfrac{1}{C^* - 1} \ln\left(\dfrac{\varepsilon - 1}{C^*\varepsilon - 1}\right)$
Parallel flow	$\varepsilon = \dfrac{1 - \exp[-NTU(1 + C^*)]}{1 + C^*}$	$NTU = -\dfrac{\ln[1 - \varepsilon(1 + C^*)]}{1 + C^*}$
Shell-and-tube One-shell pass; 2, 4, 6, ... tube passes (TEMA E shell)	$\varepsilon_1 = \dfrac{2}{(1 + C^*) + (1 + C^{*2})^{0.5} \dfrac{\left(1 + \exp[-NTU \sqrt{1 + C^{*2}}]\right)}{\left(1 - \exp[-NTU \sqrt{1 + C^{*2}}]\right)}}$	$NTU = -\dfrac{1}{(1 + C^{*2})^{0.5}} \ln\left(\dfrac{E - 1}{E + 1}\right)$ $E = \dfrac{2/\varepsilon_2 - (1 + C^*)}{(1 + C^{*2})^{0.5}}$
n-shell passes; 2*n*, 4*n*, 6*n*, ... tube passes	$\varepsilon = \left[\left(\dfrac{1 - \varepsilon_1 C^*}{1 - \varepsilon_1}\right)^n - 1\right]\left[\left(\dfrac{1 - \varepsilon_1 C^*}{1 - \varepsilon_1}\right)^n - C^*\right]^{-1}$	Use above two equations with $\varepsilon_2 = \dfrac{F - 1}{F - C^*} \qquad F = \left(\dfrac{\varepsilon C^* - 1}{\varepsilon - 1}\right)^{1/n}$
Crossflow Both fluids unmixed	$\varepsilon = 1 - \exp\left[\dfrac{NTU^{0.22}}{C^*}\left(\exp[-C^* NTU^{0.78}] - 1\right)\right]$	
Both fluids mixed	$\varepsilon = \dfrac{1}{\dfrac{1}{1 - \exp(-NTU)} + \dfrac{C^*}{1 - \exp(-C^* NTU)} - \dfrac{1}{NTU}}$	
C_{min} mixed, C_{max} unmixed	$\varepsilon = 1 - \exp\left[-\dfrac{1}{C^*}\left(1 - \exp[-C^* NTU]\right)\right]$	$NTU = -\dfrac{\ln\left(C^* \ln[1 - \varepsilon] + 1\right)}{C^*}$
C_{max} mixed, C_{min} unmixed	$\varepsilon = \dfrac{1}{C^*}\left\{1 - \exp\left[-C^*\left(1 - \exp[-NTU]\right)\right]\right\}$	$NTU = -\ln\left(1 + \dfrac{\ln[1 - \varepsilon C^*]}{C^*}\right)$

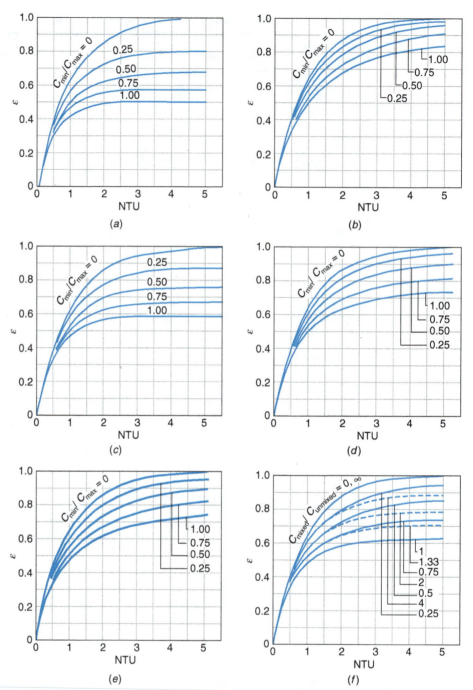

FIGURE 13-8 Effectiveness curves for several heat exchanger types. (a) Parallel flow. (b) Counterflow. (c) Shell-and-tube with one shell pass and any multiple of two tube passes (two, four, etc. tubes passes). (d) Shell-and-tube with two shell passes and any multiple of four tube passes (four, eight, etc. tube passes). (e) Single-pass, crossflow with both fluids unmixed. (f) Single-pass, crossflow with one fluid mixed and the other unmixed.
(*Source*: F. P. Incropera and D. P. DeWitt, *Introduction to Heat Transfer*, 3rd ed., Wiley New York, 1996. Used with permission.)

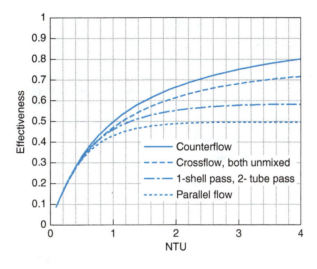

FIGURE 13-9 Comparison of effectiveness for counterflow, crossflow, multipass, and parallel flow heat exchangers ($C^* = 1.0$).

TABLE 13-4 Designing or Rating a Heat Exchanger using the ε-*NTU* Method

Design	Rating
Known information: the type of heat exchanger and basic configuration (e.g., diameter and wall thickness of tubes); the two fluids and their flow rates; the inlet temperatures of the two fluids; the required heat duty or the two outlet temperatures	**Known information:** the geometry (number, size, spacing, and layout of tubes, fin geometry, shell geometry, etc.) and type of heat exchanger (shell-and-tube, plate-fin, fin-tube, etc.); the two fluids and their flow rates; the two inlet temperatures
Objective: determine the area needed	**Objective:** determine the overall heat transfer rate or the two outlet temperatures

<table>
<tr><th>Steps to follow</th><th>Steps to follow</th></tr>
</table>

Steps to follow (Design)	Steps to follow (Rating)
1. Calculate ε from given data using Eq. 13-25.	1. Calculate the heat transfer coefficients on each side of the heat exchanger using the given geometry, fluid, flow rates, and appropriate correlations.
2. Calculate C^* with Eq. 13-33.	
3. Calculate wall resistance and estimate fouling resistances if required.	2. Calculate wall resistance and estimate fouling resistances if required.
4. Assume number of tubes or channels needed in heat exchanger.	3. Calculate the overall heat transfer coefficient using Eq. 13-3.
5. Calculate the heat transfer coefficients on each side of the heat exchanger using the given and assumed geometric parameters, fluid, and flow rates and appropriate correlations.	4. Calculate *NTU* (Eq. 13-32), C^* (Eq. 13-33), and $\dot{Q}_{max}$ (Eq. 13-27).
	5. Evaluate ε using appropriate equation or figure.
6. Calculate the overall heat transfer coefficient using Eq. 13-3.	6. Calculate actual heat duty using Eq. 13-24 and outlet temperatures using Eq. 13-7 or Eq. 13-8.
7. Evaluate *NTU* for the heat exchanger geometry.	
8. Calculate $A = (NTU)\, C_{min}/U$.	
9. Repeat steps 4 through 8 until solution converges. (Use the area calculated in step 8 to estimate the tubes/channels in step 4.)	

EXAMPLE 13-4 Effectiveness-*NTU* rating problem

The designers of a new car need the heat transfer rate of the crossflow heat exchanger to be used for the radiator; single passes are used on each side. The cross-sectional frontal area of the exchanger is 1.5 ft². The air side is finned with a heat transfer area of 40 ft². The engine coolant (use water properties) flows inside tubes and enters the exchanger at 240°F with a flow rate of 50 gal/min. The air enters at 100°F and 1 atm pressure. The car speed is 55 mph. From previous investigations, you estimate the overall heat transfer coefficient (based on the air side area) to be 87 Btu/h·ft²·°F.

a) Determine the heat transfer rate (in Btu/s).

b) Determine the water outlet temperature (in °F).

Approach:

The schematic shows all the given information.

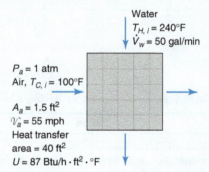

We are given information about the two fluids (flow rates and inlet temperatures) and the complete heat exchanger geometry. Since we are asked to determine the heat transfer rate of this existing heat exchanger, this is a heat exchanger rating problem, and the preferred analysis approach is the ε-*NTU* method. Once the heat transfer rate is determined, the outlet water temperature can be calculated from conservation of energy.

Assumptions:

Solution:

The governing equation for the radiator heat transfer rate, $\dot{Q}_R$, where the subscript R indicates the heat transfer rate calculated with the rate equation, is

$$\dot{Q}_R = \varepsilon C_{min} \left(T_{H,i} - T_{C,i} \right)$$

The two inlet temperatures are given in the problem statement. We need to determine the minimum heat capacity rate, C_{min}, and the heat exchanger effectiveness, ε.

The heat capacity rate is defined as $C = \dot{m}c_p = \rho\dot{V}c_p = \rho\mathcal{V}A c_p$. The water volume flow rate is known, and for the air the cross-sectional flow area and velocity are given. We need the density of both fluids. For water at 240°F, Table B-6 gives us 59.1 lbm/ft³. The ideal gas equation is used

A1. Air is an ideal gas.

[A1] for the air density at the inlet temperature because that is where the velocity is known:

$$\rho_a = \frac{P_a M}{\bar{R}T} = \frac{14.7\,\text{psia}\left(28.97\,\frac{\text{lbm}}{\text{lbmol}}\right)}{\left(10.73\,\frac{\text{psia·ft}^3}{\text{lbmol·R}}\right)(100+460)\,\text{R}} = 0.0709\,\frac{\text{lbm}}{\text{ft}^3}$$

$$\dot{m}_w = \rho_w \dot{V}_w = \left(59.1\,\frac{\text{lbm}}{\text{ft}^3}\right)\left(50\,\frac{\text{gal}}{\text{min}}\right)\left(\frac{0.1337\,\text{ft}^3}{1\,\text{gal}}\right)\left(\frac{1\,\text{min}}{60\,\text{s}}\right) = 6.58\,\frac{\text{lbm}}{\text{s}}$$

$$\dot{m}_a = \rho_a \mathcal{V}_a A_a = \left(0.0709\,\frac{\text{lbm}}{\text{ft}^3}\right)\left(55\,\frac{\text{mi}}{\text{h}}\right)(1.5\,\text{ft}^2)\left(\frac{5280\,\text{ft}}{1\,\text{mi}}\right)\left(\frac{1\,\text{h}}{3600\,\text{s}}\right) = 8.58\,\frac{\text{lbm}}{\text{s}}$$

A2. Specific heat is constant.

The specific heats should be at the average temperature of each fluid [A2]. Because we do not know the outlet temperatures, first we estimate those temperatures, calculate the average temperatures, look up the specific heats at those temperatures, and then calculate the outlet temperatures. If the calculated and estimated values are close, then the problem is finished; if they are not close, then we must iterate. We estimate the water outlet temperature to be 220°F and the average water temperature to be 230°F; likewise, the estimated air outlet temperature is 140°F and its average temperature is 120°F. From Table B-6, the water specific heat is $c_{p,w} = 1.006$ Btu/lbm·°F and from Table B-8, the air specific heat is $c_{p,a} = 0.24$ Btu/lbm·°F, so the heat capacity rates are:

$$C_w = \dot{m}_w c_{p,w} = \left(6.58 \, \frac{\text{lbm}}{\text{s}}\right)\left(1.006 \, \frac{\text{Btu}}{\text{lbm·°F}}\right) = 6.62 \, \frac{\text{Btu}}{\text{s·°F}}$$

$$C_a = \dot{m}_a c_{p,a} = \left(8.58 \, \frac{\text{lbm}}{\text{s}}\right)\left(0.24 \, \frac{\text{Btu}}{\text{lbm·°F}}\right) = 2.06 \, \frac{\text{Btu}}{\text{s·°F}}$$

Hence, $C_{min} = C_a$.

This is a single-pass crossflow heat exchanger, finned on one side with tubes on the other, so the configuration is similar to that shown in Figure 13-4b in which both flows are unmixed flows. The effectiveness can be determined from Figure 13-8e or with the appropriate equation from Table 13-3. We need the heat capacity ratio and $NTU = UA/C_{min}$. The heat capacity ratio is obtained from the individual heat capacity rates calculated above:

$$\frac{C_{min}}{C_{max}} = \frac{2.06 \, \text{Btu/s·°F}}{6.62 \, \text{Btu/s·°F}} = 0.31$$

The overall heat transfer coefficient and its accompanying surface area are specified in the problem statement, so that

$$NTU = \frac{UA}{C_{min}} = \frac{\left(87 \, \frac{\text{Btu}}{\text{h·ft}^2\text{·°F}}\right)(40 \, \text{ft}^2)\left(\frac{1 \, \text{h}}{3600 \, \text{s}}\right)}{2.06 \, \frac{\text{Btu}}{\text{s·°F}}} = 0.469$$

From Figure 13-8e, $\varepsilon \approx 0.37$, so that

$$\dot{Q}_R = \varepsilon C_{min}\left(T_{H,in} - T_{C,in}\right) = 0.37\left(2.06 \, \frac{\text{Btu}}{\text{s·°F}}\right)(240 - 100) \, °\text{F} = 106.7 \, \frac{\text{Btu}}{\text{s}}$$

A3. The system is steady.
A4. Potential and kinetic energy effects are negligible.
A5. No work occurs in the heat exchanger.
A6. Water is incompressible.

The outlet water temperature is obtained from conservation of mass and energy applied to a control volume around the water with assumptions [A3], [A4], and [A5]:

$$\dot{m}_A = \dot{m}_B = \dot{m}_w \qquad \text{and} \qquad \dot{Q}_E = \dot{m}_w \left(h_B - h_A\right)$$

where the subscript E represents the heat transfer rate calculated with conservation of energy. $\dot{Q}_E$, has the same magnitude as that calculated with the rate equation, $\dot{Q}_R$. However, the rate equation always gives a positive number, and for this problem conservation of energy gives a negative number (heat transfer out of the water). Hence, $\dot{Q}_R = -\dot{Q}_E$. We assume the water is an ideal liquid [A6] with a constant specific heat [A2] so that $\Delta h = c_p \Delta T$. Substituting this information into the governing energy equation, we solve for the water outlet temperature:

$$T_B = T_A - \frac{\dot{Q}_R}{\dot{m}_w c_{p,w}}$$

We have all the properties and flows, so that

$$T_B = T_A - \frac{\dot{Q}_R}{\dot{m}_w c_{p,w}} = 240°F - \frac{106.7 \frac{Btu}{s}}{\left(6.58 \frac{lbm}{s}\right)\left(1.006 \frac{Btu}{lbm \cdot °F}\right)} = 224°F$$

$$T_D = T_C + \frac{\dot{Q}_R}{\dot{m}_a c_{p,a}} = 100°F + \frac{106.7 \frac{Btu}{s}}{\left(8.58 \frac{lbm}{s}\right)\left(0.24 \frac{Btu}{lbm \cdot °F}\right)} = 152°F$$

Comments:

With such a low effectiveness, this is a rather inefficient heat exchanger design. No iteration is required because the calculated outlet temperatures are close to what we assumed.

EXAMPLE 13-5 Heat exchanger with one constant temperature fluid

Air flows through a 1.5-cm I.D./2.5-cm O.D tube ($k = 52$ W/m·K) that is 5 m long. The inlet temperature is 42°C, the required outlet temperature is 52°C, and the flow rate is 50 kg/h. We want to use condensing steam on the outside surface to heat the air. Over a range of steam pressures, the condensing heat transfer coefficient can be assumed to be 1000 W/m²·K. What steam temperature is required to obtain this outlet temperature? What is the steam pressure?

Approach:

The given information and a schematic of the problem are shown here.

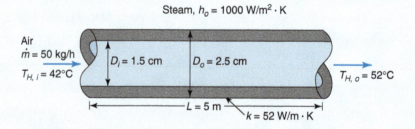

Sufficient information is given on the airflow to calculate the heat transfer rate. In the discussion of the *LMTD* method, generally when the heat transfer rate was given, the heat transfer area was sought. However, the complete heat exchanger geometry is given in this problem. In the *ε-NTU* method, generally when the heat exchanger geometry was given, the heat transfer rate was sought. This problem is different from either of those types. We want to evaluate the steam temperature (the driving temperature in the system). Because of the constant wall temperature, we can approach this problem using three different methods. However, after manipulation, we can see that the three approaches are identical.

Assumptions:

Solution:

Method 1
In Chapter 12, we developed an equation for convective heat transfer between a single-phase fluid and a constant temperature fluid:

$$T_o = T_{st} - (T_{st} - T_i) \exp\left(-UA/\dot{m}c_p\right)$$

Solving for the constant temperature steam, T_{st}:

$$T_{st} = \frac{T_o - T_i \exp\left(-UA/\dot{m}c_p\right)}{1 - \exp\left(-UA/\dot{m}c_p\right)}$$

In the problem statement, we are given the fluid inlet temperature, T_i, the fluid exit temperature, T_o, and the mass flow rate. We can calculate the surface area and look up the specific heat. When we have evaluated the heat transfer coefficient from a correlation, then we can calculate T_{st}, the steam temperature.

Method 2

The governing rate equation for the ε-*NTU* method is

$$\dot{Q} = \varepsilon C_{min}\left(T_{H,i} - T_{C,i}\right)$$

The condensing steam has an infinite heat capacity rate ($C_{max} \to \infty$), so C_{min} is that of the air and can be determined from the given information. The cold inlet temperature is that of the air. The hot inlet temperature is the constant steam temperature. To evaluate the effectiveness, we recognize that whenever one fluid remains at a constant temperature (either condensing or boiling) the effectiveness is $\varepsilon = 1 - \exp\left(-NTU\right)$, where $NTU = UA/C_{min}$ and $C_{min} = \dot{m}c_p$. Combining these expressions, the same equation for the steam temperature is obtained as in Method 1.

Method 3

In this approach we equate the governing rate equation for the *LMTD* method with conservation of energy applied to the air flowing through the heat exchanger and assume [A1], [A2], and [A3]:

A1. The system is steady.
A2. Potential and kinetic energy effects are negligible.
A3. No work occurs in the heat exchanger.

$$\dot{Q} = UA\Delta T_{LM} = \dot{m}c_p\left(T_i - T_o\right)$$

The log mean temperature difference is:

$$\Delta T_{LM} = \frac{(T_i - T_{st}) - (T_o - T_{st})}{\ln\left(\dfrac{T_i - T_{st}}{T_o - T_{st}}\right)}$$

Substituting the *LMTD* expression into the previous equation, we solve for T_{st} and obtain the same expression as in the previous two methods.

We begin by evaluating the air properties at the average air temperature of $47°C = 330$ K using Table A-7: $\rho = 1.076$ kg/m^3, $\mu = 1.99 \times 10^{-5}$ kg/m·s, $Pr = 0.708$, $c_p = 1.007$ kJ/kg·K, and $k = 0.0283$ W/m·K.

The Reynolds number is $Re = \rho \mathcal{V} D/\mu$. For a straight circular tube, we rewrite the Reynolds number as $Re = 4\dot{m}/\pi \mu D$ (see Example 13-1). Therefore,

$$Re = \frac{4\dot{m}}{\pi \mu D_i} = \frac{4\left(50\,\text{kg/h}\right)\left(\dfrac{1\,\text{h}}{3600\,\text{s}}\right)}{\pi\left(1.99 \times 10^{-5}\,\dfrac{\text{kg}}{\text{m·s}}\right)(0.015\,\text{m})} = 59,243$$

This is turbulent flow. We choose the Dittus-Boelter equation since it covers the Reynolds and Prandtl numbers in this problem (with the Pr exponent of 0.4 since we are heating the air) to calculate the heat transfer coefficient:

$$Nu = \frac{hD}{k} = 0.023 Re^{0.8} Pr^{0.4} = 0.023\,(59243)^{0.8}\,(0.708)^{0.4} = 132$$

$$h = \frac{132\left(0.0283\,\dfrac{\text{W}}{\text{m·K}}\right)}{0.015\,\text{m}} = 249\,\frac{\text{W}}{\text{m}^2\text{·K}}$$

A4. Ignore fouling.

Now we can calculate the overall heat transfer coefficient. Assume fouling resistances are negligible [A4] and no fins, so that $\eta_{o,o} = \eta_{o,i} = 1$:

$$\frac{1}{UA} = \frac{1}{h_i A_i} + \frac{\ln(r_o/r_i)}{2\pi kL} + \frac{1}{h_o A_o}$$

$$= \frac{1}{\left(249 \frac{W}{m^2 \cdot K}\right) \pi \, (0.015 \, m) \, (5 \, m)} + \frac{\ln(2.5/1.5)}{2\pi \left(52 \frac{W}{m \cdot K}\right)(5 \, m)} + \frac{1}{\left(1000 \frac{W}{m^2 \cdot K}\right) \pi \, (0.025 \, m) \, (5 \, m)}$$

$$= 0.0170 \frac{K}{W} + 0.00031 \frac{K}{W} + 0.00255 \frac{K}{W} = 0.0199 \frac{K}{W}$$

$$UA = 50.2 \frac{W}{K} \quad \rightarrow \quad \frac{UA}{(\dot{m}c_p)} = \frac{50.2 \, W/K}{(50 \, kg/h)(1 \, h/3600 \, s)(1007 \, J/kg \cdot K)} = 3.59$$

Incorporating these into our main equation,

$$T_{st} = \frac{T_o - T_i \exp(-UA/\dot{m}c_p)}{1 - \exp(-UA/\dot{m}c_p)} = \frac{52°C - (42°C) \exp(-3.59)}{1 - \exp(-3.59)} = 52.3°C$$

From Table A-10, at this temperature the saturation pressure is 13.9 kPa.

Comments:

If we use the Gnielinski correlation to calculate the air heat transfer coefficient, we would obtain $f = 0.02$, $Nu = 119$, $h = 225 \, W/m^2 \cdot K$, $UA = 46.0 \, W/m^2 \cdot K$, and $T_{st} = 52.4°C$.

EXAMPLE 13-6   **Regenerator in Brayton Cycle**

A Brayton cycle with regeneration uses air as the working fluid. The pressure ratio is 11.2, the inlet air is at 30°C, 1 atm, and the turbine inlet temperature is 1150°C. Compressor and turbine are isentropic. The regenerator is a single-pass, crossflow heat exchanger. One hundred (100) silicon carbide ceramic tubes ($k = 24 \, W/m \cdot K$, 50-mm inner diameter, 75-mm outer diameter, $L = 9.0 \, m$) are arranged such that the outside heat transfer coefficient is 35 W/m²·K. The total airflow rate inside the tubes is 1.3 kg/s; the fuel flow rate is 5% of that of the air. The inner tube surface is clean; the outer tube surface has a fouling factor of $2 \times 10^{-4} \, m^2 \cdot K/W$. Evaluate all air properties at 300 K.

a) Determine the cycle thermal efficiency.

b) Determine the net power output (in kW).

Approach:

A schematic of the system is shown on the next page.

We start with the definition of cycle thermal efficiency, $\eta_{cycle} = \dot{W}_{net}/\dot{Q}_{in}$. The net work is evaluated with the methods discussed in Chapter 8. The input heat transfer rate is determined by applying conservation of energy to the combustor; for that, we need the air inlet temperature to the combustor. This temperature must be obtained from a heat exchanger analysis applied to the regenerator. The geometry and flows of the regenerator are given. We can obtain the inlet temperatures from a cycle analysis. The outlet temperatures are obtained from a heat exchanger analysis, and the preferred approach is the ε-NTU method.

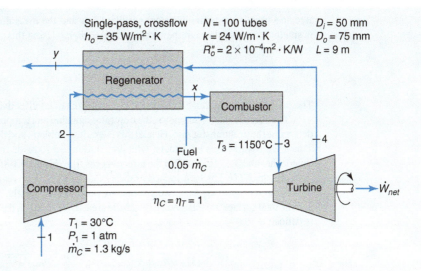

Single-pass, crossflow $\qquad$ N = 100 tubes $\qquad$ D_i = 50 mm
h_o = 35 W/m^2 · K $\qquad$ k = 24 W/m · K $\qquad$ D_o = 75 mm
$\qquad\qquad$ R_o'' = 2 × 10^{-4}m^2 · K/W $\qquad$ L = 9 m

T_3 = 1150°C

T_1 = 30°C
P_1 = 1 atm
$\dot{m}_C$ = 1.3 kg/s

Fuel 0.05 $\dot{m}_C$

$\eta_C = \eta_T = 1$

$\dot{W}_{net}$

Assumptions:

A1. The system is steady.
A2. Potential and kinetic energy effects are negligible.
A3. Air is the working fluid and is an ideal gas.
A4. The turbine and compressor are adiabatic.
A5. Specific heat is constant.
A6. The turbine and compressor are isentropic.

Solution:

For use in the cycle thermal efficiency, net power is $\dot{W}_{net} = \dot{W}_T - \dot{W}_C$. Using a cold-air-standard analysis and applying conservation of mass and energy to the turbine and compressor with assumptions [A1], [A2], [A3], [A4], and [A5], we obtain

$$\dot{W}_C = \dot{m}_C c_p (T_2 - T_1) \qquad \text{and} \qquad \dot{W}_T = \dot{m}_T c_p (T_3 - T_4)$$

Fuel mass flow rate is 5% of that of the air, so $\dot{m}_T = 1.05\dot{m}_C$. The inlet temperatures (T_1 and T_3) are given. The pressure ratio is known; for an isentropic turbine and compressor [A6], the outlet temperatures can be determined. The specific heat and the ratio of specific heats are evaluated at 300 K from Table A-8: $c_p = 1.005$ kJ/kg·K and $c_p/c_v = 1.40$.

$$T_2 = T_1 \left(\frac{P_2}{P_1} \right)^{(c_p/c_v-1)/(c_p/c_v)} = 303 \text{ K} (11.2)^{(1.40-1)/(1.4)} = 604.2 \text{ K}$$

$$T_4 = T_3 \left(\frac{P_4}{P_3} \right)^{(c_p/c_v-1)/(c_p/c_v)} = 1423 \text{ K} \left(\frac{1}{11.2} \right)^{(1.40-1)/(1.4)} = 713.6 \text{ K}$$

The net power is

$$\dot{W}_{net} = \dot{m}_T c_p (T_3 - T_4) - \dot{m}_C c_p (T_2 - T_1)$$

$$= \left(1.3 \frac{\text{kg}}{\text{s}} \right) \left(1.005 \frac{\text{kJ}}{\text{kg·K}} \right) [1.05(1423 - 713.6) - (604.2 - 303)]\text{K} \left(\frac{1 \text{ kW}}{1 \text{ kJ/s}} \right) = 580 \text{ kW}$$

A7. The air and fuel enter together.

For the input heat transfer rate, we assume the fuel and air enter together [A7], so that

$$\dot{Q}_{in} = \dot{m}_T c_p (T_3 - T_x)$$

We need to evaluate T_x. From the definition of heat exchanger effectiveness (for an ideal gas with constant specific heat),

$$\varepsilon = \frac{\dot{Q}_{act}}{\dot{Q}_{max}} = \frac{\dot{m}_C c_p (T_x - T_2)}{C_{min} (T_4 - T_2)}$$

Because we have assumed constant specific heats and the mass flow rate through the compressor is smaller than that through the turbine, $C_{min} = \dot{m}_C c_p$. Using this information in the last equation and solving for T_x:

$$T_x = T_2 + \varepsilon \left(T_4 - T_2 \right)$$

The only unknown in this expression is the heat exchanger effectiveness.

The heat exchanger geometry, the flow rates, and the inlet temperatures are specified; we want to determine the heat transfer rate. Hence, this is a rating problem, and the ε-NTU approach is preferred. This is a crossflow heat exchanger, with one fluid mixed (outside the tubes) and one fluid unmixed (inside the tubes). We can obtain the effectiveness from Figure 13-8f once we have the heat capacity ratio and the NTU. The heat capacity ratio is $C_{mixed}/C_{unmixed} = \dot{m}_T c_p / \dot{m}_C c_p = 1.05/1 = 1.05$.

From $NTU = UA/C_{min}$, we see that we need to evaluate the overall heat transfer coefficient and the total surface area. Based on the inside area, $A_i = N\pi D_i L$, the overall heat transfer coefficient is defined as

$$\frac{1}{U_i A_i} = \frac{1}{\eta_{o,i} h_i A_i} + \frac{R_i''}{\eta_{o,i} A_i} + R_w + \frac{R_o''}{\eta_{o,o} A_o} + \frac{1}{\eta_{o,o} h_o A_o}$$

There are no fins, so $\eta_{o,i} = \eta_{o,o} = 1$; the outer surface is clean, so $R_o'' = 0$ and $R_w = \ln \left(r_o/r_i \right) / 2\pi kLN$. Incorporating this information into the expression for the overall heat transfer coefficient and simplifying, we obtain

$$U_i = \left[\frac{1}{h_i} + R_i'' + \frac{D_i \ln \left(r_o/r_i \right)}{2k} + \frac{D_i}{h_o D_o} \right]^{-1}$$

We are given the outside heat transfer coefficient, the fouling resistance, and enough information to calculate the wall resistance.

To evaluate the inside heat transfer coefficient, we need the Reynolds number, $Re = \rho \mathcal{V} D_i / \mu$. For a straight circular tube, we rewrite the Reynolds number as $Re = 4\dot{m}/\pi \mu D_i$. Using Table A-7, [A8] the air properties at 300 K are: $k = 0.0261$ W/m·K, $\mu = 1.85 \times 10^{-5}$ kg/m·s, and $Pr = 0.712$. The flow rate is the total air flow, and the Reynolds number is calculated for a single tube, so:

A8. Properties are evaluated at 300 K.

$$Re = \frac{4\dot{m}}{\pi \mu D_i N} = \frac{4 \left(1.3 \text{ kg/s} \right)}{\pi \left(1.85 \times 10^{-5} \text{ kg/m·s} \right) \left(0.05 \text{ m} \right) 100} = 17,894$$

This is a turbulent flow, and the Gnielinski correlation is appropriate:

$$Nu = \frac{hD}{k} = \frac{\left(f/8 \right) \left(Re_D - 1000 \right) Pr}{1 + 12.7 \left(f/8 \right)^{1/2} \left(Pr^{2/3} - 1 \right)}$$

The friction factor is

$$f = \left(0.79 \ln Re - 1.64 \right)^{-2} = \left(0.79 \ln \left(17894 \right) - 1.64 \right)^{-2} = 0.027$$

and the Nusselt number is

$$Nu = \frac{\left(0.027/8 \right) \left(17894 - 1000 \right) \left(0.712 \right)}{1 + 12.7 \left(0.027/8 \right)^{1/2} \left(0.712^{2/3} - 1 \right)} = 47.7$$

$$h_i = \frac{Nu \, k}{D_i} = \frac{47.7 \left(0.0261 \text{ W/m·K} \right)}{0.05 \text{ m}} = 24.9 \, \frac{\text{W}}{\text{m}^2 \cdot \text{K}}$$

For the overall heat transfer coefficient,

$$U_i = \left[\frac{1}{h_i} + R_i'' + \frac{D_i \ln (r_o/r_i)}{2k} + \frac{D_i}{h_o D_o} \right]^{-1}$$

$$= \left[\frac{1}{24.9 \frac{W}{m^2 \cdot K}} + 2 \times 10^{-4} \frac{m^2 \cdot K}{W} + \frac{(0.05 \, m) \ln (0.075/0.05)}{2 \left(24 \frac{W}{m \cdot K} \right)} + \frac{0.05 \, m}{\left(35 \frac{W}{m^2 \cdot K} \right) (0.075 \, m)} \right]^{-1}$$

$$= 16.7 \frac{W}{m^2 \cdot K}$$

$$NTU = \frac{U_i A_i}{C_{min}} = \frac{(16.7 \, W/m^2 \cdot K) \, \pi \, (0.05 \, m)(9 \, m) 100 \, (1 \, J/1 \, W \cdot s)}{(1.3 \, kg/s) \, (1.005 \, kJ/kg \cdot K) \, (1000 \, J/1 \, kJ)} = 1.81$$

The heat capacity ratio is 1.05. Therefore, from Figure 13-8e, $\varepsilon \approx 0.59$.

We use T_2 and T_4 and the heat exchanger effectiveness to determine the outlet temperature from the regenerator:

$$T_x = T_2 + \varepsilon \, (T_4 - T_2) = 604.2 \, K + 0.59 \, (713.6 - 604.2) \, K = 668.7 \, K$$

Finally, the input heat transfer rate, the net power, and the cycle thermal efficiency are

$$\dot{Q}_{in} = \dot{m}_T c_p \, (T_3 - T_x) = 1.05 \left(1.3 \frac{kg}{s} \right) \left(1.005 \frac{kJ}{kg \cdot K} \right) (1423 - 668.7) \, K \left(\frac{1 \, kW}{1 \, kJ/s} \right)$$

$$= 1034 \, kW$$

Finally,

$$\eta_{cycle} = \frac{\dot{W}_{net}}{\dot{Q}_{in}} = \frac{580 \, kW}{1034 \, kW} = 0.561$$

Comments:

Heat exchangers are part of other systems, and exchanger performance will have a direct effect on the overall system performance. In this problem, the regenerator increased the cycle efficiency (as we discussed in Chapter 8), but to determine its effect we had to first evaluate the regenerator's effectiveness.

EXAMPLE 13-7   **System design problem**

You work for a mechanical engineering firm that specifies and installs heat pumps for heating in the winter and cooling in the summer. To decrease energy requirements in the summer and winter, the evaporator section of the heat pump is submerged in a large tank of water. (See the schematic.) During winter operation, air removes heat from the water and it slowly freezes, creating an ice-water bath at 0°C. During the summer, the ice-water bath is used to cool warm air.

You have been told to design a new ice-bath/heat exchanger system using standard equipment. The air/ice-water heat exchanger to be submerged has 10 tubes, each 50 mm in diameter and 1.50 m long. The air inlet temperature is 24°C. The fan characteristics are shown in the schematic. (The curve can be approximated by $\Delta P = 12.5 - 2.9 \dot{m} - 298 \dot{m}^2$ where ΔP is in Pa when $\dot{m}$ is in kg/s.) The ice-water tank has a volume of 10 m³ and initially contains 80% ice by volume. (This is the company design specification.) Assume that the fittings and other piping to and from the heat exchanger contribute a pressure drop equal to that of the heat exchanger itself. Determine how long it would take to completely melt the ice (in seconds and days). Use an ice density of 920 kg/m³ and a heat of fusion of 3.34×10^5 J/kg.

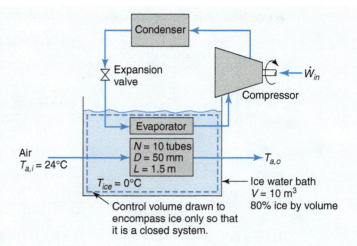

Approach:

We know from previous study that integration of the rate form of the conservation of energy equation can give us a time interval. Hence, we begin by applying conservation of energy only to the ice in the ice-water bath. But from the energy equation, we need a heat transfer rate. That suggests we need to analyze the air/ice-water heat exchanger. We have the heat exchanger geometry, fluids, and two of the temperatures. We do not have the airflow rate, but we do have the fan characteristic curve. If we balance the fan performance against the system requirements, then we can obtain the flow, which will allow us to evaluate the remaining parts of the problem.

Assumptions:

A1. Potential and kinetic energy effects are negligible.

Solution:

We begin with conservation of energy applied only to the ice in the tank. Because the ice is a closed system, and assuming [A1], the equation we need to solve is:

$$\dot{Q} - \dot{W} = \frac{dU}{dt}$$

We must be careful to not eliminate the work term. From daily experience, we know that ice and liquid water have different densities. (Ice floats.) Thus, there must be $p\,dV$ work involved in the melting of the ice, which is a constant pressure process ($P_2 = P_1$). Integrating the equation with respect to time and rearranging,

$$\int \dot{Q}\,dt = \int dU + \int \dot{W}\,dt = (U_2 - U_1) + [P(V_2 - V_1)]$$

$$= m_{ice}\{(u_2 - u_1) + (P_2v_2 - P_1v_1)\} = m_{ice}h_{sf}$$

where h_{sf} is the heat of fusion of ice. We define $X = V_{ice}/V_{tot}$ as the volume fraction of ice in the total volume, so that the mass of ice is $m_{ice} = X\rho_{ice}V_{tot}$.

To evaluate the integral involving the heat transfer term, we must know how $\dot{Q}$ varies with time, and to do that we must consider what is happening around the heat exchanger. The heat transfer coefficient on the air side will be much smaller than that on the ice-water side, so we ignore the ice-water heat transfer coefficient [A2]. (Likewise, we ignore the thermal resistance of the metal tubes [A3].) The temperature of the ice-water mixture remains constant at the freezing point of water. The air inlet temperature and flow rate are constant, so the air heat transfer coefficient is constant, too. Thus, $\dot{Q}$ does not vary with time, and the integral is easily evaluated to give $\int \dot{Q}\,dt = \dot{Q}t$ where t is the time required to melt the ice.

A2. We ignore the heat transfer coefficient on the ice-water side of the heat exchanger.
A3. Ignore the wall thermal resistance.

Combining all the terms into the energy equation, we obtain:

$$t = \frac{X\rho_{ice}V_{tot}h_{sf}}{\dot{Q}}$$

where everything except $\dot{Q}$ is known or can be evaluated from the given information.

We can calculate the heat transfer rate with the energy or rate equation:

$$\dot{Q} = hA \, \Delta T_{LM} = \dot{m}c_p \left(T_{a,i} - T_{a,o}\right)$$

However, the *LMTD* uses the air outlet temperature, $T_{a,o}$, which is unknown. To find $T_{a,o}$ we can use one of the approaches demonstrated in Example 13-5. Using the equation for convection heat transfer between a single-phase fluid and a constant temperature fluid (Chapter 12), and recognizing that for this system $U = h$,

$$T_{a,o} = T_{ice} - \left(T_{ice} - T_{a,i}\right) \exp\left(-hA/\dot{m}c_p\right)$$

The total surface area is $A = N\pi DL$, where N is the number of tubes, L their length, and D their diameter. The air specific heat can be found in the appropriate table. If we knew the value of the air mass flow rate, $\dot{m}$, we could evaluate the air heat transfer coefficient, h, and then we could calculate everything else in the problem.

We have not used the fan performance curve or the statement that we can assume the fittings and other piping to and from the heat exchanger contribute a pressure drop equal to that of the heat exchanger itself. The performance curve shows what the fan can supply to the system. For different total air mass flow rates, we can calculate the pressure drop across the heat exchanger system with

$$\Delta P_{system} = \Delta P_{heat \; exchanger} + \Delta P_{fittings} \simeq 2\Delta P_{heat \; exchanger} = 2\left[f\frac{L}{D}\rho\frac{\mathcal{V}^2}{2}\right]_{heat \; exchanger}$$

We could plot the system demand curve on the same figure as the fan performance curve or could solve the equations analytically. The intersection of the fan curve and the system curve is the operating point of the system.

Considering the evaluation of the friction factor and the heat transfer coefficient, we need the air properties, which should be evaluated at the average air temperature; with an unknown air outlet temperature, we assume an average temperature of 292 K (check this once we have calculated the outlet temperature). From Tables A-7 and A-8: $c_p = 1.007$ kJ/kg·K, $\rho = 1.193$ kg/m^3, $k = 0.0257$ W/m·K, $\mu = 1.81 \times 10^{-5}$ N·s/m^2, and $Pr = 0.709$. The density of ice is 920 kg/m^3, and the heat of fusion is 3.34×10^5 J/kg.

The system demand curve is calculated with

$$\Delta P_{system} \simeq 2\left[f\frac{L}{D}\rho\frac{\mathcal{V}^2}{2}\right]_{heat \; exchanger} \tag{1}$$

The tube length, L, and diameter, D, are given.

Mass flow rate and velocity in each tube are calculated with

$$\dot{m} = \frac{\rho \mathcal{V} A_x}{N} \tag{2}$$

$$\mathcal{V} = \frac{(\dot{m}/N)}{\rho A_x} \tag{3}$$

where $\dot{m}$ is the total mass flow through the heat exchanger, N is the number of tubes, and the cross-sectional flow area is

$$A_x = \frac{\pi D^2}{4}$$

Assuming the flow is turbulent (check this once the mass flow rate has been obtained), the friction factor is

$$f = \frac{0.184}{Re^{0.2}} \tag{4}$$

and the Reynolds number is

$$Re = \frac{\rho \mathcal{V} D}{\mu} \tag{5}$$

The fan performance curve (with ΔP_{fan} in Pa and $\dot{m}$ in kg/s) is given as

$$\Delta P_{fan} = 12.5 - 2.9\dot{m} - 298\dot{m}^2 \tag{6}$$

and, finally,

$$\Delta P_{fan} = \Delta P_{system} \tag{7}$$

This system of equations (Eqs. 1 through 7) is solved for the system mass flow rate and pressure drop. Performing the iterative solution, we obtain

$$\dot{m} = 0.0751 \,\text{kg/s} \qquad Re = 10,562 \qquad f = 0.0289 \qquad \Delta P_{system} = \Delta P_{fan} = 10.6 \,\text{Pa}$$

Above we assumed a turbulent flow; this Reynolds number validates our assumption.

Using the Dittus-Boelter equation, which covers the system's Reynolds and Prandtl numbers, with a Prandtl number exponent of 0.3, the resulting Nusselt number is

$$Nu = 0.023 Re^{0.8} Pr^{0.3} = 0.023 \, (10562)^{0.8} \, (0.709)^{0.3} = 34.4$$

and the heat transfer coefficient is

$$h = \frac{Nu \, k}{D} = \frac{34.4 \, (0.0257 \,\text{W/m·K})}{0.05 \,\text{m}} = 17.7 \, \frac{\text{W}}{\text{m}^2 \text{·K}}$$

This gives an air outlet temperature of

$$T_{a,o} = 0°\text{C} - (0°\text{C} - 24°\text{C}) \exp\left(\frac{-(17.7 \,\text{W/m}^2\text{·K}) \, 10\pi \, (0.05 \,\text{m}) \, (1.5 \,\text{m})}{(0.0751 \,\text{kg/s}) \, (1007 \,\text{J/kg·K})}\right) = 13.8°\text{C}$$

which is about what we assumed above. The heat transfer rate is

$$\dot{Q} = \dot{m} c_p \, (T_{a,i} - T_{a,o}) = (0.0751 \,\text{kg/s}) \, (1007 \,\text{J/kg·K}) (24 - 13.8)°\text{C} = 771 \,\text{W}$$

The time required to melt the ice is

$$t = \frac{X \rho_{ice} V_{tot} h_{sf}}{\dot{Q}} = \frac{0.8 \, (920 \,\text{kg/m}^3) \, (10 \,\text{m}^3) \, (3.34 \times 10^5 \,\text{J/kg}) \, (1 \,\text{W·s/1 J})}{771 \,\text{W}}$$

$$= 3.19 \times 10^6 \,\text{s} = 36.9 \,\text{days}$$

Comments:

The approach of this problem is from large scale to small scale. We began with the overall problem of how we calculate a time. Once we decided on the energy equation, then we examined each term and asked the question: Do we know this quantity or can we calculate it? If we needed to calculate it, then we decided what equation was needed. We continued doing this until we could actually calculate a quantity and used that quantity, in turn, in each preceding step.

13.5 HEAT EXCHANGER SELECTION CONSIDERATIONS

The range of heat exchanger types is large, and to choose a heat exchanger for a particular application can be a difficult decision. For a person with no experience, this could be a nearly overwhelming task. The task can be simplified by taking advantage of the experience of others who have specified heat exchangers that operate under similar operating conditions. Experience is the best guide to heat exchanger type selection, but while previous experience should be carefully considered, it should not be relied on exclusively. Changes in operating conditions, new applications, unusual design requirements, and other out-of-the-ordinary considerations may suggest that a different approach is needed. In those cases, it is important to consider several factors. Below is a brief description of some of the factors to be taken into account during the engineering and selection of a new heat exchanger.

- **Heat transfer performance**   The first and foremost requirement when selecting a heat exchanger is that it must satisfy the application's requirements. To ensure that these requirements are met, a clear statement is given of what is needed. This includes the required heat duty, the fluids, flow rates, inlet and outlet temperatures, pressure levels, and allowable pressure drops. Because design often requires trade-offs between competing factors, the relative importance of each factor should be established when possible.

- **Pressure and temperature**   Certain heat exchanger types cannot be used at high pressures. For example, tubular heat exchangers can withstand high pressures, but heat exchangers with large flat areas and thinner materials (e.g., plate-type or compact) are limited in their maximum allowable pressures. Likewise, many heat exchanger types have restrictions on their allowable temperature level because of gasketing issues and construction type.

- **Materials**   Some materials corrode in the presence of certain fluids. Hence, the materials chosen for the heat exchanger must be compatible with the heat transfer fluids. Likewise, because heat exchangers at start-up have a temperature at the nominal ambient conditions and then reach some operating temperature after the process achieves steady state, differential thermal expansion (caused by materials with different coefficients of thermal expansion) must be taken into account. The material strength must be sufficient to withstand the pressure/temperature operating conditions. Fabrication problems may also be important in the material choice, because not all materials can be soldered, brazed, and/or welded. Likewise, if a gasketed heat exchanger is used, the gasket material must be compatible with the fluids.

- **Size and weight restrictions**   Because of a particular application, there may be length, height, width, volume, or weight restrictions on a heat exchanger. Automotive and aerospace heat exchangers need small volumes and to be light to fit in a moving vehicle. Conversely, in an oil refinery, a shell-and-tube heat exchanger may require a thick steel shell to contain a high-pressure petroleum product, and weight may not be a significant issue. Or consider a large heat exchanger fabricated in a factory: Can the heat exchanger be shipped to where it will be used by truck, or must it be shipped by rail or by barge? Or will it need to be fabricated on-site? Likewise, the support structure for the heat exchanger must accommodate the size and weight of the device. If the fluids used in a heat exchanger are expensive, toxic, or flammable, then the heat exchanger may have to be designed for a small volume.

- **Heat transfer mechanisms**   Different flow paths and geometric configurations are used depending on the mode of heat transfer. Whether the flow is laminar or turbulent, single-phase gas or liquid, or two-phase flow boiling or condensing will have an effect on the choice of surface and heat exchanger type.

- **Fouling tendencies** Fouling is hard to predict, and fouling characteristics for a given application will depend on many parameters, as discussed in Section 13.2. Fluid velocity, flow distribution through the heat exchanger, channel dimensions,

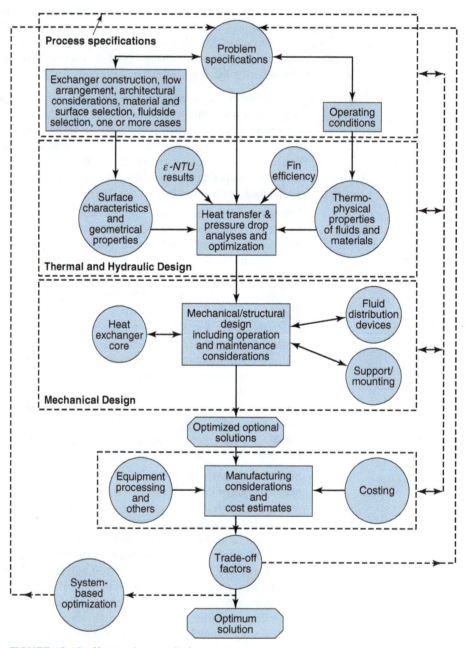

FIGURE 13-10 Heat exchanger design steps.
(*Source*: From F. Kreith, ed., *The CRC Handbook of Thermal Engineering*, CRC Press, Boca Raton, FL, 2000. Used with permission.)

and fluid type are the major contributors to whether or not fouling might become a problem. Experience with a particular fluid and heat exchanger type will give a good indication whether fouling may become a problem.

- **Maintenance**   Maintenance requirements vary widely depending on application and type of heat exchanger. Periodic cleaning and/or replacement of all or part of a heat exchanger may be required to ensure proper operation. If part or all of a heat exchanger is to be removed from a system, then provisions must be made for ease of removal (e.g., leaving enough space around the heat exchanger to remove all or part of it).

- **Safety and environmental requirements**   Issues associated with safety and environmental requirements include leakage of the process fluids into the environment, temperatures of surfaces, inventory of fluids (if toxic or flammable), and wastes from cleaning the exchanger.

- **Cost**   Economic issues always play a role in a heat exchanger specification. When you are assessing the costs for a specific application, some of the factors that may play a role include capital cost (is this an off-the-shelf heat exchanger or will the heat exchanger need to be designed and constructed from scratch?), operating costs (will pumping power be significant, and how big must the pumping equipment be?), and maintenance costs (how frequently will the heat exchanger need to be taken out of service for cleaning and/or repair, and what will be the cost of both the actual cleaning/repair and the lost production while the unit is out of service?).

In this chapter we have focused only on the thermal design of heat exchangers. Figure 13-10 shows many of the steps needed in the actual design, including the thermal-hydraulics, mechanical design, economics, and so on. Note all of the two-headed arrows. The heat exchanger design process is not linear; it involves many trade-offs and iterations. We have only touched on the basic aspects of the design process.

SUMMARY

The objective of a heat exchanger analysis generally is either (1) to determine the heat transfer rate (or *heat duty*) possible with a given heat exchanger (a *rating* problem) or (2) to design a new heat exchanger that will transfer a given heat transfer rate (a *design* or *sizing* problem).

The total thermal resistance between two fluids in a heat exchanger is evaluated by considering five thermal resistances: two convective resistances, one wall resistance, and two fouling resistances. Because these resistances are in series, we write the total thermal resistance as

$$R_{tot} = R_{conv,i} + R_{fouling,i} + R_w + R_{fouling,o} + R_{conv,o}$$

$$= \frac{1}{\eta_{o,i}h_iA_i} + \frac{R_i''}{\eta_{o,i}A_i} + R_w + \frac{R_o''}{\eta_{o,o}A_o} + \frac{1}{\eta_{o,o}h_oA_o}$$

The convective resistances require heat transfer coefficients to be given or evaluated with appropriate correlations. The wall resistance is steady, one-dimensional conduction across a wall. Fouling thermal resistance, $R_{fouling} = R''/\eta_o A$, is present whenever an unwanted substance coats a heat transfer surface and is usually expressed in terms of a *fouling factor*, R''

(see Table 13-1). Care must be exercised when using fouling factors because of large uncertainties in their values.

Instead of using total resistance, the *overall heat transfer coefficient, U,* is more often used when describing the total resistance in a heat exchanger. This quantity is defined by

$$R_{tot} = \frac{1}{UA} = \frac{1}{U_iA_i} = \frac{1}{U_oA_o}$$

$$= \frac{1}{\eta_{o,i}h_iA_i} + \frac{R_i''}{\eta_{o,i}A_i} + R_w + \frac{R_o''}{\eta_{o,o}A_o} + \frac{1}{\eta_{o,o}h_oA_o}$$

When the temperatures of the two fluids in a heat exchanger vary with position, then two methods are used to design or rate the heat exchanger. The *log mean temperature difference (LMTD)* heat exchanger thermal analysis method is governed by the equation

$$\dot{Q} = UAF\Delta T_{LM,cf}$$

where the log mean temperature difference is defined as

$$\Delta T_{LM} = \frac{\Delta T_1 - \Delta T_2}{\ln\left(\frac{\Delta T_1}{\Delta T_2}\right)} = \frac{\Delta T_2 - \Delta T_1}{\ln\left(\frac{\Delta T_2}{\Delta T_1}\right)}$$

The *LMTD* method is the preferred approach to *design* problems.

For a *counterflow heat exchanger* ($F = 1$), the temperature differences to use are

$$\Delta T_1 = T_{H,1} - T_{C,1} = T_{H,i} - T_{C,o}$$
$$\Delta T_2 = T_{H,2} - T_{C,2} = T_{H,o} - T_{C,i}$$

For a *parallel flow heat exchanger* ($F = 1$), the temperature differences to use are

$$\Delta T_1 = T_{H,1} - T_{C,1} = T_{H,i} - T_{C,i}$$
$$\Delta T_2 = T_{H,2} - T_{C,2} = T_{H,o} - T_{C,o}$$

For *all other heat exchangers*, the correction factor F is evaluated from given information about the heat exchanger, and the *LMTD* is *calculated as if the flow arrangement is counterflow.* For these situations F is less than 1, except if one of the fluids is at a constant temperature—then $F = 1$. The value of F for several heat exchanger configurations can be evaluated using Figure 13-7.

The *effectiveness-NTU* (ε-*NTU*) method of analyzing heat exchangers is governed by the equation

$$\dot{Q}_{act} = \varepsilon \dot{Q}_{max} = \varepsilon C_{min} \left(T_{H,i} - T_{C,i} \right)$$

The *effectiveness*, ε, is dependent on the heat exchanger geometry, the *number of transfer units*

$$NTU = \frac{UA}{C_{min}}$$

and the ratio of *heat capacity rates*

$$\frac{C_{min}}{C_{max}} = \frac{\dot{m}c_p\big|_{min}}{\dot{m}c_p\big|_{max}}$$

where C_{min} is the fluid with the smaller value of the product of mass flow rate and specific heat, and C_{max} is the fluid with the larger value of the product of mass flow rate and specific heat. The ε-*NTU* is the preferred method for *rating* problems. Effectiveness for commonly used heat exchangers can be evaluated with expressions from Table 13-3 or the curves in Figure 13-8.

For a heat exchanger in which one fluid remains at a constant temperature, $C_{max} \to \infty$, $C^* = 0$, the geometry becomes irrelevant, and all heat exchangers have the same expression for effectiveness:

$$\varepsilon = 1 - \exp\left(-NTU\right)$$

SELECTED REFERENCES

FRAAS, A. P., *Heat Exchanger Design*, 2nd ed., Wiley, New York, 1989.

HESSELGREAVES, J. E., *Compact Heat Exchangers: Selection, Design and Operation*, Pergamon Press, New York, 2001.

HEWITT, G. F., G. L., SHIRES, and T. R., BOTT, *Process Heat Transfer*, CRC Press, Boca Raton, FL, 1994.

INCROPERA, F. P., and D. P., DEWITT, *Fundamentals of Heat and Mass Transfer*, 5th ed., Wiley, New York, 2001.

KAKAC, S., and H., LIU, *Heat Exchangers: Selection, Rating, and Thermal Design*, 2nd ed., CRC Press, Boca Raton, FL, 2002.

KAYS, W. M., and A. L., London, *Compact Heat Exchangers*, 3rd ed., Krieger Publishing Company, Malabar, FL, 1998.

SAUNDERS, E. A. D., *Heat Exchangers: Selection, Design and Construction*, Wiley, New York, 1988.

SEKULIC, D. P., and R. K., SHAH, *Fundamentals of Heat Exchanger Design*, Wiley, New York, 2002.

SHAH, R. K, A. D, KRAUS, and D. METZGER, *Compact Heat Exchangers*, Hemisphere Publishing Corporation, New York, 1990.

Standards of Tubular Exchanger Manufacturers Association, 7th ed., Tubular Exchanger Manufacturer Association, New York, 1999.

WALKER, G., *Industrial Heat Exchangers: A Basic Guide*, 2nd ed., Hemisphere Publishing Corporation, New York, 1990.

PROBLEMS

OVERALL HEAT TRANSFER COEFFICIENTS

P13-1 Hot exhaust gases are used in the reheat section of a Rankine cycle. Consider a commercial steel tube with 5-cm outside diameter and 4.5-cm inside diameter used to convey the steam. The air side heat transfer coefficient is 85 W/m²·K, and that of the steam side is 200 W/m²·K.

a. Determine the overall heat transfer coefficient based on the inside tube area (in W/m²·K).

b. Determine the overall heat transfer coefficient based on the outside tube area (in W/m²·K).

c. Determine the overall heat transfer coefficient based on the inside area if air-side fouling is 0.0015 m²·K/W and steam-side fouling is 0.0005 m²·K/W (in W/m²·K).

P13-2 A two-shell pass, eight-tube pass heat exchanger with a surface area of 8300 ft² is used to heat 1700 lbm/min of water from 75°F to 210°F. Hot exhaust gases enter at 570°F and exit at 255°F. Assuming the exhaust gases have the same properties as air, determine

a. the overall heat transfer coefficient (in Btu/h·ft^2·°F).

b. the overall heat transfer coefficient if fouling on both sides equivalent to 0.005 h·ft^2·°F/Btu is present in the heat exchanger (in Btu/h·ft^2·°F).

P13-3   In a desalination plant, salt water is used to create pure water. Salt water is boiled, and salt concentrates in the boiler; the saltwater solution is drained from the boiler, and the pure water vapor is condensed for use. Condensing vapor at a high pressure is used to boil salt water at a lower pressure. Consider an experiment on a single tube: Condensing steam at 105°C inside the tube is used to boil salt water at 85°C. The 304 stainless-steel tube is 3 m long, has a 2.5 cm inside diameter, and is 2 mm thick. The overall heat transfer coefficient based on the inside area is 830 W/m^2·K, and the condensing coefficient is 1500 W/m^2·K. Determine the heat transfer coefficient of the boiling salt water (in W/m^2·K).

P13-4   The performance characteristics of a finned, cross-flow heat exchanger are determined in a laboratory. The heat exchanger has 100 tubes that have inside diameters of 12 mm and lengths of 2.6 m; a dense array of continuous plate fins is attached to the outside of the tubes. At one particular operating condition, measurements on the heat exchanger are: hot water inlet temperature, 174°C; hot water outlet temperature, 121°C; hot water flow rate, 0.00051 m^3/s; cold air inlet temperature, 25°C; cold air inlet pressure, 97 kPa; cold air inlet flow rate, 2.2 m^3/s. Determine the overall heat transfer coefficient based on the inside tube area (in W/m^2·K).

P13-5   Water at 200°F flows inside a 304 stainless-steel tube with a 1-in. inside diameter wall thickness of 0.05 in. Air flows over the outside surface of the tube. The water-side heat transfer coefficient is 80 Btu/h·ft^2·°F, while that of the air side is 40 Btu/h·ft^2·°F.

a. Determine the overall heat transfer coefficient based on the inside surface area (in Btu/h·ft^2·°F).

b. Determine the overall heat transfer coefficient if the air-side fouling factor is 0.0007 h·ft^2·°F/Btu and that on the water side is 0.0003 h·ft^2·°F/Btu (in Btu/h·ft^2·°F).

P13-6   A heat exchanger tube with a 25-mm outside diameter has 20 longitudinal fins with rectangular cross-sections equally spaced around the circumference of the tube. The fins are 25 mm from base to tip and 1.6 mm thick. The tube has a 2-mm wall thickness, and tube and fins are both made of plain carbon steel ($k = 60.5$ W/m·K). The inside and outside convective heat transfer coefficients are 1000 W/m^2·K and 200 W/m^2·K, respectively.

a. Determine the overall heat transfer coefficient based on the inside surface area (in W/m^2·K).

b. Determine the overall heat transfer coefficient based on the outside area (in W/m^2·K).

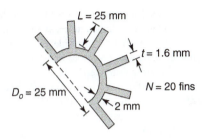

P13-7   Very thin-walled low-chromium steel ($k = 37$ W/m·K) tubes of diameter 10 mm are used in a condenser. A convection coefficient of $h_i = 5000$ W/m^2·K is associated with condensation on the inner surface of the tubes, while a coefficient of $h_o = 100$ W/m^2·K is maintained by airflow over the tubes. For a 1-m-long section of tube with 286 fins, determine

a. the overall heat transfer coefficient if the tubes are unfinned (in W/m^2·K).

b. the fin efficiency and overall heat transfer coefficient based on inner area if low-chromium-steel annular fins of thickness $t = 1.5$ mm, outer diameter $D_o = 20$ mm, and axial spacing $S = 3.5$ mm are added to the outer tube surface (in W/m^2·K).

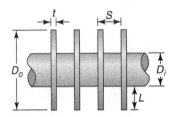

P13-8   A double-pipe heat exchanger consists of a 4-cm pipe inside a 6-cm pipe; the heat exchanger is 2 m long. The water inside the inner pipe has an average temperature of 40°C and a flow rate of 0.016 m^3/s. In the annulus (between the inner and outer pipes), unused engine oil has an average temperature of 147°C and a flow rate of 0.01 m^3/s. The inner tube has a wall thickness of 1 mm and is made of 304 stainless steel.

a. Determine the overall heat transfer coefficient based on the outside area of the inner tube (in W/m^2·K).

b. Determine the overall heat transfer coefficient based on the outside surface area of the inner tube if the water and oil sides are fouled; choose representative fouling factors from Table 13-1 (in W/m^2·K).

P13-9   Ethylene glycol enters a double-pipe heat exchanger at 17°C with flow rate of 1.5 kg/s. It is heated with water that enters the heat exchanger at 100°C with a flow rate of 0.04 kg/s. The inner pipe is 2.5 cm in diameter, the outer pipe is 3.75 cm in diameter, and the length is 3 m. Determine the overall heat transfer coefficient (in W/m^2·K) and the heat transfer rate (in W) if:

a. the water flows in the inner tube.

b. the water flows in the annular space between the two tubes.

P13-10 A heat exchanger used to heat air with hot water is constructed of individually finned tubes, as shown in the figure. The tube (1 m long with 10-mm inside diameter and 13-mm outside diameter) and fins (12 mm long and 0.5 mm thick, spaced on 5-mm centers) are constructed of brass. Air flows over the tubes with a heat transfer coefficient of 100 W/m²·K. Water with a velocity of 2 m/s enters the tube at 80°C. Determine the overall heat transfer coefficients based on the inside area, U_i, and the outside area, U_o, (in W/m²·K).

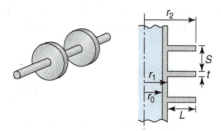

P13-11 Hot water at 100°C flows at a rate of 4.5×10^{-4} m³/s through a horizontal 316 stainless-steel pipe with a 5-cm inside diameter and a 5-mm wall thickness. Outside of the pipe is still air at 25°C and 1 atm. Determine the overall heat transfer coefficient based on the inside and outside surface areas of the pipe (in W/m²·K).

P13-12 Ethylene glycol flows inside a copper tube that has a 0.5-in. inside diameter and a 0.65-in. outside diameter. The heat transfer coefficient for the ethylene glycol is 300 Btu/h·ft²·°F. Water flows outside the tube and has a heat transfer coefficient of 550 Btu/h·ft²·°F.

a. Determine the overall heat transfer coefficient based on the outside tube area (in Btu/h·ft²·°F).

b. Determine the overall heat transfer coefficient based on the outside tube area if fouling is present on both the water and ethylene glycol sides (in Btu/h·ft²·°F). (Estimate fouling factors from Table 13-1.)

c. Discuss how much the overall heat transfer coefficient can vary depending on the choice of fouling factor.

LMTD METHOD

P13-13 To use as much energy as possible from the combustion of natural gas, heat exchangers are often placed in exhaust stacks to recover waste energy. Consider a single-pass crossflow heat exchanger. Exhaust gases (assume air properties) enter at 180°F with a flow rate of 0.31 lbm/s and exit at 130°F. Fresh air enters at 70°F with a flow rate of 0.62 lbm/s. The heat exchanger construction is such that both fluids are unmixed, and the overall heat transfer coefficient is estimated to be 35 Btu/h·ft²·°F. Determine the required area of the heat exchanger (in ft²).

P13-14 A shell-and-tube heat exchanger has tubes with 18-mm outside diameter and a wall thickness of 1.2 mm. Cold

water outside the tubes with a flow rate of 250 kg/min is heated from 30°C to 50°C with hot water that enters the heat exchanger at 105°C with a flow rate of 150 kg/min. The company design specification is to use a fluid velocity inside the tubes of about 0.4 m/s. From previous designs, the overall heat transfer coefficient based on the inside surface area is estimated to be 1800 W/m²·K. Determine the number of tubes and the required tube length if the heat exchanger is

a. counterflow.

b. parallel flow.

c. one-shell pass and two-tube passes.

d. two-shell passes and four-tube passes.

P13-15 A counterflow, concentric tube heat exchanger is designed to heat water from 20°C to 80°C using hot oil, which is supplied to the annulus at 160°C and discharged at 140°C. The thin-walled inner tube has a diameter of $D_i = 20$ mm, and the overall heat transfer coefficient is 500 W/m²·K. The design condition calls for a total heat transfer rate of 3000 W. Determine the length of the heat exchanger (in m). After three years of operation, performance is degraded by fouling on the water side of the exchanger, and the water outlet temperature is only 65°C for the same fluid flow rates and inlet temperatures. What are the corresponding values of the heat transfer rate, outlet temperature of the oil, overall heat transfer coefficient, and water-side fouling factor?

P13-16 In a cogeneration plant, the exhaust from the turbine in a Brayton cycle is used in a crossflow heat exchanger to heat pressurized liquid water inside tubes from 300°F to 400°F. The exhaust gas flow enters the heat exchanger at 850°F with a flow rate of 18 lbm/s and is considered unmixed. The overall heat transfer coefficient is 80 Btu/h·ft²·°F. The tubes are 1 in. in diameter and 16 ft long. If the heat exchanger effectiveness must be at least 75%, determine

a. the water flow rate (in lbm/s).

b. the number of tubes.

P13-17 In a refrigeration unit, R-134a at 0.18 MPa is evaporated inside in a long, thin-walled tube. The refrigerant, whose flow rate is 0.001 kg/s, enters the tube as a saturated liquid and exits as a saturated vapor, and its heat transfer coefficient is 500 W/m²·K. Air at 27°C flows with velocity of 6 m/s perpendicular to the outside of the tube. Shown in the figure is the aluminum tube ($k = 177$ W/m·K), which has eight rectangular fins inside the tube. Each fin is 5 mm long and 1 mm thick. The tube diameter is 3 cm. Determine the required tube length (in m).

P13-18 A one-shell-pass, four-tube-pass heat exchanger contains 20 thin-walled 25-mm tubes. It must be designed to heat 2.5 kg/s of water from 15°C to 85°C. The heating is to be accomplished with hot oil (c_p = 2.35 kJ/kg·K), which enters the shell side of the heat exchanger at 160°C. The oil heat transfer coefficient is 400 W/m²·K. The oil leaves the heat exchanger at 100°C. Determine the length of the shell required (in m).

P13-19 Because of its construction, the heat transfer area of a plate heat exchanger can be changed easily by adding or removing plates; in addition, counterflow can be achieved, which results in good performance. Consider the counterflow plate heat exchanger shown in the figure. The plates are 304 stainless steel 1 mm thick, 2 m wide, and 3 m long. The channels on the hot and cold sides have 5-mm gaps. Engine oil enters at 80°C with a flow rate of 0.03 m³/s and should leave at 55°C. Water, in counterflow, enters at 20°C and should leave no hotter than 30°C.

a. Determine the required water mass flow rate (in kg/s).

b. Determine the number of channels required.

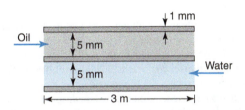

P13-20 In a counterflow heat exchanger, 3.6 kg/s of an organic fluid flows with a specific heat of 850 J/kg·K; it enters the heat exchanger at 12°C and leaves at 340°C. A high-temperature oil with a specific heat of 1900 J/kg·K enters at 650°C with a flow rate of 3 kg/s. If the outlet temperature of the cooler fluid must be increased to 450°C, with everything else remaining constant, determine the percentage increase in the heat transfer area required.

P13-21 The water flow in Problem P 13-5 is 130 lbm/min. The air enters the heat exchanger at 700°F and leaves at 500°F with a flow rate of 300 lbm/min. If no fouling is present, determine the inside heat transfer area (in ft²) if the heat exchanger is

a. counterflow.

b. parallel flow.

P13-22 The counterflow heat exchanger in Problem P 13-21 is operated for a year. All flows and inlet temperatures remain constant, but the hot fluid exits at 550°F. Determine the magnitude of the fouling factor (in h·ft²·°F/Btu).

P13-23 A closed feedwater heater is used in a Rankine cycle power plant. The feedwater (150 kg/s) is to be heated from 30°C to 90°C using steam extracted from the turbine at 200 kPa at a quality of 0.987, and the condensate should leave as a saturated liquid at 200 kPa. The overall heat transfer coefficient is estimated to be 2000 W/m²·K.

a. Determine the required heat transfer area (in m²).

b. Determine the condensate flow rate (in kg/s).

P13-24 A shell-and-tube heat exchanger is to be constructed with 0.75-in. outside diameter, 0.03-in.-thick tubes. Cold water inside the tubes has a flow rate of 500 lbm/min and is to be heated from 80°F to 110°F. Hot water with a flow rate of 350 lbm/min enters the heat exchanger at 210°F. The overall heat transfer coefficient based on the outside area is 300 Btu/h·ft²·°F. For one shell pass, tube-side water velocity of 1 ft/s, and a maximum tube length of 8 ft, determine

a. the number of tubes per pass.

b. the number of tube passes.

c. the length of the tubes (in ft).

P13-25 A crossflow heat exchanger is to be designed to heat hydrogen gas with hot water. The water is on the tube side and enters at 150°C at a flow rate of 3 kg/s with a heat transfer coefficient of 1250 W/m²·K. The hydrogen (c_p = 14.4 kJ/kg·K) is on the shell side and enters at 30°C at a flow rate of 120 kg/min with a heat transfer coefficient of 1800 W/m²·K. The required hydrogen outlet temperature is 60°C. The heat exchanger has 100 2.5-mm-thick tubes, with inside diameters of 1.5 cm, made of 347 stainless steel.

a. Determine the overall heat transfer coefficient based on the inside area (in W/m²·K).

b. Determine the required tube length (in m).

P13-26 A small oil refinery uses river water to cool some of the fluid streams in the refinery. Consider a two-shell-pass, four-tube-pass heat exchanger that uses 25 kg/s of river water at 10°C on the shell side to cool 20 kg/s of process fluid (c_p = 2300 J/kg·K) from 80°C to 25°C. If the overall heat transfer coefficient is 600 W/m²·K, determine

a. the outlet temperature of the coolant (in°C).

b. the heat transfer area required (in m²).

P13-27 Water is heated from 25°C to 80°C in a one-shell-pass, two-tube-pass shell-and-tube heat exchanger. The hot fluid is oil (c_p = 1750 J/kg·K) with a flow rate of 1 kg/s that enters the tube side of the heat exchanger at 175°C and exits at 145°C. If the overall heat transfer coefficient is 350 W/m²·K, determine

a. the heat transfer rate (in W).

b. the water flow rate (in kg/s).

c. the required heat transfer area (in m²).

P13-28 Car radiators are single-pass crossflow heat exchangers with both fluids unmixed. Water at 0.05 kg/s enters the tubes at 125°C and leaves at 55°C. Air enters the heat exchanger at 35 m³/min, 25°C, and 97 kPa. The overall heat transfer coefficient is 225 W/m²·K. Determine the required heat transfer area (in m²).

P13-29 A small Rankine cycle power plant is used in a ship. The condenser is cooled by seawater. Consider a one-shell-pass (steam-side), four-tube-pass (seawater-side) shell-and-tube heat exchanger. Steam enters the condenser at 50°C with a quality of 95% and a flow rate of 0.75 kg/s and exits as a saturated liquid; its condensing heat transfer coefficient is approximately 7,500 W/m^2·K. Seawater enters the condenser at 18°C, and its temperature at the exit should be no higher than 40°C. Assume seawater properties can be approximated with freshwater properties. The heat exchanger has 20 brass tubes of 2.5-cm inside diameter and 2.8-cm outside diameter.

a. Determine the water-side heat transfer coefficient (in W/m^2·K).

b. Determine the overall heat transfer coefficient based on the inside area (in W/m^2·K).

c. Determine the tube length required (in m).

d. Determine the tube length required if, after a long time in service, both sides of the heat exchanger have been fouled (in m).

P13-30 The oil cooler in a large diesel engine is a one-shell-pass, four-tube-pass shell-and-tube heat exchanger with 15 brass tubes of 10-mm outside diameter and 1-mm wall thickness. Oil enters the tubes at 135 °C and 0.5 kg/s and leaves at 95°C. Water enters the shell at 15°C with a flow rate of 2 kg/s and a heat transfer coefficient of 1100 W/m^2·K. Determine the shell length (in m).

P13-31 The regenerator in a small Brayton cycle is a single-pass crossflow heat exchanger with both fluids unmixed. Compressed air enters the exchanger at 300°C at 1.5 kg/s. Hot exhaust gases enter the exchanger at 850°C at 1.6 kg/s; assume the properties can be estimated as air. The overall heat transfer coefficient is 250 W/m^2·K. If we want a heat exchanger effectiveness of 75%, determine the surface area required (in m^2).

P13-32 A boiler is constructed as an unfinned crossflow heat exchanger. Hot gases at 1200°C enter the heat exchanger and flow over 400 25-mm-diameter tubes at 12 kg/s; assume the hot gas properties as those of air. Saturated liquid water enters the tubes at 8 MPa with a flow of 3.5 kg/s and leaves as a saturated vapor. The overall heat transfer coefficient is 75 W/m^2·K.

a. Determine the gas outlet temperature (in°C).

b. Determine the required tube length (in m).

P13-33 To condense 3 kg/s of saturated steam at 40°C, a shell-and-tube heat exchanger with one shell pass (steam side) and several tube passes is used. The condensing heat transfer coefficient is 11,000 W/m^2·K. Cooling water enters the 19-mm wall tubes at 15°C and exits at 24°C; the maximum velocity allowable is 1.5 m/s.

a. Determine the number of tubes required.

b. Determine the number of passes required if the maximum shell length is 2 m.

c. Determine the actual length per pass (in m).

d. Determine the percent increase in heat transfer rate if the water velocity is increased to 1.75 m/s and all other conditions remain the same as in parts a, b, and c.

P13-34 The regenerator in a Brayton cycle power plant is a crossflow heat exchanger. Air enters the regenerator at 200°C and exits at 380°C with a flow rate of 10 kg/s. Exhaust gases enter at 580°C and leave at 325°C; their properties can be approximated with those of air. The overall heat transfer coefficient is estimated to be 150 W/m^2·K.

a. Determine the required heat transfer area if both fluids are unmixed (in m^2).

b. Determine the heat transfer area if the air is unmixed and the exhaust gas is mixed (in m^2).

c. Determine the heat exchanger effectiveness in parts a and b.

P13-35 An oil cooler operates in counterflow mode. Oil (c_p = 0.5 Btu/lbm·°F) enters the heat exchanger at 195°F and leaves at 125°F with a flow rate of 400 lbm/min. Water enters at 80°F. The overall heat transfer coefficient is 100 Btu/h·ft^2·°F, and the heat transfer area is 360 ft^2. Determine the water flow rate (in lbm/min).

P13-36 A single-pass shell-and-tube heat exchanger in counterflow is to be used to heat 5000 gal/min of water from 50°F to 90°F using condensing steam on the shell side at 1 atm. The condensing heat transfer coefficient is 2000 Btu/h·ft^2·°F. The tubes are carbon steel with a 1.32-in. outside diameter and a 1.05-in. inside diameter. The maximum pressure drop through the tubes is 5 lbf/in.^{2}. Determine the required number of tubes in parallel and the tube length (in ft).

P13-37 A counterflow heat exchanger is designed to cool 2.0 kg/s of air from 70°C to 40°C. Cold air at 10°C enters on the other side with a flow rate of 2.6 kg/s. For a modified application, the basic design of the heat exchanger will remain the same, as will the two airflow rates and the cold air inlet temperature. However, the hot air now enters at 67°C and must leave at 25°C. Assume the fluid properties of the air are constant and equal on both sides. Determine the ratio of the length of the new heat exchanger to the length of the original heat exchanger.

ε-NTU METHOD

P13-38 Saturated steam at 100°C condenses in a shell-and-tube heat exchanger (one shell pass, two tube passes) with a surface area of 0.5 m^2 and an overall heat transfer coefficient of 2000 W/m^2·K. Water enters at 0.5 kg/s and 15°C.

a. Determine the outlet temperature of the water (in °C).

b. Determine the rate of steam condensation (in kg/s).

P13-39 A crossflow condenser for a two-speed air-conditioning system has both fluids unmixed. At the highest fan speed, the heat transfer rate is 35 kW and the refrigerant condenses at 65°C. The air inlet temperature is 40°C, and the air cannot have more than a 5°C temperature rise; the overall heat transfer coefficient is 150 W/m²·K. At the lower fan speed, the air velocity is half of that at the high speed, and the overall heat transfer coefficient is 125 W/m²·K. Determine the percentage decrease in heat transfer rate at the low fan speed compared to the high fan speed.

P13-40 A shell-and-tube heat exchanger with single shell and tube passes in counterflow is used to cool the oil of a large marine engine. Lake water (shell-side fluid) enters the heat exchanger at 2.0 kg/s and 15°C, while the oil enters at 1.0 kg/s and 100°C. The oil flows through 100 brass tubes, each 500 mm long and having inner and outer diameters of 6 mm and 8 mm, respectively. The shell-side heat transfer coefficient is 500 W/m²·K. Determine the oil outlet temperature (in °C).

P13-41 Water enters a heat exchanger at 70°C with a flow rate of 2 kg/s. On the other side, air enters at 25°C with a flow rate of 3 kg/s. The heat transfer area is 15 m², and the overall heat transfer coefficient is 200 W/m²·K. Determine the heat transfer rate (in kW) if the heat exchanger is

a. counterflow.

b. parallel flow.

c. crossflow with oneflow–the airflow–unmixed.

d. crossflow with both flows unmixed.

P13-42 A two-shell-pass, eight-tube-pass heat exchanger uses liquid water at 100°C to heat 2.4 kg/s of a fluid ($c_p = 2.7$ kJ/kg·K) from 25°C to 50°C. The water exits the heat exchanger at 50°C. The overall heat transfer coefficient is 700 W/m²·K. Determine the heat transfer area (in m²) using

a. the *LMTD* method.

b. the *ε-NTU* method.

P13-43 Hot air at 250°C, 100 kPa with a flow rate of 0.8 kg/s leaves a counterflow heat exchanger at 100°C. On the other side of the heat exchanger, oil ($c_p = 2100$ J/kg·K) enters at 35°C and leaves at 110°C. The overall heat transfer coefficient is estimated to be 85 W/m²·K.

a. Determine the required heat transfer area (in m²).

b. Determine the oil and air outlet temperatures if the area is increased to 25 m² (in °C).

P13-44 A shell-and-tube heat exchanger has 135 tubes (12.5-mm I.D., 0.4-mm wall thickness) in a double-pass arrangement. Each tube pass is 4.48 m long. Total inside surface area is 47.5 m². Hot exhaust gas ($c_p = 1.02$ kJ/kg·K) at 250°C flows outside of the tubes at 10 kg/s; the gas-side heat transfer coefficient is 700 W/m²·K. Boiler feedwater enters the tubes at 65°C and flows at a total flow rate of 5 kg/s. The fouling factor on the water side is 0.0002 m²·K/W. The air-side fouling factor has the same value. Ignoring wall resistance, determine the heat transfer rate (in kW).

P13-45 An engine oil cooler is made from a single tube (10 mm in diameter, 3 m long) laid out in a serpentine path with fins on the tube outside surface; both fluids are unmixed in this crossflow heat exchanger. The air-side effective area is 12 times the inside area. Air at 35°C blows perpendicular to the plane of the serpentine fin-covered tube with a flow rate of 0.6 kg/s and a heat transfer coefficient of 120 W/m²·K. Oil enters the tube at 75°C with a flow rate of 0.025 kg/s.

a. Determine the overall heat transfer coefficient based on the inside surface area assuming fully developed flow (in W/m²·K).

b. Determine the oil exit temperature (in °C).

c. Determine the oil exit temperature if entrance effects are taken into account (in °C).

P13-46 A shell-and-tube heat exchanger with 1 shell pass and 20 tube passes uses hot water on the tube side to heat unused engine oil on the shell side. The single 304 stainless-steel tube has inner and outer diameters of 20 and 24 mm, respectively, and a length per pass of 3 m. The water enters at 87°C and 0.2 kg/s. The oil enters at 7°C and 0.9 kg/s. The shell-side (oil) heat transfer coefficient is 1880 W/m²·K, and the tube-side (water) heat transfer coefficent is 3250 W/m²·K.

a. Determine the outlet temperature of the oil (in °C).

b. Determine the new outlet temperature of the oil if, over time, the oil fouls the surface such that a fouling factor of 0.003 m²·K/W can be assumed (in °C).

P13-47 Air at 27°C, 100 kPa approaches a crossflow heat exchanger with a velocity of 3.4 m/s. Hot water enters the tubes at 93°C; mass flow rate of 1.66 kg/s. The heat exchanger (mixed on the shell side) has 70 3-cm-diameter, 2-m-long tubes. (Neglect wall resistance.) The tubes are placed five deep in an in-line array with longitudinal and transverse distances between tube centers of 3.75 cm. The air-side heat transfer coefficient is 125 W/m²·K. Determine the heat transfer rate (in W).

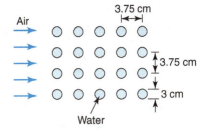

P13-48 A low-pressure boiler is a shell-and-tube heat exchanger with one shell pass and two tube passes with 100 thin-walled tubes, each with a diameter of 20 mm and a length (per pass) of 2 m. Pressurized liquid water enters the tubes at 10 kg/s and 185°C and is cooled by boiling the water at 1 atm on the outer surface of the tubes. The heat transfer coefficient of the boiling water is 4000 W/m²·K. Determine the liquid water outlet temperature (in °C).

P13-49 After the low-pressure boiler described in Problem P 13-48 has operated for six months, fouling occurs such that the fouling factor is 0.0005 $m^2 \cdot K/W$.

a. Determine the new outlet temperature (in °C).

b. Determine the percent decrease in heat transfer rate.

P13-50 Liquid R-134a ($c_p = 1260$ J/kg·K) flows inside the inner tube of a double-pipe heat exchanger at –20°C with a flow rate of 0.265 kg/s; the heat transfer coefficient is 800 $W/m^2 \cdot K$. In counterflow, water at 25°C has a flow rate of 0.14 kg/s. The thin-wall inner tube has a diameter of 2 cm, and the outer tube has a diameter of 3 cm; both are 8 m long.

a. Determine the heat transfer rate (in W).

b. Determine the water and refrigerant outlet temperatures (in °C).

c. Determine whether ice will form. (*Hint*: calculate wall temperatures).

P13-51 Saturated steam at 0.15 bar is condensed in a shell-and-tube heat exchanger with one shell pass and two tube passes with 130 brass tubes ($k = 114$ W/m·K), each with a length per pass of 2 m. The tubes have inner and outer diameters of 13.4 mm and 15.9 mm, respectively. Cooling water enters the tubes at 20°C with a total flow rate of 23.0 kg/s. The heat transfer coefficient for condensation on the outer surfaces of the tubes is 10,000 $W/m^2 \cdot K$.

a. Determine the overall heat transfer coefficient (in $W/m^2 \cdot K$) based on the outside surface area.

b. Determine the cooling water outlet temperature (in °C).

c. Determine the steam condensation rate (in kg/s).

P13-52 Milk is pasteurized in a plate heat exchanger with hot water. The two parallel passages in the heat exchanger are formed with 1-mm-thick 304 stainless-steel plates that are 3 m high and 1.2 m wide. The gap between the plates for both the water and milk flows is 5 mm. Water at 85°C enters the heat exchanger at a flow rate of 4 kg/s. The milk enters the heat exchanger at 5°C with a flow rate of 3 kg/s. The milk properties are: $\rho = 1040$ kg/m³, $\mu = 0.0021$ N·s/m², $c_p = 3900$ J/kg·K, and $k = 0.65$ W/m·K. Determine the exit temperature (in °C) of the milk if the heat exchanger is

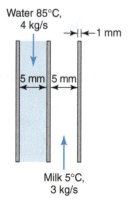

Water 85°C,
4 kg/s
→||←1 mm

5 mm | 5 mm

Milk 5°C,
3 kg/s

a. counterflow.

b. parallel flow.

P13-53 For the heat exchanger described in Problem P 13-52, if the flow length is doubled (for example, by having a 180° bend at the end of one pass so that the overall length of the heat exchanger remains 3 m), determine the milk exit temperature for a counterflow arrangement (in °C).

P13-54 In a processing plant, a single-pass heat exchanger uses condensing saturated steam at 20 psia to heat 45,000 lbm/h of air from 70°F to 170°F. The owners of the plant want to increase production; to do so, the airflow rate must be doubled. The same heat exchanger is to be used, and the outlet temperature must remain at 170°F. Assuming the overall heat transfer coefficient increases by 20% at the higher flow, determine the new required steam pressure (in psia).

P13-55 A space heater used in a university gymnasium is constructed of 60 brass tubes with 0.63-in. outside diameter, 0.48-in. inside diameter, and 3-ft length. The air blower provides 2000 ft³/min of air at 65°F and the heat transfer coefficient is 50 Btu/h·ft²·°F. Inside the tubes, 10-psig saturated steam is condensed with a heat transfer coefficient of 750 Btu/h·ft²·°F.

a. Determine the heat transfer rate (in Btu/h).

b. Determine the air exit temperature (in °F).

c. Determine the steam condensation rate (in lbm/min).

P13-56 A proposed ocean thermal energy conversion (OTEC) power plant uses ammonia as the working fluid in a Rankine cycle. Warm water from the surface of the ocean (80°F) is the heat source used to vaporize the ammonia; cold water (45°F) pumped from low ocean depths is used to condense the ammonia. Because of the small temperature difference between the warm and cold water, the cycle thermal efficiency is very low. The cycle has four identical evaporators. Each evaporator has 120,000 aluminum tubes ($k = 92$ Btu/h·ft·°F) of 2.0-in. outside diameter and 0.04-in. wall thickness that are 55.2 ft long. Water enters the evaporator at 80°F with a velocity in each tube of 5.3 ft/s. The ammonia evaporates at 72°F with a heat transfer coefficient of 1,500 Btu/h·ft²·°F. The water-side fouling factor is 0.0003 h·ft²·°F/Btu. Seawater properties are: $\rho = 64.1$ lbm/ft³, $\mu = 2.32$ lbm/h·ft, $c_p = 0.94$ Btu/lbm·°F, and $k = 0.340$ Btu/h·ft·°F.

a. Determine the heat transfer rate in one evaporator (in Btu/h).

b. Determine the maximum theoretical power output from a plant using four evaporators (in kW).

P13-57 A one-shell-pass, two-tube-pass heat exchanger uses condensing steam on the shell side with a heat transfer coefficient of 3000 $W/m^2 \cdot K$ to heat liquid water from 27°C to 68°C. The water flow rate is 5 kg/s. The 2-m-long heat exchanger has 25 tubes of 304 stainless steel, each of 2-cm inside diameter with 1-mm wall thickness. Determine the required steam pressure (in kPa).

P13-58 Hot exhaust gases are used on the shell-side of a two-shell-pass, four-tube-pass shell-and-tube heat exchanger to heat

2.5 kg/s of liquid water from 35°C to 85°C. The gases, assumed to have the properties of air, enter at 200°C and leave at 100°C. The overall heat transfer coefficient when the exchanger is clean is 180 W/m²·K. If a fouling factor of 0.0006 m²·K/W is known to exist after operating for a period of time, determine the *additional* area required in the heat exchanger to have the same heat transfer rate (in m²).

P13-59 A crossflow heat exchanger has 50 tubes made from 302 stainless steel; each tube is 4 m long and of 2.5-cm inside diameter and 2.5-mm wall thickness. Water enters the tubes at 27°C with a total flow rate of 150 kg/min. Airflow on the shell side (mixed) enters at 260°C with a flow rate of 100 kg/min; the shell-side heat transfer coefficient is 525 W/m²·K.

a. Determine the heat transfer rate (in W).

b. Determine the water outlet temperature (in °C).

P13-60 In a crossflow heat exchanger, the hot and cold sides are separated by a plate 1.0 mm thick. The hot side of the plate has straight rectangular cross-section fins 5 mm long and 0.1 mm thick, spaced 4 mm on center. The cold side of the plate also has straight rectangular cross-section fins 5 mm long, 0.1 mm thick, and spaced 3 mm on center. The hot-side fluid (c_p = 1.3 kJ/kg·K) has a flow rate of 70 kg/h, enters at 250°C, and has a heat transfer coefficient of 80 W/m²·K. The cold-side fluid (c_p = 2.1 kJ/kg·K) has a flow rate of 90 kg/h, enters at 70°C, and has a heat transfer coefficient of 80 W/m²·K. The height of both the hot- and cold-side flow passages is 5 mm. The heat exchanger length in the direction of hot flow is 1 m and that in the direction

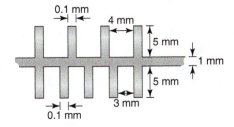

of cold flow is 0.75 m. The separating plate and fins are 2024-T6 aluminum.

a. Determine the overall heat transfer coefficient based on the hot side area (in W/m²·K).

b. Determine the fluid outlet temperatures (in °C).

P13-61 Your supervisor assigns you the task of purchasing a heat exchanger to cool 85 gal/min of oil (μ = 139× 10^{-5} lbm/ft·s, k = 0.074 Btu/h·ft·°F, c_p = 0.52 Btu/lbm·°F, ρ = 53 lbm/ft³) from 250°F using 18,750 ft³/min of air at 1 atm and 70°F. However, in the storage building you find a new single-pass crossflow heat exchanger that has a 25 × 25 array of 0.025-in. thick 304 stainless-steel tubes (k = 9.4 Btu/h·ft·°F) that are 2 ft long and of 0.5-in. outside diameter. The air-side (outside the tubes) heat transfer coefficient is 80 Btu/h· ft²·°F.

a. Determine the possible heat transfer rate (in Btu/h).

b. Determine the oil outlet temperature (in °F).

c. Determine the new heat transfer rate and oil outlet temperature if an oil fouling factor of 0.005 h·ft²·°F/Btu and an air fouling factor of 0.002 h·ft²·°F/Btu are used.

P13-62 A steam condenser is constructed of 100 brass tubes that are 8 ft long and of 1.25-in. outside diameter and 1.0-in. inside diameter. Water enters the tubes at 60°F with a total flow rate of 800 gal/min. Saturated steam at 5 psia is condensed on the shell side of the heat exchanger and has a heat transfer coefficient of 1250 Btu/h·ft²·°F.

a. Determine the heat transfer rate (in Btu/h).

b. Determine the outlet temperature of the water (in °F).

P13-63 The condenser tubes described in Problem P 13-62 are retrofitted with annular exterior fins. The brass fins are 0.25 in. long, 0.125 in. thick, and spaced 0.375 in. apart. All other dimensions remain the same, as do the heat transfer coefficients on both sides.

a. Determine the heat transfer rate (in Btu/h).

b. Determine the outlet temperature of the water (in °F).

CHAPTER 14

RADIATION HEAT TRANSFER

14.1 INTRODUCTION

Radiation is the transmission of energy by electromagnetic waves. All materials found in everyday life, such as the pages of this book, the walls of a room, tabletops, lampshades, people, and so on emit thermal radiation as long as their temperatures are above absolute zero. Heat is transferred when the radiative energy emitted by one body is absorbed by another and can occur whether or not there is a medium between the source of radiation and the body absorbing the radiation. For example, radiation from the sun travels through the vacuum of outer space and penetrates the glass of a window pane before being absorbed in a room.

Radiation is the mode of heat transfer most likely to be overlooked by a novice engineer. In high-temperature applications, radiation is usually important; however, it can also be significant in many moderate- and low-temperature applications. When other modes of heat transfer are weak, radiation remains and must be considered.

Examples of thermal radiation applications include the following:

At high temperature

- The spread of fires is largely determined by radiative effects.
- Space heaters provide warmth by radiative heating.
- The filament of a lightbulb produces not only light but also heat, which raises the temperature of the glass of the bulb.
- Gases and soot in combustion chambers exchange heat by radiation throughout their volume and with the walls of the chamber.
- In optical-fiber manufacture, molten glass is pulled into filaments inside a high-temperature furnace. The speed of processing is determined by the radiative interchange between the partially transparent glass and the walls of the furnace.
- In laser cutting and welding, a high-temperature plasma forms above the workpiece and exchanges heat by radiation with the surface below.
- In stars, thermal radiation strongly influences the temperature distribution in the stellar plasma.

At moderate temperature

- The outer casing of a motor exchanges heat with the surroundings by radiation and natural convection. Radiation is important even when the motor surface is only warm to the touch.
- When food is cooked in a conventional oven, thermal radiation plays a major role.
- High-performance thermal insulation in which the effects of conduction and convection have been minimized is still subject to unwanted radiative transfer effects. The back of the insulation is often coated with a reflective surface to minimize radiative effects.

- Humans reject some of their body heat to the environment by thermal radiation. In addition, they are warmed by the sun through a radiative transfer mechanism.
- Radiation is a major factor in climate control of buildings and automobiles. The greenhouse effect acts to warm interior spaces. In hot climates, motorists cover the windows of parked cars with silvered radiation shields to prevent overheating.
- Solar collectors gather thermal radiation to supply hot water to homes.
- Radiation on a planetary scale is cited as playing a role in global warming.

At low temperatures

- In cryogenic containers, care must be taken to minimize heat leaks from the environment, some of which are due to radiation.
- In spacecraft, the ultimate heat sink for all energy generated on board is outer space, which has a temperature of approximately 4 K. In the vacuum of space, heat transfer from a spacecraft is by radiation only.

In this chapter, we confine our attention to radiation heat transfer between solid surfaces separated by a transparent medium. Fundamental laws are introduced, and radiative energy balances are developed. An important new concept is the radiative "enclosure," which is analogous to the control volume used in previous chapters. Geometric factors that affect radiation are introduced and explained.

14.2 FUNDAMENTAL LAWS OF RADIATION

Radiation has a dual character—under some circumstances it behaves like a wave and under other circumstances it behaves like a particle. The particle model, in which radiation energy is carried by photons, is useful in thinking about emission and absorption. At an atomic level, a photon is emitted when an electron drops from a high energy level to a low energy level. The photon travels through space at the speed of light until it is absorbed by another atom. Absorption is the opposite of emission. When a photon is absorbed, an electron rises from a low to a high energy level.

Thermal radiation arises when electrons transition among vibrational and rotational energy bands within an atom or molecule. The level of electron excitation in these bands determines the temperature of the material. Thermal radiation is part of the electromagnetic spectrum (Figure 14-1) and spans the ultraviolet, visible, and infrared regions. Wavelengths of thermal radiation range from about 0.1 to 100 μm. Thermal radiation shares many features in common with other types of electromagnetic radiation; it is reflected, refracted, scattered, diffracted, and so on, as well as being emitted and absorbed.

Radiation is emitted by solids, liquids, and gases. In most solids and liquids, thermal radiation emitted at a particular location is absorbed a very short distance away. If, for example, a photon is emitted near the center of a snowball, it will be absorbed before it reaches the surface. Photons emitted very near a surface, however, can escape (Figure 14-2). The zone near the surface from which emitted photons can reach the surroundings is often very thin. For our purposes, we assume the photon is emitted at the surface.

We are interested in the case of a solid surface surrounded by a transparent gas or a vacuum. Air is transparent to thermal radiation, except when distances are very large (on the order of kilometers). Carbon dioxide and water vapor are only partially transparent to thermal radiation and will not be considered here. In Figure 14-3a, a solid surface adjacent

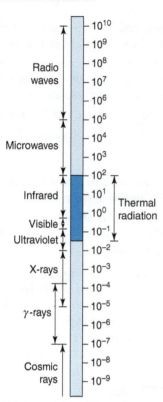

FIGURE 14-1   The electromagnetic spectrum with wavelengths of thermal radiation indicated (in microns).

to a transparent gas or vacuum emits radiation. Radiation from the environment is also incident on the surface. Part of the incident radiation is reflected and part is absorbed.

If all the radiation is absorbed, the surface is called a **black surface**. This is an important limiting case in radiation heat transfer. Real surfaces are characterized by how closely they resemble black surfaces. By definition:

A black surface absorbs all the radiation incident upon it.

A black surface is shown in Figure 14-3b; there is no reflected component from the surface.

Many surfaces that are not black in color are considered black for radiation purposes. If a surface is at a moderate temperature, we only see radiation *reflected* from the surface. The radiation *emitted* by the surface is in the infrared range and cannot be detected by the

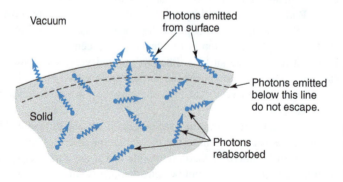

FIGURE 14-2   Photons emitted within a solid and reabsorbed or released to the surroundings.

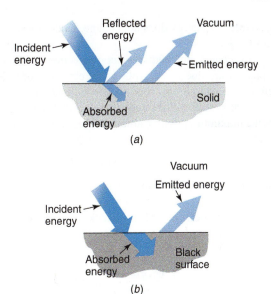

FIGURE 14-3 Radiation interaction with opaque surfaces: (a) a general surface; (b) a black surface.

human eye. (If a surface is very hot, it glows due to emitted radiation and can be seen.) If visible light strikes a surface and is completely absorbed, nothing is reflected and the surface appears black. Note, however, that we are often interested in the reflection, absorption, and emission of invisible infrared radiation. A surface that is red or white in the visible range may be black in the infrared range. Furthermore, glowing surfaces may also be "black." The sun radiates like a black surface at approximately 5800 K.

Black surfaces are perfect absorbers. They are also perfect emitters, emitting the maximum possible energy that any surface can emit at a given temperature. To demonstrate this, consider a small black object at temperature T_1 placed in an evacuated oven, as shown in Figure 14-4. The walls of the oven are also black and are maintained at temperature T_2. Experience shows that the small black object will change temperature until, after some time, it reaches equilibrium at T_2.

Consider an energy balance on the black object at equilibrium. From the first law for a closed system,

$$\frac{dE}{dt} = \dot{Q} - \dot{W}$$

No work is done on or by the object; therefore $\dot{W} = 0$. The object is stationary, and there is no change in kinetic or potential energy. The temperature of the object is steady

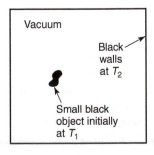

FIGURE 14-4 A small black object suspended in an evacuated oven with black walls.

at equilibrium, so internal energy does not change. Under these circumstances, the energy equation reduces to $\dot{Q} = 0$.

The oven is evacuated; no conduction or convection is present. The only heat transfer terms that are present are due to absorption and emission of radiation. The *net* heat, $\dot{Q}$, is zero. Therefore, the energy absorbed by the black object equals the energy emitted. Since a black surface absorbs all the radiation incident upon it, it follows that the black surface absorbs the maximum possible amount of radiation. Because emitted energy equals absorbed energy:

A black surface emits the maximum possible radiation at a given temperature.

All real surfaces emit less than a black surface. The black surface is the upper bound on what is possible.

The amount of radiation emitted by a black surface was first determined experimentally by Stefan in 1879 to be

$$\frac{\dot{Q}_{emitted}}{A} = E_b = \sigma T^4 \tag{14-1}$$

where E_b is the **emissive power** of a black surface, σ is the **Stefan-Boltzmann constant**, A is the area of the surface, and T is the absolute temperature (R or K). Because the emissive power depends on the fourth power of the temperature, high-temperature bodies emit much more than low-temperature ones. Emissive power has units of W/m^2 or Btu/h·ft^2. The Stefan-Boltzmann constant has values of

$$\sigma = 5.6697 \times 10^{-8} \frac{W}{m^2 \cdot K^4}$$

$$\sigma = 0.17123 \times 10^{-8} \frac{Btu}{h \cdot ft^2 \cdot R^4}$$

The temperature in Eq. 14-1 must always be expressed in absolute terms (either K or R).

Thermal energy is not emitted at a single wavelength but rather over a range of wavelengths. In 1900, Max Planck derived an equation for the energy emitted by a blackbody into vacuum as a function of wavelength. Planck's law is

$$E_{b\lambda} = \frac{C_1 \lambda^{-5}}{e^{C_2/\lambda T} - 1} \tag{14-2}$$

where $E_{b\lambda}$ is emissive power per unit area, per unit wavelength, and λ is wavelength. The constants C_1 and C_2 are

$$C_1 = 2\pi h c_o^2 = 3.742 \times 10^8 \frac{W \cdot \mu m^4}{m^2}$$

$$C_2 = \frac{h c_o}{k} = 1.439 \times 10^4 \, \mu m \cdot K$$

where h is Planck's constant, k is Boltzmann's constant, and c_o is the speed of light in vacuum. Wavelength is typically measured in microns, with $1 \, \mu m = 10^{-6}$m. To derive Eq. 14-2, Planck assumed that energy was quantized in discrete packets. His work was the beginning of quantum mechanics, which has grown into a major field in physics.

Planck's law gives emitted energy per unit wavelength. To determine the total energy emitted at all wavelengths, Eq. 14-2 is integrated as

$$E_b = \int_0^\infty \frac{C_1 \lambda^{-5}}{e^{C_2/\lambda T} - 1} d\lambda = \sigma T^4$$

where details of the integration are not shown. Through this integration, Planck showed that the Stefan-Boltzmann constant, previously obtained experimentally, could be derived from more fundamental physical constants as

$$\sigma = \frac{2\pi^5 k^4}{15 c_o^2 h^3}$$

Figure 14-5 is a plot of Planck's law and shows the spectral energy distribution from a blackbody at different temperatures. (In radiation theory, **spectral** refers to any quantity which varies with wavelength.) The total amount of radiation emitted at a given temperature is the area under the curve in Figure 14-5.

Blackbody radiation displays several important features. First, as temperature increases, the total amount of radiation emitted also increases. This is consistent with the Stefan-Boltzmann law (Eq. 14-1). Second, at moderate temperatures, most of the radiation is emitted in the infrared. Third, at higher temperatures, the peak of the distribution

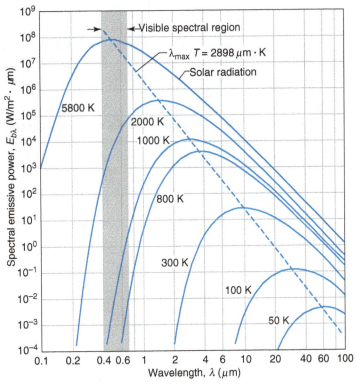

FIGURE 14-5   Blackbody emissive power as a function of wavelength.
(*Source:* F. P. Incropera and D. P. DeWitt, *Introduction to Heat Transfer*, 4th ed., Wiley, New York, 2002 p. 675. Used with permission.)

shifts to the left, and more radiation is emitted at short wavelengths. The range of visible light, which extends from 0.4 to 0.7μm, is indicated in Figure 14-5.

When a surface has a temperature near 800 K, radiation becomes visible to the naked eye and the surface glows red. Hot coals in a fire are a good example. The surface appears red because some of the emitted radiation is in the longer wavelengths of the visible spectrum, that is, near 0.7 μm. At higher temperatures, the eye sees a broader range of visible wavelengths. A white-hot surface emits all the visible wavelengths and is hotter than a red-hot surface.

At each temperature, there is a maximum in the spectral energy distribution. The wavelength of this maximum point is given by Wien's displacement law as

$$\lambda_{max}T = C_3 = 2898\mu\text{m}\cdot\text{K}$$

Wien's law may be derived by differentiating Planck's law with respect to wavelength and setting the result equal to zero.

The sun emits like a black surface at about 5800 K. It is no accident that the peak of solar radiation lies in the visible range. Human eyes have evolved to detect radiation in the part of the spectrum that has the highest intensity. If we lived in a different solar system, we would probably see in a different range of the electromagnetic spectrum, corresponding to the temperature of the star in that system.

Real surfaces emit less energy than black surfaces. In describing real surfaces, we use a quantity called emissivity, defined as

$$\varepsilon = \frac{\text{actual emitted energy}}{\text{blackbody emitted energy}} \qquad (14\text{-}3)$$

The value of emissivity varies between zero and unity. For a black surface, $\varepsilon = 1$. In general, emissivity depends on temperature, wavelength, and direction of emission.

The power emitted by a real surface as a function of wavelength is compared to that emitted by a black surface in Figure 14-6. Real surfaces emit less overall than black surfaces; furthermore, they emit less at every wavelength. The distribution for a real surface always lies within the envelope of the black surface. The variation of emissivity with wavelength is important in some applications, such as solar collectors, and will be discussed in Section 14.10. In most other circumstances, the so-called **gray** surface assumption can be used. A gray surface emits in the same pattern as a black surface, as shown in Figure 14-6. The gray emitted power is always a fixed fraction less than the blackbody power. From Eq. 14-3, this fraction is the emissivity. Thus, the emissivity of a gray surface does not depend on wavelength.

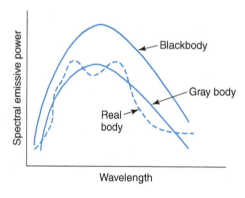

FIGURE 14-6 Spectral behavior of real surfaces, gray surfaces, and black surfaces.

The term *gray* is used by analogy to visible light. When we see a body at moderate temperatures, our eyes detect reflected light. If a body reflects all incident light by the same amount regardless of wavelength, it appears gray. If, on the other hand, the body absorbs blue light preferentially, it appears orange. Our eyes have evolved to be able to see the world either in an array of colors or in black and white. In the daytime, when light intensity from the sun is strong, everything around us is colored, but by moonlight, when light intensity is very weak, objects appear in shades of gray. Making the gray assumption in radiation calculations is similar to viewing the world in moonlight. Nevertheless, the gray assumption gives excellent results for many cases.

Emissivity, in general, also varies with angle of emission. Figure 14-7a shows a sketch of the energy emitted from a blackbody as a function of angle. This distribution can be derived from energy balance considerations, but the derivation is beyond the scope of this text. Figure 14-7b compares black emission to emission from a real surface. Real surfaces emit less overall than black surfaces; furthermore, they emit less in every direction. The distribution for a real surface always lies within the envelope of the black distribution.

The analysis of radiation from real surfaces taking account of angular variations is complex and will not be discussed here. Under most circumstances, the assumption of a **diffuse surface** produces good results. A diffuse surface is one that emits in the same pattern as a black surface, as shown in Figure 14-7c. The emissivity of a diffuse surface is not a function of angle.

Analysis of radiation heat transfer is greatly simplified by assuming surfaces are both gray and diffuse. The emissive power of a gray, diffuse surface is

$$\frac{\dot{Q}_{emitted}}{A} = E = \varepsilon \sigma T^4$$

where E is emissive power of a real surface.

In formulating an energy balance on a surface, we must consider reflected, absorbed, and transmitted energy as well as emitted energy. To characterize reflected energy, the reflectivity, ρ, is defined as

$$\rho = \frac{\text{reflected energy}}{\text{incident energy}}$$

Like emissivity, reflectivity of real surfaces depends on temperature, direction, and wavelength. Reflectivity is actually more complex than emissivity, because it varies, in general, with both angle of incidence *and* angle of reflection. There are two limiting cases of reflective behavior that are used to simplify analysis—the diffuse surface and the **specular surface** (Figure 14-8). In a diffuse reflector, the angle of reflection is independent of the angle of incidence. In a specular reflector, the angle of reflection equals the angle of incidence. Real surfaces often fall somewhere between these two behaviors, as shown in Figure 14-8c.

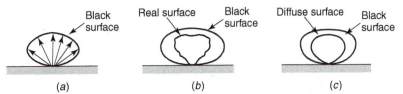

FIGURE 14-7   Emissive power as a function of direction: (a) black surface; (b) real surface; (c) diffuse surface.

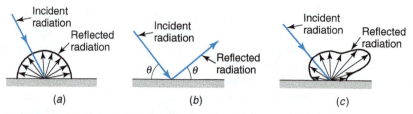

FIGURE 14-8 Reflected energy as a function of direction: (a) diffuse reflection; (b) specular reflection; (c) reflection from a real surface.

Reflection is analogous to the behavior of a ball bouncing off a surface. If the surface is smooth compared to the size of the ball (and there is no spin on the ball), the ball will depart at the same angle at which it arrived (Figure 14-9a). On the other hand, if the surface is rough compared to the size of the ball, the ball will depart at some unpredictable angle, as in Figure 14-9b. In radiation, the "ball" is a photon, and the "size" of the ball is the photon wavelength. If the surface is **optically smooth**, then the height of surface protrusions is small compared to the wavelength of the photon. In such a case, reflection is specular and angle of incidence equals angle of reflection. An optically rough surface has protrusions of comparable or greater size than the photon wavelength, and such a surface reflects diffusely.

Specular reflectors in the visible range include mirrors, shiny metal surfaces, glass sheets, and still water. The near-perfect reflection behavior allows us to see images in these surfaces. Surfaces that are specular in the visible are generally specular in the infrared as well, since infrared radiation has longer wavelengths than visible light. It is possible to perform straightforward radiation analyses assuming either perfectly diffuse reflectors or perfectly specular reflectors. Most common surfaces are more nearly diffuse than specular, and in this text, we will assume all surfaces reflect diffusely.

To characterize absorption, we define absorptivity as

$$\alpha = \frac{\text{absorbed energy}}{\text{incident energy}}$$

Absorptivity depends, in general, on temperature, wavelength, and direction. Wavelength effects are discussed in Section 14-10. In the analyses in this text, we assume that absorptivity does not depend on angle of incidence, that is, that all surfaces absorb diffusely.

The final radiative property is transmissivity, τ, defined as

$$\tau = \frac{\text{transmitted energy}}{\text{incident energy}}$$

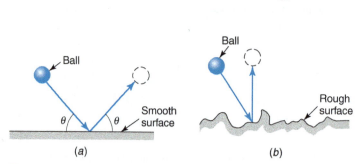

FIGURE 14-9 Behavior of a ball bouncing off a surface: (a) smooth surface; (b) rough surface.

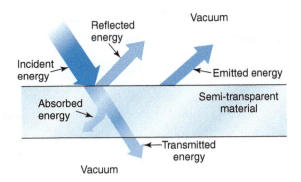

FIGURE 14-10 A semitransparent plane layer of material.

Transmissivity applies to a plane layer of material, as shown in Figure 14-10. Like the other radiative properties, transmissivity depends on temperature, wavelength, and direction. In addition, it depends on the thickness of the layer through which radiation travels. A material with a transmissivity of unity ($\tau = 1$) is perfectly **transparent**. Thin layers of some gases, such as air and oxygen, are virtually transparent to thermal radiation. A material with a transmissivity between zero and unity is **semitransparent**. Ordinary glass is semitransparent in the visible wavelength range, as is liquid water. Very few materials are semitransparent in the infrared, salt crystals being the prime example. A material with a transmissivity of zero is **opaque** ($\tau = 0$). Most solids are opaque in the visible wavelength range. Even glass and water, which transmit visible light, are opaque to infrared radiation.

The thermal radiative properties are not all independent. Referring to Figure 14-10, we note that all the incident radiation is either absorbed, reflected, or transmitted. If we use conservation of energy on the incoming energy, and then divide each term by the amount of incoming energy, the fractions absorbed, reflected, and transmitted must sum to unity, that is,

$$\alpha + \rho + \tau = 1$$

For an opaque surface, $\tau = 0$ and

$$\boxed{\alpha + \rho = 1} \tag{14-4}$$

There is also a relationship between absorptivity and emissivity. To demonstrate this, consider a small, gray, diffuse body at temperature T_1 placed in an evacuated oven, as shown in Figure 14-11. The walls of the oven are black and are maintained at temperature

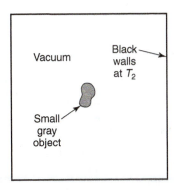

FIGURE 14-11 A small gray body in an oven with black walls.

T_2, where $T_2 > T_1$. The radiation incident on the small body comes from the oven walls. The energy absorbed by the small body is, then,

$$\dot{Q}_{abs} = \alpha \sigma T_2^4$$

After sufficient time has passed, the temperature of the small body will rise to T_2. At equilibrium, the energy emitted by the small body equals the energy absorbed. The body is gray and diffuse; therefore, it emits

$$\dot{Q}_{emit} = \varepsilon \sigma T_2^4$$

Equating absorbed and emitted energy,

$$\alpha \sigma T_2^4 = \varepsilon \sigma T_2^4$$

which reduces to

$$\boxed{\alpha = \varepsilon} \tag{14-5}$$

This relationship is called **Kirchhoff's law**. Strictly speaking, it applies only when the surface producing the incident radiation and the surface receiving the radiation are at the same temperature. We are interested in the case of radiative exchange between surfaces at different temperatures. As long as the temperature difference between surfaces is no more than a few hundred degrees, we can apply Kirchhoff's law as an approximation. Kirchhoff's law does not apply, for example, to solar radiation interacting with surfaces at moderate temperatures. The sun is thousands of degrees hotter than the surface of a solar collector, and the absorptivity of the solar collector is not generally equal to emissivity.

Values of emissivity for many materials are available in Table A-17. Emissivity depends, to some extent, on whether or not a material is an electrical conductor. Insulators usually have high values of emissivity, with a typical range from 0.8 to 0.99. Metals, on the other hand, usually have low values of emissivity, varying from 0.001 to 0.7. Surface condition has a major effect on emissivity, especially in metals. Thermal radiation is absorbed, emitted, and reflected from a thin layer near the surface; therefore, any alteration of the surface can be significant. This effect is clearly detectable in the visible wavelengths. A smudge on a mirror distorts the reflected image. A thin layer of oxide on a metal surface renders the metal dull and unpolished. A thin coat of paint changes the color of a surface. If the paint coat is very thin, the underlying color of the base material partially shows through. Similar phenomena occur in the infrared. A clean, highly polished metal surface may have an emissivity of 0.01. Dirt and oxide are dielectrics (insulators) and have emissivities near 0.9. If a metal surface is dirty or oxidized, its emissivity is somewhere between that of a clean, shiny metal and an insulator. These comments on emissivity are summarized in Table 14-1.

TABLE 14-1 Dependence of emissivity on surface condition and type of material

Material	Approximate Emissivity
Nonmetals	0.8–0.99
Heavily oxidized or very dirty metals	0.7
Lightly oxidized or dirty metals	0.5
Clean, shiny metals	0.1–0.3
Bright, polished metals	0.001–0.1

EXAMPLE 14-1   Energy balance on the planet earth

The average temperature of the earth is largely controlled by radiation from the sun. The sun radiates like a black sphere with a radius of 6.95×10^8 m and a temperature of 5800 K. The average distance between the sun and the earth is 1.5×10^{11} m. Assume the earth is black and isothermal, with no atmosphere and a radius of 6.37×10^6 m. Fission reactions within the earth's crust generate heat at the rate of 5.5×10^{16} W. Calculate the average surface temperature of the earth.

Approach:

Conservation of energy can be applied to the earth to find its temperature. For a closed, steady system, a balance is achieved among the heat generated by fission reactions, the solar radiation absorbed by the earth, and the energy emitted by the earth. To determine the energy reaching earth from the sun, draw an imaginary sphere around the sun with a radius equal to the sun–earth distance. Find the area of interception between earth and the imaginary sphere and use this to calculate the fraction of all the energy leaving the sun which is intercepted by the earth.

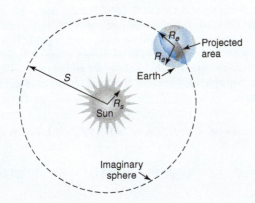

FIGURE 14-12   Projected area of the earth as viewed from the sun.

Solution:

Assumptions:

Define the earth as a closed system and use the first law in rate form, that is,

$$\frac{dE}{dt} = \dot{Q} - \dot{W}$$

A1. The temperature of the earth does not change with time.

Assume the temperature of the earth is steady [A1] (to simplify the analysis) and that earth's kinetic energy is not changing with time [A2]. Also assume the earth has a circular orbit, so that no work is done on the earth by the sun [A3]. Under these circumstances, the first law becomes $\dot{Q} = 0$. The net heat transfer is given by

A2. There is no change in kinetic energy. (The earth is not slowing down.)

$$\dot{Q} = \dot{Q}_{gen} + \dot{Q}_{abs} - \dot{Q}_{emit} = 0$$

A3. The earth's orbit is circular.

where $\dot{Q}_{abs}$, the radiation absorbed by the earth, $\dot{Q}_{emit}$, the radiation emitted by the earth, and $\dot{Q}_{gen}$, the heat generated by fission within the earth, are all taken to be positive. To find the heat absorbed, first calculate the total heat emitted by the sun using the Stefan-Boltzmann law.

A4. The sun radiates like a blackbody.

The sun radiates like a blackbody at 5800 K [A4][A5]. The total radiation emitted by the sun is [A6]:

A5. The sun is isothermal.

$$\dot{Q}_s = A_s \sigma T_s^4 = 4\pi R_s^2 \sigma T_s^4$$

A6. The sun is spherical.

where A_s is the surface area of the sun and R_s is the radius. Substituting values

$$\dot{Q}_s = 4\pi \left(6.95 \times 10^8 \text{ m}\right)^2 \left(5.67 \times 10^{-8} \frac{\text{W}}{\text{m}^2 \cdot \text{K}^4}\right) (5800 \text{ K})^4 = 3.89 \times 10^{26} \text{ W}$$

A7. The sun emits uniformly in all directions.

We assume solar radiation is emitted uniformly in all directions [A7]. To determine the amount of solar radiation impinging on the earth, draw an imaginary sphere with the sun at the center, as shown in Figure 14-12. The radius of the sphere is the distance between the sun and the earth. The heat flux due to solar radiation on the inside surface of the sphere is

$$q_s'' = \frac{\dot{Q}_s}{A_{sph}} = \frac{\dot{Q}_s}{4\pi S^2}$$

where A_{sph} is the area of the imaginary sphere and S is the distance between the sun and the earth. Inserting values,

$$q_s'' = \frac{(3.89 \times 10^{26}\ \text{W})}{4\pi\left(1.5 \times 10^{11}\ \text{m}\right)^2} = 1377\ \frac{\text{W}}{\text{m}^2}$$

A8. The sun's rays are parallel in earth orbit.
A9. The earth is spherical.

A10. The earth radiates like a blackbody.

Because the distance between the sun and earth is so large, the sun's rays essentially have parallel paths by the time they reach the earth [A8], and the rays impinge on the outer surface of the earth from a single direction. The total radiation intercepted by the earth is the heat flux at earth orbit multiplied by the area of a circle with the earth's radius [A9]. This circle is the area of the earth projected onto a flat plane perpendicular to the sun's rays. It is also the intersection of the imaginary sphere around the sun with the earth, as shown in Figure 14-12. The earth is assumed to behave like a blackbody [A10]. Therefore, it absorbs all radiation incident upon it. The energy absorbed by the earth is

$$\dot{Q}_{abs} = \pi R_e^2 q_s'' = \pi \left(6.37 \times 10^6\ \text{m}\right)^2 \left(1377\ \frac{\text{W}}{\text{m}^2}\right) = 1.76 \times 10^{17}\ \text{W}$$

where R_e is the radius of the earth. Applying the first law to the earth gives

$$\dot{Q}_{gen} + \dot{Q}_{abs} - \dot{Q}_{emit} = 0$$

Therefore, the emitted power is

$$\dot{Q}_{emit} = \dot{Q}_{gen} + \dot{Q}_{abs} = 5.5 \times 10^{16} + 1.76 \times 10^{17} = 2.31 \times 10^{17}\ \text{W}$$

From the Stefan-Boltzmann law,

$$\dot{Q}_{emit} = A_e \sigma T_e^4$$

where A_e is the (actual) surface area of the earth and T_e is the temperature of the earth. Solving for temperature gives

$$T_e = \left(\frac{\dot{Q}_{emit}}{A_e \sigma}\right)^{0.25} = \left(\frac{\dot{Q}_{emit}}{4\pi R_e^2 \sigma}\right)^{0.25} = \left(\frac{2.31 \times 10^{17}\ \text{W}}{4\pi \left(6.37 \times 10^6\ \text{m}\right)^2 \left(5.67 \times 10^{-8}\ \frac{\text{W}}{\text{m}^2 \cdot \text{K}^4}\right)}\right)^{0.25} = 299\ \text{K}$$

Comments:

The earth is not actually black and reflects a fraction of incident sunlight. In addition, the atmosphere scatters and absorbs radiation, a fact that should be included in a more advanced analysis.

14.3 VIEW FACTORS

Example 14-1 required the calculation of the fraction of energy leaving the sun and striking the earth. For this simple geometry, the calculation was relatively easy. In general, to perform an energy balance on a surface exchanging heat by radiation with another surface, the incident radiation must be known. This radiation arrives from surrounding surfaces, and the geometry is not usually as straightforward as for the sun–earth pair. To aid in calculating incident radiation, we define the **view factor**, $F_{i \to j}$, as:

> $F_{i \to j}$ *is the fraction of diffuse radiation leaving surface i that arrives at surface j by a straight-line route.*

The view factor is a geometric quantity and depends only on the size, orientation, and spacing of the surfaces involved. In Figure 14-13, radiation is shown leaving surface i and either arriving at or missing surface j. The view factor is the ratio of all the radiative contributions that strike surface j to all energy leaving surface i. It is a dimensionless quantity with a value varying between zero and unity. View factors are also sometimes called **configuration factors**, **shape factors**, or **angle factors**.

If the view factor between i and j is zero, then no radiation from i reaches j by a direct path. For example, if two surfaces lie in the same plane, as shown in Figure 14-14a, then the view factor between them is zero, and we say that the two surfaces cannot "see" each other. In Figure 14-14b, a hemispherical enclosure is shown. The base of the enclosure is a

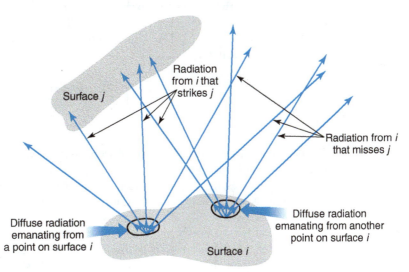

FIGURE 14-13 Diffuse radiation leaves surface i in all directions and at all locations. Some of the radiation strikes surface j.

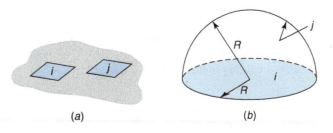

FIGURE 14-14 (a) Two flat surfaces that lie in the same plane have no view of each other. (b) The view factor between the base of a hemisphere and the inside surface of the hemisphere is unity.

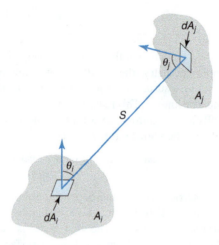

FIGURE 14-15 Geometric quantities used in finding the view factor between two arbitrary surfaces.

circular surface that radiates to the hemispherical surface surrounding it. All the radiation leaving the circular surface impinges on the inside of the hemisphere; therefore, the view factor, $F_{i \to j}$, from the circle to the hemisphere is unity.

For these simple cases, the view factor can be found by inspection, but for the general case, the view factor must be determined by integration over the two surface areas. Figure 14-15 shows two arbitrary surfaces exchanging heat by radiation. The differential areas dA_i and dA_j are joined by a straight line of length S. The quantity θ_i is the angle between the normal to dA_i and the line S. Likewise, θ_j is the angle between the normal to dA_j and the line S. The view factor is given by

$$F_{i \to j} = \frac{1}{A_i} \int_{A_i} \int_{A_j} \frac{\cos \theta_i \cos \theta_j \, dA_i \, dA_j}{\pi S^2} \tag{14-6}$$

The derivation of Eq. 14-6 is beyond the scope of this text. In some cases, the integration in Eq. 14-6 can be performed analytically to produce an algebraic formula for the view factor. In most cases, the integration must be computed numerically. View factors for several common three-dimensional geometries are shown in Figure 14-16 through Figure 14-18. Table 14-2 contains view factor for some two-dimensional (infinitely long) geometries.

On occasion, symmetry can be used to determine view factors. For example, consider a cubical enclosure. The view from an inside surface of the cube to any of the four adjacent surfaces is the same, by symmetry. The view to the opposite wall is different. As a more subtle example, consider a small spherical thermocouple bead suspended at the end of an open tube, as shown in Figure 14-19. The plane at the end of the tube passes through the center of the sphere. Radiation from the thermocouple can either escape to the environment or strike the inside of the tube. From the symmetry of the situation, each path is equally likely. Therefore, the view factor from the thermocouple to the inside wall of the tube is 0.5, as is the view factor from the thermocouple to the environment.

The view factor, $F_{i \to j}$, is related to $F_{j \to i}$. A formula for $F_{j \to i}$ may be obtained by exchanging the i and j indices in Eq. 14-6 to give

$$F_{j \to i} = \frac{1}{A_j} \int_{A_j} \int_{A_i} \frac{\cos \theta_j \cos \theta_i \, dA_j \, dA_i}{\pi S^2} \tag{14-7}$$

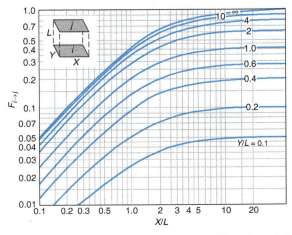

$$\overline{X} = X/L, \overline{Y} = Y/L$$

$$F_{i\to j} = \frac{2}{\pi \overline{X}\,\overline{Y}} \left\{ \ln \left[\frac{(1+\overline{X}^2)(1+\overline{Y}^2)}{1+\overline{X}^2+\overline{Y}^2} \right]^{1/2} \right.$$

$$+ \overline{X}(1+\overline{Y}^2)^{1/2} \tan^{-1} \frac{\overline{X}}{(1+\overline{Y}^2)^{1/2}}$$

$$+ \overline{Y}(1+\overline{X}^2)^{1/2} \tan^{-1} \frac{\overline{Y}}{(1+\overline{X}^2)^{1/2}}$$

$$\left. - \overline{X} \tan^{-1} \overline{X} - \overline{Y} \tan^{-1} \overline{Y} \right\}$$

FIGURE 14-16 View factor between two aligned parallel rectangles: graphical form and equation form.
(*Source:* F. P. Incropera and D. P. DeWitt, *Introduction to Heat Transfer*, 4th ed., Wiley, New York, 2002, p. 754. Used with permission.)

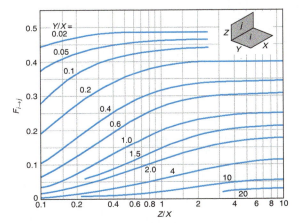

$$H = Z/X, \; W = Y/X$$

$$F_{i\to j} = \frac{1}{\pi W} \left(W \tan^{-1} \frac{1}{W} + H \tan^{-1} \frac{1}{H} \right.$$

$$- (H^2 + W^2)^{1/2} \tan^{-1} \frac{1}{(H^2 + W^2)^{1/2}}$$

$$+ \frac{1}{4} \ln \left\{ \frac{(1+W^2)(1+H^2)}{1+W^2+H^2} \left[\frac{W^2(1+W^2+H^2)}{(1+W^2)(W^2+H^2)} \right]^{W^2} \right.$$

$$\left. \left. \times \left[\frac{H^2(1+H^2+W^2)}{(1+H^2)(H^2+W^2)} \right]^{H^2} \right\} \right)$$

FIGURE 14-17 View factor between two perpendicular rectangles with a common edge: graphical form and equation form.
(*Source:* F. P. Incropera and D. P. DeWitt, *Introduction to Heat Transfer*, 4th ed., Wiley, New York, 2002, p. 754. Used with permission.)

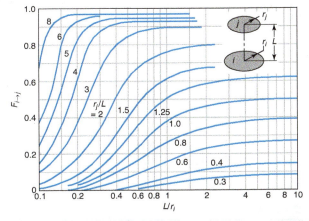

$$R_i = r_i/L, \; R_j = r_j/L$$

$$S = 1 + \frac{1+R_j^2}{R_i^2}$$

$$F_{i\to j} = \frac{1}{2}\{S - [S^2 - 4(r_j/r_i)^2]^{1/2}\}$$

FIGURE 14-18 View factor between two aligned parallel disks: graphical form and equation form.
(*Source:* F. P. Incropera and D. P. DeWitt, *Introduction to Heat Transfer*, 4th ed., Wiley, New York, 2002, p. 754. Used with permission.)

TABLE 14-2 View factors for some common two-dimensional geometries

Geometry	Relation
Parallel plates with midlines connected by perpendicular 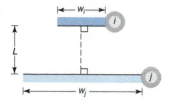	$F_{i \to j} = \dfrac{[(W_i + W_j)^2 + 4]^{1/2} - [(W_j - W_i)^2 + 4]^{1/2}}{2W_i}$ $W_i = w_i/L, \; W_j = w_j/L$
Inclined parallel plates of equal width and a common edge 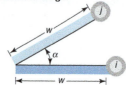	$F_{i \to j} = 1 - \sin\left(\dfrac{\alpha}{2}\right)$
Perpendicular plates with a common edge 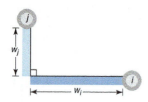	$F_{i \to j} = \dfrac{1 + (w_j / w_i) - [1 + (w_j / w_i)^2]^{1/2}}{2}$
Three–sided enclosure 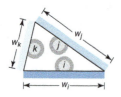	$F_{i \to j} = \dfrac{w_i + w_j - w_k}{2w_i}$
Parallel cylinders of different radii 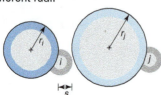	$F_{i \to j} = \dfrac{1}{2\pi}\left\{ \pi + [C^2 - (R + 1)^2]^{1/2} - [C^2 - (R - 1)^2]^{1/2} \right.$ $\left. + (R - 1)\cos^{-1}\left[\left(\dfrac{R}{C}\right) - \left(\dfrac{1}{C}\right)\right] \right.$ $\left. - (R + 1)\cos^{-1}\left[\left(\dfrac{R}{C}\right) + \left(\dfrac{1}{C}\right)\right] \right\}$ $R = r_j/r_i, \; S = s/r_j$ $C = 1 + R + S$

(*Source:* F. P. Incropera and D. P. DeWitt, *Introduction to Heat Transfer*, 4th ed., Wiley, New York, 2002, p. 751. Used with permission)

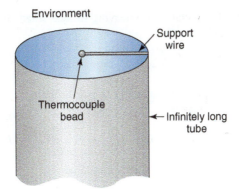

Environment

Support wire

Thermocouple bead

← Infinitely long tube

FIGURE 14-19 A thermocouple bead suspended at the center of an open tube end.

Comparing Eq. 14-6 and Eq. 14-7 reveals that

$$A_i F_{i \rightarrow j} = A_j F_{j \rightarrow i} \qquad (14\text{-}8)$$

Eq. 14-8 is called the **reciprocity relation**. It is one of the most useful formulas in radiation analysis. Sometimes it is easy to calculate $F_{i \rightarrow j}$ and difficult to calculate $F_{j \rightarrow i}$. For example, consider the disk radiating to the inside of a hemisphere shown in Figure 14-14b. By inspection, we find that $F_{i \rightarrow j} = 1$. From reciprocity,

$$F_{j \rightarrow i} = \frac{A_i F_{i \rightarrow j}}{A_j} = \frac{A_i}{A_j} = \frac{\pi R^2}{2\pi R^2} = \frac{1}{2}$$

Reciprocity allowed us to determine the view factor from the hemisphere to the disk, $F_{j \rightarrow i}$, without evaluating the integral in Eq. 14-6.

Some surfaces can "see" themselves. If a surface is concave, then radiation leaving from one location might be intercepted at another point on the same surface. For example, the inside of a hemisphere can see itself. In such a case, we may define a view factor, $F_{i \rightarrow i}$, which is the fraction of radiation leaving surface i that arrives at another location on surface i. For planar and convex surfaces, $F_{i \rightarrow i}$ is zero.

To use view factors in energy balances, we introduce the concept of a radiative **enclosure**. An enclosure is a three-dimensional region in space completely encased by bounding surfaces, as shown schematically in Figure 14-20. The surfaces may be planar, concave, or convex in shape. Consider radiation leaving surface 1 in Figure 14-20a. Conservation of energy states that we must account for all the energy leaving surface 1. Because the enclosure is closed, all the radiation leaving surface 1 must arrive at either 2, 3, or 4.

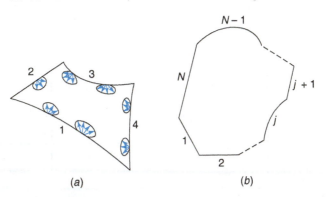

(a)

(b)

FIGURE 14-20 (a) An enclosure with four surfaces. (b) An enclosure with N surfaces.

Hence, the fractions of the total energy leaving surface 1 and arriving at each of the other surfaces must sum to unity, that is,

$$F_{1\to 2} + F_{1\to 3} + F_{1\to 4} = 1 \tag{14-9}$$

In the general case, an enclosure may have an arbitrary number of surfaces, as shown in Figure 14-20b. Then Eq. 14-9 generalizes to

$$F_{i\to 1} + F_{i\to 2} + F_{i\to 3} + \cdots + F_{i\to i} + \cdots + F_{i\to j} + \cdots + F_{i\to (N-1)} + F_{i\to N} = 1$$

or

$$\sum_{j=1}^{N} F_{i\to j} = 1 \tag{14-10}$$

Note that we have included $F_{i\to i}$ to take into account the possibility that surface i might see itself. Eq. 14-10 is called the **summation** relation. It can be used, for example, to find the last view factor in an enclosure analysis once all the other view factors are known.

In some cases, the geometry under consideration is not totally enclosed. For example, consider a small, short tube open at both ends suspended in a large room, as shown in Figure 14-21. The inside surface of the tube radiates to the walls of the room through the open ends. A photon leaving the open end is likely to be reflected many times from the walls and finally absorbed by a wall surface. Because the walls are far away, it is unlikely that the photon will ever be reflected back into the tube before being absorbed. From the point of view of the inside wall of the tube, the opening at the end of the tube absorbs all radiation incident upon it; that is to say, it behaves like a black surface. As a result, we may cover the ends of the tube by imaginary surfaces that are black at the temperature of the walls in the room. We can use this constructed enclosure in a radiation analysis inside the tube.

It is frequently possible to "plug holes" in enclosures by using black imaginary surfaces. As long as there are no exterior surfaces near the holes that can reflect radiation back into the enclosure, the assumption of an imaginary black plane is a good approximation.

A special case of an enclosure is the cavity, as shown in Figure 14-22. Radiation may enter or leave through a small opening in the cavity wall. Now take the perspective of looking into the cavity from the outside. Radiation incident on the opening in the cavity is unlikely ever to be reflected back out. An entering photon will reflect off the interior

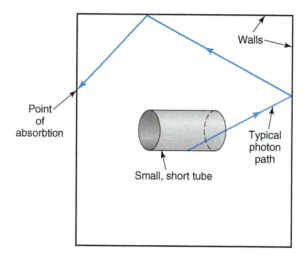

FIGURE 14-21 A tube whose inside curved surface exchanges heat by radiation with the walls of a room.

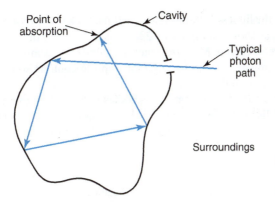

FIGURE 14-22 A photon enters a cavity and is absorbed after several bounces.

walls one or more times and then be absorbed. Because the opening is small compared to the size of the enclosure, the radiation is effectively trapped inside. Since the cavity opening absorbs all radiation incident upon it, it is black. We may cover the opening with an imaginary black surface at the temperature of the interior walls of the cavity.

We are familiar with the cavity effect in the visual wavelength range. For example, suppose you are standing on a sidewalk and looking at a building across the street. The windows appear dark, even in daytime. The surfaces of the room inside the building are not black, but they appear to be so. The room acts like a cavity, and the windows act like small openings. Visible radiation from the sun enters the window but is absorbed in the interior of the room and is not reflected back out. Another example of a cavity is the human eye. The pupil of the eye appears black; in fact, it is a transparent lens that allows light to reach the retina, as shown in Figure 14-23a. The interior of the eye is a cavity with a small opening—the pupil. Note, however, that if a photographer shines a flash directly

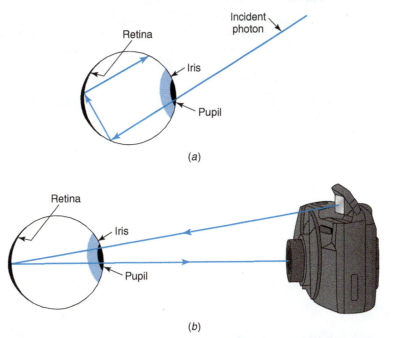

FIGURE 14-23 Appearance of the eye under different lighting conditions. (a) The inside of the eye is a cavity, with the pupil behaving like a blackbody. (b) When a photographic flash reflects directly off the retina into the camera lens, the people in the picture appear to have red pupils.

into the eye, the light sometimes reflects off the back of the retina and returns directly to the camera lens, as shown in Figure 14-23b. This is the cause of pictures with "red-eye," since, in fact, the retina is red. Modern cameras minimize or eliminate red-eye by triggering one or more preliminary flashes to cause the pupil to contract so there is less chance of light being reflected from the retina.

Cavities are used to create blackbodies in laboratory experiments. For that purpose, the inside walls of the cavity are made as black as possible, using lampblack or black paint. Such a cavity is more nearly like a perfect black surface than any real surface.

EXAMPLE 14-2   View factors in a room

Radiation occurs between the floor, walls, and ceiling of a room with a floor area of 8 ft by 12 ft and a ceiling height of 8 ft. Find the view factor from an end wall of the room to each of the other five surfaces (Figure 14-24).

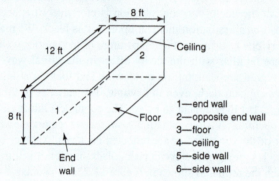

1—end wall
2—opposite end wall
3—floor
4—ceiling
5—side wall
6—side walll

FIGURE 14-24   Geometry of the room.

Approach:

First find the view factor from an end wall to the opposite end wall, $F_{1\rightarrow2}$, using the plot or equation in Figure 14-16. By symmetry, $F_{1\rightarrow3} = F_{1\rightarrow4} = F_{1\rightarrow5} = F_{1\rightarrow6}$. Using this fact and the summation relation (Eq. 14-10), we can determine all remaining view factors.

Solution:

To find the view factor, $F_{1\rightarrow2}$, refer to Figure 14-16. Let $X = Y = 8$ and $L = 12$. With these definitions,

$$\frac{X}{L} = \frac{Y}{L} = \frac{8}{12} = 0.667$$

From the plot in Figure 14-16, $F_{1\rightarrow2} \approx 0.11$. To find the remaining view factors, apply the summation relation (Eq. 14-10) in the form

$$F_{1\rightarrow2} + F_{1\rightarrow3} + F_{1\rightarrow4} + F_{1\rightarrow5} + F_{1\rightarrow6} = 1$$

By symmetry,

$$F_{1\rightarrow3} = F_{1\rightarrow4} = F_{1\rightarrow5} = F_{1\rightarrow6}$$

Combining the last two equations gives

$$F_{1\rightarrow2} + 4F_{1\rightarrow3} = 1$$

Solving for $F_{1\rightarrow3}$,

$$F_{1\rightarrow3} = \frac{(1 - F_{1\rightarrow2})}{4} = \frac{(1 - 0.11)}{4} = 0.2225$$

The remaining view factors are

$$F_{1\rightarrow4} = F_{1\rightarrow5} = F_{1\rightarrow6} = 0.2225$$

Four significant digits were retained so that all view factors would sum exactly to unity; however, in reality, this result is not accurate to more than 2 significant figures. At this point, all five view factors are known.

Comments:

- In winter, the outside wall of a room is cooler than the other surfaces, hence the need for a radiation analysis.
- In an actual room, the contents (furniture, people, etc.) may partially block the view from one surface to another.
- One could also have used Figure 14-17 to find the first view factor and then determined the other factors from summation and symmetry.
- Greater accuracy could have been obtained by using the equation in Figure 14-16 instead of the graph.

14.4   SHAPE DECOMPOSITION

The technique of **shape decomposition** greatly expands the number of view factors that can be evaluated. A major advantage is that shape decomposition allows calculation of view factors without resorting to integration. The fundamental idea is illustrated in Figure 14-25. The arbitrary surface 1 radiates to arbitrary surface 2, which is composed of two subsurfaces, 2a and 2b. All the radiation striking 2 must strike either 2a or 2b; therefore,

$$F_{1\rightarrow2} = F_{1\rightarrow2a} + F_{1\rightarrow2b} \tag{14-11}$$

This equation is useful if two of the view factors are known and one is to be calculated. For example, consider two disks in parallel planes, as shown in Figure 14-26. The view factors $F_{1\rightarrow2}$ and $F_{1\rightarrow2b}$ can be calculated using Figure 14-18. Therefore, from Eq. 14-11, $F_{1\rightarrow2a}$ can be determined. The view factor $F_{1\rightarrow2a}$ is the fraction of radiation leaving surface 1 that arrives at a ring-shaped surface in a parallel plane. Shape decomposition is a simple and effective method for calculating this view factor.

In the shape decomposition technique, surface 2 can be divided into any arbitrary number of subsurfaces and is not limited to two. In the general case, surface 2 is subdivided

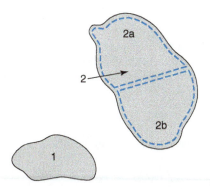

FIGURE 14-25 Surface 1 radiates to surface 2, which is divided into two parts.

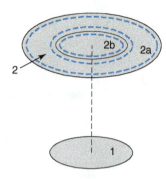

FIGURE 14-26 Construction for finding the view factor $F_{1 \to 2a}$.

into N subsurfaces, and Eq. 14-11 becomes

$$F_{1 \to 2} = F_{1 \to 2a} + F_{1 \to 2b} + F_{1 \to 2c} + \cdots + F_{1 \to 2N} = \sum_{i=1}^{N} F_{1 \to 2i} \qquad (14\text{-}12)$$

Shape decomposition is particularly powerful when used in conjunction with reciprocity and symmetry, as illustrated in the next example.

EXAMPLE 14-3 **View factors in an oven**

A heat lamp in an oven is used to cure green ceramic parts. As part of the analysis of radiative transfer in the oven, the view factors between a side wall surface and a bottom wall surface are needed. The figure shows one section of the interior of the oven, where surface 1 is on the bottom and surface 2 is on one side of the oven. Find the view factor $F_{1b \to 2b}$.

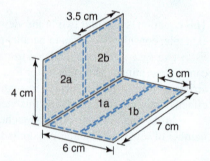

Approach:

Use Figure 14-17 to find the view factors from 1 to 2 and from 1a to 2. Take advantage of symmetry; surface 1 has the same view of 2a and 2b. In addition, 1a has the same view of 2a and 2b. Decompose surface 1 into 1a and 1b and surface 2 into 2a and 2b, as needed. Some view factors can be found by reciprocity.

Assumptions: **Solution:**

Begin by decomposing the view factor from 1 to 2 to get

$$F_{1 \to 2}^{*} = F_{1 \to 2a} + F_{1 \to 2b}$$

A1. Surface 1 and 2 emit and reflect diffusely.

The asterisk on the view factor from 1 to 2 indicates that this view factor can be determined from known formulas or plots. In this case, $F_{1\to2}^*$ can be calculated from Figure 14-17 [A1]. By symmetry, $F_{1\to2a} = F_{1\to2b}$, therefore

$$F_{1\to2}^* = 2F_{1\to2b} \tag{14-13}$$

By reciprocity:

$$F_{2b\to1} = \frac{A_1 F_{1\to2b}}{A_{2b}} \tag{14-14}$$

Solving Eq. 14-13 for $F_{1\to2b}$ and substituting into Eq. 14-14 gives

$$F_{2b\to1} = \frac{A_1 F_{1\to2}^*}{2A_{2b}} \tag{14-15}$$

We want to calculate $F_{1b\to2b}$. To make further progress, decompose $F_{2b\to1}$:

$$F_{2b\to1} = F_{2b\to1a} + F_{2b\to1b} \tag{14-16}$$

If we knew $F_{2b\to1a}$, we could calculate $F_{2b\to1b}$, from this equation. Reciprocity would then allow us to find $F_{1b\to2b}$. In order to determine $F_{2b\to1a}$, we start from

$$F_{1a\to2}^* = F_{1a\to2a} + F_{1a\to2b} = 2F_{1a\to2b} \tag{14-17}$$

where we have used the fact that $F_{1a\to2a} = F_{1a\to2b}$ by symmetry. From reciprocity

$$F_{2b\to1a} = \frac{A_{1a} F_{1a\to2b}}{A_{2b}} \tag{14-18}$$

Solving Eq. 14-17 for $F_{1a\to2b}$ and substituting in Eq. 14-18 produces

$$F_{2b\to1a} = \frac{A_{1a} F_{1a\to2}^*}{2A_{2b}} \tag{14-19}$$

At this point, we solve Eq. 14-16 for $F_{2b\to1b}$ and substitute Eq. 14-19 to get

$$F_{2b\to1b} = F_{2b\to1} - F_{2b\to1a} = F_{2b\to1} - \frac{A_{1a} F_{1a\to2}^*}{2A_{2b}}$$

Next, substitute Eq. 14-15 into this expression

$$F_{2b\to1b} = \frac{A_1 F_{1\to2}^*}{2A_{2b}} - \frac{A_{1a} F_{1a\to2}^*}{2A_{2b}} = \frac{A_1 F_{1\to2}^* - A_{1a} F_{1a\to2}^*}{2A_{2b}}$$

Finally, use reciprocity to obtain

$$F_{1b\to2b} = \frac{A_1 F_{1\to2}^* - A_{1a} F_{1a\to2}^*}{2A_{1b}}$$

Since $A_{1a} = A_{1b} = A_1/2$, this becomes

$$F_{1b\to2b} = F_{1\to2}^* - \frac{F_{1a\to2}^*}{2}$$

To find $F_{1\to2}^*$, use Figure 14-17 with

$$H = \frac{Z}{X} = \frac{4}{7} = 0.571 \qquad W = \frac{Y}{X} = \frac{6}{7} = 0.857$$

$$F_{1\to2}^* \approx 0.18$$

To find $F_{1a\rightarrow 2}^{*}$ use Figure 14-17 with

$$H = \frac{Z}{X} = \frac{4}{7} = 0.571 \qquad W = \frac{Y}{X} = \frac{3}{7} = 0.428$$

$$F_{1a\rightarrow 2}^{*} \approx 0.27$$

Finally

$$F_{1a\rightarrow 2b} = 0.18 - \frac{0.27}{2} = 0.045$$

Comments:

The final view factor is rather small; only about one-twentieth of the radiation from 1b strikes 2b. When surfaces are not close together and facing each other, view factors tend to be small. Greater accuracy could have been obtained by using the equation in Figure 14-17 rather than the graph.

14.5 RADIATIVE EXCHANGE BETWEEN BLACK SURFACES

If all surfaces in an enclosure can be idealized as black, the radiative analysis is relatively simple. Consider two arbitrary black surfaces that exchange heat by radiation, as shown in Figure 14-27. Surface 1 emits radiation, and some fraction of this radiation strikes surface 2. In addition, surface 2 emits radiation, some of which strikes surface 1. If the two surfaces are at different temperatures, there is a net transfer of energy between them given by

$$\left\{ \begin{array}{c} \text{net radiative transfer} \\ \text{between 1 and 2} \end{array} \right\} = \left\{ \begin{array}{c} \text{radiation leaving 1} \\ \text{and arriving at 2} \end{array} \right\} - \left\{ \begin{array}{c} \text{radiation leaving 2} \\ \text{and arriving at 1} \end{array} \right\}$$

(14-20)

The energy emitted by surface 1 per unit area is $E_{b1} = \sigma T_1^4$; therefore, the energy emitted by the surface is $E_{b1}A_1$, where A_1 is the surface area. The fraction of this energy that strikes 2 is $F_{1\rightarrow 2}$. With these considerations, Eq. 14-20 becomes

$$\dot{Q}_{1\rightarrow 2} = E_{b1}A_1 F_{1\rightarrow 2} - E_{b2}A_2 F_{2\rightarrow 1}$$

(14-21)

where $\dot{Q}_{1\rightarrow 2}$ is the net radiative transfer between 1 and 2. From reciprocity, $A_1 F_{1\rightarrow 2} = A_2 F_{2\rightarrow 1}$. Using this in the second term on the right-hand side of Eq. 14-21 gives

$$\dot{Q}_{1\rightarrow 2} = E_{b1}A_1 F_{1\rightarrow 2} - E_{b2}A_1 F_{1\rightarrow 2} = A_1 F_{1\rightarrow 2}(E_{b1} - E_{b2})$$

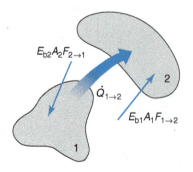

FIGURE 14-27 Two isothermal black surfaces exchanging heat by radiation.

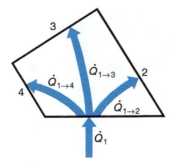

FIGURE 14-28 An enclosure of four isothermal black surfaces.

Substituting the Stefan-Boltzman law, $E_b = \sigma T^4$,

$$\dot{Q}_{1\rightarrow 2} = A_1 F_{1\rightarrow 2}\sigma \left(T_1^4 - T_2^4\right) \tag{14-22}$$

If $T_1 > T_2$, the net radiation is from 1 to 2 and $\dot{Q}_{1\rightarrow 2}$ is positive. Conversely, if $T_2 > T_1$, the net radiation is from 2 to 1 and $\dot{Q}_{1\rightarrow 2}$ is negative.

Eq. 14-22 can be applied within an enclosure of black surfaces. For example, Figure 14-28 shows an enclosure with four black surfaces. The net heat leaving surface 1 by radiation is given by

$$\dot{Q}_1 = \dot{Q}_{1\rightarrow 2} + \dot{Q}_{1\rightarrow 3} + \dot{Q}_{1\rightarrow 4}$$

More generally, in an enclosure of N surfaces, the net heat leaving surface i by radiation is

$$\dot{Q}_i = \sum_{j=1}^{N} \dot{Q}_{i\rightarrow j} = \sum_{j=1}^{N} A_i F_{i\rightarrow j}\sigma \left(T_i^4 - T_j^4\right) \tag{14-23}$$

If a surface is maintained at a constant temperature, then the net heat that leaves by radiation must be balanced by conduction, convection, heat generation, or some other form of energy. In real applications, radiation often occurs in combination with these other modes of heat transfer.

The net radiant heat that leaves one surface must arrive at one or more other surfaces in the enclosure. As defined by Eq. 14-23, net radiant heat leaving is a positive quantity and net radiant heat arriving is a negative quantity. Conservation of energy requires that, in steady state,

$$\sum_{i=1}^{N} \dot{Q}_i = 0 \tag{14-24}$$

EXAMPLE 14-4 Radiation from a black groove

An engineer suggests that radiative transfer from a black surface can be improved by cutting grooves in the surface. In Figure 14-29, a very long triangular groove exchanges heat by radiation with the surroundings, which may be considered black and at 20°C. The walls of the groove, which are maintained at 160°C, are perpendicular at the bottom, as shown in Figure 14-29. Calculate the heat transfer to the surroundings from the groove per meter of length and compare this to the heat that would be transferred if there were no groove.

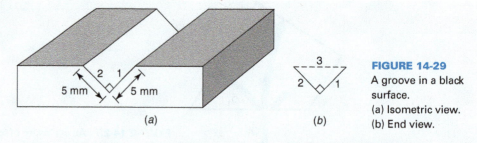

FIGURE 14-29
A groove in a black surface.
(a) Isometric view.
(b) End view.

Approach:

Cover the top of the groove with an imaginary black surface at the temperature of the surroundings to form an enclosure. Use Eq. 14-23 to find the net heat transferred by radiation from one side of the groove. Calculate the necessary view factors using Table 14-2. To find the total radiant energy leaving the groove, double the amount leaving one wall. Finally, calculate the energy that would be transferred by an infinitely long black strip with the same width as the groove, for comparison.

Assumptions:

A1. The surroundings are black at 20°C.

A2. The groove is black.

A3. The groove walls are isothermal.

A4. The groove is two-dimensional (infinitely long).

Solution:

We begin by creating an enclosure for analysis. Any radiation that leaves the groove must cross an imaginary black surface covering the groove (surface 3 on Figure 14-29b) and be absorbed by the surroundings. Radiation from the surroundings that enters the groove is black radiation at 20°C [A1]. In effect, the surroundings act exactly like a black surface covering the groove.

To determine the net heat leaving surface 1 by radiation, apply Eq. 14-23 to the three-surface enclosure to get [A2][A3]

$$\dot{Q}_1 = A_1 F_{1\to1}\sigma \left(T_1^4 - T_1^4\right) + A_1 F_{1\to2}\sigma \left(T_1^4 - T_2^4\right) + A_1 F_{1\to3}\sigma \left(T_1^4 - T_3^4\right)$$

Since $T_1 = T_2$, this reduces to

$$\dot{Q}_1 = A_1 F_{1\to3}\sigma \left(T_1^4 - T_3^4\right)$$

To find the view factor, $F_{1\to3}$, use the formula for a three-sided enclosure in Table 14-2 to get [A4]

$$F_{1\to3} = \frac{w_1 + w_3 - w_2}{2w_1}$$

Noting that the groove is a right triangle with surface 3 as the hypotenuse, we find

$$F_{1\to3} = \frac{5 + \sqrt{5^2 + 5^2} - 5}{2 \times 5} = 0.707$$

The net heat leaving surface 1 per meter may now be calculated as

$$\dot{Q}_1 = (5\,\text{mm}) \left(\frac{1\,\text{m}}{1000\,\text{mm}}\right)(1\,\text{m})(0.707)\left(5.67 \times 10^{-8}\frac{\text{W}}{\text{m}^2 \cdot \text{K}^4}\right)\left[(160 + 273)^4 - (20 + 273)^4\right]\text{K}^4$$

$$\dot{Q}_1 = 5.57\,\text{W}$$

In fact, there is no net heat transfer between surface 1 and itself or between surface 1 and surface 2, which are at the same temperature. Therefore $\dot{Q}_1 = \dot{Q}_{1\to3}$. By symmetry, $\dot{Q}_1 = \dot{Q}_2 = \dot{Q}_{2\to3}$. The total heat leaving the groove is

$$\dot{Q}_{groove} = \dot{Q}_1 + \dot{Q}_2 = 5.57 + 5.57 = 11.1\,\text{W}$$

For comparison, we calculate the total radiant heat that would leave a black strip 1 m long with the width of surface 3 and be radiated to the surroundings. It is

$$\dot{Q}_{strip} = A_{strip}\sigma \left(T_{strip}^4 - T_{surr}^4\right)$$

$$\dot{Q}_{strip} = (7.07 \text{ mm})\left(\frac{1 \text{ m}}{1000 \text{ mm}}\right)(1 \text{ m})\left(5.67 \times 10^{-8} \frac{W}{m^2 \cdot K^4}\right)\left[(160 + 273)^4 - (20 + 273)^4\right] K^4 = 11.1 \text{ W}$$

The groove radiates the same amount as the equivalent flat surface.

Comments:

Adding grooves to the surface did not improve heat transfer at all. The extra surface area created was exactly canceled by the reduced view from the groove to the surroundings. Note that if you are inside the groove and looking out, you see an equivalent black surface at the temperature of the surroundings. If you are outside the groove and looking in, you see an equivalent black surface at the temperature of the walls of the groove.

The energy leaving the groove can be found from conservation of energy. Using $\sum\limits_{i=1}^{N} \dot{Q}_i = 0$, we see that $\dot{Q}_1 + \dot{Q}_2 + \dot{Q}_3 = 0$. The energy leaving the groove is $\dot{Q}_{groove} = -\dot{Q}_3 = \dot{Q}_1 + \dot{Q}_2$. We could also have found the view factor by using Table 14-2.

14.6 RADIATIVE EXCHANGE BETWEEN DIFFUSE, GRAY SURFACES

When all surfaces are black, it is not necessary to deal with reflected radiation. In most applications, however, surfaces are not perfectly black and reflections must be considered. In general, reflection from a surface depends on the wavelength of the incoming radiation and on the incident angle. Taking these factors into account complicates the analysis and often leads to a computer simulation. We can simplify the analysis by making two assumptions: that the surfaces in the enclosure are gray and that they are diffuse.

Using the gray assumption is analogous to representing a scene in black and white instead of color. Most of the information is preserved, although important facets might be lost in the black and white representation. In many instances, a gray radiation analysis will produce results close to the actual spectral analysis. We also assume that all surfaces in the enclosure are diffuse. This implies that the angle of reflection is independent of the angle of incidence. If the enclosure contains shiny metallic surfaces, this assumption may be inaccurate. Even in that case, however, the multiple reflections within the enclosure often mask the effects of specular reflection and the diffuse results are reasonably accurate.

In this section we discuss enclosures with surfaces that are gray, diffuse, and isothermal. As in the black surface enclosures, we assume that the enclosure is filled with a transparent medium, either vacuum or a gas that does not absorb or emit radiation. We account for reflection with a new concept – radiosity. The **radiosity**, J, is defined as all the radiation leaving a surface and includes both emitted and reflected components.

In Figure 14-31, incident radiation, G, strikes a gray, diffuse surface and is partially reflected. In addition, the surface emits radiation εE_b. Radiosity, J, is the sum of the emitted and reflected radiation, that is,

$$J = \varepsilon E_b + \rho G \qquad (14\text{-}25)$$

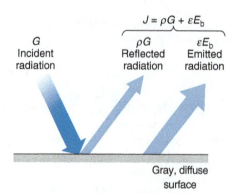

FIGURE 14-31 Radiation to and from an opaque, gray, diffuse surface.

Both J and G are heat fluxes; that is, they are heat transfer per unit area. The units of J and G are the same as those of emissive power, either W/m² in the SI system or Btu/h·ft² in the British system.

For an opaque surface, from Eq. 14-4,

$$\rho = 1 - \alpha$$

Using this in Eq. 14-25 produces

$$J = \varepsilon E_b + (1 - \alpha)G$$

Substituting Kirchhoff's law (Eq. 14-5) gives

$$\boxed{J = \varepsilon E_b + (1 - \varepsilon)G} \tag{14-26}$$

The net heat flux leaving surface i by radiation is the difference between outgoing and incoming radiation (see Figure 14-31), that is,

$$\frac{\dot{Q}_i}{A_i} = J_i - G_i \tag{14-27}$$

where net heat transfer, $\dot{Q}_i$ has been divided by area A_i, because J_i and G_i are heat *fluxes*, while $\dot{Q}_i$ is simply heat transfer. Solving Eq. 14-26 for G and substituting into Eq. 14-27 produces

$$\frac{\dot{Q}_i}{A_i} = J_i - \left[\frac{J_i - \varepsilon_i E_{bi}}{(1 - \varepsilon_i)}\right]$$

simplifying and solving for $\dot{Q}_i$ gives

$$\boxed{\dot{Q}_i = \frac{A_i \varepsilon_i}{1 - \varepsilon_i}(E_{bi} - J_i) \qquad \text{gray, diffuse surface}} \tag{14-28}$$

Eq. 14-28 does not apply to a black surface. For a black surface, $\varepsilon = 1$, and the denominator of Eq. 14-28 becomes zero. Instead, use Eq. 14-26 for a black surface to get

$$\boxed{J = E_b = \sigma T^4 \qquad \text{black surface}} \tag{14-29}$$

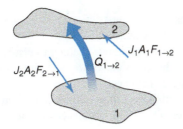

FIGURE 14-32 Two isothermal, gray, diffuse surfaces exchanging heat by radiation.

Physically, this equation states that only emitted energy leaves the black surface and there is no reflected component.

With these preliminaries, we are now ready to consider radiation in an enclosure of diffuse, gray surfaces. Figure 14-32 shows two gray, diffuse surfaces that exchange heat by radiation. The net radiant heat transfer between the two surfaces is

$$\left\{ \begin{array}{c} \text{net radiative transfer} \\ \text{between 1 and 2} \end{array} \right\} = \left\{ \begin{array}{c} \text{radiation leaving 1} \\ \text{and arriving at 2} \end{array} \right\} - \left\{ \begin{array}{c} \text{radiation leaving 2} \\ \text{and arriving at 1} \end{array} \right\}$$

$$(14\text{-}30)$$

The total radiant energy leaving surface 1 per unit area is J_1; therefore, the total energy leaving the surface is $J_1 A_1$, where A_1 is the surface area. The fraction of this energy that strikes 2 is $F_{1 \to 2}$. With these considerations, Eq. 14-30 becomes

$$\dot{Q}_{1 \to 2} = J_1 A_1 F_{1 \to 2} - J_2 A_2 F_{2 \to 1} \qquad (14\text{-}31)$$

From reciprocity, $A_1 F_{1 \to 2} = A_2 F_{2 \to 1}$. Using this in the second term on the right-hand side gives

$$\boxed{\dot{Q}_{1 \to 2} = A_1 F_{1 \to 2}(J_1 - J_2)} \qquad (14\text{-}32)$$

Furthermore, exchanging indices in Eq. 14-31 produces

$$\dot{Q}_{2 \to 1} = J_2 A_2 F_{2 \to 1} - J_1 A_1 F_{1 \to 2}$$

Comparing this with Eq. 14-31 reveals that

$$\dot{Q}_{i \to j} = -\dot{Q}_{j \to i} \qquad (14\text{-}33)$$

In an enclosure of N surfaces, the net heat leaving surface i by radiation is the sum of the net radiative transfers between surface i and each of the other surfaces in the enclosure. In equation form,

$$\boxed{\dot{Q}_i = \sum_{j=1}^{N} \dot{Q}_{i \to j}} \qquad (14\text{-}34)$$

As in the case of an enclosure with black surfaces, conservation of energy requires that

$$\sum_{i=1}^{N} \dot{Q}_i = 0$$

It should be noted that Eq. 14-31 and Eq. 14-32 apply only if radiosity is uniform over the surface. In other words, the surface must be isothermal and must reflect the same amount of energy at every location. This restriction is not apparent from the derivation but is needed to derive Eq. 14-6 for view factors.

It is possible to use an electric resistance analogy to visualize the effects of radiation in an enclosure. Figure 14-33 shows an enclosure with three surfaces and the associated resistance network. The "voltages" in the network are represented by either emissive power, E_b, or radiosity, J. The "currents" are either net radiation leaving a surface, $\dot{Q}_i$, or net radiation between two surfaces, $\dot{Q}_{i \to j}$. The resistances in the network are determined by analogy to those in an electric circuit, where

$$\text{current} = \frac{\text{voltage}}{\text{resistance}}$$

Applying this idea to the resistance, R_1, in Figure 14-33 gives

$$\dot{Q}_1 = \frac{E_{b1} - J_1}{R_1} \tag{14-35}$$

Comparing Eq. 14-35 with Eq. 14-28 reveals that

$$R_1 = \frac{1 - \varepsilon_1}{A_1 \varepsilon_1}$$

Similar expressions can be obtained for R_2 and R_3. In general, the **surface resistance to radiation, R_i,** is

$$\boxed{R_i = \frac{1 - \varepsilon_i}{A_i \varepsilon_i}} \tag{14-36}$$

The surface resistance depends only on the area and emissivity of surface i and does not depend on the placement, size, or properties of any other surface in the enclosure. For a black surface, $R_i = 0$ and $J_i = E_{bi}$.

Referring again to Figure 14-33, the resistance $R_{1 \to 2}$ is

$$\dot{Q}_{1 \to 2} = \frac{J_1 - J_2}{R_{1 \to 2}}$$

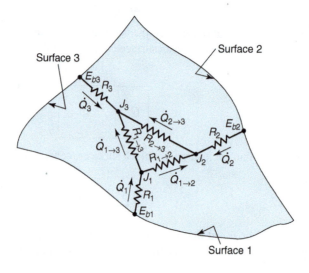

FIGURE 14-33 The resistance analogy in a three-surface enclosure.

Comparing this with Eq. 14-32 shows that

$$R_{1\rightarrow 2} = \frac{1}{A_1 F_{1\rightarrow 2}}$$

More generally, for any two surfaces in the enclosure, the **space resistance to radiation** is

$$\boxed{R_{i\rightarrow j} = \frac{1}{A_i F_{i\rightarrow j}}} \tag{14-37}$$

Figure 14-33 includes three space resistances.

Finally, in electrical circuits, the total current into a nodal point must equal the total current out of that point (this can also be viewed as a steady energy balance). Applying this idea to the node point, J_1, in Figure 14-33 produces

$$\dot{Q}_1 = \dot{Q}_{1\rightarrow 2} + \dot{Q}_{1\rightarrow 3}$$

This equation is none other than Eq. 14-34 with $i = 1$ and $N = 3$ ($\dot{Q}_{1\rightarrow 1}$ is identically zero). Eq. 14-34 applies at each of the three nodal points: J_1, J_2, and J_3. The relation $\dot{Q}_{i\rightarrow j} = -\dot{Q}_{j\rightarrow i}$ (see Eq. 14-31) can be used to change the directions of arrows representing $\dot{Q}_{1\rightarrow 2}, \dot{Q}_{1\rightarrow 3}$, and $\dot{Q}_{2\rightarrow 3}$ in Figure 14-33. For example, at node J_2,

$$\dot{Q}_2 = \dot{Q}_{2\rightarrow 1} + \dot{Q}_{2\rightarrow 3}$$

The utility of the resistance analogy is demonstrated in the next section, where it is applied to the simplest case possible: an enclosure with only two surfaces.

14.7 TWO-SURFACE ENCLOSURES

When an enclosure contains only two surfaces, the resistance network assumes a particularly simple form, as shown in Figure 14-34. The energy exchange between surface 1 and surface 2 depends on three resistances in series. The total resistance between the two surfaces is

$$R_{tot} = R_1 + R_{1\rightarrow 2} + R_2$$

Substituting Eq. 14-36 and Eq. 14-37 gives

$$R_{tot} = \frac{1 - \varepsilon_1}{A_1 \varepsilon_1} + \frac{1}{A_1 F_{1\rightarrow 2}} + \frac{1 - \varepsilon_2}{A_2 \varepsilon_2}$$

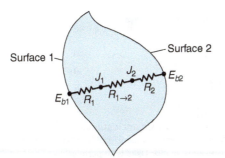

FIGURE 14-34 The resistance analogy for two surfaces.

The net heat transfer by radiation between surface 1 and 2 is

$$\dot{Q}_{1\rightarrow2} = \frac{E_{b1} - E_{b2}}{R_{tot}} = \frac{\sigma(T_1^4 - T_2^4)}{\dfrac{1 - \varepsilon_1}{A_1\varepsilon_1} + \dfrac{1}{A_1 F_{1\rightarrow2}} + \dfrac{1 - \varepsilon_2}{A_2\varepsilon_2}} \tag{14-38}$$

From Eq. 14-33 and Eq. 14-34,

$$\dot{Q}_1 = \dot{Q}_{1\rightarrow2}$$

$$\dot{Q}_2 = \dot{Q}_{2\rightarrow1} = -\dot{Q}_{1\rightarrow2} = -\dot{Q}_1$$

To solve problems with two surface enclosures, the surface temperature must be known at one or both surfaces. If the surface temperature is not known at one of the surfaces, then the net radiation leaving the surface, $\dot{Q}_i$, must be known. Figure 14-35 gives equations for some common two-surface enclosures.

In Chapter 3, the net radiative heat transfer from a small, gray, diffuse surface at T_1 to surroundings at T_2 was expressed as

$$\dot{Q}_{1\rightarrow2} = \varepsilon_1 \sigma A_1 \left(T_1^4 - T_2^4\right)$$

This equation applies if the surrounding surfaces are large compared to surface 1 and all radiation from surface 1 reaches the surroundings. In that case, $A_2 \gg A_1$ and

Large (infinite) parallel planes

A_1, T_1, ε_1
A_2, T_2, ε_2

$\begin{aligned} A_1 &= A_2 = A \\ F_{1\rightarrow2} &= 1 \end{aligned}$

$\dot{Q}_{1\rightarrow2} = \dfrac{A\sigma(T_1^4 - T_2^4)}{\dfrac{1}{\varepsilon_1} + \dfrac{1}{\varepsilon_2} - 1}$

Long (infinite) concentric cylinders

r_1
r_2

$\dfrac{A_1}{A_2} = \dfrac{r_1}{r_2}$

$F_{1\rightarrow2} = 1$

$\dot{Q}_{1\rightarrow2} = \dfrac{\sigma A_1 (T_1^4 - T_2^4)}{\dfrac{1}{\varepsilon_1} + \dfrac{1 - \varepsilon_2}{\varepsilon_2}\left(\dfrac{r_1}{r_2}\right)}$

Concentric spheres

r_1
r_2

$\dfrac{A_1}{A_2} = \dfrac{r_1^2}{r_2^2}$

$F_{1\rightarrow2} = 1$

$\dot{Q}_{1\rightarrow2} = \dfrac{\sigma A_1 (T_1^4 - T_2^4)}{\dfrac{1}{\varepsilon_1} + \dfrac{1 - \varepsilon_2}{\varepsilon_2}\left(\dfrac{r_1}{r_2}\right)^2}$

Small convex object in large surroundings

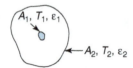

A_1, T_1, ε_1
A_2, T_2, ε_2

$\dfrac{A_1}{A_2} \approx 0$

$F_{1\rightarrow2} = 1$

$\dot{Q}_{1\rightarrow2} = \sigma A_1 \varepsilon_1 (T_1^4 - T_2^4)$

FIGURE 14-35 Common two-surface enclosures.
(*Source:* M. J. Moran, H. N. Shapiro, B. R. Munson, and D. P. Dewitt, *Introduction to Thermal Systems Engineering,* Wiley, New York, 2003, p. 498. Used with permission.)

$F_{1\rightarrow2} = 1$, and Eq. 14-38 reduces to

$$\dot{Q}_{1\rightarrow2} = \frac{\sigma\left(T_1^4 - T_2^4\right)}{\dfrac{1 - \varepsilon_1}{A_1\varepsilon_1} + \dfrac{1}{A_1}}$$

which simplifies to $\dot{Q}_{1\rightarrow2} = \varepsilon_1\,\sigma A_1\left(T_1^4 - T_2^4\right)$. This demonstrates the consistency between the radiation equations in this chapter with those presented in earlier chapters.

EXAMPLE 14-5   Heat loss from a vacuum bottle

A vacuum bottle with a height of 10 in. contains hot coffee at 160°F. The container consists of an inner bottle centered within an outer casing that is at 35°F. The space between the inner bottle and the casing is evacuated, and the walls are coated with aluminum to minimize radiative heat transfer losses. There is negligible heat transfer at the ends of the container. In a new vacuum bottle, the emissivity of all surfaces is 0.05, but in an older container, the finish becomes dull and the emissivity rises to 0.25. Calculate the rate of heat loss from the coffee for both a new and an old vacuum bottle. Use data in the schematic.

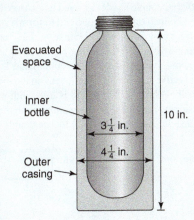

Approach:

There is no net radiation in the axial direction out the ends of the evacuated space. The view factor from the inner bottle to the casing is unity by inspection. The rate of heat transfer may be determined from Eq. 14-38.

Assumptions:

A1. The walls are gray and diffuse.
A2. The walls are isothermal.
A3. Radiosity on each surface is constant.
A4. There is no net radiation out the end of the evacuated space.

Solution:

From Eq. 14-38, the net rate of heat transfer between the inner bottle (surface 1) and the casing (surface 2) is [A1][A2][A3]

$$\dot{Q}_{1\rightarrow2} = \frac{E_{b1} - E_{b2}}{R_{tot}} = \frac{\sigma(T_1^4 - T_2^4)}{\dfrac{1 - \varepsilon_1}{A_1\varepsilon_1} + \dfrac{1}{A_1 F_{1\rightarrow2}} + \dfrac{1 - \varepsilon_2}{A_2\varepsilon_2}}$$

If we neglect the ends of the container, all the radiation leaving the bottle arrives at the casing [A4]; therefore, $F_{1\rightarrow2} = 1$. The areas of the two cylinders are

$$A_1 = \pi(3.25\ \text{in.})(10\ \text{in.}) = 102\ \text{in.}^2$$

$$A_2 = \pi(4.25\ \text{in.})(10\ \text{in.}) = 134\ \text{in.}^2$$

When the vacuum bottle is new, the emissivity of both surfaces is 0.05 and the rate of heat loss is calculated as

$$\dot{Q}_{1\to2} = \frac{\left(0.1714 \times 10^{-8} \dfrac{\text{Btu}}{\text{h}\cdot\text{ft}^2\cdot\text{R}^4}\right)\left[(160+460)^4 - (35+460)^4\right]\text{R}^4}{\left[\dfrac{1-0.05}{(102)\,\text{in.}^2\,(0.05)} + \dfrac{1}{(102)\,\text{in.}^2\,(1)} + \dfrac{1-0.05}{(134)\,\text{in.}^2\,(0.05)}\right]\left(\dfrac{144\,\text{in.}^2}{1\,\text{ft}^2}\right)} = 3.09\frac{\text{Btu}}{\text{h}}$$

For an old vacuum bottle, the calculation is repeated with $\varepsilon = 0.25$, giving

$$\dot{Q}_{1\to2} = 17.0\frac{\text{Btu}}{\text{h}}$$

Comments:

Aging of the coating on a vacuum bottle can seriously affect performance. Coffee will cool more rapidly in the older container.

EXAMPLE 14-6   Measurement error in a thermocouple

A thermocouple is a device used to measure temperature. It is constructed of two wires of dissimilar metal joined together at one end to form a small bead. In the schematic, a thermocouple is inserted in a gas flow. A cylindrical radiation shield open on both ends encloses the thermocouple, as shown. The gas is at 500°C and the pipe wall is at 140°C. The shield has an emissivity of 0.09, while the emissivity of the thermocouple bead is 0.7 and the emissivity of the pipe wall is 0.93. The convective heat transfer coefficient on the thermocouple is 45 W/m²·°C and on the radiation shield, it is 30 W/m²·°C. The shield is long with a diameter of 3 cm, while the pipe diameter is 5.6 cm. Assuming the shield is large compared to the size of the thermocouple, calculate the thermocouple temperature. What would the thermocouple read if the radiation shield were removed?

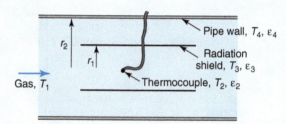

Approach:

Write an energy balance on the thermocouple bead including both convection and radiation to the shield. Because the thermocouple is a small body in a large enclosure (the shield), the shield may be modeled as a black surface from the point of view of the thermocouple. Also develop an energy balance on the radiation shield. The radiative exchange between the shield and the pipe wall may be modeled as that between two infinitely long cylinders using Figure 14-35. In this case, use the actual emissivity of the shield in the calculation. Solve the two energy equations for the unknown thermocouple and shield temperatures.

Assumptions:

A1. The thermocouple is very small compared to the shield.

A2. There is no conduction along the thermocouple wires.

Solution:

We begin with an energy balance on the thermocouple bead. From the perspective of the thermocouple, the shield is a large enclosure and appears black [A1]. The net radiative heat transfer leaving the thermocouple equals the convective heat transfer to the thermocouple [A2]; therefore,

$$A_2\varepsilon_2\sigma\left(T_2^4 - T_3^4\right) = h_2A_2\left(T_1 - T_2\right)$$

where h_2 is the heat transfer coefficient at the thermocouple surface and A_2 is the area of the bead. Other variables are indicated on the schematic.

We also perform an energy balance on the radiation shield. Here convection occurs on both sides of the shield. The shield and the pipe wall act approximately like two infinitely long cylinders [A3], so the equation for net radiative transfer from Figure 14-35 may be used. The resulting energy balance is [A4][A5]

A3. The shield and pipe wall act like two infinitely long cylinders.
A4. The shield is gray and diffuse.
A5. The shield is isothermal.

$$\frac{A_3 \sigma \left(T_3^4 - T_4^4\right)}{\frac{1}{\varepsilon_3} + \frac{1 - \varepsilon_4}{\varepsilon_4} \left(\frac{r_3}{r_4}\right)} = 2 h_3 A_3 \left(T_1 - T_3\right)$$

where the shield is surface 3 and the pipe wall is surface 4. Notice that the shield temperature is independent of the thermocouple temperature. From the perspective of the shield, the thermocouple is too small to be of significance. In the energy balance on the shield, the *actual* emissivity of the pipe wall is used. From the perspective of the pipe wall, the shield is large enough to be significant.

The last two equations are simultaneous equations in two unknowns: T_2 and T_3. The known parameters are

$$T_1 = 500°C = 773\,K \quad \varepsilon_2 = 0.7 \quad h_2 = 45\frac{W}{m^2 \cdot K} \quad r_1 = 1.5\,cm \quad \sigma = 5.67 \times 10^{-8}\frac{W}{m^2 \cdot K^4}$$

$$T_4 = 140°C = 413\,K \quad \varepsilon_3 = 0.09 \quad h_3 = 30\frac{W}{m^2 \cdot K} \quad r_2 = 2.8\,cm$$

$$\varepsilon_4 = 0.93$$

Note that the areas in each equation cancel. The temperatures must be expressed in K. The two equations are nonlinear, and it is not possible to solve for T_2 and T_3 directly. This is often the case when radiation is combined with convection and/or conduction. The solution may be found using equation-solving software to give the result:

$$T_3 = 476°C$$

$$T_2 = 485°C$$

The final thermocouple temperature is 485°C. If there is no radiation shield, the energy balance on the thermocouple becomes

$$A_2 \varepsilon_2 \sigma \left(T_2^4 - T_4^4\right) = h_2 A_2 \left(T_1 - T_2\right)$$

where the only difference is that the thermocouple exchanges heat by radiation with the pipe wall instead of the shield. The thermocouple temperature in this case is 372°C.

Comments:

With the shield in place, the thermocouple reads a temperature of 485°C, which is close to the gas temperature of 500°C. The shield prevents the thermocouple from "seeing" the cold pipe wall and cooling as a result. Since the shield is reflective, it looses little heat by radiation and takes on a temperature close to the gas temperature (shield temperature was 476°C). The thermocouple sees only the shield, which is at a relatively high temperature.

When the shield is absent, the thermocouple radiates directly to the cold pipe wall and reads an erroneous value for gas temperature of 372°C.

14.8 THREE-SURFACE ENCLOSURES

In Figure 14-33, the resistance analogy for an enclosure formed from three diffuse, gray surfaces is given. Each surface is at a uniform temperature and radiosity. At each of the three node points, J_1, J_2, and J_3, the sum of the incoming heat fluxes must equal the sum of the outgoing heat fluxes (see Eq. 14-34). This leads to the following three equations:

$$
\begin{aligned}
\dot{Q}_1 &= \frac{J_1 - J_2}{R_{1 \to 2}} + \frac{J_1 - J_3}{R_{1 \to 3}} \\
\dot{Q}_2 &= \frac{J_2 - J_3}{R_{2 \to 3}} + \frac{J_2 - J_1}{R_{2 \to 1}} \\
\dot{Q}_3 &= \frac{J_3 - J_1}{R_{3 \to 1}} + \frac{J_3 - J_2}{R_{3 \to 2}}
\end{aligned}
\qquad (14\text{-}39)
$$

The net heat transfer at each surface is related to the surface resistance by

$$
\dot{Q}_1 = \frac{E_{b1} - J_1}{R_1} = \frac{\sigma T_1^4 - J_1}{R_1}
$$

$$
\dot{Q}_2 = \frac{E_{b2} - J_2}{R_2} = \frac{\sigma T_2^4 - J_2}{R_2}
$$

$$
\dot{Q}_3 = \frac{E_{b3} - J_3}{R_3} = \frac{\sigma T_3^4 - J_3}{R_3}
$$

Rearranging these gives

$$
\begin{aligned}
J_1 &= \sigma T_1^4 - R_1 \dot{Q}_1 \\
J_2 &= \sigma T_2^4 - R_2 \dot{Q}_2 \\
J_3 &= \sigma T_3^4 - R_3 \dot{Q}_3
\end{aligned}
\qquad (14\text{-}40)
$$

Eq. 14-39 and Eq. 14-40 are six equations in six unknowns. Three of the unknowns are the radiosities, J_1, J_2, and J_3. The other three are temperatures or net heat transfer at a surface. At each surface, either T_i or $\dot{Q}_i$ must be known. Furthermore, the temperature must be known for at least one surface. For example, the six unknowns might be $J_1, J_2, J_3, T_1, \dot{Q}_2, \dot{Q}_3$. Alternatively, the six unknowns might be $J_1, J_2, J_3, T_1, T_2, T_3$. However, the six unknowns *cannot be* $J_1, J_2, J_3, \dot{Q}_1, \dot{Q}_2, \dot{Q}_3$. In practical radiation applications, the fundamental quantities of interest are surface temperature and net heat transfer at a surface. The radiosity is merely an intermediate variable needed to determine radiative transport.

Eq. 14-39 and Eq. 14-40 were written for three surfaces. The analysis can easily be extended to any finite number of surfaces. For an analysis of N surfaces, the number of view factors to be calculated, in the general case, is N^2. Commercial codes exist to aid in determining these view factors and solving the resulting system of equations.

In some cases, the system of equations can be simplified. If a surface is black, then, from Eq. 14-36, $R_i = 0$. From Eq. 14-40, if surface resistance is zero, the radiosity is equal to the emissive power or $J_i = \sigma T_i^4$.

Another special case is the **reradiating surface**. A reradiating surface receives no heat by conduction or convection and, therefore, $\dot{Q}_i = 0$. A typical reradiating surface is perfectly insulated on the back side. The front side, which radiates, is not exposed to convection. For a reradiating surface, all radiation incident on the surface leaves by radiation. Since $\dot{Q}_i = 0$ for a reradiating surface, from Eq. 14-40, $J_i = \sigma T_i^4$.

In addition to the six equations in Eq. 14-39 and Eq. 14-40, we may also write (see Eq. 14-24)

$$\dot{Q}_1 + \dot{Q}_2 + \dot{Q}_3 = 0 \qquad (14\text{-}41)$$

This equation is not linearly independent of the other six. It may be used in place of one of the six equations in the system. Note that all of the above equations apply to steady-state systems.

EXAMPLE 14-7   Radiation heat leak from an oven

A view port in the side of a furnace consists of a quartz window set at the end of a cylindrical hole in the furnace wall, as shown in the schematic. The quartz has an emissivity of 0.89, a temperature of 550°C, and is opaque to infrared radiation. Conduction and convection at the side wall of the hole are negligible in comparison to radiation. The surroundings are at 20°C, and there are no reflecting surfaces near the open end of the view port. Calculate the amount of heat transfer that leaks from the furnace through the view port.

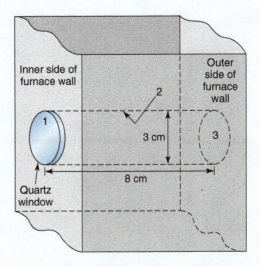

Approach:

The exit of the view port may be covered with an imaginary black surface at the temperature of the surroundings. The view port can then be modeled as a three-surface enclosure. Eq. 14-39 and Eq. 14-40 are used to find the temperature and the rate of heat loss from the furnace. The equations can be simplified by noting that one end of the hole is black ($J_i = \sigma T_i^4$) and that the side wall of the hole is a reradiating surface for which $\dot{Q}_i = 0$. To compute the resistances, view factors are needed. The view factors for two parallel disks are available in Figure 14-18. All remaining view factors in the enclosure can be calculated using the summation and reciprocity relationships.

Assumptions:

Solution:

We define the quartz window as surface 1, the side wall of the hole as surface 2, and the exposed end of the hole as surface 3. Surface 3 is an imaginary black surface at the temperature of the surroundings; therefore, $\varepsilon_3 = 1$ [A1]. The surface resistance for surface 3 is (see Eq. 14-36)

A1. There are no reflecting surfaces near the exit of the hole.

$$R_3 = \frac{1 - \varepsilon_3}{A_3 \varepsilon_3} = 0$$

From Eq. 14-40, the radiosity of surface 3 is

$$J_3 = \sigma T_3^4$$

Substituting values

$$J_3 = (5.67 \times 10^{-8}) \frac{W}{m^2 \cdot K^4} (20 + 273)^4 \, K^4 = 418 \frac{W}{m^2}$$

A2. Radiosity is uniform over each surface.
A3. The quartz and the sides of the hole are gray and diffuse.
A4. All surfaces are isothermal.

From Eq. 14-39 applied to surfaces 1 and 2 [A2][A3][A4],

$$\dot{Q}_1 = \frac{J_1 - J_2}{R_{1 \to 2}} + \frac{J_1 - J_3}{R_{1 \to 3}} \tag{14-42}$$

$$\dot{Q}_2 = \frac{J_2 - J_3}{R_{2 \to 3}} + \frac{J_2 - J_1}{R_{2 \to 1}} \tag{14-43}$$

From Eq. 14-40 applied to surface 1

$$J_1 = \sigma T_1^4 - R_1 \dot{Q}_1 \tag{14-44}$$

A5. The sides of the hole are insulated.

In the last three equations, J_3 and T_1 are known. Because there is no conduction or convection at surface 2, $\dot{Q}_2 = 0$ [A5]. The remaining three unknowns are J_1, J_2, and $\dot{Q}_1$. Before we can calculate the unknowns, we need the resistances, which depend on view factors.

Surface 1 and surface 3 are two disks of equal size in parallel planes. To find $F_{1 \to 3}$, use the equation in Figure 14-18, which is

$$F_{1 \to 3} = \frac{1}{2} \left\{ S - \left[S^2 - 4 \left(\frac{r_3}{r_1} \right)^2 \right]^{0.5} \right\}$$

where r_1 and r_3 are the radii of the disks and S is given by

$$S = 1 + \frac{1 + (r_3/L)^2}{(r_1/L)^2}$$

In this equation, L is the depth of the hole. Substituting values

$$S = 1 + \frac{1 + (1.5/8)^2}{(1.5/8)^2} = 30.4$$

which leads to

$$F_{1 \to 3} = \frac{1}{2} \left\{ 30.4 - \left[(30.4)^2 - 4 \left(\frac{1.5}{1.5} \right)^2 \right]^{0.5} \right\} = 0.0329$$

The view factor from the quartz window (surface 1) to the side of the hole (surface 2) may be determined using the summation relation:

$$F_{1 \to 2} + F_{1 \to 3} = 1$$

$$F_{1 \to 2} = 1 - F_{1 \to 3} = 1 - 0.0329 = 0.967$$

By symmetry,

$$F_{3 \to 2} = F_{1 \to 2} = 0.967$$

To find $F_{2\to3}$, use reciprocity, that is,

$$F_{2\to3} = \frac{A_3 F_{3\to2}}{A_2} = \frac{\pi r_3^2 F_{3\to2}}{2\pi r_3 L} = \frac{\pi (1.5)^2 (0.967)}{2\pi (1.5)(8)} = 0.0907$$

The last view factor is found by symmetry

$$F_{2\to1} = F_{2\to3} = 0.0907$$

All the relevant space resistances may now be calculated from Eq. 14-36 as

$$R_{1\to2} = \frac{1}{A_1 F_{1\to2}} = \frac{1}{\pi (0.015)^2 \text{ m}^2 (0.967)} = 1463 \text{ m}^{-2}$$

$$R_{1\to3} = \frac{1}{A_1 F_{1\to3}} = \frac{1}{\pi (0.015)^2 \text{ m}^2 (0.0329)} = 4.31 \times 10^4 \text{ m}^{-2}$$

Using reciprocity,

$$R_{2\to1} = \frac{1}{A_2 F_{2\to1}} = \frac{1}{A_1 F_{1\to2}} = R_{1\to2} = 1463 \text{ m}^{-2}$$

Using symmetry,

$$R_{2\to3} = \frac{1}{A_2 F_{2\to3}} = \frac{1}{A_2 F_{2\to1}} = R_{2\to1} = 1463 \text{ m}^{-2}$$

The surface resistance for surface 1 is, from Eq. 14-37,

$$R_1 = \frac{(1 - \varepsilon_1)}{A_1 \varepsilon_1} = \frac{(1 - 0.89)}{\pi (0.015)^2 \text{ m}^2 (0.89)} = 175 \text{ m}^{-2}$$

We now have all we need to solve Eq. 14-42 through Eq. 14-44. Substituting values into these equations gives

$$\dot{Q}_1 = \frac{J_1 - J_2}{1463 \text{ m}^{-2}} + \frac{J_1 - 418 \left(\text{W/m}^2 \right)}{4.31 \times 10^4 \text{ m}^{-2}}$$

$$0 = \frac{J_2 - 418 \left(\text{W/m}^2 \right)}{1463 \text{ m}^{-2}} + \frac{J_2 - J_1}{1463 \text{ m}^{-2}}$$

$$J_1 = \left(5.67 \times 10^{-8} \right) \frac{\text{W}}{\text{m}^2 \cdot \text{K}^4} (550 + 273) \text{ K}^4 - \left(175 \text{ m}^{-2} \right) \dot{Q}_1$$

Solving these three equations simultaneously yields

$$J_1 = 2.44 \times 10^4 \frac{\text{W}}{\text{m}^2}$$

$$J_2 = 1.24 \times 10^4 \frac{\text{W}}{\text{m}^2}$$

$$\dot{Q}_1 = 8.78 \text{ W}$$

The net heat transfer that leaks from the furnace is the heat transfer leaving surface 1, or 8.78 W.

Comments:

The heat leak is negligible because the window is small and the view factor between 1 and 3 is also small. Some of the radiation leaving the quartz window is reflected back to the window by the sides of the hole.

14.9 VARIATION OF THERMOPHYSICAL PROPERTIES WITH WAVELENGTH AND DIRECTION

(Go to www.wiley.com/college/kaminski)

SUMMARY

By definition, a black surface absorbs all the radiation incident upon it. A black surface also emits the maximum possible radiation that any surface at that temperature can emit. The total amount of radiation emitted from a black surface is

$$\frac{\dot{Q}_{emit}}{A} = E_b = \sigma T^4$$

where E_b is the emissive power of a black surface, σ is the Stefan-Boltzmann constant, A is the area of the surface, and T is the absolute temperature (R or K). The Stefan-Boltzmann constant has values of

$$\sigma = 5.6697 \times 10^{-8} \frac{W}{m^2 \cdot K^4}$$

$$\sigma = 0.17123 \times 10^{-8} \frac{Btu}{h \cdot ft^2 \cdot R^4}$$

The variation of emissive power with wavelength is given by Planck's law:

$$E_{b\lambda} = \frac{C_1 \lambda^{-5}}{e^{C_2/\lambda T} - 1}$$

where $E_{b\lambda}$ is emissive power per unit area, per unit wavelength, and λ is wavelength. The constants C_1 and C_2 are

$$C_1 = 2\pi h c_o^2 = 3.742 \times 10^8 \frac{W \cdot \mu m^4}{m^2}$$

$$C_2 = \frac{h c_o}{k} = 1.439 \times 10^4 \mu m \cdot K$$

At each temperature, there is a maximum in the spectral energy distribution. The wavelength of this maximum point is given by Wien's displacement law as

$$\lambda_{max} T = C_3 = 2898 \mu m \cdot K$$

Real surfaces are characterized by the emissivity, which is

$$\varepsilon = \frac{actual\ emitted\ energy}{blackbody\ emitted\ energy}$$

For a gray surface, emissivity does not depend on wavelength, and for a diffuse surface, emissivity does not depend on angle. The emissive power of a gray, diffuse surface is

$$\frac{\dot{Q}_{emit}}{A} = E = \varepsilon \sigma T^4$$

where E is emissive power of a real surface.

To characterize reflected energy, the reflectivity, ρ, is defined as

$$\rho = \frac{reflected\ energy}{incident\ energy}$$

To characterize absorption, we define absorptivity as

$$\alpha = \frac{absorbed\ energy}{incident\ energy}$$

To characterize transmission, we define transmissivity as

$$\tau = \frac{transmitted\ energy}{incident\ energy}$$

Emissivity, reflectivity, absorptivity, and transmissivity depend, in general, on temperature, wavelength, and direction. For gray, diffuse surfaces, these properties depend only on temperature. By conservation of energy, the fractions absorbed, reflected, and transmitted must sum to unity:

$$\alpha + \rho + \tau = 1$$

For an opaque surface, $\tau = 0$, and

$$\alpha + \rho = 1$$

Kirchhoff's law relates absorptivity to emissivity as

$$\alpha = \varepsilon$$

This equation is approximate and typically applies when the temperature of the source of radiation is within a few hundred degrees of the temperature of the surface receiving the radiation.

We define the view factor, $F_{i \rightarrow j}$, as the fraction of the total diffuse radiation leaving surface i that arrives at surface j by a straight-line route. View factors are subject to the reciprocity relation:

$$A_i F_{i \rightarrow j} = A_j F_{j \rightarrow i}$$

and the summation relation:

$$\sum_{j=1}^{N} F_{i \rightarrow j} = 1$$

Shape decomposition allows us to write

$$F_{1 \rightarrow 2} = F_{1 \rightarrow 2a} + F_{1 \rightarrow 2b} + F_{1 \rightarrow 2c} + \cdots + F_{1 \rightarrow 2N} = \sum_{i=1}^{N} F_{1 \rightarrow 2i}$$

In an enclosure of N black surfaces, the net heat leaving surface i by radiation is

$$\dot{Q}_i = \sum_{j=1}^{N} \dot{Q}_{i \to j} = \sum_{j=1}^{N} A_i F_{i \to j} \sigma (T_i^4 - T_j^4)$$

In any enclosure, conservation of energy requires that, in steady state,

$$\sum_{i=1}^{N} \dot{Q}_i = 0$$

Radiosity, J, is the sum of emitted and reflected radiation, that is,

$$J = \varepsilon E_b + \rho G$$

The units of J and G are the same as those of emissive power, either W/m^2 in the SI system or Btu/h·ft^2 in the British system. For a black surface,

$$J = E_b = \sigma T^4$$

The surface resistance to radiation, R_i, for a gray, diffuse surface is

$$R_i = \frac{1 - \varepsilon_i}{A_i \varepsilon_i}$$

For a black surface, $R_i = 0$ and $J_i = E_{bi}$. The space resistance to radiation is

$$R_{i \to j} = \frac{1}{A_i F_{i \to j}}$$

In a two-surface enclosure, the net heat transfer by radiation between surface 1 and 2 is

$$\dot{Q}_{1 \to 2} = \frac{E_{b1} - E_{b2}}{R_{tot}} = \frac{\sigma (T_1^4 - T_2^4)}{\dfrac{1 - \varepsilon_1}{A_1 \varepsilon_1} + \dfrac{1}{A_1 F_{1 \to 2}} + \dfrac{1 - \varepsilon_2}{A_2 \varepsilon_2}}$$

In a three-surface enclosure, the following six equations are solved simultaneously:

$$\dot{Q}_1 = \frac{J_1 - J_2}{R_{1 \to 2}} + \frac{J_1 - J_3}{R_{1 \to 3}}$$

$$\dot{Q}_2 = \frac{J_2 - J_3}{R_{2 \to 3}} + \frac{J_2 - J_1}{R_{2 \to 1}}$$

$$J_1 = \sigma T_1^4 - R_1 \dot{Q}_1$$
$$J_2 = \sigma T_2^4 - R_2 \dot{Q}_2$$
$$J_3 = \sigma T_3^4 - R_3 \dot{Q}_3$$

$$\dot{Q}_3 = \frac{J_3 - J_1}{R_{3 \to 1}} + \frac{J_3 - J_2}{R_{3 \to 2}}$$

This approach can be generalized to any number of surfaces. A reradiating surface is adiabatic with $\dot{Q}_i = 0$ and $J_i = \sigma T_i^4$.

The fraction of radiation in a wavelength band from λ_1 to λ_2 is given by

$$E_{b, \lambda_1 - \lambda_2} = \left(F_{0 - \lambda_2 T} - F_{0 - \lambda_1 T} \right) \sigma T^4$$

Hemispherical, total emissivity is related to hemispherical, spectral emissivity by

$$\varepsilon = \frac{E}{E_b} = \frac{\displaystyle\int_0^\infty \varepsilon_\lambda E_{b\lambda} d\lambda}{\sigma T^4}$$

SELECTED REFERENCES

CENGEL, Y. A., *Introduction to Thermodynamics and Heat Transfer*, McGraw-Hill, New York, 1997.

INCROPERA, F. P., and D. P. DEWITT, *Introduction to Heat Transfer*, 4th ed., Wiley, New York, 2002.

KREITH, F., and M. S. BOHN, *Principles of Heat Transfer*, 6th ed., Brooks/Cole, Pacific Grove, CA, 2001.

MILLS, A. F., *Heat Transfer*, Irwin, Boston, 1992.

SIEGEL, R., and J. R. HOWELL, *Thermal Radiation Heat Transfer*, 3rd ed., Hemisphere, Washington, DC, 1992.

SURYANARAYANA, N. V., *Engineering Heat Transfer*, West, New York, 1995.

THOMAS, L. C., *Heat Transfer*, Prentice Hall, Englewood Cliffs, NJ, 1992.

PROBLEMS

Problems designated with WEB refer to material available at www.wiley.com/college/kaminski.

FUNDAMENTAL RELATIONS

P14-1 The tungsten filament of a lightbulb is at 2600 K. Assuming the filament is black, calculate the wavelength at which the maximum power is emitted. In what range of the electromagnetic spectrum does this wavelength fall?

P14-2 Measurements of the spectrum of the star Vega show that radiation peaks at 0.29μm. Neglecting the Doppler shift of the star, estimate Vega's effective surface temperature.

P14-3 The planet Mars has an average radius of 3380 km and an average distance from the sun of 2.28×10^8 km. The sun radiates like a black sphere with a radius of 6.95×10^8 m and a temperature of 5800 K. Assume Mars is black and isothermal, with no atmosphere and no internal heat generation and calculate its average surface temperature.

P14-4 Radiation from the sun incident on the earth's outer atmosphere has been measured as 1353 W/m². The average distance between the sun and the earth is 1.5×10^{11} m. If the planet Pluto is at a distance from the sun of 5.87×10^9 km, find the solar flux incident on Pluto.

P14-5 A furnace has an inside area of 320 ft² and black walls. A radiant power of 95 W issues from a rectangular opening in the furnace wall. If the opening measures 5 in. by 6 in., determine the interior wall temperature. If the emissivity of the walls is 0.88, what is the wall temperature?

VIEW FACTORS

P14-6 A square window of side length 12 cm is located in the center of an oven wall measuring 35 cm by 45 cm. The depth of the oven is 50 cm. Considering all the oven walls as one surface and the glass window as a second surface, find the view factor from the oven walls to the window.

P14-7 Two square plates of size 8 cm by 8 cm are directly opposite each other in parallel planes. What should the spacing between the plates be so that the view factor from one to the other is 0.5?

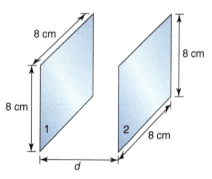

P14-8 A cryogenic dewer consists of two concentric cylinders. The inner cylinder has a length of 2.2 ft and a diameter of 8 in. The outer cylinder has a length of 2.6 ft and a diameter of 10 in. The space between the cylinders is evacuated. Find the view factor from the outer to the inner cylinder. Include the end area in the calculation.

P14-9 The inside of a sphere is divided into two hemispherical surfaces. Find the view factor from one hemisphere to the other.

P14-10 A room in an art gallery has a floor area of 16 ft by 16 ft and 10-ft-high ceilings. One wall is part of the exterior of the building and the other three are interior walls. In winter, the exterior wall is cooler than the other three and looses heat to the environment by radiation and convection. Calculate the view factor from the floor to the exterior wall and from each interior wall to the exterior wall.

P14-11 During a chemical-vapor deposition process, a disk-shaped silicon substrate 2.3 cm in diameter is placed in a reaction chamber. The chamber is cylindrical with a height of 13 cm and a diameter of 5 cm. The silicon substrate rests on the bottom surface, as shown. Find the view factor from the curved side wall of the chamber (surface 1) to the substrate (surface 2).

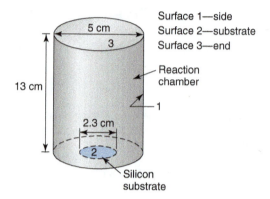

P14-12 Find the view factors between all the surfaces within a cubical enclosure.

SHAPE DECOMPOSITION

P14-13 A flat panel heater is suspended over a warming tray, as shown. Calculate the view factor from the heater to the tray.

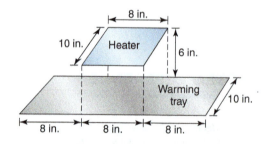

P14-14 Using data on the figure, find the view factor, $F_{3 \to 1}$. Surfaces 2 and 3 are perpendicular at their common edge.

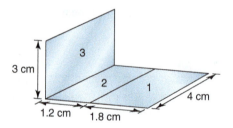

P14-15 Two flat rings of equal size in parallel planes exchange heat by radiation. Using data in the figure, calculate the view factor from one ring to the other.

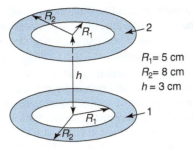

$R_1 = 5$ cm
$R_2 = 8$ cm
$h = 3$ cm

P14-16 A tubular enclosure is divided into four surfaces for the purpose of a radiation analysis. Surfaces 1 and 2 are on the inner wall of the tube and surfaces 3 and 4 are on the ends. Find the view factor $F_{1\rightarrow 2}$.

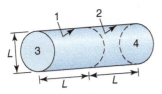

P14-17 Two rectangular surfaces face each other in parallel planes, as shown in the figure. The top surface is aligned directly above the bottom surface, and both surfaces have the same size. Each surface is divided into two equal parts, so that $A_1 = A_2 = A_3 = A_4$. Find the view factor $F_{1\rightarrow 4}$.

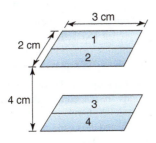

P14-18 Three radiation zones are used in an oven analysis. Surfaces 1 and 2 are placed on the side wall and surface 3 is placed on the bottom of the oven. Using dimensions in the figure, calculate $F_{1\rightarrow 3}$.

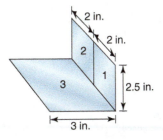

TWO-SURFACE ENCLOSURES

P14-19 A planar black surface at 850°C is parallel to a second planar black surface at 300°C. The two surfaces are separated by a small gap. If a third black surface is inserted in the gap, calculate

a. the net radiative heat flux between the outer surfaces with and without the insert.

b. the temperature of the inserted surface.

P14-20 A wire coated with insulation runs through the center of an evacuated cylindrical channel in a space station. Both the wire and the channel wall may be approximated as black surfaces. The channel wall is maintained at 50 K, and the wire dissipates 75 mW per cm of length. The outer diameter of the wire is 0.3 cm and the inner diameter of the channel is 1.5 cm. Calculate the temperature of the wire surface.

P14-21 A generator is constructed of an inner rotating cylinder (the rotor) and an outer, stationary cylindrical shell (the stator). During testing, the rotor develops a thermally sensitive vibration and the engineer suggests that one side of the rotor is hotter than the other. To correct the problem, a stripe of black paint is applied to the hot side. The unpainted rotor surface has an emissivity of 0.2 and the painted surface has an emissivity of 0.98. The rotor outer surface is at 160°C and the stator inner surface is at 95°C. The diameter of the rotor is 1.2 m and the gap between rotor and stator is 5 cm. Calculate the net radiative flux from the rotor on the painted portion and on the unpainted portion.

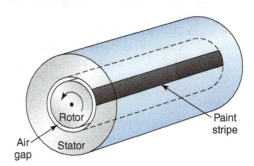

P14-22 Two large flat plates in parallel planes are at 60°F and 420°F. The cold plate has an emissivity of 0.6 and the hot plate has an emissivity of 0.2. Find the net rate of heat transfer between the plates per unit surface area. If the emissivities of the two plates are switched, how does the net rate of heat transfer change?

P14-23 A thermocouple is used to measure the temperature of a hot gas flowing in a pipe whose wall is at 350°C. A cylindrical radiation shield, large compared to the size of the thermocouple, encloses the thermocouple, as shown. The shield has an emissivity of 0.13. The emissivity of the thermocouple bead is 0.68 and the emissivity of the pipe wall is 0.94. The convective heat transfer coefficient on the thermocouple is 70 W/m²·°C, and the thermocouple reads 500°C. The convective heat transfer coefficient on the shield is 35 W/m²·°C. The shield has a diameter of 3.5 cm and the pipe has a diameter of 6 cm. Calculate the actual gas temperature.

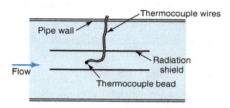

P14-24 A hot potato with a surface temperature of 375°F and emissivity of 0.93 radiates to a large room with walls at 70°F. To keep the potato warm, someone covers it loosely with a single layer of aluminum foil, which has an emissivity of 0.2. Calculate the net rate of radiation loss from the potato with and without the foil. Assume the potato is spherical with a diameter of 4.5 in. and the foil forms a spherical shell around the potato.

P14-25 A motorist leaves a car parked on a driveway with the engine on. The driveway is covered with a layer of ice 0.5 in. thick at 32°F. The undercarriage of the car in the vicinity of the catalytic converter is at a temperature of 190°F. Assume the undercarriage radiates like a black surface and the ice has an emissivity of 0.75. Ignoring heat transfer by convection, calculate the time required to melt the ice. The latent heat of fusion and density of ice are 143.5 Btu/lbm and 62.4 lbm/ft³, respectively.

P14-26 An astronaut with a surface area of 1.8 m² generates 150 W of body heat during a space walk. The exterior of the spacesuit is exposed to outer space, which is black at 4 K. Both exterior and interior surfaces of the suit are silvered, with an emissivity of 0.4. It is necessary to keep the surface of the astronaut's body, which has an emissivity of 0.85, at no less than 16°C in steady state. Should the suit be made of one layer of silvered material or two layers?

P14-27 A cryogenic dewar is constructed of two concentric spheres 45 cm and 53 cm in diameter. Liquid nitrogen at 100 K is stored in the inner sphere, and the space between the spheres is evacuated. The outer sphere has a temperature of 220 K, and the emissivity of both spheres is 0.023. If the latent heat of vaporization of the nitrogen is 210 kJ/kg, determine the number of kilograms of nitrogen evaporated per hour.

THREE-SURFACE ENCLOSURES

P14-28 During a deposition process, a long strip heater is centered above a long, well-insulated plate, as shown in the figure.

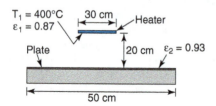

The heater is maintained at 400°C and the surroundings are black at 25°C. The geometry is effectively two-dimensional and radiation is the only mode of heat transfer. Using data on the figure, find the plate temperature.

P14-29 A chemical reaction chamber is in the shape of a cylinder with a height of 2.6 ft and a diameter of 0.6 ft. A disk-shaped heater with an emissivity of 0.92 entirely covers the bottom of the chamber and generates 1.3 kW of heat. The side wall is at 86°F and the top end is at 65°F. Both the side wall and top have an emissivity of 0.73. The chamber is partially evacuated, and the only significant mode of heat transfer present is radiation. The back of the heater is well-insulated, so that all the heater power is removed by radiation into the chamber. Find the heater temperature.

P14-30 A very long enclosure is formed from two perpendicular, equal-width plates and a slanted cover plate, as shown. The cross-section of the enclosure is in the shape of an isosceles right triangle. Assuming gray and diffuse surfaces, calculate the net radiative heat transfer from the hottest plate.

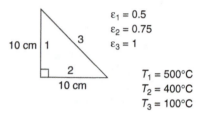

P14-31 A short, thin-walled tube of length 2.5 in. and diameter 0.75 in. is open at both ends. The tube is made of copper ($\varepsilon = 0.25$) and is maintained at 155°F. The surroundings surfaces are at 65°F. Including radiation from both the outside and the inside of the tube, find the net rate of radiative transfer.

P14-32 A ceramic cooktop has a heating element 9 in. in diameter. A person's hands are placed over the heating element at a distance of 4.5 in. Model the hands as a disk with an emissivity of 0.90 and radius of 9 in. and the heating element as a parallel disk with an emissivity of 0.97. The heating element is at 350°F, and the walls of the kitchen are at 70°F. If the hands are reradiating surfaces, what is their temperature?

P14-33 An attic space is in the shape of an isosceles triangle, with dimensions as shown. The floor of the attic is at 50°F, the two ceiling surfaces under the roof are at 30°F, and the air in the attic is at 40°F. The convective heat transfer coefficient is 1.4 Btu/h·ft²·°F on the floor and 0.66 Btu/h·ft²·°F on the ceiling surfaces. The emissivity of the ceiling surfaces is 0.93.

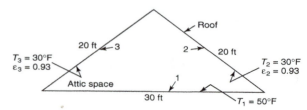

Calculate the heat transferred from the floor by radiation and by convection per foot of depth if the floor is covered with

a. unsilvered insulation ($\varepsilon_1 = 0.87$).

b. silver-backed insulation ($\varepsilon_1 = 0.19$).

P14-34 Two diffuse, gray, flat plates of size 60 cm by 45 cm face each other in parallel planes. One plate has an emissivity of 0.66 and a temperature of 460°C, while the other has an emissivity of 0.31 and a temperature of 120°C. The spacing between the plates is 14 cm. The surroundings are black at 20°C. Calculate the net rate of radiation heat transfer from the hot plate.

P14-35 In the middle of a blizzard, a homeowner's furnace fails. In desperation, the homeowner turns on the oven of the electric range and opens the oven door. The thermostat for the oven is set at 450°F, and the inside walls of the oven are black. Assume the oven door acts like a reradiating surface (i.e., neglect convection) with an emissivity of 0.89. The walls of the kitchen are at 55°F. Calculate the temperature of the oven door and the electric power consumed by the oven.

P14-36 An evacuated enclosure is in the shape of a cube with a side length of 12 cm. The top has an emissivity of 0.85 and a temperature of 800°C, while the bottom has an emissivity of 0.47 and a temperature of 210°C. The remaining four sides are perfectly insulated. Find the side wall temperature.

WAVELENGTH AND ANGULAR VARIATIONS

P14-37 (WEB) A selective surface has a spectral, hemispherical absorptivity that varies with wavelength, as shown in the figure. Find the total hemispherical absorptivity at 185°C.

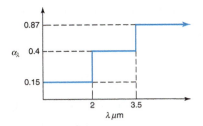

P14-38 (WEB) The spectral emissivity of a polished aluminum plate is 0.7 for $\lambda < 6.6\mu m$ and 0.4 for $\lambda > 6.6\mu m$. Calculate the power emitted by this plate at 300 K and at 900 K.

P14-39 (WEB) The tungsten filament of an incandescent light-bulb is heated to 2600 K. What fraction of the energy emitted is in the visible range (0.4 to $0.7\ \mu m$)? What is the wavelength of maximum emission?

P14-40 (WEB) A glowing coal can be seen in a darkened room at the Draper point, which is 798 K. What fraction of the energy emitted by the coal at this temperature is in the visible range (0.4 to $0.7\ \mu m$)?

P14-41 (WEB) A roughly spherical satellite with a diameter of 1.6 m is in orbit around the sun at an average distance of 6.7×10^8 km. The sun radiates like a black sphere with a radius of 6.95×10^8 and a temperature of 5800 K. The surface of the satellite is coated with a material having an absorptivity of 0.88 for $\lambda < 2.5\mu m$ and an absorptivity of 0.12 for $\lambda > 2.5\mu m$. The satellite rotates and may be assumed to be isothermal at 310 K. Determine the amount of internal heat generation in the satellite.

P14-42 (WEB) The windshield of an automobile is made of glass with a transmissivity of 0.91 between 0.3 and 3 μm. Outside this range, the transmissivity is virtually zero. Calculate the total transmissivity for the windshield for solar radiation ($T_{sun} \approx 5800$ K) and for radiation from the car seats, which are at 20°C.

P14-43 (WEB) An engineer sends an oxidized metal test sample to the laboratory to determine the emissivity. The lab reports a value of 0.72 for the total, normal emissivity. What value of total, hemispherical emissivity can the engineer expect?

P14-44 (WEB) The normal, spectral absorptivity of silicon oxide on aluminum is 0.96 for $\lambda < 1.4\mu m$ and 0.04 for $\lambda > 1.4\mu m$. Calculate the total, hemispherical emissivity at 300°C for this surface.

TABLES IN SI UNITS

TABLE A-1 Molecular weight and critical-point properties

Substance	Formula	Molecular Weight M kg/kmol	Ratio of Gas Constant to Molecular Weight $\bar{R}/M$ kJ/kg·K	Critical-Point Properties T_c K	P_c MPa	$\bar{v}_c$ m³/kmol
Air	—	28.97	0.2870	132.5	3.77	0.0883
Ammonia	NH_3	17.03	0.4882	405.5	11.28	0.0724
Argon	Ar	39.948	0.2081	151	4.86	0.0749
Benzene	C_6H_6	78.115	0.1064	562	4.92	0.2603
Bromine	Br_2	159.808	0.0520	584	10.34	0.1355
n-Butane	C_4H_{12}	58.124	0.1430	425.2	3.80	0.2547
Carbon dioxide	CO_2	44.01	0.1889	304.2	7.39	0.0943
Carbon monoxide	CO	28.011	0.2968	133	3.50	0.0930
Carbon tetrachloride	CCl_{12}	153.82	0.05405	556.4	4.56	0.2759
Chlorine	Cl_2	70.906	0.1173	417	7.71	0.1242
Chloroform	$CHCl_3$	119.38	0.06964	536.6	5.47	0.2403
Dichlorodifluoromethane (R-12)	CCl_2F_2	120.91	0.06876	384.7	4.01	0.2179
Dichlorofluoromethane (R-21)	$CHCl_2F$	102.92	0.08078	451.7	5.17	0.1973
Ethane	C_2H_6	30.020	0.2765	305.5	4.48	0.1480
Ethyl alcohol	C_2H_5OH	46.07	0.1805	516	6.38	0.1673
Ethylene	C_2H_4	28.054	0.2964	282.4	5.12	0.1242
Helium	He	4.003	2.0769	5.3	0.23	0.0578
n-Hexane	C_6H_{14}	86.178	0.09647	507.9	3.03	0.3677
Hydrogen (normal)	H_2	2.016	4.1240	33.3	1.30	0.0649
Methane	CH_4	16.043	0.5182	191.1	4.64	0.0993
Methyl alcohol	CH_3OH	32.042	0.2595	513.2	7.95	0.1180
Methyl chloride	CH_3Cl	50.488	0.1647	416.3	6.68	0.1430
Nitrogen	N_2	28.013	0.2968	126.2	3.39	0.0899
Nitrous oxide	N_2O	44.013	0.1889	309.7	7.27	0.0961
Oxygen	O_2	31.999	0.2598	154.8	5.08	0.0780
Propane	C_3H_8	44.097	0.1885	370	4.26	0.1998
Propylene	C_3H_6	42.081	0.1976	365	4.62	0.1810
Sulfur dioxide	SO_2	64.063	0.1298	430.7	7.88	0.1217
Tetrafluoroethane (R-134a)	CF_3CH_2F	102.03	0.08149	374.3	4.067	0.1847
Trichlorofluoromethane (R-11)	CCl_3F	137.37	0.06052	471.2	4.38	0.2478
Water	H_2O	18.015	0.4615	647.3	22.09	0.0568

Source: Kobe, K.A. and R.E. Lynn Jr., *Chemical Review* 52(1953), pp. 117–236; and ASHRAE, *Handbook of Fundamentals*, Atlanta, GA: ASHRAE, Inc., 1993, pp. 16.4 and 36.1.

TABLE A-2 Thermophysical Properties of Solid Metals

Composition	Melting Point (K)	ρ (kg/m³)	c_p (J/kg·K)	k (W/m·K)	α·10⁶ (m²/s)	100	200	400	600	800	1000	1200	1500	2000	2500
						...									

The property columns at various temperatures are reported as $k\,(\mathrm{W/m\cdot K})\,/\,c_p\,(\mathrm{J/kg\cdot K})$.

Composition	Melting Point (K)	ρ (kg/m³)	c_p (J/kg·K)	k (W/m·K)	α·10⁶ (m²/s)	100	200	400	600	800	1000	1200	1500	2000	2500
Aluminum Pure	933	2702	903	237	97.1	302/482	237/798	240/949	231/1033	218/1146					
Alloy 2024-T6 (4.5% Cu, 1.5% Mg, 0.6% Mn)	775	2770	875	177	73.0	65/473	163/787	186/925	186/1042						
Alloy 195, Cast (4.5% Cu)		2790	883	168	68.2			174	185						
Beryllium	1550	1850	1825	200	59.2	990/203	301/1114	161/2191	126/2604	106/2823	90.8/3018	78.7/3227	/3519		
Bismuth	545	9780	122	7.86	6.59	16.5/112	9.69/120	7.04/127							
Boron	2573	2500	1107	27.0	9.76	190/128	55.5/600	16.8/1463	10.6/1892	9.60/2160	9.85/2338				
Cadmium	594	8650	231	96.8	48.4	203/198	99.3/222	94.7/242							
Chromium	2118	7160	449	93.7	29.1	159/192	111/384	90.9/484	80.7/542	71.3/581	65.4/616	61.9/682	57.2/779	49.4/937	
Cobalt	1769	8862	421	99.2	26.6	167/236	122/379	85.4/450	67.4/503	58.2/550	52.1/628	49.3/733	42.5/674		
Copper Pure	1358	8933	385	401	117	482/252	413/356	393/397	379/417	366/433	352/451				
Commercial bronze (90% Cu, 10% Al)	1293	8800	420	52	14		42/785	52/460	59/545						
Phosphor gear bronze (89% Cu, 11% Sn)	1104	8780	355	54	17		41	65	74						
Cartridge brass (70% Cu, 30% Zn)	1188	8530	380	110	33.9	75	95/360	137/395	149/425						
Constantan (55% Cu, 45% Ni)	1493	8920	384	23	6.71	17/237	19/362								
Germanium	1211	5360	322	59.9	34.7	232/190	96.8/290	43.2/337	27.3/348	19.8/357	17.4/375	17.4/395			
Gold	1336	19300	129	317	127	327/109	323/124	311/131	298/135	284/140	270/145	255/155			
Iridium	2720	22500	130	147	50.3	172/90	153/122	144/133	138/138	132/144	126/153	120/161	111/172		
Iron Pure	1810	7870	447	80.2	23.1	134/216	94.0/384	69.5/490	54.7/574	43.3/680	32.8/975	28.3/609	32.1/654		

Table continued. Thermophysical properties of selected metallic solids. Columns: Composition | Melting Point (K) | ρ (kg/m³) | cₚ (J/kg·K) | k (W/m·K) | α·10⁶ (m²/s) | then Properties at Various Temperatures (K), k (W/m·K) / cₚ (J/kg·K).

Composition	Melting Point (K)	ρ (kg/m³)	c_p (J/kg·K)	k (W/m·K)	$\alpha\cdot10^6$ (m²/s)	100	200	400	600	800	1000	1200	1500	2000	2500
Armco (99.75% pure)		7870	447	72.7	20.7	95.6/215	80.6/384	65.7/490	53.1/574	42.2/680	32.3/975	28.7/609	31.4/654		
Carbon steels															
Plain carbon (Mn ≤ 1%, Si ≤ 0.1%)		7854	434	60.5	17.7			56.7/487	48.0/559	39.2/685	30.0/1169				
AISI 1010		7832	434	63.9	18.8			58.7/487	48.8/559	39.2/685	31.3/1168				
Carbon–silicon (Mn ≤ 1%, 0.1% < Si ≤ 0.6%)		7817	446	51.9	14.9			49.8/501	44.0/582	37.4/699	29.3/971				
Carbon–manganese–silicon (1% < Mn ≤ 1.65%, 0.1% < Si ≤ 0.6%)		8131	434	41.0	11.6			42.2/487	39.7/559	35.0/685	27.6/1090				
Chromium (low) steels															
½Cr-¼Mo-Si (0.18% C, 0.65% Cr, 0.23% Mo, 0.6% Si)		7822	444	37.7	10.9			38.2/492	36.7/575	33.3/688	26.9/969				
1 Cr-½Mo (0.16% C, 1% Cr, 0.54% Mo, 0.39% Si)		7858	442	42.3	12.2			42.0/492	39.1/575	34.5/688	27.4/969				
1 Cr-V (0.2% C, 1.02% Cr, 0.15% V)		7836	443	48.9	14.1			46.8/492	42.1/575	36.3/688	28.2/969				
Stainless steels															
AISI 302		8055	480	15.1	3.91			17.3/512	20.0/559	22.8/585	25.4/606				
AISI 304	1670	7900	477	14.9	3.95	9.2/272	12.6/402	16.6/515	19.8/557	22.6/582	25.4/611	28.0/640	31.7/682		
AISI 316		8238	468	13.4	3.48			15.2/504	18.3/550	21.3/576	24.2/602				
AISI 347		7978	480	14.2	3.71			15.8/513	18.9/559	21.9/585	24.7/606				
Lead	601	11340	129	35.3	24.1	39.7/118	36.7/125	34.0/132	31.4/142						
Magnesium	923	1740	1024	156	87.6	169/649	159/934	153/1074	149/1170	146/1267					
Molybdenum	2894	10240	251	138	53.7	179/141	143/224	134/261	126/275	118/285	112/295	105/308	98/330	90/380	86/459
Nickel Pure	1728	8900	444	90.7	23.0	164/232	107/383	80.2/485	65.6/592	67.6/530	71.8/562	76.2/594	82.6/616		

(Continued)

TABLE A-2 (Continued)

Properties at Various Temperatures (K) — each cell shows k (W/m·K) / c_p (J/kg·K)

Composition	Melting Point (K)	ρ (kg/m³)	c_p (J/kg·K)	k (W/m·K)	$\alpha \cdot 10^6$ (m²/s)	100	200	400	600	800	1000	1200	1500	2000	2500
Nichrome (80% Ni, 20% Cr)	1672	8400	420	12	3.4			14 / 480	16 / 525	21 / 545					
Inconel X750 (73% Ni, 15% Cr, 6.7% Fe)	1665	8510	439	11.7	3.1	8.7 / –	10.3 / 372	13.5 / 473	17.0 / 510	20.5 / 546	24.0 / 626	27.6 / –	33.0 / –		
Niobium	2741	8570	265	53.7	23.6	55.2 / 188	52.6 / 249	55.2 / 274	58.2 / 283	61.3 / 292	64.4 / 301	67.5 / 310	72.1 / 324	79.1 / 347	
Palladium	1827	12020	244	71.8	24.5	76.5 / 168	71.6 / 227	73.6 / 251	79.7 / 261	86.9 / 271	94.2 / 281	102 / 291	110 / 307		
Platinum Pure	2045	21450	133	71.6	25.1	77.5 / 100	72.6 / 125	71.8 / 136	73.2 / 141	75.6 / 146	78.7 / 152	82.6 / 157	89.5 / 165	99.4 / 179	
Alloy 60Pt–40Rh (60% Pt, 40% Rh)	1800	16630	162	47	17.4	–	52	52	59	65	69	73	76		
Rhenium	3453	21100	136	47.9	16.7	58.9 / 97	51.0 / 127	46.1 / 139	44.2 / 145	44.1 / 151	44.6 / 156	45.7 / 162	47.8 / 171	51.9 / 186	
Rhodium	2236	12450	243	150	49.6	186 / 147	154 / 220	146 / 253	136 / 274	127 / 293	121 / 311	116 / 327	110 / 349	112 / 376	
Silicon	1685	2330	712	148	89.2	884 / 259	264 / 556	98.9 / 790	61.9 / 867	42.2 / 913	31.2 / 946	25.7 / 967	22.7 / 992		
Silver	1235	10500	235	429	174	444 / 187	430 / 225	425 / 239	412 / 250	396 / 262	379 / 277	361 / 292			
Tantalum	3269	16600	140	57.5	24.7	59.2 / 110	57.5 / 133	57.8 / 144	58.6 / 146	59.4 / 149	60.2 / 152	61.0 / 155	62.2 / 160	64.1 / 172	65.6 / 189
Thorium	2023	11700	118	54.0	39.1	59.8 / 99	54.6 / 112	54.5 / 124	55.8 / 134	56.9 / 145	56.9 / 156	58.7 / 167			
Tin	505	7310	227	66.6	40.1	85.2 / 118	73.3 / 215	62.2 / 243							
Titanium	1953	4500	522	21.9	9.32	30.5 / 300	24.5 / 465	20.4 / 551	19.4 / 591	19.7 / 633	20.7 / 675	22.0 / 620	24.5 / 686		
Tungsten	3660	19300	132	174	68.3	208 / 87	186 / 122	159 / 137	137 / 142	125 / 145	118 / 148	113 / 152	107 / 157	100 / 167	95 / 176
Uranium	1406	19070	116	27.6	12.5	21.7 / 94	25.1 / 108	29.6 / 125	34.0 / 146	38.8 / 176	43.9 / 180	49.0 / 161			
Vanadium	2192	6100	489	30.7	10.3	35.8 / 258	31.3 / 430	31.3 / 515	33.3 / 540	35.7 / 563	38.2 / 597	40.8 / 645	44.6 / 714	50.9 / 867	
Zinc	693	7140	389	116	41.8	117 / 297	118 / 367	111 / 402	103 / 436						
Zirconium	2125	6570	278	22.7	12.4	33.2 / 205	25.2 / 264	21.6 / 300	20.7 / 322	21.6 / 342	23.7 / 362	26.0 / 344	28.8 / 344	33.0 / 344	

Source: Incropera, F.P., and D.P. Dewitt, *Introduction to Heat Transfer*, 4th ed., Wiley, New York, 2002.

TABLE A-3 Thermophysical Properties of Solid Nonmetals

	Melting Point (K)	Nonmetallic compounds Properties at 300 K				Properties at Various Temperatures (K) k (W/m·K)/c_p (J/kg·K)									
Composition		ρ (kg/m³)	c_p (J/kg·K)	k (W/m·K)	$\alpha \cdot 10^6$ (m²/s)	100	200	400	600	800	1000	1200	1500	2000	2500
Aluminum oxide, sapphire	2323	3970	765	46	15.1	450	82	32.4	18.9	13.0	10.5				
								940	1110	1180	1225				
Aluminum oxide, polycrystalline	2323	3970	765	36.0	11.9	133	55	26.4	15.8	10.4	7.85	6.55	5.66	6.00	
								940	1110	1180	1225				
Beryllium oxide	2725	3000	1030	272	88.0			196	111	70	47	33	21.5	15	
								1350	1690	1865	1975	2055	2145	2750	
Boron	2573	2500	1105	27.6	9.99	190	52.5	18.7	11.3	8.1	6.3	5.2			
								1490	1880	2135	2350	2555			
Boron fiber epoxy (30% vol) composite	590	2080													
k, ∥ to fibers				2.29		2.10	2.23	2.28							
k, ⊥ to fibers				0.59		0.37	0.49	0.60							
c_p			1122			364	757	1431							
Carbon Amorphous	1500	1950	—	1.60	—	0.67	1.18	1.89	2.19	2.37	2.53	2.84	3.48		
Diamond, type IIa insulator	—	3500	509	2300	—	10,000	4000	1540							
						21	194	853							
Graphite, pyrolytic	2273	2210	709	1950											
k, ∥ to layers						4970	3230	1390	892	667	534	448	357	262	
k, ⊥ to layers				5.70		16.8	9.23	4.09	2.68	2.01	1.60	1.34	1.08	0.81	
c_p						136	411	992	1406	1650	1793	1890	1974	2043	
Graphite fiber epoxy (25% vol) composite	450	1400													
k, heat flow ∥ to fibers				11.1		5.7	8.7	13.0							
k, heat flow ⊥ to fibers				0.87		0.46	0.68	1.1							
c_p			935			337	642	1216							
Pyroceram, Corning 9606	1623	2600	808	3.98	1.89	5.25	4.78	3.64	3.28	3.08	2.96	2.87	2.79		
								908	1038	1122	1197	1264	1498		

(Continued)

TABLE A-3 (Continued)

Nonmetallic compounds

Composition	Melting Point (K)	ρ (kg/m³)	c_p (J/kg·K)	k (W/m·K)	$\alpha \cdot 10^6$ (m²/s)	100	200	400	600	800	1000	1200	1500	2000	2500

Properties at 300 K — columns: ρ (kg/m³), c_p (J/kg·K), k (W/m·K), $\alpha \cdot 10^6$ (m²/s)

Properties at Various Temperatures (K) — k (W/m·K) / c_p (J/kg·K)

Composition	Melting Point (K)	ρ (kg/m³)	c_p (J/kg·K)	k (W/m·K)	$\alpha \cdot 10^6$ (m²/s)	100	200	400	600	800	1000	1200	1500	2000	2500
Silicon carbide — k	3100	3160	675	490	230			—	—	—	87	58	30		
Silicon carbide — c_p								880	1050	1135	1195	1243	1310		
Silicon dioxide, crystalline (quartz)	1883	2650													
k, ∥ to c axis				10.4		39	16.4	7.6	5.0	4.2					
k, ⊥ to c axis				6.21		20.8	9.5	4.70	3.4	3.1					
c_p			745			—	—	885	1075	1250					
Silicon dioxide, polycrystalline (fused silica) — k	1883	2220	745	1.38	0.834	0.69	1.14	1.51	1.75	2.17	2.87	4.00			
— c_p						—	—	905	1040	1105	1155	1195			
Silicon nitride — k	2173	2400	691	16.0	9.65	—	—	13.9	11.3	9.88	8.76	8.00	7.16	6.20	
— c_p							578	778	937	1063	1155	1226	1306	1377	
Sulfur — k	392	2070	708	0.206	0.141	0.165	0.185								
— c_p						403	606								
Thorium dioxide — k	3573	9110	235	13	6.1			10.2	6.6	4.7	3.68	3.12	2.73	2.5	
— c_p								255	274	285	295	303	315	330	
Titanium dioxide, polycrystalline — k	2133	4157	710	8.4	2.8			7.01	5.02	3.94	3.46	3.28			
— c_p								805	880	910	930	945			

(Continued)

TABLE A-3 Thermophysical Properties of Solid Nonmetals (continued)

Common Materials

Description/Composition	Temperature (K)	Density, ρ (kg/m³)	Thermal Conductivity, k (W/m·K)	Specific Heat, c_p (J/kg·K)
Asphalt	300	2115	0.062	920
Bakelite	300	1300	1.4	1465
Brick, refractory				
Carborundum	872	—	18.5	—
	1672	—	11.0	—
Chrome brick	473	3010	2.3	835
	823		2.5	
	1173		2.0	
Diatomaceous	478	—	0.25	—
silica, fired	1145	—	0.30	
Fire clay, burnt 1600 K	773	2050	1.0	960
	1073	—	1.1	
	1373	—	1.1	
Fire clay, burnt 1725 K	773	2325	1.3	960
	1073		1.4	
	1373		1.4	
Fire clay brick	478	2645	1.0	960
	922		1.5	
	1478		1.8	
Magnesite	478	—	3.8	1130
	922	—	2.8	
	1478		1.9	
Clay	300	1460	1.3	880
Coal, anthracite	300	1350	0.26	1260
Concrete (stone mix)	300	2300	1.4	880
Cotton	300	80	0.06	1300
Foodstuffs				
Banana (75.7% water content)	300	980	0.481	3350
Apple, red (75% water content)	300	840	0.513	3600
Cake, batter	300	720	0.223	—
Cake, fully baked	300	280	0.121	—
Chicken meat, white	198	—	1.60	—
(74.4% water content)	233	—	1.49	
	253		1.35	
	263		1.20	
	273		0.476	
	283		0.480	
	293		0.489	
Glass				
Plate (soda lime)	300	2500	1.4	750
Pyrex	300	2225	1.4	835
Ice	273	920	1.88	2040
	253	—	2.03	1945
Leather (sole)	300	998	0.159	—
Paper	300	930	0.180	1340
Paraffin	300	900	0.240	2890
Rock				
Granite, Barre	300	2630	2.79	775
Limestone, Salem	300	2320	2.15	810
Marble, Halston	300	2680	2.80	830
Quartzite, Sioux	300	2640	5.38	1105
Sandstone, Berea	300	2150	2.90	745

(Continued)

TABLE A-3 (Continued)

Common Materials

Description/Composition	Temperature (K)	Density, ρ (kg/m^3)	Thermal Conductivity, k (W/m · K)	Specific Heat, c_p (J/kg · K)
Rubber, vulcanized				
Soft	300	1100	0.13	2010
Hard	300	1190	0.16	—
Sand	300	1515	0.27	800
Soil	300	2050	0.52	1840
Snow	273	110	0.049	—
		500	0.190	—
Teflon	300	2200	0.35	—
	400		0.45	—
Tissue, human				
Skin	300	—	0.37	—
Fat layer (adipose)	300	—	0.2	—
Muscle	300	—	0.41	—
Wood, cross grain				
Balsa	300	140	0.055	—
Cypress	300	465	0.097	—
Fir	300	415	0.11	2720
Oak	300	545	0.17	2385
Yellow pine	300	640	0.15	2805
White pine	300	435	0.11	—
Wood, radial				
Oak	300	545	0.19	2385
Fir	300	420	0.14	2720

Source: Incropera, F.P., and D.P. DeWitt, *Introduction to Heat Transfer*, 4th ed., Wiley, New York, 2002.

TABLE A-4 Thermophysical Properties of Solid Insulating Materials

Description/Composition	Typical Properties at 300 K		
	Density, ρ (kg/m^3)	Thermal Conductivity, k (W/m · K)	Specific Heat, c_p (J/kg · K)
Blanket and Batt			
Glass fiber, paper faced	16	0.046	—
	28	0.038	—
	40	0.035	—
Glass fiber, coated; duct liner	32	0.038	835
Board and Slab			
Cellular glass	145	0.058	1000
Glass fiber, organic bonded	105	0.036	795
Polystyrene, expanded			
Extruded (R-12)	55	0.027	1210
Molded beads	16	0.040	1210
Mineral fiberboard; roofing material	265	0.049	—
Wood, shredded/cemented	350	0.087	1590
Cork	120	0.039	1800
Loose Fill			
Cork, granulated	160	0.045	—
Diatomaceous silica, coarse	350	0.069	—
Powder	400	0.091	—
Diatomaceous silica, fine powder	200	0.052	—
	275	0.061	—
Glass fiber, poured or blown	16	0.043	835
Vermiculite, flakes	80	0.068	835
	160	0.063	1000

(*Continued*)

TABLE A-4   (Continued)

Description/Composition	Typical Properties at 300 K		
	Density, ρ (kg/m³)	Thermal Conductivity, k (W/m·K)	Specific Heat, c_p (J/kg·K)
Formed/Foamed-in-Place			
Mineral wool granules with asbestos/inorganic binders, sprayed	190	0.046	—
Polyvinyl acetate cork mastic; sprayed or troweled	—	0.100	—
Urethane, two-part mixture; rigid foam	70	0.026	1045
Reflective			
Aluminum foil separating fluffy glass mats; 10–12 layers, evacuated; for cryogenic applications (150 K)	40	0.00016	—
Aluminum foil and glass paper laminate; 75–150 layers; evacuated; for cryogenic application (150 K)	120	0.000017	—
Typical silica powder, evacuated	160	0.0017	—

Source: Incropera, F.P., and D.P. DeWitt, *Introduction to Heat Transfer*, 4th ed., Wiley, New York, 2002.

TABLE A-5   Thermophysical Properties of Solid Building Materials

Description/Composition	Typical Properties at 300 K		
	Density, ρ (kg/m³)	Thermal Conductivity, k (W/m·K)	Specific Heat, c_p (J/kg·K)
Building Boards			
Asbestos–cement board	1920	0.58	—
Gypsum or plaster board	800	0.17	—
Plywood	545	0.12	1215
Sheathing, regular density	290	0.055	1300
Acoustic tile	290	0.058	1340
Hardboard, siding	640	0.094	1170
Hardboard, high density	1010	0.15	1380
Particle board, low density	590	0.078	1300
Particle board, high density	1000	0.170	1300
Woods			
Hardwoods (oak, maple)	720	0.16	1255
Softwoods (fir, pine)	510	0.12	1380
Masonry Materials			
Cement mortar	1860	0.72	780
Brick, common	1920	0.72	835
Brick, face	2083	1.3	—
Clay tile, hollow			
1 cell deep, 10 cm thick	—	0.52	—
3 cells deep, 30 cm thick	—	0.69	—
Concrete block, 3 oval cores			
Sand/gravel, 20 cm thick	—	1.0	—
Cinder aggregate, 20 cm thick	—	0.67	—
Concrete block, rectangular core			
2 cores, 20 cm thick, 16 kg	—	1.1	—
Same with filled cores	—	0.60	—
Plastering Materials			
Cement plaster, sand aggregate	1860	0.72	—
Gypsum plaster, sand aggregate	1680	0.22	1085
Gypsum plaster, vermiculite aggregate	720	0.25	—

Source: Incropera, F.P., and D.P. DeWitt, *Introduction to Heat Transfer*, 4th ed., Wiley, New York, 2002.

TABLE A-6 Thermophysical Properties of Liquids

Water (saturated liquid)

T °C	ρ kg/m^3	c_p kJ/kg·K	k W/m·K	$\mu \times 10^{4*}$ N·s/m^2	$\nu \times 10^6$ m^2/s	$\alpha \times 10^7$ m^2/s	Pr	$\beta \times 10^4$ 1/K
0.01	999.8	4.226	0.569	17.6	1.76	1.35	13.07	−0.448
5	999.9	4.206	0.578	15.0	1.50	1.37	10.91	0.244
10	999.6	4.195	0.587	12.9	1.29	1.40	9.23	0.894
15	999.0	4.187	0.595	11.2	1.12	1.42	7.90	1.50
20	998.2	4.182	0.603	9.85	0.987	1.44	6.83	2.06
25	997.0	4.178	0.611	8.72	0.875	1.47	5.97	2.58
30	995.6	4.176	0.618	7.79	0.782	1.49	5.26	3.06
35	993.9	4.175	0.625	7.00	0.704	1.51	4.68	3.49
40	992.2	4.175	0.631	6.34	0.639	1.52	4.19	3.88
45	990.2	4.176	0.637	5.77	0.583	1.54	3.78	4.23
50	988.0	4.178	0.643	5.29	0.535	1.56	3.44	4.66
55	985.7	4.179	0.648	4.88	0.495	1.57	3.15	4.95
60	983.2	4.181	0.653	4.52	0.460	1.59	2.89	5.24
65	980.5	4.184	0.658	4.21	0.429	1.60	2.67	5.53
70	977.7	4.187	0.662	3.93	0.402	1.62	2.49	5.82
75	974.9	4.190	0.666	3.69	0.379	1.63	2.32	6.10
80	971.8	4.194	0.670	3.48	0.358	1.64	2.18	6.39
85	968.6	4.198	0.673	3.25	0.336	1.66	2.03	6.67
90	965.3	4.202	0.676	3.07	0.318	1.67	1.91	6.95
95	961.9	4.206	0.679	2.91	0.303	1.68	1.80	7.23
100	958.3	4.211	0.681	2.76	0.288	1.69	1.71	7.51
110	951.0	4.224	0.685	2.50	0.263	1.71	1.54	8.07
120	943.1	4.232	0.687	2.28	0.242	1.72	1.41	8.61
130	934.8	4.250	0.688	2.10	0.225	1.73	1.29	9.61
140	926.1	4.257	0.688	1.94	0.209	1.75	1.20	9.69
150	916.9	4.270	0.687	1.80	0.196	1.75	1.12	10.2
160	907.4	4.285	0.685	1.68	0.185	1.76	1.05	10.7
170	897.3	4.340	0.681	1.58	0.176	1.75	1.00	11.3
180	886.9	4.396	0.677	1.49	0.168	1.74	0.96	11.8
190	876.0	4.480	0.671	1.40	0.160	1.71	0.94	12.3
200	864.7	4.501	0.665	1.33	0.154	1.71	0.90	12.8
250	799.2	4.857	0.616	1.08	0.135	1.59	0.85	15.2
300	712.5	5.694	0.540	0.883	0.124	1.33	0.93	17.5

*For example, at 5°C, $\mu \times 10^4 = 15.0$ N·s/m$^2 \rightarrow \mu = 15.0 \times 10^{-4}$ N·s/m^2.

Engine oil (unused)

T K	ρ kg/m^3	c_p kJ/kg·K	k W/m·K	$\mu \times 10^{4*}$ N·s/m^2	$\nu \times 10^6$ m^2/s	$\alpha \times 10^7$ m^2/s	Pr	$\beta \times 10^4$ 1/K
273	899.1	1.796	0.147	38500	4282	0.91	47000	7.0
280	895.3	1.827	0.146	21700	2424	0.88	27500	7.0
290	890.0	1.868	0.145	9990	1122	0.872	12900	7.0
300	884.1	1.909	0.145	4860	549.7	0.859	6400	7.0
310	877.9	1.951	0.145	2530	288.2	0.847	3400	7.0
320	871.8	1.993	0.143	1410	161.7	0.823	1965	7.0
330	865.8	2.035	0.141	836	96.56	0.800	1205	7.0
340	859.9	2.076	0.139	531	61.75	0.779	793	7.0
350	853.9	2.118	0.138	356	41.69	0.763	546	7.0
360	847.8	2.161	0.138	252	29.72	0.753	395	7.0
370	841.8	2.206	0.137	186	22.10	0.738	300	7.0
380	836.0	2.250	0.136	141	16.87	0.723	233	7.0

(Continued)

TABLE A-6   (Continued)

T K	ρ kg/m³	c_p kJ/kg·K	k W/m·K	$\mu \times 10^{4*}$ N·s/m²	$\upsilon \times 10^6$ m²/s	$\alpha \times 10^7$ m²/s	Pr	$\beta \times 10^4$ 1/K
390	830.6	2.294	0.135	110	13.24	0.709	187	7.0
400	825.1	2.337	0.134	87.4	10.59	0.695	152	7.0
410	818.9	2.381	0.133	69.8	8.524	0.682	125	7.0
420	812.1	2.427	0.133	56.4	6.945	0.675	103	7.0
430	806.5	2.471	0.132	47.0	5.828	0.662	88	7.0

*For example, at 300 K, $\mu \times 10^4 = 4860$ N·s/m² $\rightarrow \mu = 4860 \times 10^{-4}$ N·s/m².

Ethylene glycol

T K	ρ kg/m³	c_p kJ/kg·K	k W/m·K	$\mu \times 10^{4*}$ N·s/m²	$\upsilon \times 10^6$ m²/s	$\alpha \times 10^7$ m²/s	Pr	$\beta \times 10^4$ 1/K
273	1130.8	2.294	0.242	651	57.6	0.933	617	6.5
280	1125.8	2.323	0.244	420	37.3	0.933	400	6.5
290	1118.8	2.368	0.248	247	22.1	0.936	236	6.5
300	1114.4	2.415	0.252	157	14.1	0.939	151	6.5
310	1103.7	2.460	0.255	107	9.69	0.939	103	6.5
320	1096.2	2.505	0.258	75.7	6.91	0.940	73.5	6.5
330	1089.5	2.549	0.260	56.1	5.15	0.936	55.0	6.5
340	1083.8	2.592	0.261	43.1	3.98	0.929	42.8	6.5
350	1079.0	2.637	0.261	34.2	3.17	0.917	34.6	6.5
360	1074.0	2.682	0.261	27.8	2.59	0.906	28.6	6.5
370	1066.7	2.728	0.262	22.8	2.14	0.900	23.7	6.5
373	1058.5	2.742	0.263	21.5	2.03	0.906	22.4	6.5

*For example, at 300 K, $\mu \times 10^4 = 157$ N·s/m² $\rightarrow \mu = 157 \times 10^{-4}$ N·s/m².

Glycerin

T K	ρ kg/m³	c_p kJ/kg·K	k W/m·K	$\mu \times 10^{4*}$ N·s/m²	$\upsilon \times 10^6$ m²/s	$\alpha \times 10^7$ m²/s	Pr	$\beta \times 10^4$ 1/K
273	1276.0	2.261	0.282	106000	8310	0.977	85000	4.7
280	1271.9	2.298	0.284	53400	4200	0.972	43200	4.7
290	1265.8	2.367	0.286	18500	1460	0.955	15300	4.8
300	1259.9	2.427	0.286	7990	634	0.935	6780	4.8
310	1253.9	2.490	0.286	3520	281	0.916	3060	4.9
320	1247.2	2.564	0.287	2100	168	0.897	1870	5.0

*For example, at 300 K, $\mu \times 10^4 = 7990$ N·s/m² $\rightarrow \mu = 7990 \times 10^{-4}$ N·s/m².

R-12 (saturated liquid)

T K	ρ kg/m³	c_p kJ/kg·K	k W/m·K	$\mu \times 10^{4*}$ N·s/m²	$\upsilon \times 10^6$ m²/s	$\alpha \times 10^7$ m²/s	Pr	$\beta \times 10^4$ 1/K
230	1528.4	0.8816	0.068	4.57	0.299	0.505	5.9	18.5
240	1498.0	0.8923	0.069	3.85	0.257	0.516	5.0	19.0
250	1469.5	0.9037	0.070	3.54	0.241	0.527	4.6	20.0
260	1439.0	0.9163	0.073	3.22	0.224	0.554	4.0	21.0
270	1407.2	0.9301	0.073	3.04	0.216	0.558	3.9	22.5
280	1374.4	0.9450	0.073	2.83	0.206	0.562	3.7	23.5
290	1340.5	0.9609	0.073	2.65	0.198	0.567	3.5	25.5
300	1305.8	0.9781	0.072	2.54	0.195	0.564	3.5	27.5
310	1268.9	0.9963	0.069	2.44	0.192	0.546	3.4	30.5
320	1228.6	1.0155	0.068	2.33	0.190	0.545	3.5	35.0

*For example, at 300 K, $\mu \times 10^4 = 2.54$ N·s/m² $\rightarrow \mu = 2.54 \times 10^{-4}$ N·s/m².

TABLE A-6   (Continued)

Mercury

T °C	ρ kg/m³	c_p kJ/kg·K	k W/m·K	$\mu \times 10^{4*}$ N·s/m²	$\upsilon \times 10^6$ m²/s	$\alpha \times 10^7$ m²/s	Pr	$\beta \times 10^4$ 1/K
0	13628	0.1403	8.20	16.90	0.124	42.9	0.028	
20	13579	0.1394	8.69	15.48	0.114	45.9	0.024	1.82
50	13506	0.1386	9.40	14.05	0.104	50.2	0.020	
100	13385	0.1373	10.51	12.42	0.0928	57.2	0.016	
150	13264	0.1365	11.49	11.31	0.0853	63.5	0.013	
200	13145	0.1358	12.34	10.54	0.0802	69.1	0.011	
250	13025	0.1357	13.07	9.96	0.0765	73.9	0.010	
316	12847	0.1340	14.02	8.65	0.0673	81.4	0.008	

*For example, at 200°C, $\mu \times 10^4 = 10.54$ N·s/m² $\rightarrow \mu = 10.54 \times 10^{-4}$ N·s/m².

Sources: adapted from N.V. Suryanarayana, *Engineering Heat Transfer*, West Publishing Minneapolis/St.Paul, 1995; L.C. Thomas, *Heat Transfer*, Prentice-Hall, Englewood Cliffs, NJ, 1992; F. P. Incropera and D. P. DeWitt, *Introduction to Heat Transfer*, 3rd ed., Wiley, New York, 1996.

TABLE A-7   Thermophysical Properties of Gases

Air (at 1 atm)

T K	ρ kg/m³	c_p kJ/kg·K	k W/m·K	$\mu \times 10^{5*}$ N·s/m²	$\upsilon \times 10^6$ m²/s	$\alpha \times 10^4$ m²/s	Pr
100	3.6010	1.0266	0.009246	0.6924	1.923	0.02501	0.770
150	2.3675	1.0099	0.013735	1.0283	4.343	0.05745	0.753
200	1.7684	1.0061	0.01809	1.3289	7.490	0.10165	0.739
250	1.4128	1.0053	0.02227	1.488	9.490	0.13161	0.722
300	1.1774	1.0057	0.02624	1.846	15.68	0.22160	0.708
350	0.9980	1.0090	0.03003	2.075	20.76	0.2983	0.697
400	0.8826	1.0140	0.03365	2.286	25.90	0.3760	0.689
450	0.7833	1.0207	0.03707	2.484	28.86	0.4222	0.683
500	0.7048	1.0295	0.04038	2.671	37.90	0.5564	0.680
550	0.6423	1.0392	0.04360	2.848	44.34	0.6532	0.680
600	0.5879	1.0551	0.04659	3.018	51.34	0.7512	0.680
650	0.5430	1.0635	0.04953	3.177	58.51	0.8578	0.682
700	0.5030	1.0752	0.05230	3.332	66.25	0.9672	0.684
750	0.4709	1.0856	0.05509	3.481	73.91	1.0774	0.686
800	0.4405	1.0978	0.05779	3.625	82.29	1.1951	0.689
850	0.4149	1.1095	0.06028	3.765	90.75	1.3097	0.692
900	0.3925	1.1212	0.06279	3.899	99.3	1.4271	0.696
950	0.3716	1.1321	0.06525	4.023	108.2	1.5510	0.699
1000	0.3524	1.1417	0.06752	4.152	117.8	1.6779	0.702
1100	0.3204	1.160	0.0732	4.44	138.6	1.969	0.704
1200	0.2947	1.179	0.0782	4.69	159.1	2.251	0.707
1300	0.2707	1.197	0.0837	4.93	182.1	2.583	0.705
1400	0.2515	1.214	0.0891	5.17	205.5	2.920	0.705
1500	0.2355	1.230	0.0946	5.40	229.1	3.262	0.705
1600	0.2211	1.248	0.100	5.63	254.5	3.609	0.705
1700	0.2082	1.267	0.105	5.85	280.5	3.977	0.705
1800	0.1970	1.287	0.111	6.07	308.1	4.379	0.704
1900	0.1858	1.309	0.117	6.29	338.5	4.811	0.704
2000	0.1762	1.338	0.124	6.50	369.0	5.260	0.702
2100	0.1682	1.372	0.131	6.72	399.6	5.715	0.700
2200	0.1602	1.419	0.139	6.93	432.6	6.120	0.707
2300	0.1538	1.482	0.149	7.14	464.0	6.540	0.710
2400	0.1458	1.574	0.161	7.35	504.0	7.020	0.718
2500	0.1394	1.688	0.175	7.57	543.5	7.441	0.730

*For example, at 100 K, $\mu \times 10^5 = 0.6924$ N·s/m² $\rightarrow \mu = 0.6924 \times 10^{-5}$ N·s/m².

TABLE A-7　(Continued)

Nitrogen (at 1 atm)

T K	ρ kg/m³	c_p kJ/kg·K	k W/m·K	$\mu \times 10^{5*}$ N·s/m²	$\nu \times 10^6$ m²/s	$\alpha \times 10^4$ m²/s	Pr
100	3.4388	1.070	0.00958	0.688	2.00	0.0260	0.768
150	2.2594	1.050	0.0139	1.006	4.45	0.0586	0.759
200	1.6883	1.043	0.0183	1.292	7.65	0.104	0.736
250	1.3488	1.042	0.0222	1.549	11.48	0.158	0.727
300	1.1233	1.041	0.0259	1.782	15.86	0.221	0.716
350	0.9625	1.042	0.0293	2.000	20.78	0.292	0.711
400	0.8425	1.045	0.0327	2.204	26.16	0.371	0.704
450	0.7485	1.050	0.0358	2.396	32.01	0.456	0.703
500	0.6739	1.056	0.0389	2.577	38.24	0.547	0.700
550	0.6124	1.065	0.0417	2.747	44.86	0.639	0.702
600	0.5615	1.075	0.0446	2.908	51.79	0.739	0.701
700	0.4812	1.098	0.0499	3.210	66.71	0.944	0.706
800	0.4211	1.220	0.0548	3.491	82.90	1.16	0.715
900	0.3743	1.146	0.0597	3.753	100.3	1.39	0.721
1000	0.3368	1.167	0.0647	3.999	118.7	1.65	0.721
1100	0.3062	1.187	0.0700	4.232	138.2	1.93	0.718
1200	0.2807	1.204	0.0758	4.453	158.6	2.24	0.707
1300	0.2591	1.219	0.0810	4.662	179.9	2.56	0.701

*For example, at 100 K, $\mu \times 10^5 = 0.688$ N·s/m² → $\mu = 0.688 \times 10^{-5}$ N·s/m².

Oxygen (at 1 atm)

T K	ρ kg/m³	c_p kJ/kg·K	k W/m·K	$\mu \times 10^{5*}$ N·s/m²	$\nu \times 10^6$ m²/s	$\alpha \times 10^4$ m²/s	Pr
100	3.9450	0.962	0.00925	0.764	1.94	0.0244	0.796
150	2.5850	0.921	0.0138	1.148	4.44	0.0580	0.766
200	1.9300	0.915	0.0183	1.475	7.64	0.104	0.737
250	1.5420	0.915	0.0226	1.786	11.58	0.160	0.723
300	1.2840	0.920	0.0268	2.072	16.14	0.227	0.711
350	1.1000	0.929	0.0296	2.335	21.23	0.290	0.733
400	0.9620	0.942	0.0330	2.582	26.84	0.364	0.737
450	0.8554	0.956	0.0363	2.814	32.90	0.444	0.741
500	0.7698	0.972	0.0412	3.033	39.40	0.551	0.716
550	0.6998	0.988	0.0441	3.240	46.30	0.638	0.726
600	0.6414	1.003	0.0473	3.437	53.59	0.735	0.729
700	0.5498	1.031	0.0528	3.808	69.26	0.931	0.744
800	0.4810	1.054	0.0589	4.152	86.32	1.16	0.743
900	0.4275	1.074	0.0649	4.472	104.6	1.41	0.740
1000	0.3848	1.090	0.0710	4.770	124.0	1.69	0.733
1100	0.3498	1.103	0.0758	5.055	144.5	1.96	0.736
1200	0.3206	1.115	0.0819	5.325	166.1	2.29	0.725
1300	0.2960	1.125	0.0871	5.884	188.6	2.62	0.721

*For example, at 100 K, $\mu \times 10^5 = 0.764$ N·s/m² → $\mu = 0.764 \times 10^{-5}$ N·s/m².

Carbon dioxide (at 1 atm)

T K	ρ kg/m³	c_p kJ/kg·K	k W/m·K	$\mu \times 10^{5*}$ N·s/m²	$\nu \times 10^6$ m²/s	$\alpha \times 10^4$ m²/s	Pr
280	1.9022	0.830	0.0152	1.40	7.36	0.0963	0.765
300	1.7730	0.851	0.0166	1.49	8.40	0.110	0.766
320	1.6609	0.872	0.0181	1.56	9.39	0.125	0.754
340	1.5618	0.891	0.0197	1.65	10.6	0.142	0.746

(Continued)

TABLE A-7 (Continued)

Carbon dioxide (at 1 atm)

T K	ρ kg/m³	c_p kJ/kg · K	k W/m · K	$\mu \times 10^{5*}$ N · s/m²	$\upsilon \times 10^6$ m²/s	$\alpha \times 10^4$ m²/s	Pr
360	1.4743	0.908	0.0212	1.73	11.7	0.158	0.741
380	1.3961	0.926	0.0228	1.81	13.0	0.176	0.737
400	1.3257	0.942	0.0243	1.90	14.3	0.195	0.737
450	1.1782	0.981	0.0283	2.10	17.8	0.245	0.728
500	1.0594	1.02	0.0325	2.31	21.8	0.301	0.725
550	0.9625	1.05	0.0366	2.51	26.1	0.362	0.721
600	0.8826	1.08	0.0407	2.70	30.6	0.427	0.717
650	0.8143	1.10	0.0445	2.88	35.4	0.497	0.712
700	0.7564	1.13	0.0481	3.05	40.3	0.563	0.717
750	0.7057	1.15	0.0517	3.21	45.5	0.637	0.714
800	0.6614	1.17	0.0551	3.37	51.0	0.712	0.716

*For example, at 300 K, $\mu \times 10^5 = 1.49$ N · s/m² → $\mu = 1.49 \times 10^{-5}$ N · s/m².

Hydrogen (at 1 atm)

T K	ρ kg/m³	c_p kJ/kg · K	k W/m · K	$\mu \times 10^{5*}$ N · s/m²	$\upsilon \times 10^6$ m²/s	$\alpha \times 10^4$ m²/s	Pr
100	0.24255	11.23	0.0670	0.421	17.4	0.246	0.707
150	0.16156	12.60	0.101	0.560	34.7	0.496	0.699
200	0.12115	13.54	0.131	0.681	56.2	0.799	0.704
250	0.09693	14.06	0.157	0.789	81.4	1.15	0.707
300	0.08078	14.31	0.183	0.896	111	1.58	0.701
350	0.06924	14.43	0.204	0.988	143	2.04	0.700
400	0.06059	14.48	0.226	1.082	179	2.58	0.695
450	0.05386	14.50	0.247	1.172	218	3.16	0.689
500	0.04848	14.52	0.266	1.264	261	3.78	0.691
550	0.04407	14.53	0.285	1.343	305	4.45	0.685
600	0.04040	14.55	0.305	1.424	352	5.19	0.678
700	0.03463	14.61	0.342	1.578	456	6.76	0.675
800	0.03030	14.70	0.378	1.724	569	8.49	0.670
900	0.02694	14.83	0.412	1.865	692	10.30	0.671
1000	0.02424	14.99	0.448	2.013	830	12.30	0.673
1100	0.02204	15.17	0.488	2.130	966	14.60	0.662
1200	0.02020	15.37	0.528	2.262	1120	17.00	0.659
1300	0.01865	15.59	0.568	2.385	1279	19.55	0.655
1400	0.01732	15.81	0.610	2.507	1447	22.30	0.650
1500	0.01616	16.02	0.655	2.627	1626	25.30	0.643
1600	0.01520	16.28	0.697	2.737	1801	28.15	0.639
1700	0.01430	16.58	0.742	2.849	1992	31.30	0.637
1800	0.01350	16.96	0.786	2.961	2193	34.35	0.639
1900	0.01280	17.49	0.835	3.072	2400	37.30	0.643
2000	0.01210	18.25	0.878	3.182	2630	39.75	0.661

*For example, at 100 K, $\mu \times 10^5 = 0.421$ N · s/m² → $\mu = 0.421 \times 10^{-5}$ N · s/m².

Water vapor (at 1 atm)

T K	ρ kg/m³	c_p kJ/kg · K	k W/m · K	$\mu \times 10^{5*}$ N · s/m²	$\upsilon \times 10^6$ m²/s	$\alpha \times 10^4$ m²/s	Pr
380	0.5863	2.060	0.0246	1.271	21.68	0.204	1.06
400	0.5542	2.014	0.0261	1.344	24.25	0.234	1.04
450	0.4902	1.980	0.0299	1.525	31.11	0.308	1.01
500	0.4405	1.985	0.0339	1.704	38.68	0.388	0.998
550	0.4005	1.997	0.0379	1.884	47.04	0.474	0.993
600	0.3652	2.026	0.0422	2.067	56.60	0.570	0.993
650	0.3380	2.056	0.0464	2.247	66.48	0.668	0.996
700	0.3140	2.085	0.0505	2.426	77.26	0.771	1.00

(Continued)

TABLE A-7   **(Continued)**

Water vapor (at 1 atm)

T K	ρ kg/m³	c_p kJ/kg·K	k W/m·K	$\mu \times 10^{5*}$ N·s/m²	$\upsilon \times 10^6$ m²/s	$\alpha \times 10^4$ m²/s	Pr
750	0.2931	2.119	0.0549	2.604	88.84	0.884	1.00
800	0.2739	2.152	0.0592	2.786	101.7	1.00	1.01
850	0.2579	2.186	0.0637	2.969	115.1	1.13	1.02

*For example, at 400 K, $\mu \times 10^5 = 1.344$ N·s/m² → $\mu = 1.344 \times 10^{-5}$ N·s/m².

Steam (at saturation pressure)

T K	P kPa	ρ kg/m³	c_p kJ/kg·K	k W/m·K	$\mu \times 10^{5*}$ N·s/m²	$\upsilon \times 10^6$ m²/s	$\alpha \times 10^4$ m²/s	Pr
273.15	0.611	0.00485	1.854	0.0182	0.802	1654	20.24	0.817
280	0.990	0.00767	1.858	0.0186	0.829	1081	13.05	0.828
290	1.917	0.01435	1.864	0.0193	0.869	605.6	7.215	0.839
300	3.531	0.02556	1.872	0.0196	0.909	355.6	4.096	0.868
310	6.221	0.04361	1.882	0.0204	0.949	217.6	2.486	0.875
320	10.53	0.07153	1.895	0.0210	0.989	138.3	1.549	0.892
330	17.19	0.1134	1.911	0.0217	1.029	90.74	1.001	0.906
340	27.13	0.1742	1.930	0.0223	1.069	61.37	0.6633	0.925
350	41.63	0.2600	1.954	0.0230	1.109	42.65	0.4527	0.942
360	62.09	0.3781	1.983	0.0237	1.149	30.39	0.3161	0.961
370	90.40	0.5373	2.017	0.0245	1.189	22.13	0.2261	0.979
373.15	101.33	0.5956	2.029	0.0248	1.202	20.18	0.2052	0.983
380	128.69	0.7479	2.057	0.0254	1.229	16.43	0.1651	0.995
390	179.4	1.020	2.104	0.0263	1.269	12.44	0.1225	1.02
400	245.5	1.368	2.158	0.0272	1.305	9.539	0.09214	1.04
420	437.0	2.353	2.291	0.0298	1.379	5.861	0.05528	1.06
440	733.3	3.831	2.46	0.0317	1.450	3.785	0.03364	1.13
460	1171	5.988	2.68	0.0346	1.519	2.537	0.02156	1.18
480	1790	9.009	2.94	0.0381	1.588	1.763	0.01438	1.23
500	2640	13.05	3.27	0.0423	1.659	1.271	0.009912	1.28
520	3770	19.05	3.70	0.0475	1.733	0.9097	0.006739	1.35
540	5238	26.67	4.27	0.0540	1.81	0.6787	0.004742	1.43
560	7108	37.17	5.09	0.0637	1.91	0.5139	0.003367	1.53
580	9451	51.81	6.40	0.0767	2.04	0.3937	0.002313	1.70
600	12350	72.99	8.75	0.0929	2.27	0.3110	0.001455	2.14
620	15910	106.4	15.4	0.114	2.59	0.2434	0.0006957	3.50
630	17970	133.3	22.1	0.130	2.80	0.2101	0.0004413	4.76
640	20270	175.4	42	0.155	3.20	0.1824	0.0002104	8.67
645	21520	222.2		0.178	3.70	0.1665		
647.3	22120	312.5		0.238	4.50	0.1440		

*For example, at 300 K, $\mu \times 10^5 = 0.909$ N·s/m² → $\mu = 0.909 \times 10^{-5}$ N·s/m².
Source: Adapted from L.C. Thomas, *Heat Transfer*, Prentice-Hall, Englewood Cliffs, NJ, 1992; F. P. Incropera and D. P. DeWitt, *Introduction to Heat Transfer*, 3rd ed., Wiley, New York, 1996.

TABLE A-8   **Ideal Gas Specific Heats**

Ideal Gas Specific Heats in Tabular Form (c_p, c_v, kJ/kg · K; $k = c_p/c_v$)

Temp. K	c_p	c_v	k	c_p	c_v	k	c_p	c_v	k	Temp. K
	Air			Nitrogen, N₂			Oxygen, O₂			
250	1.003	0.716	1.401	1.039	0.742	1.400	0.913	0.653	1.398	250
300	1.005	0.718	1.400	1.039	0.743	1.400	0.918	0.658	1.395	300
350	1.008	0.721	1.398	1.041	0.744	1.399	0.928	0.668	1.389	350

(Continued)

TABLE A-8 (Continued)

Ideal Gas Specific Heats in Tabular Form (c_p, c_v, kJ/kg · K; $k = c_p/c_v$)

Temp. K	c_p	c_v	k	c_p	c_v	k	c_p	c_v	k	Temp. K
	Air			Nitrogen, N_2			Oxygen, O_2			
400	1.013	0.726	1.395	1.044	0.747	1.397	0.941	0.681	1.382	400
450	1.020	0.733	1.391	1.049	0.752	1.395	0.956	0.696	1.373	450
500	1.029	0.742	1.387	1.056	0.759	1.391	0.972	0.712	1.365	500
550	1.040	0.753	1.381	1.065	0.768	1.387	0.988	0.728	1.358	550
600	1.051	0.764	1.376	1.075	0.778	1.382	1.003	0.743	1.350	600
650	1.063	0.776	1.370	1.086	0.789	1.376	1.017	0.758	1.343	650
700	1.075	0.788	1.364	1.098	0.801	1.371	1.031	0.771	1.337	700
750	1.087	0.800	1.359	1.110	0.813	1.365	1.043	0.783	1.332	750
800	1.099	0.812	1.354	1.121	0.825	1.360	1.054	0.794	1.327	800
900	1.121	0.834	1.344	1.145	0.849	1.349	1.074	0.814	1.319	900
1000	1.142	0.855	1.336	1.167	0.870	1.341	1.090	0.830	1.313	1000

Temp. K										Temp. K
	Carbon dioxide, CO_2			Carbon monoxide, CO			Hydrogen, H_2			
250	0.791	0.602	1.314	1.039	0.743	1.400	14.051	9.927	1.416	250
300	0.846	0.657	1.288	1.040	0.744	1.399	14.307	10.183	1.405	300
350	0.895	0.706	1.268	1.043	0.746	1.398	14.427	10.302	1.400	350
400	0.939	0.750	1.252	1.047	0.751	1.395	14.476	10.352	1.398	400
450	0.978	0.790	1.239	1.054	0.757	1.392	14.501	10.377	1.398	450
500	1.014	0.825	1.229	1.063	0.767	1.387	14.513	10.389	1.397	500
550	1.046	0.857	1.220	1.075	0.778	1.382	14.530	10.405	1.396	550
600	1.075	0.886	1.213	1.087	0.790	1.376	14.546	10.422	1.396	600
650	1.102	0.913	1.207	1.100	0.803	1.370	14.571	10.447	1.395	650
700	1.126	0.937	1.202	1.113	0.816	1.364	14.604	10.480	1.394	700
750	1.148	0.959	1.197	1.126	0.829	1.358	14.645	10.521	1.392	750
800	1.169	0.980	1.193	1.139	0.842	1.353	14.695	10.570	1.390	800
900	1.204	1.015	1.186	1.163	0.866	1.343	14.822	10.698	1.385	900
1000	1.234	1.045	1.181	1.185	0.888	1.335	14.983	10.859	1.380	1000

Source: Adapted from K. Wark, *Thermodynamics*, 4th ed., McGraw-Hill, New York, 1983, as based on "Tables of Thermal Properties of Gases," NBS Circular 564, 1955.

Ideal Gas Specific Heats in Equation Form (kJ/kmol · K)

$$\frac{\bar{c}_p}{R} = \alpha + \beta T + \gamma T^2 + \delta T^3 + \varepsilon T^4$$

T is in K, equations valid from 300 to 1000 K

Gas	α	$\beta \times 10^3$	$\gamma \times 10^6$	$\delta \times 10^9$	$\varepsilon \times 10^{12}$
CO	3.710	−1.619	3.692	−2.032	0.240
CO_2	2.401	8.735	−6.607	2.002	0
H_2	3.057	2.677	−5.810	5.521	−1.812
H_2O	4.070	−1.108	4.152	−2.964	0.807
O_2	3.626	−1.878	7.055	−6.764	2.156
N_2	3.675	−1.208	2.324	−0.632	−0.226
Air	3.653	−1.337	3.294	−1.913	0.2763
SO_2	3.267	5.324	0.684	−5.281	2.559
CH_4	3.826	−3.979	24.558	−22.733	6.963
C_2H_2	1.410	19.057	−24.501	16.391	−4.135
C_2H_4	1.426	11.383	7.989	−16.254	6.749
Monatomic gases*	2.5	0	0	0	0

*For monatomic gases, such as He, Ne, and Ar, $\bar{c}_p$ is constant over a wide temperature range and is very nearly equal to 5/2 $\bar{R}$.

Source: Adapted from K. Wark, *Thermodynamics*, 4th ed., McGraw-Hill, New York, 1983, as based on NASA SP-273, U.S. Government Printing Office, Washington, DC, 1971.

TABLE A-9   Ideal Gas Properties of Air

T(K), h and u (kJ/kg), s° (kJ/kg · K)			$\Delta s = 0$		
T	h	u	P_r	v_r	s°
200	199.97	142.56	0.3363	1707.	1.29559
210	209.97	149.69	0.3987	1512.	1.34444
220	219.97	156.82	0.4690	1346.	1.39105
230	230.02	164.00	0.5477	1205.	1.43557
240	240.02	171.13	0.6355	1084.	1.47824
250	250.05	178.28	0.7329	979.	1.51917
260	260.09	185.45	0.8405	887.8	1.55848
270	270.11	192.60	0.9590	808.0	1.59634
280	280.13	199.75	1.0889	738.0	1.63279
285	285.14	203.33	1.1584	706.1	1.65055
290	290.16	206.91	1.2311	676.1	1.66802
295	295.17	210.49	1.3068	647.9	1.68515
300	300.19	214.07	1.3860	621.2	1.70203
305	305.22	217.67	1.4686	596.0	1.71865
310	310.24	221.25	1.5546	572.3	1.73498
315	315.27	224.85	1.6442	549.8	1.75106
320	320.29	228.42	1.7375	528.6	1.76690
325	325.31	232.02	1.8345	508.4	1.78249
330	330.34	235.61	1.9352	489.4	1.79783
340	340.42	242.82	2.149	454.1	1.82790
350	350.49	250.02	2.379	422.2	1.85708
360	360.58	257.24	2.626	393.4	1.88543
370	370.67	264.46	2.892	367.2	1.91313
380	380.77	271.69	3.176	343.4	1.94001
390	390.88	278.93	3.481	321.5	1.96633
400	400.98	286.16	3.806	301.6	1.99194
410	411.12	293.43	4.153	283.3	2.01699
420	421.26	300.69	4.522	266.6	2.04142
430	431.43	307.99	4.915	251.1	2.06533
440	441.61	315.30	5.332	236.8	2.08870
450	451.80	322.62	5.775	223.6	2.11161
460	462.02	329.97	6.245	211.4	2.13407
470	472.24	337.32	6.742	200.1	2.15604
480	482.49	344.70	7.268	189.5	2.17760
490	492.74	352.08	7.824	179.7	2.19876
500	503.02	359.49	8.411	170.6	2.21952
510	513.32	366.92	9.031	162.1	2.23993
520	523.63	374.36	9.684	154.1	2.25997
530	533.98	381.84	10.37	146.7	2.27967
540	544.35	389.34	11.10	139.7	2.29906
550	554.74	396.86	11.86	133.1	2.31809
560	565.17	404.42	12.66	127.0	2.33685
570	575.59	411.97	13.50	121.2	2.35531
580	586.04	419.55	14.38	115.7	2.37348
590	596.52	427.15	15.31	110.6	2.39140
600	607.02	434.78	16.28	105.8	2.40902
610	617.53	442.42	17.30	101.2	2.42644
620	628.07	450.09	18.36	96.92	2.44356
630	638.63	457.78	19.84	92.84	2.46048
640	649.22	465.50	20.64	88.99	2.47716
650	659.84	473.25	21.86	85.34	2.49364
660	670.47	481.01	23.13	81.89	2.50985
670	681.14	488.81	24.46	78.61	2.52589
680	691.82	496.62	25.85	75.50	2.54175

(Continued)

TABLE A-9 Ideal Gas Properties of Air

T(K), h and u (kJ/kg), s° (kJ/kg · K)			Δs = 0		
T	h	u	P_r	v_r	s°
690	702.52	504.45	27.29	72.56	2.55731
700	713.27	512.33	28.80	69.76	2.57277
710	724.04	520.23	30.38	67.07	2.58810
720	734.82	528.14	32.02	64.53	2.60319
730	745.62	536.07	33.72	62.13	2.61803
740	756.44	544.02	35.50	59.82	2.63280
750	767.29	551.99	37.35	57.63	2.64737
760	778.18	560.01	39.27	55.54	2.66176
770	789.11	568.07	41.31	53.39	2.67595
780	800.03	576.12	43.35	51.64	2.69013
790	810.99	584.21	45.55	49.86	2.70400
800	821.95	592.30	47.75	48.08	2.71787
820	843.98	608.59	52.59	44.84	2.74504
840	866.08	624.95	57.60	41.85	2.77170
860	888.27	641.40	63.09	39.12	2.79783
880	910.56	657.95	68.98	36.61	2.82344
900	932.93	674.58	75.29	34.31	2.84856
920	955.38	691.28	82.05	32.18	2.87324
940	977.92	708.08	89.28	30.22	2.89748
960	1000.55	725.02	97.00	28.40	2.92128
980	1023.25	741.98	105.2	26.73	2.94468
1000	1046.04	758.94	114.0	25.17	2.96770
1020	1068.89	776.10	123.4	23.72	2.99034
1040	1091.85	793.36	133.3	22.39	3.01260
1060	1114.86	810.62	143.9	21.14	3.03449
1080	1137.89	827.88	155.2	19.98	3.05608
1100	1161.07	845.33	167.1	18.896	3.07732
1120	1184.28	862.79	179.7	17.886	3.09825
1140	1207.57	880.35	193.1	16.946	3.11883
1160	1230.92	897.91	207.2	16.064	3.13916
1180	1254.34	915.57	222.2	15.241	3.15916
1200	1277.79	933.33	238.0	14.470	3.17888
1220	1301.31	951.09	254.7	13.747	3.19834
1240	1324.93	968.95	272.3	13.069	3.21751
1260	1348.55	986.90	290.8	12.435	3.23638
1280	1372.24	1004.76	310.4	11.835	3.25510
1300	1395.97	1022.82	330.9	11.275	3.27345
1320	1419.76	1040.88	352.5	10.747	3.29160
1340	1443.60	1058.94	375.3	10.247	3.30959
1360	1467.49	1077.10	399.1	9.780	3.32724
1380	1491.44	1095.26	424.2	9.337	3.34474
1400	1515.42	1113.52	450.5	8.919	3.36200
1420	1539.44	1131.77	478.0	8.526	3.37901
1440	1563.51	1150.13	506.9	8.153	3.39586
1460	1587.63	1168.49	537.1	7.801	3.41247
1480	1611.79	1186.95	568.8	7.468	3.42892
1500	1635.97	1205.41	601.9	7.152	3.44516
1520	1660.23	1223.87	636.5	6.854	3.46120
1540	1684.51	1242.43	672.8	6.569	3.47712
1560	1708.82	1260.99	710.5	6.301	3.49276
1580	1733.17	1279.65	750.0	6.046	3.50829
1600	1757.57	1298.30	791.2	5.804	3.52364
1620	1782.00	1316.96	834.1	5.574	3.53879
1640	1806.46	1335.72	878.9	5.355	3.55381

(Continued)

TABLE A-9 (Continued)

$T(K)$, h and u (kJ/kg), $s°$ (kJ/kg · K)			$\Delta s = 0$		
T	h	u	P_r	v_r	$s°$
1660	1830.96	1354.48	925.6	5.147	3.56867
1680	1855.50	1373.24	974.2	4.949	3.58335
1700	1880.1	1392.7	1025	4.761	3.5979
1750	1941.6	1439.8	1161	4.328	3.6336
1800	2003.3	1487.2	1310	3.944	3.6684
1850	2065.3	1534.9	1475	3.601	3.7023
1900	2127.4	1582.6	1655	3.295	3.7354
1950	2189.7	1630.6	1852	3.022	3.7677
2000	2252.1	1678.7	2068	2.776	3.7994
2050	2314.6	1726.8	2303	2.555	3.8303
2100	2377.4	1775.3	2559	2.356	3.8605
2150	2440.3	1823.8	2837	2.175	3.8901
2200	2503.2	1872.4	3138	2.012	3.9191
2250	2566.4	1921.3	3464	1.864	3.9474

Source: Adapted from K. Wark, *Thermodynamics*, 4th ed., McGraw-Hill, New York, 1983, as based on J. H. Keenan and J. Kaye, *Gas Tables*, Wiley, New York, 1945.

TABLE A-10 Thermodynamic Properties of Saturated Steam–Water (Temperature Table)

Temp. °C T	Press. kPa P	Specific Volume m³/kg		Internal Energy kJ/kg			Enthalpy kJ/kg			Entropy kJ/kg·K		
		Sat. Liquid v_f	Sat. Vapor v_g	Sat. Liquid u_f	Evap. u_{fg}	Sat. Vapor u_g	Sat. Liquid h_f	Evap. h_{fg}	Sat. Vapor h_g	Sat. Liquid s_f	Evap. s_{fg}	Sat. Vapor s_g
0.01	0.6113	0.001 000	206.14	0.00	2375.3	2375.3	0.01	2501.3	2501.4	0.0000	9.1562	9.1562
5	0.8721	0.001 000	147.12	20.97	2361.3	2382.3	20.98	2489.6	2510.6	0.0761	8.9496	9.0257
10	1.2276	0.001 000	106.38	42.00	2347.2	2389.2	42.01	2477.7	2519.8	0.1510	8.7498	8.9008
15	1.7051	0.001 001	77.93	62.99	2333.1	2396.1	62.99	2465.9	2528.9	0.2245	8.5569	8.7814
20	2.339	0.001 002	57.79	83.95	2319.0	2402.9	83.96	2454.1	2538.1	0.2966	8.3706	8.6672
25	3.169	0.001 003	43.36	104.88	2304.9	2409.8	104.89	2442.3	2547.2	0.3674	8.1905	8.5580
30	4.246	0.001 004	32.89	125.78	2290.8	2416.6	125.79	2430.5	2556.3	0.4369	8.0164	8.4533
35	5.628	0.001 006	25.22	146.67	2276.7	2423.4	146.68	2418.6	2565.3	0.5053	7.8478	8.3531
40	7.384	0.001 008	19.52	167.56	2262.6	2430.1	167.57	2406.7	2574.3	0.5725	7.6845	8.2570
45	9.593	0.001 010	15.26	188.44	2248.4	2436.8	188.45	2394.8	2583.2	0.6387	7.5261	8.1648
50	12.349	0.001 012	12.03	209.32	2234.2	2443.5	209.33	2382.7	2592.1	0.7038	7.3725	8.0763
55	15.758	0.001 015	9.568	230.21	2219.9	2450.1	230.23	2370.7	2600.9	0.7679	7.2234	7.9913
60	19.940	0.001 017	7.671	251.11	2205.5	2456.6	251.13	2358.5	2609.6	0.8312	7.0784	7.9096
65	25.03	0.001 020	6.197	272.02	2191.1	2463.1	272.06	2346.2	2618.3	0.8935	6.9375	7.8310
70	31.19	0.001 023	5.042	292.95	2176.6	2469.6	292.98	2333.8	2626.8	0.9549	6.8004	7.7553
75	38.58	0.001 026	4.131	313.90	2162.0	2475.9	313.93	2321.4	2635.3	1.0155	6.6669	7.6824
80	47.39	0.001 029	3.407	334.86	2147.4	2482.2	334.91	2308.8	2643.7	1.0753	6.5369	7.6122
85	57.83	0.001 033	2.828	355.84	2132.6	2488.4	355.90	2296.0	2651.9	1.1343	6.4102	7.5445
90	70.14	0.001 036	2.361	376.85	2117.7	2494.5	376.92	2283.2	2660.1	1.1925	6.2866	7.4791
95	84.55	0.001 040	1.982	397.88	2102.7	2500.6	397.96	2270.2	2668.1	1.2500	6.1659	7.4159
MPa												
100	0.101 35	0.001 044	1.6729	418.94	2087.6	2506.5	419.04	2257.0	2676.1	1.3069	6.0480	7.3549
105	0.120 82	0.001 048	1.4194	440.02	2072.3	2512.4	440.15	2243.7	2683.8	1.3630	5.9328	7.2958
110	0.143 27	0.001 052	1.2102	461.14	2057.0	2518.1	461.30	2230.2	2691.5	1.4185	5.8202	7.2387
115	0.169 06	0.001 056	1.0366	482.30	2041.4	2523.7	482.48	2216.5	2699.0	1.4734	5.7100	7.1833
120	0.198 53	0.001 060	0.8919	503.50	2025.8	2529.3	503.71	2202.6	2706.3	1.5276	5.6020	7.1296
125	0.2321	0.001 065	0.7706	524.74	2009.9	2534.6	524.99	2188.5	2713.5	1.5813	5.4962	7.0775

(Continued)

TABLE A-10 (Continued)

Temp. °C T	Press. MPa P	Specific Volume m³/kg Sat. Liquid v_f	Sat. Vapor v_g	Internal Energy kJ/kg Sat. Liquid u_f	Evap. u_{fg}	Sat. Vapor u_g	Enthalpy kJ/kg Sat. Liquid h_f	Evap. h_{fg}	Sat. Vapor h_g	Entropy kJ/kg·K Sat. Liquid s_f	Evap. s_{fg}	Sat. Vapor s_g
130	0.2701	0.001 070	0.6685	546.02	1993.9	2539.9	546.31	2174.2	2720.5	1.6344	5.3925	7.0269
135	0.3130	0.001 075	0.5822	567.35	1977.7	2545.0	567.69	2159.6	2727.3	1.6870	5.2907	6.9777
140	0.3613	0.001 080	0.5089	588.74	1961.3	2550.0	589.13	2144.7	2733.9	1.7391	5.1908	6.9299
145	0.4154	0.001 085	0.4463	610.18	1944.7	2554.9	610.63	2129.6	2740.3	1.7907	5.0926	6.8833
150	0.4758	0.001 091	0.3928	631.68	1927.9	2559.5	632.20	2114.3	2746.5	1.8418	4.9960	6.8379
155	0.5431	0.001 096	0.3468	653.24	1910.8	2564.1	653.84	2098.6	2752.4	1.8925	4.9010	6.7935
160	0.6178	0.001 102	0.3071	674.87	1893.5	2568.4	675.55	2082.6	2758.1	1.9427	4.8075	6.7502
165	0.7005	0.001 108	0.2727	696.56	1876.0	2572.5	697.34	2066.2	2763.5	1.9925	4.7153	6.7078
170	0.7917	0.001 114	0.2428	718.33	1858.1	2576.5	719.21	2049.5	2768.7	2.0419	4.6244	6.6663
175	0.8920	0.001 121	0.2168	740.17	1840.0	2580.2	741.17	2032.4	2773.6	2.0909	4.5347	6.6256
180	1.0021	0.001 127	0.194 05	762.09	1821.6	2583.7	763.22	2015.0	2778.2	2.1396	4.4461	6.5857
185	1.1227	0.001 134	0.174 09	784.10	1802.9	2587.0	785.37	1997.1	2782.4	2.1879	4.3586	6.5465
190	1.2544	0.001 141	0.156 54	806.19	1783.8	2590.0	807.62	1978.8	2786.4	2.2359	4.2720	6.5079
195	1.3978	0.001 149	0.141 05	828.37	1764.4	2592.8	829.98	1960.0	2790.0	2.2835	4.1863	6.4698
200	1.5538	0.001 157	0.127 36	850.65	1744.7	2595.3	852.45	1940.7	2793.2	2.3309	4.1014	6.4323
205	1.7230	0.001 164	0.115 21	873.04	1724.5	2597.5	875.04	1921.0	2796.0	2.3780	4.0172	6.3952
210	1.9062	0.001 173	0.104 41	895.53	1703.9	2599.5	897.76	1900.7	2798.5	2.4248	3.9337	6.3585
215	2.104	0.001 181	0.094 79	918.14	1682.9	2601.1	920.62	1879.9	2800.5	2.4714	3.8507	6.3221
220	2.318	0.001 190	0.086 19	940.87	1661.5	2602.4	943.62	1858.5	2802.1	2.5178	3.7683	6.2861
225	2.548	0.001 199	0.078 49	963.73	1639.6	2603.3	966.78	1836.5	2803.3	2.5639	3.6863	6.2503
230	2.795	0.001 209	0.071 58	986.74	1617.2	2603.9	990.12	1813.8	2804.0	2.6099	3.6047	6.2146
235	3.060	0.001 219	0.065 37	1009.89	1594.2	2604.1	1013.62	1790.5	2804.2	2.6558	3.5233	6.1791
240	3.344	0.001 229	0.059 76	1033.21	1570.8	2604.0	1037.32	1766.5	2803.8	2.7015	3.4422	6.1437
245	3.648	0.001 240	0.054 71	1056.71	1546.7	2603.4	1061.23	1741.7	2803.0	2.7472	3.3612	6.1083
250	3.973	0.001 251	0.050 13	1080.39	1522.0	2602.4	1085.36	1716.2	2801.5	2.7927	3.2802	6.0730
255	4.319	0.001 263	0.045 98	1104.28	1496.7	2600.9	1109.73	1689.8	2799.5	2.8383	3.1992	6.0375
260	4.688	0.001 276	0.042 21	1128.39	1470.6	2599.0	1134.37	1662.5	2796.9	2.8838	3.1181	6.0019
265	5.081	0.001 289	0.038 77	1152.74	1443.9	2596.6	1159.28	1634.4	2793.6	2.9294	3.0368	5.9662
270	5.499	0.001 302	0.035 64	1177.36	1416.3	2593.7	1184.51	1605.2	2789.7	2.9751	2.9551	5.9301
275	5.942	0.001 317	0.032 79	1202.25	1387.9	2590.2	1210.07	1574.9	2785.0	3.0208	2.8730	5.8938
280	6.412	0.001 332	0.030 17	1227.46	1358.7	2586.1	1235.99	1543.6	2779.6	3.0668	2.7903	5.8571
285	6.909	0.001 348	0.027 77	1253.00	1328.4	2581.4	1262.31	1511.0	2773.3	3.1130	2.7070	5.8199
290	7.436	0.001 366	0.025 57	1278.92	1297.1	2576.0	1289.07	1477.1	2766.2	3.1594	2.6227	5.7821
295	7.993	0.001 384	0.023 54	1305.2	1264.7	2569.9	1316.3	1441.8	2758.1	3.2062	2.5375	5.7437
300	8.581	0.001 404	0.021 67	1332.0	1231.0	2563.0	1344.0	1404.9	2749.0	3.2534	2.4511	5.7045
305	9.202	0.001 425	0.019 948	1359.3	1195.9	2555.2	1372.4	1366.4	2738.7	3.3010	2.3633	5.6643
310	9.856	0.001 447	0.018 350	1387.1	1159.4	2546.4	1401.3	1326.0	2727.3	3.3493	2.2737	5.6230
315	10.547	0.001 472	0.016 867	1415.5	1121.1	2536.6	1431.0	1283.5	2714.5	3.3982	2.1821	5.5804
320	11.274	0.001 499	0.015 488	1444.6	1080.9	2525.5	1461.5	1238.6	2700.1	3.4480	2.0882	5.5362
330	12.845	0.001 561	0.012 996	1505.3	993.7	2498.9	1525.3	1140.6	2665.9	3.5507	1.8909	5.4417
340	14.586	0.001 638	0.010 797	1570.3	894.3	2464.6	1594.2	1027.9	2622.0	3.6594	1.6763	5.3357
350	16.513	0.001 740	0.008 813	1641.9	776.6	2418.4	1670.6	893.4	2563.9	3.7777	1.4335	5.2112
360	18.651	0.001 893	0.006 945	1725.2	626.3	2351.5	1760.5	720.5	2481.0	3.9147	1.1379	5.0526
370	21.03	0.002 213	0.004 925	1844.0	384.5	2228.5	1890.5	441.6	2332.1	4.1106	0.6865	4.7971
374.14	22.09	0.003 155	0.003 155	2029.6	0	2029.6	2099.3	0	2099.3	4.4298	0	4.4298

Source: Tables A-10 to A-13 adapted from Van Wylen, G. J., R. E. Sonntag, and C. Borgnakke, *Fundamentals of Classical Thermodynamics,* 4th ed; Wiley, New York, 1994.

TABLE A-11 Thermodynamic Properties of Saturated Steam–Water (Pressure Table)

Press. kPa P	Temp. °C T	Specific Volume m³/kg		Internal Energy kJ/kg			Enthalpy kJ/kg			Entropy kJ/kg·K		
		Sat. Liquid v_f	Sat. Vapor v_g	Sat. Liquid u_f	Evap. u_{fg}	Sat. Vapor u_g	Sat. Liquid h_f	Evap. h_{fg}	Sat. Vapor h_g	Sat. Liquid s_f	Evap. s_{fg}	Sat. Vapor s_g
0.6113	0.01	0.001 000	206.14	0.00	2375.3	2375.3	0.01	2501.3	2501.4	0.0000	9.1562	9.1562
1.0	6.98	0.001 000	129.21	29.30	2355.7	2385.0	29.30	2484.9	2514.2	0.1059	8.8697	8.9756
1.5	13.03	0.001 001	87.98	54.71	2338.6	2393.3	54.71	2470.6	2525.3	0.1957	8.6322	8.8279
2.0	17.50	0.001 001	67.00	73.48	2326.0	2399.5	73.48	2460.0	2533.5	0.2607	8.4629	8.7237
2.5	21.08	0.001 002	54.25	88.48	2315.9	2404.4	88.49	2451.6	2540.0	0.3120	8.3311	8.6432
3.0	24.08	0.001 003	45.67	101.04	2307.5	2408.5	101.05	2444.5	2545.5	0.3545	8.2231	8.5776
4.0	28.96	0.001 004	34.80	121.45	2293.7	2415.2	121.46	2432.9	2554.4	0.4226	8.0520	8.4746
5.0	32.88	0.001 005	28.19	137.81	2282.7	2420.5	137.82	2423.7	2561.5	0.4764	7.9187	8.3951
7.5	40.29	0.001 008	19.24	168.78	2261.7	2430.5	168.79	2406.0	2574.8	0.5764	7.6750	8.2515
10	45.81	0.001 010	14.67	191.82	2246.1	2437.9	191.83	2392.8	2584.7	0.6493	7.5009	8.1502
15	53.97	0.001 014	10.02	225.92	2222.8	2448.7	225.94	2373.1	2599.1	0.7549	7.2536	8.0085
20	60.06	0.001 017	7.649	251.38	2205.4	2456.7	251.40	2358.3	2609.7	0.8320	7.0766	7.9085
25	64.97	0.001 020	6.204	271.90	2191.2	2463.1	271.93	2346.3	2618.2	0.8931	6.9383	7.8314
30	69.10	0.001 022	5.229	289.20	2179.2	2468.4	289.23	2336.1	2625.3	0.9439	6.8247	7.7686
40	75.87	0.001 027	3.993	317.53	2159.5	2477.0	317.58	2319.2	2636.8	1.0259	6.6441	7.6700
50	81.33	0.001 030	3.240	340.44	2143.4	2483.9	340.49	2305.4	2645.9	1.0910	6.5029	7.5939
75	91.78	0.001 037	2.217	384.31	2112.4	2496.7	384.39	2278.6	2663.0	1.2130	6.2434	7.4564
MPa												
0.100	99.63	0.001 043	1.6940	417.36	2088.7	2506.1	417.46	2258.0	2675.5	1.3026	6.0568	7.3594
0.125	105.99	0.001 048	1.3749	444.19	2069.3	2513.5	444.32	2241.0	2685.4	1.3740	5.9104	7.2844
0.150	111.37	0.001 053	1.1593	466.94	2052.7	2519.7	467.11	2226.5	2693.6	1.4336	5.7897	7.2233
0.175	116.06	0.001 057	1.0036	486.80	2038.1	2524.9	486.99	2213.6	2700.6	1.4849	5.6868	7.1717
0.200	120.23	0.001 061	0.8857	504.49	2025.0	2529.5	504.70	2201.9	2706.7	1.5301	5.5970	7.1271
0.225	124.00	0.001 064	0.7933	520.47	2013.1	2533.6	520.72	2191.3	2712.1	1.5706	5.5173	7.0878
0.250	127.44	0.001 067	0.7187	535.10	2002.1	2537.2	535.37	2181.5	2716.9	1.6072	5.4455	7.0527
0.275	130.60	0.001 070	0.6573	548.59	1991.9	2540.5	548.89	2172.4	2721.3	1.6408	5.3801	7.0209
0.300	133.55	0.001 073	0.6058	561.15	1982.4	2543.6	561.47	2163.8	2725.3	1.6718	5.3201	6.9919
0.325	136.30	0.001 076	0.5620	572.90	1973.5	2546.4	573.25	2155.8	2729.0	1.7006	5.2646	6.9652
0.350	138.88	0.001 079	0.5243	583.95	1965.0	2548.9	584.33	2148.1	2732.4	1.7275	5.2130	6.9405
0.375	141.32	0.001 081	0.4914	594.40	1956.9	2551.3	594.81	2140.8	2735.6	1.7528	5.1647	6.9175
0.40	143.63	0.001 084	0.4625	604.31	1949.3	2553.6	604.74	2133.8	2738.6	1.7766	5.1193	6.8959
0.45	147.93	0.001 088	0.4140	622.77	1934.9	2557.6	623.25	2120.7	2743.9	1.8207	5.0359	6.8565
0.50	151.86	0.001 093	0.3749	639.68	1921.6	2561.2	640.23	2108.5	2748.7	1.8607	4.9606	6.8213
0.55	155.48	0.001 097	0.3427	655.32	1909.2	2564.5	655.93	2097.0	2753.0	1.8973	4.8920	6.7893
0.60	158.85	0.001 101	0.3157	669.90	1897.5	2567.4	670.56	2086.3	2756.8	1.9312	4.8288	6.7600
0.65	162.01	0.001 104	0.2927	683.56	1886.5	2570.1	684.28	2076.0	2760.3	1.9627	4.7703	6.7331
0.70	164.97	0.001 108	0.2729	696.44	1876.1	2572.5	697.22	2066.3	2763.5	1.9922	4.7158	6.7080
0.75	167.78	0.001 112	0.2556	708.64	1866.1	2574.7	709.47	2057.0	2766.4	2.0200	4.6647	6.6847
0.80	170.43	0.001 115	0.2404	720.22	1856.6	2576.8	721.11	2048.0	2769.1	2.0462	4.6166	6.6628
0.85	172.96	0.001 118	0.2270	731.27	1847.4	2578.7	732.22	2039.4	2771.6	2.0710	4.5711	6.6421
0.90	175.38	0.001 121	0.2150	741.83	1838.6	2580.5	742.83	2031.1	2773.9	2.0946	4.5280	6.6226
0.95	177.69	0.001 124	0.2042	751.95	1830.2	2582.1	753.02	2023.1	2776.1	2.1172	4.4869	6.6041
1.00	179.91	0.001 127	0.194 44	761.68	1822.0	2583.6	762.81	2015.3	2778.1	2.1387	4.4478	6.5865
1.10	184.09	0.001 133	0.177 53	780.09	1806.3	2586.4	781.34	2000.4	2781.7	2.1792	4.3744	6.5536
1.20	187.99	0.001 139	0.163 33	797.29	1791.5	2588.8	798.65	1986.2	2784.8	2.2166	4.3067	6.5233
1.30	191.64	0.001 144	0.151 25	813.44	1777.5	2591.0	814.93	1972.7	2787.6	2.2515	4.2438	6.4953
1.40	195.07	0.001 149	0.140 84	828.70	1764.1	2592.8	830.30	1959.7	2790.0	2.2842	4.1850	6.4693
1.50	198.32	0.001 154	0.131 77	843.16	1751.3	2594.5	844.89	1947.3	2792.2	2.3150	4.1298	6.4448

(Continued)

TABLE A-11   (Continued)

Press. MPa P	Temp. °C T	Specific Volume m³/kg Sat. Liquid v_f	Sat. Vapor v_g	Internal Energy kJ/kg Sat. Liquid u_f	Evap. u_{fg}	Sat. Vapor u_g	Enthalpy kJ/kg Sat. Liquid h_f	Evap. h_{fg}	Sat. Vapor h_g	Entropy kJ/kg·K Sat. Liquid s_f	Evap. s_{fg}	Sat. Vapor s_g
1.75	205.76	0.001 166	0.113 49	876.46	1721.4	2597.8	878.50	1917.9	2796.4	2.3851	4.0044	6.3896
2.00	212.42	0.001 177	0.099 63	906.44	1693.8	2600.3	908.79	1890.7	2799.5	2.4474	3.8935	6.3409
2.25	218.45	0.001 187	0.088 75	933.83	1668.2	2602.0	936.49	1865.2	2801.7	2.5035	3.7937	6.2972
2.5	223.99	0.001 197	0.079 98	959.11	1644.0	2603.1	962.11	1841.0	2803.1	2.5547	3.7028	6.2575
3.0	233.90	0.001 217	0.066 68	1004.78	1599.3	2604.1	1008.42	1795.7	2804.2	2.6457	3.5412	6.1869
3.5	242.60	0.001 235	0.057 07	1045.43	1558.3	2603.7	1049.75	1753.7	2803.4	2.7253	3.4000	6.1253
4	250.40	0.001 252	0.049 78	1082.31	1520.0	2602.3	1087.31	1714.1	2801.4	2.7964	3.2737	6.0701
5	263.99	0.001 286	0.039 44	1147.81	1449.3	2597.1	1154.23	1640.1	2794.3	2.9202	3.0532	5.9734
6	275.64	0.001 319	0.032 44	1205.44	1384.3	2589.7	1213.35	1571.0	2784.3	3.0267	2.8625	5.8892
7	285.88	0.001 351	0.027 37	1257.55	1323.0	2580.5	1267.00	1505.1	2772.1	3.1211	2.6922	5.8133
8	295.06	0.001 384	0.023 52	1305.57	1264.2	2569.8	1316.64	1441.3	2758.0	3.2068	2.5364	5.7432
9	303.40	0.001 418	0.020 48	1350.51	1207.3	2557.8	1363.26	1378.9	2742.1	3.2858	2.3915	5.6772
10	311.06	0.001 452	0.018 026	1393.04	1151.4	2544.4	1407.56	1317.1	2724.7	3.3596	2.2544	5.6141
11	318.15	0.001 489	0.015 987	1433.7	1096.0	2529.8	1450.1	1255.5	2705.6	3.4295	2.1233	5.5527
12	324.75	0.001 527	0.014 263	1473.0	1040.7	2513.7	1491.3	1193.6	2684.9	3.4962	1.9962	5.4924
13	330.93	0.001 567	0.012 780	1511.1	985.0	2496.1	1531.5	1130.7	2662.2	3.5606	1.8718	5.4323
14	336.75	0.001 611	0.011 485	1548.6	928.2	2476.8	1571.1	1066.5	2637.6	3.6232	1.7485	5.3717
15	342.24	0.001 658	0.010 337	1585.6	869.8	2455.5	1610.5	1000.0	2610.5	3.6848	1.6249	5.3098
16	347.44	0.001 711	0.009 306	1622.7	809.0	2431.7	1650.1	930.6	2580.6	3.7461	1.4994	5.2455
17	352.37	0.001 770	0.008 364	1660.2	744.8	2405.0	1690.3	856.9	2547.2	3.8079	1.3698	5.1777
18	357.06	0.001 840	0.007 489	1698.9	675.4	2374.3	1732.0	777.1	2509.1	3.8715	1.2329	5.1044
19	361.54	0.001 924	0.006 657	1739.9	598.1	2338.1	1776.5	688.0	2464.5	3.9388	1.0839	5.0228
20	365.81	0.002 036	0.005 834	1785.6	507.5	2293.0	1826.3	583.4	2409.7	4.0139	0.9130	4.9269
21	369.89	0.002 207	0.004 952	1842.1	388.5	2230.6	1888.4	446.2	2334.6	4.1075	0.6938	4.8013
22	373.80	0.002 742	0.003 568	1961.9	125.2	2087.1	2022.2	143.4	2165.6	4.3110	0.2216	4.5327
22.09	374.14	0.003 155	0.003 155	2029.6	0	2029.6	2099.3	0	2099.3	4.4298	0	4.4298

TABLE A-12   Thermodynamic Properties of Steam (Superheated Vapor)

T °C	v m³/kg	u kJ/kg	h kJ/kg	s kJ/kg·K	v m³/kg	u kJ/kg	h kJ/kg	s kJ/kg·K	v m³/kg	u kJ/kg	h kJ/kg	s kJ/kg·K
	$P = 0.010$ MPa ($T_{sat} = 45.81$°C)				$P = 0.050$ MPa ($T_{sat} = 81.33$°C)				$P = 0.10$ MPa ($T_{sat} = 99.63$°C)			
Sat.	14.674	2437.9	2584.7	8.1502	3.240	2483.9	2645.9	7.5939	1.6940	2506.1	2675.5	7.3594
50	14.869	2443.9	2592.6	8.1749								
100	17.196	2515.5	2687.5	8.4479	3.418	2511.6	2682.5	7.6947	1.6958	2506.7	2676.2	7.3614
150	19.512	2587.9	2783.0	8.6882	3.889	2585.6	2780.1	7.9401	1.9364	2582.8	2776.4	7.6134
200	21.825	2661.3	2879.5	8.9038	4.356	2659.9	2877.7	8.1580	2.172	2658.1	2875.3	7.8343
250	24.136	2736.0	2977.3	9.1002	4.820	2735.0	2976.0	8.3556	2.406	2733.7	2974.3	8.0333
300	26.445	2812.1	3076.5	9.2813	5.284	2811.3	3075.5	8.5373	2.639	2810.4	3074.3	8.2158
400	31.063	2968.9	3279.6	9.6077	6.209	2968.5	3278.9	8.8642	3.103	2967.9	3278.2	8.5435
500	35.679	3132.3	3489.1	9.8978	7.134	3132.0	3488.7	9.1546	3.565	3131.6	3488.1	8.8342
600	40.295	3302.5	3705.4	10.1608	8.057	3302.2	3705.1	9.4178	4.028	3301.9	3704.7	9.0976
700	44.911	3479.6	3928.7	10.4028	8.981	3479.4	3928.5	9.6599	4.490	3479.2	3928.2	9.3398
800	49.526	3663.8	4159.0	10.6281	9.904	3663.6	4158.9	9.8852	4.952	3663.5	4158.6	9.5652
900	54.141	3855.0	4396.4	10.8396	10.828	3854.9	4396.3	10.0967	5.414	3854.8	4396.1	9.7767
1000	58.757	4053.0	4640.6	11.0393	11.751	4052.9	4640.5	10.2964	5.875	4052.8	4640.3	9.9764
1100	63.372	4257.5	4891.2	11.2287	12.674	4257.4	4891.1	10.4859	6.337	4257.3	4891.0	10.1659
1200	67.987	4467.9	5147.8	11.4091	13.597	4467.8	5147.7	10.6662	6.799	4467.7	5147.6	10.3463
1300	72.602	4683.7	5409.7	11.5811	14.521	4683.6	5409.6	10.8382	7.260	4683.5	5409.5	10.5183

(Continued)

TABLE A-12 (Continued)

T °C	v m³/kg	u kJ/kg	h kJ/kg	s kJ/kg·K	v m³/kg	u kJ/kg	h kJ/kg	s kJ/kg·K	v m³/kg	u kJ/kg	h kJ/kg	s kJ/kg·K
	P = 0.20 MPa (T_{sat} = 120.23°C)				**P = 0.30 MPa (T_{sat} = 133.55°C)**				**P = 0.40 MPa (T_{sat} = 143.63°C)**			
Sat.	0.8857	2529.5	2706.7	7.1272	0.6058	2543.6	2725.3	6.9919	0.4625	2553.6	2738.6	6.8959
150	0.9596	2576.9	2768.8	7.2795	0.6339	2570.8	2761.0	7.0778	0.4708	2564.5	2752.8	6.9299
200	1.0803	2654.4	2870.5	7.5066	0.7163	2650.7	2865.6	7.3115	0.5342	2646.8	2860.5	7.1706
250	1.1988	2731.2	2971.0	7.7086	0.7964	2728.7	2967.6	7.5166	0.5951	2726.1	2964.2	7.3789
300	1.3162	2808.6	3071.8	7.8926	0.8753	2806.7	3069.3	7.7022	0.6548	2804.8	3066.8	7.5662
400	1.5493	2966.7	3276.6	8.2218	1.0315	2965.6	3275.0	8.0330	0.7726	2964.4	3273.4	7.8985
500	1.7814	3130.8	3487.1	8.5133	1.1867	3130.0	3486.0	8.3251	0.8893	3129.2	3484.9	8.1913
600	2.013	3301.4	3704.0	8.7770	1.3414	3300.8	3703.2	8.5892	1.0055	3300.2	3702.4	8.4558
700	2.244	3478.8	3927.6	9.0194	1.4957	3478.4	3927.1	8.8319	1.1215	3477.9	3926.5	8.6987
800	2.475	3663.1	4158.2	9.2449	1.6499	3662.9	4157.8	9.0576	1.2372	3662.4	4157.3	8.9244
900	2.706	3854.5	4395.8	9.4566	1.8041	3854.2	4395.4	9.2692	1.3529	3853.9	4395.1	9.1362
1000	2.937	4052.5	4640.0	9.6563	1.9581	4052.3	4639.7	9.4690	1.4685	4052.0	4639.4	9.3360
1100	3.168	4257.0	4890.7	9.8458	2.1121	4256.8	4890.4	9.6585	1.5840	4256.5	4890.2	9.5256
1200	3.399	4467.5	5147.3	10.0262	2.2661	4467.2	5147.1	9.8389	1.6996	4467.0	5146.8	9.7060
1300	3.630	4683.2	5409.3	10.1982	2.4201	4683.0	5409.0	10.0110	1.8151	4682.8	5408.8	9.8780
	P = 0.50 MPa (T_{sat} = 151.86°C)				**P = 0.60 MPa (T_{sat} = 158.85°C)**				**P = 0.80 MPa (T_{sat} = 170.43°C)**			
Sat.	0.3749	2561.2	2748.7	6.8213	0.3157	2567.4	2756.8	6.7600	0.2404	2576.8	2769.1	6.6628
200	0.4249	2642.9	2855.4	7.0592	0.3520	2638.9	2850.1	6.9665	0.2608	2630.6	2839.3	6.8158
250	0.4744	2723.5	2960.7	7.2709	0.3938	2720.9	2957.2	7.1816	0.2931	2715.5	2950.0	7.0384
300	0.5226	2802.9	3064.2	7.4599	0.4344	2801.0	3061.6	7.3724	0.3241	2797.2	3056.5	7.2328
350	0.5701	2882.6	3167.7	7.6329	0.4742	2881.2	3165.7	7.5464	0.3544	2878.2	3161.7	7.4089
400	0.6173	2963.2	3271.9	7.7938	0.5137	2962.1	3270.3	7.7079	0.3843	2959.7	3267.1	7.5716
500	0.7109	3128.4	3483.9	8.0873	0.5920	3127.6	3482.8	8.0021	0.4433	3126.0	3480.6	7.8673
600	0.8041	3299.6	3701.7	7.3522	0.6697	3299.1	3700.9	8.2674	0.5018	3297.9	3699.4	8.1333
700	0.8969	3477.5	3925.9	8.5952	0.7472	3477.0	3925.3	8.5107	0.5601	3476.2	3924.2	8.3770
800	0.9896	3662.1	4156.9	8.8211	0.8245	3661.8	4156.5	8.7367	0.6181	3661.1	4155.6	8.6033
900	1.0822	3853.6	4394.7	9.0329	0.9017	3853.4	4394.4	8.9486	0.6761	3852.8	4393.7	8.8153
1000	1.1747	4051.8	4639.1	9.2328	0.9788	4051.5	4638.8	9.1485	0.7340	4051.0	4638.2	9.0153
1100	1.2672	4256.3	4889.9	9.4224	1.0559	4256.1	4889.6	9.3381	0.7919	4255.6	4889.1	9.2050
1200	1.3596	4466.8	5146.6	9.6029	1.1330	4466.5	5146.3	9.5185	0.8497	4466.1	5145.9	9.3855
1300	1.4521	4682.5	5408.6	9.7749	1.2101	4682.3	5408.3	9.6906	0.9076	4681.8	5407.9	9.5575
	P = 1.00 MPa (T_{sat} = 179.91°C)				**P = 1.20 MPa (T_{sat} = 187.99°C)**				**P = 1.40 MPa (T_{sat} = 195.07°C)**			
Sat.	0.194 44	2583.6	2778.1	6.5865	0.163 33	2588.8	2784.8	6.5233	0.140 84	2592.8	2790.0	6.4693
200	0.2060	2621.9	2827.9	6.6940	0.169 30	2612.8	2815.9	6.5898	0.143 02	2603.1	2803.3	6.4975
250	0.2327	2709.9	2942.6	6.9247	0.192 34	2704.2	2935.0	6.8294	0.163 50	2698.3	2927.2	6.7467
300	0.2579	2793.2	3051.2	7.1229	0.2138	2789.2	3045.8	7.0317	0.182 28	2785.2	3040.4	6.9534
350	0.2825	2875.2	3157.7	7.3011	0.2345	2872.2	3153.6	7.2121	0.2003	2869.2	3149.5	7.1360
400	0.3066	2957.3	3263.9	7.4651	0.2548	2954.9	3260.7	7.3774	0.2178	2952.5	3257.5	7.3026
500	0.3541	3124.4	3478.5	7.7622	0.2946	3122.8	3476.3	7.6759	0.2521	3121.1	3474.1	7.6027
600	0.4011	3296.8	3697.9	8.0290	0.3339	3295.6	3696.3	7.9435	0.2860	3294.4	3694.8	7.8710
700	0.4478	3475.3	3923.1	8.2731	0.3729	3474.4	3922.0	8.1881	0.3195	3473.6	3920.8	8.1160
800	0.4943	3660.4	4154.7	8.4996	0.4118	3659.7	4153.8	8.4148	0.3528	3659.0	4153.0	8.3431
900	0.5407	3852.2	4392.9	8.7118	0.4505	3851.6	4392.2	8.6272	0.3861	3851.1	4391.5	8.5556
1000	0.5871	4050.5	4637.6	8.9119	0.4892	4050.0	4637.0	8.8274	0.4192	4049.5	4636.4	8.7559
1100	0.6335	4255.1	4888.6	9.1017	0.5278	4254.6	4888.0	9.0172	0.4524	4254.1	4887.5	8.9457
1200	0.6798	4465.6	5145.4	9.2822	0.5665	4465.1	5144.9	9.1977	0.4855	4464.7	5144.4	9.1262
1300	0.7261	4681.3	5407.4	9.4543	0.6051	4680.9	5407.0	9.3698	0.5186	4680.4	5406.5	9.2984

(Continued)

TABLE A-12 (Continued)

T °C	v m³/kg	u kJ/kg	h kJ/kg	s kJ/kg·K	v m³/kg	u kJ/kg	h kJ/kg	s kJ/kg·K	v m³/kg	u kJ/kg	h kJ/kg	s kJ/kg·K
	$P = 1.60$ MPa ($T_{sat} = 201.41°$C)				$P = 1.80$ MPa ($T_{sat} = 207.15°$C)				$P = 2.00$ MPa ($T_{sat} = 212.42°$C)			
Sat.	0.123 80	2596.0	2794.0	6.4218	0.110 42	2598.4	2797.1	6.3794	0.099 63	2600.3	2799.5	6.3409
225	0.132 87	2644.7	2857.3	6.5518	0.116 73	2636.6	2846.7	6.4808	0.103 77	2628.3	2835.8	6.4147
250	0.141 84	2692.3	2919.2	6.6732	0.124 97	2686.0	2911.0	6.6066	0.111 44	2679.6	2902.5	6.5453
300	0.158 62	2781.1	3034.8	6.8844	0.140 21	2776.9	3029.2	6.8226	0.125 47	2772.6	3023.5	6.7664
350	0.174 56	2866.1	3145.4	7.0694	0.154 57	2863.0	3141.2	7.0100	0.138 57	2859.8	3137.0	6.9563
400	0.190 05	2950.1	3254.2	7.2374	0.168 47	2947.7	3250.9	7.1794	0.151 20	2945.2	3247.6	7.1271
500	0.2203	3119.5	3472.0	7.5390	0.195 50	3117.9	3469.8	7.4825	0.175 68	3116.2	3467.6	7.4317
600	0.2500	3293.3	3693.2	7.8080	0.2220	3292.1	3691.7	7.7523	0.199 60	3290.9	3690.1	7.7024
700	0.2794	3472.7	3919.7	8.0535	0.2482	3471.8	3918.5	7.9983	0.2232	3470.9	3917.4	7.9487
800	0.3086	3658.3	4152.1	8.2808	0.2742	3657.6	4151.2	8.2258	0.2467	3657.0	4150.3	8.1765
900	0.3377	3850.5	4390.8	8.4935	0.3001	3849.9	4390.1	8.4386	0.2700	3849.3	4389.4	8.3895
1000	0.3668	4049.0	4635.8	8.6938	0.3260	4048.5	4635.2	8.6391	0.2933	4048.0	4634.6	8.5901
1100	0.3958	4253.7	4887.0	8.8837	0.3518	4253.2	4886.4	8.8290	0.3166	4252.7	4885.9	8.7800
1200	0.4248	4464.2	5143.9	9.0643	0.3776	4463.7	5143.4	9.0096	0.3398	4463.3	5142.9	8.9607
1300	0.4538	4679.9	5406.0	9.2364	0.4034	4679.5	5405.6	9.1818	0.3631	4679.0	5405.1	9.1329
	$P = 2.50$ MPa ($T_{sat} = 223.99°$C)				$P = 3.00$ MPa ($T_{sat} = 233.90°$C)				$P = 3.50$ MPa ($T_{sat} = 242.60°$C)			
Sat.	0.079 98	2603.1	2803.1	6.2575	0.066 68	2604.1	2804.2	6.1869	0.057 07	2603.7	2803.4	6.1253
225	0.080 27	2605.6	2806.3	6.2639								
250	0.087 00	2662.6	2880.1	6.4085	0.070 58	2644.0	2855.8	6.2872	0.058 72	2623.7	2829.2	6.1749
300	0.098 90	2761.6	3008.8	6.6438	0.081 14	2750.1	2993.5	6.5390	0.068 42	2738.0	2977.5	6.4461
350	0.109 76	2851.9	3126.3	6.8403	0.090 53	2843.7	3115.3	6.7428	0.076 78	2835.3	3104.0	6.6579
400	0.120 10	2939.1	3239.3	7.0148	0.099 36	2932.8	3230.9	6.9212	0.084 53	2926.4	3222.3	6.8405
450	0.130 14	3025.5	3350.8	7.1746	0.107 87	3020.4	3344.0	7.0834	0.091 96	3015.3	3337.2	7.0052
500	0.139 98	3112.1	3462.1	7.3234	0.116 19	3108.0	3456.5	7.2338	0.099 18	3103.0	3450.9	7.1572
600	0.159 30	3288.0	3686.3	7.5960	0.132 43	3285.0	3682.3	7.5085	0.113 24	3282.1	3678.4	7.4339
700	0.178 32	3468.7	3914.5	7.8435	0.148 38	3466.5	3911.7	7.7571	0.126 99	3464.3	3908.8	7.6837
800	0.197 16	3655.3	4148.2	8.0720	0.164 14	3653.5	4145.9	7.9862	0.140 56	3651.8	4143.7	7.9134
900	0.215 90	3847.9	4387.6	8.2853	0.179 80	3846.5	4385.9	8.1999	0.154 02	3845.0	4384.1	8.1276
1000	0.2346	4046.7	4633.1	8.4861	0.195 41	4045.4	4631.6	8.4009	0.167 43	4044.1	4630.1	8.3288
1100	0.2532	4251.5	4884.6	8.6762	0.210 98	4250.3	4883.3	8.5912	0.180 80	4249.2	4881.9	8.5192
1200	0.2718	4462.1	5141.7	8.8569	0.226 52	4460.9	5140.5	8.7720	0.194 15	4459.8	5139.3	8.7000
1300	0.2905	4677.8	5404.0	9.0291	0.242 06	4676.6	5402.8	8.9442	0.207 49	4675.5	5401.7	8.8723
	$P = 4.0$ MPa ($T_{sat} = 250.40°$C)				$P = 4.5$ MPa ($T_{sat} = 257.49°$C)				$P = 5.0$ MPa ($T_{sat} = 263.99°$C)			
Sat.	0.049 78	2602.3	2801.4	6.0701	0.044 06	2600.1	2798.3	6.0198	0.039 44	2597.1	2794.3	5.9734
275	0.054 57	2667.9	2886.2	6.2285	0.047 30	2650.3	2863.2	6.1401	0.041 41	2631.3	2838.3	6.0544
300	0.058 84	2725.3	2960.7	6.3615	0.051 35	2712.0	2943.1	6.2828	0.045 32	2698.0	2924.5	6.2084
350	0.066 45	2826.7	3092.5	6.5821	0.058 40	2817.8	3080.6	6.5131	0.051 94	2808.7	3068.4	6.4493
400	0.073 41	2919.9	3213.6	6.7690	0.064 75	2913.3	3204.7	6.7047	0.057 81	2906.6	3195.7	6.6459
450	0.080 02	3010.2	3330.3	6.9363	0.070 74	3005.0	3323.3	6.8746	0.063 30	2999.7	3316.2	6.8186
500	0.086 43	3099.5	3445.3	7.0901	0.076 51	3095.3	3439.6	7.0301	0.068 57	3091.0	3433.8	6.9759
600	0.098 85	3279.1	3674.4	7.3688	0.087 65	3276.0	3670.5	7.3110	0.078 69	3273.0	3666.5	7.2589
700	0.110 95	3462.1	3905.9	7.6198	0.098 47	3459.9	3903.0	7.5631	0.088 49	3457.6	3900.1	7.5122
800	0.122 87	3650.0	4141.5	7.8502	0.109 11	3648.3	4139.3	7.7942	0.098 11	3646.6	4137.1	7.7440
900	0.134 69	3843.6	4382.3	8.0647	0.119 65	3842.2	4380.6	8.0091	0.107 62	3840.7	4378.8	7.9593
1000	0.146 45	4042.9	4628.7	8.2662	0.130 13	4041.6	4627.2	8.2108	0.117 07	4040.4	4625.7	8.1612
1100	0.158 17	4248.0	4880.6	8.4567	0.140 56	4246.8	4879.3	8.4015	0.126 48	4245.6	4878.0	8.3520
1200	0.169 87	4458.6	5138.1	8.6376	0.150 98	4457.5	5136.9	8.5825	0.135 87	4456.3	5135.7	8.5331
1300	0.181 56	4674.3	5400.5	8.8100	0.161 39	4673.1	5399.4	8.7549	0.145 26	4672.0	5398.2	8.7055

(Continued)

TABLE A-12   (Continued)

T °C	v m³/kg	u kJ/kg	h kJ/kg	s kJ/kg·K	v m³/kg	u kJ/kg	h kJ/kg	s kJ/kg·K	v m³/kg	u kJ/kg	h kJ/kg	s kJ/kg·K
	P = 6.0 MPa (T_{sat} = 275.64°C)				P = 7.0 MPa (T_{sat} = 285.88°C)				P = 8.0 MPa (T_{sat} = 295.06°C)			
Sat.	0.032 44	2589.7	2784.3	5.8892	0.027 37	2580.5	2772.1	5.8133	0.023 52	2569.8	2758.0	5.7432
300	0.036 16	2667.2	2884.2	6.0674	0.029 47	2632.2	2838.4	5.9305	0.024 26	2590.9	2785.0	5.7906
350	0.042 23	2789.6	3043.0	6.3335	0.035 24	2769.4	3016.0	6.2283	0.029 95	2747.7	2987.3	6.1301
400	0.047 39	2892.9	3177.2	6.5408	0.039 93	2878.6	3158.1	6.4478	0.034 32	2863.8	3138.3	6.3634
450	0.052 14	2988.9	3301.8	6.7193	0.044 16	2978.0	3287.1	6.6327	0.038 17	2966.7	3272.0.	6.5551
500	0.056 65	3082.2	3422.2	6.8803	0.048 14	3073.4	3410.3	6.7975	0.041 75	3064.3	3398.3	6.7240
550	0.061 01	3174.6	3540.6	7.0288	0.051 95	3167.2	3530.9	6.9486	0.045 16	3159.8	3521.0	6.8778
600	0.065 25	3266.9	3658.4	7.1677	0.055 65	3260.7	3650.3	7.0894	0.048 45	3254.4	3642.0	7.0206
700	0.073 52	3453.1	3894.2	7.4234	0.062 83	3448.5	3888.3	7.3476	0.054 81	3443.9	3882.4	7.2812
800	0.081 60	3643.1	4132.7	7.6566	0.069 81	3639.5	4128.2	7.5822	0.060 97	3636.0	4123.8	7.5173
900	0.089 58	3837.8	4375.3	7.8727	0.076 69	3835.0	4371.8	7.7991	0.067 02	3832.1	4368.3	7.7351
1000	0.097 49	4037.8	4622.7	8.0751	0.083 50	4035.3	4619.8	8.0020	0.073 01	4032.8	4616.9	7.9384
1100	0.105 36	4243.3	4875.4	8.2661	0.090 27	4240.9	4872.8	8.1933	0.078 96	4238.6	4870.3	8.1300
1200	0.113 21	4454.0	5133.3	8.4474	0.097 03	4451.7	5130.9	8.3747	0.084 89	4449.5	5128.5	8.3115
1300	0.121 06	4669.6	5396.0	8.6199	0.103 77	4667.3	5393.7	8.5473	0.090 80	4665.0	5391.5	8.4842
	P = 9.0 MPa (T_{sat} = 303.40°C)				P = 10.0 MPa (T_{sat} = 311.06°C)				P = 12.5 MPa (T_{sat} = 327.89°C)			
Sat.	0.020 48	2557.8	2742.1	5.6772	0.018 026	2544.4	2724.7	5.6141	0.013 495	2505.1	2673.8	5.4624
325	0.023 27	2646.6	2856.0	5.8712	0.019 861	2610.4	2809.1	5.7568				
350	0.025 80	2724.4	2956.6	6.0361	0.022 42	2699.2	2923.4	5.9443	0.016 126	2624.6	2826.2	5.7118
400	0.029 93	2848.4	3117.8	6.2854	0.026 41	2832.4	3096.5	6.2120	0.020 00	2789.3	3039.3	6.0417
450	0.033 50	2955.2	3256.6	6.4844	0.029 75	2943.4	3240.9	6.4190	0.022 99	2912.5	3199.8	6.2719
500	0.036 77	3055.2	3386.1	6.6576	0.032 79	3045.8	3373.7	6.5966	0.025 60	3021.7	3341.8	6.4618
550	0.039 87	3152.2	3511.0	6.8142	0.035 64	3144.6	3500.9	6.7561	0.028 01	3125.0	3475.2	6.6290
600	0.042 85	3248.1	3633.7	6.9589	0.038 37	3241.7	3625.3	6.9029	0.030 29	3225.4	3604.0	6.7810
650	0.045 74	3343.6	3755.3	7.0943	0.041 01	3338.2	3748.2	7.0398	0.032 48	3324.4	3730.4	6.9218
700	0.048 57	3439.3	3876.5	7.2221	0.043 58	3434.7	3870.5	7.1687	0.034 60	3422.9	3855.3	7.0536
800	0.054 09	3632.5	4119.3	7.4596	0.048 59	3628.9	4114.8	7.4077	0.038 69	3620.0	4103.6	7.2965
900	0.059 50	3829.2	4364.8	7.6783	0.053 49	3826.3	4361.2	7.6272	0.042 67	3819.1	4352.5	7.5182
1000	0.064 85	4030.3	4614.0	7.8821	0.058 32	4027.8	4611.0	7.8315	0.046 58	4021.6	4603.8	7.7237
1100	0.070 16	4236.3	4867.7	8.0740	0.063 12	4234.0	4865.1	8.0237	0.050 45	4228.2	4858.8	7.9165
1200	0.075 44	4447.2	5126.2	8.2556	0.067 89	4444.9	5123.8	8.2055	0.054 30	4439.3	5118.0	8.0987
1300	0.080 72	4662.7	5389.2	8.4284	0.072 65	4460.5	5387.0	8.3783	0.058 13	4654.8	5381.4	8.2717
	P = 15.0 MPa (T_{sat} = 342.24°C)				P = 17.5 MPa (T_{sat} = 354.75°C)				P = 20.0 MPa (T_{sat} = 365.81°C)			
Sat.	0.010 337	2455.5	2610.5	5.3098	0.007 920	2390.2	2528.8	5.1419	0.005 834	2293.0	2409.7	4.9269
350	0.011 470	2520.4	2692.4	5.4421								
400	0.015 649	2740.7	2975.5	5.8811	0.012 447	2685.0	2902.9	5.7213	0.009 942	2619.3	2818.1	5.5540
450	0.018 445	2879.5	3156.2	6.1404	0.015 174	2844.2	3109.7	6.0184	0.012 695	2806.2	3060.1	5.9017
500	0.020 80	2996.6	3308.6	6.3443	0.017 358	2970.3	3274.1	6.2383	0.014 768	2942.9	3238.2	6.1401
550	0.022 93	3104.7	3448.6	6.5199	0.019 288	3083.9	3421.4	6.4230	0.016 555	3062.4	3393.5	6.3348
600	0.024 91	3208.6	3582.3	6.6776	0.021 06	3191.5	3560.1	6.5866	0.018 178	3174.0	3537.6	6.5048
650	0.026 80	3310.3	3712.3	6.8224	0.022 74	3296.0	3693.9	6.7357	0.019 693	3281.4	3675.3	6.6582
700	0.028 61	3410.9	3840.1	6.9572	0.024 34	3398.7	3824.6	6.8736	0.021 13	3386.4	3809.0	6.7993
800	0.032 10	3610.9	4092.4	7.2040	0.027 38	3601.8	4081.1	7.1244	0.023 85	3592.7	4069.7	7.0544
900	0.035 46	3811.9	4343.8	7.4279	0.030 31	3804.7	4335.1	7.3507	0.026 45	3797.5	4326.4	7.2830
1000	0.038 75	4015.4	4596.6	7.6348	0.033 16	4009.3	4589.5	7.5589	0.028 97	4003.1	4582.5	7.4925
1100	0.042 00	4222.6	4852.6	7.8283	0.035 97	4216.9	4846.4	7.7531	0.031 45	4211.3	4840.2	7.6874
1200	0.045 23	4433.8	5112.3	8.0108	0.038 76	4428.3	5106.6	7.9360	0.033 91	4422.8	5101.0	7.8707
1300	0.048 45	4649.1	5376.0	8.1840	0.041 54	4643.5	5370.5	8.1093	0.036 36	4638.0	5365.1	8.0442

(Continued)

TABLE A-12 (Continued)

T °C	v m³/kg	u kJ/kg	h kJ/kg	s kJ/kg·K	v m³/kg	u kJ/kg	h kJ/kg	s kJ/kg·K	v m³/kg	u kJ/kg	h kJ/kg	s kJ/kg·K
	P = 25.0 MPa				**P = 30.0 MPa**				**P = 35.0 MPa**			
375	0.001 973 1	1798.7	1848.0	4.0320	0.001 789 2	1737.8	1791.5	3.9305	0.001 700 3	1702.9	1762.4	3.8722
400	0.006 004	2430.1	2580.2	5.1418	0.002 790	2067.4	2151.1	4.4728	0.002 100	1914.1	1987.6	4.2126
425	0.007 881	2609.2	2806.3	5.4723	0.005 303	2455.1	2614.2	5.1504	0.003 428	2253.4	2373.4	4.7747
450	0.009 162	2720.7	2949.7	5.6744	0.006 735	2619.3	2821.4	5.4424	0.004 961	2498.7	2672.4	5.1962
500	0.011 123	2884.3	3162.4	5.9592	0.008 678	2820.7	3081.1	5.7905	0.006 927	2751.9	2994.4	5.6282
550	0.012 724	3017.5	3335.6	6.1765	0.010 168	2970.3	3275.4	6.0342	0.008 345	2921.0	3213.0	5.9026
600	0.014 137	3137.9	3491.4	6.3602	0.011 446	3100.5	3443.9	6.2331	0.009 527	3062.0	3395.5	6.1179
650	0.015 433	3251.6	3637.4	6.5229	0.012 596	3221.0	3598.9	6.4058	0.010 575	3189.8	3559.9	6.3010
700	0.016 646	3361.3	3777.5	6.6707	0.013 661	3335.8	3745.6	6.5606	0.011 533	3309.8	3713.5	6.4631
800	0.018 912	3574.3	4047.1	6.9345	0.015 623	3555.5	4024.2	6.8332	0.013 278	3536.7	4001.5	6.7450
900	0.021 045	3783.0	4309.1	7.1680	0.017 448	3768.5	4291.9	7.0718	0.014 883	3754.0	4274.9	6.9886
1000	0.023 10	3990.9	4568.5	7.3802	0.019 196	3978.8	4554.7	7.2867	0.016 410	3966.7	4541.1	7.2064
1100	0.025 12	4200.2	4828.2	7.5765	0.020 903	4189.2	4816.3	7.4845	0.017 895	4178.3	4804.6	7.4057
1200	0.027 11	4412.0	5089.9	7.7605	0.022 589	4401.3	5079.0	7.6692	0.019 360	4390.7	5068.3	7.5910
1300	0.029 10	4626.9	5354.4	7.9342	0.024 266	4616.0	5344.0	7.8432	0.020 815	4605.1	5333.6	7.7653
	P = 40.0 MPa				**P = 50.0 MPa**				**P = 60.0 MPa**			
375	0.001 640 7	1677.1	1742.8	3.8290	0.001 559 4	1638.6	1716.6	3.7639	0.001 502 8	1609.4	1699.5	3.7141
400	0.001 907 7	1854.6	1930.9	4.1135	0.001 730 9	1788.1	1874.6	4.0031	0.001 633 5	1745.4	1843.4	3.9318
425	0.002 532	2096.9	2198.1	4.5029	0.002 007	1959.7	2060.0	4.2734	0.001 816 5	1892.7	2001.7	4.1626
450	0.003 693	2365.1	2512.8	4.9459	0.002 486	2159.6	2284.0	4.5884	0.002 085	2053.9	2179.0	4.4121
500	0.005 622	2678.4	2903.3	5.4700	0.003 892	2525.5	2720.1	5.1726	0.002 956	2390.6	2567.9	4.9321
550	0.006 984	2869.7	3149.1	5.7785	0.005 118	2763.6	3019.5	5.5485	0.003 956	2658.8	2896.2	5.3441
600	0.008 094	3022.6	3346.4	6.0114	0.006 112	2942.0	3247.6	5.8178	0.004 834	2861.1	3151.2	5.6452
650	0.009 063	3158.0	3520.6	6.2054	0.006 966	3093.5	3441.8	6.0342	0.005 595	3028.8	3364.5	5.8829
700	0.009 941	3283.6	3681.2	6.3750	0.007 727	3230.5	3616.8	6.2189	0.006 272	3177.2	3553.5	6.0824
800	0.011 523	3517.8	3978.7	6.6662	0.009 076	3479.8	3933.6	6.5290	0.007 459	3441.5	3889.1	6.4109
900	0.012 962	3739.4	4257.9	6.9150	0.010 283	3710.3	4224.4	6.7882	0.008 508	3681.0	4191.5	6.6805
1000	0.014 324	3954.6	4527.6	7.1356	0.011 411	3930.5	4501.1	7.0146	0.009 480	3906.4	4475.2	6.9127
1100	0.015 642	4167.4	4793.1	7.3364	0.012 496	4145.7	4770.5	7.2184	0.010 409	4124.1	4748.6	7.1195
1200	0.016 940	4380.1	5057.7	7.5224	0.013 561	4359.1	5037.2	7.4058	0.011 317	4338.2	5017.2	7.3083
1300	0.018 229	4594.3	5323.5	7.6969	0.014 616	4572.8	5303.6	7.5808	0.012 215	4551.4	5284.3	7.4837

TABLE A-13 Thermodynamic Properties of Compressed Liquid Water

T °C	v m³/kg	u kJ/kg	h kJ/kg	s kJ/kg·K	v m³/kg	u kJ/kg	h kJ/kg	s kJ/kg·K	v m³/kg	u kJ/kg	h kJ/kg	s kJ/kg·K
	P = 5 MPa (T_{sat} = 263.99°C)				**P = 10 MPa (T_{sat} = 311.06°C)**				**P = 15 MPa (T_{sat} = 342.24°C)**			
Sat.	0.001 285 9	1147.8	1154.2	2.9202	0.001 452 4	1393.0	1407.6	3.3596	0.001 658 1	1585.6	1610.5	3.6848
0	0.000 997 7	0.04	5.04	0.0001	0.000 995 2	0.09	10.04	0.0002	0.000 992 8	0.15	15.05	0.0004
20	0.000 999 5	83.65	88.65	0.2956	0.000 997 2	83.36	93.33	0.2945	0.000 995 0	83.06	97.99	0.2934
40	0.001 005 6	166.95	171.97	0.5705	0.001 003 4	166.35	176.38	0.5686	0.001 001 3	165.76	180.78	0.5666
60	0.001 014 9	250.23	255.30	0.8285	0.001 012 7	249.36	259.49	0.8258	0.001 010 5	248.51	263.67	0.8232
80	0.001 026 8	333.72	338.85	1.0720	0.001 024 5	332.59	342.83	1.0688	0.001 022 2	331.48	346.81	1.0656
100	0.001 041 0	417.52	422.72	1.3030	0.001 038 5	416.12	426.50	1.2992	0.001 036 1	414.74	430.28	1.2955
120	0.001 057 6	501.80	507.09	1.5233	0.001 054 9	500.08	510.64	1.5189	0.001 052 2	498.40	514.19	1.5145
140	0.001 076 8	586.76	592.15	1.7343	0.001 073 7	584.68	595.42	1.7292	0.001 070 7	582.66	598.72	1.7242
160	0.001 098 8	672.62	678.12	1.9375	0.001 095 3	670.13	681.08	1.9317	0.001 091 8	667.71	684.09	1.9260
180	0.001 124 0	759.63	765.25	2.1341	0.001 119 9	756.65	767.84	2.1275	0.001 115 9	753.76	770.50	2.1210
200	0.001 153 0	848.1	853.9	2.3255	0.001 148 0	844.5	856.0	2.3178	0.001 143 3	841.0	858.2	2.3104
220	0.001 186 6	938.4	944.4	2.5128	0.001 180 5	934.1	945.9	2.5039	0.001 174 8	929.9	947.5	2.4953

(Continued)

TABLE A-13   (Continued)

T °C	v m³/kg	u kJ/kg	h kJ/kg	s kJ/kg·K	v m³/kg	u kJ/kg	h kJ/kg	s kJ/kg·K	v m³/kg	u kJ/kg	h kJ/kg	s kJ/kg·K
	$P=5$ MPa ($T_{sat}=263.99$°C)				$P=10$ MPa ($T_{sat}=311.06$°C)				$P=15$ MPa ($T_{sat}=342.24$°C)			
240	0.001 226 4	1031.4	1037.5	2.6979	0.001 218 7	1026.0	1038.1	2.6872	0.001 211 4	1020.8	1039.0	2.6771
260	0.001 274 9	1127.9	1134.3	2.8830	0.001 264 5	1121.1	1133.7	2.8699	0.001 255 0	1114.6	1133.4	2.8576
280					0.001 321 6	1220.9	1234.1	3.0548	0.001 308 4	1212.5	1232.1	3.0393
300					0.001 397 2	1328.4	1342.3	3.2469	0.001 377 0	1316.6	1337.3	3.2260
320									0.001 472 4	1431.1	1453.2	3.4247
340									0.001 631 1	1567.5	1591.9	3.6546
	$P=20$ MPa ($T_{sat}=365.81$°C)				$P=30$ MPa				$P=50$ MPa			
Sat.	0.002 036	1785.6	1826.3	4.0139								
0	0.000 990 4	0.19	20.01	0.0004	0.000 985 6	0.25	29.82	0.0001	0.000 976 6	0.20	49.03	−0.0014
20	0.000 992 8	82.77	102.62	0.2923	0.000 988 6	82.17	111.84	0.2899	0.000 980 4	81.00	130.02	0.2848
40	0.000 999 2	165.17	185.16	0.5646	0.000 995 1	164.04	193.89	0.5607	0.000 987 2	161.86	211.21	0.5527
60	0.001 008 4	247.68	267.85	0.8206	0.001 004 2	246.06	276.19	0.8154	0.000 996 2	242.98	292.79	0.8052
80	0.001 019 9	330.40	350.80	1.0624	0.001 015 6	328.30	358.77	1.0561	0.001 007 3	324.34	374.70	1.0440
100	0.001 033 7	413.39	434.06	1.2917	0.001 029 0	410.78	441.66	1.2844	0.001 020 1	405.88	456.89	1.2703
120	0.001 049 6	496.76	517.76	1.5102	0.001 044 5	493.59	524.93	1.5018	0.001 034 8	487.65	539.39	1.4857
140	0.001 067 8	580.69	602.04	1.7193	0.001 062 1	576.88	608.75	1.7098	0.001 051 5	569.77	622.35	1.6915
160	0.001 088 5	665.35	687.12	1.9204	0.001 082 1	660.82	693.28	1.9096	0.001 070 3	652.41	705.92	1.8891
180	0.001 112 0	750.95	773.20	2.1147	0.001 104 7	745.59	778.73	2.1024	0.001 091 2	735.69	790.25	2.0794
200	0.001 138 8	837.7	860.5	2.3031	0.001 130 2	831.4	865.3	2.2893	0.001 114 6	819.7	875.5	2.2634
220	0.001 169 3	925.9	949.3	2.4870	0.001 159 0	918.3	953.1	2.4711	0.001 140 8	904.7	961.7	2.4419
240	0.001 204 6	1016.0	1040.0	2.6674	0.001 192 0	1006.9	1042.6	2.6490	0.001 170 2	990.7	1049.2	2.6158
260	0.001 246 2	1108.6	1133.5	2.8459	0.001 230 3	1097.4	1134.3	2.8243	0.001 203 4	1078.1	1138.2	2.7860
280	0.001 296 5	1204.7	1230.6	3.0248	0.001 275 5	1190.7	1229.0	2.9986	0.001 241 5	1167.2	1229.3	2.9537
300	0.001 359 6	1306.1	1333.3	3.2071	0.001 330 4	1287.9	1327.8	3.1741	0.001 286 0	1258.7	1323.0	3.1200
320	0.001 443 7	1415.7	1444.6	3.3979	0.001 399 7	1390.7	1432.7	3.3539	0.001 338 8	1353.3	1420.2	3.2868
340	0.001 568 4	1539.7	1571.0	3.6075	0.001 492 0	1501.7	1546.5	3.5426	0.001 403 2	1452.0	1522.1	3.4557
360	0.001 822 6	1702.8	1739.3	3.8772	0.001 626 5	1626.6	1675.4	3.7494	0.001 483 8	1556.0	1630.2	3.6291
380					0.001 869 1	1781.4	1837.5	4.0012	0.001 588 4	1667.2	1746.6	3.8101

TABLE A-14   Thermodynamic Properties of Saturated Refrigerant 134a (Temperature Table)

Temp. T °C	Press. P_{sat} MPa	Specific volume, m³/kg		Internal energy, kJ/kg		Enthalpy, kJ/kg			Entropy, kJ/kg·K	
		Sat. liquid v_f	Sat. vapor v_g	Sat. liquid u_f	Sat. vapor u_g	Sat. liquid h_f	Evap. h_{fg}	Sat. vapor h_g	Sat. liquid s_f	Sat. vapor s_g
−40	0.05164	0.0007055	0.3569	−0.04	204.45	0.00	222.88	222.88	0.0000	0.9560
−36	0.06332	0.0007113	0.2947	4.68	206.73	4.73	220.67	225.40	0.0201	0.9506
−32	0.07704	0.0007172	0.2451	9.47	209.01	9.52	218.37	227.90	0.0401	0.9456
−28	0.09305	0.0007233	0.2052	14.31	211.29	14.37	216.01	230.38	0.0600	0.9411
−26	0.10199	0.0007265	0.1882	16.75	212.43	16.82	214.80	231.62	0.0699	0.9390
−24	0.11160	0.0007296	0.1728	19.21	213.57	19.29	213.57	232.85	0.0798	0.9370
−22	0.12192	0.0007328	0.1590	21.68	214.70	21.77	212.32	234.08	0.0897	0.9351
−20	0.13299	0.0007361	0.1464	24.17	215.84	24.26	211.05	235.31	0.0996	0.9332
−18	0.14483	0.0007395	0.1350	26.67	216.97	26.77	209.76	236.53	0.1094	0.9315
−16	0.15748	0.0007428	0.1247	29.18	218.10	29.30	208.45	237.74	0.1192	0.9298
−12	0.18540	0.0007498	0.1068	34.25	220.36	34.39	205.77	240.15	0.1388	0.9267
−8	0.21704	0.0007569	0.0919	39.38	222.60	39.54	203.00	242.54	0.1583	0.9239
−4	0.25274	0.0007644	0.0794	44.56	224.84	44.75	200.15	244.90	0.1777	0.9213

(Continued)

TABLE A-14 (Continued)

Temp. T °C	Press. P_{sat} MPa	Specific volume, m³/kg		Internal energy, kJ/kg		Enthalpy, kJ/kg			Entropy, kJ/kg · K	
		Sat. liquid v_f	Sat. vapor v_g	Sat. liquid u_f	Sat. vapor u_g	Sat. liquid h_f	Evap. h_{fg}	Sat. vapor h_g	Sat. liquid s_f	Sat. vapor s_g
0	0.29282	0.0007721	0.0689	49.79	227.06	50.02	197.21	247.23	0.1970	0.9190
4	0.33765	0.0007801	0.0600	55.08	229.27	55.35	194.19	249.53	0.2162	0.9169
8	0.38756	0.0007884	0.0525	60.43	231.46	60.73	191.07	251.80	0.2354	0.9150
12	0.44294	0.0007971	0.0460	65.83	233.63	66.18	187.85	254.03	0.2545	0.9132
16	0.50416	0.0008062	0.0405	71.29	235.78	71.69	184.52	256.22	0.2735	0.9116
20	0.57160	0.0008157	0.0358	76.80	237.91	77.26	181.09	258.35	0.2924	0.9102
24	0.64566	0.0008257	0.0317	82.37	240.01	82.90	177.55	260.45	0.3113	0.9089
26	0.68530	0.0008309	0.0298	85.18	241.05	85.75	175.73	261.48	0.3208	0.9082
28	0.72675	0.0008362	0.0281	88.00	242.08	88.61	173.89	262.50	0.3302	0.9076
30	0.77006	0.0008417	0.0265	90.84	243.10	91.49	172.00	263.50	0.3396	0.9070
32	0.81528	0.0008473	0.0250	93.70	244.12	94.39	170.09	264.48	0.3490	0.9064
34	0.86247	0.0008530	0.0236	96.58	245.12	97.31	168.14	265.45	0.3584	0.9058
36	0.91168	0.0008590	0.0223	99.47	246.11	100.25	166.15	266.40	0.3678	0.9053
38	0.96298	0.0008651	0.0210	102.38	247.09	103.21	164.12	267.33	0.3772	0.9047
40	1.0164	0.0008714	0.0199	105.30	248.06	106.19	162.05	268.24	0.3866	0.9041
42	1.0720	0.0008780	0.0188	108.25	249.02	109.19	159.94	269.14	0.3960	0.9035
44	1.1299	0.0008847	0.0177	111.22	249.96	112.22	157.79	270.01	0.4054	0.9030
48	1.2526	0.0008989	0.0159	117.22	251.79	118.35	153.33	271.68	0.4243	0.9017
52	1.3851	0.0009142	0.0142	123.31	253.55	124.58	148.66	273.24	0.4432	0.9004
56	1.5278	0.0009308	0.0127	129.51	255.23	130.93	143.75	274.68	0.4622	0.8990
60	1.6813	0.0009488	0.0114	135.82	256.81	137.42	138.57	275.99	0.4814	0.8973
70	2.1162	0.0010027	0.0086	152.22	260.15	154.34	124.08	278.43	0.5302	0.8918
80	2.6324	0.0010766	0.0064	169.88	262.14	172.71	106.41	279.12	0.5814	0.8827
90	3.2435	0.0011949	0.0046	189.82	261.34	193.69	82.63	276.32	0.6380	0.8655
100	3.9742	0.0015443	0.0027	218.60	248.49	224.74	34.40	259.13	0.7196	0.8117

Source for Tables A-14 through A-16: M. J. Moran and H. N. Shapiro, *Fundamentals of Engineering Thermodynamics*, 2nd ed. (New York: Wiley, 1992), pp. 710–15. Originally based on equations from D. P. Wilson and R. S. Basu, "Thermodynamic Properties of a New Stratospherically Safe Working Fluid—Refrigerant-134a," *ASHRAE Trans.* 94, Pt. 2 (1988), pp. 2095–118.

TABLE A-15 Thermodynamic Properties of Saturated Refrigerant 134a (Pressure Table)

Press. P_{sat} MPa	Temp. T °C	Specific volume, m³/kg		Internal energy, kJ/kg		Enthalpy, kJ/kg			Entropy, kJ/kg · K	
		Sat. liquid v_f	Sat. vapor v_g	Sat. liquid u_f	Sat. vapor u_g	Sat. liquid h_f	Evap. h_{fg}	Sat. vapor h_g	Sat. liquid s_f	Sat. vapor s_g
0.06	−37.07	0.0007097	0.3100	3.41	206.12	3.46	221.27	224.72	0.0147	0.9520
0.08	−31.21	0.0007184	0.2366	10.41	209.46	10.47	217.92	228.39	0.0440	0.9447
0.10	−26.43	0.0007258	0.1917	16.22	212.18	16.29	215.06	231.35	0.0678	0.9395
0.12	−22.36	0.0007323	0.1614	21.23	214.50	21.32	212.54	233.86	0.0879	0.9354
0.14	−18.80	0.0007381	0.1395	25.66	216.52	25.77	210.27	236.04	0.1055	0.9322
0.16	−15.62	0.0007435	0.1229	29.66	218.32	29.78	208.18	237.97	0.1211	0.9295
0.18	−12.73	0.0007485	0.1098	33.31	219.94	33.45	206.26	239.71	0.1352	0.9273
0.20	−10.09	0.0007532	0.0993	36.69	221.43	36.84	204.46	241.30	0.1481	0.9253
0.24	−5.37	0.0007618	0.0834	42.77	224.07	42.95	201.14	244.09	0.1710	0.9222
0.28	−1.23	0.0007697	0.0719	48.18	226.38	48.39	198.13	246.52	0.1911	0.9197
0.32	2.48	0.0007770	0.0632	53.06	228.43	53.31	195.35	248.66	0.2089	0.9177
0.36	5.84	0.0007839	0.0564	57.54	230.28	57.82	192.76	250.58	0.2251	0.9160
0.4	8.93	0.0007904	0.0509	61.69	231.97	62.00	190.32	252.32	0.2399	0.9145

(Continued)

TABLE A-15 (Continued)

Press. P_{sat} MPa	Temp. T °C	Specific volume, m³/kg Sat. liquid v_f	Sat. vapor v_g	Internal energy, kJ/kg Sat. liquid u_f	Sat. vapor u_g	Enthalpy, kJ/kg Sat. liquid h_f	Evap. h_{fg}	Sat. vapor h_g	Entropy, kJ/kg · K Sat. liquid s_f	Sat. vapor s_g
0.5	15.74	0.0008056	0.0409	70.93	235.64	71.33	184.74	256.07	0.2723	0.9117
0.6	21.58	0.0008196	0.0341	78.99	238.74	79.48	179.71	259.19	0.2999	0.9097
0.7	26.72	0.0008328	0.0292	86.19	241.42	86.78	175.07	261.85	0.3242	0.9080
0.8	31.33	0.0008454	0.0255	92.75	243.78	93.42	170.73	264.15	0.3459	0.9066
0.9	35.53	0.0008576	0.0226	98.79	245.88	99.56	166.62	266.18	0.3656	0.9054
1.0	39.39	0.0008695	0.0202	104.42	247.77	105.29	162.68	267.97	0.3838	0.9043
1.2	46.32	0.0008928	0.0166	114.69	251.03	115.76	155.23	270.99	0.4164	0.9023
1.4	52.43	0.0009159	0.0140	123.98	253.74	125.26	148.14	273.40	0.4453	0.9003
1.6	57.92	0.0009392	0.0121	132.52	256.00	134.02	141.31	275.33	0.4714	0.8982
1.8	62.91	0.0009631	0.0105	140.49	257.88	142.22	134.60	276.83	0.4954	0.8959
2.0	67.49	0.0009878	0.0093	148.02	259.41	149.99	127.95	277.94	0.5178	0.8934
2.5	77.59	0.0010562	0.0069	165.48	261.84	168.12	111.06	279.17	0.5687	0.8854
3.0	86.22	0.0011416	0.0053	181.88	262.16	185.30	92.71	278.01	0.6156	0.8735

TABLE A-16 **Thermodynamic Properties of Superheated Refrigerant 134a Vapor**

T °C	v m³/kg	u kJ/kg	h kJ/kg	s kJ/kg · K	v m³/kg	u kJ/kg	h kJ/kg	s kJ/kg · K	v m³/kg	u kJ/kg	h kJ/kg	s kJ/kg · K
	$P = 0.06$ MPa ($T_{sat} = -37.07$°C)				$P = 0.10$ MPa ($T_{sat} = -26.43$°C)				$P = 0.14$ MPa ($T_{sat} = -18.80$°C)			
Sat.	0.31003	206.12	224.72	0.9520	0.19170	212.18	231.35	0.9395	0.13945	216.52	236.04	0.9322
−20	0.33536	217.86	237.98	1.0062	0.19770	216.77	236.54	0.9602				
−10	0.34992	224.97	245.96	1.0371	0.20686	224.01	244.70	0.9918	0.14549	223.03	243.40	0.9606
0	0.36433	232.24	254.10	1.0675	0.21587	231.41	252.99	1.0227	0.15219	230.55	251.86	0.9922
10	0.37861	239.69	262.41	1.0973	0.22473	238.96	261.43	1.0531	0.15875	238.21	260.43	1.0230
20	0.39279	247.32	270.89	1.1267	0.23349	246.67	270.02	1.0829	0.16520	246.01	269.13	1.0532
30	0.40688	255.12	279.53	1.1557	0.24216	254.54	278.76	1.1122	0.17155	253.96	277.97	1.0828
40	0.42091	263.10	288.35	1.1844	0.25076	262.58	287.66	1.1411	0.17783	262.06	286.96	1.1120
50	0.43487	271.25	297.34	1.2126	0.25930	270.79	296.72	1.1696	0.18404	270.32	296.09	1.1407
60	0.44879	279.58	306.51	1.2405	0.26779	279.16	305.94	1.1977	0.19020	278.74	305.37	1.1690
70	0.46266	288.08	315.84	1.2681	0.27623	287.70	315.32	1.2254	0.19633	287.32	314.80	1.1969
80	0.47650	296.75	325.34	1.2954	0.28464	296.40	324.87	1.2528	0.20241	296.06	324.39	1.2244
90	0.49031	305.58	335.00	1.3224	0.29302	305.27	334.57	1.2799	0.20846	304.95	334.14	1.2516
100									0.21449	314.01	344.04	1.2785
	$P = 0.18$ MPa ($T_{sat} = -12.73$° C)				$P = 0.20$ MPa ($T_{sat} = -10.09$° C)				$P = 0.24$ MPa ($T_{sat} = -5.37$° C)			
Sat.	0.10983	219.94	239.71	0.9273	0.09933	221.43	241.30	0.9253	0.08343	224.07	244.09	0.9222
−10	0.11135	222.02	242.06	0.9362	0.09938	221.50	241.38	0.9256				
0	0.11678	229.67	250.69	0.9684	0.10438	229.23	250.10	0.9582	0.08574	228.31	248.89	0.9399
10	0.12207	237.44	259.41	0.9998	0.10922	237.05	258.89	0.9898	0.08993	236.26	257.84	0.9721
20	0.12723	245.33	268.23	1.0304	0.11394	244.99	267.78	1.0206	0.09339	244.30	266.85	1.0034
30	0.13230	253.36	277.17	1.0604	0.11856	253.06	276.77	1.0508	0.09794	252.45	275.95	1.0339
40	0.13730	261.53	286.24	1.0898	0.12311	261.26	285.88	1.0804	0.10181	260.72	285.16	1.0637
50	0.14222	269.85	295.45	1.1187	0.12758	269.61	295.12	1.1094	0.10562	269.12	294.47	1.0930
60	0.14710	278.31	304.79	1.1472	0.13201	278.10	304.50	1.1380	0.10937	277.67	303.91	1.1218
70	0.15193	286.93	314.28	1.1753	0.13639	286.74	314.02	1.1661	0.11307	286.35	313.49	1.1501
80	0.15672	295.71	323.92	1.2030	0.14073	295.53	323.68	1.1939	0.11674	295.18	323.19	1.1780
90	0.16148	304.63	333.70	1.2303	0.14504	304.47	333.48	1.2212	0.12037	304.15	333.04	1.2055
100	0.16622	313.72	343.63	1.2573	0.14932	313.57	343.43	1.2483	0.12398	313.27	343.03	1.2326

(Continued)

TABLE A-16 (Continued)

T °C	v m³/kg	u kJ/kg	h kJ/kg	s kJ/kg·K	v m³/kg	u kJ/kg	h kJ/kg	s kJ/kg·K	v m³/kg	u kJ/kg	h kJ/kg	s kJ/kg·K
	$P = 0.28$ MPa ($T_{sat} = -1.23°C$)				$P = 0.32$ MPa ($T_{sat} = 2.48°C$)				$P = 0.40$ MPa ($T_{sat} = 8.93°C$)			
Sat.	0.07193	226.38	246.52	0.9197	0.06322	228.43	248.66	0.9177	0.05089	231.97	252.32	0.9145
0	0.07240	227.37	247.64	0.9238								
10	0.07613	235.44	256.76	0.9566	0.06576	234.61	255.65	0.9427	0.05119	232.87	253.35	0.9182
20	0.07972	243.59	265.91	0.9883	0.06901	242.87	264.95	0.9749	0.05397	241.37	262.96	0.9515
30	0.08320	251.83	275.12	1.0192	0.07214	251.19	274.28	1.0062	0.05662	249.89	272.54	0.9837
40	0.08660	260.17	284.42	1.0494	0.07518	259.61	283.67	1.0367	0.05917	258.47	282.14	1.0148
50	0.08992	268.64	293.81	1.0789	0.07815	268.14	293.15	1.0665	0.06164	267.13	291.79	1.0452
60	0.09319	277.23	303.32	1.1079	0.08106	276.79	302.72	1.0957	0.06405	275.89	301.51	1.0748
70	0.09641	285.96	312.95	1.1364	0.08392	285.56	312.41	1.1243	0.06641	284.75	311.32	1.1038
80	0.09960	294.82	322.71	1.1644	0.08674	294.46	322.22	1.1525	0.06873	293.73	321.23	1.1322
90	0.10275	303.83	332.60	1.1920	0.08953	303.50	332.15	1.1802	0.07102	302.84	331.25	1.1602
100	0.10587	312.98	342.62	1.2193	0.09229	312.68	342.21	1.2076	0.07327	312.07	341.38	1.1878
110	0.10897	322.27	352.78	1.2461	0.09503	322.00	352.40	1.2345	0.07550	321.44	351.64	1.2149
120	0.11205	331.71	363.08	1.2727	0.09774	331.45	362.73	1.2611	0.07771	330.94	362.03	1.2417
130									0.07991	340.58	372.54	1.2681
140									0.08208	350.35	383.18	1.2941
	$P = 0.50$ MPa ($T_{sat} = 15.74°C$)				$P = 0.60$ MPa ($T_{sat} = 21.58°C$)				$P = 0.70$ MPa ($T_{sat} = 26.72°C$)			
Sat.	0.04086	253.64	256.07	0.9117	0.03408	238.74	259.19	0.9097	0.02918	241.42	261.85	0.9080
20	0.04188	239.40	260.34	0.9264								
30	0.04416	248.20	270.28	0.9597	0.03581	246.41	267.89	0.9388	0.02979	244.51	265.37	0.9197
40	0.04633	256.99	280.16	0.9918	0.03774	255.45	278.09	0.9719	0.03157	253.83	275.93	0.9539
50	0.04842	265.83	290.04	1.0229	0.03958	264.48	288.23	1.0037	0.03324	263.08	286.35	0.9867
60	0.05043	274.73	299.95	1.0531	0.04134	273.54	298.35	1.0346	0.03482	272.31	296.69	1.0182
70	0.05240	283.72	309.92	1.0825	0.04304	282.66	308.48	1.0645	0.03634	281.57	307.01	1.0487
80	0.05432	292.80	319.96	1.1114	0.04469	291.86	318.67	1.0938	0.03781	290.88	317.35	1.0784
90	0.05620	302.00	330.10	1.1397	0.04631	301.14	328.93	1.1225	0.03924	300.27	327.74	1.1074
100	0.05805	311.31	340.33	1.1675	0.04790	310.53	339.27	1.1505	0.04064	309.74	338.19	1.1358
110	0.05988	320.74	350.68	1.1949	0.04946	320.03	349.70	1.1781	0.04201	319.31	348.71	1.1637
120	0.06168	330.30	361.14	1.2218	0.05099	329.64	360.24	1.2053	0.04335	328.98	359.33	1.1910
130	0.06347	339.98	371.72	1.2484	0.05251	339.38	370.88	1.2320	0.04468	338.76	370.04	1.2179
140	0.06524	349.79	382.42	1.2746	0.05402	349.23	381.64	1.2584	0.04599	348.66	380.86	1.2444
150					0.05550	359.21	392.52	1.2844	0.04729	358.68	391.79	1.2706
160					0.05698	369.32	403.51	1.3100	0.04857	368.82	402.82	1.2963
	$P = 0.80$ MPa ($T_{sat} = 31.33°C$)				$P = 0.90$ MPa ($T_{sat} = 35.53°C$)				$P = 1.00$ MPa ($T_{sat} = 39.39°C$)			
Sat.	0.02547	243.78	264.15	0.9066	0.02255	245.88	266.18	0.9054	0.02020	247.77	267.97	0.9043
40	0.02691	252.13	273.66	0.9374	0.02325	250.32	271.25	0.9217	0.02029	248.39	268.68	0.9066
50	0.02846	261.62	284.39	0.9711	0.02472	260.09	282.34	0.9566	0.02171	258.48	280.19	0.9428
60	0.02992	271.04	294.98	1.0034	0.02609	269.72	293.21	0.9897	0.02301	268.35	291.36	0.9768
70	0.03131	280.45	305.50	1.0345	0.02738	279.30	303.94	1.0214	0.02423	278.11	302.34	1.0093
80	0.03264	289.89	316.00	1.0647	0.02861	288.87	314.62	1.0521	0.02538	287.82	313.20	1.0405
90	0.03393	299.37	326.52	1.0940	0.02980	298.46	325.28	1.0819	0.02649	297.53	324.01	1.0707
100	0.03519	308.93	337.08	1.1227	0.03095	308.11	335.96	1.1109	0.02755	307.27	334.82	1.1000
110	0.03642	318.57	347.71	1.1508	0.03207	317.82	346.68	1.1392	0.02858	317.06	345.65	1.1286
120	0.03762	328.31	358.40	1.1784	0.03316	327.62	357.47	1.1670	0.02959	326.93	356.52	1.1567
130	0.03881	338.14	369.19	1.2055	0.03423	337.52	368.33	1.1943	0.03058	336.88	367.46	1.1841
140	0.03997	348.09	380.07	1.2321	0.03529	347.51	379.27	1.2211	0.03154	346.92	378.46	1.2111
150	0.04113	358.15	391.05	1.2584	0.03633	357.61	390.31	1.2475	0.03250	357.06	389.56	1.2376
160	0.04227	368.32	402.14	1.2843	0.03736	367.82	401.44	1.2735	0.03344	367.31	400.74	1.2638
170	0.04340	378.61	413.33	1.3098	0.03838	378.14	412.68	1.2992	0.03436	377.66	412.02	1.2895
180	0.04452	389.02	424.63	1.3351	0.03939	388.57	424.02	1.3245	0.03528	388.12	423.40	1.3149

(Continued)

TABLE A-16   (Continued)

T °C	v m³/kg	u kJ/kg	h kJ/kg	s kJ/kg·K	v m³/kg	u kJ/kg	h kJ/kg	s kJ/kg·K	v m³/kg	u kJ/kg	h kJ/kg	s kJ/kg·K
	$P = 1.20$ MPa ($T_{sat} = 46.32°C$)				$P = 1.40$ MPa ($T_{sat} = 52.43°C$)				$P = 1.60$ MPa ($T_{sat} = 57.92°C$)			
Sat.	0.01663	251.03	270.99	0.9023	0.01405	253.74	273.40	0.9003	0.01208	256.00	275.33	0.8982
50	0.01712	254.98	275.52	0.9164								
60	0.01835	265.42	287.44	0.9527	0.01495	262.17	283.10	0.9297	0.01233	258.48	278.20	0.9069
70	0.01947	275.59	298.96	0.9868	0.01603	272.87	295.31	0.9658	0.01340	269.89	291.33	0.9457
80	0.02051	285.62	310.24	1.0192	0.01701	283.29	307.10	0.9997	0.01435	280.78	303.74	0.9813
90	0.02150	295.59	321.39	1.0503	0.01792	293.55	318.63	1.0319	0.01521	291.39	315.72	1.0148
100	0.02244	305.54	332.47	1.0804	0.01878	303.73	330.02	1.0628	0.01601	301.84	327.46	1.0467
110	0.02335	315.50	343.52	1.1096	0.01960	313.88	341.32	1.0927	0.01677	312.20	339.04	1.0773
120	0.02423	325.51	354.58	1.1381	0.02039	324.05	352.59	1.1218	0.01750	322.53	350.53	1.1069
130	0.02508	335.58	365.68	1.1660	0.02115	334.25	363.86	1.1501	0.01820	332.87	361.99	1.1357
140	0.02592	345.73	376.83	1.1933	0.02189	344.50	375.15	1.1777	0.01887	343.24	373.44	1.1638
150	0.02674	355.95	388.04	1.2201	0.02262	354.82	386.49	1.2048	0.01953	353.66	384.91	1.1912
160	0.02754	366.27	399.33	1.2465	0.02333	365.22	397.89	1.2315	0.02017	364.15	396.43	1.2181
170	0.02834	376.69	410.70	1.2724	0.02403	375.71	409.36	1.2576	0.02080	374.71	407.99	1.2445
180	0.02912	387.21	422.16	1.2980	0.02472	386.29	420.90	1.2834	0.02142	385.35	419.62	1.2704
190					0.02541	396.96	432.53	1.3088	0.02203	396.08	431.33	1.2960
200					0.02608	407.73	444.24	1.3338	0.02263	406.90	443.11	1.3212

TABLE A-17   Total Emissivity of Various Surfaces

Metallic Solids and Their Oxides

Description/Composition		Emissivity, ε_n or ε_h, at Various Temperatures (K)										
		100	200	300	400	600	800	1000	1200	1500	2000	2500
Aluminum												
Highly polished, film	(h)*	0.02	0.03	0.04	0.05	0.06						
Foil, bright	(h)	0.06	0.06	0.07								
Anodized	(h)			0.82	0.76							
Chromium												
Polished or plated	(n)	0.05	0.07	0.10	0.12	0.14						
Copper												
Highly polished	(h)			0.03	0.03	0.04	0.04	0.04				
Stably oxidized	(h)					0.50	0.58	0.80				
Gold												
Highly polished or film	(h)	0.01	0.02	0.03	0.03	0.04	0.05	0.06				
Foil, bright	(h)	0.06	0.07	0.07								
Molybdenum												
Polished	(h)					0.06	0.08	0.10	0.12	0.15	0.21	0.26
Shot-blasted, rough	(h)					0.25	0.28	0.31	0.35	0.42		
Stably oxidized	(h)					0.80	0.82					
Nickel												
Polished	(h)					0.09	0.11	0.14	0.17			
Stably oxidized	(h)					0.40	0.49	0.57				
Platinum												
Polished	(h)					0.10	0.13	0.15	0.18			
Silver												
Polished	(h)			0.02	0.02	0.03	0.05	0.08				
Stainless steels												
Typical, polished	(n)			0.17	0.17	0.19	0.23	0.30				
Typical, cleaned	(n)			0.22	0.22	0.24	0.28	0.35				
Typical, lightly oxidized	(n)						0.33	0.40				
Typical, highly oxidized	(n)						0.67	0.70	0.76			
AISI 347, stably oxidized	(n)					0.87	0.88	0.89	0.90			
Tantalum												
Polished	(h)								0.11	0.17	0.23	0.28
Tungsten												
Polished	(h)							0.10	0.13	0.18	0.25	0.29

(Continued)

TABLE A-17 (Continued)

Nonmetallic Substances

Description/Composition		Temperature (K)	Emissivity ε
Aluminum oxide	(n)	600	0.69
		1000	0.55
		1500	0.41
Asphalt pavement	(h)	300	0.85–0.93
Building materials			
Asbestos sheet	(h)	300	0.93–0.96
Brick, red	(h)	300	0.93–0.96
Gypsum or plaster board	(h)	300	0.90–0.92
Wood	(h)	300	0.82–0.92
Cloth	(h)	300	0.75–0.90
Concrete	(h)	300	0.88–0.93
Glass, window	(h)	300	0.90–0.95
Ice	(h)	273	0.95–0.98
Paints			
Black (Parsons)	(h)	300	0.98
White, acrylic	(h)	300	0.90
White, zinc oxide	(h)	300	0.92
Paper, white	(h)	300	0.92–0.97
Pyrex	(n)	300	0.82
		600	0.80
		1000	0.71
		1200	0.62
Pyroceram	(n)	300	0.85
		600	0.78
		1000	0.69
		1500	0.57
Refractories (furnace liners)			
Alumina brick	(n)	800	0.40
		1000	0.33
		1400	0.28
		1600	0.33
Magnesia brick	(n)	800	0.45
		1000	0.36
		1400	0.31
		1600	0.40
Kaolin insulating brick	(n)	800	0.70
		1200	0.57
		1400	0.47
		1600	0.53
Sand	(h)	300	0.90
Silicon carbide	(n)	600	0.87
		1000	0.87
		1500	0.85
Skin	(h)	300	0.95
Snow	(h)	273	0.82–0.90
Soil	(h)	300	0.93–0.96
Rocks	(h)	300	0.88–0.95
Teflon	(h)	300	0.85
		400	0.87
		500	0.92
Vegetation	(h)	300	0.92–0.96
Water	(h)	300	0.96

*h-hemispherical.
n-normal.

Source: Incropera, F. P., and D. P. DeWitt, *Introduction to Heat Transfer*, 4th ed., Wiley, New York, 2002.

TABLES IN BRITISH UNITS
APPENDIX B

TABLE B-1 **Molecular Weight and Critical-Point Properties**

Substance	Formula	Molecular Weight M lbm/lbmol	Ratio of Gas Constant to Molecular Weight $\overline{R}/M$		Critical-Point Properties		
			Btu/lbm · R	psia · ft³/ lbm · R	T_c R	P_c psia	$\overline{v}_c$ ft³/lbmol
Air	—	28.97	0.06855	0.3704	238.5	547	1.41
Ammonia	NH_3	17.03	0.1166	0.6301	729.8	1636	1.16
Argon	Ar	39.948	0.04971	0.2686	272	705	1.20
Benzene	C_6H_6	78.115	0.02542	0.1374	1012	714	4.17
Bromine	Br_2	159.808	0.01243	0.06714	1052	1500	2.17
n-Butane	C_4H_{12}	58.124	0.03417	0.1846	765.2	551	4.08
Carbon dioxide	CO_2	44.01	0.04513	0.2438	547.5	1071	1.51
Carbon monoxide	CO	28.011	0.07090	0.3831	240	507	1.49
Carbon tetrachloride	CCl_{12}	153.82	0.01291	0.06976	1001.5	661	4.42
Chlorine	Cl_2	70.906	0.02801	0.1517	751	1120	1.99
Chloroform	$CHCl_3$	119.38	0.01664	0.08988	965.8	794	3.85
Dichlorodifluoromethane (R-12)	CCl_2F_2	120.91	0.01643	0.08874	692.4	582	3.49
Dichlorofluoromethane (R-21)	$CHCl_2F$	102.92	0.01930	0.1043	813.0	749	3.16
Ethane	C_2H_6	30.020	0.06616	0.3574	549.8	708	2.37
Ethyl alcohol	C_2H_5OH	46.07	0.04311	0.2329	929.0	926	2.68
Ethylene	C_2H_4	28.054	0.07079	0.3825	508.3	742	1.99
Helium	He	4.003	0.4961	2.6805	9.5	33.2	0.926
n-Hexane	C_6H_{14}	86.178	0.02305	0.1245	914.2	439	5.89
Hydrogen (normal)	H_2	2.016	0.9851	5.3224	59.9	188.1	1.04
Methane	CH_4	16.043	0.1238	0.6688	343.9	673	1.59
Methyl alcohol	CH_3OH	32.042	0.06198	0.3349	923.7	1154	1.89
Methyl chloride	CH_3Cl	50.488	0.03934	0.2125	749.3	968	2.29
Nitrogen	N_2	28.013	0.07090	0.3830	227.1	492	1.44
Nitrous oxide	N_2O	44.013	0.04512	0.2438	557.4	1054	1.54
Oxygen	O_2	31.999	0.06206	0.3353	278.6	736	1.25
Propane	C_3H_8	44.097	0.04504	0.2433	665.9	617	3.20
Propylene	C_3H_6	42.081	0.04719	0.2550	656.9	670	2.90
Sulfur dioxide	SO_2	64.063	0.03100	1.1675	775.2	1143	1.95
Tetrafluoroethane (R-134a)	CF_3CH_2F	102.03	0.01946	0.1052	673.7	589.9	2.96
Trichlorofluoromethane (R-11)	CCl_3F	137.37	0.01446	0.07811	848.1	635	3.97
Water	H_2O	18.015	0.1102	0.5956	1165.3	3204	0.90

Source: Kobe, K.A. and R.E. Lynn Jr., *Chemical Review* 52, 1953, pp. 117–236; and ASHRAE, *Handbook of Fundamentals*, Atlanta, GA: ASHRAE, Inc., 1993, pp. 16.4 and 36.1.

TABLE B-2 Thermophysical Properties of Solid Metals

Composition	Melting Point R	Properties at 540 R				Properties at various temperatures (R) k [Btu/h·ft·R] / c_p [Btu/lbm·R]						
		ρ lbm/ft³	c_p Btu/lbm·R	k Btu/h·ft·R	$\alpha \times 10^6$ ft²/s	180	360	720	1080	1440	1800	2160
Aluminum Pure	1679	168	0.216	137	1045	174.5 / 0.115	137 / 0.191	138.6 / 0.226	133.4 / 0.246	126 / 0.273		
Alloy 2024-T6 (4.5% Cu, 1.5% Mg, 0.6% Mn)	1395	173	0.209	102.3	785.8	37.6 / 0.113	94.2 / 0.188	107.5 / 0.22	107.5 / 0.249			
Alloy 195, Cast (4.5% Cu)		174.2	0.211	97	734			100.5 / —	106.9 / —			
Beryllium	2790	115.5	0.436	115.6	637.2	572 / 0.048	174 / 0.266	93 / 0.523	72.8 / 0.621	61.3 / 0.624	52.5 / 0.72	45.6 / 0.77
Bismuth	981	610.5	0.029	4.6	71	9.5 / 0.026	5.6 / 0.028	4.06 / 0.03				
Boron	4631	156	0.264	15.6	105	109.7 / 0.03	32.06 / 0.143	9.7 / 0.349	6.1 / 0.451	5.5 / 0.515	5.7 / 0.558	
Cadmium	1069	540	0.055	55.6	521	117.3 / 0.047	57.4 / 0.053	54.7 / 0.057				
Chromium	3812	447	0.107	54.1	313.2	91.9 / 0.045	64.1 / 0.091	52.5 / 0.115	46.6 / 0.129	41.2 / 0.138	37.8 / 0.147	35.8 / 0.162
Cobalt	3184	553.2	0.101	57.3	286.3	96.5 / 0.056	70.5 / 0.09	49.3 / 0.107	39 / 0.12	33.6 / 0.131	30.1 / 0.145	28.5 / 0.175
Copper Pure	2445	559	0.092	231.7	1259.3	278.5 / 0.06	238.6 / 0.085	227.07 / 0.094	219 / 0.01	212 / 0.103	203.4 / 0.107	196 / 0.114
Commercial bronze (90% Cu, 10% Al)	2328	550	0.1	30	150.7		24.3 / 0.187	30 / 0.109	34 / 0.130			
Phosphor gear bronze (89% Cu, 11% Sn)	1987	548.1	0.084	31.2	183		23.7	37.6	42.8			
Cartridge brass (70% Cu, 30% Zn)	2139	532.5	0.09	63.6	364.9	43.3	54.9 / 0.09	79.2 / 0.09	86.0 / 0.101			
Constantan (55% Cu, 45% Ni)	2687	557	0.092	13.3	72.3	9.8 / 0.06	1.1 / 0.09					
Germanium	2180	334.6	0.08	34.6	373.5	134 / 0.045	56 / 0.069	25 / 0.08	15.7 / 0.083	11.4 / 0.085	10.05 / 0.089	10.05 / 0.094
Gold	2405	1205	0.03	183.2	1367	189 / 0.026	186.6 / 0.029	179.7 / 0.031	172.2 / 0.032	164.09 / 0.033	156 / 0.034	147.3 / 0.037
Iridium	4896	1404.6	0.031	85	541.4	99.4 / 0.021	88.4 / 0.029	83.2 / 0.031	79.7 / 0.032	76.3 / 0.034	72.8 / 0.036	69.3 / 0.038
Iron Pure	3258	491.3	0.106	46.4	248.6	77.4 / 0.051	54.3 / 0.091	40.2 / 0.117	31.6 / 0.137	25.01 / 0.162	19 / 0.232	16.4 / 0.14
Armco (99.75% pure)		491.3	0.106	42	222.8	55.2 / 0.051	46.6 / 0.091	38 / 0.117	30.7 / 0.137	24.4 / 0.162	18.7 / 0.233	16.6 / 0.145

Carbon steels Plain carbon (Mn ≤ 1%, Si ≤ 0.1%)		490.3	0.103	35	190.6			32.8 0.116	27.7 0.113	22.7 0.163	17.4 0.279	
AISI 1010		489	0.103	37	202.4			33.9 0.116	28.2 0.133	22.7 0.163	18 0.278	
Carbon-silicon (Mn ≤ 1%, 0.1% < Si ≤ 0.6%)		488	0.106	30	160.4			28.8 0.119	25.4 0.139	21.6 0.166	17 0.231	
Carbon-manganese-silicon (1% < Mn ≤ 1.65%, 0.1% < Si ≤ 0.6%)		508	0.104	23.7	125			24.4 0.116	23 0.133	20.2 0.163	16 0.260	
Chromium (low) steels ½ Cr-¼Mo-Si (0.18% C, 0.65% Cr, 0.23% Mo, 0.6% Si)		488.3	0.106	21.8	117.4			22 0.117	21.2 0.137	19.3 0.164	15.6 0.231	
1Cr-½Mo (0.16% C, 1% Cr, 0.54% Mo, 0.39% Si)		490.6	0.106	24.5	131.3			24.3 0.117	22.6 0.137	20 0.164	15.8 0.231	
1 Cr-V (0.2% C, 1.02% Cr, 0.15% V)		489.2	0.106	28.3	151.8			27.0 0.117	24.3 0.137	21 0.164	16.3 0.231	
Stainless steels AISI 302		503	0.114	8.7	42			10 0.122	11.6 0.133	13.2 0.140	14.7 0.144	
AISI 304	3006	493.2	0.114	8.6	42.5	5.31 0.064	7.3 0.096	9.6 0.123	11.5 0.133	13 0.139	14.7 0.145	16.2 0.152
AISI 316		514.3	0.111	7.8	37.5			8.8 0.12	10.6 0.131	12.3 0.137	14 0.143	
AISI 347		498	0.114	8.2	40			9.1 0.122	1.1 0.133	12.7 0.14	14.3 0.144	
Lead	1082	708	0.03	20.4	259.4	23 0.028	21.2 0.029	19.7 0.031	18.1 0.034			
Magnesium	1661	109	0.245	90.2	943	87.9 0.155	91.9 0.223	88.4 0.256	86.0 0.279	84.4 0.302		
Molybdenum	5209	639.3	0.06	79.7	578	1034 0.033	82.6 0.053	77.4 0.062	72.8 0.065	68.2 0.068	64.7 0.070	60.7 0.073

(Continued)

TABLE B-2 (Continued)

Composition	Melting Point R	Properties at 540 R				Properties at various temperatures (R) $k\,[Btu/h\cdot ft\cdot R]/c_p\,[Btu/lbm\cdot R]$						
		ρ lbm/ft^3	c_p Btu/lbm·R	k Btu/h·ft·R	$\alpha \times 10^6$ ft^2/s	180	360	720	1080	1440	1800	2160
Nickel Pure	3110	555.6	0.106	52.4	247.6	94.8 / 0.055	61.8 / 0.091	46.3 / 0.115	37.9 / 0.141	39 / 0.126	41.4 / 0.134	44.0 / 0.141
Nichrome (80% Ni, 20% Cr)	3010	524.4	0.1	6.9	36.6			8.0 / 0.114	9.3 / 0.125	12.2 / 0.130		
Inconel X-750 (7.3% Ni, 15% Cr, 6.7% Fe)	2997	531.3	0.104	6.8	33.4	5 / —	5.9 / 0.088	7.8 / 0.112	9.8 / 0.121	11.8 / 0.13	13.9 / 0.149	16.0
Niobium	4934	535	0.063	31	254	31.9 / 0.044	30.4 / 0.059	32 / 0.065	33.6 / 0.067	35.4 / 0.069	32.2 / 0.071	39.0 / 0.074
Palladium	3289	750.4	0.058	41.5	263.7	44.2 / 0.04	41.4 / 0.054	42.5 / 0.059	46 / 0.062	50 / 0.064	54.4 / 0.067	59.0 / 0.069
Platinum Pure	3681	1339	0.031	41.4	270	44.7 / 0.024	42 / 0.03	41.5 / 0.032	42.3 / 0.034	43.7 / 0.035	45.5 / 0.036	47.7 / 0.037
Alloy 60Pt-40Rh (60% Pt, 40% Rh)	3240	1038.2	0.038	27.2	187.3			30	34	37.5	40	42.2
Rhenium	6215	1317.2	0.032	27.7	180	34 / 0.023	30 / 0.03	26.6 / 0.033	25.5 / 0.034	25.4 / 0.036	25.8 / 0.037	26.6 / 0.038
Rhodium	4025	777.2	0.058	86.7	534	107.5 / 0.035	89 / 0.052	84.3 / 0.06	78.5 / 0.065	73.4 / 0.069	70 / 0.074	67.0 / 0.078
Silicon	3033	145.5	0.17	85.5	960.2	510.8 / 0.061	152.5 / 0.132	57.2 / 0.189	35.8 / 0.207	24.4 / 0.218	18.0 / 0.226	15.0 / 0.230
Silver	2223	656	0.056	248	1873	257 / 0.044	248.4 / 0.053	245.5 / 0.057	238 / 0.059	228.8 / 0.062	219 / 0.066	208.6 / 0.069
Tantalum	5884	1036.3	0.033	33.2	266	34.2 / 0.026	33.2 / 0.031	33.4 / 0.034	34 / 0.035	34.3 / 0.036	34.8 / 0.036	35.3 / 0.037
Thorium	3641	730.4	0.028	31.2	420.9	34.6 / 0.024	31.5 / 0.027	31.4 / 0.029	32.2 / 0.032	32.9 / 0.035	32.9 / 0.037	33.9 / 0.04
Tin	909	456.3	0.054	38.5	431.6	49.2 / 0.044	42.4 / 0.051	35.9 / 0.058				
Titanium	3515	281	0.013	12.7	100.3	17.6 / 0.071	14.2 / 0.111	11.8 / 0.131	11.2 / 0.141	11.4 / 0.151	12 / 0.161	12.7 / 0.148
Tungsten	6588	1204.9	0.031	100.5	735.2	120.2 / 0.020	107.5 / 0.029	92 / 0.032	79.2 / 0.033	72.2 / 0.034	68.2 / 0.035	65.3 / 0.036
Uranium	2531	1190.5	0.027	16	134.5	12.5 / 0.022	14.5 / 0.026	17.1 / 0.029	19.6 / 0.035	22.4 / 0.042	25.4 / 0.043	28.3 / 0.038
Vanadium	3946	381	0.117	17.7	110.9	20.7 / 0.061	18 / 0.102	18 / 0.123	19.3 / 0.128	20.6 / 0.134	22.0 / 0.142	23.6 / 0.154
Zinc	1247	445.7	0.093	67	450	67.6 / 0.07	68.2 / 0.087	64.1 / 0.096	59.5 / 0.104			
Zirconium	3825	410.2	0.067	13.1	133.5	19.2 / 0.049	14.6 / 0.063	12.5 / 0.072	12 / 0.77	12.5 / 0.082	13.7 / 0.087	15.0 / 0.083

Tables B-2 through B-5 were obtained by converting quantities in Tables A-2 through A-5, respectively.

TABLE B-3 Thermophysical Properties of Solid Nonmetals

Composition	Melting Point R	Properties at 540 R				Properties at various temperatures (R) k [Btu/h·ft·R] / c_p [Btu/lbm·R]						
		ρ lbm/ft³	c_p Btu/lbm·R	k Btu/h·ft·R	$\alpha \times 10^6$ ft²/s	180	360	720	1080	1440	1800	2160
Aluminum oxide, sapphire	4181	247.8	0.182	26.6	162.5	260 / —	47.4 / —	18.7 / 0.224	11 / 0.265	7.5 / 0.281	6 / 0.293	
Aluminum oxide, polycrystalline	4181	247.8	0.182	20.8	128	76.8 / —	31.7 / —	15.3 / 0.244	9.3 / 0.265	6 / 0.281	4.5 / 0.293	3.8
Beryllium oxide	4905	187.3	0.246	157.2	947.3	109.8	30.3	113.2 / 0.322	64.2 / 0.40	40.4 / 0.44	27.2 / 0.459	19 / 0.490
Boron	4631	156	0.264	16	107.5			10.8 / 0.355	6.5 / 0.445	4.6 / 0.509	3.6 / 0.561	3 / 0.610
Boron fiber epoxy (30% vol) composite	1062	130										
$k_\parallel$ to fibers				1.3		1.2	1.3	1.31				
$k_\perp$ to fibers				0.34		0.21	0.28	0.34				
c_p			0.268			0.086	0.18	0.34				
Carbon Amorphous	2700	121.7	—	0.92	—	0.38	0.68	1.09	1.26	1.36	1.46	1.64
Diamond, type IIa insulator	—	219	0.121	1329	—	5778	2311.2 / 0.005	889.8 / 0.046	0.203			
Graphite, pyrolytic	4091	138										
$k_\parallel$ to layers				1126.7		2871.6	1866.3	803.2	515.4	385.4	308.5	258.9
$k_\perp$ to layers				3.3		9.7	5.3	2.4	1.5	1.16	0.92	0.77
c_p			0.169			0.032	0.098	0.236	0.335	0.394	0.428	0.45
Graphite fiber epoxy (25% vol) composite	810	87.4										
k, heat flow $\parallel$ to fibres				6.4		3.3	5.0	7.5				
k, heat flow $\perp$ to fibres				0.5		0.4	0.63					
c_p			0.223		5	0.08	0.153	0.29				
Pyroceram, Corning 9606	2921	162.3	0.193	2.3	20.3	3.0	2.3	2.1 / 0.29	1.9	1.7	1.7	1.7
Silicon carbide	5580	197.3	0.161	283.1	2475.7			— / 0.210	— / 0.25	— / 0.27	50.3 / 0.285	33.5 / 0.296

(Continued)

TABLE B-3 (Continued)

Composition	Melting Point R	Properties at 540 R				Properties at various temperatures (R) k [Btu/h · ft · R]/c_p [Btu/lbm · R]						
		ρ lbm/ft³	c_p Btu/lbm · R	k Btu/h · ft · R	$\alpha \times 10^6$ ft²/s	180	360	720	1080	1440	1800	2160
Silicon dioxide, crystalline (quartz)	3389	165.4										
$k_\parallel$ to c axis				6		22.5	9.5	4.4	2.9	2.4		
$k_\perp$ to c axis				3.6		12.0	5.9	2.7	2	1.8		
c_p			0.177			—	—	0.211	0.256	0.298		
Silicon dioxide, polycrystalline (fused silica)	3389	138.6	0.177	0.79	9	0.4	0.65	0.87	1.01	1.25	1.65	2.31
						—	—	0.216	0.248	0.264	0.276	0.286
Silicon nitride	3911	150	0.165	9.2	104	—	—	8.0	6.5	5.7	5.0	4.6
						—	0.138	0.185	0.223	0.253	0.275	0.292
Sulfur	706	130	0.169	0.1	1.51	0.095	0.1					
						0.962	0.144					
Thorium dioxide	6431	568.7	0.561	7.5	65.7			5.9	3.8	2.7	2.12	1.8
								0.609	0.654	0.680	0.704	0.723
Titanium dioxide, polycrystalline	3840	259.5	0.170	4.9	30.1			4.0	2.9	2.3	2	1.9
								0.192	0.210	0.217	0.222	0.225

(Continued)

TABLE B-3   (Continued)

Description/composition	Temperature R	Density ρ lbm/ft³	Thermal conductivity k Btu/h · ft · R	Specific heat c_p Btu/lbm · R
Asphalt	540	132	0.035	0.219
Bakelite	540	81.2	0.808	0.349
Brick, refractory				
Carborundum	1569	—	10.7	—
	3009	—	6.4	—
Chrome brick	851	187.9	1.3	0.199
	1481		1.4	
	2111		1.2	
Diatomaceous	860	—	0.14	—
silica, fired	2061	—	0.17	
Fire clay, burnt 2880 R	1391	128	0.57	0.229
	1931	—	0.63	
	2471	—	0.63	
Fire clay, burnt 3105 R	1391	145.2	0.75	0.229
	1931		0.8	
	2471		0.8	
Fire clay brick	860	165.1	0.57	0.229
	1660		0.86	
	2660		1.04	
Magnesite	860	—	2.2	0.269
	1660	—	1.6	
	2660		1.09	
Clay	540	91.1	0.75	0.210
Coal, anthracite	540	84.3	0.15	0.3
Concrete (stone mix)	540	143.6	0.8	0.210
Cotton	540	5	0.034	0.310
Foodstuffs				
Banana (75.7% water content)	540	61.2	0.27	0.8
Apple, red (75% water content)	540	52.4	0.29	0.859
Cake, batter	540	45	0.128	—
Cake, fully baked	540	17.5	0.069	—
Chicken meat, white	356	—	0.924	—
(74.4% water content)	419	—	0.86	
	455		0.78	
	473		0.69	
	491		0.275	
	329		0.277	
	527		0.282	
Glass				
Plate (soda lime)	540	156	0.8	0.179
Pyrex	540	139	0.8	0.199
Ice	491	57.4	1.08	0.487
	455	—	1.17	0.464
Leather (sole)	540	62.3	0.091	—
Paper	540	58	0.104	0.320
Paraffin	540	56.2	0.138	0.690
Rock				
Granite, Barre	540	164.2	1.61	0.185
Limestone, Salem	540	144.8	1.24	0.193
Marble, Halston	540	167.3	1.61	0.198
Quartzite, Sioux	540	164.8	3.10	0.263
Sandstone, Berea	540	134.2	1.67	0.178

(Continued)

TABLE B-3 (Continued)

Description/composition	Temperature R	Density ρ lbm/ft^3	Thermal conductivity k Btu/h · ft · R	Specific heat c_p Btu/lbm · R
Rubber, vulcanized				
Soft	540	68.6	0.075	0.48
Hard	540	74.3	0.092	—
Sand	540	94.6	0.156	0.191
Soil	540	128	0.3	0.439
Snow	540	6.9	0.028	—
	720	31.2	0.109	—
Teflon		137.3	0.202	—
				—
Tissue, human				
Skin	540	—	0.213	—
Fat layer (adipose)	540	—	0.115	—
Muscle	540	—	0.236	—
Wood, cross-grain				
Balsa	540	8.7	0.031	—
Cypress	540	29	0.056	—
Fir	540	25.9	0.063	0.649
Oak	540	34	0.098	0.569
Yellow pine	540	40	0.086	0.669
White pine	540	27.2	0.063	—
Wood, radial				
Oak	540	34	0.109	0.569
Fir	540	26.2	0.08	0.649

TABLE B-4 Thermophysical Properties of Solid Insulating Materials

Description/composition	Typical properties at 540 R		
	Density ρ lbm/ft^3	Thermal conductivity k Btu/h · ft · R	Specific heat c_p Btu/lbm · R
Blanket and batt			
Glass fiber, paper faced	1.0	0.026	—
	1.7	0.022	—
	2.5	0.02	—
Glass fiber, coated; duct liner	1.9	0.022	0.199
Board and slab			
Cellular glass	9.05	0.033	0.238
Glass fiber, organic bonded	6.55	0.02	0.189
Polystyrene, expanded			
extruded (R-12)	3.4	0.015	0.289
molded beads	1.0	0.023	0.289
Mineral fiberboard; roofing material	16.5	0.028	—
Wood, shredded/cemented	21.8	0.05	0.379
Cork	7.49	0.022	0.429
Loose fill			
Cork, granulated	9.9	0.026	—
Diatomaceous silica, coarse	21.8	0.039	—
powder	24.9	0.052	—
Diatomaceous silica, fine powder	12.5	0.03	—
	17.1	0.035	—

(Continued)

TABLE B-4 (Continued)

Description/composition	Typical properties at 540 R		
	Density ρ lbm/ft^3	Thermal conductivity k Btu/h · ft · R	Specific heat c_p Btu/lbm · R
Glass fiber, poured or blown	1.0	0.024	0.199
Vermiculite, flakes	5	0.039	0.199
	9.9	0.036	0.238
Formed/foamed-in-place			
Mineral wool granules with asbestos/inorganic binders, sprayed	11.8	0.026	—
Polyvinyl acetate cork mastic; sprayed or troweled	—	0.057	—
Urethane, two-part mixture; rigid foam	4.3	0.015	0.249
Reflective			
Aluminum foil separating fluffy glass mats; 10–12 layers; evacuated; for cryogenic applications (150 K)	2.5	0.000 092	—
Aluminum foil and glass paper laminate; 75–150 layers; evacuated; for cryogenic application (150 K)	7.5	0.000 0098	—
Typical silica powder, evacuated	9.9	0.000 98	—

TABLE B-5 Thermophysical Properties of Solid Building Materials

Description/composition	Typical properties at 540 R		
	Density ρ lbm/ft^3	Thermal conductivity k Btu/h · ft · R	Specific heat c_p Btu/lbm · R
Building boards			
Asbestos-cement board	119.8	0.33	—
Gypsum or plaster board	50	0.098	—
Plywood	34.0	0.07	0.290
Sheathing, regular-density	18.1	0.031	0.310
Acoustic tile	18.1	0.034	0.32
Hardboard, siding	39.9	0.054	0.279
Hardboard, high-density	63.0	0.086	0.329
Particle board, low-density	36.8	0.045	0.310
Particle board, high-density	62.4	0.098	0.310
Woods			
Hardwoods (oak, maple)	44.9	0.092	0.299
Softwoods (fir, pine)	31.8	0.069	0.329
Masonry materials			
Cement mortar	116.1	0.41	0.186
Brick, common	119.8	0.41	0.199
Brick, face	130	0.75	—
Clay tile, hollow			
1 cell deep, 10-cm-thick	—	0.30	—
3 cells deep, 30-cm-thick	—	0.39	—
Concrete block, 3 oval cores			
sand/gravel, 20-cm-thick	—	0.57	—
cinder aggregate, 20-cm-thick	—	0.38	—

(*Continued*)

TABLE B-5 (Continued)

Description/composition	Typical properties at 540 R		
	Density ρ lbm/ft^3	Thermal conductivity k Btu/h · ft · R	Specific heat c_p Btu/lbm · R
Concrete block, rectangular core			
2 cores, 20-cm-thick, 16-kg	—	0.63	—
same with filled cores	—	0.34	—
Plastering materials			
Cement plaster, sand aggregate	116.1	0.41	—
Gypsum plaster, sand aggregate	104.8	0.12	0.259
Gypsum plaster, vermiculite aggregate	44.9	0.14	—

TABLE B-6 Thermophysical Properties of Liquids

Water (saturated liquid)

T °F	ρ lbm/ft^3	c_p Btu/lbm · °F	k Btu/h · ft · °F	$\mu \times 10^{5*}$ lbm/ft · s	$\upsilon \times 10^3$ ft^2/h	$\alpha \times 10^3$ ft^2/h	Pr	$\beta \times 10^4$ 1/R
32	62.4	1.01	0.319	120	69.2	5.07	13.7	−0.37
40	62.4	1.00	0.325	104	60.0	5.21	11.6	0.20
50	62.4	1.00	0.332	88.0	50.8	5.33	9.55	0.49
60	62.3	1.00	0.340	76.0	43.9	5.47	8.03	0.85
70	62.2	1.00	0.347	65.8	38.1	5.57	6.82	1.2
80	62.2	1.00	0.353	57.8	33.5	5.68	5.89	1.5
90	62.1	1.00	0.359	51.4	29.8	5.79	5.13	1.8
100	62.0	1.00	0.364	45.8	26.6	5.88	4.52	2.0
150	61.2	1.00	0.384	29.2	17.2	6.27	2.74	3.1
200	60.1	1.00	0.394	20.5	12.3	6.55	1.88	4.0
250	58.8	1.01	0.396	15.8	9.67	6.69	1.45	4.8
300	57.3	1.03	0.395	12.6	7.92	6.70	1.18	6.0
350	55.6	1.05	0.391	10.5	6.80	6.69	1.02	6.9
400	53.6	1.08	0.381	9.1	6.11	6.57	0.927	8.0
450	51.6	1.12	0.367	8.0	5.58	6.34	0.876	9.0
500	49.0	1.19	0.349	7.1	5.22	5.99	0.870	10
550	45.9	1.31	0.325	6.4	5.02	5.05	0.930	11
600	42.4	1.51	0.292	5.8	4.92	4.57	1.09	12

*For example, at 50 °F, $\mu \times 10^5 = 88.0$ lbm/ft · s $\rightarrow \mu = 88.0 \times 10^{-5}$ lbm/ft · s.

Light oil

T °F	ρ lbm/ft^3	c_p Btu/lbm · °F	k Btu/h · ft · °F	$\mu \times 10^{5*}$ lbm/ft · s	$\upsilon \times 10^3$ ft^2/h	$\alpha \times 10^3$ ft^2/h	Pr	$\beta \times 10^4$ 1/R
60	57.0	0.43	0.077	5820	3676	3.14	1170	3.8
80	56.8	0.44	0.077	2780	1762	3.09	570	3.8
100	56.0	0.46	0.076	1530	983.6	2.95	340	3.9
150	54.3	0.48	0.075	530	351.4	2.88	122	4.0
200	54.0	0.51	0.074	250	166.7	2.69	62	4.2
250	53.0	0.52	0.074	139	94.4	2.67	35	4.4
300	51.8	0.54	0.073	83	57.7	2.62	22	4.5

*For example, at 100 °F, $\mu \times 10^5 = 1530$ lbm/ft · s $\rightarrow \mu = 1530 \times 10^{-5}$ lbm/ft · s.

Glycerin

T °F	ρ lbm/ft^3	c_p Btu/lbm · °F	k Btu/h · ft · °F	$\mu \times 10^{5*}$ lbm/ft · s	$\upsilon \times 10^3$ ft^2/h	$\alpha \times 10^3$ ft^2/h	Pr	$\beta \times 10^4$ 1/R
50	79.3	0.554	0.165	256000	116217	3.76	31000	
70	78.9	0.570	0.165	100000	45627	3.67	12500	2.8
85	78.5	0.584	0.164	42400	19445	3.58	5400	3.0
100	78.2	0.600	0.163	18800	8655	3.45	2500	
120	77.7	0.617		12400	5745		1600	

*For example, at 100 °F, $\mu \times 10^5 = 18800$ lbm/ft · s $\rightarrow \mu = 18800 \times 10^{-5}$ lbm/ft · s.

(Continued)

TABLE B-6   (Continued)

R-12 (saturated liquid)

T °F	ρ lbm/ft^3	c_p Btu/lbm · °F	k Btu/h · ft · °F	$\mu \times 10^{5*}$ lbm/ft · s	$\upsilon \times 10^3$ ft^2/h	$\alpha \times 10^3$ ft^2/h	Pr	$\beta \times 10^4$ 1/R
−40	94.8	0.211	0.040	28.4	10.8	2.00	5.4	
−20	93.0	0.214	0.040	25.0	9.68	2.01	4.8	10.3
0	91.2	0.217	0.041	23.1	9.12	2.07	4.4	10.5
20	89.2	0.220	0.042	21.0	8.48	2.14	4.0	13.4
32	87.2	0.223	0.042	20.0	8.26	2.16	3.8	17.2
60	83.0	0.231	0.042	18.0	7.81	2.19	3.5	21
100	78.5	0.240	0.040	16.0	7.34	2.12	3.5	25
120	75.9	0.244	0.039	15.5	7.35	2.12	3.5	

*For example, at 100 °F, $\mu \times 10^5 = 16.0$ lbm/ft · s → $\mu = 16.0 \times 10^{-5}$ lbm/ft · s.

Mercury

T °F	ρ lbm/ft^3	c_p Btu/lbm · °F	k Btu/h · ft · °F	$\mu \times 10^{5*}$ lbm/ft · s	$\upsilon \times 10^3$ ft^2/h	$\alpha \times 10^3$ ft^2/h	Pr	$\beta \times 10^4$ 1/R
40	848	0.0334	4.55	111	4.71	161	0.0292	
60	847	0.0333	4.64	105	4.46	165	0.0270	
80	845	0.0332	4.72	100	4.26	169	0.0252	
100	843	0.0331	4.80	96.0	4.10	172	0.0239	
150	839	0.0330	5.03	89.3	3.83	182	0.0210	
200	835	0.0328	5.25	85.0	3.66	192	0.0191	
250	831	0.0328	5.45	80.6	3.49	200	0.0175	
300	827	0.0328	5.65	76.6	3.33	209	0.0160	
400	819	0.0328	6.05	70.0	3.08	225	0.0137	840
500	811	0.0328	6.43	65.0	2.89	243	0.0119	
600	804	0.0328	6.80	60.6	2.71	259	0.0105	
800	789	0.0329	7.45	55.0	2.51	289	0.0087	

*For example, at 100 °F, $\mu \times 10^5 = 96.0$ lbm/ft · s → $\mu = 96.0 \times 10^{-5}$ lbm/ft · s.

Hydraulic Fluid

T °F	ρ lbm/ft^3	c_p Btu/lbm · °F	k Btu/h · ft · °F	$\mu \times 10^{5*}$ lbm/ft · s	$\upsilon \times 10^3$ ft^2/h	$\alpha \times 10^3$ ft^2/h	Pr	$\beta \times 10^4$ 1/R
0	55.0	0.400	0.0780	5550	3633	3.54	1030	7.6
30	54.0	0.420	0.0755	2220	1480	3.32	446	6.8
60	53.0	0.439	0.0732	1110	754	3.14	239	6.0
80	52.5	0.453	0.0710	695	477	3.07	155	5.2
100	52.0	0.467	0.0690	556	385	2.84	136	4.7
150	51.0	0.499	0.0645	278	196	2.44	80.5	3.2
200	50.0	0.530	0.0600	250	180	2.27	79.4	2.0

*For example, at 100 °F, $\mu \times 10^5 = 556.0$ lbm/ft · s → $\mu = 556.0 \times 10^{-5}$ lbm/ft · s.

Sources: Data adapted from F. Kreith, *Principles of Heat Transfer*, 3rd ed. Intext Press New York, 1973; J. R. Welty, C. E. Wicks, and R. E. Wilson, *Fundamentals of Momentum, Heat, and Mass Transfer*, 2nd ed., Wiley New York, 1976.

TABLE B-7   **Thermophysical Properties of Gases**

Air (at 1 atm)

T °F	ρ lbm/ft^3	c_p Btu/lbm ·°F	k Btu/h · ft ·°F	$\mu \times 10^{5*}$ lbm/ft · s	υ ft^2/h	α ft^2/h	Pr
0	0.086	0.239	0.0133	1.110	0.465	0.646	0.73
32	0.081	0.240	0.0140	1.165	0.518	0.720	0.72
60	0.077	0.240	0.0146	1.214	0.568	0.796	0.72
80	0.074	0.240	0.0150	1.250	0.608	0.851	0.72

(Continued)

TABLE B-7 **(Continued)**

Air (at 1 atm)

T °F	ρ lbm/ft^3	c_p Btu/lbm · °F	k Btu/h · ft · °F	$\mu \times 10^{5*}$ lbm/ft · s	υ ft^2/h	α ft^2/h	Pr
100	0.071	0.240	0.0154	1.285	0.652	0.905	0.72
120	0.069	0.240	0.0158	1.316	0.687	0.964	0.72
140	0.067	0.241	0.0162	1.347	0.724	1.023	0.72
160	0.064	0.241	0.0166	1.378	0.775	1.082	0.72
180	0.062	0.241	0.0170	1.409	0.818	1.141	0.72
200	0.060	0.241	0.0174	1.44	0.864	1.20	0.72
300	0.052	0.243	0.0193	1.61	1.115	1.53	0.71
400	0.046	0.245	0.0212	1.75	1.370	1.88	0.689
500	0.0412	0.247	0.0231	1.89	1.651	2.27	0.683
600	0.0373	0.250	0.0250	2.00	1.930	2.68	0.685
700	0.0341	0.253	0.0268	2.14	2.259	3.10	0.690
800	0.0314	0.256	0.0286	2.25	2.580	3.56	0.697
900	0.0291	0.259	0.0303	2.36	2.920	4.02	0.705
1000	0.0271	0.262	0.0319	2.47	3.281	4.50	0.713
1500	0.0202	0.276	0.0400	3.00	5.347	7.19	0.739
2000	0.0161	0.286	0.0471	3.46	7.737	10.2	0.753

*For example, at 100 °F, $\mu \times 10^5 = 1.285$ lbm/ft · s $\rightarrow \mu = 1.285 \times 10^{-5}$ lbm/ft · s.

Nitrogen (at 1 atm)

T °F	ρ lbm/ft^3	c_p Btu/lbm · °F	k Btu/h · ft · °F	$\mu \times 10^{5*}$ lbm/ft · s	υ ft^2/h	α ft^2/h	Pr
0	0.0837	0.249	0.0132	1.06	0.457	0.633	0.719
30	0.0786	0.249	0.0139	1.12	0.511	0.710	0.719
60	0.0740	0.249	0.0146	1.17	0.569	0.800	0.716
80	0.0711	0.249	0.0151	1.20	0.608	0.853	0.712
100	0.0685	0.249	0.0154	1.23	0.648	0.915	0.708
150	0.0630	0.249	0.0168	1.32	0.752	1.07	0.702
200	0.0580	0.249	0.0174	1.39	0.864	1.25	0.690
250	0.0540	0.249	0.0192	1.47	0.976	1.42	0.687
300	0.0502	0.250	0.0202	1.53	1.098	1.62	0.685
400	0.0443	0.250	0.0212	1.67	1.357	2.02	0.684
500	0.0397	0.253	0.0244	1.80	1.631	2.43	0.683
600	0.0363	0.256	0.0252	1.93	1.915	2.81	0.686
800	0.0304	0.262	0.0291	2.16	2.556	3.71	0.691
1000	0.0263	0.269	0.0336	2.37	3.244	4.64	0.700
1500	0.0195	0.283	0.0423	2.82	5.220	7.14	0.732

*For example, at 100 °F, $\mu \times 10^5 = 1.23$ lbm/ft · s $\rightarrow \mu = 1.23 \times 10^{-5}$ lbm/ft · s.

Oxygen (at 1 atm)

T °F	ρ lbm/ft^3	c_p Btu/lbm · °F	k Btu/h · ft · °F	$\mu \times 10^{5*}$ lbm/ft · s	υ ft^2/h	α ft^2/h	Pr
0	0.0955	0.219	0.0134	1.22	0.461	0.641	0.718
30	0.0897	0.219	0.0141	1.28	0.515	0.718	0.716
60	0.0845	0.219	0.0149	1.35	0.576	0.806	0.713
80	0.0814	0.220	0.0155	1.40	0.619	0.866	0.713
100	0.0785	0.220	0.0160	1.43	0.655	0.925	0.708
150	0.0720	0.221	0.0172	1.52	0.760	1.08	0.703
200	0.0665	0.223	0.0185	1.62	0.878	1.25	0.703
250	0.0618	0.225	0.0197	1.70	0.994	1.42	0.700
300	0.0578	0.227	0.0209	1.79	1.116	1.60	0.700
400	0.0511	0.230	0.0233	1.95	1.372	1.97	0.698
500	0.0458	0.234	0.0254	2.10	1.649	2.37	0.696

(Continued)

TABLE B-7   (Continued)

Oxygen (at 1 atm)

T °F	ρ lbm/ft^3	c_p Btu/lbm·°F	k Btu/h·ft·°F	$\mu \times 10^{5*}$ lbm/ft·s	v ft^2/h	α ft^2/h	Pr
600	0.0414	0.239	0.0281	2.25	1.955	2.84	0.688
800	0.0349	0.246	0.0324	2.52	2.603	3.77	0.680
1000	0.0300	0.252	0.0366	2.79	3.348	4.85	0.691
1500	0.0224	0.264	0.0465	3.39	5.472	7.86	0.696

*For example, at 100 °F, $\mu \times 10^5 = 1.43$ lbm/ft·s → $\mu = 1.43 \times 10^{-5}$ lbm/ft·s.

Carbon dioxide (at 1 atm)

T °F	ρ lbm/ft^3	c_p Btu/lbm·°F	k Btu/h·ft·°F	$\mu \times 10^{5*}$ lbm/ft·s	v ft^2/h	α ft^2/h	Pr
0	0.132	0.193	0.0076	0.865	0.236	0.298	0.792
30	0.124	0.198	0.0083	0.915	0.266	0.339	0.787
60	0.117	0.202	0.0091	0.965	0.298	0.387	0.773
80	0.112	0.204	0.0096	1.00	0.321	0.421	0.760
100	0.108	0.207	0.0102	1.03	0.343	0.455	0.758
150	0.100	0.213	0.0115	1.12	0.407	0.539	0.755
200	0.092	0.219	0.0130	1.20	0.472	0.646	0.730
250	0.0850	0.225	0.0148	1.32	0.558	0.777	0.717
300	0.0800	0.230	0.0160	1.36	0.616	0.878	0.704
400	0.0740	0.239	0.0180	1.45	0.706	1.02	0.695
500	0.0630	0.248	0.0210	1.65	0.947	1.36	0.700
600	0.0570	0.256	0.0235	1.78	1.123	1.61	0.700
800	0.0480	0.269	0.0278	2.02	1.512	2.15	0.702
1000	0.0416	0.280	0.0324	2.25	1.944	2.78	0.703
1500	0.0306	0.301	0.0340	2.80	3.287	4.67	0.704

*For example, at 100 °F, $\mu \times 10^5 = 1.03$ lbm/ft·s → $\mu = 1.03 \times 10^{-5}$ lbm/ft·s.

Hydrogen (at 1 atm)

T °F	ρ lbm/ft^3	c_p Btu/lbm·°F	k Btu/h·ft·°F	$\mu \times 10^{5*}$ lbm/ft·s	v ft^2/h	α ft^2/h	Pr
0	0.00597	3.37	0.092	0.537	3.240	4.59	0.713
30	0.00562	3.39	0.097	0.562	3.600	5.09	0.709
60	0.00530	3.41	0.102	0.587	3.996	5.65	0.707
80	0.00510	3.42	0.105	0.602	4.248	6.04	0.705
100	0.00492	3.42	0.108	0.617	4.500	6.42	0.700
150	0.00450	3.44	0.116	0.653	5.220	7.50	0.696
200	0.00412	3.45	0.123	0.688	6.012	8.64	0.696
250	0.00382	3.46	0.130	0.723	6.804	9.85	0.690
300	0.00357	3.46	0.137	0.756	7.632	11.1	0.687
400	0.00315	3.47	0.151	0.822	9.396	13.8	0.681
500	0.00285	3.47	0.165	0.890	11.232	16.7	0.675
600	0.00260	3.47	0.179	0.952	13.176	19.8	0.667
800	0.00219	3.49	0.205	1.07	17.532	26.8	0.654
1000	0.00189	3.52	0.224	1.18	22.356	33.7	0.664
1500	0.00141	3.62	0.265	1.44	36.720	51.9	0.708

*For example, at 100 °F, $\mu \times 10^5 = 0.617$ lbm/ft·s → $\mu = 0.617 \times 10^{-5}$ lbm/ft·s.

Water vapor (at 1 atm)

T °F	ρ lbm/ft^3	c_p Btu/lbm·°F	k Btu/h·ft·°F	$\mu \times 10^{5*}$ lbm/ft·s	v ft^2/h	α ft^2/h	Pr
212	0.0372	0.451	0.0145	0.870	0.842	0.864	0.96
300	0.0328	0.456	0.0171	1.000	1.09	1.14	0.95

(Continued)

TABLE B-7 (Continued)

Water vapor (at 1 atm)

T °F	ρ lbm/ft^3	c_p Btu/lbm · °F	k Btu/h · ft · °F	$\mu \times 10^{5*}$ lbm/ft · s	v ft^2/h	α ft^2/h	Pr
400	0.0288	0.462	0.0200	1.130	1.42	1.50	0.94
500	0.0258	0.470	0.0228	1.265	1.76	1.88	0.94
600	0.0233	0.477	0.0257	1.420	2.20	2.31	0.94
700	0.0213	0.485	0.0288	1.555	2.61	2.79	0.93
800	0.0196	0.494	0.0321	1.700	3.08	3.32	0.92
900	0.0181	0.50	0.0355	1.810	3.55	3.93	0.91
1000	0.0169	0.51	0.0388	1.920	4.07	4.50	0.91
1200	0.0149	0.53	0.0457	2.14	5.18	5.80	0.88
1400	0.0133	0.55	0.053	2.36	6.41	7.25	0.87
1600	0.0120	0.56	0.061	2.58	7.70	9.07	0.87
1800	0.0109	0.58	0.068	2.81	9.29	10.8	0.87
2000	0.0100	0.60	0.076	3.03	10.9	12.7	0.86
2500	0.0083	0.64	0.096	3.58	15.5	18.1	0.86
3000	0.0071	0.67	0.114	4.00	20.7	24.0	0.86

*For example, at 300 ° F, $\mu \times 10^5 = 1.000$ lbm/ft · s → $\mu = 1.000 \times 10^{-5}$ lbm/ft · s.

Steam (at saturation pressure)

T °F	P psia	ρ lbm/ft^3	c_p Btu/lbm · °F	k Btu/h · ft · °F	$\mu \times 10^{5*}$ lbm/ft · s	$v \times 10^{3*}$ ft^2/h	α ft^2/h	Pr
32	0.088	0.000300	0.443	0.0105	0.540	64734	79328	0.816
40	0.121	0.000405	0.443	0.0107	0.551	48942	59448	0.823
60	0.255	0.00082	0.445	0.0110	0.580	25457	30249	0.842
80	0.507	0.00156	0.447	0.0114	0.610	14038	16324	0.860
100	0.950	0.00283	0.450	0.0118	0.640	8148	9278	0.878
120	1.695	0.00487	0.453	0.0122	0.670	4949	5521	0.896
140	2.892	0.0081	0.458	0.0126	0.700	3129	3421	0.915
160	4.745	0.0128	0.463	0.0131	0.731	2050	2198	0.933
180	7.513	0.0197	0.470	0.0135	0.761	1387	1458	0.951
200	11.52	0.0292	0.479	0.0140	0.790	973.4	1003	0.970
212	14.69	0.0368	0.485	0.0144	0.808	790.1	804.7	0.982
220	17.18	0.0427	0.489	0.0146	0.819	690.6	697.8	0.990
240	24.95	0.061	0.501	0.0151	0.848	501.7	496.6	1.01
280	49.17	0.115	0.532	0.0164	0.904	282.5	267.7	1.06
320	89.63	0.203	0.573	0.0179	0.958	169.9	153.6	1.11
360	153.0	0.338	0.627	0.0196	1.01	107.7	92.6	1.16
400	247.2	0.536	0.695	0.0217	1.06	71.3	58.4	1.22
440	382.8	0.82	0.780	0.0244	1.11	49.2	38.3	1.28
480	567.1	1.19	0.894	0.0278	1.17	35.3	26.0	1.36
520	812.4	1.70	1.06	0.0324	1.23	26.1	18.1	1.44
560	1131.9	2.62	1.30	0.0391	1.31	17.9	11.4	1.57
600	1541.4	3.74	1.73	0.0481	1.43	13.8	7.46	1.85
640	2059.8	5.56	2.83	0.0600	1.63	10.6	3.82	2.76
680	2710.2	8.98	6.24	0.0780	1.95	7.81	1.39	5.61
700	3093.7	13.2		0.102	2.41	6.57		
705.5	3206.8	19.6		0.137	3.03	5.56		

*For example, at 100 °F, $\mu \times 10^5 = 0.640$ lbm/ft · s → $\mu = 0.640 \times 10^{-5}$ lbm/ft · s.

Sources: Data adapted from F. Kreith, *Principles of Heat Transfer*, 3rd ed., Intext Press New York, 1973; J. R. Welty, C. E. Wicks, and R. E. Wilson, *Fundamentals of Momentum, Heat, and Mass Transfer*, 2nd ed., Wiley New York, 1976.

TABLE B-8   Ideal Gas Specific Heats in Tabular Form in (c_p, c_v, Btu/lbm · R; $k = c_p/c_v$)

Temp. °F	c_p	c_v	k	c_p	c_v	k	c_p	c_v	k	Temp. °F
	Air			Nitrogen, N_2			Oxygen, O_2			
40	0.240	0.171	1.401	0.248	0.177	1.400	0.219	0.156	1.397	40
100	0.240	0.172	1.400	0.248	0.178	1.399	0.220	0.158	1.394	100
200	0.241	0.173	1.397	0.249	0.178	1.398	0.223	0.161	1.387	200
300	0.243	0.174	1.394	0.250	0.179	1.396	0.226	0.164	1.378	300
400	0.245	0.176	1.389	0.251	0.180	1.393	0.230	0.168	1.368	400
500	0.248	0.179	1.383	0.254	0.183	1.388	0.235	0.173	1.360	500
600	0.250	0.182	1.377	0.256	0.185	1.383	0.239	0.177	1.352	600
700	0.254	0.185	1.371	0.260	0.189	1.377	0.242	0.181	1.344	700
800	0.257	0.188	1.365	0.262	0.191	1.371	0.246	0.184	1.337	800
900	0.259	0.191	1.358	0.265	0.194	1.364	0.249	0.187	1.331	900
1000	0.263	0.195	1.353	0.269	0.198	1.359	0.252	0.190	1.326	1000
1500	0.276	0.208	1.330	0.283	0.212	1.334	0.263	0.201	1.309	1500
2000	0.286	0.217	1.312	0.293	0.222	1.319	0.270	0.208	1.298	2000

Temp. °F	c_p	c_v	k	c_p	c_v	k	c_p	c_v	k	Temp. °F
	Carbon dioxide, CO_2			Carbon monoxide, CO			Hydrogen, H_2			
40	0.195	0.150	1.300	0.248	0.177	1.400	3.397	2.412	1.409	40
100	0.205	0.160	1.283	0.249	0.178	1.399	3.426	2.441	1.404	100
200	0.217	0.172	1.262	0.249	0.179	1.397	3.451	2.466	1.399	200
300	0.229	0.184	1.246	0.251	0.180	1.394	3.461	2.476	1.398	300
400	0.239	0.193	1.233	0.253	0.182	1.389	3.466	2.480	1.397	400
500	0.247	0.202	1.223	0.256	0.185	1.384	3.469	2.484	1.397	500
600	0.255	0.210	1.215	0.259	0.188	1.377	3.473	2.488	1.396	600
700	0.262	0.217	1.208	0.262	0.191	1.371	3.477	2.492	1.395	700
800	0.269	0.224	1.202	0.266	0.195	1.364	3.494	2.509	1.393	800
900	0.275	0.230	1.197	0.269	0.198	1.357	3.502	2.519	1.392	900
1000	0.280	0.235	1.192	0.273	0.202	1.351	3.513	2.528	1.390	1000
1500	0.298	0.253	1.178	0.287	0.216	1.328	3.618	2.633	1.374	1500
2000	0.312	0.267	1.169	0.297	0.226	1.314	3.758	2.773	1.355	2000

Source: Adapted from K. Wark, *Thermodynamics*, 4th ed., McGraw-Hill, New York, 1983, as based on "Tables of Thermal Properties of Gases," NBS Circular 564, 1955.

Ideal Gas Specific Heats in Equation Form (Btu/lbmol · R)

$$\frac{\bar{c}_p}{\bar{R}} = \alpha + \beta T + \gamma T^2 + \delta T^3 + \varepsilon T^4$$

T is in R, equations valid from 540 to 1800 R

Gas	α	$\beta \times 10^3$	$\gamma \times 10^6$	$\delta \times 10^9$	$\varepsilon \times 10^{12}$
CO	3.710	−0.899	1.140	−0.348	0.0228
CO_2	2.401	4.853	−2.039	0.343	0
H_2	3.057	1.487	−1.793	0.947	−0.1726
H_2O	4.070	−0.616	1.281	−0.508	0.0769
O_2	3.626	−1.043	2.178	−1.160	0.2053
N_2	3.675	−0.671	0.717	−0.108	−0.0215
Air	3.653	−0.7428	1.017	−0.328	0.02632
NH_3	3.591	0.274	2.576	−1.437	0.2601
NO	4.046	−1.899	2.464	−1.048	0.1517
NO_2	3.459	1.147	2.064	−1.639	0.3448
SO_2	3.267	2.958	0.211	−0.906	0.2438
SO_3	2.578	8.087	−2.832	−0.136	0.1878
CH_4	3.826	−2.211	7.580	−3.898	0.6633
C_2H_2	1.410	10.587	−7.562	2.811	−0.3939
C_2H_4	1.426	6.324	2.466	−2.787	0.6429
Monatomic gases*	2.5	0	0	0	0

*For monatomic gases, such as He, Ne, and Ar, $\bar{c}_p$ is constant over a wide temperature range and is very nearly equal to 5/2 $\bar{R}$.

Source: Adapted from K. Wark, *Thermodynamics*, 4th ed., McGraw-Hill, New York, 1983, as based on NASA SP-273, U.S. Government Printing Office, Washington, DC, 1971.

TABLE B-9   Ideal Gas Properties of Air

T(R), h and u (Btu/lbm), $s°$ (Btu/lbm · R)			$\Delta s = 0$		
T	h	u	P_r	v_r	$s°$
360	85.97	61.29	0.3363	396.6	0.50369
380	90.75	64.70	0.4061	346.6	0.51663
400	95.53	68.11	0.4858	305.0	0.52890
420	100.32	71.52	0.5760	270.1	0.54058
440	105.11	74.93	0.6776	240.6	0.55172
460	109.90	78.36	0.7913	215.33	0.56235
480	114.69	81.77	0.9182	193.65	0.57255
500	119.48	85.20	1.0590	174.90	0.58233
520	124.27	88.62	1.2147	158.58	0.59172
537	128.34	91.53	1.3593	146.34	0.59945
540	129.06	92.04	1.3860	144.32	0.60078
560	133.86	95.47	1.5742	131.78	0.60950
580	138.66	98.90	1.7800	120.70	0.61793
600	143.47	102.34	2.005	110.88	0.62607
620	148.28	105.78	2.249	102.12	0.63395
640	153.09	109.21	2.514	94.30	0.64159
660	157.92	112.67	2.801	87.27	0.64902
680	162.73	116.12	3.111	80.96	0.65621
700	167.56	119.58	3.446	75.25	0.66321
720	172.39	123.04	3.806	70.07	0.67002
740	177.23	126.51	4.193	65.38	0.67665
760	182.08	129.99	4.607	61.10	0.68312
780	186.94	133.47	5.051	57.20	0.68942
800	191.81	136.97	5.526	53.63	0.69558
820	196.69	140.47	6.033	50.35	0.70160
840	201.56	143.98	6.573	47.34	0.70747
860	206.46	147.50	7.149	44.57	0.71323
880	211.35	151.02	7.761	42.01	0.71886
900	216.26	154.57	8.411	39.64	0.72438
920	221.18	158.12	9.102	37.44	0.72979
940	226.11	161.68	9.834	35.41	0.73509
960	231.06	165.26	10.61	33.52	0.74030
980	236.02	168.83	11.43	31.76	0.74540
1000	240.98	172.43	12.30	30.12	0.75042
1040	250.95	179.66	14.18	27.17	0.76019
1080	260.97	186.93	16.28	24.58	0.76964
1120	271.03	194.25	18.60	22.30	0.77880
1160	281.14	201.63	21.18	20.29	0.78767
1200	291.30	209.05	24.01	18.51	0.79628
1240	301.52	216.53	27.13	16.93	0.80466
1280	311.79	224.05	30.55	15.52	0.81280
1320	322.11	231.63	34.31	14.25	0.82075
1360	332.48	239.25	38.41	13.12	0.82848
1400	342.90	246.93	42.88	12.10	0.83604
1440	353.37	254.66	47.75	11.17	0.84341
1480	363.89	262.44	53.04	10.34	0.85062
1520	374.47	270.26	58.78	9.578	0.85767
1560	385.08	278.13	65.00	8.890	0.86456
1600	395.74	286.06	71.73	8.263	0.87130
1650	409.13	296.03	80.89	7.556	0.87954

(*Continued*)

TABLE B-9 (Continued)

T(R), h and u (Btu/lbm), s° (Btu/lbm · R)			Δs = 0		
T	h	u	P_r	v_r	s°
1700	422.59	306.06	90.95	6.924	0.88758
1750	436.12	316.16	101.98	6.357	0.89542
1800	449.71	326.32	114.0	5.847	0.90308
1850	463.37	336.55	127.2	5.388	0.91056
1900	477.09	346.85	141.5	4.974	0.91788
1950	490.88	357.20	157.1	4.598	0.92504
2000	504.71	367.61	174.0	4.258	0.93205
2050	518.61	378.08	192.3	3.949	0.93891
2100	532.55	388.60	212.1	3.667	0.94564
2150	546.54	399.17	233.5	3.410	0.95222
2200	560.59	409.78	256.6	3.176	0.95868
2250	574.69	420.46	281.4	2.961	0.96501
2300	588.82	431.16	308.1	2.765	0.97123
2350	603.00	441.91	336.8	2.585	0.97732
2400	617.22	452.70	367.6	2.419	0.98331
2450	631.48	463.54	400.5	2.266	0.98919
2500	645.78	474.40	435.7	2.125	0.99497
2550	660.12	485.31	473.3	1.996	1.00064
2600	674.49	496.26	513.5	1.876	1.00623
2650	688.90	507.25	556.3	1.765	1.01172
2700	703.35	518.26	601.9	1.662	1.01712
2750	717.83	529.31	650.4	1.566	1.02244
2800	732.33	540.40	702.0	1.478	1.02767
2850	746.88	551.52	756.7	1.395	1.03282
2900	761.45	562.66	814.8	1.318	1.03788
2950	776.05	573.84	876.4	1.247	1.04288
3000	790.68	585.04	941.4	1.180	1.04779
3050	805.34	596.28	1011	1.118	1.05264
3100	820.03	607.53	1083	1.060	1.05741
3150	834.75	618.82	1161	1.006	1.06212
3200	849.48	630.12	1242	0.9546	1.06676
3250	864.24	641.46	1328	0.9069	1.07134
3300	879.02	652.81	1418	0.8621	1.07585
3350	893.83	664.20	1513	0.8202	1.08031
3400	908.66	675.60	1613	0.7807	1.08470
3450	923.52	687.04	1719	0.7436	1.08904
3500	938.40	698.48	1829	0.7087	1.09332
3550	953.30	709.95	1946	0.6759	1.09755
3600	968.21	721.44	2068	0.6449	1.10172
3650	983.15	732.95	2196	0.6157	1.10584
3700	998.11	744.48	2330	0.5882	1.10991
3750	1013.1	756.04	2471	0.5621	1.11393
3800	1028.1	767.60	2618	0.5376	1.11791
3850	1043.1	779.19	2773	0.5143	1.12183
3900	1058.1	790.80	2934	0.4923	1.12571
3950	1073.2	802.43	3103	0.4715	1.12955
4000	1088.3	814.06	3280	0.4518	1.13334
4050	1103.4	825.72	3464	0.4331	1.13709
4100	1118.5	837.40	3656	0.4154	1.14079
4150	1133.6	849.09	3858	0.3985	1.14446

(Continued)

TABLE B-9 (Continued)

T(R), h and u (Btu/lbm), $s°$ (Btu/lbm · R)			$\Delta s = 0$		
T	h	u	P_r	v_r	$s°$
4200	1148.7	860.81	4067	0.3826	1.14809
4300	1179.0	884.28	4513	0.3529	1.15522
4400	1209.4	907.81	4997	0.3262	1.16221
4500	1239.9	931.39	5521	0.3019	1.16905
4600	1270.4	955.04	6089	0.2799	1.17575
4700	1300.9	978.73	6701	0.2598	1.18232
4800	1331.5	1002.5	7362	0.2415	1.18876
4900	1362.2	1026.3	8073	0.2248	1.19508
5000	1392.9	1050.1	8837	0.2096	1.20129
5100	1423.6	1074.0	9658	0.1956	1.20738
5200	1454.4	1098.0	10539	0.1828	1.21336
5300	1485.3	1122.0	11481	0.1710	1.21923

Source: Adapted from K. Wark, *Thermodynamics*, 4th ed., McGraw-Hill, New York, 1983, as extracted from J. H. Keenan and J. Kaye, *Gas Tables*, Wiley, New York, 1945.

TABLE B-10 **Thermodynamic Properties of Saturated Steam–Water (Temperature Table)**

Temp. °F T	Press. lbf/in.² P_{sat}	Specific Volume ft³/lbm		Internal Energy Btu/lbm			Enthalpy Btu/lbm			Entropy Btu/lbm · R		
		Sat. Liquid v_f	Sat. Vapor v_g	Sat. Liquid u_f	Evap. u_{fg}	Sat. Vapor u_g	Sat. Liquid h_f	Evap. h_{fg}	Sat. Vapor h_g	Sat. Liquid s_f	Evap. s_{fg}	Sat. Vapor s_g
32.018	0.088 66	0.016 022	3302	0.00	1021.2	1021.2	0.01	1075.4	1075.4	0.000 00	2.1869	2.1869
35	0.099 92	0.016 021	2948	2.99	1019.2	1022.2	3.00	1073.7	1076.7	0.006 07	2.1704	2.1764
40	0.121 66	0.016 020	2445	8.02	1015.8	1023.9	8.02	1070.9	1078.9	0.016 17	2.1430	2.1592
45	0.147 48	0.016 021	2037	13.04	1012.5	1025.5	13.04	1068.1	1081.1	0.026 18	2.1162	2.1423
50	0.178 03	0.016 024	1704.2	18.06	1009.1	1027.2	18.06	1065.2	1083.3	0.036 07	2.0899	2.1259
60	0.2563	0.016 035	1206.9	28.08	1002.4	1030.4	28.08	1059.6	1087.7	0.055 55	2.0388	2.0943
70	0.3632	0.016 051	867.7	38.09	995.6	1033.7	38.09	1054.0	1092.0	0.074 63	1.9896	2.0642
80	0.5073	0.016 073	632.8	48.08	988.9	1037.0	48.09	1048.3	1096.4	0.093 32	1.9423	2.0356
90	0.6988	0.016 099	467.7	58.07	982.2	1040.2	58.07	1042.7	1100.7	0.111 65	1.8966	2.0083
100	0.9503	0.016 130	350.0	68.04	975.4	1043.5	68.05	1037.0	1105.0	0.129 63	1.8526	1.9822
110	1.2763	0.016 166	265.1	78.02	968.7	1046.7	78.02	1031.3	1109.3	0.147 30	1.8101	1.9574
120	1.6945	0.016 205	203.0	87.99	961.9	1049.9	88.00	1025.5	1113.5	0.164 65	1.7690	1.9336
130	2.225	0.016 247	157.17	97.97	955.1	1053.0	97.98	1019.8	1117.8	0.181 72	1.7292	1.9109
140	2.892	0.016 293	122.88	107.95	948.2	1056.2	107.96	1014.0	1121.9	0.198 51	1.6907	1.8892
150	3.722	0.016 343	96.99	117.95	941.3	1059.3	117.96	1008.1	1126.1	0.215 03	1.6533	1.8684
160	4.745	0.016 395	77.23	127.94	934.4	1062.3	127.96	1002.2	1130.1	0.231 30	1.6171	1.8484
170	5.996	0.016 450	62.02	137.95	927.4	1065.4	137.97	996.2	1134.2	0.247 32	1.5819	1.8293
180	7.515	0.016 509	50.20	147.97	920.4	1068.3	147.99	990.2	1138.2	0.263 11	1.5478	1.8109
190	9.343	0.016 570	40.95	158.00	913.3	1071.3	158.03	984.1	1142.1	0.278 66	1.5146	1.7932
200	11.529	0.016 634	33.63	168.04	906.2	1074.2	168.07	977.9	1145.9	0.294 00	1.4822	1.7762
210	14.125	0.016 702	27.82	178.10	898.9	1077.0	178.14	971.6	1149.7	0.309 13	1.4508	1.7599
212	14.698	0.016 716	26.80	180.11	897.5	1077.6	180.16	970.3	1150.5	0.312 13	1.4446	1.7567
220	17.188	0.016 772	23.15	188.17	891.7	1079.8	188.22	965.3	1153.5	0.324 06	1.4201	1.7441
230	20.78	0.016 845	19.386	198.26	884.3	1082.6	198.32	958.8	1157.1	0.338 80	1.3901	1.7289
240	24.97	0.016 922	16.327	208.36	876.9	1085.3	208.44	952.3	1160.7	0.353 35	1.3609	1.7143
250	29.82	0.017 001	13.826	218.49	869.4	1087.9	218.59	945.6	1164.2	0.367 72	1.3324	1.7001
260	35.42	0.017 084	11.768	228.64	861.8	1090.5	228.76	938.8	1167.6	0.381 93	1.3044	1.6864
270	41.85	0.017 170	10.066	238.82	854.1	1093.0	238.95	932.0	1170.9	0.395 97	1.2771	1.6731

(Continued)

TABLE B-10 (Continued)

Temp. °F T	Press. lbf/in.^{2} P_{sat}	Specific Volume ft^3/lbm		Internal Energy Btu/lbm			Enthalpy Btu/lbm			Entropy Btu/lbm · R		
		Sat. Liquid v_f	Sat. Vapor v_g	Sat. Liquid u_f	Evap. u_{fg}	Sat. Vapor u_g	Sat. Liquid h_f	Evap. h_{fg}	Sat. Vapor h_g	Sat. Liquid s_f	Evap. s_{fg}	Sat. Vapor s_g
280	49.18	0.017 259	8.650	249.02	846.3	1095.4	249.18	924.9	1174.1	0.409 86	1.2504	1.6602
290	57.53	0.017 352	7.467	259.25	838.5	1097.7	259.44	917.8	1177.2	0.423 60	1.2241	1.6477
300	66.98	0.017 448	6.472	269.52	830.5	1100.0	269.73	910.4	1180.2	0.437 20	1.1984	1.6356
310	77.64	0.017 548	5.632	279.81	822.3	1102.1	280.06	903.0	1183.0	0.450 67	1.1731	1.6238
320	89.60	0.017 652	4.919	290.14	814.1	1104.2	290.43	895.3	1185.8	0.464 00	1.1483	1.6123
330	103.00	0.017 760	4.312	300.51	805.7	1106.2	300.84	887.5	1188.4	0.477 22	1.1238	1.6010
340	117.93	0.017 872	3.792	310.91	797.1	1108.0	311.30	879.5	1190.8	0.490 31	1.0997	1.5901
350	134.53	0.017 988	3.346	321.35	788.4	1109.8	321.80	871.3	1193.1	0.503 29	1.0760	1.5793
360	152.92	0.018 108	2.961	331.84	779.6	1111.4	332.35	862.9	1195.2	0.516 17	1.0526	1.5688
370	173.23	0.018 233	2.628	342.37	770.6	1112.9	342.96	854.2	1197.2	0.528 94	1.0295	1.5585
380	195.60	0.018 363	2.339	352.95	761.4	1114.3	353.62	845.4	1199.0	0.541 63	1.0067	1.5483
390	220.2	0.018 498	2.087	363.58	752.0	1115.6	364.34	836.2	1200.6	0.554 22	0.9841	1.5383
400	247.1	0.018 638	1.8661	374.27	742.4	1116.6	375.12	826.8	1202.0	0.566 72	0.9617	1.5284
420	308.5	0.018 936	1.5024	395.81	722.5	1118.3	396.89	807.2	1204.1	0.591 52	0.9175	1.5091
430	343.3	0.019 094	1.3521	406.68	712.2	1118.9	407.89	796.9	1204.8	0.603 81	0.8957	1.4995
440	381.2	0.019 260	1.2192	417.62	701.7	1119.3	418.98	786.3	1205.3	0.616 05	0.8740	1.4900
450	422.1	0.019 433	1.1011	428.6	690.9	1119.5	430.2	775.4	1205.6	0.6282	0.8523	1.4806
460	466.3	0.019 614	0.9961	439.7	679.8	1119.6	441.4	764.1	1205.5	0.6404	0.8308	1.4712
470	514.1	0.019 803	0.9025	450.9	668.4	1119.4	452.8	752.4	1205.2	0.6525	0.8093	1.4618
480	565.5	0.020 002	0.8187	462.2	656.7	1118.9	464.3	740.3	1204.6	0.6646	0.7878	1.4524
490	620.7	0.020 211	0.7436	473.6	644.7	1118.3	475.9	727.8	1203.7	0.6767	0.7663	1.4430
500	680.0	0.020 43	0.6761	485.1	632.3	1117.4	487.7	714.8	1202.5	0.6888	0.7448	1.4335
520	811.4	0.020 91	0.5605	508.5	606.2	1114.8	511.7	687.3	1198.9	0.7130	0.7015	1.4145
540	961.5	0.021 45	0.4658	532.6	578.4	1111.0	536.4	657.5	1193.8	0.7374	0.6576	1.3950
560	1131.8	0.022 07	0.3877	557.4	548.4	1105.8	562.0	625.0	1187.0	0.7620	0.6129	1.3749
580	1324.3	0.022 78	0.3225	583.1	515.9	1098.9	588.6	589.3	1178.0	0.7872	0.5668	1.3540
600	1541.0	0.023 63	0.2677	609.9	480.1	1090.0	616.7	549.7	1166.4	0.8130	0.5187	1.3317
620	1784.4	0.024 65	0.2209	638.3	440.2	1078.5	646.4	505.0	1151.4	0.8398	0.4677	1.3075
640	2057.1	0.025 93	0.1805	668.7	394.5	1063.2	678.6	453.4	1131.9	0.8681	0.4122	1.2803
660	2362	0.027 67	0.144 59	702.3	340.0	1042.3	714.4	391.1	1105.5	0.8990	0.3493	1.2483
680	2705	0.030 32	0.111 27	741.7	269.3	1011.0	756.9	309.8	1066.7	0.9350	0.2718	1.2068
700	3090	0.036 66	0.074 38	801.7	145.9	947.7	822.7	167.5	990.2	0.9902	0.1444	1.1346
705.44	3204	0.050 53	0.050 53	872.6	0	872.6	902.5	0	902.5	1.0580	0	1.0580

Source: Tables B-10 to B-13 adapted from Van Wylen, G. J., R. E. Sonntag, and C. Borgnakke, *Fundamentals of Classical Thermodynamics*, 4th ed., Wiley, New York, 1994.

TABLE B-11 Thermodynamic Properties of Saturated Steam–Water (Pressure Table)

Press. lbf/in.^{2} P	Temp. °F T_{sat}	Specific Volume ft^3/lbm		Internal Energy Btu/lbm			Enthalpy Btu/lbm			Entropy Btu/lbm · R		
		Sat. Liquid v_f	Sat. Vapor v_g	Sat. Liquid u_f	Evap. u_{fg}	Sat. Vapor u_g	Sat. Liquid h_f	Evap. h_{fg}	Sat. Vapor h_g	Sat. Liquid s_f	Evap. s_{fg}	Sat. Vapor s_g
1.0	101.70	0.016 136	333.6	69.74	974.3	1044.0	69.74	1036.0	1105.8	0.132 66	1.8453	1.9779
2.0	126.04	0.016 230	173.75	94.02	957.8	1051.8	94.02	1022.1	1116.1	0.174 99	1.7448	1.9198
3.0	141.43	0.016 300	118.72	109.38	947.2	1056.6	109.39	1013.1	1122.5	0.200 89	1.6852	1.8861
4.0	152.93	0.016 358	90.64	120.88	939.3	1060.2	120.89	1006.4	1127.3	0.219 83	1.6426	1.8624
5.0	162.21	0.016 407	73.53	130.15	932.9	1063.0	130.17	1000.9	1131.0	0.234 86	1.6093	1.8441

(*Continued*)

TABLE B-11 (Continued)

Press. lbf/in.² P	Temp. °F T_{sat}	Specific Volume ft³/lbm		Internal Energy Btu/lbm			Enthalpy Btu/lbm			Entropy Btu/lbm · R		
		Sat. Liquid v_f	Sat. Vapor v_g	Sat. Liquid u_f	Evap. u_{fg}	Sat. Vapor u_g	Sat. Liquid h_f	Evap. h_{fg}	Sat. Vapor h_g	Sat. Liquid s_f	Evap. s_{fg}	Sat. Vapor s_g
6.0	170.03	0.016 451	61.98	137.98	927.4	1065.4	138.00	996.2	1134.2	0.247 36	1.5819	1.8292
8.0	182.84	0.016 526	47.35	150.81	918.4	1069.2	150.84	988.4	1139.3	0.267 54	1.5383	1.8058
10	193.19	0.016 590	38.42	161.20	911.0	1072.2	161.23	982.1	1143.3	0.283 58	1.5041	1.7877
14.696	211.99	0.016 715	26.80	180.10	897.5	1077.6	180.15	970.4	1150.5	0.312 12	1.4446	1.7567
15	213.03	0.016 723	26.29	181.14	896.8	1077.9	181.19	969.7	1150.9	0.313 67	1.4414	1.7551
20	227.96	0.016 830	20.09	196.19	885.8	1082.0	196.26	960.1	1156.4	0.335 80	1.3962	1.7320
25	240.08	0.016 922	16.306	208.44	876.9	1085.3	208.52	952.2	1160.7	0.353 45	1.3607	1.7142
30	250.34	0.017 004	13.748	218.84	869.2	1088.0	218.93	945.4	1164.3	0.368 21	1.3314	1.6996
35	259.30	0.017 073	11.900	227.93	862.4	1090.3	228.04	939.3	1167.4	0.380 93	1.3064	1.6873
40	267.26	0.017 146	10.501	236.03	856.2	1092.3	236.16	933.8	1170.0	0.392 14	1.2845	1.6767
45	274.46	0.017 209	9.403	243.37	850.7	1094.0	243.51	928.8	1172.3	0.402 18	1.2651	1.6673
50	281.03	0.017 269	8.518	250.08	845.5	1095.6	250.24	924.2	1174.4	0.411 29	1.2476	1.6589
55	287.10	0.017 325	7.789	256.28	840.8	1097.0	256.46	919.9	1176.3	0.419 63	1.2317	1.6513
60	292.73	0.017 378	7.177	262.06	836.3	1098.3	262.25	915.8	1178.0	0.427 33	1.2170	1.6444
65	298.00	0.017 429	6.657	267.46	832.1	1099.5	267.67	911.9	1179.6	0.434 50	1.2035	1.6380
70	302.96	0.017 478	6.209	272.56	828.1	1100.6	272.79	908.3	1181.0	0.441 20	1.1909	1.6321
75	307.63	0.017 524	5.818	277.37	824.3	1101.6	277.61	904.8	1182.4	0.447 49	1.1790	1.6265
80	312.07	0.017 570	5.474	281.95	820.6	1102.6	282.21	901.4	1183.6	0.453 44	1.1679	1.6214
85	316.29	0.017 613	5.170	286.30	817.1	1103.5	286.58	898.2	1184.8	0.459 07	1.1574	1.6165
90	320.31	0.017 655	4.898	290.46	813.8	1104.3	290.76	895.1	1185.9	0.464 42	1.1475	1.6119
95	324.16	0.017 696	4.654	294.45	810.6	1105.0	294.76	892.1	1186.9	0.469 52	1.1380	1.6076
100	327.86	0.017 736	4.434	298.28	807.5	1105.8	298.61	889.2	1187.8	0.474 39	1.1290	1.6034
110	334.82	0.017 813	4.051	305.52	801.6	1107.1	305.88	883.7	1189.6	0.483 55	1.1122	1.5957
120	341.30	0.017 886	3.730	312.27	796.0	1108.3	312.67	878.5	1191.1	0.492 01	1.0966	1.5886
130	347.37	0.017 957	3.457	318.61	790.7	1109.4	319.04	873.5	1192.5	0.499 89	1.0822	1.5821
140	353.08	0.018 024	3.221	324.58	785.7	1110.3	325.05	868.7	1193.8	0.507 27	1.0688	1.5761
150	358.48	0.018 089	3.016	330.24	781.0	1111.2	330.75	864.2	1194.9	0.514 22	1.0562	1.5704
160	363.60	0.018 152	2.836	335.63	776.4	1112.0	336.16	859.8	1196.0	0.520 78	1.0443	1.5651
170	368.47	0.018 214	2.676	340.76	772.0	1112.7	341.33	855.6	1196.9	0.527 00	1.0330	1.5600
180	373.13	0.018 273	2.533	345.68	767.7	1113.4	346.29	851.5	1197.8	0.532 92	1.0223	1.5553
190	377.59	0.018 331	2.405	350.39	763.6	1114.0	351.04	847.5	1198.6	0.538 57	1.0122	1.5507
200	381.86	0.018 387	2.289	354.9	759.6	1114.6	355.6	843.7	1199.3	0.5440	1.0025	1.5464
250	401.04	0.018 653	1.8448	375.4	741.4	1116.7	376.2	825.8	1202.1	0.5680	0.9594	1.5274
300	417.43	0.018 896	1.5442	393.0	725.1	1118.2	394.1	809.8	1203.9	0.5883	0.9232	1.5115
350	431.82	0.019 124	1.3267	408.7	710.3	1119.0	409.9	795.0	1204.9	0.6060	0.8917	1.4978
400	444.70	0.019 340	1.1620	422.8	696.7	1119.5	424.2	781.2	1205.5	0.6218	0.8638	1.4856
450	456.39	0.019 547	1.0326	435.7	683.9	1119.6	437.4	768.2	1205.6	0.6360	0.8385	1.4746
500	467.13	0.019 748	0.9283	447.7	671.7	1119.4	449.5	755.8	1205.3	0.6490	0.8154	1.4645
550	477.07	0.019 943	0.8423	458.9	660.2	1119.1	460.9	743.9	1204.8	0.6611	0.7941	1.4551
600	486.33	0.020 13	0.7702	469.4	649.1	1118.6	471.7	732.4	1204.1	0.6723	0.7742	1.4464
700	503.23	0.020 51	0.6558	488.9	628.2	1117.0	491.5	710.5	1202.0	0.6927	0.7378	1.4305
800	518.36	0.020 87	0.5691	506.6	608.4	1115.0	509.7	689.6	1199.3	0.7110	0.7050	1.4160
900	532.12	0.021 23	0.5009	523.0	589.6	1112.6	526.6	669.5	1196.0	0.7277	0.6750	1.4027
1000	544.75	0.021 59	0.4459	538.4	571.5	1109.9	542.4	650.0	1192.4	0.7432	0.6471	1.3903
1200	567.37	0.022 32	0.3623	566.7	536.8	1103.5	571.7	612.3	1183.9	0.7712	0.5961	1.3673
1400	587.25	0.023 07	0.3016	592.7	503.3	1096.0	598.6	575.5	1174.1	0.7964	0.5497	1.3461
1600	605.06	0.023 86	0.2552	616.9	470.5	1087.4	624.0	538.9	1162.9	0.8196	0.5062	1.3258
1800	621.21	0.024 72	0.2183	640.0	437.6	1077.7	648.3	502.1	1150.4	0.8414	0.4645	1.3060
2000	636.00	0.025 65	0.18813	662.4	404.2	1066.6	671.9	464.4	1136.3	0.8623	0.4238	1.2861
2500	668.31	0.028 60	0.13059	717.7	313.4	1031.0	730.9	360.5	1091.4	0.9131	0.3196	1.2327
3000	695.52	0.034 31	0.08404	783.4	185.4	968.8	802.5	213.0	1015.5	0.9732	0.1843	1.1575
3203.6	705.44	0.050 53	0.05053	872.6	0	872.6	902.5	0	902.5	1.0580	0	1.0580

TABLE B-12 Thermodynamic Properties of Steam (Superheated Vapor)

T °F	v ft³/lbm	u Btu/lbm	h Btu/lbm	s Btu/lbm·R	v ft³/lbm	u Btu/lbm	h Btu/lbm	s Btu/lbm·R	v ft³/lbm	u Btu/lbm	h Btu/lbm	s Btu/lbm·R
	P = 1.0 psia (T_{sat} = 101.7°F)				P = 5.0 psia (T_{sat} = 162.21°F)				P = 10.0 psia (T_{sat} = 193.19°F)			
Sat	333.6	1044.0	1105.8	1.9779	73.53	1063.0	1131.0	1.8441	38.42	1072.2	1143.3	1.7877
200	392.5	1077.5	1150.1	2.0508	78.15	1076.3	1148.6	1.8715	38.85	1074.7	1146.6	1.7927
240	416.4	1091.2	1168.3	2.0775	83.00	1090.3	1167.1	1.8987	41.32	1089.0	1165.5	1.8205
280	440.3	1105.0	1186.5	2.1028	87.83	1104.3	1185.5	1.9244	43.77	1103.3	1184.3	1.8467
320	464.2	1118.9	1204.8	2.1269	92.64	1118.3	1204.0	1.9487	46.20	1117.6	1203.1	1.8714
360	488.1	1132.9	1223.2	2.1500	97.45	1132.4	1222.6	1.9719	48.62	1131.8	1221.8	1.8948
400	511.9	1147.0	1241.8	2.1720	102.24	1146.6	1241.2	1.9941	51.03	1146.1	1240.5	1.9171
440	535.8	1161.2	1260.4	2.1932	107.03	1160.9	1259.9	2.0154	53.44	1160.5	1259.3	1.9385
500	571.5	1182.8	1288.5	2.2235	114.20	1182.5	1288.2	2.0458	57.04	1182.2	1287.7	1.9690
600	631.1	1219.3	1336.1	2.2706	126.15	1219.1	1335.8	2.0930	63.03	1218.9	1335.5	2.0164
700	690.7	1256.7	1384.5	2.3142	138.08	1256.5	1384.3	2.1367	69.01	1256.3	1384.0	2.0601
800	750.3	1294.9	1433.7	2.3550	150.01	1294.7	1433.5	2.1775	74.98	1294.6	1433.3	2.1009
1000	869.5	1373.9	1534.8	2.4294	173.86	1373.9	1534.7	2.2520	86.91	1373.8	1534.6	2.1755
1200	988.6	1456.7	1639.6	2.4967	197.70	1456.6	1639.5	2.3192	98.84	1456.5	1639.4	2.2428
1400	1107.7	1543.1	1748.1	2.5584	221.54	1543.1	1748.1	2.3810	110.76	1543.0	1748.0	2.3045
	P = 14.696 psia (T_{sat} = 211.99°F)				P = 20.0 psia (T_{sat} = 227.96°F)				P = 40.0 psia (T_{sat} = 267.26°F)			
Sat	26.80	1077.6	1150.5	1.7567	20.09	1082.0	1156.4	1.7320	10.501	1092.3	1170.0	1.6767
240	28.00	1087.9	1164.0	1.7764	20.47	1086.5	1162.3	1.7405				
280	29.69	1102.4	1183.1	1.8030	21.73	1101.4	1181.8	1.7676	10.711	1097.3	1176.6	1.6857
320	31.36	1116.8	1202.1	1.8280	22.98	1116.0	1201.0	1.7930	11.360	1112.8	1196.9	1.7124
360	33.02	1131.2	1221.0	1.8516	24.21	1130.6	1220.1	1.8168	11.996	1128.0	1216.8	1.7373
400	34.67	1145.6	1239.9	1.8741	25.43	1145.1	1239.2	1.8395	12.623	1143.0	1236.4	1.7606
440	36.31	1160.1	1258.8	1.8956	26.64	1159.6	1258.2	1.8611	13.243	1157.8	1255.8	1.7828
500	38.77	1181.8	1287.3	1.9263	28.46	1181.5	1286.8	1.8919	14.164	1180.1	1284.9	1.8140
600	42.86	1218.6	1335.2	1.9737	31.47	1218.4	1334.8	1.9395	15.685	1217.3	1333.4	1.8621
700	46.93	1256.1	1383.8	2.0175	34.47	1255.9	1383.5	1.9834	17.196	1255.1	1382.4	1.9063
800	51.00	1294.4	1433.1	2.0584	37.46	1294.3	1432.9	2.0243	18.701	1293.7	1432.1	1.9474
1000	59.13	1373.7	1534.5	2.1330	43.44	1373.5	1534.3	2.0989	21.70	1373.1	1533.8	2.0223
1200	67.25	1456.5	1639.3	2.2003	49.41	1456.4	1639.2	2.1663	24.69	1456.1	1638.9	2.0897
1400	75.36	1543.0	1747.9	2.2621	55.37	1542.9	1747.9	2.2281	27.68	1542.7	1747.6	2.1515
1600	83.47	1633.2	1860.2	2.3194	61.33	1633.2	1860.1	2.2854	30.66	1633.0	1859.9	2.2089
	P = 60.0 psia (T_{sat} = 292.73°F)				P = 80.0 psia (T_{sat} = 312.07°F)				P = 100.0 psia (T_{sat} = 327.86°F)			
Sat	7.177	1098.3	1178.0	1.6444	5.474	1102.6	1183.6	1.6214	4.434	1105.8	1187.8	1.6034
320	7.485	1109.5	1192.6	1.6634	5.544	1106.0	1188.0	1.6271				
360	7.924	1125.3	1213.3	1.6893	5.886	1122.5	1209.7	1.6541	4.662	1119.7	1205.9	1.6259
400	8.353	1140.8	1233.5	1.7134	6.217	1138.5	1230.6	1.6790	4.934	1136.2	1227.5	1.6517
440	8.775	1156.0	1253.4	1.7360	6.541	1154.2	1251.0	1.7022	5.199	1152.3	1248.5	1.6755
500	9.399	1178.6	1283.0	1.7678	7.017	1177.2	1281.1	1.7346	5.587	1175.7	1279.1	1.7085
600	10.425	1216.3	1332.1	1.8165	7.794	1215.3	1330.7	1.7838	6.216	1214.2	1329.3	1.7582
700	11.440	1254.4	1381.4	1.8609	8.561	1253.6	1380.3	1.8285	6.834	1252.8	1379.2	1.8033
800	12.448	1293.0	1431.2	1.9022	9.321	1292.4	1430.4	1.8700	7.445	1291.8	1429.6	1.8449
1000	14.454	1372.7	1533.2	1.9773	10.831	1372.3	1532.6	1.9453	8.657	1371.9	1532.1	1.9204
1200	16.452	1455.8	1638.5	2.0448	12.333	1455.5	1638.1	2.0130	9.861	1455.2	1637.7	1.9882
1400	18.445	1542.5	1747.3	2.1067	13.830	1542.3	1747.0	2.0749	11.060	1542.0	1746.7	2.0502
1600	20.44	1632.8	1859.7	2.1641	15.324	1632.6	1859.5	2.1323	12.257	1632.4	1859.3	2.1076
1800	22.43	1726.7	1975.7	2.2179	16.818	1726.5	1975.5	2.1861	13.452	1726.4	1975.3	2.1614
2000	24.41	1824.0	2095.1	2.2685	18.310	1823.9	2094.9	2.2367	14.647	1823.7	2094.8	2.2121
	P = 120.0 psia (T_{sat} = 341.30°F)				P = 140.0 psia (T_{sat} = 353.08°F)				P = 160.0 psia (T_{sat} = 363.60°F)			
Sat	3.730	1108.3	1191.1	1.5886	3.221	1110.3	1193.8	1.5761	2.836	1112.0	1196.0	1.5651
360	3.844	1116.7	1202.0	1.6021	3.259	1113.5	1198.0	1.5812				

(Continued)

TABLE B-12 (Continued)

T °F	v ft³/lbm	u Btu/lbm	h Btu/lbm	s Btu/lbm·R	v ft³/lbm	u Btu/lbm	h Btu/lbm	s Btu/lbm·R	v ft³/lbm	u Btu/lbm	h Btu/lbm	s Btu/lbm·R
	$P = 120.0$ psia ($T_{sat} = 341.30°F$)				$P = 140.0$ psia ($T_{sat} = 353.08°F$)				$P = 160.0$ psia ($T_{sat} = 363.60°F$)			
400	4.079	1133.8	1224.4	1.6288	3.466	1131.4	1221.2	1.6088	3.007	1128.8	1217.8	1.5911
450	4.360	1154.3	1251.2	1.6590	3.713	1152.4	1248.6	1.6399	3.228	1150.5	1246.1	1.6230
500	4.633	1174.2	1277.1	1.6868	3.952	1172.7	1275.1	1.6682	3.440	1171.2	1273.0	1.6518
550	4.900	1193.8	1302.6	1.7127	4.184	1192.6	1300.9	1.6944	3.646	1191.3	1299.2	1.6784
600	5.164	1213.2	1327.8	1.7371	4.412	1212.1	1326.4	1.7191	3.848	1211.1	1325.0	1.7034
700	5.682	1252.0	1378.2	1.7825	4.860	1251.2	1377.1	1.7648	4.243	1250.4	1376.0	1.7494
800	6.195	1291.2	1428.7	1.8243	5.301	1290.5	1427.9	1.8068	4.631	1289.9	1427.0	1.7916
1000	7.208	1371.5	1531.5	1.9000	6.173	1371.0	1531.0	1.8827	5.397	1370.6	1530.4	1.8677
1200	8.213	1454.9	1637.3	1.9679	7.036	1454.6	1636.9	1.9507	6.154	1454.3	1636.5	1.9358
1400	9.214	1541.8	1746.4	2.0300	7.895	1541.6	1746.1	2.0129	6.906	1541.4	1745.9	1.9980
1600	10.212	1632.3	1859.0	2.0875	8.752	1632.1	1858.8	2.0704	7.656	1631.9	1858.6	2.0556
1800	11.209	1726.2	1975.1	2.1413	9.607	1726.1	1975.0	2.1242	8.405	1725.9	1974.8	2.1094
2000	12.205	1823.6	2094.6	2.1919	10.461	1823.5	2094.5	2.1749	9.153	1823.3	2094.3	2.1601
	$P = 180.0$ psia ($T_{sat} = 373.13°F$)				$P = 200.0$ psia ($T_{sat} = 381.86°F$)				$P = 225.0$ psia ($T_{sat} = 391.87°F$)			
Sat	2.533	1113.4	1197.8	1.5553	2.289	1114.6	1199.3	1.5464	2.043	1115.8	1200.8	1.5365
400	2.648	1126.2	1214.4	1.5749	2.361	1123.5	1210.8	1.5600	2.073	1119.9	1206.2	1.5427
450	2.850	1148.5	1243.4	1.6078	2.548	1146.4	1240.7	1.5938	2.245	1143.8	1237.3	1.5779
500	3.042	1169.6	1270.9	1.6372	2.724	1168.0	1268.8	1.6239	2.405	1165.9	1266.1	1.6087
550	3.228	1190.0	1297.5	1.6642	2.893	1188.7	1295.7	1.6512	2.558	1187.0	1293.5	1.6366
600	3.409	1210.0	1323.5	1.6893	3.058	1208.9	1322.1	1.6767	2.707	1207.5	1320.2	1.6624
700	3.763	1249.6	1374.9	1.7357	3.379	1248.8	1373.8	1.7234	2.995	1247.7	1372.4	1.7095
800	4.110	1289.3	1426.2	1.7781	3.693	1288.6	1425.3	1.7660	3.276	1287.8	1424.2	1.7523
900	4.453	1329.4	1477.7	1.8175	4.003	1328.9	1477.1	1.8055	3.553	1328.3	1476.2	1.7920
1000	4.793	1370.2	1529.8	1.8545	4.310	1369.8	1529.3	1.8425	3.827	1369.3	1528.6	1.8292
1200	5.467	1454.0	1636.1	1.9227	4.918	1453.7	1635.7	1.9109	4.369	1453.4	1635.3	1.8977
1400	6.137	1541.2	1745.6	1.9849	5.521	1540.9	1745.3	1.9732	4.906	1540.7	1744.9	1.9600
1600	6.804	1631.7	1858.4	2.0425	6.123	1631.6	1858.2	2.0308	5.441	1631.3	1857.9	2.0177
1800	7.470	1725.8	1974.6	2.0964	6.722	1725.6	1974.4	2.0847	5.975	1725.4	1974.2	2.0716
2000	8.135	1823.2	2094.2	2.1470	7.321	1823.0	2094.0	2.1354	6.507	1822.9	2093.8	2.1223
	$P = 250.0$ psia ($T_{sat} = 401.04°F$)				$P = 275.0$ psia ($T_{sat} = 409.52°F$)				$P = 300.0$ psia ($T_{sat} = 417.43°F$)			
Sat	1.8448	1116.7	1202.1	1.5274	1.6813	1117.5	1203.1	1.5192	1.5442	1118.2	1203.9	1.5115
450	2.002	1141.1	1233.7	1.5632	1.8026	1138.3	1230.0	1.5495	1.6361	1135.4	1226.2	1.5365
500	2.150	1163.8	1263.3	1.5948	1.9407	1161.7	1260.4	1.5820	1.7662	1159.5	1257.5	1.5701
550	2.290	1185.3	1291.3	1.6233	2.071	1183.6	1289.0	1.6110	1.8878	1181.9	1286.7	1.5997
600	2.426	1206.1	1318.3	1.6494	2.196	1204.7	1316.4	1.6376	2.004	1203.2	1314.5	1.6266
650	2.558	1226.5	1344.9	1.6739	2.317	1225.3	1343.2	1.6623	2.117	1224.1	1341.6	1.6516
700	2.688	1246.7	1371.1	1.6970	2.436	1245.7	1369.7	1.6856	2.227	1244.6	1368.3	1.6751
800	2.943	1287.0	1423.2	1.7401	2.670	1286.2	1422.1	1.7289	2.442	1285.4	1421.0	1.7187
900	3.193	1327.6	1475.3	1.7799	2.898	1327.0	1474.5	1.7689	2.653	1326.3	1473.6	1.7589
1000	3.440	1368.7	1527.9	1.8172	3.124	1368.2	1527.2	1.8064	2.860	1367.7	1526.5	1.7964
1200	3.929	1453.0	1634.8	1.8858	3.570	1452.6	1634.3	1.8751	3.270	1452.2	1633.8	1.8653
1400	4.414	1540.4	1744.6	1.9483	4.011	1540.1	1744.2	1.9376	3.675	1539.8	1743.8	1.9279
1600	4.896	1631.1	1857.6	2.0060	4.450	1630.9	1857.3	1.9954	4.078	1630.7	1857.0	1.9857
1800	5.376	1725.2	1974.0	2.0599	4.887	1725.0	1973.7	2.0493	4.479	1724.9	1973.5	2.0396
2000	5.856	1822.7	2093.6	2.1106	5.323	1822.5	2093.4	2.1000	4.879	1822.3	2093.2	2.0904
	$P = 350.0$ psia ($T_{sat} = 431.82°F$)				$P = 400.0$ psia ($T_{sat} = 444.70°F$)				$P = 450.0$ psia ($T_{sat} = 456.39°F$)			
Sat	1.3267	1119.0	1204.9	1.4978	1.1620	1119.5	1205.5	1.4856	1.0326	1119.6	1205.6	1.4746
450	1.3733	1129.2	1218.2	1.5125	1.1745	1122.6	1209.6	1.4901				
500	1.4913	1154.9	1251.5	1.5482	1.2843	1150.1	1245.2	1.5282	1.1226	1145.1	1238.5	1.5097
550	1.5998	1178.3	1281.9	1.5790	1.3833	1174.6	1277.0	1.5605	1.2146	1170.7	1271.9	1.5436
600	1.7025	1200.3	1310.6	1.6068	1.4760	1197.3	1306.6	1.5892	1.2996	1194.3	1302.5	1.5732
650	1.8013	1221.6	1338.3	1.6323	1.5645	1219.1	1334.9	1.6153	1.3803	1216.6	1331.5	1.6000
700	1.8975	1242.5	1365.4	1.6562	1.6503	1240.4	1362.5	1.6397	1.4580	1238.2	1359.6	1.6248
800	2.085	1283.8	1418.8	1.7004	1.8163	1282.1	1416.6	1.6844	1.6077	1280.5	1414.4	1.6701

(Continued)

TABLE B-12 (Continued)

T °F	v ft³/lbm	u Btu/lbm	h Btu/lbm	s Btu/lbm·R	v ft³/lbm	u Btu/lbm	h Btu/lbm	s Btu/lbm·R	v ft³/lbm	u Btu/lbm	h Btu/lbm	s Btu/lbm·R
	$P = 350.0$ psia ($T_{sat} = 431.82$ °F)				$P = 400.0$ psia ($T_{sat} = 444.70$ °F)				$P = 450.0$ psia ($T_{sat} = 456.39$ °F)			
900	2.267	1325.0	1471.8	1.7409	1.9776	1323.7	1470.1	1.7252	1.7524	1322.4	1468.3	1.7113
1000	2.446	1366.6	1525.0	1.7787	2.136	1365.5	1523.6	1.7632	1.8941	1364.4	1522.2	1.7495
1200	2.799	1451.5	1632.8	1.8478	2.446	1450.7	1631.8	1.8327	2.172	1450.0	1630.8	1.8192
1400	3.148	1539.3	1743.1	1.9106	2.752	1538.7	1742.4	1.8956	2.444	1538.1	1741.7	1.8823
1600	3.494	1630.2	1856.5	1.9685	3.055	1629.8	1855.9	1.9535	2.715	1629.3	1855.4	1.9403
1800	3.838	1724.5	1973.1	2.0225	3.357	1724.1	1972.6	2.0076	2.983	1723.7	1972.1	1.9944
2000	4.182	1822.0	2092.8	2.0733	3.658	1821.6	2092.4	2.0584	3.251	1821.3	2092.0	2.0453
	$P = 500.0$ psia ($T_{sat} = 467.13$ °F)				$P = 600.0$ psia ($T_{sat} = 486.33$ °F)				$P = 700.0$ psia ($T_{sat} = 503.23$ °F)			
Sat	0.9283	1119.4	1205.3	1.4645	0.7702	1118.6	1204.1	1.4464	0.6558	1117.0	1202.0	1.4305
500	0.9924	1139.7	1231.5	1.4923	0.7947	1128.0	1216.2	1.4592				
550	1.0792	1166.7	1266.6	1.5279	0.8749	1158.2	1255.4	1.4990	0.7275	1149.0	1243.2	1.4723
600	1.1583	1191.1	1298.3	1.5585	0.9456	1184.5	1289.5	1.5320	0.7929	1177.5	1280.2	1.5081
650	1.2327	1214.0	1328.0	1.5860	1.0109	1208.6	1320.9	1.5609	0.8520	1203.1	1313.4	1.5387
700	1.3040	1236.0	1356.7	1.6112	1.0727	1231.5	1350.6	1.5872	0.9073	1226.9	1344.4	1.5661
800	1.4407	1278.8	1412.1	1.6571	1.1900	1275.4	1407.6	1.6343	1.0109	1272.0	1402.9	1.6145
900	1.5723	1321.0	1466.5	1.6987	1.3021	1318.4	1462.9	1.6766	1.1089	1315.6	1459.3	1.6576
1000	1.7008	1363.3	1520.7	1.7371	1.4108	1361.2	1517.8	1.7155	1.2036	1358.9	1514.9	1.6970
1100	1.8271	1406.0	1575.1	1.7731	1.5173	1404.2	1572.7	1.7519	1.2960	1402.4	1570.2	1.7337
1200	1.9518	1449.2	1629.8	1.8072	1.6222	1447.7	1627.8	1.7861	1.3868	1446.2	1625.8	1.7682
1400	2.198	1537.6	1741.0	1.8704	1.8289	1536.5	1739.5	1.8497	1.5652	1535.3	1738.1	1.8321
1600	2.442	1628.9	1854.8	1.9285	2.033	1628.0	1853.7	1.9080	1.7409	1627.1	1852.6	1.8906
1800	2.684	1723.3	1971.7	1.9827	2.236	1722.6	1970.8	1.9622	1.9152	1721.8	1969.9	1.9449
2000	2.926	1820.9	2091.6	2.0335	2.438	1820.2	2090.8	2.0131	2.0887	1819.5	2090.1	1.9958
	$P = 800.0$ psia ($T_{sat} = 518.36$ °F)				$P = 1000.0$ psia ($T_{sat} = 544.75$ °F)				$P = 1250.0$ psia ($T_{sat} = 572.56$ °F)			
Sat	0.5691	1115.0	1199.3	1.4160	0.4459	1109.9	1192.4	1.3903	0.3454	1101.7	1181.6	1.3619
550	0.6154	1138.8	1229.9	1.4469	0.4534	1114.8	1198.7	1.3966				
600	0.6776	1170.1	1270.4	1.4861	0.5140	1153.7	1248.8	1.4450	0.3786	1129.0	1216.6	1.3954
650	0.7324	1197.2	1305.6	1.5186	0.5637	1184.7	1289.1	1.4822	0.4267	1167.2	1266.0	1.4410
700	0.7829	1222.1	1338.0	1.5471	0.6080	1212.0	1324.6	1.5135	0.4670	1198.4	1306.4	1.4767
750	0.8306	1245.7	1368.6	1.5730	0.6490	1237.2	1357.3	1.5412	0.5030	1226.1	1342.4	1.5070
800	0.8764	1268.5	1398.2	1.5969	0.6878	1261.2	1388.5	1.5664	0.5364	1251.8	1375.8	1.5341
900	0.9640	1312.9	1455.6	1.6408	0.7610	1307.3	1448.1	1.6120	0.5984	1300.0	1438.4	1.5820
1000	1.0482	1356.7	1511.9	1.6807	0.8305	1352.2	1505.9	1.6530	0.6563	1346.4	1498.2	1.6244
1100	1.1300	1400.5	1567.8	1.7178	0.8976	1396.8	1562.9	1.6908	0.7116	1392.0	1556.6	1.6631
1200	1.2102	1444.6	1623.8	1.7526	0.9630	1441.5	1619.7	1.7261	0.7652	1437.5	1614.5	1.6991
1400	1.3674	1534.2	1736.6	1.8167	1.0905	1531.9	1733.7	1.7909	0.8689	1529.0	1730.0	1.7648
1600	1.5218	1626.2	1851.5	1.8754	1.2152	1624.4	1849.3	1.8499	0.9699	1622.2	1846.5	1.8243
1800	1.6749	1721.0	1969.0	1.9298	1.3384	1719.5	1967.2	1.9046	1.0693	1717.6	1965.0	1.8791
2000	1.8271	1818.8	2089.3	1.9808	1.4608	1817.4	2087.7	1.9557	1.1678	1815.7	2085.8	1.9304
	$P = 1500.0$ psia ($T_{sat} = 596.39$°F)				$P = 1750.0$ psia ($T_{sat} = 617.31$°F)				$P = 2000.0$ psia ($T_{sat} = 636.00$°F)			
Sat	0.2769	1091.8	1168.7	1.3359	0.2268	1080.2	1153.7	1.3109	0.18813	1066.6	1136.3	1.2861
600	0.2816	1096.6	1174.8	1.3416								
650	0.3329	1147.0	1239.4	1.4012	0.2627	1122.5	1207.6	1.3603	0.2057	1091.1	1167.2	1.3141
700	0.3716	1183.4	1286.6	1.4429	0.3022	1166.7	1264.6	1.4106	0.2487	1147.7	1239.8	1.3782
750	0.4049	1214.1	1326.5	1.4767	0.3341	1201.3	1309.5	1.4485	0.2803	1187.3	1291.1	1.4216
800	0.4350	1241.8	1362.5	1.5058	0.3622	1231.3	1348.6	1.4802	0.3071	1220.1	1333.8	1.4562
850	0.4631	1267.7	1396.2	1.5320	0.3878	1258.8	1384.4	1.5081	0.3312	1249.5	1372.0	1.4860
900	0.4897	1292.5	1428.5	1.5562	0.4119	1284.8	1418.2	1.5334	0.3534	1276.8	1407.6	1.5126
1000	0.5400	1340.4	1490.3	1.6001	0.4569	1334.3	1482.3	1.5789	0.3945	1328.1	1474.1	1.5598
1100	0.5876	1387.2	1550.3	1.6399	0.4990	1382.2	1543.8	1.6197	0.4325	1377.2	1537.2	1.6017
1200	0.6334	1433.5	1609.3	1.6765	0.5392	1429.4	1604.0	1.6571	0.4685	1425.2	1598.6	1.6398
1400	0.7213	1526.1	1726.3	1.7431	0.6158	1523.1	1722.6	1.7245	0.5368	1520.2	1718.8	1.7082
1600	0.8064	1619.9	1843.7	1.8031	0.6896	1617.6	1841.0	1.7850	0.6020	1615.4	1838.2	1.7692

(Continued)

TABLE B-12 (Continued)

T °F	v ft³/lbm	u Btu/lbm	h Btu/lbm	s Btu/lbm·R	v ft³/lbm	u Btu/lbm	h Btu/lbm	s Btu/lbm·R	v ft³/lbm	u Btu/lbm	h Btu/lbm	s Btu/lbm·R
	P = 120.0 psia (T_sat = 341.3 °F)				P = 140.0 psia (T_sat = 353.08 °F)				P = 160.0 psia (T_sat = 363.60 °F)			
1800	0.8899	1715.7	1962.7	1.8582	0.7617	1713.9	1960.5	1.8404	0.6656	1712.0	1958.3	1.8249
2000	0.9725	1814.0	2083.9	1.9096	0.8330	1812.3	2082.0	1.8919	0.7284	1810.6	2080.2	1.8765
	P = 2500.0 psia (T_sat = 668.31 °F)				P = 3000.0 psia (T_sat = 695.52 °F)				P = 3500.0 psia			
Sat	0.13059	1031.0	1091.4	1.2327	0.08404	968.8	1015.5	1.1575				
650									0.02491	663.5	679.7	0.8630
700	0.16839	1098.7	1176.6	1.3073	0.09771	1003.9	1058.1	1.1944	0.03058	759.5	779.3	0.9506
750	0.2030	1155.2	1249.1	1.3686	0.14831	1114.7	1197.1	1.3122	0.10460	1058.1	1126.1	1.2440
800	0.2291	1195.7	1301.7	1.4112	0.17572	1167.6	1265.2	1.3675	0.13626	1134.7	1223.0	1.3226
850	0.2513	1229.5	1345.8	1.4456	0.19731	1207.7	1317.2	1.4080	0.15818	1183.4	1285.9	1.3716
900	0.2712	1259.9	1385.4	1.4752	0.2160	1241.8	1361.7	1.4414	0.17625	1222.4	1336.5	1.4096
950	0.2896	1288.2	1422.2	1.5018	0.2328	1272.7	1402.0	1.4705	0.19214	1256.4	1380.8	1.4416
1000	0.3069	1315.2	1457.2	1.5262	0.2485	1301.7	1439.6	1.4967	0.2066	1287.6	1421.4	1.4699
1100	0.3393	1366.8	1523.8	1.5704	0.2772	1356.2	1510.1	1.5434	0.2328	1345.2	1496.0	1.5193
1200	0.3696	1416.7	1587.7	1.6101	0.3036	1408.0	1576.6	1.5848	0.2566	1399.2	1565.3	1.5624
1400	0.4261	1514.2	1711.3	1.6804	0.3524	1508.1	1703.7	1.6571	0.2997	1501.9	1696.1	1.6368
1600	0.4795	1610.2	1832.6	1.7424	0.3978	1606.3	1827.1	1.7201	0.3395	1601.7	1821.6	1.7010
1800	0.5312	1708.2	1954.0	1.7986	0.4416	1704.5	1949.6	1.7769	0.3776	1700.8	1945.4	1.7583
2000	0.5820	1807.2	2076.4	1.8506	0.4844	1803.9	2072.8	1.8291	0.4147	1800.6	2069.2	1.8108
	P = 4000 psia				P = 5000 psia				P = 6000 psia			
650	0.02447	657.7	675.8	0.8574	0.02377	648.0	670.0	0.8482	0.02322	640.0	665.8	0.8405
700	0.02867	742.1	763.4	0.9345	0.02676	721.8	746.6	0.9156	0.02563	708.1	736.5	0.9028
750	0.06331	960.7	1007.5	1.1395	0.03364	821.4	852.6	1.0049	0.02978	788.6	821.7	0.9746
800	0.10522	1095.0	1172.9	1.2740	0.05932	987.2	1042.1	1.1583	0.03942	896.9	940.7	1.0708
850	0.12833	1156.5	1251.5	1.3352	0.08556	1092.7	1171.9	1.2596	0.05818	1018.8	1083.4	1.1820
900	0.14622	1201.5	1309.7	1.3789	0.10385	1155.1	1251.1	1.3190	0.07588	1102.9	1187.2	1.2599
950	0.16151	1239.2	1358.8	1.4144	0.11853	1202.2	1311.9	1.3629	0.09008	1162.0	1262.0	1.3140
1000	0.17520	1272.9	1402.6	1.4449	0.13120	1242.0	1363.4	1.3988	0.10207	1209.1	1322.4	1.3561
1100	0.19954	1333.9	1481.6	1.4973	0.15302	1310.6	1452.2	1.4577	0.12218	1286.4	1422.1	1.4222
1200	0.2213	1390.1	1553.9	1.5423	0.17199	1371.6	1530.8	1.5066	0.13927	1352.7	1507.3	1.4752
1300	0.2414	1443.7	1622.4	1.5823	0.18918	1428.6	1603.7	1.5493	0.15453	1413.3	1584.9	1.5206
1400	0.2603	1495.7	1688.4	1.6188	0.20517	1483.2	1673.0	1.5876	0.16854	1470.5	1657.6	1.5608
1600	0.2959	1597.1	1816.1	1.6841	0.2348	1587.9	1805.2	1.6551	0.19420	1578.7	1794.3	1.6307
1800	0.3296	1697.1	1941.1	1.7420	0.2626	1689.8	1932.7	1.7142	0.21801	1682.4	1924.5	1.6910
2000	0.3625	1797.3	2065.6	1.7948	0.2895	1790.8	2058.6	1.7676	0.24087	1784.3	2051.7	1.7450

TABLE B-13 Thermodynamic Properties of Compressed Liquid Water

T °F	v ft³/lbm	u Btu/lbm	h Btu/lbm	s Btu/lbm·R	v ft³/lbm	u Btu/lbm	h Btu/lbm	s Btu/lbm·R	v ft³/lbm	u Btu/lbm	h Btu/lbm	s Btu/lbm·R
	P = 500.0 psia (T_sat = 467.13 °F)				P = 1000.0 psia (T_sat = 544.75 °F)				P = 1500.0 psia (T_sat = 596.39 °F)			
Sat	0.019 748	447.70	449.53	0.649 04	0.021 591	538.39	542.38	0.743 20	0.023 461	604.97	611.48	0.808 24
32	0.015 994	0.00	1.49	0.000 00	0.015 967	0.03	2.99	0.000 05	0.015 939	0.05	4.47	0.000 07
50	0.015 998	18.02	19.50	0.035 99	0.015 972	17.99	20.94	0.035 92	0.015 946	17.95	22.38	0.035 84
100	0.016 106	67.87	69.36	0.129 32	0.016 082	67.70	70.68	0.129 01	0.016 058	67.53	71.99	0.128 70
150	0.016 318	117.66	119.17	0.214 57	0.016 293	117.38	120.40	0.214 10	0.016 268	117.10	121.62	0.213 64
200	0.016 608	167.65	169.19	0.293 41	0.016 580	167.26	170.32	0.292 81	0.016 554	166.87	171.46	0.292 21
250	0.016 972	217.99	219.56	0.367 02	0.016 941	217.47	220.61	0.366 28	0.016 910	216.96	221.65	0.365 54
300	0.017 416	268.92	270.53	0.436 41	0.017 379	268.24	271.46	0.435 52	0.017 343	267.58	272.39	0.434 63
350	0.017 954	320.71	322.37	0.502 49	0.017 909	319.83	323.15	0.501 40	0.017 865	318.98	323.94	0.500 34
400	0.018 608	373.68	375.40	0.566 04	0.018 550	372.55	375.98	0.564 72	0.018 493	371.45	376.59	0.563 43
450	0.019 420	428.40	430.19	0.627 98	0.019 340	426.89	430.47	0.62 632	0.019 264	425.44	430.79	0.624 70
500					0.020 36	483.8	487.5	0.6874	0.020 24	481.8	487.4	0.6853
550									0.021 58	542.1	548.1	0.7469

(Continued)

TABLE B-13 (Continued)

T °F	v ft³/lbm	u Btu/lbm	h Btu/lbm	s Btu/lbm · R	v ft³/lbm	u Btu/lbm	h Btu/lbm	s Btu/lbm · R	v ft³/lbm	u Btu/lbm	h Btu/lbm	s Btu/lbm · R
	P = 2000.0 psia (T_sat = 636.00 °F)				P = 3000.0 psia (T_sat = 695.52 °F)				P = 5000.0 psia			
Sat	0.025 649	662.40	671.89	0.862 27	0.034 310	783.45	802.50	0.973 20				
32	0.015 912	0.06	5.95	0.000 08	0.015 859	0.09	8.90	0.000 09	0.015 755	0.11	14.70	−0.000 01
50	0.015 920	17.91	23.81	0.035 75	0.015 870	17.84	26.65	0.035 55	0.015 773	17.67	32.26	0.035 08
100	0.016 034	67.37	73.30	0.128 39	0.015 987	67.04	75.91	0.127 77	0.015 897	66.40	81.11	0.126 51
200	0.016 527	166.49	172.60	0.291 62	0.016 476	165.74	174.89	0.290 46	0.016 376	164.32	179.47	0.288 18
300	0.017 308	266.93	273.33	0.433 76	0.017 240	265.66	275.23	0.432 05	0.017 110	263.25	279.08	0.428 75
400	0.018 439	370.38	377.21	0.562 16	0.018 334	368.32	378.50	0.559 70	0.018 141	364.47	381.25	0.555 06
450	0.019 191	424.04	431.14	0.623 13	0.019 053	421.36	431.93	0.620 11	0.018 803	416.44	433.84	0.614 51
500	0.020 14	479.8	487.3	0.6832	0.019 944	476.2	487.3	0.6794	0.019603	469.8	487.9	0.6724
560	0.021 72	551.8	559.8	0.7565	0.021 382	546.2	558.0	0.750 8	0.020 835	536.7	556.0	0.7411

TABLE B-14 Thermodynamic Properties of Saturated Refrigerant 134a (Liquid–Vapor): Temperature Table

Temp. T °F	Press. P_sat lbf/in.²	Specific Volume ft³/lbm Sat. Liquid v_f	Sat. Vapor v_g	Internal Energy Btu/lbm Sat. Liquid u_f	Sat. Vapor u_g	Enthalpy Btu/lbm Sat. Liquid h_f	Evap. h_fg	Sat. Vapor h_g	Entropy Btu/lbm · R Sat. Liquid s_f	Sat. Vapor s_g	Temp. T °F
−40	7.490	0.01130	5.7173	−0.02	87.90	0.00	95.82	95.82	0.0000	0.2283	−40
−30	9.920	0.01143	4.3911	2.81	89.26	2.83	94.49	97.32	0.0067	0.2266	−30
−20	12.949	0.01156	3.4173	5.69	90.62	5.71	93.10	98.81	0.0133	0.2250	−20
−15	14.718	0.01163	3.0286	7.14	91.30	7.17	92.38	99.55	0.0166	0.2243	−15
−10	16.674	0.01170	2.6918	8.61	91.98	8.65	91.64	100.29	0.0199	0.2236	−10
−5	18.831	0.01178	2.3992	10.09	92.66	10.13	90.89	101.02	0.0231	0.2230	−5
0	21.203	0.01185	2.1440	11.58	93.33	11.63	90.12	101.75	0.0264	0.2224	0
5	23.805	0.01193	1.9208	13.09	94.01	13.14	89.33	102.47	0.0296	0.2219	5
10	26.651	0.01200	1.7251	14.60	94.68	14.66	88.53	103.19	0.0329	0.2214	10
15	29.756	0.01208	1.5529	16.13	95.35	16.20	87.71	103.90	0.0361	0.2209	15
20	33.137	0.01216	1.4009	17.67	96.02	17.74	86.87	104.61	0.0393	0.2205	20
25	36.809	0.01225	1.2666	19.22	96.69	19.30	86.02	105.32	0.0426	0.2200	25
30	40.788	0.01233	1.1474	20.78	97.35	20.87	85.14	106.01	0.0458	0.2196	30
40	49.738	0.01251	0.9470	23.94	98.67	24.05	83.34	107.39	0.0522	0.2189	40
50	60.125	0.01270	0.7871	27.14	99.98	27.28	81.46	108.74	0.0585	0.2183	50
60	72.092	0.01290	0.6584	30.39	101.27	30.56	79.49	110.05	0.0648	0.2178	60
70	85.788	0.01311	0.5538	33.68	102.54	33.89	77.44	111.33	0.0711	0.2173	70
80	101.37	0.01334	0.4682	37.02	103.78	37.27	75.29	112.56	0.0774	0.2169	80
85	109.92	0.01346	0.4312	38.72	104.39	38.99	74.17	113.16	0.0805	0.2167	85
90	118.99	0.01358	0.3975	40.42	105.00	40.72	73.03	113.75	0.0836	0.2165	90
95	128.62	0.01371	0.3668	42.14	105.60	42.47	71.86	114.33	0.0867	0.2163	95
100	138.83	0.01385	0.3388	43.87	106.18	44.23	70.66	114.89	0.0898	0.2161	100
105	149.63	0.01399	0.3131	45.62	106.76	46.01	69.42	115.43	0.0930	0.2159	105
110	161.04	0.01414	0.2896	47.39	107.33	47.81	68.15	115.96	0.0961	0.2157	110
115	173.10	0.01429	0.2680	49.17	107.88	49.63	66.84	116.47	0.0992	0.2155	115
120	185.82	0.01445	0.2481	50.97	108.42	51.47	65.48	116.95	0.1023	0.2153	120
140	243.86	0.01520	0.1827	58.39	110.41	59.08	59.57	118.65	0.1150	0.2143	140
160	314.63	0.01617	0.1341	66.26	111.97	67.20	52.58	119.78	0.1280	0.2128	160
180	400.22	0.01758	0.0964	74.83	112.77	76.13	43.78	119.91	0.1417	0.2101	180
200	503.52	0.02014	0.0647	84.90	111.66	86.77	30.92	117.69	0.1575	0.2044	200
210	563.51	0.02329	0.0476	91.84	108.48	94.27	19.18	113.45	0.1684	0.1971	210

Source: Tables B-14 through B-16 are calculated based on equations from D. P. Wilson and R. S. Basu, "Thermodynamic Properties of a New Stratospherically Safe Working Fluid—Refrigerant 134a," *ASHRAE Trans.*, Vol. 94, Pt. 2, 1988, pp. 2095–2118.

TABLE B-15 Thermodynamic Properties of Saturated Refrigerant 134a (Liquid–Vapor): Pressure Table

Press. P lbf/in.²	Temp T_{sat} °F	Specific Volume ft³/lbm		Internal Energy Btu/lbm		Enthalpy Btu/lbm			Entropy Btu/lbm · R		Press. P lbf/in.²
		Sat. Liquid v_f	Sat. Vapor v_g	Sat. Liquid u_f	Sat. Vapor u_g	Sat. Liquid h_f	Evap. h_{fg}	Sat. Vapor h_g	Sat. Liquid s_f	Sat. Vapor s_g	
5	−53.48	0.01113	8.3508	−3.74	86.07	−3.73	97.53	93.79	−0.0090	0.2311	5
10	−29.71	0.01143	4.3581	2.89	89.30	2.91	94.45	97.37	0.0068	0.2265	10
15	−14.25	0.01164	2.9747	7.36	91.40	7.40	92.27	99.66	0.0171	0.2242	15
20	−2.48	0.01181	2.2661	10.84	93.00	10.89	90.50	101.39	0.0248	0.2227	20
30	15.38	0.01209	1.5408	16.24	95.40	16.31	87.65	103.96	0.0364	0.2209	30
40	29.04	0.01232	1.1692	20.48	97.23	20.57	85.31	105.88	0.0452	0.2197	40
50	40.27	0.01252	0.9422	24.02	98.71	24.14	83.29	107.43	0.0523	0.2189	50
60	49.89	0.01270	0.7887	27.10	99.96	27.24	81.48	108.72	0.0584	0.2183	60
70	58.35	0.01286	0.6778	29.85	101.05	30.01	79.82	109.83	0.0638	0.2179	70
80	65.93	0.01302	0.5938	32.33	102.02	32.53	78.28	110.81	0.0686	0.2175	80
90	72.83	0.01317	0.5278	34.62	102.89	34.84	76.84	111.68	0.0729	0.2172	90
100	79.17	0.01332	0.4747	36.75	103.68	36.99	75.47	112.46	0.0768	0.2169	100
120	90.54	0.01360	0.3941	40.61	105.06	40.91	72.91	113.82	0.0839	0.2165	120
140	100.56	0.01386	0.3358	44.07	106.25	44.43	70.52	114.95	0.0902	0.2161	140
160	109.56	0.01412	0.2916	47.23	107.28	47.65	68.26	115.91	0.0958	0.2157	160
180	117.74	0.01438	0.2569	50.16	108.18	50.64	66.10	116.74	0.1009	0.2154	180
200	125.28	0.01463	0.2288	52.90	108.98	53.44	64.01	117.44	0.1057	0.2151	200
220	132.27	0.01489	0.2056	55.48	109.68	56.09	61.96	118.05	0.1101	0.2147	220
240	138.79	0.01515	0.1861	57.93	110.30	58.61	59.96	118.56	0.1142	0.2144	240
260	144.92	0.01541	0.1695	60.28	110.84	61.02	57.97	118.99	0.1181	0.2140	260
280	150.70	0.01568	0.1550	62.53	111.31	63.34	56.00	119.35	0.1219	0.2136	280
300	156.17	0.01596	0.1424	64.71	111.72	65.59	54.03	119.62	0.1254	0.2132	300
350	168.72	0.01671	0.1166	69.88	112.45	70.97	49.03	120.00	0.1338	0.2118	350
400	179.95	0.01758	0.0965	74.81	112.77	76.11	43.80	119.91	0.1417	0.2102	400
450	190.12	0.01863	0.0800	79.63	112.60	81.18	38.08	119.26	0.1493	0.2079	450
500	199.38	0.02002	0.0657	84.54	111.76	86.39	31.44	117.83	0.1570	0.2047	500

TABLE B-16 Thermodynamic Properties of Superheated Refrigerant 134a Vapor

T °F	v ft³/lbm	u Btu/lbm	h Btu/lbm	s Btu/lbm · R	v ft³/lbm	u Btu/lbm	h Btu/lbm	s Btu/lbm · R
	$P = 10$ lbf/in.² ($T_{sat} = -29.71$ °F)				$P = 15$ lbf/in.² ($T_{sat} = -14.25$ °F)			
Sat.	4.3581	89.30	97.37	0.2265	2.9747	91.40	99.66	0.2242
−20	4.4718	90.89	99.17	0.2307				
0	4.7026	94.24	102.94	0.2391	3.0893	93.84	102.42	0.2303
20	4.9297	97.67	106.79	0.2472	3.2468	97.33	106.34	0.2386
40	5.1539	101.19	110.72	0.2553	3.4012	100.89	110.33	0.2468
60	5.3758	104.80	114.74	0.2632	3.5533	104.54	114.40	0.2548
80	5.5959	108.50	118.85	0.2709	3.7034	108.28	118.56	0.2626
100	5.8145	112.29	123.05	0.2786	3.8520	112.10	122.79	0.2703
120	6.0318	116.18	127.34	0.2861	3.9993	116.01	127.11	0.2779
140	6.2482	120.16	131.72	0.2935	4.1456	120.00	131.51	0.2854
160	6.4638	124.23	136.19	0.3009	4.2911	124.09	136.00	0.2927
180	6.6786	128.38	140.74	0.3081	4.4359	128.26	140.57	0.3000
200	6.8929	132.63	145.39	0.3152	4.5801	132.52	145.23	0.3072

(Continued)

TABLE B-16   Continued

T °F	v ft³/lbm	u Btu/lbm	h Btu/lbm	s Btu/lbm · R	v ft³/lbm	u Btu/lbm	h Btu/lbm	s Btu/lbm · R
		$P = 20$ lbf/in.² $(T_{sat} = -2.48$ °F)				$P = 30$ lbf/in.² $(T_{sat} = 15.38$ °F)		
Sat.	2.2661	93.00	101.39	0.2227	1.5408	95.40	103.96	0.2209
0	2.2816	93.43	101.88	0.2238				
20	2.4046	96.98	105.88	0.2323	1.5611	96.26	104.92	0.2229
40	2.5244	100.59	109.94	0.2406	1.6465	99.98	109.12	0.2315
60	2.6416	104.28	114.06	0.2487	1.7293	103.75	113.35	0.2398
80	2.7569	108.05	118.25	0.2566	1.8098	107.59	117.63	0.2478
100	2.8705	111.90	122.52	0.2644	1.8887	111.49	121.98	0.2558
120	2.9829	115.83	126.87	0.2720	1.9662	115.47	126.39	0.2635
140	3.0942	119.85	131.30	0.2795	2.0426	119.53	130.87	0.2711
160	3.2047	123.95	135.81	0.2869	2.1181	123.66	135.42	0.2786
180	3.3144	128.13	140.40	0.2922	2.1929	127.88	140.05	0.2859
200	3.4236	132.40	145.07	0.3014	2.2671	132.17	144.76	0.2932
220	3.5323	136.76	149.83	0.3085	2.3407	136.55	149.54	0.3003
		$P = 40$ lbf/in.² $(T_{sat} = 29.04$ °F)				$P = 50$ lbf/in.² $(T_{sat} = 40.27$ °F)		
Sat.	1.1692	97.23	105.88	0.2197	0.9422	98.71	107.43	0.2189
40	1.2065	99.33	108.26	0.2245				
60	1.2723	103.20	112.62	0.2331	0.9974	102.62	111.85	0.2276
80	1.3357	107.11	117.00	0.2414	1.0508	106.62	116.34	0.2361
100	1.3973	111.08	121.42	0.2494	1.1022	110.65	120.85	0.2443
120	1.4575	115.11	125.90	0.2573	1.1520	114.74	125.39	0.2523
140	1.5165	119.21	130.43	0.2650	1.2007	118.88	129.99	0.2601
160	1.5746	123.38	135.03	0.2725	1.2484	123.08	134.64	0.2677
180	1.6319	127.62	139.70	0.2799	1.2953	127.36	139.34	0.2752
200	1.6887	131.94	144.44	0.2872	1.3415	131.71	144.12	0.2825
220	1.7449	136.34	149.25	0.2944	1.3873	136.12	148.96	0.2897
240	1.8006	140.81	154.14	0.3015	1.4326	140.61	153.87	0.2969
260	1.8561	145.36	159.10	0.3085	1.4775	145.18	158.85	0.3039
280	1.9112	149.98	164.13	0.3154	1.5221	149.82	163.90	0.3108
		$P = 60$ lbf/in.² $(T_{sat} = 49.89$ °F)				$P = 70$ lbf/in.² $(T_{sat} = 58.35$ °F)		
Sat.	0.7887	99.96	108.72	0.2183	0.6778	101.05	109.83	0.2179
60	0.8135	102.03	111.06	0.2229	0.6814	101.40	110.23	0.2186
80	0.8604	106.11	115.66	0.2316	0.7239	105.58	114.96	0.2276
100	0.9051	110.21	120.26	0.2399	0.7640	109.76	119.66	0.2361
120	0.9482	114.35	124.88	0.2480	0.8023	113.96	124.36	0.2444
140	0.9900	118.54	129.53	0.2559	0.8393	118.20	129.07	0.2524
160	1.0308	122.79	134.23	0.2636	0.8752	122.49	133.82	0.2601
180	1.0707	127.10	138.98	0.2712	0.9103	126.83	138.62	0.2678
200	1.1100	131.47	143.79	0.2786	0.9446	131.23	143.46	0.2752
220	1.1488	135.91	148.66	0.2859	0.9784	135.69	148.36	0.2825
240	1.1871	140.42	153.60	0.2930	1.0118	140.22	153.33	0.2897
260	1.2251	145.00	158.60	0.3001	1.0448	144.82	158.35	0.2968
280	1.2627	149.65	163.67	0.3070	1.0774	149.48	163.44	0.3038
300	1.3001	154.38	168.81	0.3139	1.1098	154.22	168.60	0.3107
		$P = 80$ lbf/in.² $(T_{sat} = 65.93$ °F)				$P = 90$ lbf/in.² $(T_{sat} = 72.83$ °F)		
Sat.	0.5938	102.02	110.81	0.2175	0.5278	102.89	111.68	0.2172
80	0.6211	105.03	114.23	0.2239	0.5408	104.46	113.47	0.2205
100	0.6579	109.30	119.04	0.2327	0.5751	108.82	118.39	0.2295
120	0.6927	113.56	123.82	0.2411	0.6073	113.15	123.27	0.2380
140	0.7261	117.85	128.60	0.2492	0.6380	117.50	128.12	0.2463

(Continued)

TABLE B-16 Continued

T °F	v ft³/lbm	u Btu/lbm	h Btu/lbm	s Btu/lbm · R	v ft³/lbm	u Btu/lbm	h Btu/lbm	s Btu/lbm · R
		P = 80 lbf/in.² (*T*$_{sat}$ = 65.93 °F)				*P* = 90 lbf/in.² (*T*$_{sat}$ = 72.83 °F)		
160	0.7584	122.18	133.41	0.2570	0.6675	121.87	132.98	0.2542
180	0.7898	126.55	138.25	0.2647	0.6961	126.28	137.87	0.2620
200	0.8205	130.98	143.13	0.2722	0.7239	130.73	142.79	0.2696
220	0.8506	135.47	148.06	0.2796	0.7512	135.25	147.76	0.2770
240	0.8803	140.02	153.05	0.2868	0.7779	139.82	152.77	0.2843
260	0.9095	144.63	158.10	0.2940	0.8043	144.45	157.84	0.2914
280	0.9384	149.32	163.21	0.3010	0.8303	149.15	162.97	0.2984
300	0.9671	154.06	168.38	0.3079	0.8561	153.91	168.16	0.3054
320	0.9955	158.88	173.62	0.3147	0.8816	158.73	173.42	0.3122
		P = 100 lbf/in.² (*T*$_{sat}$ = 79.17 °F)				*P* = 120 lbf/in.² (*T*$_{sat}$ = 90.54 °F)		
Sat.	0.4747	103.68	112.46	0.2169	0.3941	105.06	113.82	0.2165
80	0.4761	103.87	112.68	0.2173				
100	0.5086	108.32	117.73	0.2265	0.4080	107.26	116.32	0.2210
120	0.5388	112.73	122.70	0.2352	0.4355	111.84	121.52	0.2301
140	0.5674	117.13	127.63	0.2436	0.4610	116.37	126.61	0.2387
160	0.5947	121.55	132.55	0.2517	0.4852	120.89	131.66	0.2470
180	0.6210	125.99	137.49	0.2595	0.5082	125.42	136.70	0.2550
200	0.6466	130.48	142.45	0.2671	0.5305	129.97	141.75	0.2628
220	0.6716	135.02	147.45	0.2746	0.5520	134.56	146.82	0.2704
240	0.6960	139.61	152.49	0.2819	0.5731	139.20	151.92	0.2778
260	0.7201	144.26	157.59	0.2891	0.5937	143.89	157.07	0.2850
280	0.7438	148.98	162.74	0.2962	0.6140	148.63	162.26	0.2921
300	0.7672	153.75	167.95	0.3031	0.6339	153.43	167.51	0.2991
320	0.7904	158.59	173.21	0.3099	0.6537	158.29	172.81	0.3060
		P = 140 lbf/in.² (*T*$_{sat}$ = 100.56 °F)				*P* = 160 lbf/in.² (*T*$_{sat}$ = 109.55 °F)		
Sat.	0.3358	106.25	114.95	0.2161	0.2916	107.28	115.91	0.2157
120	0.3610	110.90	120.25	0.2254	0.3044	109.88	118.89	0.2209
140	0.3846	115.58	125.54	0.2344	0.3269	114.73	124.41	0.2303
160	0.4066	120.21	130.74	0.2429	0.3474	119.49	129.78	0.2391
180	0.4274	124.82	135.89	0.2511	0.3666	124.20	135.06	0.2475
200	0.4474	129.44	141.03	0.2590	0.3849	128.90	140.29	0.2555
220	0.4666	134.09	146.18	0.2667	0.4023	133.61	145.52	0.2633
240	0.4852	138.77	151.34	0.2742	0.4192	138.34	150.75	0.2709
260	0.5034	143.50	156.54	0.2815	0.4356	143.11	156.00	0.2783
280	0.5212	148.28	161.78	0.2887	0.4516	147.92	161.29	0.2856
300	0.5387	153.11	167.06	0.2957	0.4672	152.78	166.61	0.2927
320	0.5559	157.99	172.39	0.3026	0.4826	157.69	171.98	0.2996
340	0.5730	162.93	177.78	0.3094	0.4978	162.65	177.39	0.3065
360	0.5898	167.93	183.21	0.3162	0.5128	167.67	182.85	0.3132
		P = 180 lbf/in.² (*T*$_{sat}$ = 117.74 °F)				*P* = 200 lbf/in.² (*T*$_{sat}$ = 125.28 °F)		
Sat.	0.2569	108.18	116.74	0.2154	0.2288	108.98	117.44	0.2151
120	0.2595	108.77	117.41	0.2166				
140	0.2814	113.83	123.21	0.2264	0.2446	112.87	121.92	0.2226
160	0.3011	118.74	128.77	0.2355	0.2636	117.94	127.70	0.2321
180	0.3191	123.56	134.19	0.2441	0.2809	122.88	133.28	0.2410
200	0.3361	128.34	139.53	0.2524	0.2970	127.76	138.75	0.2494
220	0.3523	133.11	144.84	0.2603	0.3121	132.60	144.15	0.2575
240	0.3678	137.90	150.15	0.2680	0.3266	137.44	149.53	0.2653
260	0.3828	142.71	155.46	0.2755	0.3405	142.30	154.90	0.2728
280	0.3974	147.55	160.79	0.2828	0.3540	147.18	160.28	0.2802

(Continued)

TABLE B-16 Continued

T °F	v ft³/lbm	u Btu/lbm	h Btu/lbm	s Btu/lbm · R	v ft³/lbm	u Btu/lbm	h Btu/lbm	s Btu/lbm · R
	$P = 180$ lbf/in.² ($T_{sat} = 117.74$ °F)				$P = 200$ lbf/in.² ($T_{sat} = 125.28$ °F)			
300	0.4116	152.44	166.15	0.2899	0.3671	152.10	165.69	0.2874
320	0.4256	157.38	171.55	0.2969	0.3799	157.07	171.13	0.2945
340	0.4393	162.36	177.00	0.3038	0.3926	162.07	176.60	0.3014
360	0.4529	167.40	182.49	0.3106	0.4050	167.13	182.12	0.3082
	$P = 300$ lbf/in.² ($T_{sat} = 156.17$ °F)				$P = 400$ lbf/in.² ($T_{sat} = 179.95$ °F)			
Sat.	0.1424	111.72	119.62	0.2132	0.0965	112.77	119.91	0.2102
160	0.1462	112.95	121.07	0.2155				
180	0.1633	118.93	128.00	0.2265	0.0965	112.79	119.93	0.2102
200	0.1777	124.47	134.34	0.2363	0.1143	120.14	128.60	0.2235
220	0.1905	129.79	140.36	0.2453	0.1275	126.35	135.79	0.2343
240	0.2021	134.99	146.21	0.2537	0.1386	132.12	142.38	0.2438
260	0.2130	140.12	151.95	0.2618	0.1484	137.65	148.64	0.2527
280	0.2234	145.23	157.63	0.2696	0.1575	143.06	154.72	0.2610
300	0.2333	150.33	163.28	0.2772	0.1660	148.39	160.67	0.2689
320	0.2428	155.44	168.92	0.2845	0.1740	153.69	166.57	0.2766
340	0.2521	160.57	174.56	0.2916	0.1816	158.97	172.42	0.2840
360	0.2611	165.74	180.23	0.2986	0.1890	164.26	178.26	0.2912
380	0.2699	170.94	185.92	0.3055	0.1962	169.57	184.09	0.2983
400	0.2786	176.18	191.64	0.3122	0.2032	174.90	189.94	0.3051

ANSWERS TO SELECTED PROBLEMS

Chapter 2

2-1	$W = 1240$ kJ
2-4	$\Delta U = 116$ Btu
2-7	$T = 70.3°F$
2-10	$a = 70.8$ ft/s^2 up
2-13	$P = 0.49$ psig
2-16	$P = 3.16$ torr
2-19	$W = 8.31$ J
2-22	$W = 2037$ Btu
2-25	$Q = 28.3$ kJ
2-28	$P = 65$ kPa
2-31	a) $T = 290°C$, b) $W = 300$ kJ
2-34	$W = 4.09$ J
2-37	$Q = 25.8$ kJ
2-40	$W = 90.9$ ft $\cdot$ lbf
2-43	a) $n = 1.21$, b) $W = -195$ kJ, c) $Q = -92.3$ kJ
2-46	$T = -7°C$
2-49	a) $T = 353°F$, b) $P = 67.7$ psia, c) $W = -248$ ft $\cdot$ lbf
2-53	$Q = 126$ Btu

Chapter 3

3-1	$t = 27$ min
3-4	a) $m = 709$ kg, b) $T = 29.6°C$
3-7	$\dot{Q} = 57,600$ Btu/h
3-10	$L = 30.8$ cm, $T = 3040$ K
3-13	$T = 87°F$
3-16	$\dot{W} = 39.3$ W
3-19	$T = 38.1°C$
3-22	$T_2 = 61°C$
3-24	old cost = \$5.87, new cost = \$4.19
3-28	$\dot{Q} = 8.79 \times 10^4$ Btu/h
3-31	$r_{opt} = k/h$
3-34	$\dot{Q} = -1811$ W, $T = 4.95°C$
3-37	$t = 1.33$ h
3-40	$T = 97.3°F$
3-43	$t = 7.13$ min
3-46	$\dot{Q} = 210$ Btu/h
3-49	$t = 23.1$ min

Chapter 4

4-1	$P = 19$ lbf/in^2
4-4	$P = 119$ kPa
4-7	$P = 118$ kPa
4-10	$\rho = 0.104$ lbm/ft^3
4-13	$l = 10.1$ cm
4-16	$\rho = 572$ kg/m^3
4-19	$F = 3.87 \times 10^5$ N
4-22	$F = 16,529$ lbf
4-25	$h = 0.188$ m
4-28	$x = 4.66$ cm
4-31	4.66 in/h
4-34	$L = 55.1$ m
4-37	$\dot{q} = 63.2$ W/cm^2
4-40	$P = 99.2$ kPa
4-43	$\dot{m} = 90.8$ lbm/s
4-46	$\mathcal{V} = 80.8$ m/s
4-49	$\mathcal{V}_{in} = 6.9$ m/s, $\mathcal{V}_{out} = 4.4$ m/s
4-52	$F = -2147$ N

Chapter 5

5-1	$P = 0.7917$ MPa
5-4	a) $T = 259.3°F$, b) $T = 60°F$
5-7	$m = 0.036$ lbm
5-10	a) $m = 0.198$ lbm, b) $m = 0.190$ lbm
5-13	$P = 45$ MPa
5-16	$x = 0.016$
5-19	$m = 0.019$ g
5-22	a) $V = 0.2207$ m^3, b) $V = 0.0543$ m^3 c) $V = 2.502 \times 10^{-4}$ m^3
5-26	$Q = -2074$ kJ
5-28	a) $P = 30$ psia, b) $Q = 70.74$ Btu
5-31	a) $m_{liq} = 0.59$ lbm, b) $Q = -526.7$ Btu
5-34	$W = -3.51$ Btu
5-37	$T = 156°C$
5-40	$Q = -9108$ Btu
5-43	a) $T = 93.2°F$, b) $\dot{V} = 99.7$ fl.oz/h
5-45	$\dot{Q} = 1002$ W
5-48	a) $V = 1.48$ ft^3, b) $V = 1.47$ ft^3

Chapter 6

6-1 $T = 320°F$
6-4 $T = -19.6°C$
6-7   a) $\mathcal{V} = 110\,m/s$, b) $D = 8.78\,cm$
6-10   a) $\dot{m} = 1.88\,kg/s$, b) $D = 0.303\,m$
6-13 $D = 0.203\,m$
6-16 $T = 533°F$, $\dot{V} = 2.78\,ft^3/min$
6-19 $\dot{Q} = -76.0\,Btu/h$
6-22 $P = 4.45\,MPa$
6-25 $\mathcal{V} = 14.4\,m/s$
6-28 $x = 0.956$
6-31 $\dot{m} = 0.0116\,kg/s$
6-34 $x = 0.63$
6-37 $D = 2.21\,ft$
6-40 $\dot{m} = 0.0441\,kg/s$
6-43 $\dot{m} = 43.5\,lbm/s$
6-46 $t = 35\,min$
6-49 $V = 0.0127\,m^3$

Chapter 7

7-1 $\dot{W}_{net} = 7.76 \times 10^5\,W$
7-4 $\dot{W}_{in} = 193\,W$
7-7 $\dot{W}_{net} = 3.025\,kW$, $\dot{W}_{Carnot} = 0.185\,kW$
7-10 $\eta = 0.0707$
7-13 $\eta_{actual} = 0.406$, $\eta_{Carnot} = 0.572$, irreversible
7-16 $cost/month = \$4,330,000$
7-19   a) $x = 0.93$, b) $W = 0.17\,kJ$
7-22   yes, it would be possible if the process were reversible
7-24 $W = -46\,Btu$
7-28 $S_{gen} = 0$
7-31 $\dot{S}_{gen} = 0.0208\,Btu/s \cdot R$
7-34 $\mathcal{V} = 2068\,ft/s$
7-37 $P_2 = 12.3\,kPa$, $\dot{W} = 7.02\,MW$
7-40 $T = 200°F$
7-43 $P = 180\,psia$
7-46   a) $T_2 = 189°C$, b) $\dot{W} = 4060\,kW$
7-49 $\dot{W}_{max} = 163\,kW$

Chapter 8

8-1   a) $\dot{W}_{in} = 7.11\,kW$, b) $\dot{Q} = 23.2\,kW$,
       c) $COP_{Ref} = 3.26$
8-4   a) $\dot{m}_R = 0.301\,lbm/s$, b) $\dot{W}_{in} = 7.14\,hp$,
       c) $COP_{Ref} = 3.84$
8-7 $\dot{W}_{in} = 1310\,W$

8-10   a) $\dot{W}_{in} = 5.35\,kW$, b) $COP_{HP} = 6.58$,
       c) $COP_{Ref} = 5.58$, d) $COP_{\underset{HP}{CARNOT}} = 13.4$
8-13   a) $\dot{Q}_{in} = 22,033\,kW$, b) $\dot{W}_{net} = 6136\,kW$,
       c) $\dot{Q}_{out} = 15,879\,kW$, d) $\eta_T = 0.786$,
       e) $\eta_P = 0.56$, f) $\eta_{cycle} = 0.278$
8-16   a) $\eta_{overall} = 0.286$, b) $\dot{m}_{coal} = 10,707\,tons/day$
8-19   a) $\dot{W}_{act} = 6992.7\,kW$, b) $\dot{W}_{act} = 67.1\,kW$,
       c) $\dot{Q}_{in} = 21,650\,kW$, d) $\eta_{cycle} = 0.320$
8-22   a) $\dot{W}_{net} = 1,339,000\,kW$, b) $\eta_{cycle} = 0.463$
8-25   a) $\eta_{cycle} = 0.354$, b) $\eta_{cycle} = 0.365$
8-28   a) $\dot{W}_{net} = 3300\,kW$, b) $\dot{Q}_{boiler} = 7060\,kW$,
       c) $\dot{Q}_{reheater} = 1170\,kW$, d) $\eta_{cycle} = 0.401$
8-31   a) $y = 0.192$, b) $\eta_{cycle} = 0.394$,
       c) $\eta_{cycle} = 0.369$
8-34   a) $y = 0.180$, b) $w_{T,1} = 461.7\,kJ/kg$,
       c) $w_{T,2} = 692.2\,kJ/kg$, d) $w_{P,2} = 5.1\,kJ/kg$,
       e) $q_{in,boiler} = 2782\,kJ/kg$, f) f $\eta = 0.370$
8-37   a) $\eta_{cycle} = 0.465$, b) $BWR = 0.477$,
       c) $T_4 = 615.7°F$
8-40   a) $\dot{W}_{net} = 74.6\,kW$, b) $\dot{Q}_{in} = 139.0\,kW$,
       c) $\eta_{cycle} = 0.537$, d) $BWR = 0.497$
8-43   a) $\dot{W}_{net} = 3233\,hp$, b) $\eta_{cycle} = 0.318$
8-46   a) $BWR = T_1/T_{4s}$,
       b) $BWR = T_1/\eta_T\eta_C T_{4s}$
8-49   a) $\eta_T = 0.855$, b) $\dot{W}_{net} = 5961.5\,kW$,
       c) $\eta_{cycle} = 0.392$, d) $\varepsilon = 0.435$
8-52   a) $\eta_{cycle} = 0.422$, b) $\varepsilon_{reg} = 0.722$,
       c) $\dot{V} = 37,900\,ft^3/min$
8-55   a) $w = 514.7\,kJ/kg$, b) $q = 1172\,kJ/kg$,
       c) $\eta_{cycle} = 0.439$, d) $\eta_{cycle} = 0.327$
8-58   a) $\dot{W}_C = 139\,kW$, b) $\dot{W}_C = 169\,kW$
8-61 $\eta_{cycle} = 1 - (T_1/T_3)\,r_P^{(k-1)/k}$
8-64   a) $\dot{m}_w/\dot{m}_a = 0.140\,kg\,H_2O/kg\,Air$,
       b) $\dot{m}_a = 2066\,kg/s$, $\dot{m}_w = 289\,kg/s$,
       c) $\dot{Q}_{in} = 1,908,300\,kW$, d) $\eta_{cycle} = 0.524$
8-67   a) $Q_{23} = 0.581\,kJ$, b) $Q_{41} = 0.247\,kJ$,
       c) $W_{net} = 0.334\,kJ$, d) $\eta_{cycle} = 0.576$
8-70   a) $q_{23} = 239.4\,Btu/lbm$,
       b) $q_{41} = 103.7\,Btu/lbm$,
       c) $w_{net} = 135.7\,Btu/lbm$,
       d) $\eta_{cycle} = 0.567$, e) $\eta_{Carnot} = 0.785$
8-73   a) $Q_{23} = 1.003\,kJ$, b) $\eta_{cycle} = 0.473$,
       c) $\dot{W}_{net} = 95\,kW$
8-76   a) $Q_{23} = 17.1\,Btu$, b) $T_3 = 2878\,°F$,
       c) $Q_{41} = 7.87\,Btu$, d) $\eta_{cycle} = 0.540$
8-79   a) $r_V = 20.29$, b) $T_3 = 1942\,K$,
       c) $\eta_{cycle} = 0.651$, d) $r_C = 1.942$
8-82   a) $\dot{W}_{net} = 122.3\,hp$, b) $\eta_{cycle} = 0.57$
       c) $\dot{m}_{fuel} = 28.2\,lbm/hr$

Chapter 9

9-1 $\dot{V}/W = 2.59 \times 10^{-6}\,\text{m}^3/\text{ms}$

9-4 $\Delta h = 50.1\,\text{ft}$

9-7 $\dot{V}/D = -\rho g h^3/3\mu + V_B h$

9-10 a) $\mu = 1.122 \times 10^{-3}\,\text{Ns}/\text{m}^2$,
b) $\mu = 0.953 \times 10^{-4}\,\text{Ns}/\text{m}^2$

9-13 $\mathcal{V} = \dfrac{H^2}{2\mu}\left[\dfrac{P_1 - P_2}{L} - \rho g \sin\theta\right]\left[1 - \left(\dfrac{y}{H}\right)^2\right]$

9-16 $\Delta P = 0.0457\,\text{kPa}$, $h_P = 4.66\,\text{mm}$ of water

9-19 $D = 6.2\,\text{in.}$ (next larger standard size)

9-22 a) $D = 1.70\,\text{ft}$ (next larger standard size)
b) $\dot{W} = 2.96\,\text{hp}$

9-25 a) $t = 635\,\text{min}$, b) $t = 432\,\text{min}$,

9-28 $D = 1.83\,\text{ft}$ (next larger standard size)

9-31 $t = 6922\,\text{min}$

9-34 $z_1 - z_2 = 1.97\,\text{ft}$

9-37 $\dot{W}_P = 33.3\,\text{hp}$

9-40 $P_1 = 2181\,\text{kPa}$ (absolute) $= 2080\,\text{kPa}$ (gage)

9-43 $\dot{W} = 2,213\,\text{kW}$

9-46 $D = 2.04\,\text{in.}$ (next larger standard size)

9-49 a) $P_2 = 100.04\,\text{kPa}$, b) $\dot{W} = -19.9\,\text{W}$

9-52 $P_{1,\,gage} = 620\text{kPa}$ (gage)

9-55 $L = 1260\,\text{ft}$

9-58 $\dot{W}_P = 17,960\,\text{W}$

9-61 a) $\dot{V} = 285\,\text{m}^3/\text{min}$, b) $\dot{V} = 198\,\text{m}^3/\text{min}$

9-64 $\dot{V}_1 = 0.00699\,\text{m}^3/\text{s}$

Chapter 10

10-1 $x = 0.304\,\text{m}$

10-4 $\mathcal{V} = 108\,\text{km}/\text{hr}$

10-7 $\dot{W} = 0.0265$, W, $\dot{W} = 0.15\,\text{W}$, $\dot{W} = 0.413\,\text{W}$

10-10 $\mathcal{V} = 7.21\,\text{ft}/\text{s}$

10-13 a) $\mathcal{V} = 22\text{m}/\text{s}$, b) $\mathcal{V} = 17\,\text{m}/\text{s}$,
c) $\mathcal{V} = 27\,\text{m}/\text{s}$

10-16 a) $\mathcal{V} = \left(\dfrac{2\rho_B t g \sin\theta}{C_D \rho_a \cos^2\theta}\right)^{1/2}$, b) $t = 4.21\,\text{mm}$

10-19 a) $\mathcal{V} = 37.6\,\text{m}/\text{s}$, $V = 109\,\text{m}/\text{s}$,
b) $t = 7.02\,\text{s}$, $t = 20.3\,\text{s}$, $y = 72.8\,\text{m}$, $y = 610\,\text{m}$

10-22 a) $F_{tot} = 307,600\,\text{N}$,
b) $t = 3.93\,\text{s}$, $x = 227\,\text{m}$

10-25 $\mathcal{V} = 23.1\,\text{km}/\text{hr}$

10-28 $t = 12.5\text{min}$

10-31 $D \approx 19.8\,\text{ft}$, $\theta = 19.7°$

10-34 $F_D = 209\,\text{lbf}$

10-37 $F_{tot} = 2.198 \times 10^6\,\text{N}$

10-40 2% increase

10-43 35% increase

10-46 a) $L = 50.5\,\text{km}$, b) $t = 65.1\,\text{min}$

10-49 a) $\alpha \approx 8°$, b) $L = 808\,\text{mi}$

10-52 a) $\mathcal{V}_2/\mathcal{V}_1 = 1.72$, b) $F_{D,2}/F_{D,1} = 1$

10-55 a) $F_R = 4994\,\text{N}$, b) $\theta_R = 13.3°$,
c) $\dot{W} = 3450\,\text{W}$, d) $\dot{W} = 4.14\,\text{W}$

10-58 $\delta/x = 0.351/Re_x^{1/5}$

Chapter 11

11-1 $k = 24.0\,\text{W}/\text{m K}$

11-4 a) $t = 4\,\text{cm}$, b) savings $= \$543.97$

11-7 $k_B = 54.1\,\text{W}/\text{m K}$

11-10 $T_{CL} = 178.7°\text{C}$

11-13 $T = 26.8°\text{C}$

11-16 $T(x) = \dfrac{-q''\,x^2}{2kt} + \dfrac{q''\,Lx}{kt} + T_b = 55.6°\text{C}$

11-19 $\dot{Q} = 266\,\text{W}$

11-22 a) $T_1 = 106.9°\text{C}$, b) $T_2 = 96.6°\text{C}$,
c) $T_3 = 95.1°\text{C}$

11-25 a) $\dot{Q} = 711\,\text{W}$, b) $\dot{m} = 3.58 \times 10^{-3}\,\text{kg}/\text{s}$

11-28 a) $\dot{Q} = 4,480\,\text{Btu}/\text{hr}$, b) $\dot{Q} = 1,650\,\text{Btu}/\text{hr}$

11-31 $t = 4.84\,\text{hr}$

11-34 $t = 1440\,\text{s}$

11-37 a) $t = 8.1\,\text{s}$, b) $t = 47.5\,\text{s}$

11-40 a) $t = 38.7\,\text{min}$, b) $T = 119.2°\text{C}$,
c) $T_{avg} = 101.1°\text{C}$

11-43 a) $t = 1.71\,\text{hr}$, b) $t = 0.34\,\text{hr}$

11-46 $k = 1.44\,\text{W}/\text{m K}$

11-49 a) $2L = 0.200\,\text{m}$, b) $t = 37.2\,\text{min}$

11-52 a) $T = 161.8°\text{C}$, b) $T = 33.5°\text{C}$

11-55 $t = 232.5 = 3.86\,\text{min}$

11-58 a) $q_1'' = 271.6\,\text{kW}/\text{m}^2$, $q_2'' = 496.4\,\text{kW}/\text{m}^2$,
b) 54.3% decrease, 16.5% decrease

Chapter 12

12-1 a) $\dot{Q} = 69.2\,\text{W}$, b) $T_w = 43.3°\text{C}$

12-4 $T_{out} = 997°\text{C}$

12-7 a) $\dot{Q} = 32,500\,\text{W}$, b) $\dot{Q} = 72,300\,\text{W}$

12-10 $T_{PV} = 273.6\,\text{K}$

12-13 $\dot{Q} = 120\,\text{W}$, $I = 14.7\,\text{A}$

12-16 $T_s = 513.4\,\text{K} = 240.4°\text{C}$

12-19 $T_f = 583\,\text{K} = 310°\text{C}$

12-22 $\dot{Q}/L = 103\,\text{W}/\text{m}$

12-25 $T_{out} = 39.4°\text{C}$, $\dot{Q} = 7.93\,\text{kW}$

12-28 $T_{out} = 17.6°\text{C}$, $\dot{Q} = 5320\,\text{W}$

12-31 $T_w = 272.7\,\text{K} = -0.3°\text{C}$

12-34 $\dot{Q}_{out} = 1381\,\text{MW},\ T_{out} = 39.5°\text{C},\ L = 9.85\,\text{m}$

12-37 $L = 38.6\,\text{m}$

12-40 $L = 8.48\,\text{m}$

12-43 $T = 230°\text{C}, P = 2795,\ \text{kPa}$

12-46 $h = 2.83\,\text{W/m}^2\,\text{K}, h = 2.09\,\text{W/m}^2\,\text{K}$

12-49 $\dot{Q} = 355\,\text{W}$

12-52 $\dot{Q} = 328,000\,\text{W}, t = 85.6\,\text{s} = 1.43\,\text{min.}$

12-55 $T_s = 59°\text{C}$

12-58 $t = 2.81\,\text{days}$

12-61 $\dot{Q}_{tot} = 49.7\,\text{W}, \dot{Q}_{tot} = 32.9\,\text{W}$

12-64 $\dot{Q}_{in} = 31,450\,\text{W}$

12-67 $I = 1150\,\text{A}$

12-70 $\dot{Q} = 582\,\text{W}, \dot{Q} = 500\,\text{W}$

14-7 $d = 3.04\,\text{cm}$

14-11 $F_{1\rightarrow2} = 0.0196$

14-13 $F \approx 0.5$

14-16 $F_{1\rightarrow2} = 0.178$

14-20 $T = 344\,\text{K}$

14-22 $\dot{Q}/A = 159\,\text{Btu/h}\cdot\text{ft}^2,\ \text{no change}$

14-25 $t = 2.43\,\text{h}$

14-28 $T = 272°\text{C}$

14-31 $\dot{Q}_{tot} = 1.65\,\text{Btu/h}$

14-34 $\dot{Q} = 2300\,\text{W}$

14-37 $\alpha = 0.861$

14-40 $F_{\lambda_1 T - \lambda_2 T} = 1.93 \times 10^{-8}$

14-43 $\varepsilon = 0.67$

Chapter 13

13-1   a) $U_i = 64.0\,\text{W/m}^2\text{K}$,   b) $U_o = 57.6\,\text{W/m}^2\text{K}$,
     c) $U_i = 60.3\,\text{W/m}^2\,\text{K}$

13-4 $U_i = 121.3\,\text{W/m}^2\,\text{K}$

13-7   a) $U_i = 98.0\,\text{W/m}^2\text{K}$,   b) $U_i = 481\,\text{W/m}^2,\ \text{K}$

13-10 $U_i = 903\,\text{W/m}^2\,\text{K}, U_o = 73.3\,\text{W/m}^2\,\text{K}$

13-13 $A = 5.59\,\text{ft}^2$

13-16   a) $\dot{m}_w = 17.75\,\text{lbm/s}$,   b) $N = 88\,\text{tubes}$

13-19   a) $\dot{m}_w = 32.0\,\text{kg/s}$,   b) $N = 140$

13-22 $R'' = 0.0175\,\text{hrft}^2\text{R/Btu}$

13-25   a) $U_i = 731\,\text{W/m}^2\text{K}$,   b) $l = 3.90\text{m}$

13-28 $A = 1.32\,\text{m}^2$

13-31 $A = 26.9\text{m}^2$

13-34   a) $A = 86.9\,\text{m}^2$,   b) $A = 98.2\,\text{m}^2$,
     c) $\varepsilon = 0.671$

13-37 $L_{new}/L_{old} = 1.224$

13-40 $T_{H,out} = 93.6°\text{C}$

13-43   a) $A = 14.7\text{m}^2$,   b) $T_{C,out} = 125.6°\text{C}$,
     $T_{H,out} = 68.9°\text{C}$

13-46   a) $T_{C,out} = 37.0°\text{C}$,   b) $T_{C,out} = 30.7°\text{C}$

13-49   a) $T_{out} = 147.9°\text{C}$,   b) $35.2\%\ \text{decrease}$

13-52   a) counterflow, $T_{C,out} = 12.9°\text{C}$
     b) parallel, $T_{C,out} = 9.65°\text{C}$

13-55   a) $\dot{Q} = 1.77 \times 10^5\ \text{Btu/hr}$,
     b) $T_{C,out} = 146°\text{F}$,   c) $\dot{m} = 3.10\text{lbm/min}$

13-58 $A_{clean} = 34.2\,\text{m}^2, A_{dirty} = 37.9\,\text{m}^2$

13-61   a) $\dot{Q} = 5630\,\text{Btu/min}$,   b) $T_{H,out} = 232°\text{F}$,
     c) same

Chapter 14

14-1 $\lambda_{max} = 1.12\,\mu\text{m},\ \text{infrared}$

14-4 $q'' = 0.884\,\text{W/m}^2$

INDEX